Primates

Primates

The Road to Self-Sustaining Populations

Edited by
Kurt Benirschke

With a Foreword by Jared Diamond

With 164 Figures

Proceedings of a conference held in San Diego, California, June 24–28, 1985, sponsored by The Morris Animal Foundation, Englewood, Colorado, and The Zoological Society of San Diego, San Diego, California

Springer-Verlag
New York Berlin Heidelberg
London Paris Tokyo

Kurt Benirschke, M.D. Trustee, Zoological Society of San Diego, San Diego, California 92112, and Professor of Pathology and Reproductive Medicine, University of California, La Jolla, California 92093

The cover illustration is a partial reproduction of the photograph by Frans de Waal, appearing on p. 342 of this volume.

Library of Congress Cataloging in Publication Data
Main entry under title:
Primates: the road to self-sustaining populations.
Includes bibliographies and index.
1. Primates—Congresses. 2. Wildlife conservation—Congresses. I. Benirschke, Kurt. II. Series.
QL737.P9P69 1986 333.95′9 85-27766

Typeset by Publishers Service, Bozeman, Montana.
Printed and bound by Arcata Graphics/Halliday, West Hanover, Massachusetts.
Printed in the United States of America.

9 8 7 6 5 4 3 2 1

ISBN 0-387-96270-0 Springer-Verlag New York Berlin Heidelberg
ISBN 3-540-96270-0 Springer-Verlag Berlin Heidelberg New York

Foreword

This conference represents the first time in my life when I felt it was a misfortune, rather than a major cause of my happiness, that I do conservation work in New Guinea. Yes, it is true that New Guinea is a fascinating microcosm, it has fascinating birds and people, and it has large expanses of undisturbed rainforest. In the course of my work there, helping the Indonesian government and World Wildlife Fund set up a comprehensive national park system, I have been able to study animals in areas without any human population. But New Guinea has one serious drawback: it has no primates, except for humans. Thus, I come to this conference on primate conservation as an underprivileged and emotionally deprived observer, rather than as an involved participant.

Nevertheless, it is easy for anyone to become interested in primate conservation. The public cares about primates. More specifically, to state things more realistically, many people care some of the time about some primates. Primates are rivaled only by birds, pandas, and the big cats in their public appeal. For some other groups of animals, the best we can say is that few people care about them, infrequently. For most groups of animals, no one cares about them, ever. Thus, I do not believe the rumor I heard that Kurt Benirschke is already at work on an even bigger conference to be held next year, and to be entitled "Endangered Worms and Slime Molds: The Road to Self-Sustaining Populations."

The reason for the public's interest in primates is of course their similarity to us and their relationship with us: The public can identify with primates. Even those of us concerned with conservation of other animals and plants profit from this public interest in primates. The reason is conservation biology's equivalent of the economic policy that our national administration pursues and that it terms the "trickle-down effect." Conservation of primates produces benefits that trickle down to other species. Marmosets and orangutans and other wild primates require forest, and that means conserving plant species in the forest, and that in turn means conserving birds, mammals, and insects that pollinate those plants and disperse their fruit. Thus, primate conservation saves many other species.

There are also several good scientific reasons for emphasizing primate conservation:

First, primates are ecologically important animals. They are often among the dominant animals, especially in pristine habitats where they are not hunted. For example, in Peru's Manu National Park, an area of pristine Amazonia rainforest, John Terborgh and his colleagues found that primates constitute about 40% of the frugivore biomass.

A second compelling scientific reason for emphasizing primate conservation is their close taxonomic relationship with humans, as a result of which primates offer the best comparison species for studying problems of human biology. Recent work of Charles Sibley and John Ahlquist has shown that we share about 75% of our single-copy DNA with lorises and tarsiers, about 76% with lemurs, 85% with New World monkeys, 92% with Old World monkeys, 95% with gibbons, 96% with orangutans, 98% with gorillas, and 98.5% with chimpanzees. In species groups about whose classification we have no personal prejudices, 97% sharing of single-copy DNA between two species generally means that they are ones we would consider on other grounds as congeneric. Hence a truly liberated cladistic taxonomist visiting the earth from Outer Space would surely lump humans, common chimpanzees, pygmy chimpanzees, and gorillas together in genus *Homo*. Chimpanzees and humans are genetically more similar than are red-eyed vireos and white-eyed vireos. From this genetic similarity between humans and other primates stems the importance of other primates for understanding human biology. For instance, although I have not personally done research specifically on primates, nevertheless in my research on human physiology and evolution I have had to use primate comparisons five times in the last couple of years: for understanding lithium distribution in human muscle in manic–depressive illness, for understanding the physiology and evolution of human aging, for tracing the origins of human aggression against conspecifics, for understanding the function and evolution of human sexual physiology and anatomy, and for placing human mating systems in context.

The remaining scientific reason for stressing primate conservation is that, quite simply, primates are an unusually interesting group of animals. They exhibit an unusually wide range of body sizes, a diversity of social systems, a diversity of feeding habits and digestive adaptations, an exceptional capacity for learning, and a capacity for vocal transmission of information.

Despite this scientific importance of primates, we are still remarkably ignorant about their biology. Among other things, we know practically nothing about their intestinal absorption of nutrients, little about their pheromones, nothing about the learning abilities of many species, and little about the social systems of many species. Numerous primate species are involved in multispecies foraging associations with other primates, but in almost no case do we understand the selective advantages of these foraging associations: e.g., whether both species derive benefits, and, if so, what benefits, or whether one species gets all the benefits and the other species is passive or even victimized. There have been few multigenerational studies of wild primate communities of which all members are individually recognized. This list of our areas of ignorance about primates could be extended indefinitely.

The scientific study of primates has a short history, but *Homo sapiens* has had a long history of exterminating or endangering other primates. Within the first thousand years of the settlement of Madagascar, humans exterminated about 12 lemur species, including all the largest lemurs. Much earlier, Acheulean paleolithic cultures in Africa hunted Gelada baboons. We do not know whether humans also exterminated the mysterious monkey *Xenothrix* of the West Indies or the robust australopithecines. I am not aware of any primate having been exterminated since the mass extinctions on Madagascar about a thousand years ago, but many primate species are now endangered.

For us to continue with the scientific study of primates, we need to have live primates left in the world. We have two strategies for accomplishing this goal.

One strategy is to preserve primates in the wild. This strategy has many obvious advantages. It may be the cheapest way to save primates. It preserves many other species as well through the trickle-down effect, as a side-benefit of primate conservation. It preserves the local cultures of primate populations, their local communications systems, and the culturally transmitted heritage of information communicated by these systems. The advantages of preserving wild primate communities over preserving captured individuals thrown together in zoos are analogous to the advantages of preserving intact human populations, with their shared language and libraries and music, as compared to putting six humans from the six different continents together in a cage, stripped of any shared knowledge, and making them reinvent Pidgin English, just as agricultural laborers imported from many countries to Hawaii were forced to do in the nineteenth century.

The other strategy of primate conservation is to preserve them in captivity. To obtain perspective on why we need this strategy as an option, consider the recent history of the forest birds of Guam, a tropical island in the Pacific Ocean. In August 1984 I published a short article on Guam's forest birds, drawing on a recent study that documented how 12 of the 13 native forest bird species were seriously endangered by pesticides, diseases, and introduced snakes. Bird numbers were declining rapidly, and there were no captive breeding programs. Hence I suggested in my article that there should be captive breeding programs. As of June 1985, nearly a year later, there now are captive breeding programs, but for only two of the 12 species. Several of the other species are already extinct, and the rest are expected to become extinct soon. Those ten extinct or doomed species might have been saved by captive breeding programs; they certainly were not saved in time in the wild.

For another perspective on the importance of captive breeding, think of all the animal species that were actually brought into captivity before breeding programs were well developed, that subsequently went extinct in the wild as well as in zoos, and that are therefore now extinct as species. These victims include the thylacine (Tasmanian wolf), passenger pigeon, quagga, Schomburg's deer, and possibly the wild ox of Cambodia (the kouprey). Individuals of all these species were at some point held in zoos. If captive breeding had been as successful in the last century, or even a mere 50 years ago, as it is today, those species would still be in existence.

But we have a recipe for trouble, and for internecine warfare among ourselves, if we postpone captive breeding efforts until a species is on the verge of extinction, as in the case of the California condor. At that point, captive breeding competes directly with preservation in the wild, and one is stuck with controversial and agonizing decisions. Surely, given the uncertainty in the world, one should pursue both preservation in the wild and captive breeding, and one should start the captive breeding *before* a species becomes endangered. Personally, I feel that a prudent goal would be to try to establish breeding colonies of all of the world's approximately 180 primate species. To view captive breeding as an either/or alternative to habitat preservation is like viewing having a fire extinguisher in your house as an alternative to buying a fire insurance policy: Any prudent person does both.

Unfortunately, little time is left to consummate either strategy of primate conservation. Conservation biology is the sole field of biological study with a time limit, because many of the species that we want to conserve will be extinct within 5 years. We speak of a race to cure cancer, which has no time limit: We could still cure it later if we suspended all cancer research until the year 2050. Yet there is no race, and little support, to cure extinction. The National Science Foundation has no panel, no research grants, no traineeships in conservation biology; universities have no departments and few courses in conservation biology. The basic sciences of population biology on which conservation biology depends are starved for support. The NSF panel on ecology and systematics has an annual grants budget of about $3 million, compared to the NIH annual grants budget of over $4 billion for medical research. It is as if cancer deaths irreversibly destroyed the victim's whole family, yet we approached human health problems by abolishing medical schools, cutting back support for basic research in molecular and cellular biology to $3 million annually, and abolishing organizations to foster communication between the few remaining molecular biologists and physicians.

In effect, this is the nonsensical approach that funding agencies are taking toward the field of biology that really is in a race: conservation biology. Hence, for those of us at this conference, our most pressing problem is to lobby for far larger funding of efforts in primate conservation.

Jared Diamond

Preface

Natural resources are undeniably shrinking, everyone would agree. Because of a burgeoning population with ever-expanding expectations, the primeval forests disappear with alarming rapidity. And with it die the refuges of a vast number of species. E. O. Wilson has declared that "the one process ongoing in the 1980s that will take millions of years to correct is the loss of genetic and species diversity by the destruction of natural habitat. This is the folly our descendants are least likely to forgive us."

Primates are but one component of this array of species, perhaps most visibly declaring the loss of their habitat because humans are a member of this order of some 200 species. Not only do we feel a certain kinship to our relatives from the forest, but we have also depended on primates heavily in freeing ourselves of many diseases and continue to employ their relationship toward this end. It is time to pay back a debt owed.

Many primates face certain extinction and will go unstudied, largely unknown, as the many species that vanished in Madagascar since human discovered that singular island. To arouse more widespread concern and to consider avenues of approaches that might stem this tide of extinction, an international meeting was convened 24–28 June 1985 in San Diego, and this volume brings the scientific presentations delivered then.

The intent to foster interdisciplinary communication was admirably achieved at the conference. Field scientists, primatologists, and laboratory scientists of many persuasions talked with curators of zoos, students found new direction, and behaviorists' ideas were heard by virologists from whose cautious approaches they themselves profited. Irrespective of their origin or the nature of their work, all were conservationists at heart and accepted the need for self-sustaining populations. The heartening meeting of minds went far beyond the narrow outlook of traditional disciplines. In situ as well as ex situ conservation was accepted as modes for different taxa, above all, much greater collaborative efforts should result from this first of many, one hopes, conferences on immediate conservation needs.

The volume loosely follows the theme of the meeting with a review of the status of primates in situ, conservation aspects, ex situ breeding results and methodologies, considerations of reproductive phenomena, infections, and other pathology. It considers behavioral needs to be met in captivity, genetic aspects, and relocation and reintroduction into the wild. Above all, the challenge to the scientists was to delineate immediately researchable problems that might serve as guideposts for funding agencies and to enhance the long-term survival of the most endangered. The meeting was instructive, and may the proceedings be helpful to the international community in the care of vanishing primates.

I wish to express my deep gratitude to Mary Byrd, Jane Wagner, and Donald Lindburg without whose constant support this conference would not have been so successful.

Kurt Benirschke

Contents

Errata

Primates: The Road to Self-Sustaining Populations

Edited by Kurt Benirschke

P. 895: Lines 1 and 2, instead of: "While NORs are found in one chromosome pair in *A. belzebul belzebul* they are present in three different chromosome pairs in *A. belzebul nigerrima*" the sentence should read: "While NORs are found in one chromosome pair in *A. belzebul nigerrima* they are present in three different chromosome pairs in *A. belzebul belzebul.*"

P. 896: The figure legend should read: Karyotype of *Alouatta belzebul nigerrima* with G-banding (2n = 50). NOR staining occurs in chromosome 8, in two different regions.

P. 898: The figure legend should read: Karyotype of *Alouatta belzebul belzebul* with G-banding (2n = 50).

P. 900: The figure legend should read: Chromosomes of *Alouatta belzebul belzebul* with NOR staining in three chromosome pairs.

Contributors

Daniel C. Anderson, Division of Pathobiology and Immunobiology, Yerkes Regional Primate Research Center, Emory University, Atlanta, GA 30322, USA

John H. Anderson, Associate Veterinarian, California Primate Research Center, University of California, Davis, CA 95616, USA

John Aspinall, Director/Owner of Howletts and Port Lympne Wildlife Parks, The Curzon, London W1Y 7AD, England

Jonathan D. Ballou, National Zoological Park, Smithsonian Institution, Washington, DC 20008, USA

Benjamin B. Beck, National Zoological Park, Smithsonian Institution, Washington, DC 20008, USA

Mary E. Beland, New England Regional Primate Research Center, Harvard Medical School, Southborough, MA 01772, USA

Kurt Benirschke, Trustee, Zoological Society of San Diego, San Diego, CA 92112; and Departments of Pathology and Reproductive Medicine, University of California, La Jolla, CA 92093, USA

Raoul E. Benveniste, Laboratory of Viral Carcinogenesis, National Cancer Institute, Frederick Cancer Research Facility, Frederick, MD 21701, USA

Donald S. Boy, Department of Anthropology, University of Massachusetts, Amherst, MA 01002, USA

Anne R. Brodie, Division of Pathobiology and Immunology, Yerkes Regional Primate Research Center, Emory University, Atlanta, GA 30322, USA

H. Sheldon Campbell, Formerly, President, Zoological Society of San Diego, San Diego, CA 92112. Deceased.

Laura V. Chalifoux, New England Regional Primate Research Center, Harvard Medical School, Southborough, MA 01772, USA

David J. Chivers, Subdepartment of Veterinary Anatomy, University of Cambridge, Cambridge CB2 1QS, United Kingdom

Adelmar F. Coimbra-Filho, Centro do Primatologia de Rio de Janeiro (FEEMA), Rio de Janeiro, Brazil

Jared M. Diamond Department of Physiology, 53–238, Center for Health Sciences, UCLA, Los Angeles, CA 90024, USA

James M. Dietz, National Zoological Park, Smithsonian Institution, Washington, DC 20008, USA

Lou Ann Dietz, National Zoological Park, Smithsonian Institution, Washington, DC 20008, USA

W. Richard Dukelow, Director, Endocrine Research Unit, Michigan State University, East Lansing, MI 48824, USA

Joseph Erwin, Associate Editor, National Geographic Research, Washington, DC 20036, USA

John E. Fa, Departamento de Ecologia, Instituto de Biologia, Universidad Nacional Autonoma de Mexico, 04510 Mexico, DF

Ted W. Finlay, Research Associate, Atlanta Zoological Park, Atlanta, GA 30315, USA

Nathan Flesness, International Species Inventory System, Minnesota Zoological Gardens, Apple Valley, MN 55124, USA

Jo Fritz, Administrative Director, Primate Foundation of Arizona, Tempe, AZ 85281, USA

William I. Gay, Director, Animal Resources Program, Division of Research Resources, NIH, Bethesda, MD 20205, USA

William J. Goodwin, Director, Department of Laboratory Animal Medicine, Southwest Foundation for Biomedical Research, San Antonio, TX 78284, USA

Kenneth G. Gould, Chief, Reproductive Biology, Yerkes Regional Primate Research Center, Emory University, Atlanta, GA 30322, USA

Tine Griede, Apenheul Sanctuary, J.C. Wilslaan 31, 7313 HK Apeldoorn, The Netherlands

Alexander H. Harcourt, Department of Applied Biology, University of Cambridge, Cambridge CV2 3DX, England

John P. Hearn, Director, The Institute of Zoology, The Zoological Society of London, Regent's Park, London NW1 4RY, England

Gary D. Hodgen, Scientific Director, Jones Institute for Reproductive Medicine, Department of Obstetrics and Gynecology, Eastern Virginia Medical School, Norfolk, VA 23507, USA

William V. Holt, The Institute of Zoology, The Zoological Society of London, Regent's Park, London NW1 4RY, England

Jan A.R.A.M. van Hooff, Laboratorium voor vergelijkende fysiologie, Rijksuniversiteit Utrecht, 3572 LA Utrecht, The Netherlands

Ronald D. Hunt, Director, New England Regional Primate Research Center, Harvard Medical School, Southborough, MA 01772, USA

Dennis O. Johnsen, Director, Primate Research Centers Program, Division of Research Resources, NIH, Bethesda, MD 20205, USA

Lorna D. Johnson, New England Primate Research Center, Harvard University, Southborough, MA 01772, USA

Alison Jolly, The Rockefeller University, New York, NY 10021-6399, USA

Marvin L. Jones, Registrar, Zoological Society of San Diego, San Diego, CA 92112, USA

Seymour S. Kalter, Director, NIH and WHO Collaborating Center for Reference and Research in Simian Viruses, Southwest Foundation for Medical Research, San Antonio, TX 78284, USA

Norval W. King, Jr., Associate Director for Collaborative Research, New England Regional Primate Research Center, Harvard Medical School, One Pine Hill Drive, Southborough, MA 01772, USA

Devra G. Kleiman, Assistant Director for Research and Educational Activities, National Zoological Park, Smithsonian Institution, Washington, DC 20008, USA

William R. Konstant, Wildlife Preservation Trust International, Philadelphia, PA 19104, USA

Donald G. Lindburg, Behaviorist, Research Department, Zoological Society of San Diego, San Diego Zoo, San Diego, CA 92112, USA

Linda J. Lowenstine, Primate Research Center, University of California, Davis, CA 95616, USA

Kathleen MacKinnon, Haddenham, Cambridge, CB6 3PN, United Kingdom

Wim B. Mager, Director, Apenheul Sanctuary, Natuurpark Berg and Bos, 7313 HK Apeldoorn, The Netherlands

Iqbal Malik, Department of Biology, Institute of Home Economics, University of Delhi, New Delhi, India 110011

Jeremy J.C. Mallinson, Jersey Wildlife Preservation Trust, Les Augres Manor, Trinity, Jersey, Channel Islands, British Isles

Terry L. Maple, School of Psychology, Georgia Institute of Technology, Atlanta, GA 30332, USA

Hal Markowitz, Department of Biological Sciences, San Francisco State University, San Francisco, CA 94132, USA

David E. Martin, Yerkes Regional Primate Research Center, Emory University, Atlanta, GA 30322, USA

William A. Mason, Primate Research Center, University of California, Davis, CA 95616, USA

Harold M. McClure, Associate Director for Scientific Programs and Chief, Division of Pathobiology and Immunobiology, Yerkes Regional Primate Research Center, Emory University, Atlanta, GA 30322, USA

Dennis A. Meritt, Jr., Assistant Director, Lincoln Park Zoo, Chicago, IL 60614, USA

George Migaki, Chief Pathologist, The Registry of Comparative Pathology, Armed Forces Institute of Pathology, Washington, DC 20306, USA

Katharine Milton, Department of Anthropology, University of California, Berkeley, CA 94720, USA

Russell A. Mittermeier, Chairman, IUCN/SSC Primate Specialist Group and Director, WWF-US Primate Program, World Wildlife Fund/The Conservation Foundation, Washington, DC 20037, USA

Donald E. Moore, Curator of Mammals, Burnet Park Zoo, Liverpool, NY 13088, USA

Jan Moor-Jankowski, LEMSIP, New York University Medical Center, New York, NY 10016, USA

John F. Oates, Department of Anthropology, Hunter College of CUNY, New York, NY 10021, USA

Francine Patterson, 17820 Skyline Boulevard, Woodside, CA 94062, USA

Andrew J. Petto, Department of Anthropology, University of Massachusetts, Amherst, MA 01002, USA

Jonathan I. Pollock, Duke University Center for the Study of Primate Biology and History, Durham, NC 27705, USA

Jing-Fen Qi, Beijing Natural History Museum, Beijing, People's Republic of China

Guoqiang Quan, Endangered Species Scientific Commission, The People's Republic of China, Haidian, Beijing, PRC

George Rabb, Director, Brookfield Zoo, Brookfield, IL 60513, USA

Harvey Rabin, DuPont, New England Nuclear, N. Billerica, MA 01862, USA

Herman D. Rijksen, School of Environmental Conservation Management, Bogor, Indonesia

Oliver A. Ryder, Research Department, San Diego Zoo, San Diego, CA 92112, USA

Ulysses S. Seal, Research Service, V.A. Medical Center, Minneapolis, MN 55417 and Departments of Biochemistry and Fisheries and Wildlife, University of Minnesota, St. Paul, MN 55108, USA

Charles J. Sedgwick, Department of Environmental Studies, School of Veterinary Medicine, Tufts University, North Grafton, MA 01536, USA

Prabhat K. Sehgal, New England Regional Primate Research Center, Harvard Medical School, Southborough, MA 01772, USA

Héctor N. Seuánez, Laboratorio de Citogenetica, Departamento de Genetica, Instituto de Biologia, Universidade Federal do Rio de Janeiro, Cidade Universitaria, 21.941 Rio de Janeiro, J.J., Brasil

Roger V. Short, Department of Anatomy and Physiology, Monash University, Clayton, Victoria, 3168, Australia

Joseph P. Skorupa, Department of Anthropology, University of California, Davis, CA 95616, USA

David G. Smith, Department of Anthropology, University of California, Davis, CA 95616, USA

Wladyslaw W. Socha, LEMSIP, New York University Medical Center, New York, NY 10016, USA

Charles H. Southwick, Environmental, Population and Organismic Biology, University of Colorado, Boulder, CO 80309, USA

Joseph S. Spinelli, Animal Care Facility, University of California, San Francisco, CA 94143, USA

R. Brent Swenson, Division of Pathobiology and Immunology, Yerkes Regional Primate Research Center, Emory University, Atlanta, GA 30322, USA

Shirley C. Strum, Department of Anthropology, University of California, La Jolla, CA 92093, USA

Stephen J. Suomi, National Institutes of Health, Bethesda, MD 20205, USA

Warren D. Thomas, Director, Los Angeles Zoo, Los Angeles, CA 90027, USA

Ronald L. Tilson, Curator of Research, Minnesota Zoological Gardens, Apple Valley, MN 55124, USA

John D. Toft, II, Pathology and Animal Resources Section, Battelle Columbus Laboratories, Columbus, OH 43201-2693, USA

Duane E. Ullrey, Department of Animal Husbandry, Michigan State University, East Lansing, MI 48824, USA

P. N. Vengesa, Endocrine Research Unit, Michigan State University, East Lansing, MI 48824, USA

James H. Vickers, Director of Product Quality Control, Pathobiology and Primatology Branch, DHEW PHS, Food and Drug Administration, Bethesda, MD 20205, USA

Frans de Waal, Assistant Scientist, Wisconsin Regional Primate Research Center, University of Wisconsin, Madison, WI 53715-1299, USA

Sung Wang, Endangered Species Scientific Commission, The People's Republic of China, Haidian, Beijing, PRC

David Western, Animal Research and Conservation Center, Embassy House, Harambee Avenue, Nairobi, Kenya; and University of California, La Jolla, CA 92093, USA

Leo A. Whitehair, American Red Cross, Biomedical Research and Development, Bethesda, MD 20814, USA

Robert F. Williams, Department of Obstetrics and Gynecology, University of Puerto Rico, Medical Sciences Campus, San Juan, Puerto Rico 00936

Chang Yi, Beijing Construction Worker's Hospital, Beijing, People's Republic of China

1
Primate Ethics

ROGER V. SHORT

In this presentation about the vexed question of primate ethics, I would like to appeal both to your intellect and to your emotions. How do we cope with the ethical problem posed by primate experimentation? What should we do about those primates whose populations are most endangered in the wild? Are they already too precious to do research upon, or does the very fact that they are so endangered mean that we should redouble our research efforts? If so, what types of research should we be undertaking?

If we look at a map of the world, we can see the distribution of the tropical rainforests of Central and South America, Africa, and southern Asia. In all these areas, the forest is being destroyed at an alarming rate by the ever-increasing human population, and yet these are the very areas of the world that are the homes of most of the primates. I well remember a few years ago talking to Sir Peter Scott of the World Wildlife Fund, and he said to me, in a moment of despair, "You know, perhaps if the World Wildlife Fund cashed in all its assets and bought condoms, we would do more to save endangered species than all our current activities." Alas, there is an element of truth in what he said. At the end of 1984, the population of the world was estimated to be 4.7 billion, and even the most optimistic United Nations projections suggest that we will not be able to call a halt to world population growth until our numbers have exceeded 10 billion, in about the year 2075. While populations are reaching equilibrium in Europe, North America, and the U.S.S.R., rapid growth will continue in Latin America, Africa, and southern Asia, the homes of those primates that are most endangered. Although more and more people in the developing countries of these regions will move into towns and cities, this will do little to ease the pressure on forest habitats. More trees will have to be cut down to meet the insatiable and ever-growing demands of villages, towns, and cities for wood, to use as fuel for domestic cooking and to support a host of small-scale local industries. This "Second Energy Crisis" of the Third World is sending tentacles of destruction deep into forests that are themselves far removed from sites of human habitation. For example, the Indian city of Bangalore is already obtaining some of its firewood from as far away as 640 km. To this urban pressure on our forests we must also add the considerable rural pressure from hunger for more arable land, coupled with the destruction of trees to provide browse for the ever-increasing numbers of domestic livestock.

Faced with all these demands, the future prospects for our tropical rainforests, and the primates they contain, seem grim indeed. The situation can only deteriorate still further as the human population inhabiting these regions doubles or trebles in the years ahead. How can we make forest and wildlife preservation an economically viable alternative to the immediate cash returns that can be obtained from selling the timber, the animals, and the land? That is our greatest challenge.

With such forbidding prospects for the survival of many of our primates in the wild, what should we be doing with those primates that we already hold in captivity? In many cases, zoos and primate research laboratories have literally become life insurance policies for the survival of the species; for some of the smaller primates, such as the golden lion tamarin, the numbers in captivity already exceed the total population surviving in the wild. Clearly, we have a duty to do all within our power to improve the breeding success of these captive colonies, and this will require a continuing and sustained research effort by all concerned.

I therefore find it incredible that this conference should have become a focus of attack by the Animal Liberation movement, which has picketed our meetings and criticized us in the press and on radio and television for daring to bring together field biologists, zoo keepers, and biomedical research workers who share a common interest in primates. Let us pause for a moment to consider the aims and objectives of these misguided Liberationists.

The Animal Liberation movement is becoming more vocal in Europe, in North America, and even in Australia. One of their gurus is Peter Singer, a Professor of Philosophy at Monash University. In a recent article in *The New York Times Review of Books*, Singer (1985) says, "The rise of the animal liberation movement, in the view of a number of commentators, is to be traced back to the publication of my essay Animal Liberation in these pages just over a decade ago. That essay was followed by the book of the same title, which was also published by The New York Book Review." So what does Singer's book have to say?

Singer's basic tenet, summarized in the title of his opening chapter, is that all animals are equal (Singer, 1976). He goes on to qualify this by saying that "the basic principle of equality does not require equal or identical *treatment*; it requires equal *consideration*." If we accept his premise that all animals are the equals of humans; it naturally follows that "the principle of equal consideration of interests requires us to be vegetarians," a major theme that runs throughout the book. But Singer runs into difficulties when trying to decide where to draw the vegetarian line:

> How far down the evolutionary scale shall we go? Shall we eat fish? What about shrimps? Oysters?. . . . Those who want to be absolutely certain that they are not causing suffering will not eat mollusks either; but somewhere between a shrimp and an oyster seems as good a place to draw the line as any, and better than most.

Those of you who have cut open a fresh oyster and squeezed lemon juice upon it will have seen it squirm. So why draw the line at oysters? All forms of animal life show aversive behavior in response to a painful stimulus. In 1904, H. S. Jennings had this to say about the lowly ameba:

> If amoeba were a large animal, so as to come within the every-day experience of human beings, its behaviour would at once call forth the attribution to it of states of pleasure and

pain, of hunger, desire, and the like, on precisely the same basis as we attribute these things to the dog.

So perhaps Singer should reconsider the position of the oyster and the ameba?

I find it difficult to accept that we should now do an about turn, reject the last 45,000 years of our evolutionary history as meat-eating hunter-gathers (Pilbeam, 1984), and on a whim become strict vegetarians. Would Singer propose a similar course of action for our carnivorous domestic pets, the dog and the cat? How does he propose that we feed endangered carnivores in captivity, such as the Siberian tiger or the clouded leopard? Would he have us forsake him because of their reprehensible carnivorous habits, or should we be doing research to develop vegetarian diets for all carnivores?

I reject out of hand Singer's assertion that all animals are equal. Some animals are *more* equal than others and therefore deserving of particular concern and compassion at the hands of humans, either because of their greater intelligence or because of their endangered status in the wild.

Singer would probably condemn such an attitude as "speciesist." For him, "Speciesism . . . is a prejudice or attitude of bias towards the interests of members of one's own species and against the members of other species." The obverse of speciesism is presumably altruism, and as any sociobiologist would hasten to point out to Singer, altruism toward another species is not an evolutionarily stable strategy. Speciesism is a basic property of all species; otherwise, species would not survive.

Singer rejects intelligence as a character that should be taken into account in our dealings with animals: "So the limit of sentience (using the term as a convenient if not strictly accurate shorthand for the capacity to suffer and/or experience enjoyment) is the only defensible boundary of concern for the interests of others. To mark this boundary by some other characteristic like intelligence or rationality would be to mark it in an arbitrary manner." Admittedly, it is difficult to compare intelligence between species, but it is not impossible. Jerison (1985) has recently addressed this very issue. He points out the futility of earlier attempts to measure intelligence on the basis of differences in learning ability. We now know that almost all species can learn equally well, so this is a particularly poor measure of intelligence. However, encephalization—the relative size of the brain in relation to the rest of the body—is a fundamental trait that is a direct measure of an animal's information processing capacity and hence is directly correlated with intelligence. The highest grades of encephalization are shared by humans, dolphins, and killer whales. Next come the apes and monkeys, whose degree of encephalization is twice that of "average" mammals such as deer, or wolves, which are on a par with lemurs, and with crows. Encephalization would seem to reflect a number of different intelligences and indicate the animal's knowledge of reality in relation to the information received by the brain. The large size of the human brain can be attributed to our linguistic ability, which gives us a new dimension to reality. If we are genuinely concerned about minimizing the pain and suffering of animals in captivity, it would seem essential to take encephalization into account, and thus to accord primates particular consideration and respect. All animals are manifestly *not* equal in this regard.

Singer goes on to develop his arguments about the equality of animals *ridiculo ad absurdum*. He claims that there is "a reason for preferring to use human infants—orphans perhaps—or retarded humans for experiments, rather than adults. So far as this argument is concerned nonhuman animals and infants and retarded humans are in the same category . . . If we make a distinction between animals and these humans, on what basis can we do it, other than a bare-faced—and morally indefensible—preference for members of our own species . . . The belief that human life, and only human life, is sacrosanct is a form of speciesism. . . . There will surely be some nonhuman animals whose lives, by any standards, are more valuable than the lives of some humans. A chimpanzee, dog or pig, for instance, will have a higher degree of self-awareness and a greater capacity for meaningful relations with others than a severely retarded infant or someone in an advanced state of senility." Under certain circumstances, Singer would even condone human torture, the end justifying the means: "Torturing a human being is almost always wrong, but it is not absolutely wrong."

If Singer is being serious, it seems that he would have our primate research laboratories populated instead by humans, those less fortunate than ourselves, such as orphans, epileptics perhaps, or cases of Down's syndrome or cerebral palsy, or deaf-mutes, and the aged and infirm. Is this the logic of human bioethics, the discipline that Singer professes? I have always believed in informed consent as a *sine qua non* for human experimentation; if an individual cannot be meaningfully informed about the nature of the proposed experiment, as is the case with a child, then he or she cannot be said to consent, and the experiment cannot be undertaken. Most people, whether Theists or Atheists, believe in a certain sanctity of human life. I wonder why Animal Liberationists of Singer's persuasion have not come into head-on conflict with the Right-to-Life movement?

Singer evidently has little interest in or love for animals, and he admits as much in his Preface. He goes on to complain that "to date environmentalists have been more concerned with wildlife and endangered species than with animals in general. . . . The newspapers do little better. Their coverage of nonhuman animals is dominated by human interest events like the birth of a baby gorilla at the zoo, or by threats to endangered species."

Singer then attacks the aims and objectives of biomedical research: "Experiments are performed on animals that inflict severe pain without the remotest prospect of significant benefits for humans or any other animals. . . . Experiments serving no direct and urgent purpose should stop immediately. . . . The general label medical research can be used to cover research which is not directed toward the reduction of suffering, but is motivated by a general goalless curiosity that may be acceptable as part of a basic search for knowledge when it involves no suffering, but should not be tolerated if it causes pain." He also talks of "the arrogance of the research worker who justifies everything on the grounds of increasing knowledge."

My response to this line of argument is that the lesson of history teaches us that we cannot know what we need to know until we know what there is to know. Serendipity is often our handmaiden in research, but fortunately, as Pasteur pointed out, chance favors the prepared mind. It is almost impossible to anticipate or predict the ultimate benefits of an experimental voyage of discovery into the unknown. In a free society, the pursuit of natural knowledge is one of the most precious of freedoms; witness the destructive effects of the Dark Ages on the course of European scientific understanding, when dogma ruled the day and new knowledge was often regarded as heresy, punishable by death. The Chinese

Cultural Revolution was a chilling recent reminder of that same antiscientific attitude. And as far as wildlife is concerned, fundamental research is absolutely essential, since we cannot hope to conserve until we comprehend.

In Australia, Animal Liberationists have recently made threats against my life, causing me to seek police protection, because I have defended fundamental wildlife research in public. The Royal Society for the Prevention of Cruelty to Animals has also stated that it is completely opposed to all forms of fundamental research on wildlife and has invited the general public to protest. In Singer's 1985 *New York Review of Books* article, he says, "I can imagine bizarre circumstances in which it would be right to kill and eat not only nonhuman animals but also human ones—for instance, if we were starving on a desert island and it was the only means of saving any of us." Such individuals and organizations surely deserve their North American epithet of "Humaniacs."

The concept that no animal experimentation is justified, that all humans and animals are equal, and that all humans should be vegetarians, I find ridiculous. My attitude to such Animal Liberationists is encapsulated in the resounding words of Oliver Cromwell, in his letter of 3 August 1650 to the General Assembly of the Church of Scotland: "I beseech you, in the bowels of Christ, think it possible you may be mistaken."

Nevertheless, there is justifiable public concern about the nature and intent of the animal experiments that are performed in laboratories, and since the public are also our paymasters, we must ultimately be publicly responsible for our actions. All is not well in our house. Three measures need to be taken to improve the situation. First, each country needs to draw up an agreed Code of Practice that sets down in some detail the terms and conditions under which animal experiments can be performed. Animal welfare organizations should have an input in the drafting of this code. Second, each organization or institution where experiments are performed needs to establish an Ethical Committee, composed of scientists and lay people, some of whom are not in the employ of the organization; this committee should review all proposed animal experimentation to see that it conforms with the guidelines laid down in the Code of Practice. And finally, to reassure the public that nefarious deeds are not going on behind closed laboratory doors, there is a need to establish a Federal Inspectorate of suitably qualified individuals who have freedom of entry at all times to all laboratories where animal experiments are undertaken. They should have the power to terminate experiments and to prosecute the experimenter and his or her institution for breaches of the Code of Practice. But equally, where experiments are being carried out in an acceptable manner, the government must protect the scientist from unwarranted public interference and harassment. At the end of the day, intellectual freedom is the hallmark of a civilized society, and in many branches of science, progress is simply not possible without some animal experimentation.

So let us return to the reason that has brought us here—primates, and their future survival. We need to adopt a "speciesist" approach to primates for two very good reasons: their endangered status in the wild, and their high level of intelligence. I would like to concentrate particularly on those species that are closest to humans and are among the most endangered, namely, the great apes of Africa, the mountain gorilla, the eastern and western forms of the lowland gorilla, the chimpanzee, and the pygmy chimpanzee. What can we do to ensure their continued survival?

In his Opening Address to this Conference, Jared Diamond touched on the important topic of cultural evolution, which I would like to develop further. Consider the fate of a human infant, taken from this Earth in the next space shuttle mission and deposited on some other planet, to be cared for by kindly and considerate Troglodytes. What could that child tell those Troglodytes about our culture? Nothing. All the elaborate panoply of civilization, our gestures, speech, writing, to say nothing of our more sophisticated means of communication, our vast store of information acquired over centuries, the whole of our social and material evolution, would be lost. Those Troglodytes would study the development of this nonsensical, babbling human child as it grew up and would be forced to conclude that it was a very primitive being.

Might we not be making the same mistake when we take a young great ape from the wild, rear it in a cage, and study its intellectual development? How can we know anything of the rich cultural life from which it came, which would all have been lost in the transition of one so young from the wild to captivity? If we have learned one thing from the studies of the Gardners, Patterson, and the Rumbaughs, it is that apes can communicate with us and with one another. They have a language. As Savage-Rumbaugh (1985) states: "We wish to assert here that apes are capable of intentionally telling one another things, and probably do so more in the wild than we yet realize."

In the light of these observations, surely our *most urgent* task must be to describe those wild societies in which our great apes live, so that we can really begin to appreciate the totality of their culture. Once we have destroyed their forest habitats and are forced to conserve them in captivity, what will we know of the richness and diversity of their lifestyles? Furthermore, if we are to attempt to reintroduce captive-bred apes to sanctuaries that we may be able to create for them in the wild, how well will they be able to survive if captive rearing has destroyed their culture?

So the underlying theme of my talk is "return to the wild." First and foremost, it is only by studying animals in their natural habitat in the wild that we can begin to understand why they are built in the way that they are and why they behave in the way that they do. As we continue to destroy the habitats of so many of our primates at an ever-increasing rate, the opportunities for this type of field research are fast disappearing. And if the ultimate objective of keeping any endangered species in captivity is "return to the wild," we need to know much more about the social life and culture of the species in question, so that the captive environment can at least begin to prepare the animal for its eventual reintroduction to the wild.

Where does this line of argument lead us when applied to the great apes of Africa?

Albert Schweitzer, who won the Nobel Peace Prize for his work in the village of Lambaréné, Gabon, West Africa, describes in his book *On the Edge of the Primeval Forest* what it was that motivated him to spend most of his life trying to bring primary health care to African villagers; he said that one of his chief desires was to have intercourse with these children of Nature. Although he made little attempt to adapt his lifestyle to Gabonese culture, preferring to try and impose his own Christian beliefs and Western ways upon the natives, as was the fashion of the times, Schweitzer nevertheless had a point, and perhaps we can learn from it. Should not our chief desire today be to have intercourse with these other children of Nature, the great apes of Africa, to find out how our closest relatives live, and move, and have their being, on the edge of the primeval forest?

There could be no more appropriate location for such studies than Gabon itself. It covers an area of 267,667 km^2, straddling the Equator, and yet it has a human population of only 0.6 million. As a result, 85% of the country is still covered by a tropical rainforest, much of it totally uninhabited by humans. Gabon was where the gorilla first came to the notice of Western man when a visiting American missionary, Savage, was given the skulls of a male and a female ape collected by local natives. When he subsequently showed them to Wyman, the Professor of Anatomy at Harvard, the latter immediately recognized them as belonging to a new species of great ape, which they named "gorilla" in an article published in the Boston Journal of Natural History in 1847 (Short, 1980).

Within the last 5 years, Dr. Caroline Tutin and Michel Fernandez have carried out a most thorough and painstaking survey of the gorilla and chimpanzee populations of Gabon. They have walked on line transects through all the habitat types, counting the number of gorilla and chimpanzee nests. Since each individual builds a new nest every night, and since it is possible to establish for how many weeks a nest will remain recognizable in the forest canopy, they could arrive at a reasonably objective estimate of the total population of each species. They have concluded that Gabon contains 35,000 ± 7000 gorillas and 64,000 ± 13,000 chimpanzees; these estimates of the gorilla population in particular are far higher than earlier figures, and this study has proved conclusively that gorillas are not just confined to the edge of the primeval forest, as we had previously imagined, but that they also inhabit its very depths (Tutin and Fernandez, 1984).

Tutin and Fernandez, with the help of the French Centre International de Recherches Médicales de Franceville (CIRMF), have been able to establish a new field station in the Lope-Okanda reserve, the Station d'Etudes des Gorilles et Chimpanzes (SEGC), B.P. 74, Booué, Gabon, where for the first time it is possible to observe the natural behavior of partially habituated troops of western lowland gorillas as they interact with chimpanzees. It would be impossible to overemphasize the potential significance of this discovery. Nobody previously has ever observed gorillas and chimpanzees in an area where the two species interact; what a priceless opportunity we now have to record the relative skills and attributes of our two closest living relatives as they compete with one another for space and food.

Hitherto, our knowledge of gorilla behavior in the wild has been almost totally dependent on the observations of Dian Fossey, Sandy Harcourt, Kelly Stewart, and others who have studied the small, isolated, relict population of mountain gorillas in Rwanda. We do not know how representative these animals are of the main population of western lowland gorillas in Gabon. We do not even know whether the mountain gorilla has been geographically isolated for so long that it has already become a separate species. A similar story could be told for the chimpanzee. Jane Goodall and all her students, and the Japanese working in the Mahale Mountains, have studied the behavior of one subspecies, *Pan troglodytes schweinfurthi*, at the extreme eastern end of its range in Tanzania, but we know little or nothing about the behavior of the type species, *Pan troglodytes troglodytes*, living in a tropical rainforest environment. There is even a suggestion that Gabon might contain another subspecies or species of chimpanzee, the koolookamba. Its vocalization appears to be quite distinct, and the one skull in existence differs in a number of important respects from that of the type species (Short, 1980). Since the tropical rainforests of West Africa are bisected by so many deep, wide and fast-flowing river systems, and since chimpanzees have an aversion to swimming, West Africa is ideally suited to speciation of chimpanzees as a result of geographical isolation, hence perhaps the origins of the pygmy chimpanzee.

I would like to dramatize the fate of the lowland gorilla by showing you the skull of an adult male that was found by Sandy Harcourt, Kelly Stewart, Richard Wrangham, and myself in December 1979 in the manager's office of a deserted iron ore prospecting camp formerly run by Somifer, a consortium including Bethlehem Steel Corporation, at Belinga, Gabon. You can easily recognize the bullet hole through the left ramus of the mandible (see Fig. 1.1). I showed this skull to Professor J. K. Mason, the Professor of Forensic Medicine in the University of Edinburgh, and I would like to quote from his report:

Skull of Gorilla
Submitted 4th February, 1982

The skull is that of an adult male gorilla showing a number of defects which are compatible with damage due to bullets.

In my opinion, the animal has been shot twice and possibly three times. The bullet tracks are as follows:

1. The bullet enters the ascending ramus of the left mandible, passes through the pterygoid plate destroying the ethmoid bone. The exit is through the floor of the right orbit passing out through the front of the eye. The track runs very slightly forward from left to right.

2. The entry is through the left cheek fracturing the coronoid process of the left mandible. The bullet track passes almost directly across the head, destroying the anterior part of the left frontal bone and penetrating the lower aspect of the medial wall of the right orbit. The exit is through the superior lateral wall of the right orbit where the fronto-malar suture is fractured.

3. The third track is less certain but I believe it enters the left eye and penetrates the medial wall of the orbit, enters the right orbit through the upper medial wall and exits through the anterior part of the frontal bone from which a fracture radiates to the parietal region. The track runs slightly backwards from left to right.

In addition, there is what I take to be a secondary track caused by a fragment of bullet or of bone which penetrates the sphenoid wing from left to right and ends within the right zygoma.

The arrangement of these injuries suggests to me that they were inflicted from relatively close range with the animal lying on its right side. I suggest they were delivered in the order detailed with the assailant possibly moving towards the head as he fired.

The calibre of the bullet is notoriously difficult to judge because the interposition of skin affects the size of the bony defect and we have here only one oblique entry wound to help. Taking everything into account I would suggest a .38.

In an effort to obtain more accurate information about the calibre of weapon used, the skull was subsequently submitted to ballistics experts from the local police force in Scotland, who concluded that either a .38 or a 9 mm hand gun had been used. It is highly unlikely that a Gabonese native would have had access to such a high-powered and sophisticated weapon. Who killed Cock Robin?

In law, there is a phrase "to purge your contempt." I would like to ask each of you to purge our past contempt of the gorilla and of the other great apes, by contributing to a "Great Apes of Africa Fund," which we will establish here and now

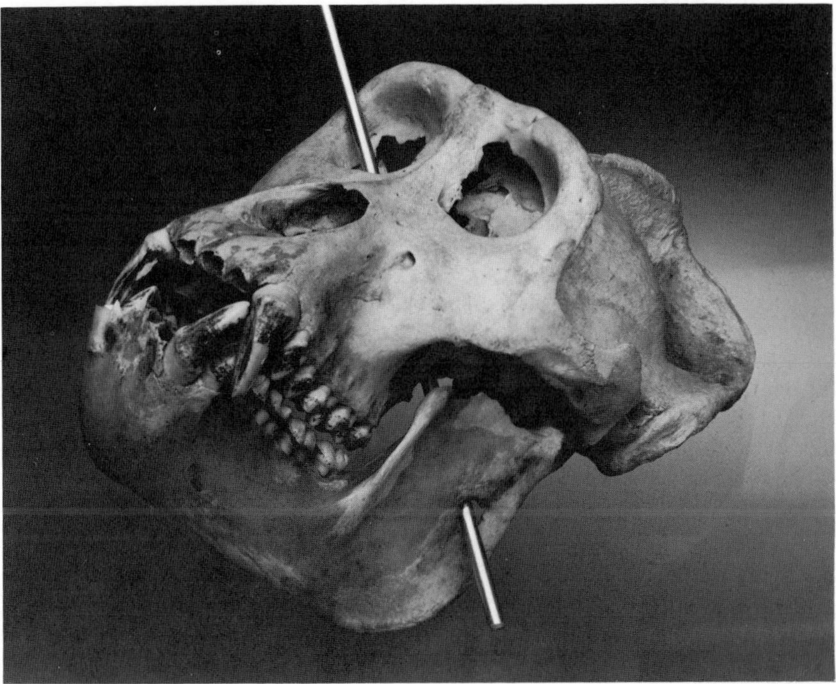

FIGURE 1.1. Skull of an adult male gorilla recently shot in Gabon with either a .38 or 9 mm hand-gun.

specifically for the advancement of knowledge about the great apes of Africa in their natural habitat in the wild.

Earlier in this Conference, Marvin Jones shamed us all when he asked those of us who had ever contributed financially to wildlife conservation to raise our hands. Pitifully few hands went up, and mine was not among them. If we, the research workers, who perceive the need for this type of research, cannot make even a token contribution, we should be ashamed of ourselves. This new fund is to be administered by Dr. Kurt Benirschke on behalf of the Trustees of the Zoological Society of San Diego, in recognition of the fact that the San Diego Zoo is unquestionably the queen of all zoos as far as conservation is concerned. I would like to open the fund by donating my cheque from the Morris Animal Foundation for my traveling expenses to bring me from Australia to San Diego to attend this meeting. It would be my hope that some of the money collected might go to support the work of Caroline Tutin and Michel Fernandez in Gabon. Let us use this as an example and a challenge to all zoos around the world, in the hope that they, too, will follow Alison Jolly's suggestion of "returning money to the wild" by adopting field research projects on wild animals in their countries of origin.

Is there anything else we could do to ensure a more secure future for the great apes of Africa, or any of the world's endangered primates? If we are realists, there is little hope that we are going to be able to stem the tide of increasing human population and hence an increasing pressure on habitats within the foreseeable future. Although habitat preservation is the ideal way of conserving endangered primates, it may be unattainable.

Are there any other areas of the world that are politically stable, with a low human population, and abundant areas of tropical rainforest, that might at least serve as temporary sanctuaries for some of our most highly endangered species? How about Australia? Could I suggest that the Primate Group of IUCN might at least consider this possibility, and initiate a botanical survey of Australian tropical rainforest, as well as looking at the palatability of the plants in question by feeding them to endangered primates already in captivity. Although the Australian ecosystem and its wildlife in particular have suffered heavily as a result of the introduction of carnivores such as the dog, cat, and fox, and herbivores such as the cow, sheep, pig, goat, horse, donkey, camel, water buffalo, and rabbit, the introduction of small numbers of a large, slowly reproducing, K-selected primate such as the gorilla, chimpanzee, or orang would be unlikely to pose a comparable threat, even to an admittedly fragile rain forest environment. Potential returns from tourism would probably finance such a scheme, and it is at least worth considering. Our chances of preserving the larger endangered primates in conventional zoos or primate research laboratories are certainly not high, and we need to consider all possible alternatives.

In conclusion, it is interesting to consider the ebb and flow of human attitudes to primates in the last hundred years or so. The early explorers were obsessed by shooting animals and sending their skins and skulls to fill our museums and private collections. Perhaps this served to salve the consciences of those who hunted gorillas for sport. This hunting took its toll; gradually it gave way to the capture of live animals to populate the growing number of zoos. The mortality rates at the hands of the professional trappers, during subsequent shipment, and in the zoological collections themselves were horrendous, but we knew no better. Breeding success in captivity came much later in the day, with the first chimpanzee birth in 1915, the first orang birth in 1928, and the first gorilla birth not until 1956. It is only within the last decade that primate reproduction has become a science, thanks to the activities of those in primate research laboratories and the few zoos with research facilities. There has also been an invaluable spin-off from human and animal in-vitro fertilization and embryo transfer research programs. So today, at a time when we see wild populations of all primates at their lowest ebb, there is some new hope. The increasing success of captive breeding programs should enable us to call a halt to further recruitment from the wild, and perhaps the ebb can be converted to a flood as we become increasingly able to return captive-bred animals to the wild—if there is any "wild" left.

In the meantime, let us always remember that we owe a enormous debt to the primates of this world; many of us owe our health, or even our very lives, to the fruits of primate research. The rhesus monkey provided the clue to our understanding of rhesus sensitization as a cause of human fetal and neonatal mortality, and hence the development of effective prophylactic measures. Rhesus monkey kidneys have made possible the large-scale commercial production of the Salk polio vaccine. Cotton-top tamarins are playing a key role in our understanding of the E-B virus, which is associated with glandular fever, Burkitt's lymphoma, or nasopharyngeal carcinoma, depending on where in the world you happen to live. A variety of old-world primates have been indispensible for the safety testing of a whole range of steroidal contraceptives, and perhaps we can gain some satisfaction from the fact that primates are continuing to play a vital role in the development of radically new approaches to human contraception such as the development of vaccines against fertilization and against pregnancy.

It was Albert Schweitzer who said "ethics are pity." Let us therefore demonstrate our compassion for primates and try by every means at our command to make "return to the wild" not just a dream, but a reality. I can think of no more fitting conclusion to this address on primate ethics than the words of Henry Beston (1928):

> We patronise them for their incompleteness, for their tragic fate of having taken form so far below ourselves, and therein we err, and greatly err. For the animal shall not be measured by man. In a world older and more complete than ours they move finished and complete, gifted with extensions of the senses we have lost or never attained, living by voices we shall never hear. They are not brethren, they are not underlings, they are other Nations, caught with ourselves in the net of life and time, fellow prisoners of the splendour and travail of the earth.

References

Beston H (1928) The outermost house. Penguin, London.

Jennings, HS (1904) Behavior of the lower organism. Indiana University Press, Bloomington.

Jerison HJ (1985) Animal intelligence as encephalization. Phil Trans Roy Soc London B 308:21–35.

Pilbeam D (1984) The descent of hominoids and hominids. Sci Amer 150:84–96.

Savage TS, Wyman J (1847) Notice of the external characters and habits of *Troglodytes gorilla*, a new species of Orang from the Gaboon River. Boston J Nat Hist 5:28–43.

Savage-Rumbaugh ES, Sevick RA, Rumbaugh DM, Rubert E (1985) The capacity of animals to acquire language: do species differences have anything to say to us? Phil Trans Roy Soc London B 308:177–185.

Schweitzer A (1948) On the edge of the primeval forest. A & C Black, London.

Short RV (1980) The great apes of Africa. J Reprod Fert Suppl 28:3–11.

Singer P (1976) Animal liberation. Jonathan Cape, London.

Singer P (1985) Ten years of animal liberation. NY Rev Books Jan 17th: 46–52.

Tutin CEG, Fernandez M (1984) Nationwide census of gorilla (*Gorilla g. gorilla*) and chimpanzee (*Pan t. troglodytes*) populations in Gabon. Amer J Primat 6:313–336.

2
The Role of Captive Populations in Global Conservation

DAVID WESTERN

Introduction

Zoos, like national parks, are torn between two conflicting goals: public service and biological conservation (Forster, 1973; Western and Henry, 1979). Both institutions arose primarily for public entertainment and recreation and survive largely because of it. Because these motivations help justify conservation (Miller, 1982), we cannot dismiss them. But, on the other hand, we cannot leave conservation to the whim of human entertainment.

If exhibition is the primary zoo purpose, one set of goals for self-sustaining populations will apply; if conservation is the primary purpose, then another set will be more appropriate. The first is dictated more by lure and profitability, the second by ecological circumstances and criteria.

There can be, and increasingly is, a link between the two, bridged by zoo education programs. However, for the purpose of identifying what role captive populations can play in global conservation, I will consider zoos from a conservation and Third World perspective, rather than from the viewpoint of public entertainment.

Therefore, we should ask if there is a clear conservation role for captive populations, and if so, how can zoos, especially in the wealthier Development World, add most to the conservation of nature. We must be quite clear about that conservation role, for it will determine the specific captive management guidelines of zoos, much as national conservation policies dictate a country's species and ecosystem management guidelines, whether within national parks or elsewhere.

Compared with the growing sophistication of captive-management techniques (Benirschke et al., 1980; see also articles in this volume), zoo conservation policies are rudimentary (Foose, 1983; Conway et al., 1984) and, because of conflicting goals and biological uncertainties (see below), frequently contradictory. Perhaps that is inevitable, but more attention to the global role of captive populations, which means a closer link between zoo and national and international conservation policies, would help clarify zoo policies. A recent example of how zoo conservation policies can benefit comes from the close working relationship between AAZPA's SSP and IUCN's SSC rhino programs (Foose, 1983). Here, the global status of rhinos, assessed by IUCN's conservation advisory body, suggested what must be done to conserve the five species and additional important subspecies. The SSP program subsequently established captive management priorities that will complement weaknesses in field conservation programs.

Captive populations, as in the case of endangered rhinos, clearly have an important role in global conservation. But not all captive populations are necessarily good for the ultimate survival of a species in the wild. Nor is it clear that self-sustaining captive populations are preferable to those dependent on replenishment from wild stocks. It all depends on the purpose of captive stocks, and whether they help or hinder the ultimate survival of free-living populations. So, what is the purpose of captive populations?

Captive self-sustaining populations are good, it is widely assumed, because they help maintain species diversity (Foose, 1983), the touchstone of modern biological conservation. Yet neither the singular goal of maximizing biological diversity, or its consequences (Western, in press), nor the establishment of self-sustaining populations as a means of achieving that goal are beyond question. And, even if they were, the imminent prospect of an extinction spasm (Myers, 1984a) and the limited zoo slots for captive animals (Conway, 1983, in press) presuppose maintaining captive species.

The term *self-sustaining population* needs clarification before a conservation role can be discussed. It has come to mean self-sustaining captive populations. But captive populations are not necessarily self-sustaining; they are maintained by humans. A genuine self-sustaining population propagates and supports itself, independent of humanity. Simply put, a self-sustaining population is a wild, natural population. We must not confuse the two, lest the captive, artificial population become a substitute for the wild, natural population.

I will distinguish between the two by referring to self-sustaining captive populations (SCPs) and self-sustaining natural population (SNPs). A gradient between the two, from zoo to wild populations, does exist. In primates, the gap is bridged by large breeding colonies such as those at Yerkes Research Primate Center, release populations such as the Cayo Santiago macaques, and human-induced isolates, such as the Virunga mountain gorillas. A bimodal distribution (a few SCPs and numerous SNPs with little in between) typifies most species populations at present. However, with range fragmentation casting more and more species into the lap of human dependency, SCPs and the mid-curve will grow at the expense of large SNPs.

We can draw two important conclusions. First, if we cannot afford to save all refugee species, then we must let some go to the wall in the interests of others. How do we decide which (Conway, in press)? Second, saving a species genome does not necessarily add to its natural evolutionary prospects. With habitat obliteration, what ecological prospects will there be for the thousands of future dodos? And if there is to be no natural place for a species, and ultimately perhaps millions of species, should we not concentrate on those that can survive independently of humanity? These larger questions, though still intractable, have a direct bearing on the role of captive propagation.

So what role can and should SCPs play in global conservation? That begs another more basic question: Why conserve at all? If we are confused over the second, we cannot be clear about the first. And, in reality, utility, whether consumptive or intangible, is a far more cogent reason to conserve than any biological argument (Western and Henry, 1979). Utility provides the means of achieving conservation even where the primary motive is biological. Urging tropical forest conservation as a pharmaceutical investment (Myers, 1984b) for example, is a utilitarian, not biological argument.

Although we must be conscious of human utility, we must be quite clear about biological priorities. Only then can we see the choices available and know what concessions to accept when necessary.

There is a largely unrecognized difference in the conservation goals of most Third World nations and modern conservation biology, a difference that complicates the identification of an SCP role. Third World preserves were primarily established to protect faunal assemblages and their habitats (an ecosystem approach, in modern parlance), not single species, whereas the preoccupation of modern conservation biology is with biological diversity, the preservation of species, and minimum viable population sizes (MVPs).

Given the limited land available for conservation in the Third World, where most biological diversity resides, those goals are not synonymous and can lead to quite different design criteria for protected areas (Western, in press). Large reserves are essential for ecological integrity, many small ones for maximizing biological diversity (Western and Ssemakula, 1981). Therefore, the SCP goals will differ depending on whether one supports ecological integrity or biotic diversity as the primary conservation goal.

If, as I believe, SCPs should help conserve nature by complementing SNPs and shadowing their genetic trajectory, then that can only be accomplished by continued genetic flow between SCPs and SNPs. This is the antithesis of present trends to establish self-sustaining captive populations. In a broader conservation view, saving a species means more than saving its genome; it means maintaining the ecosystem in which a species maintains itself, competes, adapts, or is otherwise exterminated (Frankel, 1983). That involves maintaining a species evolutionary potential on the one hand and accepting its natural ecological fate on the other.

SCP Contributions to Conservation

How do SCPs help conservation? In a variety of ways, from ultimately safeguarding a species from extinction to gathering the know-how for species and ecosystem preservation (Foose, 1983).

The role that SCPs play can be given in order of urgency and overall SNP contribution as:

1. *Last resort conservation*. Several species such as the Arabian oryx, European bison, and Przewalski's horse have survived only because they are, or were, maintained entirely in captive programs.
2. *Back-stop conservation*. SCPs act as fail-safe populations for several critically endangered species, such as the Siberian tiger and golden lion tamarin.
3. *Complementary conservation*. SCPs add to or complement the genome of several species by preserving endangered races, such as the northern white rhino.
4. *Representative conservation*. Includes a cross section of species, selected for a variety of reasons (taxonomic, ecological, etc.), which do not necessarily reflect their status in nature, e.g., baboons, vervet monkeys.

5. *Investment conservation*. We know too little to maintain SCPs or to correct imbalances caused by humans to wild populations. We can learn the necessary techniques, without interfering with wild populations, by studying factors, such as behavior, nutrition, physiology, and genetics, which affect their survival and propagation. Such research, which helps us anticipate and avert conservation crises, also offers hope that genetic banking will ultimately provide a fail-safe, in the event that a species becomes extinct.
6. *Conservation awareness and support*. Zoos have become increasingly active in parlaying public awareness raised by live exhibits and education programs into active conservation. The New York and Frankfurt Zoological Societies play a major role in international conservation.

SCP Threats to Conservation

SCPs threaten conservation in two ways, nonbiologically and biologically.

Nonbiological Threats

1. Arguing that captivity is the only hope for a species can rob SNPs of support. Attention, enthusiasm, money and know-how tend to track critically endangered species in a continual series of fire-fights. Conservation financing thrives on crisis, and crisis generates interest, including tourism and national concern. Ethiopia, in the case of mountain nyala, and Zaire, in the case of the northern white rhino, are both against SCPs, fearing they will divert SNP interest and investment.
2. Arguing that biological diversity is the primary conservation goal can weaken the justification for ecological integrity. Collecting for the sake of maintaining a little of everything, irrespective of its potential in nature, is rather like collecting antiques. It may save a natural treasure, but only because it has curiosity value, not because it has a valid ecological role.

Biological Threats

1. An SCP genome can never precisely match its parent genome or mimic the selective forces acting on it. Depending on the size of an SCP relative to the SNP and the amount and direction of genetic flow, SCPs can be complementary, irrelevant, or disruptive. An SCP without genetic flow from the SNP will diverge continuously from it (Franklin, 1980). With interchange, the impact differs. An SCP can shadow an SNP if it is small relative to it and receives sufficient genetic inflow. The reverse ratio and flow, a large SCP genome flowing into a small SNP genome, can genetically swamp the natural evolutionary processes. The same ratio with opposite flow could, on the other hand, rob the SNP genome and prejudice its independence.
2. Similar arguments can be applied to nongenetic factors, including learned behavior such as migration, feeding, reproduction, and nurturing. Loss of learned behavior is likely to be far more critical in the short term than genome loss in rupturing social and ecological systems (see the chapter by Kleiman

et al. in this volume). Such losses could destroy the prospects for reestablishing an SNP.

Criteria for SCP Management

Limited zoo space and the prospect of an extinction surge presupposes SCP selection. But how do we select?

Here we need to forge links between species (and ecosystem) conservation, ecological theory, and captive management constraints. Do we select species on the basis of endangerment, taxonomy, ecology (rarity, endemicity, ecological role, etc.), or practicability (cost and feasibility)?

I suggest that ecological criteria for captive candidates include returnability, ecological importance, representation, and taxonomy, in a way that approximates a species' role in maintaining ecosystem integrity and complexity. This approach promotes integrity at the expense of diversity, in step with the objectives of most national parks (Western, in press).

Within groups we also need criteria for selecting species, which could well run along an axis from endangerment to conservation investment, that is, the degree of urgency involved.

MVP principles are fast becoming the guideline for SCP size (see Schoenwald-Cox, 1983). If, however, SCPs are to aid nature conservation, they must be the handmaiden of SNPs and should in no way prejudice them. That requires SCPs to be an integral part of SNPs, not stand-alone populations. In turn, for SCPs to track SNPs requires genetic flow, the amount and direction depending on the status of a species in the wild. Ideally, an SCP should always be small relative to an SNP, should always receive just enough genetic inflow to track it, and should not prejudice it through reverse flow. However, the more endangered the SNP becomes, the greater the need for genetic flow and perhaps interchange (Templeton and Read, 1983). Criteria along these lines can be developed and elaborated to ensure maximum contribution of SCPs to SNP conservation.

Finally, SCPs should be reintroduced to the wild as soon as possible after a species has become extinct, or severely depleted in the wild, with a view to reestablishing SNPs as the dominant genome. That, surely, is the proper role of SCPs in global conservation?

Discussion

We should not delude ourselves that SCPs can or should present a stand-alone conservation strategy. At best, they offer a small and selectively applied supplement to conservation. The economics and technicalities of propagating more than a small proportion of refugee species is simply too formidable (Foose, 1983).

Conway (in press) calculates that the 635 IZY–AAZPA listed captive collections could, at existing carrying capacity, support no more than 330 mammal species with enough individuals (250–300) to maintain 90% of heterozygosity for 200 years. At a maintenance cost of around $1,000 per capita, zoos would need to spend \$50 million annually to support SCPs for all 200 primate species. That would leave few slots and little money for thousands of other mammals, let alone the millions of other invertebrates.

In contrast, the annual cost of maintaining huge populations of thousands of species of animals and plants in an area the size of Serengeti National Park (15,000 km^2) costs less than \$0.5 million (IUCN, 1984). That amounts to less than 1 cent per individual large mammal, four orders of magnitude less than zoo costs. We could potentially support 200 Serengeti-sized ecosystems and millions of species of all taxa, for the cost of maintaining 200 primate SCPs.

This argument is admittedly simplified, but it does underscore the need for humility when considering the role of SCPs, and the enormous potential of conserving at source, in the Third World in particular, where biological diversity could become a huge asset of truly international value. SCPs should in every respect enhance that potential, not diminish it.

We should, therefore, ensure that SCPs complement our primary goal of conserving natural areas and species. The two conservation strategies cannot be independent; nor should they be competitive. SNPs should provide the cues for fragmentary population management, whether in the wild or in captivity, and whether for biological or genetic guidelines. A great deal more genetic analysis of wild populations is required before we can assess the characteristics of natural genomes (O'Brien et al., 1985) and how to mimic them in captive programs.

Aligning SCP goals with primary conservation and defining selection criteria will help avert the first-come-first-served approach to endangered species, an approach more humanitarian (and profitable, since as in the collection of antiques, it is based on uniqueness) than ecologically sound. Zoo slots filled by the earliest refugees favor naturally rare species, endemics, and specialists. We run the risk of overloading zoos with the ecologically sensitive species to the detriment of the more important late-comers.

We presently view SCPs as a way of severing the dependence on SNPs, both to minimize the economic costs to zoos and the ecological costs to wild populations. However, although an independent captive gene pool is unlikely to jeopardize a natural population in the short-term, it will suffer progressive genetic drift (Franklin, 1980). The divergence rate will vary, depending on the initial captive genome, and the subsequent selection in both SCPs and SNPs.

I suggest that SCPs must mimic SNPs as closely as possible and reconstitute or complement them when necessary. This requires a set of guidelines that ally SCP policies to those of SNP conservation. In addition to those already identified by SSP (Conway et al., 1984), these would encompass the following:

1. *Conservation policy.* SCPs should ideally mimic and bolster SNPs and the ability of natural ecosystems to maintain themselves.
2. *Ecological criteria.* In conjunction with IUCN's SSC, draw up criteria and identify species that, if saved, could ultimately ensure minimum viable functioning for the major biomes and ecosystems.
3. *Genetic criteria.* An SCP genetic policy, based on an analysis of wild genomes, which would ensure that, whenever possible, SCPs were managed as ancillary genomes to SNPs.
4. *Behavioral criteria.* Guidelines for the maintenance of learned behaviors

essential for the reintroduction of animals to the wild (see the chapter by Kleiman et al. in this volume).

That captive populations have a major conservation role to play is beyond dispute. What is troublesome is that, despite the rapid advances in the conservation role played by zoos (Conway, 1980, 1983), many curators regard captive management as an end in itself and not as a way of keeping open our options for the conservation of nature (Conway, 1980). The stamp-collecting approach has little to contribute to nature conservation and runs the risk of being viewed by the Third World in much the same way as the western museums that plundered and still hoard their antiquities.

Fortunately, zoos are playing an ever-increasing role in conservation. Their most difficult task arises in striking a balance between what is in the best interests of conserving natural systems and what is economically and technically feasible. Only a close link between zoos, conservation biologists, and field conservationists can ensure a reasonable compromise.

Acknowledgements. I gratefully acknowledge the sponsorship of the New York Zoological Society and Shirley Strum's helpful comments.

References

Benirschke K, Lasley B, Ryder O (1980) The technology of captive propagation. In: Soulé ME, Wilcox BA (eds) Conservation biology. Sinauer, Sunderland, MA.

Conway W (1983) Captive birds and conservation. In: Perspectives in ornithology. Cambridge University Press, Cambridge.

Conway W (In press) The practical difficulties and financial implications of endangered species breeding programs.

Conway W, Foose TJ, Wagner RO (1984) Species survival plan of the American Association of Zoological Parks and Aquariums. Wheeling, West Virginia.

Frankel OH (1983) The place of management in conservation. In: Schonewald-Cox CM, Chambers SM, MacBryde B, Thomas L (eds) Genetics and conservation. Benjamin/Cumming, London.

Franklin IR (1980) Evolutionary change in small populations. In: Soulé ME, Wilcox BA (eds) Conservation biology. Sinauer, Sunderland, MA.

Foose TJ (1983) The relevance of captive populations to the conservation of biotic diversity. In: Schonewald-Cox CM, Chambers SM, MacBryde B, Thomas L (eds) Genetics and conservation. Benjamin/Cumming, London.

Forster RR (1973) Planning for man and nature in national parks. IUCN Publications New Series No. 26 Morges.

IUCN (1984) IUCN directory of afrotropical protected areas. International Union for the Conservation of Nature, Gland, Switzerland.

Miller KR (1982) Parks and protected areas: considerations for the future. Ambio 11(5):315–317.

Myers N (1984a) Genetic resources in jeopardy. Ambio 13(3):171–174.

Myers N (1984b) The primary source: tropical forests and our future. Norton, New York.

O'Brien SJ, Roelke ME, Marker L, Newman A, Winkler CA, Meltzer D, Colly L, Evermann JF, Bush M, Wildt DE (1985) Genetic basis for species vulnerability in the cheetahs. Science 227(4693):1428–1434.

Schonewald-Cox CM (1983) Preface. In: Schonewald-Cox CM, Chambers SM, MacBryde B, Thomas L (eds) Genetics and conservation. Benjamin/Cumming, London.

Templeton A, Read B (1983) The elimination of inbreeding depression in a captive herd of Speke's gazelle. In: Schonewald-Cox CM, Chambers SM, MacBryde B, Thomas L (eds) Genetics and conservation. Benjamin/Cumming, London.

Western D (In press) Primate conservation in the broader realm. In: Else J, Lee P (eds) Proceedings of the Tenth International Congress of Primatology, Primate Ecology and Conservation. Cambridge University Press.

Western D, Henry W (1979) Economics and conservation in Third World national parks. Bioscience 29(7):414–418.

Western D, Ssemakula J (1981) The future of savannah ecosystems: ecological islands or faunal enclaves? Afr J Ecol 19:7–19.

3
African Primate Conservation: General Needs and Specific Priorities

JOHN F. OATES

Introduction

Any plan for the conservation of the world's nonhuman primates must give major attention to Africa. Most primates live in the tropics, and Africa has by far the largest amount of land within the tropical zone of any continent (about 24 million km^2, compared to 14 million km^2 for South America and 9 million km^2 for Asia; UNESCO, 1978). About one-third of all living primate species occur on the African continent; depending on the classification used, some 50–60 African species are usually recognized today out of a world total of 150–180; these species are regarded as belonging to some 15–18 genera and 5 distinct subfamilies (Wolfheim, 1983; Oates, in press). About 80% of the species are monkeys, members of the family Cercopithecidae. Among the remaining lorisids and apes are our closest living relatives, the chimpanzees and gorilla, and much evidence points to Africa as the evolutionary home of human primates. This gives the African primate fauna a special significance to us.

Threats to African Primates

Although a huge area of Africa lies within the tropics, much of this area has a low or highly seasonal rainfall and supports desert, scrub, and savanna vegetation in which only a few primate species live. However, these species (such as the baboons, vervets, patas monkeys, and Senegal bushbabies) have very wide distributions and are relatively well protected in numerous national parks and other reserved areas. Although comprehensive surveys across time are lacking, available evidence suggests that these primates are not, in general, seriously threatened at present.

Most of Africa's primates (about 80% of species) live in forest habitats, particularly moist lowland rain forest. Such forest does not cover a particularly large area. It has been estimated that in 1980 there were 1.5–2 million km^2 of moist lowland forest in Africa, compared to 3 million km^2 in Asia and 5 million km^2 in South and Central America (Myers, 1980; National Research Council, 1980; UNESCO, 1984). Most forest primate species only occupy a small geographical part of the forest zone, and some species occupy very small areas indeed. For instance, the island of Fernando Po (Bioko) in the Gulf of Guinea has seven endemic forms of primate, but the whole island is only 2000 km² in size.

Not only is the area inhabited by most of Africa's primates relatively small, but much of it is also under great threat from expanding agriculture and commercial lumbering, as well as through conversion to plantations of one or a few tree species (which are often exotics). It has been estimated, for instance, that of 47,900 km² of forest in southern Ivory Coast in 1966, only 33,000 km² remained in 1974, a loss of 31% in only 8 years (Myers, 1980). Losses on this scale are likely to continue in many areas, or even to intensify. On one hand, the annual rate of human population growth in Africa for 1980–1990 has been projected as in excess of 3%, greater than in tropical America or Asia (Barr, 1981). On the other hand, environmental factors, inefficient techniques, and a lack of capital resources are causing rates of increase in food production to lag behind those in other tropical areas. In 1970–1980 the estimated annual increase in food output in Africa was 1.6%, or −1.1% per capita, and of 40 sub-Saharan countries in Africa, 14 may not have enough arable land even to support their present populations by subsistence agriculture on a sustainable basis (World Bank, 1984). Not surprisingly, one of these 14 is Ethiopia, and others include Burundi, Kenya, Nigeria, Rwanda, and Uganda. In such countries, where there is obviously great agricultural pressure on forests, are found some of the most threatened of Africa's primates. Even when pressures on forest land are not yet so great, such as Gabon, the Congo Republic, and parts of Zaire, monkeys and apes are often hunted for food (e.g., Harcourt and Stewart, 1980; Zeeve, 1985).

Consequences of Threats

As a result of these habitat pressures and hunting, and of long-term environmental changes that have confined some unique primates to very isolated areas, 13 species or very distinct local populations of African primate could be extinct or on the verge of extinction by the end of the century if significant action is not taken to reverse threats currently acting on them. These forms are listed in Table 3.1. They are mostly limited to small areas of forest where habitat destruction and/or hunting are serious problems and are estimated to have total populations numbering less than 25,000. In some cases, a recent census indicates that only a few hundred individuals remain (e.g., Tana River red colobus and mountain gorilla), while the *bouvieri* and *pennanti* forms of red colobus may already be extinct. It should be stressed, however, that robust evidence is lacking on the status of most of these populations.*

Another 10 forms will probably be in danger in 25 years time if current trends continue. These are listed in Table 3.2. Two forest *Cercopithecus* monkeys are so poorly known that they have not been included in either table. These are *Cercopithecus* "salongo," known only from a single skin collected in central Zaire (Thys van den Audenaerde, 1977), and a population of monkeys resembling *Cercopithecus preussi* recently discovered in Gabon by Harrison (Harrison, 1984). More information is needed on both these monkeys before their status can be even tentatively estimated.

The longer term future is very hard to predict, but an informed guess is that in 50 years time, in the absence of major conservation action, about half the African

TABLE 3.1. African primates in danger of extinction.

Species/subspecies	Common name	Location	Estimated size of remaining population (with source of datum)	
Macaca sylvanus	Barbary macaque	North Africa	23,000	(Fa, 1983)
Cercocebus galeritus galeritus	Tana River mangabey	Tana R., Kenya	800–1100	(Marsh, 1985)
Cercocebus galeritus subsp.	Sanje mangabey	Uzungwa Mts., Tanzania	1800–3000	(Homewood and Rodgers, 1981)
Mandrillus leucophaeus	Drill	Cameroon, Fernando Po	?	
Cercopithecus preussi	Preuss's guenon	Cameroon, Fernando Po	?	
Cercopithecus erythrogaster	White-throated guenon	Southwest Nigeria	?	
pennanti form of red colobus	Pennant's red colobus	Fernando Po	?	
preussi form of red colobus	Preuss's red colobus	Cameroon	8000	(Conservation Monitoring Centre, Cambridge, UK)
bouvieri form of red colobus	Bouvier's red colobus	Congo Republic	?	
rufomitratus form of red colobus	Tana River red colobus	Tana R., Kenya	200–300	(Marsh, 1985)
gordonorum form of red colobus	Uhehe red colobus	Uzungwa Mts., Tanzania	10,000	(Rodgers and Homewood, 1982)
kirkii form of red colobus	Zanzibar red colobus	Zanzibar	1500	(Silkiluwasha, 1981)
Gorilla gorilla beringei	Mountain gorilla	Virunga Volcanoes, and Bwindi Forest	360–370	(Wilson, 1984)

primate fauna could be either extinct or verging on extinction. This is not, of course, a unique problem. Primates and forests are under threat all around the world, and vast numbers of plant and animal species are faced with imminent extinction (National Research Council, 1980). But the loss of Africa's primates would be particularly tragic because they represent a significant fraction of our own mammalian order and live in the continent in which our own roots seem to lie.

General Conservation Needs

Can we, as scientists with a special interest in primates, do anything effective to mitigate these threats to the survival of Africa's primates and their habitats? The answer is surely yes, but a qualified yes. We must proceed thoughtfully, we must work in close collaboration with African countries, and we must constantly keep in mind the need to demonstrate to the generally poor residents of rural Africa

TABLE 3.2. Highly vulnerable African primates.

Species/subspecies	Common name	Location
Theropithecus gelada	Gelada baboon	Ethiopian Highlands
Cercopithecus diana	Diana monkey	Coastal West Africa
Cercopithecus hamlyni	Hamlyn's owl-faced monkey	Eastern Zaire
Cercopithecus erythrotis	Red-eared guenon	Fernando Po, Nigeria, Cameroon
tephrosceles form of red colobus	Western Rift red colobus	Uganda, Tanzania, ?Burundi
Colobus satanas	Black colobus	Fernando Po and Western Equatorial Africa
Gorilla gorilla gorilla	Western lowland gorilla	Western Equatorial Africa
Gorilla gorilla graueri	Eastern lowland gorilla	Eastern Zaire
Pan paniscus	Pygmy chimpanzee or bonobo	Congo Basin of Zaire
Pan troglodytes verus	Western chimpanzee	Coastal West Africa

that conservation can provide them with tangible benefits. We are unlikely to persuade these people to forego apparent opportunities to increase their prosperity (or just to maintain current subsistence levels) by simply arguing that it is morally wrong to destroy forests and to let primates go extinct.

One obvious and low-cost action that those of us who have studied African primates in the wild and who are familiar with the African environment can take is to pool our knowledge to identify as objectively as possible the most urgent conservation priorities. In particular, we need to determine the location of populations and habitats on which international attention and funding should be focused in the near future if we are to maintain the current diversity of African primates.

After carefully identifying where conservation efforts are most urgently needed, we can try to assist in putting these efforts into effect. Two kinds of action are particularly needed in Africa: more education in the meaning, value, and practice of primate (and other resource) conservation; and the establishment of more effective wildlife reserves in forest areas. There are still very few national parks or other strict wildlife sanctuaries in the forested regions of Africa, and those that do exist often do so only on paper.

National parks (which cannot be regarded as totally secure anywhere in the world) are particularly insecure in the African rainforest zone. Several of the few rainforest national parks in Africa have been demarcated on maps in areas where there is little apparent pressure at present on the environment from the human population. Very often these parks have no proper staff, have not been demarcated on the ground, and are not patrolled. When pressures grow, such parks can be readily abandoned to major encroachment and hunting.

The many forest reserves and *forêts classée* in Africa do not usually provide good protection for either their plants or their animals. Rather, they are areas reserved for government-controlled exploitation of timber, with only weak controls on hunting.

Specific Action Required

The action needed to identify African primate conservation priorities is already being undertaken by the African section of the IUCN/SSC Primate Specialist

Group. An Action Plan for African Primate Conservation for the 1986–1990 period is being compiled (Oates, in press). It proceeds as follows:

1. Species-conservation priorities are identified by rating each African primate species on the basis of (1) the degree of threat that it faces (judged on a 1–6 scale from "not especially threatened" to "highly endangered"), (2) its taxonomic uniqueness (rated on a 1–3 scale from "member of a large group of closely related species" to "monotypic genus"), and (3) its association with other threatened primates (rated 1 or 2). These ratings are then summed to give a total rating of between 3 and 11.
2. Distinct regional communities of African primates are identified (each of which contains a different set of species and subspecies). One of these is the savanna zone community, which is not yet under special threat. Eleven other distinct communities can be recognized (several of which can be subdivided further). Their geographical location is shown in Figure 3.1. Four occupy large blocks of lowland rainforest in Upper Guinea, West Equatorial Africa, the Congo Basin, and Eastern Zaire. Two are areas adjacent to these lowland

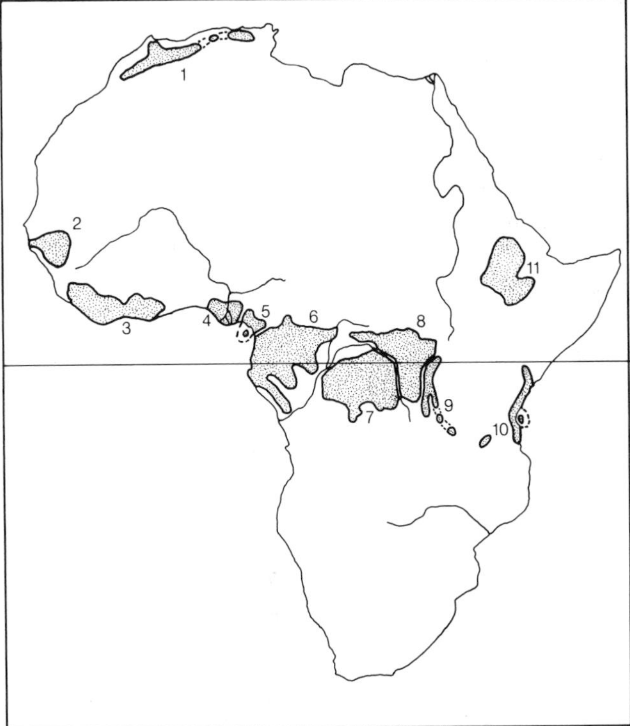

FIGURE 3.1. Regional communities of African primates. Each community has at least one endemic species or several endemic subspecies. 1, Maghreb (North Africa); 2, Casamance; 3, Upper Guinea; 4, Southern Nigeria; 5, Cameroon; 6, West Equatorial Africa; 7, Congo Basin; 8, Eastern Zaire; 9, Western Rift; 10, Coastal East Africa; 11, Ethiopian Highlands.

blocks where there are major areas of medium and high-altitude forest; these are the Mt. Cameroon area including the island of Fernando Po, and the forests on high ground along the Western Rift Valley in East Africa. The remaining five areas with special primate communities are the woodlands of the Atlas Mountains in North Africa (home of the barbary macaque), the woodlands of the Casamance area of West Africa (home of the Guinea baboon), the remnant forests of southern Nigeria (home of the white-throated guenon), the Ethiopian Highlands (home of the gelada baboon), and forests in the coastal regions of East Africa (home of several endangered populations of red colobus monkeys and crested mangabeys).

If each of these distinctive communities of primates could be adequately protected, the diversity of the African primate fauna would be maintained.

3. The Action Plan, while recognizing the need for conservation education programs, national conservation strategies, and better controls on hunting, focuses on the establishment of reserve areas. It identifies where surveys are needed to assess the status of wild primates and locate suitable reserves, and it identifies where already located reserve sites require further development. The priority of these various projects is rated on the basis of (1) the number of species in the project area that have high individual ratings for conservation action (rated on a 1–4 scale), (2) the imminence of threat to the climax ecosystem characteristic of the area (also rated on a 1–4 scale), (3) the overall primate species diversity in the area (on a 1–3 scale), and (4) the number of endemic primate forms present (1–3 scale). Ratings on each parameter are again summed, so that the highest possible project rating is 14 and the lowest 4.

Table 3.3 displays some of the results of this rating effort, showing which projects the IUCN/SSC Primate Specialist Group has identified as having the highest priority for action in 1986–1990. While putting the finishing touches to this plan, we are also bringing it to the attention of major conservation funding agencies, and IUCN itself is using the plan in its advisory dealings with African governments.

Some Practical Considerations

We certainly need to develop better reserves for primates and the ecosystems of which they are a part, and this need is particularly great in forest areas. But these reserves will not work unless they are integrated into more general resource development and management programs. People living in the areas where conservation efforts are located have to be convinced that forest conservation can bring them tangible benefits, such as more reliable water supplies and lower soil erosion. Although the complete protection of very large forest areas in representative biogeographic regions may be our ideal, we have to recognize that such reserves will often not be possible in practice. In these cases, we must argue for the benefits of careful, sustained-yield management, and for the preservation within such a managed area of sanctuaries that *are* strictly protected against any exploitation, both to act as references against which the effects of different forms of land use can be measured and to act as reservoirs of animal and plant species that may recolonize exploited land (see Harris, 1984).

TABLE 3.3. Highest priority projects for African primate conservation, 1986–1990.

Project area	Total rating
Survey projects	
Between Mamfe, Cameroon, and Obudu, Nigeria	13
Bioko (Fernando Po), Equatorial Guinea	13
East and Central Ivory Coast	12
Oban Forests, Eastern Nigeria	12
Republic of Congo	12
Lofa-Mano area, Liberia	11
Wamba area, Zaire	11
Ituri Forest, Zaire	10
Maiko National Park, Zaire	10
Ondo and Ogun States, Nigeria	10
Niger-Cross Interfluvium, Nigeria	10
Reserve development projects	
Tai National Park, Ivory Coast	11
Dja National Park, Cameroon	11
Lomako Forest, Zaire	11
Bwindi Forest Reserve, Uganda	11
Tana River National Reserve, Kenya	11
Sapo National Park, Liberia	10
Gola Forest, Sierra Leone	10
West Ghana Parks	10
Korup National Park, Cameroon	10
Lopé Reserve, Gabon	10
Northeast Gabon	10
Southern Central African Republic	10
Okomu Forest Reserve, Nigeria	10
Kibale Forest Reserve, Uganda	10

[1] Ratings are provisional. See text.
Source: Oates (in press).

It is, of course, valuable to promote the idea that wildlife is a part of a country's heritage, as worthy of protection as human artifacts, and to demonstrate the recreational value of viewing wildlife and visiting wilderness. However, a more utilitarian appeal may be more likely to pay off in the short term. For instance, a survey carried out in 1982 of 2600 students in Tanzania, a country with a fine record of wildlife conservation, found that none of the students expressed a love of wild places or a personal vested interest as reasons for maintaining national parks. Rather, they saw the parks as existing primarily to attract tourists. The author of this report, Hilary Pennington (1984), argues that environmental education must be incorporated into the standard school biology curriculum. This would seem to be especially important in countries where the environment is not particularly appealing to foreign tourists.

Captive Breeding

In no part of Africa are current threats to primate populations from habitat destruction and hunting so severe and so hard to mitigate that we should abandon efforts to conserve wild populations. Indeed, the protection of functioning ecosystems should be our main priority, since once habitats are lost, captive populations cannot be reintroduced to the wild and lose much of their significance as anything but museum pieces. However, where captive individuals of threatened African primates already exist in captivity, every effort should be made to establish a self-sustaining breeding population, through nutritional and reproductive research, and a program of genetic exchange based on a central studbook (one is being established at the Hannover Zoo, for instance, for the drill). African zoo collections should not be ignored, and efforts should be made to find the resources to allow African zoos to collaborate more effectively in international breeding programs.

The Value of Long-Term Field Research Projects

I would like to end this paper by stressing the valuable role that long-term field research projects can play in conservation efforts in Africa and other parts of the tropics, especially in the creation of protected forest ecosystems. Much of the most effective primate conservation action that has occurred in Africa in the last 20 years has come about through the existence of behavioral or ecological research projects based in threatened ecosystems. Examples are the Kibale Forest Project in Uganda, the Mountain Gorilla Project in Rwanda, and the chimpanzee studies in the Mahale Mountains of Tanzania.

Why is this so? First, primatologists obviously want their study sites and study animals to remain as undisturbed as possible. Second, their training often gives them a special awareness of local conservation problems and of the need for action to mitigate threats to the environment (of course, the field projects themselves often gather data essential for the implementation of effective conservation). Third, the primatologists (generally foreigners) usually have financial and material resources that are in short supply in rural Africa. Fourth, foreign scientists involved in these projects are relatively apolitical in an African context and do not have to pay quite the same attention to local social and political pressures as do most nationals in the project countries (in fact, governments will often listen carefully to foreign scientists' recommendations, regarding them as relatively unbiased). Fifth, long-term foreign-sponsored projects often provide more continuity of effort than is possible at present in efforts sponsored by local governments, which are highly susceptible to radical upheavals and economic crises.

There is an obvious danger here of a form of neocolonialism, of foreigners attempting to control African events without properly understanding the workings of African societies. To avoid this, field projects must work in close collaboration with local organizations and help to train local personnel. The more they become integrated with the workings of the country in which they are based, the more likely they are to generate programs that will be locally maintained for a long time.

In several places, primate field research projects are adapting and developing their base camps and study sites into field stations affiliated with local organizations (particularly universities). These stations can then be used in the long term by both nationals and foreigners for research and training. This is a welcome trend, and one that funding agencies should be strongly encouraged to support. However, these stations must as far as possible be constructed and run in such a maintenance of sophisticated equipment. Several research stations in savanna Africa have failed in the past or not fulfilled their early promise, partly because they were too costly and difficult to operate at their planned level with local resources.

In summary, primatologists can do much to assist the survival of the current diversity of the world's primates. Field workers in particular can help to draw up objective strategies for action, and they can help to establish low-cost research stations in areas identified as having high conservation priority.

References

Barr TN (1981) The world food situation and global grain prospects. Science 214:1087–1095.

Fa JE (1983) The Barbary macaque—the future. Oryx 17:62–67.

Harcourt HA, Stewart KJ (1980) Gorilla-eaters of Gabon. Oryx 15:248–251.

Harris LD (1984) The fragmented forest. University of Chicago Press, Chicago.

Harrison MJH (1984) Poster presentation, Xth Congress, International Primatological Society, Nairobi.

Homewood KM, Rodgers WA (1981) A previously undescribed mangabey from southern Tanzania. Int J Primatol 2:47–55.

Marsh CW (1985) A resurvey of Tana River primates. Report to Institute of Primate Research and Department of Wildlife Conservation and Management, Kenya.

Myers N (1980) Conversion of tropical moist forests. National Academy of Sciences, Washington, DC.

National Research Council, Committee on Research Priorities in Tropical Biology (1980) Research priorities in tropical biology. National Academy of Sciences, Washington, DC.

Oates JF (in press) Action plan for African primate conservation. IUCN/SSC Primate Specialist Group.

Pennington H (1984) Conservation awareness in Tanzania. Oryx 18:125–127.

Rodgers WA, Homewood KM (1982) Biological values and conservation prospects for the forests and primate populations of the Uzungwa Mountains, Tanzania. Biol Conserv 24:285–304.

Silkiluwasha F (1981) The distribution and conservation status of the Zanzibar red colobus. Afr J Ecol 19:187–194.

Thys van den Audenaerde, DFE (1977) Description of a monkey-skin from east-central Zaire as a probable new monkey species (Mammalia, Cercopithecidae). Rev Zool Afr 91:1000–1010.

UNESCO (1978) Tropical forest ecosystems: a state-of-knowledge report. UNESCO, Paris.

Wilson R (1984) The mountain gorilla project: progress report no. 6. Oryx 18:223–229.

Wolfheim JH (1983) Primates of the world: distribution, abundance and conservation. University of Washington Press, Seattle.

World Bank (1984) World development report 1984. Oxford University Press, New York, for the World Bank, Washington, DC.

Zeeve SR (1985) Swamp monkeys of the Lomako Forest, Central Zaire. Prim Conserv 5:32–33.

---

*_Note added in proof_: During January–March 1986 Dr. Thomas Butynski (stimulated in part by a draft of the African Primate Conservation Action Plan) surveyed the primates of Bioko Island (Fernando Po). He reports that, due to a recent decline in hunting and a regrowth of the forest, population of all species are at present in a relatively healthy state (T.M. Butynski, personal communication).

4
Gorilla Conservation: Anatomy of a Campaign

ALEXANDER H. HARCOURT

Introduction

Different species, habitats, and countries have different requirements and priorities for conservation. The conservation program for the gorilla population in Rwanda has been described as a model project, hence its inclusion in this section of the symposium. It is certainly an unusual project in several respects, including the structure of its funding, its broad approach to conservation of the region, the close monitoring of its results, its long duration, and perhaps also in the extent of its success. However, before it is used as a blueprint for other projects, the context of the program and the nature of its success need to be clearly understood. I therefore first sketch the biological, socioeconomic, and administrative setting of the program and outline the reasoning behind the form that it took; then I describe its results; and finally discuss some implications of the program for conservation. Particular emphasis is placed on the necessity for research aimed at the management of conservation areas as integral parts of a wider ecosystem that includes humankind, instead of as sacrosanct islands in a sea of hostile human influence.

Background

Gorilla Conservation in Africa

In terms of the conservation of the gorilla (*Gorilla gorilla*) in Africa as a whole, the Rwandan gorilla conservation program is a small contribution. It concerns one part of one population of one subspecies in one country. In fact, the gorilla occurs as three subspecies in many populations in nine countries of Africa (Fig. 4.1). The bulk of the population is in West Africa, where the western lowland gorilla (*Gorilla g. gorilla*) numbers about 40,000 animals, 35,000 of them in Gabon (Tutin and Fernandez, 1984). The population size of the eastern lowland gorilla (*Gorilla gorilla graueri*) of Zaire is not well known but is probably not less than 5000, judging from average population densities (Harcourt et al., 1981; Tutin and Fernandez, 1984) and the area of available habitat. Known populations of the mountain gorilla (*Gorilla g. beringei*) of Zaire, Uganda, and Rwanda

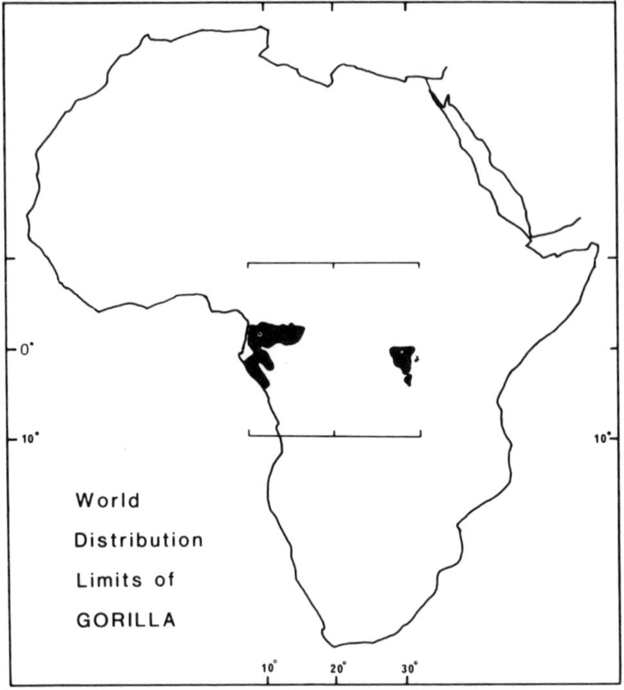

FIGURE 4.1. Distribution of *Gorilla gorilla* in Africa.

number less than 400, 260 of them in the 375 km² Virunga conservation area (Harcourt et al., 1983) and 115 in the 310 km² Bwindi Forest Reserve of Uganda (Harcourt, 1981). Most of the known mountain gorilla population lies in officially protected areas, but less than 20% of the eastern lowland and only about 6% of the western lowland populations are in parks or reserves. The Rwandan gorilla conservation program concerns only the population of the Rwandan sector, the Parc National des Volcans, of the Virunga conservation area (Fig. 4.2), i.e., less than 0.5% of the total gorilla population in Africa.

Rwandan Socioeconomics

Rwanda is one of the poorest, most densely populated countries in Africa (Prioul and Sirven, 1981). Both imports and foreign aid exceed exports. The density of Rwanda's population, 95% of which is rural, is around 250/km² but exceeds 400/km² around the Parc National des Volcans. Although over the country as a whole it is increasing at over 2.5% per year, in the region of the Park, it dropped between 1978 and 1981, because of emigration. The fact that no other area of the country is agriculturally richer and yet people are leaving the Park region indicates severe overcrowding around the Virunga conservation area. Food production has kept pace with the country's population increase, largely because of

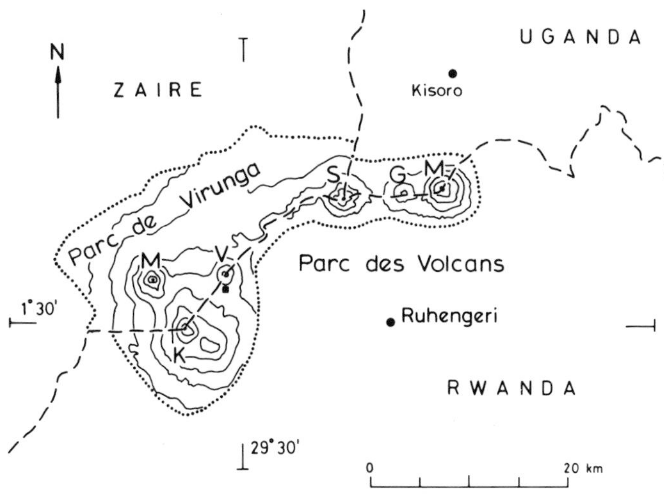

FIGURE 4.2. The Virunga conservation area. Single letters refer to names of the six main peaks. Contour lines at 400 m intervals, peaking at 4500 m at K.

immigration to previously uncultivated land. However, by 1978, 70% of theoretically arable land was taken, and Prefol and Delepierre (1973) have calculated that all potential arable land will be under cultivation by 1995.

The Parc National des Volcans (PNV) and Its Gorillas

Despite such pressures, Rwanda has a higher proportion of its area as national park—10%— than any other country in Africa, indeed than most countries of the world (IUCN, 1980). A further 5% is forest reserve. The 120 km² PNV is one-twentieth of the country's national park area and less than 0.5% of Rwanda's total land area. However, with a mean yearly rainfall of about 1800 mm, the PNV is 10% of the country's water catchment area (Spinage, 1972). Its wide altitudinal range, from 2500 m to 4500 m, incorporates a variety of montane vegetation types, including Afro-alpine zones (Lind and Morrison, 1974; Schaller, 1963), but no endemic species as far as is known.

The PNV has been under threat for some time. In 1968, a European-funded pyrethrum scheme resulted in removal of 40% of what was then a 220-km² national park. Up to the mid-seventies, herds of hundreds of cattle grazed in the Park, and in 1978 the Ministry of Agriculture planned to acquire another 40% of the Park for cattle ranching. There was also evidence of small-scale appropriation for cultivation. In addition, hunting occurred throughout the Virunga conservation area. Although gorillas were rarely sought directly, they were caught and killed in snares. The drop in the Virunga gorilla population from about 450 in the early sixties to 275 in the early seventies reflects these chronic threats (Harcourt and Fossey, 1981; Weber and Vedder, 1983).

The Rwandan Gorilla Conservation Program

Origins

It was a perceived crisis—the commencement of hunting of gorillas for trophies—and the resultant publicity that precipitated the "Mountain Gorilla Project" (MGP) (Harcourt and Curry-Lindahl, 1979). the Project was initiated in Rwanda, rather than Zaire or Uganda, more for reasons of administrative and logistical efficiency than biological necessity. The MGP was an internationally financed conservation program whose aims were to provide the Rwandan Office of Tourism and National Parks with the funds, equipment, and expertise necessary for long-term protection of its gorilla population. Gorilla conservation did not of course start in Rwanda in 1979 with the MGP. The Project augmented the existing work of the Office of National Parks and complemented the role of the Belgian government. However, the MGP is treated separately here because it was an integrated project that aimed to alter fundamentally the conduct of conservation in the PNV, indeed in the country as a whole.

Requirements

Research

Effective conservation management depends on detailed knowledge of the species and ecosystem concerned. Information about the PNV is supplied to the Office of National Parks largely through the work of an independently funded research station, the Karisoke Research Centre, established since 1967. Studies at the Centre have concentrated on the gorilla, and thus there was available at the start of the program extensive information on the demography, ecology, reproduction, and social behavior of this species (see references in Stewart and Harcourt, in press). Information on other aspects of the Virunga ecosystem is largely lacking, however.

The MGP identified and initially responded to three main requirements for the PNV in 1978: park security, conservation awareness, and tourism.

Park Security

Gorilla conservation in Rwanda was improved in 1976 by removal of all livestock from the PNV and in 1978 by tripling of the Guard force to 30 men. Nevertheless, in 1979 the Guards were still unequipped and ill-trained for foot patrols in the cold, wet montane forest.

Conservation Awareness

Any program whose reasons are at least understood, or better still supported, is going to be more successful than one where they are not. A general ignorance existed of the conservation issues affecting the PNV in particular and Rwanda in general. For example, one survey showed that 25% of farmers living on the edge of the PNV could not describe a gorilla (Weber, 1981, in press).

Tourism

An organization that is financially self-sufficient is more likely to survive than one that is not. Tourism (game viewing) was the only obvious means by which the Park could earn money, and yet in 1978 tourism was at such a low level and so negligently organized that revenue did not cover Park costs. However, financial profitability was never and still is not the goal of the PNV. Rather, the aim is to protect for the forseeable future a valued wilderness area. The tourism project was implemented to help toward this goal, not to replace it.

Implementation

Funding

External funding for development of the three described programs of the MGP after 1978 came mainly from a consortium of four international conservation agencies (see Acknowledgments). The cooperative working of the consortium is of interest in that for publicity purposes, international conservation organizations usually prefer to be sole funding agencies. The consortium functioned well in its role perhaps largely because within the overall Project, each organization was responsible for financing separate programs (cf. the multiorganization funding of the lion tamarin project; Kleiman et al., this volume).

Park Security

Guard numbers were increased by 50% ; the Guard Force was properly equipped; training and selection procedures were improved and the system of patrolling made more efficient and effective.

Conservation Awareness

The program was conducted at all levels of the population and through all forms of media (cf. the lion tamarin project; Kleiman et al., this volume). Posters, calendars, newspaper articles, and radio broadcasts publicized the endangered status of gorillas; booklets and pamphlets explained to visitors the biology of the Park and the work of conserving it; Wildlife Clubs were established and educational visits to the Park were organized for teachers and pupils; seminars on conservation were held at the National University; a vehicle specially equipped for audiovisual presentations visited villages and schools throughout the country and showed films that covered general conservation issues, e.g., deforestation, as well as the topic of gorillas and Rwandan Parks; and perhaps most encouragingly of all, a conservation program was established with the help of the Belgian government and the U.S. Peace Corps in the secondary school curriculum under the auspices of the Ministry of Education. Thus, conservation became a subject for organizations other than conservation agencies, and the Park itself and its flora and fauna became an educational resource for the country.

Tourism Development

The tourism program encouraged visits to the Park for its scenery, but concentrated on viewing of gorillas. There are no roads in the PNV, and the vegetation is very dense. Therefore, the program necessitated that tourists arrive on foot within a few meters of wild gorillas, which were to remain totally unrestrained. Guides had to be trained for the specialized work involved and gorilla groups habituated to being visited by tourists. An important aspect of the project is the tight control exerted over visitor numbers. To avoid disturbance of the gorillas and the vegetation, to ensure ease of control of tourists, and to enable the most rewarding experience possible, visits are limited to one per day per gorilla group by parties of not more than six people. Reception, booking, and information facilities were also improved.

Research

In recognition of the value of research, management of the tourism program was designed to minimize disturbance of the Research Centre's study area and study groups. In effect, a zonation system of use was established, with tourism separated from the research area.

Costs

The costs of these three programs to the consortium plus the Belgian government was around U.S.$50,000 per year. The Office of National Parks paid a further $100,000. Each gorilla thus costs approximately $1,250 per year. To maintain them in a zoo would cost at least $1,500 annually (cf. Cummins et al., 1978), without the advantage of all the other plants and animals, as well as a watershed, protected at the same time.

Results of the Mountain Gorilla Project

Protection of the PNV and Its Gorillas

The impact of the MGP on the PNV can be judged by demographic comparison of the Rwandan and Zaire sectors of the Virunga gorilla population before and after the start of improved conservation measures. Zaire, where no improvement of protection occurred, effectively acts as the control area to the "experimental" manipulation of the Rwandan population. Censuses conducted in the early seventies and then in 1981, 5 years after livestock were removed from the PNV, provide the data base (Aveling and Harcourt, 1984; Harcourt et al., 1983; Weber and Vedder, 1983).

The results demonstrate that from 1976 and continuing thereafter, conservation efforts in Rwanda have been associated with a significant increase in recruitment to the gorilla population (Table 4.1). Changes in neither total population size nor in numbers of adults were significant. However, numbers of immatures per adult female changed markedly (Figs. 4.3 and 4.4). The results clearly demonstrate that the entire difference between Rwanda and Zaire in immature recruitment was due to animals born after 1976, i.e., juveniles and infants in the

TABLE 4.1. Size and composition of Zaire (ZR) and Rwanda (RW) populations in early seventies (E70) and in 1981.

	Total	FAM[1]	AF	YAM	SA	J	I	AD	IMM
ZR E70	146.5	28	46.5	17.5	25	12	17.5	92	54.5
ZR 1981	129.5	26	41.5	19.5	18.5	13	11	87	42.5
RW E70	111.5	30	35.5	5.5	13	15	12.5	71	40.5
RW 1981	107.5	14	36.5	9.5	10.5	20	17	60	47.5

[1]Abbreviations: FAM, fully adult male; AF, adult female; YAM, young adult male (9–12 yr); SA, subadult (6–9 yr); J, juvenile (3–6 yr); I, infant (1–3 yr). AD, adult; IMM, immature (< 9 yr). Numbers are fractions because groups sharing the two regions (< 10% of total) were split between them.

1981 census, not those born before it (subadults) (Figs. 4.3 and 4.4). Numbers of subadults per adult female dropped between the censuses in both countries. However, whereas numbers of juveniles and infants per adult female also decreased in Zaire, they rose in Rwanda. Thus, in neither census did the number of subadults per adult female per group differ significantly between the countries; nor in the early seventies did number of juveniles and infants; yet in 1981, Rwanda contained twice as many juveniles and infants per adult female per group (median of 1.0) as did Zaire (median of 0.5) (U=30, n=16,14, P=0.01, Mann-Whitney U Test, one-tailed). The difference between the countries was due to both juvenile and infant numbers (Fig. 4.4), demonstrating that the improved recruitment from 1976 (detectable as juveniles in 1981) was continued in the subsequent years (infant numbers in 1981).

The changes could have been caused in several ways. The removal of livestock in 1976 must have been associated with a reduction in direct disturbance and improvement of habitat quality. More frequent and efficient patrols were associated with a lower extent of poaching in Rwanda than in Zaire (Fig. 4.5). And the daily monitoring of the groups visited by tourists increased the chance

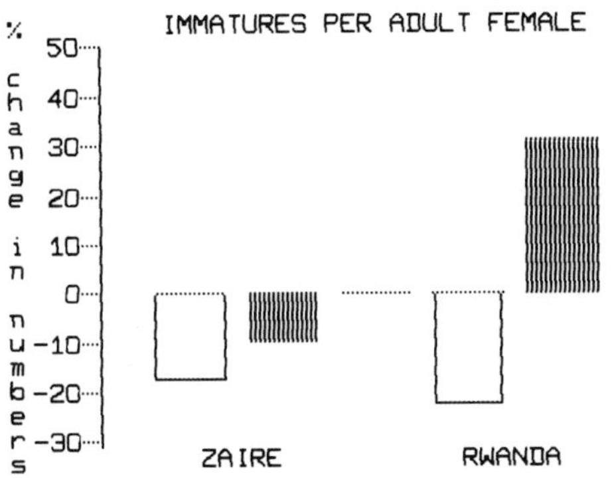

FIGURE 4.3. Percent change in numbers of immatures per adult female in Rwanda and Zaire between the early seventies (before the start of improvement of protection in Rwanda) and 1981 (after it). Open bars, subadults (6–9 yr); shaded bar, juveniles and infants (1–6 yr).

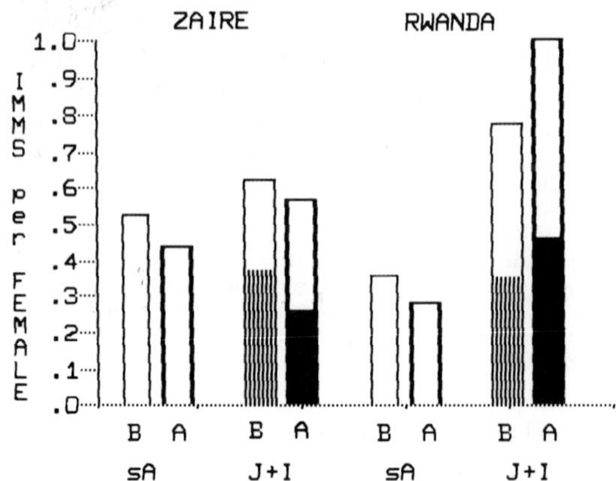

FIGURE 4.4. Numbers of immatures per adult female in Zaire and Rwanda in the early seventies and in 1981, i.e., before (B) and after (A) the start of improvement of protection in Rwanda. sA, subadults; J+I, juveniles and infants; shaded sections of histograms, infants; open sections, juveniles.

of gorillas being freed from snares: Before 1979 no gorilla had ever been released from a snare; since then, 15 have been.

Attitudes to the Park

On a rather subjective level, an evolution of pride in the gorillas and the PNV has been detectable in, for example, a greater readiness to prosecute serious infringements of the country's wildlife laws. More quantifiable, thanks to the interview

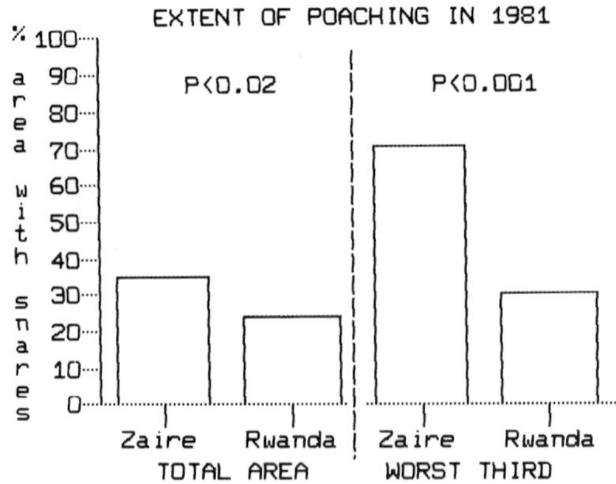

FIGURE 4.5. Percent of 1 km² squares in which snares were found in Zaire and Rwanda during 1981 census. "Worst kind" is most heavily hunted region of each country. χ^2 test corrected for continuity.

surveys directed by Weber (1981, in press), is the beneficial change in attitude to the Park and rise in knowledge of it among farmers in the region of the PNV (Table 4.2). Both the conservation awareness program and the tourism program seem to have been instrumental in influencing attitudes to the PNV. The impact of the latter is seen in the details of the answers to questions 3 and 4 of Table 4.2, which showed that by 1984, tourism was perceived as an important utility of the PNV. It is only the conservation awareness program, however, that could have produced the change evident in response to question 2.

Obviously, results from interviews have to be interpreted with caution, yet some confidence appears to be warranted. The interviews were conducted by sociology students who were Rwandan nationals, not by the expatriates known to be associated with the project; regions less intensively covered by the project produced less favorable responses; and knowledge of the Park, which compared to opinion probably cannot be biased by interviewers, improved at the same time as did opinion of it.

Tourism

The data in this section on tourist numbers and revenue come from Monfort (1980, 1985) or by calculations from data therein.

Development of tourism has been the most overtly successful part of the MGP. A total of 1352 people paid to visit the PNV in 1978, the year before the tourism program started, but 5790 in 1984, a 430% rise, even though only 70% of available bookings were taken in 1984. The increase in revenue has been far greater,

TABLE 4.2. Farmers' attitudes to conservation.

Questions	1980 $n = 72$ (%)	1984 $n = 120$ (%)
1. Convert PNV to agriculture?[1]		
Yes	51	18
	(G = 20.95; df = 1; $P < 0.001$)	
2. Forest is beneficial to water supply?		
Yes	49	86
	(G = 28.65; df = 1; $P < 0.001$)	
3. Utility of forest?		
None	17	12
Other	49	81
Not know	38	8
	(G = 28.61; df = 2; $P < 0.001$)	
4. Utility of wildlife?		
None	14	14
Other	41	63
Not know	44	18
	(G = 14.36; df = 2; $P < 0.001$)	

[1]Questions in full are as follows: 1. Should the Park be converted to agriculture? Question 2. Does the forest have a beneficial impact on water supply to your fields? Question 3. What utility does the forest have, given that wood cannot be cut or collected? Question 4. What utility does the wildlife have, given that hunting is illegal? Statistics are G tests (Sokal and Rolf, 1969).
Source: Data from Weber (1981, in press).

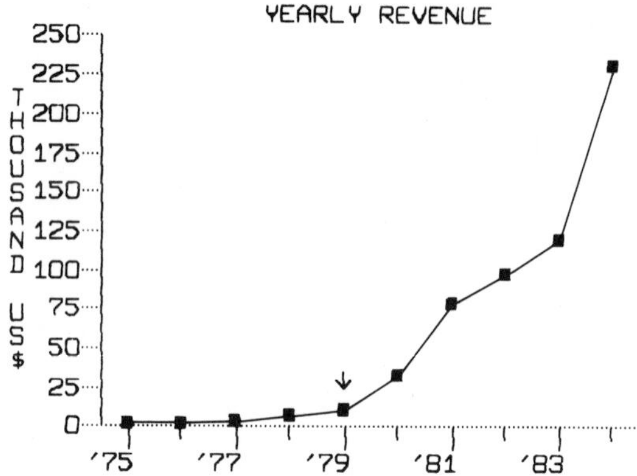

FIGURE 4.6. Yearly revenue of the Parc National des Volcans in thousands of U.S. dollars = year of start of the Mountain Gorilla Project.

3200% over the same period (Fig. 4.6), because improved facilities, especially for gorilla viewing, allowed higher Park entry fees. In 1978, an attempted visit to gorillas cost $5; by 1983 a guaranteed viewing cost $45, no more than a meal with drinks at a tourist restaurant. Tourists from abroad benefit the country's economy more than do residents because the former use more of the country's facilities and pay in foreign currency: For every dollar spent in the PNV, foreign tourists probably spend at least ten elsewhere in the country (calculated from data in Monfort, 1980). In 1984, 79% of PNV visitors were from abroad, compared to 44% in 1978.

A common complaint about international tourism to national parks in nonindustrialized countries is that entry fees reflect the income of the tourist, not the citizen, who consequently becomes excluded from the parks of his own country (Myers, 1975). Gorilla viewing and entry fees to the PNV for Rwandans are one-tenth those for expatriates, and Rwandan students can enter the Park for free. Nevertheless, Rwandan nationals accounted for only 2.3% of visits in 1984.

The gorillas are apparently not adversely affected by tourism since recruitment rates, for example, are as high in the tourist groups as the research groups (Aveling and Harcourt, 1984). Numbers of hikers are not limited, however, and damage of paths and campsites is abundantly evident. Nevertheless, the destruction is small compared to that caused by soldiers on exercise in the Park, and in fact it affects a very small proportion of the Park. The present 70 km of trails with a maximum average swathe of damage of 10 m covers less than 1% of the Park's area.

Discussion

The Mountain Gorilla Project

With the limited funding available for conservation work, the justification for a large and long-term investment, such as the MGP, needs to be examined. In fact, the PNV and the gorillas match many of the requirements on most classification lists of conservation priorities. For example, the PNV as an Afromontane forest and a former Pleistocene refugia is biologically interesting (e.g., Hamilton, 1981; White, 1981); it is a sample of a highly threatened ecosystem (Myers, 1980); and its conservation helps the welfare of the people that live around it because of its role as a water catchment area (Spinage, 1972). The gorilla is one of only four species in its taxonomic family; it is aesthetically attractive (a significant practical consideration); it has high economic potential from tourism; and its protection necessitates conservation of a rare and threatened ecosystem. Finally, the very success of the Project is perhaps justification in itself. It resulted in increased recruitment to the gorilla population, a drop in hunting, an improvement in conservation awareness in the country, and the conversion of a financial loss to a profit. Insofar as these were its general goals, the Rwandan gorilla conservation program has been successful: At a total cost of approximately $0.75 million, 120 km² of Afromontane forest and 120 gorillas are safer now than they were 10 years ago.

The Economics of Tourism

Two issues are considered here: Is the tourist industry itself profitable, and is tourism a more profitable form of land use than any other? It has been suggested that some national parks in Africa would probably not exist were it not for the perceived financial gains from tourism (Myers, 1972). Nevertheless, it is by no means the case that tourism is profitable, even for the conservation authorities, let alone the country (Mitchell, 1970; Pullan, 1984). Whether the country benefits or not probably depends on the stage of development of the tourist industry and the proportion of its requirements that can be produced within the country. Kenya, which is a relatively industrialized African country and which has a developed tourist industry, was in the early seventies keeping within the country 75% of tourist revenue and gaining three times as much from wildlife as it spent on it (Myers, 1975). Zambia, by contrast, was making a considerable loss in the same period (Pullan, 1984).

Rwanda by the late seventies was just starting to develop a tourist industry and in 1979 spent roughly $2.35 million to gain $1.85 million (Monfort, 1980). However, 68% of expenditure was capital investment. If this were reduced to mimic what might be expected of a developed industry, the present increase in tourist numbers to Rwanda indicate that a profit could be made. Tourism is developing faster than any other industry in Rwanda and in 1984 reaped $8.5 million to become the fourth major earner of foreign exchange (Monfort, 1985).

If conservation areas are to use the financial returns from tourism as part of an argument for their existence, it would help if they could demonstrate that tourism was more profitable than other forms of land use. For example, in Kenya the calculations showing that game viewing was worth $40/ha compared to $0.8 for the most profitable agricultural form of land use, pastoralism, were an important argument in the gazetting of the Amboseli National Park (Western and Henry, 1979). Similar calculations for the PNV show that when used for game viewing, it was worth in 1984 about $200/ha, if expenditure by tourists outside the Park is incorporated in the calculations. Cattle ranching by contrast would earn only about $15/ha (data from Weber, 1981). The Ministry of Agriculture's cattle scheme that was threatening the Park in 1979 could thus be conclusively rejected by the Office of National Parks as a viable concern. However, the PNV region is enormously more fertile than the Amboseli region of southern Kenya, and as arable land, the PNV is worth upwards of $300/ha (data from Monfort, 1980). Obviously, the economics are immensely more complicated than the simple sums presented here. Nevertheless, they indicate that in some circumstances, game viewing could be a feasible financial alternative to other forms of commercial use of the land.

Conservation Research: The Future

The Virunga gorilla population numbers less than 300 individuals in an elongated reserve of varied habitat totally surrounded by cultivation. The ecosystem as a whole has suffered from loss of almost entire vegetation zones and is 200 km from the next nearest region of similar altitude. Research relevant to its conservation would therefore include the genetics of small populations (Berry, 1983; Franklin, 1980; Soulé, 1980); integration of ecology with island biogeographic theory (Diamond, 1975; Simberloff and Abele, 1982; Western and Ssemekula, 1981); and of course the general ecology of the conservation area.

This list of research topics exemplifies most conservation-oriented research at present: it concentrates almost entirely on processes occurring within the conservation area. However, threats to wildlife arise largely from outside the natural habitat, not within it, and exist independently of its biological nature. Biological research within the conservation area could therefore approach a mere intellectual exercise unless the threats are fully understood and means to alleviate them are investigated and implemented. The Rwandan program described here suggests two other, usually neglected areas of research in the field of conservation. The first concerns means of mitigating the threats, and the second, of identifying them.

Multiple Use of Conservation Areas

National parks are often seen as places where animals are protected at the expense of humans, in some cases with justification (Brotoisworo, 1978). However, the increasing human population, especially in Africa where food supply has not kept pace with needs (Lele, 1981), is surely going to force the authorities into allowing exploitation of conservation areas if they are to survive (Myers, 1972): Wildlife will have to coexist with people. The drawback is lack of data on the effects that coexistence has on the wildlife.

In the gorilla program in Rwanda, for example, it was assumed that tourism was the only possible option for managed exploitation. This was partly an aesthetic judgment but also partly a technical one, despite the almost total lack of information on the effects of other forms of controlled use. The same is largely true of most national parks, except perhaps for game cropping as a means of utilization (Eltringham, 1984). Nevertheless, the level of illegal harvest from most conservation areas and yet the survival of wildlife there suggests the possibility of a greater extent of multiple use than is formally practiced at present. Occasionally, multiple utilization has been incorporated in management plans, even if it does have problems. Examples in Africa are the pastoralism that occurs in Amboseli National Park in Kenya (Western, 1982) and the Ngorongoro Conservation Area of Tanzania (Mascarenhas, 1983); the fishing industry in the Ruwenzori National Park of Uganda (Eltringham, 1984) and Akagera National Park of Rwanda; and in India, the pastoralism in the Gir forest (Berwick, 1976).

These programs are good illustrations of the sort of management of conservation areas required for the future, and of the difficulties that will be encountered. To implement them successfully, further research is needed on, for example, the size and siting of intact refuge areas for recolonization of buffer zones or zones of shifting human use; the causes of differences between used and unused areas (Johns, 1985; Skorupa, this volume); the effects of different types of usage on the ecosystem in relation to intensity, timing, and spacing of exploitation (Adams, 1975); and, very importantly, on the economic and political forces that influence conservation practice (Western, 1982).

Tourism is an example of one form of use that needs far more study. It should be a truism that financially beneficial as tourism might be, it can inflict substantial environmental costs on the conservation area (Budowski, 1976). Numerous studies in temperate regions have shown the detrimental effects of visitors on wilderness areas (Anderson and Keith, 1980; Liddle, 1975). Yet few parks anywhere limit visitor numbers, and hardly any outside of Europe and the United States conduct research on the effects of visitors on the conservation area (Western and Henry, 1979; Ralph and Maxwell, 1984), let alone on the impact of other forms of usage.

Interactions of the Conservation Area with Its Environs

Conservation biologists in large part have yet to realize that wilderness areas are part of a two-way system. The mass of data that they collect on the workings of one-half of the system, the conservation area, are quite out of proportion to the lack of knowledge both of the other part, the conservation area's environs, and of the interaction between the two. To put the dichotomy at its extreme, and keeping only to biological analysis, surveys of human demography and habitat use could be better bases for siting of reserves than studies of the biology of the endangered species itself (Eltringham, 1984).

The previous section considered some of the research required on the conservation area and the influence of the environs on it. The conservation area influences the surroundings also, and examples of the sort of research needed on this part of the interaction are studies on the role of the conservation areas in erosion control and water catchment (e.g., AIDR-Rwanda, 1981; Borman et al., 1968; Gentry and Lopez-Parodi, 1980) and as refuges for crop pests or grazing competitors. Of course, biological analysis alone is not sufficient. Conservation areas and their environs have economic as well as ecological influences on one another (see the section on the Economics of Tourism). With respect to the environs and the threat they pose, examples of the nature of the research needed are Kurji's (1976) study of the demographic structure of the human populations around Tanzania's conservation areas and their consequent ranking in relation to the threat faced from population increase, and Weber's (1981, in press) analysis of attitudes to the Parc National des Volcans in Rwanda.

In general, biologists concerned with conservation need to accept that the wildernesses that interest them are part of a larger ecosystem. With the same intensity that they at present study interrelations within the conservation areas in order to aid their management, so they must start to examine the relations of the wilderness area with the broader ecosystem for the same end.

Summary

A case study is presented of a conservation program for a small population of a threatened primate species, *Gorilla gorilla*, in a threatened and rare ecosystem, Afromontane forest, in a national park that is an important water catchment area in a poor and densely populated country, Rwanda, in east central Africa. The program had three branches, namely protection of the conservation area from illegal use, publicizing the need for conservation, and ensuring financial self-sufficiency of the Park through tourism. The program was successful: Immature recruitment to the gorilla population increased 30%; attitudes to the Park improved, e.g., a drop from 50% to 18% of farmers who thought it ought to be cultivated; and Park revenue rose by 3200%. However, tourism as a means of conservation exacts costs both on the Park and on the country. Conservation-oriented research in nonindustrialized countries at present concentrates on processes within conservation areas, thus ignoring the fact that the threats they face are largely external to the area. To enable precise identification of the threats and their mitigation, far more research is needed on (1) possibilities for multiple use of conservation areas, and (2) the nature of environmental and socioeconomic interaction between conservation areas and their environs.

Acknowledgments. Conservation in the PNV has been funded directly and indirectly by a variety of funding organizations, the major ones of which are: the Rwandan Office of Tourism and National Parks; the Belgian government; the Consortium members—the African Wildlife Foundation, the Fauna and Flora Preservation Society, the Peoples' Trust for Endangered Species and the World Wildlife Fund; the National Geographic Society (largely responsible for funding the Karisoke Research Centre); and the New York Zoological Society.

References

Adams SN (1975) Sheep and cattle grazing in forests: a review. J Appl Ecol 12:143–152.

AIDR-Rwanda (1981) Evolution entre 1958 et 1979 du couvert forestier et du debit des sources dans certaines regions du Rwanda. Geomines-Somirwa, Brussels.

Anderson DW, Keith JO (1980) The human influence on seabird nesting success: conservation implications. Biol Conserv 18:65–80.

Aveling C, Harcourt AH (1984) A census of the Virunga gorillas. Oryx 19:8–14.

Berry RJ (1983) Genetics and conservation. In: Warren A, Goldsmith FB (eds) Conservation in perspective. John Wiley & Sons, London, p 141.

Berwick S (1976) The Gir forest: an endangered ecosystem. Amer Sci 64:28–40.

Bormann FH, Likens GE, Fisher DW, Pierce PS (1968) Nutrient loss accelerated by clear cutting of a forest ecosystem. Science 159:882–884.

Brotoisworo E (1978) Nature conservation in Indonesia and its problems with special reference to primates. In: Chivers DJ, Lane-Petter W (eds) Recent advances in primatology, vol 2. Academic Press, London, p 31.

Budowski G (1976) Tourism and environmental conservation: conflict, coexistence, or symbiosis. Environ Conserv 3:27–31.

Cummins LB, Moore GT, Kalter SS (1978) The economics of non-human primate conservation. In: Chivers DJ, Lane-Petter W (eds) Recent advances in primatology, vol 2. Academic Press, London, p 293.
Diamond JM (1975) The island dilemma: lessons of modern biogeographic studies for the design of natural reserves. Biol Conserv 7:129–146.
Eltringham SK (1984) Wildlife resources and economic development. John Wiley & Sons, London.
Franklin IR (1980) Evolutionary change in small populations. In: Soulé ME, Wilcox BA (eds) Conservation biology. Sinauer Associates Inc, Sunderland, p 135.
Gentry AH, Lopez-Parodi J (1980) Deforestation and increased flooding of the Upper Amazon. Science 210:1354–1356.
Hamilton AC (1981) The quaternary history of African forests: its relevance to conservation. Afr J Ecol 19:1–6.
Harcourt AH (1981) Can Uganda's gorillas survive? A survey of the Bwindi Forest Reserve. Biol Conserv 19:85–101.
Harcourt AH, Curry-Lindahl K (1979) Conservation of the mountain gorilla and its habitat in Rwanda. Environ Conserv 6:143–147.
Harcourt AH, Fossey D (1981) The Virunga gorillas: decline of an 'island' population. Afr J Ecol 19:83–97.
Harcourt AH, Fossey D, Sabater Pi J (1981) Demography of *Gorilla gorilla*. J Zool London 195:215–233.
Harcourt AH, Kineman J, Campbell G, Redmond I, Aveling C, Condiotti M (1983) Conservation and the Virunga gorilla population. Afr J Ecol 21:139–142.
IUCN (1980) 1980 United Nations list of national parks and equivalent reserves. IUCN Commission on National Parks, IUCN, Gland, Switzerland.
Johns A (1985) Selective logging and wildlife conservation in tropical rain-forest: problems and recommendations. Biol Conserv 31:355–375.
Kurji F (1976) Conservation areas and their demographic settings in Tanzania. Res Rep No 18 (NS), Bureau of Land Use Practice, University of Dar es Salaam, Dar es Salaam.
Lele U (1981) Rural Africa: modernization, equity, and long-term development. Science 211:547–553.
Liddle MJ (1975) A selective review of the ecological effects of human trampling on natural ecosystems. Biol Conserv 7:17–36.
Lind EM, Morrison MES (1974) East African vegetation. Longman, London.
Lugo AE, Brown S (1981) Ecological monitoring in the Luquillo Forest Reserve. Ambio 10:102–107.
Mascarenhas A (1983) Ngorongoro: a challenge to conservation and development. Ambio 12:146–152.
Mitchell F (1970) The value of tourism in East Africa. E Afr Econ Rev 1970:1–21.
Monfort A (1980) Evaluation du Projet Belgo-Rwandais: Tourisme et Parcs Nationaux. Administration Générale de la Coopération au Developpement, Brussels.
Monfort A (1985) Projet Belgo-Rwandais: Tourisme et Parcs Nationaux, Rapport Annuel 1984. Administration Générale de la Coopération au Developpement, Brussels.
Myers N (1972) National parks in savannah Africa. Science 178:1255–1263.
Myers N (1975) The tourist as an agent for development and wildlife conservation: the case of Kenya. Int J Soc Econ 2:26–42.
Myers N (1980) The present status and future prospects of tropical moist forests. Environ Conserv 7:101–114.
Prefol B, Delepierre G (1973) Disponibilité et utilisation des terres au Rwanda. Institut des Sciences Agronomiques au Rwanda, Butare.
Prioul C, Sirven P (1981) Atlas du Rwanda. Ministère de la Coopération de la République Française, Kigali, Paris, Nantes.
Pullan RA (1984) The use of wildlife as a resource in the development of Zambia. In: Bee OJ (ed) Natural resources in tropical countries. Singapore University Press, Singapore, p 267.

Ralph CJ, Maxwell BD (1984) Relative effects of human and feral hog disturbance on a wet forest in Hawaii. Biol Conserv 30:291–304.

Schaller GB (1963) The mountain gorilla. Ecology and behavior. University of Chicago Press, Chicago.

Simberloff D, Abele LG (1982) Refuge design and island biogeographic theory: effects of fragmentation. Amer Nat 120:41–50.

Sokal RR, Rolf FJ (1969) Biometry. WH Freeman & Co, San Francisco.

Soulé ME (1980) Thresholds for survival: maintaining fitness. In: Soulé ME, Wilcox BA (eds) Conservation biology. Sinauer Associates Inc, Sunderland, MA, p 151.

Spinage C (1972) The ecology and problems of the Volcanoes National Park, Rwanda. Biol Conserv 4:194–204.

Stewart KJ, Harcourt AH (In press) Gorillas: variation in female relationships. In: Smuts B, Cheney D, Seyfarth R, Wrangham R (eds) Primate societies. University of Chicago Press, Chicago.

Tutin CEG, Fernandez M (1984) Nationwide census of gorilla (*Gorilla g. gorilla*) and chimpanzee (*Pan t. troglodytes*) populations in Gabon. Amer J Primatol 6:313–336.

Weber AW (1981) Conservation of the Virunga gorillas: a socio-economic perspective on habitat and wildlife preservation in Rwanda. Thesis, University of Wisconsin, Madison.

Weber AW (In press) Socioecological factors in the conservation of afromontane forest reserves. In: Gartlan JS, Marsh CW, Mittermeier RA (eds) Primate conservation in tropical rain forest. Alan R Liss, New York.

Weber AW, Vedder A (1983) Population dynamics of the Virunga gorillas: 1959–1978. Biol Conserv 26:341–366.

Western D (1982) Amboseli National Park: enlisting landowners to conserve migratory wildlife. Ambio 11:302–308.

Western D, Henry W (1979) Economics and conservation in Third World national parks. BioScience 29:414–418.

Western D, Ssemakula J (1981) The future of the savannah ecosystems: ecological islands or faunal enclaves. Afr J Ecol 19:7–19.

White F (1981) The history of the Afromontane archipelago and the scientific need for its conservation. Afr J Ecol 19:33–54.

5 Captive Chimpanzee Populations—Past, Present, and Future

ULYSSES S. SEAL and NATHAN R. FLESNESS

Introduction

Egyptian decorative art suggests that the chimpanzee may have been held in captivity since the time of the Pharaohs. Serious use of chimpanzees as research animals in North America appears to date to the work of Yerkes and the establishment of the research colony in 1930. This colony offers the oldest and longest continuous records following young chimpanzees from their arrival in North America until death. The Yerkes colony also has provided some of the earliest documented captive breeding records; several of these first offspring are still alive in the colony. However, until 1973, replenishment of captive chimpanzee stock essentially depended upon continued recruitment of animals from the wild. There were few recorded successes with second-generation breeding and no effort made to establish a self-sustaining captive population. A preliminary demographic analysis (Seal, unpublished observations) indicated that the captive chimpanzee population was not self-sustaining as then managed and that a fall in productivity would occur as the breeding population of wild-caught animals declined.

The advent of the Endangered Species Act in 1973 and of CITES in 1978 led to the abrupt termination of imports from the wild. This became dramatically apparent with the denial in 1978 of a request to import additional wild-caught animals from Africa for hepatitis vaccine development and testing. The high priority of the hepatitis projects and the unique role of the chimpanzee as a model for this and other human diseases and physiological processes led to a concerted effort to identify and evaluate the available chimpanzee resources in North America and to exploration of strategies for providing a continuing supply of captive-bred animals.

Strategy for Analysis

Four central questions guided the collection of data, its analysis, and the development of a proposed management plan for establishing a self-sustaining captive chimpanzee population for biomedical purposes in North America.

1. What are the objectives of the captive breeding programs?
   a. Is the population to be self-sustaining with no further recruitment from the wild?
   b. Is the population to produce an annual surplus that will not be returned to the captive breeding population?
   c. Is it desirable that the captive population be managed to retain the genetic diversity present in its founder stock or should active selection for captivity-adapted or domesticated research stock be initiated?
   d. Is the captive population intended to contribute directly to the conservation of the chimpanzee as a species?
   e. What time period should be encompassed in the planning?
2. What are the origins (geographic and pedigree), age and sex composition, demographic characteristics (historical and recent), and genetic status (founders, variance in family size, sex ratio of breeders, inbreeding) of the captive chimpanzee population?
   a. Is the distinction between named subspecies important since most captive animals are of uncertain geographic origins? Is it possible to identify the geographic origins of the founder stock?
   b. How many founders have contributed to the present population? Has inbreeding occurred?
   c. What is the estimated effective population size as an index of the breeding strategies that have been employed and of the loss of genetic diversity?
   d. Are new breeders being recruited from the captive-born stock?
   e. Is more than one captive population currently being managed? Are the zoo and biomedical populations effectively separate? Are there differences in demographic characteristics among institutions or groups?
3. What size population, what age and sex composition, and what annual rate of increase (birth rate) are necessary to meet the objectives of a self-sustaining population producing a specified annual surplus?
   a. Should more than one population be managed with only limited genetic movement between the populations?
   b. What is the impact of the hepatitis vaccine research programs on the biomedical chimpanzee populations?
4. What are the costs of a program to meet objective 3 and how do they compare with the costs of current management practices?

Related questions considered include:

1. What is the role of the biomedical research population in the possible conservation of the chimpanzee?
2. What are the costs and benefits of the several possible strategies for conservation and for research use?
3. What are some possible futures for the chimpanzee in the wild and over what time span?
4. Should an AAZPA/SSP plan and an official regional or international studbook program be recommended and implemented?
5. What are the implications of the NIH National Chimpanzee Management Plan for the conservation of the chimpanzee?
6. Can the zoo and biomedical communities, acting in their enlightened self-interest as cultural institutions for recreation and research, respectively, contribute to the conservation of the chimpanzee as a species in the wild?

Population Biology Guidelines

What are the genetic and demographic requirements for a captive population designed to be self-sustaining, produce an annual surplus, and to assist the conservation of the chimpanzee as a species? Can they be met with available captive animals and resources?

Recent discussions at workshops on the conservation and population biology of small populations suggest that values for the variables of (1) founder numbers and (2) generation time provide the information necessary to preserve a specified proportion of the initial genetic diversity (for example, 90%) present in the founder stocks for a defined time interval (for example, 200 years) (Soulé and Gilpin, 1986). Additional conditions for use of these solutions to the equations relating these variables are (1) rapid (one-generation) expansion of the population to the effective population size with (2) equal sex ratio of breeders and (3) equal numbers of offspring from each founder. Satisfaction of these conditions is necessary to minimize loss of diversity in the initial expansion of the population.

Given an approximate generation time of 15 years for the captive chimpanzee, then an effective population size of 100 is required to maintain 90% of starting genetic diversity for about 200 years (approximately 13 generations). This could be accomplished with about 20 reproductive founders and a census population of about 200 animals. Note that the effective population size of a generation may be estimated as a first approximation in terms of the number and sex ratio of the breeding adults in the population (as is discussed by Flesness, this volume). Distortions in sex ratio or progeny numbers distribution will reduce Ne and increase the number of founders required to reach the goal.

Can we meet this requirement for the captive populations of chimpanzees? Assessment of the resource depends upon the availability of census and pedigree information tracing each animal back to wild-caught ancestry of known geographic origin. There has been no international studbook. The ISIS data base for zoos was initiated in 1974 and did not then include the primate centers or other biomedical facilities that hold the majority of chimpanzees in North America. We began assembling the information on the biomedical animals in 1978 and have now collected as complete a historical data base as we think can be achieved. The database includes 3426 animals, living and dead, with 1770 animals living as of 31 December 1984. There are 1612 living in North America of which 394 are in zoos (24.5%) and the remainder (1218, 75.5%) in research institutions.

The Present Chimpanzee Population

The total living population of 1770 animals includes 742 known captive born, 425 known wild born, and 603 of uncertain origin but presumed to be wild born. Data on the location of capture of most of these animals is unknown, and thus it is impossible to make any clear contribution to subspecies designations.

There was a steady increase in the number of captive births in the research community from 1967 to 1973 and then in 1974 a sharp increase in captive births occurred in the zoo and biomedical populations (Fig. 5.1). Births in the zoo population have currently stabilized at about 20 per year and births in the research population at about 70–80 per year for the last 5 years. The age pyramid

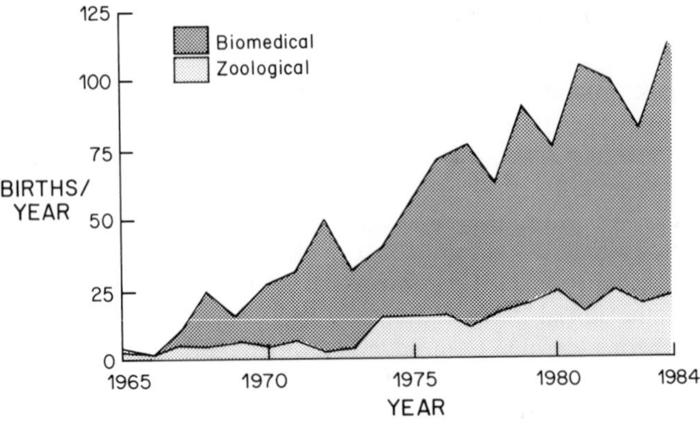

FIGURE 5.1. Chimpanzee births in the biomedical and zoo populations by year from 1965 through 1984.

of the total living North American research population encompassing 619 males and 599 females superficially has the general shape appropriate for a stable age distribution with more animals in the younger age classes at the bottom of the pyramid (Fig. 5.2). This age structure reflects the history of the shift from dependence on wild-caught animals to captive production. There is a deficit in the population of animals born in 1973–1974. If we separate out the approximately 850 living wild-caught animals, we see an older population of median age around 18 years that is now passing through its most productive years as a breeding stock.

The age pyramid of the approximately 750 captive-born animals in the research and zoo populations shows the small number of 10-year and older animals that will provide our potential stock for recruitment of captive-born animals as breeders. The numbers in this age pyramid overestimate the apparent number of animals that appear available for the captive breeding population. Some of these animals have been used for hepatitis research and in principle are not considered eligible for reintroduction into the breeding population since they can serve as potential carriers of the disease.

The age structure for the total zoo population reflects a similar history of animal acquisition. There is a striking deficiency in numbers of animals in the age classes following the year imports were curtailed. This was followed by a gradual increase in the number of captive births. The age structure of the zoo captive-born animals makes this apparent, and a small number of animals have entered the breeding age classes.

What has happened to the 103 animals born in an average year during the last 5 years? Zoos produced about 24 births per year of which 20 (80%) survived to the age of 18 months. The research community produced about 79 young of which 61 survived to the age of 18 months. Of the survivors, at least 25 (24.3%) are used for hepatitis research and are not in principle available for return to the breeding population.

There currently are 205 animals with a known history of exposure to Non-A, Non-B hepatitis in the population, which should be considered in the demo-

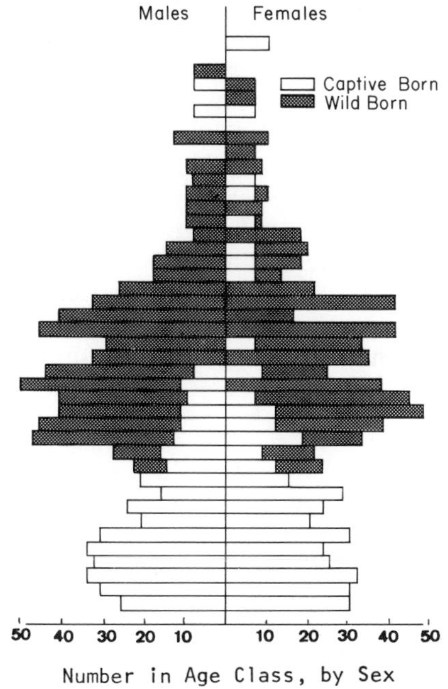

FIGURE 5.2. Age structure by sex of the living chimpanzee in the biomedical research population.

graphic analysis as nonparticipants in the breeding population. This number is still increasing since these are young animals, have a long life expectancy, and require maintenance in separate facilities after the hepatitis studies are completed (about age 5). Management plans for the future require estimates of the sex- and age-specific mortality and fecundity of the captive chimps with an assessment of the variability in these measures as a function of management objectives and practices (comparison of institutions), changes in management techniques, and of the performance of wild-caught versus captive-born animals.

Mortality

Mortality of chimpanzees is high during the first year of life, ranging around 21% of births since 1950 when data first became available (Seal et al., 1985). These rates are not a consequence of inbreeding because virtually all production is from wild-caught animals. Nearly 50% of the starting population is gone by the age of 10 and only 25% remains by the age of 25 years. This decline appears steeper during the adolescent and young adult years than is considered typical for other long-lived mammal species. The neonatal mortality rate is reminiscent of human populations without access to modern medical prenatal care. However, as described by Flesness (this volume), it may be typical of captive primate populations.

The neonatal mortality rate is of special interest because of its detrimental impact on the growth rate of the population. This mortality increases the size of the population that must be maintained for an effective population size target for preservation of genetic diversity or to produce a desired number of animals for research. Thus, analysis of neonatal deaths is one important research objective for future management of chimpanzees and other primate species.

Follow-up of 1311 captive-born chimpanzee births yielded 254 deaths during the first year. Day 0 deaths accounted for 126 or 50% of these year 1 deaths. One-half or 62 of the day 0 deaths were stillborn and 22 more were premature births. The fraction of day 0 and year 1 deaths has not noticeably changed over the past 20 years suggesting a lack of clear information on etiologies. There are similar problems with neonatal mortality in captive-born gorillas and orangutans (Seal et al., 1985).

Fecundity

The historical age-specific fecundity of females peaks at about 0.25 female offspring per breeding female per year between the ages of 12 and 22. This is about one young (male and female) every other year per breeding female during the peak breeding years. The earliest recorded age of first reproduction is about 5 years, but fecundity is very low in these young females and rises at age 8–10. Fecundity also drops off rapidly after the age of 30. It is not clear if the chimpanzee reaches an age of menopause, but there is a clear decline in fecundity that, in combination with the steady decline in numbers, means they make a small contribution to the productivity of the population.

Recruitment

The successful multiple generation management of a captive population depends upon the recruitment of captive-bred animals into the breeding population. Long-lived animals, with 10- to 20-year generation times, are a particular challenge. The generation time of the captive chimpanzee is about 12–15 years. It has been simplest for colony managers to continue the use of successful males and avoid the risk of unproductive pairings of fertile females with inexperienced or unknown males. This management bias is reflected in the skewed distribution of male family sizes in the captive chimpanzee population. Thus, in a series of 566 births of known male parentage in which 113 males sired young, 263 offspring were sired by 12 males, each of whom sired 10 or more offspring. In contrast, 252 females produced 685 young, and 4 females, with more than 10 births each, produced 53 young. The earliest recorded age of first breeding with production of young by male chimpanzees is 6 years, and 82 of 112 breeders had sired their first young by the age of 15 years.

The population of captive-born males eligible for recruitment as breeders includes 80 living animals 15 years or older. Thirteen of these captive-born males have sired 25 surviving offspring in 39 births. Thus, about 16% have become breeders in contrast to 80% of the wild-caught males. There has been no recruitment of breeders from the 90 males in the 10–14 year age group. Demographic projection of the future of the captive chimpanzee population is critically dependent upon the recruitment of captive-born male breeders. The loss of genetic diversity will be accelerated by increasing use of fewer males. Thus, an increase in the ratio of breeding males to females from 1:3 to 1:10 will result in a 60% decrease in Ne.

The Future Chimpanzee Resource

Continuation of the current trends of age-specific fecundity and mortality will in 1993 yield a population that is 70% captive born with the remaining animals being surviving wild-caught or presumed wild-caught animals.

The current mortality rates for wild-caught females indicate that 95% will be dead in 21 years. Their contribution to the productivity of the captive population will start dropping over the next 5 years, and the contribution of recruited captive-born breeders will become critical. A concerted effort to recruit the young adult and adolescent chimpanzees into the breeding population must be made now if the population is to remain stable or continue to yield replacement animals.

The problem is made particularly acute by the continued utilization of up to 50% of the biomedical productivity for hepatitis research and their loss to a clean breeding population. Thus the dilemma: How to continue current utilization and provide for future research production in the face of current high productivity of the breeders, but low recruitment of their replacements? Indeed, with any long-lived species (i.e., condors, Sumatran rhinos, northern white rhinos, etc.), this sort of population problem can be denied until demographic catastrophe strikes. Continuing to rely on the wild-caught females is possible only for another 5 years since they are aging and their numbers are declining. The number of births will begin falling rapidly in about 5 years if the status quo is maintained. Continuation of the status quo in the research population also will result in a steady decline in the number of available uncompromised animals.

The future of the North American research population will depend upon the establishment of dedicated breeding groups of animals excluded from general research use and management to recruit new captive-bred breeders on a continuing basis. A review of the living biomedical population by the managers indicates that there are 173 male and 224 female chimpanzees that could form the nucleus of clean dedicated breeding groups (Fig. 5.3). Their age structure, similar to that of the total population, indicates that the group includes wild-caught and captive-born animals. Most of the wild-caught animals in this group are currently part of the breeding population. The pool of captive-born animals that are potential breeders is much smaller. It would appear appropriate to concentrate efforts on recruitment of these animals into the breeding population.

The zoo population also faces the long-term prospect of a decline resulting from the failure to recruit new breeders because of separate management of multiple small groups of animals rather than management of the captive population as a biological unit.

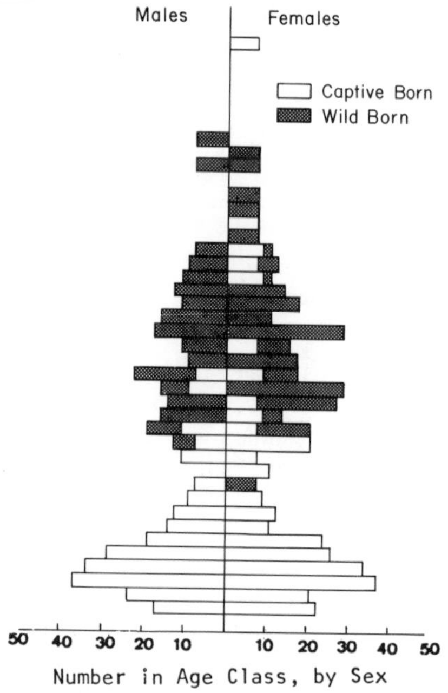

FIGURE 5.3. Age structure by sex of the chimpanzees in the biomedical research population that are considered suitable candidates for inclusion in a clean breeding population.

Stable Population

If recruitment of captive-born breeders is accomplished from the clean research population and if fecundity rates comparable to those obtained in the most productive colonies during the past 7 years are maintained, then it will be possible to maintain a stable population of 370 animals, comprised of 86 males and 284 females with an annual birth rate of about 75 animals, allowing the removal of about 40 animals per year. Sufficient potential breeding stock for this purpose has been identified, and this is the thrust of the proposed national chimpanzee management plan. In addition, the approximately 250 chimpanzees in North American exhibit collections can be managed as a self-sustaining population separately from the research collections.

Recommendations

1. Evaluate the validity as evolutionary significant units of the named subspecies using animals of unquestioned geographic origin. Utilize modern chromosomal banding and molecular genetic techniques as well as museum materials. This needs to be done immediately as a guide for expenditure of both in situ and ex situ conservation resources.

2. Establish a North American SSP with regional studbook program for the chimpanzee.
3. Consider management of the zoo and research populations as two separate populations with carefully chosen genetic exchange each generation.
4. If the taxonomic analysis establishes the existence of distinct evolutionary units within *Pan troglodytes*, then begin the establishment of separate captive breeding populations of each utilizing animals of known origin.
5. If 4 above is the case, then undertake the classification of the living captive population now while most of the captive population is in the wild-caught or Fl generation.
6. Establish a program for analysis, diagnosis, and prevention of neonatal mortality in the great apes.
7. Establish a deliberate program of recruitment of captive-born males and females as new breeding stock. Plan for recruitment of F3 animals now.
8. Within the research population determine the feasibility for maintaining at least two or more separate breeding populations as a protection against catastrophe.
9. Obtain funding for the establishment of a self-sustaining biomedical research population as a national resource.
10. Under current circumstances there is no need to import additional wild-caught animals for either research or exhibit use. The establishment of a captive population of a newly defined evolutionary significant unit now poorly represented in captivity might provide a basis for selective imports to found such a population.

Summary

Despite current optimism major problems remain to be resolved for both the zoo and research community to successfully achieve self-sustaining captive breeding chimpanzee populations.

Acknowledgments. We thank Larry Grahn for his assistance in assembly of the data and preparation of the graphics. This study was supported in part by the Division of Research Services, NIH, and the V. A. Medical Research Service.

References

Flesness NR, Garnatz G, Seal US (1984) ISIS—an international specimen information system. In: Allkin R, Bisby FAS (eds) Databases in systematics. Academic Press, New York, pp 103–112.

Seal US, Flesness NR, Foose TJ (1985) Neonatal and infant mortality in captive-born great apes. In: Graham C (ed) Clinical management of infant great apes. Alan R. Liss, Inc., New York, pp 193–203.

Soulé M, Gilpin M, Conway W, Foose T (1986) The millenium ark: how long the voyage, how many staterooms, how many passengers? Zoo Biology.

6
Responses of Rainforest Primates to Selective Logging in Kibale Forest, Uganda: A Summary Report

JOSEPH P. SKORUPA

Introduction

Approximately 80–90% of all extant nonhuman primate species occur primarily in tropical rainforest environments (Bourliere, 1973; Napier, 1962; Napier and Napier, 1967; Wolfheim, 1983). Destruction of the world's tropical rainforests has been occurring at an alarming rate over the past two to three decades (cf. Myers, 1984) and is by far the most important threat to global primate conservation (Mittermeier and Cheney, in press). Currently, only 3.6–4.4% of all tropical broadleaf forests are legally protected from exploitation (Myers, 1984:46,311; Steinlin, 1982). This figure will increase to 7% if all pending land set-aside proposals are ultimately implemented (Mabberley, 1983). It is unlikely, however, that the area of protected rainforest will ever exceed 9% of the world's presently existing total (Wadsworth cited in Hartshorn, 1984). Given the high degree of diversification and endemicity typical of tropical rainforest biota, it is clear that the conservation of rainforest primates will be difficult to ensure solely within the confines of land set-asides (e.g. Rylands, 1985).

Of the most common alternatives for rainforest exploitation, land managed for selective harvesting of timber offers the most promising hope for retaining large areas of economically productive land in a natural state that might also provide reasonable primate habitat. Already, the area of selectively logged rainforest exceeds protected rainforest by a ratio of 4:1 (Brown and Lugo, 1984), and the ratio is likely to increase over time. In addition, as economic pressures mount, it may be inevitable that selective logging will become a widespread multiple-use management option even in formerly protected "parks."

It is therefore apparent that assessments of a primate species' capacity to survive in logged rainforest should be an integral component of any long-term conservation plan. Nonetheless, few intensive studies have focused primarily on this topic (Johns, 1981, 1983a, 1983b, 1983c; Skorupa, in preparation), though many studies have given the issue brief or secondary attention (for a summary see Skorupa, in preparation).

In this report I will attempt to summarize a representative selection of the major findings from a 2-year study of the effects of selective logging on seven primate species in Kibale Forest, Uganda.

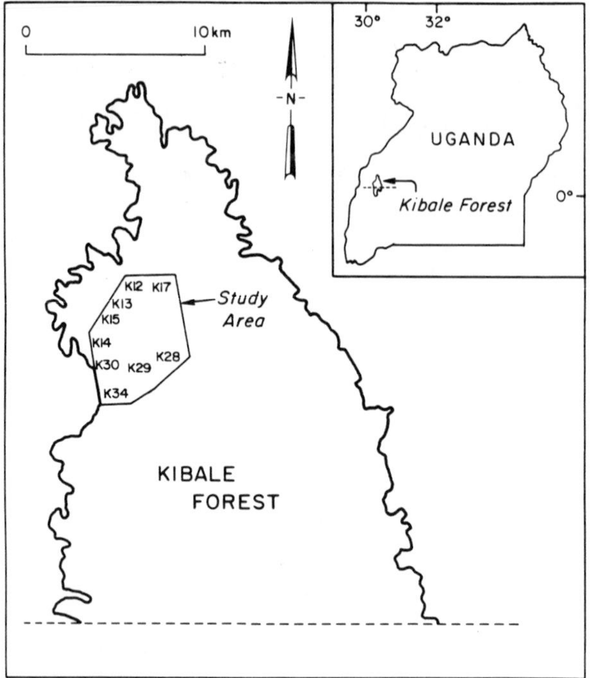

FIGURE 6.1. Map of the study area showing locations of nine timber compartments included in the study. Each compartment contains about 300 ha of forest.

Study Site and Methods

Kibale Forest is located in western Uganda (Fig. 6.1) and is best described as a medium-altitude moist evergreen forest, being distinguished from true lowland tropical rainforest by its higher altitude, lower temperature, and lower rainfall, but otherwise possessing most of the same typical features (Langdale-Brown et al., 1964). For a detailed description of the study site see Struhsaker (1975).

From 1980–1982, I studied the vegetation and primate populations found in a series of five study plots ranging from an unlogged control plot to a plot that had been heavily damaged by logging and by postlogging arboricide treatments. However, many of the results reported here will be restricted to the three most comparable plots: (1) an unlogged control plot (K-30), (2) a lightly damaged logged plot (K-14), and (3) a heavily damaged logged plot (K-15). The heavily damaged logged plot actually represents the typical intensity of disturbance caused by mechanized selective logging (i.e., ~50% destruction of the original forest stand; e.g., Ewel and Conde, 1976; Johns, 1983a, 1983b).

Trees ≥ ~35 cm circumference at breast height were surveyed in each plot using both plot sampling and point-quadrant sampling techniques. Post hoc comparisons of logged and unlogged forest are employed to infer logging impact on vegetative structure. Primate populations were censused using line-transect techniques (cf. Burnham et al., 1980). Relative abundances of primate social groups were estimated from replicate transects, correcting the data for differential visibility (by species and plot) and for differential forest access (e.g., footpaths versus logging roads). Encounter rates conformed well with the Poisson distribution, and therefore all parametric statistical procedures were performed on square-root transformed data.

To examine the relations between vegetative parameters and primate abundances, univariate correlation analysis was employed as a screening procedure to isolate suites of potentially important botanical variables, and then partial correlation analysis is employed to further reduce the variable set. The fundamental assumptions underlying these analyses are: (1) that primate densities can be expected to respond monotonically to variation in important botanical variables, and (2) that the form of the response can be adequately approximated as linear over the range of values sampled.

For a detailed explanation and evalution of my methodology see Skorupa (in preparation).

Vegetative Resources

Selective logging altered both the quantity and the quality of vegetative resources available to primates. Total basal area was 25% lower in the lightly logged plot than in the control plot and 50% lower in the heavily logged plot. Similarly, large stem density, percent canopy cover, and aggregate tree dispersion were all maximized in the unlogged control plot (Table 6.1). While total stem density and *Ficus* density were maximized in the lightly logged plot, even these variables were still significantly lower in heavily logged forest than in unlogged forest. Average patch size is the only variable that appeared to be completely unrelated to a plot's management history (Table 6.1). In general, heavily logged forest showed distinct reductions in nearly every measure of resource quantity, while lightly logged forest yielded mixed results.

Four measures of tree species diversity consistently showed that the quality of the forest resource base declined in rough proportion to the intensity of logging

TABLE 6.1. Composite botanical variables for three study plots in Kibale Forest, Uganda.

Variable	Unlogged	Lightly logged	Heavily logged
Total basal area (m^2/ha)	35.52[1]	26.66[2]	18.98[3]
Total stem density (stems/ha)	255.9[1]	267.2[1]	125.1[2]
Large stem density (stems/ha)	25.2[1]	12.0[2]	8.0[2]
Ficus density (stems/ha)	4.07[2]	6.51[1]	0.59[3]
Percent canopy cover $\geq$ 9 m	87%[1]	65%[2]	45%[3]
Percent canopy cover $\geq$ 15 m	72%[1]	50%[2]	32%[3]
Aggregate tree dispersion ($s^2/\bar{x}$)	1.51[1]	1.72[2]	1.91[3]
Mean patch size (basal area/stem)	0.14	0.10	0.15

[1,2,3]These superscripts indicate for each row which values show significant statistical differences. See Skorupa (in preparation) for details of statistical analyses.

TABLE 6.2. Comparative values for four measures of tree species diversity.

Diversity measure	Unlogged	Lightly logged	Heavily logged
Species richness[4]	25.6[1]	23.0[2]	18.2[3]
Species density[5]	25.3[1]	22.7[2]	14.3[3]
Species equitability[6]	0.620[1]	0.527[2]	0.501[3]
Diversity (H′)[7]	2.760[1]	2.484[2]	2.206[3]

[1,2,3]These superscripts indicate for each row which values show significant statistical differences. See Skorupa (in preparation) for details of statistical analyses.
[4]Average number of species encountered per 100-stem sample.
[5]Average number of species enumerated per 20 randomly located 5×50 m plots.
[6]Hill's (1973) evenness measure.
[7]Shannon-Wiener index (Wilson and Bossert, 1971).

disturbance (Table 6.2). Based on species-area curves, equal-sized primate home ranges in lightly and heavily logged forest at Kibale would contain only 85% and 48% as many tree species when compared to an equivalent unlogged home range. Contrary to what might be expected, based on the intermediate disturbance principle (Connell, 1978), disturbed *Parinari* forest at Kibale does not support a greater species diversity of adult trees than undisturbed forest (Table 6.2).

Despite the fairly unambiguous influence of logging on composite measures of forest structure (i.e., measures that lump all species of trees), between-plot variation in the abundance of trees known to provide foods useful to primates was not found to be a simple function of harvest intensity. For example, the total basal area of tree species contributing 5% or more to the annual diet of *Cercocebus albigena* (in unlogged forest) was virtually unaltered in both lightly and heavily logged forest (Table 6.3); however, this category of trees only includes the top three to five food species. Total basal areas of food trees that cumulatively provided 80% of *Cercocebus albigena*'s annual diet (in unlogged forest) were reduced by 25% in heavily logged forest (Table 6.4). Generally, between-plot variation in the abundance of food trees depends on each primate species' feeding ecology (frugivore versus folivore, generalist versus specialist, etc.) and utilization of commercially important tree species.

TABLE 6.3. Comparative total basal areas (m^2/ha) of tree species that contributed 5% or more to the annual diet of primate species in unlogged forest.[1]

Primate species	Unlogged	Lightly logged	Heavily logged
Cercocebus albigena	4.36	4.20	4.63
Cercopithecus mitis	5.73	6.11	5.38
Colobus badius	11.21	12.13	5.88
Colobus guereza	6.14	12.86	4.50

[1]Based on dietary data published by Struhsaker (1975), Waser (1975), Oates (1977), and Rudran (1978). Detailed dietary data for the other primate species in unlogged *Parinari* forest has not been published.

TABLE 6.4. Comparative total basal areas (m²/ha) of tree species cumulatively providing 80% of a primate species' annual diet in unlogged forest.[1]

Primate species	Unlogged	Lightly logged	Heavily logged
Cercocebus albigena	17.27	19.00	12.97
Cercopithecus mitis	17.49	17.14	8.30
Colobus badius	17.18	13.75	7.22
Colobus guereza	9.20	17.18	8.71

[1]Based on dietary data published by Struhsaker (1975), Waser (1975), Oates (1977), and Rudran (1978). Detailed dietary data for the other primate species in unlogged *Parinari* forest has not been published.

Primate Populations

On average, the sensitivity of census data collected during this study allowed statistical discrimination of plots whose primate abundance varied by as much or more than ±45% from the control plot. The best resolution was achieved for *Colobus badius* (±23%), and the poorest resolution was achieved for *Pan troglodytes* (±59%). Census results revealed that only one of seven primate species showed a statistically significant population decline in lightly logged forest, while five species showed statistically significant declines in heavily logged forest (Fig. 6.2). However, while the decline of *Pan troglodytes* in lightly logged forest was not statistically significant, it may nonetheless represent a biologically meaningful response. One species (*Colobus guereza*) responded positively to logging, and one species (*Cercopithecus mitis*) occurred at similar densities in all plots (Fig. 6.2).

Since there were several different patterns of primate population responses to logging (Fig. 6.2), the net effect of logging on the conservation value of a forest is best assessed using an index of total primate abundance that weights each species equally. To derive such an index I calculated, by species, what percent (0–100%) of the maximum observed abundance (in all five study plots) each study plot's value represented. I then summed these proportions by study plot and calculated the average, which serves as my *total primate index*. For example, on average, primates in unlogged forest (K-30) achieved 82.5% of their maximum observed abundance in any study plot (Table 6.5). The total primate index for lightly logged forest is encouragingly comparable to unlogged forest, while heavily logged forest appears to have had its conservation value substantially compromised (Table 6.5). Forest that had been heavily logged and subsequently treated with arboricides (to kill remnant stems of commercially undesirable tree species) exhibited the lowest total primate index.

Relations between Primate Populations and Vegetative Variables

My correlation analyses revealed that the strongest botanical correlates of primate species' abundances were fairly unique for each species (Table 6.6). *Colobus badius* was the only species whose abundance was strongly linked with a

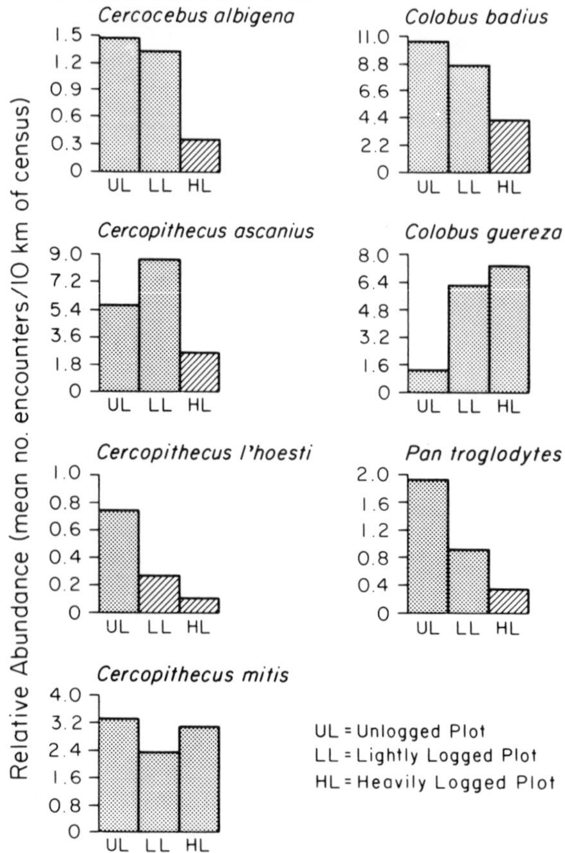

FIGURE 6.2. Relative abundance of primates in unlogged (UL), lightly logged (LL), and heavily logged (HL) forest. Abundances that are significantly lower than the values for unlogged forest are indicated by diagonal hatching.

particular set of food trees. This result is consistent with the fact that *Colobus badius* possesses a highly specialized ruminantlike digestive system (Struhsaker, 1975). *Cercopithecus ascanius* was the only species whose numbers covaried closely with *Ficus* densities, although *Cercocebus albigena* also showed a strong trend in the same direction. However, most species' abundances were best predicted by composite measures of forest structure (e.g., basal area, canopy cover, etc.), perhaps indicating a wide degree of dietary flexibility (Table 6.6). *Cercopithecus mitis* abundances did not consistently covary with any of the botanical variables I measured. Such a result is consistent with the extreme generalist status of *C. mitis*. *Cercopithecus mitis* is seemingly well adapted to a wide variety of vegetative communities over the widest latitudinal and altitudinal ranges of any diurnal forest primate in Africa (Dorst and Dandelot, 1969).

At a community level of analysis, both primate species richness and primate species diversity significantly covaried with corresponding measures of tree

TABLE 6.5. Total primate indices (TPIs) for four study plots with different management histories.[1]

Study plot	TPIs
Unlogged control (K-30)	82.5
Lightly logged (K-14)	73.2
Heavily logged (K-15)	45.9
Heavily logged and poisoned (K-13/12/17)	35.1

[1]TPIs can range from 0–100 with higher scores reflecting higher net primate conservation value. See the text for the derivation of TPIs.

species richness and tree species diversity (Fig. 6.3). Primate species equitability, however, was not significantly correlated with tree species equitability indicating that rare species of trees might be disproportionately important to primates (for a discussion of the theoretical importance of this result see Rapport, 1980). Unfortunately, selective logging commonly leads to the extirpation of rare tree species (Queensland Forestry Department, 1983; Payne, 1984). The strongest botanical predictors of total primate indices proved to be percent canopy cover at or above 15 m and tree species richness. These results suggest that both the quantity and quality of postlogging remnant forest are important for primate conservation.

Mature-Forest Core Species

Botanical variables more strongly influenced patterns of primate abundance than did interspecific interactions between primate species, a result consistent with Rodman's (1973) analysis of a Bornean primate community. Therefore, I hypothesized that densities of primate species closely coevolved with mature

TABLE 6.6. Strongest botanical predictors of primate abundance in Kibale Forest, Uganda.

Primate species	LSD (1)[1]	BA (2)	CC-15 (3)	TSR (4)	STD (5)	FD (6)	MPS (7)	TOP-80 (8)
Cercocebus albigena				*[2]	*			
Cercopithecus ascanius						*	(*)[3]	
Cercopithecus l'hoesti			*					
Cercopithecus mitis								
Colobus badius		*						*
Colobus guereza	(*)							
Pan troglodytes			*					
Total primate index			*	*				

[1]Botanical variables in order of appearance are: (1) large stem density; (2) total basal area; (3) percent canopy cover ≥ 15 m; (4) tree species richness; (5) total stem density; (6) *Ficus* density; (7) mean patch size; and (8) total basal area of trees making up the top 80% of the annual diet.
[2]*, a positive correlation between the botanical variable and primate abundance.
[3](*), a negative correlation between the botanical variable and primate abundance.

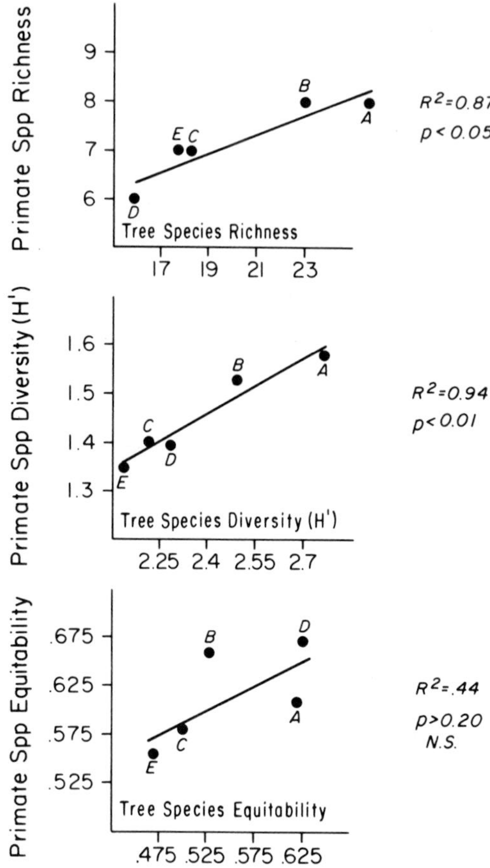

FIGURE 6.3. Relationship between primate species diversity and tree species diversity. Each lettered point represents one study plot.

forest should show strong positive covariation between plots. Consequently, dendrogrammatic analysis (Cody, 1974) of an interspecific correlation matrix should reveal whether there is a mature-forest core of species in any given primate community. Such an analysis of my Kibale data suggests that there are four species particularly dependent on mature forest: (1) *Cercocebus albigena*, (2) *Cercopithecus l'hoesti*, (3) *Colobus badius*, and (4) *Pan troglodytes* (Fig. 6.4). Accordingly, these are the species that would be threatened most by extensive logging of Kibale Forest (i.e., the mature-forest core species).

I performed a similar analysis on data provided by Marsh and Wilson (1981:56) for a peninsular Malaysian primate community, and the studies suggest that there are no mature-forest core species in the community (Fig. 6.5). Independently of Marsh and Wilson's (1981) study, Johns (1983a, 1983b, 1983c) intensively studied the effects of selective logging on the same Malaysian primate community and found few, if any, long-term detrimental influences. Johns' findings,

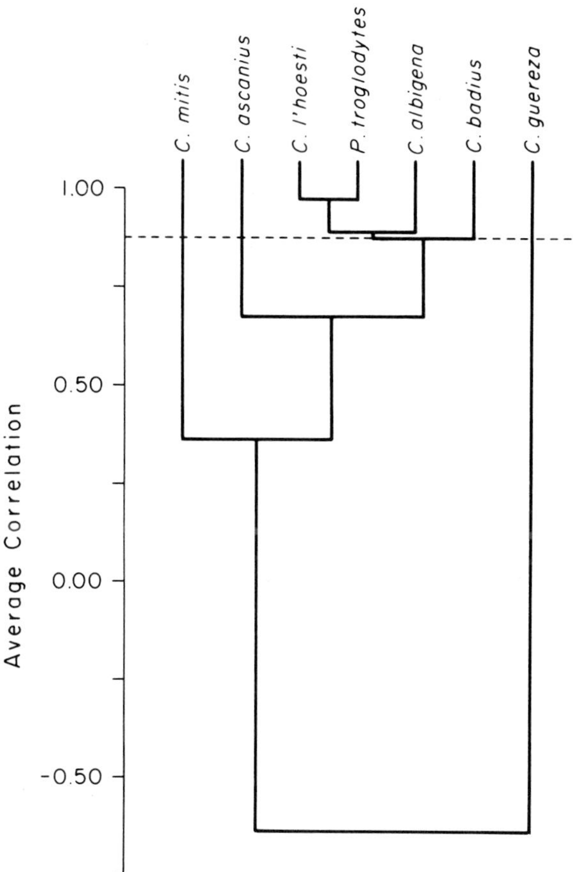

FIGURE 6.4. Dendrogrammatic representation of primate density interspecific correlation matrix for five study plots in Kibale Forest, Uganda. The dashed line indicates the 5% probability level for *r* (product–moment correlation coefficient). Species closely coevolved with mature forest should branch off at or above the dashed line. Species that benefit from logging should branch off in negative correlation space.

therefore, appear to corroborate the biological validity of my dendrogrammatic analyses. A comparison of the results for the African and Malaysian communities also cautions against extrapolating the results of Asian logging studies (where the greatest amount of research has been conducted) to other tropical rainforest regions.

Socioecological Correlates of Primate Vulnerability

Since the socioecology of Kibale primates is relatively well known (Struhsaker and Leland, 1979; Skorupa, in preparation), my comparative census data provide a unique opportunity for discerning socioecological correlates of primate

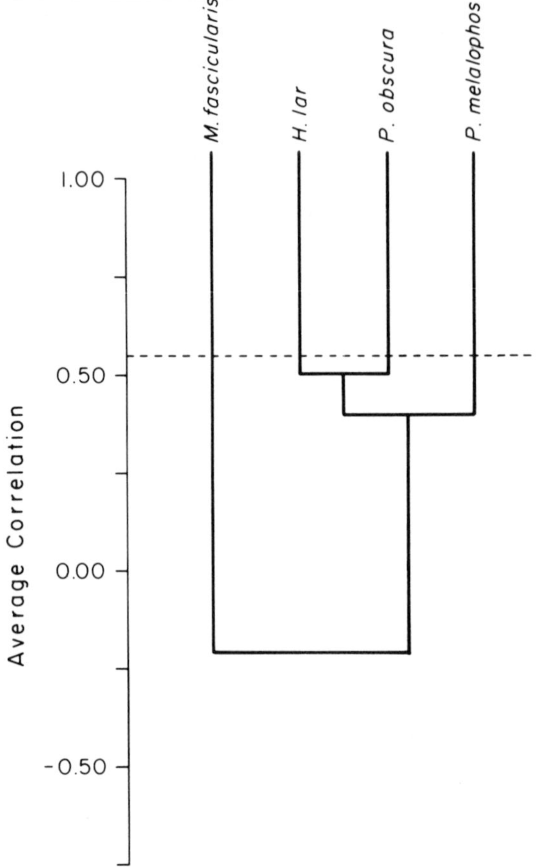

FIGURE 6.5. Dendrogrammatic representation of primate density interspecific correlation matrix for 11 study plots in peninsular Malaysia (data from Marsh and Wilson, 1981). The dashed line indicates the 5% probability level for *r* (product–moment correlation coefficient). Species closely coevolved with mature forest should branch off at or above the dashed line. Species that benefit from logging should branch off in negative correlation space. The species illustrated here are from the genera *Hylobates*, *Macaca*, and *Presbytis*.

response to selective logging. The data analyzed here are unique because both the socioecological parameters and the measures of primate response to habitat disturbance were assessed at one locality, in the same forest type (*Parinari* forest), and by standardized methods. Thus, many sources of confounding variability have been minimized (cf. Clutton-Brock and Harvey, 1977; Johns, 1983c).

Of 13 variables tested, three were significantly correlated with primate vulnerability to habitat disturbance. Group home range size, average group spread, and percent fruit, seeds, and flowers in the diet were all positively correlated with vulnerability (Table 6.7). These three variables are all related to a group's foraging ecology and, when jointly maximized, constitute a suite of socioecological

TABLE 6.7. Rank correlations between socioecological variables and primate vulnerability to habitat disturbance in Kibale Forest, Uganda.[1,2]

Socioecological variable	Spearman rank correlation	Sample size
Group size (indiv.)	−0.216	7
Group biomass (kg)	0.286	7
Day range length (m)	0.200	5
Home range area (ha)	0.786[3]	7
Group spread (m)	1.000	5
Degree of territoriality	−0.714	6
Adult sex ratio	−0.200	7
Dietetic diversity (H′)	0.100	5
Percent fruit, seeds, flowers	0.829	6
Percent mature leaves	−0.600	6
Percent all aged foliage	−0.771	6
Percent animal matter	0.257	6
Diversity of food classes	−0.543	6

[1]Parameter values for socioecological variables are presented in Skorupa (in preparation).
[2]Vulnerability to habitat disturbance is measured as the proportional difference between a primate species' abundance in undisturbed and selectively logged forest.
[3]Underlined correlation coefficients are statistically significant at the 5% probability level (one-tailed test).

characters related to the exploitation of widely dispersed, relatively rare foods. On this basis, a strong negative correlation between degree of territoriality and vulnerability should also be expected. While not quite statistically significant at the 5% level, the results do show a strong trend in the expected direction (Table 6.7).

The above suite of characters is most fully expressed in species adapted to exploit the large animal-dispersed fruits of trees endemic to mature forests (cf. Opler et al., 1977). Because trees endemic to mature forests typically occur at low densities (e.g., Hartshorn, 1978), the aforementioned strategy requires that a large resource supply area be monitored. Therefore, species adopting such a strategy are likely to be close to the limits of their resource monitoring capabilities, which would explain why the strategy is so narrowly optimized in mature undisturbed forest.

Concluding Discussion

The evidence presented here firmly suggests that the typical level of damage associated with capital-intensive, mechanized selective logging (i.e., destruction of ~50% of the prelogging forest stand) seriously compromises the primate conservation value of Kibale Forest. The results from lightly damaged forest, however, provide an optimistic basis for promoting a synergistic relationship between tropical foresters and primate conservationists. It appears that selective timber harvesting can be compatible with primate conservation in particular and biological conservation in general, if levels of damage are strictly limited. Studies of treefall dynamics in Kibale also support this conclusion (Skorupa and Kasenene, 1984).

There are at least two major policies by which damage could be substantially reduced: (1) by moving toward a greater reliance on labor-intensive logging techniques and away from capital-intensive mechanized techniques, and (2) by making a concerted effort to establish markets for a broader range of tree species, thereby increasing damage efficiency and lowering overall levels of damage associated with any particular level of production (cf. Queensland Forestry Department, 1983). The first option would seem more desirable on a number of grounds. First, labor-intensive techniques would tap an underutilized resource of most tropical nations, the large pool of unemployed labor, and thereby also provide direct local benefit. Second, labor-intensive techniques would also maximize the ratio of foreign exchange earned over foreign exchange consumed by the forestry sector. Third, the broader marketability option would only be compatible with biological conservation if the temptation to increase production goals simultaneously were resisted. Recent history argues that such an expectation is unrealistic.

In summary, if the results from Kibale are fairly representative of what would be found in other African rainforests, then the prospects for combining selective logging with primate conservation appear to be much more narrowly circumscribed in Africa than has been reported for Asian rainforests.

Acknowledgments. I'm indebted to all Kibale researchers whose studies preceded mine and provided such a useful baseline data set for unlogged forest. I thank T. M. Butynski and L. A. Isbell for sharing with me the results of their own vegetative sampling of unlogged forest (K-30). Without the guidance and support provided by T. T. Struhsaker in Uganda and my major professor, P. S. Rodman, in Davis, this study would not have been possible. I thank the World Wildlife Fund-US (project no. 1969 through T. T. Struhsaker), New York Zoological Society, and California Primate Research Center (through P. S. Rodman) for their generous financial support. I am also grateful to the President's Office of Uganda, the Ugandan National Research Council, and to the Uganda Forests Department for permission to work in Kibale Forest.

References

Bourliere F (1973) The comparative ecology of rain forest mammals in Africa and Tropical America: some introductory remarks. In: Meggars BJ, Ayensu ES, Duckworth WD (eds) Tropical forest ecosystems in Africa and South America: a comparative review. Smithsonian Institution Press, Washington, DC.

Brown S, Lugo AE (1984) Biomass of tropical forests: a new estimate based on forest volumes. Science 223:1290–1293.

Burnham KP, Anderson DR, Laake JL (1980) Estimation of density from line-transect sampling of biological populations. Wildl Monogr No 72:1–202.

Clutton-Brock TH, Harvey PH (1977) Species differences in feeding and ranging behaviour in primates. In: Clutton-Brock TH (ed) Primate ecology. Academic Press, New York.

Cody ML (1974) Competition and the structure of bird communities. Princeton University Press, New Jersey.

Connell JH (1978) Diversity in tropical forests and coral reefs. Science 199:1302–1310.

Dorst J, Dandelot P (1969) A field guide to the larger mammals of Africa. Collins, London.

Ewel J, Conde L (1976) Potential ecological impact of increased intensity of tropical forest utilisation. US Forest Service, Madison, Wisconsin.

Hartshorn GS (1978) Tree falls and tropical forest dynamics. In: Tomlinson PB, Zimmerman MH (eds) Tropical trees as living systems. Cambridge University Press, London.

Hartshorn GS (1984) Tropical forest ecosystems—a misnomer. Ecology 65:334–335.

Hill MO (1973) Diversity and evenness: a unifying notation and its consequences. Ecology 54:427–432.

Johns AD (1981) The effects of selective logging on the social structure of resident primates. Malays Appl Biol 10:221–226.

Johns AD (1983a) Wildlife can live with logging. New Sci 99:206–211.

Johns AD (1983b) Tropical forest primates and logging—can they coexist? Oryx 17:114–118.

Johns AD (1983c) Ecological effects of selective logging in a West Malaysian rain-forest. Thesis, Cambridge University, Cambridge.

Langdale-Brown I, Osmaston HA, Wilson JG (1964) The vegetation of Uganda. Government Printing Office, Entebbe, Uganda.

Mabberley DJ (1983) Tropical rain forest ecology. Chapman and Hall, New York.

Marsh CL, Wilson WL (1981) A survey of primates in peninsular Malaysian forests. Universiti Kebangsaan, Malaysia.

Mittermeier RA, Cheney DL (In press) Conservation of primates and their habitats. In: Primate societies. University of Chicago Press, Chicago.

Myers N (1984) The primary source. WW Norton and Co, New York.

Napier J (1962) Monkeys and their habitat. New Sci 15:88–92.

Napier J, Napier P (1967) A handbook of living primates. Academic Press, New York.

Oates JF (1977) The guereza and its food. In: Clutton-Brock TH (ed) Primate ecology. Academic Press, New York.

Opler PA, Baker HG, Frankie GW (1977) Recovery of tropical lowland forest ecosystems. In: Cairns J, Dickson KL, Herricks EE (eds) Recovery and restoration of damaged ecosystems. University of Virginia Press, Virginia.

Payne J (1984) Species conservation priorities in the tropical forests of Sabah, East Malaysia. In: Mittermeier RA, Konstant WR (eds) Species conservation priorities in the tropical forests of Southeast Asia. Occasional papers of the IUCN Species Survival Commission (SSC), No. 1 Gland, Switzerland.

Queensland Forestry Department (1983) Rainforest research in North Queensland. Government Printer, Queensland, Australia.

Rapport DJ (1980) Optimal foraging for complementary resources. Amer Nat 116:324–346.

Rodman PS (1973) Synecology of Bornean primates: a test for interspecific interactions in spatial distribution of five species. Amer J Phys Anthropol 38:655–660.

Rudran R (1978) Socio-ecology of the blue monkeys (*Cercopithecus mitis stuhlmanni*) of the Kibale Forest, Uganda. Smithson Contrib Zool 249:1–88.

Rylands AB (1985) Conservation areas protecting primates in Brazilian Amazonia. Primate Conservation 5:24–27.

Skorupa JP (In preparation) The effects of selective timber felling on rain-forest primates in Kibale Forest, Uganda. Thesis, University of California, Davis.

Skorupa JP, Kasenene JM (1984) Tropical forest management: can rates of natural treefalls help guide us? Oryx 18:96–101.

Steinlin HJ (1982) Monitoring the world's tropical forests. Unasylva 34:2–8.

Struhsaker TT (1975) The red colobus monkey. University of Chicago Press, Chicago.

Struhsaker TT, Leland L (1979) Socioecology of five sympatric monkey species in the

Kibale Forest, Uganda. In: Rosenblatt J, Hinde R, Beer C, Busnel M (eds) Advances in the study of behavior, vol 9. Academic Press, New York.

Waser PM (1975) Monthly variations in feeding and activity patterns of the mangabey, *Cercocebus albigena* (Lyddeker). E Afr Wildl J 13:249–263.

Wilson EO, Bossert WH (1971) A primer of population biology. Sinauer Associates Inc, Sunderland, MA.

Wolfheim JH (1983) Primates of the world: distribution, abundance, and conservation. University of Washington Press, Seattle.

7
Lemur Survival

ALISON JOLLY

Introduction

There is no good reason why any lemur species should go extinct. Present-day lemurs, those that survived the first human onslaught on Madagascar, are small, unthreatening creatues. They are not major crop raiders; they don't even offer much meat. They can live in small patches of forest, at population densities more like squirrels than apes or monkeys. Most are surprisingly adaptable. Even the ones we have not learned to breed in captivity live in a variety of forest types in the wild.

However, if present trends continue, most lemurs have no future. They are forest animals, and Madagascar's native forest is inexorably turning to barren savanna cut by livid gashes of erosion gullies.

So we have a paradox. It would take less effort, less land, less money to save viable populations of lemurs than almost any other primates, yet they are among the most threatened. Oddly, the very peril contains the seeds of hope. On the Malagasy side, their ecological and economic crisis has provoked official adoption of a "Strategie de la Conservation et du Développement Durable," a policy aimed at permanent, rational use of their land. On the international side, WWF and IUCN are at last accepting that the flora and fauna of Madagascar are an international responsibility. In this new political climate, I believe it may prove surprisingly easy to make permanent strongholds of conservation, and even evolution, for the lemurs of Madagascar.

What Is There and What Was There

Madagascar in many ways is more like a continent than an island. It is 1600 km long, the distance from San Diego to Seattle, or, to choose equivalent latitudes, Miami to the far side of Nicaragua. Its surface area of 587,000 km^2 is about that of all the eastern states from Maine to North Carolina. Those eastern states have about seven times Madagascar's population, and two and a half times as much forest, as of the last survey finished in 1950 (Jolly, 1980).

It is composed of continental rock—the ancient granites and gneisses of Gondwanaland on the east, sedimentary sandstone and limestone on the west. The trade winds of the Indian Ocean clothe the eastern escarpment in true rainforest, with maximum rainfall of 3612 mm/year on the Ile St. Marie. To the west, the climate grows progressively drier from north to south. Deciduous woodland grows in northern latitudes giving way to semiarid "spiny desert" in the south (Fig. 7.1). The lowest rainfall recorded is at Anakao, only 310 mm/year. The porous,

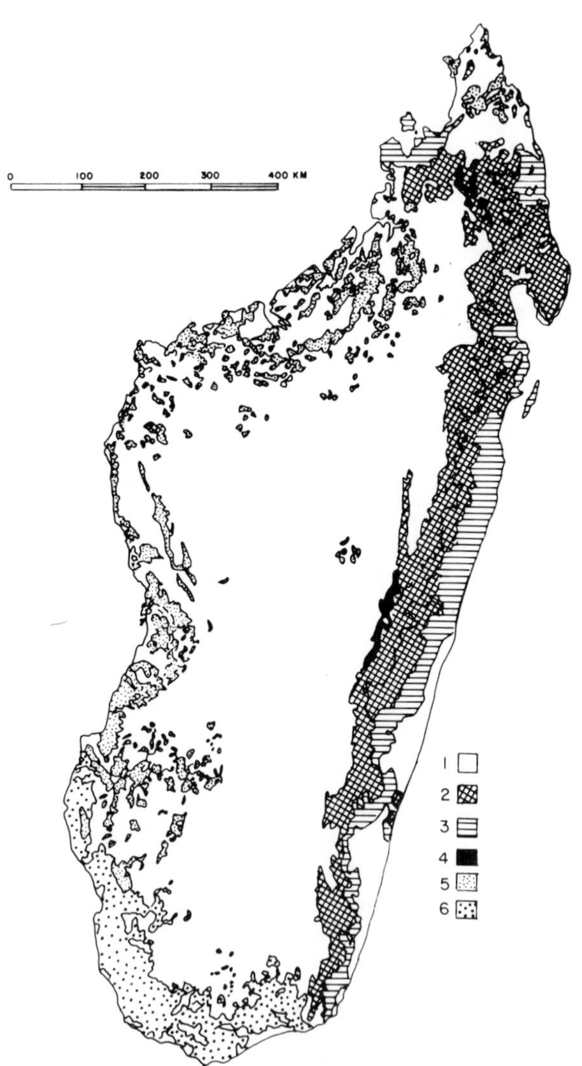

FIGURE 7.1. Existing vegetal formations in Madagascar (modified after Humbert and Cours Darne, 1965). 1: savanna and steppe formations, grassland; 2: dense rain forest; 3: savoka (heliophilous humid secondary forest formation); 4: montane forests; 5: deciduous and sclerophyllous forests; 6: xerophilous bush and thicket formations. Tattersall 1982. Figures 7.1 through 7.10 © 1982, Columbia University Press. By permission.

calcareous substrate of the southwest, which drains away even what little rain there is, supports an endemic flora that resembles the cactus forests of Mexico (Paulian, 1984).

Madagascar has an extraordinary rate of endemicity. In the rainforest, 90% of flowering plant species are unique to the island; in the spiny desert, 95%. All the amphibians and terrestrial mammals are endemic, nearly all the reptiles, and half the birds and bats. Several families exist only there; its lemurs compose four or five of the 12 or 13 families of living primates.

Geologists are still in dispute over the origins of Madagascar. It may have broken away from Africa 100 million years ago, or 200 million, or even longer (Paulian, 1984). What seems clear is that it acquired a fairly complete Gondwanaland fauna—ancestral frogs, boa constrictors, iguanid lizards, and the flightless elephant birds, the Aepyornithidae. Mammals evolved later. The mammalian fauna, including lemurs, could all have plausibly rafted to Madagascar at a period of lowered sea level, at a time when the island had not drifted too far from Africa, during the Eocene. Since 40 million years ago, there have been no further invasions of mammals until people brought their bush pigs (now forest boars), zebu cattle, dogs, cats, rats, and goats.

The indigenous Malagasy mammals are cricetid rodents, tenrecoid insectivores, viverrid carnivores, bats, and lemurs. The extraordinary radiation of lemurs was not hindered by competing monkeys or even ungulates, nor by the predation of large carnivores. A now extinct large fossa (*Crypoprocta spelea*) probably hunted with leopard-like stealth, but the ocelot-like *C. ferox* is the largest remaining viverrid.

The primates occupy all remaining forest, from humid east to dry south. We do not usually think of primates in semiarid lands, until reminded of hanuman langurs near Jodhpur or baboons on the Tibesti Plateau. In Madagascar, white sifaka (*Propithecus verreauxi*) and *Lepilemur* species live in the spiny desert, apparently obtaining all their water from dew and the leaves they eat. Madagascar has the smallest living primate, *Microcebus murinus*, the mouselemur. Until a thousand years ago, its primates also ranked with the largest. *Megaladapis* and *Archaeoindris* were in the size range of modern great apes. Madagascar is the fourth country in the world in total number of primate species, and the first in number of endemic species (Mittermeier and Oates, 1985). If one adds the 14 species extinct in the last thousand years following human settlement, it does indeed look more like a continent than a country.

Tables 7.1 and 7.2 give the array of dead and living lemur species, with their sizes, principal means of locomotion, and diet type.

Distribution in the Wild

It is safe to say that no western scientist knows the present distribution of any lemur species. There may be Malagasy foresters who have walked through the entire habitat of a species, but even they are probably not sure if it exists or not in other forests.

What determines our current knowledge is the existence of roads. Where a reserve is touched by a road, as in the Ankarafantsika that lies athwart the

TABLE 7.1. Living lemurs.

		Mean body weight kg (n)	Mean cranial length mm (n)	Karyotype (2N)	Locomotion[1]	Diet[2]
Family Lemuridae[3]						
Lemur catta	ringtailed l.	2.8 (11)	84 (27)	56	QT	Fr
L. mongoz	mongoose l.	2.4 (5)	85 (31)	60	Q	Fr
L. macaco	black l.	2.4 (11)	91 (39)	44	Q	Fr
L. fulvus fulvus	brown l.	2.6 (9)	90 (22)	60	Q	Fr
L. f. albifrons	white fronted l.	2.3 (8)	90 (8)	60	Q	Fr
L. f. rufus	redfronted l.	2.7 (15)	89 (44)	60	Q	Fr
L. f. collaris	redcollared l.	2.5 (5)	92 (16)	50,52	Q	Fr
L. f. mayottensis	Mayotte l.		91 (14)	60	Q	Fr
L. f. sanfordi	Sanford's l.	2.2 (1)	89 (11)	60	Q	Fr
L. f. albocollaris	white collared l.		88 (1)	48	Q	Fr
L. coronatus	crowned l.	~2	80 (16)	46	Q	Fr
L. rubriventer	redbellied l.		87 (52)	50	Q	Fr
Varecia variegata variegata	black and white ruffed l.	3.8 (5)	105 (33)	46	Q	Fr
V. v. rubra	red ruffed l.	3.9 (2)	104 (13)	46	Q	Fr
Family Lepilemuridae						
Lepilemur mustelinus mustelinus	sportive or weasel l., lepilemur		59 (24)	34	VL	Fol
L. m. ruficaudatus		0.915	56 (10)	20	VL	Fol
L. m. dorsalis			53 (2)	26	VL	Fol
L. m. leucopus		0.544 (10)	51	26	VL	Fol
L. m. edwardsi			58 (21)	22	VL	Fol
L. m. septentrionalis			53 (5)	34,36, 38	VL	Fol
Hapalemur griseus griseus	eastern gentlel. eastern hapalemur	0.830 (4)	65 (24)	54	VL	Bamboo
H. g. alaotrensis	Lake Alaotra h.		70 (4)	54	VL	Reeds
H. g. occidentalis	western h.		62 (7)	58	VL	Bamboo
H. simus	broad-nosed h.		81 (4)	60	VL	
Family Indriidae						
Indri indri	indri		103 (33)	40	VL	Fol
Avahi laniger laniger	eastern woolly l.	0.600–0.700	55 (19)	66	VL	Fol
A. l. occidentalis	western woolly l.	0.859 (4)	50 (3)		VL	Fol
Propithecus diadema diadema	diademed sifaka	6.5 (2)	92 (15)	42	VL	Fol
P. d. candidus			91 (7)		VL	Fol
P. d. edwardsi			88 (16)		VL	Fol
P. d. holomelas			86 (4)		VL	Fol
P. d. perrieri			87 (2)	42	VL	Fol
P. verreauxi verreauxi	white sifaka		81 (31)	48	VL	Fol
P. v. coquereli		3.9 (4)	82 (23)	48	VL	Fol
P. v. deckeni			84 (11)	48	VL	Fol
P. v. coronatus			84 (7)	48	VL	Fol
Daubentonia madagascariensis	aye-aye	2.8 (1)	87 (14)	30	Q	I

TABLE 7.1. (Continued).

		Mean body weight kg (n)	Mean cranial length mm (n)	Karyo-type (2N)	Loco-motion[1]	Diet[2]
Family Cheirogaleinae						
C. major	greater dwarf l.	0.450	55 (30)	66	Q	
C. medius	fat tailed dwarf l.	0.333 (11)	41 (41)	66	Q	Fr,I
Microcebus murinus	grey mousel.	0.060	32 (193)	66	Q	Fr,I
M. rufus	red mousel.	0.045–0.080	33 (40)	66	Q	Fr,I
Mirza coquereli	Coquerel's mousel.	~0.300	50 (20)	66	Q	I
Allocebus trichotis	Hairy-eared dwarf l.		37 (2)		Q	
Phaner furcifur	forked l.	0.350–0.500	54 (25)	48	Q	G

[1]Locomotion: Q, quadrapedal; VL, vertical bodied leaper; QT, partly terrestrial quadrapedal.
[2]Diet: Fr, mainly frugivorous; I, largely insectivorous; Fol, mainly folivorous; G, largely gummivorous.
[3]Taxonomy after Tattersall (1982). I have no personal opinions about taxonomy. *Lepilemur* and *Hapalemur* may be lemuridae. *Lepilemur* may have seven species, not six subspecies. *Mirza* may be a species of *Microcebus*, and *Allocebus* may be a species of *Cheirogaleus*. Tattersall's taxonomy is used in this paper, since the size data and distribution maps are from his book.
Source: Data from Tattersall (1982).

TABLE 7.2. Subfossil (recently extinct) lemurs.

	Cranial length (N)	Locomotion[1]	Diet[2]	Region[3]
Family Lemuridae				
Varecia (or *Lemur*) *insignis*	115	Q	Fr	Pl,NW,W,SW,SE
V. jullyi	122 (11)	Q	Fr	Pl
Family Lepilemuridae				
Hapalemur gallieni = *simus*?		VL		Pl
Megaladapis madagascariensis	241 (3)	VC	Fol	W,SW,SE
M. edwardsi	296 (10)	VC	Fol	W,SW,SE
M. grandidieri	289 (3)	VC	Fol	Pl
Family Indriidae				
Mesopropithecus pithecoides	98 (4)		Fol	Pl
M. globiceps	92 (4)		Fol	W
Archaeolemur majori	130 (17)	QT	Fr	W,SW,SE
A. edwardsi	147 (17)	QT	Fr	Pl
Hadropithecus stenognathus	136 (2)	QT	grass, seeds	Pl,W,SW,SE
Paleopropithecus ingens	194 (8)	Sus	Fol	Pl,W,SW,SE
Archaeoindris fontoynonti	269 (1)		Fol	Pl
Family Daubentoniidae				
Daubentonia robusta		Q		W

[1]Locomotion: Q, quadrupedal; QT, partly terrestrial quadrupedal; VL, vertical leaper; VC, koala-like vertical clinger; Sus, suspensory.
[2]Diet presumed from tooth structure. Diet: Fr, mainly frugivorous; Fol, mainly folivorous.
[3]"Plateau" species found at Ampasambazimba and other sites now 100 km from remaining forest; 14 species, surviving or extinct, coexisted at Ampasambazimba.
Source: Data from Tattersall (1982), Dewar (1984).

"highway" from Anatananarivo to Mahajunga, there we know a great deal about the fauna (Attenborough, 1961; Petter, 1962; Petter-Rousseaux, 1962; Richard, 1978; Albignac, 1981). Some reserves, let alone unnoted forests, which have no roads, may have remained virtually undocumented since the 1930s.

The remedies are now to hand. Satellite mapping is being attempted by several groups of scientists, to compare with the map of Humbert and Cours-Darne (1965). Even crude comparison, without extensive "ground truth," should tell us in the course of this year how many forest patches remain in roughly the same configuration and how many have been flattened into savanna. This will allow at least maximum estimates of the amount of habitat remaining in the original range of each species.

After this, we must turn from high-tech to primeval technology. Satellites can show us where the forests are but only walking through the forests and talking with local people will teach us where animals have been hunted out or failed to adapt to changed conditions. There is, as yet, no real substitute for feet.

There have been, I believe, recent attempts to survey the possible range of only four species. The aye-aye seems to have occurred early in this century in both western and eastern Madagascar and the Sambirano (Tattersall, 1982; Petter et al., 1977). However, when Petter and Peyrieras (1970) made their survey, they only heard reports of a few animals seen or killed long before in the eastern escarpment forests and the Sambirano regions. On the other hand, about 20 aye-ayes were located in the rapidly dwindling eastern coastal forests, near villages where the people believed that proximity of an aye-aye meant certain evil luck or death. Nine of these animals were captured and released on the 5 km^2 island reserve of Nosy Mangabe in 1966 (Petter and Peyrieras, 1969). Nosy Mangabe has at least two aye-aye today (Constable et al., 1985), and there continue to be rare reports of living animals on the mainland (Vaohita, personal communication). Aye-aye may, in fact, still live in both eastern and western zones (O'Connor and Pidgeon, personal communication; Dokobé, personal communication).

Petter's attempts to relocate the elusive *Hapalemur simus*, described a century earlier by Gray, led first to the identification of the large lake Alaotra subspecies of *H. griseus*, *H. g. alaotrensis*. Subsequently, Randrianasolo recognized and Albignac and Rumpler named the western form, *H. griseus olivaceus*. In this climate of interest in *Hapalemur*, the eventual rediscovery by Peyrieras and Petter of *H. simus* in 1972 in the eastern forest near Ranomafana may mean that *simus* is indeed highly localized to that forest region (Petter et al., 1977).

Propithecus diadema perrieri, probably the rarest of the sifaka subspecies, was discovered by Lavauden in 1931, photographed by Petter (Petter et al., 1977) and the subject of two *diplomes de fins d'études* by foresters of the Ecole d'Enseignement Superieur des Sciences de l'Agriculture. All found their subjects in the dry, calcareous area of Analamerana south of Diego Suarez, though the particular forest patches studied were cleared by slash and burn cultivators between each field trip, so each of the scientists located different populations in new patches.

Finally, Petter's 1975 attempt to find the habitat of *Allocebus trichotis*, after Peyrieras caught one living specimen near Mananara in 1965, did not succeed.

These four attempts, three of which end in statements such as "at least it exists just *here* and for the moment that is the known range" are exactly the form of statements we have for the more widely distributed lemurs. In fact, we have no reports from presumptive boundary zones or areas where closely related species

TABLE 7.3. Population status.[1]

	Region[2]	IUCN Status[3]	Reserves where occurs[4]	Comments
Family Lemuridae				
Lemur catta	S,SW	NT	5,10,11 Berenty	Abundant locally, semiterrestrial, and tolerates secondary forest; needs standing water or very juicy vegetation
L. mongoz	N,W,C	E/V	7	
L. macaco	Sa	E/V	4,6	A few lemur troops and small forests are sacred to Sakalava villagers
L. fulvus fulvus	E,NW	NT/V?	3?,7 Perinet	Disjunct distribution: a former plateau species?
L. f. abifrons	NE		1,3?	
L. f. rufus	W,SE		8,9,5	Most widely distributed subsp.; disjunct distribution, former plateau species?
L. f. collaris	SE		11	
L. f. mayottensis	C			
L. f. sanfordi	N	E		Restricted range
L. f. albocollaris	SE			Restricted range
L. coronatus	N	IK	Mt. d'Ambre	Restricted range
L. rubriventer	E	E?	1,2,4?,5,12	Restricted range
Varecia variegata variegata	E	E/V	1,3,5 Nosy Mangabe	
V. v. rubra	E			Restricted range
Family Lepilemuridae				
Lepilemur mustelinus mustelinus	E	IK	1,3,5,11 Perinet	All lepilemur abundant in local populations
L. m. ruficaudatus	W	E?	7	
L. m. dorsalis	Sa	V?	6	
L. m. leucopus	S	E/V	10,11, Berenty	
L. m. edwardsi	NW	IK	7,8,9	
L. m. septentrionalis	N	IK	Mt. d'Ambre	
Hapalemur griseus grieus	E	IK	1,3,4,12, Perinet	Specialized bamboo feeder; all subsp. vulnerable
H. g. alaotrensis	E	IK/E		Restricted range, habitat burned
H. g. occidentalis	W			Disjunct, restricted range
H. simus	E	E		Punctate range
Family Indriidae				
Indri indri	E	E	1,3,Perinet	
Avahi laniger laniger	E	IK	1,2,11, Perinet	
A. l. occidentalis	NW	IK	7	
Propithecus diadema diadema	E	E	1?,3	
P. d. candidus	NE			
P. d. edwardsi	I		5?	
P. d. holomelas	SE			Restricted range
P. d. perrieri	NE		Analamera	Punctate range
P. verreauxi verreauxi	S,SW	IK/NT?	11,Berenty	

TABLE 7.3. (Continued).

	Region[2]	IUCN Status[3]	Reserves where occurs[4]	Comments
P. v. coquereli	NW		7	
P. v. deckeni	W			
P. v. coronatus	Pl,NW			
Family Daubentoniidae				
Daubentonia madagascariensis	E	E	Nosy Mangabe	Extremely rare
Family Cheirogaleinae				
C. major	E	NT	1,3,4, Nosy Mangabe	
C. medius	W	IK	7,9,10 Berenty	
Microcebus murinus	W	NT	7,8,9,10,11 Berenty	Abundant in primary and secondary forest
M. rufus	E	NT	1,3,11,12, Perinet, Nosy Mangabe	Abundant in primary and secondary forest
Mirza coquereli	W	E?	9, Analabe	Disjunct range
Allocebus trichotis	E	E		Punctate range or extinct
Phaner furcifur	E,W	IK/V?	11, Analabe, Mt. d'Ambre	Disjunct but widespread range

[1]All wild lemur populations are declining because of habitat destruction. This table indicates relative urgency, not comfort.
[2]From Tattersall (1984). See figures Sa, Sambirano, C, Comores.
[3]From IUCN (1984). NT, not threatened; V, vulnerable; E, endangered; IK, insufficiently known.
[4]From IUCN (1984) and Andriamampianina (1984). Numbers indicate Reserves Integrales as in map. Names are a few well-known special reserves.

or subspecies might meet, just as we have no measures of the actual amount of forest left in any species' presumed range.

All this said, of course, it is clear that some lemurs are more threatened than others. Table 7.3 gives qualitative, relative descriptions of current populations, and Figures 7.2 through 7.11, all from Tattersall (1982), give approximate distributions.

Population Density

When we turn from extensive to intensive studies, the picture becomes suddenly much more hopeful. Richard and Sussman (1975) drew some criticism when they suggested that fairly small reserves could support viable populations of lemurs in the long term. In fact, the observed population densities of lemurs in favorable habitat commonly exceed 100 animals/km² (Table 7.4). There are exceptions: *Indri* and *P. diadema*, for instance, occur at lower density where they exist at all and eastern rainforest mammal densities are generally lower than in the west (Pollock, 1979). There is also a strong caveat. On the sedimentary rocks of western Madagascar, there is a very marked difference between upland, dry

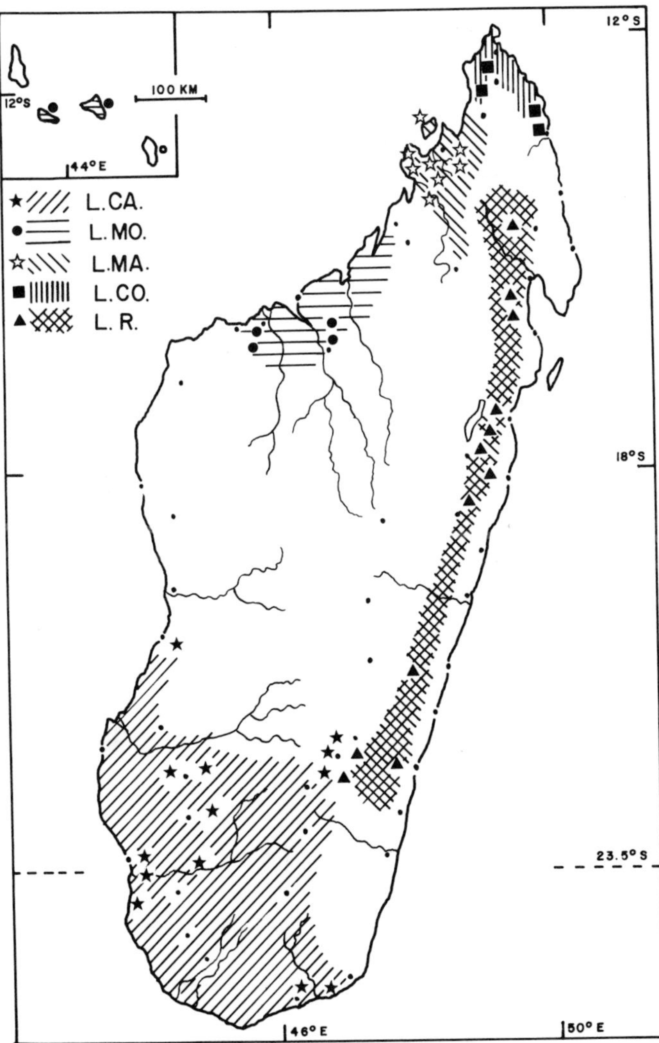

FIGURE 7.2. Distributions of various species of *Lemur*. Shaded areas represent approximate limits of distribution, symbols represent localities of museum specimens. L.CA, *Lemur catta*; L.MO., *L. mongoz*; L.MA., *L. macaco*; L.CO, *L. coronatus*; L.R., *L. rubriventer*. Tattersall 1982.

areas, and lower lying pond or streamside vegetation, even though they are only a few meters apart in height. The high reported densities are from gallery forest, or near ponds. Only Albignac, at Ampijoroa, has worked with the sparser upper level on dry sand (Table 7.4). However, densities of 100/km² are an order of magnitude greater than those of most forest monkeys and a hundred times greater than most reported densities of great apes. Our concern with the perils of

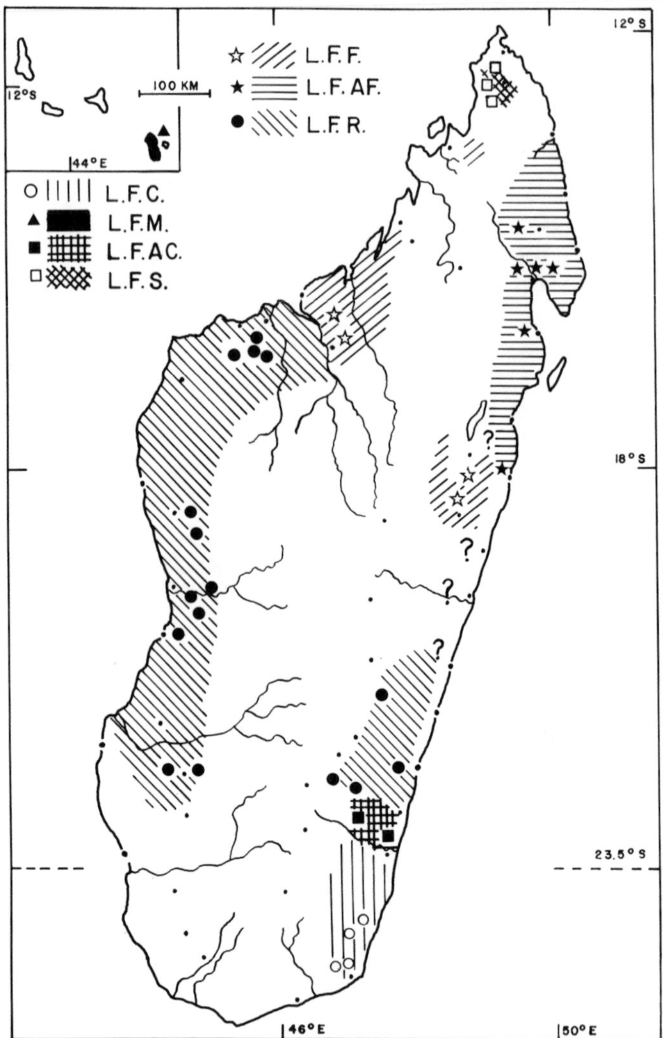

FIGURE 7.3. Distributions of the subspecies of *Lemur fulvus*. Shaded areas represent approximate distribution limits; symbols denote localities of museum specimens. L.F.F., *Lemur fulvus fulvus*; L.F.AF., *L. f. albifrons*; L.F.R., *L. f. rufus*; L.F.C., *L. f. colaris*; L.F.M., *L. f. mayottensis*; L.F.AC., *L. f. albocollaris*; L.F.S., *L. f. sanfordi*. Tattersall 1982.

inbreeding in small populations postdates Richard and Sussman's review (1975), but their conclusion stands: A reserve of even a few square kilometers may hold viable lemur populations, if it contains some well-watered habitat. We should not write off even isolated forest patches in the western savanna as valueless, before someone censuses their fauna. The large nature reserves, even though currently unprotected, surely still contain riches of both fauna and flora.

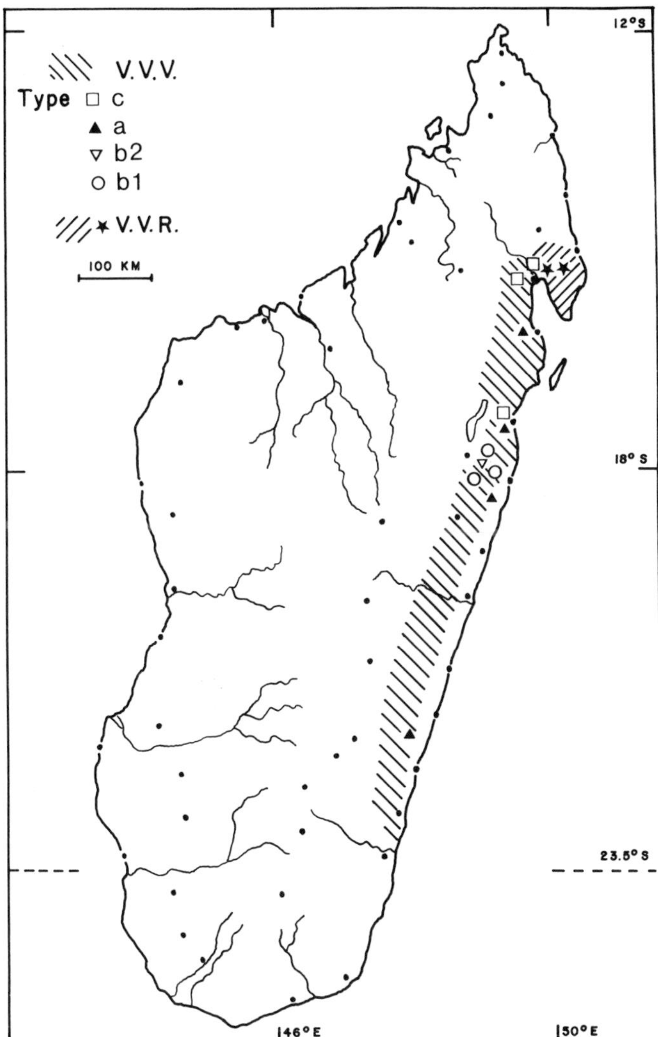

FIGURE 7.4. Distribution of *Varecia*. Shaded areas represent approximate range limits; symbols denote localities of museum specimens. V.V.V., *Varecia variegata variegata*; V.V.R., *V. v. rubra*. Tattersall 1982.

Our experience at Berenty suggests that unless there is intensive hunting, we should worry about the trees, not the lemurs. I used to argue that lemurs, as the largest mammals, would be sensitive indicators of the health of the environment. The *Lemur catta* population has been roughly stable at about 150/km^2 since 1972, the *Propithecus verreauxi* population may have increased from 100/km^2 to a significantly larger figure, and introduced *Lemur fulvus rufus* have grown from about eight animals in 1974 to about 18 in 1984, apparently creating a niche for themselves in competition with the other two species (O'Connor and Jolly, in

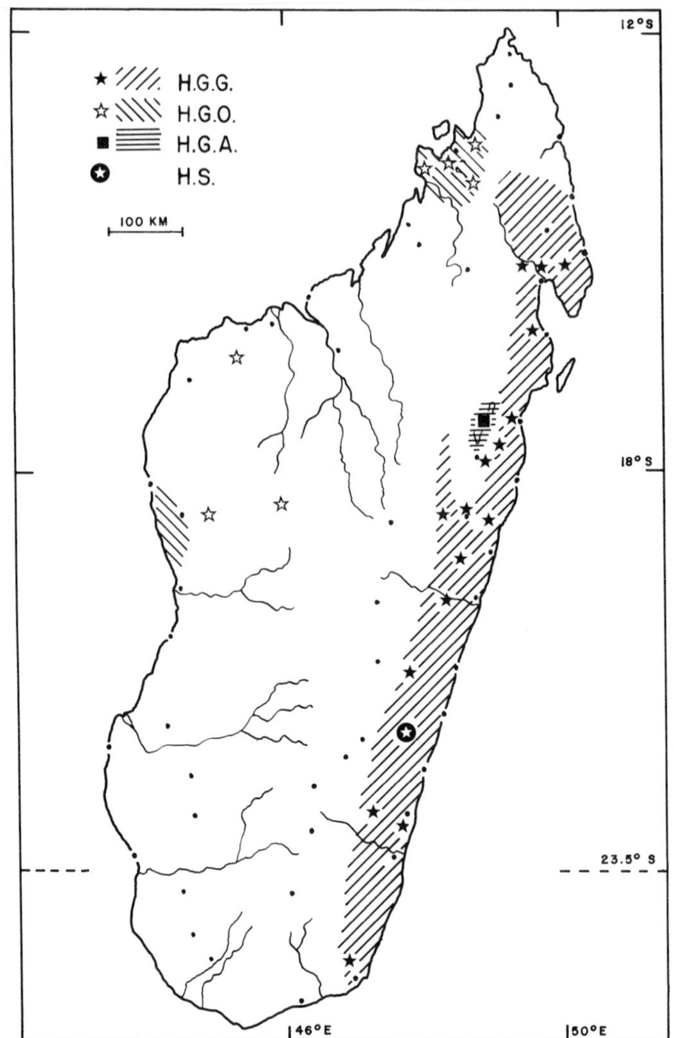

FIGURE 7.5. Distributions of *Hapalemur* species and subspecies. Shaded areas denote approximate range limits; symbols represent localities of museum specimens except in the case of *H. simus*, where the locality is one of capture and observation. H.G.G., *Hapalemur griseus griseus*; H.G.O., *H. g. occidentalis*; H.G.A., *H. g. alaotrensis*; H.S., *Hapalemur simus*. Tattersall 1982.

preparation). Meanwhile, the number of kily trees has markedly decreased (de Heaulme, personal communication; O'Connor, in preparation). The kily, *Tamarindus indica*, is the dominant species of the closed canopy forest and the major food of all three species of lemur.

The causes of the decline are as yet unknown. The forest is being invaded by a sun-loving vine, *Cissus quadralangularis*, which grows over trees and smothers

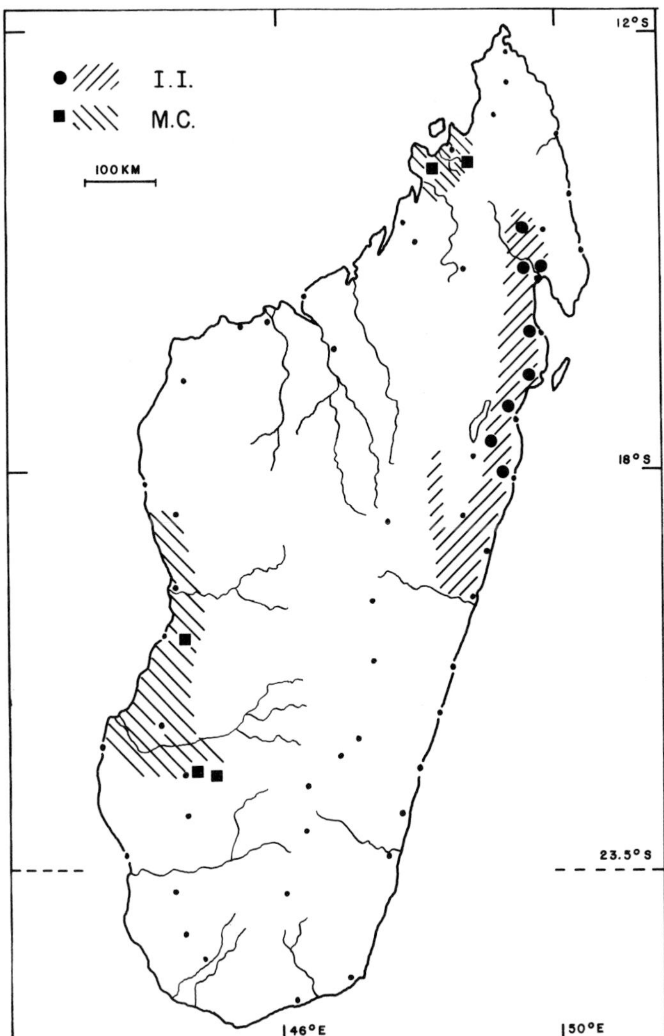

FIGURE 7.6. Distributions of *Indri indri* (I.I.) and *Mirza coquereli* (M.C.). Shaded areas represent approximate range limits; symbols denote localities of museum specimens. Tattersall 1982.

them. The *Cissus* may, paradoxically, benefit from the exclusion of goats and zebu as well as from the human consumption of tortoises that might have browsed on its stems. It may be benefiting from an edge effect, invading from cleared or degraded areas into the closed forest. There may be much more profound causes, such as lowering of the water table as the headwaters of the Mandare River are progressively deforested.

This is a particular case of a much more general moral. In Lovejoy's study of isolated forest stands in the Amazon, the populations of many animals actually

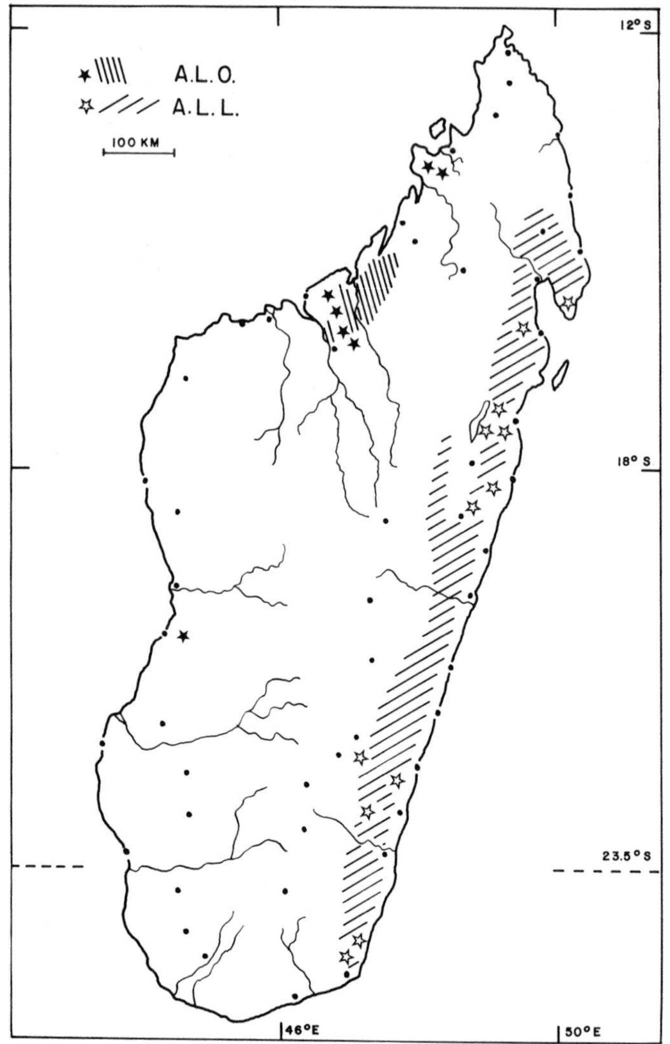

FIGURE 7.7. Distributions of *Avahi laniger* subspecies. Shaded areas represent approximate range limits; symbols denote localities of museum specimens. A.L.O., *Avahi laniger occidentalis*; A.L.L., *A. l. laniger*. Tattersall 1982.

rose just after clearing, as forest species took refuge in the remaining stands, while others benefited from the new mosaic of sunny edge. Again, though, the trees proved more quickly vulnerable, blowing over as they faced the unaccustomed phenomenon of wind (Lovejoy et al., in press).

It seems, then, that primatologists who deal with small species, such as lemurs, must willy nilly turn foresters. Barring intensive hunting for food, if we take care of the trees the lemurs may take care of themselves.

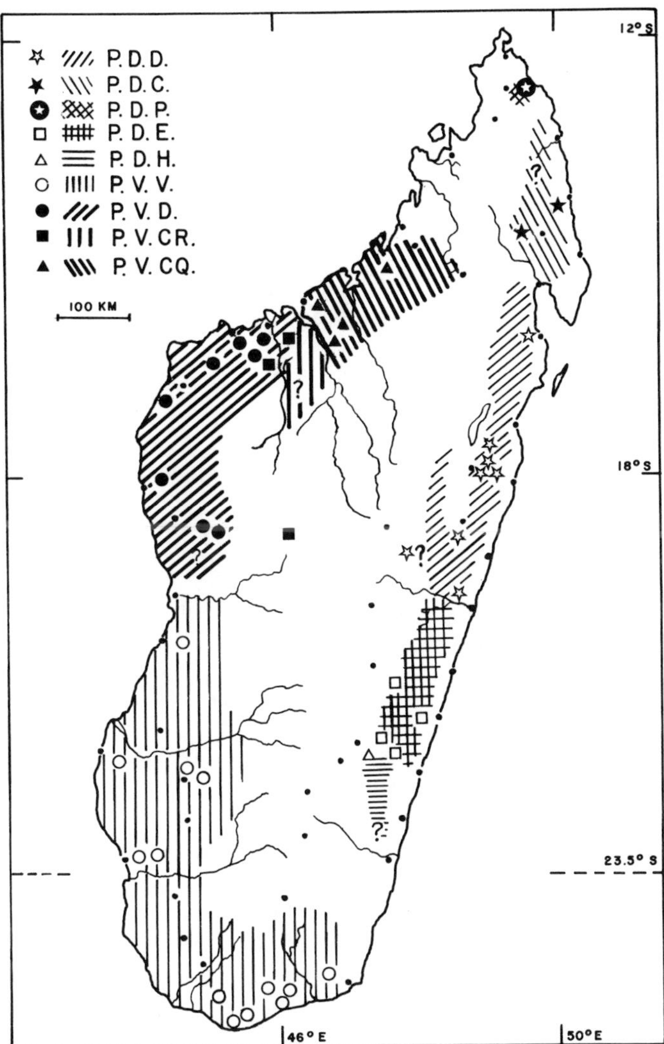

FIGURE 7.8. Distributions of *Propithecus* species and subspecies. Shaded areas represent approximate range limits; symbols denote the localities of museum specimens. P.D.D., *Propithecus diadema diadema*; P.D.C., *P. d. candidus*; P.D.P., *P. d. perrieri*; P.D.E., *P. d. edwardsi*; P.D.H., *P. d. holomelas*; P.V.V., *Propithecus verreauxi verreauxi*; P.V.D., *P. v. deckeni*; P.V.CR., *P. v. coronatus*; P.V.CQ, *P. v. coquereli*. Tattersall 1982.

Adaptability to Captivity and to Novel Situations

Only about half of the species and subspecies of lemurs are currently in captivity. Some, particularly the indriids and lepilemur, that is the folivores, have an extremely poor record of survival in captivity, let alone breeding. Very few, that

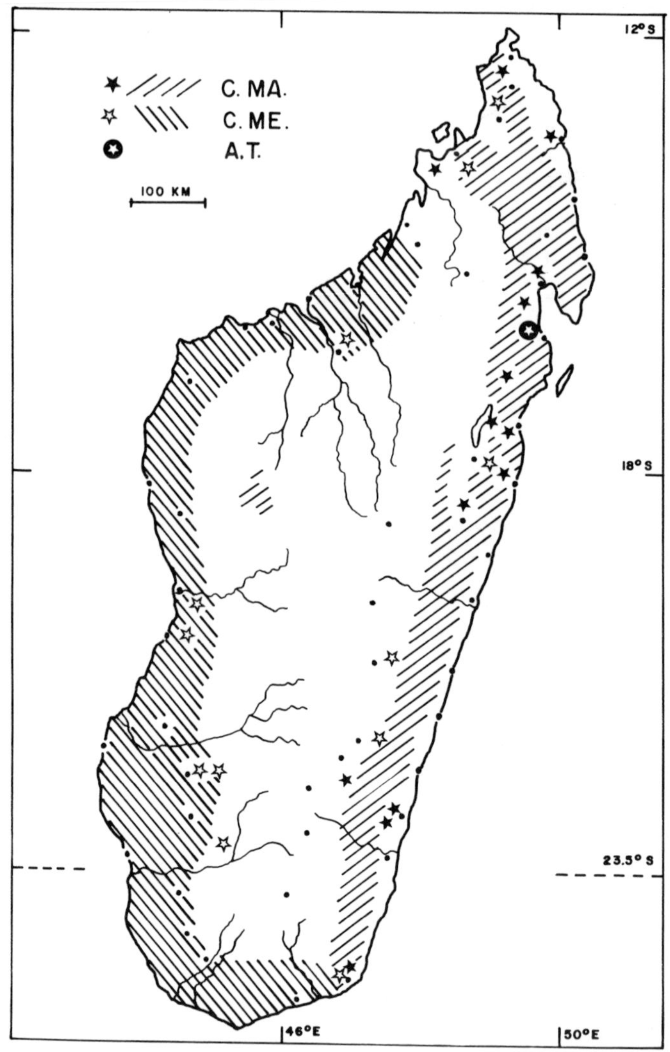

FIGURE 7.9. Distributions of the species of *Cheirogaleus* and of *Allacebus*. Shaded areas represent approximate range limits; symbols denote localities of museum specimens. C.M.A., *Cheirogaleus major*; C.ME, *C. medius*; A.T., *Allocebus trichotis*. Tattersall 1982.

is some *Lemur*, *Varecia*, and *Microcebus*, are currently breeding well enough to increase their populations. On the other hand, these few could increase almost to the point of embarrassment. Looking around the San Diego Zoo, one can imagine a lobby developing for the regazetting of Reserve #2, on the Masoala Peninsula, if only to have someplace to repatriate extra red ruffed lemur twins and triplets. Clearly, the zoos of the world are far from saturated with these beau-

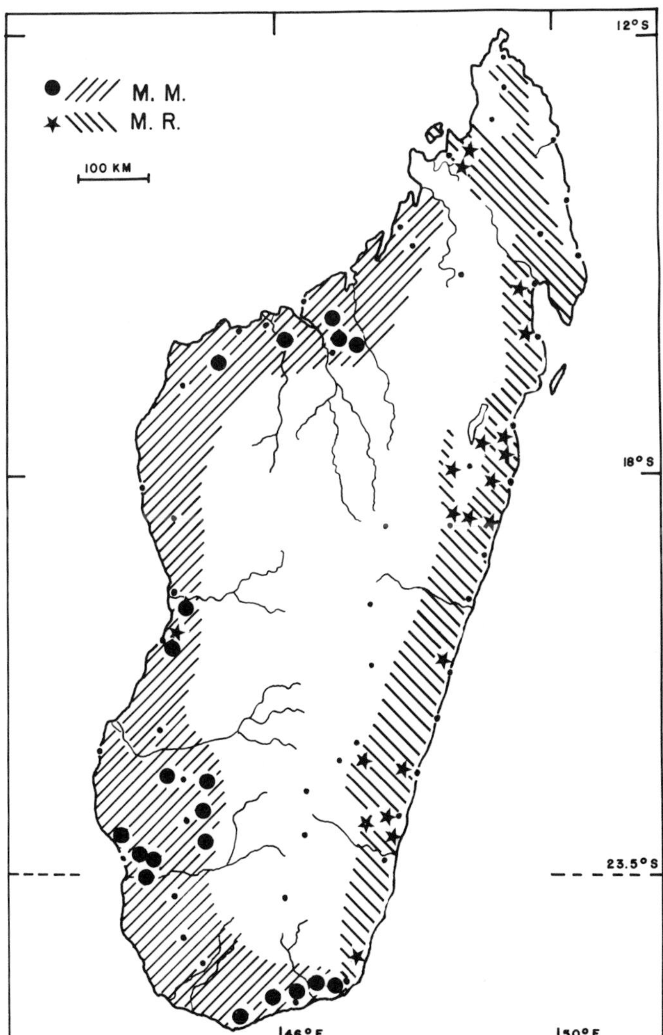

FIGURE 7.10. Distributions of *Microcebus murinus* (M.M.), and *Microcebus rufus* (M.R.). Shaded areas represent approximate range limits; symbols denote localities of museum specimens. Tattersall 1982.

tiful bellowing beasts, nor is the species yet out of danger—but Malthus stalks in the corridors.

Oddly, at least two of the "difficult" folivores seem to me to be good candidates for captive breeding. *Lepilemur* has dense populations in every site it has been studied. It eats fairly mature leaves (Albignac, 1981) and survives without standing water—indeed, without coming to the ground. In Russell's (in preparation) study, the lepilemur actually fed from few of the apparently suitable trees of their range, which leads Russell to argue that adult animals are behaviorally inflexible,

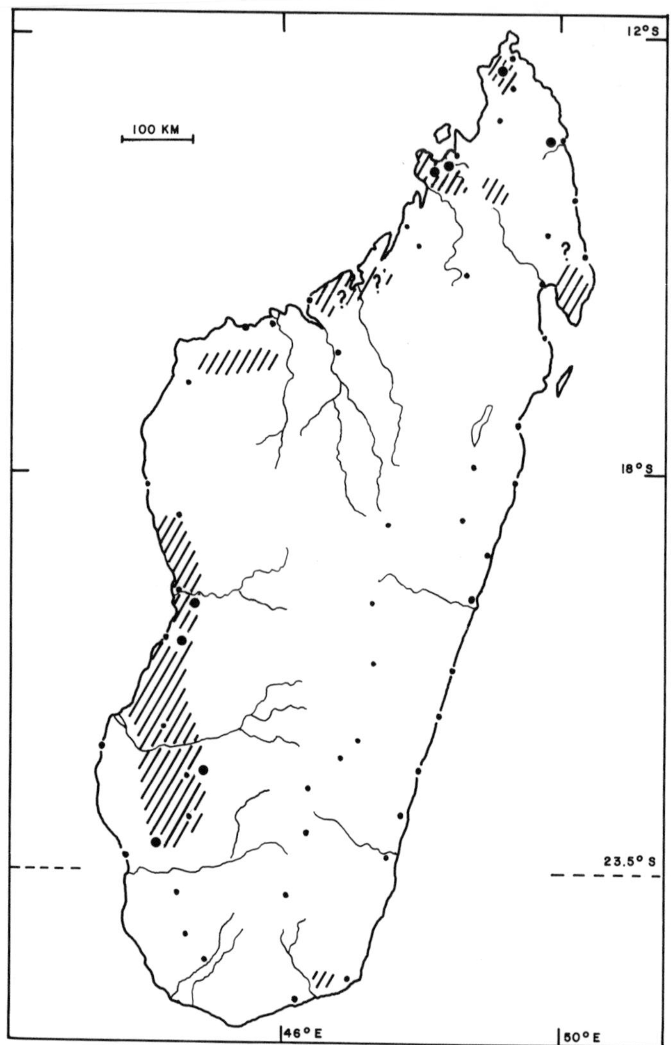

FIGURE 7.11. Distribution of populations of *Phaner furcifer*. Shaded areas represent approximate range limits; symbols denote localities of museum specimens. Tattersall 1982.

using only those food trees they learn as infants or migrating juveniles. However, if their physiological tolerance is as robust as their distribution in the wild suggests, there seems every hope of keeping them in zoos.

Similarly, the white sifaka, *Propithecus verreauxi*, feeds on different foods and with differing ranging and territorial behavior, in various parts of its range (Richard, 1978). At Berenty, where the forest edge meets human habitation, the sifaka bounce off the picnic tables or corrugated roof tops to travel to the flowers and new leaves of introduced ornamental trees. I have seen sifaka siesta-ing in

TABLE 7.4. Some reported population densities.

	Est. Pop. Dens/km²		Biomass g/ha
Analabe: Deciduous western woodland, 50 km north of Morondava, dominated by *Adansonia grandideri* (Hladik et al., 1980, from 30-ha study area near seasonal pond).			
Microcebus murinus	400		200
Cheirogaleus medius	350		500–700
Mirza coquereli	30		100
Phaner furcifur	50–60		200
Lepilemur ruficaudatus	250		2000
Lemur fulvus rufus	present		present
Propithecus v. verreauxi	present		present
Ampijoroa: Deciduous western woodland, dry sandy area not near pond, in Reserve #7, Ankarafantsika (Albignac, 1961, 16-ha study area).			
Microcebus murinus	75		
Cheirogaleus medius	12		
Lepilemur edwardsi	250		
Avahi laniger occidentalis	60–80		
Lemur fulvus fulvus	3–12		
Lemur mongoz	0	(4/km² near water)	
Propithecus verreauxi coquereli	75		
Berenty: Deciduous western gallery forest west of Ft. Dauphin, closed canopy dominated by *Tamarindus indica* and more open scrub. (Diurnal, 1 km² census. Jolly et al., 1982a, 1982b; Mertl-Millhollen et al., 1979; nocturnal transects in closed canopy area, Russell, in preparation.)			
Microcebus murinus	39		
Cheirogaleus medius	37		
Lepilemur leucopus	450–520		
Lemur catta	150		
Lemur fulvus rufus	15	(recently introduced)	
Propithecus verreauxi	87–128		
Berenty and Hazafotsy: Southern "spiny desert" dominated by *Alluaudia* (Russell, in preparation; Richard, 1978).			
Microcebus murinus	123		
Lepilemur leucopus	230		
Propithecus verreauxi	160		
Antseranomby: Closed canopy forest dominated by *Tamarindus* (10-ha study area; Sussman, 1974).			
Microcebus murinus	present		
Cheirogaleus medius	present		
Phaner furcifer	present		
Lepilemur m. ruficaudatus	200		
Lemur fulvus rufus	1227		
Lemur catta	215		
Propithecus verreauxi verreauxi	present		
Perinet: Eastern mid-altitude rainforest (Pollock, 1979).			
Microcebus rufus	present		
Cheirogaleus major	present		
Lepilemur	present		
Hapelemur g. griseus	47–62		
Lemur f. fulvus	present		
Lemur rubriventer	present		
Varecia variegata	seen 1958		
Avahi l. laniger	present		
Propithecus d. diadema	present 1970s		
Indri indri	9–16		

the roof trusses of the airplane hanger. This does not seem like behavior of a totally unadaptable animal.

We are far from understanding the needs of such lemur species in captivity, but this may be due more to lack of knowledge of their life in the wild, or to lack of imagination, than to inherent impossibility.

Threats and Hopes for the Malagasy Forests

The future Madagascar has been called "a tragedy without villains" (Jolly and Jolly, 1984). There are no big bad logging companies, no ill-conceived international highway schemes. Instead, millions of peasant farmers practice slash-and-burn cultivation in the rainforest. They set fire to western grassland, and the dry forest with it, for a "green bite" to tide cattle over the dry season. Increasingly, people cut fuelwood for the towns. Everyone in Madagascar—farmers, townspeople, luxury hotels—cooks on wood or charcoal. The economic recession means that imported goods are expensive or simply cannot be obtained, which includes kerosene for cooking.

There have been repeated attempts to reverse the trend. The last precolonial Malagasy government forbade felling virgin forest, in the Code of 305 articles of 1881. Perrier de la Bathie in 1921 presented a cogent botanist's view that the Malagasy savannas are man-made. (Almost all the endemic plants live in forest or rock formations; the savannas contain mostly common grassland species of Africa.) It may be that Madagascar was originally more of a woodland-savanna mosaic than Perrier de la Bathie believed, but even the mosaic version supposes that the human population's zebu and goats have wholly changed the flora once grazed by giant tortoises and elephant birds (Dewar, 1984; McPhee et al., 1985).

Such considerations led to the founding, in 1927, of an extraordinary set of wilderness reserves, "Reserves Naturelles Integrales" (Fig. 7.12; Table 7.5). These were among the earliest reserves in Africa. They embodied a concept we have reached more recently in America: that some irreplaceable biological areas should be protected even from causal recreation, let alone development. Legal entry is by permit, obtainable only in Antananarivo from the Direction des Eaux et Forêts. These wilderness reserves are supplemented by two national parks, meant for general wonder and enjoyment: The Montagne d'Ambre in the north and the Isalo, a rock wilderness of sandstone castles, in the south. There are also 21 smaller special reserves such as the one for indri at Perinet and that for aye-aye at Nosy Mangabe. The reserves were chosen by botanists, in the preroad era when scientists walked to exciting places, without constantly checking the date on their return ticket. They are thus as fine a selection as one could hope to make of the areas that should be preserved.

However, the system was set up by French biologists; the people to be excluded were Malagasy. When a French logging company proposed degazetting Reserve #2, the Masoalala peninsula, this was done. Only the fact that logging proved uneconomic saved the habitat of red ruffed lemur and Madagascar serpent-eagle. Meanwhile, as independence came (de jure 1960, de facto 1972–1975), foresters found it impossible to forbid peasant access to forest lands in the reserves or

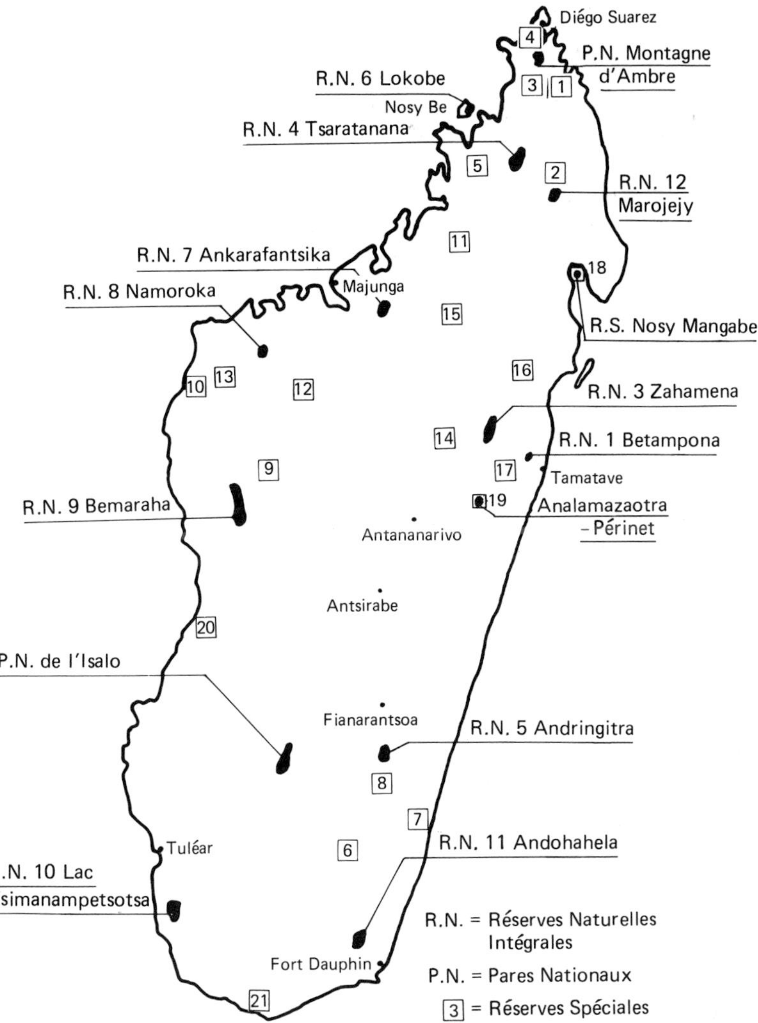

FIGURE 7.12. Map of Natural Reserves (Andriamampianina, 1984).

elsewhere. The IUCN International Congress of 1970 in Antananarivo had welcome publicity but few practical results.

Now, however, the situation may be changing. Madagascar has its own representative of WWF, a University-affiliated reserve at Bezaha Mahafaly, and plans for a new reserve at Analabé near Morondava (Sussman et al., 1985).

A top-level interministerial committee in December 1984 has debated and signed a major document: "Stratégie Malgache pour la Conservation et le Développement Durable." This strategy declares in the strongest terms that Madagascar must depend upon itself: its own soil, its own water. This means checking the erosion that silts up the fields of paddy rice or strips the topsoil from

TABLE 7.5. Malagasy reserves.[1]

Reserves Naturelles Integrales: Wilderness Reserves
1. Betampona, 2228 ha (1927)
2. Masoalala Penninsula, 30,000 ha (1927, degazetted, 1964)
3. Zahamena, 73, 160 ha (1967)
4. Tsaratanana Massif, 48,622 ha (1927)
5. Andringitra Massif, 31,160 ha (1927)
6. Lokobe, 740 ha (1927)
7. Ankarafantsika Plateau, 60,520 ha (1927)
8. Tsingy (calcareous plateau) of Namoroka, 21,742 ha (1927)
9. Tsingy of Bemaraha, 152,000 ha (1927)
10. Lake Tsimanampetsotsa, 43,200 ha (1927)
11. Andohahela, 76,020 ha (1929)
12. Marojejy Massif, 60,150 ha (1929)

National Parks
1. Isalo Massif, 81,540 ha
2. Montagne d'Ambre, 18,200 ha

Special Reserves

Province d'Antsiranana (Diego Suarez)
1. Analamerana, 34,700 ha (1956)
2. Anjanaharibe-Sud, 23,100 ha (1956)
3. Ankara, 18,220 ha (1956)
4. Foret d'Ambre, 4810 ha (1958)
5. Manongarivo, 35,250 ha (1956)

Province de Fianarantsoa

6. Kalambatritra, 28,250 ha (1959)
7. Monombo, 5020 ha (1962)
8. Pic d'Ivohibe, 3450 ha (1964)

Province de Mahajanga (Majunga)

9. Ambohijanahary, 24,750 ha (1958)
10. Bemarivo, 11,570 ha (1956)
11. Bora, 4780 ha (1965)
12. Kasijy, 18,800 ha (1956)
13. Maningozo, 7900 ha (1956)
14. Morotandrano, 42,400 ha (1956)
15. Tampoketsa d'Analamaitso, 17,150 ha (1958)

Province de Toamasina (Tamatave)

16. Ambatovaky, 60,050 ha (1965)
17. Mangerivola, 800 ha (1958)
18. Nosy Mangabe, 520 ha (1965)
19. Analamazaotra-Perinet, 810 ha (1970)

Province de Toliary (Tulear)

20. Andranomena, 6420 ha (1958)
21. Cap Sainte Marie, 1750 ha (1962)

University Reserves
1. Bezaha Mahafaly, 600 ha (1980)

Private Reserves
1. Berenty, 200 ha (1930s)
2. Analabe, 4000 ha (1930s)

[1]After Andriamampianina (1984). Founding data given in parentheses.

dryland farms. It means guarding forest on the watersheds, to preserve river runoff. It urgently means fuelwood plantations and fuelwood stands near every village. That, in turn, demands control of the wildfires and encouragement of fodder crops. In short, the government has declared that Malagasy must abandon their frontier mentality and conserve both their forest and their agricultural lands for an enduring future.

One may demur that this is utopian fantasy. Governments have been trying to stem the tide since the Code of 305 articles; will yet another document change people's ways? I can only argue that people do change, when they believe they will gain. The dustbowl, the depression, and the New Deal changed American attitudes to our land. Malagasy face very similar, and equally desperate problems today. As in America, conservation will never be won once and for all—the cowboy mentality is always ready to reemerge. But there are signs that this is a time in Madagascar when the conservation ethic will gain ground.

What can the international community do, at this point? Right now we can send delegates to the International Conference on Conservation and Enduring Development, to be held in Antananarivo 4–12 November 1985. Such delegates should understand that the conference, like the Strategie, is not primarily about saving fuzzy lemurs or serpent-eagles or plowshare tortoises: It is about saving people, agriculture, soil, and forest all together. A Malagasy view of conservation must start from human needs and human pride.

Second, we can send money. The entire recurrent budget for the National Reserve System last year, exclusive of salaries, came to $900. Any increase within the limits of ability to spend usefully would help. International organizations that already have commitments to conservation in Madagascar are listed in Table 7.6.

Finally, we can send expertise. Scientists wishing to apply for research visas should consult the International Advisory Group of Scientists, whose addresses are also given in the appendix. The role and functions of the IAGS may well change after the November conference. In any future version, the provisions will remain that visiting scientists must offer transport and collaboration to Malagasy colleagues and that conservation must form part of any scientific visitor's goals.

Summary and Conclusion

Madagascar is uniquely rich in primate fauna, with four or five of the 12 or 13 primate families, 20 species, and 40 subspecies, all endemic to the island continent. It is uniquely threatened: 14 species of giant lemur have become extinct in the past thousand years, as at least 80% of the forests have been cleared. The second wave of extinction is taking place today. However, the remaining lemurs can live at high density in favorable habitat and may be more adaptable than usually credited.

The country boasts on paper, and partially in fact, a fine system of national wilderness reserves. It will take national political will and international finance to protect this rich heritage. There are signs of hope, in the formal adoption by the Malagasy government of the "Stratégie Malgache pour la Conservation et le Développement Durable," and the convening of an international conference on

TABLE 7.6. Foreign organizations currently aiding Malagasy conservation.

Duke University Primate Center 3705 Erwin road, Durham, NC 27706
Jersey Wildlife Preservation Trust Les Augres Manor, Jersey, C.I., United Kingdom
Missouri Botanical Garden P.O. Box 299, St. Louis, MO 63166
Museum Nationale d'Histoire Naturelle Laboratoire d'Ecologie Generale 4 Ave de Petit Chateau, 91800 Brunoy, France
Wildlife Preservation Trust International 34th Street and Girard Ave, Philadelphia, PA 19104
World Wildlife Fund, Intl. Ave du Mont Blanc, CH-1196, Gland, Switzerland
World Wildlife Fund, U.S. 1601 Connecticut Ave., NW, Washington, DC 20009
World Wildlife Fund, U.K. Panda House, 29 Greville Street, London EC1, UK

implementation of the Strategie in 1985. Given the political will, it would be relatively easy to offer every species of Malagasy lemur the chance of an enduring future. It must come, however, hand in hand with the wisdom to guard forest, water, and soil not just for the lemurs' sake, but for the Malagasy people.

Appendix: Information Sheet for Foreigners Wishing to Do Biological Research in Madagascar

Biological Research in Madagascar

Basic biological research by foreigners in Madagascar is under the aegis of the Direction de la Formation Post-Universitaire (DFPU) in the Ministère de l'Enseignement Supérieur (MES). The MES supervises the Université de Madagascar and three Départements of the Centre National de Recherches de Tsimbazaza (CNRT): Histoire Naturelle (which runs the Parc Zoologique et Botanique), Entomologie Faunistique, and Sciences Humaines et Sociales.

Administration of the natural and protected areas of Madagascar (the Réserves Naturelles Integrales, Réserves Speciales, and Parcs Nationaux) is by the Direction des Eaux et Forêts (DEF) in the Ministère de la Production Animale et Les Eaux et Forêts (MPAEF). The DEF controls access to and activities (collecting/trapping) in these areas. Permits from the DEF are obtained only in Antananarivo.

Research by foreigners is of great interest to the Ministère de la Recherche Scientifique et Technologique pour le Développement (MRSTD), particularly its Direction de la Recherche sur l'Environment (DRE), and to the Représentation du "World Wildlife Fund" à Madagascar (WWF à Madagascar).

The WWF à Madagascar develops its own conservation projects with the cooperation of the MES, MPAEF, and MRSTD, and these become official government programs.

The major Réserves and Parcs are listed here, with asterisks beside sites of the conservation projects of WWF à Madagascar:

Betampona	Namoroka (Soalala)	*Nosy Mangabe
Zahamena	Bemaraha (Antsalova)	*Montagne d'Ambre
Tsaratanana	Tsimanampetsotsa	Isalo Massif
Andringitra	Andohahela	*Amboitantely
*Lokobe	Marojejy	*Vohidrazana
*Ankarafantsika	Analamazaotra (Perinet)	*Ankazobe

On 1 March 1983, the Malagasy authorities requested that research be relevant to nature conservation and that proposals be screened by an International Advisory Group of Scientists (IAGS), the composition of which was approved by them. The authorities favorably consider research proposed for the sites listed above, especially where conservation efforts are already underway, but they consider research proposed for other sites as well.

Members of International Advisory Group of Scientists (IAGS), 1 March 1983–28 February 1986

Dr. Roland Albignac (ecology and behavior of Malagasy carnivores), Faculté des Sciences, Laboratoire de Zoologie et Ecologie Animale, Université de Besançon, 25000 Besançon, France

Dr. Lee Durrell, Chairman (vocal behavior of Malagasy birds and mammals) Jersey Wildlife Preservation Trust, Les Augrès Manor, Trinity, Jersey, Channel Islands (via UK)

Dr. Alison Jolly (behavior and conservation of primates) The Rockefeller University, 1230 York Avenue, New York, NY 10021, USA

Dr. Bernd-Ulrich Meyburg (ecology and conservation of birds of prey) World Working Group of Birds of Prey (ICBP), Herbertstr. 14, D-1000 Berlin 33, West Germany

Dr. Jean-Jacques Petter (behavior and ecology of prosimians) Museum National d'Histoire Naturelle, Ecologie Générale, 4 Ave du Petit Château, 91800 Brunoy (Essonne), France

Dr. Peter Raven (plant biology) Director, Missouri Botanical Garden, P.O. Box 299, St. Louis, Missouri 63166, USA

Dr. Alison Richard (evolution and behavioral ecology of Malagasy primates) Dept. of Anthropology, 2114 Yale Station, Yale University, New Haven, CT 06520, USA

Process of Applying to Conduct Research in Madagascar

Prepare a proposal according to the guidelines given below and send it to the appropriate member of IAGS, who becomes your Advisor and sends it to the other members of IAGS and usually to a specialist consultant(s) for their comments. Your Advisor may then suggest modification to the proposal according to the comments. (If there are substantial modifications, your Advisor will request from you further fair copies of the proposal.) On favorable recommendation from your Advisor, the proposal will be sent by the Chairman of IAGS to the Director of DFPU, the President of WWF à Madagascar, the Director of DRE, and the Chef du Service de la Conservation de la Nature (DEF). Comments on the proposal will be made by them and any other authorities they may consult. The Director of DFPU will notify you (and the Chairman of IAGS) of the final decision. Reasons for rejection will be given. The successful applicant should attach the letter of approval from the DFPU to the forms requesting an entry visa for Madagascar, which are obtained from your nearest Malagasy Embassy.

It is hoped that this procedure will expedite permission to conduct research in Madagascar, but processing a proposal usually takes several months. If any applicant feels that an unreasonable delay has occurred he or she may write directly to the Chairman of IAGS (Dr. Lee Durrell).

NOTE: Any proposal sent by the applicant directly to Madagascar without going through IAGS will normally not be considered by the authorities. (Exceptions are projects sponsored through institutions having existing formal agreements with the government of Madagascar.)

Guidelines for Writing Research Proposals for Madagascar[1]

A non-French-speaking applicant should submit three copies in French and five copies in English and a French speaker should submit seven copies in French of:

1. A cover page, providing:
   a. name, address, and phone number
   b. title of project
   c. abstract of research objectives
   d. name, address, and phone number of three references (at least one of whom knows him or her very well)
   e. previous experience (and that of any other project participants) that indicates ability to carry out the research
   f. indication of ability to speak French and Malagasy (include any other project participants)
2. Not more than three single-spaced typed pages describing:
   a. the scientific significance of the research
   b. how the research will benefit nature conservation (e.g., providing faunal or floral inventories of little studied area, identifying factors that endanger species and biotic communities)
   c. potential for collaboration with Malagasy colleagues/government personnel/educators (e.g., joint publications, field teams) and potential for Malagasy student training (e.g., field assistants, imparting new techniques in experimental design and field methods)
   d. how the project can be of immediate use to education in Madagascar (e.g., lectures/tours/slides/films offered by the applicant)

[1]Information compiled by Lee Durrell, Chairman of IAGS, 13/4/84.

e. briefly, the practical aspects of the project, i.e., methods, duration, budget, transport, and potential sources of funding
f. statement that the applicant will undertake "... à fournir un nombre d'exemplaires suffisants de leurs rapports et travaux au MES qui se chargera de la diffusion aux differentes organisations nationales concernées." ("to provide copies of reports and published work to MES in sufficient numbers to permit the Ministry to disseminate them to appropriate bodies in Madagascar").

References

Albignac R (1981) Lemurine social and territorial organization in a northwestern Malagasy forest (Restricted Area of Ampijoroa). In: Chiarelli AB, Corruccini RS (eds) Primate behavior and sociobiology. Springer Verlag, Berlin, pp 25–29.

Andriamampianina J (1984) Nature reserves and nature conservation in Madagascar. In: Jolly A, Oberle P, Albignac RC (eds) Madagascar, Pergamon Press, Oxford, pp 329–228.

Attenborough D (1961) Zoo quest to Madagascar. Lutterworth: London (U.S. title: Bridge to the past).

Constable ID, Pollock JI, Ratsirarson J, Simons H (1985) Sightings of aye-ayes and red ruffed lemurs on Nosy Mangabe and the Masoala Peninsula. Prim Conserv (WWF-US Primate Program) 5:59–61.

Dewar RE (1984) Recent extinctions in Madagascar: the loss of the subfossil fauna. In: Martin P, Klein RG (eds) Quaternary extinctions. University of Arizona Press, Tucson, AZ, pp 574–599.

Humbert H, Cours-Darne G (1965) Carte internationale du tapis vegetale. French Institute of Pondicherry.

IUCN Conservation Monitoring Unit (1984) In: Jenkins M (ed) An environmental profile of Madagascar. IUCN, Cambridge, UK.

Jolly A (1980) A world like our own. Yale University Press, New Haven, CT.

Jolly A, Jolly R (1984) Malagasy economics and conservation: a tragedy without villains. In: Jolly A, Oberle P, Albignac R (eds) Madagascar. Pergamon, Oxford, pp 211–218.

Jolly A, Oliver WLR, O'Connor SM (1982a) Population and troop ranges of *Lemur catta* and *Lemur fulvis* at Berenty, Madagascar: 1980 census. Folia Primat 39:115–123.

Jolly A, Oliver WLR, O'Connor SM (1982b) *Propithecus verreauxi* population and ranging at Berenty, Madagascar, 1975 and 1980. Folia Primat 39:124–144.

Lovejoy TE, Rankin JM, Bierregaard RO, Jr, Brown KS, Emmons LH, Van der Voort ME (In press) Ecosystem decay of Amazon forest remnants. In: Proceedings of the Sixth Spring Systematics Symposium on Extinctions, Field Museum. University of Chicago Press, Chicago.

McPhee RDE, Burney DA, Wells NA (1985). Early Holocene chronology and environment of Ampasambazimba, a Malagasy subfossil lemur site. Int J Primatol.

Mertl-Millhollen A, Gustafson HL, Budnitz N, Dainis K, Jolly A (1979) Population and territory stability of the *Lemur catta* at Berenty, Madagascar. Folia Primat 31:106–122.

Mittermeier RA, Oates JF (1985) Primate diversity: the world's top countries. Primat Conserv (WWF-US Primate Program) 5:41–48.

Paulian F (1984) Madagascar: a microcontinent between Africa and Asia. In: Jolly A, Oberle P, Albignac R (eds). Madagascar. Pergamon, Oxford, pp 1–26.

Perrier de la Bathie H (1921) La vegetation Malgache. Extr Ann du Musee Coloniale de Marseille, 29 an 3 Ser, 9 Vol. Challemel, Paris.

Petter J-J (1962) Recherches sur l'ecologie et l'ethologie des lemuriens malgaches. Mem Mus Nat Hist Naturelle ns 27:1–146.

Petter J-J, Peyrieras A (1969) Nouvelle contribution à l'étude d'un lemurien malgache, le aye-aye (*Daubentonia madagascariensis* E. Geoffroy). Mammalia 34:167–193.
Petter J-J, Peyrieras A (1970) Observations éco-éthologiques sur les lémuriens malgaches due genre Hapalemur. Terre et Vie 17:356–383.
Petter J-J, Peyrieras A (1975) Preliminary notes on the behavior and ecology of *Hapalemur griseus*. In: Tattersall I, Sussman RW (eds) Lemur biology. Plenum, New York, pp 281-286.
Petter J-J, Albignac R, Rumpler Y (1977) Mammifères Lemuriens (Primate Prosimiens). Faune de Madagascar. ORSTOM, Paris, vol 44.
Petter-Rousseaux A (1962) Récherches sur la biologie de la réproduction des primates inferieurs. Thèse de l'Univ. de Paris. Mammalia.
Pollock JI (1979) Spatial distribution and ranging behavior in lemurs. In: Doyle GA, Martin RD (eds) The study of prosimian behavior. Academic Press, New York, pp 359–410.
Richard AJ (1978) Behavioral variation. Bucknell University Press, Lewisburg, PA.
Richard AF, Sussman RW (1975) Future of the Malagasy lemurs: conservation or extinction? In: Tattersall I, Sussman RW (eds) Lemur biology. Plenum, New York, pp 313–334.
Sussman RW (1974) Ecological distinctions in sympatic species of *Lemur*. In: Martin RD, Doyle GA, Walker AC (eds) Prosimian biology. Duckworth, London, pp 75–108.
Sussman RW, Richard AF, Ravelojaona G (1985) Madagascar: current projects and problems in conservation. Prim Conserv (WWF-US Primate Program) 5:53–58.
Tattersall I (1982) The primates of Madagascar. Columbia, New York.

8
The Conservation Status of Nonhuman Primates in Indonesia

KATHLEEN MACKINNON

Introduction

There are at least 28 species of nonhuman primates in Indonesia (Table 8.1) and, according to some authorities, as many as 31, if we give specific status to all the Sulawesi macaques (Fooden, 1969). Sixteen of these primates (or 19 species if we accept the higher number) are island endemics. Primates are found as far east as Sulawesi and Timor, the easternmost island of the Lesser Sundas (Nusa Tenggara), though the long-tailed macaques that occur here are believed to have been introduced originally by humans. The distributions of the various species are shown in Figure 8.1, with some illustrated in Figure 8.2.

Estimating Primate Densities

To estimate the numbers of the various species of primates remaining in Indonesia, three kinds of data are required:

1. area of remaining habitat for each species
2. group densities of each species in different habitats
3. average group sizes in different habitats

Numbers 2 and 3 multiplied together give population densities.

Table 8.2 shows the areas of original and remaining natural habitat within each species range. Habitat areas are derived from the National Conservation Plan for Indonesia (FAO, 1981/1982). Some measure of the ongoing threat to each species is given by comparing areas of original with remaining habitat to estimate percentage loss of habitat. A species' long-term security is reflected by the areas of protected habitat found within its range. For detailed data for one species, the Javan gibbon, *Hylobates moloch*, see Appendix.

Three methods are commonly used to estimate densities of forest primates:

1. *Intensive study* of a small area over a long period of time (cf. MacKinnon, 1974; Rijksen, 1978; Rodman, 1978; Chivers, 1977; Gittins, 1979)
2. *Sweep survey* method where several observers systematically search an area. This requires an extensive network of trails and good coordination between observers (cf. MacKinnon and MacKinnon, 1980a, West Malaysia)

TABLE 8.1. Primates of Indonesia.

Species	Vernacular names		Distribution in Indonesia	Extralimital	Status
	Indonesian	English			
Suborder Prosimii					
Family Lorisidae					Protected 1973
1. *Nycticebus coucang*	Kukang	Slow Loris	Sumatra, Kalimantan, Java, Bangka, Riau archipelago	Assam to Mindanao	
Family Tarsiidae					
2. *Tarsius bancanus*	Binatang hantu, Singapuar	Western tarsier	Kalimantan, Sumatra	Borneo	Protected 1931
3. *T. spectrum*	Tangkasi	Spectral tarsier	Sulawesi	—	Protected 1931
Suborder Anthropoidea					
Family Cercopithecidae					
Subfamily Colobinae					
4. *Nasalis larvatus*	Bekantan	Proboscis monkey	Kalimantan	Borneo	Protected 1931
5. *Simias concolor*	Simakobu	Snub-nosed langur, pig-tailed langur	Mentawai Islands	—	Protected 1972
6. *Presbytis aygula*	Surili	Javan leaf monkey	Java	—	Protected 1979
7. *P. cristata*	Lutung, hirangan	Silvered leaf monkey	Sumatra, Java, Bali, Kalimantan, Lombok	Burma, Indochina, Thailand, Malaya, Borneo	
8. *P. femoralis*[1]	Koka	Banded leaf monkey	Riau, Kalimantan	—	—
9. *P. frontata*	Lutung dahi putih	White-fronted leaf monkey	East Kalimantan	Borneo	Protected 1979
10. *P. hosei*	Banggat	Bornean leaf monkey	East Kalimantan	Borneo	—
11. *P. melalophos*	Chi-cha, simpai	Banded leaf monkey	Sumatra	Thailand, Malaya	—
12. *P. potenziani*	Joja	Mentawai leaf monkey	Mentawai Islands	—	Protected 1977
13. *P. rubicunda*	Kelasi	Maroon leaf monkey	Kalimantan, Karimata Islands	Borneo	Protected 1977
14. *P. thomasi*	Rungka	Thomas' leaf monkey	Sumatra, in Aceh	—	Protected 1977

Subfamily Cercopithecinae					
15. *Macaca fascicularis*	Kera Warik	Crab-eating macaque, long-tailed macaque	Sumatra, Kalimantan, Java, Bali, Nusa Tenggara east to Timor	Burma, Indochina, Thailand, Malaya, Philippines, Borneo	—
16. *M. nemestrina*	Beruk	Pig-tailed macaque	Sumatra, Kalimantan	NE India, Burma, Thailand, Malaya, Borneo	—
17. *M. maura*	Dare	Moor macaque	SW Sulawesi	—	Protected 1977
18. *M. nigra/nigrescens*	Yaki, dihe	Crested macaque	North Sulawesi	—	Protected 1970
19. *M. ochreata/brunnescens*	Hada, ndoke	Ochreate black ape, Buton macaque	SE Sulawesi, Muna and Buton	—	Protected 1977
20. *M. tonkeana/hecki*	Digo	Black macaque, Celebes macaque	NW and C Sulawesi	—	
21. *M. pagensis*	Bokoi	Mentawai macaque	Siberut, Pagai Islands (Mentawai Is.)	—	Protected 1977
Family Hylobatidae					
22. *Hylobates syndactylus*	Siamang	Siamang	Sumatra	Malaya	Protected 1931
23. *Hylobates agilis*	Ungko	Dark-handed gibbon, agile gibbon	Sumatra south of Toba lake SW Kalimantan	Malaya, South Thailand	Protected 1931
24. *H. klossii*	Bilou	Kloss gibbon	Mentawai Islands	—	Protected 1931
25. *H. lar*	Wau-wau	White-handed gibbon, Lar gibbon	Northern Sumatra	Burma, Laos, Thailand, Malaya	Protected 1931
26. *H. moloch*	Owa-owa	Javan gibbon, grey gibbon	Java	—	Protected 1931
27. *H. muelleri*	Wa-wa	Bornean gibbon	Kalimantan	Borneo	Protected 1931
Family Pongidae					
28. *Pongo pygmaeus*	Mawas	Orangutan	Kalimantan, North Sumatra	Borneo	Protected 1931

[1]*P. femoralis* may be a subspecies of *P. melalophos*.

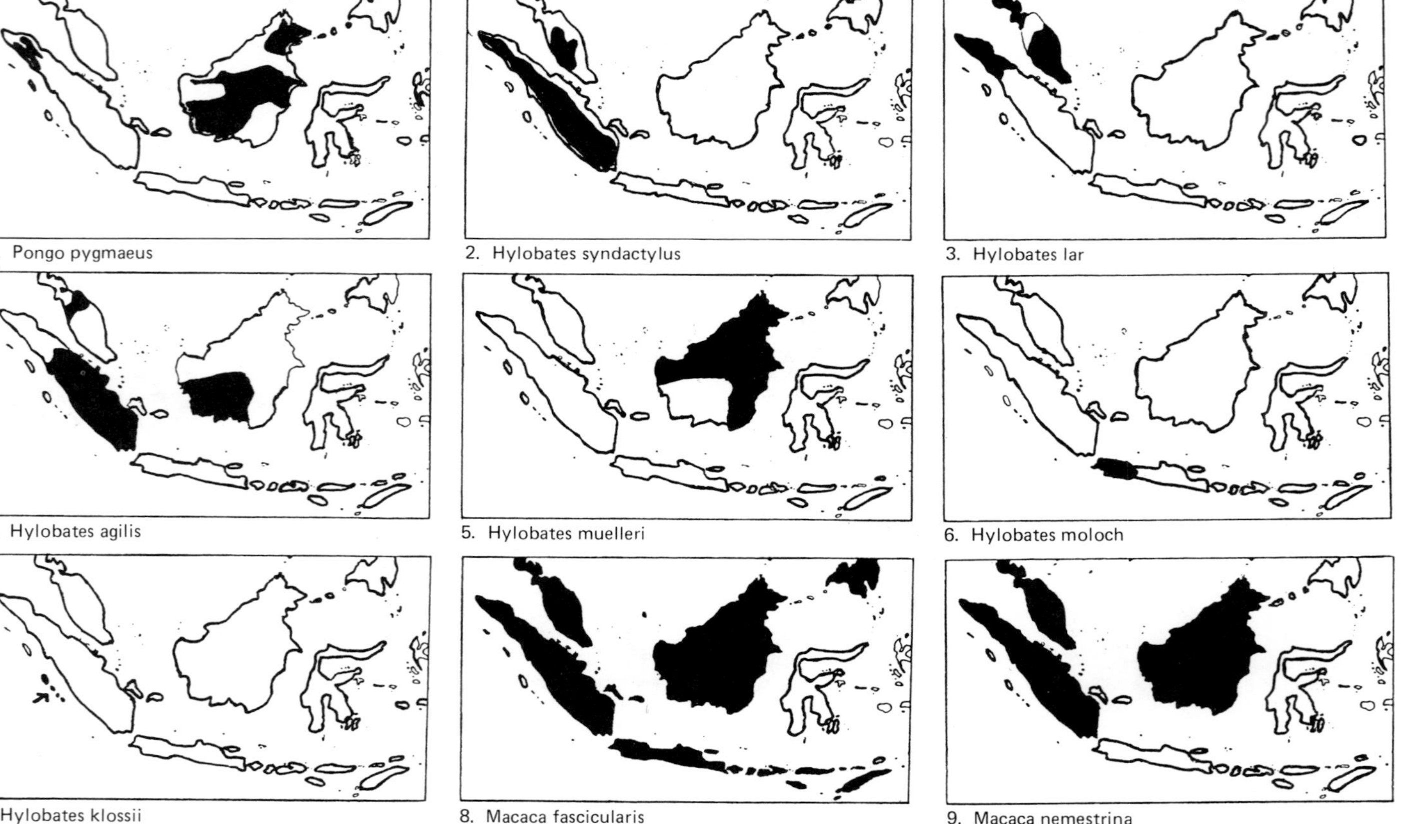

FIGURE 8.1a. Distribution of all primate species within Indonesia (and Malaysia).

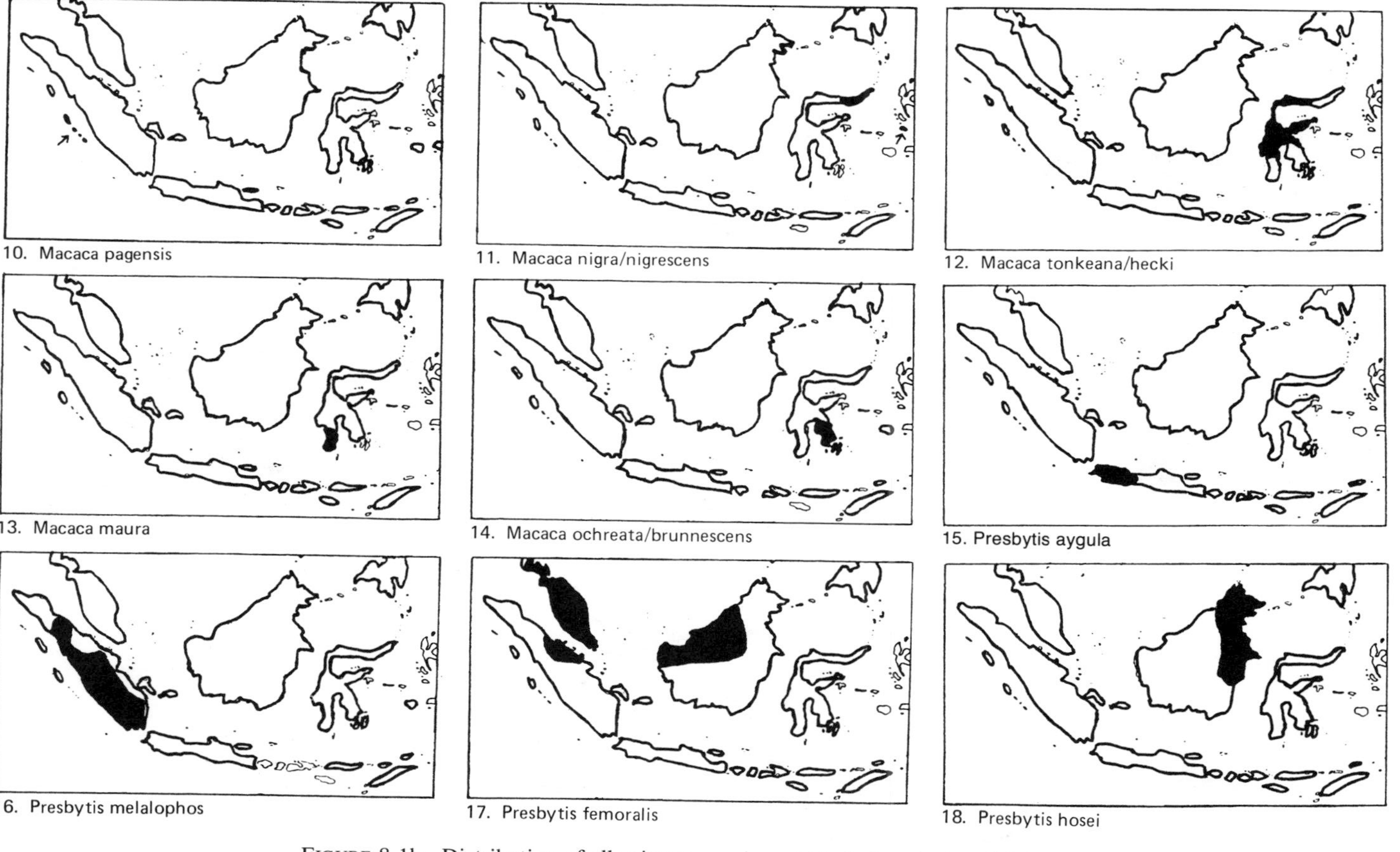

FIGURE 8.1b. Distribution of all primate species within Indonesia (and Malaysia).

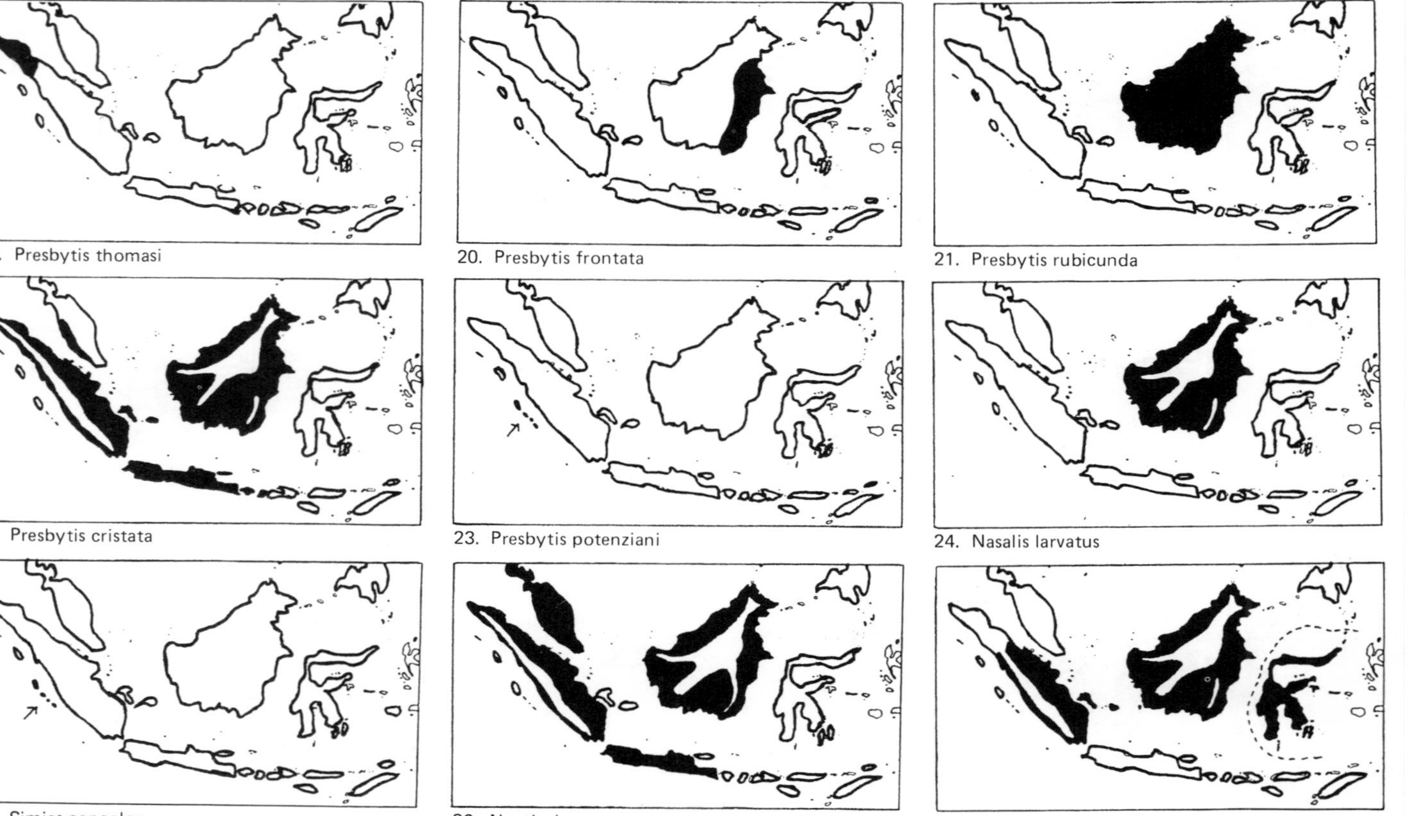

FIGURE 8.1c. Distribution of all primate species within Indonesia (and Malaysia).

(a) Slow loris, *Nycticebus coucang*, occurs in Sumatra, Kalimantan, Java.

(b) Spectral tarsier, *Tarsius spectrum*, endemic to Sulawesi.

(c) Silvered leaf monkey, *Presbytis cristata*, from Java.

(d) Crested macaque, *Macaca nigra*, endemic to north Sulawesi.

FIGURE 8.2. Some Indonesian primates.

(e) Grey gibbon, *Hylobates moloch*, endemic to West Java.

(f) Lar gibbon, *Hylobates lar*, occurs in North Sumatra.

(g) Orangutan, *Pongo pygmaeus*, from Kalimantan.

FIGURE 8.2. *Continued.*

3. *Line transect* method where the observer travels slowly along a measured line or trail recording all animals seen. This method has the advantage that large areas can be surveyed in a relatively short time, allowing one or two observers to cover many sites. Consequently, this method is often used for broad surveys, e.g., in Malaya by Southwick and Cadigan (1972) and Marsh and Wilson (1981a), in Sumatra at Wilson and Wilson (1975), and in Bangladesh by Gittins (1980).

The *line transect* method was the principal method used in this survey of Indonesian primates. During 1983, the author visited various sites throughout the archipelago to collect data on primate distribution and abundance. Transect counts were made at Pulau Kaget, the Sikonyer river and Tanjung Puting (Kalimantan), Way Kambas and Barisan Selatan (South Sumatra), Pangandaran (West Java), and Tangkoko Dua Saudara and Dumoga-Bone (North Sulawesi). For details of survey methods, see MacKinnon (1983). Several of the survey sites were selected because they had not been well studied previously and the primate data from them were compared with those from earlier field studies (see Table 8.4).

Primate densities vary in different habitats. In general, total group densities are lower in swamp forest than in lowland forest, but the density of one species *Macaca fascicularis* is markedly higher (cf. Marsh and Wilson, 1981b). Densities of *M. fascicularis*, *Presbytis cristata*, and maroon leaf monkeys *P. rubicunda* show similar densities in primary and selectively logged forest (Wilson and Wilson, 1975). Changes in density may be due to changes in number of groups/km^2 or because of changes in group size. Group sizes for certain species, e.g., *M. fascicularis* and *P. cristata*, are much higher in secondary and disturbed forest, though in very disturbed areas groups may fragment and split into several smaller groups or lone animals. *Presbytis cristata* in Way Kambas, Sumatra, were often found in small groups of two to three individuals (MacKinnon, 1983), although average group size for this species in primary forest is 15 individuals/group. *Presbytis melalophos* group counts are larger in primary than logged forest.

Several primate species have been studied intensively during long-term field projects in Indonesia and in similar habitats elsewhere in South-East Asia (Aldrich-Blake, 1980; Barrett, 1981; Bismark, 1982; Caldecott, 1981; Curtin, 1980; Galdikas, 1979; Jeffrey, 1982; MacKinnon, 1977; MacKinnon and MacKinnon, 1980b; Niemitz, 1979; Raemaekers and Chivers, 1980; Robertson, 1982; Van Schaik, 1983; Watanabe, 1981; Watanabe and Brotoisworo, 1982; Whitten, 1982; Whitten and Whitten, 1982; WWF, 1980b; Kappeler, 1984). Population data from these studies have been compared with data collected by the author and data from other field studies in different habitats to establish a conservative density estimate for each species. The conservative estimate multiplied by the area of remaining habitat for each species gives the total population estimate for Indonesia (Table 8.3).

Although the *conservative working density estimate* for each species is based on the results of intensive field studies on certain species plus census data collected during this consultancy (see Table 8.4), each estimate is qualified by subjective judgments based on a knowledge of the species' biology and distribution. For instance, the working density estimate for long-tailed macaques in primary habitat in Table 8.3 (30 animals/km^2) is much lower than the overall figure of 54.5

TABLE 8.2. Areas of original and remaining natural habitat for nonhuman primates in Indonesia.

Species	Islands	Habitat types[1]	Original area of habitat (km^2)	Area of habitat remaining (km^2)	% of habitat loss	Protected area	Reserves within species range
Orangutan *Pongo pygmaeus*	Kalimantan	Lowland and hill forest, swamp and heath forest	K) 415,000	K) 156,000	64	5800	Kutai, Tanjung Puting, Gn. Palung, Bukit Raya, Gn. Leuser, Bukit Baka
	North Sumatra		S) 89,000	S) 23,000		3500	
			504,000	179,000		9300	
Siamang *Hylobates syndactylus*	Sumatra	All forests up to 1500 m excluding mangrove and peat swamp	340,000	120,000	66	20,000	Way Kambas, Barisan Selatan (SSI), Kerinci, Gn. Leuser
Agile gibbon, dark-handed gibbon *Hylobates agilis*	C and S Sumatra SW Kalimantan	Lowland forest excluding mangrove	500,000	170,000	66	19,700	Way Kambas, Barisan Selatan (SSI), Kerinci, Kerumutan, Gn. Palung, Kendawangan, Karimun, Gn. Bentuang
White-handed gibbon *Hylobates lar*	North Sumatra	Lowland forests up to 1000 m excluding mangrove	68,000	30,800	55	4000	Gn. Leuser
Bornean gibbon *Hylobates muelleri*	NE and S Kalimantan	Lowland forest, excluding mangrove and kerangas	237,000	146,000	38	13,500	Kutai, Ulu Kayan, Gn. Becapa
Kloss gibbon *Hylobates klossii*	Mentawai Is.	All forests excluding mangrove	6500	4500	31	1490	Taitaibatti
Javan gibbon, grey gibbon *Hylobates moloch*	West Java, Gn. Slamet	Lowland and hill forests to 1500 m excluding mangrove	43,274	1608	96	600	Ujung Kulon, Gn. Halimun
Long-tailed macaque *Macaca fascicularis*	Sumatra, Java, Kalimantan, Bali, Lesser Sundas east to Timor	Coastal and riverine forest up to 1000 m disturbed forest	Sum: 87,644	29,906	66	3542	Gn. Leuser, Kerinci, Way Kambas, Kutai, Tanjung Puting, Ujung Kulon, Pangandaran, Baluran, Bali Barat
			Kal: .80,641	36,807	54	3474	
			Java + Bali: 17,177	758	96	179	
			NT: 32,519	5900	82	330	
			217,981	73,371	66	7525	

Pig-tailed macaque *Macaca nemestrina*	Sumatra, Kalimantan	Primary and secondary lowland and hill forest, raid farmland	354,115	179,140	49	27,095	Gn. Leuser, Kerinci, Way Kambas, Kutai, Tanjung Puting
Mentawai macaque *Macaca pagensis*	Mentawai Is.	Lowland forest	6500	4500	31	1490	Taitaibatti
Crested macaque *Macaca nigra/ nigrescens*	North Sulawesi	Lowland and hill forests	12,000	4800	60	2750	Tangkoko-Dua Saudara, Gn. Ambang, Dumoga-Bone
Celebes macaque *Macaca tonkeana/hecki*	Central Sulawesi	Lowland and hill forests	67,000	38,500	33	1055	Lore Kalamanta, Morowali, Panua Lempako Mampie
Moor macaque *Macaca maura*	SW Sulawesi	Lowland and hill forests	23,000	2800	88	495	Bantimurong, Karaenta
Ochreate black ape *Macaca ochreata/ brunnescens*	SE Sulawesi	Lowland and hill forests	29,500	18,500	37	1420	Rawa Opa
Proboscis monkey *Nasalis larvatus*	Kalimantan	Estuarine and riverine forest, often far inland	19,622	10,438	47	1025	Tanjung Puting, Kutai, Gn. Palung
Pig-tailed langur, snub-nosed langur *Simias concolor*	Mentawai Is.	Lowland rainforest below 450 m	6500	4500	31	1490	Taitaibatti
Javan leaf monkey *Presbytis aygula*	West Java, Gn. Slamet	Lowland and hill forest to 1500 m	43,274	1608	96	730	Ujung Kulon, Gn. Halimun
Silvered leaf monkey *Presbytis cristata*	Sumatra, Java, Kalimantan, Bali, Lombok	Coastal and riverine forest, plus on Java lowland and hill forest	322,610	133,167	59	12,585	Kutai, Gn. Palung, Gn. Leuser, SS1, Way Kambas, Berbak, Ujung Kulon, Gn. Halimun, Gn. Gede–Pangrango, Baluran, Meru Betiri, Pangandaran
White-fronted leaf monkey *Presbytis frontata*	Kalimantan, R. Kayan south to R. Barito	Primary forest up to 300 m, excluding swamp forest	90,000	35,000	61	5330	Kutai, Pleihari Martapura

TABLE 8.2. (Continued).

Species	Islands	Habitat types[1]	Original area of habitat (km^2)	Area of habitat remaining (km^2)	% of habitat loss	Protected area	Reserves within species range
Bornean or grey leaf monkey *Presbytis hosei*	Eastern Kalimantan, N of S Mahakam	Primary lowland and lower montane forest up to 1000 m	54,000	29,000	48	4220	Ulu Kayan, Mentarang, Kutai
Banded leaf monkey *Presbytis melalophos*	S Sumatra to north of Lake Toba	Primary and lower montane forest up to 1500 m	174,340	50,960	71	11,120	Kerinci, Way Kambas, Barisan Selatan
Mentawai langur *Presbytis potenziani*	Mentawai Is.	Primary lowland rain-forest	6500	4500	31	1490	Taitaibatti
Maroon leaf monkey *Presbytis rubicunda*	Kalimantan	Dry lowland and montane forest up to 1500 m, secondary forest	415,000	266,000	36	19,670	Kutai, Tanjung Puting, Gn. Palung, Pleihari Martapura
Thomas' leaf monkey *Presbytis thomasi*	Sumatra north of Lake Toba	Lowland rainforest lower montane forest	68,000	30,800	55	4000	Gn. Leuser
Banded leaf monkey *Presbytis femoralis*	East Central Sumatra and NW Kalimantan	Swamp forest and lowland forest	83,300	42,700	49	2200	Gn. Becapa, Sambas, Danau Sentarum, Gn. Bentuang/Karimun
Western tarsier *Tarsius bancanus*	Sumatra, Kalimantan	Lowland primary and secondary rainforests	450,730	198,250	56	26,280	Kutai, SSl, Kerumutan, Kerinci
Eastern or spectral tarsier *Tarsius spectrum*	Sulawesi and offshore islands	Lowland and secondary rainforest, agricultural land	154,000	70,730	54	5852	Tangkoko-Dua Saudara Dumoga Bone, Panua Lore Kalamanta
Slow loris *Nycticebus coucang*	Sumatra, Java, Kalimantan	Primary and secondary rainforest	610,570	227,883	63	31,596	Gn. Leuser, Kutai

[1]Data on habitat areas from National Conservation Plan for Indonesia, FAO-INS/78/061.

TABLE 8.3. Conservative population estimates for all nonhuman primates within Indonesia.

Species	Area of suitable habitats remaining (km²)	% of habitat protected	Conservative working density animals/km²	Total population estimate	Estimated protected population in reserves
P. pygmaeus	179,000	5.2	1	179,000	9300[1]
H. syndactylus	120,000	16.7	3	360,000	60,000
H. agilis	170,000	11.6	5	850,000	98,500
H. lar	30,800	13.0	5	154,000	20,000
H. muelleri	146,000	9.2	8	1,168,000	108,000
H. klossii	4500	33.1	8	36,000	11,920[1]
H. moloch	1608	37.3	3	4824	1800[2]
M. fascicularis:					
Primary habitat	73,371	10.2	30	2,176,860	220,040
Secondary forest	38,750	5.8	40	1,550,000	90,000
M. nemestrina	179,140	15.1	5	895,700	135,475
M. pagensis	4500	33.1	2	9000	2980[2]
M. nigra/nigrescens	4800	57.3	30	144,000	82,500
M. tonkeana/hecki	38,500	2.7	10	385,000	10,550[1]
M. maura	2800	17.7	20	56,000	9900[1]
M. ochreata/brunnescens	18,500	7.7	15	277,500	21,300
N. larvatus	10,438	9.8	25	260,950	25,625
S. concolor	4500	33.1	7	31,500	10,430[1]
P. aygula	1608	45.5	5	8040	3650[2]
P. cristata	133,167	9.5	15	1,997,505	188,775
P. frontata	35,000	15.2	2	70,000	10,660[1]
P. hosei	29,000	14.6	10	290,000	42,200
P. melalophos	50,960	21.8	20	1,019,200	222,400
P. potenziani	4500	33.1	10	45,000	14,900[1]
P. rubicunda	266,000	7.4	10	2,660,000	196,700
P. thomasi	30,800	13.0	30	924,000	120,000
P. femoralis	42,700	5.2	20	854,000	44,408
T. bancanus	198,250	13.3	50	9,912,500	1,314,000
T. spectrum	70,730	8.3	200	14,146,000	1,170,400
N. coucang	227,883	13.9	5	1,139,415	157,980

[1]Species of vulnerable status.
[2]Estimated total populations of less than 10,000 animals.

used by the Wilsons (1976) and deliberately so. Long-tailed macaques occur in various types of habitat from beach vegetation and mangrove to lowland rain-forest or secondary forest, but it is important to remember that these macaques are not distributed evenly throughout each habitat type. The long-tailed macaque is a species that throughout much of its range does not travel far from rivers, so even within a habitat such as moist lowland forest densities will be greatest near rivers and less inland. Surveys along or near rivers will give a higher density estimate than is true for the whole area. The conservative working density estimates for each species (Table 8.3), subjective as they are, are meant to give a "safe" figure applicable to the whole area of habitat considered.

TABLE 8.4. Primate species densities in different Sundaland communities (animals/km²).

	Borneo					Sumatra				West Malaysia	
	Segama	Kutai	Tanjung Puting	Kalimantan Sepaku	P. Kaget	Ranun	Ketambe	Way Kambas	Barisan Selatan (SSI)	Krau	West Malaysia
Pongo pygmaeus	2.0	4.0	2.0	—	—	2.0	5.0	—	—	—	—
Hylobates syndactylus	—	—	—	—	—	7.0	15.0	2.6	2.9	4.5	2.8
Hylobates muelleri	11.5	14.6	—	22.2	—	—	—	—	—	—	—
Hylobates lar	—	—	—	—	—	7.0	11.0	—	—	6.1	19.2
Hylobates agilis	—	—	7.0	—	—	—	—	3.5	4.3	—	—
Macaca fascicularis	7.0	5.8	47.0	19.0	35.0	50.0	48.0	22.2	present	39.0	32.2
Macaca nemestrina	9.0	5.5	present	—	—	53.0	21.0	present	—	0.5	4.0
(lowland hill forest)							6.0				
Presbytis frontata	—	present	—	5.7	—	—	—	—	50.0	—	—
Presbytis rubicunda	33.0	present	—	11.4	—	—	—	—	—	—	—
Presbytis thomasi	—	—	—	—	—	64.0	27.0	—	—	—	—
Presbytis melalophos	—	—	—	—	—	—	—	3.9	—	74.0	90.0
Presbytis hosei	5.0	20.4	—	—	—	—	—	—	34.3	—	—
Presbytis cristata	1.5	—	present	—	present	—	—	15.2	—	—	5.0
Nasalis larvatus	9.0[1]	present[1]	62.5	—	120.0	—	—	—	present	—	—

[1]Rainforest areas known not to be preferred habitat for *Nasalis*.

Source: Data from MacKinnon and MacKinnon (1980a), Wilson and Wilson (1975), Marsh and Wilson (1981a), Rijksen (1978), Rodman (1978), Galdikas (1979), ongoing research, and MacKinnon (1983).

Density Estimates for Indonesian Primates

Total population estimates for each primate species in Indonesia are presented in Table 8.3. These estimates are inevitably crude. However, they do give a working figure on which to base some conclusions about a species status. At the crudest level, these figures give some idea of order of magnitude, which species are common, which rare, and which endangered.

Orangutan

There are an estimated 179,000 km² of apparently suitable habitat remaining in Sumatra and Kalimantan, and with a conservative density of 1 animal/km² this gives a total population estimate of 179,000 animals. Rijksen (1978) published a figure of 30,000 animals for Sumatra (23,000 km² remaining habitat), and the 1979 Red Data Book lists "guesstimates" as high as 145,000 animals. The total population estimate in Table 8.3 may, however, be an overestimate owing to the species' rather patchy occurrence, especially in Kalimantan, and its vulnerability to habitat destruction and hunting pressure. The only orangutans that can be regarded as secure are those in protected areas. In several reserves, animals are known to occur at higher densities (Table 8.4) than the conservative 1/km² used to determine total population estimates, so that the number of protected animals is considerably higher than 9300, more likely 20,000 animals.

Gibbons

Table 8.3 shows that two (*moloch* and *klossii*) of the five gibbon species are rather rare, especially in protected areas. The total population estimate for Kloss' gibbon is the same as that assessed independently by Whitten et al. (WWF, 1980b) on the basis of an intensive field study. Predictions for the siamang and agile gibbon are very similar to those given by Chivers (1977) and Wilson and Wilson (1976); the density estimate for white-handed gibbon, *H. lar*, is twice as high, but only 20,000 individuals are believed to occur in reserves.

Cynomolgus or Long-Tailed Macaque, *M. fascicularis*

According to Table 8.3 there are at least some 3 million long-tailed macaques in natural habitat throughout Indonesia. This macaque is one of the few primates to show higher densities in secondary forest, and this habitat accounts for another 1.5 million animals.

The long-tailed macaque appears to be plentiful. However, it has already lost an average 60% of its natural habitat. This macaque lives in riverine habitat, often on the alluvial plains preferred for agricultural expansion, so it is particularly vulnerable to habitat loss. Although macaques can live in secondary forest, the areas of this habitat in no way make up for population losses resulting from destruction of original habitat. Wilson (1975) gave a total population estimate of over 14 million for long-tailed macaques in Sumatra, compared to only 1.4 million estimated in this survey (MacKinnon, 1983). Even allowing for Wilson's much higher mean density, the comparison still suggests a considerable population decline in the last 10 years alone. If *M. fascicularis* is to continue to be an important export species, then wild population levels should be carefully monitored.

Pig-Tailed Macaque, *M. nemestrina*

The pig-tailed macaque is much rarer than the long-tailed macaque, although it has a much larger habitat area because it lives at lower density and does not utilize all the apparently available habitat. The estimated 895,000 may seem secure, but because this species is a habitual crop raider and most of the population is outside reserves, the number does not seem so safe. This estimate is much greater than that given by Wilson (1975), who has used an impossibly small area of habitat for Sumatra of only 2316 km² for this species, compared with my own estimate of 72,500 km² for Sumatra and an additional 107,000 km² for Kalimantan. My population estimates are not so different from those of Robertson (1982), who estimated a population of 11,000–13,000 pig-tails in Gn. Leuser Reserve. My estimate for Gn. Leuser using a mean density of 5 animals/km² is 16,400.

Mentawai Monkeys

All of these monkeys can be considered endangered because of their very limited distribution and small populations protected in reserves. Whitten et al. (WWF, 1980b) give a much higher figure (39,000) for *M. pagensis*, a lower estimate for *S. concolor* (19,000), but a similar figure for *P. potenziani*. These primates are endemic to the Mentawai Islands and only secure in the Taitaibatti reserve, Siberut.

Sulawesi Macaques

The Sulawesi macaques are all island endemics with limited distributions. Not all species are rare, but some need special consideration, especially *M. maura*, which is both rare and underprotected.

The total estimate for *M. tonkeana/hecki* masks the fact that very little of the species habitat is protected within reserves. *Macaca tonkeana* only occurs at low densities within reserves such as Lore Kalamanta, but *hecki*, which is quite distinct in appearance, is even less well protected and much of its remaining habitat is being opened up for transmigration schemes. A survey on the status of these two species should be carried out as soon as possible.

Proboscis Monkeys, *N. larvatus*

Where this species occurs it is common and conspicuous, with densities of 120 animals/km² on Pulau Kaget in the Barito estuary (MacKinnon, 1983). However, it inhabits riverine forests, and these are the first areas to be cleared in Kalimantan where waterways are the main routes of transport so village settlements naturally cluster on riverbanks. The Barito river, for instance, is cleared for many hundreds of kilometers upstream to its headwaters. Some attempt should be made to monitor populations because the proboscis monkey is endemic to Borneo, and only an estimated 25,000 animals are protected in reserves in Indonesia.

Leaf Monkeys, *Presbytis spp.*

The Javan leaf monkey, *P. aygula*, is rare, having suffered, like the Javan gibbon, the destruction of much of its habitat. Little is known about *P. frontata*, the other *Presbytis* species having only 10,000 individuals inside reserves, other than that it is a rare monkey occurring at low densities throughout its range. It would be interesting to have more data on this species. *Presbytis cristata* seems relatively numerous, reflecting its wide occurrence. This monkey is common along river banks and in secondary habitats, e.g., rubber plantations and kampong lands in Kalimantan and Sumatra. The figure of two million or so animals given in Table 8.3 is considerably lower than Wilson's 5 million for Sumatra alone (Wilson, 1975), but this is partially explained by the much higher mean estimate she has used. For *P. melalophos-femoralis-thomasi*, Wilson's estimate for Sumatra alone is almost twice my estimate for Sumatra plus West Kalimantan, even using a similar density/km^2.

From the preceding discussion and the figures presented in Table 8.3, it can be seen that certain species have such low overall densities or such low numbers protected in reserves that they give cause for concern, e.g., the Javan gibbon. Other species such as long-tailed macaque and silvered leaf monkey appear to be numerous. Any species with less than 10% of its habitat protected is not secure. Nor is any species with a total population of less than 10,000 animals.

However, if we look at Table 8.2, one of the most interesting figures is the percentage habitat loss already suffered by each species. Original area of habitat was determined from physiographic and climatic maps; remaining areas are based on LANDSAT imagery and Agrarian land-use maps (FAO, 1981/1982). The latest imagery and maps are often several years old, but deforestation continues all the while, so even these figures tend to overestimate the area of remaining forest. Some species have already lost more than 90% of their original habitat; this includes the long-tailed macaque on Java, a species that everyone has so far considered to be so common that it did not matter how intensively it was cropped. The message is clear. Even species that at present are common will not be so for much longer if the present rate of deforestation continues. To safeguard future populations action must be taken now.

Threats to Indonesian Primates

The main threat facing primates in Indonesia, as elsewhere in the tropics, is loss of habitat as a result of forest clearance.

Clearing Forest for Agriculture

In order to sustain its increasing human population, Indonesia must increase the amount of land under agriculture. Every year vast areas of forest are cleared in Sumatra, Sulawesi, and Kalimantan for ladang lands by individual farmers. In overpopulated Java this process has gone on so long that the island now has no more than 8% of its original forest cover left. Indeed, since colonial times the government has encouraged farmers to migrate from overcrowded Java and Bali to the outer islands. These transmigration schemes open up new areas of forest.

The forest is felled and burned, then the land is set with crops such as hill rice, corn, and tapioca. Often the initial burning process destroys much of the humus layer so that the land is fertile and able to sustain crops for only a few years. It is then abandoned and alang-alang grassland becomes established while new areas are cleared for agriculture. Every year approximately 500,000 ha of forest are lost to agriculture with displacement and loss of primates. Some species move to adjacent forest and raid the crops; the macaques *M. fascicularis*, *M. nemestrina*, *M. nigra*, and *M. maura* in certain areas, are particularly serious agricultural pests.

Logging

Logging is a major industry in Indonesia, the second most important source of revenue after oil. There are two main logging procedures:

1. Clear cutting where *all* trees of a suitable size are taken with consequent complete destruction of the habitat and disastrous consequences for the resident primates.
2. Selective logging (8–12 trees removed/ha) where only certain species of a certain size are taken. Even with selective logging there is serious disturbance and destruction of the forest (up to 65% of the canopy is often destroyed) as a result of falling trees dragging down their neighbors, construction of logging roads, etc. Selective logging does have one major benefit over clear-felling; the fruit trees important to primates are usually not good timber trees and are often left by the loggers.

In Malaya, Marsh and Wilson (1981a) found that *logging reduced primate densities at all the sites studied*, but the effects differed with the age of the secondary stand and also between primate species. Gibbons showed little or no reduction in group density at recently logged sites but were less than half as common in 5-year-old secondary forest. Banded leaf monkey, *Presbytis melalophos*, are generally less common in *recently* logged forest than in primary forest, sometimes drastically so; this is probably also true for *M. fascicularis* (Marsh and Wilson, 1981b). In forest that has been recently logged, then left to regenerate, there seems to be an initial decline in total primate densities (Wilson and Johns, 1982; Johns, 1981). After a lapse of 5 years or so, certain primate species may make a comeback and achieve higher population densities in the disturbed secondary forest (which because of its open nature has a high level of new leaf growth) than in primary forest. The long-tailed macaque, *M. fascicularis*, an adaptable opportunist, is one such species.

Previously logged-over areas of forest may have potential as breeding areas for "ranching" *M. fascicularis* under wild conditions (MacKinnon, 1983).

Is the Indonesian Reserve System Adequate to Protect and Maintain Primate Populations?

Since pressures on forest for both agricultural land and timber are unlikely to abate in the foreseeable future, it is vital to determine if the Indonesian primates are adequately represented in the country's system of national parks and other protected areas. Ultimately, these may be the only areas where certain species will survive.

With the help of the FAO Development of National Parks Programme INS/78/061, the Department of Conservation (PHPA) has prepared a Conservation Master Plan for the whole of Indonesia reviewing the conservation priorities of each reserve on an island by island basis. Some 299 areas of natural beauty or richness have already been declared reserves. Sixteen of these are national parks and more are in process. Figure 8.3 shows existing and proposed major reserves within Indonesia.

Four types of reserves are currently defined under the forestry laws of Indonesia:

1. Cagar Alam—strict nature reserves where no management or human interference with the environment is permitted.
2. Suaka Margasatwa—game reserves where the natural balance of the environment must not be disturbed but low levels of management, visitor use, and utilization are permitted.
3. Taman Buru—hunting reserves managed specifically for hunting and fishing.
4. Taman Wisata—recreation parks maintained for outdoor recreation.

New legislation is in process for the establishment of national parks and conservation buffer zones. Furthermore, an additional 30 million ha of natural forest on steeper slopes is designated as *hutan lindung* protected for its hydrological value.

From the point of view of primate protection, the Indonesian reserve system must fulfill certain criteria:

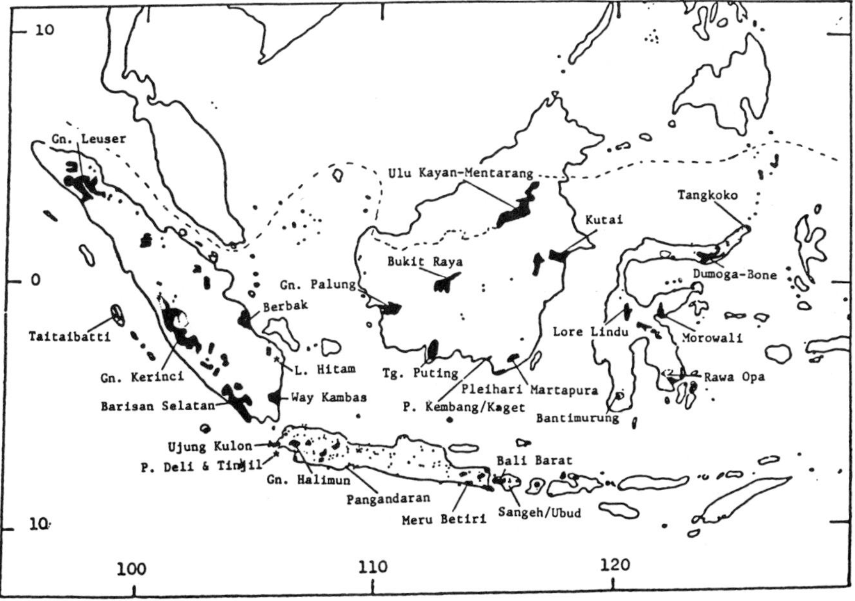

FIGURE 8.3. Major reserves (black) harboring primate populations in Indonesia and proposed locations for pilot primate ranching projects (*).

1. Reserves must be large enough to maintain genetic diversity and protect viable populations of each species. In small reserves cut off from similar habitat local extinctions will occur; e.g., Barro Colorado Island, Panama (1500 ha), has lost some 20% of its original resident bird species (Willis, 1974). Geneticists believe that breeding populations of at least 5000 individuals are necessary to maintain genetic diversity. The FAO/PHPA team planning the reserve system in Indonesia has tried to protect at least 10% of all major habitats, and larger areas where population size of certain animal or plant species is critically small (FAO, 1981/1982).
2. Primate populations within reserves are only secure if they are adequately protected. Although the list of nature reserves in Indonesia is impressive and most Indonesian primates seem to have adequate areas of habitat included within the existing reserve system, not all of these areas can be considered secure or well protected. Many have suffered considerable abuse in the past, e.g., logging, collection of forest products, extensive burning, and many areas still lack an auequate or well-trained guard force so that abuses still occur. Management problems are especially serious in regions of high human population density, e.g., in Way Kambas reserve in Lampung poachers still enter to hunt, fish, and steal timber in spite of the presence of a large guard force. Any primate species that does not have at least 10% of its remaining habitat protected or a freely interbreeding population of at least 5,000 animals should be considered endangered.

From Table 8.3 we can see that several species have less than 10% of their remaining habitat included within reserves, but of these some have large populations outside protected areas and healthy populations within reserves. Of special concern are the orangutan *Pongo pygmaeus*, the Sulawesi macaques *Macaca tonkeana/hecki*, *M. ochreata/brunnescens*, and proboscis monkey *Nasalis larvatus*, which all have rather limited distributions.

Three species—the Javan gibbon, *H. moloch*, Javan leaf monkey, *P. aygula*, and the Mentawai macaque, *Macaca pagensis*—have total population estimates of less than 10,000 animals with less than 5000 in reserves. In all cases, more than one-third of their remaining habitat lies within reserves, but the populations there are so small that these species must be regarded as seriously threatened.

Another seven species should be regarded as vulnerable since their estimated protected populations in reserves is of the order of 10,000 individuals or less. These species include the orangutan, the other three Mentawai primates, the white-fronted leaf monkey, *Presbytis frontata*, and three Sulawesi macaques—*M. tonkeana/hecki*, *M. ochreata/brunnescens*, and *M. maura*.

For the Javan gibbon and leaf monkey, although the area protected seems inadequate, there is no further room for reserve expansion on crowded Java. It is therefore inexcusable that such a valuable reserve as Gn. Halimun (40,000 ha), which probably harbors the largest populations of these endemic species, as yet has no permanent guard force.

The Mentawai monkeys, like the Kloss' gibbon, are all endemic to the Mentawai Islands, and their status must be regarded as precarious, even though 33% of their remaining habitat is protected in the Taitaibatti reserve. It is essential that PHPA go ahead with the proposed extension of the Taitaibatti reserve, the only protected area for these species.

Of the Sulawesi macaques *M. maura* is probably the most threatened. Current proposals to expand Karaenta to join Bantimurung to form a national park and for a new reserve Gn. Lompobatang should receive immediate attention. For *M. tonkeana/hecki* the picture is confused by the fact that there appears to be a large population outside the reserve system. Several large reserves have been established in Central Sulawesi, e.g., Morowali, Lore Kalamanta, but much of the habitat included within their boundaries is unsuitable for *tonkeana*. Moreover, even if *tonkeana* and *hecki* are the same species, as Groves (1980) suggests, they are strikingly different in appearance and *hecki* is worth protecting as a distinct form. *Macaca hecki* is found only on the northern peninsula, and much of its habitat is being destroyed by land clearance for transmigration schemes, e.g., at Paguyaman. *Macaca hecki* from this area are common pets even 100 km to the east at Gorontalo; this is some reflection of the number of animals being displaced. *Macaca hecki* is protected only in very small reserves such as Tangale, Panua and Tanjung Panjang; more areas should be set aside to protect this subspecies.

While it would be desirable to extend the area of reserves protecting the orangutan and its lowland rainforest habitat, the existing reserve system already protects healthy populations. Known densities of this ape are considerably higher in Gn. Leuser, Kutai, and Tanjung Puting than the conservative density estimate of 1 animal/km^2 used in Table 8.3 so that the protected population can be considered to be at least 20,000 animals. However, the orangutan is particularly vulnerable to habitat destruction and hunting pressure so that the only orangutans that can be regarded as secure are those in reserves. This vulnerability was well illustrated when fire swept through Kutai reserve in 1983 destroying much of the forest and depleting the resident wildlife.

Little is known about the leaf monkey, *Presbytis frontata*, so it is impossible to determine whether it is adequately protected within the existing reserve system until further population data are collected.

For the proboscis monkey, *N. larvatus*, the main threat is habitat loss. Where proboscis monkeys do occur they are conspicuous, occurring at high densities and in large troops. However, like long-tailed macaques they frequent some of the riverine habitats most favored by humans for logging, village settlement, and agriculture so that much of their lowland forest habitat is threatened. Less than 10% of their remaining habitat is secure in reserves; if possible, more areas should be set aside to protect this unique species.

With Indonesia's ever-increasing human population, the reserve system must be recognized as vital to the long-term survival of all Indonesian primates and perhaps, ultimately, with areas of protection forest, their final refuge. Certainly the primate populations within the reserve system must be considered as a natural reservoir for restocking surrounding areas of suitable habitat outside the reserves. Government policy to set aside some 10% of all habitats as reserves is highly commendable and should be adequate to protect most primate species apart from those that are already seriously reduced in numbers and/or confined to very small areas.

The Primate Trade in Indonesia

Table 8.1 lists the nonhuman primates found in Indonesia, their distribution, and current status. Of the 28 species found in Indonesia, 22 are classified as protected, but the legal status of this protection is not always clear. Lists of protected animals declared during Dutch Colonial Rule enjoy full legal status, but those declared since 1970 by Ministerial decree (Surat Keputusan Menteri) are not ratified by parliament. In practice, this means that while PHPA can confiscate certain animals kept as pets they probably cannot prosecute their owners.

New legislation on hunting and the animal trade that has already been under consideration for several years should be passed as soon as possible to clear this confusion.

Illegal trade in endangered and protected animals is common within Indonesia. There are many animal markets, and species such as the slow loris, *Nycticebus coucang* (protected since 1931), are openly offered for sale. Monkeys, many of them protected species, are commonly kept as pets; there are probably 100 crested macaques, *Macaca nigra/nigrescens*, held as household pets in the Dumoga valley (North Sulawesi) alone.

If PHPA confiscates these illegal pets, they are faced with the problem of what to do with them. The internal illegal trade in orangutans has been almost completely stopped, partly by developing orangutan rehabilitation centers where the confiscated animals are trained to return to the wild. Such schemes are costly in terms of money and effort, however, and contribute little to the species' overall chances of survival when the main threat is habitat loss. Their most useful function is to serve as centers for conservation education, as at Bohorok, North Sumatra.

There is a small legal trade in protected species both within the country and overseas, e.g., to provide breeding animals for zoos, but this is much better controlled. The Directorate of Forest Protection and Nature Conservation (PHPA, formerly PPA) issues permits for the transport of nonprotected animals, e.g., *Macaca fascicularis*, within Indonesia and separate permits for their export. Permits for the export or local transport of protected species, e.g., orangutan, are issued and must be signed by the Minister of Forestry (previously Minister of Agriculture).

On 28 December 1978 Indonesia ratified the Convention on International Trade in Endangered Species of Flora and Fauna (CITES). LBN (Lembaga Biologi Nasional) is the scientific authority and PHPA the management authority responsible for the implementation of CITES. The main primate trade in Indonesia involves the provision of live animals for biomedical and other research laboratories, most of which are in Europe, North America, and Japan. After 1978, when India imposed an outright ban on exports of rhesus macaques, until then the most popular primates for biomedical research, Indonesia together with Malaysia, the Philippines, and Bolivia accounted for well over half the world's exports of primates. Since 1978, Indonesia has been the biggest primate supplier with long-tailed macaques, *M. fascicularis*, accounting for most of the exports but some trade in pig-tailed macaques, *M. nemestrina*, and silvered leaf monkeys, *P. cristata*. Malaysia in 1984 also placed a total ban on all primate exports, including *M. fascicularis*. Long-term projections indicate that Indonesia will continue to be the main supplier of this species as one would expect since Indonesia has much larger remaining areas of this species habitat than the Philippines.

However one feels about the primate trade, the demand for live animals for biomedical research is unlikely to disappear in the near future, particularly while live monkeys are still used for testing batches of vaccine prior to its release for mass immunization programs. In fact, there is a world-wide trend of diminishing primate exports, and this is reflected in Indonesia. Whereas in both 1972 and 1973 Indonesia exported over 14,000 long-tailed macaques to Japan alone (Kavanagh, 1984), 10 years later, in 1982, the total export of long-tailed macaques was less than 15,000 and the quota for 1983 was reduced to 13,000 animals. The world demand for pig-tailed macaques and silvered leaf monkeys is only a few hundred animals per year, but the quotas set are considerably higher.

There are still substantial wild populations of long-tailed macaques in Indonesia, and there is no evidence that the species cannot sustain the present level of harvesting. The situation for pig-tailed macaques may be more serious, however (see below). Another cause for concern is the considerable number of monkeys "wasted" during forest clearance schemes and in the trapping programs themselves. Loss of animals as a result of death, injury, or unsuitability (wrong age/sex class) varies between 32% and 71% (Darsono, personal communication). Primates for trade are derived mainly from wild populations with the main dealers employing their own teams of highly skilled trappers. They should be encouraged to trap and trade animals displaced by land clearance schemes (or in forest about to be logged) thus reducing the pressure on healthy wild populations in undisturbed habitat.

Live-caught long-tailed macaques currently fetch ridiculously low prices, between U.S. $50 to 100 an animal to the main dealer and as little as $1–5 *per suitable animal* to the actual trapper. So long as the price per wild-caught animal remains at the current low level, wastage is likely to remain high. The higher the price a monkey fetches for the trapper or dealer, the more valuable it becomes and the better the treatment it will receive. By increasing demand for "cleaner," healthier animals, the overseas buyer could encourage responsible dealers to establish alternative supply sources for long-tailed macaques, either ranching animals on a semiwild basis in previously logged forest (see MacKinnon, 1983) or by maintaining captive breeding colonies.

In 1980, the World Wildlife Fund Program produced a comprehensive report on the animal trade, including primates, in Indonesia. The main findings, that the legal trade in protected species appears well controlled but that trade in other species needs better control and monitoring, still hold good today. Permits are required only for the transport of animals, not for their capture, so that the reported volume of trade may be only a small fraction of the number of animals taken from the wild. Moreover, export quotas set by PHPA are not based on sufficient information of the animal's biology and status in the wild.

The status of all primate species should be regularly reviewed. Certain species listed as protected because they are endemic to a relatively small area may in fact reach very high population densities within those areas and be a serious pest in some localities. *Macaca nigra/nigrescens* falls into this category. Although found only on the eastern half of the northern peninsula of Sulawesi, this animal occurs at very high densities of 3 animals/ha in Tangkoko reserve (MacKinnon and MacKinnon, in preparation), is well protected in several reserves and large areas of protection forest, and is a serious agricultural pest in certain parts of North Sulawesi and on Bacan Island, Maluku. On the other hand, the pig-tailed macaque is widely believed to be so common that the annual quota for exports has been around 10,000 a year with 7200 proposed for 1983. In fact, this animal is probably very rare in some regions and should therefore have protected status outside as well as within reserve boundaries in these areas.

Conservation Action Needed for Indonesian Primates

1. Further surveys should be carried out to monitor certain primate populations, especially for proboscis monkeys, Sulawesi macaques, and *Presbytis frontata*. Comprehensive surveys should be carried out to census and monitor wild populations of *M. fascicularis* and *M. nemestrina*, the two species most traded; the latter may be much less common than previously believed. Data for Sumatra should be compared with those of the Wilsons' 1973 survey to compare estimates and get some measure of population decline.
2. There should be no trading in those 12 primate species with an estimated 10,000 or fewer animals in reserves or less than 10% of their habitat protected. The 12 species regarded as endangered or threatened are: orangutan, proboscis monkey, the Mentawai primates, *Presbytis frontata*, Javan gibbon, and leaf monkey, and three of the Sulawesi macaques—*M. maura*, *M. tonkeana hecki*, and *M. ochreata/brunnescens*. Populations of these species in reserves should be monitored and active management implemented if necessary to maintain adequate populations.
3. Improved management and guarding of all reserves with as a first priority:
   a. implementation of a management plan and establishment of a guard force in Gn. Halimum (*H. moloch* and *P. aygula*)
   b. improved guarding in Gn. Honje (*H. moloch* and *P. aygula*).
4. Extension of existing reserves and new reserves established within a threatened species range whenever possible including:
   a. extension of the Taitaibatti reserve (Mentawai primates) as in process
   b. establishment of Gn. Lompobatang reserve (*M. maura*) as proposed
   c. establishment of a new, preferably large, reserve to protect habitat of *M. tonkeana hecki*; the rejected proposal for Randangan should be reconsidered.
5. For species with a wide geographical range, several reserves should be established to maintain regional variation. For instance, some of the island forms of *M. fascicularis* are quite distinct and well worth special protection. To be specific, the proposed Simeulue Island reserve in Aceh is a high priority.
6. Improved law enforcement to prevent hunting of protected species or their capture for pets, especially relevant to Mentawai primates, orangutan, and *M. hecki*.
7. Procedures for the control of trade and export of primates need tightening up. The Department of Forest Protection and Conservation of Nature (PHPA) as the management authority for CITES should be the only authority empowered to issue capture permits. Permits granted should not exceed the quota for each species.
8. So long as Indonesia maintains a trade in wild-caught primates, primarily long-tailed and pig-tailed macaques, supplies should, whenever possible, be drawn from animals displaced when their forest habitat is converted to agricultural land, in accordance with IUCN and WHO guidelines (1981).

Captive breeding colonies of much used species, e.g., *M. fascicularis*, should be established in both user and producer countries.

Summary

Indonesia has the widest range of primate species of any Asian country, with many endemic forms. Data were collected on distribution of Indonesian primates and areas of their original, remaining, and protected habitat. Percentage habitat loss gives some measure of the species long-term security. Total population estimates were calculated for each species.

Most primate species in Indonesia are still present in large numbers with eight species including long-tailed macaques, *M. fascicularis*, estimated at populations exceeding 1 million animals. Twelve species can be regarded as endangered or threatened, including the orangutan and several island endemics, e.g., proboscis monkeys, Mentawai primates, Javan gibbon, and Javan leaf monkey. Populations of all these species are protected within the reserve system. As elsewhere in the wet tropics, the main threat facing primate populations is forest clearance and habitat loss. The present trade in primates for biomedical research, particularly long-tailed macaques, is not regarded as a threat to the species. Recommendations are given for conservation action needed in Indonesia, with particular regard to extension of the existing reserve system and improved management and guarding of reserves.

Acknowledgments. The data in this paper were collected while the author was employed as a consultant for the World Health Organization on the Primate Resources Program. I am grateful to WHO and to the late Dr. F. Perkins for his support and encouragement. Many people and organizations provided help and information during the consultancy. I would like to thank the Directors and staff of the Indonesian Department of Forest Protection and Conservation of Nature (PHPA) for their assistance, particularly Ir Toga Siallagan and his staff (Balai V PHPA, Kalimantan), Ir Widodo, Ir Sjarif Bastaman, Ir Kurnia Rauf, and Ir Priyono (Balai II, Sumatra), Ir Tarmudji and Ir Subidio (Manado), and Pak Ubus Wardju Maskar (Pangandaran). I received considerable help from the World Wildlife Fund Indonesia Programme, especially from the headquarters staff and Jan Wind, project leader for Dumoga, and Dr. Nengah Wirawan, project leader for Kalimantan. I very much enjoyed working with my counterparts Nengah Wirawan and Ir Puspa Dewi Liman who proved excellent travel companions. I should like to thank all those other fieldworkers on whose data I have drawn, particularly those who made available unpublished data. Finally, special thanks are due to my husband Dr. John MacKinnon for his support and help with all aspects of the project.

Specimen Data Sheet: Species Distribution and Population Estimate

Family: Hylobatidae
Species: *Hylobates moloch*
English Name: Javan gibbon, grey gibbon
Indonesian Name: Owa-owa

Distribution in
Indonesia: Western Java to Gn. Slamet
Habitats: Lowland forests to 1500 m excluding mangrove and swamp
Habitat status:

	Type	Original area (km^2)	Remaining area (km^2)	Protected area (km^2)
Lowland forests:	LMa	5910	—	—
	LMv	3370	—	—
	LMo	1610	—	—
	HMv	730	—	—
	LWl	2470	44	—
	LWa	2930	165	125
	LWo	11,090	469	176
	LWv	7340	20	10
Hill forests:	HWl	300	—	—
	HWo	1530	400	72
	HWa	150	—	—
	HWv	1910	510	217
		43,274	1608	600

Figures from National Conservation Plan, vol 3 (FAO, 1982)
% habitat loss 96%
% remaining habitat protected 37.3%

Expected occurrence in following protected areas:
Gn. Honje ++ Gn. Tilu ++
Gn. Halimun ++ Gn. Pangrango-Gede ++
++, presence confirmed

Ecological category: frugivore
Conservative mean density estimate: 3 animals/km^2
Estimated total population: 4824
Threats: habitat destruction
Status: endangered

References

Aldrich-Blake FPG (1980) Long-tailed macaques. In: Chivers DJ (ed) Malayan forest primates. Plenum, New York, pp 147–165.

Barrett E (1981) The present distribution and status of the slow loris in Peninsular Malaysia. Malays Appl Biol 10(2):205–211.

Bismark M (1982) Ecology and behaviour of *Macaca nigrescens* in Dumoga Game Reserve, North Sulawesi. Balai Penelitian Hutan Report No. 392.

Caldecott JO (1981) Findings on the behavioural ecology of the pig-tailed macaque. Malays Appl Biol 10(2):213–220.

Chivers DJ (1977) The lesser apes. In: Prince Rainier, Bourne GH (eds) Primate conservation. Academic, New York, pp 539–598.

Curtin SH (1980) Dusky and banded leaf-monkeys. In: Chivers DJ (ed) Malayan forest primates. Plenum, New York, pp 107–145.

FAO (1981/1982) National Conservation Plan for Indonesia, vols 2, 3, 4, 5, 8.

Fooden J (1969) Taxonomy and evolution of the monkeys of Celebes. Bibl Primatol 10:1–148.

Galdikas BMF (1979) Orangutan adaptation at Tanjung Puting reserve, Central Borneo. Thesis, University of California, Los Angeles.

Gittins SP (1979) The behaviour and ecology of the agile gibbon (*Hylobates agilis*). Thesis, University of Cambridge.

Gittins SP (1980) A survey of the primates of Bangladesh. Project report to Forest Department, Bangladesh.

Groves CP (1980) Speciation in *Macaca*: the view from Sulawesi. In: Lindburg DG (ed) The macaques: studies in ecology, behaviour and evolution. Van Nostrand Reinhold, New York, pp 84–124.

Jeffrey S (1982) Threats to the proboscis monkey. Oryx 16(4):337–339.

Johns AD (1981) The effects of selective logging on the social structure of resident primates. Malays Appl Biol 10(2):221–226.

Kappeler M (1984) The gibbon in Java. In: Preuschoft H, Chivers DJ, Brockelman WY, Creel N (eds) The lesser apes: evolutionary and behavioural biology. Edinburgh University Press, pp 19–31.

Kavanagh M (1984) A review of the international primate trade. In: Mack D, Mittermeier RA (eds) International trade in primates. Traffic (USA), Washington, DC, pp 49–90.

MacKinnon JR (1974) Behaviour and ecology of orang-utans. Anim Behav 22:3–74.

MacKinnon JR (1977) A comparative ecology of Asian apes. Primates 18:747–772.

MacKinnon JR, MacKinnon KS (1980a) Niche differentiation in a primate community. In: Chivers DJ (ed) Malayan forest primates. Plenum, New York, pp 167–190.

MacKinnon J, MacKinnon K (1980b) The behaviour of wild spectral tarsiers. Int J Primatol 1:361–379.

MacKinnon K (1983) Report of a World Health Organisation (WHO) Consultancy to Indonesia to determine population estimates of the Cynomolgus or long-tailed macaque *Macaca fascicularis* (and other primates) and the feasibility of semi-wild breeding projects of this species. WHO Primate Resources Programme Feasibility Study: Phase II.

Marsh CW, Wilson WL (1981a) A survey of primates in Peninsular Malaysian forests. Universiti Kebangsaan Malaysia and University of Cambridge, UK.

Marsh CW, Wilson WL (1981b) Effects of natural habitat differences and disturbance on the abundance of Malaysian primates. Malays Appl Biol 10:227–249.

Niemitz C (1979) Outline on the behaviour of *Tarsius bancanus*. In: Doyle GA, Martin RD (eds) The study of prosimian behaviour. Academic Press, New York, pp 631–660.

Raemaekers JJ, Chivers DJ (1980) Socio-ecology of Malayan forest primates. In: Chivers DJ (ed) Malayan forest primates. Plenum, New York, pp 279–316.

Rijksen HD (1978) A field study on Sumatran orang-utans (*Pongo pygmaeus abelii* Lesson 1827): ecology, behaviour and conservation. H Veenman and Zonern BV, Wageningen, Netherlands.

Robertson JMY (1982) Behaviour ecology of *Macaca n. nemestrina*. Unpublished final report.

Rodman PS (1978) Diets, densities and distributions of Bornean primates. In: Montgomery GG (ed) The ecology of arboreal folivores. Smithsonian Institution Press, Washington, DC, pp 465–478.

Southwick CH, Cadigan FC (1972) Population studies of Malaysian primates. Primates 13:1–19.

Van Schaik CP (1983) The effect of group size on time budgets and social behaviour in wild long-tailed macaques (*Macaca fascicularis*). Behav Ecol Sociobiol 13:173–181.

Watanabe K (1981) Variations in group composition and population density of the two sympatric Mentawaian leaf monkeys. Primates 22:145–160.

Watanabe K, Brotoisworo E (1982) Field observation of Sulawesi macaques. Kyoto University Overseas Research Report of Studies on Non-human Primates 2:3–9.

Whitten AJ (1982) Home range use by Kloss Gibbon (*Hylobates klossii*) on Siberut Island, Indonesia. Anim Behav 30:182–198.
Whitten AJ, Whitten JEJ (1982) Preliminary observations of the Mentawai macaques on Siberut Island, Indonesia. Int J Primatol 3:445–459.
Willis EO (1974) Populations and local extinctions of birds on Barro Colorado Island, Panama. Ecol Mongr 44:153–169.
Wilson CC (1975) An estimate of primates in Sumatra excluding prosimians and orangutan. Unpublished report.
Wilson CC, Wilson WL (1975) The influence of selective logging on primates and some other animals in East Kalimantan. Folia Primatol 23:245–274.
Wilson CC, Wilson WL (1976) Behavioural and morphological variation among primate populations in Sumatra. Yearb Phys Anthrop 20:207–233.
Wilson WL, Johns AD (1982) Diversity and abundance of selected animal species in undisturbed forest, selectively logged forest and plantations in East Kalimantan, Indonesia. Biol Conserv 24:205–218.
World Wildlife Fund (1980a) The animal trade in Indonesia. WWF Indonesia Programme.
World Wildlife Fund (1980b) Saving Siberut: a conservation master plan. WWF Indonesia Programme.

9
Southeast Asian Primates

DAVID J. CHIVERS

Introduction

The Oriental Region (or Indo-Malayan Realm) comprises four subregions: Indian, Indo-Chinese, Sundaic, and Wallacean (Fig. 9.1). The main subject of this paper is the Sundaic subregion: the Malay Peninsula and the islands of Borneo, Java, and Sumatra and their outliers; this is the wettest part of the region, and hence the part most densely forested. The evergreen rainforests here constitute ecosystems that are among the most complex in the world in terms of plant forms and species diversity, and of animal diversity and abundance. The conservation of these ecosystems is consequently complex, but their importance is global as well as local.

Because the Sunda Shelf was uplifted from the sea by volcanic activity as recently as the late Miocene, with its fauna migrating first from India and then from China (Verstappen, 1975; Medway, 1972), and because the same kinds of primates that now occur there are also found in the Indo-Chinese subregion, reference will be made to the conservation of primates at Thailand, Burma, Bangladesh, Assam, and the countries of Indo-China, as well as in Malaysia and Indonesia—the two main political components of the Sundaic subregion. The few primates that have crossed eastward into the Philippines and the Indonesian islands of Sulawesi and the Lesser Sundas will also be mentioned. The habitats and primates of the Indian subcontinent are more distinctive and are the subject of a separate discussion (Southwick and Lindburg, this volume).

There is a distinct vegetation change at the Isthmus of Kra (southern Thailand), with the evergreen rainforests of the Sundaic subregion giving way to the more seasonal monsoon (semideciduous) forests of the Indo-Chinese subregion (Whitmore, 1975). The conservation situation also differs as one goes from the "islands" to the "mainland." While most of the Sunda Shelf was forested until very recently, the tropical forests of mainland Asia have been cleared continuously during historic times, so that they are now very fragmented.

The conservation situation and approach thus differ markedly between the Sundaic and Indo-Chinese subregions. While the forests of the Sunda Shelf are now being cleared rapidly, especially in the fertile lowlands (where they are most diverse), there are still extensive forests rich in wildlife. Urgent conservation is necessary, however, not just to slow the trend and conserve the great variety of

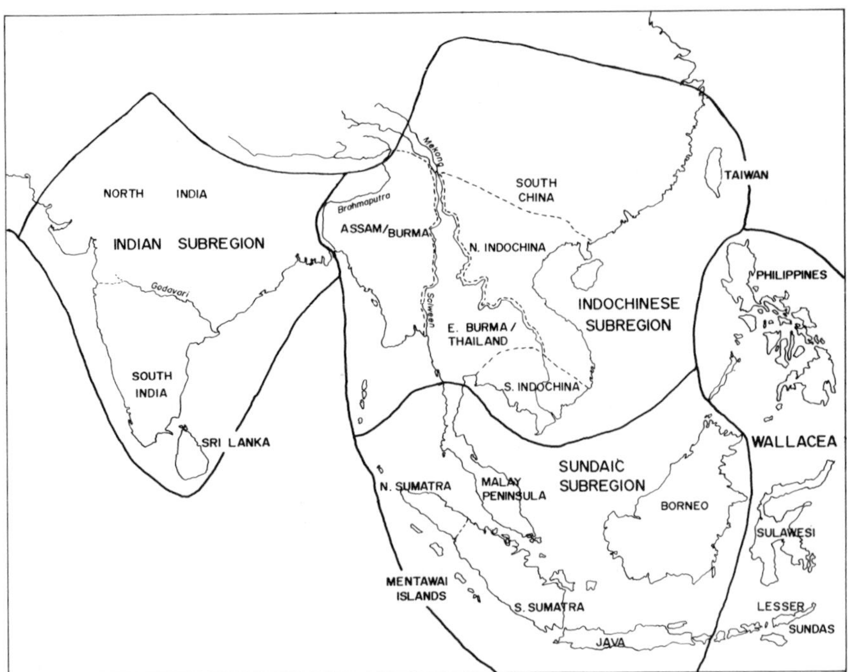

FIGURE 9.1. An outline biogeographical scheme for primate conservation in the Oriental Region, based on four subregions and 19 provinces (Marsh, in press).

primates and other animals, but to prevent the increasing threats to human life and livelihood that result from excessive forest clearance, especially climatic change, soil erosion, and flooding. The survival prospects of humans and wildlife are closely interwined, and they are enhanced by adopting a longer term approach than is currently done.

It is the history of volcanic activity, along with the dramatic climatic fluctuations of the Pleistocene, that make the Sunda Shelf of great biological interest. In particular, the recurrent changes in sea level that alternately separated and rejoined the constituent "islands" of the Sunda Shelf enhanced evolutionary processes (particularly in promoting speciation), through this sequence of isolation followed by rejoining. The consequences of these enforced evolutionary changes among animal and plant populations can still be observed. Nowhere else can one study so well the evolutionary biology of our closest relatives the apes as with the gibbons of Southeast Asia (Preuschoft et al., 1984).

There are four main kinds of primates (Fig. 9.2) in the Sundaic subregion: (1) six of the nine species of gibbon or lesser ape and, in Borneo and north Sumatra, the orangutan, the only Asian great ape; (2) about ten species of langur or leaf monkey; (3) just two species of macaque; and (4) the slow loris and, on certain islands, the tarsier. In addition, there are the "odd-nosed" colobine monkeys of coastal Borneo and the Mentawai Islands (off the west coast of Sumatra), which also have their own species of macaque. Apart from the tarsier and orangutan, all these primates have close relatives in the more seasonal forests of the Indo-Chinese subregion. Indeed, it is a matter of intense ecological interest, as yet unresolved, how they adapt to these different habitats.

Thus, the approach adopted here to the conservation of Southeast Asian primates is, first, a zoogeological one. One has to recognize the greater diversity of species and subspecies, particularly of gibbons and colobine monkeys, and seek to project as many as possible. Second, socioecological data, particularly on population structure, density, and dynamics and on habitat use, are very relevant. More practically, there is a need to recognize that conservation has to be achieved within an ecosystem approach, whereby forest protection and management bring sustained benefits to the long-term economies of the countries concerned. Apart from their relevance to human biology, one can justify a focus on primates in this approach, since they are among the largest of the arboreal forest inhabitants. Conserve them, and one has conserved adequate rainforest for all purposes and for most of its inhabitants.

Zoogeography—Distribution and Classification

The distribution of Southeast Asian primates is related in part to the distribution of forest types. In the Sundaic subregion, forests are particularly diverse (Table 9.1; Whitmore, 1975), with evergreen rainforest (mostly mixed dipterocarp) predominating, but with distinctive montane and other dry-land formations relating to the underlying soil and rocks, and smaller areas of wet-land formations that are particularly threatened. Eastern Java is the exception, being part of the more seasonal forest belt that extends northward through the Philippines. The eastern block of evergreen rainforest of the Sahul Shelf that extends eastwards to

TABLE 9.1. Forest formations of the tropical Far East.

Climate	Soil water	Forest formation
Ever-wet	Dry land	Topical lowland evergreen rainforest, e.g., lowland and hill mixed dipterocarp forests
		Tropical lower montane rainforest, e.g., upper dipterocarp and oak–laurel forests
		Tropical upper montane rainforests, e.g., ericaceous forests
		Tropical subalpine forest
		Heath forest
		Forest over limestone
		Forest over ultrabasic rocks
	Water table high (at least periodically)	Mangrove forest
		Brackish-water forest
		Peat swamp forest
		Freshwater swamp forest
Seasonally dry	Moderate annual shortage	Tropical semievergreen rainforest
	Marked annual shortage	Tropical moist deciduous forest other

Source: Whitmore (1975).

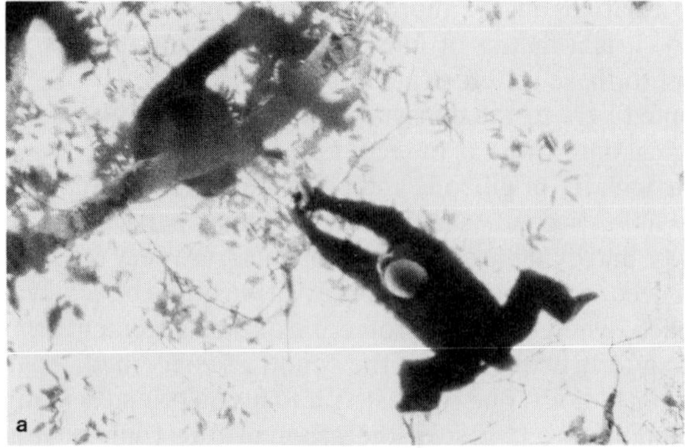

(a) Siamang (*Hylobates syndactylus*), adult pair duetting in Peninsular Malaysia.

(b) Lar gibbon (*H. lar lar*), father and juvenile daughter resting in Peninsular Malaysia.

(c) Hoolock gibbon (*H. hoolock*), adult male in Assam (R. L. Tilson).

FIGURE 9.2. Some South-east Asian rainforest primates.

(d) Hoolock gibbon (*H. hoolock*), adult female in Assam (R. L. Tilson).

(e) Dusky langur (*Presbytis obscura*) in Peninsular Malaysia (J. O. Caldecott).

(f) Banded langur (*P. melalophos*) in Peninsular Malaysia (J. O. Caldecott).

(g) Long-tailed macaque (*Macaca fascicularis*), adult female in Peninsular Malaysia.

(h) Pig-tailed macaque (*M. nemestrina*), adult male in North Sumatra (J. M. Y. Robertson).

Papua/New Guinea and northeast Australia occurs as far west as Sulawesi (Whitmore, 1975). Thus, Wallacea contains both major forest types. North of the Sunda Shelf, in the Indo-Chinese subregion, rainforest is reduced to relicts and replaced by a variety of monsoon or semideciduous forests.

The distribution of primates is Southeast Asia is considered within these three subregions of the Oriental Region, province by province to account for habitat differences (Table 9.2). Tarsiers (*Tarsius* spp.) are restricted to the forests of Borneo (and South Sumatra), Sulawesi, and the Philippines, where they have evolved into three species in isolation; the orangutan (*Pongo pygmaeus*) occurs only in Borneo and North Sumatra, as two subspecies. In contrast, the slow loris (*Nycticebus coucang*), long-tailed and pig-tailed macaques (*Macaca fascicularis* and *M. nemestrina*), and a variety of langurs (*Presbytis* spp.) and gibbons (*Hylobates* spp.) occur throughout at least two of the three subregions. While lorises and macaques show less taxonomic variations across the region, langurs and gibbons are each quite variable and distinctive, at least superficially. Herein lies a key part of the complexity of the biological, hence conservation, situation, especially as distinctive races or subspecies can be recognized in most of these primates.

There are also the "special" colobine monkeys—the "odd-nosed" monkeys—in subtropical China and north Indo-China (golden or snub-nosed monkeys, *Rhinopithecus* spp., and douc monkeys, *Pygathrix nemaeus*), in Borneo (proboscis monkey, *Nasalis larvatus*), and in the Mentawai Islands (*Nasalis concolor*).

The widespread slow loris varies in body size, pelage color and markings, the craniodental features (Barrett, 1984), with a pygmy slow loris of Indo-China (mean body weight 216 g) perhaps a separate species (Groves, 1971). The largest and smallest are otherwise found in different parts of Thailand (mean body weight 1474 and 647 g, respectively).

The macaques are considered to comprise three species groups or subgenera (Table 9.2), although Fooden (1980) identifies the stump-tailed macaque, *Macaca arctoides*, as a fourth. Morphological and some behavioral features group the pig-tailed and Mentawai Island macaque with the four or more species from Sulawesi in the *silenus* group; the two subspecies of *M. nemestrina* meet at the border of the Indo-Chinese and Sundaic subregions. The long-tailed macaque is in the *fascicularis* group, with the rhesus, Taiwan, and, according to some authorities (see Caldecott, 1983), the stump-tailed macaques; it shows little geographic variation, except perhaps in the smaller islands on the eastern edge of the Sunda Shelf. The *sinica* group is represented outside the Indian subregion by the Assamese and Tibetan macaques across the north of the Indo-Chinese subregion.

The langurs are traditionally grouped into four subgenera (Pocock, 1939), of which two are Indian (with *Presbytis entellus*, the sole species in one, just extending into the northwest of the Indo-Chinese subregion) and two are widely distributed throughout the Indo-Chinese and Sundaic subregions (Table 9.2). (*Presbytis*) or the *aygula* species group and (*Trachypithecus*) or the *cristata* species group are usually considered to be each represented by six species, many of them with subspecies.

The gibbons are divided into three subgenera: the siamang (*Symphalangus*), the concolor (*Nomascus*), and the rest (*Hylobates*) (Groves, 1972), with molecular evidence now supporting the construction of a fourth, *Bunopithecus*, for the hoolock (Prouty et al., 1983). Morphologists recognize only five species, and field biologists recognize nine, with the members of the *lar* species group readily separable by pelage and song (Preuschoft et al., 1984). The three species of the Indo-Chinese subregion are sexually dichromatic (*H. hoolock*, *H. pileatus*, *H. concolor*), with the recent addition from the south—*H. lar entelloides*—asexually dichromatic (Chivers, 1977). On the Sunda Shelf, gibbons are black in the southwest—*H. syndactylus*, H. klossi—grey or greyish in the east and southeast—*H. moloch*, *H. muelleri*, *H. agilis* (on Borneo)—and polychromatic in the center—*H. lar* and *H. agilis* in Malaya and Sumatra. The siamang is twice the size of the rest and the hoolock intermediate.

It is argued that the aridity of the Pleistocene at intervals reduced the climax rainforest of the Sunda Shelf to refugia, principally in North Indo-China, northeast Borneo, west Java, the Mentawai Islands, and North Sumatra (Brandon-Jones, 1978). This is supported by fossil evidence of savanna fauna on the central Sunda Shelf during the Pleistocene (Medway, 1972), with certain ungulate species still present today, secondarily adapted to forest. This isolation of the forest both fragmented widespread genera, such as *Hylobates* (Chivers, 1977) and *Presbytis* (Medway, 1970), promoting speciation, and isolated others, such as *Rhinopithecus*, *Pygathrix* (Groves, 1970), and *Pongo*, under the maritime influence of the edge of the Sunda Shelf.

The plant species diversity and hence (probably) the mammalian species diversity, is often greatest in these refugia; certainly they are centers of endemicity. As Whitten (1980) points out, the Mentawai Islands, separated from the rest of the Sunda Shelf for at least 500,000 years, shows the highest endemicity (species/area) in the world—greater even than in Madagascar for primates; those parts of Borneo relatively isolated from the recent incursions from Sumatra (i.e., north and east of the Kapuas and Barito rivers) come second (in Asia at least). Lorises and macaques have not been so restricted by these vegetation changes, probably spreading into the Sunda Shelf subsequently; *Tarsius* is an example of an isolated (and fragmented) genus, *Nasalis* of a fragmented one.

The species richness across the north of the Indo-Chinese subregion is explained by the admixture of Indian and Chinese faunas across the Ganges/Brahmaputra river system (and the headwaters of the Salween and Mekong rivers), by the barriers offered by the Salween and Mekong rivers, and by successive reinvasion from the Sunda Shelf in times of climatic amelioration.

Such evolutionary and zoogeographic considerations are essential for understanding the present distribution and status of primate taxa, and thus for their conservation.

Socioecology—Ecological and Behavioral Needs

The next essential for primate conservation is a consideration of the socioecology of each primate species and its place in the rainforest community, in terms of habit, habitat preferences, social structure, biomass, ranging patterns, diet, and interactions between species (Table 9.3).

Lorises and tarsiers are nocturnal and faunivorous (partly or wholly, respectively), hence they are small, living in small territories at low biomass density; lorises are "solitary," tarsiers monogamous (Barrett, 1984; MacKinnon and MacKinnon, 1980). Tarsiers are much more numerous than slow lorises, but their much smaller size results in a comparable biomass (Table 9.3).

TABLE 9.2. Classification and distribution of Southeast Asian primates.

Order → suborder → superfamily → (sub)family → genus →	Primates Strepsirhini Lemuroidea Lorisidae *Nycticebus*	Haplorhini Tarsioidea Tarsiidae *Tarsius*	Cercopithecoidea Cerocopithecinae *Macaca*	*Presbytis*	*Rhinopithecus* *Pygathrix* *Nasalis*	Hominoidea Hylobatidae *Hylobates*	Pongidae *Pongo*	
Common name →	Loris	Tarsier	Macaque	Langur		Gibbon	Orangutan	No. of species (endemic)
Indo-Chinese								
Assam/Burma	*N. coucang*	—	*M. nemestrina* *M. assamensis* *M. mulatta* *M. arctoides*	*P. entellus* *P. pileata* *P. geei* *P. phayrei*	—	*H. hoolock*	—	10 (3)
E. Burma/Thailand	*N. coucang*	—	as above and *M. fascicularis*	*P. cristata* *P. obscura* *P. phayrei*	—	*H. lar*	—	10 (0)
Subtropical China	—	—	*M. assamensis* *M. thibetana* *M. mulatta* *M. arctoides*	—	*R. roxellanae*	—	—	5 (2)
North Indochina	*N. coucang*	—	*M. assamensis* *M. mulatta* *M. arctoides*	*P. phayrei* *P. francoisi*	*R. avunculus* *P. nemaeus*	*H. concolor*	—	8 (3)
South Indochina	*N. coucang*	—	*M. assamensis* *M. fascicularis* *M. arctoides*	*P. cristata*	—	*H. pileatus*	—	5 (1)
Taiwan	—	—	*M. cyclopis*	—	—	—	—	1 (1)
Sundaic								
Malaya	*N. coucang*	—	*M. nemestrina* *M. fascicularis*	*P. melalophos* *P. obscura* *P. cristata*	—	*H. lar* *H. agilis* *H. syndactylus*	—	9 (0)
North Sumatra	N. coucang	—	*M. nemestrina* *M. fascicularis*	*P. thomasi* *P. cristata*	—	*H. lar* *H. syndactylus*	*P. pygmaeus*	8 (2)

South Sumatra	*N. coucang*	*T. bancanus*	*M. nemestrina* *M. fascicularis*	*P. melalophos* *P. cristata*	—	*H. agilis* *H. syndactylus*	—	8 (0)
Mentawai Islands	—	—	*M. pagensis*	*P. potenziani*	*N. concolor*	*H. klossi*	—	4 (4)
Java	*N. coucang*	—	*M. fascicularis*	*P. aygula* *P. cristata*	—	*H. moloch*	—	5 (2)
Borneo	*N. coucang*	*T. bancanus*	*M. nemestrina* *M. fascicularis*	*P. hosei* *P. melalophos* *P. frontata* *P. rubicunda* *P. cristata*	*N. larvatus*	*H. muelleri* *H. agilis*	*P. pygmaeus*	13 (6)
Wallacea								
Sulawesi	—	*T. spectrum*	*M. ochreata* *M. tonkeana* *M. maura* *M. nigra*	—	—	—	—	5 (5)
Philippines	—	*T. syrichta*	*M. fascicularis*	—	—	—	—	2 (1)
Lesser Sundas	—	—	*M. fascicularis*	—	—	—	—	1 (0)
Species groups (subgenera)		*silenus*	*M. nemestrina* *M. pagensis* *M. nigra* et al.	*P. aygula* *P. hosei* *P. melalophos*	*aygula* (*Semnopithecus*)			
		sinica	*M. assamensis* *M. thibetana*	*P. thomasi* *P. frontata* *P. rubicunda*				
		fascicularis	*M. fascicularis* *M. cyclopis* *M. mulatta* *M. arctoides*	*P. cristata* *P. pileata* *P. geei* *P. phayrei* *P. francoisi* *P. potenziani*	*cristata* (*Trachypithecus*)			

Source: Adapted from Marsh (in press).

TABLE 9.3. Socioecology of representative Southeast Asian forest primates—a summary.[1]

	Loris	Tarsier	Macaque		Langur		Gibbon		Orangutan
	Nycticebus coucang	*Tarsius spectrum*	*Macaca fascicularis*	*Macaca nemestrina*	*Presbytis obscura*	*Presbytis melalophos*	*Hylobates lar*	*Hylobates syndactylus*	*Pongo pygmaeus*
Habit	Nocturnal	Nocturnal	Diurnal		Diurnal		Diurnal		Diurnal
Habitat	Forest edge	Forest + edge	Forest edge	forest	Forest	forest + edge	Forest		Forest
Positional behavior	Slow climber	Vertical cling and leap	Quadrupedal—run,walk trees + ground		Quadrupedal/leaping		Suspensor—hang,climb, brachiate		Quadrumanual climb + swing
Social organization	"Solitary"	Monogamous territorial	Multimale polygyny multilevel		One-male polygyny occ. territorial		Monogamous territorial		"Solitary"
Group size	1	4	23	33	14	12	4	4	1.5
Body weight adult female (kg)	0.7	0.1	3.5	7	6.5	6.5	5.5	11	40
Group wt. (kg)	0.7	0.3	73	74	72	60	16	31	60
Biomass (kg/km²)	15	23	180	45	240	286	29	97	100
Diet	Frugivore (faunivore)	Faunivore	Frugivore/(faunivore)		Folivore/frugivore		Frugivore/folivore		Frugivore
leaves (%)	0	0	20	13	56	39	30	48	28
fruit (%)	71	0	63	74	43	58	61	44	58
animals (%)	29	100	17	13	1	3	8	8	14
Day range (km)	0.49	0.20[2]	1.08	~3.0	0.76	0.95	1.67	0.87	0.64
Home range (km²)	0.05	0.01[2]	0.40	~8.3	0.30	0.21	0.55	0.32	1.50
DR / BW	0.7	2.0	0.3	0.4	0.1	0.2	0.3	0.08	0.02
HR / BW	0.07	0.10	0.11	1.19	0.05	0.03	0.10	0.03	0.04

[1]While mean values are given—for comparative simplicity—such behavioral scores can be very variable over an annual cycle and between groups of the same species.
[2]Estimate.
Source: Raemaekers and Chivers (1980); also Barrett (1984); Caldecott (1983); MacKinnon and MacKinnon (1980); Rijksen (1978); J. M. Y. Robertson (personal communication).

The remaining primates are diurnal and frugivorous to varying extents, occurring at different densities (mainly because of diet; Hladik and Chivers, 1978), and with differing social structures.

The macaques are opportunistic frugivores living in multimale social groups; since their food is more abundant they live at higher biomass. The long-tailed macaque (*M. fascicularis*) is a forest-edge species, thriving in rural, and even urban, areas (Aldrich-Blake, 1980; Mah, 1980; Lindburg, 1980). It is replaced in the forest interior by the pig-tailed macaque (*Macaca nemestrina*), which is seen occasionally at the forest edge, and which is the most terrestrial of all these forest primates; its large, complex (multilevel) social groups live in vast home ranges (Caldecott, 1983; Robertson, in preparation). Caldecott (in press) identifies two social strategies among macaques: (1) adult males that are single-mount ejaculators, showing much paternal care, with equal sex ratios and much inbreeding, associated with richer habitats, e.g., *M. arctoides* (and *M. radiata* in India); (2) adult males that are multiple-mount ejaculators, showing no paternal care, with highly skewed sex ratios, tense social relations, and much outbreeding, associated with poorer habitats, e.g., *M. nemestrina* (and *M. fuscata* in Japan). *Macaca fascicularis* seems to exhibit both strategies, but this may be a subspecific difference, indicative of habitat state.

Since langurs are adapted to folivory, through their sacculated stomachs in which bacterial fermentation takes place, they exploit the most abundant food and occur at the highest biomasses. Social groups are smaller than in macaques, almost always based around one adult male. There are usually two species present in any Sundaic forest, with one being more frugivorous and tending toward the lower canopy levels and forest edge, and the other more folivorous. This dietary dichotomy can be equated with the *aygula* (more frugivorous, seed predators) and *cristata* (more folivorous) species groups (see also Curtin, 1980; Bennett, 1984; Davies, 1984; Davies et al., 1984).

Gibbons are monogamous, territorial, and frugivorous (particularly exploiting small sources), with elaborate suspensory and singing behavior. They occur at low biomass and, like the colobines, live in small territories. They are allopatric only when competition is reduced by one being much larger, hence more folivorous and sedentary—the siamang with the lar gibbon in Malaya, and with the lar gibbon in North Sumatra and the agile gibbon over the rest of Sumatra (Preuschoft et al., 1984).

Orangutans are essentially solitary, because of sexual dimorphism and incompatible foraging requirements (Wrangham, 1979)—females have to seek out high-quality foods; males' main needs are for quantity. They are frugivorous and relatively wide-ranging, so, despite their large size, their population density is low (MacKinnon, 1974; Rijksen, 1978). They are discussed more fully elsewhere (Rijksen, this volume).

Little is known about the ecology and behavior of the "odd-nosed" monkeys, except that they are typical folivorous colobines. The proboscis monkey is now being studied in the northwest (E. L. Bennett) and south (C. Yeager) of Borneo; while it is classically associated with mangrove forest, which is floristically impoverished, it thrives better in drier lowland forests (E. L. Bennett, personal communication).

In summary, while some species live in small groups in small home ranges (thus at high population density/biomass) because foliovorous (*Presbytis*), the

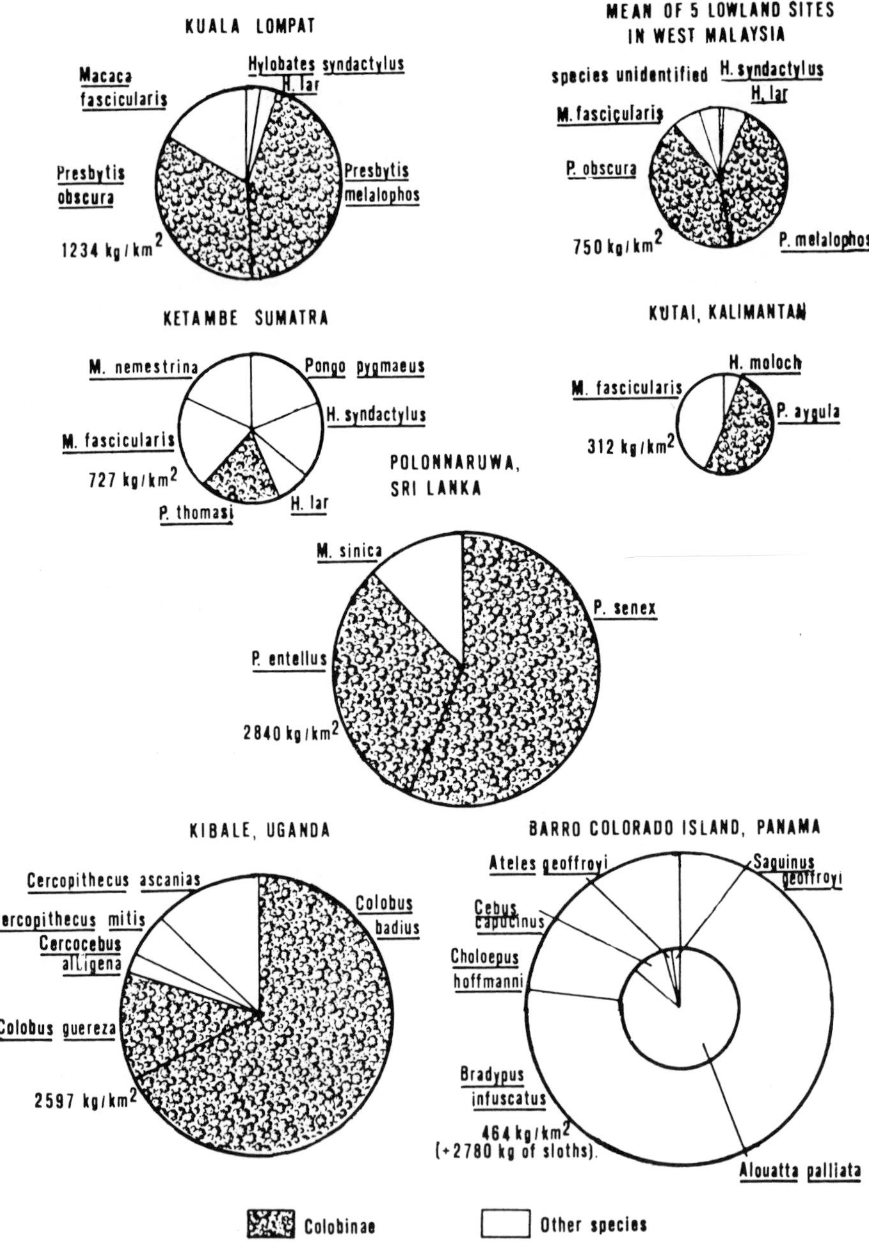

FIGURE 9.3. Comparison of primate biomasses (kg/km²) among sites in Malaysia, Indonesia, Sri Lanka, Uganda, and Panama (Marsh and Wilson, 1981).

rest occur at low biomass, whether they be small faunivores in small groups and very small home ranges (*Loris*, *Tarsius*); medium-sized frugivores in small, localized groups (*Hylobates*) or in large, wide-ranging groups (*Macaca*); or large, wide-ranging frugivores in small groups (*Pongo*). *Macaca fascicularis* is at intermediate biomass, because groups are large and home ranges small (Table 9.3); this is what would be expected for frugivores (Hladik and Chivers, 1978). The other frugivores must deviate because of the relative scarcity of suitable foods in Sundaic forests, perhaps because of competition between the numerous frugivores.

Marsh and Wilson (1981), from survey data collected on 19 routes at 20 sites in seven States of West Malaysia, regressed group density data for each primate species on various features of the forest. They show that gibbons seem to prefer forest with a high proportion of trees of the family Dipterocarpaceae (implying a high canopy and good supports for travel, sleeping, and singing) and a high diversity of tree families (implying a high dietary diversity, of fruits in particular). Langurs seem to prefer forest rich in the family Leguminosae (the major food source), with large trees (not specifically dipterocarps) and a high diversity of tree families (again for dietary reasons). For the long-tailed macaque the only significant relationship was an avoidance of dipterocarps, which accords with their preference for riverine forest and other edge habitats, where these trees are less common.

Marsh and Wilson (1981) also quantify the biomass of primates in the Malay Peninsula, which is not as reliable as group density because of the problems of accurately determining group size and individual body weights in the correct proportions for age and sex; biomass has much more ecological meaning, however. They compare these data with those from Borneo and North Sumatra, illustrating the smaller communities in the floristically poorer east of Borneo compared with Malaysia and Sumatra, and the predominance of folivorous colobines in all except North Sumatra, where fruiting trees are superabundant (at least, in the Gunung Leuser Reserve) (Fig. 9.3). These Southeast Asian communities are greatly exceeded, however, by primate densities at sites in Sri Lanka, East Africa, and Central America (if one includes sloths, along with howling monkeys, as the main folivores).

Thus, we have made good progress in understanding how the variety of macaques, langurs, and gibbons, in particular, coexist, and abundant quantitative information is available (summarized in Table 9.3). It is perhaps the nocturnal species, and the specialized and restricted colobines, that are most endangered, as much by our ignorance of them as by their biology and the status of their habitats. While gibbons are ecologically very similar (apart from the larger siamang), with certain exceptions that can be related to habitat (Chivers, 1984), the specific and subspecific populations of langurs and, to a lesser extent, of macaques merit special consideration because of their socioecological differences. Finally, the conservation of these primates and their tropical forest habitat must be considered.

Conservation of Tropical Forests and Primates

Values, Pressures, Solutions

The long-term value of tropical rainforests has been publicized widely (e.g., IUCN/WWF campaigns, Myers, 1979, 1983). It is perhaps not widely appreciated that the persistence of large areas of tropical forest is as crucial to the survival of humans as it is to the survival of animals such as monkeys and apes.

TABLE 9.4. Conservation of tropical rainforests.

Values (Long-term)	Pressures	Solutions
Water and soil balance	Hunting	Total protection of watersheds and significant representatives of each ecosystem, especially those with high plant/animal diversity
Climate rainfall pattern atmospheric gas balance	Harvesting Farming	
40–50% world's plant and animal species genetic diversity pivotal plant/animal links	Pet trade Power water oil	Wide-ranging management of buffer zones to reserves for sustained yields Agro-forestry and agriculture in areas cleared of forest, with improved efficiency
Sustained yields timber, canes, fibers, gums, waxes, resins, dyes, foods—plant and animal medicines	Selective logging Clear-felling for timber, fuel, agriculture	
Education and research		
Recreation		

These values center on both environmental and economic features and are global as well as local (Table 9.4). There are increasingly devastating examples of the effects of excessive forest clearance on water, soil, and climate. The soil is poor enough for cultivation, without being washed off the slopes to clog up urban centers and fertile alluvial land, with serious loss of life and damage to property.

It is a matter of immediate economic necessity versus long-term stability, the transition from one to the other being delicately balanced. The crisis is growing as developing countries seek to escape from the spiral of large incomes from the sale of timber and other forest products in the short term with much uncertainty about the future, into a long-term strategy of survival and prosperity from sustained yields. The forests can contribute so much more to a country's economy in the long term, than from a one-off contribution through clear felling. However much income for development can be derived from such sacrifice of the forest, and whatever one's beliefs about the viability of monocultures in a monsoonal or ever-wet region, the costs in terms of environmental catastrophes and poor or lost yields are likely to be prohibitive (e.g., Myers, 1979, 1983; Furtado, 1980).

Pressure on forests is increasing on all fronts—from local needs for fuel, materials, and food, to national needs for power, agriculture, and trade, to international needs for forest products. It is often the external pressures on a tropical forest country, which, while being so immediately welcome, in the long term are so damaging. All will suffer if tropical forests do not remain central to national (and international) economies, especially in tropical regions where the forests play such a critical role in environmental stability.

There is a tripartite solution (Table 9.4) that is already being implemented in various countries:

1. The establishment of national parks and reserves to give total protection to watersheds, thereby protecting the ecosystems at higher altitudes, and to adequate (for survival and breeding or resident flora and fauna) areas repre-

sentative of each forest ecosystem (dry-land and wet-land) in the lowlands of each country.
2. The efficient management of substantial buffer zones to these sanctuaries, and of other available tracts of forests, for sustained yields of the great variety of plant and animal products that the forest has to offer.
3. The more efficient use of land already deforested for agriculture, with the development of agro-forestry where possible.

While it may not be economically viable to protect totally more than 10% of the natural habitat in most countries, environmental problems seem to escalate when the forest cover of a country is much less than 50% of the land area (e.g., Raemaekers and Chivers, 1980; cf. Myers, 1979, 1983). This means that at least 35% of the country outside sanctuaries should be maintained in the long term as production forest.

In previous sections, the situation has been discussed from a regional and subregional viewpoint, but even the provinces do not often coincide with national boundaries. Hence the need for cooperation and coordination through international agencies, such as the Association of Southeast Asian Nations (ASEAN), the International Union for the Conservation of Nature (IUCN), the United Nations Development Program (UNDP), the Food and Agriculture Organization (FAO), and the World Bank, so that each country can act with maximum effect, since it is on a national basis that conservation (environmental management) will operate.

Primate Conservation in Southeast Asia

Having outlined the general situation, attention should now be focused on the primates. Considerable information has been accumulated on the status of primate species in the Oriental Region—23 in the Indo-Chineses subregion, 23 in the Sundaic, 7 in Wallacea, as well as 10 in the Indian subregion—and of many of the subspecies that have been identified. Some have been identified as particularly endangered (see below), and others are in urgent need of survey to resolve their status (Table 9.5). Malaysia (East and West), Indonesia, and Thailand are well surveyed, and key reserves have been identified (Table 9.6), particularly in southeast Thailand, north Sumatra, western Java, the Mentawai Islands, and parts of Borneo (Malaysian and Indonesian) and Sulawesi.

While the Sundaic subregion has been well covered, most Indo-Chinese countries have not been well surveyed, and their resident primates—the exotic colobine monkeys and the concolor and hoolock gibbons, in particular—are in urgent need of attention, as are the forests they inhabit. This situation is already being remedied in Vietnam (J. R. MacKinnon, personal communication), Bangladesh (Gittins, 1980), and subtropical China (Zhang et al., 1981; Haimoff, personal communication).

Even on the well-forested Sunda Shelf, however, the rate at which forest is disappearing is alarming, especially in the lowlands, where plant diversity is highest and wildlife at its most abundant. It is also alarming in view of the flooding, soil erosion, droughts, and fires that follow; recently fires raged for months in East Kalimantan (Berenstain, in press; Leighton, in preparation). It is estimated that lowland forests outside sanctuaries will have disappeared by the end of

TABLE 9.5. Zoogeographic provinces and species needing status surveys.

Subregion[1] Province	Countries	Species	
Indo-Chinese			
Assam/Burma	India Bangladesh Burma	*Nycticebus coucang* *Macaca assamensis* *Rhinopithecus roxellanae* *Hylobates hoolock*	*Presbytis geei* *P. pileata*
Thailand/E. Burma	Burma Thailand Kampuchea	*Macaca* spp., esp. *M. arctoides* *Presbytis phayrei*	
Subtropical China	China	*Rhinopithecus roxellanae* *Pygathrix nemaeus*	*Macaca thibetana*
North Indo-China	China Laos Vietnam	*Rhinopithecus avunculus* *Pygathrix nemaeus* *Presbytis francoisi* *Hylobates concolor*	*Macaca arctoides*
South Indo-China	Vietnam Kampuchea	*Hylobates pileatus* *Pygathrix nemaeus*	*H. concolor*
Sundaic			
Borneo	E. Malaysia Brunei Indonesia	*Nasalis larvatus* *Presbytis frontata* *Hylobates muelleri/H. agilis* border	*Pongo pygmaeus*
Java	Indonesia	*Presbytis aygula*	*Hylobates moloch*
Mentawai Islands	Indonesia	all species (except on Siberut)	
Wallacea			
Philippines	Philippines	*Tarsius syrichta*	
Sulawesi	Indonesia	*Macaca maura*	

[1]Subregion of Indo-Malayan realm.
Source: from Marsh (in press).

the century; Marsh and Wilson (1981) point out that the lowland peat-swamp and semievergreen forests of the Malay Peninsula are already very seriously threatened.

In 1975, the total gibbon population (of nine species across Southeast Asia) was estimated in order to predict, from the rate of forest clearance at that time, the population level of each species by 1990 (assuming that by then little forest would remain outside protected areas). The results led to the prediction of an 84% reduction in numbers, with four species (Kloss, moloch, pileated, and concolor) close to extinction and with the position of the other five species only slightly better (Chivers, 1977). The same applies to all the other primates, especially those confined to lowland forest formations. In 1974, about 74% of Borneo was forested, with only 2.5% protected (Chivers, 1977); 7 years later only 60% was forested, but the protected area had doubled, although about half the area in reserves was subject to shifting cultivation (Davies, 1983). Current figures (Table 9.7) indicate that 18% of the island's forests are now protected, with a further 6% proposed. Aken and Kavanagh (1983) describe 76% of Sarawak as forested, with 24% to be maintained as production forest and 2% in parks and sanctuaries (with a further 2% proposed, according to the data used in Table 9.7).

TABLE 9.6. Some important reserves for the conservation of Southeast Asian forest primates.

Subregion Province	Country	Reserve	Species
Indo-Chinese			
Assam/Burma	Burma	Alaungdaw Kathapa Tamanthi	*Presbytis pileata* *Hylobates hoolock*
Subtropical China	China	Wolong et al.	*Rhinopithecus roxellanae* *Macaca* spp.
Thailand/E. Burma	Thailand	Huay Kha Khaeng Khao Yai	5 sympatric *Macaca* spp. *Hylobates lar/H. pileatus* border zone
		Khao Soi Dao	*H. pileatus*
Sundaic			
Malaya	W. Malaysia	Krau	Malayan community on granite-derived soils
		Taman Negara	Malayan community on shale-derived soils
North Sumatra	Indonesia	Gunung Leuser	*Pongo pygmaeus, Hylobates lar, H. syndactylus, Presbytis thomasi*
South Sumatra	Indonesia	Kerinci-Seblat	S Sumatran community
Mentawai Islands	Indonesia	Siberut	*H. klossi, P. potenziani, M. pagensis, Nasalis concolor*—4 endemics
Java	Indonesia	Gunung Halimun	*Hylobates moloch* *Presbytis aygula*
Borneo	Indonesia	Tanjung Puting	Swamp forest community incl. *Nasalis larvatus*
	E Malaysia	Lanjak-Entimau	*Pongo pygmaeus* *Presbytis frontata*
		Samunsam	*Nasalis larvatus* in coastal forest
		Bako	*Nasalis larvatus* in kerangas forest
Wallacea			
Sulawesi	Indonesia	Dumoga-Bone	*Macaca nigra* *Tarsius spectrum*
		Lore-Lindu	*Macaca tonkeana* *Tarsius spectrum*

Source: from Marsh (in press).

For West Malaysia alone, Marsh and Wilson (1981) estimated, from their survey data and information on forest clearance in the late 1970s, an *annual* loss of 3400 siamang, 31,000 gibbons (lar and agile), 171,000 banded langurs, 175,000 dusky langurs, and 45,000 long-tailed macaques. Extrapolation to the rest of the Sundaic subregion produces a projection of devastation on a horrific scale. In emotional terms, this is an inexcusable loss of life; in economic terms, it is a terrible waste of resources.

TABLE 9.7. Protected forests in some Southeast Asian countries.

	Total land area	1975[1]				1985[2]		
		Forested area		Protected forests		Protected forests: Actual + proposed land area		
	km²	km²	%	km²	%	km²	%	(Proposed only)
"Assam"	121,900	47,900	39	235	0.2	4937	4	(0)
Bangladesh	142,776					4498	3	(1)
Burma	678,033					11,886	2	(1)
Thailand	514,910	94,452	18			41,484	8	(0)
West Malaysia	128,013	66,950	52	8150	6	39,138	31	(5)
East Malaysia	201,727	128,700	64	1740	1	71,297	35	(2)
Indonesia								
Sumatra	473,970	260,000	55	18,280	4	76,597	16	(3)
Java	126,501	28,000	22	2422	2	12,294	10	(4)
Kalimantan	539,500	419,000	78	11,410	2	105,724	20	(11)

[1]From Chivers (1977).

[2]From Protected Areas Data Unit, Conservation Monitoring Centre, Cambridge.

In a thorough analysis of the status of primates in Indonesia (Sumatra, Java, Kalimantan, and Sulawesi) MacKinnon (1983, this volume) shows that, of the 28 species occurring there, three are highly endangered (moloch gibbon, Mentawai macaque, and Javan langur) and seven are vulnerable (orangutan, Kloss gibbon, two of the Sulawesi macaques, Mentawai pig-tailed "langur" and langur, and the Bornean white-fronted langur). The lar gibbon, another Sulawesi macaque, the proboscis monkey, two more Bornean langurs, and the siamang come next, with Müller's gibbon, long-tailed and pig-tailed macaques, silvered, red, and banded langurs; the two tarsiers are the most abundant, and thus probably the least threatened, particularly in view of the protected populations in reserves, which are increasingly forming an extensive network.

Thus, recent figures on the areas of protected forests are superficially encouraging (Table 9.7); they show a significant increase for most countries over the last 10 years (including proposals for the future). It is not necessarily so encouraging for the primates, however. There are two problems obscured by these optimistic figures:

1. Accurate data on forest areas are difficult to come by, and they do not reflect accurately the protection given to an area, e.g., 10% of Java is proposed as protected forest area (Table 9.7), and yet only 8% is currently forested (K. S. MacKinnon, personal communication)—the worthy statements now burgeoning on paper are not necessarily followed up on the ground.
2. There are different sorts of protection, since some countries include in these figures areas protected for timber production; while such long-term (if they are such) forest estates are of great value to many animals, they are unsuitable for others (see below).

Such issues need clarifying and the desired procedures need to be implemented. It is important that totally protected forests should cover all forest types and be large enough to sustain populations of the larger animals residing therein (e.g., Medway and Wells, 1971), as well as being considered in relation to the areas of forests managed for sustained yields. Once such strategies are developed, labor for protection and management should not be a problem in these expanding nations, it being more a matter of tapping the right financial sources. Given the intrinsic value of these forests, however, and the debt that the international community owes these forested nations, finance should not be a problem either, but it probably will be, and much pressure and support will be required by those seeking to implement viable long-term strategies for forest protection and management.

Special Problems

Finally, reference must be made to the role of managed forests in primate conservation and to the fate of primates made homeless by forest clearance.

The influence of selective logging on primate populations has been studied in East Kalimantan by Wilson and Wilson (1975) and in West Malaysia by Johns (1981, 1983a, 1983b), with a useful discussion based on surveys by Marsh and Wilson (1981). Even though extraction may be light (ten trees/ha or 4% of trees), 45% of the total stand (68% of biomass) is damaged in the process of access, felling, and extraction (Johns, 1983b). While orangutans and proboscis monkeys are intolerant of logging (Wilson and Wilson, 1975), gibbons and langurs (and macaques) are remarkably tolerant, at least of low levels of extraction, although their behavior is affected in the short term by the disturbance and disruption of food supplies, and population dynamics is upset for some time. Once access has been gained to these forests, however, it is vital that hunting and cultivation are prevented, if the primate populations are to recover.

Both gibbons and langurs adjust their foraging strategies by eating more foliage as fruit availability declines in newly logged forest (Johns, 1983b). Gibbons maintain their territories, but the stress affects their breeding so that there is predicted to be a latent decline in population size, which takes more than 20 years to recover fully (Fig. 9.4). Langurs emigrate temporarily from the disturbed area, and there is increased mortality of immature monkeys, so that populations decline markedly (from this and the future breeding loss), again taking at least 20 years to recover to the original level, it is surmized (Marsh and Wilson, 1981).

The effects of light selective logging are not so drastic to primates and other rainforest mammals as might be suspected, because it enhances the diversity of microhabitats characteristic of the mosaic of successional stages of climax forest (Johns, 1983b). Thus, squirrels shift from fruit to bark and sap, and browsing mammals and small predators become more abundant because of increased food supply. Bird communities maintain much the same trophic structure, but species composition may be changed markedly; dietary generalists survive better than faunivores and frugivores, whose food supply may be very disrupted. In contrast, amphibians and other cold-blooded animals cannot survive the increased temperature, sunshine, and decreased humidity.

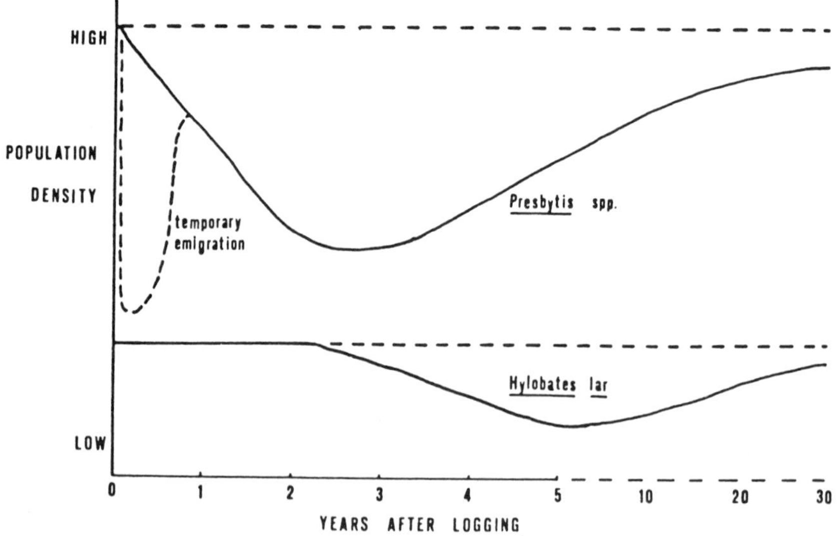

FIGURE 9.4. Hypothesized effects of logging on populations of langurs and gibbons (Marsh and Wilson, 1981).

Selective logging that is light and carefully controlled, therefore, can produce a habitat with a diverse and abundant fauna, but such conditions must be adhered to rigidly and the fauna will be significantly different from that in undisturbed climax forest. Current selective logging techniques, however, present problems other than those already mentioned (Johns, 1983b); the cutting of climbers prior to felling reduces food supply, although it does reduce tree-fall damage. More seriously, postfelling poison girdling to eliminate defective and noncommercial trees affects food supply as well as sites for resting and supports for travel. At present, the least damaging extraction techniques are prohibitively expensive, but, in view of the environmental benefits, ways should be found to make them economically viable.

Clear felling results in primates either remaining and perishing (augmented by trapping and hunting) or escaping and disrupting the adjacent population (as well as being disrupted ecologically and socially themselves). Hunting is not as much of a threat to primates in West Malaysia (where by 1969 more than 40% of the forest peoples had [been] moved out of the forest, abandoning their traditional ways) as it is in Borneo (Chivers and Davies, 1979).

Much concern has been expressed internationally, and quite rightly, at the illegal trade in wild animals, especially primates; demand is often greatest for those that are most endangered. Pressure on nations exporting monkeys for biomedical research are such that bans have been imposed, even for less endangered species and for monkeys bred in captivity in their habitat country. Much time and money is spent reintroducing confiscated primates, especially apes (e.g., Aveling, 1982), to their natural habitat. Little thought and effort is given, however, to the majority of primates in distress, to the vast numbers made homeless by clear felling for agriculture.

It has been stressed that clear felling should cease in the foreseeable future, for a whole variety of reasons, including economic ones; but, until it does, horrific death toll (see above) should be reduced. The ideal solution is to move social groups to suitable unpopulated habitat, but apart from forest that has been overhunted, forest that still stands has its own healthy populations. Furthermore, we have not yet developed the appropriate translocation techniques for forest primates (but see Strum, this volume, for savanna ones)—trapping, transport, release, and monitoring—with the minimum of physical and social harm, even though guidelines have been produced (e.g., Caldecott and Kavanagh, 1983). Some experience has been gained with great apes, but the value has been more in conservation education than in saving significant numbers (e.g., Aveling, 1982). It is an elaborate and expensive technique, which merits full investigation because of its potential for conservation (in the broadest sense).

The obvious solution is rescue and "use." The reduction of suffering, saving of lives, and contribution to primate biology and conservation and human welfare would seem to be indisputable, far outweighing the ethical problems posed. Biomedical needs could be supplied from captive-bred animals, or from animals "ranched" along the lines developed elsewhere in the tropics, as proposed for long-tailed macaques in Indonesia by MacKinnon (1983); this would save the wild populations from further depradation. Research on captive social groups yields information invaluable for the conservation of species in the wild, particularly on nutrition and reproduction. The management of social groups can provide primates for a variety of purposes, including a nucleus for return to the

TABLE 9.8. IUCN/SSC global strategy for primate conservation: draft proposals for Southeast Asia, summary.

Indonesia	
Java	*Hylobates moloch, Presbytis aygula*
Kalimantan	*Pongo pygmaeus, Nasalis larvatus, H. agilis/H. muelleri* border zone
Mentawai Is.	4 rare, endemic primates
Sumatra	Gunung Leuser and Kerinci-Seblat Reserves
Sulawesi	Macaques
	Tarsius spectrum
Philippines	*Macaca fascicularis*
	Tarsius syrichta
East Malaysia	*Nasalis larvatus*
	Coastal and riverine forests
	Pongo pygmaeus, Lanjak-Entimau Reserve
Thailand	Sympatric macaques
	H. lar/H. pileatus border zone
Kampuchea	
Laos	Status surveys
Vietnam	Park development
South China	

Southeast Asian primates currently listed in IUCN Red Data Book (1982).

Species	Status
Tarsiidae	
Tarsius bancanus	Indeterminate
Tarsius syrichta	Endangered
Cercopithecinae	
Macaca pagensis	Indeterminate
Colobinae	
Presbytis potenziani	Indeterminate
Nasalis larvatus	Endangered
Nasalis (Simias) concolor	Endangered
Rhinopithecus roxellanae	Rare
Pygathrix nemaeus	Endangered
Hylobatidae	
Hylobates klossi	Vulnerable
Hylobates pileatus	Endangered
Hylobates concolor	Indeterminate
Hylobates moloch	Endangered
Pongidae	
Pongo pygmaeus	Endangered

Source: A. A. Eudey (personal communication).

wild, especially for the most endangered species; it also offers unique educational opportunities, both for local people and tourists, especially if done under seminatural conditions, with concomitant economic benefits. There are many opportunities, but they are not being developed as rapidly or widely as they should be, and the main losers are the primates themselves, with humans coming a close second.

Conclusions

The complex variety of Southeast Asian primates and their distribution in relation to the equally complex variety of tropical forest habitats has been described and explained; emphasis has been placed on the Sundaic subregion of the Oriental Region. The behavior and ecological needs of lorises, tarsiers, macaques, langurs, gibbons, and orangutans have been described. The status of each primate, in relation to the rapid clearance of its forest habitat, has been discussed. The importance of protecting sufficiently large areas of each forest type in each Southeast Asian country has been stressed, in relation to the long-term environmental and economic benefits that they will provide. Recognizing that only a small proportion of the forests can be left undisturbed, full reference has been made to the long-term values and ways of careful management of large areas of forest, both for wildlife and humans. Concern has been expressed about the current, alarming waste of primate resources.

In the light of the endangered species listed in the IUCN *Red Data Book* and the collation of data now available on primate distribution and abundance, in relation to forest type, extent and, status, the Primate Specialist Group (Asian Section) of the Species Survival Commission of IUCN has redrafted its contribution to the Global Strategy for Primate Conservation (key points summarized in Table 9.8). Adopting an ecosystems and zoogeographical approach, it represents a blend of action to save the most endangered species (and subspecies) and to survey the lesser known areas and resident species (often restricted and believed to be endangered). The approach is justified since primates, because of their large size and arboreal habits, are among the best indicators of the health of the forests concerned, and it is the whole ecosystem and its diversity that we are seeking to conserve—for humans as well as primates, and all the other animals and plants on which they depend.

Despite the gloomy situations described herein, there are also signs of significant progress—but will it be enough, and in time?

Acknowledgments. I am grateful to Kathy MacKinnon, Eamonn Barrett, and Yarrow Robertson for their comments on the draft of this paper.

References

Aken K, Kavanagh M (1983) Species conservation priorities in the tropical forests of Sarawak, Malaysia. In: Species conservation priorities in the tropical forests of Southeast Asia. IUCN Species Survival Commission, Gland, Switzerland.

Aldrich-Blake FPG (1980) Long-tailed macaques. In: Chivers DJ (ed) Malayan forest primates. Plenum, New York, pp 147–165.

Aveling R (1982) Orang utan conservation in Sumatra, by habitat protection and conservation education. In: de Boer LEM (ed) The orang utan: its biology and conservation. Dr W Junk, The Hague, pp 299–315.

Barrett EBM (1984) The ecology of some nocturnal arboreal mammals in the rain forest of Peninsular Malaysia. Thesis, University of Cambridge.

Bennett EL (1984) The banded langur: ecology of a colobine in West Malaysian rainforest. Thesis, University of Cambridge.

Berenstain L (In press) Responses of long-tailed macaques to drought and fire in eastern Borneo. Biotropica.

Brandon-Jones D (1978) The evolution of recent Asian Colobinae. In: Chivers DJ, Joysey KA (eds) Recent advances in primatology, vol 3. Evolution. Academic, London, pp 323–325.

Caldecott JO (1983) An ecological study of the pig-tailed macaque in Peninsular Malaysia. Thesis, University of Cambridge.

Caldecott JO (In press) Sexual behaviour, societies and the ecogeography of macaques. Anim Behav.

Caldecott JO, Kavanagh M (1983) Guidelines for the use of translocation in the management of wild primate populations. Oryx 17:135–139.

Caldecott DJ (1977) The lesser apes. In: Prince Rainier, Bourne GH (eds) Primate conservation. Academic, New York, pp 539–598.

Chivers DJ (1984) Feeding and ranging: a summary. In: Preuschoft H, Chivers DJ, Brockelman WY, Creel N (eds) The lesser apes: evolutionary and behavioural biology. Edinburgh University Press, pp 267–281.

Chivers DJ, Davies AG (1979) The abundance of primates in the Krau Game Reserve, Peninsular Malaysia. In: Marshall AG (ed) The abundance of mammals in Malesian rain forests. University of Hull, pp 9–32 (Department of Geography, Miscell Series, no 22).

Curtin SH (1980) Dusky and banded leaf monkeys. In: Chivers DJ (ed) Malayan forest primates. Plenum, New York, pp 107–145.

Davies AG (1983) Distribution, abundance and conservation of simian primates in Borneo. In: Harper D (ed) Symposium on the conservation of primates and their habitats. University of Leicester, pp 122–148 (Vaughan Paper no 31).

Davies AG (1984) An ecological study of the red leaf monkey (*Presbytis rubicunda*) in the dipterocarp forests of northern Borneo. Thesis, University of Cambridge.

Davies AG, Caldecott JO, Chivers DJ (1984) Natural foods as a guide to nutrition of Old World primates. In: UFAW (ed) Standards in laboratory animal management. Universities Federation for Animal Welfare, Potters Bar, England, pp 225–244.

Fooden J (1980) Classification and distribution of living primates (*Macaca* Lacépède, 1799). In: Lindburg DG (ed) The macaques. Van Nostrand Reinhold, New York, pp 1–9.

Furtado JI (ed) (1980) Tropical ecology and development, proceedings of the Vth international symposium of tropical ecology, April 1979, Malaysia. The International Society of Tropical Ecology, Kuala Lumpur.

Gittins SP (1980) A survey of the primates of Bangladesh. Project report to Forest Department, Bangladesh.

Groves CP (1970) The forgotten leaf-eaters, and the phylogeny of the Colobinae. In: Napier JR, Napier PH (eds) Old World monkeys: evolution, systematics and behaviour. Academic, New York, pp 555–587.

Groves CP (1971) Systematics of the genus *Nycticebus*. In: Proceedings of the 3rd Int Congr Primatol (Zurich, 1970). Karger, Basel, pp 44–53.

Groves CP (1972) Systematics and phylogeny of gibbons. In: Rumbaugh DM (ed) Gibbon and siamang, vol 1. Karger, Basel, pp 1–89.

Hladik CM, Chivers DJ (1978) Ecological factors and specific behavioural patterns determining primate diet. In: Chivers DJ, Herbert J (eds) Recent advances in primatology, vol 1. Behaviour. Academic, London, pp 433–444.

Johns AD (1981) The effects of selective logging on the social structure of resident primates. Malays Appl Biol 10:221–226.

Johns AD (1983a) Selective logging and primates: an overview. In: Harper D (ed) Symposium on the conservation of primates and their habitats. University of Leicester, pp 86–100 (Vaughan Paper no 31).

Johns AD (1983b) Ecological effects of selective logging in a West Malaysian rain-forest. Thesis, University of Cambridge.

Lindburg DG (ed) (1980) The macaques: studies in ecology, behavior and evolution. Van Nostrand Reinhold, New York.

MacKinnon JR (1974) The behaviour and ecology of wild orang-utans (*Pongo pygmaeus*). Anim Behav 22:3–74.

MacKinnon JR, MacKinnon KS (1980) The behaviour of wild spectral tarsiers. Intern J Primatol 1:361–379.

MacKinnon KS (1983) To determine population estimates of *Macaca fascicularis* (and other primates) and the feasibility of semi-wild breeding projects of this species. WHO Primate Resources Programme, feasibility study.

Mah YL (1980) Ecology and behaviour of *Macaca fascicularis*. Thesis, University of Malaya, Kuala Lumpur.

Marsh CW (In press) A framework for primate conservation priorities in Asian moist tropical forests. In: Marsh CW, Mittermeier RA, Gartlan JS (eds) Conservation of primates and tropical forests. Allan R Liss, New York.

Marsh CW, Wilson WL (1981) A survey of primates in Peninsular Malaysian forests. Universiti Kebangsaan Malaysia and University of Cambridge.

Medway Lord (1970) The monkeys of Sundaland. In: Napier JR, Napier PH (eds) Old World monkeys: evolution, systematics and behaviour. Academic, New York, pp 513–553.

Medway Lord (1972) The Quaternary mammals of Malesia: a review. In: Ashton P, Ashton M (eds) The Quaternary era in Malesia, University of Hull, pp 63–83 (Department of Geography, Miscell Series, no 13).

Medway Lord, Wells DR (1971) Diversity and density of birds and mammals at Kuala Lompat, Pahang. Malay Nat J 24:238–247.

Myers N (1979) The sinking ark. Pergamon, Oxford.

Myers N (1983) A wealth of wild species: storehouse for human welfare. Westview, Boulder, CO.

Pocock RI (1939) The fauna of British India, including Ceylon and Burma, vol I. Mammalia. Taylor and Francis, London.

Preuschoft H, Chivers DJ, Brockelman WY, Creel N (eds) (1984) The lesser apes: evolutionary and behavioural biology. Edinburgh University Press.

Prouty LA, Buchanan PD, Pollitzer WS, Mootnick AR (1983) *Bunopithecus*: a genus-level taxon for the hoolock gibbon (*Hylobates hoolock*). Amer J Primatol 5:83–87.

Raemaekers JJ, Chivers DJ (1980) Socio-ecology of Malayan forest primates. In: Chivers DJ (ed) Malayan forest primates. Plenum, New York, pp 279–316.

Rijksen HD (1978) A field study on Sumatran orang-utans (*Pongo pygmaeus abelii* Lesson 1827): ecology, behaviour and conservation. H. Veeman and Zonen BV, Wageningen, The Netherlands.

Verstappen HT (1975) On palaeoclimates and land form development in Malesia. In: Bartstra G-J, Casparie WA (eds) Modern Quaternary research in South-east Asia. Balkema, Rotterdam, pp 3–36.

Whitmore TC (1975) Tropical rain forests of the Far East. Clarendon, Oxford.

Whitten AJ (1980) The Kloss gibbon in Siberut rain forest. Thesis, University of Cambridge.

Wilson CC, Wilson WL (1975) The influences of selective logging on primates and some other animals in East Kalimantan. Folia Primatol 23:245–274.

Wrangham RW (1979) On the evolution of ape social systems. Soc Sci Infor 18:335–368.

Zhang Y-Z, Wang S, Quan G-Q (1981) On the geographical distribution of primates in China. J Hum Evol 10:215–226.

10
Conservation of Orangutans: A Status Report, 1985

HERMAN D. RIJKSEN

Introduction

The "red-haired man of the rainforest," or orangutan, is among the most spectacular creatures on Earth, and although representing the closest among human mammal relatives in Southeast Asia, he is in grave danger of extinction. Ironically also "cultivated" man is seriously endangered, but while humans choke the world with their biomass, aggravated by disproportional exploitation and pollution, they supplant and eradicate their closest relative as they exploit and destroy his habitat so as to serve so called "development." In this time of Western self-scorn with respect of the qualitative aspects of what is commonly understood as "development," it may be enlightening to notice that the existence of the red-haired man of the rainforest has been endangered whenever he happened to come into contact with "cultivated" man, even long before any European set a foot in his habitat: He has been hunted to extinction in all the regions where the earliest agricultural migrants from the Asian mainland settled.

When the first Europeans arrived to colonize and convert the Indomalayans to their technocratic civilization pattern, the orangutan's distribution range had shrunk already to relic populations. Confined to the upper half of Sumatra and some stretches of Borneo, none were to be found any more on the Asian mainland, the Southeast Asian peninsula, Java, the southern half of Sumatra, and those regions in Borneo where neolithic agriculturalists had settled (i.e., near the mouths of the largest rivers). Drawing from the ethnic histories of the peoples of Sumatra, Malaysia, and Borneo, it is clear that the demise of the red-haired "man," in liaison with the "negrito-type" man (and the so-called "punan" ethnic group) has been a direct consequence of the (agri-)cultural urge for headhunting—a uniquely human form of competition. And although the introduction of "Western" civilization has effectively stopped headhunting of human subjects, it in fact boosted the eradication of (at least) the orangutan—albeit mainly in an indirect way, by engendering a human population explosion and overexploitation of its forest habitat. Nevertheless, it also came to foster some awareness for the sorrowful plight of man's nearest relative.

Not many people have had the privilege to see and come to know the orangutan. Yet it is largely owing to the concern and the ensuing efforts of those few privileged individuals that world-wide measures have been taken in order to seriously try to conserve the remaining relic populations of apes as effectively as is feasible under the adverse circumstances. Thus, in the 1940s the (then still colonial) government of Indonesia was persuaded to decree a total ban on hunting and trade in "orang utan" and/or preserved parts thereof. In addition, it set aside, after due deliberation and after the local authorities had voluntarily given their full consent, some major "reserves" in which the ape might be conserved. They comprise the Gunung Leuser/Sikundur/Langkat Utara group of wildlife reserves (currently Gunung Leuser National park; 10,000 km^2), the Kotawaringin Sampit group of wildlife reserves (currently Tanjung Puting S.M.; some 350,000 ha), and the Kutai wildlife reserve (currently National Park; 306,000 ha).

The areas were uninhabited by people at the time of their establishment as reserves, and great care was taken by the government that only those areas were included where any prospect for profitable cultivation or exploitation was absent, either because of adverse soil and terrain conditions, such as in the Gunung Leuser area and Tanjung Puting, or because of excessive remoteness. At that time the reserves covered about 20% of the distribution range of the Sumatran and less than 3% of the Bornean populations of apes. Fortunately, most of its distribution range was covered by the territory of Indonesia, because British colonial rule never installed any sanctuaries in the Bornean territories. Only quite recently has the Sarawak government set aside major reserves for its unique rainforest types including the orangutan, almost reluctantly followed by the Sabah government.

In the meantime, at the instigation of B. Harrisson, a world-wide ban on orangutan trade has been enforced during the 1970s, a move that effectively curbed at least one aspect of the steady decline of apes.

Biological Aspects

The orangutan is a frugivore, although its diet comprises all sorts of vegetable produce (leaves, growthlayers, wood, and flowers) from a wide variety of rainforest flora, as well as animal protein, notably insects (e.g., ants and termites), and even mineral soil. With reference to its food requirements, the rainforest should be seen as a "patchy" habitat, with an irregular distribution, not just in space and quantities of food, but also in terms of quality and availability in time. Yet, within the vast stretches of seemingly uniform rainforest that used to cover all tropical Southeast Asia, the ape appears to inhabit mainly the alluvial plains and valleys and the adjacent slopes of surrounding hills and mountain complexes, patches where, in general, the conditions for its existence seem to be most suitable. For there the forest, as a result of regularly occurring disturbances of a minor, "adaptable" scale, cannot attain a "mature" stage and consequently supports the highest relative incidence of suitable fruit trees and lianas. The extensive, mature rainforests of the flat lowlands, dominated by Dipterocarp species, offer far less suitable conditions.

Not surprisingly, albeit rather unfortunate for the ape, humans have the same preference for such patches in the rainforest as they offer the most fertile soil, relative to their surroundings while the rivers and streams assure easy access. As a consequence, both close relatives wherever their distribution range overlap are bound to engage in a conflict situation in which man invariably exerts his superior powers at the cost of the ape's existence. It is not that the apes are invariably destroyed in the conflict, but the harassment to which they are being exposed by a human invasion into what used to be their habitat causes "stress" in their ecological and social relationships patterns, and if not terminating their lives, it certainly affects their reproduction.

The orangutan's social organization can best be characterized as a dynamic pattern of limited gregariousness, in which temporary associations can be engendered by the occurrence of large fruit sources. The formation of real, albeit temporary, social groupings, as well as the alternative, antagonistic encounters, are based on long-standing relationships among the individuals who "know" each other more or less intimately from their adolescent (and subadult) stage of relatively high sociability.

Interestingly, the subadult males develop a strategy of "raping" lower ranking adult and subadult females, which appears to have very little, if any, immediate reproductive relevance, but seems to serve as status advertisement and the establishment of a "choice parameter" with respect to the female victim. Consequently, when a female comes into estrus she actively engages a particular male while avoiding others. It is worthy of notice that orangutans in captivity fail to develop (or maintain) a "normal" courtship pattern (triggered by the female's proceptive behavior, rather than her presence), but maintain such "rape-behavior" all their lives. The extremely restricted possibilities with respect to varied social contacts and subsistence problems under captive conditions apparently prevent the development of the complex mental capacities of orangutans to the normal standard. One would expect that a normal, mentally healthy orangutan requires an ontogeny characterized by options for the establishment and maintenance of a variety of dynamic relationship patterns including all aspects of social life and the problems of subsistence. This implies that apes born and raised under captive conditions inevitably become mentally retarded, while captive conditions for a wild-raised ape should be considered as mental torture.

Be that as it may, the reproductive output of the orangutan is rather limited; it is not so much restricted by social and environmental constraints as by the inherent capacity of the female. A female orangutan, under wild conditions, attains maturity around her twelfth year and may not conceive before an age of 13–15 years. Then the birth interval is exceptionally long, and in the case of a viable offspring extends for at least 5 years, implying that with an estimated life-expectancy of 35 years, a female cannot produce more than at least five offspring. Yet more important, it also implies that recruitment in case of vacant niches (due to hunting pressure, for example) is terribly slow, even under ideal conditions. It will be clear that stress caused by human impact, either upon the ape population itself or upon its habitat, can only aggravate the problem of low reproductive success.

In this respect, it should be realized that the orangutan commonly occupies a more or less permanent home range, rather than living a nomadic life-style, as has been suggested quite often. In that home range the complex distribution pattern of resources appears well known. Consequently, it may be expected that forcing an ape out of his home range, into unfamiliar areas, with unfamiliar and therefore hostile conspecifics, exposes it to fatal risks: the common idea of forest exploiters that destruction of the forest without physically harming the animals—"only" driving them into adjacent regions of (still) intact forest—may be nearly as fatal as killing them outright.

Current Status: Distribution Pattern

From the knowledge that the orangutan mainly occupies alluvial plains and valleys and their adjacent water catchments in rainforest, combined with the findings of a common density of population in such patches, one could, with the aid of an accurate topographic map that shows the extent of current land use easily arrive at a rather accurate figure for the total population sizes of apes in Sumatra and Borneo. Unfortunately, however, such maps, if possibly existent, as well as recent aerial photographs are suspiciously guarded by a few privileged authorities and are not available even to the government authorities concerned with forest protection and nature conservation in Sumatra and Borneo. As a consequence, the precise extent of forest conversion and destruction in these two major islands is shrouded in a veil of ignorance, while the official data for the extent of forest are gross overestimates.

Nonetheless, with some effort it is possible to make a rough estimate of the extent of remaining rainforest and even the types within from crude timber concession and land use maps available at the departments concerned. Thus for Kalimantan, the official figure for forest coverage varies between 50 million ha and 37 million ha, whereas an estimate from the maps amount to some 32 million ha at most, including secondary scrub as well as mangrove. It may be an interesting detail to note that the policy of the Indonesian government is to keep 66% of the total land area of Kalimantan under forest cover (i.e., 36 million ha). For the Northern part of Sumatra (Aceh and Sumatra Utara province), the official figures come closer to the guesses from the maps, namely 6.6 million ha of forest against 6.3 million ha.

However, as these figures lump all forest types together, a more meaningful approach is to make an estimate of the suitable habitat for the apes, employing from whatever crude maps are currently available for the Sumatran and Bornean region the extent of water courses and their catchments. Such an approach yields some 2000 km² in Sumatra and some 12,000 km² in Borneo within the currently known distribution range. These figures exclude possible relic populations of extremely small size, such as the recently discovered population in the tiny patch of remnant protection forest on Gunung Talamau in West Sumatra province, some 100 km north of Padang and well south of earlier reports of relic populations. They also exclude the suitable patches of habitat that were recently destroyed (1982–1983) in the gargantuan forest firest that raged over large parts of Kalimantan.

Taking into consideration the finding that in alluvial habitat the orangutan may occur in densities of three to five individuals/km² and supposing that an alluvial belt of suitable habitat does not exceed 3 km from the river in the mountainous terrain of Sumatra and some 20 km in the flat, undulating terrain of Borneo while subtracting the stretches of alluvial area known to be destroyed and/or occupied by people, either for agriculture or timber exploitation, one arrives at a figure of some 6000 individuals for the Sumatran and some 37,000 individuals for the current Bornean population. It is noteworthy how these figures contrast with the numbers of "civilized" man in the same regions: Borneo is at present occupied by a population of some 10 million humans increasing at a rate of 3% per year; the population of northern Sumatra amounts to some 12 million individuals increasing at a rate of 3.4% per year.

In view of the fact that the suitable habitat for apes in Kalimantan has dwindled between 88% and 95% during the last two decades, and this process continues with amplified vigor, these rough guesses should be taken as little more than an exercise of mental gymnastics should be taken as little more than an exercise of mental gymnastics without relevance; an indication rather than a fact, because the figure is almost certainly a gross overestimate already at the time when this line is being read. The significant fact is that the orangutan is critically endangered, many being scattered throughout small isolated forest patches that are unlikely to sustain their system functions. And with a recruitment rate so despairingly low, while under continuous persecution, it is doubtful whether the ape will be allowed to survive by "civilized" man.

Current Status: Protection

The government of Indonesia has recently increased the number of conservation areas from some 4 million ha (in 1970) to more than 11 million ha. Of these, a total of some 3.5 million ha overlap with the distribution range of the orangutan in Kalimantan and little less than 1 million ha in Sumatra; i.e., some 30% and 40%, respectively. In Sarawak recently two significant conservation areas have been established, covering some 3000 km^2 or about 25% of the distribution range. In Sabah the areas for the conservation of the red ape are still in the process of establishment and may come to cover some 2000 km^2 or about 10% of the distribution range in the state.

On paper these figures look as formidable as the strenuous effort that was necessary to wrestle the areas from other land-use allocations and get them legally established. However, from the point of view of the ape, only the effectiveness of the administrative measures in real life count. Also, an area of gigantic size in administrative terms, such as, for example, the Gunung Leuser National Park with its close to 1 million ha for an orangutan offers not more than some 10,000–12,000 ha of suitable habitat, since more than 65% of the area is so mountainous as to lie above some 2000 m altitude, the absolute limit of ape comfort. Unfortunately, the same applies to all the other regions where the conservation area overlaps with the distribution range of the ape, although the adverse conditions in Borneo usually concern uniform "maturity" of the forest system with a dominance of nonfood species and a paucity of travel accommodation (i.e., lianas).

Moreover, the "paper" figure may suggest an undisturbed, if not "pristine," condition, which in most cases is a fallacy; because of the authoritive ignorance with respect to the actual situation in the field, many conservation areas have been or are still being damaged through not seldom legal incursions of timber exploitation and local land-use programs. Because conservation and exploitation are handled by two departments of one ministry, it is not rare that a newly established "reserve" meant to conserve rainforest systems in fact comprises little more than an extensive grass plains with scattered remnants of tree vegetation. Proposed conservation areas may first be logged over before being established as sanctuaries; but on a map they still look reassuringly green.

Yet more serious from the viewpoint of the ape is the fact that the paperwork has rarely been backed up with actions designed to protect the extremely vulnerable rainforest systems and to curb encroachment, whether from organized exploiters or from local squatters. It seems as though measures of law enforcement with respect to nature and natural resources cannot easily be effected in a development context. It is true that the establishment of rehabilitation stations has, to some extent, curbed the hunting pressure supplying a pet market. Yet, an average of ten young orangutans are still being taken into the stations in Sumatra and Central Kalimantan, indicating a yearly loss of at least 30 individuals as a result of illegal capture and/or hunting activities. The same applies to Sabah, where an annual loss of at least 20 apes has been noted. Still these figures are probably dwarfed by the real numbers perishing in the remoter areas of Borneo (among many ethnic communities, e.g., the Iban, the orangutan is still hunted for supposed magical properties of its body parts) and Aceh province, notably in timber concessions. For there the assault on the ape population may be both direct, due to recreation hunting, and (inevitably) indirect following the massive habitat destruction, but is invariably hidden from the public view.

It must be clear that this is not meant as a reproach to authorities governing the regions where the orangutan still occurs: People from so called advanced technocratic societies can hardly understand that such measures as would effectively conserve significant areas of rainforest bear upon all aspects of the demographic and socialeconomic situation and cannot simply be enforced. It is a situation in which people in general, and authorities in particular, are being trapped in a fundamental ambiguity between conservation and development; a situation emerging from on the one hand the fear of losing one's cultural identity and integrity (perhaps amplified with a fundamental inability to cope with novelties), and on the other hand the obligatory pressure for development, in what is essentially still a medieval society.

As a consequence, the authorities, invariably endowed with much more political than ecological insight, try to muddle up the concept of conservation and what is commonly understood as development and readily come to adhere to false ideas about the supposed value of conservation areas. Thus, they consider that such seemingly robust systems as are called jungle, full of valuable wood, could as well be exploited for those aspects that do not immediately seem vital for the maintenance of the few spectacular organisms that appear to have offered the major justification for establishment of the area's status. It is as though leaving some possibly marketable commodity or cultivable soil unexploited is a mortal sin, only to be compensated by at least some kind of yield. Hence an eager acceptance of the consumptive National Parks concept and great interest for the buffer-zone concept in developing countries because they represent socioeconomically positive symptoms treatment, albeit at the expense of the integrity of the system to be conserved. Yet, a serious break-down will probably become evident only during a successor's term of office. Thus, in National Parks recreational attractions involving "rehabilitant" orangutans can be developed; the rare nesting beaches of sea turtles can be made into more or less commerical "breeding-stations"; and some conservation areas may soon suffer the development of deer farming.

It must be considered as a fatal failure of modern ecology that the International Union for the Conservation of Nature (IUCN) has been allowed to officially sanction the muddle of concepts concerning conservation and development in a widely advertized document entitled the World Conservation Strategy. Rather than recognizing how all physical development is achieved at the expense of natural systems, the Strategy advocates how the concept of conservation should be interpreted so as to become almost identical to utilization. So far as can be judged from the current state of affairs, notably in the Indo-Malaysian Realm, the Strategy has effectively supplanted the long-standing, frustratingly strenuous attempts to protect the rainforest systems of which the orangutan is a component, for an emphasis on far more lucrative (administrative) activities concerned with physical development. Whereas protection was executed in the field, where the conflict of interests between the orangutan and civilized man obsessed by development does occur in real life, the Strategy has directed conservation to the industrial manager's office where the money should enter.

References

Galdikas BMF (1978) Orangutan adaptation at Tanjung Puting Reserve, Central Borneo. Thesis.

MacKinnon JR (1981–1982) National Conservation Plan for Indonesia, vol. 2 and 5. UNDP/FAO Report.

Rijksen HD (1978) A field-study on Sumatran orang utans. Wageningen.

Rijksen HD (1982) How to save the man of the rainforest. In: de Boer LEM (ed) The orang utan. The Hague.

11
The Natural Breeding Strategy of Gibbons (*Hylobates lar*): Are We Managing the Captive Population by Design or by Default?

DONALD E. MOORE

Introduction

Gibbons and siamangs (*Hylobates* spp.) are among the most appealing of all primates; their active arm-swinging mode of locomotion through the trees, human-like facial expressions and family organization, and beautiful songs make them very popular with zoo visitors of all ages.

All gibbons and siamangs are currently listed as Endangered under the U.S. Endangered Species Act and are on Appendix I of CITES, the Convention on International Trade in Endangered Species (Hill et al., 1979). The total captive population of *Hylobates* recorded by the International Species Inventory System (ISIS) as of December 1984 was 450 animals, 96 (21%) of which were wild born. Of all *Hylobates* listed by ISIS, 228 are white-handed or "lar" gibbons (*H. lar*), 32 (14%) of which were wild born (ISIS, 1984). The captive poulation appears to be self-sustaining; despite a negative net population change in 1984, the population grew by 14 animals during the 5-year period 1980–1984 (ISIS, 1984). Gibbons seem to be freely traded among zoos, but more weight is apparently given by curators and directors to coat color and age than to relatedness or possible subspecific differences of animals to be paired for breeding. Only 15 of the 228 *H. lar* in ISIS are listed by subspecies (4 *H. l. lar* and 11 *H. l. carpenteri*) (ISIS, 1984). In fact this paper is a result of the Burnet Park Zoo, Syracuse, New York, receiving a large, long-haired, blonde-phase male and a small, relatively short-haired, dark-phase female white-handed gibbon; these animals were to be paired for breeding, but their extreme phenotypic differences prompted us to question the advisability of this action.

The purpose of this paper is to examine current classification of white-handed gibbon subspecies, our knowledge of gibbon sociobiology in nature, and the relationship of this information to genetic considerations for management of captive populations. The white-handed gibbon may serve as a model for the captive management of other species of gibbon and siamang and possibly for other primates that exhibit philopatric dispersal in nature.

Classification

The diagnostic features of all *Hylobates* are summarized by Chivers and Gittins (1978) and Haimoff (1983) to assist managers in accurate identification of each species; a useful key to distinguish between species is provided. In Chivers and Gittins (1978), illustrations also detail conspicuous key characteristics. Most species appear to be geographically, but not necessarily reproductively, isolated; seas, rivers, and mountains form boundaries between isolated populations (Chivers and Gittins, 1978).

Groves (1972) presents a review of the systematics of gibbons and taxonomic revision of the group. Historic distribution, dispersal, and subspecies formation of the four currently recognized subspecies of white-handed gibbons are discussed. Range maps and discussions of distributions are also provided. A synthesis of range and taxons discussed by Groves (1972), Chivers and Gittins (1978), and Haimoff (1983) is developed for lar gibbons in Figure 11.1.

The Sumatran white-handed gibbon (*H. lar vestitus*) is distinguished from mainland subspecies by its lack of any dark phase, its tendency to be dark on cap and venter, and its relatively small size and high cranium (Groves, 1972). Additionally, Frisch (1967) found that it was more similar in dental features to *H. agilis* than to mainland races of *H. lar* because it has more reduced third molars and more buccal hypoconulid. Withers hair length is 44–60 mm (Groves, 1972).

The Malayan white-handed gibbon (*H. l. lar*) has two color phases: a creamy pale phase and a relatively light dark phase. In dark-phase individuals, the hair base (except on crown hairs) is clearly grayer and lighter than the tip for one-half to two-thirds the length of the hair. Withers hair length is 29–55 mm (Groves, 1972).

The Tenasserim white-handed gibbon (*H. l. entelloides*) also has pale and dark color phases. In the dark phase, only one-third of the basal portion of the hair length is lighter than the shaft; hair length is 29–56 mm (Groves, 1972). Note that this subspecies and *H. l. lar* are very similar in these particular key characteristics.

The Chiengmai white-handed gibbon (*H. l. carpenteri*) is the northernmost *H. lar*, found in the monsoon forest region of northern Thailand; the dark phase is very dark chocolate brown, and hair bases are lighter for up to one-half the total length. The hair of this subspecies is very long, 79–103 mm (Groves, 1972).

Face ring characteristics may also be useful in identifying these four subspecies (Haimoff, 1983).

Capture location of founders is the best criterion for identification of individuals to their subspecies, since there is little overlap in distribution (Fig. 11.1); however, animals supplied from the wild are often accompanied by incorrect capture-site information as a result of efforts to circumvent export restrictions in the true country of origin (Chivers and Gittins, 1978). The four races have fairly distinct physical attributes, though, which should allow managers to correctly identify wild-caught individuals. As we noted, the *entelloides* and *lar* races are most similar to one another, but dark-phase animals should be identifiable by measuring relative length of hair base discoloration. Correct subspecific identification of individuals may be very important for long-term management of self-sustaining gibbon population; we elaborate on this below (and see Benirschke, 1983).

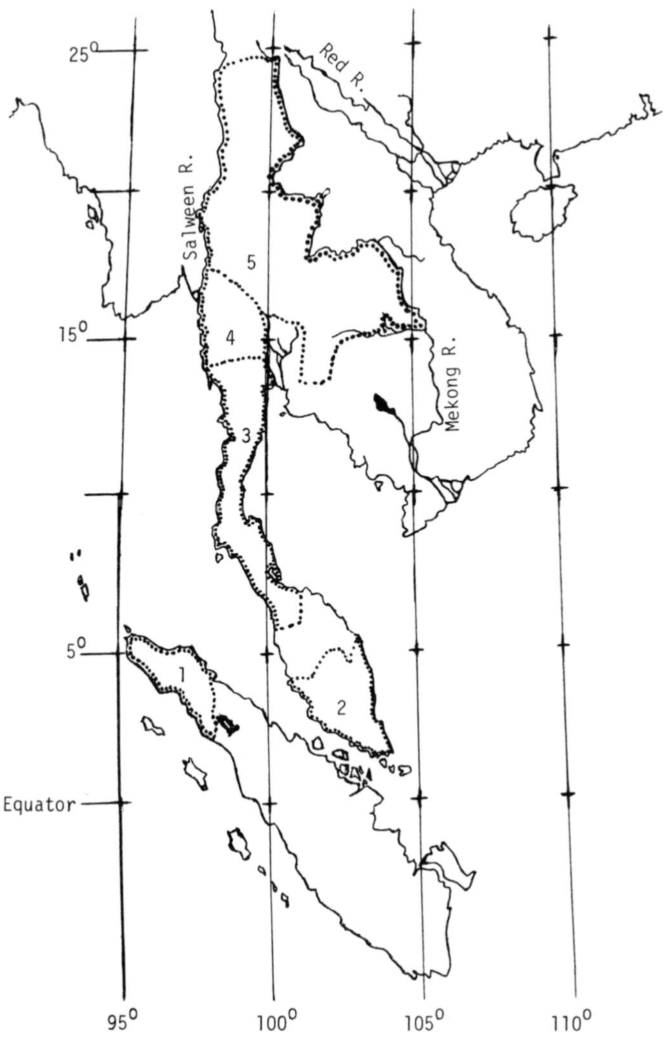

FIGURE 11.1. Distributions of the *H. lar* group. 1, *H. l. vestitus*; 2, *H. l. lar*; 3, *H. l. entelloides*; 4, Intergrades; 5, *H. l. carpenteri*. (After Groves, 1972; Chivers and Gittins, 1978; Haimoff, 1983).

Social Behavior in the Wild

Gibbons and siamangs are the only apes with monogamous mating systems. Gibbon family units are made up of an adult pair and up to four offspring (Ellefson, 1974; Tilson, 1981). Individuals within a group can be classified by age (see Ellefson, 1974). Infants are in age class 0 to 2 or 2.5 years; this stage ends when the infant stops sleeping with its mother, usually around the time a sibling is born. The juvenile stage extends through the next 2 years; it is characterized by great interest in play, complete independence from the mother, and increased agonistic behavior between the juvenile and parents over food. The juvenile stage ends about the time a second sibling is born. When youngsters are between 4 and 6 years of age, they are classed as adolescents, and begin to feed and forage on the periphery of the family group; the tendency to become increasingly peripheral is probably influenced by both the adults' increasing aggression toward the adolescent and by the adolescent's own motivation to feed by him/herself. The subadult period extends from the age of 6 years until the animal finds a mate, at about 8 years of age (Ellefson, 1974). During the subadult period, the peripheral, mature offspring are evicted from the natal territory and forced to compete with solitary gibbons, and occasionally groups, for territory space and a mate (Tilson, 1981). Acquisition and defense of a territory by a mature male gibbon probably improves his chance of acquiring a mate and may play a major role in the subsequent reproductive success of the pair (Brockelman et al., 1973; Tilson, 1981). Once a gibbon is finally pair-bonded, it is considered an adult (Ellefson, 1974; Tilson, 1981).

Tilson (1981) studied Kloss's gibbon (*H. klossi*) family group formation in the Indonesian tropical rainforest. His field studies involved describing intra- and intergroup social dynamics, particularly the transition of individuals from unmated to mated status; Tilson's work presents the only convincing evidence so far presented for mechanisms of family formation in a *Hylobates* species. Individuals and groups were readily identified in the field (Tilson, 1981), so this study provides accurate information on natural gibbon dispersal and mating systems; a review of Tilson's results is critical if we are to begin to understand the possible consequences of current gibbon management in captivity.

Gibbon family groups partition suitable forest habitat into a mosaic of almost contiguous territories averaging 11 ha (Tilson, 1981) to 46 ha (Chivers, 1972). Territories are aggressively defended against trespass by other adults (Chivers, 1972; Ellefson, 1974; Tilson, 1981). In addition to territorial family groups of Kloss's gibbon in Tilson's study area, there are solitary males without territories peripheral to territorial gibbons. Presumably, these animals are floating males that characterize populations in which all suitable breeding habitat is held by territorial animals (Tilson, 1981). Tilson was able to study family formation strategies of a number of individuals from different natal groups.

One means by which Kloss's gibbon forms a family is for a solitary male to establish a territory and attract a female from an adjacent territory. The male and female in this instance were together intermittently for 12 months, then continually for about 4 months before successful mating and conception occurred (Tilson, 1981).

One territory on Tilson's study site was abandoned, then contested by three solitary males. One of the males was a peripheral subadult from an adjacent territory; over a 2-month period, this male's family left their own territory and accompanied him into the adjacent territory. Tilson heard a number of conflict vocalizations which indicated that the family encountered one or more males on each excursion. After each excursion, the family would return to its own territory, but the subadult male remained behind for 3–4 days before rejoining them. After the fourth family excursion, the subadult male remained behind permanently; within weeks Tilson first observed contact between the subadult and a solitary female, and within 3 months of that sighting the pair was together continuously. Thirteen months after this date of pair-bond formation, an infant was born.

One family, discussed by Tilson at length, accompanied its subadult daughter into a subadult male's adjacent territory by easing into the territory over a period of months. The female finally left her family group and remained with the male during a morning chorus; at that time, the family group immediately returned to its original adjacent territory.

One group tried to usurp a slice of an adjacent group's territory for its son. The son occupied the area for awhile, but 6 months later the displaced group had reoccupied the area, forcing the subadult male to (presumably) become a transient male.

Another family formation strategy observed by Tilson was the replacement of an adult male by his son. After 25 days of successful territorial defense, the son copulated with his mother. Six months later, copulation was successful; observations made 8 months after that indicated the female and her son were fully pair-bonded mates with an apparently healthy infant. The same relationship may have developed in another mother–son pair, but the study was terminated before Tilson could observed mating. Another female, accompanied by an adolescent male and apparently widowed, successfully defended her space for about 2 weeks against trespassing subadult males; at least one of the trespassers was from an adjacent territory (Tilson, 1981).

Long lifespan, low birth rate, slow maturation of young, long-term parental care, and peripheralization of subadults is common to all gibbon species; Kloss's gibbons extend their parental care to assist subadult offspring in achieving reproductive success (Tilson, 1981). Once a subadult is peripheralized and subsequently separated from its group, it can either stay near its natal territory or leave and search elsewhere for space and a mate (Tilson, 1981). Tilson lists four possible options available to the subadult for establishment of a territory and subsequent reproductive success. The best option is to establish themselves in a neighboring space. They are under pressure from their same-sex parent to leave, and, as shown by Tilson's data, they are supported by their parents in their attempts to secure adjacent space. In three or four cases in which successful establishment of a breeding pair was observed, at least one of the pair was an offspring of inhabitants of an adjacent or nearby territory (Tilson, 1981). Hence most Kloss's gibbons, possibly most gibbons, do not disperse very far from their natal territories before settling down to mate and reproduce. The possible consequences of this behavior are discussed below.

Dispersal and Genetic Consequences

Soulé and Wilcox (1980) provided the impetus for wildlife managers to focus more on the genetic considerations of conservation biology. Genetic management of rare and endangered species was more practically presented for applied scientists and managers in *Genetics and Conservation*, by Schonewald-Cox et al. (1983). In both Schonewald-Cox et al. and Soulé and Wilcox, genetic and evolutionary theories are explored in attempts to derive implications that might be useful in population management; genetic variation, inbreeding depression, and extinction are discussed at length. Alternative views (Shields, 1982, 1983) to suggested management for relatively large population size and panmixis are not discussed or are only discussed briefly (e.g., Cade, 1983). These alternative interpretations may be applied in the management of gibbons and other rare and endangered primates.

"Inbreeding" is viewed as anathema by many zoo curators and staff, but Shields (1982) presents inbreeding as a continuum, from mild (effective population size [Ne] greater than 100 but less than 1000), through intense ($2 < Ne < 100$) to extreme inbreeding or incest ($Ne = 2$, e.g., a nuclear family with sib-, half-sib, or parent-offspring mating). Although Shields' "inbreeding" occurs even in large but finite demes, he expects its intensity to be weak enough that its consequences may be considered negligible. Shields assumes that transmission of exact duplicates of a successful ancestral genome (= inbreeding) is advantageous for the individuals concerned: animals that survive under local environmental conditions pass their survival characteristics on to offspring that successfully survive and reproduce under the same environmental conditions, etc. Matings between genetically very different individuals (i.e., with different genomes) can disrupt favorable coadapted gene complexes because of recombinational load and can lead to outbreeding depression (Price and Waser, 1979). If inbreeding is adaptive, Shields (1982) theorized, mechanisms promoting its occurrence should be found in nature.

Dispersal patterns observed as animals move from their natal areas to breeding areas often promote inbreeding. The most common pattern of dispersal observed in nature, Shields found, is philopatry, or dispersal to a site near the organism's birthplace. Philopatry does, indeed, increase inbreeding intensity (Shields, 1982).

Rather than analyze dispersal as absolute distances moved by individuals, which may be impressively long in some mammals, Shields (1982) uses effective dispersal, defined as a group's median dispersal divided by average home range diameter (diameter of the area occupied exclusively by sedentary individuals). Effective dispersal basically indicates how many home range units away from its natal area an individual disperses. Philopatry is then defined as a median effective dispersal of less than 10 home range units. Philopatric species often show age- and sex-specific differences in dispersal, and adults usually show site-tenacity to areas where they have bred successfully (Shields, 1982). Shields (1982, 1983) gives examples of many vertebrate species which, though dispersing individuals move impressive absolute distances, move fewer than 10 home range units away from natal areas before settling down to breed, hence they are categorized as philopatric. Tilson's (1981) limited data, which show dispersal to adjacent or nearby spaces, suggest that effective dispersal in gibbons is also philopatric.

Philopatry in low-fecundity vertebrates such as gibbons results in small, semi-isolated and relatively inbred demes in nature (Shields, 1983). An ecogenetic hypothesis developed to explain philopatry suggests that philopatry both permits and maintains local adaptations, thereby benefitting individuals or even populations (Shields, 1982). Philopatry minimizes migrational genetic load and increases the level of inbreeding. Thus, inbreeding maintains locally adapted genomes and protects a local gene complex by preventing its dilution by "unrelated" alleles from different areas with different selective properties; "outbreeding depression" resulting from fertility declines in hybrids will continuously favor inbreeding in natural populations (Shields, 1982). Shields (1982, 1983) cites empirical evidence consistent with this hypothesis.

Discussion and Conclusion

Inbred demes may be expected to have reduced genetic variability and evolutionary potential, but each isolated deme will tend to fix alleles at different loci, so that genetic variability of the megapopulation may be maintained through low-level interdeme migration (Shields, 1982, 1983). Thus, a high level of genetic variability may be maintained in a relatively small megapopulation (say, Ne $<$ 500) if it is managed by division into isolated demes and by controlled migration between demes. This genetic variability is expected to be greater than the variability from random panmixis of the megapopulation (Shields, personal communication).

Subspecific taxonomy of lar gibbons has been based upon careful study of geographic variation (Groves, 1972); distinctive subspecific taxonomy is a good indicator of genetic diversity within a species (Chambers and Bayless, 1983). If we assume (1) that Shields (1982, 1983) is correct about the benefits of relative degrees of inbreeding and the costs of outbreeding in naturally inbreeding species, such as gibbons, and (2) that our objective is to conserve the genetic diversity of managed species such as *Hylobates lar*, then management of populations at the subspecies level is indicated. We have special concern that lar gibbon subspecies may have been isolated long enough that hybridization may result in outbreeding depression in hybrids and/or their offspring, in the form of reproductive failure or inability to survive if reintroduced into ancestral environments.

Price and Waser (1979) found that physiological compatibility between parental genotypes was important in determining offspring fitness *even in* environments not similar to those of the parents. Managers faced with eventual reintroduction of endangered species would benefit from experiments on the reintroduction into the wild of nonendangered, captively inbred and outbred mammals to ascertain whether or not Price and Waser's results are applicable to mammalian species.

We conclude that studbooks should be initiated and maintained for gibbons and siamangs, that a Species Survival Plan (SSP) should be considered for at least one species of lesser ape, and that the SSP Propagation Group should seriously consider a moratorium on subspecies hybridization until the isolated wild populations of gibbons and siamangs have been subjected to more study (see Benirschke, 1983; Chambers and Bayless, 1983; Maple, 1985). A moratorium on hybridization might result in a dramatic decrease in gibbon reproduction for a period of 2 years while animals are shifted between institutions and pair-bonded. The status of hybrid adults might become an issue. Managers should pay equal attention to possible outbreeding depression (see Shields, 1982), and to inbreeding depression and demographics in the captive gibbon population (see Foose, 1983). Once subspecies are identified, the populations can maintain genetic variability if zoo regions (western and eastern North America, Europe, etc.) are managed as demes with low-level interdeme migration (and see Foose, 1983). These are only a few suggestions.

We should stop "managing" these endangered, enchanting lesser apes by default and begin managing their populations by some logical design. When we preserve the natural integrity of unique taxonomic units, such as gibbon subspecies, we will preserve the species' future evolutionary potential.

Acknowledgments. Thanks to Jill Tripp for critical review of this manuscript, and to Bill Shields for many thought-provoking discussions—I hope I have adequately represented his theory.

References

Benirschke K (1983) The impact of research on the propagation of endangered species in zoos. In: Schonewald-Cox C, Chambers S, MacBryde B, Thomas W (eds) Genetics and conservation: a reference guide for managing wild animal and plant populations. Benjamin/Cummings, Menlo Park, CA, pp 402–413.

Brockelman WY, Ross BA, Pantuwatana S (1973) Social correlates of reproductive success in the gibbon colony on Ko Klet Kaeo, Thailand. Amer J Phys Anthrop 38:637–640.

Cade TJ (1983) Hybridization and gene exchange among birds in relation to conservation. In: Schonewald-Cox C, Chambers S, MacBryde B, Thomas W (eds) Genetics and conservation: a reference for managing wild animal and plant populations. Benjamin/Cummings, Menlo Park, CA, p 304.

Chambers SM, Bayless JW (1983) Systematics, conservation, and the measurement of genetic diversity. In: Schonewald-Cox C, Chambers S, MacBryde B, Thomas W (eds) Genetics and conservation: a reference for managing wild animal and plant populations. Benjamin/Cummings, Menlo Park, CA, pp 349–363.

Chivers DJ (1972) The siamang and the gibbon in the Malay Peninsula. Gibbon and Siamang 1:103–134.

Chivers DJ, Gittins SP (1978) Diagnostic features of gibbon species. Intern Zoo Yrbk 18:157–164.

Ellefson JO (1974) A natural history of white-handed gibbons in the Malayan Peninsula. Gibbon and Siamang 3:1–136.

Foose TJ (1983) The relevance of captive populations to the conservation of biotic diversity. In: Schonewald-Cox C, Chambers S, MacBryde B, Thomas W (eds) Genetics and conservation: a reference for managing wild animal and plant populations. Benjamin/Cummings, Menlo Park, CA, pp 374–401.

Frisch JE (1967) The gibbons of the Malay Peninsula and of Sumatra. A comparative odontological study. Primates 8:297–310.

Gittins SP (1978) The species range of the gibbon *Hylobates agilis*. In: Chivers DJ, Joysey KA (eds) Recent advances in primatology, vol 3. Evolution. Academic, New York, pp 319–321.

Groves CP (1972) Systematics and phylogeny of gibbons. Gibbon and Siamang 2:1–89.

Haimoff EH (1983) Gibbon systematics for zoological parks. AAZPA Reg Conf Proc: 69–78.

Hill CA, Warren WC, Wolfe EE (1979) AAZPA manual of federal wildlife regulations. Amer Assoc Zool Parks and Aquar, Wheeling, WV.

ISIS (1984) ISIS species distribution reports 31 December 1984. Microfiche, International Species Inventory System, Apple Valley, MN.

Maple T (1985) SSP report: orang-utan. AAZPA Newslet 26(4):7–8.

Price MV, Waser NM (1979) Pollen dispersal and optimal outcrossing in *Delphinium nelsoni*. Nature 277:294–297.

Schonewald-Cox C, Chambers S, MacBryde B, Thomas W (eds) (1983) Genetics and conservation: a reference for managing wild animal and plant populations. Benjamin/Cummings, Menlo Park, CA.

Shields WM (1982) Philopatry, inbreeding, and the evolution of sex. State University of New York Press, Albany.

Shields WM (1983) Genetic considerations in the management of the wolf and other large vertebrates: an alternate view. In: Carbyn, LN (ed) Wolves in Canada and Alaska: their status, biology, and management. Canadian Wildlife Service Report Series No. 45, Ministry of Supplies and Services, Ottawa, Ontario, pp 90–92.

Soulé ME, Wilcox BA (1980) Conservation biology: an evolutionary-ecological perspective. Sinauer, Sunderland, MA.

Tilson RL (1981) Family formation strategies of Kloss's gibbons. Folia Primatol 35:259–287.

12
The Primates of India: Status, Trends, and Conservation

CHARLES H. SOUTHWICK and DONALD G. LINDBURG

Introduction

India has long been known as one of the rich primate areas of the world, both in species diversity and population abundance. Fourteen species of nonhuman primates occur in India—six species of macaques, five of langurs, two of lorises, and one species of gibbon (Table 12.1). If the nations immediately east of India, Burma and Sri Lanka, are considered, two more species of macaques are included, one more species of langur, and an additional gibbon species.

In population abundance, the rhesus macaque and the Hanuman langur have also been considered among the largest populations of any nonhuman primate, sharing this distinction with the baboons of Africa. In the 1950s, monkey populations in northern India were variously estimated from 10 to 20 million (Corbett, 1953; G. C. Pandit, 1959, personal communication), although these estimates were based on the judgments of natural historians and not on scientific surveys. Actual field surveys, first completed in 1960, arrived at lower figures, estimating the rhesus population of India to be less than 2 million (Southwick et al., 1965).

These same rhesus populations were, of course, the source of the greatest utilization of any primate species in the world for biomedical research, pharmaceutical testing, and vaccine production. In the 1950s and early 1960s, the United States alone was importing as many as 100,000 rhesus monkeys per year, and India's world-wide exports were often twice this number. India's primate exports declined throughout the late 1960s and early 1970s to slightly more than 20,000 per year, and they ceased entirely in April 1978 with India's ban on exports.

Throughout this time, India has shown a remarkable growth in primatology. Field and laboratory studies of primates have increased markedly, as evidenced by a series of symposia, conferences, and books on primate research. Some of these have emphasized primates in biomedical research (Prasad and Anand Kumar, 1977), whereas others have focused on field studies (Roonwal and Mohnot, 1977; Seth, 1983; Roonwal et al., 1984), and all have had major papers on conservation. The expansion of both scientific interest and public concern about primate biology and conservation in India has been very encouraging.

In this paper, we consider what is known about the primates of India in regard to their distribution, population status, trends, and conservation prospects.

TABLE 12.1. Nonhuman primates of India.

Species	Common name	General status	IUCN status[1]	CITES status[2]
Presbytis entellus	Hanuman langur or common langur	Common throughout India, but declining	—	I
P. johnii	Nilgiri langur	Common, but limited to small areas in south	V	II
P. geei	Golden langur	Rare, endangered	R	I
P. phayrei	Phayre's leaf monkey	Generally rare, but common in limited area	—	II
P. pileata	Capped langur	Generally rare, but common in limited area	—	I
Macaca mulatta	Rhesus macaque	Reduced, but still common in north	—	II
M. radiata	Bonnet macaque	Common in south	—	II
M. assamensis	Assamese macaque	Unknown	—	II
M. arctoides	Stump-tailed macaque	Rare, endangered	—	II
M. nemestrina	Pig-tailed macaque	Rare, endangered	In	II
M. silenus	Lion-tailed macaque	Rare, endangered	E	I
Hylobates hoolock	Hoolock gibbon	Rare, endangered	—	I
Loris tardigradus	Slender loris	Unknown	—	II
Nycticebus coucang	Slow loris	Unknown	—	II

[1]IUCN ratings: R, rare; V, vulnerable; In, interdeterminate; —, not rated; E, endangered.
[2]CITES ratings: I, Appendix I, rare and endangered; II, Appendix II, threatened and vulnerable.

Common Langurs

The most widely distributed species of nonhuman primate in India is the Hanuman langur, or common langur, *Presbytis entellus*. It occurs throughout the country except in the westernmost sections of the Rajasthan desert and the easternmost areas of India bordering Burma (Fig. 12.1, from Wolfheim, 1983). It is currently recorded from all provinces of India except Tripura and Manipur, east of Bangladesh (Tiwari, 1983). The Hanuman langur is the truly sacred monkey of India, though all primates share this distinction to some extent. Despite its widespread distribution and sacred status, there are two points of concern. First, a recent survey shows that it is not so abundant as commonly thought, and second, several studies indicate that its populations are declining.

The nationwide primate survey conducted by the Zoological Survey of India from 1978 to 1983 showed the common langur to be the most abundant monkey species in India, but still its population was estimated to be only 233,800 individuals (Tiwari, 1983). There were no previous surveys with which this may be compared, but we can say without a doubt that this is a surprisingly small figure considering the dimensions of its range. The largest provincial pop-

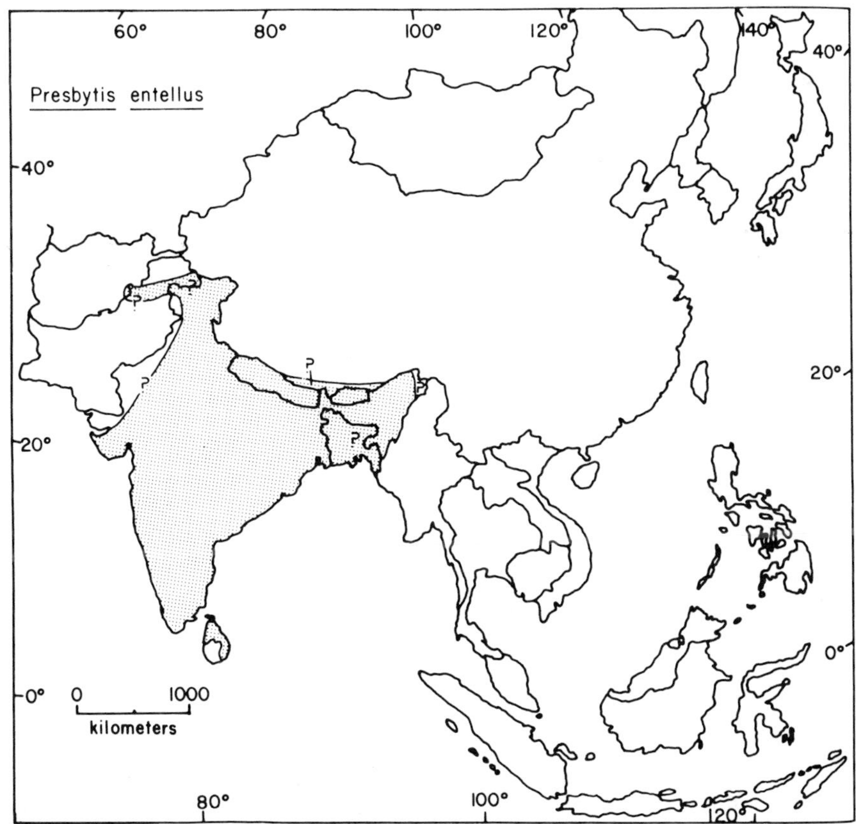

FIGURE 12.1. Distribution of the common or Hanuman langur (from Wolfheim, 1983). Primates of the World—Distribution, Abundance, and Conservation. Copyright © Jaclyn H. Wolfheim. By permission, University of Washington Press, Seattle, WA.

ulations were found in Madhya Pradesh (77,800), Uttar Pradesh (44,100), and Rajasthan (25,200).

Kurup (1984a, 1984b) studied langur populations in south India and concluded that their population structure showed evidence of decline. This conclusion was based on the small average group size of 16.7±1.1, a depressed ratio of young to adults, and evidence of a low birth rate. Kurup's census data showed that only 28.6% of the adult females had infants, and only 38% of the population was immature. Neither was considered adequate for population recruitment or replenishment.

Mohnot (1971) and Hrdy (1974, 1977) have also shown population changes and depletion affecting the young and immature age groups in the langur populations of Rajasthan. During Hrdy's 5-year study period at Mt. Abu, about half the langurs born died in infancy. One troop had an 83% infant mortality, and two of the Mt. Abu troops showed 10–33% reduction in troop size over the 4-year period.

Sugiyama and Parthasarathy (1978) studied the population changes of common langurs in the Dharwar district of southern India from 1961 to 1976 and found that the population decreased 47.8% in this 15-year period. Groups declined from 43 to 22, and the total population decreased from 626 to 327.

We do not fully understand why a sacred monkey, not subject to commercial trapping and highly revered in a cultural sense, should show these declines. Sugiyama and Parthasarathy concluded that "the significant decline of the langur population in cultivated fields (open land) shows an increased human impact on langurs as being responsible for this decrease." They and Oppenheimer (1977) noticed frequent harassment by villagers: Chasing, throwing of sticks and stones, and attacks by village dogs were signs of serious disturbance and competition. The demand for food in India is so great that any monkeys which raid village crops are under considerable pressure. Furthermore, many environmental changes, such as the replacement of natural forest with eucalyptus, teak, and casuarina plantations, were also considered detrimental to langur populations.

The conservation prospects for the Hanuman langur depend upon several aspects of India's ecology and economy—forest conservation, agricultural production, education, and attitudes. These are all so similar to the needs of other species that they will be considered jointly at the end of this paper.

The other species of langurs in India are much less abundant and more restricted in distribution than the Hanuman langur.

Nilgiri Langurs

The Nilgiri langur, *Presbytis johnii*, is restricted to the southern portion of the western Ghat mountains of peninsular India (Fig. 12.2). It is limited to evergreen monsoon forests in only three provinces (Kerala, Karnataka, and western Tamil Nadu). Within this limited area, it is locally abundant, and its total population has been variously estimated from 5000 to 15,000 (Wolfheim, 1983). The greatest threat to its numbers is forest clearing for timber production, teak plantations, and firewood. It is also subject to some hunting pressure for blood and organs to produce health cures and aphrodisiacs (Poirier, 1970).

Golden, Phayre's, and Capped Langurs

The golden langur, *Presbytis geei*, the rarest langur in India, is limited to a very small area of central Assam, bordering Bhutan (Mukherjee and Saha, 1974; Mukherjee, 1980). It occurs in deciduous and evergreen tropical and subtropical forests, and its population is estimated from 540 to 6000 individuals. Golden langurs are protected in only one forest reserve in India and one wildlife sanctuary in Bhutan. Outside of these areas, golden langurs are subject to hunting. This species is obviously in need of more study and greater protection.

Phayre's leaf monkey, *Presbytis phayrei*, is limited to a small portion of eastern India in Tripura (Fig. 12.2). Most of its geographic range extends eastward into Burma, Thailand, Laos, and Vietnam. In Tripura, Mukherjee (1982a,b), found a total of 36 groups containing 409 individuals in mixed deciduous and evergreen

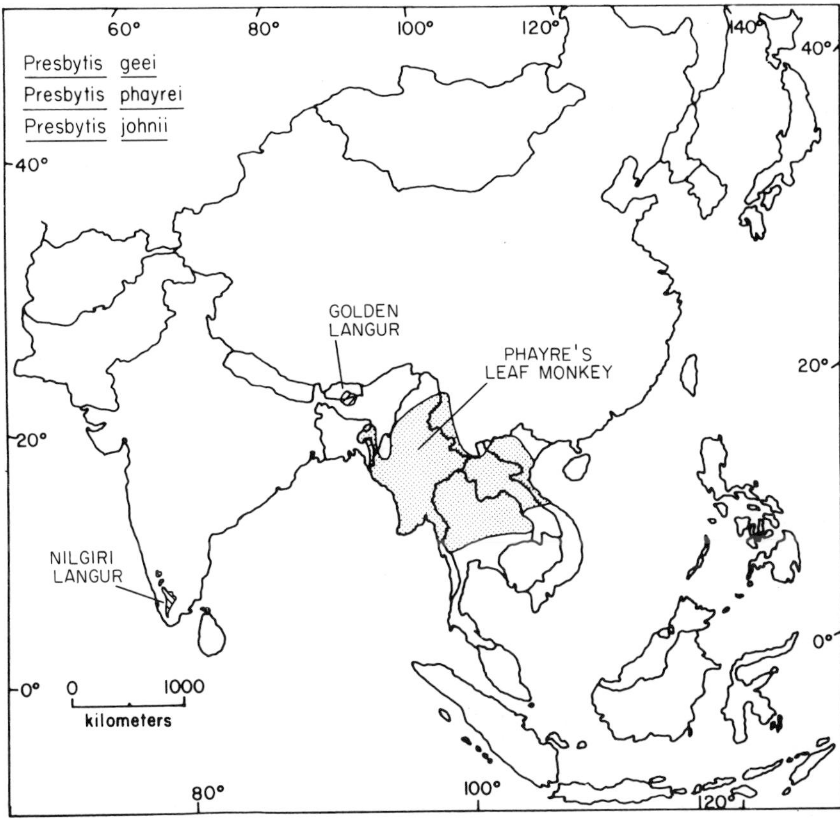

FIGURE 12.2. Distribution of the golden langur, Phayre's leaf monkey, and the Nilgiri langur (from Wolfheim, 1983). Primates of the World—Distribution, Abundance, and Conservation. Copyright © Jacyln H. Wolfheim. By permission, University of Washington Press, Seattle, WA.

forests. These were in a survey area of 3945 km^2, representing one of the richest primate communities of India—the same forests contained rhesus and stump-tailed macaques, capped langurs, hoolock gibbons, and slow lorises. No total population estimates of Phayre's leaf monkeys are available for India, but these forests of Tripura are logical areas for further study and conservation action.

The capped langur, *Presbytis pileata*, occupies a broader geographic range in eastern India than either Phayre's or the golden langur, occurring in Assam, Tripura, and Manipur, but its total range extends no further than Bangladesh and western Burma. No population estimates are available. Although the capped langur has been described as abundant in one forest in Bangladesh (Green, 1978) and common in the Garo Hills of Assam, it is probably subject to hunting and habitat alteration throughout its full geographic range. Its presence in Tripura, in close association with Phayre's leaf monkey and the hoolock gibbon, provides another strong reason for focusing conservation attention in Tripura.

Rhesus Macaques

Throughout Asia, rhesus monkeys, *Macaca mulatta*, have the widest geographic distribution of any species of nonhuman primate in the world. Rhesus occur from Afghanistan in the west to Vietnam, Hong Kong, and eastern China, as far north as Beijing (Bangjie, 1985). In India, rhesus are found throughout northern and central India to an overlapping boundary with the bonnet macaque south of Hyderabad (Fooden et al., 1981).

Ecologically, rhesus have the widest environmental and habitat ranges of any nonhuman primate. They inhabit pine and oak forests over 12,000 feet elevation in the Himalayas, lowland mangrove forests in the Gangetic delta, and tropical rainforests in Burma and Thailand. Furthermore, they are among the most commensal of all primates, occurring in villages, towns, cities, temples, and other areas of crowded human activity.

Despite this exceptionally wide geographic and ecologic distribution, rhesus populations have declined markedly throughout much of their range. In India, their numbers have decreased from many millions, often described as 5 to 10 million, sometimes as high as 20 million, in the 1940s and 1950s, to an estimated population of less than 2 million in 1960, with further declines to estimates of less than 200,000 in 1983 (Tiwari, 1983; Southwick and Siddiqi, 1983). Figures 12.3 and 12.4 show the relative decline in rhesus populations between 1960 and 1980 in roadside, canal bank, and village habitats in northern India. These declines were extensively documented over the past 25 years and have been attributed to a variety of economic, ecologic, and social factors. Among these were: (1) the growth of the human population of India from 350 to over 700 million on the last 30 years, and a current net increase of more than 1 million people per month; (2) expanding agricultural demands, increasing competition for food and space, unfavorable habitat changes, and more frequent disturbances on all animals that consume crops; (3) excessive trapping of rhesus for commercial utilization throughout the 1950s and 1960s; (4) deforestation of 8.4 million acres of forest land in India from 1951 to 1973 (Bowander, 1982); (5) reforestation, when it occurred, with monocultures of eucalyptus, and (6) generally less habitat and less protection for rhesus, until 1978 when the export ban was placed in effect.

Rhesus population declines were also illustrated by comparisons of the work of Dolhinow and Lindburg (1983) with the more recent surveys of the Zoological Survey of India (Tiwari, 1983). In 1964 and 1965, Dolhinow and Lindburg (1983) estimated the rhesus populations in the forests of Uttar Pradesh at 200,000 and those of Himachal Pradesh at 70,000. The Zoological Survey of India did not divide their population estimates by habitat, since forest areas in the late seventies were so frequently invaded by agriculture and village sites, but they did provide a total estimate for all of Uttar Pradesh, including the Gangetic plains, of only 67,200 rhesus and for all of Himachal Pradesh of only 19,200. These figures in themselves represent declines of 66.4% and 72.5%, but actually forest populations would be only a fraction of the total Z.S.I. figures, so the real decline in forest populations would be greater than these percentages.

Despite all of these forces acting against rhesus populations, rhesus numbers have shown some recovery in local areas in recent years. Several events of the past decade have favored rhesus populations. First of all, the government of India banned all export of monkeys in April 1978, an action that did not have an immediate impact on the monkey population (Southwick et al., 1982), but that now seems to be aiding considerably. Second, the Green Revolution has greatly improved India's food production, eased the threat of food shortages and famine, and made India self-sufficient in food grains production. Third, the conservation movement made positive strides under the leadership of Prime Minister Indira Gandhi, and Prime Minister Rajiv Gandhi apparently shares this interest and concern. Fourth, there is evidence of some return to more traditional religious values in India, for we now observe more people feeding monkeys in India as an act of respect and religious celebration. At least in certain locations, these factors are producing a more favorable situation for nonhuman primates in India.

Several examples of locally increasing rhesus populations may be given. Figure 12.5 shows a rural rhesus population in Aligarh District, 130 km southeast of Delhi. This population, censused three times annually by Southwick and Siddiqi since 1959, declined significantly in the 1960s, and then leveled off in the early to mid-1970s. Although the number of groups declined from 22 to 8, several of these groups have become very large, one tripling in size over this 25-year period. As a result, the total population is now at approximately the same level as in 1960, though it might be more vulnerable to habitat changes since nearly 50% of the population is now in one social group. Nonetheless, the recovery has been remarkable, and the hope is that the largest groups will give rise to smaller groups that may take up separate home ranges. This has already occurred once.

Southwick and Siddiqi have also demonstrated the possibility of trapping subgroups from excessively large groups, especially from areas in which they are

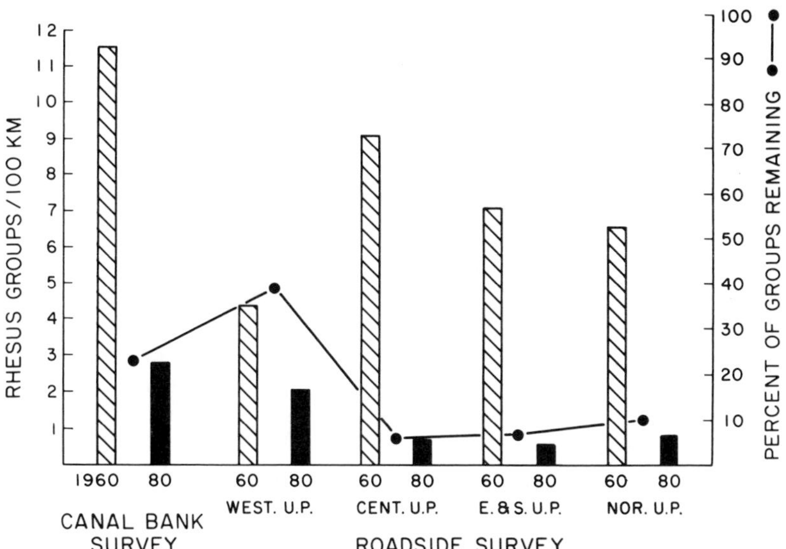

FIGURE 12.3. Population changes of rhesus monkeys in canal bank and roadside habitats, 1960–1980. U.P. = Uttar Pradesh.

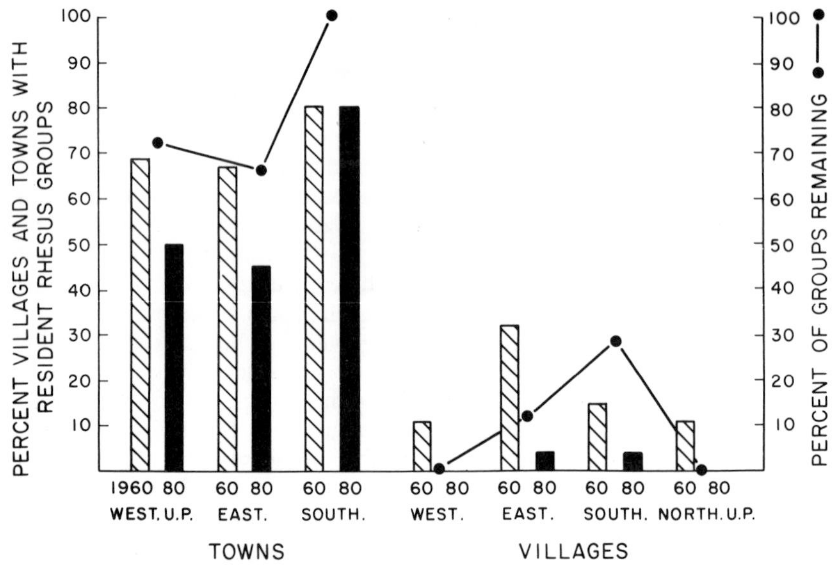

FIGURE 12.4. Population changes of rhesus monkeys in village and town habitats, 1960–1980.

causing crop damage, and reestablishing them in suitable habitats that monkeys had formerly occupied. By this means, they established a new group along a canal bank in Aligarh District in the fall of 1983 (Southwick et al., 1984).

Siddiqi has also studied a small rhesus population in Banda District of southern U.P. that has more than doubled in the last 25 years. When first censused in 1960, it consisted of 109 monkeys in five groups. By 1980, this population, on a sacred hill with Hindu ashrams and temples, had increased to 267 monkeys in 11 groups.

Seth and his students have been studying a number of rhesus groups in western and northern U.P., Haryana, and eastern Rajasthan that increased 66% in a 5-year period, from a total of 400 monkeys in nine groups in 1975 to 665 in the same nine groups in 1980 (Seth, 1983a,b).

One of the most remarkable examples of rhesus population growth has been recorded at the archaeological site of Tughlaqabad on the southern fringe of New Delhi. A well-protected population with an improving habitat and supplemental food source increased more than 20% annually for at least three consecutive years in the early 1980s. The population expanded from two groups of 160 monkeys in 1980 to five groups of 286 monkeys in 1983 (Malik et al., 1984).

These examples show that rhesus populations can respond quickly to the proper combination of protection and habitat. They demonstrate how various conservation measures can be most effective in restoring depleted primate populations.

In the forests of northern India not subject to deforestation and agricultural invasion, rhesus populations seem to be relatively stable and in ecological balance with their environment. In the Asarori forest of Dehra Dun, forest populations first studied in the early 1960s by Lindburg and subsequently by Makwana (1978) and Pirta (1978) in the 1970s have shown substantial populations of approximately 500 to 600 rhesus monkeys in 32 km², a density of 16–18

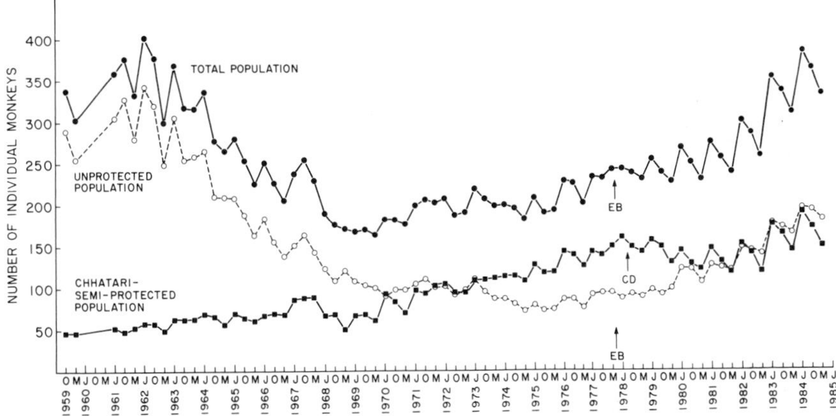

FIGURE 12.5. Population changes of rhesus monkeys in Aligarh District, northern India. EB = Export ban; CD = Habitat displacement of Chhatari Group.

monkeys/km². These figures are comparable to those of Southwick and Siddiqi in Corbett National Park, where populations were stable between 1960 and 1971, showing a density of 13.5 rhesus/km².

Bonnet Macaques

The bonnet monkey, *Macaca radiata*, is the common macaque of south India, occupying essentially the same niches as the rhesus in the north. It is abundant in a variety of forest and commensal habitats, and two recent studies of its population ecology have been completed. Kurup (1984a,b) surveyed bonnet populations in the four southern provinces, which represent its areas of greater abundance (although not its total range). In Andhra Pradesh, Tamil Nadu, Kerala, and Karnataka, Kurup estimated a totel of 9450 bonnet groups with an average group size of 17.73, or a total population of 167,500. With bonnets known to occur in Maharashtra (Fooden et al., 1981), the total population of bonnet macaques in India may equal that of rhesus monkeys.

Demographic studies of bonnet monkeys in a more limited area between Mysore and Ooctacamund have been made by Singh et al. (1984) who found an average group size of 21.0. Both Kurup and Singh et al. conclude that the age distributions indicated that the populations were "reproductively vigorous" and were holding their own. Although village, town, and roadside groups are often in conflict with the human population, general population pressures in south India are not so intense as in the north (with the possible exception of Kerala).

Assamese Macaques

Assamese macaques, *Macaca assamensis*, are limited to the Himalayan range of northern India. They only occur in montane forests over a wide elevational range of 300 m to 3500 m. Although they have been considered abundant in limited areas of India (Mohnot, 1978), no populations surveys have been conducted. Southwick et al. (1964) observed two groups in Darjeeling district, and they are known to occur in Kathmandu Valley. Schaller and Bertrand, however, spent several weeks in Assam looking for Assamese macaques and found only rare and scattered groups. The species is very much in need of field study and conservation attention.

Stump-Tailed and Pig-Tailed Macaques

Both of these species, *Macaca arctoides* and *Macaca nemestrina*, are rare in India and limited to forest regions of eastern Assam, Manipur, and Tripura. Mukherjee (1977) has observed both species in Tripura, but no field studies have been done, nor are any population estimates available. Bertrand and Lahiri spent several weeks in central Assam in 1964 looking for stump-tailed macaques, but found none. Certainly both species are in dire need of field study and strict conservation measures in India. The main geographic ranges of the two species occur in southeast Asia, so the world populations are not dependent upon those in India.

Lion-Tailed Macaques

The lion-tailed macaque, *Macaca silenus*, is a seriously endangered species, limited to small areas in the southern half of the western Ghat mountains of Kerala, Tamil Nadu, and Karnataka (Fig. 12.6). Its total world population in nature has been variously estimated from 405 (Green and Minkowski, 1977), to over 1200 (Kurup, personal communication). Ali (1985) estimates its current minimum population at 915. Recent discoveries of several populations in Karnataka may increase the known numbers of lion-tailed macaques in the wild by as many as 3,000 according to Karanth (1985). The lion-tailed macaque lives only in the evergreen monsoon forests and does not adapt to commensal habitats. Hence, it is highly susceptible to logging, disturbance, and the conversion of natural forests to teak plantations. Although it seemed several years ago that lion-tailed macaques were on an inevitable course toward extinction, several recent developments provide more encouragement.

First of all, the primatological community has drawn both national and international attention to the plight of lion-tails, and this has resulted in definite conservation measures. The late Prime Minister Indira Gandhi, acutely aware of its precarious status, enlarged Anamalai Wildlife Sanctuary and relocated several villages to provide more security for groups in that area. Even more significantly, she delayed a large-scale hydroelectric scheme for Silent Valley, a primary forest area on the Kerala Border where lion-tails were known to occur but had never been studied. It is hoped that this area will be preserved despite the economic pressures for development.

The other major improvement in the conservation of lion-tailed macaques involves the development of captive breeding programs. These programs may play a critical role in conservation efforts of species that have a limited geographical distribution and that are subject to extensive habitat decline. Konstant and Mittermeier (1982) have summarized reintroductions of neotropical primates and Kleiman et al. (this volume) report on the recent reintroduction of the highly

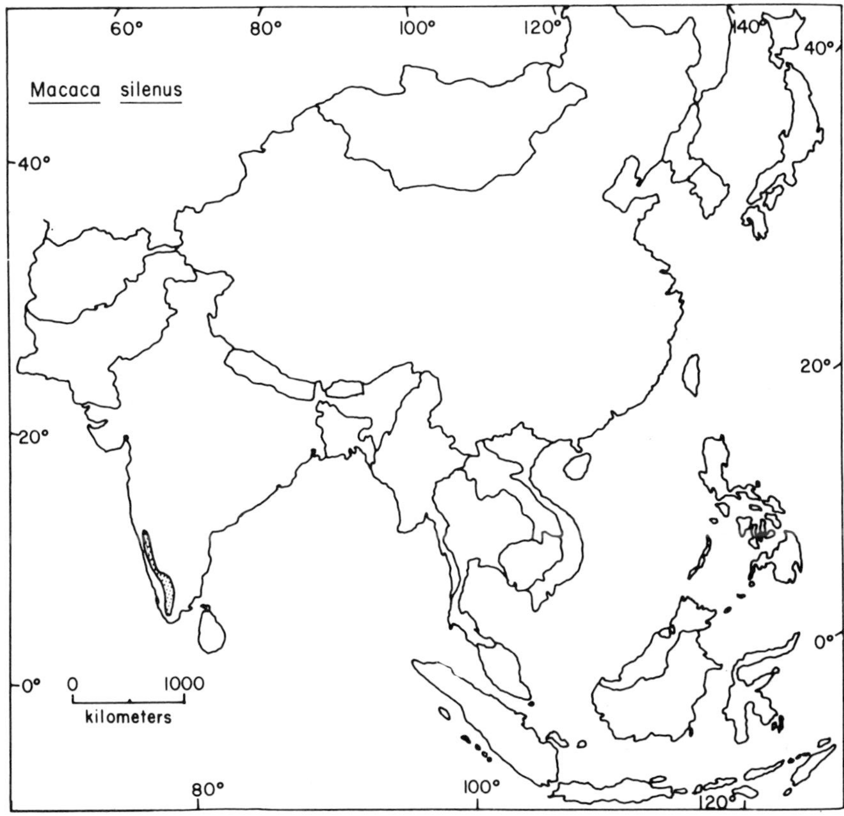

FIGURE 12.6. Distribution of the lion-tailed macaque (from Wolfheim, 1983). Primates of the World—Distribution, Abundance, and Conservation. Copyright © Jacyln H. Wolfheim. By permission, University of Washington Press, Seattle, WA.

endangered golden lion tamarin. For India's rarer primates, a captive population that could play a future role in restocking depleted wild populations exists only for the pig-tailed, stump-tailed, and lion-tailed macaques. Of these three, only the latter is at present classified as endangered.

The captive population of lion-tails is estimated at about 350 individuals (Lindburg, 1980a), held in some 30 zoos around the world. For the North American population a management plan is under development and a regional studbook has been published (Gledhill, 1983). One of the major problems in the conservation of this species has been the low rate of reproduction in both natural and captive populations (Green and Minkowski, 1977; Lindburg, 1980a,b). Studbook records indicate, however, that in North American zoos births have generally increased from approximately ten per year in the 1960s and early 1970s to over 20 per year in recent years. Figure 12.7 shows the cumulative birth success for this population, and this curve has a strong exponential increase. Despite improvement in the birth rate for at least a segment of the captive world population, too many lion-tailed macaques in captivity still remain in small nonbreeding assemblages. The genetic and behavioral suitability of captive-born individuals for reintroduction schemes is also an issue of major concern, if these individuals are to contribute to the maintenance of wild populations. A concerted international effort is needed to address these and other problems related to a total conservation effort. A first step in this direction is seen in results of an international symposium (Heltne, 1985) held in Baltimore in 1982.

Hoolock Gibbons

The hoolock gibbon, *Hylobates hoolock*, occurs in eastern India in the provinces of Assam, Tripura, and Manipur, and also in the neighboring countries of Bangladesh and Burma (Fig. 12.8). Chivers (1977) estimated total hoolock populations by 1980 of 91,000 individuals, based on an average of four gibbons/km^2 in all available forest in Bangladesh, India, and Burma. Tilson (1979) also used an average density figure from one study area, to extrapolate to all forest areas in Assam and provided an estimate of 24,000 for India alone. It is doubtful whether the procedure of using a density figure from a favorable study area as a basis for estimating a broad regional population is valid in view of the extent of hunting and agricultural pressure now occurring. Of the total number of 91,000 hoolock gibbons estimated by Chivers, however, he estimated only 1000 for India.

Brockelman and Chivers (1984) have recently considered the hoolock gibbon highly endangered in India, existing only in restricted isolated pockets. Gittins (1984) provided a total population estimate of hoolock gibbons for Bangladesh of 3000 and confirmed that they exist only in small isolated numbers, primarily in

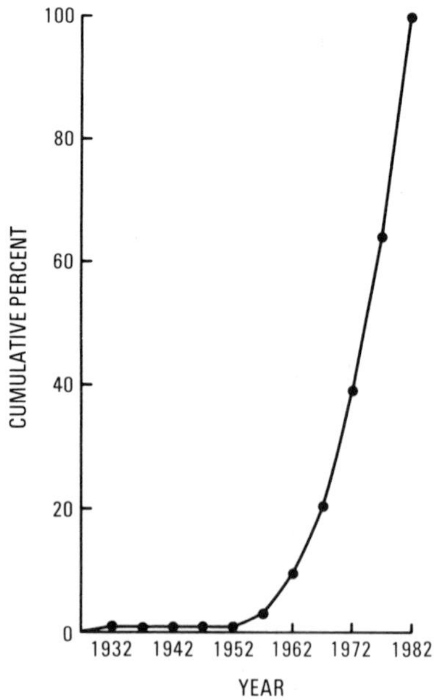

FIGURE 12.7. Cumulative percentages of total lion-tailed macaque births in captivity.

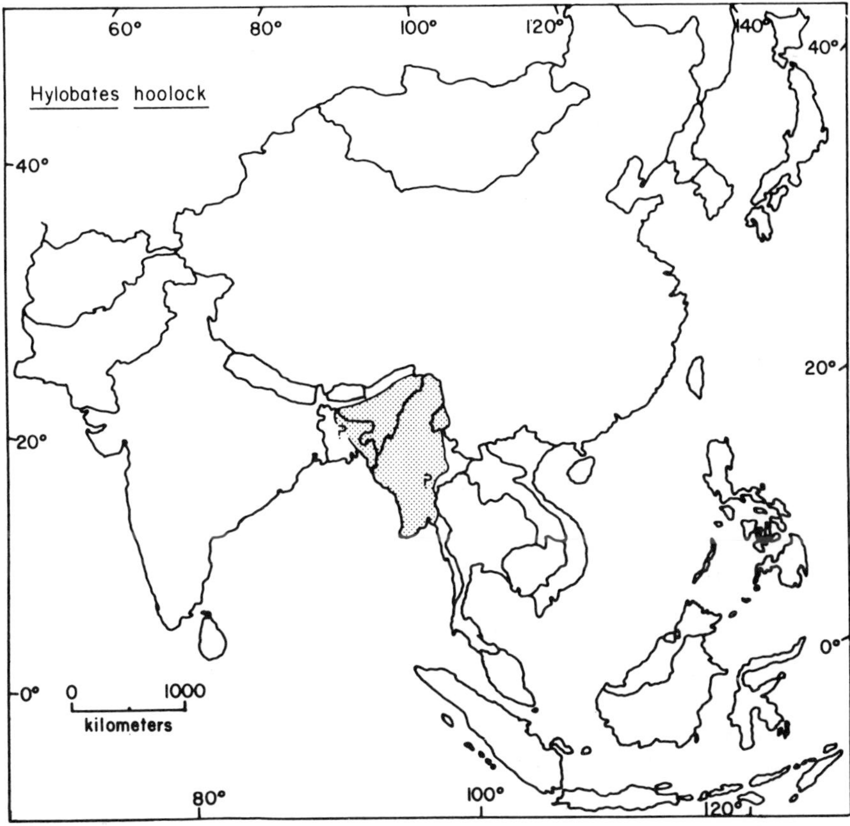

FIGURE 12.8. Distribution of hoolock gibbon (from Wolfheim, 1983). Primates of the World—Distribution, Abundance, and Conservation. Copyright © Jaclyn H. Wolfheim. By permission, University of Washington Press, Seattle, WA.

southeastern Bangladesh bordering Tripura. Mukherjee (1982a) surveyed the hoolock gibbon in Tripura and found that only 1360 km² out of a total of 4080 km² of forest provided suitable gibbon habitat. He estimated a gibbon population in Tripura of 340–453 groups with an average group size of 3.2; hence, a total population of 1088–1449 individuals in this province. Mohnot (1978) has also noted the hoolock gibbon in Meghalaya and Nagaland, in eastern Assam, but did not provide a population estimate. The gibbon in India definitely deserves more study and conservation attention.

Slender and Slow Loris

Two species of loris occur in India, the slender loris, *Loris tardigradus*, in southern peninsular India, and the slow loris, *Nycticebus coucang*, in eastern India in approximately the range of the hoolock gibbon. Both are forest-dwelling nocturnal species with no population studies available to provide data on numbers. The slender loris has been considered abundant in southern India (Wolfheim, 1983), but the basis for this is unknown. Petter and Hladik (1970) observed average population densities in Sri Lanka of only one individual per 100 km^2.

Like the hoolock gibbon, the fates of the lorises are intimately tied to tropical forests, and hence they are in considerable danger.

Conclusions on Conservation Problems

The problems of primate conservation in India fall into two broad categories: The first is the overall problem of forest conservation, and the second is primate management in commensal and agricultural environments.

Most of the species we have reviewed, in fact 11 out of 14, are dependent upon forest environments. Certainly all possible priority must be given in India to maintaining forest areas throughout the country. Tropical forests are necessary for the survival of all langurs except the Hanuman langur, for the gibbon, the lorises, and the stump-tailed, pig-tailed, and lion-tailed macaques. Temperate forests are necessary for the survival of the Assamese macaque. Furthermore, forests are an important, if not totally necessary, habitat for the three main commensal species in India, the rhesus, the Hanuman langur, and the bonnet macaques.

These latter species, however, can survive in domestic habitats if allowed to do so. Here the problem is one of agricultural competition and human harassment. With increasing pressures of human population, all three species can suffer considerably from disturbance—even the sacred Hanuman langur seems to be faring poorly from excessive human competition and harassment. Several independent studies report declining populations.

The best solutions for commensal species seem to lie in programs of management and education. Locally, primate populations can become too large for their own good. If they cause serious and intolerable crop losses, they may lose the support of local people. Villagers can literally harass monkeys out of existence by stoning and chasing them, breaking up groups, and denying them food. The extent and frequency of this behavior depends in part upon broad economic and social conditions in India.

The best counteraction against these kinds of events are for India to increase its agricultural output and to improve the economic welfare of all its citizens; for conservationists to emphasize the cultural and scientific value of primates. and for primatologists and wildlife biologists to recognize local areas where monkeys are too abundant and cause excessive crop damage. Managed relocations can then reduce local densities and simultaneously restore primates to areas from which they have been depleted.

Fortunately, India is a country with a substantial scientific and educational base and a strong development in the areas of conservation and primatology. India is also a country with a heritage and sensitivity to the values of its animal life. If these qualities are not snuffed out by the demands of human population growth and the realities of economics, primate conservation may have a bright future in India.

Acknowledgments. We are grateful for the cooperation and assistance of the late Dr. M. Babar Mirza and Drs. M. F. Siddiqi, K. K. Tiwari, S. M. Alam, M. R. Siddiqi, M. A. Beg, T. N. Ananthakrishnan, M. L. Roonwal, R. P. Mukherjee, R. K. Lahiri, J. A. Khan, M. Neville, P. Dolhinow, M. Bertrand, M. S. Rai, the late F. B. Bang, C. Wallace, T. W. Simpson, B. C. Pal, J. R. Oppenheimer, M. Y. Farooqui, J. A. Cohen, S. W. Ashraf, B. H. H. Hingorani, T. T. Srivastava, I. Quereshi, R. N. Chatterji, and Mrs. Susan Chopra. The field work of CHS has been supported since 1959 by U.S. Public Health Service Grants RG 6262 to Ohio University, RO7-AI-10048, MH-18440, and RR-00910 to the Johns Hopkins University, and RR-01245 to the University of Colorado. The field work was initially begun when CHS was a Fulbright Research Fellow at Aligarh University with support from the U.S. Educational Foundation in India. The field work of DGL in India was supported by U.S.P.H.S. Grant FR-00169 to the California Primate Research Center, Davis, California.

References

Ali R (1985) An overview of the status and distribution of the lion-tailed macaque. In: Heltne P (ed) The lion-tailed macaque: status and conservation. Alan R Liss, New York, pp 13–25.

Bangjie T (1985) The status of primates in China. Prim Conserv 5:63–81.

Bowander B (1982) Deforestation in India. Intern J Environ Stud 18:223–236.

Brockelman WY, Chivers DJ (1984) Gibbon conservation: looking to the future. In: Preuschoft H, Chivers DJ, Brockelman WY, Creel N (eds) The lesser apes: evolutionary and behavioural biology. Edinburgh University Press, Edinburgh, pp 3–12.

Chivers DJ (1977) The lesser apes. In: Prince Rainier, Bourne GH (eds) Primate conservation. Academic, New York: pp 539–598.

Corbett J (1953) Jungle lore. Oxford University Press, London, p 168.

Dolhinow P, Lindburg DG (1983) A population survey of forest-dwelling rhesus monkeys in north India. In: Seth PK (ed) Perspectives in primate biology. Today and Tomorrow, New Delhi, pp 201–209.

Edwin C.J., Chopra SRK (1984) Group characteristics of *Macaca mulatta* inhabiting submontane and montane forests of the Indian subcontinent: a comparative review. In: Roonwall ML, Mohnot SM, Rathore RS (eds) Current primate researches. University of Jodhpur, Jodhpur, pp 307–313.

Fooden J (1981) Taxonomy and evolution of the *sinica* group of Macaques. 2. Species and subspecies accounts of the Indian bonnet macaque, *Macaca rradiata*. Fieldiana, New Series No 9:1–52.

Fooden J, Mahabal A, Saha SS (1981) Redefinition of the rhesus macaque-bonnet macaque boundary in peninsular India (Primates: *Macaca mulatta, M. radiata*). J Bombay Nat Hist Soc 78:463–474.

Gittins SP (1984) The distribution and status of the hoolock gibbon in Bangladesh. In: Preuschoft H, Chivers DJ, Brockelman WY, Creel N (eds) The lesser apes: evolutionary and behavioural biology. Edinburgh Univerity Press, Edinburgh, pp 13–18.

Gledhill L (1983) Lion-tailed macaque, *Macaca silenus*, North American regional studbook. Woodland Park Zoological Garden, Seattle.

Green KM (1978) Primates of Bangladesh: a preliminary survey of population and habitat. Biol Conserv 13:141–160.

Green S, Minkowski K (1977) The lion-tailed monkey and its south Indian rain forest habitat. In: Prince Rainier, Bourne GH (eds) Primate conservation. Academic, New York, pp 289–337.

Heltne PG (ed) (1985) The lion-tailed macaque: status and conservation. Monographs in primatology, vol 7. Alan R Liss, New York, p 411.

Hrdy S (1974) Male-male competition and infanticide among the langurs (*Presbytis entellus*) of Abu, Rajasthan. Folia Primatol 22:19–58.

Hrdy S (1977) The langurs of Abu. Harvard University Press, Cambridge, p 361.

Karanth KU (1985) Ecological status of the lion-tailed macaque and its rainforest habitats in Karnataka, India. Prim Conserv 6:73–78.

Konstant WR, Mittermeier RA (1982) Introduction, reintroduction and translocation of neotropical primates: past experiences and future possibilities. Intern Zoo Yrbk 22:69–77.

Kurup GU (1984a) Nonhuman primate census surveys in southern India. In: Roonwal ML, Mohnot SM, Rathore NS (eds) Current primate researches. University of Jodhpur, Jodhpur, pp 57–65.

Kurup GU (1984b) Census survey and population ecology of the Hanuman langur, *Presbytis entellus* (Dufresne) in South India. Proc Indian Natl Sci Acad B50(3):245–256.

Lindburg DG (1971) The rhesus monkey in North India: an ecological and behavioral study. In: Rosenblum LA (ed) Primate behavior: developments in field and laboratory research, vol 2. Academic, New York, pp 1–106.

Lindburg DG (1980a) Status and captive reproduction of the lion-tailed macaque, *Macaca silenus*. Intern Zoo Yrbk 20:60–64.

Lindburg DG (ed) (1980b) The macaques: studies in ecology, behavior and evolution. Van Nostrand Reinhold, New York, p 384.

Makwana SC (1978) Field ecology and behavior of the rhesus macaque (*Macaca mulatta*): group composition, home range, roosting sites, and foraging routes in the Asarori Forest. Primates 19:483–492.

Malik I, Seth PK, Southwick CH (1984) Population growth of free-ranging rhesus monkeys at Tughlaqabad. Amer J Primatol 7:311–321.

Mohnot SM (1971) Ecology and behavior of the Hanuman langur, *Presbytis entellus* (Primates: Cercopithecidae) invading fields, gardens and orchards around Jodhpur, western India. Trop Ecol 12:237–249.

Mohnot SM (1978) The conservation of non-human primates in India. In: Chivers DJ, Lane-Petter W (eds) Recent advances in primatology, vol 2. Academic, London, pp 47–53.

Mohnot SM, Gadgil M., Makwana SC (1981) On the dynamics of the Hanuman langur populations of Jodhpur (Rajasthan, India). Primates 22:182–191.

Mukherjee RP (1977) Rhesus and other monkeys of Tripura. Newslet Zool Surv India 3:111.

Mukherjee RP (1980) Distribution and present status of the golden langur, *Presbytis geei* Khajuria in some forests of Assam, India. Proc Wild Life Workshop 129–132.

Mukherjee RP (1982a) Survey of non-human primates of Tripura, India. Zool Soc India 34:70–81.

Mukherjee RP (1982b) Phayre's leaf monkey (*Presbytis phayrie* Blyth, 1847) of Tripura. J Bombay Nat Hist Soc 79:47–56.

Mukherjee RP, Saha SS (1974) The golden langurs (*Presbytis geei* Khajuria, 1956) of Assam. Primates 15:327–340.

Oppenheimer JR (1977) *Presbytis entellus*, the Hanuman langur. In: Prince Rainier, Bourne GH (eds) Primate conservation. Academic, New York, pp 469–512.

Petter JJ, Hladik CM (1970) Observations sur le domaine vital et la densite population de *Loris tardigradus* dans les forets de Ceylan. Mammalia 34:394–409.

Pirta RS (1978) Observations on group size, group composition and home range of rhesus monkeys in Asarori Forest, North India. Sci Res Banaras Hindu Univ 28:123–135.

Poirier FE (1970) The Nilgiri langur (*Presbytis johnii*) of south India. In: Rosenblum LA (ed) Primate behavior, vol 1. Academic, New York, pp 251–283.

Prasad MRN, Anand Kumar TC (eds) (1977) Use of non-human primates in biomedical research. Indian Natl Sci Acad, New Delhi, p 426.

Roonwal ML, Mohnot SM (1977) Primates of South Asia: ecology, sociobiology, and behavior. Harvard University Press, Cambridge, p 421.

Roonwal ML, Mohnot SM, Rathore NS (eds) (1984) Current primate researches. University of Jodhpur Press, Jodhpur, p 627.

Saha SS (1984) The present southern limit of the rhesus macaque (*Macaca mulatta*) in peninsular India, especially the Godavari and Krishna river basins. In: Roonwal ML, Mohnot SM, Rathore NS (eds) Current primate researches. University of Jodhpur, Jodhpur, pp 153–165.

Seth PK (ed) (1983a) Perspectives in primate biology. Today and Tomorrow, New Delhi, p 242.

Seth PK, Seth S (1983b) Population dynamics of free-ranging rhesus monkeys in different ecological conditions in India. Amer J Primatol 5:61–67.

Singh M, Akram N, Pirta RS (1984) Evolution of demographic patterns in the bonnet monkey (*Macaca radiata*). In: Roonwal ML, Mohnot SM, Rathore NS (eds) Current primate researches. University of Jodhpur, Jodhpur, pp 7–16.

Southwick CH (1985) Some encouraging signs from India's rhesus populations. Amer J Primatol 8:191.

Southwick CH, Beg MA, Siddiqi MR (1965) Rhesus monkeys in North India. In: DeVore I (ed) Primate behavior—field studies of monkeys and apes. Holt, Rinehart and Winston, New York, pp 111–159.

Southwick CH, Ghosh A, Louch CD (1964) A roadside survey of rhesus monkeys in Bengal. J Mammal 45:443–448.

Southwick CH, Johnson R, Siddiqi MF (1984) Subgroup relocation of rhesus monkeys in India as a conservation measure. Amer J Primatol 6:423.

Southwick CH, Richie T, Taylor H, Teas HJ, Siddiqi MF (1980) Rhesus monkey populations in India and Nepal: patterns of growth, decline, and natural regulation. In: Cohen MN, Malpass RS, Klein HG (eds) Biosocial mechanisms of population regulation. Yale University Press, New Haven, pp 151–170.

Southwick CH, Siddiqi MF (1977) Population dynamics of rhesus monkeys in northern India. In: Prince Rainier, Bourne GH (eds) Primate conservation. Academic, New York, pp 339–362.

Southwick CH, Siddiqi MF (1983) Status and conservation of rhesus monkeys in India. In: Seth PK (ed) Perspectives in primate biology. Today and Tomorrow, New Delhi, pp 227–236.

Southwick CH, Siddiqi MF, Cohen J, Oppenheimer JR, Khan J, Ashraf SW (1982) Further declines in rhesus populations in India. In: Chiarelli AB, Corruccini RS (eds) Advanced views in primate biology. Springer-Verlag, Berlin, pp 128–137.

Southwick CH, Siddiqi MF, Oppenheimer JR (1983) Twenty-year changes in rhesus monkey populations in agricultural areas of northern India. Ecology 64:434–439.

Sugiyama Y, Parthasarathy MD (1978) Population change of the Hanuman langur (*Presbytis entellus*) 1961–76, in Dharwar area, India. J Bombay Nat Hist Soc 75:860–867.

Tilson RL (1979) On the behavior of hoolock gibbons (*Hylobates hoolock*) during different seasons in Assam, India. J Bombay Nat Hist Soc 76:1–16.

Tiwari KK (1983) Report on census of rhesus macaque and Hanuman langur of India. Mimeo Report, National Primate Survey, Zool Soc of India, Calcutta, p 68.

Wolfheim J (1983) Primates of the world: distribution, abundance and conservation. University of Washington Press, Seattle, p 831.

13
Increased Home Range for a Self-Sustaining Free-Ranging Rhesus Population at Tughlaqabad, India

IQBAL MALIK

Introduction

In northern India two reasons for the decline of free-ranging rhesus population could be a change in the habitat, or constraints on expansion of habitat.

During a long-term study of the free-ranging rhesus population of Tughlaqabad it was observed that an increase in the home range would make it possible to have a self-sustaining and ever-growing population.

The most dominant group of the area made a preemptive move to check any possible decline in their population by first expanding its home range and then changing its core area. Thus, a positive correlation was witnessed in the population growth and the potentialities of the habitat.

Field Methods

This study is a longitudinal survey begun in 1980 to observe selected behavioral aspects of the rhesus monkey of the area with emphasis on population dynamics. Related counts (Malik et al., 1984) were made in March before the birth season, in July and August immediately after the birth season, and in October and November following monsoon and just prior to winter. This method provided data on the minimum March and maximum July and August populations of the year and a transitional period from monsoon to winter. A record of births, deaths, disappearances, accidents, and injuries was also kept. Hence, when the largest group of the area increased its home range and then changed their core area, this was immediately noted and hence a constant watch was made to observe the movements of this group. Observations have been made from March 1980 to January 1985.

Study Area

Tughlaqabad is an ancient city site and fourteenth century fort, situated on the southern edge of New Delhi at 30°21′ north latitude and 78°76′ east longitude. The fort was built of massive stones on a rocky hill with the outer ramparts integrated into the hill so that the entire structure rises 50–90 feet above the surrounding plain. The flat and fertile area surrounding the fort contains croplands, pasture, two forested areas, and encroaching suburban development. A road runs through the southern part of the area. Across the road, to the south of the fort, is the tomb of Giasuddin Tughlaq who died in A.D. 1325.

Tughlaqabad has a subtropical climate with marked seasonal changes. During the months of May and June, daytime temperatures often reach 40–45 °C. In the months of December and January, temperatures fall to 7–9 °C. Monsoon occurs from the end of June to early July until mid-September. The area receives an annual average of 567 cm of rain; rain in winter and spring is light and sporadic.

Results

Initial Home Range

From March 1980 to November 1984 the home range of rhesus monkey groups under study extended throughout the fort and surrounding areas covering approximately 5 km² (2.5 × 2.0 km). The fort constituted one-fourth of the total area, the forest plantations occupied another one-fourth, and the surrounding farms constituted the remaining half (Fig. 13.1).

The natural vegetation inside the fort is nerophytic, generally grasses and arid forbs and shrubs. Outside the fort, vegetation is more mesophytic, and better groundwater supports trees and crops, primarily wheat and pulses. The main trees present are Indian jujube (*Zizyphus jujuba*), neem or margosa (*Azadirachta indica*), shersham or sissoo (*Dalbergia sissoo*), oak (*Quercus incana*), acacia (*Acacia arabica*), pipal (*Ficus religiosa*), and date palm (*Phoenix dactylifera*). Other than people, the dominant fauna include rhesus monkeys, cattle, buffalo, donkeys, goats, dogs, jackals, mongoose, lizards, and a great variety of birds, both migratory and resident. Peacocks, partridges, pigeons, crows, sparrows, vultures, mynahs, and kites are all common (Malik et al., 1984).

Extended Home Range

In 1984 after the breeding (May–July) but before the onset of winter group A extended its home range. Adjoining the southern end of their initial home range is a walled Air Force enclosure. A part of this establishment, a completely restricted area, was included in the home range (Fig. 13.1).

Group A started frequenting it and then used it for sleeping and resting; eventually they spent most of their time inside this Air Force compound, which is now their core area. The canteen inside this area seems to meet a major portion of their diet. At times they come out to feed at the roadside, but it is never quite certain that the whole group has converged upon the visitor as they would do before. The rhesus have been seen on the northern wall of this compound and at times on the eastern and western but never on the southern wall. The extent of utilization of this space is not exactly known, but a vague estimate is that the extended home range of Group A is 1 km².

The vegetation of the Air Force area is similar to that on the outside, with the exception that cultivation would be minimal, limited to kitchen gardens that

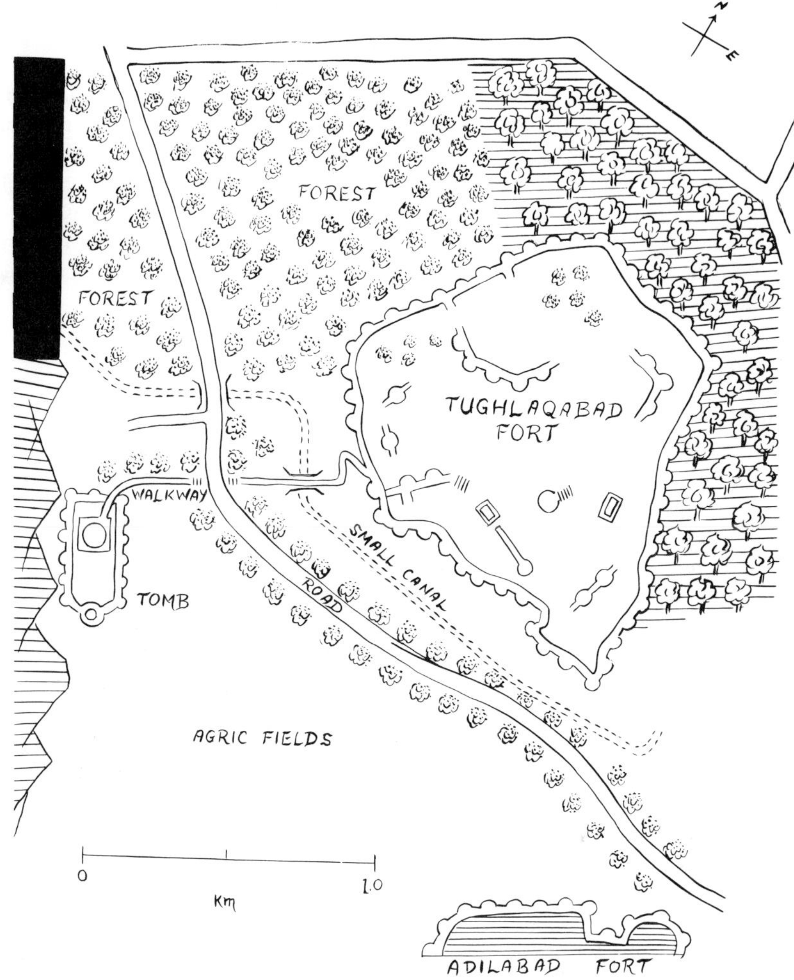

FIGURE 13.1. Home range of Tughlaqabad monkeys showing Tughlaqabad fort, tomb, forest plantations, agricultural fields, canal, hills, surrounding roads, and Adilabad. ▭, initial home range; ■, extended home range; ▥, scope for expansion of home range.

would be zealously guarded against raids by the rhesus. Trees visible from the outside are sheesham or sissoo (*Dalbergia sissoo*), neem or margosa (*Azadirachta indica*). It would be safe to presume that there would be other trees, some perhaps even providing fruit consumed by human populations. The fauna would differ slightly from the outside because dogs, goats, donkeys, buffalo, and cattle would not be allowed inside although mongoose, lizards, jackals, and snakes would be difficult to restrict. The birds would be similar to the outside.

Scope of Further Expansion of Home Range

The habitat provides a vast scope for the further expansion of the home range of rhesus of the area (Fig. 13.1). On the southeast side of their territory is Adilabad. This is a considerably smaller fort than the one presently used, but it provides similar facilities. To date the monkeys have only visited the boundaries of this fort for water, but may in future spend more time there. Further south are rocky hills with xerophylic vegetation and few or no predators. Toward the north beyond the fort are patches of forest with a road that is busy and could be an excellent source of food for them. Toward the east beyond the home range are more forests, which can provide good cover. Rhesus have so far not visited these areas.

Population Growth and Expansion of Home Range

At Tughlaqabad the rhesus population has increased in 5 years 119.38%, from 160 monkeys in July 1980 to 351 in July 1984, an average increase per year of 22.70%. It is felt that this increase is due to the habitat, which provides protection, abundance of food and water, and good cover with scope for expansion. With the right combination of ecological and behavioral factors, rhesus not only sustained themselves, but the population more than doubled in 5 years. The rhesus of the area are preadapted to the environment and could capitalize on increased habitat.

In 1980 the number of rhesus per square kilometer of home range was 32. With the increase of population every year the number per square kilometer kept increasing (Table 13.1), reaching a maximum in July 1984 with 70.2 rhesus/km^2 of home range. This congestion probably led to the expansion of their initial home range. In November 1984 the number came down to 57.4 monkeys/km^2, almost that of 1983 (57.2 rhesus/km^2). It seems that the maximum number of rhesus that the initial home range could sustain is around 286.

Fission and Expanded Home Range

The social behavior interacting with environmental parameters determines the number of rhesus that may exist in a group. At the beginning of this study in January 1980 the Tughlaqabad rhesus population consisted of two groups: A, consisting of 92 monkeys, and B, consisting of 28. By the summer of 1983, the population had grown to 286 monkeys, and the number of groups had increased

TABLE 13.1. Number of rhesus per square kilometer of home range, July 1980–November 1984.

Date of census	Total rhesus population of the area	Home range (km^2)	No. of rhesus/km^2
July 1980	160	5	32
July 1981	201	5	40.2
July 1982	244	5	48.8
July 1983	286	5	57.2
July 1984	351	5	70.2
November 1984	351	6	57.4

TABLE 13.2. Effect of expansion of home range on fission in group A.

Date of census	Home range (km^2)	No. of rhesus of parent group	Formation of splinter ranks from group A
July 1980	5	123	—
July 1981	5	154	123 (A) 31 (C)
July 1982	5	148	133 (A) 15 (D)
July 1983	5	152	123 (A) 29 (E)
July 1984	5	155	—
November 1984	6	155	—

to five (Malik et al., 1985). Groups C, D, and E were splinter groups of A, which remained the largest group, none of the members of group B joined either C, D, or E, nor did groups C and D contribute to each other or to group E. Group B remained intact throughout the study period.

The first fission of group A took place in December 1980, toward the end of the rainy season. The group size was 123, and a subgroup of 21 separated to form group C. Fifteen months later, in March 1982, at the beginning of the birth season, the size of group A was 120, and the second fission occurred when 11 individuals left to form group D. Group A was reduced to 109 individuals, but after the birth season of 1982 in June, group A numbered 133. The third fissioning occurred in the spring of 1983 when group A numbered 137 individuals and 29 left to form group E. By July of 1983, group A had been restored to a level of 123 through births, and by July 1984, the number went up to 155 but no fissioning took place. Table 13.2 shows the effect of expansion of home range on fission. In the initial home range group A apparently could contain only a certain number of individuals, approximately 120, and still maintain coordinated activities as a social unit. Once the number exceeded this general limit, a splinter group was formed. The reason for the lack of fission in 1984 (when the number was 155) could be the increased home range of group A. Expanded home range, it seems, can contain a large number of individuals. In the coming years a record of population fissioning of groups might reveal the optimum number of rhesus that can be contained in group A in this expanded home range.

There was a clear dominance pattern in intergroup encounters. Group A remained the most powerful group, as well as the largest, at the termination of this study in January 1985, as it had been since 1980.

Discussion

It has been proposed that in northern India a reason for the decline in the rhesus population could be the changing beliefs of the rural people who no longer consider them sacred (Seth and Seth, 1983). But at Tughlaqabad, people consider them sacred. They are given protection against undue harassment. One instance that never fails to come to mind is that of a lorry driver who had accidentally run over a monkey and was given a severe beating. It is still true that humans in large numbers come from great distances to feed rhesus. If it was a change in beliefs of rural populations then it would conform with that of the urban population, which is still that of worship/reverence. Another reason could be the changes in habitat caused by deforestation, overgrazing, commercial development, and spread of cities (Seth and Seth, 1983). The changes in habitat were observed at Tughlaqabad. For example in 1982 a shooting range was constructed on the southern side of their territory that precipitated increased activity. The noise of the firing added to their insecurity and fear as a result of which they would take to the fort. The road that runs through the area has an ever-increasing amount of traffic. With the increase in tourists, related facilities have started, for instance tea stalls and vendors. But if commercialization or development were reasons then the transformation would have been made a long time ago. But for a long time after the construction of the shooting ranges, rhesus were still using the same sleeping quarters and were not dislodged. On the other hand, the monkeys depend on and obtain a major portion of their diet from humans and a preference for a location secluded from human interference (Fa, 1983) would not be applicable here as the move made is one to a place that is not totally free of human beings. True, a civilian population does not have access to the place, but the canteen from which they get their provisions would be a much used area constantly bustling with activity.

Brennan and Else (1984) in their study of the De Brazza monkey (*Cercopithecus neglectus*) suggested that the remaining population (just over 100 in Kenya) be translocated as the first step in attempts to save them. It is felt that if the present urbanization continues then the rate of growth would fall unless they are translocated. A point to be noted here in favor of the habitat would be the availability of a suitable locality in the same habitat. Fa (1983) noted that there is a positive correlation with numbers of monkeys and the available habitat. It is therefore my contention that though human interference was not a factor, it was a preemptive move by the monkeys to check a decline in their population. Since it is known that the number had risen from 120 animals in 1980 to 339 in 1984 (Malik et al., 1984) and the number of groups from two to five, a frequent confrontation between groups could not have been avoided for long. This in turn would have led to more aggression. Aggression, as it has been observed, is an inevitable consequence of congestion. Southwick (1967) and Alexander and Roth (1971) observed that the aggression in the captive groups of rhesus and Japanese macaques, respectively, increased under crowding conditions. Southwick et al. (1965) reported that adult males attacked other members of a group including the infants at feeding time. Mukherjee (1976) observed that males of the Mahabali temple attacked group members during feeding and nonfeeding times and even unprovoked. This was the result of the population having increased with no scope for expansion of their territory. The aggression may have further increased the mortality. Another point to be noted is that it was group A that made the move and not any of the less dominant groups. This would indicate that the move was one of choice rather than compulsion. It would seem that the new locality had some aspects that were attractive to the most dominant of the groups. It is away from the road and hence more peaceful and the animals are less prone to accidents. Provisioning by humans would be readily at hand. Food from nature would be in abundance because it is said that trees are denser and more diverse than the trees outside the compound, and hence more useful to the monkeys. Probably, too, harmful interspecies interaction would be lessened if not totally absent because this area is a walled compound where dogs would not be able to enter.

To sum up it can be said that it is not human pressures or threat from any other species of animals or lack of food or pressure from their own kind that has prompted the most dominant group to choose a locale of the same habitat to move to. It is felt that they chose to expand their home range because a certain amount of congestion was felt and now that the locality meets all their needs, they choose to stay there.

References

Brennan E, Else JG (1984). The status of Kenya's De Brazza monkey population. A need for immediate conservation efforts. Intern J Primatol 5:303–410.

Fa JE (1983) Habitat distribution and habitat preference in Barbary Macacques (*Macaca sylvanus*) Intern J Primatol 5:273–286.

Malik I, Seth PK, Southwick CH (1984) Population growth of free ranging rhesus monkeys in Tughlaqabad. Amer J Primatol 7:311–321.

Malik I, Seth PK, Southwick CH (1985) Group fission in free ranging rhesus monkeys at Tughlaqabad; Northern India. Int J Primatol 6:411–421.

Mukherjee RP (1976) Effects of crowding on temple rhesus monkeys of Imphal, Manipur. J Bombay Nat His Soc 74:275–281.

Seth PK, Seth S (1983) Population dynamics of free ranging rhesus monkeys in different ecological conditions in India. Amer J Primatol 5:61–67.

Southwick CH (1967) An experimental study of intra group agonistic behaviour in rhesus monkeys. Behaviour 28:182–209.

Southwick CH, Beg MA, Siddiqui MR (1965) Rhesus monkeys in north India. In: De Vore I (ed) Primate behaviour: field studies of monkeys and apes. Holt, Rinehart and Winston, New York, pp. 111–159.

14
Balancing the Wild/Captive Equation—The Case of the Barbary Macaque (*Macaca sylvanus* L.)

JOHN E. FA

Introduction

Hitherto, the struggle to secure self-sustaining primate populations has been two-pronged, with efforts to perpetuate a species concentrated either in captivity or in the wild. The ultimate goals of zoo captive breeding programs and of conservation agencies (with their desire to preserve genetic diversity) are essentially one and the same—to protect the species in its natural state. Their approach, however, differs: One works toward the eventual introduction of captive-bred animals into wild habitats, the other for the protection of natural habitats. Yet, if there is to be any measure of success in conserving an endangered species, the two approaches must be integrated. Unfortunately, in most cases, the two fronts are largely unconnected, and many conservation measures remain merely "on paper" intentions of goodwill. Two fundamental problems with a strong scientific basis operate against pursuing such a divorced approach. On the one hand, captive propagation cannot be carried on indefinitely—after a finite number of generations (depending on population size) the stock will show the effects of domestication, random change, and inbreeding depression. On the other hand, by the time a good number of animals is available for reintroduction there may no longer be any wild habitat to release them into. The survival of an endangered primate must be secured by producing viable release animals before deleterious captivity effects appear and before all available habitats disappear in the wild. Such a fine target requires complementary planning in captive and wild environments. Successful conservation measures can be seen as similar to balancing an algebraic equation in which captive propagation is a *function* of the protection of the species in the wild. The Barbary macaque (*Macaca sylvanus* L.) represents a prime example of a vulnerable species whose preservation in natural conditions is still possible through habitat protection reinforcement from already existing captive-breeding programs. Success depends on the direction and fine tuning of both efforts. This paper briefly considers the state of the Barbary macaque both in the wild and in captivity and summarizes the possible dangers facing the animal in both situations.

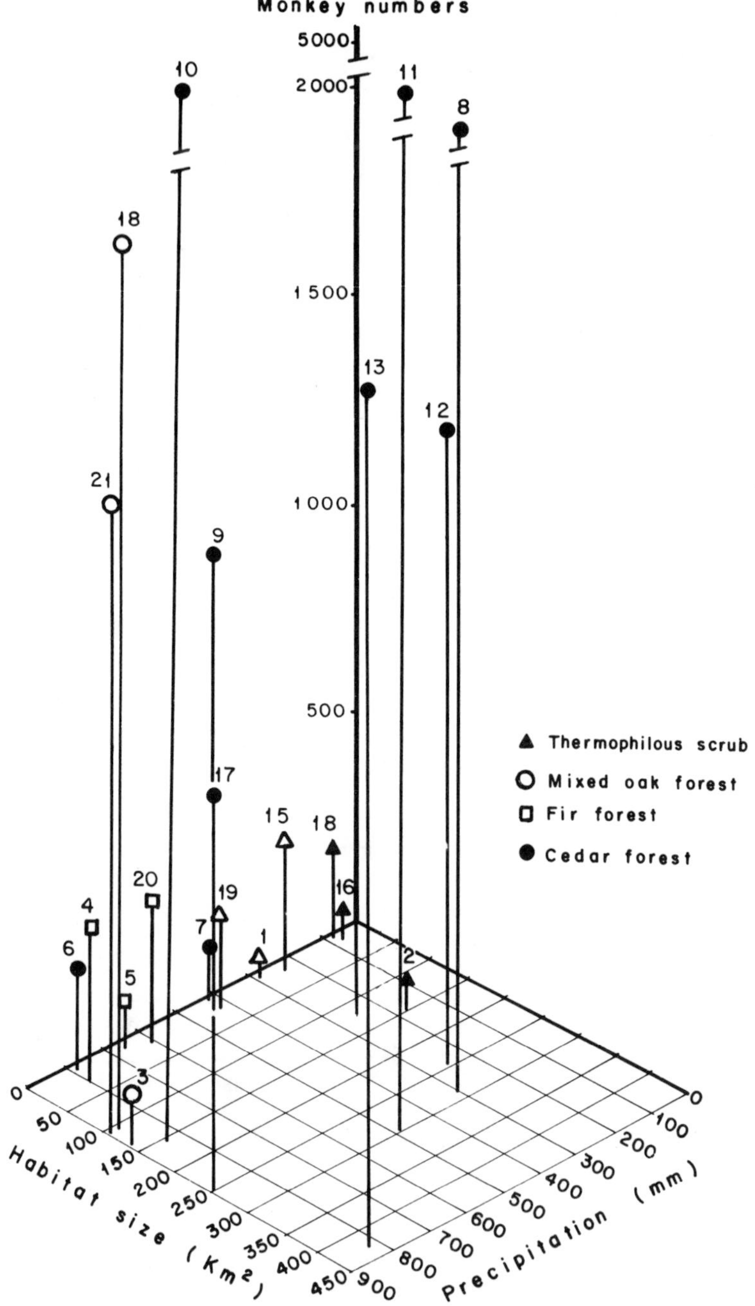

FIGURE 14.1. Three-dimensional plot of precipitation (as an indirect measure of primary production), habitat size, and monkey population size for the Barbary macaque areas now found in Morocco and Algeria. Further descriptions of location of each habitat are given in Fa et al. (1984). Morocco, *Rif*: 1, Djebel Moussa; 2, Djebel Kelti, Kaiat, Sidi-Salah;

Distribution and Abundance in the Wild

Present Distribution and Status of Barbary Macaque Populations

The Barbara macaque in the wild is now confined to only a few isolated areas in Morocco and Algeria. A more extensive description of the distribution sites and their numerical importance can be found in Fa et al. (1984). A brief description for the purposes of this paper is given below. Data on numbers, habitat size, and precipitation for each locality are summarized in Figure 14.1.

In Morocco, the species exists in large numbers only in the Moyen Atlas where an estimated maximum of 16,000 animals (Taub, 1978) live primarily in the high-altitude mixed cedar (*Cedrus atlantica* [Endl.] Carriere/*Quercus ilex* L.) forests of the Central Zone. Estimated densities range from 44 (Deag, 1974) to 60 (Taub, 1978) animals/km². In the Rif Mountains, in north Morocco, numbers are lower, with a maximum of perhaps 600 animals at densities ranging from one to eight animals/km² (Fa, 1982). The main ideas of distribution here are the fir (*Abies pinsapo* Boiss.) forests of Djebel Lakraa, Tissouka, and Tazoute where approximately 400 macaques live (Fa, 1982; Mehlman, 1984). In the Rif, the monkey is also found in oak (*Quercus faginea* Lamk/*Q. pyrenaica* Willd/*Q. suber* L.) forests at Djebel Bouhassim at densities of about one animal/km² (total estimated population of 70–100 monkeys) and in lowland scrub habitats, but at considerably lower numbers and densities. Reports of monkeys in the Ourika valley, the only site in which they are found in the Haut Atlas, have also been recently substantiated (Deag and Crook, 1971), but only three groups were observed by Drucker (quoted in Fa et al., 1984).

In Algeria, the two mountain ranges of the Grande and Petite Kabylies provide the main refuge sites for the Barbary macaque. Both localities were recently surveyed by Menard (in Fa et al., 1984) and about 23% (5500 animals) of the total Barbary macaque numbers are estimated to occur there. High densities and large populations are still present in the oak (*Quercus faginea*/*Q. afares*/*Q. suber*) forests of Agkfadou (2100) and Guerrouch (1500) and in the cedar forests in the Djurdjura (1750). A smaller population of around 300 monkeys may be found in the fir/cedar forests (*Abies numidica* de Lannoy/*Cedrus atlantica*) on Djebel Babor. Three other small populations (maximum of 50 animals each locality) are reported in scrub at the Chiffa gorge, south of Alger, at Kerrata in the Chabet Akra, and on the Pic des Singes in Bejaia. Monkeys in all of these last sites are well habituated to humans and readily accept food offered to them. As a result, these monkeys tend to be concentrated around sites frequented by tourists.

Habitat Isolates—Patches in Space and Time

Of the 21 habitat isolates identified in which the Barbary macaque is still found, the largest blocks are those of cedar (m±S.D.: 160.79±142.72 km²) and mixed oak forests (114.0±24.25 km²) and the smallest those of thermophilous scrub

---

Djebel Bouhassim; 4, Djebels Tissouka, Talassemtane, Lakraa; 5, Djebel Tazoute; 6, Djebel Tizirane. *Moyen Atlas*: 7, Fes/Taza; 8, Azrou/Ifrane/Mischlifene; 9, Ain Leuh/El Hamman; 10, Seheb; 11, Ajdir; 12, Itzere; 13; Midelt. *Haut Atlas*: 14, Ourika Valley. Algeria, *Blida*: 15, Chiffa. *Grande Kabylie*: 16, Pic des Singes, Bejaia; 17, Djurdjura; 18, Agkfadou. *Petite Kabylie*: 19, Chabet el Akra, Kerrata; 20, Djebel Babor; 21, Djebel Guerrouch.

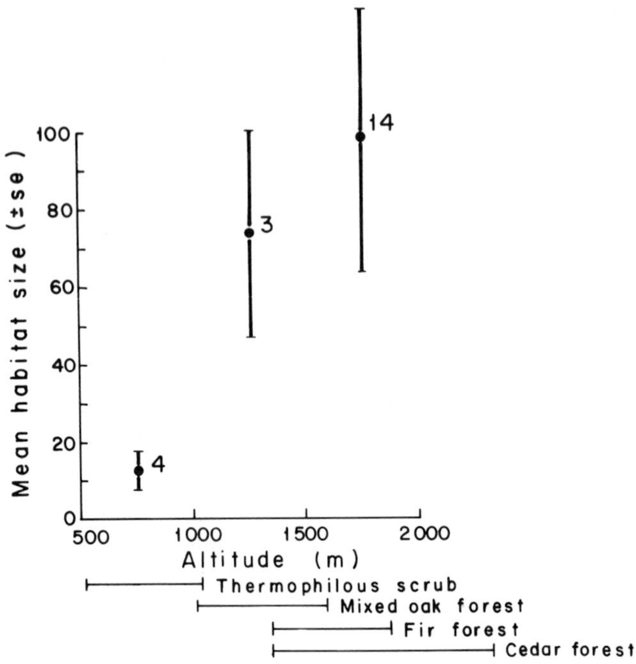

FIGURE 14.2. Relation between habitat size (in km^2 $\pm$ S.E.) and altitude (m.a.s.l.) for all Barbary macaque habitats in Morocco and Algeria.

(13.17$\pm$5.83 km^2) and fir (14.20$\pm$1.92 km^2) with the largest discrete area, the Moyen Atlas cedar forests, of 1116 km^2. There is a general tendency for the largest blocks to be found at higher altitudes (Fig. 14.2). From the inferred primordial vegetation cover of both countries, Fa (1984a) has shown that cedar forests in fact represent only about 4% of the entire land surface area where cedar can grow, but about 78% of the total monkey populations live in this habitat. Although the number of macaques living in the cedar forests in relation to the potential habitat areas indicates a concentration of monkeys in cedar, when this is weighed against the actual availability of habitats, the presence of higher monkey populations in cedar can be explained as an artifact of this relationship. That cedar is now the more abundant habitat type is a result of human destruction and modification of most of the lowlands and the consequent restriction of forests to the highland regions. For example, Fa (1983a) has calculated that for North Morocco almost all of the oak forests have been lost since prehistoric times, while there has been virtually no change in the area covered by cedar and fir forests. The extent of the high-mountain conifers relative to the lower altitude forests is now disproportionately biased in favor of the former though the potential area available for the others is far larger.

For over a thousand years, the forest has provided fuel, timber, pasture, and land for clearing and cultivation (Thirgood, 1984). The vast influx of Bedouin peoples from the east (A.D. 643–698) was the greatest single influence to mold the North African landscape. In all lands under Arab domination, pastoral nomadism substituted the sedentary agriculture that had developed during the previous Roman period. In fact, traditional Islamic concepts of land ownership and tenancy brought the natural vegetation, especially lowland forests, under severe pressure but had little influence on the mountains (see Mikesell, 1961). Because of their inaccessibility the forests went largely untouched.

Population Sizes, Density, and Habitat Quality

It is an established axiom in ecology that if isolates are censused, those of greater area will have more species (Wilcox, 1980). This relationship occurs primarily because larger areas have more habitat and sometimes greater diversity. The quantitative relationship between species and area can be shown to follow from properties of ecological communities that have practical consequences to conservation. This power function, which produces a straight line when variates are log transformed, applies also to single species in different-sized habitat isolates. For the Barbary macaque, this has already been demonstrated by Fa (1984a); a modified plot of this is shown in Figure 14.3. Here, the Moyen Atlas cedar forest blocks, which were considered as separate samples in Fa (1984b), are included as one continuous habitat isolate for the sake of clarity. Indeed, this cedar forest area in the Moyen Atlas represents a single geographic zone where monkey populations are likely to experience gene flow from all its parts.

In primates, intraspecific variations in group size, home range area, or population density are usually related to habitat quality (Freeland, 1979; Caldecott,

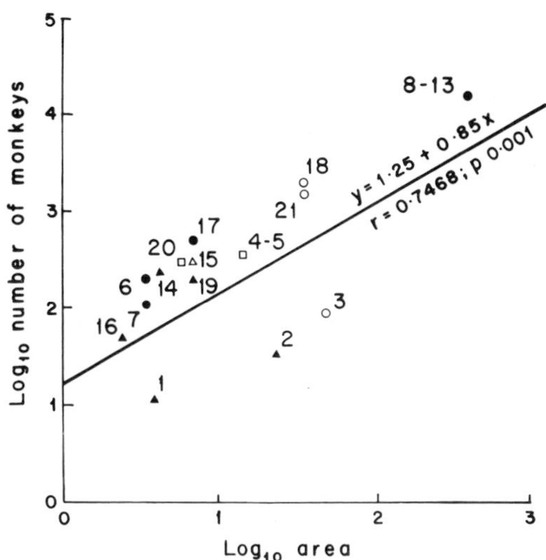

FIGURE 14.3. Numbers–area relation for Barbary macaque populations in the different habitat blocks throughout Morocco and Algeria. Data from Table 1 in Fa (1984). Numbers alongside symbols refer to localities mentioned in Figure 14.1. Habitat symbols as in Figure 14.1.

1980; Takasaki, 1981). More specifically, home range area is usually a good measure of habitat quality since it correlates negatively with habitat quality (Neville, 1968; Yoshiba, 1968; Kano, 1972). The latter has been defined by Takasaki (1981) as type of habitat (coniferous or deciduous) and extent of human disturbance. For those Barbara macaque areas for which there is information on home range sizes, habitat quality has been established by using a number of measures proposed by Bishop et al. (1981), as shown in Fa et al. (1984). The results obtained, together with data on home range size and monkey densities, are shown in Table 14.1. A log plot of habitat quality and home range size shows that there is a significant positive correlation between both parameters (Fig. 14.4)—monkey groups in the less disturbed areas, those in the Moyen Atlas cedar forests, have the smallest home range sizes. The implication is that the cedar forests have the highest populations as a direct consequence of a relatively better food supply. The macaque is an eclectic feeder capable of using a wide variety of food items in all habitat types. As a corollary to this, a plot of monkey density against home range size in Figure 14.5 shows that there is, as expected, a significant relationship between high monkey density in an area and a small home range size. Whether this inferred higher productivity in cedar forest is a result of habitat type or of different levels of human disturbance is important. Fa (1983a, 1984a) demonstrated that, in the North Moroccan habitats at least, potential primary productivity is not a good predictor of monkey population size, but that primary production figures for some areas are lower purely because of human action. Indeed, areas of different habitat types fall on the same regression line in the

TABLE 14.1. Data used in the analysis of the relationship between habitat quality and Barbary macaque populations in seven study sites in North Africa.

Locality	Habitat type	Group size	Habitat quality[1]	Home range size (ha)	Density (animals/km²)	Source
1. Bou Jirrir, Moyen Atlas	Cedar forest	40	9	25.00	50.00	Drucker (1984)
2. Ain Kahla, Moyen Atlas	Cedar forest	39	8	18.59	60.00	Taub (1978)
3. Ain Kahla, Moyen Atlas	Cedar forest	25	7	11.70	44.00	Deag (1974)
4. Djurdjura, Grande Kabylie	Cedar forest	41	11	166.60	13.80	Fa et al. (1984)
5. Talassemtane, Rif	Fir forest	48	12	400.00	8.30	Mehlman (1984)
6. Agkfadou, Grande Kabylie	Mixed oak forest	32	10	333.00	9.60	Fa et al. (1984)
7. Djebel Bouhassim, Rif	Mixed oak forest	27	13	1200.00	0.37	Fa (unpublished data)

[1]Habitat quality is the total rating of the sum of ranks (1–5, good to poor) given to (1) description of home range, (2) harassment of animals, (3) habituation to humans, (4) presence of predators, as defined by Bishop et al. (1981) and used by Fa et al. (1984). A fifth category, a measure of primary productivity of the habitat given as a relative rank denoting the condition of the habitat's primary production, is also used here.

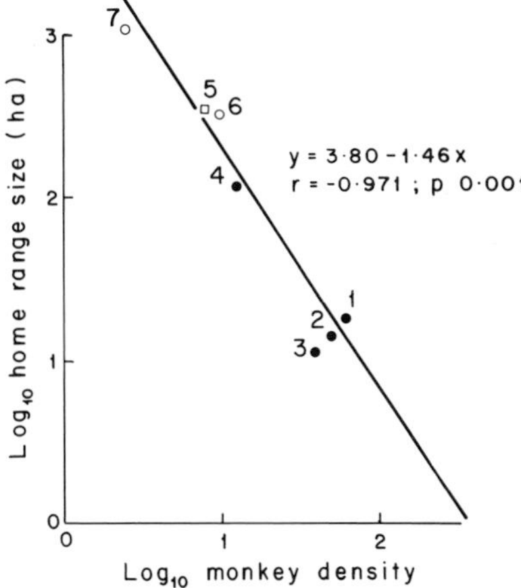

FIGURE 14.4. Log plot of habitat quality against home range size of Barbary macaque groups in the seven studied localities, shown in Table 14.1. Symbols are the same used in Figure 14.1.

area–numbers plot in Figure 14.3, and, as shown by Fa (1984a), no habitat type can then be implicated. Thus, the present condition of most Barbary macaque populations in both Morocco and Algeria seems to reflect the level of habitat disturbance (overgrazing by livestock, tree-felling, and burning) as is in fact demonstrated in Figure 14.4 (see Drucker, 1984, and Mehlman, 1984). Figure 14.6 illustrates the interrelationship between the main factors promoting the survival of larger population sizes in the cedar habitats.

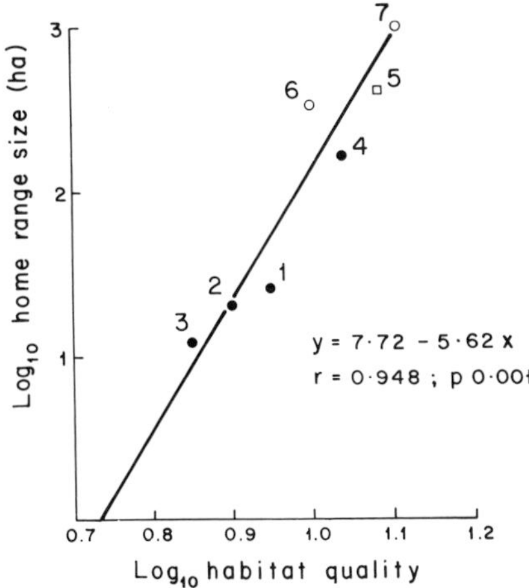

FIGURE 14.5. Log plot of density of monkeys against home range size for Barbary macaque groups in the seven studied habitats shown in Table 14.1. Symbols are the same used in Figure 14.1.

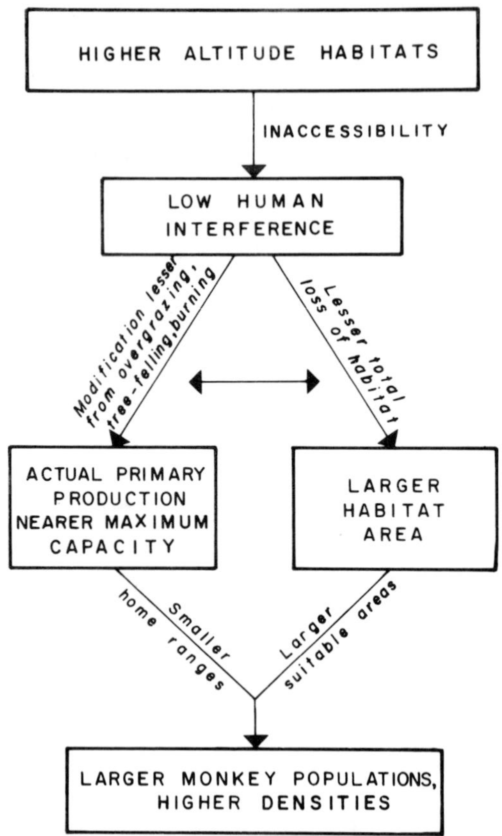

FIGURE 14.6. Effects of altitude and human interference levels on density and population size of Barbary macaques in North Africa.

Pressures and the Fate of Barbara Macaque Habitats

There is already a significant volume of literatuare underlining the pressures affecting the Barbary macaque habitats in Morocco and Algeria. As Thirgood (1984) has pointed out, the problem of the Mediterranean forest is a sociological one, though it has ecological consequences. In the Mediterranean basin, humans are not superimposed on the ecosystem but are an integral part of it and, most often, have been the ecological dominant. Both Deag (1974, 1977) and later Taub (1978, 1984) concluded that legal and illegal commercial woodcutting by Berber tribesmen in the Moyen Atlas were the principal threats to the continued survival of Barbary macaque habitat. Deforestation, through cutting and burning, is also seen by Mehlman (1984) as a major danger in the Rif fir forests. Perhaps more seriously, Fa (1983a) and Drucker (1984) have shown the monkey habitats to be also subject to depression and sometimes depletion of herbage production from overgrazing by livestock. Intensive grazing in forests is usually accompanied by tree-felling to provide fodder for goats during times of shortage (Mehlman, 1984), thus accelerating the deterioration of the habitat.

It is almost certain that grazing in the Mediterranean contributes as much as fires to the maintenance of paraclimaxes and eventually to barren areas (Quezel, 1980). Drucker's (1984) comparison of a group of monkeys' ($n = 40$) food intake with that of livestock (livestock and macaques feed on largely the same herbage plants) in a Moyen Atlas cedar forest points out the enormous offtake of vegetation by livestock and stresses the consequent negative effects on the monkeys through habitat decline (Fig. 14.7). Calorific intake values for a single livestock are up to 16 times greater than the observed mean calorific intake for a single macaque. Despite the common assertion by North African forestry officials that the monkeys cause damage to the forests there is little evidence of overcropping by macaques. The main danger to the forest resides in excessive grazing and trampling by livestock. The combination of tree-felling, burning, and overpasturing, in that order generally, is disastrous for the monkey habitats since regeneration cannot take place.

Although fires have already constituted a natural ecological phenomenon in Mediterranean forests, in the last decades their impact has been greater as a result of deliberate or accidental starts perpetrated by humans.

Rates of area loss caused by fire for existing habitats are difficult to assess. However, in France, as an example of a Mediterranean habitat for which there is information, littoral forest areas lose an annual average of 4% of their surface area through fire (Quezel, 1980). As a result, it is calculated that the life expectancy of reafforestation in this region does not exceed 25 years. Large areas of the Moroccan and Algerian forests are also annually burned. According to Mahlman (1984), 1.5% of the Talassemtane fir forest was burned in 10 years, but no other figures are available for other parts. But using this figure as a minimum potential loss rate of habitat and 4% as a maximum (from the French data and assuming that these are realistic figures to apply to the different habitats, i.e., that there is

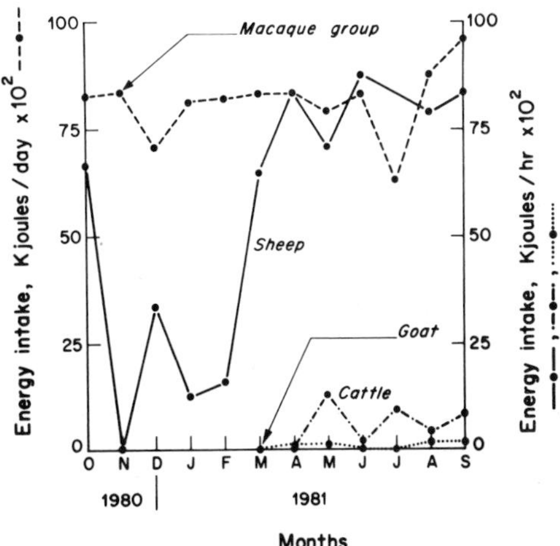

FIGURE 14.7. Comparison of monthly energy intake of a Barbary macaque troop (40 members) and by livestock inhabiting the Bou Jirrir forest in the Moroccan Moyen Atlas. (From Drucker, 1984).

no difference in loss rate between habitats), one can estimate the collapse process of the Barbary macaque habitats on the basis of the regression equation in Figure 14.3. Application of the collapse theory, as used by Soulé et al. (1979), to Barbary macaque habitats is appropriate because insularization is complete. It is thus possible to obtain theoretical extinction curves using the equation 1 ln $(A_0) = T$, where a is the rate of habitat loss with time (years), A_0 is an initial area, A_p is the present area, a constant derived from the numbers–area regression, and T the estimated time from A_0 to A_p. The largest macaque habitat is used as A_0 whereas A_p values are taken from the habitat size data shown above. The model predicts an annual exponential loss of habitat. The main assumption underlying this exponential model (like Soulé's et al., 1979) is that areas lost to fire never recover. The implication for the macaques is that their populations are progressively reduced because of the shrinking primary habitats. Figure 14.8 shows the decline in numbers for both rates of habitat loss. Based on these assumptions, it would take around 360 years for the entire macaque population in the Moyen Atlas to disappear at a loss rate of 0.15% and 180 years at 4%.

Captive Propagation

Demographic Vigor of Captive Barbary Macaque Groups

According to records of the International Zoo Yearbooks, since 1964 the number of Barbary macaques in captivity has increased progressively from an average of 2 to 168 animals per collection. The number of collections keeping the monkey has also grown from 9 to 28 in the same period with 70% being European. Of these, 9 are British, 7 German, 6 are in the United States; the rest are in Africa (2), Canada (1), Poland (1), Czechoslovakia (1), and Gibralter (1). The creation of two large enclosures (11–20 ha) in France (Kintzheim and Rocamadour) and

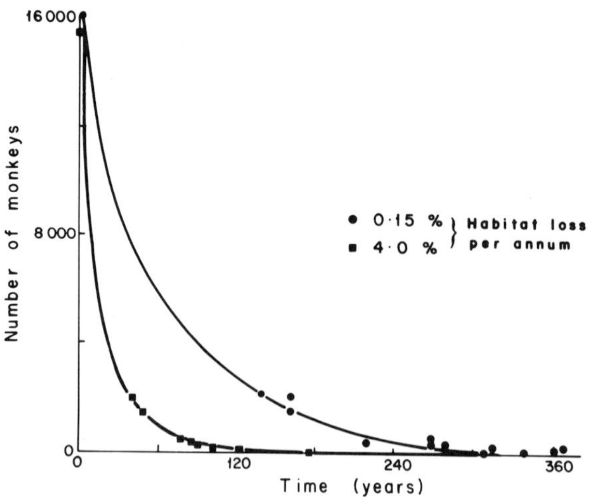

FIGURE 14.8. Plots of depletion rates for all Barbary macaque habitats in North Africa assuming an annual exponential decay rate (see text for explanation).

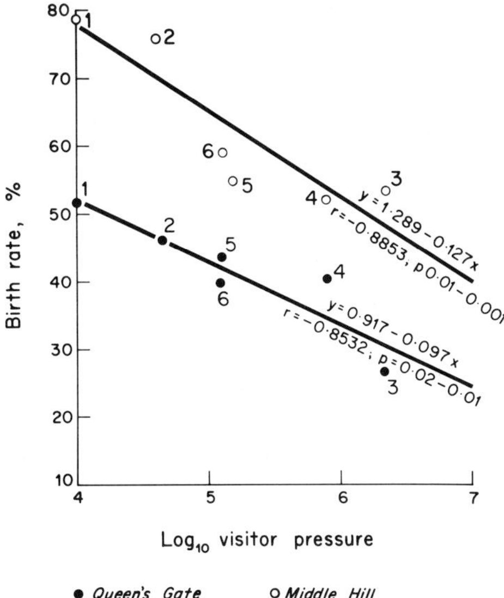

FIGURE 14.9. Relation between visitor pressure and birth rates for the two troops of Barbary macaques in Gibraltar. Each point on the plot refers to a half decade period from 1950. 1, 1950–1955; 2, 1955–1960; 3, 1960–1965; 4, 1965–1970; 5, 1970–1975; 6, 1975–1980.

one in West Germany (Salem) by Baron Gilbert de Turckheim accounts for the increase in captive stock after 1974. About 200 monkeys are born in these enclosures each year and the total number of surviving captive young rose from 11 between 1966 and 1973 to 108 from 1974 to 1980. Although these parks originally imported animals from the wild they are now completely self-supporting (Deag, 1977; Turckheim and Merz, 1984); indeed, they have produced a surplus that led to a successful reintroduction of 232 monkeys in 1980 (Merz, 1984).

The impact of Turckheim's parks on the status of the species in captivity is significant and merits discussion. These parks annually produce a disproportionate number of the total captive stock—of the 124 young that survived in 1980, 98 (or 79%) were from these institutions (Stevenson, 1984). Approximately 68% of all sexually mature females in these parks gave birth each year, and infant mortality is merely 9.1% (Turckheim and Merz, 1984; Paul and Thommen, 1984). There is no doubt that the conditions of semicaptivity and of restricted human interference (even though over half a million people annually visit each park) are crucial in determining their success. The level of human interference, in particular, is critical. It has been shown for the Gibraltar Barbary macaques that uncontrolled human interference has detrimental effects on birth rate (Fa, 1984b; Fig. 14.9).

The reservoir potential of the Turckheim parks is obvious. The lamda value calculated from survival probabilities taken from Gibraltar (Fa, 1984b) is 1.133; the population can double itself each year. As a corollary, the estimated exponential growth curves are steep for all Turckheim's groups (Fig. 14.9). This means that under unchecked growth and no space limitations to set the populations at

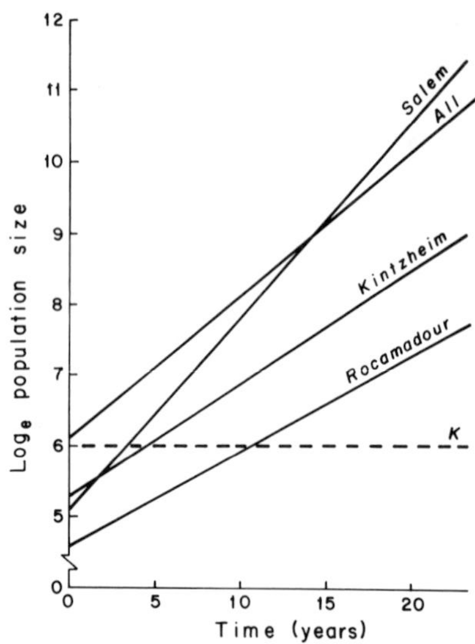

FIGURE 14.10. Exponential rates of increases for the three main Barbary macaque enclosures owned by Baron Gilbert de Turckheim. The carrying capacity (*K*) of the parks is determined from the total number of members achieved at Salem by 1984. This figure of 200 monkeys, according to W. Angst (personal communication) represent the maximum number that can be kept in a park to avoid an increase in aggression between monkeys and between monkeys and visitors.

carrying capacity *K* in Figure 14.10, a total of over 48,000 (double the present wild population) could be bred in 24 years.

Captive and Wild Barbary Macaques—Balancing the Equation

The above sections have summarized the situation of the Barbary macaque in the wild and in captivity. A deliberate emphasis has been placed on the wild situation largely because conservation in the wild is seen as fundamental and captive propagation as an essential backup.

While there have been significant advances in the captive propagation of species in recent years, they have been due more to the manipulative approaches of species preservation than to natural reproduction (Conway, 1978, 1980; Benirschke et al., 1980). Pinder and Barkham (1978) for mammals and Stevenson (1984) for primates have clearly shown that, despite the claims of some zoos and wildlife parks that rare animals are kept for conservation purposes (i.e., captive propagation), there is still little evidence of the successful breeding of large numbers of release. As Conway (1980) has stated, though the concept of captive propagation and release into the wild is encouraged by recent advances in animal care, the existing inadequate facilities, failing socioeconomic mechanisms, and shrinking wild habitats all work against efforts in this direction. It may be that successful conservation attempts are those that concentrate on practicable compromises between in situ and ex situ preservation (sensu Frankel and Soulé, 1984); as in the case of the Barbary macaque whose numbers make the species still redeemable. This is not to suggest that efforts to save the most endangered species should be discouraged but rather that given the limited world-wide resources to save all species, it may be more realistic to concentrate on less precarious species. AAZPA and ISIS's work (Seal et al., 1977) may represent the kind of concerted action that could make sense of captive propagation attempts world-wide but that should be geared at the preservation of the species in their natural state.

The Barbary macaque, perhaps prematurely, has reached a self-sustaining level in captivity in the seminatural enclosures. This effectively means that, while there is an accelerated growth of animals in captivity, conservation of wild populations and their habitats is still lagging. Once again, efforts in the wild and in captive environments are out of phase. The dilemma now posed is for how long can the monkeys be kept in captive conditions (and the consequences of this action are known) in the hope that some day they may be used. A likely solution is, perhaps, to use surplus captive animals to repopulate depleted areas in the wild. Yet any rash decision to reintroduce animals into natural habitats that are unprotected will lead inevitably to their disappearance once again. It would be totally unrealistic to presume that all habitats are viable monkey localities. Any reintroduction must be preceded by an in-depth study of the viability of the proposed release site. Since the captive situation is now such as to afford a stock of animals for release into the wild, the priority has to be to ensure that certain areas are receptive to such an introduction. A first step in this direction is to help promote the establishment of natural parks in Morocco and to afford help to Algeria in managing existing ones (Fa, 1983b, 1984c). These are the habitat islands likely to survive in the long term, and through proper management they should be able to maintain self-sustaining wild populations. It is imperative that all efforts should be concentrated in a well-balanced and coordinated attempt to establish permanent refuge areas for the Barbary macaque. Unilateral action (in the form of an ill-timed reintroduction) would lead to the loss of valuable surplus animals and would do little to alter the status of the animal in the wild. In the case of the Barbary macaque, the time has come for the two faces of conservation (reintroduction of captive stock and protection of wild habitats) to confront each other and work together. The failure to do so will lead to yet another doomed attempt to save a vulnerable species.

Acknowledgments. I am most grateful to D. Piñero and J. Soberon for their help in computing and to them and C. Macias and R. Bye for reading the manuscript. I would like to acknowledge my gratitude to Tere Medina for typing the manuscript and to Felipe Villegas for the drawings. My warmest thanks again go to my wife, Monique, for her never-ending support and encouragement. I thank G. R. Drucker and Plenum Publishing Corporation for permission to reproduce Figure 14.6.

References

Benirschke K, Lasley B, Ryder O (1980) The technology of captive propagation. In: Soulé ME, Wilcox BA (eds) Conservation biology, an evolutionary-ecological perspective. Sinauer, Sunderland, MA, pp 225–242.

Bishop N, Blaffer-Hrdy S, Teas J, Moore J (1981) Measures of human influence in habitats in South Asian monkeys. Intern J Primatol 2:153–167.

Caldecott JO (1980) Habitat quality and populations of two sympatric gibbons (Hylobatidae) on a mountain in Malaya. Folia Primatol 33:291–309.

Conway WG (1978) A different kind of captivity. Anim King 81:4–9.

Conway WG (1980) An overview of captive propagation. In: Soulé ME, Wilcox BA (eds) Conservation biology, an evolutionary-ecological perspective. Sinauer, Sunderland, MA, pp 199–208.

Deag JM (1974) A study of the behaviour and ecology of the wild Barbary macaque (*Macaca sylvanus* L.). Unpublished Ph.D. Thesis, University of Bristol.

Deag JM (1977) The status of the Barbary macaque (*Macaca sylvanus* L.) in captivity and factors influencing its distribution in the wild. In: Prince Rainier, Bourne G (eds) Primate conservation. Academic, New York, pp 267–287.

Deag JM, Crook JM (1971) Social behaviour and 'agonistic buffering' in the wild Barbary macaque *Macaca sylvanus* L. Folia Primatol 15:183–200.

Drucker GR (1984) The feeding ecology of the Barbary macaque and cedar forest conservation in the Moroccan Moyen Atlas. In: Fa JE (ed) The Barbary macaque—a case study in conservation. Plenum, New York/London, pp 135–164.

Fa JE (1982) A survey of population and habitat of the Barbary macaque (*Macaca sylvanus* L.) in North Morocco. Biol Conserv 14:45–67.

Fa JE (1983a) An analysis of the status of the Barbary macaque (*Macaca sylvanus* L.) in the wild—steps towards its conservation. In: Harper D (ed) Symposium on the conservation of primates and their habitats. Vaughan Papers, University of Leicester, Leicester, pp 58–85.

Fa JE (1983b) The Barbary macaque—the future. Oryx 17(2):62–67.

Fa JE (1984a) Habitat distribution and habitat preference in Barbary macaques (*Macaca sylvanus*). Intern J Primatol 5(3):273–286.

Fa JE (1984b) Structure and dynamics of the Barbary macaque population in Gibraltar. In: Fa JE (ed) The Barbary macaque—a case study in conservation. Plenum, New York/London, pp 263–306.

Fa JE (1984c) Conclusions and recommendations. In: Fa JE (ed) The Barbary macaque—a case study in conservation. Plenum, New York/London, pp 319–334.

Fa JE, Taub DM, Menard N, Stewart PJ (1984) The distribution and current status of the Barbary macaque in North Africa. In: Fa JE (ed) The Barbary macaque—a case study in conservation. Plenum, New York/London, pp 79–112.

Frankel OH, Soulé ME (1984) Conservation and evolution. Cambridge University Press, Cambridge.

Freeland WJ (1979) Mangabey (*Cercocebus albigena*) social organization and population density in relation to food use and availability. Folia Primatol 32:108–124.

Kano T (1972) Distribution and adaptation of the chimpanzee on the eastern shore of Lake Tanganyika. Kyoto Univ Afr Stud 7:37–129.

Mehlman PT (1984) Aspects of the ecology and conservation of the Barbary macaque in the fir forest habitat of the Moroccan Rif mountains. In: Fa JE (ed) The Barbary macaque—a case study in conservation. Plenum, New York/London, pp 165–202.

Merz E (1984) Report on a reintroduction of Barbary macaques of North Africa. Unpublished report. 53 pp.

Mikesell MW (1961) Northern Morocco; a cultural geography. Publs in Geography 14, University of California Press, Berkeley/Los Angeles.

Neville MK (1968) Ecology and activity of Himalayan foothill rhesus monkeys (*Macaca mulatta*). Ecology 40:110–123.

Paul A, Thommen D (1984) Timing of birth, female reproductive success and infant sex ratios in semifree-ranging Barbary macaques (*Macaca sylvanus*). Folia Primatol 42(1):2–16.

Pinder NJ, Barkham JP (1978) An assessment of the contribution of captive breeding to the conservation of rare mammals. Biol Conserv 13:187–245.

Quezel P (1980) Biogéographie et écologie des conifères sur le pourtour meditérranéen. In: Pesson A (ed) Ecologie forestière. Bordas, Dunod and Gauthier-Villars, Paris, pp 207–225.

Seal US, Makeyt DG, Bridgwater D, Simmons L, Murtfeldt L (1977) ISIS: a computerized record system for the management of wild animals in captivity. Intern Zoo Yrbk 17:68–70.

Soulé ME, Wilcox BA, Moltby C (1979) Benign neglect: a model of faunal collapse in the game reserves of East Africa. Biol Conserv (15):259–272.

Stevenson MF (1984) The sense and direction of captive breeding programmes—the position of the Barbary macaque. In: Fa JE (ed) The Barbary macaque—a case study in conservation. Plenum, New York/London, pp 203–219.

Takasaki M (1981) Troop size, habitat quality and home range area in Japanese macaques. Behav Ecol Sociobiol 9:277–281.

Taub DM (1978) Geographic distribution and habitat diversity of the Barbary macaque *Macaca sylvanus* L. Folia Primatol 27:108–133.

Taub DM (1984) A brief historical account of the decline in the geographic distribution of the Barbary macaque in North Africa. In: Fa JE (ed) The Barbary macaque—a case study in conservation. Plenum, New York/London, pp 71–78.

Thirgood JV (1984) The demise of Barbary macaque habitat—past and present forest cover of the Maghreb. In: Fa JE (ed) The Barbary macaque—a case study in conservation. Plenum, New York/London, pp 19–69.

Turckheim G de, Merz E (1984) Breeding Barbary macaques in outdoor open enclosures. In: Fa JE (ed) The Barbary macaque—a case study in conservation. Plenum, New York/London, pp 241–262.

Wilcox BA (1980) Insular ecology and conservation. In: Soule ME, Wilcox BA (eds) Conservation biology—an evolutionary-ecological perspective, Sinauer, Sunderland, MA, pp 95–118.

Yoshiba K (1968) Local and intertrooop variability in ecological behavior of common Indian langurs. In: Jay PC (ed) Primates: studies in adaptation and variability. Holt, Rinehart and Winston, New York, pp 217–242.

15
Primate Status and Conservation in China

SUNG WANG and GUOQIANG QUAN

Introduction

Sixteen species or 25 subspecies of primates representing three families and six genera are recorded in China (Quan et al., 1981a, 1981b; Zhang et al., 1981; Honacki et al., 1982; Li and Lin, 1983; Fooden et al., 1985). Among them, the Taiwan macaque (*Macaca cyclopis*), the Tibetan stump-tailed macaque (*M. thibetana*), and the golden monkey (*Rhinopithecus roxellanae*) are species endemic to China.

China is situated in the eastern part of the Eurasian continent, covering the cold temperate, temperate, warm temperate, subtropical, and the tropical zones from north to south. Various types of vegetation can be found throughout the country, and those inhabited by primates are the temperate coniferous forest, mixed coniferous and broadleaf forest, deciduous broadleaf forest, subtropical evergreen broadleaf forest, tropical monsoon forest and rain forest, and the subalpine coniferous forest (Zhang et al., 1981). Most of the species occur in the southern parts of the country, especially in Yunnan Province where as many as 12 species are recorded. The rhesus monkey is the only species that has extended its distributional range as far as the northern part of Hebei Province, reaching approximately 41° north latitude.

Limiting factors influencing the dispersal of primates are primarily the temperature and the food supply in relation to the vegetation types and its growing season. However, human activities should also be considered as an important threat to the present status of primates. Long-term human activities, including deforestation and overexploitation, have greatly diminished or extirpated populations of primates such as gibbons in the vast area over which they once ranged (Gao et al., 1981). Similarly, as indicated in some investigations (Wen et al., 1981), the rhesus monkey, which had been abundant over a wide range in the northwestern and northern parts of the country in ancient times, now survives only in small, isolated areas in the north, with its population size already on the verge of extinction, although it is still abundant in other parts of its range.

With the rapid development of the Chinese economy, wildlife conservation has been of much concern to the authorities as well as to scientists (Wang, 1984). Numerous field surveys and investigations have been conducted in recent years. Special attention was paid to endemic and rare species, including those of primates. As an extension to our previous brief introduction to the situation of primate conservation in China (Quan et al., 1981a, 1981b), the authors here attempt to make a further systematic review of the primate status and conservation of the country based on the information available.

Species Accounts

Family Lorisidae

NYCTICEBUS COUCANG BENGALENSIS SLOW LORIS

Status: Rare. Its primary distribution is restricted to the extreme western and southern parts of Yunnan, though it has also been reported in southwestern Guangxi. It inhabits tropical monsoon forest and tropical rainforest. Its population size is estimated at no more than 1000 and is rare even in Guangxi. The species is still threatened by habitat destruction and hunting although it is already listed as protected wildlife and involved in six of the established nature reserves. Further strict measures for controlling its hunting should be taken.

Family Cercopithecidae

MACACA ARCTOIDES STUMP-TAILED MACAQUE

Status: Threatened. This species is distributed in Yunnan, Guangdong, and Damingshan Mt., southern Guangxi. Its habitats include tropical rainforest, tropical monsoon forest, and southern subtropical evergreen broadleaf forest with rainforest component, up to an altitude of 2500 m above sea level. Population size is estimated at 3000–4000 in Guangxi (Wu, 1983), 6000 in Yunnan, and very rare in Guangdong. The species is threatened by hunting and by its restricted range although involved in two of the established nature reserves.

MACACA ASSAMENSIS ASSAM MACAQUE

Status: Threatened. *Macaca a. assamensis* occurs in southern and western Yunnan and extends westward to the extreme southeastern corner of Tibet (Feng et al., 1984) as well as eastward to the southern border of Guangxi. *Macaca a. pelops* is only found in Buochiu Valley and Jilongbu Valley on the south slope of Mt. Himalaya, Tibet. It inhabits tropical rainforest and monsoon forest and also subtropical evergreen broadleaf forest. Population size of both subspecies is estimated at 3000-5000 in Guangxi (Wu, 1983), approximately 10,000 in Yunnan, and rare in Tibet. The species suffers from hunting and habitat destruction and is involved in two of the established nature reserves.

MACACA CYCLOPIS TAIWAN MACAQUE

Status: Undetermined, but probably endangered. Its distribution is confined to Taiwan. Its habitats mainly include temperate forest, cypress forest, conifer-hardwood forest, and tropical hardwood forest. No estimation of its population size is available, but numbers have declined because of human interference with its habitat (specifically logging) and trapping. A possible low estimate of

1000-2000 animals was given annually. The species is protected in one of the established national parks. There is a clear need for conservation policies to preserve these monkeys (Chen, 1969; Poirier and Davidson, 1979).

Macaca Mulatta Rhesus Monkey

Status: Abundant, but its northernmost population is extremely endangered. *Macaca m. mulatta* occurs in Yunnan, Guangxi, Guangdong, and northward to Shanxi and Hebei in north China, represented by isolated small population in Xinrong, Hebei (lat. 40°40′), Taihanshan Mt., Henan, and Zhongtiaoshan Mt., southeast Shanxi. *Macaca m. vestita* occurs in southwestern Yunnan, southern Tibet (Feng et al., 1984), western Sichuan, and southern Qinghai (Chang and Wang, 1963). *Macaca m. brachyurus* is confined to Hainan Island.

The habitat of the rhesus monkey includes various types of vegetation: tropical forest, subtropical forest, and temperate forest. In the western part of its range, it inhabits subtropical forest, temperate mixed coniferous and broadleaf forest, and subalpine coniferous forest up to 3400 m.

Estimation of population size is not fully available. It was reported that in Guangxi, the number is estimated at about 36,000-50,000 (Wu, 1983), 5000-6000 in Guizhou, and 50,000 in Yunnan. In Guangdong, 27–31 troops inhabit some isles at the mouth of the Zhujiang River, with their total number estimated at 900. It was estimated that the northernmost population in Hebei consisted of around 200 members in the 1950s, but that number has been reduced in the past decades to merely a few individuals. Total number estimated is about 150,000. It is estimated that not fewer than 10,000 individuals are killed and trapped annually. Exportation of the rhesus monkey and other species is now regulated according to the provisions of the Endangered Species Convention (CITES) as well as to the domestic regulations concerned. The species is involved in at least 29 of the established nature reserves.

Macaca nemestrina leonina Pig-Tailed Macaque

Status: Rare. This species only occurs in extreme southern Yunnan and inhabits both tropical monsoon forest and tropical rainforest. Estimation of population size is unknown but is thought to be very low and declining because of habitat destruction and uncontrolled hunting. The species should be protected in at least one of the established nature reserves.

Macaca thibetana Tibetan Stump-Tailed Macaque

Status: Threatened. *Macaca thibetana* occurs in the central and eastern parts of China, ranging east from Fujian, 109°21′ east longitude and westward as far as to Baoxin in Suchuan, 102°40′ east longitude; south from Tiange, Gongcheng, 25° north latitude, in Guangxi, and northward as far as to Qingling Mt. in Shaanxi and Wenxien, Wudu, 32°50′ north latitude, in Gansu (Fooden, 1983; Fooden et al., 1985).

Its habitat includes northern subtropical deciduous broadleaf forest with evergreen species, middle subtropical evergreen broadleaf forest, evergreen deciduous broadleaf mixed forest, and southern subtropical evergreen broadleaf forest with rainforest component. The monkey also occurs at higher altitudes in bare, rocky, mountainous regions (Xiong, 1984; Fooden et al., 1985). Population is estimated at about 1500 in Guangxi (Wu, 1983), 2000-3000 in Guizhou, and 240 in Wenxien in Gansu. The total estimated number is 50,000. The species is still threatened by hunting and trapping in some areas. The animals are to be protected in 11 established nature reserves.

Pygathrix nemaeus nemaeus Douc Langur

Status: Extirpated. This langur has been reported only once from Hainan, but it seems doubtful whether it was really on the island.

Rhinopithecus roxellanae Golden Monkey, Snub-Nosed Monkey

Rhinopithecus r. roxellanae Sichuan Golden Monkey

Status: Endangered. This subspecies occurs in Minshan Mountain ranges and Quonglaishan Mountain ranges in central and northern Sichuan, extending to southern Gansu, and also in Qingling Mt. in Shaanxi (Chen et al., 1982) as well as Shennongjia Mt. in Hubei, inhabiting the northern subtropical deciduous broadleaf forest with evergreen species as well as the middle subtropical evergreen broadleaf deciduous broadleaf mixed forest. The total population size is estimated at not less than 10,000 and is expected to be increased through long-term protection. The species is to be involved in 15 nature reserves established mainly for panda, golden monkey, and takin.

Rhinopithecus r. bieti Yunnan Golden Monkey

Status: Endangered. The subspecies is restricted to Yunling Mountain Ranges, western Yunnan, within the area of 98°40′–100°00′ east longitude, and 26°30′–31°00′ north latitude. It inhabits the alpine dark coniferous forest, with an elevation of 3350–4000 m (Yang, 1984; Li et al., 1981). Population size is estimated at 5000 individuals; the subspecies is threatened by hunting and trapping as well as by deforestation. It is to be involved in four established nature reserves.

Rhinopithecus r. brelichi Guizhou Golden Monkey

Status: Critically endangered. This subspecies is found only at Fanjingshan Mt. in northeastern Guizhou, 108°48′30″–108°55′55″ east longitude, 27°46′50″–28°01′30° north latitude, a total area of 180 km^2. Inhabiting the middle subtropical evergreen broadleaf and deciduous broadleaf mixed forest, with an elevation of 1400–1800 m. Recent survey reported only eight groups to exist, about 500–670 individuals (Quan and Xie, 1981; Xie et al., 1982). The whole inhabited area has been set up as a nature reserve for its ecosystem as well as for the monkey. The species is threatened by its extremely narrow distributional range and small population.

Presbytis entellus Grey Langur

Status: Rare. *Presbytis e. schistaceus* occurs in Buoqiu Valley, Jilong Valley, and Nilamu, southwestern Tibet; *P. e. lania* was once recorded from Chunpeitang at the southern border of Tibet; however, no further information was obtained. The langurs are found in the tropical rainforest, monsoon forest, and subtropical evergreen broadleaf forest, as well as in warm temperate coniferous and broadleaf mixed forest. No nature reserve has been established to protect the species yet (Feng et al., 1984).

Presbytis francoisi Francois' Monkey

Status: Endangered. *Presbytis f. francoisi* is distributed in southern Guangxi and northeastern and southwestern Guizhou. *Presbytis f. leucocephalus* is only found in a small area (200–500 km^2) in south Guangxi. Its habitat appears to be middle subtropical evergreen broadleaf forest and southern subtropical evergreen broadleaf forest with rainforest component. Population size is estimated at 4000–5000 for *P. f. francoisi* and about 400 for *P. f. leucocephalus* only (Wu, 1983). The species is threatened by both deforestation and overhunting. Strict measures should be taken for protecting the species from extinction. It is to be involved in five established nature reserves.

Presbytis phayrei Phayre's Leaf Monkey

Status: Rare and endangered. *Presbytis p. crepusculus* occurs in Xishuanbanna area of the extreme southern Yunnan; *Presbytis p. shanicus* is found in western Yunnan. Its habitat appears to be tropical monsoon forest and rainforest. Formerly, it was rather abundant within its range. However, the number has drastically declined during the past decades as a result of deforestation and trapping. The population size is estimated at approximately 5000, and it is to be involved in one of the established nature reserves.

Presbytis pileatus shortridgei Capped Langur

Status: Rare. Its distribution is confined to Gongshan and Dulong, extreme northwestern Yunnan. It inhabits southern subtropical evergreen broadleaf forest with rainforest component. It was first reported in 1974 (Li and Lin, 1983); however, no further data are available.

Family Hylobatidae

Hylobates concolor Black Gibbon

Status: Critically endangered. *Hylobates c. concolor* occurs in southeastern Yunnan and Hainan Island, while *H. c. leucogenys* is found in extreme southern Yunnan. It inhabits tropical rain and monsoon forest and southern subtropical evergreen broadleaf forest with rainforest component. Estimation of its total number is not available. However, it has been reported that there are fewer than 30 individuals still surviving in some isolated forest areas in Hainan, although the total number was estimated at 2000 in the 1950s. The Yunnan population has also been severely diminished because of deforestation and hunting. The tremendous change in the status of the black gibbon could be regarded as a typical example of a primate species being threatened to extinction by habitat destruction. The species is to be protected in seven established nature reserves.

HYLOBATES HOOLOCK HOOLOCK GIBBON

Status: Critically endangered. It is distributed in southwestern Yunnan, inhabiting tropical monsoon and tropical rainforest. *Hylobates hoolock* is thought to be threatened with extirpation because of its restricted range and heavy habitat destruction, though it is involved in one of the nature reserves so far established.

HYLOBATES LAR CARPENTERI LAR GIBBON

Status: Extremely rare. It is found in the extreme southwestern part of Yunnan, inhabiting the tropical monsoon and rainforest. The species has seldom been encountered in recent years; however, it is to be protected in one of the nature reserves so far established.

Conservation Problems and Recommendations

As early as 1962, an announcement was published by the State Council of China regulating wildlife protection. Most of the primate species inhabiting China have been listed since that time in either the first or the second category, and their hunting has been controlled by the management authorities at the state or provincial level. Hunting of the slow loris, leaf monkeys, golden monkeys, and the gibbons is prohibited. Nevertheless, the situation of wildlife conservation including primates up to now is far from satisfactory. A number of precious or rare species of primates have drastically declined. Several species are on the brink of extinction, although some individual species such as the Sichuan golden monkey have increased to some extent.

There are several major threats to primates. Habitat destruction, however, should be regarded as the main threat, especially the deforestation in southern parts of the country where most primate species are concentrated. Logging and cultivation have decreased the extent of natural forested area. For example, in Hainan Island the natural forested area has decreased from 25.7% to 8.5% of the total land area in 25 years. A similar figure also occurs in southern Yunnan. Thus, the tropical and subtropical species are facing a critical situation of lost habitat.

Another important threat to primates is overexploitation and killing for various purposes. Export for trade has caused the macaque species, mainly the rhesus monkey, to decline rapidly in population since the 1950s. Capturing monkeys for medical use, scientific experiments, traditional medicine preparation, and even food use has caused monkey populations to suffer seriously. It is estimated that 10,000–20,000 monkey skins are purchased annually.

Another primary threat that should be mentioned here is that monkeys are often killed as agricultural pests. These cases happen mainly in the mountainous regions. In addition, monkeys are also captured for pets or street displays.

In order to improve the present situation and promote primate conservation, the following recommendations and suggestions are made:

1. Strengthen primate survey and investigation. A primate conservation strategy should be worked out on the basis of a series of well-planned surveys.
2. Enforce domestic regulations and acts and emphasize international conventions by means of public education.
3. Further strengthen the management authorities at both the state and local levels.
4. Set up more primate nature reserves for the multiple purposes of gene pool, science, education, tourism, etc.
5. Further international exchange and collaboration should be considered, including cosponsoring training courses on primate conservation, joint surveys or research, joint symposia, scholar exchanges, joint-sponsored primate center, etc.

References

Chang C, Wang T-Y (1963) Faunistic studies of mammals of the Chinghai Province. Acta Zool Sinica 15:125–138.

Chen, Min Z, Gan Q, Luo S, Xie W (1982) The golden monkey of the Qing Mountain and its protection. Wildlife (2):7–10.

Cheng JZ (1969) Vertebrate fauna of Taiwan. Taiwan Commercial, Taipei.

Feng Z, Zheng C, Cai G (1984) A checklist of mammals of Xizang (Tibet). Acta Theriol Sinica 4:341–358.

Fooden J (1983) Taxonomy and evolution of the *Sinica* group of macaques. 4. Species account of *Macaca thibetana*. Fieldiana: Zoology, New Series, No 17.

Fooden J, Quan G, Wang Z, Wang Y (1985) The stumptail macaques of China. Amer J Primat 8:11–30.

Gao Y-T, Wen H-R, He Y-H (1981) The change of historical distribution of Chinese gibbons (*Hylobates*). Zool Res 2:1–8.

Honacki JH, Kinman KE, Koeppl JW (1982) Mammal species of the world. Allen & ASC, Lawrence, KA.

Kao Y-T, Lu C-K, Chang C, Wang S (1962) Mammals of the Hsi-Shuan-Pan-Na area in southern Yunnan. Acta Zool Sinica 14:180–196.

Li Z, Ma S, Hua C, Wang Y (1981) The distribution and habit of Yunnan snub-nosed monkey. Zool Res 2:9–16.

Li Z, Zhengyu L (1983) Classification and distribution of living primates in Yunnan, China. Zool Res 4:111–120.

Poirier FE, Davidson DM (1979) A preliminary study of the Taiwan macaque. Quart J Taiwan Mus 32:123–191.

Quan G, Xie J (1981) Notes on *Rhinopithecus roxellanae brelichi* Thomas. Acta Theriol Sinica 1:113–116.

Quan G, Wang S, Zhang Y (1981a) On the recent status and conservation of primates in China. Acta Theriol Sinica 1:99–104.

Quan G, Wang S, Zhang Y (1981b) The classification and distribution of Primates in China. Wildlife 3:7–14.

Wang S (1984) Species conservation. In:National conservation strategy. Unpublished.

Wen H-R, He Y-H (1981) The historical status of the rhesus monkey in north China. Acta Henan Normal Univ 1:37–44.

Wu M (1983) On the distribution and number estimation of primates in Guangxi Province. Acta Theriol Sinica 3:16.

Xie J-H, Liu Y, Yang Y (1982) Preliminary survey of ecological conditions of the Guizhou golden monkey. Sci Survey Fanjingshan Mountain Preserve, pp 215–221.

Xiong C (1984) Ecological studies of the stump-tailed macaque. Acta Theriol Sinica 4:1–10.

Yang D (1984) Locomotory characters of *Rhinopithecus bieti*. Acta Theriol Sinica 4:34.

Zhang Y, Wang S, Quan G (1981) On the geographical distribution of Primates in China. J Hum Evol 10:215–225.

16
Primate Conservation Priorities in the Neotropical Region

Russell A. Mittermeier

Introduction

Neotropical primates are found from southern Mexico, through central America and northern South America south as far as southern Brazil and northern Argentina (Fig. 16.1) and, with the exception of Chile and Uruguay, occur in all mainland South and Central American countries and on a few Caribbean islands as well. Sixteen genera and some 65 species are currently recognized, which means that approximately one-third of all living primate species occur in this part of the world. Of these 16 genera and 65 species, we consider two genera, 12 species, and a total of 28 taxa to be already endangered, and several of these are literally on the verge of extinction (Table 16.1). In this paper, we analyze the primate conservation priorities for the Neotropical region by major biogeographic region and by country and then discuss some of the most endangered species in more detail. For the purposes of this paper, we will not consider either *vulnerable* or *rare* species (as defined in the IUCN *Red Data Book*), but only those that are considered to be already endangered. These are the animals of most immediate concern, and they serve quite well to focus conservation activities over the next 5 years.

Threats to Neotropical Primates

The reasons for the disappearance of primates in the Neotropical region are much the same as in the rest of the world (e.g., see Mittermeier and Coimbra-Filho, 1977; Mittermeier et al., in press) and will not be discussed in any detail here. Suffice it to say that the primary cause is tropical forest destruction (Fig. 16.2), as it is almost everywhere else. All Neotropical primates are arboreal, and they simply disappear in the face of the kind of widespread forest clearance that is taking place in many parts of South and Central America.

The next most important threat is hunting of primates as a source of food for our own species. The effects of such hunting vary greatly from region to region and from species to species, but it can be a very important factor in some areas, especially in parts of Amazonia (Mittermeier and Coimbra-Filho, 1977; Mittermeier, 1977). Larger species such as spider monkeys (*Ateles* spp.) and woolly monkeys (*Lagothrix lagotricha* sspp.) are heavily hunted in most parts of Amazonia and have already disappeared in many parts of their former ranges, even in areas of otherwise suitable primary forest habitat (Fig. 16.3).

FIGURE 16.1. Map of Central and South America showing the distribution of Neotropical primates (drawing by S. Nash).

TABLE 16.1. The most endangered Neotropical primate genera, species and subspecies.

Endangered genera	Endangered species	Endangered subspecies
Brachyteles	*Callithrix aurita*	*Callithrix argentata leucippe*
Brachyteles arachnoides	*Callithrix flaviceps*	*Saguinus bicolor bicolor*
Leontopithecus	*Callithrix geoffroyi*	*Chiropotes satanas satanas*
Leontopithecus rosalia	*Saguinus oedipus*	*Cacajao calvus calvus*
Leontopithecus chrysomelas	*Callicebus personatus*	*Cebus apella xanthosternos*
Leontopithecus chrysopygus	*C. p. personatus*	*Alouatta belzebul ululata*
	C. p. nigrifrons	*Ateles geoffroyi azuerensis*
	C. p. melanochir	*Ateles geoffroyi panamensis*
	Saimiri oerstedi	*Ateles geoffroyi grisescens*
	S. o. oerstedi	*Ateles fusciceps fusciceps*
	S. o. citrinellus	*Ateles belzebuth hybridus*
	Alouatta fusca	*Ateles belzebuth marginatus*
	A. f. fusca	
	A. f. clamitans	
	Lagothrix flavicauda	

FIGURE 16.2. Forest destruction in Peruvian Amazonia (photo by A. Young).

FIGURE 16.3. White-bellied spider monkey (*Ateles b. belzebuth*) shot for food in Colombia (photo by F. Medem).

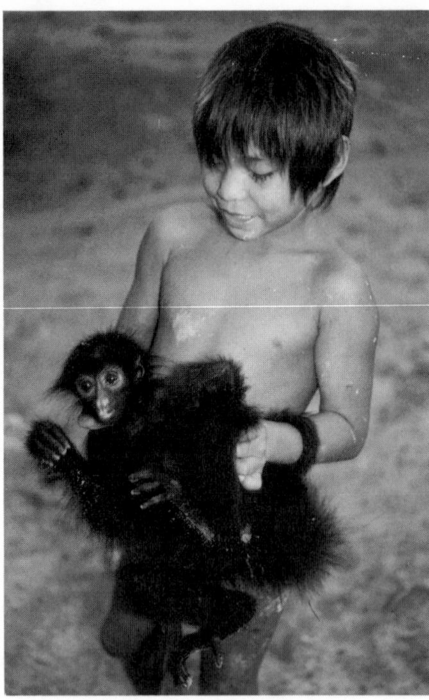

FIGURE 16.4. Juvenile black spider monkey (*Ateles paniscus paniscus*) being kept as a pet by Trio Indian boy in Surinam (photo by M. J. Plotkin).

The third major threat is live capture of primates, either for export or to serve a local pet trade (Fig. 16.4). For the most part, this is a minor factor compared to habitat destruction and food hunting, but for certain endangered species that happen to be in heavy demand, it can be quite serious. Species that have been hurt by the trade include the cotton-top tamarin (*Saguinus oedipus*) from northern Colombia, which was trapped by the thousands for export to research labs (Hernandez-Camacho and Cooper, 1976), the woolly monkey (*Lagothrix lagotricha* sspp.), which were and still are very popular as pets for local people in Amazonia (Mittermeier and Coimbra-Filho, 1977), and the golden-headed lion tamarin (*Leontopithecus chrysomelas*), a highly endangered species from southern Bahia in Brazil that has recently entered the illegal trade both within Brazil and internationally (Mallinson, 1984; Mittermeier, unpublished observation).

Priority Regions for the Conservation of Neotropical Primates

In order to determine the highest priority regions for the conservation of Neotropical primates, we have analyzed the distribution of these animals by major *phytogeographical* regions. These regions are based on recent work by Gentry (1982, and personal communication) and include Amazonia (in the broad Hylaean sense that covers the Guianas as well as the Amazon drainage), the Atlantic forest region of eastern Brazil, the Cerrado/Caatinga/Chaco regions of central Brazil, Paraguay, and Bolivia, the northern Andes (north of the Peru–

Ecuador border and including the Chocó), the southern Andes, northern Colombia/Venezuela, and Middle America, including all of Central America and Mexico (Fig. 16.5).

As indicated in Figure 16.6, the highest primate diversity occurs in Amazonia, which as 12 genera and 39 species, followed by the Atlantic forest region of eastern Brazil, with 6 genera and 15 species. The relatively high numbers for the northern and southern Andes are somewhat misleading, since they result mainly from the extension of certain Amazonian primate species a short way into the Andean foothills above the 1000-m contour line that we have taken as the boundary between the two Andean regions and Amazonia. Once above 1000 m in the Andes proper, there are very few primates, and indeed there appears to be only one primate species endemic to the Andes, the Peruvian yellow-tailed woolly monkey (*Lagothrix flavicauda*).

Endemic species serve to highlight the importance of Amazonia and the Atlantic forest even further (Fig. 16.6). Although most of the regions have few if any endemic species, the Atlantic forest has two endemic genera and nine endemic species and Amazonia has five endemic genera and 26 endemic species.

Priority Countries for the Conservation of Neotropical Primates

Since most conservation programs are developed by country and not by broad biogeographical region, we have also analyzed primate distribution for each country in the Neotropics. Figure 16.7 shows that Brazil is by far the richest

FIGURE 16.5. Major phytogeographic regions in tropical America (modified from Gentry, 1982).

FIGURE 16.6. Map of Central and South America showing the number of Neotropical primate species and genera in each of the major phytogeographic regions shown in Figure 16.5. The figures above the line are total numbers of species and genera; those below are numbers of endemic species and genera (drawing by S. Nash).

country in the region, with 16 genera and 51 species, and it is followed by Colombia and Peru, each with 12 genera and 27 species. As it turns out, these three countries are also at the top of the world list of primate diversity, with Brazil being the single richest country on earth for primates and Peru and Colombia being fifth and sixth (Table 16.2).

Brazil also stands out in primate endemism, with one in every three species being found only in Brazil and nowhere else. Peru and Colombia have five endemic species between them, but no other South or Central American country has any (Fig. 16.7). Again, on a global basis, Brazil's level of endemism is quite high, being surpassed only by Madagascar, with its almost total primate endemism, and Indonesia, with roughly 50% (Table 16.3).

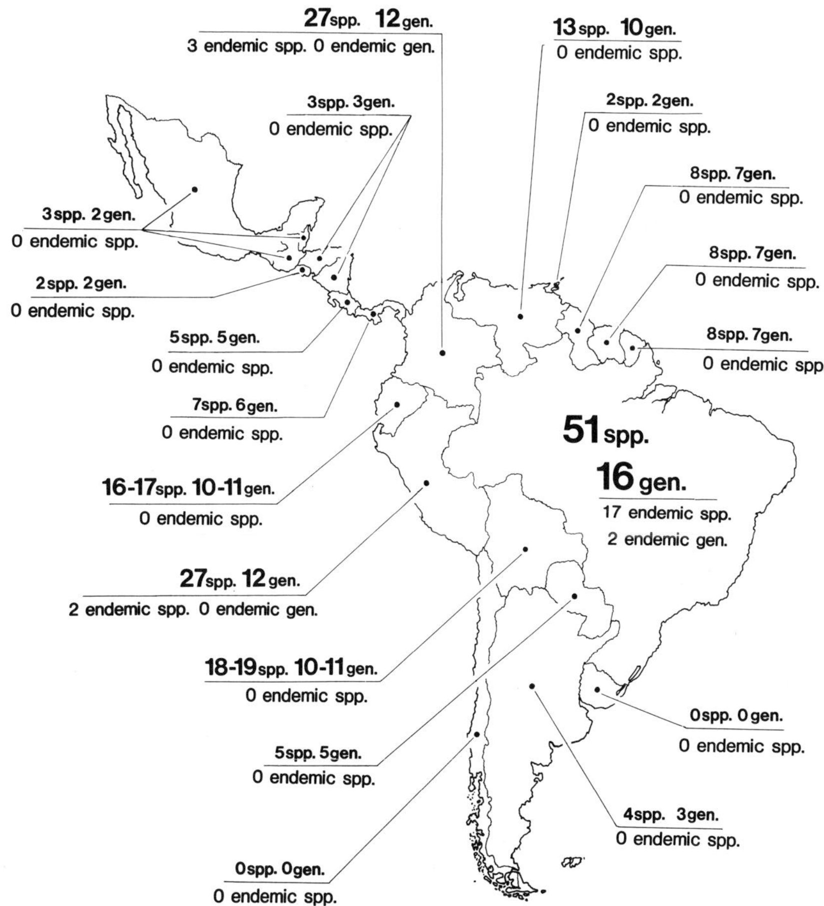

FIGURE 16.7. Map of Central and South America showing the number of Neotropical primate species and genera in each country. The figures above the line are total numbers of species and genera; those below are numbers of endemic species and genera (drawing by S. Nash).

Distribution of Endangered Neotropical Primates by Region and by Country

Table 16.1 lists the Neotropical primate genera, species, and subspecies that we consider endangered and updates the list provided in the latest IUCN *Red Data Book* (Thornback and Jenkins, 1982). It is important to note that the 28 taxa in this table represent a minimum estimate of the number of endangered primates in this region, since we have included only those animals for which we have some data or for which an educated guess can be made as to status. As data on other

TABLE 16.2. The top countries in the world for primate diversity.

Country	No. of species	No. of genera
Brazil	51	16
Zaire	29–32	13–15
Cameroon	28–29	14
Madagascar	28	13
Peru	27	12
Colombia	27	12
Indonesia	27–30	8
Nigeria	23	13
Congo	22	14
Equatorial Guinea	21–22	12
Central African Republic	19–20	11–12
Gabon	19	11
Uganda	19	11
Bolivia	17–18	11–12
Angola	18–19	10–11

Source: from Mittermeier and Oates (1985).

little-known or unknown species are gathered, it will undoubtedly be necessary to add to the endangered list.

Looking at the distribution of endangered taxa by region, by far the largest number (two genera, nine species) occur in the Atlantic forest region of Brazil, followed by Amazonia and Middle America (Fig. 16.8), and looking at their distribution by country, it is clear that Brazil, with 19 endangered taxa (18 of them endemic), is again right at the top of the list (Fig. 16.9).

It is obvious from this analysis that if we want to maintain the current diversity of Neotropical primates, it will be necessary to place special emphasis on Brazil, and particularly on the Atlantic forest region of Brazil and the Brazilian portion of Amazonia (Fig. 16.10), and secondarily on Colombia, Peru, and parts of Middle America. In the remainder of this paper, we discuss the status of several of the most endangered Neotropical primates and some of the conservation programs that are either planned or already underway to ensure their survival.

What is Being Done to Ensure the Survival of Endangered Neotropical Primates

The Atlantic Forest Region of Eastern Brazil

Since the Atlantic forest region of eastern Brazil has a major portion of the Neotropical primates now considered endangered, it is a good place to begin. The Atlantic forest is a unique series of ecosystems that are quite distinct from the much more extensive Amazonian forests to the northwest (Fig. 16.10). They once stretched fairly continuously from the state of Rio Grande do Norte in northeastern Brazil south as far as Rio Grande do Sul, the southernmost Brazilian state, and had southwestern outliers in extreme northern Argentina and eastern Paraguay (Fig. 16.10). In all, the Atlantic forests once covered about 1,000,000 km^2 or about 12% of Brazil's total land surface and included some of

FIGURE 16.8. Map of Central and South America showing the number of endangered primates (from Table 16.1) in each major phytogeographic region (drawing by S. Nash).

the most impressive forest formations anywhere on earth. However, this region was the first part of Brazil to be colonized, it has developed into the agricultural and industrial center of the country, and it has within its borders two of the largest cities in South America, Rio de Janeiro and São Paulo. The result has been large-scale forest destruction, especially in the last two decades of rapid economic development, and we estimate that only about 1–5% of the original forest cover remains in the 13 different Brazilian states that make up this region.

Since 1979, World Wildlife Fund–U.S. has supported a broad conservation program in the Atlantic forest region, with 15 major projects on various aspects of conservation. Primate conservation activities have been a major component of the Atlantic forest program from the outset, and the whole program grew out of what was initially a small primate survey. The bulk of the work in the program is being carried out by a team of some 20 Brazilian students and researchers, with some American and British participation as well.

The primate survey project has now investigated some 35 different parks, reserves, and privately protected forests, mainly in the southeast but also in the

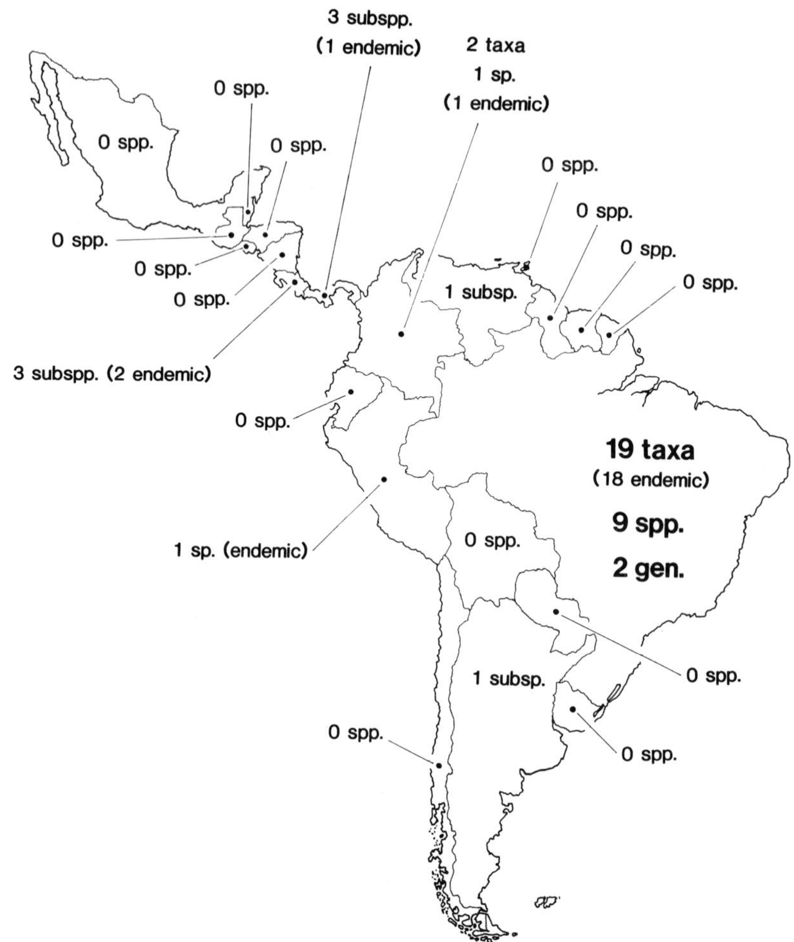

FIGURE 16.9. Map of Central and South America showing the number of endangered primates (from Table 16.1) in each country (drawing by S. Nash).

northeast and the south. Although this project is still underway and will continue for some time to come, it has already given a good indication of the status and distribution of the primates found in this region.

Of the 21 species and subspecies of primates that occur in the Atlantic forest, it appears that 15 are endemic and 13 of these endemics are already endangered (Table 16.4). Using the criterion of taxonomic uniqueness, two endangered genera of the Atlantic forest must be considered the highest primate conservation priorities in the Neotropics, since they are the only primates in the entire region that are actually in danger of becoming extinct as genera. These are *Brachyteles*, a monotypic genus with just one species, the muriqui (*Brachyteles arachnoides*), and *Leontopithecus*, the lion tamarins, with three species: the golden lion tamarin (*Leontopithecus rosalia*), the golden-headed lion tamarin

FIGURE 16.10. Map of Central and South America showing the location of Amazonia and the Atlantic forest region of eastern Brazil in relation to country borders (drawing by S. Nash).

(*Leontopithecus chrysomelas*), and the golden-rumped lion tamarin (*Leontopithecus chrysopygus*).

The muriqui is worthy of special mention, since it is the largest of the New World monkeys and also one of the most endangered. It was once found through much of southeastern Brazil, from the southern part of the state of Bahia south as far as São Paulo and perhaps Paraná (Fig. 16.11), and the original population has been estimated at about 400,000 individuals by the Brazilian conservationist Alvaro Aguirre (1971). Indeed, the muriqui was once so abundant that several of the early expeditions, like that of the German explorer-naturalist Prince Maximilian zu Wied, at times lived almost exclusively from muriqui meat. Unfortunately, the muriqui has suffered tremendously from habitat destruction and has always been a prime target for hunters. As a result, it has declined to the point where it is now on the verge of extinction.

As part of our Atlantic forest conservation program, we have been conducting surveys of this animal and other eastern Brazilian primates since 1979 and have

TABLE 16.3. Primate endemism in the world's top 15 countries for primates.

Country	No. species (No. endemic species)	No. genera (No. endemic genera)	% of endemic species	% of endemic genera
Madagascar	28 (26)	13 (12)	93	92
Indonesia	27–30 (12–15)	8 (1)	44–50	12.5
Brazil	51 (18)	16 (2)	35	12.5
Colombia	27 (3)	12 (0)	11	0
Peru	27 (2)	12 (0)	7	0
Zaire	29–32 (2)	13–15 (0)	6–7	0
Nigeria	23 (1)	13 (0)	4	0
Cameroon	28–29 (0)	14 (0)	0	0
Congo	22 (0)	14 (0)	0	0
Equatorial Guinea	21–22 (0)	12 (0)	0	0
Central African Republic	19–20 (0)	11–12 (0)	0	0
Gabon	19 (0)	11 (0)	0	0
Uganda	19 (0)	11 (0)	0	0
Bolivia	17–18 (0)	11–12 (0)	0	0
Angola	18–19 (0)	10–11 (0)	0	0

succeeded in accounting for only 240 muriquis in ten widely separated sites located in four different states. These populations range from a low of four to five individuals to about 100 individuals, and several of them are still threatened by poaching.

It is necessary to point that these are minimum estimates and that we have a number of reports that may lead to discoveries of new populations, especially in the Serra do Mar region of coastal São Paulo. However, we have adopted the very conservative policy of counting only those animals whose existence we have actually been able to document with certainty. With a highly endangered species such as muriqui that has often been exterminated from areas of apparently suitable habitat, extrapolation from one area to another is a dangerous procedure that can give completely false indications of remaining numbers.

It is obvious that the future of this species will depend on careful management of wild populations and on development of a captive breeding program as well, in part to hold animals that have been confiscated from the illegal trade. The kind of research needed for the development of management programs has been gathered in part by studies such as those of Strier of Harvard University and Milton of the University of California at Berkeley (Milton, this volume) and by ongoing studies by a number of Brazilian researchers, most notably the team headed by Celio Valle of the University of Minas Gerais (e.g., Fonseca, 1985). A captive

TABLE 16.4. Monkeys of the Atlantic forest region.

Species and subspecies	21
Endemic	15
Endangered	14
Endemic	13
Vulnerable	2
Not yet in danger	5

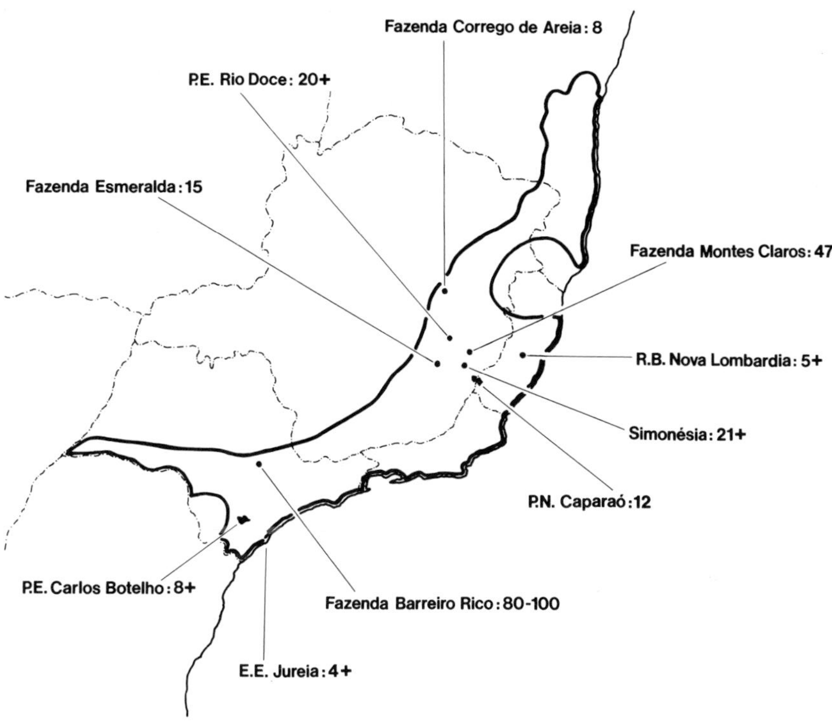

FIGURE 16.11. Map of southeastern Brazil, showing the original range of the muriqui (heavy dark line), and the ten locations in which it is known to still survive. The numbers given are minimum estimates for each area.

breeding project is planned as well, and a large cage for the muriqui is now being built at the Rio de Janeiro Primate Center, with support from Wildlife Preservation Trust International.

The muriqui is an important "flagship species" for the Atlantic forest region as well, and provides an excellent example of how primates can be used to sell the cause of tropical forest conservation as a whole. It is not only the largest of all New World monkeys, it is also the largest mammal endemic to Brazil. As such, it is most appropriate as a symbol for the Atlantic forest conservation program, and indeed for the Brazilian conservation movement as a whole, and could very well become for Brazil what the giant panda (*Ailuropoda melanoleuca*) is for China. To publicize the plight of this species and its Atlantic forest habitat, World Wildlife Fund launched a Muriqui Campaign in 1981 in conjunction with the Brazilian Conservation Foundation (FBCN), the Rio de Janeiro Primate Center (CPRJ-FEEMA), and the University of Minas Gerais (UFMG). This campaign has included lecture programs in the cities and in the interior, development of museum exhibits, and distribution of t-shirts, posters, stickers, and other educational materials, and has succeeded in getting a great deal of media coverage for the muriqui. World Wildlife Fund also produced a film on the animal, entitled *Cry of the Muriqui*, by Andy Young, and this has been translated into Portuguese for use on Brazilian television. It has already been shown hundreds of times in Brazil, especially in small towns and villages in the immediate vicinity of key protected areas for the muriqui. The muriqui has now become so popular that it even graces the cover of a Brazilian phone book and appears on two new Brazilian postage stamps. All of this activity has led to a general increase in conservation

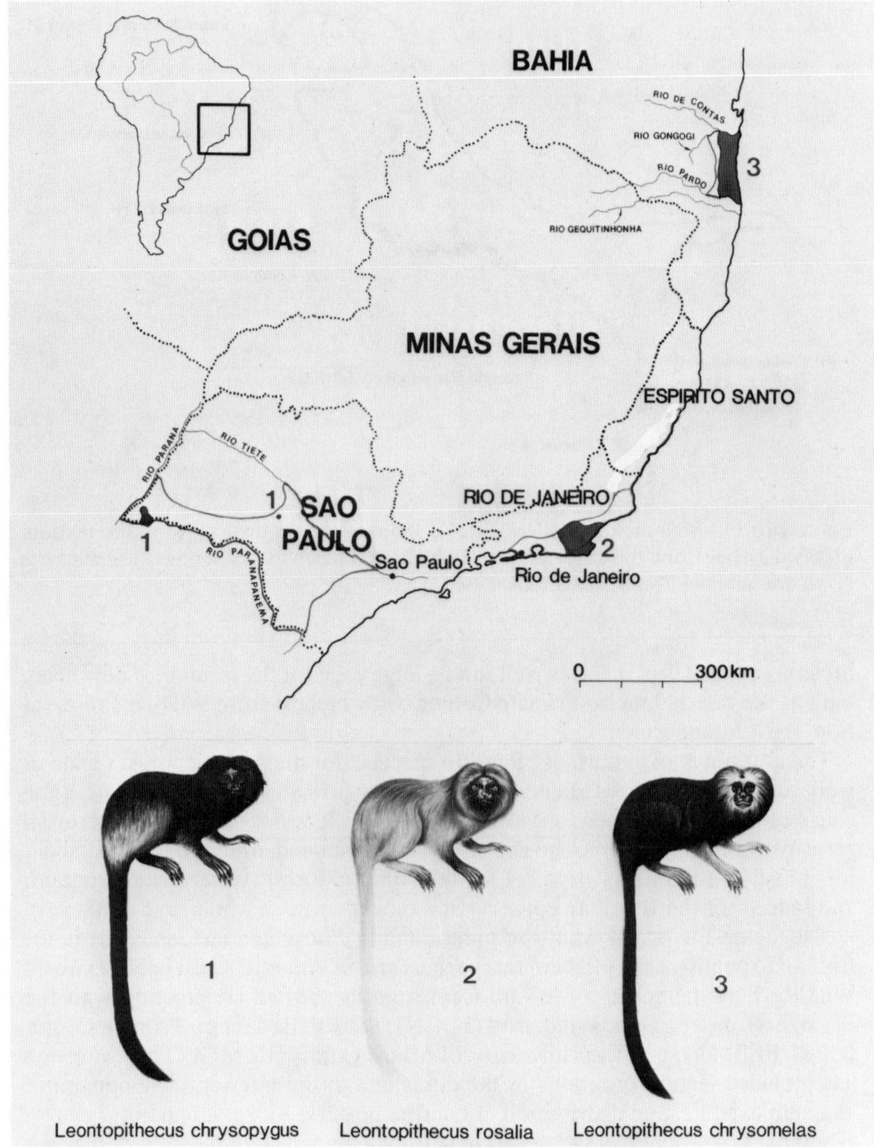

FIGURE 16.12. Past and present distribution of the three lion tamarins. The original distribution of each species is indicated by a continuous solid line; present distribution is shaded (drawing by S. Nash).

awareness in southeastern Brazil, and it is hoped that this heightened awareness will be instrumental in helping to save what remains of the Atlantic forest and its spectacular fauna.

The three lion tamarin species are also restricted to the Atlantic forest, and their original range is similar to, but less extensive than, that of the muriqui. The ranges of all three species were already quite small when the first Europeans arrived in Brazil, and they have now been reduced to tiny, isolated fragments by widespread forest destruction (Fig. 16.12).

The golden lion tamarin (*Leontopithecus rosalia*) is found only in the state of Rio de Janeiro, and its hope for survival in the wild now rests mainly with a single small (5000 ha) protected area, the Poço das Antas Biological Reserve. A major international program is now underway to improve the condition of this reserve and to ensure that the lion tamarin population is sufficiently large and well-protected to permit the survival of the species. The program is a joint effort involving the National Zoo in Washington, DC, the Rio de Janeiro Primate Center (CPRJ-FEEMA), the Brazilian Forestry Development Institute (IBDF), which manages the reserve, and World Wildlife Fund-U.S. It includes ecological studies, restoration of the habitat, and reintroduction of captive-born animals under the direction of James Dietz of the National Zoo. In addition, an outstanding education project is being carried out by Lou Ann Dietz, and it has succeeded in getting a great deal of press coverage and in creating great enthusiasm for golden lion tamarin conservation among the local people. As a result of these efforts over the past 3 years, prospects for the survival of wild populations of this spectacular and uniquely Brazilian primate are better than they have been at any time during the last 15 years—although a great deal of work remains to be done. Fortunately, the golden lion tamarin appears to be secure in captivity, with a large captive population thriving in U.S. and European zoos (see Kleiman et al., this volume).

The golden-headed lion tamarin (*Leontopithecus chrysomelas*) is found only in the southern part of the state of Bahia (Fig. 16.12), and though its range is also quite small, it still has more habitat left than either of its two congeners. Unfortunately, forest destruction has been especially heavy within its range over the past 5 years, and the single protected area that exists in this region, the 5000-ha Una Biological Reserve, remains totally inadequate in its present state. An education campaign is now planned for this region as well, and the University of Minas Gerais and the University of Paraiba will be undertaking a field program there that will use the Golden Lion Tamarin Project in Rio de Janeiro as a model. This work will be supported by World Wildlife Fund and the Brazilian Conservation Foundation (FBCN).

The area in which the golden-headed lion tamarin occurs is also home to several other endangered primates that occur nowhere else, among them the buff-headed tufted capuchin (*Cebus apella xanthosternos*), the northern masked titi (*Callicebus personatus melanochir*), and Wied's marmoset (*Callithrix kuhli*, Fig. 16.13), and it may still harbor a few groups of muriqui (*Brachyteles arachnoides*) and northern brown howler monkey (*Alouatta fusca fusca*). Action is needed there as soon as possible, since the remaining forest is being destroyed very rapidly, and an illegal trade is developing as well. During the past 2 years, at least 60 golden-headed lion tamarins have been smuggled into Beglium, France, Japan, and Hong Kong, and an unknown number (possibly exceeding that entering the international trade) has found its way into the illegal internal

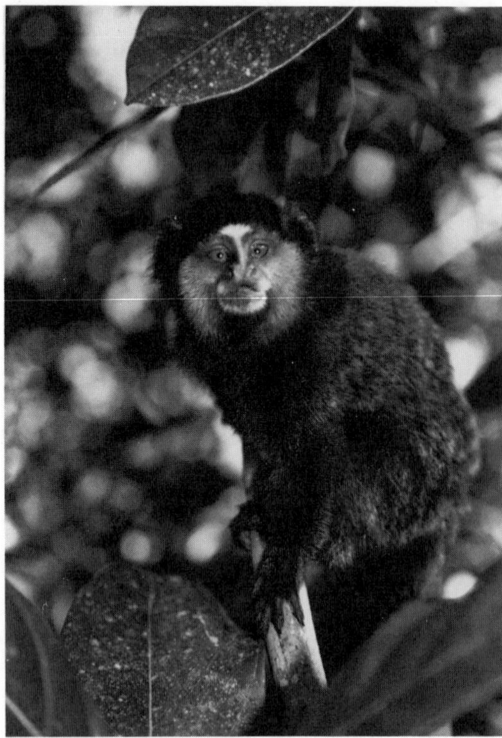

FIGURE 16.13. Wied's marmoset (*Callithrix kuhli*), another species found only in southern Bahia (photo by R. A. Mittermeier).

market within Brazil (Mallinson, 1984; Mittermeier, unpublished observations). Only one legal breeding colony exists at this time at the Rio de Janeiro Primate Center.

The third of the lion tamarins, the golden-rumped or black lion tamarin (*Leontopithecus chrysopygus*), is certainly the rarest and probably the most endangered of all New World monkeys. Always restricted to the interior of the state of São Paulo, Brazil's most developed state, it now occurs with certainty in only two widely separated reserves, the 37,157-ha Morro do Diabo State Reserve in the extreme southwestern part of the state and the 2178-ha Caitetus Reserve in central São Paulo (Fig. 16.12). It is unlikely that more than 100 individuals remain in the wild, and a project is now underway to determine exactly how many still exist and how they can best be conserved. The state of São Paulo has taken an interest in the future of this species, and a collaborative effort involving the State Forestry Institute (IF), the Energy Company of São Paulo (CESP), the Brazilian Conservation Foundation (FBCN), the Rio de Janeiro Primate Center (CPRJ-FEEMA), the University of Minas Gerais, and World Wildlife Fund–U.S. is now underway. Claudio de Padua of the Rio de Janeiro Primate Center is conducting survey work and ecological studies of this species, and his findings should provide the foundation for a management plan for both reserves. Again, there is only one captive breeding colony in the world, at the Rio de Janeiro Primate Center.

Amazonia

Amazonia is by far the richest area for primates in the Neotropics (Fig. 16.6) and certainly one of the richest in the world. At the generic level, the primates of greatest concern in Amazonia are *Ateles* and *Lagothrix*, which we consider the next most endangered New World primates at this taxonomic level. Neither of these are in nearly as critical a situation as *Brachyteles* and *Leontopithecus*, which are really in danger of becoming extinct as genera, but both *Ateles* and *Lagothrix* are disappearing rapidly or are already gone in much of Amazonia. They are heavily hunted as food almost everywhere, and *Lagothrix* is usually the preferred monkey food species and the most sought-after pet in the region. Both genera have slow reproductive rates as well and are thus unable to bounce back in the face of exploitation the way that some of the smaller monkeys such as the callitrichids can. Finally, they are more demanding in choice of habitat and less able to adapt to human-modified environments. The result is that these monkeys are usually the first to disappear when humans move into a previously undisturbed area, and they have already been exterminated from large areas of Amazonia where they once occurred. Market hunting has been an especially serious threat and has resulted in the disappearance of these animals even in areas of otherwise pristine primary forest habitat (Hernandez-Camacho and Cooper, 1976; Mittermeier and Coimbra-Filho, 1977; Mittermeier et al., in press).

World Wildlife Fund–U.S. has a special program focusing on *Ateles* both in captivity and in the wild, and it is discussed in greater detail by Konstant (this volume; Konstant et al., 1985).

Aside from the spider monkeys and woolly monkeys, the most endangered Amazonian monkeys are several subspecies with very restricted ranges in areas of heavy development. These include the southern bearded saki (*Chiropotes satanas satanas*), found in the lower Amazonian region south of the city of Belem, the white uakari (*Cacajao calvus calvus*), from a tiny area between the Rio Solimões and the Rio Japurá in upper Amazonia, the pied tamarin (*Saguinus bicolor bicolor*), known only from the general vicinity of the city of Manaus, and the white marmoset (*Callithrix argentata leucippe*), which occurs only between two small tributaries of the Rio Tapajos in an area cut by the Transamazonian Highway. All four of these animals are endemic to Brazilian Amazonia.

Endangered Primates from Other Regions

Although the majority of endangered primates in the Neotropics occur in the Atlantic forest and Amazonia, several are endemic to other regions. Three of the best examples are the yellow-tailed woolly monkey (*Lagothrix flavicauda*), the Central American squirrel monkey (*Saimiri oerstedi*), and the cotton-top tamarin (*Saguinus oedipus*, Fig. 16.14).

The yellow-tailed woolly monkey is the only truly Andean primate species and is restricted to a small area in the Andes of northern Peru. Rediscovered in 1974 after a lapse of almost 50 years (Mittermeier et al., 1975, 1977), it is threatened by habitat destruction and hunting in its once isolated cloud forest range. It is the most endangered Peruvian primate and, like the muriqui in Brazil, is also the largest mammal endemic to its country. World Wildlife Fund–U.S. is supporting

FIGURE 16.14. Sticker and t-shirt design depicting the cotton-top tamarin (*Saguinus oedipus*), to be used in a public awareness campaign in Colombia. The text reads: "White titi monkey in danger of extinction. Help us to save it. The forest is my home, don't destroy it." (Drawing by S. Nash, text by J. Ramirez Cerquera.)

several different projects on behalf of this species. An education and public awareness campaign is now underway in the region in which the species occurs. It is being carried out by Peruvian researcher, Mariella Leo Luna, and involves the same kind of campaign materials used in Brazil (e.g., stickers, posters, t-shirts, etc.), together with a film produced as a special World Wildlife Fund project. Leo Luna is also investigating two potential reserve sites in the region to determine appropriate boundaries for new protected areas, and another World Wildlife Fund project is gathering the data needed to develop a management plan for the Abiseo National Park, a 565,000-ha protected area established in 1983 (Leo Luna, 1982).

The Central American squirrel monkey (*Saimiri oerstedi*) is found only in western Panama and adjacent parts of southeastern Costa Rica. Although squirrel monkeys are usually among the most abundant of Neotropical primates in South America, this is certainly not the case with the isolated, remnant populations of the Central American representative. *Saimiri oerstedi* is highly endangered, and Hershkovitz (1984) has recently demonstrated that the species can be divided into two distinct subspecies, *S. o. oerstedi* and *S. o. citrinellus*. A recent World Wildlife Fund-supported ecological study and population survey of this monkey in Costa Rica revealed that there are no more than 3000 *S. o. oerstedi* in Costa Rica and less than 300 *S. o. citrinellus* (Boinski, 1986). Since the species may already be extinct in Panama, it is possible that these Costa Rican animals are all that remain. Fortunately, each subspecies is represented in one of Costa Rica's national parks, Corcovado for *S. o. oerstedi* and Manuel Antonio for *S. o. citrinellus*.

The cotton-top tamarin from northern Colombia is another highly endangered Neotropical monkey. It is found only in northwestern Colombia, an area that has been subjected to heavy forest destruction in the past two decades. The animal was also heavily persecuted for the primate trade in the 1960s and early 1970s, and it is estimated that some 20,000–30,000 were exported during the peak years of the trade (Hernandez-Camacho and Cooper, 1976). *Saguinus oedipus* is still well represented in captivity, with some 700–1000 being found in zoos and research labs in the United States and Europe (S. Tardiff, personal communication), and the development of a studbook and a species survial plan (SSP) for this species should be considered a high conservation priority for the zoo community.

Unfortunately, there is very little up-to-date information on the status of the cotton-top tamarin in the wild. However, it was already very depleted when the last survey took place in 1975 (Neyman, 1977). The situation is certain to have deteriorated since then, and it is likely that this species is in a position similar to that of the golden lion tamarin in Brazil. World Wildlife Fund has plans for a new survey in this region, which would have as its primary aims assessing the current status of the species and locating an appropriate protected area. A public awareness campaign is planned as well, with the goal of developing this attractive monkey as a symbol (Fig. 16.14) along the lines of the muriqui and the yellow-tailed woolly monkey.

The Action Plan for Neotropical Primate Conservation

Finally, it important to point out that all of the conservation efforts in which World Wildlife Fund and other conservation organizations are involved are not being carried out in piecemeal fashion, but rather as part of a broad international effort being coordinated by the IUCN/SSC Primate Specialist Group. The goal of this group is *to maintain the current diversity of the Order Primates*, with dual emphasis on:

1. ensuring the survival of particular endangered species wherever they happen to occur
2. protecting large numbers of primates in areas of high primate diversity and abundance.

To achieve this goal, the Primate Specialist Group is developing special Action Plans for Primate Conservation for each of the four major regions of the world in which primates occur: Asia, Africa, Madagascar, and the Neotropics. The Neotropical Action Plan is in the process of being completed, and it includes projects for all the species discussed in this paper, together with a number of training and education efforts aimed at enhancing the capacity of people from the tropical countries to carry out the conservation efforts needed from within. In Brazil, this has already been largely achieved, with 90% of the projects that World Wildlife Fund is supporting there being carried out by Brazilians, and it is hoped that a similar position will be reached in the rest of the Neotropical region within the next 10 years. If we can develop in the tropical countries both an awareness of the need for primate conservation and the local capacity to carry it out, we will be that much closer to ensuring that all of the primate species alive today will still be with us as we enter the next century.

References

Aguirre AC (1971) O mono *Brachyteles arachnoides*. Brazilian Academy of Sciences, Rio de Janeiro.

Boinski S (1986) Evaluation of the status of the squirrel monkey, *Saimiri oerstedi*, in Costa Rica. Prim Conserv 6:15–16.

Fonseca G da (1985) Observations on the ecology of the muriqui (*Brachyteles arachnoides* E. Geoffroy 1806): implications for its conservation. Prim Conserv 5:48–52.

Gentry A (1982) Neotropical floristic diversity: phytogeographical connections between Central and South America, Pleistocene climatic fluctuations, or an accident of the Andean orogeny? Ann Miss Bot Gard 69:557–593.

Hernandez-Camacho J, Cooper RW (1976) The nonhuman primates of Colombia. In: Thorington RW, Heltne PG (eds) Neotropical primates: field studies and conservation. National Academy of Sciences, Washington, DC.

Hershkovitz P (1984) Taxonomy of squirrel monkeys genus *Saimiri* (Cebidae, Platyrrhini): a preliminary report with description of a hitherto unnamed form. Amer J Primatol 7:155–210.

Konstant WR, Mittermeier RA, Nash SD (1985) Spider monkeys in captivity and in the wild. Prim Conserv 5:82–109.

Leo Luna M (1982) Conservation of the yellow-tailed woolly monkey, *Lagothrix flavicauda*, in Peru. Intern zoo Yrbk 22:47–52.

Mallinson JC (1984) Golden-headed lion tamarin contraband—a major conservation problem. IUCN/SSC Primate Specialist Group Newsletter 4:23–25.

Mittermeier RA (1977) Distribution, synecology and conservation of Surinam monkeys. Thesis, Harvard University, Cambridge, MA.

Mittermeier RA, Coimbra-Filho AF (1977) Primate conservation in Brazilian Amazonia. In: Prince Rainier, Bourne GH (eds) Primate conservation. Academic, New York, pp 117–166.

Mittermeier RA, Oates JF (1985) Primate diversity: the world's top countries. Prim Conserv 5:41–48.

Mittermeier RA, Macedo H de, Luscombe BA (1975) A woolly monkey rediscovered in Peru. Oryx 13:41–46.

Mittermeier RA, Macedo H de, Luscombe BA, Cassidy J (1977) Rediscovery and conservation of the Peruvian yellow-tailed woolly monkey (*Lagothrix flavicauda*. In: Prince Rainier, Bourne GH (eds) Primate conservation. Academic, New York, pp 95–115.

Mittermeier RA, Oates JF, Eudey AE, Thornback J (In press) Primate conservation. In: Mitchell G (ed) Comparative primate biology. Alan R Liss, New York.

Neyman PF (1977) Aspects of the ecology and social organization of free-ranging cotton-top tamarins (*Saguinus oedipus*) and the conservation status of the species. In: Kleiman DG (ed) The biology and conservation of the Callitrichidae. Smithsonian Institution, Washington, DC, pp 39–71.

Thornback J, Jenkins M (1982) The IUCN Mammal Red Data Book, Part 1: threatened mammalian taxa of the Americas and the Australasian zoogeographic region (excluding Cetacea). IUCN, Gland, Switzerland.

17 Ecological Background and Conservation Priorities for Woolly Spider Monkeys (*Brachyteles arachnoides*)

KATHARINE MILTON

Introduction

A number of excellent papers have discussed factors that must be considered to maintain and breed endangered species successfully in captivity. It is clear that what primates do easily in the wild becomes far more problematic under captive conditions and that the range of problems which must be resolved in each case is often both complex and unexpected. In this paper, I will consider the value of field observations—both in terms of providing essential information for the implementation of viable conservation programs in the wild and to illustrate how field studies can identify problems that could occur in trying to maintain and breed a particular endangered species in captivity. Clearly, the more information we have on the natural behavior of particular primate species, the less likely we are to make some careless mistake with respect to their conservation. Now is the proper time to collect this essential information base—while we are not yet under critical pressure with respect to the imminent extinction of most primate species. Then, should conditions alter in the natural environment, we will be able to act with reference to facts rather than our best guesses or instincts. Though I will be focusing my attention here on the endangered primate species *Brachyteles arachnoides*, I should stress at the onset that I view our major goal as the conservation of ecosystems—ecological and evolutionary processes—rather than the preservation of single species. I don't think we should lose sight of this major objective in our consideration of these special cases, as important as they are, since by conserving the ecosystem we also conserve endangered species.

Woolly Spider Monkeys

The highly endangered primate that I have been studying in the wild is the woolly spider monkey, *Brachyteles arachnoides* E. Geoffroy 1806. With an adult body weight of some 12–15 kg, woolly spider monkeys are the largest endemic mammals of Brazil and the largest naturally occurring primates in the New World. This genus is traditionally placed in the subfamily Atelinae, along with spider monkeys (*Ateles*) and woolly monkeys (*Lagothrix*) (but see also Rosenberger, 1981). Unlike these latter genera, which are polytypic and have a wide geographical distribution, *Brachyteles* is a rare, monotypic genus, entirely confined to one restricted area of southeastern Brazil. Its geographical range is largely encompassed by the states of Minas Gerais, Espirito Santo, Rio de Janeiro, and São Paulo. This is not the best locale for any monkey, let alone a large and somewhat specialized monkey such as *Brachyteles*, because most of this region of Brazil encompasses the area with by far the greatest human population density. This is the area of huge cities, including São Paulo, one of the largest cities in the world today, the area of big business and industry, the area of intensive farming acitivity, and the area of huge cattle ranches. Thus, from a precontact population estimated at some 300,000 or more animals, the present-day woolly spider monkey population may well consist of no more than 200–300 animals, living in remote montane regions or in small forest pockets, often on private land. We can view the great decline in numbers of this species as due almost entirely to habitat destruction. It is important, however, to stress the fact that the large size of woolly spider monkeys alone makes them unusually vulnerable. Further, the monotypic status of the genus and its restricted geographical range give us only this one area of Brazil to work in with respect to safe, viable reserves and parks for the perpetuation of the species in its natural environment.

Woolly spider monkeys are buff gold in color with black faces and extremities. In some other locales, the face or extremities may be pink or mottled. Animals have a fully prehensile tail and use the brachiating locomotor mode. Brazilians know this monkey by the common name *mono* or *monocarvoeiro*. Few, if any, Brazilians use the term *muriqui* in referring to woolly spider monkeys, and in scholarly papers in Portuguese, when the scientific name is not employed, the word *mono* is used (e.g., Aguirre, 1971; Valle et al., 1983). Woolly spider monkeys are notoriously difficult to maintain in captivity. At the present time, there is only one woolly spider monkey in captivity, an adult female in the São Paulo Zoological Garden. Certainly one major problem in keeping woolly spider monkeys under captive conditions has been the mistaken notion, until very recently, that animals were strongly frugivorous. In fact, woolly spider monkeys are very strongly folivorous and may in fact be the most folivorous New World primate. It would be difficult indeed to get sufficient natural food to feed even one of these large monkeys in captivity, which certainly argues caution in trying to confine or breed them until more is known both about the content of the wild diet and the digestive physiology of the species.

Ecology and Behavior: An Overview

My field study of woolly spider monkeys took place over an 11-month period at a private ranch, Fazenda Barreiro Rico, in São Paulo State. Barreiro Rico is located some 2 hours' drive from the city of São Paulo and is situated 480 m above sea level in a juncture between two rivers, the Piricicaba and Tiete (Fig. 17.1). This region receives an annual average of 1200 mm of rainfall. Most rain falls in a single 6-month period and the other 6 months of the year are notably drier (see Milton, 1984, for a full description of the habitat). In data collection, I was greatly aided by a Brazilian biology student, Carlos de Lucca, from the University of Rio Claro.

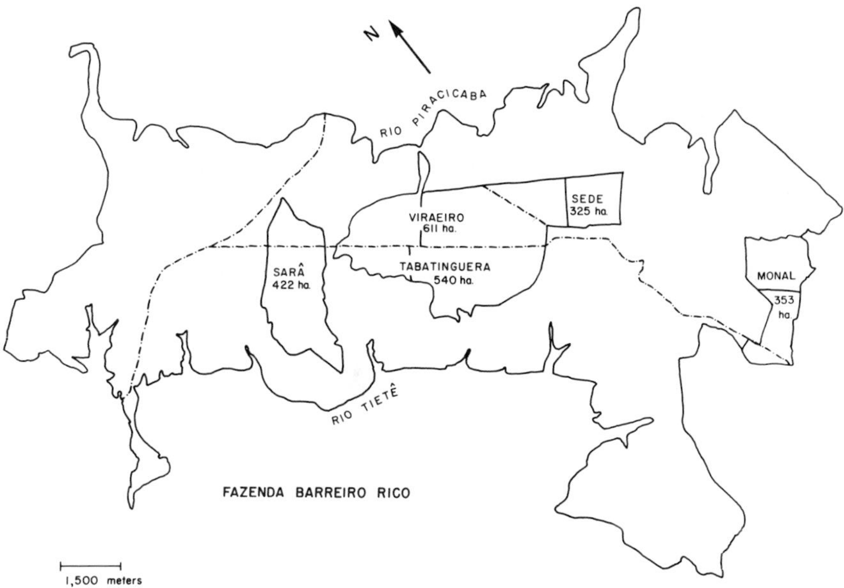

FIGURE 17.1. Map of Fazenda Barreiro Rico in São Paulo State, Brazil, showing locations and size in hectares of the five forested areas on this ranch. Hatched lines represent roads.

Barreiro Rico has over 3000 ha of forest, divided into five patches (Fig. 17.1). Surveys located woolly spider monkeys in three of these forests: Monal with an estimated 12–14 woolly spider monkeys, Veraeiro with some 23–28, and Sarâ, my main study site, with some 43–45 (Milton and de Lucca, 1984) (Fig. 17.2). The total woolly spider monkey population for this ranch thus numbers some 80 to perhaps well over 100 individuals, making it by far the largest single concentration of these rare monkeys discovered to date. All three subpopulations contain adult males, adult females, immature animals, and tiny infants; they are all viable and show no evidence of any inbreeding problems at present. As has been suggested (e.g., U. Seal, personal communication), it is important to get information on the approximate ages of these different forest patches and the amount of time each has been separated in order to better understand the genetic structure of the present-day woolly spider monkey population at Barreiro Rico.

In carrying out field work, I was particularly interested in obtaining quantitative data both on forest composition and on the phenological production patterns of tree species. The type of forest found at Barreiro Rico once covered huge areas of this region of Brazil. Today, however, almost all of this forest type has been totally destroyed with perhaps no more than some 2–4% of the original precontact forest cover still remaining (Veja, 1982). In essence then, this forest at Barreiro Rico is a living museum that preserves, at least in part, the basic pattern of this forest type, which has by now been almost entirely exterminated. In my 1984 paper, I present a complete list of all tree species encountered in a 3-ha sample of

FIGURE 17.2. Adult female woolly spider monkey. Animals spend greater than 50% of their daylight hours quietly resting, a behavior that should help to conserve energy for these monogastric leaf-eaters. (Courtesy of R. Mittermeier.)

this forest as well as phenological data on leaf, flower, and fruit production. Here, however, I wish to make two important points. One is that the forest in this area of Brazil is somewhat different than that in other areas of tropical forest with which I am familiar in that it contains relatively few tree species. There is an average of only 30 tree species per hectare at Barreiro Rico, and in a total of 3 ha only 46 species were encountered. Contrast this, for example, with 1 ha of tropical forest outside the city of Manaus in Amazonas, where you might encounter some 170 different tree species. Further, the Barreiro Rico forest is characterized by a high degree of species dominance. One species alone makes up more than 24% of the total larger stems (≥ 60 cm CBH) in the forest and seven species account for more than 65%. In my work I found that woolly spider monkeys did not eat leaves from most of the dominant tree species in the forest but rather concentrated leaf-eating on other rarer tree species and on vine leaves (Milton, 1984).

Dietary data were collected in 12 sample sets, spaced over the 11-month collection period. Each sample set lasted from 4 to 9 continuous days. Data show that at Barreiro Rico, woolly spider monkeys are extremely folivorous, spending some 41% to more than 93% of total feeding time eating only leaves. Animals also eat some fruits and flowers, and there are notable fluctuations between sample sets in time spent feeding on foods from these different dietary categories. Woolly spider monkeys routinely eat mature as well as young leaves. Indeed, mature leaves from a few rare tree species along with vine leaves are staple foods for the species and are relied on heavily when other more seasonal plant foods are in scant supply (Milton, 1984). Not only do animals have a specialized diet, they also appear to have a specialized digestive tract. Calm habituated woolly spider monkeys, both males and females, defecate copious quantities of fecal matter some 12 to 14 times per 12-hour observational day. I hypothesize that animals have a hindgut fermentation strategy predicated on the rapid fermentation of more degradable structural components in the diet. The more refractory lignified portion of the diet appears to be expelled from the gut in fairly regular pulses throughout the day (Milton, 1984).

Fecal collections from woolly spider monkeys in all three forests revealed little evidence of any internal helminths or protozoa. In a sample of 23 individuals, only two had any intestinal parasites—in both cases the same amoeba species, *Entamoeba hartmannii*, that in humans is considered nonpathogenic. It has occurred to me that perhaps natural chemical compounds in the foliage which the animals eat or perhaps the rather continuous nature of the defecation pattern might in part relate to this lack of internal parasites. *Alouatta fusca* at Barreiro Rico likewise have very few intestinal parasites (Milton, in preparation).

In terms of social organization, woolly spider monkeys were somewhat surprising in that they showed a fusion–fission pattern of social organization very similar to that described for two highly frugivorous primates, the spider monkey(*Ateles* ssp.) and the common chimpanzee (*Pan troglodytes*). Individual woolly spider monkeys come together and drift apart in no set pattern, and small groups of two to four individuals are the norm, though, on occasion, woolly spider monkeys come together in groups of 20 or more individuals. During the study, individually recognizable females remained in discrete areas of the forest while males appeared to be itinerant, traveling through the forest alone or in all-male bands of up to eight individuals. I believe that males travel extensively relative to females to maximize their contacts with potential sexual partners.

Mating in this species was also surprising and highly dramatic. Males can detect (apparently through olfactory cues) when a female is sexually receptive, and within a few hours as many as nine males may be congregated in a mating aggregation around a single sexually receptive female (Milton, 1984, 1985a, 1985b). These males follow the female continuously for a period of some 24–28 hours, attempting to copulate with her whenever possible. I have observed as many as 11 copulations with a single female in a 12-hour period by four or more different males (Milton, 1985b). Though woolly spider monkeys are strongly polygynous, they lack sexual dimorphism in body size and both males and females have canine teeth scarcely larger than incisors (Hill, 1962; Zingeser, 1973; Milton, 1985c). Males in a mating aggregation show remarkably little aggression toward one another even though it is clear that they are all intensely excited. Male woolly spider monkeys lack large canines, but they do have extremely large testes, certainly among the largest or the largest relative to body size of any extant anthropoid. This large testis size suggests that much intermale competition for reproductive success may occur at a postcopulatory rather than precopulatory level, perhaps by sperm competition (Milton, 1985b).

Woolly spider monkeys are interesting primates from any point of view—they are interesting in terms of their phylogeny, their geographical distribution, their morphology, their diet, their mating pattern, and their social organization. One intriguing fact to emerge from recent field work is that woolly spider monkeys at another study site, Montes Claros in Minas Gerais, do not show a fusion–fission pattern of social organization (Valle et al., 1983; Strier, in press). Rather, at Montes Claros, woolly spider monkeys are organized into cohesive and discrete troops (Valle et al., 1983; Strier, in press). Theoretical work predicts that the distribution patterns of key items of diet may exert a strong influence on particular patterns of primate social organization (Clutton-Brock and Harvey, 1977; Wrangham, 1980). With woolly spider monkeys we appear to have the opportunity to determine firsthand just how the patterning of key items of diet may affect the different patterns of social organization we've observed and perhaps also to tease apart the relative influences of phylogeny and the immediate environment.

Conservation Measures

Certainly, the most pressing and immediate concern is to ensure that all existing areas still known to contain woolly spider monkeys are rigorously monitored and protected. The World Wildlife Fund has located woolly spider monkeys now in some ten locales (W. Konstant, personal communication). Four of these sites, including Barreiro Rico, are on private land while the others are largely biological reserves or state parks. What is striking, indeed amazing, about these localities is the extremely low population estimates that often have been obtained. For example, the 35,000-ha Rio Doce State Park shows an estimated population of 20 woolly spider monkeys while the 4500-ha Nova Lombardia Reserve shows a total woolly spider monkey population of only about ±five animals. Similar estimates come from other parks—some 12 woolly spider monkeys in the Caparao National Park, some eight woolly spider monkeys in the Carlos Botelho State Park, and so on (W. Konstant, personal communication). Contrast these low estimates with my estimate of some 42–45 woolly spider monkeys in the 422-ha forest of Sarâ at Barreiro Rico—a forest that, as we have seen, does not provide a high degree of species diversity in terms of diet but that nonetheless can support a good-sized population of woolly spider monkeys as well as perhaps 100 *Alouatta fusca*, perhaps 80 or more *Callicebus personatus*, some 30–40 *Cebus apella*, and some three to six *Callithrix aurita*. Why are these estimates from other sites often so low? And what can be done to increase population size at these sites?

Here we have a somewhat novel situation in that we have the parks and reserves already declared in most cases, and yet we don't have the population of woolly spider monkeys we would predict and expect living in them. It would be useful to get an estimate of how much area in these parks and reserves is suitable habitat for woolly spider monkeys. This would permit realistic estimates for future population growth. If suitable habitat is extensive, efforts should be concentrated on determining why population size in these parks and reserves is so low. It is possible that hunting is taking place in these areas, even though this is in fact illegal. If hunting of woolly spider monkeys is occurring, immediate measures should be taken to end it. In this case, the best allocation of funds would be to set up a patrol system through existing agencies in these respective Brazilian states to protect parks and reserves such that animals are no longer shot. If this is indeed the principal factor responsible for these amazingly low population estimates, then once the hunting is stopped, the situation should improve dramatically. It may be that these low estimates are not valid. We do need to carry out further work to better ascertain the population size of woolly spider monkeys in areas where they are now known to exist as well as continue to survey any remaining forested areas in this geographical region for evidence of woolly spider monkeys. Further, and most important, in these state parks and biological reserves, where there is potential for a great increase in the size of woolly spider monkey populations, careful attention should be given to details of forest composition and the actual diet of woolly spider monkeys in these respective forests. In this way, we can better understand the interrelationship between the monkeys and their food sources and make certain that no immediate environmental problem is hindering population growth.

In my view, the first priority for woolly spider monkey conservation is to implement a program along the lines outlined above—that is, to establish a clear set of conservation objectives, an action plan, that revolves around the known populations of woolly spider monkeys now living in established parks and reserves and that seeks to increase total population size. At some point, if those populations now living on private land are threatened by habitat destruction, translocation or reintroduction may become necessary. The literature is replete with cautionary tales regarding the delicacy of such projects and the high failure rate (Kakuyo, 1980). Certainly we need to know more about the ecology of the existing populations, for translocation or reintroduction are more likely to succeed when there is a thorough knowledge of the behavioral and physiological adaptations animals possess to cope with their natural environment and an understanding of how these can best be accommodated in the new environment (Kakuyo, 1980).

Ultimately, it would be ideal to have this and other species common to this geographical region protected by a three-tiered conservation system that incorporates (1) the largest areas still remaining with natural vegetation as protected parks and reserves, (2) smaller protected areas wherever suitable forest cover still exists or can be extended through natural reforestation, and (3) rigorous protection of the highly viable populations currently found on private land, such as those at Barreiro Rico or Montes Claros. Long-term leasing might help to accomplish this latter objective. Fund-raising efforts and publicity campaigns are an essential part of any conservation effort, but it is important to realize that they must be accompanied by a solid program of conservation objectives. Such objectives should be formalized in consultation with the Brazilian officials involved in conservation programs for these respective areas.

At the present time there seems little need for any type of formal breeding program with respect to woolly spider monkey conservation—rather, funds and efforts should be directed toward increasing population size in the wild. In particular, no woolly spider monkeys should be removed from the wild for breeding experiments, and severe penalties should be enforced to make sure this does not occur. If woolly spider monkeys are seized in the illegal animal trade in Brazil, I suggest that they be placed back in the natural environment by the age of 2.5 to 3 years. By this age, they should be able to deal adequately with wild foods and should not require supplementary nursing (Milton, unpublished observation). This conference has provided ample documentation of the difficulties of maintaining many primate leaf-eaters under captive conditions and has further noted that immature rather than adult animals are far more likely to explore the habitat and try novel foods if released into the wild from captivity (e.g., Kleiman et al., this volume). Once the wild populations in these larger parks and reserves show a documented increase in numbers, it might then be feasible to plan and then to implement a captive breeding program. I regret that the cage constructed near

Rio de Janeiro to hold animals seized in the illegal animal trade is being referred to as a "breeding facility." Rather, I think the term "holding facility" or "maintenance facility" would be more appropriate.

Conclusion

I will conclude by reminding you first, that we ourselves are Primates and second, that our Primate order arose in a tropical forest environment. Even today, most primate species are found in tropical forests. In essence then, we owe our order to a particular set of challenges and opportunities posed to our ancestors by life in this environment. We should realize that the trees may underlie much of our own evolution in terms of many important primate traits. One particularly outstanding trait of the Primates order is the fact that all primates have relatively large brains in relation to body mass. We don't as yet totally understand the significance of this large brain. Perhaps it is in some manner related to diet and food location, perhaps it is related to a high degree of sociality, another striking characteristic of most higher primates. Whatever the factors involved, it seems clear that the large brain of primates gives them considerable behavioral plasticity in comparison with animals from many other orders and that higher primates apparently have to learn a great deal about their environment (both physical and social) as they mature. Given the fact that learning is such a key feature of primate life in the wild, I wonder if many species of higher primate can in fact be successfully introduced into the natural environment if bred in captivity.

For example, as I have discussed, the woolly spider monkeys at Barreiro Rico do not eat leaves from most of the more common tree species in their environment. How do they know to avoid these leaves and seek out leaves from rarer species? We simply do not know the answer. How do woolly spider monkeys know where these rare trees are? How do animals know when to visit these trees? There is a tremendous amount of learning of some type involved in this pattern of leaf-eating, regardless of whatever innate components are involved. Primates raised in captivity, no matter how benign and delightful their conditions, simply are not going to grow up learning this type of essential information. And what happens often when captive-bred animals are introduced into the wild or when animals actually born in the wild are simply translocated from one locale to another? We will consider the reintroduction of black-handed spider monkeys (*Ateles geoffroyi*) into the nature preserve of Barro Colorado Island in the Republic of Panama.

Barro Colorado Island has been a nature preserve since the early 1920s. There is no human interference in its ecology. Most of the island is now covered in tall tropical forest, some primary forest, and other sections, areas of older second growth (Milton, 1980). Black-handed spider monkeys at one time were found naturally in this area, but prior to the 1920s, they had been completely eliminated by hunters. Between 1959–1961 some 15–20 immature black-handed spider monkeys, captured in other areas, were purchased in the market in Panama City and released, with provisioning, onto Barro Colorado. It was planned that animals would gradually move into the forest and become established and that then provisioning could be halted. Of this number (15–20), all but five disappeared and certainly died. Thus, if this had been all of the black-handed spider monkeys still extant, we would have lost 75% of the remaining population in this one reintroduction experiment. Luckily, one male survived, along with four females, and these five animals did become established in the forest (Milton, 1979). But today (1985), we find that the total population of black-handed spider monkeys on Barro Colorado Island, some 25 years after reintroduction, is just about, but perhaps not quite, equal to the original number of spider monkeys that were released. Today there are some 15 black-handed spider monkeys living on Barro Colorado Island, and this number includes three of the original introducees. It takes a long time for such populations to build up in the wild once they have been reduced to extremely low numbers, even under the best of conditions.

Habitat preservation and protected large reserves make the best sense, including monetary sense, in terms of just about any aspect of primate conservation. And, in the case of primates, my position is that such reserves are far more critical and far more important than may be the case for many other animal species from other orders. Our efforts should be directed at ecosystem conservation through protected reserves and parks, and, if we are successful, the primate species in these ecosystems will doubtless be able to perpetuate themselves.

References

Aguirre AC (1971) O mono *Brachyteles arachnoides* (E. Geoffroy). Academia Brasileira de Ciências, Rio de Janeiro.

Clutton-Brock TH, Harvey PH (1977) Primate ecology and social organization. J Zool London 183:1–39.

Hill WCO (1962) Primates: comparative anatomy and taxonomy, vol. 4. Interscience, New York.

Kakuyo K (1980) The effects of translocation of the Rothschild's giraffes (*Giraffa camelopardalis rothschildi* Lydekker) from Lewa Downs Farm to Lake Nakuru National Park. Unpublished Master's thesis, University of Nairobi.

Milton K (1979) Factors influencing leaf choice by howler monkeys: a test of some hypotheses of herbivore leaf choice. Amer Nat 114:362–378.

Milton K (1980) The foraging strategy of howler monkeys: a study in primate economics. Columbia University Press, New York.

Milton K (1984) Habitat, diet and activity patterns of free-ranging woolly spider monkeys. Intern J Primatol 5:491–514.

Milton K (1985a) Urine washing behavior in the woolly spider monkey. Z Tierpsychol 67:154–160.

Milton K (1985b) Mating patterns of woolly spider monkeys: implications for female choice. Behav Ecol Sociobiol 17:53–59.

Milton K (1985c) Multimale mating and absence of canine tooth dimorphism in woolly spider monkeys (*Brachyteles arachnoides*). Amer J Phys Anthrop 68:519–523.

Milton K, de Lucca C (1984) Population estimate for *Brachyteles* at Fazenda Barreiro Rico. IUCN/SSC Newslet 4:27–28.

Rosenberger AL (1981) Systematics: the higher taxa. In: Coimbra-filho AF, Mittermeier RA (eds) Ecology and behavior of neotropical primates, vol. I. Academia Brasileira de Ciências, Rio de Janeiro.

Strier K (In press) Reprodução de *Brachyteles arachnoides* (Primates, Cebidae). In: A primatologia no Brasil.

Valle C, Alves MC, Santos IB, Pinto CAM, Mittermeier RA (1983) Observações sobre o comportamento do mono (*Brachyteles arachnoides*) na Fazenda Montes Claros, Caratinga, Minas Gerais, Brasil. I. População. II. Reação ao observador. III. Interação entre grupos. IV. Uso da agua. Resumos X Congresso de Zoologia. Imprensa Universitaria, Belo Horizonte. Universidade Federal de Minas Gerais, Belo Horizonte, Brazil.

Veja (1982) Vol 708, p 82.

Wrangham RW (1980) An ecological model of female-bonded primate groups. Behaviour 75:262–300.

Zingeser MR (1973) Dentition of *Brachyteles arachnoides* with reference to Alouattine and Atelinine affinities. Folia Primatol 20:351–390.

18
Successes and Failures of Captive Breeding

MARVIN L. JONES

For the past 35 years, I have been collecting data on the history of mammals in captivity, with particular emphasis on those species that have reproduced and in the determination of life spans. This information has been gleaned from the available literature in a number of languages including annual reports and inventories of zoological gardens, which may have limited local distribution. I also have personally examined the archives of scores of zoological collections around the world. Much of the archival material has never before been published and is used extensively in Tables 18.2 and 18.3 of this chapter.

The zoo professional has available the yearly publication of the Zoological Society of London, the International Zoo Yearbook, which presents articles on captive breeding and detailed lists of primates born in specific collections each year, as well as censuses of rare and endangered species. For the research scientist, there is the news bulletin of the Institute of Laboratory Animal Resources and the Laboratory Primate Newsletter; the latter have often published surveys of the numbers of primates born in captivity in responding collections but rarely offer specifics by institution.

Nevertheless, all three publications serve a useful purpose and are readily available. Thus, it is not my intention to repeat what they already have offered, but rather to take the historical approach and offer some personal comments on the many factors that affect captive breeding of primates.

There have been many symposia convened on the subject of the use of primates in biomedical research, with overall details of the problems of such breeding and the numbers of specimens born. Many of these have been published and are also readily available, so I also will not go into detail on these aspects of captive breeding.

Table 18.1 shows those species of nonhuman primates that have so far failed to bear young in captivity, to the best of my knowledge. These are the outright failures in captive breeding. Fortunately, the list is not very long, only 51 species, which grows shorter with each passing year. Several of the species have never been brought into captivity, such as *Allocebus trichotis* or *Cercopithecus dryas*. Some were collected in limited numbers or may have survived for only a brief time, and still others may currently be living in a number of widely scattered collections, with little probability of reproduction taking place. On the other hand, there are some that could have bred but were exhibited in the years gone by, and what records remain do not explain why breeding did not take place.

TABLE 18.1. Species of nonhuman primates that have not borne young in captivity.

Allocebus trichotis (Gunther, 1875)
Phaner furcifer (Blainville, 1841)
Hapalemur simus Gray, 1870
Lepilemur dorsalis Gray, 1870
Lepilemur edwardsi Forbes, 1894
Lepilemur leucopus Major, 1894
Lepilemur microdon Forbes, 1894
Lepilemur mustelinus I. Geoffroy, 1851
Lepilemur septentrionalis Rumpler and Albignac, 1975
Indri indri (Gmelin, 1788)
Lichanotus laniger (Gmelin, 1788)
Propithecus diadema Bennett, 1832
Daubentonia madagascariensis (Gmelin, 1788)
Nycticebus pygmaeus Bonhote, 1907
Galago elegantulus LeConte, 1857
Galago granti Thomas and Wroughton, 1907
Galago inustus Schwartz, 1930
Tarsius spectrum (Pallas, 1779)
Saguinus inustus Schwartz, 1951
Alouatta belzebul (Linnaeus, 1766)
Alouatta fusca (E. Geoffroy, 1812)
Alouatta pigra Lawrence, 1933
Cacajao melanocephalus (Humboldt, 1812)
Callicebus personatus (E. Geoffroy, 1812)
Callicebus torquatus (Hoffmannsegg, 1807)
Chiropotes albinasus (I. Geoffroy and Deville, 1848)
Chiropotes satanas (Hoffmannsegg, 1807)
Lagothrix flavicauda (Humboldt, 1812)
Brachyteles arachnoides (E. Geoffroy, 1806)
Pithecia albicans Gray, 1860
Pithecia hirsuta Spix, 1873
Cercocebus galeritus Peters, 1879
Cercopithecus dryas Schwartz, 1932
Cercopithecus erythrogaster Gray, 1866
Cercopithecus salongo Thys van den Audenzerde, 1977
Cercopithecus wolfi Meyer, 1891
Colobus badius (Kerr, 1792)
Colobus pennanti Waterhouse, 1838
Colobus pruessi (Matschie, 1900)
Colobus rufomitratus Peters, 1879
Colobus satanas Waterhouse, 1838
Colobus verus Van Beneden, 1838
Macaca thibetana (Milne-Edwards, 1870)
Nasalis concolor Miller, 1903
Presbytis frontata (Muller, 1838)
Presbytis hosei (Thomas, 1889)
Presbytis potenziani (Bonaparte, 1856)
Presbytis rubicunda (Muller, 1838)
Presbytis thomasi (Collett, 1863)
Pygathrix avunculus (Dollman, 1912)
Pygathrix brelichi (Thomas, 1903)

Table 18.2 offers the complete captive history of one such species, the almost legendary aye-aye, *Daubentonia madagascariensis*, of Madagascar. It is known, for instance, that in 1884 35 individuals were captured for exhibition in European zoos, but only one survived and arrived at the Menagerie of the Jardin des Plantes

TABLE 18.2. Captive specimens of *Daubentonia madagascariensis*.

Zoo	Sex code[1]	Received	Died
Natura Artis Magistra	0.0.1	22 Apr 1905	11 Nov 1912
Amsterdam, Netherlands	0.0.1	3 Jun 1914	15 Sep 1937
Zoological Gardens	0.0.2	1910	1910
Antwerp, Belgium	0.0.1	1910	1912
Zoological Gardens	0.0.1	10 Jan 1902	1903
Berlin, Germany	1.0	15 Nov 1907	1916
Zoological Gardens	0.0.1	1913	1913
Dublin, Ireland			
Zoological Gardens	0.0.1	1887	?
Hamburg, Germany	0.0.1	1909	?
Zoological Gardens	0.0.1	1911	?
Frankfurt, Germany			
Zoological Gardens	0.0.1	1930	?
Leipzig, Germany			
Zoological Gardens	0.0.1	12 Aug 1862	1864
London, England	0.0.1	28 Oct 1887	22 Sep 1896
	0.0.2	12 Sep 1908	?
	0.0.1	1911	?
	0.0.1	28 Sep 1929	26 Aug 1931
	0.0.1	28 Sep 1929	26 Sep 1932
Menagerie Jardin	0.0.1	3 Nov 1880	30 Apr 1882
des Plantes	0.0.1	3 Nov 1880	1 Sep 1883
Paris, France	0.0.1	28 Mar 1881	19 Jun 1882
	0.0.1	12 Sep 1883	20 May 1884
	0.0.1	12 Sep 1883	6 Dec 1888
	0.0.1	14 Apr 1884	15 Sep 1884
	0.1	28 Aug 1907	28 Oct 1907
	0.0.1	28 Aug 1907	16 Jun 1915
	0.0.1	30 Dec 1909	27 Jan 1910
	0.0.1	28 Dec 1921	4 Jun 1928
	1.0	20 Sep 1927	19 Nov 1930
	0.0.1	20 Sep 1927	4 Dec 1929
	0.0.1	20 Sep 1927	5 Mar 1932

[1]Sex code: 0.0.0, male/female/sex undetermined.

in Paris, where it lived for 5 months. Thirty other individuals are known to have been shown at a few other zoos in Europe; the species is never known to have arrived alive in North America. It will be noted that one specimen lived at the Artis zoo in Amsterdam from 1914 to 1937, and it is somewhat ironic that this zoo, famous for being the home of one of the last living quaggas (*Equus quagga*), also was the home of the last known aye-aye to live outside of Madagascar in captivity. A glance at the list will show that the Paris zoo often had more than one specimen at a time, but why reproduction never took place cannot be determined from the records available, for despite its great rarity and uniqueness, very little was recorded about its habits, the food it ate, or its behavior. One of the rarest of living primates may have been saved through captive reproduction if zoos then had had the same interest we have today in trying to preserve the rare species of the world.

Table 18.3 does offer some promise of hope, for here are 127 species of primates that have borne young in captivity. At the suggestion of Flesness, the species have been listed by the year of first known captive birth and the location of

TABLE 18.3. Species of nonhuman primates that have borne young in captivity, by first year of birth and location.

Species	Year	Place
Macaca fascicularis (Raffles, 1821)	1824	Paris JdP
Papio papio (Desmarest, 1820)	1824	Paris JdP
Macaca mulatta (Zimmerman, 1780)	1825	Paris JdP
Loris tardigradus (Linnaeus, 1758)	1832	London
Callithrix jacchus (Linnaeus, 1758)	1832	London
Cercopithecus aethiops (Linnaeus, 1758)	1835	London
Macaca radiata (E. Geoffroy, 1812)	1843	London
Cercocebus torquatus (Kerr, 1792)	1844	London
Lemur fulvus E. Geoffroy, 1796	1846	London
Cebus capucinus (Linnaeus, 1758)	1849	London
Macaca nemestrina (Linnaeus, 1766)	1854	Amsterdam
Galago senegalensis E. Geoffroy, 1796	1855	London
Lemur catta Linnaeus, 1758	1858	London
Macaca sylvanus (Linnaeus, 1758)	1859	Dresden
Macaca sinica (Linnaeus, 1771)	1861	Amsterdam
Lemur macaco Linnaeus, 1766	1865	Hamburg
Leontopithecus rosalia (Linnaeus, 1766)	1872	London
Mandrillus sphinx (Linnaeus, 1758)	1876	Dublin
Lemur mongoz Linnaeus, 1766	1883	Paris JdP
Macaca maura F. Schinz, 1825	1884	Philadelphia
Microcebus coquerali A Grandidier, 1867	1885	London
Cercopithecus mona (Schreber, 1774)	1885	Berlin
Galago demidovi (Fischer, 1806)	1891	Lilford-England
Lemur rubriventer I. Geoffroy, 1850	1901	Berlin
Presbytis entellus (Dufresne, 1797)	1892	Calcutta
Macaca fuscata (Blyth, 1875)	1894	Berlin
Cebus apella (Linnaeus, 1758)	1895	London
Cercopithecus talapoin (Schreber, 1774)	1896	Rotterdam
Nycticebus coucang (Boddaert, 1785)	1899	Rotterdam
Lemur coronatus (Gray, 1842)	1899	London
Cheirogaleus major (E. Geoffroy, 1812)	1901	Dublin
Theropithecus gelada (Ruppell, 1831)	1904	Paris JdP
Varecia variegata (Kerr, 1792)	1906	Berlin
Cercopithecus campbelli (Waterhouse, 1838)	1909	Amsterdam
Mandrillus leucophaeus (F. Cuvier, 1807)	1910	Berlin
Macaca silenus (Linnaeus, 1758)	1911	Rotterdam
Pan troglodytes (Gmelin, 1788)	1915	Abreu-Havana
Macaca nigra (Desmarest, 1822)	1928	San Diego
Pongo pygmaeus (Linnaeus, 1760)	1928	Berlin
Otolemur crassicaudatus (E. Geoffroy, 1812)	1929	London
Macaca cyclopsis (Swinhoe, 1863)	1930	Berlin
Hapalemur griseus (Link, 1795)	1931	Basel
Ateles belzebuth E. Geoffroy, 1806	1934	Vienna
Cercopithecus mitis Wolf, 1822	1934	San Diego
Macaca arctoides (I. Geoffroy, 1831)	1935	Berlin
Presbytis obscura (Reid, 1837)	1935	Frankfurt
Cebuella pygmaea (Spix, 1823)	1936	Frankfurt
Saguinus bicolor Spix, 1823)	1936	Frankfurt
Aotus trivirgatus (Humboldt, 1811)	1937	San Diego
Cercocebus albigena (Gray, 1850)	1938	San Diego
Hylobates muelleri Martin, 1841	1938	San Diego
Saimiri sciureus (Linnaeus, 1758)	1940	Kluver-Chicago
Hylobates lar (Linnaeus, 1771)	1940	Philadelphia
Presbytis cristata (Raffles, 1821)	1941	San Diego
Hylobates concolor (Harlan, 1826)	1941	Cleres

TABLE 18.3. Continued.

Species	Year	Place
Eythrocebus patas (Schreber, 1775)	1942	Chicago
Presbytis pileata (Blyth, 1843)	1942	San Diego
Cercopithecus diana (Linnaeus, 1758)	1943	Chicago
Cercopithecus neglectus Schlegel, 1876	1945	San Diego
Perodicticus potto (Muller, 1766)	1947	Washington
Cercopithecus petaurista (Schreber, 1774)	1947	San Diego
Saguinus oedipus (Linnaeus, 1758)	1948	Bronx
Ateles geoffroyi Kuhl, 1820	1948	Washington
Tarsius syrichta (Linnaeus, 1758)	1949	London
Pithecia pithecia (Linnaeus, 1766)	1950	San Diego
Saguinus midas (Linnaeus, 1758)	1951	London
Colobus angolensis Sclater, 1860	1952	Bronx
Colobus guereza Ruppell, 1835	1952	Bronx
Propithecus verreauxi A. Grandidier, 1867	1954	Tananarive
Saguinus imperator (Goeldi, 1907)	1954	Bronx
Lagothrix lagothricha (Humboldt, 1812)	1954	Baltimore
Cercopithecus cephus (Linnaeus, 1758)	1955	Washington
Gorilla gorilla (Savage and Wyman, 1847)	1956	Columbus
Microcebus murinus (J. F. Miller, 1777)	1957	Petter-Brunoy
Cercopithecus hamlyni Pocock, 1907	1958	Rotterdam
Cercopithecus l'hoesti Sclater, 1899	1958	Antwerp
Cebus albifrons (Humboldt, 1812)	1959	Houston
Cebus olivaceus (Linnaeus, 1758)	1959	Dallas
Allenopithecus nigroviridis (Pocock, 1907)	1959	San Diego
Cercocebus agilis Milne-Edwards, 1886	1959	Liberec
Macaca assamensis (M'Clelland, 1840)	1959	Berlin Tierpark
Presbytis senex (Erxleben, 1777)	1959	Colombo
Hylobates pileatus (Gray, 1861)	1959	Columbus
Galago alleni Waterhouse, 1838	1960	Chicago
Callithrix argentata (Linnaeus, 1766)	1960	Tel Aviv
Ateles paniscus (Linnaeus, 1758)	1960	Dallas
Presbytis francoisi (Pousargues, 1898)	1960	JMC-Inuyama
Ateles fusciceps Gray, 1866	1961	Dallas
Cercopithecus nictitans (Linnaeus, 1766)	1961	Baltimore
Arctocebus calabarensis (J. A. Smith, 1860)	1962	Amsterdam
Hylobates syndactylus (Raffles, 1821)	1962	Milwaukee
Pan paniscus Schwartz, 1929	1962	Frankfurt
Cacajao calvus (I. Geoffroy, 1847)	1963	MoJ-Miami
Presbytis phayrei Blyth, 1847	1963	Milwaukee
Pygathrix roxellana (Milne-Edwards, 1870)	1963	Shanghai
Galago zanzibaricus Matschie, 1893	1964	Wroclaw
Callimico goeldi (Thomas, 1904)	1964	Heinemann-Wiesbaden
Callicebus moloch (Hoffmannsegg, 1807)	1964	Barro Colorado
Cercopithecus pogonias Bennett, 1833	1964	Cincinnati
Presbytis johni (Fischer, 1829)	1964	Delhi
Cercopithecus ascanius (Audebert, 1799)	1964	Antwerp
Saguinus fuscicollis (Spix, 1823)	1965	San Diego
Saguinus mystax (Spix, 1823)	1965	Washington
Saguinus nigricollis (Spix, 1823)	1965	San Diego
Cercopithecus erythrotis Waterhouse, 1838	1965	Cincinnati
Nasalis larvatus (Wurmb, 1787)	1965	San Diego
Hylobates moloch (Audebert, 1798)	1966	Winnipeg
Alouatta seniculus (Linnaeus, 1766)	1967	MoJ-Miami
Colobus polykomos (Zimmermann, 1780)	1967	Jersey
Callithrix humeralifer (E. Geoffroy, 1812)	1968	Adelaide
Pithecia monachus (E. Geoffroy, 1812)	1968	Frankfurt

TABLE 18.3. Continued.

Species	Year	Place
Cheirogaleus medius E. Geoffroy, 1812	1969	Duke-Durham
Leiplemur ruficaudatus A. Grandidier, 1867	1969	Tananarive
Saguinus labiatus (E. Geoffroy, 1812)	1969	London
Saguinus leucopus (Gunther, 1877)	1969	Medellin
Pygathrix nemaeus (Linnaeus, 1771)	1969	Memphis
Hylobates hoolock (Harlan, 1834)	1969	Vancouver
Macaca ochreata (Ogilby, 1841)	1970	Bronx
Hylobates agilis F. Cuvier, 1821	1970	JMC-Inuyama
Alouatta palliata (Gray, 1849)	1971	Washington
Saimiri oerstedi Reinhardt, 1872	1972	Amsterdam
Presbytis geei Khajuria, 1956	1973	Gauhati
Presbytis melalophos (Raffles, 1821)	1974	Jakarta
Alouatta caraya (Humboldt, 1812)	1975	Columbia
Macaca tonkeana (Meyer, 1899)	1975	Rotterdam
Hylobates klossi (Miller, 1903)	1975	Basel
Presbytis aygula (Linnaeus, 1758)	1977	Jakarta
Tarsius bancanus Horsfield, 1821	1983	Washington
Microcebus rufus E. Geoffroy, 1834	1969	Petter-Brunoy

[1]Full names of cited institutions:
Private collection of Madame Abreu, Havana, Cuba (extinct). Zoological Gardens, Adelaide, Australia. Natura Artis Magistra, Zoological Gardens, Amsterdam, Netherlands. Societe d'Zoologie 'Anvers, Antwerp, Belgium. The Baltimore Zoo, Baltimore, Maryland, USA. Smithsonian Institution Research Station, Barro Colorado, Panama. Zoological Gardens, Basel, Switzerland. Zoological Gardens, Berlin, Federal Republic of Germany. Tierpark Berlin, Berlin, German Democratic Republic. New York Zoological Park, Bronx, New York, USA. Zoological Gardens, Calcutta, India. Lincoln Park Zoological Gardens, Chicago, Illinois, USA. Zoological Gardens, Cincinnati, Ohio, USA. Zoological Park, Cleres, France. Riverbanks Zoo, Columbia, South Carolina, USA. Zoological Gardens, Colombo, Sri Lanka. Columbus Zoo, Powell, Ohio, USA. Dallas Zoo, Dallas, Texas, USA.Zoological Gardens, Delhi, India. Zoological Gardens, Dresden, German Democratic Republic. Phoenix Park Zoo, Dublin, Ireland. Duke University Primate Center, Durham, North Carolina, USA. Zoological Gardens, Frankfurt, Federal Republic of Germany. Assam State Zoo, Gauhati, India. Zoological Gardens, Hamburg, Germany (extinct). Private collection of Herr Heinemann, Wiesbadan, Federal Republic of Germany. Zoological Gardens, Houston, Texas, USA. Zoological Gardens, Jakarta, Indonesia. Jersey Wildlife Preservation Trust, Jersey, Channel Islands. Japan Monkey Center, Inuyama, Japan. Research collection of Dr. Kluver, University of Chicago, Chicago, Illinois, USA. Zoological Gardens, Liberec, Czechoslovakia. Private collection of Lord Lilford, England (extinct). Zoological Gardens, London, England. Santa Fe Zoological Park, Medellin, Colombia. Zoological Garden, Memphis, Tennessee, USA. Zoological Park, Milwaukee, Wisconsin, USA. Monkey Jungle, Miami, Florida, USA. Jardin des Plantes, Menagerie (Museum National d'Histoire Naturelle), Paris, France. Primate colony of Prof. Petter, Brunoy, France. Zoological Gardens, Philadelphia, Pennsylvania, USA. Zoological Gardens, Rotterdam, Netherlands. Zoological Gardens, San Diego, California, USA. Zoological Garden, Shanghai, People's Republic of China. Botanical and Zoological Park Tsimbazaza, Tananarive, Madagascar. Zoological Garden, Tel Aviv, Israel (extinct). Stanley Park Zoo, Vancouver, British Columbia, Canada. Schonabrunn Zoological Gardens, Vienna, Austria. National Zoological Park, Washington, District of Columbia, USA. Assiniboine Park Zoo, Winnipeg, Manitoba, Canada. Zoological Garden, Wroclaw, Poland.

the collection where this took place. Most were the result of captive conception, but this cannot be guaranteed for all. Zoos are cited more than private collections because records from this sector are more available. Many have come from the already mentioned zoo archives, especially for the Berlin Zoological Gardens in Germany, where good records exist going back to about 1845, and for the Menagerie of the Jardin des Plantes in Paris, France. Records for this zoo from at least 1808 are available, and I had the pleasure of examining them for several weeks in 1982.

Examination of the list may offer some surprises because unfortunately there is little material available in the English language on the history of animals in captivity. Allow me now to discuss some of the factors that affect what species have bred in the past and which are being bred today.

In the early years of captive exhibition, from about 1752 to 1875, when zoos began to appear on the scene in the United States, specimens were the end result of exploration of the unknown parts of the world, frequently single individuals brought home by returning diplomats, soldiers of fortune, and ship captains.

These were chiefly the hardy species that could withstand the rigors of a long sea voyage and rudimentary care. Macaques, baboons, some lemuroids, and a handful of guenons and tropical American primates survived long enough to be donated to the major collections of the day, where they were eagerly examined by zoologists and frequently given scientific names on the spot. On infrequent occasions some breeding took place, but it was rarely sustained breeding, and both the parents and young invariably survived for very short periods of time, a few weeks or months, sometimes a year or more.

As the numbers of fixed zoological collections began to increase, there appeared on the scene a new entrepreneur, the animal dealer, a man much maligned today, but who served a useful purpose at the time, and still does, although today he is more or less a broker, dealing primarily in captive-bred specimens. These dealers either would have agents who plied the major ports of the world buying whatever was available, or sent out their own collecting expeditions. Hagenbeck is the best known. Warren Buck, Wilfrid Frost, William Cross, the Jamrach family, Otto Fockelmann, Ellis Joseph, and Reiche were major figures in the history of primate breeding, as was John Daniel Hamlyn, for whom Pocock in 1907 named a new guenon species from the Congo that Hamlyn had sold to Walter Rothschild and that in turn was shown at the London zoo and wound up at the British Museum.

With the establishment of zoological gardens in the United States and in those colonial lands where the British were dominant, as well as a growing number in Europe, captive breeding of primates began to accelerate. The number of species bred for the first time began to rise as did longevity of primates in general, human as well as nonhuman. Diseases that affected the lives of humans frequently also decimated the nonhuman primate as well. With the advent of the animal dealer, specimens were better cared for en route to their new homes—after all, this represented an investment of money and time. Once tuberculosis was conquered and zoos began to exhibit primates behind glass, in the late 1920s, and specimens in the temperate climates were given fresh air, the overall picture improved. Gorillas that Hornaday once felt would never live to any length of time in captivity now grew to adulthood but did not breed. The Dutch exporter van Geuns was able to export family groups of orangutans, which did start to breed in 1928. Madame Abreu in Havana raised the first chimpanzee in 1915 and was soon followed by several major zoos. Private collections also were being established by Robert Yerkes, and the Sukhumi station came into being in the U.S.S.R. By 1946, most of the problems attendant on the keeping of primates were being overcome, and, as Table 18.3 shows, primate breeding took a major spurt forward. The advent of the jet plane reduced transit time, and more enlightened dealers appeared on the scene.

Research into a number of human diseases and the space program brought about large-scale importations of primates for the laboratory, thousands of individuals, many more than all of the zoos of the world had ever shown.

In 1966, the International Union for the Conservation of Nature and Natural Resources, IUCN, published what it called the *Red Data Book* for mammals, and a new term came into use, *endangered species*. This was widely accepted by the media as well as the zoological community and is the most common term used today, often incorrectly, when we wish to attract attention to captive-breeding programs. Three years later, the United States passed the Endangered Species Act, which has been amended several times since, a major regulation that brought considerable change to the way zoos and research collections acquired primates. A few years later under the prodding of the United States, the Convention in Trade of Endangered Species (CITES) came into being, which even further restricted the movement of primates, both wild born and captive born. All of these regulations and publications resulted in a shift of what species would be bred and kept. Capuchins were the first to disappear from zoo inventories, followed by the more common macaques and baboons. Only the lion-tailed macaque, barbary ape, the Celebesian macaques, and gelada remained as part of zoo collections, although in Europe the hamadryas continues to populate monkey islands, and there has been a recent trend both in Europe and the United States to import the Japanese macaque for exhibition purposes.

They were replaced by the rare species of tropical America and Asia, such as uakari, howlers, saki, titi, and langurs. Fortunately, many of these bred, some only in recent years, with sustained breeding in a few collections such as Douc langur at Cologne, Basel, San Diego, and Stuttgart, but overall there was limited success, and for many species only a few individuals remain, in a few collections.

Today the major factors affecting the breeding of primates in captivity are more limited in scope. For one, there is the subject of money, financing for all of the many projects that all of us are vitally concerned with; where will the money come from, and how will it be distributed? Who will be responsible, the human factor. Who will volunteer to spend long and arduous hours in the field to make those observations so necessary if we are to succeed with those 51 species that have yet to be bred in captivity? Who will apply for the grants to study them in captivity or conduct further research be it behavioral or biomedical, and where will this be done and in what setting? Which species will be favored with inclusion into one of the many Species Survival Programs of the American Association of Zoological Parks and Aquariums, a bold and imaginative plan for the future, or one of the many consortia being developed by the more progressive zoos of Europe and Australia. These, too, are important determinants of what species will be bred and how and where.

The already mentioned regulations such as CITES and the U.S. Endangered Species Act brought into prominence another factor in captive breeding of primates, which may at times seem to be an irritant to many of us here today, but which must be given careful consideration and thought, the rise of what CITES calls the nongovernmental organizations. There are many. Some are violently against any and all use of primate species in biomedical research and, as we have seen have adopted violent tactics to, as they say, liberate animals from laboratories with which they do not agree, both in Europe and in the United States. Others are opposed to the removal of any primates from the wild for any purpose whatsoever, be it captive breeding in the zoo setting or for research. Many have gained the ear of the mass media, and all must be considered, for the public perception of primates has changed considerably in recent years, especially in Europe and the United States. They are no longer looked upon as freaks in a sideshow but as fellow creatures inhabiting a shrinking world that deserve to be kept in better surroundings. I will not attempt to judge the merits of biomedical research, but only wish to point out that there is another major factor which should not be totally discounted as the ravings of a small vocal minority.

Now it is time to turn to a more happy series of events, the current state of captive breeding. Once CITES and the other regulations came into being and those of us in the zoo and research community began to realize that we would not be able to continue to replace lost specimens by vast infusions of new replacements, captive breeding took on a new importance. As Table 18.3 shows, from 1946 to 1969 almost 60 species had bred in captivity for the first time, which means that breeding had already commenced on a large scale, and with considerable success, but now our major efforts focused on expanding captive colonies of as many species as possible. The common species that zoos no longer maintained often moved into the research sector; this is where all of the many macaque and baboon species and some of the mangabeys are currently living and breeding by the thousands, as well as squirrel monkeys and many marmosets. Duke University established a colony of leumoroids that has outpaced many zoos, often today living in large acre-sized compounds. All of the large anthropoid apes became commonplace exhibits at most zoos, and Yerkes expanded its colony from just chimpanzees to also include the gorilla and orangutan, with which it has had considerable success. One of the two oldest living chimpanzees, Bula, born at the Abreu colony in 1930, still is living at Yerkes center. The other is a male, Jimmy, brought to Rochester Zoo, New York, in 1931. Many new longevity records were and are being established in all sectors, zoo and research.

Later at this conference we will hear about one of the most highly successful captive programs of all, that for the golden lion tamarin, *Leontopithecus rosalia*, so ably conducted by Devra Kleiman and her enthusiastic co-workers at the National Zoological Park, at both the Washington, DC, facility and the new research center at Front Royal, Virginia. Scores of zoos around the world now belong to this program, and hundreds of individuals have been born and reared, and as we will hear have even been reintroduced back into the native habitat where in this year they, too, have reproduced. Those who may question the feasibility of saving many other very endangered forms of primate life, such as mountain gorilla, need only to study this program to see that it can be done, if there is complete cooperation between all sectors of the zoological community, field, zoo, and research.

The first gorilla came into captivity in the 1850s, and it was not until a hundred years later that the first was born in captivity. By the time that Ronald Nadler and I published our listing of captive births in 1974, over 100 had been born, and today that number closely approaches 300. Yes, there are problems at some sites, but the species is still breeding in many locations around the globe. In 1984, breeding took place at a center in Gabon, ably assisted by the return to Africa of a male from the collection of John Aspinall in England, a major gorilla breeding center, where the animals live in family groups.

The first orangutan was born in 1928 at Berlin, and while breeding at first took place only at those zoos that had imported fully adult animals, today the species is represented in over 160 collections world-wide and the number born in captivity now approaches the 900 mark. Rotterdam bred the first full second generation orangutan in 1963, and there have been many more since, although in an attempt to maximize founder representation we continue to breed from the surviving wild-bred specimens rather than just add to the numbers of full second generation born animals. There no longer is a market for wild-born orangutans, and there has not been for over 10 years.

The Zoological Society of San Diego has concentrated its efforts in recent years on a number of species, and one group in particular, lemuroids. We acquired one male and two female ring-tailed lemurs, *Lemur catta*, in 1962, which were probably wild born, and a second male from Madagascar in 1969. From 1964 to 1984 more than 95 have been reared and distributed among some 28 other zoological collections, where many have in turn reproduced. This is now the most common lemur in capivity, bred in scores of facilities around the world.

Our greatest success at San Diego has been with the ruffed lemurs. We legally imported from Madagascar two pairs of black and white ruffed lemur, *Varecia variegata*, in 1965 and 1967 and have procured some outcross males from other zoos and collections in 1978, 1979, and 1982. From 1970 to 1984, 87 specimens have been born and reared and many distributed out to some 27 other collections. In 1966 and 1970 we secured two pairs of the more endangered red phase of this species, *Varecia variegata rubra*, which we believe is a full subspecies and not just a color variant. Breeding commenced in 1973, and to date 70 have been born and reared and, much like the black and white form, have been sent to some 19 collections. The majority of these have been placed out on breeding loan, and some of both have in turn already reproduced at their new homes. I should add that San Diego was the first collection to breed these in North America. Miss Diane Brockman, Curator of Mammals at the San Diego Zoo, maintains the studbook for ruffed lemurs that was published this year. I keep the studbook for orangutans, and we published our first edition of this revised book in 1980 and will have a second edition ready in 1986. Other staff members of San Diego Zoo also maintain studbooks, Arabian oryx by Dr. James Dolan and slender-horned gazelles by Carmi Penny. This is only part of our commitment to the preservation of endangered species.

In recent years, zoos and research centers alike have begun to give attention to a relatively new area of interest in the role of captive breeding, genetics, but as this conference will so amply demonstrate, we are willing to learn and listen and profit from the mistakes we have made in the past.

With the help of the aforementioned breeding programs of zoo organizations in the United States and Europe, the ever-increasing use of studbooks, and the inclusion of individual specimens no matter where they may be living into the International Species Inventory System, ISIS, there is no reason not to believe that the future of primate breeding in captivity is a bright one.

I would recommend that we endeavor in the future to be as open and communicative with one another as possible; there must be more and better methods of exchanging ideas, problems and solutions, and even animals between the zoo and research community, and we must both listen to the wise counsel of the field zoologist. Working together as a team, we can assure the future for all primate species.

References

Crandall L (1964) Management of wild mammals in captivity. University of Chicago Press, Chicago.

Honacki J, Kinman K, Koeppl J (1982) Mammal species of the world. Allen Press, Lawrence, KS.

19
History of Geoffroy's Tamarins, *Saguinus geoffroyi*, at Lincoln Park Zoological Gardens, 1974–1985

DENNIS A. MERITT, JR.

The early to mid-seventies were an unsettled time in the Republic of Panama. National parks were not yet fully established, and those areas designated as reserves or protected areas provided little more than paper protection. RENARE, Direccion Nacional de Recursos Naturales Renovables, the Panamanian governmental agency responsible for wildlife and natural resources, was badly understaffed and poorly funded. The need for conservation efforts was well recognized, the mechanism to start it was established, but the resources to carry it out were essentially nonexistent. Animal dealers were legally operating in the Republic, some with good credentials and excellent facilities, others with no objective other than to make the most money in the shortest possible time. This latter group showed no regard for the animals in their care; the animals were treated as a commodity to be exploited whenever and however possible. The sale of birds and small mammals was open business in the public market. Roadside vendors of small parrots and young mammals, as well as iguanas, were not an uncommon site. There was a well-established system of animal collectors, most operating with their base in Panama and others operating both in Colombia and Panama. Animals from Colombia not uncommonly found their way into Panama and from there into the United States, western Europe, and Japan. The export of mammal hides and large quantities of crocodilian skins was well known and clearly visible, even to an untrained biologist. Wildlife was being exploited at every turn and usually not by Panamanian nationals. The handwriting was on the wall, for if this exploitation continued, Panama would lose forever some of its most valuable living resources. This situation in Panama was not unlike that in other Latin American countries at the same time in history, but in Panama it was not as well organized, and it was clearly reversible.

It was during an extended field study in the Republic of Panama in 1974 that I saw my first Geoffroy's tamarin, *Saguinus geoffroyi*, in the wild. At the same time, living tamarins were being offered for sale in the *mercado publico*, as pets to unsuspecting shoppers. This was during an era in Panama when animal sales in the market were open, commonplace, unregulated, and legal. Young of the previous year were offered for sale during the late dry season, yearlings and juveniles at other times of the year. Adults were not usually sold because they adjusted poorly to the captive conditions under which they were usually held. Even casual shoppers for live animals realized that there was little hope in trying to tame adults and therefore did not buy them.

Geoffroy's tamarin, *Saguinus geoffroyi*, is the northernmost form of tamarin and is found in Panama but not Costa Rica. This species may range into Colombia but clearly not farther than areas adjacent to Panama where the tamarin is common. In Panama, it is called "mono titi" and "bichichi" by the Choco Indians. Moynihan (1970), Mendez (1970), and more recently Hershkovitz (1977) have described the animal in detail, and it is sufficient to say that this species is not easily confused with any other marmoset or tamarin. The closest similar form is the cotton-top tamarin, *Saguinus oedipus*, of Colombia, which superficially may resemble Geoffroy's tamarin to the untrained eye. There is no sexual dimorphism in the latter species, although some older adult females are noticeably larger than most males. (For the purposes of this paper and to avoid confusion, I will continue to use the common name Geoffroy's tamarin and the taxonomic designation, *Saguinus geoffroyi*. I have no intention of entering into the taxonomic discussion surrounding these primates.)

This tamarin is diurnal and arboreal and in nature is shy and retiring even when approached silently and cautiously. It is usually found in family units or in small groups. Preferred habitat is secondary to scrubby forest, rarely primary forest. The animals show a preference for vine tangles and dense but not high tree forms. They easily clear forest gaps, tree falls, and dirt roadways by jumping across the openings from tree to tree. I have seen family units, i.e., adult male and female, twin offspring nearly grown and newborn young, in close association with human settlement and light agricultural activity. As long as they are not hunted or in any other way molested, they appear to be able to survive and thrive in and around areas of urbanization, when provided with suitable habitat. Food in nature consists of vegetation including buds and fruits, insects, and available suitable vertebrates such as small lizards and geckos. Other and larger vertebrate forms may be eaten when the opportunity presents itself.

Since most of the tamarins in the *mercado publico* were being purchased by U.S. military personnel stationed in Panama, their chances for survival and a productive life were limited. Through the cooperation of RENARE, the Panamanian governmental agency responsible for wildlife protection and exportation, I was given permission to obtain and export tamarins. This was prior to the international protection ultimately afforded Geoffroy's tamarins by CITES (the Convention on International Trade in Endangered Species) when Panama became a signatory nation. This was also prior to its taxonomic reassignment by the U.S. Department of the Interior so that it became listed in federal regulations as a subspecies of cotton-top tamarin, *Saguinus oedipus*, and thus was officially declared an endangered species by the United States. Since these particular animals had already been removed from the wild, their chance for survival as pets was limited, and since they could not easily be returned to the wild, their best hope for the future appeared to be in a captive propagation program at the Zoological Gardens. With this in mind, during annual trips to the Republic of Panama, I obtained additional tamarins that were paired with animals already at the Zoological Gardens. This represented the start of the first major and coordinated captive propagation program in the United States for this species. Animals were exported from Panama in 1975, 1976, 1977, 1979, and 1981, totaling 14 individuals of mixed ages. The sex ratio of this combined group was five males and nine females. Knowing that Geoffroy's tamarins are pair-bonded and that multiple reproducing female groups with single males were unlikely, an attempt was made to obtain sufficient males and females to initiate several potential breeding pairs. These were housed as single pairs after a suitable introduction period and our own in-house, mandatory quarantine period. During this time they were acclimated to captive conditions, diets, screened for and treated for parasites if present, and tested for tuberculosis. Parasite loads were minimal and easily treated, and there were no positive TB reactions.

An additional stimulus to obtain and initiate a captive breeding colony of these tamarins was the then heavy pressure they were under as a species in their country of origin. Continuing land reform, while playing a small role in their long-term survival in nature, was really insignificant overall compared to other human factors. Geoffroy's tamarin had for years been used by researchers in the Republic of Panama in studies of leishmaniasis, trypanosomiasis, and other tropical parasitic diseases. Their use was restricted to occasional animals and to blood or skin sampling in these studies. At the completion of these necessary and important screening studies, most of the animals were intact and living, essentially none the worse for their contribution to human medicine. Two major events in the early to mid-seventies directly and adversely affected wild populations of these tamarins. In the first case, in the course of field work undertaken by a graduate student from the United States, substantial numbers of Geoffroy's tamarins were wild caught and sacrificed to undertake standard measurements, stomach content analysis, and to determine their reproductive state (Dawson, 1976, 1977). These animals, as well as additional specimens, eventually were deposited in museum collections so that they were not entirely wasted; more properly in this case, they were put to the best use considering the circumstances.

The second event that affected wild tamarin populations negatively, but perhaps not as dramatically, was their use in biomedical studies in the United States and Europe. This was at a time when the search was on for a suitable primate model for studies of colon cancer, leukemia, and other forms of cancer. Geoffroy's tamarin was desirable for a number of reasons and was exported to biomedical laboratories to be evaluated for use in these studies. Fortunately for the species, other models were developed and soon available and Geoffroy's tamarin fell out of fashion rather quickly. Even more fortunately, some of these tamarins found their way from the pharmaceutical research laboratories into the hands of a small number of concerned facilities once they were declared nonusable and surplus to biomedical research needs. Many of these thrived and reproduced in their new surroundings. Small groups were scattered in a few locations in North America but not a significant number of animals overall. Nevertheless, there was some hope for the long-term survival of the captive population.

Births in the captive propagation program, the result of captive breeding, began in 1974 and continue to the present (Fig. 19.1). A total of 111 offspring have been produced thus far. There have been 17 single births, 38 twin births, and 6 sets of triplets (Fig. 19.2). Singletons account for 28% of all births, twins 62%, and triplets 10% (Fig. 19.3). Births have occurred in every month of the year (Fig. 19.4). The least productive months have been May and November. The most productive months have been January, March, April, July, and December. The remaining months are intermediate. If we turn from birth events, whether single or multiple, and look at individuals born per month, we find that the bulk

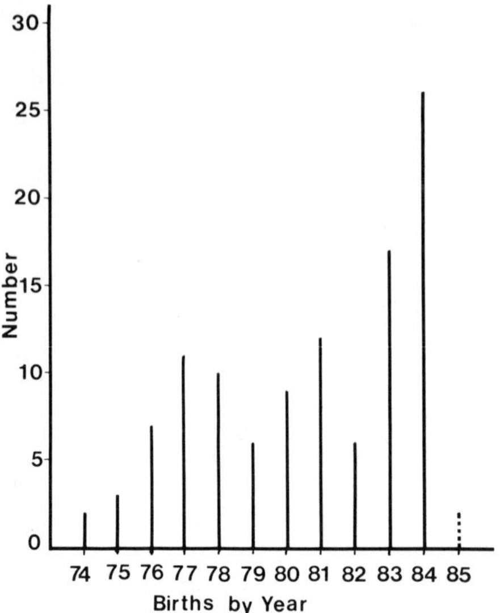

FIGURE 19.1. Births by year in the Geoffroy's tamarin breeding program from 1974 to the present.

TYPE OF BIRTH:

single	twin	triplet
17	38	6

(total birth events : 61)

FIGURE 19.2. Distribution of single, twin, and triplet births in Geoffroy's tamarin breeding program, from 1974 to the present.

INDIVIDUALS BORN:

single	twin	triplet
17	76	18

(total births : 111)

FIGURE 19.3. Number of individuals produced by each type of birth event.

of individuals have been born in the months of January through April (Fig. 19.5). These 4 months account for 62 out of 111 births, or 56%. The least productive month for individuals born is November.

Captive-born tamarins have produced young themselves as well. A female born 23 March 1978 gave birth to her first offspring at 24 months of age. A male captive born on 13 May 1981 sired offspring that were born when he was 26 months old (Fig. 19.6). These two examples represent the earliest age at which a male or female produced young. Additional captive-born tamarins were routinely siring or giving birth to offspring by 36 and 33 months of age, respectively.

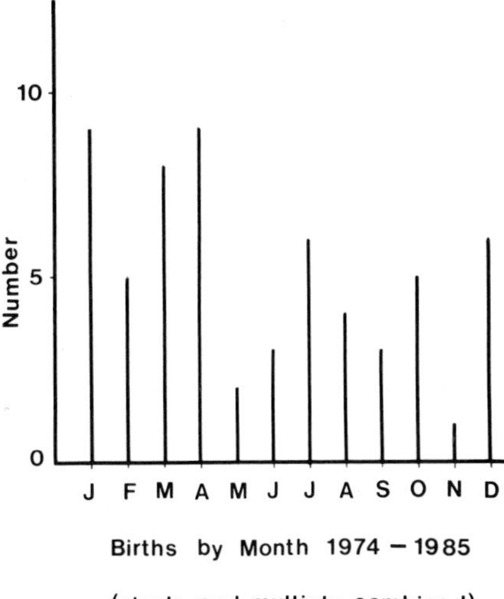

FIGURE 19.4. Total number of births by month, 1974–1985.

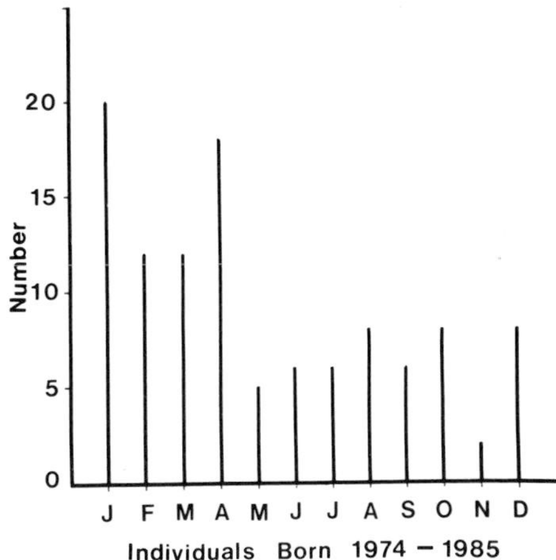

FIGURE 19.5. Individuals born per month, from 1974 to the present.

FIRST BIRTHS TO CAPTIVE BORN TAMARINS :

female:
born 23 March 78
first litter: 24 months old

male:
born 13 May 81
first litter: 26 months old

FIGURE 19.6. Age of first births to captive-born tamarins.

	Reproductive Interval:	Offspring Born:
females:	23 Mar 76 – 14 Jan 81	13
	24 Apr 77 – 28 Sept 81	13
	02 Dec 80 – 09 Jan 85	11
males:	02 Sept 74 – 24 Aug 77	13
	23 Mar 76 – 28 Sept 81	15
	10 Apr 82 – 21 Jan 84	7

FIGURE 19.7. Reproductive capabilities of a representative group of tamarins.

Representative of the reproductive capabilities of these tamarins are the following birth records of selected males and females (Fig. 19.7). A female produced 13 offspring in a span of 5 years; another, 11 in the same length of time; and a third produced 11 offspring in just under 4.5 years. A male sired seven offspring in less than 2 years; a second male sired 13 offspring in 35 months; and a third, 15 offspring in 5.5 years. This represents 2.4 offspring produced per year per female and 3.5 offspring sired per year per male. In 45 births, the sex ratio was 20 males and 25 females (0.8:1.0). Two litters per year per pair was not uncommon in experienced, mature pairs. Interbirth intervals ranged from 154 days to 265 days (n = 7:154; 191; 212; 220; 253; 264; 265).

The minimal longevity of wild-born tamarins received as young adults is 8 years and 5 months for a female, 10 years, and still living, for a male. For captive-born tamarins: female 8 years and 2 months (living), male, 9 years and 3 months (living).

In the 11 years since my first trip to Panama, the situation has changed dramatically for Geoffroy's tamarin. The outlook is vastly improved for this small primate. National parks have not only been established in the Republic, but protected areas are indeed protected. The animal dealers are no longer operating, and RENARE closely and effectively controls the exportation of any wildlife. Only those living animals to be used for scientific, educational, or propagation purposes are allowed to be exported, and not until full justification for need has been proven. Animals are no longer offered for sale in the public market, and no living native animal can be legally maintained without the permission of RENARE.

With each passing year, as I conduct my own field studies in Panama, I have been able to see Geoffroy's tamarin in more locations, in larger numbers, in viable, reproducing groups. These localized populations are growing, in some cases spreading, and appear secure in many locations. Areas that formerly had substantial numbers of these tamarins are now being repopulated by the spread of natural groups. New areas are being inhabited, areas suitable for these tamarins, but up until now without the population numbers to spread into them. Tamarins are now found in suitable habitats adjacent to sizeable human populations, living in apparent harmony with *Homo sapiens*. Geoffroy's tamarin is not endangered nor threatened in the Republic of Panama. In fact, in some locales it is quite common, being the most frequently seen diurnal mammal.

This spring, during the middle of the dry season, as I sat and observed a group of Central American agoutis, *Dasyprocta punctata*, at my study site on the Pacific side of the Isthmus, I was struck by the obvious increase in the actual number of individuals as well as the number of species of wildlife in the forest around me. Late one afternoon, after some days of intense field work, four small forms made their way across my field of vision. At first, I dismissed them as the ever present variegated squirrel, *Sciurus variegatoides*. It registered in my brain that their tails didn't quite look right and forced me to respond. There, in a tree in the clearing, were two adult and two juvenile Geoffroy tamarins casually feeding above the agoutis. As I watched them in their various activities, I recorded in my field notes, "10 March 1985, new species at study site, Geoffroy's tamarin." It was then that I realized that in 11 years of sitting and watching in this particular place, I had never seen "mono titi" before. There is every reason to believe that this primate's survival in the Republic of Panama is assured, given the programs and protection that it, as well as other wildlife, is afforded at this time.

Acknowledgments. I am grateful to the Lincoln Park Zoological Gardens, Lester E. Fisher, D.V.M., Director, for allowing me time to conduct these field studies. The American Philosophical Society (Philadelphia) partially funded portions of some of these studies. G. F. Meritt assisted in various phases of the field work. The staff of Gorgas Memorial Laboratory cooperated in many ways. Dr. and Mrs. N. B. Gale assisted and encouraged field work in innumerable ways. The Directors of RENARE, past and present, as well as E. Vallester, Y. Hidalgo, and D. Botello, all of RENARE, allowed field work and cooperated in the issuance of permits. I am indebted to these individuals and to these agencies for their assistance and cooperation.

References

Dawson GA (1976) Behavioral ecology of the Panamanian tamarin, *Saguinus oedipus* (Callitrichidae, Primates). Ph.D. dissertation, Michigan State University, East Lansing, MI.

Dawson GA (1977) Composition and stability of social groups of the tamarin, *Saguinus oedipus geoffroyi*, in Panama: ecological and behavioral implications. In: Kleiman DG (ed) The biology and conservation of the Callitrichidae. Smithsonian Institution Press, Washington, DC, pp 23–37.

Hershkovitz P (1977) Living New World monkeys (Platyrrhini) Volume I. University of Chicago Press, Chicago, pp 753–806.

Mendez E (1970) Los principales mamiferos silvestres de Panama. Edicion Privada, Panama, pp 1–283.

Moynihan M (1970) Some behavior patterns of Platyrrhine monkeys. II. *Saguinus geoffroyi* and some other tamarins. Smithsonian Institution Press, Washington, DC, pp 1–77.

Moynihan M (1976) The New World primates. Princeton University Press, New Jersey, pp 1–262.

20
The Management of Prosimians in Captivity for Conservation and Research

JONATHAN I. POLLOCK

Introduction

At a recent international conference on primates, only four out of 156 papers presented concerned prosimians although more than one-fifth of the primate species and over one-third of the genera are prosimians. Why do ethologists, physiologists, and comparative zoologists persist in typifying prosimians—the "one point on the curve" approach—when 104 of the 507 extant primate taxa are members of this suborder? Ignorance about diversity within the group is one reason, but belief of uniformity accompanies some of our prejudices: Prosimians are not glamorous to look at; they are mostly small-bodied and nocturnal, cryptically colored, living in small groups or alone, and have limited intelligence or behavioral flexibility. In short, our approach is colored by characteristics of the order that we attribute to the monkeys and apes. By way of outlining some of the differences and similarities of the two suborders, this paper aims to redress the balance in a small way using the principles and problems of colony management by means of illustration.

Approximately one-half the prosimian species are kept in captivity at present, but of these 22 species, ten are in reproductive difficulties (Table 20.1), and their survival in captivity is in doubt. From captive population sizes and information on fecundity, only ten prosimian species can at present be regarded as "self-sustaining" in captivity. The majority of species not held in captivity and, unfortunately, most of those faring less well there, are endangered or severely threatened in the wild.

In this paper I shall discuss briefly three topics on prosimian husbandry that are appropriate to the conservation and research-oriented framework of this conference: nutrition, reproduction, and population biology.

Nutrition

Despite the fact that prosimians are a diverse group, a substantial proportion can be generalized—for management purposes—as relatively small, nocturnal, insect-eating forms, interacting socially mostly with individuals they do not move about with (Table 20.2). Species possessing all these characteristics account for 42% of the 43 extant species. The remainder, however, occupy a wide

TABLE 20.1. Population sizes of prosimians kept in captivity in the world.[1]

Lorisoidea		Lemuroidea		Tarsioidea	
A. calabarensis	"A few"	*M. murinus*	147	*T. syrichta*	28
P. potto	"A few"	*C. medius*	45	*T. bancanus*	9
L. tardigradus[2]	75	*L. macaco*	193		
N. coucang	75	*L. fulvus*	796		
G. senegalensis[3]	353	*L. catta*	694		
G. crassicaudatus[3]	162	*V. variegata*	462		
G. garnetti	17	*L. mongoz*	130		
G. demidovii	40	*L. coronatus*	29		
		M. coquereli	15		
		P. verreauxi	11		
		H. griseus	10		
		C. major	4		

[1]These values are estimates based on information obtained from International Zoo Yearbooks, studbooks, and Duke Primate Center records. The figures represent 1985 values assuming a *net* increase of 5% per year since the last full census available.
[2]Underlined species are those most severely threatened at present.
[3]Excludes research institution holdings.

array of niches, locomotor specializations, and dietary adaptations. Because prosimians are almost exclusively arboreal and have limited opposability of the digits, the seed, grass, and rhizome-eating baboon ecotype is not represented in the suborder. There is also no specialized mature-leaf eating homolog to *Colobus guereza*. The group includes, however, its own primate stars: for example, the woodpecking *Daubentonia* and the totally carnivorous *Tarsius*.

A number of experimental studies have examined prosimian diets in captivity (King, 1974; Petter-Rousseaux and Hladik, 1980; Poliak, 1981). In general, these show that prosimians adapt readily to artificial diets and that ad libitum

TABLE 20.2. Prosimian size, activity patterns, diet, and group size.

	Number of species
Weight (g)	
$<$ 1000	21
$\simeq$ 1000	10
$>$ 1000	12
Activity	
Nocturnal	31
Crepuscular or nocturnal and diurnal	5–6
Diurnal	6–7
Diet	
Insects or insects and fruit	20
Frugivore/folivore	8
Folivore	13
Other (e.g., gum)	2
Group size	
"Solitary"	30
"Family" groups	8
"Larger" groups	5

presentation regimes permitting voluntary food consumption to be measured can separate species reflecting in part tendencies of food selection variation in the wild (Hladik, 1979). One finding is that voluntary food intake variation in some prosimians reproduces natural seasonal patterns of food ingestion in the wild (Petter-Rousseaux and Hladik, 1980; Hladik et al., 1980). In some species there are associated fluctuations in body weight (Petter-Rousseaux, 1980; McCormick, 1981).

As small-bodied primates, prosimians are expected to require a diet richer in energy and protein than monkeys and apes. However, experimental measurements of basal metabolic rates and the calorific content of food consumed on an adlibitum basis suggest that weight for weight this is not the case (Table 20.3). Data on 13 species were recovered from published papers including the computation of the number of calories consumed in the Brunoy study (Petter-Rousseaux and Hladik, 1980) by five prosimian species, using Ministry of Agriculture food tables (MAFF, 1976). By converting values of oxygen consumption and voluntary food intake to the number of Kcals used or consumed per metabolic body size (weight in kg to the power 3/4), the values can be compared across species. All the values of basal metabolic rate (BMR) are slightly or substantially lower than expected for the class Mammalia as a whole. The BMR figure reported by McCormick for *Cheirogaleus medius* is artificially high because the animals were allowed to move around the experimental chamber. It should be noted that BMR values for the ponderous quadrumanal *Perodicticus potto* and *Nycticebus coucang* are respectively 63% and 41% of the expected value under thermoneutral condi-

TABLE 20.3. Basal metabolic rates and voluntary food intake in prosimians.[1]

	BMR	VFI
Mammals	70	105–140
G. demidovii	69;55	
G. elegantulus	65;56	
G. senegalensis	55	
G. crassicaudatus	55;47–51	
A. calabarensis	61	
P. potto	44;41;36;36	
N. coucang	29;28	
M. murinus	45–61	67;67[2]
C. medius	47;60–138	63
L. fulvus	20–46	90;74[2]
L. mongoz		90
Lemur sp.		98
M. coquereli		137
P. furcifer		94
Lepilemur mustelinus		35[3]

[1] Values are expressed in Kcal/weight (kg)$^{0.75}$ per day. Each value represents a single individual. Where a number of individuals are involved, the range is given. The mammalian values for basal metabolic rate (BMR) and voluntary food intake (VFI) are given.
[2] Animals experiencing weight loss over the experimental period.
[3] Field study measurement.
Source: Data recovered and adapted from King (1974, 1978); Poliak (1981); Hladik (1979); Petter-Rousseaux and Hladik (1980); McCormick (1981); Hildewein (1972); Hildewein and Goffart (1975); Daniels (1984); Palacio (1977); Müller (1975, 1977, 1979, 1983); Whittow et al. (1977); Dobler (1982); Müller and Jakshe (1980); Chevillard (1976); Charles-Dominique and Hladik (1971).

tions. (It is not, perhaps, irrelevant that pottos are notoriously difficult to anesthetize for surgical treatment.) The voluntary food intake (VFI) values in Table 20.3 are, in some cases, annual means, created by averaging the results from four seasons. With the exception of *Phaner furcifer*, winter and autumn values are, in these cases, between 25% and 56% below spring and summer values. In *C. medius*, a winter VFI value of only 30% of the expected maintenance metablic rate no doubt reflects the extreme physiological rhythm of this species, which in the wild may enter a state of torpor for up to 6 months of the year utilizing fat reserves stored in the tail. The aseasonal pattern of VFI in *Phaner* is interesting with respect to its gum-eating habits. Gum is probably available as a food source more reliably throughout the year than any other food type in the forests of western Madagascar.

Measures of protein consumption alone are of limited value in assessing dietary sufficiency because amino acid composition can have a major influence on requirements (Samonds and Hegsted, 1973). However, values of 2–3 g protein/kg body weight per day have been experimentally obtained for adult *Macaca*, with infants and New World primates needing substantially more (National Research Council, 1978). Although a wide range of protein-containing foods were not supplied in the adlibitum study at Brunoy, the values obtained (Table 20.4) appear to be on the low side considering that protein requirements are normally inversely related to body size allometrically. Protein:energy ratios for gross energy intake in the VFI studies ranged from 16% to 19.5%. Considering that Old World primates and humans adequately maintain themselves on 10–13% ratios (King, 1978) and that prosimians may require less protein per unit body weight, it is likely that many zoo diets for prosimians are unnecessarily rich in protein.

Very little is known of the vitamin and mineral needs of prosimians. Standard vitamin/mineral supplements are commonly given and prove adequate in terms of weight maintenance growth and general health. Some information has recently, however, become available on the vitamin D and vitamin C requirements of prosimians, which is of both practical and academic importance.

Reports of impaired mobility, reluctance to move, and bone deformities in *Lemur catta* and *Varecia variegata* at Cincinnati Zoo were linked by Tompson and

TABLE 20.4. Protein consumption in prosimians computed by measures of voluntary food intake.

Species	Mean protein intake g protein/day	Source
M. murinus	3.3 (1.5–4.6)[1]	Adapted from Petter-Rousseaux and Hladik (1980)
C. medius	1.5 (0.9–2.6)	Adapted from Petter-Rousseaux and Hladik (1980)
M. coquereli	3.2 (2.3–4.6)	Adapted from Petter-Rousseaux and Hladik (1980)
P. furcifer	2.8 (2.7–2.9)	Adapted from Petter-Rousseaux and Hladik (1980)
L. fulvus	0.6 (0.3–0.9)	Poliak (1981)
	1.9	King (1974)
L. mongoz	1.9	King (1974)

[1] The range shown refers either to variation between individuals or to variation across the year. Poliak's animals suffered severe or moderate weight loss because of an unsatisfactory diet.

Lotshaw (1978) to low serum calcium levels (less than 5 mg/dl, which is 40–50% below normal mammalian values). Together with mild hyperphosphatemia (9–11 mg/dl) this produced low calcium:phosphorus ratios of 0.3–0.6 compared to the normally recommended 1–2. Poor skeletal mineralization and thin bone cortices, as well as gross deformities in some cases, were said to have resulted from a diet whose sole source of vitamin D was three hard-boiled eggs for five lemurs each day. The condition was corrected by vitamin D supplementation and improving the calcium to phosphorus ratio of their diet. It is interesting to note that the adult female ring-tailed lemur failed to show pathological signs on the unimproved diet. It is likely that this resulted from her preferential access to desirable food items, in this case the eggs. Females are dominant to other group members in *L. catta* (Jolly, 1966) and in other lemurs (Pollock, 1979), a feature that commonly influences food consumption by individuals in group feeding situations.

Calcium and vitamin D concentrations in commercially available formula food for primates are high enough to alleviate such problems, especially since the high vitamin D_3 requirement of New World primates has received such publicity. But could they be too high for prosimians? At Duke, evidence of hypercalcemia and great variation in 1,25-dihydroxycholecalciferol—the principal serum metabolite of vitamin D_3—were associated with soft tissue calcium deposits found on necropsy in *Hapalemur griseus* (Gray et al., 1982). Monkey chow has since been rationed to this species. A study recently completed in North Carolina examined bone density changes, serum calcium, and vitamin D metabolites in *L. fulvus* fed a vitamin D-free monkey chow (Purina Batch No. R5038) for 4 months (Pollock et al., in preparation). Although no pathological symptoms appeared, a small reduction in humerus bone mineral content was observed in all animals. The lemurs in this case were maintained inside in cages lit by cool, white fluorescent lights. Many species of prosimian are thickly fur-covered, nocturnal species spending the daytime in nests or tree-holes well shaded from light. In these cases opportunities for the photochemcal conversion of 7-dehydroxycholesterol in the skin tissues to the vitamin itself may be severely restricted. It has been recently suggested that the heavy coating of wool and lanolin on unsheared sheep limit this reaction and cause low vitamin D_3 serum concentrations compared with sheared sheep (Horst and Littledike, 1982). Furthermore, some prosimian species—notably *Lepilemur*—are said to be entirely folivorous (Charles-Dominique and Hladik, 1971), and almost all plant material unless cut and cured in the sun is thought to be very poor in this vitamin (Horst et al., 1984).

It is possible, therefore, that prosimians may be acutely sensitive to this vitamin, that some adequately maintain themselves solely on low dietary doses of ergocholesterol, and that the present-day quantities of D_3 captive *Lemur* are ingesting in monkey chow (500–700 IU/day) are far too great. Until more scientific data are available it is probably wise to maintain prosimians in conditions where an adequate light source of the appropriate spectral intensities (i.e., including wavelength components less than 310 nm) are provided. Simultaneously, a diet including vitamin D should be given equivalent to no more than the NRC recommended levels (2000 IU/kg diet). The values suggested by Bourne (1975) for *Macaca mulatta* may be more appropriate for prosimians (25 IU/kg body weight/day).

Although most mammals synthesize ascorbic acid, it has long been recognized that vitamin C supplementation is necessary to avoid scurvy in humans and a number of simian primates. From the point of view of the colony manager, this is of significance because the chemical instability of ascorbic acid makes the provision of fresh sources of this vitamin, such as citrus fruits and green leafy vegetables, very important.

In 1966, it was reported that the liver of tree-shrews (*Tupaia glis*) and slow lorises (*N. coucang*) synthesized ascorbic acid in vitro (Elliot et al., 1966). A few years later evidence was published to suggest that pottos (*P. potto*) and and thick-tailed galagos (*Galago crassicaudatus*) were also synthesizers (Nakajima et al., 1969). Recently these observations were extended to include most of the prosimians kept at Duke (Table 20.5). All species tested, with the exception of *Tarsius bancanus*, possess the enzyme *L*-gulonolactone oxidase, which is responsible for vitamin C synthesis (Pollock and Mullin, in preparation). It is probable that ascorbic acid is not, therefore, a "vitamin" for lemurs, lorises, and galagos although without more quantitative information on rates of synthesis it would be unwise to recommend omitting ascorbic acid completely from their diet. For *Tarsius*, however, the result is of some significance. Because the diet of *Tarsius* is totally carnivorous, its vitamin C needs must be met by the ascorbate content of the invertebrate prey on which it spends most of its time feeding (Niemitz, 1984). At Duke, tarsiers consume between 25 and 35 crickets daily (Wright, personal communication) which, at a concentration of 75 μg ascorbate/g crickets wet weight, would supply a minimum of 1.0 mg vitamin C daily. It is remarkable that the National Research Council recommended requirement for simians is 10 mg/kg body weight per day, a figure very similar to that observed for the 120 g tarsier.

This finding is also of academic importance because it provides further support for a phylogenetically closer association for *Tarsius* with the Anthropoidea than with the prosimian suborder (Pollock and Mullin, in preparation).

Reproduction

Although many female primates, especially in the Old World, cycle reproductively only during certain times of the year (Lancaster and Lee, 1965), it is not known in most cases whether these "breeding seasons" result from adaptation to local ecological conditions or responsiveness to higher level cues such as photoperiod. Since many seasonal conditions correlate with daylength, this can only be examined by means of the geographical transfer of animals or careful scrutiny of reproductive timing in different groups of the same species occupying different latitudes (Lancaster and Lee, 1965; Vandenbergh, 1973) and different habitats. An analysis by Van Horn (1980) attempted to provide evidence of a photoperiodic control of breeding seasons in *M. mulatta* and *M. fuscata*. While demonstrating latitudinal clines of birth seasons in both species (interestingly in opposite directions), this study failed to account for significant outliers. This is an important reservation because nonphotoperiodic influences *can* be responsible for shifts in birth peaks (Varley and Vessey, 1977).

Among prosimians, many species exhibit a marked seasonality in many aspects of their behavior and physiology, including reproduction (Doyle, 1974; Jolly, 1966; Tattersall, 1982; Petter-Rousseaux, 1980). All the Malagasy lemurs for which there is information breed seasonally with specific, short breeding periods followed by a lengthy quiescent phase. In these cases, strong evidence exists for

TABLE 20.5. Summary of *L*-gulonolactone oxidase activity in the livers and kidneys of simian and prosimian primates.

Species	Liver	Kidney	Source
M. mulatta	–[1]	–	Nakajima et al. (1969)
P. troglodytes	–	–	Nakajima et al. (1969)
A. geoffroyi	–	–	Nakajima et al. (1969)
S. sciureus	–	–	Nakajima et al. (1969)
S. oedipus	–	–	Unpublished results reported in Yess and Hegsted (1967)
S. mystax	–	–	
T. bancanus	–	–	Pollock and Mullin (in preparation)
T. glis	+	–	Nakajima et al. (1969)
N. coucang	+	–	Nakajima et al. (1969)
P. potto	+	–	Nakajima et al. (1969)
G. crassicaudatus	+	–	Nakajima et al. (1969)
G. senegalensis	+	–	Pollock and Mullin (in preparation)
G. garnetti	+	–	Pollock and Mullin (in preparation)
L. tardigradus	+	–	Pollock and Mullin (in preparation)
M. murinus	+	–	Pollock and Mullin (in preparation)
C. medius	+	–	Pollock and Mullin (in preparation)
H. griseus	+	–	Pollock and Mullin (in preparation)
L. fulvus	+	–	Pollock and Mullin (in preparation)
L. macaco	+	–	Pollock and Mullin (in preparation)
V. variegata	+	–	Pollock and Mullin (in preparation)

[1] +, enzyme present; –, enzyme absent.

a direct photoperiodic regulation of reproductive activity that is amenable to experimental manipulation (Petter-Rousseaux, 1970, 1972, 1975; Martin, 1972; Van Horn, 1975, 1980). Transferring lemurs between two hemispheres, for example, shifts the breeding and birth seasons by 6 months, and artificially changing the rate of daylength changes in captivity alters the timing of reproduction. By doubling this rate, mouse lemurs can be brought twice into reproductive condition within one year.

On the mainland of Africa and Asia, *G. senegalensis*, *G. crassicaudatus*, and *Nycticebus* spp. can be found in nontropical latitudes experiencing substantial seasonality in photoperiod, temperature, rainfall, and plant and animal productivity. A very sharp birth peak in *G. crassicaudatus* was determined by Bearder (1974) at 24° south latitude in northeastern Transvaal where all infants were born over a period of 3 weeks. For *G. senegalensis*, two well-defined birth peaks have been observed (Bearder, 1969; Butler, 1967; Doyle et al., 1971) possibly corresponding to two pregnancies a year for at least some females. Postpartum conceptions are probably more dependent on small body size than specific environmental conditions in the wild since evidence has accumulated for its occurrence in all small-bodied galagos.

When these species are brought into captivity, seasonality in breeding is either lost naturally (Doyle, 1974) or can be suppressed by exposure to a constant photoperiod. At Duke University, a reduction of the interbirth interval has been obtained by allowing galagos to "free-run" reproductively in respect to photoperiod (Izard, personal communication). This result implies that certain light regimes might be inhibitory rather than others stimulatory. There is some evidence that *reducing* the duration of daylength in "short-day" breeding species, or *increasing* it in "long-day" species stimulates reproductive activity more than recreating specific breeding season photoperiods constantly (see Van Horn, 1980). Photoinhibition provides a fascinating tool for research in prosimian reproductive physiology, and its significance is enhanced by observations by Van Horn (1980) that social stimuli from male *L. catta* can cause photoinhibited females to start reproductive cycling.

The maintenance in captivity of breeding seasonality in these species and the consequences of attempts to interfere with it are of some importance: First, the evolutionary entrainment of animals to natural cycles in the environment such as light, water, food, and temperature may be reflected in deep, interrelated physiological rhythms that require simulation under artificial conditions. For example, mean testicular length, which is an index of plasma testosterone concentration in normal male lemurs (Van Horn et al., 1976; Petter-Rousseaux and Picon, 1981) and spermatogenesis in *Macaca* (Sade, 1964), varies with photoperiod. Under constant photoperiod, testicular dimensions of *L. catta* continue to fluctuate for approximately one year before settling down to a level *below* that of the previous seasonal maximum (Van Horn et al., 1976), which might reflect reduced fecundity.

Second, most seasonally breeding female prosimians cycle only two or three times in the year. Receptivity forms an extremely small proportion of the duration of the cycle—usually between 10 and 24 hours—and consequently matings may not be observed. It is, therefore, important to manage and monitor the female cycle closely by observation of vaginal "opening" (in those species that seal up outside the breeding period), swelling, and color changes, and by regular vaginal smears where necessary (see D'Souza, 1978). Frequent, even daily, vaginal smears have no discernible effect on fertility in the *Microcebus* and *Cheirogaleus* colonies at Duke University. In males, testicular development and sex-associated behavior should be followed. Because of the very short period of sexual activity, it is of particular importance, for purposes of equalizing the genetic contribution of small numbers of founders, that information about *male* fecundity be unambiguous.

Third, because cycle lengths plus gestation periods are long in relation to the period in the year over which females are in reproductive condition, seasonally breeding prosimians are not well adapted to second pregnancies within the year, except for the smallest species of *Galago* and *Microcebus*. In many species, the gestation period is approximately equal to the cycle length multiplied by the maximum number of cycles occurring in the year. This adaptation to a single birth peak is reflected in the difficulty of obtaining in captivity second pregnancies in cases where females lose full-term infants or where early postnatal mortality occurs (Table 20.6). In *C. medius* short estrous cycles (18–23 days) are probably necessary because of the time constraints affecting females: they have between 4 and 6 months to enter breeding condition, conceive, give birth, and grow their litters to an adult size with sufficient fat reserves, before the period of winter torpor begins. Consequently, "second chances" have a poor prognosis, and the role of the manager is to reduce infant mortality because of the long time period separating births.

From the perspective of captive management, breeding seasonality can be advantageous. Benefits include limitations in the duration of aggressive episodes, the relatively easy mating of females to males from other social groups (thereby

TABLE 20.6. Instances of second full-term pregnancies by lemurs in the same year at Duke University Primate Center since 1977 (seven breeding seasons).

Species	Number of postpartum conceptions	Number of second births	Cycle length	Gestation length
M. murinus	4	0	40–60 d ($\bar{x}$ + 50; Glatston, 1979)	60–64 d (Glatston, 1979) DUPC data
C. medius	13	5(1)[1]	18–23 d ($\bar{x}$ = 19.7 + 1.58; Foerg, 1982a)	62 d (Foerg, 1982a)
L. macaco	9	0	~ 30 d (DUPC)	120 d (DUPC)
L. fulvus fulvus	16	3(1)[1]	~ 30 d (DUPC)	120 d (DUPC)
L. catta	4	0	39.3 d (Evans and Goy, 1968)	130–135 d (DUPC)
V. variegata	11	0	40–42 d (Foerg, 1982b)	102.5 ± 1.5 d (Foerg, 1982b)

[1] Second pregnancy occurring *without* first infant(s) death.

integrating social and genetic management), and the opportunities presented for cross-fostering infants when females giving birth fall ill, require surgery, produce very large litters, or simply fail to care for their young (Pollock and Katz, in preparation).

Some prosimian species, while appearing healthy and maintaining weight in captivity, fail to reproduce successfully. Species prone to this problem include pottos, slow lorises, crowned lemurs, gentle lemurs, sifakas, and mongoose lemurs (see Schaaf and Stuart, 1983). Typically a period of activity is followed by reproductive quiescence, although some species appear to be "turned off" by the trauma of capture, transport, or captivity itself. Two second-generation *H. griseus* have been born at Duke, but in the course of 29 adult female breeding-years involving seven individual females only 11 offspring have been born, and the population in captivity is declining. Since 1962, 26 adult female breeding-years in *Propithecus verreauxi* have produced ten infants, six of them from a single female. Attempts to keep both other indriid genera, *Indri* and *Avahi*, in captivity have been spectacularly unsuccessful, and there is, therefore, an urgent need for careful reflection and intensive research on all these exotic species. The general activity, low incidence of health problems, and stable, satisfactory body weights of these species, with the possible exception of *Propithecus*, testify to the success of captive maintenance of these species, and we must turn to reproductive physiology as our starting point for investigation.

The endocrinological characterization of the prosimian female reproductive cycle is in its initial stages. Of particular concern are the events surrounding ovulation. Profiles of estrogen and progesterone plasma concentration changes that have been determined for *G. crassicaudatus* (Eaton et al., 1973), *L. catta* (Van Horn and Resko, 1977; Bogart et al., 1977), and *L. macaco* (Bogart et al., 1977) bear a marked similarity to each other and to those of simian species. The pattern of gonadotropin release is, however, uncertain. Normal et al. (1978) obtained binding between plasma from *L. catta* females and antiovine LH using an RIA technique, but recent attempts have failed to detect any biological activity of LH (assessed by both rat and mouse Leydig cell assays for testosterone) in serum collected in time-progressive samples from LH–RH-treated male *L. fulvus* in the breeding season (Dunaif, personal communication). Because of the perhaps subtle influences of light on the system, it is likely that both an understanding of the control of the female cycle in prosimians and possible therapeutic interventions will only be successful when the chemical basis of the photoinduction of breeding cycles is understood and can be simulated accurately. Using regimes of FSH and HCG injection developed in studies on simians, however, *G. crassicaudatus* at Duke have recently conceived infants and carried them to term (Izard, personal communication).

Although Sade's (1964) rather complete analysis of the relationships between testicular dimensions, plasma testosterone, and spermatogenesis in *M. mulatta* have not been repeated for any prosimian species, there is abundant evidence that several male reproductive features change dramatically with the seasons. Close relationships exist between testicle size, plasma testosterone, and photoperiod in *L. catta* (Van Horn, 1980), between plasma testosterone concentration and stages of the estrous cycle of females in *L. catta* (Van Horn et al., 1976), between testicular dimensions and plasma testosterone in *Microcebus murinus* (Petter-Rousseaux and Picon, 1981), *L. catta* (Van Horn et al., 1976), *V. variegata* (Bogart et al., 1977), and *L. mongoz* (Pollock, unpublished observation), and between testicular sizes and stages of the female estrous cycle in *V. variegata* (Foerg, 1982b); see Table 20.7.

Testicular dimensions are probably the most appropriate way to monitor male prosimian fecundity at present. Assays of serum testosterone from male prosimians during the breeding season show much variation between individuals and species (Table 20.8). This may be due to the pulselike secretion of the hormone into the blood stream and an additive memory capability of the relevant receptors; previous researchers have generally taken only single samples. Unfortunately, multiple-sampling protocols present technical difficulties for the many prosimians weighing less than 500 g.

TABLE 20.7. Studies on male prosimian reproductive function.[1]

	Plasma testosterone	Breeding season	Stage of ♀ cycle	Spermatogenesis
Testicle size	*L. catta* *L. mongoz* *V. variegata* *M. murinus* *L. macaco*	Many species	*V. variegata*	?
	Plasma testosterone	*M. murinus* *L. catta*	?	?
		Breeding season	–	?
			Stage of ♀ cycle	?

[1] Each box reflects, for named species, a determined relationship between two of five parameters of male reproductive function.

TABLE 20.8. Plasma testosterone concentrations in prosimian primates.

Species	Time of sampling	Value[1]	Reference
M. murinus	Peak breeding season	52.4 (range ?–120)	Petter-Rousseaux and Picon (1981)
L. catta	Breeding season	27.6 ± 7.3	Van Horn (1980)
L. catta	Breeding season	6.2 (range 1.2–12.3) *n* = 5	Bogart et al. (1977)
L. catta	Breeding season	10.1–28.9	Evans and Goy (1968)
L. macaco	Breeding season	6.1 (*n* = 1)	Bogart et al. (1977)
V. variegata	Breeding season	6.6 (range 6.3–6.9) (*n* = 2)	Bogart et al. (1977)
V. variegata	Breeding season	40–100	Foerg (personal communication)
L. mongoz	Breeding season	10.7	Duke data Unpublished
L. catta	Diestrus period	Below 5	Van Horn (1980)
L. catta	Nonbreeding season	3.6 (range 0.2–20.7) *n* = 16	Bogart et al. (1977)
L. catta	Nonbreeding season	1.8	Evans and Goy (1968)
V. variegata	Nonbreeding season	0.5–1.0	Foerg (personal communication)
L. macaco	Nonbreeding season	1.56 (range 0.8–2.0) *n* = 3	Bogart et al. (1977)

[1] All values are in nanograms/ml.

Caution must be observed, however, in interpreting changes in testicular size. In many small mammals, 20% changes in rates of spermatogenesis have been reported without a corresponding change in testicular volume, especially early and late in the season (Woodall, 1985) when this information is so urgently needed.

Furthermore, factors other than endogenous physiological rhythms are involved. For example, testicle sizes in male *L. catta* housed separately from females fluctuate seasonally with a lower amplitude than those in heterosexual groups (Van Horn et al., 1976). In ruffed lemurs, not only does male testicle size within a social group peak on the actual day of mating, the only day of behavioral estrous, but these changes are suppressed in young fertile males housed with an adult male in the same group (Foerg, 1982b). Such socially mediated influences provide opportunities for increasing productivity through management intervention.

Population Biology

With only 22 of the 43 prosimian species held in captivity, many of these with small population sizes, careful management of the breeding and birth period is essential. Infant prosimians surviving the first 72 hours have a good prognosis for survival to maturity. Records from Duke University show that until the end of

1977, 82% of the deaths across all taxa within 2 months of birth occurred in this period, with 58% in the first 24 hours.

Because infant death is sometimes associated with mutilation or cannibalism—although their involvement as a *cause* of death is questionable—valuable females considered to be at risk have been routinely separated from their social groups prior to giving birth. Analysis of *Varecia* records at Duke, for example, show that infants born to primiparous mothers have a higher 24-hour mortality rate than those born to multiparous mothers (Fig. 20.1). Furthermore, this difference is accounted for by those primiparous mothers that, before management changes, remained with their social groups. Consequently, it is primiparous females that now attract most attention in this species during the birth season.

The closely managed *Galago* colony at Duke has been subject to intensive reproductive research in recent years. For all three species held, 10-day infant mortality is lower in mothers isolated before birth (Fig. 20.2). Recent data have also demonstrated a higher incidence of infant mortality in primigravid than in multigravid females (Izard and Simons, in press a).

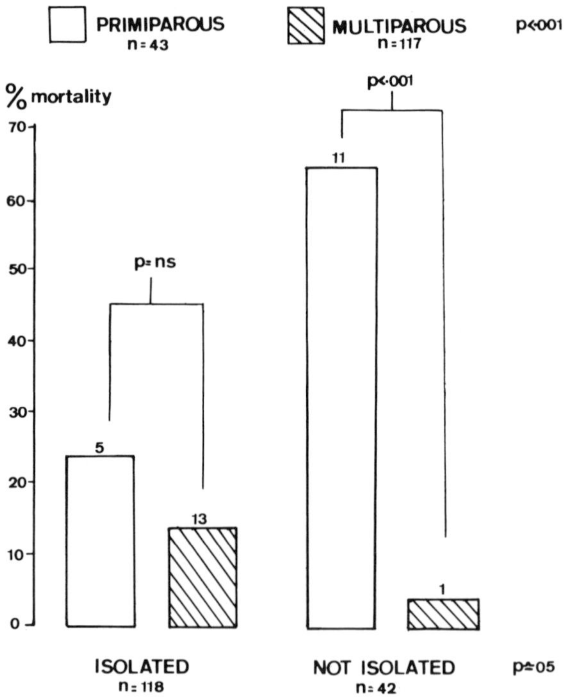

FIGURE 20.1. Mortality of infant *Varecia* within 24 hours of birth according to female gravidity and isolation management. Sample sizes and the results of statistical comparisons are shown (G_{1df} tests of mortality rates between primiparous vs. multiparous, $G = 11.96$, $P < 0.001$; isolated vs. not isolated, $G = 3.4$, $P \simeq 0.05$; isolated primiparous vs. isolated multiparous, $G = 0.39$, NS; not isolated primiparous vs. not isolated multiparous $G = 19.8$, $P < 0.001$). Data from Duke University 1972–1984.

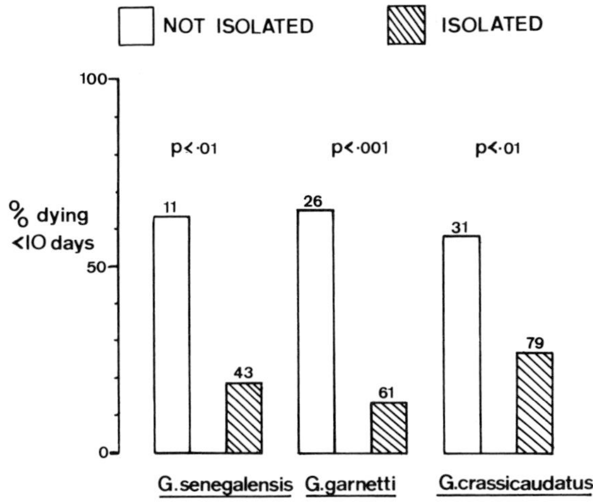

FIGURE 20.2. Mortality rates in three species of *Galago* within 10 days of birth at Duke University Primate Center. Sample sizes and the results of statistical comparisons between infants born to isolated and not isolated mothers are shown (G_{ldf} tests for *G. senegalensis* $G = 8.07$, $P < 0.01$; *G. garnetti* $G = 23.4$, $P < 0.001$; *G. crassicaudatus* $G = 9.39$, $P < 0.01$). Data from Izard and Simons (in press b).

By way of illustrating the critical importance of infant mortality to population growth in a prosimian species, it is instructive to examine the reproductive performance of one of the species still at risk in captivity: *Loris tardigradus.* This species, which weighs only 200 g, can, in captivity, produce a single infant every 9.5 months from 18 months of age (Izard and Rasmussen, 1985). Assuming that Bowden's (1984) parameters of life-span (13 years) and postreproductive life-span fraction (0.3) are accurate, a maximum lifetime reproductive potential for a female of 10.0 offspring is possible. Estimating a modest mortality rate of 25%, this figure is reduced to 7.5. Although dying infants will probably shorten the delay to the next birth, this is unlikely to be significant because of the very long gestation period (5.5 months) in this species (Izard and Rasmussen, 1985). By the end of the first 5 years with the *L. tardigradus* pair brought into captivity and reproducing at the optimal rate, the total colony size will number 12. Clearly this species is adapted to minimize infant mortality in the wild, a feature of their natural history that the captive breeder must reproduce for population survival.

In the above example it was assumed that the first infant born was a female and that the sex ratio at all times was 1:1. If by chance the first two surviving offspring had been *male*, the colony size at the end of the same period would be 7. Stochastic influences on population growth can dominate the structure of colonies of little-known prosimians with *k*-selected reproductive strategies. For this reason, analysis of sex ratios and sex-dependent mortality achieves great significance. Records from a variety of sources provide evidence for a modest male-favored sex ratio bias of about 10% in many prosimians and a partially compensating male-disfavored selective mortality over the first year of life (Fig. 20.3).

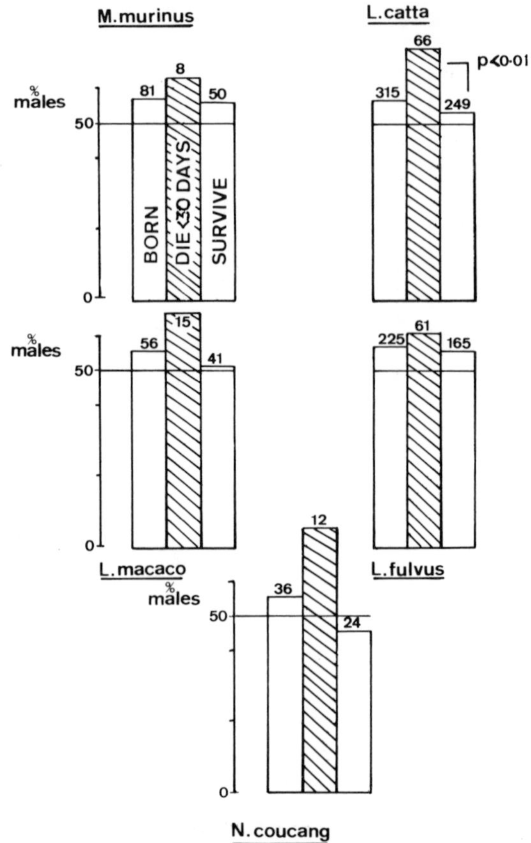

FIGURE 20.3. Sex ratio at birth and early infant (≤ 30 days) mortality in five prosimian species. The percentage males born, dying within 30 days, and surviving past this period are shown. No significant statistical differences between the sex ratio of dying and surviving prosimian young are present except in *Lemur catta* ($G_{ldf} = 8.79$, $P < 0.01$), although the same trend is evident in many species. Data were extracted from International Zoo Yearbooks 1979–1982.

In a recent survey of this topic in captive and wild populations of *G. crassicaudatus*, it was claimed that the preponderance of male births was not accompanied by "differential death before maturity" (Clark, 1978). However, statistical analysis of the data presented in Clark's paper shows that the sex ratio of *G. crassicaudatus* surviving to maturity does not differ significantly from equality ($G_{ldf} = 0.43$; NS) whereas that of infants born does ($\chi^2 = 4.05$; $P < 0.05$; Clark, 1978). This result agrees with records kept at Duke University where both the percentage males born and the percentage infants dying within 10 days that are males is slightly higher than for females in *G. senegalensis*, *G. crassicaudatus*, and *G. garnetti* (Izard and Simons, in press a).

An apparent exception is provided by *V. variegata*, where early female mortality is the greater. In this case, the difference can be attributed to poor survivability of female infants to 24 hours born to primiparous mothers (Fig. 20.4). Overall sex ratio at birth is 52% in this species, increasing to 56% after 1 day at Duke University. In this case, it is possible that a high plane of nutrition is accelerating development and causing females to give birth prematurely in captivity: litter size in young primiparous females is smaller than in "old" primiparous females (Pollock and Katz, in preparation). Pregnancy failure in young *Varecia* is higher than in old females, but the sex ratio of abortions is not known. This may be an artefact of captivity, but *Varecia* is a nonconformist in many ways, most especially as a primate with multiple litters, three pairs of mammae, and oral transport of young to and from a constructed nest, contrasting with its large body size and occupation of apparently stable, less productive rainforest habitats.

With well-established, large, captive populations such as *V. variegata*, *L. fulvus* and *L. catta*, and some *Galago* spp., sex-ratio imbalances are probably of small consequence. In many taxa, however, small captive population sizes (see Table 20.1) are threatened by sex-ratio fluctuations through biological design or by chance. The sex of a dying infant, in such cases, can seriously alter the structure of the subsequent breeding colony.

Small founder populations and few opportunities for subsequently broadening the genetic base of prosimians in captivity has inevitably led to consanguinous pairings. This has exposed two conditions, both in *Varecia*: "funnel-chested," and "hairlessness," the first being of no great importance either for the health or reproductive potential of the animal, the second being a deleterious condition caused by homozygosity in an autosomal recessive gene brought into captivity by

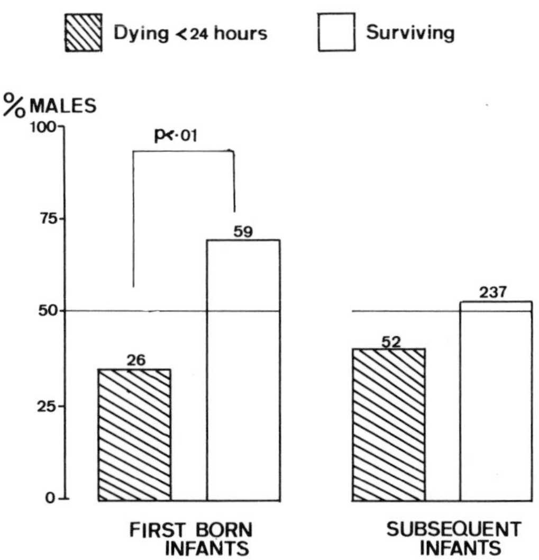

FIGURE 20.4. Sex ratio and infant mortality in *Varecia*. Fewer females are born, but more die young in this species in captivity. The sex ratio of surviving infants born to primiparous females is significantly different from dying infants (G_{Idf} = 9.05, $P < 0.01$). This is not the case for multiparous females (G_{idf} = 2.44, NS). Data from Duke University records and the *Varecia* studbook compiled by Dr. D. Brockman, San Diego Zoo.

a wild-caught individual. No simple genetic defects at present are known to pose a threat to captive populations of prosimians.

A widespread effect of inbreeding depression, however, has been documented recently for 16 groups of captive primates, including four species of prosimian (Ralls and Ballou, 1982). A higher inbred infant and juvenile mortality rate was determined for the sample as a whole and for five of the captive groups individually. It may be, however, that the effects of inbreeding—which cannot always be avoided by improved genetic management—is amenable to improved husbandry. For example, in the above paper, groups can be divided into those that have a significantly higher inbred young mortality rate and those that have not. Not only does the latter group have almost identical death rates for *outbred* young (37.0%) and inbred young (38.0%), but the overall *outbred* young death rates in the two groups are significantly different. The proportion of outbred young dying in those groups in which there is significantly higher *inbred* young mortality (43.6%) is higher than in those groups where inbred young are not especially at risk (37.0%; $G_{ldf} = 14.3, P < 0.001$). It might be, therefore, that certain groups of captive primates experience inbreeding depression mostly at high overall mortality rates, suggestive of difficulties they are experiencing in captivity. If this is the case, and many deleterious effects of homozygosity will be dependent on the environment, the effects of inbreeding should in some situations be controlled by management improvements.

A more detailed examination of the effects of inbreeding on prosimian viability has been conducted at Duke University on *L. fulvus* (Reinhartz, 1981). No differences in prereproductive mortality, sex ratio at birth, fertility, or longevity could be related to inbreeding coefficients. However, infants born to inbred females did suffer a higher mortality, perhaps supporting the suboptimal environment concept for such species. It was probably improved animal care that reduced *L. fulvus* 12-month mortality rates from 34.8% prior to 1980 to 16.0% after this date at Duke University. The highly outbred *L. macaco* figures are, respectively, 41.4% and 27% for the same periods. It should be noted that management mitigation of inbreeding depression does not necessarily apply to all taxa. It is known, for example, that for ungulates high overall mortality rates are not implicated in inbreeding depression effects (data reanalyzed from Ralls et al., 1979) and that the inbreeding coefficient of the dam is not significant (Ballou and Ralls, 1982).

For captive species founded by only a few individuals, such as most prosimian species, great care must be taken to genetically manage their reproduction. The studbooks now prepared or being prepared for *V. variegata* and *L. macaco* are an important step in this direction, and more studbooks should now be planned. Obvious practical methods such as equalizing founder contributions, monitoring sex ratios, and, most importantly, enhancing the reproductive opportunities for individuals with apparently reduced fecundity, should be pursued. Positive methods such as the controlled elimination of inbreeding depression should be considered (Templeton and Read, 1983). However, it should be noted that inbreeding depression must not be confused in direction or magnitude with poor adaptability to captive conditions that may have a genetic basis and persist for some time. Intentionally reducing genetic variability may have dangerous results in exposing a defenseless population to a rapidly changing pathogenic environment and, furthermore, probably severely compromises reintroduction schemes.

Conclusion

While basic research on general husbandry, reproductive physiology, and nutrition is still urgently needed for certain problem species, present management skills are adequate for vastly improving the status of prosimians kept in captivity. A good case can be made for increasing the number of species in captivity by careful selection. Examples of species with a good chance of success now are *T. spectrum*, *N. pygmaea*, *G. alleni*, *M. rufus*, *C. major*, *L. rubriventer*, and *Daubentonia madagascariensis*. Each species should be held so that concentrations of individuals are at least initially close to each other to facilitate breeding transfers. Breeding institutions should limit their holdings of different species and increase their population sizes wherever possible.

Because of poor habitat protection in host countries, reintroduction programs for prosimians are not currently a sensible objective. Captive management goals should therefore be set with this in mind—"a long-term, self-sustaining captive population" approach. Identification of those species at risk in the wild that have captive potential should succeed some consensus of opinion on the taxonomy of the suborder—especially in relation to *Galago*, *Lepilemur*, and *Microcebus*.

This is an urgent task. In Madagascar alone, 34% of the primate species have become recently extinct. At a present-day average of 1.8 taxa per species, this is equivalent to a loss of about 25 prosimian taxa in the last 1000 years. By observation and research, those working with prosimians in captivity have an obligation, together with biologists in the field, to provide both expertise and material support to the authorities responsible for protecting these species in the wild.

Acknowledgments. I would like to thank Professor E.L. Simons, A. Katz, and the staff at Duke University Primate Center for all they have taught me about prosimians.

Research mentioned in this paper was funded by the following grants: BSR 83-00653, BNS 81-20529, 40RR01302-04, and BNS 8407570.

This is publication number 274 of the Duke University Primate Center.

References

Ballou J, Ralls K (1982) Inbreeding and juvenile mortality in small populations of ungulates. Biol Cons 24:239–272.

Bearder SK (1969) Territorial and intergroup behaviour of the lesser bushbaby, *Galago senegalensis moholi* (A. Smith), in semi-natural conditions and in the field. Unpublished MSc Thesis, Witwatersrand University, Johannesburg.

Bearder SK (1974) Aspects of the ecology and behaviour of the thick-tailed bushbaby *Galago crassicaudatus*. PhD Thesis, Witwatersrand University, Johannesburg.

Bogart MH, Kumamoto AT, Lasley BL (1977) A comparison of the reproductive cycle of three species of lemur. Folia Primat 28:134–143.

Bourne GH (1975) The rhesus monkey. Academic, New York.

Bowden DM (1984) Aging. Adv Vet Sci Comp Med 28:305–341.

Butler H (1967) Seasonal breeding of the Senegal galago in the Nulba mountains. Folia Primat 5:167–175.

Charles-Dominique P, Hladik M (1971) Le Lepilemur du sud de Madagascar: ecologie, alimentation et vie sociale. Terre Vie 25:3–66.

Chevillard M-Ch (1976) Capacites thermoregulatrices d'un lemurien malgache, *Microcebus murinus* (Miller 1777). Thesis, University of Paris.

Clark AB (1978) Sex ratio and local resource competition in a prosimian primate. Science 201:163–165.

Daniels HL (1984) Oxygen consumption in *Lemur fulvus*: deviation from the ideal model. J Mammal 65(4):584–592.

Dobler H-J (1982) Temperaturregulation und Sauerstoffverbrauch beim Senegal und Zwerggalago (*Galago senegalensis, Galago(Galagoides) demidovii*). Bonn Zool Beitr 33(1):33–59.

Doyle GS (1974) Behavior of prosimians. In: Schrier AM, Stollnitz F (eds) Behavior of nonhuman primates, 5:50–123.

Doyle GA, Anderson A, Bearder SK (1971) Reproduction in the lesser bushbaby (*Galago senegalensis moholi*) under semi-natural conditions Folia Primat 14:15–22.

D'Souza F (1978) The detection of oestrus. In: Watson PF (ed) Artificial breeding of non-domestic animals. Symp Zoo Soc Lond 43:175–194.

Eaton GG, Slab A, Resko AJ (1973) Cycles of mating behaviour, oestrogen and progesterone in the thick-tailed bushbaby (*Galago crassicaudatus crassicaudatus*) under laboratory conditions. Anim Behav 21:309–315.

Elliot O, Yess NJ, Hegsted DM (1966) Biosynthesis of ascorbic acid in the tree-shrew and slow loris. Nature 212:739–740.

Evans CS, Goy RW (1968) Social behaviour and reproductive cycles in captive ring-tailed lemurs (*Lemur catta*). J Zool 156:181–197.

Foerg R (1982a) Reproduction in *Cheirogaleus medius*. Folia Primatol 39:49–62.

Foerg R (1982b) Reproductive behavior in *Varecia variegata*. Folia Primatol 38:108–121.

Glatston ARH (1979) Reproduction and behaviour of the lesser mouse lemur (*Microcebus murinus*, Miller 1777) in captivity. Thesis, University College, London.

Gray TK, Lester G, Moore G, Crews D, Simons EL, Stuart M (1982) Serum concentrations of calcium and vitamin D metabolites in prosimians. J Med Primat 11:85–90.

Hildewein G (1972) Metabolisme énergetique de quelques mammifères et oiseaux de la forêt equatoriale. Arch Sci Physiol 26:387–400.

Hildewein G, Goffart M (1975) Standard metabolism and thermoregulation in a prosimian *Perodicticus potto*. Comp Biochem Physiol 50A:201–213.

Hladik M (1979) Diet and ecology in prosimians. In: Doyle GA, Martin RD (eds) The Study of prosimian behaviour. Academic, New York, pp 307–358.

Hladik CM, Charles-Dominique P, Petter JJ (1980) Feeding strategies of five nocturnal prosimians in the dry forest of the west coast of Madagascar. In: Nocturnal Malagasy primates. Academic, New York, pp 41–74.

Horst RL, Littledike ET (1982) Comparison of plasma concentrations of vitamin D and its metabolites in young and aged animals. Comp Biochem Physiol 73(B):485–489.

Horst RL, Reinhardt TA, Russell JR, Napoli JL (1984) The isolation and identification of vitamin D_2 and vitamin D_3 from *Medicago sativa* (Alfalfa plant). Arch Biochem Biophys 231(1):67–71.

Izard MK, Rasmussen TD (1985) Reproduction in the slender Loris (*Loris tardigradus malabaricus*). Amer J Primatol 8:153–165.

Izard MK, Simons EL (In press a) Infant survival and litter size in primigravid and multigravid galagos. J Med Primatol.

Izard MK, Simons EL (In press b) Management of reproduction in a breeding colony of bushbabies. Proc Xth Intern Primatol Conf, Cambridge University Press.

Jolly A (1966) Lemur behavior. Chicago University Press, Chicago.

King GJ (1974) The feeding and nutrition of lemurs (*Lemur* spp.) at Jersey Zoological Park. Ann Rep Jersey Wildl Pres Trust 8:81–96.

King GJ (1978) Comparative feeding and nutrition in captive, non-human primates. Brit J Nutr 40:55–62.

Lancaster JB, Lee RB (1965) The annual reproductive cycle in monkeys and apes. In: DeVore I (ed) Primate behavior. Holt, Rinehart and Winston, New York, pp 486–513.
MAFF (Ministry of Agriculture and Food) (1976) Manual of nutrition. HMSO, London.
Martin RD (1972) A laboratory breeding colony of the lesser mouse lemur. Breed Prim, pp 161–171.
McCormick SA (1981) Oxygen consumption and torpor in the fat-tailed dwarf lemur (*Cheirogaleus medius*): rethinking prosimian metabolism. Comp Biochem Physiol 68A:605–610.
Müller EF (1975) Temperature regulation in the slow loris. Naturwissenschaften 62:140–141.
Müller EF (1977) Energiestoffwechsel, Temperaturregulation und Wasserhaushalt beim Plumplori (*Nycticebus coucang*, Boddaert 1785). Thesis, Tubingen.
Müller EF (1979) Energy metabolism, thermoregulation and water budget in the slow loris, *Nycticebus coucang* (Boddaert 1785). Comp Biochem Physiol 64A:109–199.
Müller EF (1983) Wärme und Energiehaushalt bei Halbaffen (Prosimiae). Bonn Zool Beitr 34(1–3):29–71.
Müller EF, Jaksche H (1980) Thermoregulation, oxygen consumption, heart rate and evaporative water loss in the thick-tailed bushbaby (*Galago crassicaudatus*, Geoffroy 1812). Z Säugetierk 45:269–278.
Nakajima Y, Shantha TR, Bourne GH (1969) Histochemical detection of *L*-gulonolactone: phenazine methosulfate oxidoreductase activity in several mammals with special reference to synthesis of vitamin C in primates. Histochemie 18:293–301.
National Research Council (1978) Nutrient requirements of non-human primates. Nutrient Requirements of Domestic Animals No. 14 NAS.
Niemitz C (1984) Synecological relationships and feeding behaviour of the genus *Tarsius*. In: Niemitz C (ed) Biology of tarsiers. Gustav Fischer Verlag, Berlin, pp 59–76.
Norman RL, Brandt H, Van Horn RN (1978) Radioimmunoassay for luteinizing hormone (LH) in the ring-tailed lemur (*Lemur catta*) with antiovine LH and ovine ^{125}I-LH. Biol Reprod 19:1119–1124.
Palacio C (1977) Standard metabolism and thermoregulation in three species of lorisoid primates. MSc Thesis, University of Florida.
Petter-Rousseaux A (1970) Observations sur l'influence de la photopériode sur l'activité sexuelle chez *Microcebus murinus* (Miller 1777) en captivité. Annals Biol Anim Biochim Biophys 10:203–208.
Petter-Rousseaux A (1972) Application d'un système semestriel de variation de la photopériode chez *Microcebus murinus* (Miller 1777). Annals Biol Anim Biochim Biophys 12:367–375.
Petter-Rousseaux A (1975) Activité sexuelle de *Microcebus murinus* (Miller 1777) soumis à des régimes photopériodiques expérimentaux. Annals Biol Anim Biochim Biophys 15:503–508.
Petter-Rousseaux A (1980) Seasonal activity rhythms, reproduction and body weight variations in five sympatric nocturnal prosimians, in simulated light and climatic conditions. In: Nocturnal Malagasy primates. Academic, New York, pp 137–152.
Petter-Rousseaux A, Hladik CM (1980) A comparative study of food intake in five nocturnal prosimians in simulated climatic conditions. In: Nocturnal Malagasy primates. Academic, New York, pp 169–180.
Petter-Rousseaux A, Picon R (1981) Annual variation in the plasma testosterone in *Microcebus murinus*. Folia Primatol 36:183–190.
Poliak SC (1981) L'alimentation des lemuriens en captivite. Thesis, Université Paul Sabatier de Toulouse.
Pollock JI (1979) Female dominance in *Indri indri*. Folia Primatol 31:143–164.
Ralls K, Ballou J (1982) Effects of inbreeding on infant mortality in captive primates. Intern J Primatol 3(4):491–505.

Ralls K, Brugger K, Ballou J (1979) Inbreeding and juvenile mortality in small populations of ungulates. Science 206:1101–1103.

Reinhartz GE (1981) The effects of inbreeding in a captive population of a prosimian primate. MSc Thesis, Duke University.

Sade DS (1964) Seasonal cycle in size of testes of free-ranging *Macaca mulatta*. Folia Primatol 2:171–180.

Samonds KW, Hegsted DM (1973) Protein requirements of young cebus monkeys (*Cebus albifrons* and *apella*). Amer J Clin Nutr 26:30–40.

Schaaf CD, Stuart MD (1983) Reproduction of the mongoose lemur (*Lemur mongoz*) in captivity. Zoo Biol 2:23–38.

Tattersall I (1982) The primates of Madagascar. Columbia University Press.

Templeton AR, Read B (1983) The elimination of inbreeding depression in a captive herd of Speke's gazelle. In: Schonewald-Cox CM, Chambers SM, MacBryde B, Lawrence Thomas W (eds) Genetics and conservation. Benjamin/Cummings, Menlo Park, CA, pp 241–262.

Tompson FN, Lotshaw RR (1978) Hyperphosphatemia and hypocalcemia in lemurs. J Amer Vet Med Assoc 173(9):1103–1106.

Vandenbergh JG (1973) Environmental influences on breeding in rhesus monkeys. Symp 4th Intern Congr Primatol:Primate Reproductive Behaviour 2:1–17.

Van Horn RN (1975) Primate breeding season: photoperiodic regulation in captive *Lemur catta*. Folia Primatol 24:203–220.

Van Horn RN (1980) Seasonal reproductive patterns in primates. Prog Reprod Biol 5:181–221.

Van Horn RN, Resko JA (1977) The reproductive cycle of the ring-tailed lemur (*Lemur catta*): sex steroid levels and sexual receptivity under controlled photoperiods. Endocrinology 101(3):1579–1586.

Van Horn RN, Beamer NB, Dixson AF (1976) Diurnal variations of plasma testosterone in two prosimian primates (*Galago crassicaudatus crassicaudatus* and *Lemur catta*). Biol Reprod 15:523–528.

Varley MA, Vessey SH (1977) Effects of geographic transfer on the timing of seasonal breeding of rhesus monkeys. Folia Primatol 28:52–59.

Whittow GC, Scammell CA, Manuel JK, Rand D, Leong M (1977) Temperature regulation in a hypometabolic primate, the slow loris. Arch Int Physiol Biochim 85:139–151.

Woodall PF (1985) An evaluation of some methods for measuring male fecundity in small mammals. J Zool 206:263–267.

Yess NJ, Hegsted DM (1967) Biosynthesis of ascorbic acid in the acouchi and agouti. J Nutr 92:331–333.

21
Corral Breeding of Nonhuman Primates

WILLIAM J. GOODWIN

Introduction

Nonhuman primates are maintained in a variety of ways, from the indoor single-cage system to large enclosures and islands. The housing method selected is usually governed by the intended use of the animals. Climatic conditions are also a major factor when selecting the type of housing to be employed. The single-cage system is usually employed when frequent access to the animal is required. A variety of systems are used to house small groups of primates, including indoor pens in climate-controlled facilities, indoor-outdoor pens, corn cribs, and outdoor gang cages. These facilities are used for housing small breeding colonies, usually one male and several females, and for the long-term holding of small groups of animals. Larger colonies of primates are housed in outdoor enclosures, commonly referred to as corrals, and on islands.

This paper describes the various corral systems for the long-term holding and production of several species of nonhuman primates. The species of primates being maintained, the various types of construction, and the management techniques employed will be discussed.

The original corral for housing simians was constructed at the Oregon Primate Research Center in 1966 (Alexander et al., 1969). This corral was designed to house *Macaca fuscata* and is still in use after nearly 20 years. The original corral facility for maintaining great apes was the one constructed for chimpanzees at Holloman Air Force Base, New Mexico, in 1965 (Van Riper et al., 1967). This facility is no longer in use and will not be described further in this paper.

Species

A number of biomedically important primate species belonging to the family Cercopithecidae are being maintained under corral conditions. These include the rhesus monkey, other macaques, and the baboon. The simian species that are being bred successfully in corrals are the ones that live naturally in social units or troops and are adapted to living in temperate climates in the wild. Corrals are also being used successfully to maintain chimpanzees. The species of primates currently being maintained in corrals and the location of these facilities are shown in Table 21.1.

TABLE 21.1. Primates maintained in corrals (1985).

Species	Facility
M. mulatta	Oregon Primate Research Center, Beaverton, OR
	Yerkes Primate Research Center, Lawrenceville, GA
	Delta Primate Research Center, Covington, LA
	California Primate Research Center, Davis, CA
	Caribbean Primate Research Center, San Juan, PR
P. cynocephalus	Southwest Foundation for Biomedical Research, San Antonio, TX
	Washington Primate Research Center, Medical Lake, WA
M. fuscata	Oregon Primate Research Center, Beaverton, OR
M. radiata	California Primate Research Center, Davis, CA
M. nemestrina	Yerkes Primate Research Center, Lawrenceville, GA
M. arctoides	Yerkes Primate Research Center, Lawrenceville, GA
M. niger	Yerkes Primate Research Center, Lawrenceville, GA
C. atys	Yerkes Primate Research Center, Lawrenceville, GA
P. troglogytes	University of Texas System Cancer Center, Bastrop, TX
	Yerkes Primate Research Center, Lawrenceville, GA

Physical Facilities

Corrals vary considerably in their size and type of construction. The areas enclosed vary from 0.25 to 6.0 acres, and the walls vary in height from 3.0 to 4.8 m. A variety of materials are used for wall construction as follows: (1) all galvanized steel, (2) chain-link fence with galvanized steel, (3) all chain-link fence covered with chain link, and (4) prestressed concrete. All corrals have facilities for capturing the animals for routine management practices. Some have attached shelters, and others have small structures located inside the corrals to provide shelter, shade, and visual barriers. Corrals with inclined walls also provide shelter and shade. Primate feed is usually provided from the exterior in galvanized feeders and water is provided ad libitum in automatic water valves or lixits.

Corrals are currently employed for housing primates at eight major research institutions in the United States (Table 21.1). The design of the corrals is somewhat different at each institution; therefore, they will be discussed separately.

Oregon Primate Research Center

This center has two 2-acre rectangular corrals housing *Macaca fuscata* and six 1-acre corrals housing *M. mulatta*. The basic construction of the eight corrals is essentially the same. The walls are constructed of 26-gauge corrugated galvanized steel and are 3.5 m high slanting inward at a 15° angle. The walls are supported by 5.0-cm diameter steel pipes set at 2.4-m intervals around the perimeter. Each pipe is embedded in concrete with braces slanting outward at a 60° angle. Covered areas located between the corrals contain capture facilities, feeders, and an automatic watering system. These areas also provide shelter in inclement weather. Plywood A-frame hutches are placed in the corrals providing additional shelter, vertical space, and visual barriers. Approximately 80 rhesus monkeys are housed in each 1-acre corral.

A second 2-acre corral with a connecting tunnel has been constructed adjacent to the original corral to house the increase in the *M. fuscata* population.

Delta Primate Research Center

This center has 22 0.5-acre corrals housing *M. mulatta*. Each corral has a 3.6-m vertical wall that consists of 1.2 m of 12-gauge chain-link fence and 2.4 m of 26-gauge corrugated galvanized metal sheets. The metal sheeting is bolted to the interior of the galvanized steel pipe framework. The vertical supports are 7.5-cm galvanized pipe. The galvanized pipes and the lower edge of the chain-link fence are embedded in a deep concrete footing. The horizontal members of the framework are 3.7-cm galvanized pipe. The walls are braced externally with 6.25-cm galvanized pipe set at a 45° angle from the top to the ground. The bases of these pipes are set in concrete and are spaced 6 m apart. The wire at the base of the wall reduces the wind load and permits ventilation in hot weather.

Each corral is provided with an exterior pen that serves as a safety entrance, a holding area, and a capture facility. The animals have access to this pen through a chain-link-covered, partitioned tunnel. Feeders are located inside these pens. The catch pen can be modified to provide shelter by covering the sides with panels. Additional shelter is provided in the form of plywood A-shaped hutches. In addition to shelter, these hutches provide vertical spaces, shade, and visual barriers. Each corral houses from 40 to 60 rhesus monkeys.

In addition to the 22 corrals described above, this center has a large enclosure originally designed to house chimpanzees and currently used to house rhesus monkeys. The overall dimensions of this compound are 29.7 × 121.1 m. The walls are 4.8 m high and are inclined inward at a 12° angle for the lower 3.6 m; the top 1.2 m is further inclined at a 30° angle. The first 2.4 m are chain-link fence, and the top 2.4 m are flat steel metal. One wall has a cinder block capture pen that is 2.7 × 3 m and a 2.4 × 4.8 m observation room.

Yerkes Primate Research Center

This center has 22 compounds for primate living groups, each with attached heated indoor housing for use during inclement weather. These compounds range in size from 15 × 15 m to 37.8 × 37.8 m. One compound houses a chimpanzee breeding colony, and the remaining 21 compounds house a variety of monkeys as shown in Table 21.1. The monkey compounds have basically similar construction differing only in size and the types of indoor quarters. Most of the compounds are in multiple sets with common walls. The outside vertical walls are 4.8 m high with the bottom 1.8 m being 9-gauge chain-link fence and the top 3.0 m made of 16-gauge galvanized steel sheets. The support consists of 10-cm I-beams and angle iron. Walls between adjacent compounds are 4.8 m of galvanized steel sheets fastened to both sides of the I-beams.

All of the compounds are equipped with concrete culverts and heavy-duty playground equipment to provide shelter and vertical space. Indoor units consist variously of concrete block, prefabricated metal buildings, or modified trailers. They vary in size from 54 to 121 m^2, depending on the size of the compound. The indoor units are equipped with perches to provide ample space. Feed is provided in metal feeders serviced from the outside of the compound, and water is available in automatic watering devices.

The single chimpanzee compound is 23 × 30 m with 4.8 m high walls. The bottom 1.8 m is 6-gauge chain-link fencing reinforced with steel bars, and the top

3.0 m is 16-gauge galvanized steel sheets. The compound supports are 25-cm I-beams and galvanized steel tubing. The indoor quarters for this facility are in a concrete and concrete block building 6 × 7.5 m with four indoor cages, service corridor, and three service rooms.

Caribbean Primate Research Center

This center has three 0.3-acre corrals and one 2.0-acre corral at its Sabana Seca facility housing *M. mulatta*. The walls are 4.8 m high with the bottom 1.2 m being chain-link fence and the top 3.6 m constructed of 26-gauge galvanized metal sheets. The corral walls are supported by 7.5-cm galvanized pipes, 6.0 m in length, embedded in concrete. A feeding and capture facility is located adjacent to the corral.

The 2-acre corral encloses a heavily wooded hillside; therefore, trees and shrubs provide shade and vertical space. Feed is provided by spreading it around the perimeter of the fence and in galvanized metal feeders.

California Primate Research Center

This facility has 11 unique 0.5-acre corrals. These corrals have walls constructed with a 9-gauge wire fabric, and the entire corral is covered with the same wire fabric. The wire fabric on the walls is attached to 6.25-cm diameter galvanized pipes spaced 2.4 m apart and placed in a concrete footing. The walls of these corrals are 2.0 m high. The wire fabric cover is supported by 8.75-cm diameter metal posts spaced 6 m apart in each direction. Each support post has two lateral arms supporting the wire cover. In addition, two lateral rods extend from the top of the support pole and are attached to the wire fabric. These configurations provide adequate support of the wire fabric cover. The corners of the corrals are enclosed with 26-gauge galvanized metal sheets providing protection in inclement weather. A series of perch bars for the animals are located in the protected corners.

The supporting facilities for each corral include a 6 × 3 m entrance lock and a 4.2 × 1.0 m capture pen and chute. Elevated A-frame huts provide shelter, shade, vertical space, and visual barriers. Water is provided by an automatic watering system, and the primate chow is provided in metal feeders and scattered on the ground inside the corral. Sprinklers are placed on the top of each support pole. These are used to moderate the temperature on hot days (95°F+) and in maintaining the vegetation in the corral.

Southwest Foundation for Biomedical Research

Two 6.0-acre dodecagon corrals have been constructed at this facility to house a breeding colony of baboons, *Papio cynocephalus anubis*. The original 6.0-acre corral was constructed in 1979 and has been described by Goodwin and Coelho (1982). This 12-sided configuration was chosen over the more usual rectangular corrals because it encompasses a greater area with the same number of linear meters of wall. The walls are 3.5 m high constructed of 22-gauge galvanized metal and inclined inward at a 15° angle. The metal sheets are riveted and screwed to steel cross-members attached to angle steel support braces buried 2.1m in a concrete pylon. The inclination of the wall provides approximately a 2-m area of shade and protection from inclement weather. Openings 1.2 m high and 2.4 m long are located at 30.5-m intervals around the perimeter. These provide wind relief and locations for feeders and the automatic watering system. The galvanized metal feeders are serviced from outside the walls. This corral houses approximately 500 baboons.

A second 6.0-acre 12-sided corral designed to house juvenile baboons was constructed in 1984. Each side is 45 m long constructed in 7.5-m sections of 26-gauge corrugated steel. The wall is 3 m high inclined inward at a 15° angle. Two strands of electrified wire are located at the top of the wall. Electricity is provided by two livestock fence chargers. The wall is placed on a concrete footing 15 cm wide and 45 cm underground. The metal sheets are riveted and screwed to the two horizontal cross-members that are attached to steel posts placed 7.5 m apart. Each post is supported by a steel brace bolted to the post and to a concrete footing. Each section contains a 1.8 × 1.0 m opening for feeders and the automatic watering system. The juvenile corral can also house approximately 500 baboons.

The supporting facilities include a 21.6 × 6.0 m (L × W) capture pen, four holding pens, two vehicle entrance locks, and a 7.6 × 4.9 m work area. The work area includes a hydraulic squeeze mechanism for restraining baboons, cages for separating animals, scales, a U-shaped roller conveyor system for moving sedated baboons during processing, work space for technicians, and toilet and shower facilities. A refrigerated feed room and an observation tower are located adjacent to the capture facility. A chain-link tunnel (partially portable) connects the juvenile corral with the central work area.

Washington Primate Research Center

This primate center, located at the University of Washington, has a primate field station at Medical Lake, Washington. This facility is a large three-story building that houses breeding groups of *M. nemestrina* and *P. cynocephalus* in indoor harem units. In 1979, a 0.6-acre outdoor corral was constructed to house approximately 30 baboons. This was a pilot project to determine the suitability of this type of housing for baboons under severe winter conditions. The corral area was formed by fencing between two wings of the building that are at a 90° angle. A 12 × 6 m heated building containing animal housing space, a capture area, and a work area is provided.

During a 5-year period, the morbidity and mortality was slightly less and the birth rate slightly higher than for those baboons housed in the indoor harem units. These results encouraged the expansion of this system for housing baboons.

A new corral enclosing an additional 3.6 acres and encompassing the original 0.6 acres was constructed in 1984. The fence of both corrals is constructed of heavy chain link 2.4 m in height with 2.4 m of galvanized steel metal beginning 1.8 m from the ground and extending 1.8 m above the chain-link fence. The total height of the fence is 4.2 m. The fence is supported by 5.0-cm galvanized pipe embedded in concrete and is inclined inward at a 15° angle. An 18 × 9 m insulated, heated metal building provides shelter during inclement weather. This building has a concrete floor and gutters that direct effluent to the sewer system.

This corral system with the heated shelter has proven effective and efficient for maintaining baboons in a location with severe winter weather.

University of Texas Systems Cancer Center Science Park

This institution has developed a unique system for housing a large colony of *Pan troglodytes*. This facility is designed to rehabilitate laboratory-reared, behaviorally deficient chimpanzees and to house breeding colonies. It includes quarantine facilities, single or group housing for long-term holding, and semifree-ranging compounds for established breeding groups (Riddle et al., 1982). The eight octagonal compounds (1287 m^2) surround a central service area. The walls of the compounds are constructed of prefabricated, prestressed, reinforced concrete 4.8 m high with the top 1.2 m inclined inward at a 45° angle. Ventilation and viewing windows are located around the compound walls. Each compound has an entrance gate permitting vehicle access to the inside of the compound. A heated den is located adjacent to each compound.

The service area contains two nurseries, eight treatment rooms, an isolation ward, a procedural room, diet kitchen, and rooms for storage of supplies and equipment.

A complex of concrete, steel, and wooden climbing structures of various designs are located in the center of each compound. They are designed to provide the animals with shade, shelter, escape routes, visual enrichment from high perches, and exercise. Rooftop observation decks allow surveillance of the animals.

Colony Management

The management policy usually employed in primate corral operations is to provide adequate care and to intervene only when necessary. The management system has proven to be very effective as evidenced by several successful corral breeding programs. Observations are made on a daily basis in order to identify and remove sick or injured animals and to record births. Behavioral observations are made in some corral operations as part of ongoing research programs. Animals are fed a commercial primate chow on a daily basis and water is usually available ad libitum. The animals are captured and evaluated one or more times per year. The capture methods vary with the corral designs and the primate species involved. The usual procedure is for several persons to enter to corral and herd the animals into the capture facility located adjacent to the corral. Rhesus monkeys are usually herded into a partitioned chain-link tunnel and then transferred to holding cages for processing. Some research institutions physically restrain these animals and others sedate them with ketamine hydrochloride. Baboons and larger primates are herded into a squeeze device and sedated with ketamine hydrochloride prior to processing. Routine processing includes physical examinations, tuberculin testing, tattooing, weighing, and inoculations if necessary. Biological samples and data for research purposes are collected at this time.

The primary cause of morbidity and mortality in corral colonies is related to trauma. Some of the other medical problems encountered include pneumonia, diarrhea, and hypothermia.

There are several advantages of the corral method of housing animals. The animals are confined under semifree-ranging conditions permitting normal social interactions and physical development. The cost of the construction of corrals on a per animal basis is considerably less than the individual cage or gang cage method. The cost of maintaining the animals is considerably less owing to the reduction in labor costs. It is estimated that the per diem costs for animals housed in corrals is less than one-half the cost for primates housed in gang cages and less than one-fourth the cost of maintaining primates in individual cages. The cost factor is one of major consideration when designing facilities for sustaining primate colonies over long periods of time.

Summary

The corral system has proven to be an effective, efficient, and economical method for maintaining several species of primates on a long-term basis. A review of all of the corral systems currently in use in the United States is presented. The various designs, construction materials, supporting facilities, and management procedures are described.

Acknowledgments. The author wishes to thank individuals of the various primate facilities for the information provided for this paper.

References

Alexander BK, Hall AS, Bowens JM (1969) A primate corral. J Am Vet Med Assoc 155:1144–1150.

Goodwin WJ, Coelho AM Jr (1982) Development of a large scale baboon breeding program. Lab Anim Sci 32:672–676.

Riddle KE, Keeling ME, Alford PL, Beck TF (1982) Chimpanzee holding, rehabilitation and breeding: facilities design and colony management. Lab Anim Sci 32:525–533.

Van Riper DC, Fineg J, Day PW (1967) Development of a primate resource. Lab Anim Care 17:472–478.

22
Environments for Captive Propagation of Primates: Interaction of Social and Physical Factors

JOSEPH ERWIN

Introduction

In countries of origin, nonhuman primates compete for space with rapidly expanding human populations. The loss of primary habitat as a result of the encroachment of humans on previously undeveloped areas is clearly the most serious threat to wild primates. As primate ranges become restricted, competition within species (as well as between species) intensifies. At critical levels, populations exceed carrying capacities by so much that they cannot be sustained, and rapid declines in numbers result. It is necessary that optimal sizes for wild preserves be identified if conservation in natural settings is to achieve long-term success.

In captive settings, many of the problems associated with critical sizes of home range are eliminated. Sufficient food is provided, and potential predators are absent. Health care is available, and climatic hazards are usually eliminated. Still, many factors related to carrying capacity apply to captive as well as natural settings, and these must be considered along with other practical aspects of management if captive propagation efforts are to succeed.

This report includes summaries of research projects (conducted by myself and my colleagues) that have dealt with such issues as social organization, social bonds, social roles, aggression, spatial change, social density, provision of cover, group formation, crowding, and risk of trauma or disease as related to environmental design.

Enduring Relationships

Wild macaques and baboons (as well as other kinds of primates) form social groups based on enduring social bonds among group members. Females remain with their mothers and sisters in the groups where they were born, while males leave their natal groups early in adolescence. Females are especially intolerant of unfamiliar females and usually attack them if they are encountered.

In captivity, as in the wild, many kinds of primates are intolerant of strangers. Introduction of unfamiliar animals into groups can result in violent, even lethal, attacks on the newcomers. Consequently, it is desirable to consider the nature of social bonds and roles when primate breeding groups are being formed. If the animals involved are very rare, it is especially critical that potential problems be anticipated and avoided.

Fundamental information from laboratory experiments designed to increase understanding of attachment processes can aid in prevention of unnecessary violence during breeding group formation. The information is vital to avoid the loss of animals that may represent the last prospect of saving a species from extinction.

An example is provided below of some work we did in the early 1970s. It contributed basic data that aided understanding of attachment processes and mechanisms of rhesus macaque social organization. In addition, characteristics were demonstrated that require consideration in managing primate breeding colonies.

Reunion Studies

Same-Sexed Peers

Twelve 4.5-year-old rhesus macaques, six of each sex, were reunited with peers with which they had spent the second year of their lives. Their responses were compared with those they exhibited when paired with unfamiliar peers matched for age and sex with the familiar peers. Despite the passage of more than 2 years since the animals had been together, there were clear differences in responses to familiar and unfamiliar peers. Subjects displayed less aggression, fear/submission, and disturbance indicators, and more affiliative behaviors while paired with familiar than with unfamiliar animals. In fact, females always fought if they were unfamiliar and never did so if they were familiar. Females embraced, groomed, contacted, and remained near their familiar peers. The evidence indicated that attachments among female rhesus macaques were very strong and persistent. While some males exhibited strong mutual attachment, this was not typical. Detailed results of the study are published elsewhere (Erwin et al., 1974).

The results of this study are consistent with data from field studies (e.g., Lindburg, 1971; Southwick et al., 1965) and indicate that the female–female attachment system is very basic. The existence of such consistent aggressive responses to unfamiliar animals by female rhesus macaques suggests that it is unwise to add animals to established groups. If such additions must be made, great care must be taken, because the motivation to attack strangers appears to be very deeply rooted in the biological, psychological, and social nature of these animals.

Opposite-Sexed Peers

Twelve 5.5-year-old rhesus macaques, six of each sex and the same Ss as in the previously described study, were reunited with opposite-sexed peers with which they had been housed for 6 months but from which they had been separated for 2 years. The animals gave some evidence that they recognized their former cage mates. No animals ever directed any aggression or threatened aggression toward familiar peers, but females sometimes threatened unfamiliar males, and males attacked unfamiliar females even if they had previously copulated with them. Males mounted familiar females more frequently than unfamiliar females, and females permitted familiar males to mount more quickly than unfamiliar males. Detailed results of the study were published elsewhere (Erwin and Flett, 1974).

Again, the data were consistent with field data (e.g., Neville, 1966; Lindburg, 1971) that have indicated that rhesus macaques can develop long-term heterosexual relationships in nature. The fact that some affiliative tendencies withstood a separation of about 2 years indicates that these attachments were strong. At the same time, comparison of the results of this study with the one on same-sexed pairings indicated that the female–female bonds were by far the most consistent of any relationships that were examined (Erwin et al., 1975). The strength of macaque female–female bonds cannot be ignored in setting up breeding groups. Awareness of these relationships was useful in some of the other studies described below.

Environments and Health Risks

The work described above was conducted at the California Primate Research Center at University of California, Davis, and was completed in 1974. That year I moved to the Regional Primate Research Center at University of Washington to work at the Primate Field Station at Medical Lake on a project concerned with the causes of premature birth and other poor pregnancy outcomes. This was a breeding colony, primarily for pig-tailed macaques (closely related to the rare lion-tailed macaques). At that time the mission included conservation of the species as well as propagation and use in biomedical and behavioral research. There was a commitment to housing these primates in social groups. Although many of the macaques at the California PRC were in social groups, most of the animals with which I had worked had been socially housed, but usually only in pairs. I was surprised to see that there were many serious consequences of social housing. One of those problems (and one familiar to every primate keeper and primate curator, as well as every breeding colony manager) was aggressive behavior and trauma. Fighting was common, as were serious bite wounds. The director of the center was concerned and asked me to study the problem and make some recommendations. That work is summarized below.

Background

The Primate Field Station at Medical Lake, Washington, housed more than 1000 pig-tailed macaques, nearly all in social groups indoors. The rooms in which the groups were housed were about the size one would expect to see in a traditional primate house in a zoo. At the time I arrived, each group had access to two rooms for most of each day. Once each day, the group was crowded into one room while the other was cleaned, and later shifted to the clean room while the other was cleaned. This cleaning regimen had been adopted not long before in response to concerns over enteritis (especially shigellosis) that was common among animals in the colony. The previous cleaning strategy had involved cleaning an end room, then shifting the group in the next cage to the clean room, cleaning that room and shifting the next group over, and so on down the line. The concern was that there was some overlap in room occupation, and to the extent that the cleaning was imperfect (despite use of pressure washers and disinfectants), enteritic pathogens could be transmitted. Since enteritic diseases seemed to be stress related, it had been decided that providing two rooms, rather than just one for each group, would alleviate stress caused by crowding. By the time I arrived, it was clear that treatments for enteritis were still frequent and bite wounds were still very common.

Because the new procedure involved crowding animals into one room of their two-room suite each day, I suspected that the fighting that resulted in injury might have been in response to the temporary crowding. Our first study examined this hypothesis, and the results led to a series of experiments and surveys.

Crowding Studies

Short-Term Crowding: Spatial Change

The first experiment involved observation of 98 pig-tailed macaques housed in six groups. Five groups included one adult male, and one group included no adult male. Each group was observed for 20 minutes with access to both rooms of their suite, then for the same amount of time while confined in one room, and then again with access to both rooms. We were surprised to discover that there was invariably more contact aggression (grab, push, hit, bite) and threatened aggression (open mouth, stare, head bob) when animals had access to two rooms than when they had access to only one room (with one exception involving the group that included no male). It was almost inconceivable to us that the more crowded condition would result in less aggression, so we repeated the experiment with more groups and additional controls.

Spatial Change Crowding: Replication

The replication study involved eight groups (not used in the previous study) containing 109 pig-tailed macaques (one adult male per group). Each group was observed in two experimental sequences: in two rooms, then one room, then two rooms, and in one room, two rooms, then one room. The control sequences were three consecutive observation periods with single- or double-room access. In every case, for every group, without exception, there was more contact and threatened aggression when the groups had access to two rooms than when they had access to one room. Invariably there was about twice as much aggression when the animals had twice as much space.

Reduction of Trauma and Disease

The results of these experiments were published and discussed in Anderson et al. (1977) and Erwin (1977, 1979). A subsequent analysis of records revealed that treatments for bite wounds had doubled when the two-room suite housing strategy was adopted. Thus, the short-term behavioral observations accurately indicated the long-term process in the colony. It is worth noting here that the housing strategy was changed again after this series of studies was completed. Each group had a two-room suite but occupied only one room at a time, on alternate days. An analysis of the long-term consequences of this change clearly indicated that the frequency of treatments for trauma, enteritis, and respiratory disorders all decreased dramatically after the adoption of the single-room housing strategy. The combination of group isolation and reduction of group stress resulting from aggression apparently was responsible for reduction of communicable disease in the colony.

Perhaps the most important point of these studies is that the common-sense solution to the initial problem turned out to be worse than the problem. The solution required careful and systematic measurement of behavior and good records of management and clinical procedures.

Subsequent studies provided some understanding of the social processes responsible for the environmental impact on behavior, trauma, and disease, and these studies are described below. It was necessary to examine crowding from another perspective. Although there was consistency in response to change in the amount of space available in the studies reported above, there was also an indication of more aggression in groups containing more animals.

Crowding: Social Density

A survey was conducted in which aggressive behavior in groups was measured and examined in relation to the number of adult females per group and the number of adult males per group. In this study, 92 groups were surveyed. The number of females per group was highly correlated with the amount of contact aggression that occurred. The effect was particularly strong for groups containing more than ten females. Clearly this recommended that groups of pig-tailed macaques in these conditions should contain ten or fewer adult females.

It was interesting to note that aggression in groups containing no adult males was much higher than for groups containing at least one male. The number of males made no difference. The results of this study by Erwin and Erwin (1976) along with similar results from specific surveys under similar conditions clearly implied that the presence of a male reduced aggression among females.

Social Roles

Groups With or Without Males

Aggressive behavior was surveyed in three groups containing only adult females with their offspring and three matched groups, each containing a male. Without exception, females fought more in the groups containing no adult males (Sackett et al., 1975). In a follow-up study comparing 15 one-male groups with five groups containing no males, the same relationship was found (Dazey et al., 1977). The results of the Erwin and Erwin (1976) study mentioned above were consistent with these results. Apparently the presence of males inhibited aggression among females.

Experimental Removal of Males

To determine whether the presence of a male inhibited female–female aggression in specific groups, we experimentally removed males from their groups. The removals were for only 20 minutes, and female–female aggressive behavior was measured before removal, during male absence, and after male return. In each of the six groups studied, female aggression increased markedly during the absence of the male (see Oswald and Erwin, 1976).

Experimental Removal of Females

Considering it possible that the removal of any animal might increase aggressive behavior in groups, Swenson and Bartlett (1976) performed similar studies with removal of high-, middle-, and low-ranking females. There was no increase in aggression among females associated with these removals whether or not a male was present in the groups.

Social Role Interacted with Environment

A survey of the location and duration of female–female aggressive bouts indicated that more aggressive bouts which lasted longer and included more acts of contact aggression (grab, push, hit, bite) occurred in the room of the two-room suite where the male was not present. If a female attacked another in the presence of the male, she typically only bit the victim once and no retaliation occurred. When in the room away from the male, females attacked, victims retaliated, and extended fights occurred, sometimes continuing until the male came to the opening into the second room or entered the room and threatened or attacked one of the combatants. For the most part, the mere presence of a male was sufficient to limit fighting among females (see Erwin, 1979).

Cage Furniture and Aggression

Experimental Provision of "Cover"

While it was clear that escape into a room away from the male was not escape at all, it seemed plausible that some smaller escape structure placed within a single room might result in reduced aggression. Concrete culverts were introduced into six rooms. Baseline measurements of behavior were made prior to the placement of the culverts and were compared with behavior after installation. The macaques immediately began to use the cylinders to sit on and in and to escape from conspecifics giving chase. Unfortunately, two of the groups underwent major social changes during the test phase of the study. In the most serious case, the resident male was removed and a male with his small group of females was introduced. Serious fighting ensued, and all the original females were injured or killed. Clearly, the provision of cover was not sufficient to override the serious consequences of placing unfamiliar female macaques together (Erwin et al., 1976).

Zoo Experiences

Some years after leaving the Medical Lake facility, I became Curator of Primates for the Chicago Zoological Society. It became necessary on several occasions to attempt to apply some of the earlier work. For example, when a siamang female was abusing her infant and the older sibling was fighting with the mother over the infant, we added ropes to the cage. This reduced the aggression, at least on a temporary basis. The addition of cage furniture to a cage that included talapoins and red-tailed guenons noticeably reduced aggression. Addition of shelving to baboon quarters seemed to reduce aggression. But all these and many other examples from zoos are anecdotal. There is a need for standardizing and systematizing the wisdom of primate keepers and curators in this area.

Formation of Groups

Group Merger and Other Techniques

When groups were being reconstituted for other reasons at the Medical Lake facility, we examined three basic patterns of group formation: (1) merger of two existing groups (with a male familiar to the females of one group or unfamiliar with either); (2) merger of three or four subgroups of three females (males as above); (3) all strangers, or at least, no more than two animals familiar with each other. The highest risk group was that in which two existing groups were merged, especially if a male familiar with one group was included. Fights typically continued until all females unfamiliar with the male were killed, injured, or otherwise removed. There were virtually no injuries among the groups composed of complete strangers, despite the observation of much agonistic behavior in those groups. Thus, it appears that it would be best to avoid formation of new macaque groups by merging groups. If it must be done, groups should be formed of complete strangers or animals already all familiar with each other (recall section on bonds).

Multispecies Groups

The Tropic World facility at the Chicago Zoological Park in Brookfield, Illinois, includes exhibits containing several species of primates simultaneously. The exhibit areas are spacious and each species has its own off-exhibit sleeping quarters. Without going into great detail, the species housed together are mentioned here. In the African section of Tropic World, mandrills, sooty mangabeys, Kolb's guenons, talapoins, and guerezas are together daily on exhibit. The area is large and detailed with artificial trees and vines that allow escape routes, and there has been little serious aggression in the situation.

In the Asian section of Tropic World there have been siamangs, white-handed gibbons, crab-eating macaques, and silvered leaf monkeys. Interactions between siamangs and other species have been intense, with much interaction between them and the crab-eating macaques, but with no injuries to either. Siamangs have chased gibbons, but the gibbons have virtually always eluded them, and no injuries have ensued. At times, however, the male gibbon jumped to the nearby orangutan exhibit under pressure from siamangs. Interaction between orangutans and gibbons, crab-eating macaques, and silvered leaf monkeys have all been without problem. Silvered leaf monkeys were injured by crab-eating macaques and siamangs, and an infant was killed by a gibbon. A silvered leaf monkey stole a crab-eating macaque infant but did not injure it.

In the South American area, spider monkeys, squirrel monkeys, and capuchins are exhibited together. There have been no injuries in that situation. The area is characterized by tall artificial trees and multiple vine pathways that offer abundant escape routes.

The Tropic World exhibits have demonstrated that many species can be exhibited together if a large enough environment is supplied and structures for climbing and escape are provided that are ecologically valid and within the capacities of the animals to use. Observation of the animals in the wild, if possible, provides the strongest basis for designing captive environments with the features that will be used effectively.

Conclusions

1. Many kinds of primates develop attachments or emotional bonds with conspecifics that may endure over years. These social bonds should be considered in management decisions regarding group integrity, genetic management, and formation of new breeding groups.
2. There is no substitute for careful documentation and experimentation to determine whether management strategies are accomplishing their intended goals. Social and physical environments can interact in complex ways that lead to counterintuitive results.
3. More space is not always best. It is the quality of the space that is most important, rather than the sheer quantity.
4. Cleaner is not always better. Cleanliness can be insufficient to overcome psychological and social disturbances.
5. Provision of cage furniture and cage enrichment can reduce psychological and social problems in primate groups. Enrichment usually does not supersede social stability as a contributor to tranquility.
6. Introduction of unfamiliar primate groups to each other can result in serious fighting, injury, and death.
7. Design of environments for propagation and exhibition of primates can include release of animals into multispecies enclosures. There is apparently little risk in doing so for some species. Most primates can be easily trained to enter a holding area for the night that permits access for treatment or other purposes.

Summary

Information from experiments and surveys regarding the roles of social and physical environments in captive management and propagation of primates is reported, including the following: (1) risk of violence was much higher when unfamiliar rhesus macaques were introduced than when familiar animals were reunited, even after 2 years of separation, especially for females; (2) pig-tailed macaques fought more and suffered more frequent bite wounds and disease problems when the amount of space available to them was doubled, and the effect was reversed when groups were returned to the original amount of space; (3) fighting among female pig-tailed macaques in social groups increased dramatically in groups containing more than ten females; (4) fighting was more frequent in all-female groups of pig-tailed macaques than those in which an adult male was housed; (5) experimental removal of adult males from groups of pig-tailed macaques resulted in increased aggression among females, while removal of high-, medium-, or low-ranking females did not have that effect; (6) provision of cage furniture offering cover resulted in reduced fighting in pig-tailed macaque groups but did not prevent violence in groups that underwent major social changes; (7) new groups formed by merger of existing groups suffered many more injuries than new groups formed by introduction of several subgroups or of unfamiliar animals; (8) experiences with many primate species in a zoological park emphasize the need to consider species-specific characteristics in the design and management of primates in laboratories, breeding colonies, and zoos. The need to consider temporal factors in primate management is emphasized, along with the importance of documentation of all aspects of primate health, behavior, and management.

References

Anderson B, Erwin N, Flynn D, Lewis L, Erwin J (1977) Effects of short-term crowding on aggression in captive groups of pigtail monkeys (*Macaca nemestrina*). Aggress Behav 3:33–46.

Dazey J, Kuyk K, Oswald M, Martenson J, Erwin J (1977) Effects of group composition on agonistic behavior of captive pigtail macaques, *Macaca nemestrina*. Amer J Phys Anthropol 46:73–76.

Erwin J (1977) Factors influencing aggressive behavior and risk of trauma in the pigtail macaque (*Macaca nemestrina*). Lab Anim Sci 27:541–547.

Erwin J (1979) Aggression in captive macaques: interaction of social and spatial factors. In: Erwin J, Maple TL, Mitchell G (eds) Captivity and behavior: primates in breeding colonies, laboratories, and zoos. Van Nostrand Reinhold, New York.

Erwin J, Flett M (1974) Responses of rhesus monkeys to reunion after long-term separation: cross-sex pairings. Psychol Rep 35:171–174.

Erwin J, Maple TL, Willott J, Mitchell G (1974) Persistent peer attachments of rhesus monkeys: responses to reunion after two years of separation. Psychol Rep 34:1179–1183.

Erwin J, Maple TL, Welles J (1975) Responses of rhesus monkeys to reunion: evidence for exclusive and persistent bonds between peers. Contemp Primatol Karger, Basel, pp 254–262.

Erwin J, Anderson B, Erwin N, Lewis L, Flynn D (1976) Aggression in captive groups of pigtail monkeys: effects of provision of cover. Percept Mot Skills 42:319–324.

Erwin N, Erwin J (1976) Social density and aggression in captive groups of pigtail monkeys (*Macaca nemestrina*). Appl Anim Ethol 2:265–269.

Lindburg D (1971) The rhesus monkey in North India: an ecological and behavioral study. In: Rosenblum L (ed) Primate behavior: developments in field and laboratory research, vol 2. Academic, New York, pp 1–106.

Neville MK (1966) A study of the free-ranging behavior of rhesus monkeys. Thesis, Harvard University, Cambridge.

Oswald M, Erwin J (1976) Control of intragroup aggression by male pigtail monkeys (*Macaca nemestrina*). Nature 262:686–688.

Sackett D, Oswald M, Erwin J (1975) Aggression among captive female pigtail monkeys in all-female and harem groups. J Biol Psychol 17:17–20.

Southwick C, Beg M, Siddiqui M (1965) Rhesus monkeys in North India. In: DeVore I (ed) Primate behavior: field studies of monkeys and apes. Holt, Rinehart and Winston, New York, p 111–159.

Swenson L, Bartlett L (1976) Experimental removal of female pigtail monkeys from groups. Paper presented Int Congr Primatol, Cambridge, England.

23
Behavior Requirements for Self-Sustaining Primate Populations— Some Theoretical Considerations and a Closer Look at Social Behavior

JAN A.R.A.M. VAN HOOFF

Museums may maintain specimens "for eternity" through preservative chemicals, but a Zoo's duty lies in passing on living species to posterity through appropriate breeding techniques. This process involves natural propagation, with all its characteristic behavior patterns. Because of this dimension, Zoos are "Behavior museums."

Heini Hediger (1982)

Introduction

This quotation of the pioneering advocate of species adequate systems of animal maintenance in zoos spells out succinctly the essence of our theme. During the last 15 years or so we have witnessed a strongly increasing interest in the behavioral aspects of our dealings with animals in conditions of confinement. It coincides with an almost paradigmatic change, in the Kuhnian sense, in the attitude of science and society toward animals. This is exemplified by the revived scientific consideration of animals as purposely striving aware creatures (e.g., Griffin, 1976). This change in scientific outlook undoubtedly has arisen in connection with and, in turn, feeds the more general concern about the welfare and "rights" of animals under our dominion. Among these, the primates occupy a special position. It has become obvious that their cognitive potentialities and the complexity and flexibility of their behavior have tended to be grossly underestimated in the past.

This behavioral potential undoubtedly is adaptive in that it meets the functional demands to which a species has been exposed during its evolutionary history. In other words, the ultimate factors of evolutionary selection have shaped programs of development that bring about relevant needs and the behavioral instruments to satisfy them.

In the natural situation, where populations are self-sustaining, the behavioral requirements are therefore adequate "by definition" (be it not always factually!). In confined, managed, artificial conditions such as are offered by the zoo and, especially, the laboratory and the bioindustry, the needs and the behavioral instruments often do not match the conditions; disturbed functions and stereotyped behaviors are the most obvious expressions of this (e.g., Hediger, 1942; Morris, 1964; Meyer-Holzapfel, 1968; Mitchell, 1970; Erwin et al., 1979).

Unnatural Confinement and Behavior

To evaluate the effects of unnatural confinement on behavior, we should consider present ethological views on the structure of behavior. We can conceive the behavior of an organism as a hierarchically integrated set of systems or goal-directed functions. The goals may be to get food, to achieve sexual contact, to avoid predators, to get and remain clean, etc. Each of these functions may command various routines and subroutines with their own second-order goals; they serve as instrumental actions in the realization of the major goals (e.g., Leyhausen, 1965; Baerends, 1976; Dawkins, 1976).

As a rule, an animal can do only one thing at a time; in other words, these functions mutually exclude each other's performance and have to share the available time (e.g., McFarland, 1974). They can even be said to compete for expression, the greatest need becoming dominant. The respective function acquires the hegemony, more or less monopolizing perceptive and effector mechanisms and thus inhibiting alternative functions (e.g., once an animal is set on mating, it temporarily loses its interest in food, and vice versa).

An important point is that the various functional systems do differ in the ease with which they can achieve dominance over one another. This depends not only on the relative strengths of the motivational factors of contending systems, but also on their specific nature. In this respect I find it useful to qualify different behavioral functions as more "reactive" or more "spontaneous" (van Hooff, 1974; cf. Hughes, 1980).

"Reactive" Behavior Functions

Some, such as predator avoidance or territorial defense, must be called upon immediately once the environmental releasers present themselves. Therefore, these functions must be able to interrupt the hegemony of other functions. They are primarily of a reactive nature and the "need" depends on an environmental change causing a discrepancy between, on the one hand, the animal's norm representation (goal value, expectancy) of (its relation with) the environment and, on the other hand, the actual situation.

In unnatural situations, such reactive functions can be overstimulated. Such overstimulation can pathologically disrupt the behavioral equilibrium, for example, when other functions are thwarted by the arousal associated with abnormally intense or frequent activation of the fight, flight, fright mechanisms (i.e., "stress"). Thus, Boër (1983) has suggested that the disquieting failure of male gorillas in many institutions to exhibit sexual behavior is due to stress, excessive excretion of ACTH causing spermatogenic arrest. This stress could be due to a housing system in which the animals are continuously confronted with observers at close quarters, leading to an overstimulation of the territorial defense system (Maple and Stine, 1982). Nevertheless, other explanations may apply as well (Resko, 1982).

Disruptive, stressful overstimulation may occur in several contexts in a zoo. Glatston et al. (1984) have compared the behavior of family groups of cotton-topped tamarins (*Saguinus o. oedipus*) when they were on display to the public and when they were housed off display. They suspected that the presence of the public was a disturbance, which might even be responsible for high infant mortality. Indeed, a strong difference was found. Animals on display had less body contact. In particular they avoided their young more often. Both the father and the mother had a lower percentage of the contact initiations between them and the children. In addition they attacked their persistent children more frequently.

"Spontaneous" Behavior Functions

Other behavioral functions are comparatively spontaneous in that variations of endogenous factors are important in the direct causation of the behavior. Feeding behavior is a case in point: as enteroreceptive measures of the actual state of satiation come to deviate from the relevant norm values or set points, its motivation grows and lends potency to releasing and steering factors in the environment. In this case the "need" may be generated by a deviation from a norm of some variable that fluctuates as a function of some "decay" or "renewal" process, either autonomously or under the protracted, summated influence of external factors. Another example: Exploratory behavior may provide satisfying certainty that subsequently decays again and drops below a threshold value.

We may expect that the more spontaneous behaviors have low priority, i.e., they may be easily interrupted and fill the time left free by the more dominant systems (cf. McFarland, 1974). Dependent as they are on an internally determined need, they will not easily be overstimulated by the presence of the potential releasing and directing stimuli (a satiated predator will not suffer under the continued presence of its prey animals). Yet, here also disruptive effects may occur, due, in this case, to understimulation. If the adequate stimuli are not forthcoming, the behavior may be released and directed by inadequate stimuli (Morris, 1964). In response to these, however, the behavior may fail to bring about the intended, satisfying feedback effects ("frustration"). Consequently, the behavior may be released with such frequency or duration as to disrupt adaptive behavioral integration (cf. Wood-Gush, 1973; Ödberg, 1976). Eventually, such behaviors may rigidify and develop into stereotyped habits. In situations of stimulus deprivation an animal may try to get access to or even to create stimulus situations that resemble the adequate "needed" stimuli. A relatively innocent and even amusing example is provided by *Procyon lotor*, the "washing bear." The activity that gave it its name has nothing to do with any sense of hygiene, nor does it serve to moisten its food, as Bierens de Haan (1932) could establish. A raccoon that, in a barren cage, carries food into its drinking bowl only to retrieve it with the typical "washing" motions a little later appears to have discovered a way to perform the behavior with which it would grope for small food items such as invertebrates in humus or the mud of puddles (Lyall-Watson, 1963).

As Markowitz (1982) has recently noted as well, animals often like to work for their food, even if it is freely available. In other words, performing the instrumental behaviors normally subservient in reaching a major goal may in some cases be rewarding in itself.

If the relevant stimuli and incentives fail to come forth, animals may eventually become lethargic (Fox, 1983).

Behavioral Requirements: The Minimum and the Maximum Option

The problem of behavior requirements must be considered with the different reasons in mind for which primate populations are being sustained and the environmental conditions associated with these reasons.

Apart from feral populations that in some cases, because of the destruction and fragmentation of their habitat and reduction of their numbers, are meeting rather unnatural environmental and demographic conditions, there are two other types of population living in unnatural confinement: the laboratory population and the zoo population. In both situations, instances of abnormal behavior indicative of an impairment of psychological well being are not hard to find.

In the laboratory setting, the effectiveness, efficiency, and economics of experimental procedures may foster highly unnatural situations. Even when one maintains that there are physiological experiments for which the behavioral and psychological condition of the animal is not relevant (but see Markowitz and Spinelli, this volume), such impairments will undoubtedly affect the sustainance potential of the stock from which the animals are taken, if not the well being of the animals involved. In this case, the problem is: What are the behavioral requirements that have to be fulfilled in order to obtain a self-sustaining population. One might call this the *minimum option*.

In zoo populations, there are still too many examples of species that are not self-sustaining. Notably problematic in this respect are the prosimians (see Pollock, this volume). *Lemur mongoz*, for example, reproduces poorly in captivity, and its rate of reproduction lags behind that of other species. Schaaf and Stuart (1983) found that animals imported from the wild tended to reproduce initially but that reproduction halted after a while. Captive-born animals hardly reproduced at all. The authors compared data about various aspects of the animals' ecology and ethology in the wild, as far as known, with data on environmental conditions, management routines, and the composition of groups, collected from a number of institutions keeping these animals. They related these to the occurrence of reproduction in these institutions in the hope of discovering which might be the crucial factors. They directed their attention to climatological, photoperiodic, and dietary factors, the temporal variations in these, and their influence on daily and seasonal activity patterns. They investigated social organization and group structure regimes in captivity. In this case, the investigation did not succeed in providing any definitive answers. However, it is exemplary as an exploratory approach treating the captive situation in various institutions as a variable.

In addition to this *minimum option* there is a *maximum option*. The chief commitments of a zoo are to recreative education and to conservation. It is of importance that the visitor acquire an appreciation of animals that to the fullest possible extent express their richness as functioning biological systems (Hediger, 1982). In this *maximum option*, the requirements refer to those factors that lead to an optimal and naturally full development of an animal's behavioral potentialities.

This *maximum option* should be attempted of course a fortiori, when animals are being preserved for the purpose of reintroduction into the wild; we would like to realize even the requirements for the preservation of adaptive traditions or at least of the capability to develop them easily.

This also means that we want to prevent domestication as much as possible. During this conference, some speakers have noted the very real danger of domestication. To ensure that the wanted genotypes are preserved, to be able to conduct a naturalistic artificial selection, it is important that the genetic dispositions can express themselves adequately.

The "Naturalistic" Solution

Recently, Hutchins and his co-workers, among others, pleaded that the abovementioned objectives should be achieved by trying to incorporate the complexity of the natural ecological setting of a species into its zoo environment (Hutchins et al., 1984). They emphasized increasing the spatial and temporal variation in enclosures with naturalistic elements. They furthermore advised increasing complexity by adapting feeding routines and by providing a natural social environment, where biologically adequate. Dahl (1982) urged the introduction of fluctuations in important abiotic factors such as temperature and light.

Two types of effects can be achieved in this way. For the visitor, the educational value of an enclosure that simulates the natural environment is obvious. The visitor obtains a greater understanding and appreciation of a particular biocenose and its inhabitants if various species of waterfowl, for example, are displayed in a replica of a marsh landscape instead of a void space with a basin made of bricks and concrete. The admirable new gorilla exhibit at Seattle simulates a tropical forest cleared by slash-and-burn agriculture. Its highly varied and naturalistic appearance, achieved by logs and vegetation in and around an irregularly shaped enclosure, no doubt makes a much more favorable cognitive and emotional impact on the visitor than some new glass, concrete, and plastic designs to be seen elsewhere, however architecturally pleasing these may be.

In many cases, such environmental enrichment will at the same time not fail to enhance the behavioral activity of the animals (van Hooff, 1967, 1973). It will thus further add to the fascination of the public. It may also offer conditions that are conducive to reproduction.

More important, a world offering complex and changing stimuli will stimulate the development of behavioral flexibility and generalized behavioral capacities. Through play and the rewards it entails exploration and manipulation to control become "voluntary" and autochtonously satisfying. The impact on young developing animals is particularly great and can manifest itself in brain structure and chemistry (Fagen, 1982; Mason, this volume). Of course, now the circle is closed: Animals developed under richer conditions will have greater "behavioral needs," and will, therefore, be affected more strongly in their welfare in an impoverished setting.

However, two questions remain. First, which environmental aspects are responsible for the possible beneficial effects? Naturalistic complexity in itself is not essential. Simply being an observer in a complex world does not bring about the enrichment responses, discussed above, as Ferchmin et al. (1975) could demonstrate experimentally at least in the rat.

What matters is whether the environment offers adequate opportunities for particular kinds of interaction and whether we know which are the relevant aspects to be offered (Fagen, 1982). The experience with the Seattle gorilla display offers an illustration of this point. The exhibit includes a small stream. Observations by Brown et al. (1982) had led to the expectation that it might induce activity. These observations already had shown also that individual variation might be expected as well. As it happened, the gorillas at the Seattle display were rather indifferent to the water.

What is interesting and stimulating depends on phylogenetically determined behavioral propensities and corresponding sensitivities and on the way experience and habituation interact with these. An example of the latter: After the opening of the large naturalistic chimpanzee consortium at the Arnhem Zoo, the animals were fascinated by fish, which they managed to scoop from the ditch occasionally (van Hooff, 1973). This habit, however, has not been observed by any one of my collaborators since then.

Programmed Stimulation of Behavior

A second question is whether a naturalistic setting that is realizable within the confines of a zoo exhibit can offer the relevant cues for the entire behavioral potential. The obvious example is the stimulation of predatory behaviors, which, for obvious reasons, poses a problem in zoos. Already in 1961, Morris proposed the introduction of procedures and the installation of equipment that could prompt in animals certain responses requiring them to bring into play the full potential of their behavioral capabilities. One can think not only of complex motor capabilities but also of perceptive and cognitive abilities. Such prompting would be indicated in the case of instrumental behavioral routines and appetitive behaviors that normally are insufficiently self-reinforcing, in that their specific feedbacks are not (or no longer) satisfying on their own. They normally derive their motivational impulses from higher order motivational systems. These pursuits choose their instruments on the basis of some extraneous cost-effectiveness evaluation (the lion does not need to run very hard anymore to get its food; there is an easier way).

Lately, Markowitz especially has advocated a behavior-engineering approach based on operant conditioning, in which certain animal performances are made instrumental in achieving natural satisfactions (Markowitz, 1982; Markowitz and Spinelli, this volume). This development has been criticized by Hutchins et al. (1979) on several grounds. A major objection is that such operant behaviors can come to dominate other behaviors, especially if they are the most important occupations in an otherwise barren environment. They may then become rigidified automatic routines, stereotypies of the kind one is actually out to get rid of in zoos (Morris, 1966; Meyer-Holzapfel, 1968). Moreover, experimental work with rats has shown that simply teaching an animal to perform a motor skill, even a complex one, does not entail the kind of "enrichment responses" in development that are brought about in an incitingly complex and variable environment (Ferchmin and Eterovic, 1977). Nevertheless, I agree with Forthman Quick (1984) that, like the other objections, this one can be met by building conditions for perceptive and motor variability into the program. Furthermore, special consideration should be given to the constraints in learning that manifest themselves in species-specific limitations and predispositions in learning abilities (Hinde and Stevenson-Hinde, 1973). If naturalistic relations between the releasing stimuli and the wanted response type and between the response type and the reinforcer are sought, a combination of naturalistic enrichment and behavioral engineering offers great promises.

The Social Environment

One of the most stimulating aspects of an animal's environment is formed by its conspecifics, its social companions.

An important part of its behavior is social behavior, concerned with maintaining an optimal position in the social organization. Primates show a great variety of social organizations. Whereas nocturnal prosimians tend to live rather solitary lives, most diurnal species live in groups. These may be very cohesive, such as a troop of yellow baboons or a gorilla group, they may be organized more loosely, as in a chimpanzee community, they may be large or they may be small, such as a monogamous family of gibbons, and there is even the odd man out among the diurnal primates, the orangutans, where the individuals mostly go their own way.

During evolution, natural selection has endowed animals with the emotional dispositions and the cognitive capacities that result in a particular social environment, namely in that social environment in which the functions that ultimately determine their inclusive fitness can be performed most efficiently and effectively.

Although opinions are not yet unanimous on this point with respect to primates (cf. Wrangham, 1979, 1983), there is increasing empirical evidence from eco-ethological field studies that the primary benefit of sociality in these animals is the increased protection against predators (e.g., van Schaik and van Hooff, 1983; van Schaik, 1983). In most species, this benefit is greater than the major disadvantage brought about by the increased hindrance of competition (van Schaik et al., 1983). The orangutan is the obvious exception among the diurnal primates (Galdikas, 1979; McKinnon, 1979; Sugardjito et al., in press).

Depending on how factors such as the intensity of predation and the availability and dispersion of food vary from time to time and from location to location, it might be adaptive if the social organization could be adjusted in accordance. In recent years, we have learned that the social organization of primates is not rigid in all cases. There is some flexibility within a species-specific range. This subject has recently been reviewed by Dunbar (1982).

Thus, a strict distinction between species that form one-male groups and species that form multimale groups is not always possible. Some species can do both. Cords (1984) found this for the red-tailed monkey and Tsingalia and Rowell (1984) for the blue monkey. Comparison of the ranging and fission-fusion patterns of *Macaca fascicularis* in places where they meet predators (e.g., Ketambe, Sumatra) and where they do not (Simeleu, off Sumatra) shows that these patterns differ strongly in accordance with the risk of predation (van Schaik and van Noordwijk, in press). An analogous variation may also exist in chimpanzees (Tutin et al., 1983).

This flexibility has an interesting consequence with respect to the question of what type of social organization one should try to realize in a zoo environment; should one try to create socially naturalistic exhibits? What, then, is naturalistic in this respect?

A restrictive criterion is of course whether the social interactions generated in a particular environmental and demographic setting result in stressful overstimulation with deleterious effects. Thus, multimale groups of some species may be undesirable in conventional zoo cages, even though these occur naturally in the wild and even though this prevents the expression of important aspects of their social possibilities. The strategic aspects of the social behavior that make the relationships between adult male chimpanzees so fascinating (van Lawick-Goodall, 1975; Bygott, 1979; Nishida, 1983) and that have been so well explicated by de Waal's (1982) studies on the Arnhem Zoo colony cannot express themselves in the one-male situation, which is still the only possibility in most zoos. As a rule, we will find less complex social situations in captivity than in the wild.

But the reverse is possible as well. Recently, Edwards and Snowdon (1980) studied the orangutans at Vilas Park Zoo. Their observations support the impression that can also be gained in some other zoos, namely that even *adult* orangutans can live in a harmonious one-male group. The animals developed affinitive behaviors, such as grooming, food sharing by the adult male, and paternal care, which have never been reported from the wild. Rijksen (1978), McKinnon (1979), and Galdikas (1979) have noted the occasional occurrence of groupings in the wild. Often these may simply be aggregations on a rich source of food. Recently, Sugardjito et al. (in press) have been able to demonstrate that socially coordinated groups may occur at the peak of availability of certain fruits. The conclusion is that the orangutan is normally solitary to avoid food competition and that it can do so because its size and arboreal life-style make it practically immune to predation.

For captive orangutans, the need to separate for foraging is absent, and we see that social dispositions which normally remain dormant in nature are released in captivity. In terms of the well being of the animals, this undoubtedly is a welcome enrichment of their social environment. Must we now deliberately attempt the establishment of social groups for orangutans? Or shall we house them more solitarily and call for the behavioral engineer to create some foraging problems for them? Environmental designs that allow the animals to follow their own preferences, as planned for the new orangutan exhibit at the Atlanta Zoo (Maple and Finlay, this volume), seem to be the best solution to problems such as this.

An intensification of social interaction is a common effect of the captive situation where the social partner often is an unnaturally continuous and close stimulus source. Kummer and Kurt (1967) and Rowell (1967) have shown this by comparing time budgets of baboon groups in the wild and in captivity.

Foraging had decreased and migration was absent; the "free" time had been taken by grooming, play, and agonistic conflicts. Which of these behaviors in particular increases makes all the difference. If aggressive behavior is overstimulated, patterns of redirection of aggression resulting in scapegoating may become prominent. More relaxed and affiliative patterns may be suppressed, at least in some relationships, and we see the deleterious effects of what is commonly referred to as crowding.

These effects can often be obtained by reducing available space, hence the name. It is too simplistic, however, to see this merely as a direct effect of a reduced distance to the neighbors. Whether crowding effects will occur often depends more subtly on the way spatial relations affect particular roles. Thus, a reduction of the living space that makes it more surveyable for the alpha male in a group of macaques may facilitate the exertion of his control role, so that the aggression between the females in the group is actually decreased instead of increased (Erwin et al., 1979; Erwin, this volume). For a further elaboration see also de Waal (this volume). This illustrates the importance of an insight in the social role system of a particular species (i.e., the social goals that animals in various positions within the group attempt to reach and maintain) and the way this is influenced by quantitative and qualitative aspects of the environment.

How important ethological insight in the role system of an animal species is to prevent and cure abnormal behaviors becomes clear from the following example. The Dutch zoos of Emmen and Rhenen had problems in their groups of Hamadryas baboons. Young animals were born, but they were invariably abused to death by the leading male, most likely their own father. In the wild, Hamadryas baboons live in harem communities. Each harem consists of an adult male who carefully herds one or more females and controls their contacts with animals outside the harem (e.g., Kummer, 1968). Rijksen (1981) studied the group at Emmen Zoo and found the following:

1. The animals had arrived as juveniles and had grown up together.
2. There was no clear harem structure. Even the leading male did not succeed in herding females; the latter behaved rather independently.
3. When a mother with a newly born baby reentered the group, the male stole it, carried it around, threw it about, and abused it till it succumbed.
4. The rewarding feedbacks of this behavior obviously were that he became the center of attention.

The explanation is that the animal had discovered a most effective way to reach the goal he could not attain otherwise, namely to make the females attend to him.

A similar situation existed in the Rhenen Zoo (Wiepkema, personal communication). In both zoos, the behavior developed independently and had already acquired the character of a habit, and might have developed into a very maladaptive tradition.

The solution, and therefore the managerial advice in similar cases, is as follows. When founding a colony of this species, young females should be paired with older experienced males that achieve control easily from the onset.

With these last examples we have entered the field of mating systems in primates and their implications for management in confinement. This subject is dealt with specifically by Tilson (this volume).

Reproduction and Overproduction

Once "the minimum option" has been achieved and factors that hinder mate selection and acceptance, sexual behavior, and acceptance and adequate care of young can be controlled, and once survival problems have been mastered, we

TABLE 23.1. Interbirth intervals in one captive population and four wild populations of chimpanzee.[1]

		Mean interval	N_i	$N♀$	Interval range	Median interval
Arnhem Zoo	Young stays	4.0	7	5	2.5–2.8	4.0
	Young gone	1.6	24	12	0.7–4.1	1.2
Kaskakela (Goodall, 1983)	Young stays	5.2	17	12	4.5–7.6	5.6
Bossou (Sugiyama, 1984)	Young stays	4.4	12	—	3–7	—
Mahale K (Nishida, 1977)		—	—	3	5–6$^+$	—
Mahale M (Hiraiwa-Hasegawa et al., 1984)		6.0	7	7	5.2–7.1	6.0

[1] The captive data distinguish cases where the offspring stayed with the mother and cases where the offspring died or was removed before the birth of the next infant. N_i, number of interbirth intervals measured. $N♀$, number of mothers involved.

may expect strong growth of captive populations. Not only may animals survive longer, but also reproduction may go faster.

If we compare parameters of reproduction of wild populations with those of captive populations that are representative in a naturalistic respect, both in demographic composition and in the relevant aspects of the management regime (e.g., no early removal of offspring), we can conclude that such captive colonies can have a higher rate of reproduction. Smith (1982) made such a comparison concerning *Macaca mulatta*. The same applies to *Pan troglodytes* (Table 23.1). Of the Arnhem Zoo's chimpanzee colony of 20 animals present in 1973, 12 animals and 24 offspring survive at present. The "luxury" problem of overproduction is near.

This is what Hediger (1982) has to say about it: "For the majority of surplus zoo animals, the best solution would be a kind of cropping, the humane selective elimination of those animals that can no longer be maintained. Of course, for the zoo biologist, who seeks to preserve healthy living species, this is a tragic way.... This method, however, is certainly preferable to surgical or medicinal castration, which would force the animal in an unnatural lifestyle." It is certainly true that birth control could rob primate groups of a most important form of enrichment of the social environment, the cycle of young appearing, being the group focus, growing up, and acquiring their place. We begin to appreciate the important place that young animals occupy in group life by their role, active or only passive, in triadic interactions (Dunbar, 1984; Taub, 1984).

There is a second, albeit still more remote problem. If we are going to worry about the effects of genetic change and zoo domestication, we have to think of adequate replacement of natural selection. A more active approach than solely maintaining genetic diversity may be required. In that case, we must give thought to what should be our selection criteria and whether behavioral criteria will be of any significant use. Since we shall have to select from behavioral phenotypes, we must have phenotypes to select from, allowing for a certain degree of overproduction.

We should come to terms with this sensitive problem. There is a biological viewpoint and there is an ethical one, if you want a sentimental viewpoint—and they are clearly not the same. Are there solutions to these problems that satisfy the rationalizing biologists, but that are or can be made acceptable also to the community at large?

References

Baerends GP (1976) The functional organization of behaviour. Anim Behav 24:726–738.

Bierens de Haan JA (1932) Uber das sogenannte 'Waschen' des Waschbären (*Procyon lotor*). Biol Zentralbl 52:329–342.

Boër M (1983) Several examinations on the reproductive status of lowland gorillas (*Gorilla g. gorilla*) at Hannover Zoo. Zoo Biol 2:267–280.

Brown SG, Dunlap WP, Maple TL (1982) Notes on water-contact by a captive male lowland gorilla. Zoo Biol 1:243–249.

Bygott JD (1979) Agonistic behaviour, dominance and social structure in wild chimpanzees of the Gombe National Park. In: Hamburg DA, McCown ER (eds) The great apes. Benjamin Cummings, Menlo Park, CA, pp 405–427.

Cords M (1984) Mating patterns and social structure in redtail monkeys (*Cercopithecus ascanius*). Z Tierpsychol 64:313–329.

Dahl JF (1982) The feasibility of improving the captive environments of the Pongidae. Amer J Primatol Suppl 1:77–85.

Dawkins R (1976) Hierarchical organization: a candidate principle for ethology. In: Bateson PPG, Hinde RA (eds) Growing points in ethology. Cambridge University Press, pp 7–54.

Dunbar RIM (1982) Intraspecific variations in mating strategy. In: Bateson PPG, Klopfer PH (eds) Perspectives in ethology, 5. Ontogeny. Plenum, New York, pp 385–431.

Dunbar RIM (1984) Infant-use by male Gelada in agonistic contexts: agonistic buffering, progeny protection or soliciting support. Primates 25:28–35.

Edwards SD, Snowdon CT (1980) Social behavior of captive group-living orang-utans. Intern J Primatol 1:39–62.

Erwin J (1979) Aggression in captive macaques. In: Erwin J, Maple TL, Mitchell G (eds) Captivity and behavior. Van Nostrand Reinhold, New York, pp 139–171.

Erwin J, Maple TL, Mitchell G (1979) Captivity and behavior: primates in breeding colonies, laboratories and zoos. Van Nostrand Reinhold, New York.

Fagen R (1982) Evolutionary issues in development of behavioral flexibility. In: Bateson PPG, Klopfer PH (eds) Perspectives in ethology, 5. Ontogeny. Plenum, New York, pp 365–383.

Ferchmin PA, Eterovic VE (1977) Brain plasticity and environmental complexity: role of motor skills. Physiol Behav 18:455–461.

Ferchmin PA, Bennett EL, Rosenzweig MR (1975) Direct contact with enriched environment is required to alter cerebral weight in rats. J Comp Physiol Psychol 88:360–367.

Forthman Quick DL (1984) An integrative approach to environmental engineering in zoos. Zoo Biol 3:65–77.

Fox MW (1983) Farm animals; husbandry, behavior and veterinary care, viewpoints of a critic. University Park Press, Baltimore.

Galdikas BMF (1979) Orang utan adaptation at Tanjung Puting Reserve: mating and ecology. In: Hamburg DA, McCown BER (eds) The great apes. Benjamin/Cummings, Menlo Park, CA, pp 194–233.

Glatston AR, Geilvoet-Soeteman E, Hora-Pecek E, Hooff JARAM van (1984) The influence of the zoo environment on social behavior of groups of cotton-topped tamarins, *Saguinus o. oedipus*. Zoo Biol 3:241–253.

Goodall J (1983) Population dynamics during a 15 year period in one community of free-living chimpanzees in the Gombe National Park, Tanzania. Z Tierpsychol 61:1–60.

Griffin DR (1976) The question of animal awareness. Rockefeller University Press, New York.

Hediger H (1942) Wildtiere in Gefangenschaft, ein Grundriss der Tiergartenbiologie. Schwabe, Basel.

Hediger H (1982) Zoo biology; retrospect and prospect. Zoo Biol 1:85–86.

Hinde RA, Stevenson-Hinde J (1973) Constraints on learning. Academic, London.

Hiraiwa-Hasegawa M, Hasegawa T, Nishida T (1984) Demographic study of a large-sized unit-group of chimpanzees in the Mahale mountains, Tanzania: a preliminary report. Primates 25:401–413.

Hooff JARAM van (1967) The care and management of chimpanzees with special emphasis on the ecological aspects. Aeromed Res Lab-Techn Rep 67:15.

Hooff JARAM van (1973) The Arnhem Zoo chimpanzee consortium: an attempt to create an ecologically and social acceptable habitat. Internatl Zoo Yrbk 13:195–205.

Hooff JARAM van (1974) Een ethologische beschouwing over abnormaal gedrag en dierlijk welzijn. In: Huisman GH (ed) Diergeneeskunde, ethiek, ethologie. Fac Diergeneeskunde, Utrecht, pp 2–17.

Hughes BO (1980) The assessment of behavioural needs. In: Sybesma V (ed) The laying hen and its environment. Nijhoff, Den Haag.

Hutchins M, Hancock D, Calib T (1979) Behavioral engineering in the zoo: a critique. Internatl Zoo News 25(7):18–23, 25(8):18–23, 26(1):20–27.

Hutchins M, Hancocks D, Crockett C (1984) Naturalistic solutions to the behavioral problems of captive animals. Zool Garten NF Jena 154:28–42.

Kummer H (1968) Social organization of Hamadryas baboons: a field study. University of Chicago Press, Chicago.

Kummer H, Kurt F (1967) A comparison of social behaviour in captive and wild Hamadryas baboons. In: Vagtborg BH (ed) The baboon in medical research. Texas University Press, Austin, pp 65–80.

Lawick-Goodall J van (1975) Chimpanzees of Gombe National Park: thirteen years of research. In: Kurth G, Eibl-Eibesfeldt I (eds) Hominisation und Verhalten. Fischer, Stuttgart, pp 56–100.

Leyhausen P (1965) Uber die Funktion der relativen Stimmungshierarchie. Z Tierpsychol 22:412–494.

Lyall-Watson M (1963) A critical re-examination of food 'washing' behaviour in the raccoon *Procyon lotor* L. Proc Zool Soc London 141:371–393.

Maple TL, Stine WW (1982) Environmental variables and great ape husbandry. Amer J Primatol 1:67–76.

Markowitz H (1982) Behavioral enrichment in the zoo. Van Nostrand Reinhold, New York.

McFarland DJ (1974) Time-sharing as a behavioural phenomenon. In: Bateson PPG, Hinde RA (eds) Advances in the study of behaviour, 5. Cambridge University Press, pp 55–93.

McKinnon JR (1979) Reproductive behavior in wild orangutan populations. In: Hamburg DA, McCown ER (eds) The great apes. Benjamin/Cummings, Menlo Park, CA, pp 194–233.

Meyer-Holzapfel M (1968) Abnormal behavior in zoo animals. In: Fox MW (ed) Abnormal behavior in animals. WB Saunders, Philadelphia/London/Toronto, pp 476–503.

Mitchell G (1970) Abnormal behavior in primates. In: Rosenblum L (ed) Primate behavior; developments in field and laboratory research, I. Academic, New York, pp 195–249

Morris D (1961) Active life for zoo animals. New Scient 241:773–776.

Morris D (1964) The responses of animals to a restricted environment. Symp Zool Soc London 13:99–118.

Morris D (1966) The rigidification of behaviour. Phil Trans Roy Soc London (B) 251:327–330.

Nishida T (1977) The social structure of chimpanzees of the Mahale mountains. In: Hamburg DA, McCown BER (eds) The great apes. Benjamin/Cummings, Menlo Park, CA, pp 73–121.

Nishida T (1983) Alpha status and agonistic alliance in wild chimpanzees (*Pan troglodytes schweinfurthii*). Primates 24:318–336.

Ödberg FO (1976) Problemes de comportement des animaux sauvages en captivité. Ann Med Vet 120:113–125.

Resko JA (1982) Endocrine correlates of infertility in male primates. Amer J Primatol Suppl 1:37–44.

Rijksen HD (1978) A field study on Sumatran orang utans (*Pongo pygmaeus* abelii LESSON 1827), ecology, behaviour and conservation. Meded Landbouwhogeschool Wageningen 78:1–420.

Rijksen HD (1981) Infantkilling: a possible consequence of a disputed leader role. Behaviour 78:138–168.

Rowell TE (1967) A quantitative comparison of the behaviour of a wild and a caged baboon group. Anim Behav 15:499–509.

Schaaf CD, Stuart MD (1983) Reproduction in the mongoose lemur (*Lemur mongoz*) in captivity. Zoo Biol 2:23–38.

Schaik CP van (1983) Why are diurnal primates living in groups? Behaviour 87:120–144.

Schaik CP van, Hooff JARAM van (1983) On the ultimate causes of primate social systems. Behaviour 85:91–117.

Schaik CP van, Noordwijk MA van (In press) The evolutionary effect of the absence of felids on the social organization of the Simeulue monkey (*Macaca fascicularis fusca*, MILLER 1903). Folia Primatol.

Schaik CP van, Noordwijk MA van, Boer RJ de, Tonkelaar I den (1983) The effect of group size on time budgets and social behaviour in wild long-tailed macaques (*Macaca fascicularis*). Behav Ecol Sociobiol 13:173–181.

Smith DG (1982) A comparison of the demographic structure and growth of free ranging and captive groups of rhesus monkeys (*Macaca mulatta*). Primates 23:24–30.

Sugardjito J, Boekhorst IJA te, Hooff JARAM van (In press) Ecological constraints on the grouping of wild orangutans (*Pongo pygmaeus*) in the Gunung Leuser National Park, Sumatra, Indonesia. Intern J Primatol.

Sugiyama Y (1984) Population dynamics of wild chimpanzees at Bossou, Guinea, between 1976 and 1983. Primates 25:391–400.

Taub DM (1984) Primate paternalism. Van Nostrand Reinhold, New York.

Tsingalia HM, Rowell TE (1984) The behaviour of adult male blue monkeys. Z Tierpsychol 64:253–268.

Tutin CEG, McGrew WC, Baldwin PJ (1983) Social organization of savanna-dwelling chimpanzees, *Pan troglodytes verus*, at MT Assirik, Senegal. Primates 24:154–173.

Waal FBM de (1982) Chimpanzee politics. Jonathan Cape, London.

Wood-gush DGM (1973) Animal welfare in modern agriculture. Brit Vet J 129:167–174.

Wrangham RW (1979) On the evolution of ape social systems. Soc Sci Inf 18:335–368.

Wrangham RW (1983) Ultimate factors determining social structure. In: Hinde RA (ed) Primate social relationships. Blackwell, Oxford, pp 255–262.

24
Early Socialization

WILLIAM A. MASON

Introduction

Socialization refers to the development of an individual as a social being and participant in a society or social system. This implies that development is viewed in relation to certain specified social endpoints, which, in primate research, are usually defined functionally and with reference to species-typical norms. The course of social development is obviously subject to many different influences.

The objectives of research on socialization are to describe the normal course of social development; to delineate influential organismic and situational variables; and, ultimately, to arrive at coherent, logically consistent, and comprehensive sets of scientific generalizations (principles, models, theories) that are able to predict and explain various developmental outcomes. The conceptual underpinnings of socialization research are derived mainly from developmental psychology, particularly organismic models of development (Overton and Reese, 1973; Mason, 1979).

Research on the social development of the nonhuman primates has been going on now for about three decades. The methods have included normative descriptions obtained on captive and free-ranging animals, as well as experimental modifications of rearing environments; cross-sectional as well as longitudinal data have been obtained. Although interest has focused on events occurring before reproductive maturity, it should be remembered that the entire life span, including senescence, properly falls within the purview of developmental research and theory. Socialization, as an aspect of life-span development, is included in this domain. The principal species studied has been the rhesus monkey. Although the rhesus monkey obviously cannot be considered to be fully representative of all simian primates, sufficient data have been collected on other species of Old World monkeys, on chimpanzees, and (to a lesser extent) on New World monkeys to support some broad conclusions. The major findings have been reviewed repeatedly and are widely known (see Mason, 1965, 1971, 1985; Mason et al., 1968; Sackett, 1970; Suomi, 1976). I confine myself here to generalizations and implications derived from this research that I perceive as most pertinent to problems of raising primates in captivity.

Generalizations

At every stage in its postnatal career, the individual primate is an active participant in its own development. From the moment it emerges from the birth canal, it is an ongoing enterprise with established needs and dispositions and in active commerce with its environment. It has the potential for a range of developmental outcomes; only a portion of this potential is fully realized under any circumstances. Specific outcomes are the result of an epigenetic process, the cumulative effect of the continuing interaction between the living system and its environment. From the beginning of its postnatal life, the relations of the newborn monkey or ape with its surroundings are selective and transactional. It responds differentially to the objects and events it encounters in the world around it, depending on its current organizational state, and both its organization and the environment may be altered as a consequence. This element of dynamic interchange or reciprocity between the developing primate and its environment is particularly prominent, of course, in its relations with other animals. Traditional dichotomies, such as innate/acquired and cognitive/conative, have not been helpful in describing or interpreting the phenomena of social development.

Socialization is not a unitary process, but the result of changes in many participating systems that can be approached from different levels of analysis. At the higher levels, these systems are abstractions—hypothetical constructs—based on inferences drawn from diverse lines of evidence, including the form of specific behavior patterns, information about their motivational attributes and functions (including physiological aspects), and about their changes over time and in response to different rearing environments.

Based on such evidence, it is possible to identify two major functional systems that are particularly pertinent to the socialization process. I have characterized these as the filial ("mother-directed") and exploitative ("other-directed") developmental programs (Mason, 1971). Although both systems are present throughout life, their relative prominence changes as development proceeds. They can therefore be regarded as the basis for two contrasting and complementary developmental trends (see Fig. 24.1).

The mother-directed trend is the most clearly focused and the easiest to identify and characterize. The survival of the newborn monkey or ape is critically dependent on establishing and maintaining a satisfactory relationship with its mother. This is accomplished in the beginning largely through the presence in the neonate of specific and strongly motivated patterns of sensorimotor coordination, such as clinging, rooting, and sucking. Apart from their specific functions, these behaviors share the characteristics of being most likely to occur under conditions of elevated or increasing psychological arousal and of resulting in arousal reduction. Arousal reduction appears to be an important factor in the formation of a filial attachment or infantile emotional bond, which in most species under normal circumstances is focused on the biological mother. The responses mediating the infant's adjustment to the mother are implemented and supported by complementary behaviors exhibited by the mother. Mothers, however, seem unaware (or at best vaguely aware) of the functional significance of their infant-directed activities. As might be expected, therefore, the infant nonhuman primate is better equipped than is the human infant to participate actively in the management of its relationship with its mother, and it is required to do so. The nature and extent

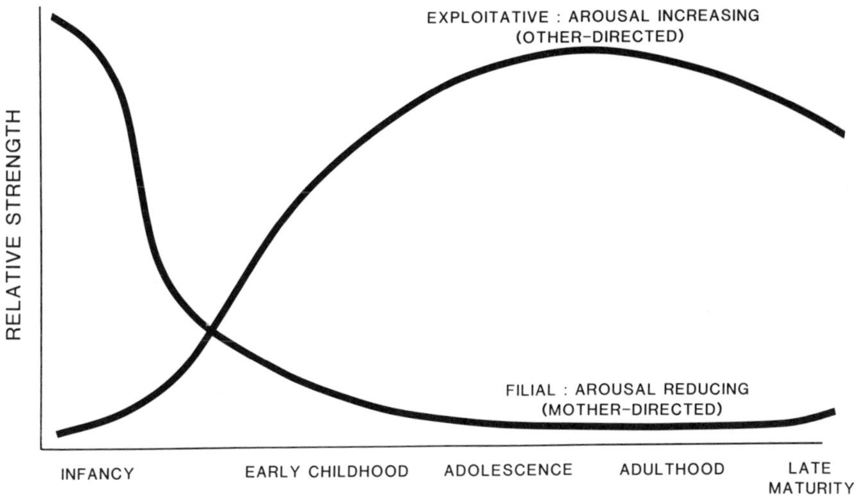

FIGURE 24.1. Schematic representation of hypothetical trends in the development of the filial and exploitative programs.

of the differences between human and nonhuman primates vary with species and are least among the great apes. Even in the great apes, however, mothers are able to compensate only within narrow limits for deficiencies in the early behavioral organization of their infants. Although filial responses and motivations wane as development proceeds, they remain in the repertoire and under appropriate conditions can be displayed by fully mature individuals.

As the developing individual gradually shifts its focus from the parent toward the surrounding environment, the exploitative trend becomes an increasingly prominent influence on its behavior. In contrast to the mother-directed behaviors, which occur under conditions of high arousal and are arousal-reducing, other-directed behaviors occur under conditions of low or moderate arousal, and they share the general characteristic of increasing arousal. Other-directed behaviors are therefore stimulation-seeking. In its earliest manifestations, the exploitative trend is expressed in visual exploration, motor play, and various investigatory activities. Moderately novel or variable objects and events elicit or sustain exploitative behaviors and are primary agents facilitating the developing individual's move away from the mother and into the larger world. Social objects and social interactions are a potent source of novelty and variability. In the ordinary course of events, therefore, the exploitative trend brings the young animal into increasing contact with other individuals differing in age, gender, social status, temperament, and so on. This trend thereby creates numerous occasions for adjustment to the varied and changing conditions of life in a social group and presents multiple opportunities for the acquisition of social knowledge and social skills.

The "mother-directed" and "other-directed" developmental programs can be modified in many of their aspects by individual experience, sometimes with bizarre consequences. Apparently, neither developmental trend can be suppressed entirely, however, and will express itself in *some form* over a very broad range of rearing conditions. Clinging and sucking, for example, may become habitually directed toward any number of objects, including the individual's own body. When this occurs, even though the behaviors no longer serve their primary biological functions of maintaining contact with the mother and providing nourishment, their primitive relationships to psychological arousal are unchanged. Similarly, behaviors expressing the exploitative or stimulation-seeking trend can assume a variety of forms, partly depending on the available options the environment affords. Stimulation seeking may be expressed in apparently aimless pacing or other "cage stereotypies," in noise-making, and in destructive actions directed toward the physical environment, or even toward the animal's own body.

Although these trends will be found in all simian primates, it is obvious that the details will vary with species. Species differ in developmental rates and in behavioral plasticity or adaptability. It is also common knowledge that they vary in many different aspects of their life modes—in the kinds of social organization they create, in the form and quality of their social relationships, in their use of space, in their levels of general activity, and in their time and energy budgets, including virtually every aspect of their feeding and foraging activities. Furthermore, it is safe to assume that some intraspecific variation will be found in all these parameters and that the extent of such variations will also differ systematically according to species. Species-specific differences in life modes reflect differences at the individual level in behavioral dispositions, temperament, and behavioral potential that are clearly pertinent to the socialization process.

Implications

Socialization is a historical process viewed with reference to certain endpoints, usually related to some aspect of adult performance. In scientific research on socialization, these endpoints are usually selected on the basis of some combination of criteria, including presumed biological relevance, methodological convenience (e.g., ease of recording, reliability of measurement), and the particular theoretical concerns of the investigator. The point to be emphasized is that the criteria used to evaluate socialization effects in research projects represent a small sample selected from a much broader set of possibilities on the basis of some fairly explicit rationale.

The questions I am principally concerned with in this section are: What can the available information on socialization contribute to the problems of establishing and managing self-sustaining populations of captive primates? What criteria will be used to assess the adequacy of early socialization experiences? And what rationales will govern the selection of these criteria? Of these questions, the selection of criteria is the most critical issue, in my view, because the criteria that have been used in studies of socialization do not cover the full spectrum of socialization effects that are relevant to the larger concerns of this conference and because the criteria that are relevant to these concerns will vary, depending on the overall objectives and specific aims of a given management program.

Two obvious and time-honored criteria for assessing the adequacy of early socialization experience are the capability of the individual to engage in biologically effective mating and the ability of the female to provide adequate postnatal care. Enough is known to indicate the sufficient conditions for meeting these criteria in captivity for a number of primate species: Ideally, individuals should be raised at least through weaning by their biological mothers and have the experience of living in a social group that provides a reasonable approximation of the size and age-sex composition of the groups in which they would develop in nature. Based on data available for rhesus monkeys and chimpanzees, it seems likely that the experience of being mothered has a positive effect on adult maternal behavior and possibly adult mating behavior in both sexes. There are also clear suggestions that social interaction of the developing individual with animals of the same age and with younger animals (e.g., siblings) may have significant positive effects on sexual and parental behavior. The achievement of adult reproductive competence is minimally dependent on other (nonsocial) environmental factors, assuming, of course, that these are adequate to maintain the individual in good health.

If, for whatever reason, it is impractical or impossible to raise an animal in a quasi-normal social setting, the obvious alternative is to arrange social experiences that are appropriate to the developmental status of the individual, as defined, in part, with reference to the developmental programs described above. Here the evidence is not sufficiently systematic to indicate specific measures, although it suggests some general guidelines: Generally speaking, the rearing environment should be arranged to permit a clear differentiation of the filial and exploitative programs in an approximation of the natural sequence. If the biological mother is the developing individual's sole companion, she will eventually become the target of playful and other exploitative behaviors, which the mother may reject or respond to punitively. Whether her reactions will have any serious deleterious effects on the subsequent reproductive competence of her offspring is not entirely clear at this point, although it is safe to assume that whatever the effects, they will not be beneficial. In species characterized by incest "taboos," it is possible that exclusive and extended housing with parents of the opposite sex will have an inhibitory effect on sexual development. The appropriate social outlet for the exploitative trend under normal circumstances is individuals other than the parents. For rhesus monkeys, the data indicate that contact with peers is important to the development of effective sexual performance. Here, too, however, such contacts should be arranged with a view to their relationship to developmental trends. Although contacts with peers may be allowed to begin fairly early in life, they should be limited to brief periods occurring at frequent intervals, rather than the result of continuous cohabitation. One argument against continuous exposure to peers from an early age is that each individual develops a strong infantile attachment to the other. In effect, each animal becomes a substitute mother for the other, and this tends to prolong the "mother-directed" trend and interferes with the transition to more mature modes of relating and interacting socially. In addition, intermittent exposures tend to increase arousal, which facilitates social play and sexual "rehearsal" in the immature animal.

It will be recognized, of course, that reproductive competence provides a minimal standard for assessing the adequacy of early socialization regimes. For many areas of biomedical research, in which the primary need is for the predictable and continuing production of healthy, captive-born subjects, reproductive competence may be all that is required. For many other areas of concern, however, the criterion of reproductive competence will be necessary but clearly not sufficient. Those who are interested in exhibiting primates before the public, in the description and analysis of individual behavior, in the study of social processes and social organization—in short, individuals who, for whatever reason, are interested in primatology or the broader aspects of primate biology—will be concerned with how "representative" their subjects are in relation to the natural populations to which they hope their findings will apply. For such persons, the issue of socialization assumes much richer connotations than is the case when the primary concern is with the production of animal models.

This obvious difference in standards of adequate socialization seems often to be overlooked in discussions concerning the design of captive environments and the management of captive populations. Another point that is often overlooked is that representativeness also implies evaluative criteria. Our current understanding of primate socialization effects, although far from complete, is sufficient to provide useful guidelines for achieving various developmental outcomes. We can suggest rearing regimes that produce individuals that are free from the more obvious and undesirable stigmata of captive rearing and that fulfill the minimal criterion of reproductive competence. We can also suggest regimes which will produce individuals that approximate their free-ranging counterparts in some specific aspects of their knowledge or performance, or that even surpass them in some respects. It is clear, however, that primates growing up in captivity are never *fully* representative of primates socialized in the wild. In any captive environment, numerous constraints exist on diet, space, predator-prey relationships, the availability of mates, social companions, and so on, which differ from those operating in the natural habitat. Directly or indirectly, they are imposed by human beings for such reasons as convenience, lack of resources or information, or a humane concern with maintaining individuals in good health by eliminating unnecessary stress or disease. Based on what we have learned from socialization research, it is reasonable to assume that the individual growing up in captivity will be "representative" of its developmental milieu. The ways in which it resembles or differs from an individual socialized in more naturalistic circumstances is an empirical matter that presumes the availability of explicit evaluative criteria based on appropriate information about a species' life history, including significant variations (if any) across habitats in important aspects of its life mode. Such information, of course, can also be used to design rearing environments.

I am not suggesting that even those concerned with generalizing from their findings on captive specimens to natural populations will hold precisely the same criteria as to what constitutes a "normal" or representative member of the species. An investigator who is measuring the performance of a captive animal on a learning task in a highly structured experimental setting will presumably be less concerned with the details of a subject's early socialization experience than is the person who is concerned with investigating some aspect of social cognition. Insofar as these investigators have something to say about the design of rearing environments, their specific requirements and recommendations are likely to be notably different. In both cases, however, the issue revolves around what the experts in tests-and-measurements call validity and the criteria by which it will be established.

What I am advocating here is that any program aimed at creating self-sustaining populations of captive primates will ideally be based on stated goals and explicit criteria relating to the selection of particular individuals of particular species and to the design of the environments in which they and their progeny will spend their lives. The range of possibilities is wide. One might, for example, propose the "domestication" of certain species as a feasible and defensible goal. In domestication, goals and criteria can be stated in terms that are frankly and narrowly anthropocentric. They have direct implications for the selection of breeding stock and for the ways in which animals are housed, fed, and maintained from the moment of birth. Changes in behavior often accompany domestication, commonly as the result of modifications in developmental processes (Price, 1984). That this will also occur in any program aimed at domestication of a primate species seems likely, and the possibility carries obvious implications for the issue of representativeness. Clearly, however, domestication is only one of many conceivable and legitimate goals that are germane to the general theme of this conference.

The critical questions are, Which goals do we seek and why? There are no simian utopias in nature and no universally ideal breeding programs and rearing regimes in captivity. Just as there are a diversity of primate species in nature, so there are a diversity of reasons for wishing to maintain them as captive populations. Not all of them will be mutually compatible, nor universally appreciated. What is viewed as a minimally adequate socialization regime from one standpoint may be considered grossly extravagant from another. Which view is correct? Socialization research will not resolve this question. Given the goals of a captive program and the appropriate evaluative criteria, however, socialization research can assess the adequacy of various rearing regimes and look for ways of making improvements. This will be its principal contribution as an applied discipline. Whether a breeding program's goals are appropriate and worthwhile is, strictly speaking, outside the purview of socialization research. This is not to say that those engaged in basic research on the socialization process will be indifferent to this issue. On the contrary, they may be expected to have a lively concern with the goals of captive breeding programs and the design of rearing regimes. Furthermore, they are likely to be among those whose recommendations are sometimes viewed as extravagant. I hope that as we proceed in our discussions of these important issues, it will be the plurality and not the single viewpoint that prevails.

Summary

1. Socialization refers to the development of an individual as a social being and participant in a society or social system, implying that development is viewed with reference to the achievement of specified endpoints.

2. The objectives of research on socialization are to describe the normal course of social development, to delineate influential organismic and situational variables, and to arrive at principles that are able to predict and explain various developmental outcomes.

3. The newborn monkey or ape is a highly organized, ongoing enterprise from the moment of birth and is an active participant in its own development. It possesses the potential for a range of developmental outcomes, and only a portion of this potential is realized under any given set of circumstances.

4. The particular outcomes that are actually achieved are the result of the dynamic interaction between the individual's existing state or organization and what it encounters in its surroundings.

5. Socialization is a process in which the whole organism is involved, and it proceeds on many levels. Diverse lines of evidence obtained at various levels have led to the inference of two major developmental systems that display different motivational attributes, serve different functions, and follow different developmental courses. These are the filial (mother-directed) and exploitative (other-directed) developmental programs.

6. The filial program is prominent early in life. It is activated under conditions of high psychological arousal, and the performance of filial responses is associated with arousal reduction. It functions to create and maintain a satisfactory adjustment to the mother and plays an important part in the formation of the filial bond.

7. The exploitative system becomes progressively more prominent as development proceeds. This system is activated by conditions of low or moderate psychological arousal, and the behaviors associated with it lead to increases in arousal. The exploitative system, being stimulation seeking, brings the individual into contact with the relatively unfamiliar or variable elements in its environment, including individuals other than the mother.

8. Although both systems are found in all species of simian primates and are present throughout life, the details of their development, including the particular behaviors in which they are expressed, will vary with species and with rearing conditions.

9. Under natural circumstances, the interaction of these developmental programs with environmental influences results in a normally socialized "species-typical" individual that can function as an effective member of the society.

10. In captive settings, the interaction of these programs with environmental variables may result in individuals that deviate markedly from the norms that characterize free-ranging populations, or in individuals that constitute a fair approximation of these norms in one or more aspects of their performance, or that even surpass them in some respects. Current knowledge is sufficient to provide guidelines as to the socialization regimes that will produce these various outcomes.

11. Owing to the inevitable presence of environmental constraints, primates growing up in captivity will never be fully representative of their free-ranging counterparts in all aspects of their experience, their biology, or their behavior. It is unrealistic, therefore, to establish complete representativeness as an ideal to be sought in captivity.

12. A more realistic objective is based on the recognition that, just as there are many conceivable and legitimate goals for establishing self-perpetuating primate populations in captivity, so there will be many criteria (not necessarily compatible) by which the adequacy of various socialization regimes will be assessed.

References

Mason WA (1965) The social development of monkeys and apes. In: DeVore I (ed) Primate behavior: field studies of monkeys and apes. Holt, Rinehart and Winston, New York, pp 514–543.

Mason WA (1971) Motivational factors in psychosocial development. In: Arnold WJ, Page MM (eds) Nebraska symposium on motivation. University of Nebraska Press, Lincoln, pp 35–67.
Mason WA (1979) Ontogeny of social behavior. In: Marler P, Vandenbergh JG (eds) Handbook of behavioral neurobiology, vol. 3. Social behavior and communication. Plenum, New York, pp 1–28.
Mason WA (1985) Experiential influences on the development of expressive behaviors in rhesus monkeys. In: Zivin G (ed) The development of expressive behavior: biology-environment interactions. Academic, New York, pp 117–152.
Mason WA, Davenport RK, Menzel EW (1968) Early experience and the social development of rhesus monkeys and chimpanzees. In: Newton G, Levine S (eds) Early experience and behavior. CC Thomas, New York, pp 440–480.
Overton WF, Reese HW (1973) Models of development: methodological implications. In: Nesselroade JR, Reese HW (eds) Life-span developmental psychology: methodological issues. Academic, New York, pp 65–86.
Price EO (1984) Behavioral aspects of animal domestication. Quart Rev Biol 59:1–32.
Sackett GP (1970) Innate mechanisms, rearing conditions, and a theory of early experience effects in primates. In: Jones MR (ed) Miami symposium on the prediction of behavior. University of Miami Press, Coral Gables, FL, pp 11–53.
Suomi SJ (1976) Mechanisms underlying social development: a reexamination of mother-infant interactions in monkeys. Minn Symp Child Psych 10:201–228.

25
Behavioral Aspects of Successful Reproduction in Primates

STEPHEN J. SUOMI

Introduction

There can be little argument that the road to self-sustaining populations of primates—especially those species that are endangered, threatened, or seemingly destined for such classification in the near future—can be greatly advanced through improvements in breeding success. It should also be clearly recognized that advances in current knowledge of primate reproductive physiology can make invaluable contributions to significant travel along that road. The promise that recently developed techniques, such as in vitro fertilization and embryo transplants, provide for expanding captive populations is both noteworthy and encouraging.

On the other hand, we now know that successful reproduction among virtually all primate species involves a great deal more than the merging of sperm and egg and the subsequent nurturing of pregnancy to term. Reproductive success in the true biological sense involves not only the production of offspring but also the survival of those offspring to reproductive age themselves and the propagation of their genes in successive generations of conspecifics (Wilson, 1975). Events that in nature precede the occurrence of copulation can be every bit as crucial for long-term reproductive success as the most basic aspects of reproductive physiology and prenatal development. The same is true for events that follow successful parturition. If an individual is inept in its sexual or parental behavior, most questions regarding reproductive physiology become moot.

This chapter focuses on advances in our knowledge of behavioral factors that promote successful reproduction—in the true biological sense—in primates raised from birth in captivity, several generations removed from feral environments. Most of the data to be discussed come from observations of rhesus monkeys (*Macaca mulatta*), among the least threatened and, biologically speaking, most successful nonhuman primate species in the world today, even considering their steep decline in population on the Indian subcontinent over the past quarter century (Southwick, 1982). Twenty-five years ago, researchers first began to appreciate the problems posed by captivity and then-prevalent animal husbandry practices for successful long-term propagation of these monkeys. A decade's worth of subsequent research fully documented the nature and extent of such problems. Solutions to these problems were discovered in the early 1970s, and implementation of new practices derived from these discoveries has been in place since then. Paying attention to behavioral aspects of reproduction has yielded valuable insights and produced some promising surprises. This chapter summarizes the lessons learned from these research endeavors.

Reproductive Behavior Problems and Causal Factors

When, back in the 1950s, the first generation of rhesus monkeys born in captivity grew to sexual maturity in laboratories and zoos, severe problems in reproductive and maternal behavior emerged in many of the captive-born subjects. Perhaps the most thoroughly documented cases of such problems came from the University of Wisconsin's Primate Laboratory, which in the early 1950s initiated a major program of captive breeding, the first of its kind in the history of primate research. While feral-born rhesus monkeys flourished reproductively in this captive environment, the initial results of efforts to breed first-generation captive-born rhesus monkeys were not entirely encouraging. Aberrant sequences of copulatory activities were commonly displayed by both young males and females, despite the fact that the males typically had species-normative sperm counts and the females exhibited regular menstrual cycles. The young males seemed especially incompetent. Even when paired with receptive, multiparous, feral-born females, these males almost universally failed to achieve intromission; obviously, they also failed to sire any offspring (Harlow, 1962). In contrast, many of the captive-born females introduced to reproductively successful feral-born males became pregnant and delivered viable infants. Unfortunately, many of these females who became mothers turned out to be deficient in their maternal behavior (Harlow et al., 1966). Thus, a large proportion of captive-born, first-generation rhesus monkeys displayed behavioral abnormalities of procreation and infant care that in the strictly biological sense overrode any advantages that captive living, with its ensured adequate nutritional supply, freedom from parasite and predation, and ready availability of specialized veterinary care, might advance potential reproductive success.

Subsequent research efforts by Harlow and his colleagues indicated that the reproductive problems ultimately could be traced to the early rearing experiences of these first-generation offspring. Almost all of the young adults, both male and female, who exhibited inappropriate or incomplete sexual behavior had not been reared by their biological mothers. Instead, they had been separated from their mothers shortly after birth and were subsequently hand-reared in the laboratory neonatal nursery. Similarly, the overwhelming majority of primiparous captive-born females who failed to care adequately for their offspring had themselves been reared in the nursery, not by their biological mothers (Harlow termed them "motherless mothers"). The original decisions to raise these captive-born offspring of feral-born parents in the nursery rather than by their real mothers had been made largely on the basis of veterinary considerations regarding hygiene and concern for the infants' health and safety. Most zoos at the time had made similar decisions regarding the care of captive-born infant primates and continued to do so for many years, some to this day.

While most cases of aberrant reproductive behavior and inadequate maternal care could be traced to the lack of a mother, not all nursery-reared infants turned out to be sexually incompetent or maternally deficient. Instead, a number of post-nursery social experiences were found to reduce greatly the incidence and severity of such behavioral problems. Foremost among these were experiences with peers, especially those who themselves had been mother reared. The earlier in life that experiences with peers began, the greater the gain in subsequent social, sexual, and parental competence. The majority of nursery-reared rhesus monkeys whose exposure to peers began within the first 1–3 months of life displayed functionally effective, if not completely normal behaviorally, patterns of reproductive activity (Harlow and Harlow, 1969; Goy et al., 1974). Nursery-reared females who had early social experience with peers were three times as likely to display normal maternal care of their first-born offspring as those females who were not physically exposed to peers prior to 6 months of age (Ruppenthal et al., 1976).

Other research, largely performed in the late 1960s and early 1970s, revealed that early social experiences with peers need not be extensive to reduce the risk of behavior problems later in life, especially with respect to copulation. For example, Rosenblum (1971) showed that peer interactions of only 20 minutes each day beginning at 1 month of age provided sufficient social stimulation to promote the normal development of most basic social behaviors characteristic of the species. Moreover, extensive early experience with conspecifics did not need to be maintained beyond the first year of life to improve nursery-reared monkeys' prognosis for later breeding and caretaking activities (rhesus monkeys typically become physiologically capable of procreating during their fourth year of life, females slightly earlier than males). Additionally, it was found that a large proportion of nursery-reared females who displayed inadequate maternal care of their first-born offspring turned out to be capable and competent mothers when they bore their second and subsequent infants (Harlow et al., 1966). More extensive analysis of an expanded data set revealed that the crucial factor for competent care of later born offspring was the amount of time spent with first-born offspring. Virtually all nursery-reared mothers who failed to care adequately for their first infants, but who remained with them for more than a week, were good mothers toward their later born offspring. In contrast, females who spent less than 2 days with their first-born offspring continued to exhibit inadequate maternal behavior toward subsequent offspring (Ruppenthal et al., 1976).

Thus, many of the problems of reproduction and infant care displayed by first-generation, laboratory-born offspring of feral-born parents brought into laboratories and zoos were attributed to early social deprivation, especially in terms of lack of a real mother and restricted access to peers. The obvious conclusion reached by researchers at the time was that social behaviors, and especially biologically crucial reproductive and parental care behaviors, were not entirely innate but instead involved a certain degree of learning through direct physical interactive experience. On the other hand, some of the studies showed that the social experience need not be extensive for the acquisition of basic reproductive and caretaking behavioral capabilities—and once they had been learned, these basic capabilities were not forgotten for the rest of the lifespan (Suomi and Ripp, 1983).

Preventive Social Interventions and Long-Term Consequences for Reproductive Success

The research summarized in the previous section clearly documented some of the long-term hazards of the once-preferred practice of nursery rearing of captive-born primate infants. Findings from these studies suggested possible solutions to situations, both in laboratories and zoos, in which nursery rearing was the *only* viable option, e.g., in cases in which a new mother rejected her neonate or experienced lactational failure and no other lactating females were available as potential foster mothers. At the Wisconsin Primate Laboratory, these findings were used to develop preventive practices involving social interventions for infants born to captive-reared parents, beginning in the early 1970s. At that time, a basic decision had been made to make the laboratory self-sufficient with respect to supply of rhesus monkey subjects, i.e., all new subjects would henceforth come from the laboratory's own breeding colony, and new members of this breeding colony would come from the existing laboratory population. One result of this decision was that no additional feral-born monkeys were brought into the colony after 1970. A second result was that increasing numbers of breeders were captive born, and many of these captive-born rhesus monkeys had been nursery reared. For such a system to work effectively, of course, the nursery-reared monkeys had to be reproductively and maternally competent. Such was the rationale for the development and implementation of preventive social intervention practices.

These intervention practices can be summarized as follows. First, whenever possible, all new infants born into the colony were reared by their biological mothers until at least the age of weaning. Thus, simply reducing the incidence of hand rearing in the nursery was a top priority. Second, in cases in which the biological mother failed to provide adequate care for her infant or was otherwise indisposed, attempts were made to find a substitute adult female who would adopt the newborn infant and thus serve as a foster mother. This practice also served to reduce the incidence of nursery rearing. Third, for those infants who, for one reason or another, required early separation from mother and subsequent rearing in the nursery, efforts were made to provide these infants with as much social stimulation as was feasible, beginning as early in life as was possible. In actual practice, this usually involved placing the infant in a peer group as soon as it left the nursery. If continuous housing with peers (or with peers and older monkeys) was not possible for certain infants, efforts were made to provide them with peer experience by introducing them to agemates in a social playroom for at least an hour several times each week for at least the first year of life.

Additional preventive intervention practices focused on social housing conditions provided to the laboratory-born monkeys when they were juveniles and adolescents and when they became parents themselves. By and large, these monkeys were maintained in stable social groups that contained, at a minimum, like-reared agemates. Whenever possible, these groups also contained older monkeys, including reproductively active adults and mothers caring for their own infants. It was felt that the nursery-reared juveniles and adolescents would benefit the most if (1) the groups were stable in composition, with few deliberate additions or subtractions, and (2) if the age range of group members was varied sufficiently to approximate the age range of members typically found in feral troops. Finally,

(3) whenever possible the groups included at least some relatives of the nursery-reared subjects, so that the social composition of any one group included two or more maternal kinship lines.

A final set of intervention practices concerned the social housing condition of nursery-reared primiparous females during pregnancy, parturition, and the post-natal period up to (at least) weaning. Previously, the practice most common at primate research facilities and zoos had been to isolate females during the latter stages of pregnancy through the immediate postpartum period. The rationale offered at the time was that group members, especially adult males, might harass the pregnant females, increasing the danger of miscarriage, or they might try to kill the newborn during or shortly after delivery. Such fears were in large part based on a few unfortunate incidents reported by a few laboratories and zoos in which primiparous females living in unstable social groups had lost their infants (cf. Dazey and Erwin, 1976; and Nadler, 1980, for some relevant examples). However, we felt that a stable social group, especially one containing maternal kin, might provide a source of social support for these new and inexperienced mothers. Thus, whenever possible, primiparous females remained in their social group throughout pregnancy and delivery, during which time the group was kept as stable as possible.

How effective were these preventive intervention strategies? The data from the Wisconsin colony are clear-cut. Figure 25.1 presents the relative incidence of

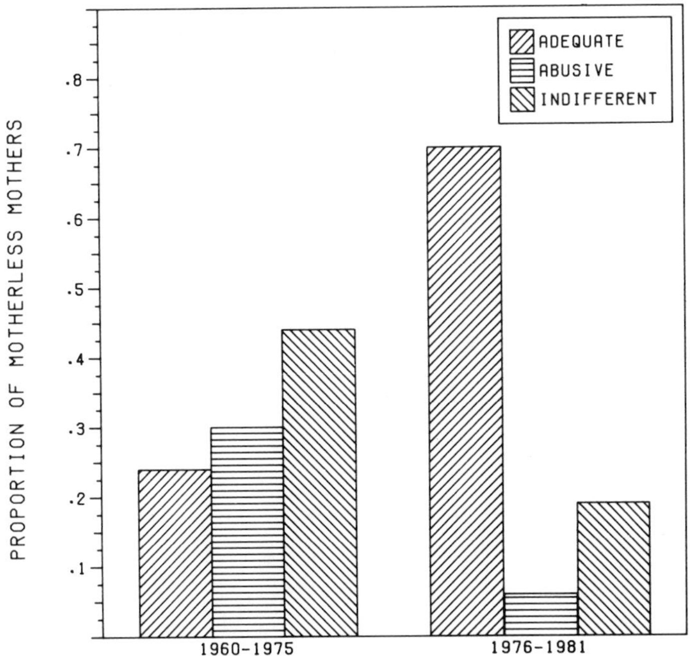

FIGURE 25.1. Relative incidence of adequate, abusive, and indifferent care of first-born offspring by nursery-reared mothers in the Wisconsin colony, 1960–1975 and 1976–1981 (from Suomi and Ripp, 1983).

adequate and inadequate care (broken down into cases of failure to nurse [indifferent] and actual infant abuse) displayed by primiparous, nursery-reared females in the Wisconsin colony from 1960 to 1975, when few of these practices were in force, with the relative incidence of adequate, indifferent, and abusive maternal behavior from 1976 to 1981, the first 5-year period when most of these practices were fully in place. As is evident in the figure, the rate of adequate mothering increased by a factor of three during that 5-year period, a highly significant change ($p < 0.001$, Fisher exact test). Correspondingly, there were proportional decreases in the incidence of abusive and indifferent maternal care among the "new" motherless mothers. These figures represent substantial improvement in the prognosis for primiparous motherless mothers in the colony; in particular, the incidence of offspring abuse among these mothers was not markedly different from that of "mothered" primiparous rhesus monkey females in colonies around the country (A. Hendrickx, 1982, personal communication). Nevertheless, the incidence of adequate care by these motherless mothers, while much improved, was still 20% below that of primiparous females in the colony who had been reared by their own mothers.

On the other hand, not only was there dramatic improvement in the maternal care by primiparous mothers, but this improvement also seemed to carry over to care of subsequent offspring. Figure 25.2 presents the relative proportions of

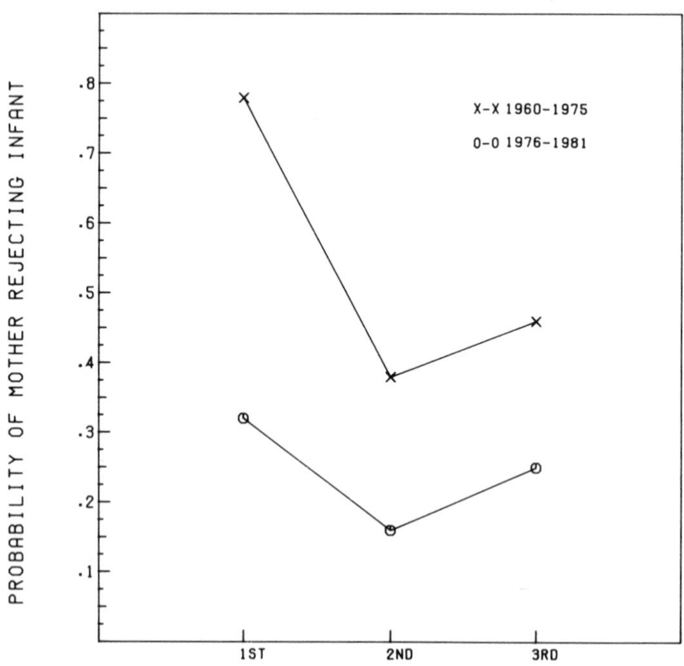

FIGURE 25.2. Relative incidence of inadequate maternal care as a function of birth order, 1960–1975 and 1976–1981 (from Suomi and Ripp, 1983).

motherless mothers who failed to provide adequate maternal care for their first, second, and third offspring in the 1960–1975 and the 1976–1981 samples, respectively.

In both samples, there was marked improvement (i.e., a decrease in probability of not caring for one's infant) in care of second- and third-born offspring over that of first borns. Thus, the apparent improvement in prognosis for displaying adequate maternal care among the 1976–1981 sample was clearly not limited to first-born offspring. Rather, the same general trend for improved care of subsequent offspring shown in the earlier sample was essentially replicated (i.e., the proportional rate of improvement with subsequent offspring did not differ between the two samples of motherless mothers, even though the "absolute" rates of incidence of inadequate care differed substantially, as shown in Figure 25.2).

These data clearly indicate that preventive social interventions such as those described above can greatly increase the reproductive success of captive-born rhesus monkeys, especially those reared in a laboratory nursery. Indeed, the rate of successful mothering among primiparous nursery-reared females in the Wisconsin colony since 1981 has continued to improve and currently exceeds 90%.

An Experiment in Outdoor Living

The preventive intervention practices initiated for nursery-reared infants described thus far all focused on social enrichment of the subjects' environments at various points in development. In the last part of this chapter, I will describe a different strategy that involves exposing nursery-reared rhesus monkeys to outdoor settings that physically resemble, at least in certain respect, a feral environment. Although the original purpose of the study was to provide a therapeutic environment in which monkeys with less than ideal early social histories might facilitate the acquisition of social skills, it is now apparent that the results have important implications for a larger issue—the extent to which laboratory-born-and-reared monkeys several generations removed from the wild can adapt to naturalistic settings with a minimum of exposure to other, more sophisticated conspecifics.

For the past 10 years a small troop of rhesus monkeys, the adults of which were born and nursery reared at the Wisconsin Primate Laboratory, have been spending their springs, summers, and falls (and this past winter) in multiacre enclosures, first in rural Wisconsin and, for the past 2 years, in rural Maryland. This project, carried out in collaboration with Peggy O'Neill, began when we inherited a group of 1-year-old monkeys who had been part of a study investigating the development of social play in nursery-reared monkeys given periodic brief exposure to peers (for details of monkeys' early rearing and test histories, see Mears and Harlow, 1975). After assembling these youngsters into a permanent peer group, we moved them to the first of several multiacre enclosures during their second year of life. Here, they were provisioned daily and provided with a shelter from the elements, but they also had unlimited access to forested areas, grassy fields filled with small bushes, and (in the Maryland enclosure) a small pond. What they did not have was any exposure to any other monkeys, either feral or laboratory born. Thus, whatever social repertoires they developed over the past decade has been acquired solely through interactions with each other, and any knowledge of how to cope with challenges provided by a field-like setting they have developed on their own.

Somewhat to our surprise, these monkeys have shown remarkable adjustment to their outdoor settings over the past decade. The core group of juveniles are now 12-year-old adults, and they have had considerable reproductive success. The group has almost tripled in size since being put outdoors, and it now contains first- and second-generation offspring of the original core group of nursery-reared peers. Furthermore, the behavioral repertoires of these first- and second-generation offspring appear to be completely normal by rhesus monkey standards (some of the original nursery-reared adults still occasionally display brief bouts of stereotypic activity).

Since the move to the new enclosure in Maryland, it has been possible to keep the troop in an outdoor setting on a year-round basis, and this has produced additional findings of interest. First, the members of the study group clearly demonstrate seasonal changes in their patterns of behavior and social organization, and these changes generally have been consistent with those reported by field workers studying rhesus monkey groups living in nontropical climates that included winters with snow. Thus, for example, the present group this past fall exhibited a clear-cut breeding season, and for the first time since the initial first-generation progeny was produced 7 years ago (in Wisconsin), all of this year's new infants were born within a 10-week "birth season."

Second, the social organization and demographic dynamics of the study group continue to reflect the same basic principles that have been reported to govern wild-living troops of rhesus monkeys. Specifically, this past year additional first-generation female progeny of the original group members produced offspring of their own (second-generation progeny), and in each case the resulting patterns of mother-infant interactions were species-normative in their constituent behavior patterns and sequence of developmental change. Moreover, each of these primiparous females raised her offspring in the close company of female kin, with relatively few interactions involving adult females who were not blood relatives. Thus, the matriarchal organization characterizing this captive-born group, as it characterizes rhesus monkey troops in the wild, has been maintained through another generation. In addition, during the past year the oldest first-generation male progeny in the group passed through puberty and was expelled by other troop members (we removed him from the 5-acre enclosure after he was repeatedly attacked by the adult females in the group, with no intervention by the adult males; this male was then placed in a new social group of unrelated juveniles). This clear case of adolescent male peripheralization is consistent with the phenomenon of male emigration reported for feral rhesus monkey troops (e.g., Lindburg, 1973). During the past year, we also successfully cross-fostered an orphaned infant from another colony to one of the present group's multiparous females who had delivered a stillborn fetus in the previous week. The successful adoption represents a procedure we plan to use for experimental purposes in future years.

The discovery that laboratory-born-and-reared rhesus monkeys readily adapt to naturalistic habitats and retain species-normative behavioral repertoires and patterns of social organization in the absence of any exposure to wild-born, ferally experienced adult conspecifics is of considerable significance in its demonstration that such complex behavioral capabilities and characteristics are clearly preserved in the genetic heritage of members of an advance nonhuman primate species, even after many generations of laboratory living. The fact that the social backgrounds of the original group members were less than ideal gives added emphasis to the significance of these monkeys' apparently successful adaptation to a naturalistic habitat, albeit a provisioned one.

Summary and Conclusions

In this chapter, the notion that successful reproduction and subsequent propagation goes beyond mere consideration of reproductive physiology was illustrated with several examples from a quarter century of research with captive-born rhesus monkeys. Findings from this research serve to emphasize the importance of appropriate social stimulation early in life for reproductive success later on. On the other hand, at least for rhesus monkeys, captive breeding and even nursery rearing need not inevitably result in socially damaged juveniles, adolescents, and adults, unable to pass their genes to succeeding generations. Instead, the adaptive flexibility of this species has been shown to be truly remarkable. Of perhaps even greater importance for the theme of this volume is the finding that even individuals with relatively deprived early social backgrounds can learn to live in naturalistic environments with no access to any other conspecifics except other monkeys reared under comparable conditions. The fact that these monkeys can produce generations of offspring who likewise thrive in provisioned outdoor environments and follow species-normative patterns of social organization and demographic dynamics provides powerful testimony to the argument that major portions of rhesus monkey social activity have deep-rooted genetic foundations. They also provide a solid basis for optimism regarding the feasibility of some day releasing captive-born rhesus monkeys into true feral environments that contain predators but lack provisions supplied by human caretakers.

Of course, it should be pointed out that rhesus monkeys represent an unusually adaptive nonhuman primate species, capable of surviving and successfully reproducing under many different environmental conditions. It may well be that the behavioral flexibility and long-term maintenance of species-normative patterns of social behavior and organization, even after many generations of captivity, are not shared by most other primate species, especially those that are currently among the most endangered (e.g., the lion-tail macaque, a species closely related to rhesus monkeys in the genetic sense but not in terms of present adaptive success). In other words, the promising picture that captive rhesus monkeys provide may not generalize to other, even closely related, primate species. Nevertheless, these findings do generate some basis for optimism, and they provide additional motivation to pursue efforts by humans to help nonhuman primates travel down the road to self-sustaining populations.

References

Dazey J, Erwin J (1976) Infant mortality in *Macaca nemestrina*: neonatal and postneonatal mortality at the Regional Primate Research Center Field Station at the University of Washington, 1967–1974. Theriogen 5:267–279.

Goy RW, Wallen K, Goldfoot DA (1974) Social factors affecting the development of mounting behavior in male rhesus monkeys. In: Montagna W, Sadler W (eds) Reproductive behavior. Plenum, New York.

Harlow HF (1962) The heterosexual affectional system in monkeys. Amer Psych 17:1–9.

Harlow HF, Harlow MK (1969) Effects of various mother-infant relationships on rhesus monkey behaviors. In: Foss BM (ed) Determinants of infant behavior, vol 4. Methuen, London.

Harlow HF, Harlow DK, Dodsworth RO, Arling GL (1966) Maternal behavior of rhesus monkeys deprived of mothering and peer associations in infancy. Proc Amer Phil Soc 110:58–66.

Lindburg DG (1973) The rhesus monkey in North India: an ecological and behavioral study. In: Rosenblum LA (ed) Primate behavior, vol 2. Academic, New York.

Mears CE, Harlow HF (1975) Play: early and eternal. Proc Nat Acad Sci 72:1878–1882.

Nadler RD (1980) Child abuse: evidence from nonhuman primates. Dev Psychobiol 13:507–512.

Rosenblum LA (1971) The ontogeny of mother-infant behavior in macaques. In: Moltz H (ed) The ontogeny of vertebrate behavior. Academic, New York.

Ruppenthal GC, Arling GL, Harlow HF, Sackett GP, Suomi SJ (1976) A 10-year perspective of motherless mother monkey behavior. J Abnorm Psych 85:341–349.

Southwick CH (1982) The status of the rhesus monkey population on the Indian subcontinent. Public Report to the American Society of Primatologists, Atlanta, GA.

Suomi SJ, Ripp C (1983) A history of motherless mother monkey mothering at the University of Wisconsin Primate Laboratory. In: Reite M, Caine NG (eds) Child abuse, the nonhuman primate data. Alan R Liss, New York.

Wilson EO (1975) Sociobiology: the new synthesis. Cambridge University Press, New York.

26
Conflict Resolution in Monkeys and Apes

FRANS B.M. DE WAAL

Introduction

Social groups of free-living animals are often compact in spite of a fair amount of internal strife. There appears to exist social homeostasis, that is, a dynamic equilibrium between cohesive and disruptive social forces. Maintenance of this equilibrium requires both restraints on the expression of aggression and mechanisms to normalize disturbed relationships. The resulting state of harmony, however, is probably not the reason—neither proximately nor ultimately—why group members buffer their aggression. Individuals decide on a day-to-day basis into which relationships they will or will not put efforts to resolve tensions. These decisions are undoubtedly guided by self-interest. Peace at the group level is, in this view, a product of selfish compromises at the individual level.

Conflict resolution is a virtually uncharted territory, not only in animals but also, in the sector of private relationships, in humans. In the past decades, the emphasis has been on the factors causing aggression, not on those ending it. Subject indices of textbooks in social psychology and animal behavior provide plenty of references to aggression, violence, and competition but few, if any at all, to the way these modes of behavior are kept under social control.

Here, I will review some recent work in primatology to illustrate the promise and problems of studying social organization from the perspective of conflict resolution. The topic is of practical interest. As explained subsequently, the capacity of primates to deescalate tensions actively causes a more variable response to environmental factors, such as available space, than generally believed on the basis of rodent studies (Calhoun, 1962; Archer, 1970).

The Concept of Reconciliation

There are numerous indications in the primate literature that body contact, especially grooming, plays an important role in the regulation of aggression (e.g., Ellefson, 1968; van Lawick-Goodall, 1968; Blurton-Jones and Trollope, 1968; Poirier, 1970; Lindburg, 1973; Ehrlich and Musicant, 1977). The function of these contacts is commonly described as "reassurance," "appeasement," and "arousal reduction." Note that this terminology stresses effects on internal states rather than on interindividual relationships. Since arousal reduction is a principal

FIGURE 26.1. Situation 10 minutes after a protracted, noisy conflict between two adult male chimpanzees at Arnhem Zoo. The challenged male (left) fled into the tree. He is now being approached by his adversary, who stretches out a hand. Just after this photograph was taken the two males had a physical reunion and climbed down together (Photo by the author; from de Waal, 1982).

component of primate sociability (Mason, 1964), we need a new conceptual framework that emphasizes the social implications.

One useful concept is *reconciliation*. It refers to friendly reunions of former adversaries not long after their conflict. The first tacit use of this concept is recognizable in work by Seyfarth (1976) and McKenna (1978). Seyfarth (1976) reports that agonistic conflicts among feral female baboons (*Papio cynocephalus*) are more often followed by proximity and affiliative behavior if one or both females carry newborn infants. He sees this as an indication of a changed quality of female antagonism caused by the presence of infants.

McKenna (1978) carefully analyzed how social grooming may help to change tense situations into peaceful ones among captive Hanuman langurs (*Presbytis entellus*). He observes: "In one context the recipient of aggression grooms to appease the aggressor and to prevent the aggression from continuing, while in another context the aggressor grooms the victim, thereby assuring the recipient that aggression has ceased and that nonaggressive behaviors can resume between them."

Calling such postconflict interactions "reconciliations," as proposed by de Waal and van Roosmalen (1979), is not merely a matter of a new label. With it comes a set of assumptions and connotations. The term implies, for instance, that there is something special about contacts between former opponents; something that is absent from contacts with bystanders to the conflict. The latter type of contacts can very well serve a reassurance function, but never the function of *repair* of the social network. Only the antagonists themselves can mend their disturbed relationship.

TABLE 26.1. Frequencies of kissing and embracing among chimpanzees in two postconflict contexts.[1]

Context	Behavior	
	Kiss	Embrace
Reconciliation	23	8
Consolation	19	57

[1]χ^2 is 20.3, $P < 0.001$.
Source: de Waal and van Roosmalen (1979).

Studies on the large chimpanzee (*Pan troglodytes*) colony of Arnhem Zoo, The Netherlands, support the above distinction by demonstrating a behavioral difference between the two types of postconflict interaction. Typically, contact between former adversaries is preceded by an invitational hand gesture (Fig. 26.1) and involves mouth-to-mouth kissing. This intensive contact pattern is less common in other contexts. If, after a conflict, one of the participants seeks comfort from a third individual, this usually involves an embrace or pat on the shoulder.

Table 26.1 compares the frequencies of kissing and embracing for reconciliations (i.e., first nonagonistic contact between former opponents) and *consolations* (i.e., contacts between the recipient of aggression and third individuals within 5 minutes following the conflict). The data stem from our first study on postconflict behavior (de Waal and van Roosmalen, 1979). Subsequent studies in Arnhem have confirmed the behavioral differences between the two interaction types. Furthermore, a comparison with control levels of behavior indicates that kissing occurs approximately three times as frequently during reconciliations and embracing six times as frequently during consolations than during normal contacts in the colony (Griede and Willemsen, unpublished observations).

If social repair is an important function of postconflict interactions, one expects not only, as demonstrated above, that the former adversary is treated differently from others, but also that he or she is frequently selected as a contact partner. Thus, another assumption associated with the concept of reconciliation is that of *attraction* between adversaries. This is a testable hypothesis, which stands in marked contrast to the idea, implicit in most of the literature, that the principal effect of aggression is mutual avoidance and dispersal. In our chimpanzee work, we found already strong evidence for the former hypothesis (de Waal and van Roosmalen, 1979), but for a conclusive verification a more carefully controlled study was needed. New observations were made on a large breeding group of rhesus monkeys (*Macaca mulatta*) at the Wisconsin Primate Center (de Waal and Yoshihara, 1983).

Nearly 600 pairs of opponents were observed twice: first, for a period immediately following their conflict; second, on the next day, at the same time, without preceding aggression. Figure 26.2 shows a comparison between the postconflict and matched-control periods with respect to nonagonistic body contact between former opponents. The percentage of pairs with contact was significantly higher after aggression. It is clear from the graph that this difference was entirely created during the first 10 minutes.

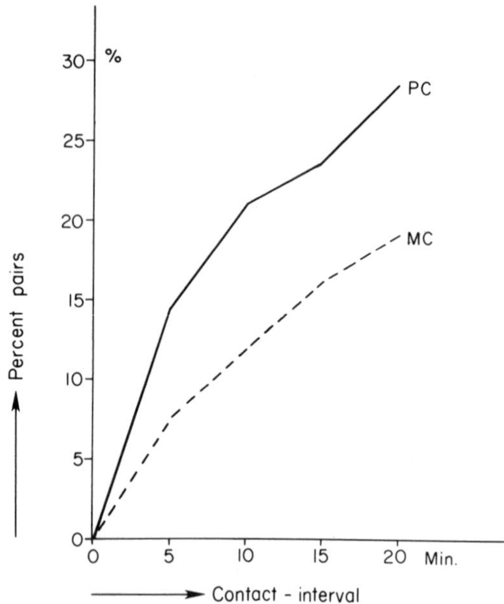

FIGURE 26.2. Cumulative percentage of opponent pairs making contact within a certain time interval in a large rhesus monkey group. PC, postconflict observations. MC, matched-control observations. From de Waal and Yoshihara, 1983.

One possible explanation of the difference is that aggression causes a general surge in contact, indiscriminately involving many group members, including former adversaries. This explanation was refuted by our data: The contact increase occurred especially between the antagonists themselves. Thus, rhesus monkeys remember with whom they have had fights, showing selective attraction to these particular individuals for a period of at least 10 minutes.

Species, Sex, and "Social Class" Differences

Our research has corroborated the basic hypotheses associated with the concept of reconciliation not only for chimpanzees and rhesus monkeys, but recently also for bonobos (*Pan paniscus*) and stump-tail macaques (*Macaca arctoides*). Actually, tension reduction and social repair seem better developed in the latter two species. The data are still being processed, but it is already evident that marked contrasts exist within the genera *Pan* and *Macaca*.

This variability is important in connection with the next research step, which is to investigate the interplay between reassurance and reconciliation behavior, on the one hand, and the nature of social relationships and the level of aggression, on the other. Comparisons can be made between species, between the sexes per species, and between categories of relationships (e.g., kin vs. nonkin). The first systematic cross-species comparisons of aggression and aggression-control can be found in Bernstein et al. (1983) and Thierry (1984a, 1984b).

Thierry observed a captive group of the rare tonkeana monkey (*Macaca tonkeana*), which possesses a remarkably rich repertoire of reassurance gestures. This species' agonistic behavior was compared to that of long-tailed (*M. fascicularis*) and rhesus macaques. Besides differences in the frequencies of postconflict appeasement and affiliation, which were highest for tonkeana monkeys, the investigator reports two further differences: (1) Biting was very rare in the tonkeana group, and (2) confrontations among tonkeanas were more symmetrical, with recipients of aggression frequently counterthreatening their opponents.

According to Thierry (1984b), there may exist a relation between effective reassurance mechanisms and a low intensity of aggression. The resulting reduction in injury risks provides an opportunity for subordinates to retaliate and protest against dominants. Possibly, then, the way conflicts are settled, the rigidity versus flexibility of dominance relations, and the tendency to make up and comfort one another are expressions of a single social/psychological complex.

For this reason, studies of conflict resolution should also pay attention to *tolerance* between dominants and subordinates. In primates, tolerance patterns range from food sharing in the apes to the less spectacular cofeeding of many monkey species. In order to study the distribution of social tolerance in our rhesus group, a variant on the classical dominance test was used. Normally, such tests involve deprivation of food or water for 24 hours or more, after which a monopolizeable source is presented to the group. In this study, water-deprivation lasted only 3 hours, and the source, a water basin, was shareable by four to eight monkeys, depending on body size. More than a thousand interactions among adults were videotaped.

Four types of encounters were distinguished: aggressive exclusion (6%); approach/retreat exclusion (49%); simultaneous drinking (25%), and one monkey drinking while another sat nearby (20%). Comparison of drinking priorities with the group's stable and linear formal hierarchy (based on the direction of submissive teeth-baring; de Waal and Luttrell, 1985) revealed a division into an upper and lower class. Precedence in drinking tests was as expected between individuals of different classes but was not clear-cut within the classes. This was due to variations in social tolerance. Kin always belonged to the same class and showed high tolerance. Sharing of the basin was also more common among unrelated females of the same class than of different classes. Interestingly, the two classes seemed to be kept apart through special intolerance between females ranking at the class borders, i.e., through competitiveness of females at the bottom of the upper class toward females at the top of the lower class. For more details see de Waal (1986).

The recent discovery of this class structure helped solve an important puzzle regarding postconflict behavior. Both in chimpanzees and rhesus monkeys, conflicts among females are less often reconciled than male–male and male–female conflicts (de Waal, 1984a). For chimpanzees, this sex difference is not surprising in view of well-documented differences in sociability and coalition formation (Wrangham, 1979; Halperin, 1979; Nishida, 1983; de Waal, 1982, 1984b, in press). Male chimpanzees associate in bands with a flexible coalition structure. Since every rival is potentially an ally, and vice versa, males cannot afford to hold grudges. Female chimpanzees live more dispersed lives, forming stable bonds with offspring and perhaps a few friends; females can be more selective in their reconciliations.

In macaques, on the other hand, females build complex, cohesive societies. It was not expected, therefore, that they would reconcile less often than males. The solution seems to be that the females in our rhesus group, which includes approximately 50 individuals, limit their peace efforts to certain spheres of social interest. These spheres overlap with the social classes. I did not realize this at the time of the original analysis. Figure 26.3 presents a reanalysis of the data collected by de Waal and Yoshihara (1983), taking the class division into account. The measure of postconflict attraction used in this analysis has a built-in correction for normal contact levels. It gives the proportion of opponent pairs that established nonagonistic contact within 10 minutes following the conflict, and no contact, or later, in a matched-control observation.

Given that postconflict attraction among kin was 30.2% ($N = 63$), the scores of three adult nonkin categories came very close: conflicts among males, conflicts of males with upper-class females, and conflicts among upper-class females. Since all adult males occupied medium to high ranks, this means that reconciliations among adults were most common within the upper regions of the hierarchy, regardless of sex. Intermediate scores were reached for conflicts of lower-class females with each other and with males. By far the lowest postconflict attraction, less than 5%, existed between females of different classes. This difference was significant: Female–female conflicts between classes were less often reconciled than those within classes (χ^2 is 8.2, $p < 0.01$).

Since associations and grooming were not particularly rare between females of different social classes, weak bonding does not seem to be the explanation for the

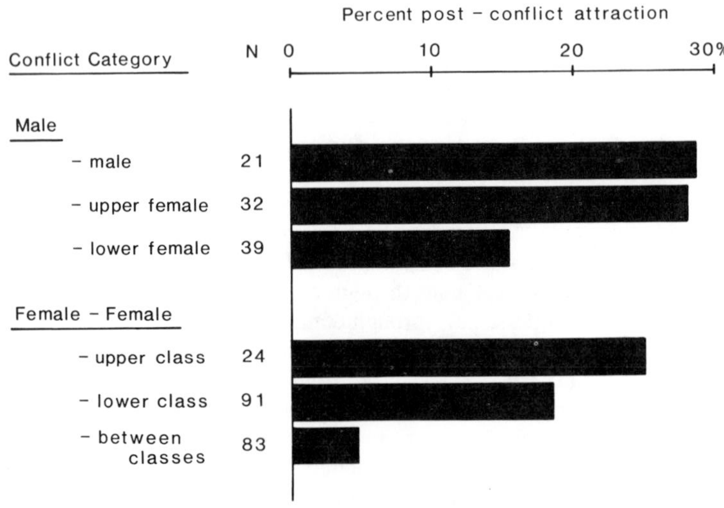

FIGURE 26.3. Reanalysis of data collected by de Waal and Yoshihara (1983), taking the female class structure into account. The figure presents postconflict attraction (defined in the text) within six categories of relationships among unrelated adult rhesus monkeys in a large group. The upper half gives relationships of males with each other and with females; the lower half gives relationships among females. *N* is the number of agonistic interactions observed.

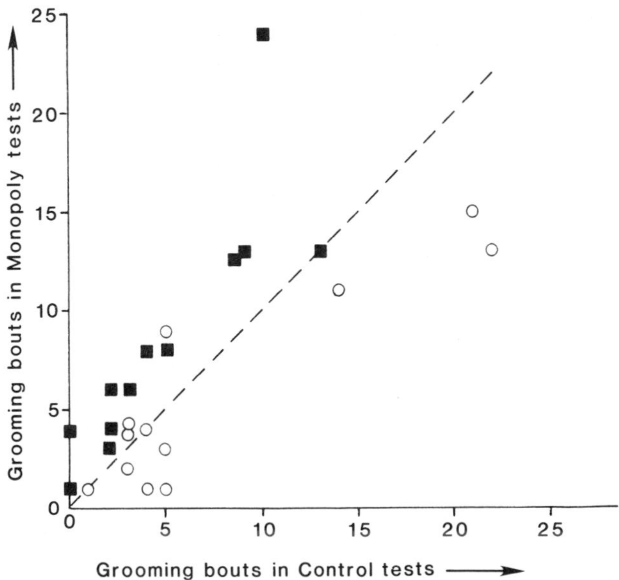

FIGURE 26.4. Comparison of number of grooming bouts performed by 3-year-old rhesus monkeys in small isosexual groups (three male and three female groups with four members per group). In monopoly tests the groups received one piece of apple; in control tests they did not receive extra food. Data for each individual are from ten 30-minute tests of both types. Solid squares represent the males, open circles the females. From de Waal, 1984a.

relative lack of reconciliations between them. I rather seek the explanation in the difference in privileges and social tolerance demonstrated in the drinking tests. Perhaps this difference goes together with solidarity and mutual support within the classes, especially within the upper class. If so, this would provide females with an important incentive to keep good relationships with classmates. This hypothesis, giving a strategical aspect to female reconciliations, is presently under investigation.

Finally, this does not mean that rhesus monkeys do not show any sex differences with regard to conflict resolution. In a series of experiments with small isosexual groups of young monkeys, a tense moment was created by means of the introduction of a single piece of apple. In the 30 minutes following the food competition, males huddled together and groomed more than in control tests; females did not (Fig. 26.4). Although the males' behavior was not identical with reconcilation (their grooming activity was not correlated with the amount and direction of aggression per test), it probably did serve some tension-resolving function. In isosexual groups of the same size, equal numbers of males and females occupy top, middle, and low ranks. Therefore, this sex difference was not, as in the large breeding group, confounded by rank or class differences (deWaal, 1984a).

Environmental Factors

Aggressive behavior is not only followed by intensive contact behavior; it sometimes seems to be replaced by it. At moments of competition, social contacts among apes increase in frequency in an apparent attempt to forestall aggression. The patterns differ per species. The bonobos of the San Diego Zoo, for example, use sociosexual behavior in conflict situations. They have sex at feeding time, or when one of them finds an object that others want to hold (de Waal, in preparation; see Kuroda, 1984, for observations in the wild). Male chimpanzees at Arnhem Zoo, on the other hand, groom each other at high rates when one of the adult females becomes sexually attractive or during the tense and dangerous periods of hierarchical change (de Waal, 1982, in press; see also Coe and Levine, 1980).

Because of these strong counterbalancing mechanisms, the influence of the environment on primate aggression is less straightforward than in many other species. Concepts that became popular in the sixties, such as "crowding" and "stress," have not proven to be of much help in understanding these influences. Personally, I am always skeptical when an outburst of violence in a captive group of monkeys or apes is ascribed to crowding; it seems too easy an explanation. Social developments within the group are more likely to underly such incidents (e.g., Samuels and Hendrickson, 1983). Except under extreme conditions, with monkeys kept like canned sardines (Elton, 1979), there is no evidence for the same dramatic crowding effects and "social pathology" in primates as reported for rodents (Erwin, 1979; Eaton et al., 1977; Nieuwenhuijsen and de Waal, 1982; McGuire et al., 1983). A similar absence of negative effects has been suggested for humans (Freedman, 1977).

In Arnhem Zoo, approximately 25 chimpanzees live in the summer on an island of 0.7 ha, but in the winter in a heated hall with a surface area 20 times smaller than that. The colony does not show any clear-cut seasonal pattern in its reproductive activity. Nieuwenhuijsen and de Waal (1982) compared behavior during two summer and three winter periods in order to determine the effect of crowding on aggression and to test the hypothesis that there might be an increase in reassurance behaviors under the crowded condition. As compared to studies on monkeys, the effects on aggression were minor in spite of the unusually large reduction in living space. The frequency of aggression during the winter period was 1.7 times the summer level, and the aggression intensity did not change.

At the same time, there was a marked increase in submissive "greeting" rituals during the winter (with a factor of 2.4), an appeasement gesture often shown to intimidating dominants. The grooming rate also increased during winters (with a factor of 2.0). This combination of changes was interpreted as an indication that the apes were trying to *cope*, rather successfully, with the increased tensions caused by spatial restriction. There was an intensification rather than a deterioration of social life, except perhaps for juveniles and infants whose play behavior decreased in frequency during the winter periods.

The effects of crowding on social behavior were not the same at the beginning and the end of the winter period of 5 months. Some parameters changed dramatically during the first weeks, after which the situation returned to summer levels. Therefore, traditional short-term experiments are perhaps of little value to understand environmental effects on the well being of captive primates. Also, the effects of a particular environment may depend on whether the animals live in it continuously or only part of the time. It is doubtful, for example, whether the Arnhem chimpanzees can be kept year-round in their winter quarters without problems. Major shifts in the social hierarchy have always taken place during outdoor periods, as if challengers of the status quo waited for this opportunity. Provocations require escape space. If this space is not provided at any time in the year, attempts at social change may be made indoors with all possible risks of dangerous escalation.

Further research is needed to determine under which conditions primates are able to absorb social tensions, and which conditions hinder this important capacity. Since "commitment" to social relationships takes time, changes in group composition are potentially the most disruptive factor (Erwin, 1979; see Kessler et al., 1985, for the shocking consequences of a management policy ignoring this basic fact). The mechanisms of conflict resolution described in this chapter seem to operate best in the context of stable, long-term relationships. The study of these buffering mechanisms, rather than of aggression per se, may give us a better understanding of the complex effects of the captive environment on primate behavior. Ultimately, this will assist improvement of the environment.

Acknowledgments. Writing of this chapter was supported by NIH grant RR00167 to the Wisconsin Regional Primate Research Center. This is publication No. 25-014 of the WRPRC. I thank Mary Schatz and Jackie Kinney for typing of the manuscript.

References

Archer J (1970) Effects of population density on behavior in rodents. In: Crook J (ed) Social behavior in birds and mammals. Academic, New York, pp 169–210.

Bernstein I, Williams R, Ramsay M (1983) The expression of aggression in Old World monkeys. Intern J Primatol 4:113–125.

Blurton-Jones NG, Trollope J (1968) Social behaviour of stump-tailed macaques in captivity. Primates 9:365–394.

Calhoun J (1962) Population density and social pathology. Sci Amer 206:139–148.

Coe C, Levine R (1980) Dominance assertion in male chimpanzees (*Pan troglodytes*). Aggress Behav 6:161–174.

Eaton G, Modahl K, Johnson D (1977) Aggressive behavior in a confined troop of Japanese macaques. Paper presented at the Annual Meeting of the Western Psychological Association, Seattle, Washington.

Ehrlich A, Musicant A (1977) Social and individual behaviors in captive slow lorises. Behaviour 60:195–220.

Ellefson JO (1968) Territorial behavior in the common white-handed gibbon, *Hylobates lar*. In: Jay P (ed) Primates: studies in adaptation and variability. Holt, Rinehart and Winston, New York, pp 180–199.

Elton R (1979) Baboon behavior under crowded conditions. In: Erwin J, Maple T, Mitchell G (eds) Captivity and behavior. Van Nostrand Reinhold, New York, pp. 125–138.

Erwin J (1979) Aggression in captive macaques: interaction of social and spatial factors. In: Erwin J, Maple T, Mitchell G (eds) Captivity and behavior. Van Nostrand Reinhold, New York, pp 139–171.

Freedman J (1977) Crowding and behavior, WH Freeman, San Francisco.

Halperin S (1979) Temporary association patterns in free-ranging chimpanzees. In: Hamburg D, McCown E (eds) The great apes. Benjamin Cummings, Menlo Park, CA, pp 491–499.

Kessler M, London W, Rawlis R, Gonzales J, Martinez H, Sanchez J (1985) Management of a harem breeding colony of rhesus monkeys to reduce trauma-related morbidity and mortality. J Med Primatol 14:91–98.

Kuroda S (1984) Interactions over food among pygmy chimpanzees. In: Sussman R (ed) The pygmy chimpanzee. Plenum, New York, pp 301–324.

Lawick-Goodall J van (1968) A preliminary report on expressive movements and communication in the Gombe Stream chimpanzees. In: Jay P (ed) Primates: studies in adaptation and variability. Holt, Rinehart and Winston, New York, pp 313–374.

Lindburg DG (1973) Grooming behavior as a regulator of social interactions in rhesus monkeys. In: Carpenter C (ed) Behavioral regulators of behavior in primates. Bucknell University, Lewisburg, PA, pp 85–105.

Mason W (1964) Sociability and social organization in monkeys and apes. In: Berkowitz L (ed) Advances in experimental social psychology 1. Academic, New York, pp 277–305.

McGuire M, Raleigh M, Johnson C (1983) Social dominance in adult male vervet monkeys: general considerations. Soc Sci Inf 22:89–123.

McKenna J (1978) Biosocial functions of grooming behavior among the common Indian langur monkey (*Presbytis entellus*). Amer J Phys Anthrop 48:503–510.

Nieuwenhuijsen C, Waal F de (1982) Effects of spatial crowding on social behavior in a chimpanzee colony. Zoo Biol 1:5–28.

Nishida T (1983) Alpha status and agonistic alliance in wild chimpanzee. Primates 24:318–336.

Poirier FE (1970) Dominance structure of the Niigiri langur (*Presbytis johnii*) of South India. Folia Primatol 12:161–186.

Samuels A, Hendrickson R (1983) Outbreak of severe aggression in captive *Macaca mulatta*. Amer J Primatol 5:277–281.

Seyfarth R (1976) Social relationships among adult female baboons. Anim Behav 24:917–938.

Thierry B (1984a) Clasping behaviour in *Macaca tonkeana*. Behaviour 89:1–28.

Thierry B (1984b) Patterns of agonistic interactions in three species of macaque (*Macaca mulatta*, *M. fascicularis*, *M. tonkeana*). Paper presented at the Xth Congress of the International Primatological Society in Nairobi, Kenya, 1984.

Waal F de (1982) Chimpanzee politics. Jonathan Cape, London.

Waal F de (1984a) Coping with social tension: sex differences in the effect of food provision to small rhesus monkey groups. Anim Behav 32:765–773.

Waal F de (1984b) Sex-differences in the formation of coalitions among chimpanzees. Ethol Sociobiol 5:239–255.

Waal F de (In press). The reconciled hierarchy; on the integration of dominance and social bonding in primates. Quart Rev Biol.

Waal F de (1986) Class structure in a rhesus monkey group; the interplay between dominance and tolerance. Anim Behav. 34:1033–1040.

Waal F de, Luttrell L (1985) The formal hierarchy of rhesus monkeys: an investigation of the bared-teeth display. Amer J Primatol 9:73–85.

Waal F de, Roosmalen A van (1979) Reconciliation and consolation among chimpanzees. Behav Ecol Sociobiol 5:55–66.

Waal F de, Yoshihara D (1983) Reconciliation and re-directed affection in rhesus monkeys. Behaviour 85:224–241.

Wrangham R (1979) Sex differences in chimpanzee dispersion. In: Hamburg D, McCown E (eds) The great apes. Benjamin Cummings, Menlo Park, CA, pp 481–490.

27
Resocialization of Asocial Chimpanzees

JO FRITZ

Introduction

If we are to have a self-sustaining, captive population of chimpanzees (*Pan troglodytes*), we must have breeders and parents, some of which can be supplied by resocialization. However, we cannot depend solely upon resocialization techniques to add breeders to the genetic pool nor to produce animals that are adequate parents.

The term *asocial* as used herein simply means that the animal has had little or no species-specific social experience. Insufficient opportunities have prevented it from acquiring the social experience necessary for compatible group living, successful reproduction, and adequate parental behaviors.

Resocialization is the process used to introduce or restore behaviorally deficient chimpanzees to a more normal state of social living. In general, this means eliminating any behaviors that prevent an individual from interacting with and relating to other group members (Fritz and Fritz, 1979). We use *resocialize* as a generic term in that a distinction is not made between the term socialization for isolation-reared animals and resocialization for restricted-reared animals.

The Primate Foundation of Arizona has resocialized 59 chimpanzees, most of which have formed the nucleus of the present breeding colony. However, our criteria for successful resocialization do not necessarily include reproduction, and successful resocialization does not necessarily produce animals that are captive duplicates of their wild relatives. Animals currently in the resocialization program are juveniles that were born at the Foundation. They have temporarily cycled into biomedical research and subsequently returned to the colony. Our criteria for successful resocialization have become more stringent for these juveniles based upon their known rearing history. We expect them all to become breeders and parents.

Methods

The methods for resocialization are the same as reported by Fritz and Fritz (1979). These methods of maneuvering the animal through increasingly complex social situations have been successfully utilized since 1969 and the reader is referred to that publication for more detail.

Resocialization programs must be planned to allow maximum social opportunity for each animal. This requires enough animals to provide a variety of ages and sexes in the social groupings. The facility should be designed to allow for ease of controlled shifting of individuals between the social groups, and well-trained observers must be available to document progress.

Focal-animal observations are conducted one time per day for a minimum of 10 consecutive work days. When the individual stabilizes or when stress behaviors, such as rocking and screaming, are reduced in frequency, the number of focal-animal observations are gradually decreased to a minimum of one time per week. With each change, for example, a new cage or the addition of a social partner, the formal observations are again increased to one time per day. Ad libitum observations are made and recorded throughout the day for the entire colony. It should be stressed that resocialization personnel must be trained observers who are intimately aware of and able to recognize individual chimpanzees behavioral patterns.

Asocial chimpanzeess should not be rapidly forced into social situations. The introduction to social living should be accomplished gradually to prevent withdrawal and possible increase in undesirable behavioral patterns. Both frequency and circumstance of stereotypies are of considerable importance. They are often the first clue that one has gone too far, too fast. It has also been our experience that the behavior of the animals toward personnel, i.e., timidity, aggression, etc., is often the behavior they will initially exhibit toward other animals. These observations will provide fairly reliable projections of the individual's behavior in a social situation and the necessary general characteristics, i.e., gentleness, playfulness, etc., of the animal's first social partner (Fritz and Fritz, 1979).

The behavioral history (when available) and the source, e.g., zoo, laboratory, etc., of the animal are also important. Slightly different resocialization techniques are required based upon the animal's source. The animals must also disregard their previous management routines and learn new ones. Because of the vast amount of individual variation the process must be tailored to each animal and its needs. Careful attention is paid to the behavior of each individual and its preference for or dislike of companions of both sexes. This requires a wide variety of animals to afford adequate selection of social partners.

Group composition is changed frequently for a variety of reasons, such as deletion of an overly dominant or extremely subordinate individual and, more recently, in our returned youngsters to eliminate what appears to be excessive pair-bonding behavior. In any case, the composition of the social groups changes as a result of the addition or deletion of individuals achieving social growth and experience.

Observations that determine the decisions to move individuals to other situations are (1) the degree of integration into a social group, (2) the status or role of the individual within a group, and (3) evidence of appropriate or inappropriate behaviors.

Chimpanzees have come to the Primate Foundation of Arizona from zoos, laboratories, circuses, and private individuals (Table 27.1). There are still too few animals in each category to be able to make any statement regarding different results based upon the source. However, source may still prove to be an important factor because slightly different resocialization techniques are required for animals from different sources.

TABLE 27.1. Percentage received from each source.

Zoo	39
Performer	27
Laboratory	22
Pet	12
	100
Males $N = 27$	Females $N = 32$

Criteria for a successful resocialization program require individualized techniques and that the animal (1) can be moved from group to group with positive interaction, (2) initiates and receives all socially acceptable behaviors, and (3) has the ability to assume different roles in different groups.

Results

During the past 15 years, only one animal failed to achieve socialization based upon our criteria (Table 27.2). Animals that met this criterion on arrival were considered to be social and are not included in these data.

Copulatory behavior is not an automatic result of resocialization as shown by the frequency with which both sexes achieved the goal of copulatory behavior (Table 27.3). We do not use it as a major criterion for successful resocialization. The number of subjects of each sex have been adjusted downward for animals that have not reached copulatory age or are not reproductively fit, i.e., are castrated. Although there is no significant difference by sex, when all individuals are considered, the trend is that males show less success. There may be an interaction of sex and source upon success, but the numbers of animals in each category are still too few for tests of significance.

Adequate mothering is also not an automatic by-product of a resocialization program (Table 27.4). The number of subjects has been adjusted downward for animals that copulate but have never conceived or that have not reached reproductive age.

As an interesting note in the Performer category, we could have listed a higher percentage of mothers than breeders. An ex-performer female had been the dominant leader of two younger females in the same act. She retained this position on arrival and was also protective of other young females in the social group.

TABLE 27.2. Percentage resocialized from each source.

Zoo	100
Performer	100
Laboratory	92
Pet	100
Total regardless of source	98
$N = 59$	

TABLE 27.3. Percentage achieved copulatory behavior.

	Males	Females
Zoo	67	69
Performer	80	50
Laboratory	17	80
Pet	100	100
Total regardless of source	58	70
	$N = 19$	$N = 30$

However, she did not copulate and led the other females into aggressive action toward any male in the group. She responded extremely well to spoken commands, and this coupled with her protective behavior toward younger animals led us to predict that given the opportunity, she would be a good mother. We moved her to a biomedical research facility (Laboratory for Experimental Medicine and Surgery in Primates, New York University Medical Center) where she entered an experimental artificial insemination program. This was successful, and she gave birth to a female infant. Paul Fritz, our Colony Director, provided the moral support and the telephone commands that were to be authoritatively spoken to the new mother. The result: She raised her infant and is regarded as a very good mother.

A great deal of allo-mothering or aunting behavior is observed in our colony. However, it has not proven to be 100% successful in teaching adequate mothering behaviors. We have seen females that frequently allo-mother but that will not accept or take care of their own infants.

We have had as much or more success in teaching adequate mothering behaviors by the use of conditioned voice control and the "respect" that the chimpanzees accord to Paul Fritz. He is the dominant member of the colony and an experienced animal trainer. Paul has entered the cage of several young mothers and emphatically informed them that they must pick up the infant and hold it. They do, and after several days of this pointed instruction, they go on to become good mothers.

Fifty infants have been born to resocialized colony animals since 1973. Thirteen of these 50 were rejected by their mothers and had to be hand reared. However, in several instances, females that had rejected their first infants successfully reared subsequent offspring. At present, there are eight mother–infant pairs in the colony. Our first second-generation birth occurred in 1983. The event was especially significant in that the sire was hand reared from 30 days of age by

TABLE 27.4. Percentage of adequate mothers.

	Copulators	Mothers
Zoo	69	67
Performer	50	50
Laboratory	80	25
Pet	100	100
Total regardless of source	70	58
	$N = 30$	$N = 26$

the author. Since that birth, he has sired three additional offspring. All infants are being successfully reared by their mothers, and we look forward to a third generation of births to adequate parents. On the negative side, there are 10 nonbreeding and three breeding but nonparental females and nine males that have never exhibited any form of copulatory behavior; all of these animals met our criteria for resocialization and adapted well to social group living.

While different source has not yet been indicative of different resocialization results, animals from different sources exhibited different behaviors upon arrival. Based upon past contact with other chimpanzees, many ex-zoo animals arrived as social animals; however, in several cases, the contact had been detrimental, e.g., a very young female had been housed with an extremely aggressive older male. On arrival, she reacted with overwhelming hysterical fear when any male approached her. These animals also required resocialization to the extent of overcoming fears or other problems associated with detrimental social contact. In any case, these social zoo animals became the initial cadre of "teachers" and social partners for the asocial animals. Many of these zoo animals also practiced coprophagy, "finger painting," and depilation (overgrooming) to the point of total baldness. We rapidly discovered that these behaviors, as well as social behaviors, are learned by others while motion stereotypies, such as rocking, are not.

The ex-laboratory subjects were mere shadows of normal chimpanzees. Some did not know that they could or should climb. This may also have been due to a lack of muscle development, such as that reported for monkeys in caged versus free-ranging environment (Turnquist, 1985). While our caging is not to be compared to a free-ranging situation, the cages are considerably larger than laboratory cages. In these animals, social skills did not develop until motor skills, such as climbing, were learned. All of these chimpanzees used vocalizations incorrectly or not at all. This also may be a direct comparison to a study with monkeys which provided evidence that while motor and vocal patterns seem to develop even in complete social isolation, the proper use of these behaviors appears to be highly dependent upon social experience (Hopf et al., 1984). Few of these ex-research animals practiced coprophagy, perhaps because in laboratory cages, the feces fall through and are not available. However, all exhibited some form of stereotypic movement, i.e., twirling, scooting, rocking, etc.

Ex-pets had few bad habits. However, unless they arrived at a very young age, they appeared to have little interest in associating with their own species. They eventually achieved sociality by our criteria, but on the whole they appeared to prefer to remain on the periphery of most groups. Those that arrived before 5 years of age were the easiest to resocialize. For them, integration into a peer group was rapid, as long as the human surrogate provided security by remaining in the cage during the daily introduction periods. In most cases, this security was no longer necessary after 2 or 3 days.

Most of the ex-performers were asocial on arrival. Interestingly, however, they were the easiest to handle, the most rapid to socialize, and had few, if any, bad habits. Dressing up and turning somersaults cannot be a prerequisite to chimpanzee social life, but the stimulation and occupation that is associated with training and learning (even circus-type tricks) may be one of the most important factors in understanding the differences seen among the singly reared, asocial chimpanzees from all four sources.

For animals of all sources, the problem of lack of opportunity for social learning is confounded by the behavioral repertoire that they learn during isolation. They learn to cling to themselves, to self-suck, and to provide their own motor feedback, which develops into stereotypic activity (Suomi et al., 1974). These stereotypic behaviors initially take precedence over social learning, and they prove to be exceedingly resistant to change.

In many cases, behaviors such as coprophagy and depilation (to the point of baldness) remain evident even when the animal is comfortable and occupied within its environment. Based upon the activity and aggression levels of the whole group, these behaviors do not appear to be related to either stress or boredom. It appears more likely that they have become part of that animal's natural repertoire of behaviors. However, the stereotypic activities do not totally prevent other normal behaviors from appearing. Animals that rock, practice excessive grooming, etc., have gone on to become breeders and parents. Frequently, however, an animal that exhibits stereotypic activity may also be deviant in another behavior. For example, one of our breeding males occasionally rocks. He is also overly aggressive toward females that are not in estrus. On the other hand, he is gentle and playful with infants and can be dominated by estrous females.

In all behaviors, males appear to be the most severely affected by the lack of sociality and are the more difficult of the two sexes to socialize (Fritz and Fritz, 1979). They either become whimpering, fearful, "quivering bowls of jelly," or overly aggressive bullies. Also, more males practice one of the most unacceptable of the stereotypic behaviors—self-mutilation. In 15 years, we have had four self-mutilators, three males and one female; two were from zoos and two from labortories. After 4 years of manipulating one 12-year-old male through every conceivable group or pair situation, we were forced to euthanize him. He had exposed bones across both cheeks and wrists and had various defects over his legs, chest, and shoulders. A 3-year-old male stopped self-mutilating within several months after being gradually introduced to a young peer group. The third male was 10 years old. He did not self-mutilate as long as he was the dominant animal in his group. This made it difficult to supply him with appropriate social partners because he was also overly aggressive. However, after 10 years, the mutilation behavior has stopped and the aggression diminished.

The single female is 6 years old and has proved to be extremely difficult. She is an ex-zoo animal that exhibits severe motion stereotypies when placed in even mildly stressful situations. At the same time, she appears to regard self-mutiliation as a normal quiet-time activity. The behavior may be replacing more normal grooming behaviors. For example, when she is integrated into a new social group or new members are introduced to her group, she sits against a solid barrier and rocks violently, literally knocking the air from her lungs. As she recovers from the social change, this behavior diminishes and her contact with group members increases. Then there is a brief period of attempted sociality, which includes inept grooming and timid play bouts. However, this too slowly diminishes, and she sits off by herself and self-mutilates. The mutilation behavior coincides with resting periods and the other normal diminished daily activity periods of the whole group. We have twice taken her back into very much younger groups and slowly brought her forward with the group. We have recently initiated this technique for the third time. This time, she is dividing her day equally between social play and withdrawn self-mutilation. However, both quality and quantity of play have increased as compared to past efforts, and we are beginning to believe that there is hope for her.

Discussion

If we want to have chimpanzees in the United States 50 years from now, captive propagation of the chimpanzee is a critical necessity (Fritz and Fritz, 1983; Fritz and Nash, 1983). A recent survey of the U.S. chimpanzee population revealed that of the 411 captive births between 1926 (the first captive birth) and 1973 (animals that reached reproductive age in 1985) only eight living captive-born males and 27 living captive-born females reproduced (Fritz and Fritz, 1982). Thus, the far greater percentage of breeders and parents of today are wild-caught animals, most of which are rapidly approaching the slowed reproductive years of old age.

We should recognize that these wild-born breeders did have prior mothering and social experience in a stimulating environment during the first 12 to 24 months of their life. Therefore, we must change current management practices of removing the infant from its mother at birth or at a very young age. If the infant must be hand reared, it must be given as much stimulation and social opportunity as possible (Fritz and Fritz, 1982, 1985). The issue is clear-cut: The more restrictive the early social environment, the more deficient is the chimpanzee's social repertoire. Early social experience may be critical to the subsequent development of social and reproductive behaviors (Riesen, 1971).

The total U.S. captive chimpanzee population numbers approximately 1400, which would suggest it to be in a good state for continued population growth and genetic diversity. However, this figure may be deceptive. Of this 1400, approximately 1200 are owned or being used by biomedical research. They may have been living in single cages for perhaps too long a time. They may socialize to the point of group living, but not to the extent of breeding or evidencing parental behaviors. Second, many of the research animals were captive born and removed from their mothers at birth. Only a few of these animals have reached puberty. It is not yet known to what extent social and reproductive behaviors will emerge, even under the best of conditions (Fritz and Nash, 1983). Only one of the resocialized animals in our colony was captive born, and she spent 12 months with her mother. Therefore, we can only hypothesize that there is more hope of successful resocialization if the animal has experienced a period of maternal reciprocity and sociality following birth. Thus, it is imperative that intense efforts be rapidly initiated to resocialize as many as possible of the asocial chimpanzees in the population. Any chimpanzee that is housed as a solitary animal should be a candidate for resocialization.

Whether or not resocialization is considered successful depends completely upon the desired end result. If you desire that all "bad habits" be completely eliminated, it is our experience that the success rate will be approximately 50%. This will depend upon the individual animal and the age at which the resocialization process is initiated. If you desire only sociality to the point that the animal can live without fear in a group, that it will exhibit social behaviors such as grooming and compatible sharing of available space, then given time, patience, and an appropriate facility and staff, one can expect to be 98% successful. The emphasis on time and patience is critical because complex social behaviors such as play, sex, and aggression do not suddenly emerge. They are acquired only through prolonged physical interaction. As they gradually develop, they are integrated into existing repertoires (Suomi et al., 1974).

Should one decide that success will be judged only upon whether or not the animals become breeders and parents, the success rate rapidly declines. Again, it will depend upon the extent of the deviant behaviors, the age of the animal at onset of resocialization (the younger the animal the higher the success rate), the availability of socially adequate animals to serve as "teachers," and the amount of time devoted to the project. The issue of parenting will still be problematic. It does not necessarily follow that good breeders are good parents. Of course, there are exceptions to these generalities. Everyone who works closely with chimpanzees knows that they spend a good portion of their day figuring out ways to prove you wrong.

Summary

Is the so-called art of resocialization necessary and worth the time, expense, and effort? Will it play a role in saving the chimpanzee? I believe the answer is an unequivocal *Yes*.

Captive breeders are desperately needed, and a 50% success rate is better than none. A few parents to assure that there are also parents in the next generation are better than no parents. Chimpanzee facilities are rapidly reaching the point of being overcrowded. We may need to relieve some of the crowded conditions by placing a large number of the present captive population of ex-research animals in a free-ranging environment or in large social groups. They will have to learn the social behaviors required for group living. Captive-born, hand-reared infants or infants that are removed at a very young age from their mothers will have to learn the social behaviors of their species. As long as there are single animals maintained in zoological gardens, laboratories, or as house pets and performers, sooner or later there will be a need for them to be integrated into a group.

The need to resocialize has been created by our uses of the chimpanzee and current management practices. These uses have produced a large population of nonsocial, nonbreeding animals. We must not depend solely upon resocialization programs to solve this problem. We must also change and improve our management procedures to limit the need for resocialization programs. However, resocialization remains as the only way to correct out previous neglects and to give every asocial chimpanzee the opportunity to become a normal, social chimpanzee. Only then will there be enough healthy, behaviorally normal chimpanzees to put us on the road to a self-sustaining captive population.

Acknowledgments. None of this would be possible without Paul Fritz, the chimpanzees, and the observers who collected the data. I am greatly indebted to Leanne T. Nash, Ph.D., for statistical assistance and valuable advice. Thomas L. Wolfle, D.V.M., also made many useful comments.

References

Fritz P, Fritz J (1979) Resocialization of chimpanzees: ten years of experience at the Primate Foundation of Arizona. J Med Primatol 8:202–221.

Fritz J, Fritz P (1982) Great ape hand-rearing with a goal of normalcy and a reproductive continuum. In: Fowler ME (ed) Proceed Amer Assoc Zoo Vet, Annual Meeting, November 1982, New Orleans, LA.

Fritz J, Fritz P (1983) Captive chimpanzee population crisis. Proceed Amer Assoc Zoos, Parks, and Aquar Regional Meeting, March 1983, Santa Barbara, CA.

Fritz J, Fritz P (1985) The hand-rearing unit: management decisions that may affect chimpanzee development. In: Graham CE, Bowen JA (eds) Clinical management of infant great apes. Alan R Liss, New York, p 1.

Fritz J, Nash LA (1983) Rehabilitation of chimpanzees: captive population crisis. Lab Prim Newslett 22:4–7.

Hopf S, Herzog M, Vogl-Kohler C (1984) Social integration of surrogate-reared infant squirrel monkeys to captive groups. Acta Paedopsychiat 50:79–95.

Riesen AH (1971) Nissen's observations on the development of sexual behavior in captive born, nursery-reared chimpanzees. In: Bourne GH (ed) The chimpanzee. Karger AG, Basel, p 1.

Suomi SJ, Harlow HF, Novak MA (1974) Reversal of social deficits produced by isolation rearing in monkeys. J Hum Evol 3:527–534.

Turnquist J (1985) Passive joint mobility in patas monkeys (*Erythrocebus patas*): rehabilitation of caged animals after release into a free-ranging environment. Amer J Phy Anth 67:1–5.

28 Primate Mating Systems and Their Consequences for Captive Management

RONALD L. TILSON

Introduction

Excluding *Homo sapiens*, the living primates account for about 182 species (after Napier and Napier, 1967; Kavanagh, 1983). They are confined to tropical and subtropical forests and savannas of Africa, Asia, and America; they live as solitary individuals or in social groups; and they employ a variety of mating strategies, each a product of selection that maximizes reproductive success. However, the mating relationship yielding maximal success is not necessarily the same for both sexes. In many species, males attempt to fertilize as many females as possible while females try to monopolize the male and the resources he can contribute to her reproductive effort. The result is often a fundamental conflict of interests (Wittenberger, 1981).

> Within any single species there is a growing realization that males and females have overlapping but nonidentical reproductive interests and the type of mating system shown by a species results from interactions between the individual interests of each sex. Optimal mating systems for promoting individual reproductive interests often differ for each sex and when it does, the interests of one sex constrain the reproductive options open to the other. Thus, mating system theory revolves around two major issues: the factors determining which sex predominates in shaping each mating system, and the factors determining which mating system is optimal for members of the 'controlling' sex (Wittenberger and Tilson, 1980, p. 197).

Here I briefly review concepts in mating system theory (Altmann et al., 1977; Bradbury and Vehrencamp, 1977; Emlen and Oring, 1977; Graul et al., 1977; Clutton-Brock and Harvey, 1978; Wittenberger, 1979, 1980; Wittenberger and Tilson, 1980; Rutberg, 1983), then characterize general primate mating systems and the selective forces that have molded them. Because the bulk of theoretical development in mating systems theory addresses the evolution of polygyny—and the models are mostly derived from avian systems—I approach this presentation from the perspective of a monogamous species. I then attempt to indicate the consequences of species-specific mating systems that are liberated from certain selective pressures typical to the natural environment but absent in the captive environment. Finally, suggestions for long-term captive management strategies for selected primate species are offered.

Definitions

Mating systems are most commonly classified according to the duration of pair-bonding and the number of mates acquired by each sex (Wittenberger, 1979, 1981). For convenience I use Gouzoules' (1984) terms in which "mating systems are distinguished by the number of breeding adults typically present in groups of a given species. *Monogamous* and *polygynous* groups are both characterized by the presence of a single breeding male. They are distinguished by the number of breeding females, with monogamous groups containing a single breeding female and polygynous groups from two to several females. Similarly, both polygynous and *promiscuous* groups contain several adult females and are distinguished by the number of breeding males, with promiscuous groups typically containing several sexually mature breeding males" (after Gouzoules, 1984, p. 102).

Typically, mating terminology has been defined with reference to the duration and quality of the pair-bond (Selander, 1972; Kleiman, 1977; Wittenberger, 1979; Wittenberger and Tilson, 1980) or to resources (Emlen and Oring, 1977). Other authors stress the influence of patterns of female sociality (Wittenberger, 1980; Wrangham, 1981; Stacey, 1982; Rutberg, 1983) or stress genetic contributions to mating system definitions (Gowaty, 1983; Gowaty and Karlin, 1984).

In general, mating systems in the primates do not covary with phylogeny; species exhibiting monogamy, polygyny, and promiscuity can be found among both the prosimians and the apes (Gouzoules, 1984). Monogamy in primates is rare (Kleiman, 1977; Wittenberger and Tilson, 1980), having evolved in only 14% of the species (Rutberg, 1983). All other primate species are polygamous: either polygynous or promiscuous.

Solitary Versus Group-Living Primates

So-called solitary primates exhibit either polygynous or promiscuous—also referred to as *overlap promiscuity* (Wittenberger, 1979)—mating systems. They are not really solitary. The term refers to animals that are usually found alone, rather than to the absolute absence of social interactions (Wasser and Jones, 1983). In general, female home ranges in most solitary species overlap considerably while those of males are usually larger, exclusive, and encompass one to several females' home ranges. Females typically depend on secretive behavior and concealment for protection and they gain little benefit from male parental care or vigilance (Wittenberger, 1981). Typical examples include bushbabies (*Galago* sp.), the lesser mouse and sportive lemurs (*Microcebus murinus* and *Lepilemur mustelinus*, respectively), and the orangutan (*Pongo pygmaeus*) (see Gouzoules, 1984, for review and references).

Promiscuity

Promiscuity results from selection against prolonged pair-bonding. Eisenberg et al. (1972) distinguished between unimale, age-graded, and multimale primate species on the basis of the degree of tolerance manifested among fully adult and similarly aged males. Multimale primate groups generally exhibit promiscuity, although the mating system is often called *hierarchical promiscuity*, which is based on male dominance hierarchies and has evolved primarily in cohesive female groups that are too large for single males to control (Wittenberger, 1981). It is especially prevalent in terrestrial or semiterrestrial primates (Crook, 1972; Jolly, 1972; Eisenberg et al., 1972), exemplified by all of the macaques (*Macaca* sp.), ring-tailed and black lemurs (*Lemur* sp.), most baboons (*Papio* sp.), Hanuman and silver-leaf monkeys (*Presbytis* sp.), the red colobus monkey (*Colobus badius*), several mangabeys (*Cercocebus albigena* and *C. atys*), vervet monkeys (*C. aethiops*), talapoin monkeys (*Miopithecus talapoin*), squirrel monkeys (*Saimiri sciureus*), howler monkeys (*Alouatta* sp.), spider monkeys (*Ateles geoffroyi*), and chimpanzee (*Pan troglodytes*) (see Gouzoules, 1984, for review and references).

Male dominance hierarchies probably result from competition for access to receptive females and to limited resources. However, the ability of dominant males to control access to females is debatable. Many studies infer a relationship between social rank and reproductive success (e.g., Hausfater, 1975; Packer, 1979). In some respects this evidence must be viewed with a certain amount of skepticism because males of high and low status are not equally visible to human observers, and many copulations of low-ranking males may be missed (e.g., rhesus monkeys, Drickamer, 1974). Other tactics that low-ranking male primates sometimes use is to form coalitions to dominate an otherwise dominant male. They may also use furtive tactics to copulate with females out of view of high-ranking males. Female primates do exercise choices when accepting mates, and they sometimes prefer subordinate males over dominant males (Dixson et al., 1973; Eaton, 1973; Hausfater, 1975; Lindburg, 1975; Seyfarth, 1978; Bachman and Kummer, 1980). To counter this, males often resort to aggressive tactics to control access to receptive females, especially during peak estrus (e.g., Hausfater, 1975).

Recent studies address the actual probabilities of offspring produced by males of differing status. Using electrophoretic and serological genetic markers to assess paternity, Duvall and co-workers (1976) found that the dominant male of one rhesus group did not father more offspring than several low-ranking males (see also Smith, 1982; Stern and Smith, 1984, for additional evidence).

Polyandry

A prolonged association and essentially exclusive mating relationship between one female and two or more males—or *polyandry*—is an unusual mating system found primarily in birds (Wittenberger, 1981) and is almost always associated with sex role reversal (Jenni, 1974). Among the primates, some tamarins and marmosets live in extended family groups composed of largely unrelated adults of both sexes. In such groups, only a single female is reproductively active, and she may mate with more than one male, who subsequently provides care for the young. It has been suggested that such breeding systems are polyandrous (Sussman, 1985), but more evidence concerning genetic paternity (e.g., Smith, 1982) will be necessary to clarify this position.

Polygynous Primate Species

The evolution of polygynous mating systems in primates is complicated by a number of environmental factors (Wittenberger, 1979) and by the fact that social structure among polygynous species varies considerably (Gouzoules, 1984). Polygynous species are characterized by social units containing one breeding male and several adult females with dependent offspring. However, within this definition, the social organization of groups ranges from those species in which only one sexually mature male is present in the group (e.g., patas monkeys, *Erythrocebus patas*; purple-faced leaf monkeys, *Presbytis senex*; Nilgiri langurs, *P. johnii*; Hanuman langurs, *P. entellus*; red-tail monkeys, *Cercopithecus ascanius*; blue monkeys, *C. mitis*; Lowe's guenons, *C. campbelli*) to those in which mature males of different ages and status may coexist—also called "age-graded structure"—represented by silver leaf monkeys, *Presbytis cristatus*; spider monkeys, *Ateles geoffroyi*, and gorillas, *Gorilla gorilla* (see Gouzoules, 1984, for references). It is believed that only the oldest alpha male has access to reproductive females in age-graded groups, but as discussed above, this is not always true, and until genetic markers are employed for assessing paternity, this question remains to be resolved.

Other polygynous species, such as gelada (*Theropithecus gelada*) and hamadryas (*Papio hamadryas*) baboons, fall somewhere in between the two above patterns. By joining social groups, females usually lose the option of forming prolonged monogamous pair-bonds with a male (Wittenberger, 1981). They nevertheless retain the ability to select mates in many ways (see Ralls, 1977). An intriguing question in social mammals concerns how males compete for mates. In some species such as hamadryas baboons (Kummer, 1968, 1971), individual males control access to entire female groups, giving rise to *harem polygyny*, while in others (e.g., baboons and macaques) males establish dominance hierarchies within them. Although patas monkeys are typically characterized as a harem-polygynous species (Hall, 1968; Olson, 1985), at least some populations switch to alternate reproductive tactics that include male guarding of individual females (Harding and Olson, 1985).

An important point by Gouzoules (1984, p. 108) is that males in polygynous species are often depicted as more intolerant of one another than are males in promiscuous species. However, male intolerance seems more to be based on context rather than absolute rules. For example, in some species (red howler monkeys, *Alouatta seniculus*; Hanuman langurs) group structure varies between one-male and multimale groups, from population to population, or even within a single population. Further, some species (e.g., Hanuman langurs, gelada baboons), males lacking females band together into "bachelor" or "all-male groups." The context of such social groupings in the wild should offer managers of captive populations important insights.

Monogamous Primate Species

The evolution of monogamy is a complex issue, because there is no single explanation for it that is universally valid for all species (Kleiman, 1977; Wittenberger and Tilson, 1980; Rutberg, 1983). "The only common factor seems to be that monogamy is likely to occur if the male can do better by staying to help the female raise their offspring than by leaving in the hopes of fertilizing more females. Yet, the reason why this should be so in any given case depends on the fine details of the biology of the species in question. Consequently, the immediate causes of monogamy are almost as diverse as the species that practice it" (Dunbar, 1984, p. 15).

Monogamous Old World primates include all nine species of gibbons (*Hylobates* sp.), Mentawai leaf monkeys (*Presbytis potenziani*), the simakobu or Mentawai pig-tailed langurs (*Nasalis concolor*), some populations of deBrazza's monkeys (*Cercopithecus neglectus*), the indris (*Indri indri*), mongoose lemurs (*Lemur mongoz*), forked-tail dwarf lemurs (*Phaner furcifer*), and tarsiers (see references below).

New World representatives include most, if not all, of the marmosets and tamarins (*Cebuella pygmaea*, *Callithrix* sp., *Saguinus* sp., and *Leontopithecus rosalia*), Goeldi's monkeys (*Callimico goeldi*), night monkeys (*Aotus trivirgatus*), titi monkeys (*Callicebus* sp.), and the sakis (*Pithecia* sp.) (after Kleiman, 1977; Wittenberger and Tilson, 1980; Rutberg, 1983). Marmoset and tamarin groups tend to be larger than the typical monogamous family group because offspring may remain in the group after maturation, giving rise to the "extended family group" (see Kleiman, 1977), although only the dominant adult pair typically breed (see below). Female marmosets and tamarins are usually very aggressive toward other females, as are most females within monogamous family groups (Wittenberger and Tilson, 1980).

Females are more likely to benefit from monogamy than males because female reproductive success is usually limited by time and energy constraints and can be increased by male parental assistance. By contrast, males are less likely to practice monogamy because their reproductive success is only limited by the number of females they can inseminate (Wittenberger and Tilson, 1980).

Most discussions of the evolution of monogamy have centered on the necessity and extent of male parental investment to the survival of young (Emlen and Oring, 1977; Kleiman, 1977; Ralls, 1977; Wittenberger, 1979; Wittenberger and Tilson, 1980). Rutberg's (1983, p. 95) version is that monogamy "evolves when females choose to live apart from other females and when males choose to accompany and defend individual females over a prolonged time period. Obligate monogamy (after Kleiman, 1977) evolves as breeding success grows increasingly dependent on male parental investment." The model, which is similar to the others, is based on the premise that predation pressure selects for group formation in diurnal primates and that food availability acts as the principal constraint on its size.

Mating Systems and Evolutionary Constraints

The role of predation in driving some animal societies toward aggregation is well established (Alexander, 1974; Triesman, 1975; Caraco, 1979a, 1979b; Caraco and Pulliam, 1984). Within groups, individuals may reduce predation on themselves by concealment (Paloheimo, 1971; Vine, 1971) or by detecting predator approaches more efficiently than solitary individuals (Siegfried and Underhill, 1975; Treisman, 1975; Caraco, 1979b) or by reducing the probability of being caught by the predator—the dilution effect (Bertram, 1978). Also, group formation may reduce the space about an individual where it is most vulnerable by interposing conspecifics between themselves and an attacking predator (Williams, 1966; Hamilton, 1971) and sometimes allows for active defense against predators (Krunk, 1964; Curio et al., 1978). Other advantages that individuals gain in large groups is that they can increase feeding rates by reducing the time spent on predation detection (Pulliam, 1973; Caraco, 1979a; Pulliam and Caraco, 1983). But these benefits are offset by numerous disadvantages. These include in particular the increased foraging and social competition experienced by groups (Murton, 1967; Fretwell, 1972; Caraco, 1979a) and increased conspicuousness to a searching predator (Pulliam and Mills, 1977).

In summary, although the empirical evidence is unavailable, it is generally accepted that predation pressure (references above) and the distribution of resources (Crook, 1972; Eisenberg et al., 1972; Altmann et al., 1977; Clutton-Brock and Harvey, 1977, 1978; Wrangham, 1981; Wittenberger, 1981) influence social organization in a number of species. Alexander (1974) suggests that sociality, particularly in primates, is solely in response to predation. More generally, Caraco and Pulliam (1984) view social organization as a set of coadapted traits, so that antipredator, foraging, and mating strategies covary dependently (e.g., Crook, 1972; Pulliam, 1973; Altmann, 1974; Caraco, 1979a; Brown, 1982). The obvious problem is that it is difficult to disentangle effects of any single above trait on sociality from influences exerted by the other traits.

In the captive environment, predation pressure and foraging competition are significantly reduced if not eliminated altogether. The emancipation from these selection processes shifts the evolutionary emphasis onto mating strategies and their consequences. Male parental investment in its various forms thus becomes a leading variable for increasing individual reproductive success for captive-held species.

Mating Systems and Parental Investment

If males contribute significantly to the care of offspring, both male and female fitness may be maximized in monogamy. High paternal care has been noted among many species of monogamous mammals, and several authors suggest that male investment in offspring is one of the principal advantages, if not the prerequisite, for the evolution of long-term monogamous pair bonds (Trivers, 1972; Clutton-Brock and Harvey, 1977, 1978; Kleiman, 1977; Ralls, 1977; Wittenberger and Tilson, 1980; Zeveloff and Boyce, 1980; Kleiman and Malcolm, 1981; Rutberg, 1983).

Parental investment is defined by Trivers (1972, p. 139) as "any investment by the parent in an individual offspring that increases the offspring's chance of surviving (and hence reproductive success) at the cost of the parent's ability to invest in other offspring." Other authors (Wittenberger and Tilson, 1980; Kleiman and Malcolm, 1981) have pointed out that there are many types of parental investment that need not be divided among the young, thereby reducing any one individual's fitness. Thus, there is "shareable parental assistance" such as predator vigilance, territory or resource maintenance, and freedom from harassment and "nonshareable" forms of parental assistance, i.e., parental investment that cannot be apportioned among several offspring without reducing the amount received by each (Wittenberger and Tilson, 1980). These terms are similar to Altmann et al.'s (1977) terms of "depreciable and non-depreciable" forms of parental investment. "A depreciable contribution is like a non-renewable or slowly renewable resource: it is reduced in availability to one individual to the extent that it is expended on or used by another" (Altmann et al., 1977, p. 409).

Kleiman and Malcolm (1981) distinguish between "direct and indirect" forms of parental investment. Direct investment includes such acts as feeding or carrying of young (i.e., nonshareable), whereas indirect investment is equated to shareable forms of parental investment. The distinction betwen shareable or nonshareable (depreciable or nondepreciable) forms of parental investment are clear in the extreme cases, but often as not, most overall patterns of parental contributions include elements of both (Kleiman and Malcolm, 1981). Identification of the various forms of nonshareable parental investment found among the living primates may be of greater significance to their long-term survival in captivity (see below).

Male parental investment in primates has been reviewed by Mitchell (1969), Mitchell and Brandt (1972), Hrdy (1976), Redican (1976), Kleiman (1977), and Wittenberger and Tilson (1980). Kleiman and Malcolm (1981) provide a qualitative assessment of the parental care data base in which "nearly 40% of the primate genera are reported to exhibit some form of direct male parental care, the highest for any individual order" (p. 357). Among monogamous primates, Rutberg (1983) notes that in 89% of the species for which data exist, males are involved either in defending territories from other pairs (shareable investment) or in carrying young (nonshareable investment, unless they are twins) or both (mixed strategies).

Carrying of young, which is well known for the callitrichids and monogamous cebids, is mostly absent in the larger monogamous primates (one exception is *Symphalangus syndactylus*, which shows a prolonged period of infant dependence (Preuschoft et al., 1985), while territoriality is almost universal among obligate monogamous species (Rutberg, 1983). Although there is little direct evidence to support the notion that male territorial defense contributes to either adult fecundity or infant survivability within the primates, interspecific comparisons, especially among avian species (Wittenberger, 1979), together with the observation of the ubiquity of territoriality among monogamous primate species (Rutberg, 1983; Wright, 1985) suggests that the correlation may be valid. Thus, any loss of male assistance—typically caused by removing the infant from the social group shortly after birth or by removing the adult male before parturition (see Rasmussen and Tilson, 1984)—could significantly reduce the female's reproductive rate or survival prospects of the young.

There are still a number of primate species for which virtually no field data are available concerning social complexities. Without such knowledge we are not in a good position to manage captive primate populations properly. Identification of the male's role within the context of parental aid is going to become increasingly more important as the demand for excellence in formulating captive primate breeding programs increases.

Consequences for Captive Management

Given a fundamental understanding of mating systems theory and a realization of how the captive environment emancipates individuals from certain selection pressures, what then are the basic issues? Kleiman (1980) has suggested three points of interest. They are:

1. The degree of contact between the mother and young during rearing.
2. The degree to which females will rear young together.
3. The degree to which the male or other nonproducing offspring participate in parental care (after Kleiman, 1980, p. 244).

Two other concerns need to be addressed. They are the process by which mates are selected and the impact of genetic and demographic management strategies that are currently being developed for selected species.

Mate choice among primates is not understood, but we do know in several cases that courtship in monogamous species can be extremely long—in gibbons 6–12 months is not unusual—and that more than one unmated male is considered by the unmated female (Tilson, 1981). By contrast, in the captive situation pairs are matched (often with little regard to age differences), given a brief period of familiarization time (usually a few weeks at best), and then are expected to be sexually and socially compatible. Not uncommonly reproduction fails. The excuse is that one or the other is incompatible. On the other hand, perhaps the fault lies in design, rather than desire. Solutions can involve longer periods of familiarization before introduction and experimental designs that include more than a single choice. Cooperative programs on a regional basis among zoos and primate facilities would be one way to approach the problem.

Another concern with mating systems manipulation in captive primates is related to the genetic demographers' dilemma. As genetic markers become more refined, more accurate identification of paternity in multimale promiscuous social groups will become a reality. The technique, however, may not be a panacea for all species at all times. For example, a preliminary study of genetic marker screening of male Japanese macaques failed to produce sufficient genetic variation to be able to expect reasonable success at paternity identification (Smith and Tilson, unpublished observations). Thus, how will the species (and presumably other similar species) be managed genetically on a long-term basis? Will separate populations now in captivity be managed as discrete entities—as groups rather than individuals within the context of the AAZPA's Species Survival Programs (see Foose, 1981, 1983; Ballou, 1984) or will they be manipulated into one-male harem polygynous groups—in which paternity can be assured? This kind of solution may be convenient, but restricting the number of breeding males in typical multimale breeding units may entail certain costs.

For example, males of some species demonstrate depressed reproductive effort when confined and denied access to other males. It is believed that male–male interaction involving the establishment and/or maintenance of social dominance may be essential for maintaining high levels of blood androgen levels, which in turn are correlated with successful performance of sexual behavior (e.g., rhesus monkeys; Rose et al., 1971). In the absence of sufficient intrasexual competition, the male libido declines and reproduction is curtailed. This kind of reproductive malaise is commonplace in zoos (Lasley et al., 1981), and it is a critical field of reproductive specialization that needs addressing.

Females may be similarly affected. Kleiman (1980) has suggested that in the absence of intrasexual male competition depressed blood androgen levels in males could result in depressed female reproductive function. At insufficient levels females' estrous cycles may be irregular or depressed, both of which result in loss of reproduction.

Other forms of reproductive suppression include depressed fecundity. Among rhesus monkeys, low-ranking females produce fewer offspring than high-ranking females. The same relationship is reported for gelada baboons, and data from a number of other baboon populations reveal that females living in groups with many adult females per adult male have lower birth rates than females in groups with relatively more males. In all cases severe competition between females is evident, and low-ranking females suffer greater stress as a consequence of their social rank and, as a result, show temporary infertility through disruption of their endocrine system (see Dunbar, 1985).

In other cases, reproductive suppression is brought about by direct effects of physical harassment. In both talapoin monkeys and gelada baboons, suppression of ovulation results when low-ranking females are subjected to excessive aggression from high-ranking females. As Dunbar (1985, p. 18) points out, "this is precisely the evolutionary consequence of their behaviour, for high-ranking females may gain a considerable reproductive advantage over less fortunate lower-ranking conspecifics. The risk of reproductive suppression may simply be one of the unavoidable consequences of living in groups."

Reproductive suppression in other forms is known as "reproductive despotism" for cases in which only one female breeds in a social group comprised of several females (Wasser and Barash, 1983). For example, among marmosets and tamarins, only one female breeds, and even though other potentially sexually mature females are present, only the behaviorally dominant female shows an ovarian cycle; the subordinate female(s) do not (e.g., Evans and Hodges, 1984).

Reproduction suppression need not be hormonal. Numerous studies have shown that social-dependent conditions cause substantial variance in female reproductive success in gorillas, chimpanzees, macaques, langurs, and baboons (see Wasser and Barash, 1983, for review and references).

Other Social Considerations

The pursuit of food, space, and mates is fundamental to an animal's survival. Here I have considered the theoretic considerations of mating systems and their consequences for captive management. Equally important are the theoretic issues that address spatial considerations (see Maple and Finlay, this volume), a subject that has often been ignored because of perceived limitations of the captive environment. More ideas are needed as to how a species and individual animal's perceived spatial needs may vary (see Stricklin et al., 1979; Innis et al., 1985).

In summary, the trade-offs for mating systems manipulation are difficult to identify because of their complexities. One juncture I advocate for decision-making policy is to maintain the species social milieu in captivity as it is understood in the wild. This orientation reflects the philosophy underlying genetic guidelines for SSP's developed for more closely managed species. Selection (in an evolutionary sense) at the captive-breeding level should not be based on arbitrary decisions for personal or institutional convenience in managing a species, but on the principle of maintaining maximal genetic diversity, achieved in part by keeping the group's social complexity intact. Natural selection rather than human selection should be the ultimate arbitrator of genetic quality.

References

Alexander RD (1974) The evolution of social behavior. Ann Rev Ecol Syst 5:325–383.

Altmann SA (1974) Baboons, space, time and energy. Amer Zool 14:221–248.

Altmann SA, Wagner SS, Lenington S (1977) Two models for the evolution of polygyny. Behav Ecol Sociolbiol 2:397–410.

Bachmann C, Kummer H (1980) Male assessment of female choice in hamadryas baboons. Behav Ecol Sociobiol 6:315–321.

Ballou JD (1984) Strategies for maintaining genetic diversity in captive populations through reproductive technology. Zool Biol 3:311–324.

Bertram BCR (1978) Living in groups: predators and prey. In: Krebs JR, Davies NB (eds) Behavioural ecology: an evolutionary approach. Blackwell, Oxford, pp 64–96.

Bradbury JR, Vehrencamp SL (1977) Social organization and foraging in emballonurid bats. III. Mating systems. Behav Ecol Sociobiol 2:1–17.

Brown JL (1982) Optimal group size in territorial animals. J Theor Biol 95:793–810.

Caraco T (1979a) Time budgeting and group size: a theory. Ecology 60:611–617.

Caraco T (1979b) Time budgeting and group size: a test of theory. Ecology 60:618–627.

Caraco T, Pulliam HR (1984) Sociality and survivorship in animals exposed to predation. In: Price PW (ed) New ecology: novel approaches to interactive systems. Wiley, New York, pp 280–309.

Clutton-Brock TH, Harvey PH (1977) Primate ecology and social organization. J Zool London 183:1–39.

Clutton-Brock TH, Harvey PH (1978) Mammals, resources and reproductive strategies. Nature 273:191–195.

Crook JH (1972) Sexual selection, dimorphism, and social organization in the primates. In: Campbell BG (ed) Sexual selection and the descent of man, 1871–1971. Aldine, Chicago, pp 231–281.

Curio E, Ernst U, Vieth W (1978) The adaptive significance of avian mobbing. Z Tierpsych 48:184–202.

Dixson AFG, Everitt GF, Herbert J, Rugman SM, Scrutton DM (1973) Hormonal and other determinants of sexual attractiveness and receptivity in rhesus and talapoin monkeys. In: Phoenix CH (ed) Primate reproductive behavior. Symposia of the 4th International Congress of Primatology, vol. 2. Karger, Basel, pp 36–63.

Drickamer LC (1974) Social rank, observability, and sexual behaviour of rhesus monkeys (*Macaca mulatta*). J Reprod Fert 37:117–120.

Dunbar R (1984) The ecology of monogamy. New Sci 103:12–15.

Dunbar R (1985) Stress is a good contraceptive. New Sci 105:16–18.

Duvall SW, Bernstein IS, Gordon TP (1976) Paternity and status in a rhesus monkey group. J Reprod Fert 47:25–31.

Eaton GG (1973) Social and endocrine determinants of sexual behavior in simian and prosimian females. In: Phoenix CH (ed) Primate reproductive behavior. Symposia of the 4th International Primatology Congress. Karger, Basel, pp 20–35.

Eisenberg JF, Muckenhirn NA, Rudran R (1972) The relation between ecology and social structure in primates. Science 176:863–874.

Emlen ST, Oring LW (1977) Ecology, sexual selection and the evolution of mating systems. Science 197:215–223.
Evans S, Hodges JK (1984) Reproductive status of adult daughters in family groups of common marmosets (*Callithrix jacchus jacchus*). Folia Primatol 42:127–133.
Foose TJ (1981) Demographic problems and management in captive populations. Proc AAZPA Annual Conf 1980:46–68.
Foose TJ (1983) The relevance of captive populations to the conservation of biotic diversity. In: Schonewald-Cox CM, Chambers SM, MacBryde B, Thomas WL (eds) Genetics and conservation: a reference for managing wild animal and plant populations. Benjamin/Cummings, Menlo Park, CA, pp 374–401.
Fretwell SD (1972) Populations in a seasonal environment. Monogr Pop Biol 5:1–219.
Gouzoules S (1984) Primate mating systems, kin associations, and cooperative behavior: Evidence for kin recognition? Yrbk Phys Anthrol 27:99–134.
Gowaty PA (1983) Male parental care and apparent monogamy among eastern bluebirds (*Sialia sialis*). Amer Nat 121:149–157.
Gowaty PA, Karlin AA (1984) Multiple maternity and paternity in single broods of apparently monogamous eastern bluebirds (*Sialia sialis*). Behav Ecol Sociobiol 15:91–95.
Graul WD, Derrickson SR, Mock D (1977) The evolution of avian polyandry. Amer Nat 111:812–816.
Hall KRL (1968) Behaviour and ecology of the wild petas monkey, *Erythrocebus patas*, in Uganda. In: Joy P (ed) Primates: studies in adaptation and variability. Holt, Rinehart and Winston, New York, pp 31–121.
Hamilton WD (1971) Geometry for the selfish herd. J Theoret Biol 31:295–311.
Harding RSO, Olson DK (1985) Male mating strategies in patas monkeys. Amer J Phys Anthro 66:179 (Abstract).
Hausfater G (1975) Dominance and reproduction in baboons (*Papio cynocephalus*). Contrib Primatol 7:1–150.
Hrdy SB (1976) Care and exploitation of nonhuman primate infants by conspecifics other than the mother. Adv Study Behav 6:101–158.
Innis GS, Balph MH, Balph DF (1985) On spatial requirements of captive social animals. Anim Behav 33(2):680–682.
Jenni DA (1974) Evolution of polyandry in birds. Amer Zool 14:129–144.
Jolly A (1972) The evolution of primate behavior. Macmillan, New York.
Kavanagh M (1983) A complete guide to monkeys, apes and other primates. Viking, New York.
Kleiman DG (1977) Monogamy in mammals. Quart Rev Biol 52:39–69.
Kleiman DG (1980) The sociobiology of captive propagation. In: Soulé ME, Wilcox BA (eds) Conservation biology: an evolutionary-ecological perspective. Sinauer, Sunderland, MA, pp 243–262.
Kleiman DG, Malcolm JR (1981) The evolution of male parental investment in mammals. In: Gubernick DJ, Klopfer PH (eds) Parental care in mammals. Plenum, New York, pp 347–387.
Kruuk H (1964) Predators and antipredator behaviour of the black-headed Gull. Behaviour Supplement 11:1–127.
Kummer H (1968) Social organization of hamadryas baboons: a field study. University of Chicago Press, Chicago.
Kummer H (1971) Primate societies. Aldine-Atherton, Chicago.
Lasley BL, Lindburg DG, Robinson PT, Bernirschke K (1981) Captive breeding of exotic species. J Zoo An Med 12:67–73.
Lindburg DG (1975) Mate selection in rhesus monkey (*Macaca mulatta*). Amer J Phys Anthro 42:315 (Abstract).
Mitchell G (1969) Paternalistic behavior in primates. Psych Bull 71:399–417.
Mitchell G, Brandt EM (1972) Paternal behavior in primates. In: Poirier FE (ed) Primate socialization. Random House, New York, pp 48–63.

Murton RK (1967) The significance of endocrine stress in population control. Ibis 109:622–623.
Napier JR, Napier PH (1967) A handbook of living primates. Academic, London.
Olson DK (1985) The importance of female choice in the mating system of wild patas monkeys. Amer J Phys Anthro 66:211 (Abstract).
Packer C (1979) Male dominance and reproductive activity in *Papio anubis*. Anim Behav 27:37–45.
Paloheimo JE (1971) A stochastic theory of search: implications for predator-prey situations. Math Biosci 12:105–132.
Preuschoft H, Chivers DJ, Brockelman WY, Creel N (1985) The lesser apes. Edinburgh University Press, Edinburgh.
Pulliam HR (1973) On the advantages of flocking. J Theoret Biol 38:419–422.
Pulliam HR, Caraco T (1983) Living in groups: is there an optimal group size? In: Krebs JR, Davies NB (eds) Behavioural ecology: an evolutionary approach, 2nd ed. Sinauer, Sunderland, MA, pp 122–147.
Pulliam HR, Mills GS (1977) The use of space by sparrows. Ecology 58:1393–1399.
Ralls K (1977) Sexual dimorphism in mammals: avian models and unanswered questions. Amer Natur 111:917–938.
Rasmussen JL, Tilson RL (1984) Food provisioning by adult maned wolves (*Chrysocyon brachyurus*). Z Tierpsychol 65:346–352.
Redican WK (1976) Adult male-infant interactions in nonhuman primates. In: Lamb ME (ed) Role of the father in child development. Wiley, New York, pp 345–385.
Rose RM, Holaday JW, Bernstein IS (1971) Plasma testosterone, dominance rank and aggressive behaviour in male rhesus monkeys. Nature 231:366–368.
Rutberg AT (1983) The evolution of monogamy in primates. J Theor Biol 104:93–112.
Selander RK (1972) Sexual selection and dimorphism in birds. In: Campbell BG (ed) Sexual selection and the descent of man, 1871–1971. Aldine, Chicago, pp 180–230.
Seyfarth RM (1978) Social relationships among adult male and female baboons. I. Behaviour during sexual consortship. Behaviour 64:204–226.
Siegfried WR, Underhill LR (1975) Flocking as an anti-predator strategy in doves. Anim Behav 23:504–508.
Smith DG (1982) Use of genetic markers in the colony management of nonhuman primates: a review. Lab Anim Sci 32:540–546.
Stacey P (1982) Female promiscuity and male reproductive success in social birds and mammals. Amer Nat 120:51–64.
Stern BR, Smith DG (1984) Sexual behaviour and paternity in three captive groups of rhesus monkeys (*Macaca mulatta*). Anim Behav 32:23–32.
Stricklin WR, Graves HB, Wilson LL (1979) Some theoretical and observed relationships of fixed and portable spacing behavior of animals. Appl Anim Ethol 5:201–214.
Sussman RW (1985) Communal breeding among the Callitrichidae and its adaptive significance. Amer J Phys Anthro 66(2):236 (Abstract).
Tilson RL (1981) Family formation strategies in Kloss's gibbons. Folia Primatol 35:259–287.
Treisman M (1975) Predation and the evolution of gregariousness. Anim Behav 23:801–825.
Trivers RL (1972) Parental investment and sexual selection. In: Campbell BG (ed) Sexual selection and the descent of man, 1871–1971. Aldine, Chicago, pp 136–179.
Vine I (1971) Risk of visual detection and pursuit by a predator and the selective advantage of flocking behavior. J Theoret Biol 30:405–422.
Wasser SK, Barash DF (1983) Reproductive suppression among female mammals: implications for biomedicine and sexual selection theory. Quart Rev Biol 58:513–537.
Wasser PM, Jones WT (1983) Natal philopatry among solitary mammals. Quart Rev Biol 58:355–390.

Williams GC (1966) Adaptation and natural selection. Princeton University Press, Princeton.

Wittenberger JF (1979) The evolution of mating systems in birds and mammals. In: Marler P, Vandenbergh J (eds) Handbook of behavioral neurobiology: social behavior and communication. Plenum, New York, pp 271–349.

Wittenberger JF (1980) Group size and polygamy in social mammals. Amer Nat 115:197–222.

Wittenberger JF (1981) Animal social behavior. Duxburg, Boston.

Wittenberger JF, Tilson RL (1980) The evolution of monogamy: hypotheses and evidence. Amer Rev Ecol Syst 11:197–232.

Wrangham R (1981) An ecological model of female-bonded primate groups. Behaviour 75:269–299.

Wright PC (1985) What do monogamous primates have in common? Amer J Phys Anthro 66:244 (Abstract).

Zeveloff SI, Boyce MSC (1980) Parental investment and mating systems in mammals. Evolution 34:973–982.

29 The Interbirth Interval in Primates: Effects of Pregnancy and Nursing

ROBERT F. WILLIAMS

Introduction

This chapter will review the endocrine events associated with the restoration of cyclic ovarian function during the postpartum interval, specifically examining the duration of anovulation in nonnursing mothers and the prolongation of infertility by nursing. The majority of the physiological data have been obtained from studies of rhesus and cynomolgus monkeys housed in individual cages under environmentally controlled conditions. These laboratory studies will be compared and contrasted to data available for monkeys in confined groups and under free-ranging conditions. Finally, data for other primate species will be summarized and related to the more detailed presentations for the macaques.

The Interbirth Interval

The components of the interbirth interval (IBI) vary between the primate species; the schematic illustration in Figure 29.1 is developed for macaques. There are three major components to the interbirth interval: anovulatory, ovulatory, and pregnancy segments. The *anovulatory* segment encompasses several physiologic situations. If nursing does not occur in rhesus and cynomolgus monkeys, the pregnancy is followed by a 3-month interval of anovulation (Goodman and Hodgen, 1978). The refractoriness of the hypothalamic–pituitary–ovarian (H–P–O) axis during this interval is a residual effect of the pregnancy. If nursing occurs, the anovulatory interval is extended (Williams and Hodgen, 1982); function of the H–P–O axis is suppressed by the suckling stimulus. Therefore, during the first 3 months of nursing, at least two mechanisms operate to suppress ovulation: (1) the pregnancy-induced refractoriness, and (2) the suckling stimulus. By the fourth postpartum month, the suckling stimulus appears to be the only inhibition of ovulation.

In monkeys housed under controlled laboratory conditions, these two types of anovulation characterize the anovulatory segment of the IBI. If the monkeys are housed under conditions that allow seasonal breeding to persist, then the anovulatory interval of the IBI is further confounded, in that the periods of pregnancy-induced refractoriness and suckling-induced anovulation may occur, but within the seasonal interval of anovulation.

INTERBIRTH INTERVAL

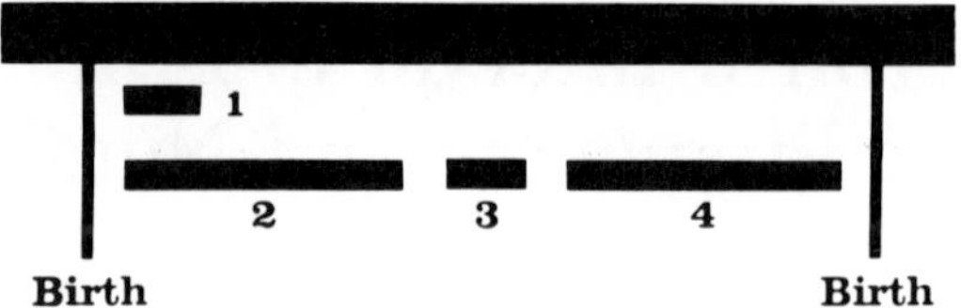

1. anovulation - non - nursing
2. anovulation - nursing
3. ovulation resumed
4. pregnancy

FIGURE 29.1. A schematic representation of the components of the interbirth interval in macaques.

Following the anovulatory segment of the IBI, ovulatory menstrual cycles are reestablished, fertility is restored, and pregnancy can occur. In monkeys being housed with fertile males, the majority of the IBI will be composed of almost equal intervals of anovulation and pregnancy, with the ovulatory interval being very short.

Postpartum Interval in Nonnursing Monkeys

Figure 29.2 characterizes the endocrine pattern for nonnursing monkeys during the first four postpartum months (Goodman and Hodgen, 1978). In the data for the one cynomolgus monkey presented in this figure, the first ovulation is presumed to have occurred at about day 98 postpartum. This conclusion is made from the increased concentrations of estradiol, LH, and FSH on day 97 and subsequent secretion of progesterone for 2 weeks. It should be noted from Figure 29.2 that there were no major changes in the daily concentrations of LH and FSH prior to the resumption of follicular growth that culminated in the observed ovulation. Therefore, the 3-month anovulatory interval is not due to a suppression of LH and FSH secretion that could be detected by once-daily blood collections. Also evident in this figure and characteristic of the monkeys in this study was the occurrence of menstruation prior to the first ovulation. Apparently, this menstruation, not preceded by ovulation, is indicative of increased follicular growth but failure of this folliculogenesis to culminate in ovulation. However, associated with this follicular development is apparently adequate steroid secretion to induce endometrial development and subsequently menstruation when ovulation fails to occur (Williams and Hodgen, 1982; Williams et al., 1982). Thus, the reinitiation of follicular growth in the nonnursing mother is a gradual process, in which initial waves of folicular development do not culminate in ovulation but do secrete adequate steroids to result in menstruation. Eventually, the endocrine

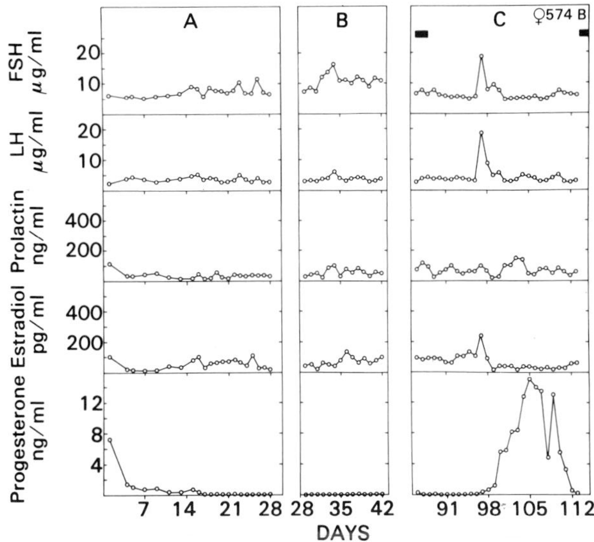

FIGURE 29.2. Composite pattern of circulating gonadotropins, prolactin, and ovarian steroids during the puerperim (the first 42 days postpartum) and during the first menstrual cycle postpartum in an individual intact cynomolgus monkey. Solid horizontal bars in top frame of panel C indicate menses. See text for additional details. Reproduced from Goodman and Hodgen, 1978, with permission.

milieu supports follicular growth that does proceed to ovulation as evidenced in panel C of Figure 29.2.

The refractory status of H–P–O axis evident during the first three postpartum months may be the result of diminished function of one or more components of this axis. Experiments have evaluated the responsiveness of the ovaries to gonadotropins (Goodman and Hodgen, 1978) and the hypothalamic–pituitary unit to estrogen (Williams et al., 1979; Plant et al., 1980) (i.e., induction of midcycle-like surge release of gonadotropins by estrogens). In anovulatory, nonnursing monkeys, treatment with exogenous gonadotropins during the second month postpartum resulted in development of multiple follicles and ovulation, thus demonstrating that the postpartum ovary is not totally refractory to gonadotropic stimulation. Since all treatments with exogenous gonadotropins are pharmacologic, the ovaries of nonnursing monkeys could be less sensitive to FSH/LH stimulation than in cycling monkeys, but this cannot be determined from this type of experiment. Therefore, concerning ovarian function, we can only conclude that the ovaries are not totally refractory to gonadotropic stimulation.

Within normal menstrual cycles, an integral component of the ovulatory process is the induction by rising estrogens of the midcycle LH/FSH surge (estrogen-positive feedback) (Yamaji et al., 1971). To confirm that the hypothalamic–pituitary unit is sensitive to estrogen-positive feedback, an estrogen-provocation test may be undertaken, in which monkeys are treated with estradiol benzoate, and within 48 hours, a surge release of LH and FSH should occur. Following estrogen-provocation tests in nonnursing monkeys, no gonadotropic responses has been observed during the first postpartum month. Early in the second month these estrogen-provocation tests result in a surge release of LH and FSH, and thereafter throughout the postpartum interval in nonnursing monkeys (Williams et al., 1979; Plant et al., 1980).

The inhibition of estrogen-positive feedback, rather than being a direct inhibition at the pituitary, may actually be secondary to a diminished secretion of gonadotropin-releasing hormone (GnRH) from the hypothalamus. In addition to directly stimulating the release of LH and FSH from the pituitary, GnRH is self-priming (i.e., increasing the pituitary response to GnRH stimulation) and primes the pituitary to be responsive to estrogen-positive feedback. Therefore, during the first postpartum month the failure of estradiol benzoate to elicit midcycle-like gonadotropin surges may reflect diminished GnRH secretion and a resultant desensitization of the pituitary to estrogen stimulation. Consistent with this hypothesis is the observation that maximum pituitary response to a GnRH provocation test is achieved coincident with the restoration of positive estrogen feedback at the beginning of the second postpartum month (di Zerega et al., 1981). Estrogen-positive feedback and increased GnRH responsiveness may result from increased hypothalamic secretion of GnRH and its priming effect on the pituitary.

Despite the inhibited estrogen-positive feedback and diminished GnRH responses during the first month of the puerperium, other mechanisms regulating ovarian function must also be inhibited, since the first postpartum ovulation does not occur in nonnursing monkeys until 2 months after restoration of maximal LH/FSH responses to estrogen and GnRH provocations. These other mechanisms may be a diminished sensitivity of the ovary to basal levels of gonadotropins, as discussed previously, or more likely an aberrant pattern of LH/FSH secretion incapable of supporting follicular growth that culminates in ovulation. Normally ovulatory monkeys become anovulatory when regions of the hypothalamus involved in GnRH secretion are destroyed (Knobil et al., 1980). Replacement therapy with GnRH secretion is unsuccessful in inducing ovulatory cycles when given in an unvarying mode of administration. In contrast, the administration of GnRH in an episodic pattern (hourly infusions) results in a corresponding pattern of gonadotropin secretion and the restoration of cyclic ovarian function. Therefore, in nonnursing monkeys, months 2 and 3 of anovulation may be due to altered pulsatile secretion of LH/FSH, secondary to aberrant GnRH secretion. This hypothesis awaits testing.

The preceding paragraphs have considered the altered endocrine milieu of the postpartum interval in nonnursing mothers. But what is the mechanism by which the pregnancy induces the lesion in the functional status of the H–P–O axis? Several studies have tested hypotheses concerning this mechanism; however, the mechanism remains enigmatic. These studies reveal that the pregnancy need not to be full term (167 days) to induce the 3-month refractoriness, but rather, no more than 120 days in length (Williams et al., 1979).

The endocrine functions of the H–P–O axis during the anovulatory interval, following abortion at 120 days, are indistinguishable from those in nonnursing mothers. In contrast, pregnancies terminated at 21 days are followed by an anovulatory interval of only 30 days, and midcycle-like LH/FSH surges are inducible immediately after the abortion (di Zerega et al., 1979).

Apparently, the increased serum concentrations of estradiol and progesterone throughout pregnancy do not induce the postpartum interval of anovulation (Williams and Hodgen, 1981). In nonpregnant monkeys, in which the pregnancy levels of these hormones have been simulated, ovulatory menstrual cycles were resumed within 6 weeks of cessation of the steroid therapy. This 6-week interval is significantly shorter than that which follows normal pregnancy; therefore, these steroids are discounted as the sole agents inducing this postpartum anovulatory interval. Chorionic gonadotropin, which is detectable in serum for approximately the first 42 days of pregnancy (Hodgen et al., 1974), is known to inhibit follicular development (Goodman and Hodgen, 1982), but it is difficult to hypothesize a mechanism for its action, since it is only detectable for the first 6 weeks of gestation. Thus, the question still remains, What is the pregnancy mechanism between 21 and 120 days of gestation that induces the inhibition of the H–P–O axis observed in nonnursing macaque mothers?

Postpartum Interval in Nursing Monkeys

If the macaque mother nurses an infant, the interval of anovulation is prolonged beyond that observed in nonnursing mothers (Williams and Hodgen, 1982).

In the laboratory setting (individual cage; controlled environment), the nursing mother remains anovulatory until the infant is permanently weaned; usually, this separation is made, managerially, by 150 days of age. However, we have observed in one study that mothers remained anovulatory through 450 days if the infants were not separated from the mothers (Aso and Williams, unpublished observations).

Figure 29.3 presents the endocrine profile of the first ovulatory cycles after weaning of the infant at 150 days of age (Williams and Hodgen, 1982). It is evident that the first ovulation occurred at approximately 40 days after weaning and was followed by a luteal phase of normal duration and a second ovulatory cycle. Uniformly, the first ovulation was not preceded by menstruation, in contrast to the observation made in nonnursing monkeys. Therefore, because of this difference in menstruation between nonnursing and nursing mothers, and the brief interval (40 days) for resumption of ovulation after weaning versus the 95 days in nonnursing mothers, previous conclusions have been that "the reestablishment of ovulatory cycles is a gradual process in non-nursing monkeys, with initial follicular growth supporting endometrial proliferation but failing to culminate in ovulation," while in nursing mothers, the inhibition of the H–P–O axis is absolute prior to weaning, but that the reacquisition of its functional integrity is rapid following weaning.

As previously described for nonnursing mothers, evaluations have been made in nursing monkeys of the ovarian responses to gonadotropin therapy and the hypothalamic–pituitary response to estrogen provocation. During the first and second postpartum weeks, nursing monkeys were treated with hMG and hCG; follicular development and ovulation occurred (di Zerega et al., 1979). Therefore, as in nonnursing mothers, the ovaries of nursing monkeys are not totally refractory to gonadotropin stimulation. Of course, since the dosage of LH and FSH in these studies was pharmacologic, the results do not exclude the possibility

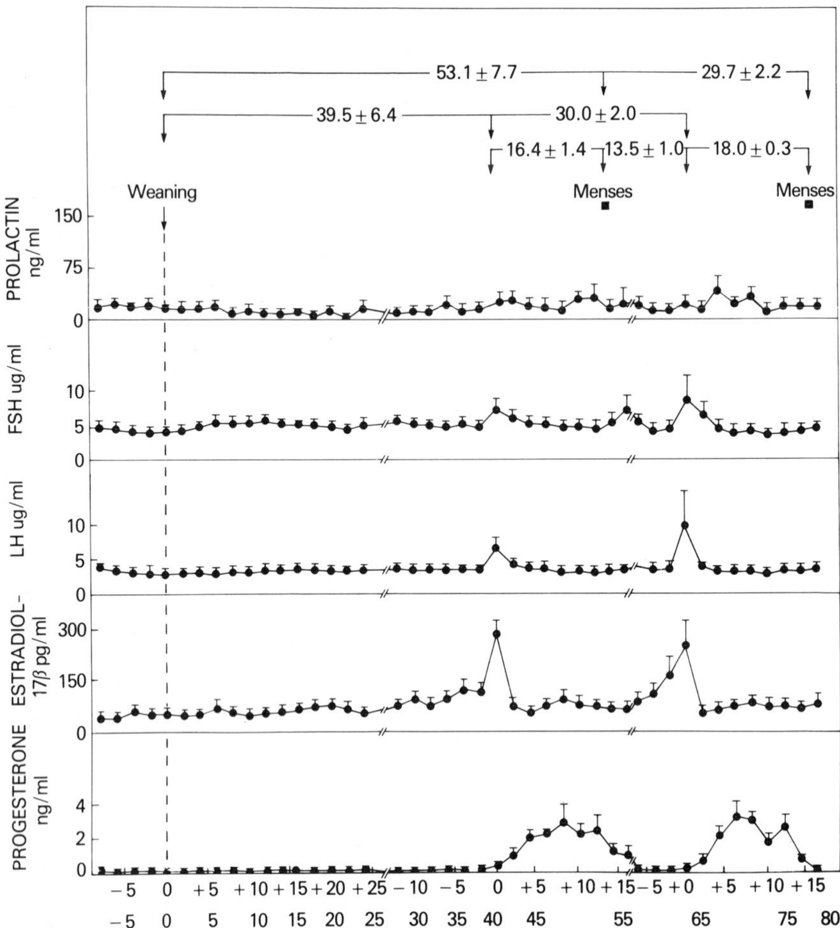

FIGURE 29.3. Composite patterns of peripheral serum concentrations of gonadotropins, prolactin, and ovarian steroids in the mother after weaning of the infant at 5 months of age. Blood samples were collected on alternate days. Data (mean ± S.E.) are synchronized to the day of weaning and subsequent gonadotropin surges, as indicated by the time scale immediately below the abscissa. The bottom time scale represents the consecutive days from weaning. Reproduced from Williams and Hodgen, 1982, with permission.

that there is a diminished ovarian sensitivity to physiologic concentrations of gonadotropins. In contrast to nonnursing monkeys, the surge mode of gonadotropin secretion is inhibited throughout the postpartum interval in nursing monkeys (Williams et al., 1979). In addition, nursing clearly inhibits the basal secretion of gonadotropins. In normally cycling females, removal of the ovaries is followed by a five-fold increase in the basal levels of gonadotropins; this is caused by the removal of negative ovarian steroid feedback. Nonnursing, postpartum monkeys demonstrate this same increased secretion of LH and FSH following ovariectomy, but nursing monkeys do not (Weiss et al., 1976). In nursing mothers, the gonadotropin levels only double following ovariectomy, thus demonstrating the suppression of basal secretion of LH/FSH by nursing. In addition, in preliminary studies, an absence of pulsatile gonadotropin secretion has been noted during nursing (Williams, unpublished observations). As for nonnursing monkeys, it is a reasonable hypothesis that the primary lesion in the H–P–O axis is the diminished pulsatile secretion of the GnRH from the hypothalamus, which would account for the other effects observed in pulsatile, basal, and surge patterns of gonadotropin secretion, and the attendant cessation of ovarian function.

In nonnursing mothers, the suppression of the H–P–O axis is by an endogenous mechanism; in contrast, in nursing mothers, this suppression is due to the interaction of mother and infant, presumably the suckling stimulus of the infant. Utilizing the surge mode of gonadotropin secretion as an endpoint, the role of mother–infant interaction in suppressing the function of the hypothalamic–pituitary unit has been clearly demonstrated (Plant et al., 1980). In this study, estrogen provocation tests were administered to monkeys who were nonnursing, nursing their own infants, and nursing foster infants. Prior to receiving the foster infants, these females had been having normal, ovulatory cycles. In the nonnursing monkeys, estrogen provocation elicited a midcycle-like LH/FSH surge by the end of the first month postpartum; in the natural mothers and foster mothers, no gonadotropin surges occurred following estrogen treatment during the first 6 months of the study, thus demonstrating the inhibitory action of nursing and further demonstrating that nursing behavior does not have to be preceded by pregnancy for the function of the hypothalamic–pituitary unit to be suppressed.

The mechanism by which the suckling stimulus inhibits the function of the H–P–O axis remains enigmatic. Clearly, the stimulation of neurons leading from the nipple increases prolactin secretion from the pituitary (Aso and Williams, 1985). Much attention has focused on the role of prolactin in anovulatory conditions (McNeilly et al., 1982). In laboratory-housed monkeys, the amount of suckling, cuddling, and cradling is apparently increased at night, and there is an attendant increase in serum prolactin concentrations at night (Aso and Williams, 1985). Whether this nocturnal increase in prolactin (specifically associated with nursing behavior) has an effect on H–P–O function remains to be determined. If prolactin is the mediator of the suckling stimulus, prolactin may induce lactational anovulation (1) by altering neurotransmitter secretion that regulates GnRH secretion, (2) by directly inhibiting GnRH and/or gonadotropin secretion, or (3) by directly reducing ovarian sensitivity to gonadotropins. However, results from other studies with monkeys, in which nursing-induced prolactin secretion was inhibited with pharmacologic agents, suggest that lactational prolactin secretion does not, or at least not totally, account for the inhibition of gonadotropin secretion (Schallenberger et al., 1981).

Further research is required to identify the mechanism by which suckling inhibits the H–P–O axis and to determine the various components of mother–infant behavior that affect the inhibitory mechanism.

Interbirth Intervals in Outdoor Housing

Under controlled laboratory conditions, rhesus and cynomolgus monkeys exhibit an anovulatory interval of about 3 months, if nursing does not occur, and a prolongation of this interval if the mother nurses an infant (Williams and Hodgen, 1982). This extension of anovulation may last long after the offspring is no longer dependent upon breastfeeding for its nutrition. After the nutritional need for nursing has passed, the nursing behavior may be a "comfort" nursing, which still results in the inhibition of ovarian function.

In an outdoor environment, the effect of nursing behavior is dependent upon the type of living condition. In free-ranging monkeys, annual births usually occur, with 85% of colony members having births each year (Koford, 1965; Carpenter, 1972; Drickamer, 1974). Therefore, the inhibitory effect of the infant's suckling lasts only about 6 months. Of course, the nursing interval corresponds to the nonbreeding season. That there is an effect of suckling on maternal reproduction, separate from the effect of seasonal breeding, is evident from matings and subsequent births for mothers who have stillbirths or early neonatal deaths (Koford, 1965; Drickamer, 1974; Small, 1983; Wilson et al., 1978). In the breeding season, following stillbirth or neonatal death, these mothers mate earlier in the breeding season and have earlier births in the birth season than do monkeys who are nursing a 6-month-old infant. Presumably, the later occurrences of mating and parturition in nursing mothers results from the suckling-induced suppression of the H–P–O axis.

In monkeys housed outdoors, but in a "harem" or "gang" cage, the effects of nursing behavior may transcend a breeding season, resulting in such colonies having an annual birth rate of less than 50% (Anderson and Simpson, 1979), in contrast to the 85% rate for free-ranging monkeys (Koford, 1965; Carpenter, 1972; Drickamer, 1974). This difference in annual birth rates may be due to differences in the interaction of the mother and infant in the two different environments. In comparing the free-ranging and gang-caged mother–infant pairs, the infants spent more "time-off" the mother, and the mother rejected a larger proportion of the infant's attempts to suckle in the free-ranging setting than in the large cages (Berman, 1980). Additionally, the infants in the free-ranging setting played a larger role in maintaining proximity to the mother than did the infants in the harem cages. These parameters may reflect a more intensive mother–infant relationship in the gang-cage setting, which results in: (1) a greater suckling stimulus, (2) a more prolonged suppression of the H–P–O axis, and (3) an annual birth rate of less than 50% (Anderson and Simpson, 1979).

The Postpartum Interval in Other Primates

In other species of primates, the postpartum interval has not been studied as thoroughly as in the macaques. For most other species (chimpanzee being an exception; Nadler et al., 1981) postpartum data primarily consist of determination of the interbirth interval. In the great apes, the interbirth interval is significantly extended by a suckling infant (gorilla 4.5 years, Nadler et al., 1981; Harcourt et al., 1980); chimpanzee 5.6 years, (Nadler et al., 1981; Teleki et al., 1976; Tutin, 1980). Likewise, in baboons (Altmann et al., 1977) a nursing infant extends the IBI (11 months following infant loss and 22 months with a surviving infant). In New World monkeys, postpartum reproduction is less uniform. The howler monkeys (Glander, 1980) demonstrate an effect of lactation on the IBI like that observed in Old World species, but, in contrast, the owl monkey (Hunter et al., 1979) has no effect of lactation on reproduction and has an IBI similar to its length of gestation. For the squirrel monkey (Wolf et al., 1975; Coe and Rosenblum, 1978; Dukelow, 1985; Coe et al., 1985), the lactation interval coincides with the nonbreeding season, with mothers having nursing infants mating immediately in the breeding season. Therefore, discerning the effect of lactation in squirrel monkeys is confounded by the coexistence of the nonbreeding season and lactation. Marmosets (Poole and Evans, 1982; Lumm and McNeilly, 1982) demonstrate no effect of lactation on reproductive function, and though reproduction of tamarins has been reported to be suppressed by lactation (French, 1983), other studies find tamarins to be like marmosets (Kirkwood et al., 1983), with no inhibitory effect of nursing.

Postpartum Physiology and the Management of Primate Populations

If a primate species exhibits lactational anovulation, management of the postpartum interval will require intervention to separate the mother and infant, in order to increase fecundity. Therefore, it will be difficult to manage the postpartum interval in noncaptive populations. In captive populations, the separation of the infant from the mother may result in the early restoration of ovulatory menstrual cycles. However, in separating the infant, consideration must be given to the infant's proper socialization. Additionally, if the species exhibits an interval of pregnancy-induced anovulation, then no increase in fertility will occur by weaning the infant prior to the completion of this interval.

A specific example of this type of management could be the application to harem breeding groups of rhesus monkeys. Since the annual birthrate is about 50% in this type of caging, intervention to reduce the amount of suckling during the breeding season should increase the fecundity of these groups. By the breeding season, the infant should be 5 to 6 months old and should not be impaired nutritionally by intervention at this age. A major concern must be the need for proper social development of the infant, and efforts should be made that the management decision does not compromise this development.

However, before management schemes for most species can be developed, further research is required to clarify the physiology of the postpartum interval and the contribution of pregnancy-induced anovulation and lactational anovulation in determining the length of the interbirth interval.

Acknowledgment. I am very appreciative of the excellent secretarial assistance given by Lourdes M. Mendoza de Romney in the preparation of this manuscript.

References

Altmann J, Altmann SA, Hausfater G, McCuskey SA (1972) Life history of yellow baboons: physical development, reproductive parameters, and infant mortality. Primates 18:315.

Anderson DM, Simpson MJA (1979) Breeding performance of a captive colony of rhesus macaques (*Macaca mulatta*). Lab Anim 13:275.

Aso T, Williams RF (1985) Lactational amenorrhea in monkeys: effects of suckling on prolactin secretion. Endocrinology 117:1727.

Berman CM (1980) Mother–infant relationships among free-ranging rhesus monkeys on Cayo Santiago: a comparison with captive pairs. Anim Behav 28:860.

Carpenter CR (1972) Breeding colonies of macaques and gibbons on Santiago Island, Puerto Rico. In: Beveridge WIB (ed) Breeding primates. S Karger, New York, p 76.

Coe CL, Rosenblum LA (1978) Annual reproductive strategy of the squirrel monkey (*Saimiri Sciureus*). Folia Primatol 29:19.

di Zerega GS, Williams RF, Morin ML, Hodgen GD (1979) Anovulation after pregnancy termination: ovarian *versus* hypothalamic–pituitary factors. J Clin Endocrinol Metab 49:594.

di Zerega GS, Williams RF, Hodgen GD (1981) Pregnancy induced anovulation in the primate ovarian cycle. In: Schwartz NB, Hunzicker-Dunn M (eds) Dynamics of ovarian function. Raven, New York, p 147.

Drickamer LC (1974) A ten-year summary of reproductive data for free ranging *Macaca mulatta*. Folia Primatol 121:61.

French JA (1983) Lactation and fertility: an examination of nursing and interbirth intervals in cotton-top tamarins (*Saguinus o. oedipus*). Folia Primatol 40:276.

Glander KE (1980) Reproduction and population growth in free-ranging mantled howling monkeys. Amer J Phys Anthro 53:25.

Goodman AL, Hodgen GD (1978) Postpartum patterns of circulating FSH, LH prolactin, estradiol, and progesterone in nonsuckling cynomolgus monkeys. Steroids 31:731.

Goodman AL, Hodgen GD (1982) Evidence for an extraluteal and antifolliculogenic action of chorionic gonadotropin in rhesus monkeys. Endocrinology 110:1315.

Harcourt AH, Fossey D, Stewart KJ, Watts DP (1980) Reproduction in wild gorillas and some comparisons with chimpanzees. J Reprod Fertil Suppl 28:59.

Hodgen GD, Tullner WW, Vaitukaitis JL, Ward DN, Ross GT (1974) Specific radioimmunoassay of chorionic gonadotropin during implantation in rhesus monkeys. J Clin Endocrinol Metab 39:457.

Hunter J, Martin RD, Dixson AF, Rudder BCC (1979) Gestation and inter-birth intervals in owl monkey (*Aotus trivirgatus griseimembra*). Folia Primatol 31:165.

Kirkwood JK, Epstein MA, Terlecki AJ (1983) Factors influencing population growth of a colony of cotton-top tamarins. Lab Anim 17:35.

Knobil E, Plant TM, Wildt L, Belchetz PE, Marshall G (1980) Control of the rhesus monkey menstrual cycle: permissive role of hypothalamic gonadotropin-releasing hormone. Science 207:1371.

Koford CB (1965) Population dynamics of rhesus monkeys on Cayo Santiago. In: Devore I (ed) Primate behavior: field studies of monkeys and apes. Holt, Rinehart and Winston, New York, p 160.

Lunn SF, McNeilly AS (1982) Failure of lactation to have a consistent effect on interbirth interval in the common marmoset, *Callithrix jacchus jacchus*. Folia Primatol 37:99.

McNeilly AS, Glasier A, Jonassen J, Howie PW (1982) Evidence for direct inhibition of ovarian function by prolactin. J Reprod Fertil 65:559.

Nadler RD, Graham CE, Collins DC, Kling OR (1981) Postpartum amenorrhea and behavior of apes. In: Graham CE (ed) Reproductive biology of the great apes. Academic, New York, p 69.

Plant TM, Schallenberger E, Hess DL, McCormack JT, Dufy-Barbe L, Knobil E (1980) Influence of suckling on gonadotropin secretion in the female rhesus monkey (*Macaca mulatta*). Biol Reprod 23:760.

Poole TB, Evans RG (1982) Reproduction, infant survival and productivity of a colony of common marmosets (*Callithrix jacchus jacchus*). Lab Anim 16:88.

Schallenberger E, Richardson DW, Knobil E (1981) Role of prolactin in the lactational amenorrhea of the rhesus monkey (*Macaca mulatta*). Biol Reprod 25:370.

Small MF (1983) Females without infants: mating strategies in two species of captive macaques. Folia Primatol 40:125.

Teleki G, Hunt EE Jr, Pfifferling JH (1976) Demographic observations (1963–1973) on the chimpanzees of Gombe National Park, Tanzania. J Hum Evol 5:559.

Tutin CEG (1980) Reproductive behavior of wild chimpanzees in the Gombe National Park, Tanzania. J Reprod Fertil Suppl 28:43.

Weiss G, Butler WR, Dierschke DJ, Knobil E (1976) Influence of suckling on gonadotropin secretion in the postpartum rhesus monkeys. Proc Soc Exp Biol Med 153:330.

Williams RF, Hodgen GD (1981) Reinitiation of ovulatory menstrual cycles in postpartum monkeys: effects of fetectomy at mid-pregnancy. In: Schwartz NB, Hunzicker-Dunn M (eds) Dynamics of ovarian function. Raven, New York, p 141.

Williams RF, Hodgen GD (1982) Initiation of the primate ovarian cycle with emphasis on perimenarchical and postpartum events. In: Greep RO (ed) Reproductive physiology IV, International Review of Physiology, vol. 27. University Park Press, Baltimore, p 1.

Williams RF, Johnson DK, Hodgen GD (1979) Resumption of estrogen-induced gonadotropin surges in postpartum monkeys. J Clin Endocrinol Metab 49:422.

Williams RF, Turner CK, Hodgen GD (1982) The late pubertal cascade in perimenarchial monkeys: onset of asymmetrical ovarian estradiol secretion and bioassayable LH release. J Clin Endocrinol Metab 55:660.

Wilson ME, Gordon TP, Bernstein IS (1978) Timing of births and reproductive success in rhesus monkey social groups. J Med Primatol 7:202.

Wolf RH, Harrison RM, Martin TW (1975) A review of reproductive patterns in new world monkeys. Lab Anim Sci. 25:814.

Yamaji T, Dierschke DJ, Hotchkiss J, Bhattacharya AN, Surve AH, Knobil E (1971) Estrogen induction of LH release in the rhesus monkey. Endocrinology 89:1034.

30
Embryonic Loss in Primates in Relation to In Vitro Fertilization and Embryo Transfer

GARY D. HODGEN

Introduction

In science, we are called upon to recognize and assimilate "the facts" as interpretations, based on data samples that we trust statistically to represent a larger body of conditions or responses, but that are not practically available for examination as a whole. Unlike the performing arts, we cannot produce useful results through creative expression alone; rather, imaginative ideas must be tested with minimal bias to ascertain elements of truth, even if they prove only partially valid. With maturity, these truths are seen to interconnect, producing the dogma of current understanding. In this context, the evidence available to offer secure interpretation on the frequency and causes of early embryonic loss among primates is severely limited. Frequently, requisite data are either incomplete or nonexistent even for humans and the more commonly studied laboratory primates. Despite such restrictions, I hope to present a few persuasive interpretations on the incidence and origins of early embryonic loss in both human and nonhuman primates, with emphasis on their relationship to in vitro fertilization and embryo transfer (IVF/ET).

Limitations on Comparisons Among Primates

Our knowledge of early in vivo or in vitro reproductive events of any primates, human and nonhuman, pales beside the comprehensive contributions of Austin, Chang, Blandau, Pincus, Biggers, Yoshinaga, Iritani, Whittingham, Yanagiachi, Brackett, Brinster, and many others who have studied fertilization and early development in nonprimate laboratory and farm mammals. Indeed, among both human and nonhuman primates, except for the original observations of Rock and Hertig in the human and Hartman in macaques, few data were available until the achievement of Edwards and Steptoe eventually brought IVF/ET to human clinical application in 1977–1978. More recently, there have been important new findings on preimplantation reproductive processes of primates from groups led by Cruxatto, Buster, and Trounson in women and by Marston, Kraemer, Dukelow, Pope, Kuehl, Bavister, and Hodgen in macaques, baboons, and squirrel monkeys; these observations include numerous studies both in vivo and in vitro. Even so, the bulk of our recent understanding of gamete maturation, fertilization, and early embryonic development is of the human primate and has derived from the collective monumental achievement of numerous IVF/ET teams who have taken this novel infertility treatment from an experimental technique to a proven therapeutic course over the past 6 years. Accordingly, because of the paucity of data among nonhuman primate species, I will employ the larger experience on human embryos as a background against which some comparisons with various nonhuman primates can be made. This matrix will be overlaid by contrasting observations on fertilization and embryogenesis in vivo and in vitro.

Of course, primates, like any other order, have diversity of reproductive functions peculiar to each group, genus, and species. Although our focal point is IVF/ET, we cannot avoid the context of differing female reproductive cycles, the seasonality of testicular function in males, and many other differentiating factors. For example, the rhesus monkey has an idealized 28-day menstrual cycle, like the woman, whereas in baboons, the menstrual cycle is typically near 35 days. In contrast, squirrel monkeys do not menstruate; they have an overt estrous cycle instead. Further, while seasonality of reproductive efficiency may be only a subtle factor in humans, marmoset fertility is strictly regulated by ambient conditions, with other nonhuman primates influenced to varying degrees even in controlled laboratory settings of photoperiod, temperature, and humidity. As well, we remain only slightly informed as to the importance of behavioral factors, such as social order and phenomena. More specifically, there are significant anatomical and functional differences: The rhesus monkey has an S-shaped cervical canal, making visualization of the cervical os and embryo transfer more difficult than in baboons; whereas most of the primates are typically monovular, the tamarins and marmosets usually bear twins or triplets spontaneously. The diversity includes a range from subtle to overt differences in implantation and placentation; whereas some achieve definitively decidualized endometrium, others do not.

So what is common to early reproductive processes among primates? Certainly, a distinguishing characteristic of primates is the production of chorionic gonadotropin by the trophoblast, although secretion patterns are unique, as are the endocrinologic effects that extend corpus luteum function in cycles of conception. Among higher adult female primates (macaques, baboons, apes, and humans), menstruation is an overt characteristic. As well, single ovulations abound, with occasional exceptions. Although there are subtle variations of function, the interactions of the hypothalamus, pituitary, and testes or ovaries are quite similar among higher primates. Importantly for IVF/ET research, from what we know, folliculogenesis and oocyte maturation among macaques is temporally and functionally very similar to that of women. Also, there is an extended interval from birth to menarche, with menopause at age 30 to 40 years in macaques. Each of these factors bears on our use of nonhuman primates for IVF/ET research.

Among the most important qualifications of laboratory primates as models for IVF/ET in humans is their inherent genetic diversity. This is to be contrasted with the confined genetic pools of inbred strains of rodents or the artificial selection preferred in domestic farm animals for generations. While the impact of this genetic differential (restricted vs. mongrel) on reproductive efficiency is difficult to isolate from environmental factors, clearly the capacity to reproduce efficiently has long been one of the high priorities among animal breeders in agriculture and for production of laboratory rodents. Accordingly, the genetic diversity among feral primates studied in the laboratory, with preservation of more individual variation of response, is similar to that of humans.

The final points to be appreciated are the issues of high cost and low availability for study of nonhuman primate reproduction. Thus, the numbers of observations are usually small. Whereas embryonic loss in cattle has been studied on a scale of hundreds of thousands, or even millions of observations, typically data on non-human primate embryos may be collected with rigorous effort just to obtain a dozen or fewer observations. Some endangered species of primates are essentially unavailable for reproduction research, even toward their own preservation. For all of the above reasons, our comparative knowledge of primate reproduction is often both highly informative and relevant to issues of human fertility, where credible data are present; however, too often the data are absent or inconclusive where natural diversity, expense, and unavailability are profoundly rate-limiting factors along the course to new understanding of gametogenesis, fertilization, early embryogenesis, and most specifically, the causes of early embryonic loss.

Embryonic Loss in Primates After Natural Conception

That human reproduction efficiency is glaringly low within a given menstrual cycle is acknowledged by reports that viable pregnancy rates in women are only about 25% even when coitus is known to occur near the time of ovulation. Paradoxically, how then is humankind experiencing the global problems of overpopulation? The answer lies in two principal factors: (1) that a normal ovulatory woman who engages in unprotected intercourse has 12 to 13 opportunities per year to initiate a viable pregnancy; and (2) that her years of reproductive potential extend over three decades. Thus, it is mostly by repetition and longevity that the human primate reproduces successfully (Smart et al., 1982). With regard to nonhuman primates studied in controlled indoor laboratory or exhibit conditions, data for evaluation of inherent reproductive efficiency are scarce, except in relatively small groups of macaques, baboons, squirrel monkeys, and perhaps certain marmosets. Given the modest sample available and the potential for influence from experimental conditions, as well as intrinsic seasonality of many nonhuman primates, conclusions about the rates of ovulation, mating, fertilization, and subsequent embryonic loss may be only slightly better than conjecture. Even so, we know subjectively from the records of breeding colonies of the more numerous laboratory primates that most so-called good breeders will become pregnant within three to five ovulatory menstrual or estrous cycles. However, we also know that the objectivity of such observations is biased by the influences of artificial selection (both males and females), including the fickle demands of investigators for timed pregnancies or managerial decisions influenced by economic considerations. Among exhibitionary (zoo) primates, including apes, the difficulties in achieving viable pregnancy are well known. But again, the absence of natural habitat and normal social order may introduce misleading interpretations. Despite these severe limitations, the preponderance of actual and inferential data point to a viable pregnancy rate in the 20–30% range for any single ovulatory cycle. From my own experience, during 15 years of timed-pregnancy matings in our program at NIH, a 50% conception rate per cycle in either rhesus or cynomolgus macaques was never achieved, even when employing females and males having successful records of reproductive performance during previous years. Although less than satisfactory data are available from both human and nonhuman primates, the chance that a given egg will be fertilized through timely copulation and achieve viable gestation may not exceed 25%.

Surely, many factors contribute to a relatively low reproductive efficiency. There is, of course, frank male and female infertility in primates of all species, including anovulation and a host of potential pathophysiologic conditions well known only in humans. But this is not the real question here. Rather, our interest is focused on the paradox of frequent (prevalent) reproductive failures in the short run among men and women as well as nonhuman primates; yet, these same individuals reproduce quite successfully overall throughout their reproductive lives. The human data clearly demonstrate that when apparent ovulation and timely coitus or artificial insemination occur, the chance for fertilization and early embryonic development is high (Buster et al., 1983). Similarly, more modest experiences in nonhuman primates point out a like situation. That is, the discrepancy between the high frequency of preimplantation embryo recovery by lavage of the fallopian tubes or uterus (> 70%) versus the relative infrequency of establishing viable pregnancy (−25%) is readily apparent. After subtraction of spontaneous abortions in the range of 8–10% for unmanipulated conceptions in both humans and several common laboratory primates following unambiguous confirmation of pregnancy, the remainder surely provides a crude estimate of natural early embryonic loss is primates (Edmonds et al., 1982). Moreover, although spontaneous ovulation and timely coitus may result in more than 70% of the eggs being fertilized and manifesting onset of cleavage, typically only about 30% of these embryos survive to achieve successful implantation and establishment of viable gestation. In other instances, fertilization itself fails. Accordingly, even though hard evidence is lacking, early embryonic loss in primates generally may exceed two of every three eggs fertilized. Indeed, this seems to be the normal course of early reproductive events (Kreitmann et al., 1980).

What are the factors depreciating the ultimate viability of early primate embryos? With regard to causal factors of embryonic loss after fertilization or conceptions achieved naturally, there are few data to guide us. Surely, gamete quality per se, lethal genetic combinations, and inadequacies of the maternal milieu (endocrine, immunoprotective, sepsis, nutrition, environmental stress, etc.) all bring a cumulative toll of embryonic atresia; statistically, these negative factors defy the hope that a plurality of embryos may survive in primates even under the best of natural conditions.

Embryonic Loss in Primates After IVF/ET

Ovarian Stimulation and Oocyte Collection

Although the initial success of human IVF/ET derived from collection of an oocyte from the spontaneous dominant follicle during the LH surge, by 1981 most protocols employed ovarian stimulation methods by administration of gonadotropins, clomiphene citrate, or a combination of them (Vargyas et al., 1984); more recently, the potential use of high-dose GnRH pulse therapy has been considered (Fig. 30.1). The purpose in all these approaches to enhancement

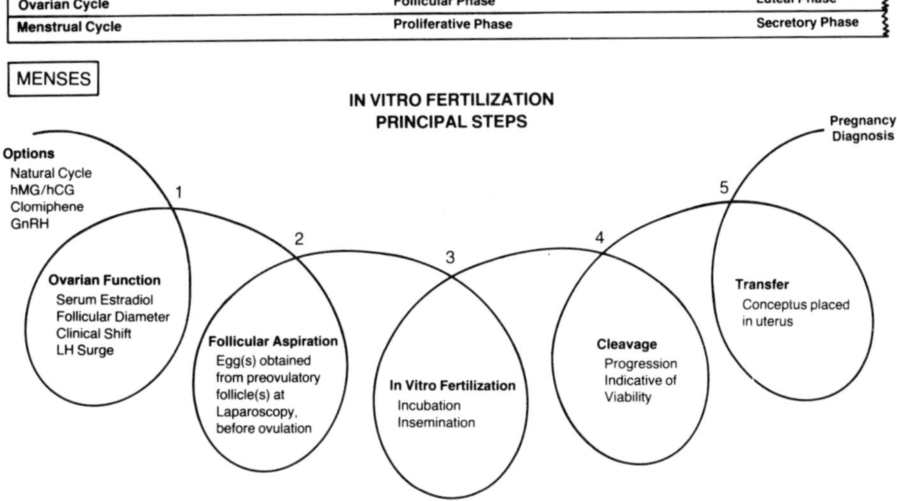

FIGURE 30.1. The five principal steps in human IVF/ET therapy.

of the ovarian cycle was to increase the numbers of fertilizable eggs collected, resulting in larger numbers of embryos transferred per patient, and ultimately, higher pregnancy rates. Indeed, the increased number of viable embryos transferred has been the single most important factor elevating pregnancy rates in women undergoing IVF/ET therapy.

Much of the physiological basis for these strategies of ovarian hyperstimulation in women derived from studies in nonhuman primates (Hodgen, 1982; Goodman and Hodgen, 1983). For example, when gonadotropins are injected into women or macaques, most females (~ 85%) do not manifest a timely LH surge (Goodman and Hodgen, 1978; Schenken and Hodgen, 1983; Garcia et al., 1983); typically hCG is given as a surrogate LH surge. This creates a special dilemma for physicians; when should the hCG be given (Fig. 30.2)? If the dose of hCG is administered optimally for only the most advanced follicles, then aspiration of the several large follicles usually present will often result in collection of numerous immature eggs, characterized by the germinal vesicle and densely packed granulosa cells. Alternatively, when hCG treatment is delayed too long, eggs aspirated from the most advanced follicles will often be degenerating from postmaturity (Hodgen, 1983a). Paradoxically, when hCG is used, it brings the advantage of regulating the time for laparoscopy (Fig. 30.2) and avoidance of frequent monitoring for the spontaneous LH surge (Edwards and Steptoe, 1983). With regard to in vitro maturation of immature eggs, prior studies showed some fertilization but low ultimate viability in squirrel monkeys (Kuehl and Dukelow, 1979), as well as rhesus (Avendano et al., 1975; Gould and Graham, 1976) and cynomolgus monkeys (Kreitmann et al., 1980). Similarly, even though most IVF/ET clinics have incorporated steps for maturation of "GV eggs" in vitro, there is little evidence to support that insemination, fertilization, and transfer of embryos so derived add very much to the pregnancy rates achieved after collec-

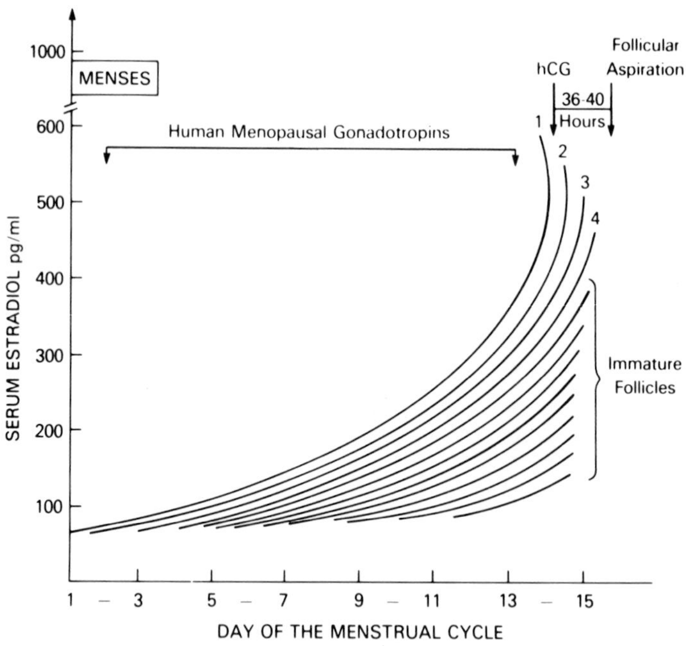

FIGURE 30.2. Illustration that multiple follicular development is only quasi-synchronous and that timely hCG treatment influences follicular and oocyte maturity.

tion of mature preovulatory oocytes. Besides nuclear maturation, it seems probable that ooplasmic and membraneous changes are also required to accommodate ultimate embryo viability. Thus, the quality of the egg is a principal factor influencing the rate of embryonic loss in primates after IVF/ET.

Sperm Capacitation and Insemination In Vitro

Perhaps no area of IVF/ET in primates is as understudied as the andrologic factors. The human sperm seems to require little special handling to acquire fertilizing capacity (Veeck et al., 1983). On the other hand, more attention is given to preparation of sperm obtained from nonhuman primates before insemination in vitro (Bavister et al., 1983; Clayton and Kuehl, 1984; Kuehl and Dukelow, 1982). Because the evidence is so sparse, it is premature to describe in vitro capacitation of primate sperm. Indeed, some investigators resist the whole concept of capacitation in higher primates.

One area of significant change in human IVF/ET therapy is the addition of fewer sperm per oocyte. Whereas only a few years ago it was common to add several hundred thousand sperm to each egg, current trends gaining acceptance include addition of as few as 20,000 sperm per oocyte, with maintenance of high fertilization rates and less polyspermy (Jones et al., 1984).

Embryo Development In Vitro and Transfer

Ideally, proper in vitro fertilization would allow differentiation of the fertilized egg in a sequence and at a rate contemporaneous with events in the maternal milieu, where some degree of ultimate harmony and synchrony must exist to support implantation after embryo transfer. To evaluate the culture conditions used for the embryo differentiation, we maintained in vivo fertilized monkey ova or embryos in culture during several days (Kreitmann and Hodgen, 1981). Among 14 eggs four were at one-cell stage, four at the first cleavage, three with four blastomeres, and three with eight blastomeres. Three (21.4%) were morphologically degenerate at the time of collection. The media used to culture the 11 "good" embryos were either Ham's F10, supplemented with glutamine, 15% fetal calf serum (heat inactivated), penicillin G (50 U/ml) and streptomycin base (50 U/ml), or TCM 199 with Hank's salts, supplemented with 15.8 mmol/L sodium bicarbonate, 15% fetal calf serum, and penicillin G. These embryos resembled those described in humans (Avendano et al., 1975; Edwards and Steptoe, 1975a, 1975b), baboons (Hendrickx and Kraemer, 1968; Kraemer and Hendrickx, 1971; Panigel et al., 1975), and rhesus monkeys (Lewis and Hartman, 1933, 1938). That is, their stage of development after ovulation (determined retrospectively from the time of the LH surge) corresponded to those described for several primate species (Kuehl and Dukelow, 1979).

The in vitro development of embryos cultured in petri dishes containing a 300 μl droplet of medium covered by mineral oil and placed in a humidified incubator at 37°C, saturated with 5% CO_2 in air, was generally slower than expected in vivo, as shown in Figure 30.3. These culture conditions probably mimicked poorly those of parallel stage of development in vivo. Accordingly, as the interval of extracorporeal embryo differentiation lengthened, the disparity (asynchrony)

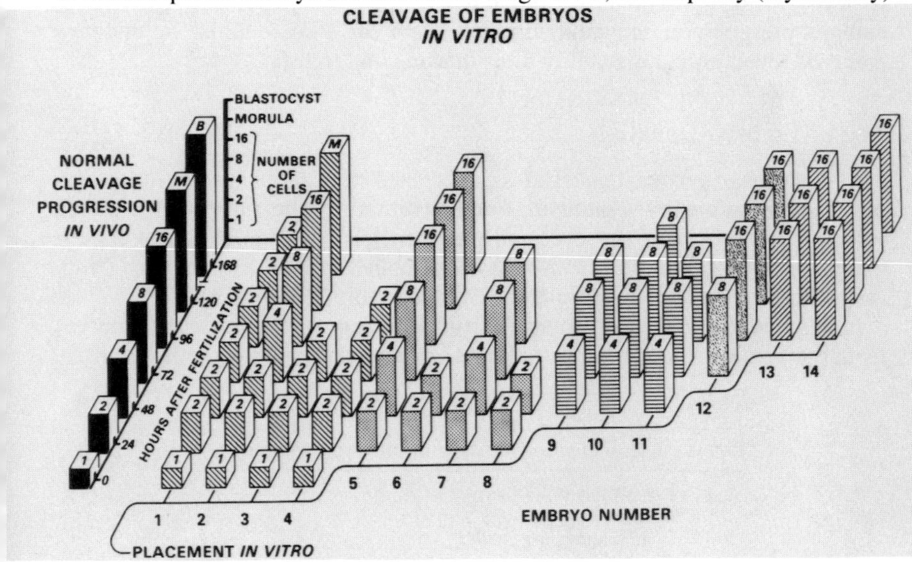

FIGURE 30.3. Demonstration of qualitative deficiencies in the culture condition in vitro. Embryos collected after fertilization in vivo were studies in vitro.

between embryonic and maternal status increased, thereby reducing the probability of achieving timely implantation with maternal recognition of pregnancy and prevention of menstruation in the fertile menstrual cycle.

Although the media developed in recent years surely is improved for human (Lopata et al., 1980), baboon (Clayton and Kuehl, 1984), and monkey (Bavister et al., 1983) embryos, in vitro conditions are still lacking qualitatively in comparison with the milieu of a healthy fallopian tube and uterine lumen. It is also noteworthy that human embryos are usually transferred at about the four-cell stage, although some may be more or less advanced (Jones et al., 1984). You may ask why current protocols do not allow for these embryos to develop into morulae or blastocysts before transfer. The single most important reason is that our in vitro incubation conditions are deficient, causing a retarded pace of embryo development and thereby allowing the maternal events to continue out of synchrony with embryo growth; this scenario risks onset of the next menstrual flow before maternal recognition of pregnancy can be established.

With regard to freezing of human embryos, the data available are limited to a few dozen cases, with the results after thawing and transfer still inconclusive. Even so, the births of a few children derived from frozen embryos have been reported (Trounson and Mohr, 1983; Zeilmaker et al., 1984). Recently, Pope et al. (1980) have attempted pregnancies in baboons after thawing embryos stored in liquid nitrogen. With the increased number of embryos produced per patient and the threat of multiple pregnancy when three or more embryos are transferred, interest in cryopreservation of human embryos is increasing, along with the ethical debate on its moral and legal appropriateness. Among endangered primates, it may be important to ensure preservation of some species by developing reliable technology for IVF/ET in association with cryopreservation. However, with regard to both human and nonhuman primates, the current methods for embryo preservation in liquid nitrogen seem far from optimal in that few embryos retain ultimate viability after thawing and transfer.

Egg or Embryo Donation

That donation of germinal material is an accepted clinical practice for many couples seeking infertility treatment is demonstrated by the success of artificial insemination by donor. However, only recently has donation of human eggs or embryos, fertilized in vitro or in vivo been applied to infertility therapy. Clinically, there are several conditions that motivate this development.

In humans, the advent of in vitro fertilization has made transfer of embryos from the petri dish into the uterus almost commonplace with perhaps over 1000

TABLE 30.1. Indications for egg or embryo donation.

1. Inaccessible ovaries
2. Genetic disease
3. Contraindications for IVF
4. Ovarian dysgenesis
5. Premature menopause
6. Surgical castration
7. Failed IVF

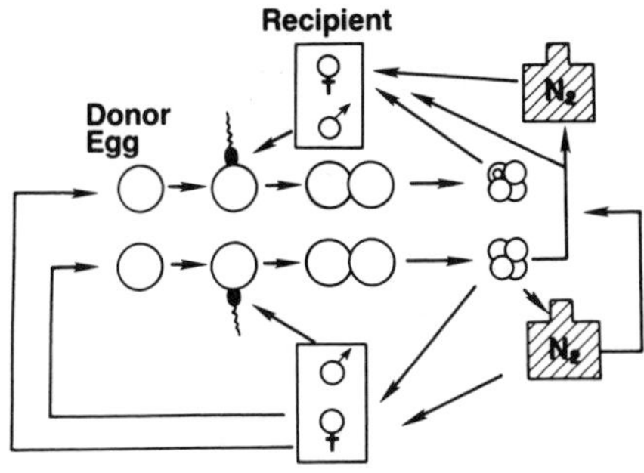

FIGURE 30.4. Scenarios for IVF/ET combined with donation of "extra" eggs or embryos.

children born world-wide by the end of 1984 (Seppala, 1984, personal communication). During this past year, human donor eggs and surrogate embryos have been transferred successfully to spontaneously cycling infertile recipients, yielding normal children. Some of the clinical indications for donor eggs or embryos are listed in Table 30.1. As with previous animal studies in many species, the first step in human donor egg or embryo transfer is synchronizing the cycle for the female providing the egg (donor) with the one receiving the egg (recipient). In this way the endometrium of the recipient woman will be in a state to implant and nourish the newly developing embryo.

Donor embryos can be obtained through various means. In vitro fertilization programs sometimes have "extra" eggs or embryos since often protocols allow no more than three or four embryos to be transferred at one time to the patient, while frequently five to ten eggs are obtained during laparoscopy. With or without cryopreservation, if the initial couple agrees to donate an egg or embryo to another infertile couple, the "extra" embryo(s) can be transferred to the recipient's uterus. Alternatively, the "extra" eggs can first be fertilized in vitro with the sperm of the recipient's husband (Fig. 30.4).

In vivo fertilized embryos can also be obtained (Bustillo et al., 1984). In this technique, the donor woman would be artificially inseminated with the sperm of the recipient's husband at the time of ovulation. Several days later, after the embryo enters the uterus, but before it implants, a lavage of the uterine cavity is performed and the embryo retrieved, similar to the preexisting technology used in collection of cattle embryos.

A specifically designed catheter is placed in the uterine cavity through the cervix. The cavity is then lavaged or gently washed, taking care that almost all the media injected are recovered. The embryo transfer is performed by injecting the embryo, in a small amount of media, into the recipient's uterine cavity.

Results of a series of embryo recoveries and transfers in spontaneously cycling women have been reported. Out of 29 recovery attempts during cycles of spontaneous ovulation, 12 embryos were obtained (41%) 5 to 7 days after insemination. All 12 embryos were transferred yielding two intrauterine pregnancies (16.7%) and one ectopic pregnancy (8.3%).

Since these techniques are in a sense nonsurgical and require no anesthesia, ovum transfer or surrogate embryo transfer has the potential for becoming a routine, relatively inexpensive office procedure. It could be offered as a therapy in most categories of female infertility when other methods have failed.

Pregnancy Without Ovarian Function

Certain types of female infertility, however, are not amenable to any of the above procedures. Within this group are women with a normal uterus but lacking ovaries, or having ovaries that fail to respond to ovulatory therapies. For them, donor egg/embryo or surrogate embryo transfer combined with a specialized sequence of estrogen and progesterone therapy can provide new potential for pregnancy. The feasibility of this approach was demonstrated in a monkey model (Table 30.2), using rhesus monkeys whose ovaries had been removed several years previously; they served as embryo recipients.

Normally, in these macaques, as in the human, a single egg is drawn into the fallopian tube after ovulation. Following natural insemination, fertilization takes place in the fallopian tube, where cell division proceeds. Typically, within 3 to 4 days the embryo enters the uterus (as an early morula now 16 to 32 cells) and implants 8 to 9 days after fertilization, when the blastocyst contains a few hundred cells. According to this schedule of events, preimplantation embryos were harvested by retrograde irrigation of the donor female (Hodgen, 1983b). The procedure for embryo recovery and transfer in monkeys required abdominal surgery, since it is technically difficult to pass instruments through the cervix of these monkeys. Note the advantage of transcervical (nonsurgical) embryo transfer into baboons, where cervical access is easier.

TABLE 30.2. Pregnancies in overiectomized monkeys after estrogen–progestin therapy and surrogate embryo transfer.

	Surrogate embryo transfer[1]		
	Observations	Transfer of surrogate embryos	Pregnancies[2]
Uterotubal irrigations[3]	23[4]		
Collection of embryos/eggs	17		
Tubal recoveries	7		
"viable" embryo[5]	4 (4–10 cells)	4	1 (8 cell)
degenerating embryo	2 (6 cells, with fragmentation)		
degenerating eggs	1 (fragmented)		
Uterine recoveries	10 (12 or more cells)		
"viable" embryos[5]	8 (> 12 cells)[6]	7	3 (morula)
degenerating embryos	2 (one 12 cells; one fragmented)		

[1] Embryos transferred from intact monkeys to estrogen–progesterone treated ovariectomized females.

[2] Four live-born infants were delivered naturally (three male, one female); in one case, a cynomolgus embryo was successfully transferred to a rhesus castrate.

[3] Twenty-eight donor monkeys were employed; in two cases no suitable recipients were available; in three instances the cycle was anovulatory, leaving 23 test subjects.

[4] One donor female became pregnant and died on day 63 of gestation; massive peritoneal hemorrhage derived from an ectopic (periovarian) implantation, with an intraabdominal abscess.

[5] The term "viable" means apparently healthy embryo as assessed subjectively by microscopy.

[6] One embryo was lost from the cannula before transfer.

For implantation to occur, the recipient endometrium must be hormonally prepared to sequential estrogen and progesterone. This can be done naturally in cycling females by spontaneously synchronizing the menstrual cycles of the donor and recipient. When the recipient lacked intrinsic ovarian function, or as in this study, if the female was without ovaries, we administered sequential estrogen and progesterone in a regimen that mimicked the natural ovarian cycle and, later, the corpus luteum of early pregnancy in the monkey. By sequential insertion of subcutaneous silastic capsules of estrogen and progesterone, we attained a steady release of these steroid hormones in order to develop an adequate endometrium for the fertile menstrual cycle.

In one study (Table 30.2), 23 embryo recoveries were attempted in the natural cycle and 17 embryos or unfertilized ova were obtained. Only 12 of these appeared to be "normal" embryos and 11 were transferred to recipients. Of these 11, four implanted and manifested a positive (macaque chorionic gonadotropin) pregnancy test at 21 days postconception (Fig. 30.5). All four of these pregnancies resulted in full-term normal infants. Interestingly, one of these was an interspecies transfer from a cynomolgus monkey donor to a rhesus monkey recipient. Notice that even when advanced embryos were transferred only four of 11 were ultimately successful.

The only complication from this series of recoveries and transfers was a ruptured tubal pregnancy in a donor at 63 days of gestation. It is presumed that this

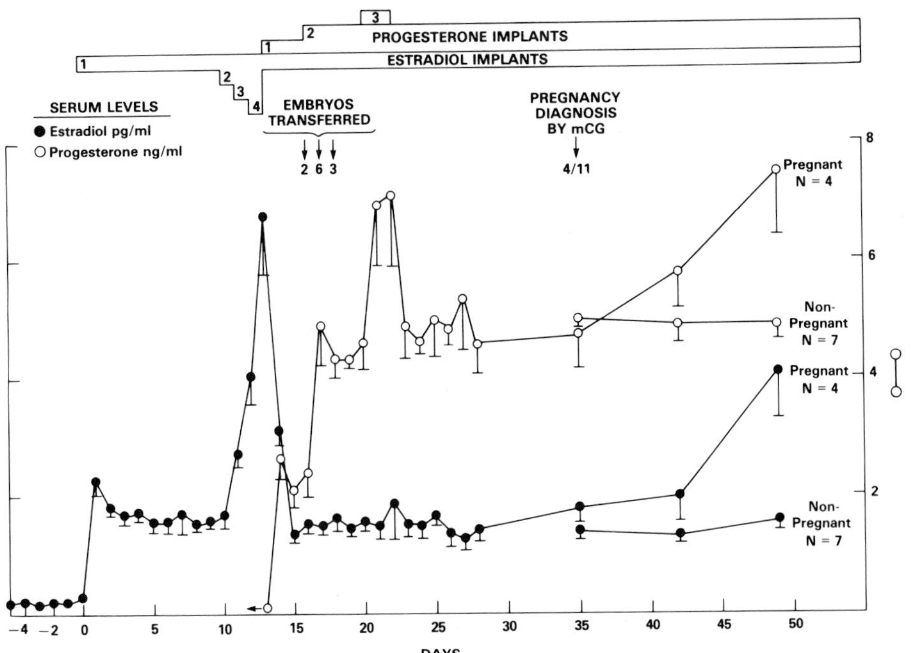

FIGURE 30.5. Pregnancy in ovariectomized monkeys.

was the result of the retrograde tubal irrigations, since in monkeys spontaneously ectopic pregnancies are infrequent.

Similarly, a substantial portion of infertile women are barren because of nonfunctioning ovaries but have a potentially functioning uterus. Women with gonadal dysgenesis (lack of development of ovaries); women who have had their ovaries removed because of ovarian tumors or endometriosis; and women with premature ovarian failure, idiopathic or secondary to cancer radiotherapy or chemotherapy, all could be offered the opportunity to attain pregnancy via surrogate embryo transfer or by donor egg or embryo IVF/ET combined with appropriate estrogen–progesterone therapy. Since these monkey studies were performed (Hodgen, 1983b), a full-term human pregnancy has, in fact, been

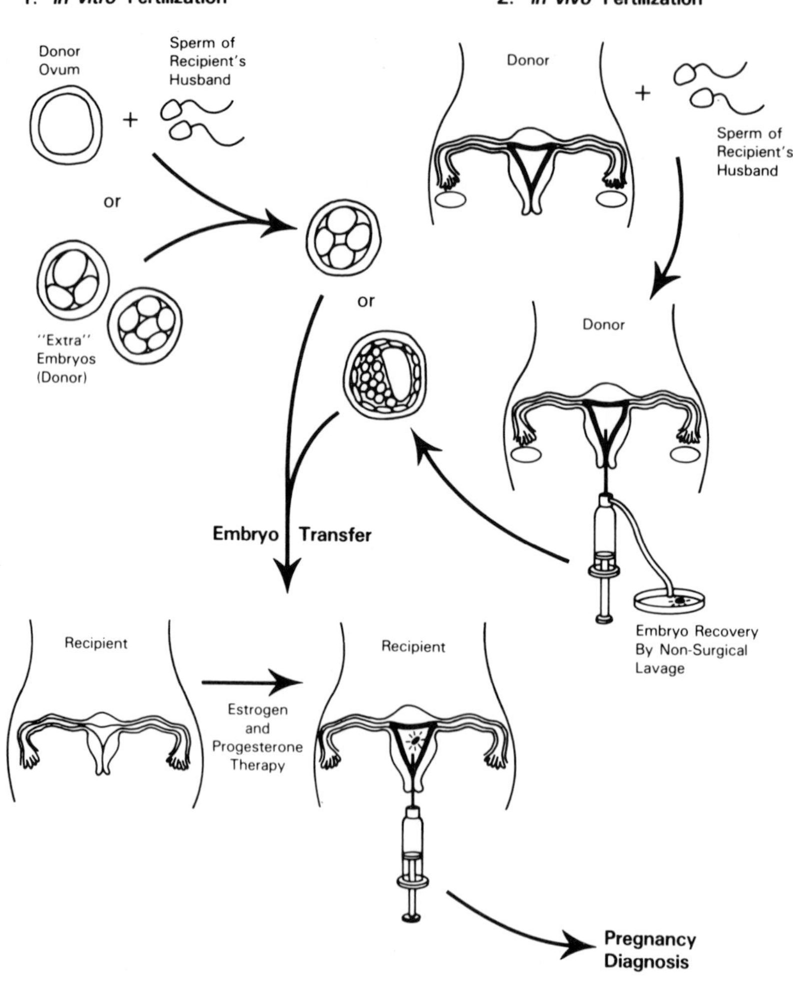

FIGURE 30.6. Infertility treatments for couples requiring embryo transfer in association with ovarian failure.

reported in a woman with premature ovarian failure, using similar replacement hormone therapy (Lutjen et al., 1984). Based on these few data from monkeys and women, surely the luteal phase "window" is fairly wide; that is, lack of exact endometrial synchrony does not seem to be a major cause of early embryonic loss in primates.

The use of donor eggs and embryo or surrogate embryos is in one respect the female counterpart to artificial insemination (Fig. 30.6). The difference lies first in the substantial theoretical risk to the woman who is an embryo donor when fertilization is accomplished in vivo. Ectopic or unwanted intrauterine pregnancy could occur if lavage of the uterine cavity is not 100% efficient at embryo recovery. Also, insemination with sperm from the recipient's husband adds the risk of venereally transmitted disease. These donor risks could be eliminated if "extra" eggs or embryos from IVF/ET programs were used, rather than recovering in vivo fertilized embryo from the donor uterine cavity. Second, there is the unknown risk to the recipient in carrying a conceptus whose genetic makeup is totally foreign (both maternal and paternal genomes) to her own, as opposed to the normal pregnancy state where half of the fetal antigens are of maternal origin. The observation that there was no problem in the monkey (cynomolgus to rhesus) even with an interspecies transfer, as well as the reported human pregnancy, seems to lessen such concern. However, ethical and legal questions must be resolved over human donor eggs or embryos and surrogate embryo transfer procedures. Society and the individual patient (couple) will have to decide in what circumstances these new reproductive technologies are acceptable for infertility treatment.

Summary

In summary, the frequency of early embryonic loss in primates generally is much higher than in inbred livestock or rodent species, whether fertilization and early development occur in vivo or in vitro. On the whole, ultimate embryo viability among primates seems to depend more on the intrinsic quality of gametes and their capacity to exert timely progression of embryogenic events. Although the data available are severely limiting, many embryos either contain lethal encodes from fertilization or lack requisite genomes to cope with early growth and differentiation. While acknowledging an essential role for the maternal milieu in early embryo survival or loss, the data may indicate that such problems are secondary to genetic aberrations inherent to high rates of early embryo loss among primates, both in vivo and after IVF/ET (Jones et al., 1983; Enders et al., 1982).

References

Avendano S, Croxatto HD, Pereda J, Croxatto HB (1975) A seven-cell human egg recovered from the oviduct. Fertil Steril 26:1167.

Bavister BD, Boatman DE, Leibfried L, Loose M, Vernon MW (1983) Fertilization and cleavage of rhesus monkey oocytes in vitro. Biol Reprod 28:983–999.

Buster JE, Bustillo M, Thorneycroft IA, Simon JA, Boyers SP, Marshall JR, Seed RG,

Louw JA (1983) Non-surgical transfer of an in vivo fertilized donated ovum to an infertility patient. Lancet 1:816.

Bustillo M, Buster JE, Cohen SW, Thorneycroft IA, Simon JA, Boyers SP, Marshall JR, Louw JA, Seed RG (1984) Nonsurgical ovum transfer as a treatment in infertile women: preliminary experience. JAMA 251:1171.

Clayton O, Kuehl TJ (1984) The first successful in vitro fertilization and embryo transfer in a non-human primate. Theriogenology 21:228.

Edmonds DK, Londsay KS, Miller JF, Williamson E, Wood PJ (1982) Early embryonic mortality in women. Fertil Steril 38:447.

Edwards RG, Steptoe PC (1975a) Induction of follicular growth, ovulation and luteinization in the human ovary. J Reprod Fertil Suppl 22:121.

Edwards RG, Steptoe PC (1975b) Physiological aspects of human embryo transfer. In: Behrman SJ, Kistner RW (eds) Progress in infertility. Little, Brown, Boston, pp 377–409.

Edwards RG, Steptoe PC (1983) Current status of in vitro fertilization and implantation of human embryos. Lancet 2:1265.

Enders AC, Hendrickx AG, Binkerd PA (1982) Abnormal development of blastocysts and blastomeres in the rhesus monkey. Biol Reprod 26:353–366.

Garcia JE, Jones GS, Acosta AA, Wright J (1983) Human menopausal gonadotropin/human chorionic gonadotropin follicular maturation for oocyte aspiration: Phase I, 1981. Fertil Steril 39:167.

Goodman AL, Hodgen GD (1978) Post partum patterns of circulating FSH, LH, prolactin, estradiol and progesterone in nonsuckling cynomolgus monkeys. Steroids 31:731–744.

Goodman AL, Hodgen GD (1983) The ovarian triad of the primate menstrual cycle. Rec Prog Horm Res 39:1–73.

Gould KG, Graham CE (1976) Maturation in vitro of oocytes recovered from prepubertal rhesus monkeys. J Reprod Fertil 46:269.

Hendrickx AG, Kraemer DC (1968) Preimplantation stages of baboon embryos. Anat Rec 162:111.

Hodgen GD (1982) The dominant ovarian follicle. Fertil Steril 38:281–300.

Hodgen GD (1983a) Oocyte transfer and fertilization in vivo. In: Crosignini P (ed) In vitro fertilization and embryo transfer. Serono Symposia 47:126–138.

Hodgen GD (1983b) Surrogate embryo transfer combined with estrogen–progesterone therapy in monkeys. JAMA 250:2167–2171.

Jones HW Jr, Acosta AA, Andrews MC, Garcia JE, Jones GS, Mantzavinos T, McDowell J, Sandow BA, Veeck L, Whibley TW, Wilkes CA, Wright GL Jr (1983) What is a pregnancy? A question for programs of in vitro fertilization. Fertil Steril 40:728.

Jones HW Jr, Acosta AA, Andrews MC, Garcia JE, Jones GS, McDowell JS, Rosenwaks Z, Sandow BA, Veeck LL, Wilkes CA (1984) Three years of in vitro fertilization at Norfolk. Fertil Steril 42:826.

Kraemer DC, Hendrickx AG (1971) Description of stages I, II, and III. In: Hendrickx AG (ed) Embryology of the baboon. University of Chicago Press, Chicago, pp 45–62.

Kreitmann O, Hodgen GD (1981) Retarded cleavage rates of preimplantation monkey embryos in vitro. JAMA 246:627–629.

Kreitmann O, Lynch A, Nixon WE, Hodgen GD (1980) Ovum collection, induced luteal dysfunction, in vitro fertilization, embryo development and low tubal ovum transfer in primates. In: Hafez ESE, Semm K (eds) In vitro fertilization and embryo transfer. MIT, vol. 30, pp 303–324.

Kuehl TJ, Dukelow WR (1979) Maturation and in vitro fertilization of follicular oocytes of the squirrel monkey (*Saimiri sciureus*). Biol Reprod 21:545.

Kuehl TJ, Dukelow WR (1982) Time relations of squirrel monkey (*Saimiri sciureus*) sperm capacitation and ovum maturation in an in vitro fertilization system. J Reprod Fertil 64:135–137.

Lewis WH, Hartman CG (1933) Early cleavage stages of the egg of the monkey (rhesus). Contracept Embryol Carnegie Inst 24:187.

Lewis, WH, Hartman CG (1938) Tubal ova of the rhesus monkey. Contrib Embryol Carnegie Inst 29:7.

Lopata A, Johnston IWH, Hoult IJ, Spiers TA (1980) Pregnancy following intrauterine implantation of an embryo obtained by in vitro fertilization of a preovulatory egg. Fertil Steril 33:117.

Lutjen P, Trounson A, Leeton J, Wood C (1984) Donation of oocytes. Nature 307:174–175.

Panigel M, Kraemer DC, Kalter SS, Smith GC, Heberline RL (1975) Ultrastructure of cleavage stages and preimplantation embryos of the baboon. Anat Embryol 147:45.

Pope CE, Pope VZ, Beck LR (1980) Nonsurgical recovery of uterine embryos in the baboon. Biol Reprod 23:657–662.

Schenken R, Hodgen GD (1983) Follicle-stimulating hormone induced ovarian hyperstimulation in monkeys: blockade of the luteinizing hormone surge. J Clin Endocrinol Metab 57:50–55.

Smart YC, Fraser IS, Roberts TK, Clancy RL, Cripps AW (1982) Fertilization and early pregnancy loss in healthy women attempting conception. Clin Reprod Fertil 1:177.

Trounson A, Mohn L (1983) Pregnancy after freezing, thawing and transfer of human embryos. Nature 305:707.

Vargyas JM, Morente C, Shangold G, Marrs RP (1984) The effect of different methods of ovarian stimulation for human in vitro fertilization and embryo replacement. Fertil 42:745–749.

Veeck LL, Wortham JWE Jr, Witmyer J, Sandow BA, Acosta AA, Garcia JE, Jones GS, Jones HW Jr (1983) Maturation and fertilization of morphologically immature human oocytes in a program of in vitro fertilization. Fertil Steril 39:594.

Zeilmaker GH, Gent A, Rijkmans C (1984) Pregnancies following transfer of intact frozen-thawed embryos. In: European Sterility Congress, Monaco, Abstract 034.

31
Artificial Acceleration of Reproduction

JOHN P. HEARN

Introduction

The human population of the world reached 1 billion as recently as 1850. Since then, it has more than quadrupled to approximately 4.5 billion today, and the increase will continue. Table 31.1 summarizes the projected increase in human population over the next 50 years, to a level of approximately 10 billion in 2050 (McNamara, 1984). The great success of the human primate has already had devastating effects on the survival of many other species of animals, including nonhuman primates. The clearance of forest environments and the expansion of human agricultural populations into marginal lands previously available to wild primate populations will undoubtedly result in more and more primate species becoming threatened, endangered, or extinct.

There is limited time in which to chart a course of action. Crucial to any effective plan that would increase the chances of survival of wild primate populations is a fundamental understanding of the limiting factors in primate reproductive physiology. It is imperative that we learn more about these factors from primate populations in captivity and then apply the knowledge gained to realistic conservation policies. In future, populations of primates in captivity and in the wild will need to be considered as one stock to be managed as a whole.

All is not yet lost. Our knowledge of the reproductive physiology of primates, developed from studies of the human and of the conventional laboratory primates, is already extensive. But although the general rules are understood, there is a great deal of species variation even within the primates that requires eludicdation for individual species before we can understand the delicate balance through which these species relate to their environmental conditions. It is ironic that research in laboratory primates has contributed significantly to improvements in human conditions, which in turn has led to a greater threat to nonhuman primate populations in the wild. In recent years, considerable advances have been made in the treatment of infertility in the human, and there are now a battery of methods available for accelerating the breeding of the human male and female (Fen et al., 1981; Rowe and Ramarozoka, 1980). These advances have yet to be translated to improve the conservation of nonhuman primates. A part of the problem may be the assumption that because methods are now developed for the human, they can be generally applied to other species. This is not so. Species var-

TABLE 31.1A. Human population projections from 1950–2050 (in millions).[1]

	1950	1980	2000	2025	2050	Total fertility rate 1982
Developing countries	1670	3298	4884	6941	8400	4.2
Developed countries	834	1137	1263	1357	1380	1.9
Total world	2504	4435	6147	8298	9780	3.6

[1] These figures are selected from the 1984 World Bank estimates and projections.

TABLE 31.1B. Human population projections from 1950–2050 (in millions) in selected countries now the homelands of major wild nonhuman primate stocks.[1]

	1950	1980	2000	2025	2050	Total fertility rate 1982
China	603	980	1196	1408	1450	2.3
India	362	687	994	1309	1513	4.8
Indonesia	77	146	212	283	330	4.3
Brazil	53	121	181	243	279	3.9
Kenya	6	17	40	83	120	8.0

[1] These figures are selected from the 1984 World Bank estimates and projections.

iation means that successful application of these methods requires a greater fundamental understanding of the other species.

Overproduction is an essential feature of reproduction in most species, and fecundity is well below the potential rate. For example, the human male produces over 40,000 sperm/hr throughout reproductive life, and a normal ejaculate may contain over 50 million sperm. Sperm counts beneath 5 million/ml are thought to indicate infertility, but pregnancies have been achieved with sperm counts of less than 1 million/ml. The human female has approximately 2 million oocytes in the ovaries at birth and most of these will be lost by follicular atresia. Only about 350 will be shed at ovulations during the reproductive life of the female, and fewer than ten will usually result in babies born and reared to maturity. In both male and female, therefore, it should be possible within a few years to recover sperm, eggs, and embryos for storage and eventual transfer. This will not only enable the establishment of sperm, egg, and embryo banks, but it will also add a dimension to the management of reproduction that will facilitate improved genetic management of stocks of small populations kept in captivity and in the wild. In tandem with these developments, advances are now being made in the development of noninvasive methods of monitoring reproduction (Hearn, 1984). For example, urinary steroid assays are becoming available that reduce the necessity for the manipulation and handling of animals (Hodges and Hearn, 1983). These methods require further development into kits that could be used in the field or in small captive collections of primates that do not have the back-up of biochemical or medical laboratories.

The encouragement of natural methods of breeding must always remain a priority. It would be counterproductive to put so much emphasis on artificial breeding that any significant numbers of animals depend solely on artificial methods. There is still a great deal to be learned in order to facilitate optimal captive breeding. This includes behavior, environmental conditions, the genetic requirements of small populations, the optimal social groupings, and the requirements for the successful rearing of young.

In all the fields mentioned above, more basic knowledge is required or there will be little progress. In this discussion, the progress made in the reproductive management of the human and other primate species is reviewed and the applications of these methods are explored, where relevant, to primate conservation both in captivity and in the wild.

The Male

Our understanding of male reproductive physiology still lags behind that of the female. This is in part due to the greater production of gametes by the male and the relative ease with which sperm from the human male may be frozen. Yet, it is only in the human male, the cockerel, and the bull that sperm can be frozen easily. Species variations in the biochemical constituents of sperm require that each species be considered separately (Holt, 1984). Table 31.2 lists some of the methods now usable or being developed for the manipulation of male reproduction, but many of these are still only feasible in laboratory rodents, domesticated farm stock, and humans.

Stimulation

It is not generally necessary to stimulate the male reproductive system to produce gametes, for the reasons given above. The natural overproduction of sperm in most mammals has made it unnecessary, other than in particular cases of infertility or to overcome seasonal effects, to use hormone injections, minipumps, or diet additives of hormones to promote reproduction. Examples of such methods include the administration of melatonin, which has yet to be proved as a hormone that can overcome seasonal periods of infertility, or the administration of gonadotropin-releasing hormone (GnRH) to increase the production rate of gametes.

Recovery

The electroejaculator is now widely used in many species in order to recover sperm. Yet it is only in a few species that we have adequate knowledge about the in vitro capacitation of sperm where its recovery is helpful in accelerating breeding. Among nonhuman primates, there are not a great many successes, and the procedures have certainly not become routine. Yet this has not deterred a number of investigators from building up what they call "sperm banks" in which sperm collected from a wide range of species has been frozen. Many of these collections may better be termed "dead sperm banks," since without the fundamental studies required to test the survival of sperm during freezing, the collections are useless. Unfortunately, there is no quick solution, and a careful scientific approach is essential if this field is to be of value more generally in primate conservation. It should be remembered that the success in preservation of sperm from the human, cockerel, and bull came only after many years and much investment in basic research. Progress in extending the field to new species may be quicker, but it cannot be done overnight, and an impatient, superficial approach will not produce results (Holt, this volume).

TABLE 31.2. Artificial methods of reproduction relevant to the management and breeding of male primates.[1]

Stimulation	Sperm maturation Electroejaculation Capacitation
Recovery/Storage	New cryoprotectants/freezing Nonfrozen storage Assessment of viability
Suppression	Steroids (estrogens, progestagens) Vaccines (GnRH, gonadotropins) Biochemical inhibitors Surgical (nonreversible)

[1] A number of these methods are currently in use or under development for the human and for domesticated animals. Few have yet been translated to routine procedures for nonhuman primates.

Suppression

The suppression of fertility in males is not a requirement for most of the species considered here. While an effective means of suppressing reproduction in the human male would be a great advantage, most nonhuman primates require acceleration of reproduction and not suppression. However, there are occasions when such systems would be useful, such as in the temporary and reversible inhibition of fertility of individual animals that are becoming too well represented in the gene pool. It may not be desirable to render such animals permanently infertile. The possible methods being developed in this field include steroidal implants, vaccines against gonadotropins or gonadotropin releaser hormone, or the development of new agents that block sperm maturation in the epididymis. None of these are yet fully tested or effective, and the problem appears to be in individual responses to treatments and the great production rate of sperm, which is difficult to inhibit without causing behavioral side effects.

TABLE 31.3. Artificial methods of reproduction relevant to the management and breeding of female primates.[1]

Stimulation	Gamete development Ovulation Synchronization of the cycle In vitro fertilization
Recovery/Storage	Embryo freezing and thawing Sexing of embryos Micromanipulation (embryo splitting) Inner cell mass transfer/Chimeras
Suppression	Long-acting steroids (synthetics) Vaccines (gonadotropins, zona antigens) Postcoital inhibitors Surgical

[1] A number of these methods are currently in use or under development for the human and for domesticated animals. Few have yet been translated to routine procedures for nonhuman primates.

The Female

The precise manipulation of the female reproductive system is to an extent easier than that of the male. Although gamete production by the female is continuous, other than in seasonally breeding species, ovulation is a discrete event and the numbers of eggs that are successfully ovulated are relatively small. Table 31.3 lists some of the ways in which the female reproductive system may be manipulated. Many of these methods are now becoming routine practice in the management of reproduction in the human and in domesticated farm stock. They have yet to be applied to nonhuman primates with any routine degree of success. As for the male, it is not unlikely that extension of these methods to new species will require greater fundamental knowledge of the species differences in the biochemistry of the female gamete and also of the embryo.

Stimulation

It is possible to stimulate follicular and oocyte development in the female using a number of methods. Ovulation can be induced in women and in domesticated stock by injections of gonadotropin-releasing hormone (GnRH) or with individual gonadotropins. In addition, pregnant mare serum gonadotropin or chorionic gonadotropin have been used extensively. Using these methods, it is possible to induce ovulation and to synchronize it to a time when artificial insemination may be carried out. However, it is important that great care be taken in the purity and the amount of gonadotropin used, because they may raise antibodies that render the individual being treated infertile in the longer term. These methods have yet to be used in a variety of nonhuman primates with any degree of success, and greater knowledge of the normal hormone profiles and the timing of oocyte development and ovulation is required.

In nonprimate species, a great advantage of the female system is that it be synchronized using prostaglandin F2α. Unfortunately, the primate corpus luteum appears to be insensitive to prostaglandin F2α, although some reexamination of this is necessary in the light of recent results in the marmoset monkey, where the corpus luteum may be controlled easily by the administration of prostaglandin F2α (Summers et al., 1985).

In domesticated species such as the sheep and cow, recent research shows that there is some possibility of overcoming the natural suppression of fertility during seasonal anestrus using melatonin implants or diet additives. These have not yet been attempted on nonhuman primate species.

Recovery

When ovulation has been induced and successful fertilization in vivo obtained, embryo recovery is a delicate but relatively straightforward procedure. At present, the surgical recovery of embryos has a greater success rate than nonsurgical flushing, but it would be of great benefit if nonsurgical methods could be further developed to reduce trauma and the possibility of adhesions in females subjected to surgery. The flushing of early embryos using nonsurgical techniques has long been a feature of management of cattle breeding, and in the past 2 years it has proved possible in the human, several nonhuman primates, and exotic equids, bovids, and ungulates (Summers, 1986).

A small proportion of attempts at in vitro fertilization in the human are now successful, but it is still impractical for the majority of nonhuman primates, and, once again, one must realize the great time and effort expended in bringing success to this procedure in the human. It is unlikely that the procedure will be widely available in the reproductive management of nonhuman primates for some years.

Once embryos have been collected or developed from in vitro fertilization techniques, they must be transferred to a female who is exactly at the right stage of the cycle. This needs careful monitoring of the cycle, particularly in those species that do not give a clear, overt sign of menstruation or estrus. The noninvasive monitoring of the reproductive cycle has advanced considerably in recent years, and the availability of urinary estrogen or pregnanediol assays should continue this advance (Hodges and Eastman, 1984).

The field of embryo manipulation is potentially of benefit to the management of primate breeding. The surgical separation of blastomeres from a four- or eight-cell embryo to produce several babies from an individual embryo is still at an early stage of development. This technology is now feasible in rodents and in sheep under closely controlled laboratory conditions, but it is unlikely to be of any general application in primate conservation for some years, other perhaps than to individual cases of rare species where such investment is essential (Hearn and Hodges, 1985). A further potential field of embryo manipulation, again only justifiable in extreme cases, is the transfer of the inner cell mass of one embryo into the trophoblast of a second. This technique allows the growth and development of the embryo of one species within the trophoblast of another and reduces the risk of rejection of the embryo by the surrogate mother. However, it is unlikely that this technique will be of any great value in conservation, since the young primate is so dependent on learning its behavioral cues from its mother that will fit it for its own future reproductive life. Consequently, although these manipulations may be of some interest in investigating basic questions to the barrier of interspecies breeding, they are unlikely to be of any great practical benefit to primate conservation in the next 10 years.

The freezing of embryos is now increasingly possible in humans, cattle, and in some nonhuman primates and ungulates (Summers, in press). Yet it is too soon to expect that this technology can be widely applied, and it is not unlikely that species differences in embryo reactions to the freezing procedure may yet become apparent. It may be necessary to take a similar approach to that needed for the freezing of sperm from various species. Therefore, it will be necessary to develop new cryoprotectants related to the surface components of embryos in order to prevent damage or destruction of the embryos during the freezing and thawing process.

Undoubtedly, one of the areas where advance would be most helpful would be in the in vitro maturation of oocytes. As noted earlier, there is a great overproduction of oocytes in the ovary, and if these could be recovered, stored frozen, and then matured in vitro before transfer at an appropriate time, it might enhance our chances of accelerating breeding in primates. Recent work shows that the process is possible, but it is going to be several years before it is developed, even under closely controlled laboratory conditions (Polge, 1985).

Suppression

New methods for the long-term suppression of reproduction in individual females are likely to be available in the next decade. These will either be through using slow release implants of steroids, vaccines against chorionic gonadotropin (Hearn, 1979) or gonadotropin-releasing hormone, or biochemical blockers that reversibly inhibit the fertility of animals that are overrepresented in the gene pool. This would give us an improved management of reproduction without permanent loss of any individual's reproductive capacity. Because the emphasis of this discussion is on the accelerated breeding of endangered primates, it is not necessary to go into further detail of reproductive suppression here.

The Young

Neonatal loss is a major waste of reproductive capacity in all animal species, both in captivity and in the wild. Most Old World primate species produce one young at a time, and the investment in parental care is great. The New World primates, often producing twins or triplets, are more r-selected and the neonatal losses are higher. In captivity, the neonatal survival of primates is not high, and frequently the loss rates are in excess of 50%. In order to reduce this waste, it is important that we know more about the constituents of primate milk and the species variation in milk composition that would allow us to produce better synthetic or substitute diets. It is well known that removing young primates from the family group can have long-term behavioral complications that reduce the potential of future breeding. Consequently, methods for collaborative rearing of young primates in captivity, whereby diet additives are given to the young without removing them permanently from the family group, would be an advantage (Hearn and Burden, 1979). There are hand-rearing procedures for many of the common laboratory primate species, but hand rearing the young of endangered or rare species is less well developed.

A further area where greater knowledge is required is in the management of young primates around the time of puberty. There is a tendency to put young male and female primates together for breeding much earlier than they would naturally pair off in the wild. This is undue haste, because although these animals are theoretically capable of breeding soon after they reach puberty, when the female is cycling regularly and the male is producing sperm, they are incapable of the maturity required to successfully rear their young. The full synchronization of the reproductive system, including the hypothalamus, the pituitary, the ovary, the uterus, and the embryonic dialogue with the mother, may take several years to develop after puberty (Hearn, 1983). It is necessary for us to understand more about sexual maturity rather than puberty, and management procedures of captive primate stocks need to be based more firmly on such knowledge.

There is a need for greater cooperation between field and laboratory scientists in improving the environmental conditions provided to primates in captivity. Wherever possible, these should be based on the environmental cues that the animals use in the wild. An enrichment of captive environments, designed to provide as natural a setting as possible, may be more time consuming, but it will result in greater long-term breeding success. All primates are intelligent, and the rearing of such species under deprived conditions will lead eventually to boredom, reduction of breeding potential, and perhaps even to loss of successful reproduction. These aspects will become increasingly important if any successful reintroduction of primates from captivity to the wild is to be carried out.

Conclusions

The advances made in the artificial acceleration of reproduction in the past 5 years mean that we are entering a decade where almost anything is possible. Yet we are still ignorant of the basic reproductive physiology of many primate species. In extending research to assist with the captive breeding or the management in the wild of rare species, little progress will be made unless the fundamental limiting factors are understood.

The first priority is improved natural breeding, requiring a greater knowledge of social and behavioral development in the species concerned. What are the cues necessary to ensure successful breeding? What are the limiting factors in genetics, disease, nutrition, reproduction, and veterinary care? What are the important features to be learned from primates in the wild that will benefit their management and breeding as the wild diminishes, as it will with the human population set to double in the next 50 years?

In the male and in the female, greater production can now be stimulated, embryos can be recovered and transferred, and neonatal loss may be reduced. What are the realistic targets to be achieved in captivity and to what extent can such work help in the wild? How can reproduction be accelerated to achieve greater production, improved genetic balance, and fitness without upsetting "natural constraints"? What are the realistic goals for transfer of gene material between captive populations to ensure outbreeding, or to transfer sperm, eggs, or embryos from captivity to the wild or vice versa? What are the priority species and the areas where an impact can be made with the limited resources available?

We have to look in the future to captive and wild populations as one stock, to be managed as a whole to improve the chances of survival for both. A lot is possible, but careful thought is necessary if efforts are not to be wasted.

References

Fen CC, Griffin D, Woolman A (1981) Recent advances in fertility regulation. World Health Organisation, Afar S.A., Geneva.

Hearn JP (1979) Immunological interference with the maternal recognition of pregnancy. Ciba Found. Symp. 64:353–376.

Hearn JP (ed) (1983) Reproduction in New World primates. MTP, Lancaster, UK.

Hearn JP (1984) Scientific report of the Zoological Society of London 1982–1983. J Zool (London) 204:1–99.

Hearn JP, Burden FJ (1979) "Collaborative" rearing of neonatal marmosets. Lab Anim 13:131–133.

Hearn JP, Hodges JK (eds) (1985) Advances in animal conservation. Symp Zool Soc London 54:1–282.

Hodges JK, Eastman SAK (1984) Monitoring ovarian function in marmosets and tamarins by the measurement of urinary estrogen metabolites. Am J Primatol 6:187–197.

Hodges JK, Hearn JP (1983) The prediction and detection of ovulation: applications to comparative medicine and conservation. In: Jeffcoate SL (ed) Ovulation. Methods for its prediction and detection. John Wiley, Chichester, UK, pp 103–122.
Holt WV (1984) Membrane heterogeneity in the mammalian spermatozoon. Int Rev Cytol 87:159–194.
McNamara RS (1984) The population problem: time bomb or myth. World Bank, Washington, DC.
Polge C (1985) Embryo manipulation. In: Hearn JP, Hodges JK (eds) Advances in animal conservation. Symp Zool Soc London 54:123–132.
Rowe PJ, Ramarozoka SRR (1980) Workshop on the diagnosis and treatment of infertility. Pittman, Bath, UK.
Summers PM (In press) Collection, storage and use of mammalian embryos. Intern Zoo Yrbk 24/25.
Summers PM, Wennink CJ, Hodges JK (1985) Cloprostenol induced luteolysis in the marmoset monkey, *Callithrix jacchus*. J Reprod Fertil 73:133–138.

32
Collection, Assessment, and Storage of Sperm

WILLIAM V. HOLT

Introduction

The development of techniques for the preservation of bull semen, which led directly to the current commercial use of artificial insemination in dairying, enjoyed a number of advantages not available for analogous studies in primate breeding and conservation. Considerable scientific effort and financial resources were invested into these research programs (see Smith, 1961), and luckily, the bull spermatozoa were a fortunate first choice for freezing studies because of their resilience during freezing and storage.

The problems of semen storage in rare and endangered species have to be approached differently, because there is often no possibility of performing even the simplest experiment to determine optimal conditions for semen preservation. When such possibilities do arise they should be exploited, but in general there is a need for the elucidation of fundamental factors that are important both for the evaluation of seman quality and for the greater understanding of processes involved in sperm preservation. With this in mind, the main aim of the present review is to point out areas that require further examination as well as to identify and discuss recent developments in sperm assessment and preservation.

Semen Collection Techniques

Specific aspects of semen collection in nonhuman primates were reviewed in some detail by Hendrickx et al. (1978), and other points of particular relevance to the great apes were more recently discussed by Gould (1983) and Bader (1983). General aspects of semen collection were also reviewed in 1978 by Watson (1978) and were again discussed by Marshall (1984). Because there have been few, if any, significant technical developments in this field since these articles were published, there is little point in covering the same literature again. Instead, a brief summary of the main methods available, together with some comments on their use, will be presented.

Electroejaculation, which involves the stimulation of nerves supplying the reproductive organs, is the most universally applicable technique for the collection of semen, having been used with a wide variety of primates ranging from tree shrews and marmosets to orangutans and gorillas. Watson (1978) published a comprehensive list of primate species from which semen has been collected by this method.

The versatility of electroejaculation can be attributed to:

1. its use in conjunction with anesthetized animals, hence minimizing risks to either the animal or to personnel
2. the availability of rectal electrodes for a wide range of body sizes
3. the lack of any requirements for previous training of the subjects

Some workers have experienced urine contamination of ejaculates obtained by electroejaculation. This problem was discussed by Watson (1978), who suggested that the application of excessive voltage or anterior displacement of electrodes might be responsible.

Perhaps the most significant effect of electroejaculation on semen characteristics is possible dilution through the overproduction of accessory gland secretions. While there is little evidence to show that this effect occurs in primates, these findings having been made in domestic species, this may only be a consequence of inadequate examination of the phenomenon. Some caution is therefore indicated when addressing the possible significance of ejaculate volume or sperm density variations in samples collected by this method.

The collection of semen by artificial vagina (AV), widely used with domestic species, cannot normally be applied to wild species for two main reasons: The potential risks of injury to personnel are too great, and there is a requirement that the animals be trained to mount the AV. Nevertheless, ingenious efforts to collect semen from conscious animals have been reported. Martin et al. (1978) regularly obtained an ejaculate from a chimpanzee that masturbated for reward. Mastroianni and Manson (1963) developed an electrical method for collecting semen from macaques, whereby an AC monophasic stimulus of 20–40 V was intermittently applied between two electrodes, one placed at the base of the penis and the other near the glans penis. This method has been used by several workers and was discussed in detail by Hendrickx et al. (1978).

In some primate species, the ejaculate gels to form a stable coagulum, which interferes with subsequent handling of the semen sample. To some extent this problem can be overcome by the use of proteolytic enzymes (Watson, 1978), but it is also possible to recover spermatozoa by allowing them to swim out into an incubation medium (H. D. M. Moore, personal communication).

Although the use of ejaculatory procedures for collecting semen may be the only ones available for use with living animals, the potential value of postmortem specimens should not be overlooked. Spermatozoa remain structurally intact and retain their motility in the epididymis for several hours after death; indeed, Graham et al. (1978) proposed that spermatozoa could be processed for freezing as much as 48 hours after death. While the fertilizing ability of spermatozoa would diminish considerably during such a long period, it is likely that useful samples could be recovered from cauda epididymides stored for short intervals. The optimal conditions for storing epididymides after death, with the intention of retaining sperm viability and fertility, remain to be established. Cooling to 0°C, as suggested by Graham et al. (1978), may cause cold shock and membrane damage in some species.

Semen Assessment

While the ultimate test of semen quality is its fertilizing ability when females are inseminated, it is clearly impossible to use this method of semen assessment except in specialized cases. The problem that faces those without this facility is, therefore, how to make a realistic appraisal of the likely fertility of an individual animal or ejaculate from the semen characteristics available for measurement.

Defining the problem in this way leaves open a second question: Which of the characteristics should we measure? The identification of semen parameters that correlate best with fertility is difficult to achieve, especially in species where there is little possibility of performing the necessary large-scale studies required to establish these relationships. Our current semen assessment techniques rely heavily, therefore, on experience gained in human medicine or agricultural practice; even here it is surprising that relatively little progress has been made in understanding the precise significance of semen parameters.

Once a semen sample has been obtained, a number of possible options are open for its assessment.

Volume and Sperm Density

These can usually be measured without difficulty, although they may present a problem if a coagulum has formed. Evidence from the human is contradictory regarding the important that should be placed upon these parameters; it is generally accepted that if sperm density in the human ejaculate falls below about 10–20 million/ml then men will tend to be infertile (McLeod and Gold, 1951), but exceptions to this trend are known to occur (Davis et al., 1979; Bostofte et al., 1984). Sperm density and semen volume also show considerable variation between ejaculates in men and probably also in nonhuman primates. Moreover, if ejaculates are collected by electrical stimulation, these inconsistencies would almost certainly be amplified.

Measurements such as semen volume and sperm density are, of course, of most value if they can be compared with previously collected data, preferably from subjects of known fertility. Unfortunately, baseline information of this nature is scarce for the nonhuman primates; a table of data pooled by Hendrickx et al. (1978) from a number of sources shows that only one study large enough to provide statistically meaningful control data on semen parameters in nonhuman primates had been published. This study of 128 rhesus monkeys revealed a wide range of semen volumes, 0.2–5.0 ml, and sperm densities, 100–1500 million/ml.

Sperm Morphology

The pleiomorphism exhibited by gorilla spermatozoa (Seuanez et al., 1977), where considerable diversity in sperm head morphology is apparent within single ejaculates, is a feature shared with human spermatozoa. Such diversity, where variable proportions of cells differ in size and shape from what is considered to be "normal," has led to the proposal that morphological examination of human semen should provide information relevant to fertility assessment. Among others, Aitken and colleagues (1982) found that the incidence of abnormal spermatozoa was one of the most revealing conventional semen parameters they examined; in this study, a group of individuals with unexplained infertility showed significantly fewer normal spermatozoa than the fertile control group.

Extending this principle to the gorilla is likely to be of value but needs further exploration in this species. Two relatively recent reports of semen analysis in infertile lowland gorillas (Platz et al., 1979, 1980) documented surprisingly large proportions of spermatozoa classified as abnormal. Of the six animals covered in these two reports, five showed greater than 92% abnormal spermatozoa in their ejaculates, while the remaining one exhibited 65%. Platz et al. (1979, 1980), in both their reports on gorilla semen, performed testicular biopsies on their subjects and found the high incidence of sperm abnormalities to be correlated with defective spermiogenesis. These findings appear to explain the apparent infertility of these animals, although the causes underlying the defective spermatogenic process remains obscure. It is relevant here to point out that gorillas, at least in captivity, have a general tendency toward testicular abnormalities or even degeneration (Dixson et al., 1980).

It is of some passing interest that the single ejaculate obtained by Platz and his colleagues (1979) from a siamang gibbon contained 98% abnormally shaped spermatozoa; at present this figure stands alone for this species but may indicate the usefulness of sperm morphology estimates for the assessment of gibbon semen.

Sperm Motility

The mammalian spermatozoon probably requires considerable thrust to penetrate the resilient investments of the ovum, and it is reasonable to suppose, therefore, that assessments of sperm motility in semen would be good indicators of fertility. Since sperm movement is one of the most obvious characteristics of an ejaculate, anyone possessing a suitable microscope can easily make a visual assessment.

In an effort to improve the subjective procedure by putting it on a semiquantitative basis, Emmens (1947) introduced a system that assigned scores of motility based upon the subjective judgment of both the proportion of motile cells and their speed of progression. Various modifications of this procedure have been used since the introduction of this system and are still widely used. It is important to realize, however, that this type of method is best applied when a large number of samples are both randomized to eliminate observer bias and when the results can be analyzed by suitable statistical techniques. Such procedures are, therefore, less than ideal for use in diagnostic work, whether clinical or veterinary, since these conditions can rarely be fulfilled.

Clearly, there is a need for the development of objective methods for the assessment of sperm motility, which by eliminating the subjective element can be applied to intermittently received samples. Interesting new developments toward this goal have concentrated upon the objective measurement of sperm swimming speed as an indicator of fertility (Milligan et al., 1980; Katz and Overstreet, 1981; Aitken et al., 1982; Holt et al., 1984, 1985).

The value of sperm speed measurements with respect to human spermatozoa has been investigated in groups of fertile and infertile individuals (Milligan et al., 1980; Aitken et al., 1982), where a high degree of correlation between speed and the fertility of individuals or individual ejaculates was demonstrated. A recent study performed in this laboratory confirmed these relationships, fertility in this case being assessed using the zona-free hamster egg sperm penetration assay (Holt et al., 1984, 1985), as well as the in vitro penetration of human oocytes.

Until recently, techniques for the measurement of sperm swimming speed have generally been so time consuming and tedious that this parameter has been of little use in routine semen assessment. Developments in computer technology have, however, permitted the development of new and easily performed procedures for speed measurement. One such technique, which was developed in this laboratory for the assessment of experimentally applied treatment effects in semen cryopreservation research, has proved immensely useful in the evaluation of semen quality in exotic animals as well as in the clinical diagnosis of human male infertility (Holt et al., 1984, 1985). This particular procedure permits the semi-automatic tracing of sperm paths, thereby providing sufficient data for the computer to calculate a mean speed of forward progression for each cell examined. Using this equipment, the speed of 50–60 individual spermatozoa can be determined within 10 minutes, thus providing considerable objective information about the semen sample in a very short time. Further relevant information can be gained by examining the frequency distributions of sperm speeds in the sample which, at least in the human, seems to be a more powerful way of assessing fertility than the simple determination of mean sperm speed (see Fig. 32.1).

The application of these principles to the assessment of nonhuman primate semen deserves investigation, but the absence of suitable background data for these species on parameters such as sperm speed rules out their diagnostic use at

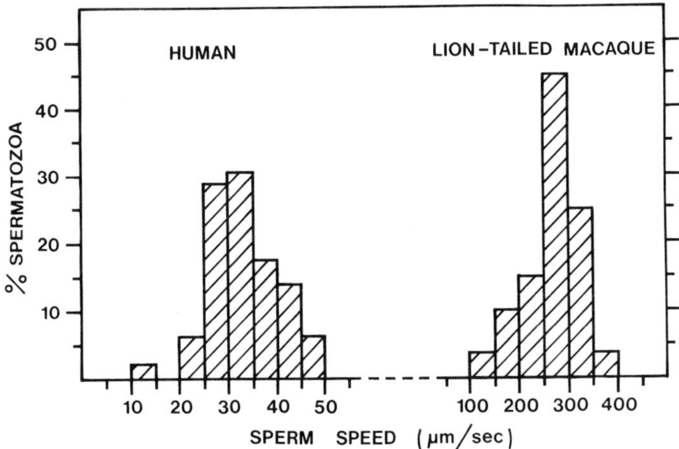

FIGURE 32.1. This figure illustrates the distribution of sperm swimming speeds in ejaculated semen from two primate species, the human and the lion-tailed macaque. Swimming speeds range widely within a given sample, but there is no overlap between the two populations, the mean speed for the macaque (266 μm/second) being almost an order of magnitude greater than than of the human (34 μm/second). Examples of this sort emphasize that assessment of semen quality through motility parameters needs careful background work for each individual species of interest.

present. That these parameters are likely to differ significantly, both from the human and from one another, is revealed by some preliminary measurements of sperm swimming speed carried out in this laboratory. Ejaculated spermatozoa from a lion-tailed macaque, after dilution in a physiological salt solution, displayed a mean speed (±S.E.M.) of 266 ± 8.6 μm/second, the range covering 102–350 μm/second (see Fig. 32.1). This would compare with human spermatozoa, which under comparable conditions in this laboratory are considerably slower, with a mean speed of 34 μm/second and a range of 12–60 μm/second (Fig. 32.1). Measurements made on a frozen/thawed sample of gorilla semen revealed a mean sperm speed (±S.E.M.) of 27 ± 3.4 μm/second, a figure more in keeping with the human values.

With respect to human spermatozoa, there is every indication that the measurement of parameters associated with the motile behavior of spermatozoa will ultimately provide more information relevant to fertility than can currently be gained by the simple measurement of speed. For example, Aitken et al. (1985) found recently that spermatozoa exhibiting inadequate lateral head displacement while swimming in seminal plasma showed poor penetration of cervical mucus, itself a correlate of fertility.

Sperm Membrane Integrity Assessment

Sperm membrane damage appears to be a major consequence of cold shock and semen freezing and probably accounts for much of the loss of fertility associated with these procedures. Techniques to monitor sperm membrane damage include the popular use of eosin/nigrosin staining as a live/dead indicator, specific Giemsa staining of acrosomal membranes (Watson, 1975), and electron microscopy.

The search for efficient methods of assessing sperm membrane integrity has recently been boosted by research into the distribution and development of sperm membrane antigens. In this and other laboratories, murine monoclonal antibodies have been generated that are specific to sperm membrane surface components but display broad species cross reactivity. These antibodies, when used in conjunction with appropriately labeled conjugates, serve as excellent markers of cellular components such as the acrosome or midpiece and will undoubtedly become more widely used to establish cellular integrity as the associated technical procedures are progressively simplified.

Heterologous Ovum, Sperm Penetration Test

Nearly a decade ago, Yanagimachi et al. (1976) reported that human spermatozoa were able to penetrate the vitellus of the zona-free hamster egg; a considerable body of literature now exists on the use of this phenomenon as an investigative test of human infertility (Yanagimachi, 1984). The unique receptivity of the hamster egg plasma membrane also allows the penetration of spermatozoa from other species, but the application of this technique to nonhuman primate fertility assessment has received little attention. It has been established, however, that spermatozoa from common marmoset (Moore, 1981) and orangutan (H. D. M. Moore, personal communication) are capable of egg penetration in this system. One of the potential problems to be overcome in the application of this technique in a range of species is that conditions permitting sperm capacitation have to be determined individually for each new species studies. Until such basic information is available, it will be difficult to judge the significance of negative penetration rates.

A further development of the zona-free hamster egg test has been the possibility of preparing and examining chromosome spreads derived from spermatozoa (Yanagimachi, 1984). Among the potential uses of this method would be the correlation of morphological sperm abnormalities with chromosomal defects, in gorilla as well as human semen. The technique would also be of use in toxicological studies applied to experimental groups of primates, where chromosomal damage inflicted by compounds affecting meiosis could be directly assessed.

Semen Freezing

Practical aspects of the cryopreservation of semen were recently reviewed by Marshall (1984); a more detailed consideration of the problem was presented by Watson (1979), who more recently also reviewed the field of human semen cryopreservation (Watson, 1985). Semen cryopreservation in nonhuman primates was also reviewed by Hendrickx et al. (1978).

From these and other reviews, it is apparent that developments in the technology of semen freezing have reached a stage where there is a requirement for a breakthrough in our understanding of the fundamental processes involved in successful sperm cryoprotection. This might eliminate the present requirement that optimal conditions for freezing have to be formulated through trial and error for each individual species.

With this in mind, it is instructive to consider the possible factors responsible for such major species differences in the response to semen cryopreservation.

The Sperm Plasma Membrane and Cryopreservation

The cell membrane is of crucial importance as an interface between the cytoplasm and the external environment, but its integrity is dependent upon its physical state, which in turn is temperature dependent. Since membrane damage is one of the most frequent observations connected with sperm cryoinjury, it is likely that membrane composition and organization will be an important determinant of cryosensitivity. Efforts to attribute species differences in cold-shock sensitivity to sperm membrane compositional differences have met with some success (White and Darin-Bennett, 1976; Darin-Bennett and White, 1977). The considerable organizational complexity of the sperm plasma membrane (Holt, 1984) means, however, that membrane composition differs between regions of the same cell. It will therefore be necessary to find ways of determining membrane domain composition before the potential of this approach can be fully realized.

Recent work in this laboratory directed toward understanding these relationships has shown that sperm plasma membranes undergo major organizational changes during the cooling procedures that precede freezing (Holt and North, 1984). Comparative studies of these effects, linked with observations of cryosensitivity, would be of value in understanding the causes of cold-induced sperm damage.

Previous reports have shown that exogenous lipids can protect spermatozoa against cold shock (Evans and Setchell, 1978; Watson, 1981), but the mechanism of this effect has never been explained. The possibility that exogenous lipids can modulate membrane fluidity, thus allowing manipulation of sperm sensitivity to cold-induced damage above freezing point, is currently under investigation in this laboratory. To date we have found that exogenous lipids such as cholesterol and phosphatidylcholine, especially when prepared in the form of multilamellar liposomes, protect spermatozoa during cooling and storage by affecting the physical properties of the sperm plasma membrane. For experimental reasons these studies were performed on ram spermatozoa, but the effect is unlikely to be limited to this species alone.

These studies correlate well with investigations into the mechanisms of cryoprotection mediated by the low-density component of egg yolk (LDF), which has been proposed by several workers (Pace and Graham, 1974; Watson, 1976; Foulkes, 1977; Watson, 1981) as the active constituent of egg yolk. Watson (1981) found that the phospholipid component of this lipoprotein was of primary importance in the protection of spermatozoa, the associative proteins possibly serving to enhance binding to the sperm surface. It is therefore likely that this substance also acts by modulating the fluidity of the sperm plasma membrane, and preliminary studies in this laboratory seem to support this suggestion.

A recent report that human spermatozoa have a highly specialized cytoskeletal organization that coincides with discrete membrane domains (Virtanen et al., 1984) raises the interesting possibility that these might be involved with the maintenance of membrane stability. Clearly, these findings ought to be pursued further in relation to sperm cryopreservation, because there is every likelihood that the extent and stability of the cytoskeletal networks will exhibit species differences.

Cryoprotectants

It is becoming increasingly clear that spermatozoa require a different type of protection during the prefreeze stages from that required once ice begins to form. This may explain the dual requirement for glycerol or dimethyl sulfoxide, the antifreeze agents, and the lipoproteins present in egg yolk or milk components.

The protective action of lipoproteins upon the cell membrane at temperatures above zero has already been discussed above; once the temperature falls to freezing point, however, it is likely that a number of different deleterious processes have to be counteracted if loss of sperm viability is to be prevented. Clearly, ice crystal formation at the time of freezing is a potential source of cellular damage, causing both mechanical damage to the cell and osmotic damage through the production of high extracellular solute concentrations. These solutes tend to draw water out of the cell, with the consequent disruption of membranes and the further generation of local pockets of high salt content.

Substances such as glycerol and dimethyl sulfoxide are believed to minimize cell damage under these conditions by maintaining water in solution during freezing. Some experiments with nonmetabolizeable sugars have indicated that these may also confer cryoprotection by dehydrating the cells to some extent prior to freezing (Nagase et al., 1968). Relatively little attention has since been paid to the possible use of sugars as sperm cryoprotectants, although their potential has been more fully explored in relation to plant cells.

Mathias et al. (1984) have recently introduced an entirely new concept in freezing technology, whereby cells are undercooled to subzero temperatures, i.e., −10°C or −15°C, dispersed in stable emulsions of paraffin oil and wax. Using this technique, these authors have preserved plant and yeast cells with good postthaw survival. Erythrocytes have also been maintained by this technique with about 50% survival after thawing. Unfortunately, this method has not yet been tried with mammalian spermatozoa, although its application to turkey spermatozoa was apparently unsuccessful (Zavos and Graham, 1981). A new principle such as this for semen preservation, which avoids the actual freezing process, may, however, be a timely stimulus for further experimentation in mammalian semen preservation.

Semen Preservation in Primates

Hendrickx et al. (1978), reviewing semen cryopreservation procedures in nonhuman primates, noted that the major obstacle to freeze-preservation of simian semen is the maintenance of fertilizing capacity upon thawing, despite the recovery of good postthaw motility. This apparent discrepancy between fertility and postthaw motility appears to be best illustrated by the recent birth of a young male gorilla in Memphis after insemination of the mother with frozen semen (Douglass, 1981). The motility of the frozen semen was described as poor, with only about 10% of the spermatozoa moving.

This case is exceptional in resulting in a successful birth, other recently reported attempts to use frozen semen in great apes having resulted in failure. Bader (1983) reported three inseminations with frozen semen in chimpanzees, and Matern (1983) described a similar attempt with one bonobo female (*Pan paniscus*). Although these authors documented their freezing and insemination protocols in detail, this type of attempt contributes very little to a better understanding of semen cryopreservation in great apes simply because they are not controlled studies.

Controlled studies are more feasible with the smaller laboratory primates, but here too this type of data is only available to a limited extent. Leverage et al. (1972) found that out of 48 inseminations carried out with frozen semen in rhesus monkeys, only one conception resulted; this was subsequently lost after 40 days gestation. Cho et al. (1975) similarly reported 2/13 conceptions in cynomolgus monkeys inseminated with frozen semen. These too terminated spontaneously within 6–8 weeks.

Whether these spontaneous terminations point to a further drawback in the use of frozen primate semen is difficult to say at present; it certainly appears that this is not a problem in the field of human artificial insemination with frozen semen (Leeton et al., 1980).

Conclusions

From the foregoing discussion, points that emerge of general relevance in the field of semen assessment and preservation are:

1. Not enough is known about the specific importance of individual attributes of sperm function. Large studies need to be set up where possible to identify the most valuable diagnostic components of semen quality.
2. The interspecific variation in sperm survival during cooling and freezing cannot be adequately explained at present. Research into this question would not only be of practical benefit but would likely be of fundamental importance in the cell biology of spermatozoa.
3. Mechanisms of cryoprotectant action must be more fully understood to be exploited optimally.

In the light of these gaps in our current knowledge of semen assessment and preservation, it is clear that the situation with particular respect to nonhuman primates is much worse. Too little data exist at present upon which to base judgments on the effectiveness of, for example, particular freezing schedules and methods. Given the appropriate opportunities, however, the technical ability to perform detailed studies of these areas, both in laboratory and field studies, now exists. We should therefore pursue these opportunities if primate breeding by artificial insemination in captivity is to be unhampered by the unknowing use of poor-quality semen or, conversely, by the failure to recognize good-quality semen.

References

Aitken JR, Best FSM, Richardson DW, Djahanbakhch O, Mortimer D, Templeton AA, Lees MM (1982) An analysis of sperm function in cases of unexplained infertility: conventional criteria, movement characteristics, and fertilizing capacity. Fertil Steril 38:212–221.

Aitken RJ, Sutton M, Warner P, Richardson DW (1985) Relationship between the movement characteristics of human spermatozoa and their ability to penetrate cervical mucus and zona-free hamster oocytes. J Reprod Fertil 73:441–449.

Bader H (1983) Electroejaculation in chimpanzees and gorillas and artificial insemination in chimpanzees. Zoo Biol 2:307–314.

Bostofte E, Serup J, Rebbe H (1984) Interrelations among the characteristics of human semen, and a new system for classification of male infertility. Fertil Steril 41:95–102.

Cho F, Honjo S, Makita T (1975) Fertility of frozen-preserved spermatozoa of cynomolgus monkeys. In: Kondo S, Kawai M, Ehara A (eds) Contemporary primatology. Int Congr Primatol, vol 5. S Karger, Basel, pp 125–133.

Darin-Bennett A, White IG (1977) Influence of cholesterol content of mammalian spermatozoa on susceptibility to cold-shock. Cryobiology 14:466–470.

Davis G, Jouannet P, Martin-Boyce A, Spira A, Schwartz D (1979) Sperm counts in fertile and infertile men. Fertil Steril 31:453–455.

Dixson AF, Moore HDM, Holt WV (1980) Testicular atrophy in captive gorillas (*Gorilla g. gorilla*). J Zool (London) 191:315–322.

Douglass EM (1981) First gorilla born using artificial insemination. Intern Zoo News 28:9–15.

Emmens CW (1947) The motility and viability of rabbit spermatozoa at different hydrogen ion concentrations. J Physiol 106:471–481.

Evans RW, Setchell BP (1978) Association of exogenous phospholipids with spermatozoa. J Reprod Fertil 53:357–362.

Foulkes JA (1977) The separation of lipoproteins from egg yolk and their effect on the motility and the integrity of bovine spermatozoa. J Reprod Fertil 49:277–284.

Gould KG (1983) Diagnosis and treatment of infertility in male great apes. Zoo Biol 2:281–293.

Graham EF, Schmehl MKL, Evensen BK, Nelson DS (1978) Semen preservation in nondomestic mammals. In: Watson PF (ed) Symp Zool Soc London, 43. Academic, London, pp 153–173.

Hendrickx AG, Thompson RS, Hess DL, Prahalada S (1978) Artificial insemination and a note on pregnancy detection in the non-human primate. In: Watson PF (ed) Symp Zool Soc London, 43. Academic, London, pp 219–240.

Holt WV (1984) Membrane heterogeneity in the mammalian spermatozoon. Intern Rev Cytol 87:159–194.

Holt WV, North RD (1984) Partially irreversible cold-induced lipid phase transitions in mammalian sperm plasma membrane domains: freeze-fracture study. J Exp Zool 230:473–483.

Holt WV, Moore HDM, Parry A, Hillier S (1984) A computerized technique for the measurement of sperm cell velocity: correlation of the results with in vitro fertilization assays. In: Thompson W, Harrison RF, Bonnar J (eds) The male factor in human infertility diagnosis and treatment. XI World Congr Fertil Steril (Dublin). MTP, Lancaster, UK, pp 3–7.

Holt WV, Moore HDM, Hillier SG (1985) Computer assisted measurement of sperm swimming speed in human semen; correlation of results with in vitro fertilization assays. Fertil Steril 44:112–119.

Katz DF, Overstreet JW (1981) Sperm motility assessment by videomicrography. Fertil Steril 35:188–193.

Leeton J, Selwood T, Trounson A, Wood C (1980) Artificial donor insemination, frozen versus fresh semen. Aust NZ J Obst Gynec 20:205–207.

Leverage WE, Valerio DA, Schultz AP, Kingsbury E, Dorey C (1972) Comparative study on the freeze-preservation of spermatozoa. Primate, bovine and human. Lab Anim Sci 22:882–889.

MacLeod J, Gold RZ (1951) The male factor in fertility and infertility. II. Spermatozoon counts in 1000 men of known fertility and 1000 cases of infertile marriage. J Urol 66:436–449.

Marshall CE (1984) Considerations for cryopreservation of semen. Zoo Biol 3:343–356.

Martin DE, Graham CE, Gould KG (1978) Successful artificial insemination in the chimpanzee. In: Watson PF (ed) Symp Zool Soc London, 43. Academic, London, pp 249–260.

Mastroianni L, Manson WA (1963) Collection of monkey semen by electroejaculation. Proc Soc Exp Biol Med 112:1025–1027.

Matern B (1983) Problems and experiences in performing artificial insemination in Bonobos (*Pan paniscus*). Zoo Biol 2:303–306.

Mathias SF, Franks F, Trafford K (1984) Nucleation and growth of ice in deeply undercooled erythrocytes. Cryobiology 21:123–132.

Milligan MP, Harris S, Dennis KJ (1980) Comparison of sperm velocity in fertile and infertile groups as measured by time-lapse photography. Fertil Steril 34:509–511.

Moore HDM (1981) An assessment of the fertilizing ability of spermatozoa in the epididymis of the marmoset monkey (*Callithrix jacchus*). Intern J Androl 4:321–330.

Nagase H, Yamashita S, Irie S (1968) Protective effects of sugars against freezing injury to bull spermatozoa. Proc VI Int Congr Anim Reprod Artif Insem (Paris), vol 2, pp 1111–1113.

Pace MM, Graham EF (1974) Components in egg yolk which protect bovine spermatozoa during freezing. J Anim Sci 39:1144–1149.

Platz C, Wildt BS, Wildt D, Bridges C, Seager S (1979) Fertility analysis in the lowland gorilla and siamang gibbon using electroejaculation. Ann Proc Amer Assoc Zoo Vet 88–90.

Platz CC, Wildt DE, Bridges CH, Seager SW, Whitlock BS (1980) Electroejaculation and semen analysis in a male lowland gorilla, *Gorilla gorilla gorilla*. Primates 21:130–132.

Seuanez HN, Carothers AD, Martin DE, Short RV (1977) Morphological abnormalities in spermatozoa of man and great apes. Nature 270:345–347.

Smith AU (1961) Biological effects of freezing and supercooling. Edward Arnold, London.

Virtanen I, Badley RA, Paasivuo R, Lehto VP (1984) Distinct cytoskeletal domains revealed in sperm cells. J Cell Biol 99:1083–1091.

Watson PF (1975) Use of a giemsa stain to detect changes in the acrosomes of frozen ram spermatozoa. Vet Rec 97:12–15.

Watson PF (1976) The protection of ram and bull spermatozoa by the low density lipoprotein fraction of egg yolk during storage at 5°C and deep-freezing. J Thermal Biol 1:137–141.

Watson PF (1978) A review of techniques of semen collection in mammals. In: Watson PF (ed) Symp Zool Soc London, 43. Academic, London, pp 97–126.

Watson PF (1979) The preservation of semen in mammals. In: Finn CA (ed) Oxford reviews of reproductive biology, vol 1. Clarendon, Oxford, pp 283–350.

Watson PF (1981) The roles of lipid and protein in the protection of ram spermatozoa at 5°C by egg-yolk lipoprotein. J Reprod Fertil 62:483–492.

Watson PF (1985) Recent advances in sperm freezing. In: Thompson W, Joyce DN, Newton JR (eds) In vitro fertilization and donor insemination. Royal College of Obstetricians and Gynaecologists, London, pp 261–267.

White IG, Darin-Bennett A (1976) The lipids of sperm in relation to cold shock. Proc VIII Int Congr Anim Reprod Artif Insem (Krakow), pp 951–954.

Yanagimachi R (1984) Zona-free hamster eggs: their use in assessing fertilizing capacity and examining chromosomes of human spermatozoa. Gam Res 10:187–232.

Yanagimachi R, Yanagimachi H, Rogers BJ (1976) The use of zona-free animal ova as a test system for the assessment of the fertilizing capacity of human spermatozoa. Biol Reprod 15:471–476.

Zavos PM, Graham EF (1981) Preservation of turkey spermatozoa by the use of emulsion and supercooling methods. Cryobiology 18:497–505.

33
Artificial Insemination of Nonhuman Primates

KENNETH G. GOULD and DAVID E. MARTIN

Introduction

There is an increasing interest in and need for artificial insemination (AI) techniques applicable to nonhuman primates. As the pressure on endangered species increases, so does the incentive for development of plans for their husbandry and maintenance. Plans are being formulated for conservation based on the best genetic data available. Such plans are frequently based on mathematical projections of factors, including minimum viable population size, breeding rate, generation time, and the need for maintenance of adequate genetic diversity. The mathematical formulations used most often involve a critical presupposition, namely, the equal representation of founder stock. Artificial insemination has a unique potential as a tool to achieve this goal, providing as it does a means for rapid and easy manipulation of the genetic pool of such species. Artificial insemination can play a potential role in increasing the breeding success of species needed for biomedical research and in increasing the availability of animals of precise age. When coupled with techniques for freeze preservation of gametes, it can provide an invaluable means for avoiding premature loss of an individual from the pool as a result of disease or death and by providing a means for future reintroduction of present genetic material as a way of controlling genetic drift.

The technique of artificial insemination is not new and has been practiced in some species since Spallanzani successfully inseminated a bitch in the late 1700s. In the early part of the twentieth century, the technique was refined and developed by Soviet scientists for use in horses (Perry, 1960). It has been most actively developed for use in the cattle industry, with a major expansion in its use during the 1950s, a fact not independent of the development in the same period of satisfactory methods for cryopreservation of bovine spermatozoa. Reviews of the development of the technique for application in this area are available (Perry, 1960; Jones, 1971). While the value of artificial insemination does not reside wholly in the use of frozen/thawed spermatozoa, the correlation of the two is of major significance in any species.

There is a wide difference in the success rate of AI in differing species. Thus, in the bovine, the success rate approaches 100% after a single insemination; while in nonhuman primates in which AI has been demonstrated, the rate is less than 50% (Table 33.1). This discrepancy may relate in part to the natural rate of

TABLE 33.1. Species in which AI has been successful.

Species	Date	Frozen	Reference
Baboon	1972	No	Kraemer et al.
Baboon	1985	No	Keuhl (unpublished observations)
Chimpanzee	1975	No	Hardin et al.
Chimpanzee	1978	No	Martin et al.
Chimpanzee	1985	No	Gould et al.
Gorilla	1983	Yes	Douglass and Gould
Gorilla	1985	No	Melbourne Zoo
Long-tail	1975	Yes	Cho et al.[1]
Rhesus	1973	No	Eisle et al.
Rhesus	1985	No	Lu et al.
Squirrel	1967	No	Bennett[1]

[1] No term pregnancy.

pregnancy maintenance in the two groups, which in turn is a reflection of differences in the endocrinology and physiology of the reproductive cycle.

An excellent review of artificial insemination in nonhuman primates was provided in 1978 by Hendrickx et al., and it is the intent of this paper to provide an update to the material then available with special reference to the role of AI in maintenance of endangered primate species.

The procedure of artificial insemination is basically simple, involving only the introduction of semen, or sperm, into the female reproductive tract by mechanical means. For the technique to succeed, however, several steps are involved: collection and preparation of sperm, timing or induction of ovulation, and placement of the prepared sperm in the optimal location within the female tract. Each of these steps offers problems unique to nonhuman primates.

The Male Component

While it is intuitively evident that semen collected after a "natural" ejaculation, i.e., after copulation or masturbation, will most closely represent the semen deposited at natural mating and therefore potentially offer the best fertility, it is not often possible to obtain such specimens.

Such specimens, however, are not produced at a predictable time for recovery when fresh (e.g., rhesus monkey) or are not easily collected because of difficulty in recovery from the female, but on occasion such samples can be recovered. Only the chimpanzee, of the nonhuman primate species, has been reported to provide masturbated ejaculates on a predictable basis. In those cases where an artificial vagina is not used, it is impossible to avoid cold shock of the sperm or, in some individuals, contamination with saliva or fecal material. Bader (1983) describes a situation in a pair of bonobos in which the behavioral pattern was such that a state of coitus interruptus existed and freshly ejaculated semen could be recovered. The conditions of such a collection may dictate the initial treatment of the semen after collection with regard to immediate washing with medium, incubation in antibiotic solution, etc. In most primate species, the semen forms some type of coagulum. The degree and rapidity of liquefaction of this coagulum varies between species; and in some, e.g., rhesus monkey and chimpanzee, complete liquefaction rarely occurs, while in other species the degree of liquefaction varies between individuals and specimens.

In species in which masturbates are not available, electrostimulation must be used to promote semen emission (Watson, 1978). Such stimulation is provided via electrodes placed directly on the penis or by electrodes placed on a rectal probe. The stimulation parameters are different for the two procedures. Penile stimulation provides direct sensory stimulation and requires that the ascending neural pathways be intact and functional. This precludes the use of anesthesia during the procedure, which is a problem in some species where physical restraint is impractical. Further, the success of the procedure requires that the animal be relatively unexcited and relaxed, a situation that usually restricts the number of subjects in a given colony and may preclude the use of any given male.

The method of penile electrostimulation originally reported by Mastroianni and Manson (1963) involves administration of square wave DC current that is applied via two metal band electrodes placed around the penis, one just below the glans, the other at the base of the shaft. Current is applied either in an intermittent surge or gradually increased over 10–15 seconds to a level shown to be effective and held there for up to 2 minutes (M.Y. Lu, 1985, personal communication; Settlage and Hendrickx, 1974; Van Pelt and Keyser, 1970). It is interesting and noteworthy that this method works only with square wave stimuli, in contradistinction to the stimulation using a rectal probe, which appears to require sine wave alternating current (Warner et al., 1974).

We have used equipment designed to provide a sine wave stimulus successfully for penile electrostimulation in the rhesus monkey and squirrel monkey, but when connected to penile electrodes the equipment used (Warner et al., in preparation) provides a sine wave with some clipping, i.e., squaring. We found that after placement of electrodes, direct stimulation with a current of 3–4 mA, at 25–27 V, 30 Hz is effective. This stimulation prompts semen emission/ejaculation in most animals in less than 25 seconds. If no semen is collected, the animal is rested for 1–2 minutes and the procedure repeated.

It is usually practical to collect a second or third ejaculate provided a rest of 5 minutes or more is provided between stimulations (Settlage, 1971; Settlage and Hendrickx, 1974).

A complication of the penile stimulation method is the incidence of trauma to the penis, which can arise as a result of incorrect or overtight connection of metal band electrodes or from application of excessive electrical stimulation, resulting in erythema and/or burning. It is to be emphasized that these complications should *not* occur if the method is used correctly. We have found that 2–3 mm wide strips of a commercially available disposable electrical ground plate (3M Scotchplate #1149) provide an effective, disposal electrode that is a soft, flexible, nontraumatic connector for this purpose.

Rectal probe electroejaculation has been used successfully in a variety of species. First reported by Gunn (1936) for use in the ram and developed by Laplaud and Cassou (1948), the method works by stimulating contraction of the male accessory glands in the pelvis, with subsequent emission (technically not ejaculation) of semen. When used in a conscious animal, it is possible for complete ejaculation to occur, but in the anesthetized animal this is not so.

Smaller animals (<2 kg) are restrained in a supine position and larger animals in lateral recumbency or, when appropriate (e.g., macaques), in a suitable

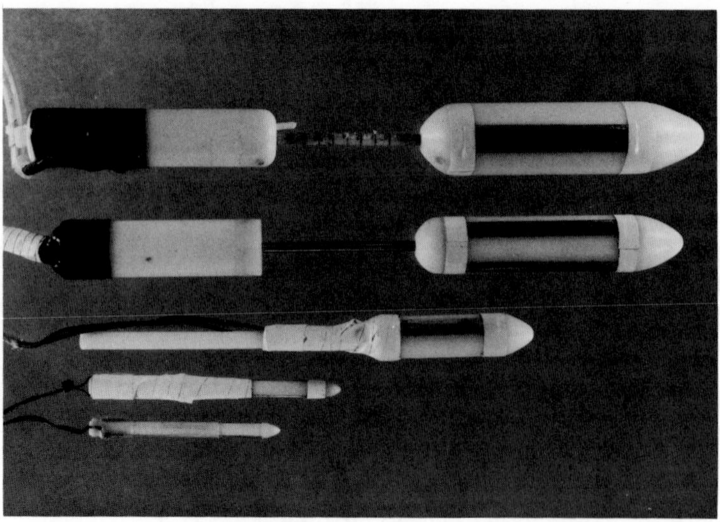

FIGURE 33.1. Rectal probes used in nonhuman primates.

restraint chair. Anesthesia is kept as light as is practical while consonant with operating safety, as better response is obtained with light sedation. It is now generally accepted that the best electrode configuration comprises two or three longitudinal electrodes, as opposed to the circular (ring) electrodes used earlier, mounted on the surface of a probe of a diameter sufficient to provide moderate distension of the rectum in order to promote contact of the electrodes with the rectal wall (Fig. 33.1).

Electrical current is applied in a series of increments, the maximum value at each stage being raised in approximately 2-V steps, with six repetitions at each voltage level. The stimulus is applied by varying a potentiometer in a sinusoidal manner, with a rapid rise, a pause of 1–2 seconds around the maximum value, and a more rapid drop. Each stimulus occupies 3–4 seconds. As higher voltages are used, the current is not reduced to zero between stimuli.

The most effective frequency for stimulation is approximately 20 Hz, with a tendency for smaller species to require a higher frequency (30–50 Hz), and stimulation parameters for response vary from 2 to 16 V and from 7.0 to 340 mA. These values do not account for various electrode areas, however, it is more meaningful to express stimuli in terms of current per unit of electrode area (mA/m^2). A mean value of 0.24 mA/m^2 appears to prompt semen emission regardless of body weight. The resistance at the electrode/rectal mucosa interface, although predictable for an individual or species, varies with electrode area between 60 and 600 ohms (Fig. 33.2). Identification of a single stimulating equipment/probe combination is valuable because it provides an objective means for comparison of techniques.

At this time, the technique of RPE is considered safe and the side effects minimal. On two occasions the repeated use of RPE in lemur species has been

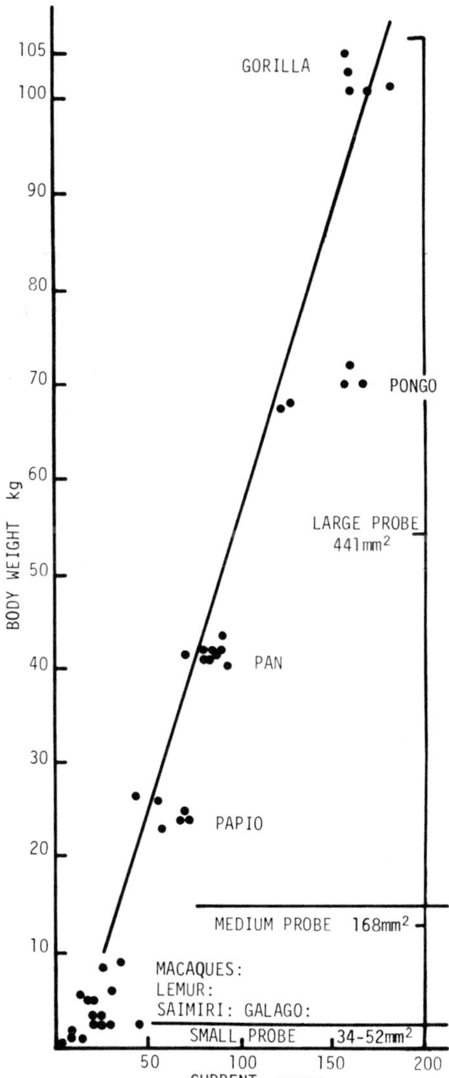

FIGURE 33.2. Relationship of delivered current, body weight, and electrode area for RPE.

correlated with the later occurrence of transient urethral blockage, attributed to coagulation and irritation by accessory gland secretions within the duct (B.L. Lasley, 1983, personal communication; Y. Rumpler, 1983, personal communication).

As a result of the fundamental difference in effector pathway between rectal and penile stimulation, there are significant differences in the volume of semen collected, with penile stimulation yielding, when effective, ejaculates of a volume equal to or exceeding the natural value. Rectal probe electroejaculation, on the other hand, provides specimens of a variable volume, always reported as lower than that for natural ejaculates. There is still no clear evidence to modify the statement made by Hendrickx et al. (1978) that penile stimulation is less predictably effective than RPE but, when effective, yields superior results. It appears that semen parameters other than volume show less variability between collection methods with significant parameters of count/ml, percent live, and percent motile sperm being similar (Gould, 1982; Table 33.2).

Anatomical differences between species place the accessory glands in physically different locations relative to the rectal stimulating electrode, resulting in variation in the distribution of the stimulating field as a result of attenuation by various tissues (Gould et al., 1978). Such anatomical differences may alter the relative contribution of each accessory gland to the recovered semen. In addition, provision of the optimum size of probe and electrode is significant. It is interesting, for example, that reported data for rectal probe electroejaculation of the baboon shows consistently higher semen volume, relative to the natural volume, than for any other species; the chimpanzee is second in this regard (Hendrickx et al., 1978) (Table 33.3). In both these species, the seminal vesicles lie most closely apposed to the stimulating electrodes, and both species are of sufficient size to permit use of relatively large electrodes on the probes.

TABLE 33.2. Semen parameters: comparison of collection method.

Species	Method[1]	Vol. mean	Count mean	% live mean	% mot. mean	N	Reference
Chimpanzee	AV	2.40	620.00	70.35	NA	15	Gould (1982)
Chimpanzee	MAST.	3.78	540.00	65.00	NA	27	Gould (1982)
Chimpanzee	RPE	1.02	280.00	84.47	NA	9	Gould (1982)
Chimpanzee	RPE	1.90	743.00	52.70	NA	11	Bader (1983)
Crab-eating monkey	PE	0.83	347.00	NA	NA	40	From Hendrickx et al. (1978)
Crab-eating monkey	RPE	1.20	458.00	NA	57.00	17	From Hendrickx et al. (1978)
Gorilla	MAST.	NA	22.00	31.00	NA	1	Rafael (1985, personal communication)
Gorilla (fertile)	RPE	0.62	121.00	62.00	51.00	12	Gould and Martin (unpublished observations)
Gorilla (fertile)	RPE	0.38	41.00	54.00	32.00	9	Gould and Kling (1982)
Rhesus monkey	PE	1.11	608.00	76.00	68.00	10	Mann and Gould (unpublished observations)
Rhesus monkey	PE	2.90	523.00	94.00	NA	50	Lu et al. (1985)
Rhesus monkey	PE	3.36	420.00	NA	76.00	128	From Hendrickx et al. (1978)
Rhesus monkey	RPE	1.10	875.00	NA	NA	33	From Hendrickx et al. (1978)
Rhesus monkey	RPE	0.75	267.00	92.00	NA	50	Lu et al. (1985)

[1] Abbreviations: AV, artificial vagina; MAST., masturbation; RPE, rectal probe electroejaculation; PE, penile electrostimulation.

It has been shown, however, that semen recovered by all three methods can be used successfully to induce pregnancy, and the relative advantage of one method over another may resolve itself to that of provision of greater total numbers of sperm for subsequent multiple insemination.

After collection of the ejaculate, various steps should be taken to evaluate and prepare the sperm either for insemination or preservation. As pointed out by Hendrickx et al. (1978) and Harrison (1975), the most important semen parameters are probably count/ml and percent progressive motility. Harrison (1975; Table 33.4) has suggested the use of a Predicted Reproductive Value Score based on a weighted evaluation of count/ml and progressive motility, using the formula:

$$\text{PRV} = \frac{2\ (\text{motility rank value}) + \text{concentration rank value}}{3}$$

This provides a method of expressing the fact that, for example, a semen sample with a count of 3×10^7/ml and 60% progressive motility is equal to or better than a sample of 6×10^7/ml with only 30% progressive motility. It remains to be demonstrated, however, that the former sample will be superior for freezing. Holt et al. (1984) have provided objective evidence of a correlation between progressive sperm velocity and potential fertilizing capacity. These data both support the contention of Harrison (1975) and provide an indicator of a valuable area for further research.

Various methods have been used to count sperm, the most frequently used being counting of a diluted sample using a hemocytometer. In recent years, a variant of the hemocytometer counting chamber has been marketed (Sefi Medical Instruments), following the design described by Makler (1980). This chamber, which uses a counting space only 10 μm deep, as opposed to 100 μm in the Neubauer chamber, permits direct counting of most semen samples and is also of value in measurement of motility and morphology. It is a most useful advance. The precise method of preparing sperm for counting varies with individual preference. Some workers count motile sperm, while others take appropriate steps to immobilize or kill sperm before counting.

The most routine method of identifying "live" or membrane-intact sperm has been measurement of the permeability of the plasma membrane to eosin. Live or membrane-intact sperm exclude the dye, while sperm with compromised membranes permit dye entry, staining red. The most frequently used method, of Swanson and Bearden (1951), has been modified by Martin and Davidson (1976) for primate sperm in an attempt to reduce the precipitation seen when nigrosin is used as the background stain. More recently the use of eosin Y (1% in sodium citrate 3%) has regained popularity and can provide readily interpretable slides.

Semen Preparation for AI

It is difficult to compare accurately the success rate of AI in those species in which it is not frequently reported with natural pregnancy rates because of variation in timing of insemination.

Although initial reports of AI in primates utilized whole semen, usually placed into the vagina (Mastroianni and Rosseau, 1965; Bennett, 1967; Dede and Plentl,

TABLE 33.3. Semen parameters.

Species	Method	Vol. min.	Vol. max.	Vol. mean	Count min.	Count max.	Count mean	% live mean	% mot. mean	$N =$	References
Ateles fusiceps	RPE			0.70			17.20				From Kraemer and Kuehl (1980)
Cebus apella	RPE	0.20	1.40	0.58	38.50	612.00	255.00	58.60	62.06	126	Bush et al. (1975)
Cebus apella	RPE	0.30	1.00	0.60	56.00	740.00	161.00		24.00	15	From Hendrickx et al. (1978)
Cebus apella	RPE	1.20	2.20	1.91	38.00	400.00	207.66			6	Nagle and Denari (1982)
Cercocebus galeritus	RPE	1.10	1.50	1.30	542.00	609.00	576.00		60.00	2	From Hendrickx et al. (1978)
Cercopithecus aethiops	RPE	0.30	2.00	0.90	166.00	811.00	440.00		39.00	23	Roussel and Austin (1968)
Erythrocebus patas	RPE	0.40	1.00	0.60	250.00	3600.00	1153.00		45.00	21	From Hendrickx et al. (1978)
Gorilla gorilla	RPE	0.10	1.85		0.00	1300.00				24	Seager et al. (1982)
Gorilla gorilla	RPE	0.11	0.50	0.31	2.50	28.00	15.20		11.00	2	Platz et al. (1980)
Gorilla gorilla (fertile)	RPE	0.20	0.70	0.38	7.50	160.00	41.00	54.00	32.00	9	Gould and Kling (1982)
Gorilla gorilla (fertile)	RPE	0.20	2.10	0.62	0.00	728.00	121.00	62.00	51.00	12	Gould and Martin (unpublished observations)
Hylobates lar	RPE	0.50	4.00	1.30	51.00	350.00	152.00		9.00	13	Roussel and Austin (1968)
Macaca irus	PE	0.05	6.00	0.83	110.00	1120.00	347.00			40	From Hendrickx et al. (1978)
Macaca irus	RPE	0.60	3.00	1.20	161.00	830.00	458.00		57.00	17	Roussel and Austin (1968)
Macaca mulatta	PE			2.90			523.00	94.00		50	Lu et al. (1985)
Macaca mulatta	PE	0.50	1.90	1.11	130.00	1370.00	608.00	76.00	68.00	10	Mann and Gould (unpublished observations)

Macaca mulatta	PE	0.20	5.00	3.36	93.00	1500.00	420.00		76.00	128	From Hendrickx et al. (1978)
Macaca mulatta	RPE	0.20	4.50	1.10	100.00	3600.00	1069.00		58.00	33	Roussel and Austin (1968)
Macaca mulatta	RPE			0.75			267.00	92.40		50	Lu et al. (1985)
Macaca radiata	PE	1.90	2.30		1105.00	1251.00			74.00	16	From Hendrickx et al. (1978)
Macaca speciosa	RPE	0.40	4.00	1.60	214.00	1268.00	468.00		49.00	22	From Hendrickx et al. (1978)
Pan paniscus	RPE	1.40	2.20	1.80	615.00	1027.00	821.00	75.00	68.00	2	Gould and Martin (unpublished observations)
Pan troglodytes	AV	0.20	5.60	2.40	9.50	910.00	620.00	70.35		15	Gould (1982)
Pan troglodytes	MAST.	1.00	8.00	3.78	11.60	2000.00	540.00	65.00		27	Gould (1982)
Pan troglodytes	RPE	0.10	3.00	1.02	8.00	1110.00	280.00	84.47		9	Gould (1982)
Pan troglodytes	RPE	0.30	2.70	1.90			743.00	52.70		11	Bader (1983)
Papio cynocephalus	RPE			3.60			71.20			42	Kraemer and Vera Cruz (1969)
Papio cynocephalus	RPE						16.35	88.00	61.70	14	Lister (1975)
Pongo pygnaeus	RPE	0.20	3.20	1.10	10.00	128.00	61.00	59.00	47.00	5	Warner et al. (1974)
Saguinus nigricollis	RPE			0.03			833.30				From Kraemer and Kuehl (1980)
Saimiri sciureus	RPE	0.21	0.46	0.38	0.20	181.00	97.00				Chen et al. (1981)
Saimiri sciureus	RPE	0.04	0.29	0.16	93.20	847.00	426.00		65.30	24	Denis et al. (1976)
Theropithecus gelada	RPE	0.50	2.00	1.00	351.00	651.00	503.00		21.00	5	From Hendrickx et al. (1978)
Tupia glis	RPE	0.10	0.10	0.10	90.00	117.00	103.00		52.00	4	From Hendrickx et al. (1978)

[1] Abbreviations: See Table 33.2.

TABLE 33.4. Rank values for semen parameters.

Rank Value	0	1	2	3	4	5	6	7	8	9	10
Motility - 1%	0	10	20	30	40	50	60[1]		70[1]		80+
Concentration	0	50	100	150	200	250	300	350	500	750	1000

[1] Samples with 60% motility are given a rank value of 6.5, those with 70% motility receive 8.5. Samples with motility values and/or concentrations between values listed in the table should receive the lower rank value.
Source: From Harrison (1975). Reprinted with permission.

1966; Hardin et al., 1975), subsequent reports suggest an improvement in success rate when intrauterine or intraperitoneal insemination is used (Valerio et al., 1971; Czaja et al., 1975; Hendrickx et al., 1978; Gould et al., 1985; Van Pelt, 1970; Kraemer, 1983, personal communication) (Table 33.1), with conception rates as high as 52% (Lu et al., 1985) or 64% (Czaja et al., 1975) resulting from intrauterine insemination.

The conception rates for natural mating reported by Dede and Plentl (1966), Valerio et al. (1971), and Settlage et al. (1973), and against which their AI results were compared, seem low relative to current rates reported by other workers. However, as indicated by Settlage et al. (1973), the females used in the trials were not selected on the basis of past breeding history, although late review of their breeding history suggested a positive correlation of parity with pregnancy rate in the study. Similarly, the success rate for intraperitoneal insemination and intrauterine insemination reported by Van Pelt (1970), Czaja et al. (1975), and Lu et al. (1985), while much higher than the natural rates previously reported, approach those expected in a well-managed colony. It is perhaps unrealistic to expect that rates of AI success would exceed those for natural mating, unless the females involved exhibited some abnormality in sperm transport, in which case suitable placement of the sperm could result in an increased fertility rate.

Seminal plasma is not a simple medium solely providing a vehicle for suspension of sperm. Although not yet fully analyzed, we know it contains many compounds in balance which, singly, act to promote or inhibit fertilizing capacity of sperm. It provides nutrients for sperm metabolism and, because it contains part of the epididymal secretion, possibly contains compounds that promote sperm motility. It also contains components that inhibit capacitation and potentially influence the stability of the membrane system of the sperm. Especially in species in which semen deposition is normally intravaginal, there will be modification of the local environment of sperm as they traverse the female tract to the site of fertilization, with the bulk of the seminal plasma remaining extrauterine.

It is logical, therefore, to proceed on the assumption that the sperm carrier fluid should be modified suitably prior to use of sperm suspensions for intrauterine insemination. At this time, it seems appropriate to prepare sperm for intrauterine insemination using a washing technique designed to (1) maximize the percentage of progressively motile sperm, based on the presumption that immotile sperm will not enter the uterus and/or contribute to the fertilizing capacity of motile sperm, (2) remove, by dilution, the antifertility compounds present in the seminal plasma, and (3) maintain such sperm in a nonhostile medium that at least provides an adequate energy source. Any further specific requirements remain to be identified. It is apparent that this approach is the same as that taken for preparation of sperm for in vitro fertilization; any method used in that procedure will probably aid in successful AI using the intrauterine route. This approach is supported by analysis of $Na^{125}I$-labeled surface components that show the surface composition of sperm to be unaltered during incubation (transit) in cervical mucus but to be rapidly modified upon exposure in vivo to uterine fluids (Young and Gould, unpublished observations).

We routinely prepare sperm by the following method: After collection, semen is allowed to liquify at 37°C for 30 minutes. A convenient method is to place the entire coagulum in a 6-ml syringe and to place the nozzle of the syringe into a 15-ml conical tube. The plunger remains in the syringe. After liquefaction occurs, the plunger is removed from the syringe, thus permitting the liquefied portion of the seminal coagulum to flow into the centrifuge tube, where its volume can be measured. Replacement of the plunger and compression of the remaining coagulum permit extrusion of additional fluid and accurate identification of the volume of coagulum. If required, sperm can be flushed from the remaining coagulum by passage of an appropriate volume, usually 1–2 ml, of a defined medium. This medium can be one of several available, including Dulbecco's salt solution, BWW, or Hams F10 with or without serum supplementation. When it is impractical to obtain recipient serum, fetal calf or human cord serum is substituted. Serum is heated to 56°C for 30–60 minutes prior to use. The liquefied portion of the semen is centrifuged gently at 150 g for approximately 5 minutes at room temperature, the supernatant is decanted, and the pellet is resuspended in approximately 1 ml of defined medium. After resuspension, the specimen is again centrifuged at 150 g for 5 minutes, and the pellet is resuspended in approximately 0.2 ml of medium. A further 1 ml of medium is cautiously overlaid on this, and the sperm are allowed to swim into the overlaying medium for a period of approximately 30 minutes. An aliquot of 0.25 to 0.5 ml of the supernatant, referred to as "washed sperm," is used for intrauterine insemination. Both after recovery and prior to insemination, the quality of semen is evaluated, using light microscopy and established methods, with regard to number of sperm, percentage of motility, and percentage of live sperm.

Ovulation Timing

In many of those primate species that, because of their endangered status, are prime candidates for AI anesthesia is needed for conduct of the procedure. This is especially true when intrauterine insemination is used. The combination of anesthesia risk and the endangered status, together with good husbandry practice, dictates that AI be performed as infrequently as possible in a given cycle. This, in turn, increases the significance of accurate timing of ovulation. Again, there is a parallel to be observed in the requirements of an AI program for nonhuman primate species and an in vitro fertilization program demanding recovery of mature oocytes. There are a variety of methods available for monitoring the natural primate cycle (Table 33.5). Of these, calendar timing is of limited use, as shown, for example, by the discrepancy in pregnancy rates obtained by Valerio et al. (1970) and Czaja et al. (1975; Table 33.1) in which studies the technique of insemination, etc., are comparable with the exception of the timing of ovulation date. Many, but not all, nonhuman primate species exhibit external markers that provide a guide to ovulation time. Perineal tumescence, with or without change

TABLE 33.5. Methods for timing ovulation.

Menstrual interval (calendar)
Perineal swelling/coloration
Vaginal cytology
Basal body temperature
Cervical mucus changes
Palpation of reproductive tract
Urinary/serum steroid hormones
Urinary/serum protein hormones

in color, is observed in the baboon, pig-tail macaque, mangabey, gibbon, and chimpanzee (Hendrickx, 1965; Bullock et al., 1972; Mann et al., 1980; Dahl and Collins, in preparation; Young and Orbison, 1944). In the gorilla it is possible to detect a cyclical change in labial swelling resulting in alteration of the length of the vulval aperture (Nadler et al., 1977). In other macaques, an altered coloration, without perineal turgescence, occurs in the perineum and other areas of the body; and these changes can be used to estimate ovulation time (Czaja et al., 1975).

Vaginal cytology has been evaluated in several species, but the consensus is that interpretation of vaginal cell smears requires a high degree of skill; and, while it is possible to identify cyclic changes, the test is of little prognostic value (De Allende et al., 1945; Gillman, 1937; Mahoney, 1970). Basal body temperature changes have been demonstrated around the time of ovulation in *Macaca mulatta* (Balin and Wan, 1969) and *Pan troglodytes* (Graham et al., 1978); but the technique requires placement of indwelling temperature telemeters and is not routinely practical or especially accurate. However, in those species in which correlation of physical markers with ovulation has been directly examined using corroborative laparoscopy, the correlation of the two events is not sufficiently precise to provide a satisfactory indicator that a single insemination will initiate pregnancy.

Other techniques for prediction of ovulation, including alteration in the NaCl content of cervical mucus and palpation of the reproductive tract, have been used in macaques (e.g., Wilson et al., 1970; Mahoney, 1970; Hartman, 1933) and reviewed by Mahoney (1976).

The changing pattern of endocrine hormones that characterizes the primate menstrual cycle has been documented for a large number of species. The two observations most pertinent to ovulation detection are the rise and peak in circulating estrogen associated with follicle development and the rapid surge of LH, which initiates the ovulation process. Detection of either of these events is of predictive value in timing of artificial insemination (Hotchkiss et al., 1971). It has been shown that measurement of estrogen in urine, when indexed to creatinine, can provide useful information on the potential fertility of a given female (Lasley et al., 1982).

Until recently, a major disadvantage of hormone assays for prediction of ovulation was the time taken to conduct the assay. It is now possible, however, to perform assays for estrogens and LH in less than 3 hours, and such assays will be increasingly valuable. A continuing problem, however, is the need to collect suitable specimens for assay; and, depending on the species to be used, urine samples may be the only practical material for assay. The most attractive assay for timing natural ovulation would appear to be a rapid urinalysis for LH. Such a test was available for the great apes (Gould and Faulkner, 1981), which used a modified method of interpreting the nonhuman primate pregnancy test developed by Hodgen et al. (1976) and supplied by the National Institutes of Health.

Alterations in that test, however, have reduced its reliability and application in this area (Bader, 1983; Bambra et al., 1984). Fortunately, there are other newer tests that appear to work adequately for the apes. We have recently tested a solid-phase enzyme-linked assay for LH (Ovustick urine LH kit, Monoclonal Antibodies, Inc., Mountain View, CA) in the chimpanzee, rhesus monkey, and squirrel monkey. The test, as provided, indicates the periovulatory rise in chimpanzee urinary LH, with ovulation occurring approximately 24 hours after the first positive test result. It does not appear to work for the monkeys tested, nor for the baboon (R. M. Eley, 1985, personal communication).

The problems and deficiencies mentioned above, which reduce the effectiveness with which the time of natural ovulation can be predicted, demonstrate the need for another method of obtaining a timed ovulation. It is logical, then, to review the practicability of ovulation induction or synchronization. There have been reports since 1962 of the use of monkey pituitary extract (Simpson and van Wagenen, 1962) or human menopausal gonadotropins for the induction of follicle development in nonhuman primates (Dede and Plentl, 1966; Bennett, 1967). The reported results are variable, with Simpson and van Wagenen (1962) reporting the occurrence of multiple ovulations in *M. mulatta* from a single ovary and Dede and Plentl (1966) reporting no difference in presumed ovulation and pregnancy rate when HMG was used. Such differences are, at least in large part, a result of different regimens of administration, coupled with variation in the potency of the products being used. More recently, studies have been reported that investigated the potential use of purified human FSH (urinary) for stimulation of follicle development and ovulation in the cynomolgus monkey and in the squirrel monkey (Schenken et al., 1984; Yano and Gould, 1985). This product (Urofollitropin/Metrodin, Serono Laboratory, Inc., Randolph, MA) has also been shown to be effective in the chimpanzee (Gould, unpublished observations) although the latter species appears to be more refractory than is the woman. It is also practical to induce follicle development in the nonhuman primate using Clomiphene citrate (Gould et al., 1985).

It is important to remember, however, that for successful artificial insemination, the most important factor is the timing of ovulation and not, at least yet, the induction of multiple pregnancy. This consideration dictates an altered strategy in the use of hormones that stimulate follicle development and prompt ovulation. In most cases in which artificial insemination is contemplated, the female is potentially fertile, as demonstrated by the presence of ovarian cyclicity and the absence of clinically evident reasons for infertility. In such individuals, artificial synchronization of ovulation requires that two criteria be fulfilled: There must be a mature follicle capable of responding to a stimulus for ovulation, and a suitable stimulus for ovulation must be provided. The latter criterion is easily met; injection of HCG in a dose appropriate to the species has been shown repeatedly to promote ovulation. The former criterion is best met by providing a modest stimulus to follicle development. Schenken et al. (1984) showed that administration of supraphysiological doses of hFSH to cynomolgus monkeys late in the follicular phase prompted development of supplementary follicles but did not prompt ovulation of more than one. We have used Clomiphene citrate during the late follicular phase in the chimpanzee and gorilla, not to prompt development of supplementary follicles, but to ensure an adequate gonadotropic stimulus for development of the already selected primary follicle (Hodgen, 1982; Gould, 1983). Injection of HCG is used to trigger ovulation at a predetermined time, which is calculated as well as possible to precede the predicted time of ovulation by 24 hours. In species in which it is possible to monitor for the onset of the spontaneous LH surge, it is advisable to conduct such a test to ensure that a spontaneous LH surge does not precede the exogenous stimulus.

If there is evidence suggesting that the manipulations and possible anesthetic episode associated with artificial insemination may delay ovulation (e.g., Lasley et al., 1982), it is advisable to delay insemination until after the predicted time of ovulation following HCG administration.

Site of Insemination

Four sites of insemination have been described in the literature: vaginal, intracervical, intrauterine, and intraperitoneal. Of these, the vaginal site is the physiologically normal one in nonhuman primates. It is the simplest route to use for insemination and is the only route that is practical in unanesthetized females. This site for deposition has the disadvantages that sperm must still traverse the entire length of the reproductive tract; it requires the highest number of sperm for success, a factor of some importance when stored semen samples may be at a premium; and has the lowest reported success rate in the literature (Czaja et al., 1975; Gould et al., 1985; Kraemer, 1985, personal communication).

The success rate of artificial insemination in macaques is improved when intracervical or intrauterine insemination is utilized. The major reason for use of the intracervical site, as opposed to the intrauterine, appears to be related to the anatomy of the cervix in several nonhuman primates, which makes it rather difficult to pass a catheter into the uterus. In those species in which a vaginal plug of coagulum is usually deposited, intrauterine insemination may be required for successful artificial insemination. At the present time, for example, intravaginal insemination has been unsuccessful in the lemur, possibly because it has not been practical to reproduce the effect of the naturally placed plug (Y. Rumpler, 1985, personal communication).

Intrauterine insemination of washed sperm appears to provide the highest success rate for artificial insemination in several species (Gould et al., 1985; M.Y. Lu, 1985, personal communication; Hendrickx et al., 1978) and has resulted in a 64% success rate in the macaque (Czaja et al., 1975) and a 21% success rate when used in the chimpanzee (Gould et al., 1985).

At this time, the only documentation of the successful use of the intraperitoneal route of insemination in nonhuman primates is that of Van Pelt (1970), which showed a 30% success rate in a small test series.

Provided the sperm sample has been adequately washed prior to insemination, the technique for intrauterine insemination is simple, and sperm can be deposited in utero via any suitable device, most usually a feeding tube or cannula of suitable size. The technique requires that the subject be anesthetized, and this may adversely influence the natural motility and activity of the pelvic organs. It was this consideration that prompted Gould et al. (1985) to use rectal probe electrostimulation of the female at the time of insemination. When electrostimulation was used in the chimpanzee, pregnancy resulted in 6 of 29 attempts. In a subsequent series without electrostimulation, no pregnancies resulted from 15 attempts (Gould, 1985, unpublished observations). Although the numbers are small, there appears to be an indication for the use of some stimulation method to ensure uterine activity at the time of insemination, presumably having an effect on sperm transport to the site of fertilization.

Future Development

The feasibility of using artificial insemination as a tool in breeding endangered primates has been clearly demonstrated. AI has been successful in a number of primate species (Table 33.1). The success rate of the procedure remains low; and yet, under optimum conditions, AI should achieve, or exceed, the success rate of timed breeding. The problems that have been identified above—use of anesthetics, timing of ovulation, and site of delivery—remain as areas that each require further work. While it is unlikely that the need for anesthesia will be eliminated, more information is required on the direct and indirect effects of anesthetics on AI. A suggested protocol for artificial insemination, which could be adapted readily to use in other nonhuman primate species, has been described by Gould et al. (1985; Fig. 33.3).

Lasley et al. (1983) suggest that use of ketamine HCl as an immobilizing agent may be associated with disruption of the normal follicular phase, which makes it more important that insemination be timed after ovulation has occurred. Other workers, using monkeys, have recorded no disruption of the menstrual cycle by repeated anesthetic (Channing et al., 1977; Koyama et al., 1977). Because the male gamete has a longer fertile life than the female, this requirement reduces the period of time in which fertilization can occur and, thus, makes timing of ovulation more critical.

	Male
Semen collection:	Rectal probe electroejaculation (RPE) or artificial vagina or automasturbation
Semen liquefaction:	30 minutes at 37°C
Semen preparation:	Centrifugation; wash with Ham's F10 + 15% serum; cleaning of sperm by swimming into overlaid medium
	Female
Ovulation synchronization:	Clomiphene citrate, 50 mg p.o. daily day 2-day 6 max swelling, HCG, 5000 IU; day 7, AI 36 hours after HCG injection
Insemination:	Intrauterine via #7 FR pediatric feeding tube, 5×10^6–5×10^8 sperm in 0.5 ml medium; coagulum placed on cervix; electrostimulation, 15 stimuli of 65 mA, 20 HZ, 3 seconds, via RPE stimulator and probe

FIGURE 33.3. Abbreviated protocol for artificial insemination in the chimpanzee.

Further development of methods for determination of the natural time of ovulation should have a high priority in future work. At this time, the most promising area for such an approach would seem to focus on rapid measurement of urinary LH, which is the most reliable single indicator available. A second area for research on this subject is the more thorough evaluation of electrolyte/conductivity changes in vaginal secretions. It is possible that measurement of such changes will be of predictive value in some species. A vital part of successful fertilization is the pickup by the oviduct fimbriae of the ovum as it leaves the follicle. There is scant information on the effect of anesthetic and hormone manipulations on this process, although such information as we have, derived from observation of the process in anesthetized laboratory animals, suggests a minimum disruption of the ovum pickup and transport. Also little studied is the effect of anesthesia on the motility of other pelvic organs, in particular the uterus and oviducts. Although uterine contraction has been reported as playing a role in sperm transport in the human and it has been suggested that electrostimulation enhances the fertilization rate by increasing sperm transport (Gould et al., 1985), such suggestions remain to be documented and proven. This, then, is also an area that requires more attention.

A recurrent problem in AI of endangered species is the fact that the selection of semen donors may be severely limited. This means that there is no selection process available for the "best" semen, which in turn suggests that methods for preparing semen must be optimized and media for sperm maintenance further perfected. In some species, including the gorilla and owl monkey (Seager et al., 1982; Dixon, 1983), semen quality from males of proven fertility can be very poor as measured by conventional means. Further investigation of the mechanism and hormonal control of spermatogenesis appears warranted, with emphasis on methods for reducing or eliminating seasonality in spermatogenesis as well as increasing the quality of semen from those species with an unusual pattern of spermatogenesis (e.g., *Aotus*).

As the techniques for successful, routine AI are further developed, their potential for use as effective tools in species management will be enhanced, and the concept of the frozen ark becomes closer to reality, providing this and future generations with a resource of inestimable value.

Acknowledgments. The research reported in this paper was supported by NIH grant RR-00992 and by NIH grant RR-00165 to the Yerkes Regional Primate Research Center, which is fully accredited by the American Association for Accreditation of Laboratory Animal Care. The assistance of Dr. A. Raphael (Dallas Zoo), Dr. Lu Ming Yi (Institute for Medical Biology, Kunming, PRC), Dr. Duane Kraemer (Texas A&M), and Dr. T. J. Kuehl (Southwest Foundation) in permitting the quotation of unpublished data is gratefully acknowledged.

References

Bader H (1983) Electroejaculation in chimpanzees and gorillas and artificial insemination in chimpanzees. Zoo Biol 2:307–314.

Balin H, Wan LS (1969) Basal body temperature as an index of ovulation in the rhesus monkey. Abst Soc Stud Reprod 2:7.

Bambra CS, Eley RM, Wall H (1984) Limitations of the nonhuman pregnancy kit for pregnancy diagnosis in baboons. J Med Primatol 13:219–227.

Bennett JP (1967) Artificial insemination of the squirrel monkey. J Endocrinol 37:473–474.

Bullock DW, Paris CA, Goy RW (1972) Sexual behaviour swelling of the sex skin and plasma progesterone in the pigtail macaque. J Reprod Fertil 31:225–236.

Bush DE, Russell LH Jr, Flowers AI, Sorensen AM Jr (1975) Semen evaluation in capuchin monkeys (*Cebus apella*). Lab Anim Sci 25:588–593.

Channing CP, Fowler S, Engel B, Vitek K (1977) Failure of daily injections of ketamine HCl to adversely alter menstrual cycle length, blood estrogen and progesterone levels in the rhesus monkey. Proc Soc Exp Biol Med 155:615–619.

Chen JJ, Smith ER, Gray GD, Davidson JM (1981) Seasonal changes in plasma testosterone and ejaculatory capacity in squirrel monkeys (*Saimiri sciureus*). Primates 22:253–260.

Cho F, Honjo S, Makita T (1975) Fertility of frozen-preserved spermatozoa of cynomolgus monkeys. Contemporary Primatology, Proceedings of the 5th International Congress of Primatology. S Karger, Basel, pp 125–133.

Czaja JA, Eisele SG, Goy RW (1975) Cyclical changes in the sexual skin of female rhesus: relationships to mating behaviour and successful artificial insemination. Fed Proc 34:1680–1684.

De Allende ILC, Shorr E, Hartmann CG (1945) A comparative study of the vaginal smear cycle of the rhesus monkey and human. Contr Embryol 31:1–26.

Dede JA, Plentl AA (1966) Induced ovulation and artificial insemination in a rhesus colony. Fertil Steril 17:757–764.

Denis LT, Poindexter AN, Ritter MAK, Seager SWJ, Deter RL (1976) Freeze preservation of squirrel monkey sperm for use in timed fertilization studies. Fertil Steril 27:723–729.

Dixon AF (1983) The owl monkey (*Aotus trivirgatus*). In: Hearn J (ed) Reproduction in New World primates: new models in medical science. MTP, Lancaster, UK, pp 69–113.

Gillman J (1937) The cyclical changes in the vaginal smear in the baboon and its relationship to the perineal swelling. S Afr J Med Sci 2:44–56.

Gould KG (1982) Ovulation detection and artificial insemination. Amer J Primatol Suppl 1:15–25.

Gould KG (1983) Ovum recovery and in vitro fertilization in the chimpanzee. Fertil Steril 40:378–383.

Gould KG, Faulkner JR (1981) Development, validation and application of a rapid method for detection of ovulation in great apes and women. Fertil Steril 35:676–682.

Gould KG, Kling OR (1982) Fertility in the male gorilla (gorilla gorilla): relationship to semen parameters and serum hormones. Amer J Primatol 2:311–316.

Gould KG, Martin DE, Warner H (1985) Improved method for artificial insemination in the great apes. Amer J Primatol 8:61–67.

Gould KG, Warner H, Martin DE (1978) Rectal probe electroejaculation of primates. J Med Primatol 7:213–222.

Graham EF, Schmehl MKL, Evensen BK, Nelson DS (1978) Semen preservation in non-domestic mammals. Symp Zool Soc London 43:153–173.

Gunn RMC (1936) Fertility in sheep. Artificial production of seminal ejaculation and the characters of the spermatozoa contained therein. Bull CSIR Aust 94:1–116.

Hardin CJ, Liebherr G, Fairchild O (1975) Artificial insemination in chimpanzees. Intern Zoo Yrbk 15:132–134.

Harrison RM (1975) Normal sperm parameters in *Macaca mulatta*. Lab Prim Newslet 14(1):10–13.

Hartman CG (1933) Pelvic (rectal) palpation of the female monkey with special reference to the ascertaining of ovulation time. Amer J Obst Gyn 26:600–608.

Hendrickx AG (1965) The menstrual cycle of the baboon as assessed by the vaginal smear, vaginal biopsy and perineal swelling. In: Austin H (ed) The baboon in medical research, vol II. University of Texas Press, Austin/London.

Hendrickx AG, Thompson RS, Hess DL, Prahalada S (1978) Artificial insemination and a note on pregnancy detection in the nonhuman primate. Symp Zool Soc London 43:219–240.

Hodgen GD (1982) The dominant ovarian follicle. Fertil Steril 38:281–287.

Hodgen GD, Niemann WH, Turner CK, Chen H-C (1976) Diagnosis of pregnancy in chimpanzees using the nonhuman primate pregnancy test kit. J Med Primatol 5:247–251.

Holt WV, Moore HDM, Parry A, Hillier S (1984) A computerized technique for the measurement of sperm cell velocity: correlation of the results with in vitro fertilization assays. In: Harrison RF, Bonner J, Thompson W (eds) The male factor in human infertility. MTP, Lancaster, UK, pp 3–7.

Hotchkiss J, Atkinson LW, Knobil E (1971) The time course of serum estrogen and luteinizing hormone concentrations during the menstrual cycle of the rhesus monkey. Endocrinology 89:177–183.

Jones RC (1971) Uses of artificial insemination. Nature 229:534–537.

Kraemer DC, Kuehl TJ (1980) Semen collection and evaluation of breeding soundness in nonhuman primates. in: Morrow DA (ed) Current therapy in theriogenology. W.B. Saunders, Philadelphia, pp 1134–1137.

Kraemer DC, Vera Cruz NC (1969) Collection, gross characteristics and freezing of baboon semen. J Reprod Fertil 20:345–348.

Kraemer DC, Vera Cruz NC (1972) Breeding baboons for laboratory use. In: Beverage WIB (ed) Breeding primates. Karger, Basel, pp 42–47.

Koyama T, de la Peria A, Magino N (1977) Plasma estrogen, progestin and luteinizing hormone during the normal menstrual cycle in the baboon: role of luteinizing hormone. Amer J Obst Gyn 127:67–72.

Laplaud M, Cassou R (1948) Recherches sur l'electro-ejaculation chez le taureau et le verrat. CR Soc Biol 142:726–727.

Lasley BL, Czekala NM, Presley S (1982) A practical approach to evaluation of fertility in the female gorilla. Amer J Primatol Suppl 1:45–50.

Lister RE (1975) Evaluation of fertility in male baboons (*Papio cynocephalus*). Breeding simians for developmental biology. Lab Anim Handbk 6:183–189.

Lu M-Y, Su Y, Wang L, Pan H, Pan B, Mu WW (1985) Artificial insemination of rhesus monkey and laparotomy and induced labor of pregnant rhesus monkey. Medical Biological Research. (In Chinese; reprint available from K Gould.)

Mahone JP, Dukelow WR (1978) Semen preservation in *Macaca fascicularis*. Lab Anim Sci 28:556–561.

Mahoney CJ (1970) A study of the menstrual cycle in *Macaca irus* with special reference to the detection of ovulation. J Reprod Fertil 21:153–163.

Mahoney CJ (1976) Breeding. In: Section 31: Primates of UFAW Handbook in the Care and Management of Lab Animals, 5th ed, pp 401–418.

Makler A (1980) The improved ten micrometer chamber for rapid sperm count and motility evaluation. Fertil Steril 33:337–338.

Mann DR, Gould KG, Collins DC, Blank MS (1980) The sooty mangabey: an alternate model for studies in primate reproduction. Abstract #270, the Endocrine Society 62nd Annual Meeting, June, 1980, Program and Abstracts, p 142.

Martin DE (1981) Breeding great apes in captivity. In : Graham CE (ed) Reproductive biology of the great apes. Academic, New York, pp 343–373.

Martin DE, Davidson MW (1976) Differential live-dead stains for bovine and primate spermatozoa. Proc 8th Int Cong Anim Reprod and AI. Krakow, Poland 4:919–922.

Martin MJ (in press) Menstrual cycle monitoring for artificial insemination in a captive lowland gorilla. Fert Soc of Australia (Abstr).

Mastroianni L, Manson WA (1963) Collection of monkey semen by electroejaculation. Proc Soc Exp Biol Med 112:1025–1027.

Mastroianni L, Rosseau CU (1965) Influence of the intrauterine coil on ovum transport and sperm distribution in the monkey. Amer J Obst Gyn 93:416–420.

Matern B (1983) Problems and experiences in performing artificial insemination in bonobos (*Pan paniscus*). Zoo Biol 2:303–306.

Nadler RD, Graham DE, Neill JD, Collins DC (1977) Genital swelling, pituitary gonadotropins and progesterone in the menstrual cycle of the lowland gorilla. In: Abstracts for the Inaugural Meeting of the American Society of Primatologists, April 1977, Seattle, WA, pp 16–19.

Nagle CA, Denari JH (1982) The reproductive biology of capuchin monkeys. Intern Zoo Yrbk 22:143–150.

Perry EJ (ed) (1960) The artificial insemination of farm animals, 3rd rev edition. Rutgers University Press, New Brunswick, NJ.

Platz CC Jr, Wildt DE, Bridges CH, Seager SW, Whitlock BS (1980) Electroejaculation and semen analysis in a male lowland gorilla, *Gorilla gorilla gorilla*. Primates 21(1):130–132.

Roussel JD, Austin CR (1968) Improved electroejaculation of primates. J Inst Anim Tech 19:22–32.

Schenken RS, Williams RF, Hodgen GD (1984) Ovulation induction using "pure" follicle-stimulating hormone in monkeys. Fertil Steril 41:629–634.

Seager SWJ, Wildt DE, Schaffer N, Platz CC (1982) Semen collection and evaluation in *Gorilla gorilla gorilla*. Amer J Primatol Suppl 1:13.

Settlage DSF (1971) Establishment of normal parameters in semen analysis of highly fertile rhesus monkeys. MS Thesis, University of California, Davis, CA.

Settlage DSF, Hendrickx AG (1974) Electroejaculation technique in *Macaca mulatta*. Fertil Steril 25:157–159.

Settlage DSF, Swan S, Hendrickx AG (1973) Comparison of artificial insemination with natural mating technique in rhesus monkeys, *M. mulatta*. J Reprod Fertil 32:129–132.

Simpson ME, van Wagenen G (1962) Induction of ovulation with human urinary gonadotropins in the monkey. Fertil Steril 13:140–152.

Swanson EW, Bearden HJ (1951) An eosin-nigrosin stain for differentiating live and dead bovine spermatozoa. J Anim Sci 10:981–987.

Valerio DA, Ellis EB, Clark ML, Thompson GE (1969) Collection of semen from macaques by electroejaculation. Lab Anim Care 19:250–252.

Valerio DA, Leverage WE, Bensenhaver JC, Thornett HD (1971) The analysis of male fertility, artificial insemination and natural matings in the laboratory breeding of macaques. In: Goldsmith EI, Moor-Jankowski J (eds) Medical primatology 1970. Karger, Basel, pp 515–526.

Valerio DA, Leverage WE, Munster JH (1970) Semen evaluation in macaques. Lab Anim Care 20:734–740.

Van Pelt LF (1970) Intraperitoneal insemination of *Macaca mulatta*. Fertil Steril 21:159–162.

Van Pelt LF, Keyser PE (1970) Observations on semen collection and quality in macaques. Lab Anim Care 20:726–733.

Warner H, Martin DE, Keeling ME (1974) Electroejaculation of the great apes. Ann Biomed Eng 2:419–432.

Watson PF (1978) A review of techniques of semen collection in mammals. Symp Zool Soc London 43:97–126.

Wilson JG, Fradkin R, Hardman A (1970) Breeding and pregnancy in rhesus monkeys used for teratological testing. Teratology 3:59–71.

Yano Y, Gould KG (1985) Induction of follicular growth in the squirrel monkey (*Saimiri sciureus*) with human urinary follicle-stimulating hormone (Metrodin). Fertil Steril 43:799–803.

Young WC, Orbison WD (1944) Changes in selected features of behaviours in pairs of oppositely sexed chimpanzees during the sexual cycle and after ovariectomy. J Comp Psych 37:107–143.

34
Primate Models for Fertilization and Early Embryogenesis

W. Richard Dukelow and P. N. Vengesa

Introduction

The overwhelming preponderance of research work on the rhesus monkey until the last decade has provided a sound basis of use of this animal as a model for humans. In recent years, the banning of exportation of rhesus monkeys from India, coupled with the major increase in the use of in vitro fertilization in a variety of species including the human, has led to the examination of the basic aspects of fertilization and early embryonic development in many primate species. The present review will examine this progress from the time of ovulation and sperm production through the time of implantation.

Oogenesis and Preovulatory Events on the Ovarian Surface

In nearly all animals, oogenesis ceases soon after birth, although an interesting exception to this pattern exists in the loris and galago, two prosiminans. In these species, active oogenesis continues into adult life. In other species that have been studied, the oocytes which will eventually yield the early embryos are all present at the time of birth.

The complex embryological events that precede ovulation are covered in other chapters. Similarly, the process of follicle selection is covered by Dr. Hodgen, and in this chapter we will deal only with the events on the ovarian surface immediately prior to ovulation. These ovarian preovulatory events have recently been summarized (Dukelow, 1983a, 1983b). Laparoscopic studies have delineated these changes, and these events have been most thoroughly described in the cynomolgus macaque (*Macaca fascicularis*) (Jewett and Dukelow, 1971a, 1973; Rawson and Dukelow, 1973; Dukelow et al., 1972). By examination of the very discrete morphological changes that occur one can predict the time of ovulation, and this has been used in timed-mating procedures or to study basic ovulation phenomena. About 2 days prior to ovulation, the exact site of the developing follicles can be identified on the surface of the ovary by a swelling and darkening of part of the ovarian surface and an increase of about 35% in the total ovarian size. About 30 hours prior to ovulation, a stellate pattern of blood vessels occurs on the developing follicles; by 8–10 hours before ovulation, the vasculature is very pronounced and the follicular cone or stigma (the point on the follicle that

that will rupture) can be identified. The increased blood flow is a result of enhanced luteinizing hormone (LH) secretion at this time. About 10 hours prior to ovulation, the oviductal fimbria moves to envelop the developing follicle and prepare for the pickup of the ovum from the ovarian surface. Prior to ovulation, a degree of luteinization can be observed through the laparoscope at the base of the follicle. In the squirrel monkey (Harrison and Dukelow, 1974), similar vascular patterns are observed, but there is more extensive hemorrhaging at the base of the follicle, which makes observation difficult. In the galago (Dukelow et al., 1973), a very pronounced protrusion of the follicle occurs. Follicular vessels can be observed at the base of the follicle and occasionally near the apex. The formation of a clear area (the stigmata) is not as evident in either the squirrel monkey or the galago as in the cynomolgus macaque. Basically similar patterns of periovulatory follicular development have been described in the Japanese macaque (*Macaca fuscata*), the baboon, and the chimpanzee (Nigi, 1977a, 1977b; Wildt et al., 1977; Graham, 1981; Graham et al., 1973). The increased vasculature on the preovulatory follicle has been confirmed by Zeleznik et al. (1981) who also showed an increase in human chorionic gonadotropin binding sites by day 9 of the cycle. The latter were particularly evident in the dominant follicle about to ovulate. This work was with the rhesus monkey. In the cebus monkey (*Cebus apella*), a stellate pattern of small blood vessels was also noted. These became more pronounced 24 hours before ovulation. A network of small vessels was uniformly spread across the follicle wall (Nagle et al., 1980). These workers found that 10 hours before ovulation the follicles showed a conical aspect with hemorrhages occurring in the follicular wall. In the cebus, ovulation time was characterized by a very tight adherence of the fimbria to the follicular surface that was maintained for at least 2 hours. The movement of the fimbria to the dominant follicle in the cynomolgus macaque is similar to that seen in the human. In the rhesus monkey, the fimbria are normally found more closely surrounding one pole of the ovary in a caplike fashion. In the pig-tailed macaque (*M. nemestrina*), the fimbrial position is intermediate between that of the rhesus and cynomolgus macaque.

The diameter of the preovulatory follicle varies according to body size in primates. In humans this is about 10 mm. Corresponding values for other primates are: gorilla, 7 mm; rhesus, 6.3–6.6 mm; baboon, 6 mm; gibbon, 5–6 mm; New World monkeys, 5–6 mm; tarsiers, 1 mm. The amount of granulosa cells is similar in all of these species with the follicular size difference primarily due to the secretion of fluid in the antrum. The size of the ovary is usually proportional to body size except in the platyrrhine species in which the ovaries are disproportionally large. The bonnet macaque (*M. radiata*) shows unusual follicular development that is characterized by invaginations or folding of the follicular wall into the antrum (Barnes et al., 1978).

Ovulation

The biochemical and biophysical changes in the follicle of the nonhuman primate have not been extensively studied relative to ovulation. However, in other species, particularly the pig and the rabbit, extensive studies have been carried out. Obviously in primates there are major changes that occur immediately prior to ovulation, and some of these are mentioned above. It is well established that the follicle wall weakens just prior to rupture. Early workers believed that the increasing size of the antrum would lead to increased pressure within the follicle that ultimately would lead to the rupture itself. Other theories of ovulation have examined the role of contractility of ovarian tissue near the base of the follicle and, particularly, the enzymatic activity in the follicle itself. Others have likened ovulation to an inflammatory response (Espey, 1980), and indeed there appear to be similarities between these two natural phenomena. The process of ovulation was reviewed some years ago by Espey (1978), and it now appears there is no significant increase in the intrafollicular pressure at ovulation. In fact, pressure at times decreases just prior to ovulation. There is a marked increase in the distensibility of the follicular wall just prior to ovulation, and this causes a thinning of the wall and greater ease of rupture. A major component of the follicular wall is the collagen, which gives structural strength to the follicle itself. At ovulation the normal bonding of collagen fibrils is disrupted and the tension reduced with stretching of the tissues. An enzymatic agent (collagenase) has been identified in follicular wall tissue and is activated at the time of ovulation. It is likely that other enzymes, hormones, or biochemical compounds are involved in the ovulation process, but more extensive studies must be carried out. When rupture does occur, the oocyte and its surrounding cumulus cells are expelled and the stigma is plugged by clotted serum as the antrum refills with plasma. Okamura et al. (1980) described the ultrastructure of the apical wall in 16 human follicles at varying stages of development. In the mature follicles, the fibroblastic cytoplasm was well developed, rich in lysosomelike granules, and contained peripheral multivesicular structures. Intercellular collagen fibers were sparse. These workers suggested that the collagen fibers were digested by the content of the liposomal granules and the multivesicular structures. In the cynomolgus macaque, a large protruding stigma appeared as ovulation approached, and the total time required for ovum expulsion was short, 30 seconds (Rawson and Dukelow, 1973). The adherence of the egg mass to the ovarian surface and the strength of this adhesion is of importance in the mechanism of transfer of the oocyte to the oviduct, and one can speculate that the sweeping motion of the cilia (lining the fimbria) over the ovarian surface occurs in primates as in other species. In the cynomolgus macaque, complete rupture of blood vessels on the follicular surface does not occur, and, in fact, some of the vessels can be seen in the newly formed corpus luteum (Jewett and Dukelow, 1971a). In the Japanese macaque, just prior to ovulation the follicular wall becomes very thin at the apex and the cytoplasm of the granulosa cells and theca interna cells is flattened (Nigi, 1978). Ovulation between the left and right ovary appears random despite occasional reports to the contrary. Dukelow (1977) reported on ovulations observed laparoscopically in 138 cycles of the cynomolgus macaque. Ovulation appeared on the left ovary 62.6% of the time, and no significant effect was noted on cycle length relative to the ovulatory status of the previous cycle or the occurrence of ovulatory cycles.

Ovulation Induction

Although timing of reproduction to natural ovulation is preferable, there are times, whether due to the stress of captivity or the seasonality of the animal, when exogenous administration of steroids or gonadotropins is necessary to induce follicular maturation and ovulation. The purpose of exogenous administration of steroids is to mimic the events that occur in the normal cycle. In the normal cycle of most species, an estrogen peak precedes an LH rise, and an increased level of progesterone occurs several hours before the LH peak but after the initiation of the LH surge. In the stump-tailed macaque (*M. arctoides*), a secondary follicle stimulating hormone (FSH) surge occurs 2–3 days after the LH peak (Wilks et al., 1980). The significance of this second FSH surge is not certain, but it may enhance postovulatory progesterone and estrogen synthesis. Normally, even with use of steroids to induce follicular development, gonadotropins are included in the regimen to stimulate ovulation itself. Dr. Gertrude van Wagenen and her associates spent over 25 years developing a method for gonadotropin induction of ovulation of the rhesus (1968). A variety of domestic animal source gonadotropins were utilized with varying degrees of success, and even with the discovery of pregnant mare serum (PMS), which has been used to trigger follicular growth in many species, this failed to provide an answer to ovulation induction in the macaque. The reports of Knobil et al. (1956) of species specificity of growth hormone in nonhuman primates led van Wagenen and her associates to speculate that a variety of FSH sources can be used to stimulate follicular growth but that there is a species-specificity requirement for the LH source to actually cause ovulation. Accordingly, most gonadotropins used in nonhuman primates (and in humans) today derive from primate sources. The most common sources of FSH are human menopausal gonadotropin (HMG) or pituitary compounds from human or nonhuman primate sources. For the induction of ovulation, human chorionic gonadotropin (HCG) is most commonly used. Additionally, clomiphene citrate, a nonsteroidal compound with FSH action, is used in both human and nonhuman primates. Controlling the exact ovulation time with this compound is more difficult. The time of administration of HCG is also important. Williams and Hodgen (1980) administered 1000 IU of HCG during the late follicular phase in rhesus monkeys, either on day 9 or day 11. If the HCG was given coincidentally with the natural gonadotropin surge, no pharmacological effects were induced. However, when administered before the initiation of spontaneous FSH and LH surges, the HCG induced apparent atresia of the dominant follicle, arrest of ovarian function lasted more than 3 weeks, and, in half the monkeys, an altered modulation of tonic FSH secretion independent of LH release where serum FSH levels persisted about threefold above normal. Ovulation induction can occur in the squirrel monkey by a wide variety of regimens and has recently been reviewed (Dukelow, 1983c). The initial work used regimens derived from studies with common laboratory and domestic species and utilized PMS supplemented with HCG to induce ovulation. Although the number of animals is small, the work of Bennett (1967a) decisively showed that multiple ovulation could be induced with PMS-HCG. Unfortunately, animals soon become refractory to PMS, and thus the use of this ovulatory regimen could not be repeatedly used in the same animal. This has important implications relative to the breeding of endangered and captive species. The refractoriness to PMS and the production of antibodies against this protein hormone greatly limit its usefulness in many species.

A repeat of the early work of Bennett and a study of the use of FSH, HMG, and HCG was reported (Dukelow, 1970). This work demonstrated that 1 mg of domestic animal source pituitary FSH given daily for 4 days with a single injection of 250–500 IU of HCG on the fourth day would result in ovulation in about 60% of the animals. In subsequent studies (Dukelow, 1979), the minimum effective dose of HCG following 4 days of FSH was found to be 100–250 IU. This treatment, then, of 4 days FSH and a single injection of 250 IU of HCG, has been used continuously in the authors' laboratory over the past 16 years with virtually no change in its efficacy. Its use has been confirmed in the squirrel monkey by Travis and Holmes (1974), and it has also been used in a variety of other laboratory, domestic, and zoo animals. The regimen can be repeated over several years with no refractoriness to the gonadotropins. Using the regimen described, squirrel monkeys have been ovulated 20 or more times (at intervals of 3 weeks or more) with no significant effect on the percentage of animals ovulating or the number of ovulations per animal.

Seasonal Responses

Seasonal responses to ovulation induction regimens have been found in a number of primate species. The rhesus shows distinctly seasonal ovulation (Riesen et al., 1971). Five of 24 menstrual cycles were ovulatory during the summer and 30 of 31 during the winter. This seasonality is not present in all rhesus monkeys, but it is of such a degree that it must be considered in captive breeding programs or studies of basic reproductive physiology. The squirrel monkey is a seasonal breeder (Harrison and Dukelow, 1973), and this seasonality is reflected in the response to the FSH-HCG induction regimen mentioned above. Kuehl and Dukelow (1975a) demonstrated that the seasonal effect could be overcome by doubling the dose of FSH from 1 to 2 mg/day or by extending the 1 mg dose to 5 days instead of 4. Varying the dose of HCG did not affect ovulation during the anovulatory season. Moreover, the seasonal effect could be overcome by the use of estradiol and progesterone in various combinations. Ovulation induced out of season results in oocytes that are capable of being fertilized in vitro (Kuehl and Dukelow, 1979) and in vivo (Jarosz et al., 1977), and the latter has resulted in normal pregnancy and birth.

Sperm Capacitation

The study of sperm capacitation and fertilization of the oocytes, as well as the early preimplantation stages of development, is critical to determining the exact development biology of primate species. Furthermore, this work has important applications relative to captive breeding and propagation of endangered species. For over 60 years, researchers attempted to combine sperm and oocytes in vitro and bring about fertilization. In 1951, Drs. C. R. Austin and M. C. Chang discovered the phenomenon that we now call capacitation, and this revolutionized approaches to in vitro fertilization research. Basically, the term capacitation implies that the sperm must incubate for a period of time (normally in the female reproductive tract) before they achieve the capacity to fertilize the oocyte. The long delay in the discovery of this basic physiological phenomenon is probably due to the fact that capacitation (and the subsequent acrosome reaction) are very short temporally and thus their discovery eluded scientists. In most common laboratory species, the time required for capacitation ranges from 1 to 15 hours. Once capacitation was discovered and its biochemical nature ascertained, it was a rather rapid development to achieve successful in vitro fertilization in laboratory animals. Today we have seen in vitro fertilization occurring in about 20 animal species, and over 1600 births have occurred in the human primate as a result of in vitro fertilization and embryo transfer. Capacitation is also assumed to occur in vitro through exposure of the sperm to female reproductive tract secretion. This includes the follicular fluid components that are enclosed in the cumulus matrix surrounding the oocyte.

The classical method for demonstrating capacitation in any species is to mate the animal naturally at a fixed interval after ovulation. Then, by recovering the oocytes or embryos at varying times and studying the temporal stages of development, one can determine if a period of delay was experienced before the sperm penetrated the egg. Such approaches are necessarily crude but have traditionally given preliminary evidence for capacitation. In 1933, Hartman demonstrated that the rhesus monkey ovulates on approximately day 13 of the menstrual cycle and van Wangenen (1945) showed that the optimum time for mating in the rhesus was on day 11 or 12, suggesting that a period of delay was required. Marston and Kelly (1968) used timed mating studies in the rhesus monkey and reported a 3–4 hour period of time for capacitation in vivo. This value corresponding to the 5–7 hour value reported for humans (Soupart and Morgenstern, 1973; Soupart and Strong, 1974). In the squirrel monkey, sperm used for in vitro fertilization require an incubation period of 2–5 hours before fertilization occurs (Kuehl and Dukelow, 1982). In addition to these direct measures of sperm capacitation, some studies have been conducted where indirect measures of capacitation were used. A variety of methods have been used usually based on the characteristics possessed by capacitated sperm but not by freshly ejaculated sperm. These include increased oxygen uptake, the presence of decapacitation factor in primate seminal plasma, and the ability to remove a tetracycline coating from the surface of the sperm after incubation in a foreign uterine environment. All of these procedures have been used with rhesus sperm (Dukelow and Chernoff, 1969) and all gave positive results relative to a requirement for sperm capacitation in this species. In more recent years, the zona-free hamster egg (with the zona pellucida removed by trypsin treatment) has been used as a test system to observe penetration of the ovum and the need for capacitation. This technique was developed by Yanagimachi et al. (1976) and has been ascribed to be a measure of capacitation and the acrosome reaction. Human sperm (Yanagimachi et al., 1976) and squirrel monkey sperm (Burke, 1979) will penetrate the zona-free hamster egg. Kreitmann and Hodgen (1980) found limited usefulness of this test with the rhesus. It is of value in assessing the fertility of sperm, but there is some question if this is a true measure of capacitation. In 1967, Bedford and Shalkovsky found a role of sperm capacitation in penetrating the cumulus layer of the cells around the oocyte, and the acrosome reaction has been implicated in the penetration of the zona pellucida. One can argue if these layers are removed from the hamster oocyte prior to exposure of the sperm one may be by-passing the capacitation process.

Sperm Enzymes

In the past 20 years, a great deal of scientific research has been devoted to determining the enzyme components of the acrosome resting on the head of the sperm beneath the plasma membrane. During the capacitation process and particularly during the acrosome reaction, various enzymes are released or activated, and these enzymes obviously play a role in allowing the sperm to penetrate the cumulus cell layer, the corona radiata, and the zona pellucida surrounding the oocyte. Basically, three enzymes have been suggested to play a major role in penetration (McRorie and Williams, 1974). The first enzyme, hyaluronidase, is released from the sperm during contact with the cumulus cells. About 50% of hyaluronidase is free within the acrosome, and the remainder is bound to the inner acrosomal membrane. This was the first enzyme discovered, and it acts to dissolve or disperse the hyaluronic acid matrix that binds the cumulus cells. Following release of this enzyme, an esterase called corona penetrating enzyme allows passage of the sperm to the area of the zona pellucida. At this point a third enzyme called acrosin allows penetration of the zona pellucida itself. The latter two enzymes are bound to the inner acrosome membrane of the sperm. Other enzyme systems have been proposed for the activation of these enzymes. Acrosin inhibitors to rhesus monkey sperm have been identified in the oviductal fluid (Stambaugh et al., 1974). The presence of acrosin appears to be universal in many species, and much more emphasis has been placed on this enzyme from human sperm (Tobias and Schumacher, 1976; Anderson et al., 1981a, 1981b). Lysosomal hydrolases isolated from sperm acrosomes and seminal plasma were studied in nine chimpanzee semen samples by Srivastava et al. (1981). Beta-*N*-acetyl-hexosaminidase and hyaluronidase showed the highest specific activity in acrosomal extracts. Other enzymes isolated included arylsulfatase, β-glucuronidase and α-L-fucosidase. Seminal plasma also showed high activity of acid and alkaline phosphatases and low specific activity of β-galactosidase. More work is needed before the exact function of each of these enzymes can be shown, and in addition there is a need for comparative studies between primate species.

Sperm Penetration of the Oocyte

Fertilization can be defined as the events that lead to the actual fusion (syngamy) of the female and male pronucleus, or it can be defined as the actual penetration process. After the sperm head has penetrated the zona pellucida, it attaches to the vitelline membrane and is engulfed by the egg plasma membrane. When this occurs, the sperm head, midpiece, and tail are drawn into the cytoplasm where the latter two structures are rapidly dismantled (within 9 hours) and the sperm head moves toward the center of the ooplasm. These events have been extensively

TABLE 34.1. Temporal development of in vitro fertilized squirrel monkey embryos.

Hours after insemination	Event
2–4.7	Capacitation
6	Condensation and swelling of sperm
10	Nucleate stage
16	First cleavage metaphase
46–52	Second cleavage
52–72	Third cleavage
96–120	Blastocyst stage

studied in laboratory species, but few studies have been carried out with the nonhuman primate. In the squirrel monkey (Dukelow, 1983c), a series of events occur relating to the condensation of the chromosomes, the fusion of the pronuclei, and early cleavage, which are indicated in Table 34.1. In the rhesus monkey, Bavister et al. (1983) found that the sperm tail remnants remain 18–24 hours after fertilization in only two of 13 fertilized oocytes. These events are the immediate preamble to the initiation of embryonic development of the preimplantation embryo.

In Vivo Fertilization and Timed Mating

Basic studies on capacitation, fertilization, and gestation length normally involve the mating of an animal at a time interval related to a known time of ovulation. Allowing for the time of capacitation, the exact moment of fertilization can then be determined. Difficulties with this procedure in various species reflect the inability of the investigators to induce or predict accurately the time of ovulation. Nevertheless, some of the earliest studies of naturally fertilized primate ova derive from research attempts to predict the time of ovulation and recovery of the oocytes or embryos relative to that time. Lewis and Hartman (1941) studied rhesus monkey eggs recovered from the one- to 16-cell stage and cultured in vitro. These were largely anecdotal studies in which the time of ovulation was estimated and with natural mating. Generally speaking, early investigators assumed that ovulation occurred at midcycle or slightly before, and this necessitated the prediction of ovulation based on the time of the previous cycle. At least in the case of the cynomolgus macaque, this technique is not particularly accurate (Yoshido et al., 1982). In a retrospective study of the stump-tailed macaque using timed matings of 20 minutes or less that resulted in pregnancy, Bruggemann and Dukelow (1980) reported the highest conception rate after matings on day 11 of the cycle. Similarly, Jewett and Dukelow (1971b) used laparoscopy in the cynomolgus macaque to produce pregnancies based on the follicular structure. Dukelow and Bruggemann (1979) summarized pregnancies that occurred in colonies of stump-tailed and cynomolgus macaques over a 10-year period with timed mating. Successful pregnancies were calculated on the basis of the day of breeding compared to the previous cycle length (DB/CL ratio), and additional pregnancies were available from a limited number of rhesus matings as well. Maximum conception occurred at a ratio of 0.40 to 0.41 with a range for successful matings from 0.39 to 0.44. In similar studies with the pig-tailed macaque where male/female exposure was 48 hours, Blakley et al. (1981) found that the optimum time for mating was a DB/CL ratio of 0.46. In a similar calculation with seven known timed matings in chimpanzees (D. Martin, personal communication), a DB/CL ratio of 0.58 was noted. Others have used variation in timed matings with daily observation of the vaginal lavage for the presence of sperm. Stolzenberg et al. (1979) used this procedure with the squirrel monkey to assess pregnancy rate in a large colony. Similarly, the time of detumescence has been used as an indicator of ovulation (Blakley et al., 1981) in the pig-tailed macaque and in baboons. These techniques are useful in breeding programs to estimate gestation length crudely, but they are of less value to time fertilization precisely as is required in capacitation–fertilization studies. Such techniques represent

TABLE 34.2. Comparative rates of primate preimplantation development.

Species	Hours after fertilization						
	Two polar bodies	2-cell	4-cell	8-cell	16-cell	Morula	Blastocyst
Squirrel monkey	6–22	16–40	46–52	52–72			96
Rhesus monkey		24–36	36–48	48–72	72–96		
Cynomolgus monkey		24	48	48–72			
Baboon	6–24	24	48	48–72	96–120	120–148	96–144
Human	12	30–38	38–46	51–72	85–96	96–135	123–147

Source: Data from: Ariga and Dukelow (1977); Kuehl and Dukelow (1979); Lewis and Hartman (1941); Bavister et al. (1983); Kreitmann et al. (1982); Kraemer and Hendrickx (1971); Pope et al. (1982); Kuehl (1983); Hertig et al. (1954); Croxatto et al. (1972); Avendano et al. (1975); Edwards and Steptoe (1975).

indirect estimates of ovulation time and suffer from normal biological variation between animals. Such studies with in vivo fertilization (and correlated with in vitro fertilization) yield developmental rates for the early preimplantation stages such as those shown in Table 34.2. By comparison between the species listed in this table, one can see that the developmental stages of nonhuman primates of both Old World and New World monkeys is comparable to that reported in the human. This verifies their value as an experimental model for such research.

In Vitro Fertilization and Embryo Transfer

Recently the subject of in vitro fertilization and embryo transfer in nonhuman primates has been reviewed (Dukelow, 1983a, 1983b, 1983c). In vitro fertilization has been achieved in five nonhuman primate species: the squirrel monkey, the marmoset, the baboon, the rhesus macaque, and the cynomolgus macaque. In 1972, two reports appeared (Johnson et al., 1972; Cline et al., 1972) announcing in vitro fertilization of squirrel monkey oocytes. The first full publication of the latter studies appeared in 1973 (Gould et al., 1973). According to this report, 22 mature oocytes were recovered of which 11 showed sperm in the perivitelline space, extrusion of the second polar body, or pronuclear formation. Six of the 11 cells cleaved to the two-cell stage. Expanded reports of in vitro fertilization from the former laboratory (that of the authors) appeared in 1975 (Kuehl and Dukelow, 1975b) with 32 of 79 oocytes fertilized in vitro. In vitro fertilization studies in the squirrel monkey had the advantage of a wide variety of background techniques for ovulation induction and semen collection in the monkey (Bennett, 1967a, 1967b). The degree of ovulation can be controlled by the FSH-HCG regimens reported earlier. Laparoscopic techniques are also available for the collection of follicular oocytes from the squirrel monkey and for semen collection by electrostimulation. Using these procedures in trials over a 5-year period, 745 oocytes were aspirated from 2168 follicles. Of these oocytes, 18.4% were atretic, and of the remaining 608 oocytes, 38% matured to metaphase II stage. Of these, 33.5% were fertilized in vitro. Oocytes were classified by the system of Soupart and Morgenstern (1973), and this was correlated with the in vitro fertilization rate. In subsequent studies, the effect of varying quantities of cumulus cells on squirrel monkey oocyte in vitro fertilization were observed (Chan et al., 1982). If cumulus cells were absent, maturation was reduced in the oocytes but even as little as one-quarter of the oocyte covered with cumulus cells resulted in 70% fertilization in vitro. It was also found that if oocytes were collected 15–16 hours after HCG and allowed to incubate for an additional 21 hours prior to sperm addition (i.e., fertilization 37 hours after HCG), a higher level of fertilization was achieved. Furthermore, if 1 or 10 μm of dibutyryl cAMP was added to the culture medium, in vitro fertilization increased from 60% to 90%. This effect was the result of stimulation of sperm capacitation, the acrosome reaction, and whiplash motility, all of which are required for penetration of the oocyte. The additive affected the sperm only and had no effect on the oocyte itself. Similarly, cAMP had no effect on the subsequent cleavage of the fertilized oocyte. Kreitmann et al. (1982) successfully fertilized 22 cynomolgus monkey oocytes after incubation with homologous sperm with some development to the morula stage. Bavister et al. (1983) utilized PMS for 12 days beginning on day 3 to 5 of the cycle in the rhesus monkey. Follicles were aspirated laparoscopically about 30 hours after HCG, and 43% of the oocytes showed signs of fertilization. The times of cleavage were comparable to earlier reports by Lewis and Hartman (1941) with in vivo fertilized rhesus oocytes but were somewhat faster than the cleavage rates reported by Kreitmann and Hodgen (1981). Kuehl (1983) reported successful in vitro fertilization of baboon oocytes recovered from ovaries surgically. Additionally, Kuehl and colleagues were the first to report a live birth resulting from in vitro fertilization and embryo transfer (Clayton and Kuehl, 1984; Kuehl et al., in press).

Fertilization is only the first step in this process, and embryo transfer is required to bring about a successful pregnancy. The first successful surgical transfer of nonhuman primates was by Kraemer et al. (1976). A 5-day-old baboon embryo was transferred to a synchronized but nonmated baboon, and 174 days later an infant was delivered by a caesarian section. In the rhesus monkey, a simple surgical flushing technique was developed by Hurst et al. (1976). Nine embryos and two unfertilized ova were recovered from 22 flushes. Marston et al. (1977) used a similar technique to recover fertilized eggs from the rhesus monkey and transferred these to the opposite oviduct or uterus of the donor animal. All eight such transfers to the oviduct resulted in pregnancy. Additionally, these workers transferred one five-cell embryo and two eight-cell embryos to the uterus without a resulting pregnancy. Subsequently, however, two six-cell embryos were transferred to the uterus and pregnancies resulted. It was these experiments that led other English workers, using human in vitro fertilized embryos, to attempt the transfer of early developmental stage eggs directly to the uterus, and this subsequently resulted in the first successful in vitro fertilization and transfer of human eggs. In 1980, Pope et al. successfully recovered uterine embryos nonsurgically from the baboon. This was accomplished with a modified Isaacs endometrial cell sampler. Thirty-seven eggs were recovered from 80 flushes on 33 baboons, and to date over 500 blastocysts have been recovered by this technique. The procedure was subsequently used to recover a four-cell embryo that was then transferred and resulted in the birth of the first nonsurgical in vitro fertilized nonhuman primate (Pope et al., 1983). The remaining major event in this series of pioneering efforts occurred on July 25, 1983, exactly 5 years to the day from the birth of the first human in vitro fertilized, nonsurgically transferred offspring. This was the work by Kuehl et al. (in press) where a female baboon was delivered after a

TABLE 34.3. Chromosomal abnormality of in vitro maturing and matured squirrel monkey oocytes.

Developmental stage	Abnormality incidence (%)	Type
Metaphase I	7/28 (25.9)	Univalents
Metaphase II	7/95 (7.4)	4 Hyperploidy; 3 hypoploidy
Metaphase III	7/50 (14.0)	6-21/23; 1-20/24

normal gestation length from a primigravida baboon. Earlier this female received four embryos that had been collected from follicles on the ovary of a female that was autopsied. After 24 hours of culture at 37°C to achieve maturation, sperm were added. At 24 hours, 19 of 22 mature ova were fertilized, and of these three zygotes were transferred into the oviduct of the recipient. Sonography was utilized to study development throughout the pregnancy. Once embryos have been produced from nonhuman primates in quantity, it is possible to study normal embryonic development of these embryos by a number of different means. These include the morphological stage of development relative to time, the chromosomal normality of the embryo, and the metabolic or biochemical normality. One of the problems in studying chromosomal normality of multicelled embryos is the tendency for chromosomes to mix following disruption of the blastomere membranes during the fixing process. Kamaguchi et al. (1976) developed a technique in Chinese hamsters that was subsequently modified for chromsomal study on squirrel monkey oocytes before and after in vitro fertilization (Mizoguchi and Dukelow, 1981). Utilizing this technique Asakawa and Dukelow (1982) and Asakawa et al. (1982) studied chromosomal number in oocytes recovered from follicles of the squirrel monkey after ovulation induction and of embryos of the same species following in vitro fertilization. The results of these studies are shown in Tables 34.3 and 34.4. By the metaphase II stage 7.4–14.0% abnormalities were observed, a value comparable to that found in other laboratory species without in vitro fertilization. In similar trials with squirrel monkey oocytes exposed to in vitro fertilization, an incidence of abnormalities from 9% to 16% occurred with the common abnormalities being missing or extra chromosomes. These levels are also comparable to those found with other species including humans. Interestingly, the incidence of triploidy or in vitro fertilized squirrel

TABLE 34.4. Chromosomal abnormality in in vitro fertilized squirrel monkey oocytes.

Developmental stage	Abnormality incidence (%)
Metaphase II	
haploidy	5/72 (8.9)
polar body	4/29 (13.8)
First cleavage metaphase	
diploidy	3/25 (12.0)
triploidy	5/30 (16.7)

TABLE 34.5. Effect of sperm concentration on fertilization in vitro and triploidy in the squirrel monkey.

Sperm concentration per culture ($\times 10^5$)	Ova fertilized / total ova	Fertilization percent	Triploidy observed
2.5–4.0	24/40	60.0	+
4.0–5.5	25/65	38.5	
5.5–7.0	39/72	54.2	+
7.0–8.5	13/24	54.2	+
8.5–10.0	21/54	42.9	
10.0–11.5	10/19	52.6	
11.5–13.0	2/5	40.0	+
13.0–14.5	3/11	27.3	
> 14.5	9/17	52.9	+
Total	146/302	48.3	

monkey oocytes was 16.7%. Triploidy is commonly encountered with in vitro fertilization in all laboratory species and has been reported by Lopata et al. (1978) in human in vitro fertilized embryos. The cause of the triploidy is not known. In studies with the squirrel monkey (Table 34.5) no effect was found between sperm concentration in the culture and the level of triploidy observed or percent fertility. While the fertilization rate in vitro with humans is high (70–90%), the percent delivery of living children is low (10–25%). This greater embryonic death loss after fertilization should reflect the incidence of triploidy that occurs, and greater emphasis of research is needed to reduce this factor. Little research has focused on the biochemistry or metabolism of in vitro fertilized primate embryos. The work of Hutz et al. (1983a) was designed to study protein synthesis, uptake of steroid hormones, oxygen consumption, and overall viability of squirrel monkey embryos produced by this method. Incorporation of labeled leucine, as an indicator of protein synthesis, declined with oocyte maturation in vitro and remained constant after in vitro fertilization. There was a nonsignificant elevation at first cleavage. Estradiol and progesterone uptake increased after in vitro fertilization in these oocytes, but there were no further changes in uptake of either steroid at first cleavage. RNA synthesis, as measured by uridine incorporation, decreased in oocytes recovered from squirrel monkeys 36 hours after HCG compared to oocytes recovered at 16 hours. There was an approximate doubling of uridine incorporation after fertilization with further increase as development progressed beyond the first cleavage division. This is in accord with other studies on early embryonic stages of development in the mouse. Uridine incorporation has also been studied on a comparative basis for nonfertilized oocytes in squirrel monkeys and humans (Hutz et al., 1983b). There was a significant decline in both uptake and incorporation of uridine into squirrel monkey oocytes 36 hours after HCG, and similarly RNA synthesis diminished in human oocytes collected 35 hours after HCG compared to 12 hours after HCG. These studies provide emphasis for the use of increased interval between HCG administration of follicular aspiration in order to recover mature oocytes for in vitro fertilization.

Conclusion

Ovulation, ovulation induction, sperm capacitation, and fertilization have been studied for over 50 years in nonhuman primates, but it has only been in the past 15 years that extensive studies on sizeable numbers of embryos have been carried out. The number of species has been relatively small, and the majority of the findings come from studies in the squirrel monkey, baboon, rhesus, and the cynomolgus macaque. Nevertheless, the fertilization process appears to be similar to that identified in other mammals.

While the increasing basic studies on reproductive phenomena in nonhuman primates continues to be novel and of significance to a better understanding of the animal, the important application is to the natural mating of the animals themselves. While studies on artificial insemination and in vitro fertilization are interesting and of basic value, they still represent efforts that can be applied only in extreme cases of infertility or with problem cases of endangered species. Efforts to preserve species should still orient to the basic natural techniques of reproduction, and the sophisticated technological approaches should be applied only in extreme cases for such species.

References

Anderson RA, Oswald C, Zaneveld LJD (1981a) Inhibition of human acrosin by monosaccharides and related compounds: structure-activity relationships. J Med Chem 24:1288–1291.

Anderson RA, Beyler SA, Mack SR, Zaneveld LJD (1981b) Characterization of a high-molecular-weight form of human acrosin. Biochem J 199:307–316.

Ariga, S, Dukelow WR (1977) Recovery of preimplantation blastocysts in the squirrel monkey by a laparoscopic technique. Fertil Steril 28:577–580.

Asakawa T, Dukelow WR (1982) Chromosomal analyses after in vitro fertilization of squirrel monkey (*Saimiri sciureus*) oocytes. Biol Reprod 26:579–583.

Asakawa, T, Chan PJ, Dukelow WR (1982) Time sequence of in vitro and chromosomal normality in metaphase I and metaphase II of the squirrel monkey (*Saimiri sciureus*) oocyte. Biol Reprod 27:118–124.

Avendano S, Croxatto HD, Pereda J, Croxatto HB (1975) A seven-cell human egg recovered from the oviduct. Fertil Steril 26:1167–1172.

Barnes RD, Lasley BL, Hendrickx AG (1978) Midcycle ovarian histology of the bonnet monkey, *Macaca radiata*. Biol Reprod 18:537–553.

Bavister BD, Boatman DC, Leibfried L, Loose M, Vernon MW (1983) Fertilization and cleavage of rhesus monkey oocytes in vitro. Biol Reprod 28:983–999.

Bedford JM, Shalkovsky S (1967) Species specificity of sperm capacitation in the rabbit. J Reprod Fertil 13:361–364.

Bennett JP (1967a) The induction of ovulation in the squirrel monkey (*Saimiri sciureus*) with pregnant mares serum (PMS) and human chorionic gonadotropin (HCG). J Reprod Fertil 13:357–459.

Bennett JP (1967b) Artificial insemination of the squirrel monkey. J Endocrinol 13:473–474.

Blakley GB, Beamer TW, Dukelow WR (1981) Characteristics of the menstrual cycle in nonhuman primates. IV. Timed mating in *Macaca nemestrina*. Lab Anim 15:351–353.

Bruggeman S, Dukelow WR (1980) Characteristics of the menstrual cycle in nonhuman primates. III. Timed mating in *Macaca arctoides*. J Med Primatol 9:213–221.

Burke DB (1979) In vitro sperm-ovum interaction utilizing golden hamster and squirrel monkey spermatozoa with hamster zona-free ova. MS Thesis, Endocrine Research Unit, Michigan State University, East Lansing, 57 pp.

Chan RJ, Hutz RJ, Dukelow WR (1982) Nonhuman primate in vitro fertilization: seasonality, cumulus cells, cyclic nucleotides, ribonucleic acid and viability assays. Fertil Steril 38:609–615.

Clayton O, Kuehl TJ (1984) The first successful in vitro fertilization and embryo transfer in a nonhuman primate. Theriogenology 21:228.

Cline EM, Gould KG, Foley CW (1972) Regulation of ovulation, recovery of mature ova and fertilization in vitro of mature ova of the squirrel monkey (*Saimiri sciureus*). Fed Proceed 31:277.

Croxatto HB, Diaz S, Fuentealga B, Croxatto HD, Carrillo D, Fabres C (1972) Studies on the duration of egg transport in the human oviduct. I. The time interval between ovulation and egg recovery from the uterus in normal women. Fertil Steril 23:477–458.

Dukelow WR (1970) Induction and timing of single and multiple ovulations in the squirrel monkey (*Saimiri sciureus*). J Reprod Fertil 22:303–309.

Dukelow WR (1977) Ovulatory cycle characteristics in *Macaca fascicularis*. J Med Primatol 6:33–42.

Dukelow WR (1979) Human chorionic gonadotropin: induction of ovulation in the squirrel monkey. Science 206:234–235.

Dukelow WR (1983a) The nonhuman primate as a reproductive model for man. In: Dukelow WR (ed) Nonhuman primate models for human diseases. CRC, Boca Raton, FL, pp 79–105.

Dukelow WR (1983b) Ovum recovery and embryo transfer in primates. In: Adams CE (ed) Mammalian egg transfer. CRC, Boca Raton, FL, pp 165–174.

Dukelow WR (1983c) The squirrel monkey (*Saimiri sciureus*). In: Hearn JP (ed) Reproduction in New World primates. MTP, Lancaster, UK, pp 149–179.

Dukelow WR (1985) Reproductive cyclicity and breeding. In: Rosenblum LA, Coe CL (eds) Handbook of squirrel monkey research. Plenum, New York.

Dukelow WR, Bruggemann S (1979) Characteristics of the menstrual cycle in nonhuman primates. II. Ovulation and optimal mating times in macaques. J Med Primatol 8:79–87.

Dukelow WR, Chernoff HN (1969) Primate sperm survival and capacitation in a foreign uterine environment. Amer J Physiol 216:682–686.

Dukelow WR, Harrison RM, Rawson JMR, Johnson MP (1972) Natural and artificial control of ovulation in nonhuman primates. Med Primatol 1:232–236.

Dukelow WR, Jewett DA, Rawson JMR (1973) Follicular development and ovulation in *Macaca fascicularis*, *Saimiri sciureus* and *Galago senegalensis*. Amer J Phys Anthro 38:207–208.

Edwards RG, Steptoe PC (1975) Physiological aspects of human embryo transfer. In: Behrman SJ, Kistner RW (eds) Progress in infertility. Little, Brown, Boston, pp 377–409.

Espey LL (1978) Ovarian contractility and its relationship to ovulation: a review. Biol Reprod 19:540–551.

Espey LL (1980) Ovulation as an inflammatory reaction—a hypothesis. Biol Reprod 22:73–106.

Gould KG, Cline EM, Williams WL (1973) Observations on the induction of ovulation and fertilization in vitro in the squirrel monkey (*Saimiri sciureus*). Fertil Steril 24:260–268.

Graham CE (1981) Menstrual cycle of the great apes. In: Graham CE (ed) Reproductive biology of the great apes. Academic, New York, pp 1–43.

Graham CE, Keeling M, Chapman C, Cummins LB, Haynie J (1973) Methods of endoscopy in the chimpanzee. Relations of ovarian anatomy, endometrial histology and sexual swelling. Amer J Phys Anthro 38:211–216.

Harrison RM, Dukelow WR (1973) Seasonal adaptation of laboratory-maintained squirrel monkeys (*Saimiri sciureus*). J Med Primatol 2:277–283.

Harrison RM, Dukelow WR (1974) Morphological changes in *Saimiri sciureus* ovarian follicles as detected by laparoscopy. Primates 15:305–309.

Hartman C (1933) Pelvic (rectal) palpation of the female monkey with special reference to the ascertainment of ovulation time. Amer J Obst Gyn 26:600–608.

Hertig AT, Rock J, Adams EC, Mulligan WJ (1954) On the preimplantation stages of the human ovum: a description of four normal and four abnormal specimens ranging from the second to the fifth day of development. Contributions to Embryology, Carnegie Institution, Washington 35:199–230.

Hurst PR, Jefferies K, Eckstein P, Wheeler AG (1976) Recovery of uterine embryos in rhesus monkeys. Biol Reprod 15:429–434.

Hutz RJ, Chan PJ, Dukelow WR (1983a) Nonhuman primate in vitro fertilization: biochemical changes associated with embryonic development. Fertil Steril 40:521–524.

Hutz RJ, Holzman GV, Dukelow WR (1983b) Synthesis of ribonucleic acid in oocytes collected from squirrel monkeys and humans following chorionic gonadotropin administration. Amer J Primatol 5:267–270.

Jarosz SJ, Kuehl TJ, Dukelow WR (1977) Vaginal cytology induced ovulation and gestation in the squirrel monkey. Biol Reprod 16:97–103.

Jewett DA, Dukelow WR (1971a) Follicular morphology in *Macaca fascicularis*. Folia Primatol 16:216–220.

Jewett DA, Dukelow WR (1971b) Laparoscopy and precise mating techniques to determine gestation length in *Macaca fascicularis*. Lab Prim Newslet 10:16–17.

Jewett DA, Dukelow WR (1973) Follicular observation and laparoscopic aspiration techniques in *Macaca fascicularis*. J Med Primatol 2:108–113.

Johnson MJ, Harrison RM, Dukelow WR (1972) Studies on oviductal fluid and in vitro fertilization in rabbits and nonhuman primates. Fed Proceed 31:278.

Kamaguchi Y, Funaki Y, Mikamo K (1976) A new technique for chomosome study of murine oocytes. Proceed Japan Acad 52:316–319.

Knobil E, Morse A, Green RO (1956) The effects of beef and monkey hypohysectomized rhesus monkey. Anat Rec 224:331–335.

Kraemer DC, Hendrickx AG (1971) Description of stages I, II, III. In: Hendrickx AG (ed) Embryology of the baboon. University of Chicago Press, Chicago, pp 45–52.

Kraemer DC, Moore GT, Kramen MA (1976) Baboon infant produced by embryo transfer. Science 192:1246–1247.

Kreitmann O, Hodgen GD (1980) Low tubal ovum transfer: an alternate to in vitro fertilization. Fertil Steril 34:375–378.

Kreitmann O, Hodgen GD (1981) Retarded cleavage rates of preimplantation monkey embryos in vitro. JAMA 246:627–630.

Kreitmann O, Lynch A, Nixon WE, Hodgen GD (1982) Ovum collection, induced luteal dysfunction, in vitro fertilization, embryo development and low tubal ovum transfer in primates. In: Hafez ESE, Semm K (eds) In vitro fertilization and embryo transfer. MTP, Lancaster, UK, pp 303–324.

Kuehl TJ (1983) In vitro and xenogenous fertilization of baboon follicular oocytes. Fertil Steril 39:422.

Kuehl TJ, Dukelow WR (1975a) Ovulation induction during the anovulatory season in *Saimiri sciureus*. J Med Primatol 4:23–31.

Kuehl TJ, Dukelow WR (1975b) Fertilization in vitro of *Saimiri sciureus* follicular oocytes. J Med Primatol 4:209–216.

Kuehl TJ, Dukelow WR (1979) The effect of a synthetic polypeptide, threonyl-prolyl-arginyl-lysin, on ovulation in the squirrel monkey (*Saimiri sciureus*). Biol Reprod 21:545–556.

Kuehl TJ, Dukelow WR (1982) Time relations of squirrel monkey (*Saimiri sciureus*) sperm capacitation and ovum maturation in an in vitro fertilization system. J Reprod Fertil 64:135–137.

Kuehl TJ, Clayton O, Reyes PS (In press) Live birth following in vitro fertilization and embryo transfer in a nonhuman primate. Amer J Primatol.

Lewis WH, Hartman CG (1941) Tubal oval of the rhesus monkey. Contributions to Embryology, Carnegie Institution, Washington 29(108):9–14.

Lopata A, Brown JB, Leaton JF, McTalbot J, Wood C (1978) In vitro fertilization of preovulatory oocytes and embryo transfer in infertile patients. Fertil Steril 30:27–35.

Marston JH, Kelly WA (1968) Time relationships of spermatozoa penetration into the egg of the rhesus monkey. Nature (London) 217:1073–1074.

Marston JH, Penn R, Sivelle PC (1977) Successful autotransfer of tubal eggs in the rhesus monkey (*Macaca mulatta*). J Reprod Fertil 49:175–176.

McRorie RA, Williams WL (1974) Biochemistry of mammalian fertilization. Ann Rev Biochem 43:777–803.

Mizoguchi H, Dukelow WR (1981) Gradual fixation method for chromosomal studies of squirrel monkey oocytes after gonadotropin treatment. J Med Primatol 10:180–186.

Nagle CA, Riarte A, Quiroga S, Azorero RM, Carril M, DeNari JH, Rosner JM (1980) Temporal relationship between hormonal profile in the capuchin monkey (*Cebus apella*). Biol Reprod 23:629–635.

Nigi H (1977a) Laparoscopic observations of ovaries before and after ovulation in Japanese monkey (*Macaca fuscata*). Primates 18:243–259.

Nigi H (1977b) Laparoscopic observations of follicular rupture in the Japanese macaque (*Macaca fuscata*). J Reprod Fertil 50:387–388.

Nigi H (1978) Histological ovarian changes in *Macaca fuscata* before and after ovulation. Jap J Vet Sci 40:297–307.

Okamara H, Takenzka A, Yajima Y, Nishimura T (1980) Ovulatory changes in the wall at the apex of the human graafian follicle. J Reprod Fertil 58:153–155.

Pope CE, Pope VZ, Beck LR (1980) Nonsurgical recovery of uterine embryos in the baboon. Biol Reprod 23:657–662.

Pope CE, Pope VZ, Beck LR (1982) Development of baboon preimplantation embryos to post-implantation stages in vitro. Biol Reprod 27:915–923.

Pope CE, Pope VZ, Beck LR (1983) Successful nonsurgical transfer of a nonsurgically recovered four-cell uterine embryo in the baboon. Teratology 19:144.

Rawson JMR, Dukelow WR (1973) Observation of ovulation in *Macaca fascicularis*. J Reprod Fertil 34:187–190.

Riesen JW, Meyer RK, Wolf RC (1971) The effect of season of occurrence of ovulation in the rhesus monkey. Biol Reprod 5:111–114.

Soupart P, Morgenstern LL (1973) Human sperm capacitation and in vitro fertilization. Fertil Steril 24:462–478.

Soupart P, Strong PA (1974) Ultrastructural observations on human oocytes fertilized in vitro. Fertil Steril 25:11–44.

Srivastava PN, Farooqui AS, Gould KG (1981) Studies on hydrolytic enzymes of chimpanzee semen. Biol Reprod 25:363–369.

Stambaugh R, Seitz HM, Mastroianni L (1974) Acrosomal proteinase inhibitors in rhesus monkey (*Macaca mulatta*) oviduct fluid. Fertil Steril 25:352–357.

Stolzenberg SJ, Jones DCL, Kaplan JN, Barth RA, Hodgen GD, Madan SM (1979) Studies with time-pregnant squirrel monkeys (*Saimiri sciureus*). J Med Primatol 8:29–38.

Tobias PS, Schumacher GFB (1976) The extraction of acrosin from human spermatozoa. Biol Reprod 15:187–194.

Travis JC, Holmes WN (1974) Some physiological and behavioral changes associated with oestrus and pregnancy in the squirrel monkey (*Saimiri sciureus*). J Zool 174:41–66.

Van Wagenen GW (1945) Mating and pregnancy in the monkey. Anat Rec 91:304.

Van Wagenen G (1968) Induction of ovulation in *Macaca mulatta*. Fertil Steril 19:15–29.

Wildt DE, Doyle LL, Stone SC, Harrison RM (1977) Correlation of perineal swelling with serum ovarian hormone levels, vaginal cytoogy and ovarian follicular development during the baboon reproductive cycle. Primates 18:261–270.

Wilks JW, Marciniak RD, Hildebrand DL, Hodgen GD (1980) Periovulatory endocrine events in the stumptailed monkey (*Macaca arctoides*). Endocrinology 107:237–244.
Williams RF, Hodgen GD (1980) Disparate effects of human chorionic gonadotropin during the late follicular phase in monkeys: normal ovulation, follicular atresia, ovarian acyclicity, and hypersecretion of follicle stimulating hormone. Fertil Steril 36:64–68.
Yanagimachi R, Yanagimachi H, Rogers BJ (1976) The use of zona-free animal ova as a test system for the assessment of the fertilizing capacity of human spermatozoa. Biol Reprod 15:471–476.
Yoshida T, Nakajima M, Hiyaoka A, Suzuki MT, Cho F, Honjo S (1982) Menstrual cycle lengths and the estimated time of ovulation in the cynomolgus monkey (*Macaca fascicularis*). J Exp Anim 31:165–174.
Zeleznik AJ, Schuler HM, Reichart LE (1981) Gonadotropin-binding sites in the rhesus monkey ovary: role of the vasculature in the selective distribution of human chorionic gonadotropin to the preovulatory follicle. Endocrinology 109:356–362.

35
Housing and Furniture

WARREN D. THOMAS

The approach to housing and furniture for primates in zoos has undergone a series of transitions representing the changes in thinking and attitude toward captive primates. We have progressed from small wire or barred cages totally barren in nature through glass-fronted, tile displays still maintaining the barren aspect, to the trend toward large, naturalistic exhibits. Through all of this, the propping or furniture of the cage and the attention to the animal's basic needs have been ongoing problems.

There are as many solutions as there are efforts. Our zeal has sometimes overridden our common sense, and our efforts have been skewed in ways guaranteed to produce failure.

The ideal, of course, is to be able to recreate the animal's natural environment to show not only the animal but the forces that shaped it. This addresses the needs of the animal and shows the animal to the public in its most natural form. This should be the ultimate goal. Sometimes we have put so much effort into the animal's needs that we have created displays that totally lose their effectiveness on the viewer; at other times we have produced esthetically pleasant exhibits that are incompatible with the needs of the animal. We have even disregarded such basic needs as for primates who tend to need horizontal branches and we have propped the exhibit with nothing but verticals and vice versa. We have experimented with plants and propping materials, sometimes disastrously.

The larger primates, certainly the anthropoid apes, have been housed in excellent displays built around an existing group of mature trees large enough to support the destructive tendencies of the animals housed therein.

More often than not, however, the exhibits have been built in areas that do not have such vegetation, and it is rare that a zoo can afford to move adult, mature trees into such a display. Protection for young vegetation is, by necessity, overbearing and ugly and usually not very effective. I remember a palm tree that was exhibited in with a group of orangutans. I was obsessed with keeping a live palm tree in their display, and the orangs, of course, had nothing to do but figure out new ways to thwart me. Over a period of 4 years I constantly had to think up new ways to preserve that tree as the orangs always figured out new ways to destroy it. It was a running battle of wits, which I finally lost.

We must strike a balance at a point where the needs of the animal are met, the exhibit is esthetically pleasing, and the display can be maintained and properly cleaned. We often tend to forget the fact that with all of our grandiose plans and clever innovations for designing the display, someone has to take care of it.

Certainly with the great apes, boredom is a consideration that must be dealt with. One of our speakers today will deal with environmental engineering, which is a way to occupy the animal and cope with boredom. Relatively simple methods can help. We tend to give our animals a considerable amount of browse strewn all over the exhibit; edible flowers, rice and sunflower seeds are set out in such a fashion as to induce the animal to spend a lot of its time moving about the exhibit searching out the little tidbits that are hidden away. The browse itself is a source of nutrition and roughage; once the green leaves and tops of the branches are eaten, the remainder becomes an artifact for them to amuse themselves with.

Summary

The basic considerations are, first and foremost, the needs of the animal and the presentation made to the public. You should always be able to walk in front of any exhibit and ask yourself, "Does this exhibit teach anyone anything?" If your answer is negative, a reappraisal and readjustment to the exhibit should be made. If the answer is positive, there should still be a reappraisal to determine how the exhibit could be improved.

Last, but certainly not least, one must never lose sight of the fact that the exhibits must be maintained by an animal keeper. There is probably no single, correct formula; like all else, everything must be tailored to the local situation and the animal in question, in order to discover the best approach possible in a rational, creative manner.

36
The Howletts Gorilla Bands

John Aspinall

In this paper I would like to expose some of the background to the work that we have done with lowland gorillas at Howletts since the colony was founded in 1958. I also bear in mind that the only information of much interest must be different from that contained in my other two papers (Aspinall, 1980, 1982). As far as the underlying factors are concerned, it seems clear to me that Howletts has enjoyed certain obvious advantages over other organizations which, to some degree, has been reflected in our performance as breeders and keepers of this remarkable ape. The fact that certain other orthodox establishments have done as well or better than we have—for instance, Lincoln Park—in no way weakens my postulate that Howletts has performed better than most.

Perhaps the most revealing approach is the negative one. Some things are not responsible for our success, and one of these is money. It is true that the two zoos, Howletts and Port Lympne, have always had enough to survive and are reasonably well placed at the moment, but the capital value of them both combined would not measure up to the cost of some of the single zoo exhibits that have recently been built in the United States and elsewhere. In terms of financial resources, we would not make the top 50 zoos in the United States, so money can be safely ruled out as anything much more than an enabling factor.

It might be thought by some that we have on call a number of accredited experts upon whose special knowledge and research we can draw at will. This simply is not and has never been the case. Nobody who works in the zoos, with the exception of the director and full-time veterinary surgeon, in management or otherwise, has any academic qualifications whatsoever—no Ph.D.'s, no degrees, diplomas, or baccalaureates. The progress of the colony of gorillas cannot then be explained by exceptional intakes of academic learning and scientific expertise.

A further negative that fails to gain admission is the false belief that the two zoos are not open to the public. In other words, some have suggested that we are singularly advantaged by not having to cater for a viewing public. The truth is that Howletts has been open to the paying public every day of the year for the last 9 years, and the 17 gorilla births have all occurred since 1975. I would add a rider here: In my view, on balance, the absence of a "public" would be beneficial to the apes. The advantages for them of stimulation and diversion created by the visitors is outweighed by the danger of contagious infection and by the time, space, and cost needed to cater for the public need, all of which could be better deployed on the animals themselves. Also on the scale is the educational factor, but I have never been wholly convinced of the positive propaganda value of zoos in relation to wildlife. What I can say is that one of the long-term aims of the Howletts and Port Lympne Foundation is to restrict public attendance at the zoos to a few weeks a year. Such a move would prove costly because revenue from the public covers somewhat less than a third of running costs. We are still some years away from achieving this objective.

To sum up, the success we have had is not really attributable to money, scientific research, qualified expertise, or privacy. It behooves us then to look elsewhere to try and find an answer to this conundrum. Why has a poorly funded zoo in Kent owned by a part-time dilettante and supported by a cast of unqualified personnel made such satisfactory progress in gorilla husbandry?

Perhaps I am both the last and the first person who should answer such a question, but at least let me attempt to explain to you what I feel and think on the matter. An advantage does accrue to us from the fact that the zoo was initially a private one from 1958 to 1976. Its philosophical origin was surprisingly simple: I wished to befriend certain wild animals—a two-way process, of course—and create conditions whereby they could reproduce and be happy. The deeper aspects of conservation grew in the rich humus of my first intent, but in 1958 I was only dimly aware of the approaching avalanche of human biomass that is now sweeping away what is left of the natural world.

The zoo was born because I believed I could cross the barrier and mix with wild creatures—get to know them in an atmosphere of mutual trust and affection. The whole concept was subjective and personal and remains so to this day. I favored out of choice certain species and certain individuals within those species. By the same mysterious process, barely understood, whereby a man finds and returns affection and delight in the company of his friends, I acted and reacted with the tigers, gorillas, elephants, and rhinos with which I surrounded myself. This subjective approach has been adopted through natural example by many of the keeping staff. It works effectively, for it raises the whole problem of custodianship to a level of personal involvement where the emotions can be mobilized to diminish much of the drudgery of work, to exile boredom and to transmute as if by magic the routines of toil to times of enjoyment and contentment.

The trick is worked by encouraging the keepers to forge personal bonds with the high mammals, thus adding a whole new perspective to their work and a behavior stimulus to those animals. The private and personal origins of the zoo concentrated our minds on thinking about the needs and comforts of the animals freed from any concern for the needs of a visiting public. Municipal zoos must look to City Hall for finance and cater for the perceived desires of a mass public, the adult portion of whom have a vote in local elections. The fact that the word "exhibit" is the common expression in institutionalized zoos for a wild animal enclosure tells the full story. Zoos cannot be blamed for this or even criticized. Zoos are for people with the emphasis on children. Beneath the weight of such a *force majeur*, the directors must bow low and bow often. They have my sympathy and many of them my admiration: my sympathy because I can well imagine the pressures exerted by parks commissioners, municipal education departments, and others; my admiration because I know what astonishing successes have been made by many zoos in spite of having to cater for huge influxes of human beings on a year-round basis.

There are several easily understood benefits to be gained by entry into the gorilla bands:

1. Learning about group behavior and relationships. There is nothing quite like being in with a family of gorillas to feel the cross currents of emotion and apprehension that constantly pass through the band.
2. Being within a group necessitates successful mood interpretation. A failure or a misreading could have painful consequences. In the case of Djoum (male, 410 pounds), a false move could invite disaster. The tone of a family is imposed by the dominant male. If he is relaxed, we are all relaxed. If he is tense, we are ill at ease along with all the rest of his family. Our behavior is markedly different from that of Dian Fossey when she mixes with wild mountain gorillas (Fossey, 1983, and personal communication). She told me that when a large male approached her she adapted a posture of total female submission. She said this came naturally to her under such a circumstance. In the case of myself, Peter Halliday (head gorilla keeper), and Ian Williams (second gorilla keeper), all of whom mix with the apes, the response is instinctively quite different. We measure up to wrestle and play as comes naturally to men. If any of the gorillas, even Djoum, play too roughly or in an ill-tempered way, we answer back and let our displeasure be known. The gorillas have long ago understood that we are not as strong as we are tall, and when they close with us, they deploy only a fraction of their power. Djoum, who is the largest primate in the United Kingdom, is the possessor of preternatural strength; this, added to his majestic appearance and compelling personality, makes each hour we spend with him and his family of three wives and two children a *tour de force*. We feel when we are in with him that "vital tension," to borrow an expression from George Schaller (1964), that is the hallmark of a wild animal—a raised consciousness of all that is going on about us. This helps us to study the band and watch for slight changes in mood or tone.
3. An ability to enter the gorilla enclosures is also useful in life-threatening situations that must occur from time to time. Once Ian Williams rushed inside to release 3-year-old (male) Jomie's head from a fork in the bars when his mother Lomie was pulling him in the wrong direction. He certainly saved his life. In another instance, Peter Halliday went in and cut the umbilical cord that was wrapped round newborn Kaja's neck before her mother JuJu strangled her with it. On many occasions, Peter and Ian have gone in to retrieve dangerous or valuable objects that have either been flung in by the public or have been grabbed from us along the service corridor. Sometimes when a gorilla is dangerously ill, the company and attention of a friendly keeper, night and day, can make a crucial difference to its survival. In the last 10 years, at least three lives may have been saved in this manner, those of JuJu (female), Bamenda (female), and Sidonie (female).
4. In the event of an escape, a close physical friendship could be the telling factor in the successful capture of the animal.
5. The presence of keepers, male and female, is most useful in the introduction and assimilation of orphaned young into a formed band. This process, which can take up to 2.5 years, is most difficult, time consuming, and absolutely essential. Our experience is that the hand-raised orphan must make friends

essential. Our experience is that the hand-raised orphan must make friends with the new family one at a time. Anything other than this piecemeal approach is too dangerous. Many gorilla young have been lost in zoos by too hasty an attempt at assimilation.

The role model of the dominant male in captivity is difficult to compare with that of the wild silverback. At Howletts with the large groups that we have at hand to study, some interesting information has come to light. The head of section, Peter Halliday, believes that a captive overlord has reduced power over the females and that their power is correspondingly increased. He argues that in the wild, the patriarch is the defender of the group against all enemies and all rival silverbacks. In captivity, there are no external threats as no zoo would have the temerity to introduce outside males as rivals, let alone outside threats. In the wild, the patriarch decides when and where to turn in for the night and when and where to forage in the morning. He is in fact the guide and navigator and may well enjoy the confidence of the females for his topographical expertise and for his botanical skills.

It is believed that only a small percentage of wild males ever rise to the patriarchate. The competition to reach this high position must be severe indeed. In zoos, however, the females do not choose their leader and seldom join him voluntarily. A zoo silverback does not have to measure his strength, character, and personality against any rivals. Circumstances place him where he is and man-made ones at that.

Many zoo silverbacks are thus ill fitted for a position that in nature they would probably have failed to reach. The zoo adult gorilla males' effective roles are thus reduced to stud matters and to adjudication in disputes. There must be considerable truth in Halliday's argument and we know from our own observations the extreme difficulty young, not exceptionally gifted males have in asserting their authority over the females. Bitam, an aspiring young silverback of 14 years, has still some way and a year or two to go before he can assume control over the females, Shamba (27 years), Mouila (25 years), Baby Doll (24 years), and JuJu (23 years). Perhaps in the wild he could by now have proved his worth to them by protective behavior, such as dispersing poachers, interlopers, or even tourists.

Whether Bitam will ever "make it" as a replacement for Kisoro, who has gone to the Denver zoo, only the future will tell. Dijoum at 17 years is a natural leader and in my view and that of all his keepers would have quite easily attracted a large following around him in the wild: that is if he had not "inherited" a band through the process of ultimogeniture, which may well be a method of preserving some family continuity among wild gorillas. We have not yet had enough experience of adult males to know whether Djoum's majestic confidence as a leader, his absolute assurance of his own position and power are contributary factors in his tolerance and enjoyment of our company. We hope that he is not a "one off" case and that others will follow his example. Halliday, Williams, and I predict that the young male Kibobo (5 years), by Kisoro out of Baby Doll, will prove as reliable and dependable as Djoum has done. Already at his tender age he shows qualities that are necessary for the overlordship of a band. He has acquired the confidence of all the adult females in his group and of their progeny as well, most of whom are his kindred. Mothers will allow him to carry their infant young with a trust not extended to his elder full brother, Kibabu (8 years). He initiates play with us and shows no fear of his human cousins. He has, like mighty Djoum, a magnetic personality laced with an undefinable charm. We have never seen him panic, and his steady nature becalms those of his age group when a quarrel begins. His intelligence is of the highest order, on a par with Djoum's. One day, perhaps, he will start a band with Juma (female, 4 years) and other unrelated females.

I would like, finally, to relate a few incidents that throw some light into the state of mind of gorillas and give us an insight into their emotional world. Gorillas, compared to chimpanzees and men, live in a psychological redoubt, aloof, remote, reserved—these adjectives are in constant use by writers on gorillas. Schaller (1964) went so far as to describe them as "transcendental." I go along with these descriptions and endorse them. Everyone almost without exception, so it seems, falls under their spell. Gorillas are emotionally very close to us. We have learned to modify our behavior to suit theirs. Humor, guilt, and punishment can be seen in their reactions. If you tickle them and make them laugh they do the same back and expect you to laugh too. If you have a quarrel with one, a rapprochement follows with back patting, conciliatory gurgles and embraces. Recently I forgot to take into Djoum's family my usual quota of chocolates and "pats de fruit," which they specially look forward to. Djoum came up to me, felt my trouser pockets, and on finding them empty gave me a punishing bite in the arm. The message was clear: You have forgotten the goodies, don't let us down again.

When Djoum was younger, we would initiate play by squaring up to him on our knees for a wrestle or by a short chase, but, as he matured, we thought it wiser to allow him to make the opening moves. After polishing off all the "bribes" we bring with us, he usually saunters over and invites us to play. Only after a prolonged play session with us will he then allow his wives and children to join in. Sometimes when facing two of us head on, his son Jamie (5 years) and daughter Juma (3 years) will tackle him from behind, creating a welcome diversion. When Peter, Ian, and I are chatting, with him next to us, we have to be careful not to make fun of him because he knows if our remarks are unflattering and responds accordingly. Sometimes he punishes us for failing to give him complete attention; at other times he shows a lazy tolerance and permits us to make a fuss of Mushie and Founa in front of his eyes. He has a strange passion for eating our eyelashes and get quite annoyed if we don't indulge him. He removes mine by clamping my head in his enormous hands and biting them out with his incisors. Peter and Ian are far more generous to him in this respect than I am.

Redirected aggression can sometimes present us with a problem. When we know a female is in oestrus we make a point of not disturbing the apes for obvious reasons. However, sometimes we don't find out until we are already in the enclosure. Then the oestrus is usually in its first stages, which is why we have not noticed it. Djoum's females sometimes tease him when they are in oestrus in an attempt to "wind him up." This can be quite awkward for us, and the worst offender is Founa, who has a naturally flirtatious nature. Her game is to come and embrace one of us to stimulate Djoum, and we don't enjoy the experience at all. Once we noticed that his head wife, Lomie, was in very early estrus and did not present to Djoum. Djoum knew she was getting interesting but suddenly, feeling frustrated, he turned on me with a sudden cough and mock bit me in the small of my back. I was down on all fours anyway and felt the pressure of his bite. Peter Halliday told me he had half my back in his mouth and looked very angry. I came out of it quite unscathed and a minute or two later Lomie had moved off and Djoum came up to me to play as usual.

The following recommendations are made for successful gorilla husbandry:

1. Keep them in large groups.
2. Large indoor playroom and extensive well-equipped outside runs with a roof open to the weather and cobwebbed with brachiating bars. The latter is essential for the development of pectoral, shoulder, and upper arm muscles.
3. Deep litter of oat straw in outside run for foraging and day nesting. Straw also for bedding inside house.
4. Numerous outside contraptions for play: chutes, paddling pools, metal trees, platforms, suspended spheres of galvanized piping, giant marker buoys, a hanging-garden of sisal ropes. A large flattened sphere with apertures for stick fishing, etc.
5. At least 200 types of food a year. Lowland gorillas are 50% fruit eaters, so a large variety of tropical fruits are required as well as numerous herbs, wild and domestic, garden vegetables, and at least eight or nine species of broad-leaved tree branches. I await with interest Caroline Tutin's study of lowland gorilla diet in the Gabon.
6. Keepers to be encouraged to stay with the gorillas all their lives and bond strengthen with them.
7. Treat gorillas at least as equal or superior to man.

The last behest is a difficult one for us, shored up as we all are by a deeply engrained belief in our divine superiority over all other living things. The latest comparative work of endocrinologists and molecular biologists reveal the closeness of the gorilla to man. I for one feel proud to hail him as a cousin, feel privileged to work for him, and to be accepted by him as an honorary member of his family.

Acknowledgments. I have drawn on the vast experience of the head of the gorilla section at Howletts, Peter Halliday, who has been with the apes for 12 years, and on Ian Williams, the second keeper, both by notes and personal comments, for the compilation of this paper, for which I thank them both.

References

Aspinall J (1976) The best of friends. Macmillan.

Aspinall J (1980) The husbandry of gorillas in captivity. J Reprod Fertil Suppl 28:71–77.

Aspinall J (1982) Some aspects of gorilla behaviour, with particular reference to adult males and their infant progeny. Hanover.

Fossey D (1983) Gorillas in the mist. Houghton Mifflin, Boston.

Schaller G (1964) The year of the gorilla. Chicago University Press, Chicago.

37 Using Outside Areas for Tropical Primates in the Northern Hemisphere: *Callitrichidae*, *Saimiri*, and *Gorilla*

WIM B. MAGER and TINE GRIEDE

Introduction

Tropical houses seem to be a new wave in zoo architecture. Almost any large zoo that can find the money will demolish the old monkey cage rows. They break down the biological desert of steel and concrete and build their jungle house, especially in the northern hemisphere. There is a lot to say against the old-fashioned monkey cages in a row, but little or nothing against a good tropical house. A small zoo, however, that must depend on entrance fees will never find the enormous amount of money needed to build a good tropical house. Moreover, in only a few jungle houses the quality of the space and furniture, available to the monkeys, equals or even approaches the quality of the nice picture offered to the visitors.

Tropical primates do not only thrive at tropical temperatures. By using outside areas it is possible even for smaller zoos to provide inexpensive high quality exhibits. In combination with good inside accommodation, an outside area can also be used in the winter in the northern hemisphere. For this reason, the general zoo-technical management that is used at Apenheul might be of interest. These aspects are illustrated with reference to the more important species of the park, i.e., *Callitrichidae*, *Saimiri*, and *Gorilla*.

Apenheul

Based on the philosophy that tropical primates can also flourish in the outdoors in the northern hemisphere, Apenheul has made a park of three islands and five areas where monkeys roam semifree and visitors are able to walk through. The different areas are separated by water moats, which had to be built artificially because of the sandy soil. To prevent the animals from drowning, the slopes of the moats have swamplike, muddy edges, which caution them to stay back. Visitors pass the water moats by crossing bridges. The monkeys are afraid of the bridges because at times when the park is closed these are electrically wired (hot). The positive and negative poled grids are separated by rubber.

A "monkey-proof" fence (Fig. 37.1) surrounds the entire park, representing an unclimbable barrier. In stress situations where electricity would not prevent the animals from leaving their enclosure, this fence has turned out to be escape-

FIGURE 37.1. The monkeys can climb halfway on the fence, but cannot pass a soft PVC overhanging flap.

proof. In this way, Apenheul keeps squirrel monkeys, white-faced sakis, several families of marmosets and tamarins, barbary macaques, gibbons, gorillas, patas, capuchins, woolly and spider monkeys, and ring-tailed lemurs.

Free Choice

Keeping monkeys or apes outside in cold climates requires suitable inside accommodations. Houses should be considered as shelters against the climate, as well as safe sleeping dens to assure a feeling of security. They must be heated, and experiments need to be done for each species to decide upon a minimum temperature at which each species remains active and feels comfortable, safe, and secure. The temperature in their inside enclosure should not be too high, as the primates should have a free choice of being inside or outside all year long and difference in temperature should be minimized as far as possible. The free choice is accomplished by plastic hatches or flaps (Fig. 37.2), a system introduced by the Frankfurt Zoo. Open holes in a building create draughts, and these plastic hatches prevent them. All species easily learn to handle these doors in the exits of their enclosures.

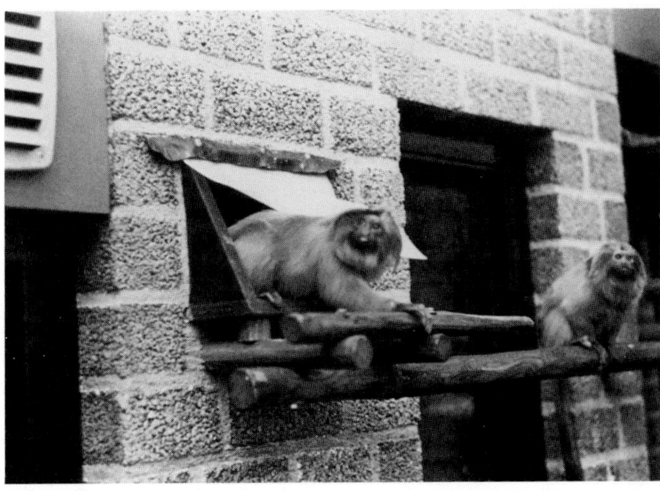

FIGURE 37.2. Because of the sloping construction of the door post, the plastic hatch closes tight and prevents draughts.

The advantages of free choice are numerous. Above all, monkeys and apes seem to know that there is a big garden at their disposal at all times. This "monkey knowledge" makes it possible to have relatively small inside houses, which saves in the cost of construction, maintenance, cleaning, and heating. Another cost-saving feature is found in the outside area itself because, if large enough, it needs no cleaning. Sun and rain are superb disinfectants.

Marmosets and Tamarins (*Callitrichidae*)

All inside rooms for marmosets have been built according to the same principle: there are two exits to the outside and a sliding door in the middle (Fig. 37.3). Each inside cage contains a nestbox with one opening. The feeding trays are about 1.5 m above ground. Since marmosets and tamarins do a lot of scent-marking (Kleiman and Mack, 1980), and disinfection actually means destroying their home, disinfectants are seldom used and even when used one compartment is cleaned at a time.

Many zoos could take advantage of captive-born *Callitrichidae* behavior, which keeps them very close to their sleeping dens and allows them to roam semifree. Captive-born *Callithrix jacchus*, *Saguinus oedipus*, *Callimico goeldi*, and *Leontopithecus rosalia*, once released, generally stay close to their homes. In fact, they hardly even use tree branches, preferring to use the vast climbing structures erected near their houses. However, if both parents are born in a semifree-ranging family, they may start to roam the entire territory. In this situation, a family of common marmosets at Apenheul uses over 1.5 ha of wooded area bordered on one side by water and the other side by the unclimbable fence.

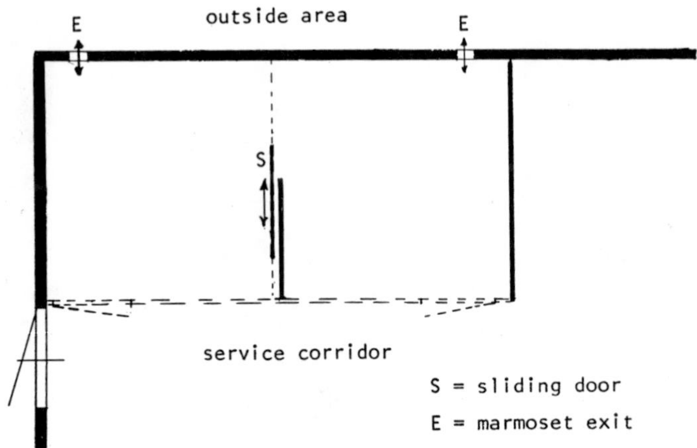

FIGURE 37.3. Plan of a marmoset cage.

In order to accustom a family group to the routine of semifree ranging during the day and returning to their bedrooms at night, during the first few days one or two animals are cut off from the rest of the family by closing the sliding door. The rest of the family is forced to enter a temporary outside cage, approximately 1 m^3 in size is sufficient. The separated individual has to be given free choice of leaving through the other exit. It keeps visual contact with the other family members, so it does not go far away and it will easily return to its inside quarters. By rotating the individuals of the family-group they all get a chance to learn to know the "free" outside-environment. After a week or so the temporary outside cage can be demolished and the then free-roaming family will each night happily return to their inside quarters.

White-faced sakis (*Pithecia pithecia*) show this same type of behavior and make a good combination with, for instance, golden lion tamarins (*Leontopithecus rosalia*). Another combination of Callitrichidae that thrives very well together is Goeldi's marmosets (*Callimico goeldi*) with pygmy marmosets (*Cebuella pygmaea*).

Squirrel Monkeys (*Saimiri s. boliviensis*)

Squirrel monkeys can live in large bands (Baldwin and Baldwin, 1981). A group home must be composed of a number of well-equipped and furnished rooms and, above all, with many tunnels, all interconnected in a "creep-through, sneak-through system" (Fig. 37.4). If the monkeys always have the opportunity to go outside, the rooms themselves can be rather small. The possibility of some privacy is far more important than the actual size of the rooms. A monkey house built according to such a system would benefit any group of Simiae.

Saimiri will stay around their enclosures for months or even more than a year and then suddenly leave without any plausible reason. Thus, an unclimbable fence or large outside cage is needed. The Apenheul group now numbers 117 animals. This number has grown from 18 wild-caught animals brought in during 1974–1975. In the summer they often sleep in the trees, as well as in their night

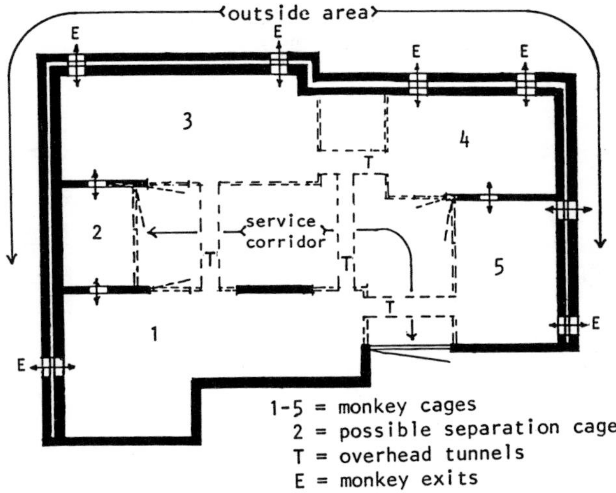

FIGURE 37.4. Many cages together, each connected through tunnels, create the proper housing for primates living in large nonfamily groups.

quarters. There are almost no veterinary consultations, and there is hardly any interference from the keeper staff. This group has shown us new ways of keeping monkeys in captivity.

Problems encountered need to be mentioned also. In the winters of 1982–83 and 1983–84 three animals died, probably because of their inability to cope with the spartan way in which they were kept. Also, there were some problems with tails, especially in offspring born the previous spring of low-ranking females. Tips of the tails would die off because of low temperatures. Low-ranking females are forced to stay outside too long and the accompanying youngsters do not curl their tails around their bodies like their mothers do. Even temperatures of +5°C influence the blood circulation in the youngsters' tails enough to cause problems. Since we have locked the squirrel monkeys inside during part of the winter, this problem has been alleviated. In the winter the animals are now only released on beautiful and windless days, choice food being used to persuade them to return to their cages at sundown. To make the animals feel comfortable, inside temperatures should be 15°C.

Gorilla (*G. g. gorilla*)

Apenheul's gorilla group consists of 12 animals, of which seven were imported from the wild in 1974 as youngsters. Five have been born at Apenheul of four different females, and all were mother-reared within the group. The present sex ratio is three males and nine females.

As well as a 2-ha island, the gorillas have a well-structured house (Mager, 1981) with a visitor area in the center (Fig. 37.5). An overhead tunnel connects the two sides of the gorilla building, which makes all rooms join, creating a circle of six interlinked rooms.

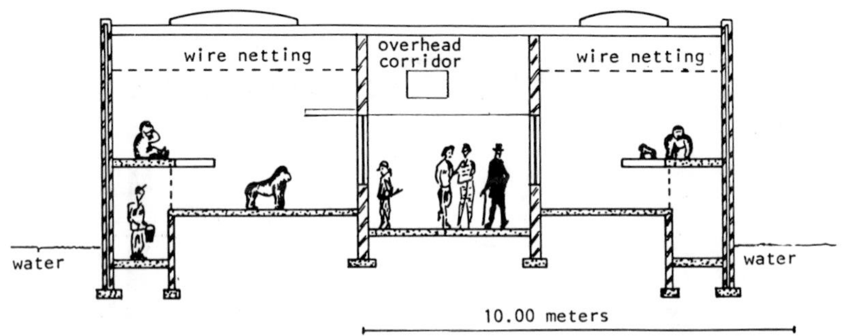

FIGURE 37.5. Cross section of the gorilla building of Apenheul.

The inside visitor area is small, but since the gorillas are only inside during very bad weather, and there are few visitors at that time, it does not present a problem. Through the windows of the visitor area the animals are able to see their mates on the other side of the building. The cost of this building was $120,000 in 1974.

The conventional bear pit exhibit is probably the worst place in which to house an ape family. Apes and monkeys want to look around and like to peep at the outside. Small holes in the walls of the gorilla building serve this purpose, as glass "bricks" function as peep-holes.

The quality of housing is of considerable influence on the social life of a group. The Apenheul group is very cohesive. They eat and sleep together, thus there are no night cages. Newborns remain with the group from birth.

By lowering the temperature degree by degree in the inside quarters, it was concluded that lowland gorillas feel comfortable at temperatures of about 11°C, with a relative humidity of about 75%. Lower temperatures cause them to have goose pimples. But at 11°C they sleep well, have a good appetite, and behave normally. Of course, in these circumstances enough material for good straw beds must be available at night. At this low temperature heating bills are considerably less, and as a result of less temperature difference between inside and outside, the animals spend more time on the island. Also, bacteria do not flourish at the lower temperatures even though the walls are not tiled.

Diet also contributes to the gorillas' health. They receive only one-half of a banana each day, some apples, a small piece of self-made ape-cake, but large amounts of green vegetables, and fresh branches for roughage. In all these years, only one gorilla has been treated once with antibiotics.

Conclusion

Keeping primates in the northern hemisphere does not necessarily require expensive tropical houses. A well-equipped inside and outside enclosure, both always accessible to the animals, is a much less expensive alternative. It also improves their resistance against diseases, thus colds are rare or, as in gorillas, completely absent. Inside accommodations may be relatively small, since nonhuman primates seem to know that on the outside there is a big garden at their disposal.

The animals must feel comfortable and safe in their houses. Multiple structured buildings with a creep-through, sneak-through system to guarantee privacy to these social animals are needed, as well as windows and peep-holes to satisfy their curiosity about life outside. Nestboxes for marmosets and nest material for gorillas must be available during the night. Buildings should be heated when the temperature is below 11°C (for gorillas) or 15°C (for squirrel monkeys and marmosets). Sharing exhibits with other primate species helps prevent boredom.

At Apenheul, many primates roam semifree. While this way of keeping animals is not applicable to every zoo or not allowed in every country, for the monkeys and apes it is not important whether they are kept in big outside cages or semifree. They do not seem to be aware of feeling caged or, as zoo visitors call it, being imprisoned.

For primate conservation, it would help a great deal to replace the old cages with well-constructed and furnished small houses combined with large (encaged) natural enclosures.

References

Baldwin JD, Baldwin JI (1981) The squirrel monkeys, Genus *Saimiri*. In: Coimbra Filho AF, Mittermeier RA (eds) Ecology and behavior of neotropical primates. Academia Brasileira de Ciencias, Rio de Janeiro, pp 277–330.

Kleiman DG, Mack DS (1980) Effects of age, sex and reproductive status on scent mark frequencies in the golden lion tamarin, *Leontopithecus rosalia*. Folia Primatol 33:1–14.

Mager WB (1981) Stimulating maternal behaviour in the lowland gorilla, *Gorilla g. gorilla*, at Apeldoorn. Intern Zoo Yrbk 21:138–143.

38
Evaluating the Environments of Captive Nonhuman Primates

TERRY L. MAPLE and TED W. FINLAY

Introduction

The architect John Zeisel (1977) recently asserted that "buildings themselves must be seen as hypotheses to be tested rather than solutions to be lived with." Whole environments, including buildings, have been designed and constructed with captive animals, caretakers, and visitors in mind. What do we know about how these environments affect users? And what techniques are currently available to test hypotheses and otherwise evaluate them in an objective and systematic fashion?

Evaluations of captive nonhuman primate environments enhance the road to self-sustaining populations in several ways. First, by identifying the variables that contribute to successful husbandry and propagation, evaluations lead directly to more effective breeding colonies. A second benefit of evaluation is that design errors (and design traps) are discovered and can be eliminated from subsequent designs. Thus, evaluations are cost effective. Finally, evaluations can appraise user satisfaction such that public attitudes and animal welfare become more objective constructs.

Our program of research, which began at Emory University in 1975, was initiated to better understand the effects of a captive life on the great apes (Fig. 38.1). To succeed it was first necessary to acquire a deeper understanding of the various pongid taxa. To this end we read the literature, carried out experiments and observations, and summarized what we had learned in a series of publications (Maple, 1979, 1980, 1982; Maple and Hoff, 1982). The senior author's move to the Georgia Institute of Technology in 1978 presented an opportunity to develop a research program in "applied primatology," which evolved and narrowed as we endeavored to blend the broader traditions of comparative and environmental psychology.

Environmental Psychology

The literature of environmental psychology, overwhelmingly a human subject data base, is frequently applicable to animal settings. In their well-known text, Bell et al. (1978, p. 3) defined the field thus: "Environmental psychology is the study of the interrelationship between behavior and the built and natural environment."

FIGURE 38.1. This inadequate ape habitat (vintage 1950s) permitted no vertical locomotion. Orangutans, the most arboreal of the apes, could climb on the one small steel apparatus as illustrated here.

This literature has been especially helpful in solving management problems. For example, it was recently necessary to select a color to repaint the outmoded feline house at the Atlanta Zoo. Since these hard cages are abhorred by the public, we wanted to do all we could in the short run to improve them. An initial effort was made to soften the environments with the addition of plants and natural material and increase space by the removal of partitions. To further "open" up the environment, we selected a light wall color that had been perceived by human subjects as "open." The EP literature provided guidelines on colors that were appropriate (Acking and Küller, 1972) and those that were not (Bennett and Rey, 1972).

In a comparative environmental psychology, findings and principles from the literature can be broadly applied. For example, there exists a well-developed literature on human spatial behavior and social density (cf. Heimstra and McFarling, 1978) that is relevant to the problems of animal housing. In fact, a recent study of chimpanzee crowding at the Arnhem Zoo suggests that chimpanzee and human adaptations to high-density living are indeed quite similar (Nieuwenhuijsen and deWaal, 1982). As the authors phrased their conclusion: "Apparently, advanced mammalian forms, such as chimpanzees and human beings, are able to cope with crowding-induced stress by engaging in alternatives to overt aggression" (p. 5).

While we appreciate the value of animal models of human behavior, the utility of human models of animal behavior is equally promising but far less appreciated.

The application of human data to solve nonhuman primate problems is particularly appropriate (Maple, 1983).

Postoccupancy Evaluation (POE)

A common methodology for evaluating environments is known as postoccupancy evaluation (O'Reilly et al., 1981; Zimring and Reizenstein, 1981). The primary goals of POE have been to (1) obtain immediate feedback on the outcome of environmental designs, and (2) develop a useful data base for future designs. These data can be used to "fine tune" an existing structure or produce a state-of-the-art design. We have had experience with both. An appropriate tool for this type of research is time-series analysis (Rotton, 1985).

In the evaluation of zoo environments, three major users must be studied—the animals, the caretakers, and the visitors. Although primate centers and laboratories do not generally cater to the public, we believe that such environments can be made aesthetically pleasing and comfortable for animals and people alike. The University of Texas' superior facility for chimpanzees in Bastrop, Texas, is an example of a laboratory that has been designed to accommodate animals, staff, and the occasional visitor (Riddle et al., 1982). A complete POE studies the effects of the environment on all identified users.

A number of recent studies have concentrated on the animal's responses to enclosure modification or translocation to a new environment. A pretest/posttest research design is typically employed.

The Animal's Response to Habitat Change

For example, Clark et al. (1982) recorded the behavior of a small group of chimpanzees before and after their translocation from a laboratory to a naturalistic island habitat. After 22 weeks on the man-made island, stereotyped and self-directed behaviors were dramatically less evident while their overall level of activity increased.

In a study conducted at the Audubon Zoo in New Orleans, the positive effects of naturalistic environments were further demonstrated (Maple and Stine, 1982). In this POE, both gorillas and orangutans exhibited a decline in aggressiveness and some improvement in their affiliative behavior. The naturalistic environment stimulated natural behavior patterns, some of which (e.g., brachiation) could not be expressed in the barren cages that they formerly occupied (Fig. 38.2).

It is sometimes the case that new environments are used by the animals in ways not contemplated by the design team. During the course of our evaluation of the Audubon ape enclosures, we discovered that the male lowland gorilla became a habitual user of the water moat, waterfall, and stream (Brown et al., 1982; Fig. 38.3). This habit was correlated with high humidity but was not affected by zoo attendance or ambient temperature. From these systematic observations, we came to regard water as a type of environmental enrichment.

Similarly, Chamove and Anderson (1979) discovered that the addition of deep wood chip litter reduced aggression and increased the activity of monkeys living

FIGURE 38.2. At the Audubon Zoo in New Orleans our research team witnessed meaningful brachiation in a male orangutan occurring for the first time in his adult life. The new naturalistic habitat is conducive to the expression of species-typical behavior patterns.

in laboratory cages. Manipulations such as these also have been labeled as environmental enrichment. According to the authors: "We recommend deep litter as one technique of enhancing conditions for captive primates. It has real potential for promoting good health and induces positive kinds of behavior among species that invest a great deal of time and energy in foraging in their natural environment" (p. 316).

In a pioneering study of European zoo environments, Wilson (1982) discovered that it was not the *amount* of space that determined activity; rather, it was the objects within space that really counted. In this study, activity was highly correlated with the number of cagemates, and the presence of playthings, especially *movable* playthings (see also Maple and Bloomstrand, 1984).

Toward an Applied Primatology

According to Monaghan (1984), behavioral knowledge is germane to solving the problems that result from our "exploitation" of animals. The application of behavioral knowledge in the primate realm we have characterized as "applied primatology." Our research has been increasingly conducted in zoo and laboratory settings where the results of our studies have provided tangible solutions to management problems.

A recent student project conducted at the Atlanta Zoo is instructive. Kenneth Blank recently evaluated the provision of a "foraging medium" (straw) in a small glass-fronted primate cage. Under our supervision, Blank recorded aggression, fear, affiliation, foraging, inactivity, and abnormal behaviors exhibited by two

FIGURE 38.3. An adult male gorilla submerged chest-high in a water moat. Water may be regarded as a form of environmental enrichment.

spot-nosed guenons in the respective barren and straw-laden environments. The enrichment condition reduced inactivity by one-third and increased foraging by a factor of four. These animals were engaged in natural behavior patterns, and we have hypothesized that the public's perception of these animals is now more positive. This hypothesis will be formally tested in the very near future.

Design Traps

In our work in applied settings, we have frequently encountered design problems which by their very nature defy solutions. For example, zoo managers may resist the provision of climbing apparatus if the floor is made of hard concrete. The threat of injury compromises enrichment in a dilemma that we have labeled a *design trap*. In one zoo we noticed that a moat was so narrow that vertical structures on the siamang island were limited by the potential for escape. Similarly, small drains that are clogged by browse must be regarded as a design trap.

The way out of design traps is to design with enrichment in mind. Naturalistic environments are especially conducive to enrichment (Coe, 1985; Hancocks, 1980, 1983; Hutchins et al., 1984; Maple, 1979).

Zoo Visitor Behavior

In a provocative essay published over a decade ago, the environmental psychologist Robert Sommer (1972) argued that dilapidated U.S. zoos frequently failed in their mission to educate the public. As Sommer suggested: "Despite excellent intentions, even the best zoos may be creating animals stereotypes that are not only incorrect but that actually work against the interests of wildlife preservation" (p. 27).

Until recently there was no evidence to support or refute this contention. However, an important study by Rhoads and Goldsworthy (1979) demonstrated that "the setting in which an animal is exhibited determines to a great extent the public's attitude toward the animal" (Fig. 38.4). By presenting slides of animals in natural, seminatural, and caged environments, these investigators discovered that the caged animals were perceived as significantly less dignified, confined, unhappy, unnatural, tame, and dependent. A factor analysis of these data further suggested that the display of animals in cages detracts from their inherent dignity.

In our laboratory at the Georgia Institute of Technology, we have become interested in separating public perceptions about enclosures from their attitudes about the subjects. To accomplish this, we developed the Zoo Environmental Description Scale (ZEDS) and the Animal Attitude Survey (AAS).

In developing the ZEDS instrument, we initially exposed 55 undergraduates to 20 slides of various and diverse great ape enclosures, ranging from barred cages to open wooded exhibits. These subjects were then provided a list of 122 bipolar adjective pairs. Each pair was rated on an 11-point descriptive scale. When the instrument is used in the zoo, respondents simply circle the number that best describes the enclosure.

The Animal Attitude Survey is a type of Thurstone scale of 22 items derived from a preliminary pool of 51 items tailored specifically to the great apes. The zoo visitors simply check the items with which they agree. After calculations for central tendencies, the resultant score represents the visitors' attitudes toward the animal.

FIGURE 38.4. Dead trees and naturalistic surrounding improve the aesthetics of zoo habitats and have a positive effect on zoo visitor attitudes.

In a pilot study that employed both instruments, visitors at the Atlanta Zoo and New Orleans' Audubon Zoo were surveyed. Data from the Animal Attitude scale revealed that visitors to both zoos held equally positive attitudes toward orangutans. However, the two enclosures were rated quite differently. The small, barren, and tiled cage in Atlanta was rated significantly lower than the open, moated enclosure in New Orleans.

To some degree the inherent popularity of an animal can "inoculate" against a negative evaluation of its surroundings. However, it is very clear to us that the zoo-going public is becoming increasingly able to discriminate among enclosures that are "good, bad, and ugly" (Maple, 1979). This emerging consensus has been corroborated by the work of Wolf and Tymitz (1981) who found: "Though the majority of visitors could not identify what should be included in the habitat of a particular specimen, one criterion was clear: captivity must be comfortable" (p. 50).

Future Trends

Our research has been recently focused on an attempt to quantitatively compare environments. While it is currently possible to compare environments by their size, the *quality* of the environment has eluded comparison.

We have gathered some data that permit us to compare environments for spatial volume which, in our opinion, is a more meaningful index of space. Furthermore, by weighing components we have achieved a measure of objective evaluation. To further objectify this process we are using independent raters. Ultimately, it should be possible to compare, rank, and fully describe the most

FIGURE 38.5. When captive gorillas live in groups, as they do in the wild, it is more likely that they will exhibit natural behaviors including reproductive behavior and parenting.

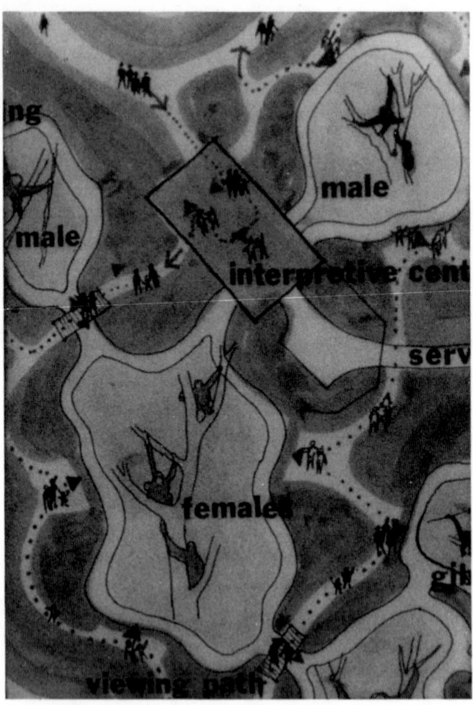

FIGURE 38.6. Preliminary concept plan for orangutans at the "new Atlanta zoo." Males are "peripheralized" and can be contacted by interested females through special one-way access gates (illustration by Jon Charles Coe).

complex environments according to unbiased standards. A similar effort to quantify space has been made by Innis et al. (1985), who remarked:

> Our analysis does not resolve the question of how much space captive animals should actually have. An animal's perceived spatial needs may vary, from central to peripheral portions of an enclosure, for example. . . or they may depend upon resource distribution or social context. However, the analysis does in our view, offer some guidelines that, if followed, will promote the welfare of captive animals. . . (p. 681)

We are on the verge of resolving the question for a few species. With a proven methodology, it should be possible to extend it to many others.

Perhaps the most exciting application of evaluation techniques and environmental psychology is when they can be used to design an innovative exhibit. In collaboration with the design firm of Coe and Lee Associates of Philadelphia, we have been engaged in the design of a major habitat for great apes.

The environment we are planning is naturalistic within and without, and it takes into account the way that the environment looks as well as the way that the animals are socially organized (Fig. 38.5). The gorilla habitat will provide enough space for several harem breeding units. Although the harems will be separated by a moat, the illusion of competition will be preserved for apes and visitors alike. This facility was inspired by the complex enclosures in San Francisco, Seattle, and Apeldoorn (Netherlands, as described by Mager, 1981; Fig. 38.6). A full explication of this unique design will be the subject of another paper (Maple and Coe, in preparation).

We are able to contemplate such a complex design because of our unique and exciting relationship to the Yerkes Primate Center. When our innovative habitats are complete, we will translocate the Yerkes gorilla population and many of their orangutans to the new Atlanta Zoo facilities. In this setting, it will be possible to educate and entertain the public, as well as continue the important Yerkes studies of great ape reproductive biology. By this combination of resources we will become partners in great ape propagation, wildlife conservation, and zoo biology.

Acknowledgments. Our research has been supported by a number of sources including NIH grant RR-00165 to the Yerkes Primate Center from the U.S. Public Health Service, Division of Research Resources; NIH Contract NO1-RS-7142 to the University of Texas System Cancer Center; and small grants to the senior author from the Wildlife Preservation Trust International, and the Atlanta Zoological Society. For the methodology to develop the ZEDS instrument we acknowledge the work of Kasmar (1970).

References

Acking CA, Küller R (1972) The perception of an interior as a function of its colour. Ergonomics 15:645–654.

Bell PA, Fisher JD, Loomis RJ (1978) Environmental psychology. WB Saunders, Philadelphia.

Bennett CA, Rey P (1972) What's so hot about red? Hum Fact 14:149–154.

Brown SG, Dunlap WP, Maple TL (1982) Notes on water-contact by a captive male lowland gorilla. Zoo Biol 1(3):243–249.

Chamove AS, Anderson JR (1979) Woodchip litter in macaque groups. J Inst Anim Tech 30(2):69–74.

Clarke AS, Juno CJ, Maple TL (1982) Behavioral effects of a change in the physical environment: a pilot study. Zoo Biol 1(4):371–380.

Coe JC (1985) Design and perception: making the zoo experience real. Zoo Biol 4(2):197–208.

Hancocks D (1980) Bringing nature into the zoo: inexpensive solutions for zoo environments. Int J Stud Anim Prob 1(3):170–177.

Hancocks D (1983) Gorillas go natural. Anim King 86(3):10–16.

Heimstra NW, McFarling LH (1978) Environmental psychology. Brooks/Cole, Monterey, CA.

Hutchins M, Hancocks D, Crockett D (1984) Naturalistic solutions to the behavioral problems of captive animals. Zool Gart 54(1/2):28–42.

Innis GS, Balph MH, Balph DF (1985) On spatial requirements of captive social animals. Anim Behav 33(2):680–682.

Kasmar JV (1970) The development of a useable lexicon of environmental descriptors. Environ Behav 12:153–169.

Mager W (1981) Stimulating maternal behaviour in the lowland gorilla at Apeldoorn. Intern Zoo Yrbk 21:138–142.

Maple TL (1979) Great apes in captivity: the good, the bad, and the ugly. In: Erwin J, Maple TL, Mitchell G (eds) Captivity and behavior. Van Nostrand Reinhold, New York, pp 239–272.

Maple TL (1980) Orang-utan behavior. Van Nostrand Reinhold, New York.

Maple TL (1982) Orang utan behavior and its management in captivity. In: deBoer LEM (ed) The orang utan: its biology and conservation. Dr W Junk, The Hague, pp 257–267.

Maple TL (1983) Environmental psychology and great ape reproduction. Intern J Stud Anim Prob 4(4):295–299.
Maple TL, Bloomstrand M (1984) Feeding, foraging and mental health. In: Proc Dr. Scholl Conf Nutrition. Lincoln Park Zoo, Chicago.
Maple TL, Hoff MP (1982) Gorilla behavior. Van Nostrand Reinhold, New York.
Maple TL, Stine WW (1982) Environmental variables and great ape husbandry. Amer J Primatol Suppl 1:67–76.
Monaghan P (1984) Applied ethology. Anim Behav 32:908–915.
Nash VJ (1982) Tool use by captive chimpanzees at an artificial termite mound. Zoo Biol 1(3):211–221.
Nieuwenhuijsen K, deWaal FBM (1982) Effects of spatial crowding on social behavior in a chimpanzee colony. Zoo Biol 1(1):5–28.
O'Reilly J, Shettel-Neuber J, Vining J (1981) The use of post-occupancy evaluations in an aviary. EDRA 12:318–325.
Rhoads DL, Goldsworthy RJ (1979) The effects of zoo environments on public attitudes toward endangered wildlife. Intern J Envir Stud 13:283–287.
Riddle KE, Keeling ME, Alford PL, Beck TF (1982) Chimpanzee holding, rehabilitation and breeding: facilities design and colony management. Lab Anim Sci 32(5):525–533.
Rotton J (1985) Time-series analysis: a tool for environmental psychologists. Pop Env Psych Newsl 12(1):15–19.
Sommer R (1972) What do we learn at the zoo? Nat Hist 81(7):26–27, 84–85.
Wilson SF (1982) Environmental influences on the activity of captive apes. Zoo Biol 1(3):201–209.
Wolf RL, Tymitz BL (1981) Studying visitor perceptions of zoo environments: a naturalistic view. Intern Zoo Yrbk 21:44–53.
Zeisal J, Epp G, Demos S (1977) Low rise housing for older people. US Dept Hous Urban Develop, Washington, DC.
Zimring CM, Reizenstein JE (1981) A primer on postoccupancy evaluation. J Am Inst Archs (Nov), 52–58.

39 Environmental Engineering for Primates

HAL MARKOWITZ and JOSEPH S. SPINELLI

Introduction

We continue to overpopulate and wantonly use unreplenishable resources at rates that promise to leave little of the current diversity of nature for our children's children to experience. Unfortunately, this means that we are likely to sustain those populations of primates that we are able to preserve in greatly reduced ranges compared with those that they originally inhabited. For those that we do manage to maintain for future generations, there are two major questions concerning "quality" that we must address. The first question is one of quality of life for those remaining primates, and the second is the quality of those primates that are maintained in less than naturally diverse environments.

In "civilized" captive environments, we can keep animals alive for considerably longer than their natural life expectancies in the wild. This is largely because we protect animals from predators, treat them for diseases, immunize them against infections, shelter them from extreme weather changes, and meet their nutritional needs. In nature, these animals are subject to forces of natural selection from which we carefully shield them in captivity. Thus, primates who have been in captivity through several generations are clearly quite different from contemporary conspecifics who live in natural habitats. There is increasing documentation of these differences in dimensions ranging from behavior through muscular and skeletal development (Gay et al., 1985). In the course of evolution, primates have been selected in considerable part based upon their abilities to gather food and to avoid predators. Depending on the species in question, the primate may have enviable abilities to leap and brachiate, to glide through the air, to inflict mortal wounds with powerful canines, etc. In the interest of promoting longevity and safely maintaining stocks of animals, we often eliminate the need or the opportunity for primates to exercise these natural characteristics.

It is only in relatively recent times that significant attention has been paid to the possibility of focusing some of our advanced technological capabilities on developing environments that are responsive to animals in ways that encourage species-appropriate activities (Markowitz, 1982, 1983, in press; Markowitz and Stevens, 1978). This relatively new work in behavioral enrichment is based upon the premise that animals that have learned to control parts of their own environments, that have learned that the environment is responsive to their actions, that do more than sit and wait for provisioning and testing, are bound to be more representative of their species than those that are cared for in ways that are only paralleled in some situations where humans are incarcerated.

For those animals that are displayed in zoos and aquariums, we are beginning to develop more adequate-appearing environments to reach the educational goals of these institutions. It is crucial that we combine these elegant appearances with opportunities for the primates to display some of their natural behaviors and to engage in species-appropriate exercise (Forthman-Quick, 1984; Foster-Turley and Markowitz, 1982; Markowitz, 1982). Unfortunately, it has sometimes been suggested that one must choose between methods which emphasize contingencies encouraging behavior and more "natural" approaches (Hutchins et al., 1978–1979). This faulty dichotomy ignores the fact that animals living in nature survive and evolution progresses as a function of their abilities to meet natural contingencies. Where animals are in our care, most of us are unwilling to subject them to many of the forces of natural selection (Markowitz, 1977).

In fairness to the animals, and in the interest of maintaining fitness, we must strive to identify the best methods of environmental engineering to simultaneously promote species-appropriate behavior and provide relative safety. The examples described below will illustrate some varying approaches to behavioral enrichment with the first descriptions used to emphasize questions of quality of life and the later ones emphasizing quality (e.g., fitness and species representativeness) of primates living under these conditions. We hope that this method of organization will not interfere with the recognition that both better lives and better animals are bound to result from properly engineered environments.

Quality of Life

In the San Francisco Zoo, there are some clear contrasts between a beautiful appearing exhibit and a simple procedure with respect to their effects on the behavior of great apes. San Francisco has one of the largest and most attractive gorilla exhibits in the zoo world. When the animals were first introduced to this exhibit, although the adult male took a while to venture forth, the youngsters and females began to explore their new environment and reveled in climbing the trees, watching limbs break under their weight, and stripping the bark. In the interest of preserving the limited number of trees in the grotto, a management decision was made to use hardware cloth around the bases of the trees in order to prevent stripping. Although it is still possible for gorillas to climb the trees, as time has progressed an unfortunate behavioral pattern has developed: They spend much of their time in a portion of the exhibit that is devoid of natural foliage. There the public encourages their presence by calling them and (when no one is monitoring the visitors) by throwing in food and objects with which they can play. Although this is a beautiful appearing exhibit, it is not a beautifully functioning environment. Providing interesting naturalistic contingencies would encourage the gorillas to use their attractive habitat more extensively. For example, since the gorillas are brought into cages each evening, it might be a partial solution to provide some interesting foods placed in the trees, or in other places where the gorillas might forage rather than beg for them.

John Alcaraz and some of the other keepers in the same zoo have initiated an interesting enrichment project for young orangutans. With the help of a number of dedicated volunteers, they take the young orangs into a wooded part of the grounds and encourage them to climb trees, swing on ropes, and generally become familiar with the forest (Markowitz, 1983). These orangutans are beautiful to behold as they climb to the tops of trees, encouraged initially by presentation of sunflower seeds and other treats that John and his colleagues provide them. Although the management of the zoo is willing to listen to suggestions that orangutans might spend happier, more active lives in an environment designed to incorporate a substantial section of the existing forest, there are some significant reasons for concern about protecting these representatives of an endangered species from the perils of a more complex environment. If adult orangutans were introduced to this forest, there is a finite possibility that they might fall from branches that have not been "worked over" by animals preceding them. Indeed, in the wild, primates do sometimes suffer or even die because of falls from high places. A second critical concern is how to provide veterinary care and appropriate routine examination of relatively free-ranging apes. Unless there is a behavioral protocol that encourages them to use regularly parts of their environment where ease of capture for examination is guaranteed, the introduction of apes to a forest will be accompanied by unacceptable reductions in our ability to care for and preserve dwindling primate species.

In spite of these and other bonafide concerns about keeper and animal safety, there is good reason to press on to find solutions that will allow us to house primates in more complex environments with naturalistic contingencies. The single most important reason is that if we really care for the animals that we are conserving, we should be striving to provide them with environments that enhance their quality of life. Inevitably, there will be some measurable increases in danger to the animals as a function of the complexity of natural environments compared with the appurtenances of more barren traditional captive environments. If quality of life for the resident primates is not sufficient reason for some to have the courage to take the risks associated with this type of enrichment, there is another critical factor that they might wish to consider. This is the question of the extent to which we are conserving animals that are truly representative of their species, a point we shall return to below.

When the animals in question are on display in zoos, there is a responsibility for visitors to be effectively educated about the need to work to conserve the habitats of these species in nature. Surely, we can do a more adequate job of this if people see the primates' behavioral beauty within a microcosm of their natural surroundings. Anyone who has seen orangutans forage, move through trees, and apparently glory in the expression of their natural abilities, cannot help but recognize that we are committing a horrible crime in eliminating the last vestiges of their homes.

Properly designed naturalistic environments that include large forested areas and appropriate behavioral protocols can greatly reduce the danger of fatal accidents. The dangers that remain are outweighed by the advantages for the animals and our increased ability to communicate the need for conservation of natural habitats. We need, in the case of orangutans, to begin with young, resilient animals who grow up in forested areas and, as they mature, learn in naturalistic ways which behaviors are safe and which are dangerous. With earnest effort and good fortune, it may be possible to develop captive colonies that are reproductive in environments that allow them to remain fit, capable, and happier representatives of their species.

While it is interesting to speculate about more ideal exhibits that may be developed in the future, it is also true that many primates will continue to live in areas decidedly unlike those they would inhabit in nature. In some cases, this is because of limited budgets for the facilities in which they live. Sometimes it is because the animals are being used in research, which necessitates that they be caged individually in environments that make them easily accessible. In a sense, these primates are subject to some of the same stress factors that humans encounter in limited urban spaces. With one species (*Homo sapiens*), we have come to recognize the necessity of providing alternative forms of physical exercise, and we have invented innumerable ways to challenge our problem-solving skills.

Although it may be more pleasant to design wonderful environments for use where they are plausible, the sheer number of animals whose lives will be affected is greater when we consider less expensive ways to enhance opportunities in limited space. Consequently, a considerable part of our effort has gone into behavioral enrichment devices that are clearly unnatural solutions to reducing the boredom of clearly unnatural living places. In one such environment in the Portland Zoo, a speed game was installed to entertain mandrills (*Mandrillus sphinx*), one of which learned to play so well that he was able to beat zoo visitors in fair contests and to compete with a computer at remarkably fast reaction times (Markowitz et al., 1982).

This game was originally designed and installed in an effort to provide some diversion that might reduce the aggression between this male mandrill and his cagemates. It accomplished this goal in a very convincing fashion, both reducing aggression and increasing the amount of space that was used by the females (Yanofsky and Markowitz, 1978). Although it taught the public nothing about the natural behaviors of this species, it did provide those who were defeated by the mandrill some sense of the relative competence of the two primate species in making rapid discriminations and quick responses.

Perhaps this point can be made most easily by thinking about changes in human attitudes about another mammalian group. Most people have ignored carefully and accurately presented data about ways in which we are endangering marine mammals. Their only observations of the skills of these animals have come from contrived shows conducted in seaquariums or from watching them as "actors" in television shows that make absurd anthropomorphic claims about our understanding of the dolphin's or killer whale's "mental" abilities. Although we might be more pleased if humans were more responsive to carefully presented data than they are to the entertainment industry, it would be foolish to ignore the fact that some people who would never have been interested in marine animals were it not for "Flipper" have significantly changed their attitudes with respect to the need for conservation of cetaceans. The speed game was designed primarily with the mandrills' welfare in mind, but one mandrill's ability to beat humans in speed games was noted in programs by all three major television networks as well as by other media. Many people who otherwise might think only of the mandrill's unusual physical appearance now recognize that it has behavioral skills as well.

The final point that we wish to address concerning quality of life resulting from environmental engineering efforts is the most philosophically difficult one. We humans tend to think of ourselves as making largely voluntary decisions about our activities, although a more careful analysis may show that a considerable part of behavior is determined by effects of factors such as advertising and economic coercion. However, there are certainly many cases in which we appear to decide spontaneously between a variety of options. We do not like to believe that our behavior is controlled by others, and, in parallel fashion, most of us would prefer that the design of environments for other primates allow them as much volition as possible. It is important for us to remember that before we are able to exercise our options we must learn how to participate effectively in various activities. Only after we have learned to play the piano, either through systematic training or laborious trial and error, are we able to choose to entertain ourselves with this particular device. Only after years of preliminary effort are we able to decide to choose to spend our time solving mathematical problems. For understandable reasons that have in large part to do with the influence of outmoded theories of instinct (Beach, 1955; Lehrman, 1953), many critics of environmental engineering approaches have equated learning with being forced to perform because of "conditioning" by researchers.

In our work, we have endeavored to develop models that present opportunities to animals rather than forcing them to respond. We have been fortunate in having a number of ways to illustrate unequivocally that some animals have chosen to respond. First, there is the fact that a number of species have worked to obtain food in the presence of identical free food (Markowitz, 1982). This is in keeping with numerous findings first highlighted by Neuringer (1969). Second, in two cases in the Portland Zoo, where primates fed themselves for a number of years by means that required some considerable movement around the cage, youngsters that were never taught how to use these opportunities by the research staff became proficient at feeding themselves as a function of watching older cagemates.

One case involved white-handed gibbons (*Hylobates lar*) who fed themselves by leaping and brachiating several meters in the air between two stations. In spite of opportunity for free feeding in the evening, these gibbons continued to exercise and feed themselves for 7 years. Animals introduced to the cage always learned how to feed themselves by observing other gibbons. Diana monkeys (*Cercopithecus diana*) also moved between widely separated stations in order to feed themselves, with the added contingency that they could deposit a token (which they had earned by exercising) in exchange for food. Juveniles that had never been trained in use of this apparatus and youngsters born after its introduction all learned this complex way to feed themselves by watching the older monkeys. These animals were never deprived. If they chose not to feed themselves, they were fed at the same time as other residents of the primate house. Given the opportunity, animals prefer to gather their own food actively.

Quality of Animals

As primates become increasingly scarce, it is important that we use these valuable animals in the most efficient manner possible. Sometimes this means that we will need to restrict their lives more in order to best accomplish educational or research goals. It is important not to lose sight of the fact that we need to maintain vigorous representatives of these species unless we are willing to settle for rather dull institutionalized specimens. Some authors (e.g., Cheney, 1978) have argued rather cogently that the only way to maintain the best animals in many species may be to introduce predator–prey interactions, even within zoo and laboratory settings. Since public opinion precludes exercising that option in most cases, we will focus instead on some methods that may be used to keep primates both safe and more fit than the average in captivity.

In the Panaewa Rainforest Zoo near Hilo, Hawaii, an exhibit was built that allowed gibbons to deliver their own food by swinging on vines, and at the same time ensured that they could be regularly examined and captured for treatment without discomfort (Markowitz, 1982). This was accomplished by installing sensors within the vines and recording when the animals swung on them. Patterns of movements between vines required to deliver food were selectable by the staff. The food was not delivered near the vines but instead was delivered to the gibbons' hut, which they could only reach by brachiating along a very long hemp-covered cable. This ensured abundant exercise, and when the gibbons entered the hut, it was easy to close the door behind them. We have also designed and briefly tested some apparatus that would allow primates to feed themselves regularly by entering enclosures with electronically controlled closing mechanisms. Using this type of mechanism, primates can be habituated to closing themselves into examination spaces with a minimum of stress. For animals such as gibbons, which often present great difficulties when it is necessary to capture them, utilization of this technique could make life easier for both animals and their keepers.

Many institutions that keep primates are not sufficiently staffed to weigh the animals as often as might be desirable. With the use of contemporary electronics and load cells, it is possible to find relatively inexpensive methods to solve this problem. In the San Francisco Zoo, we have installed an electronic scale in a hollow tree in the dwarf lemur (*Cheirogaleus medius*) exhibit. When the lemurs rest on the platform in the tree, a large fluorescent display indicates their weight. Zoo visitors are encouraged by a graphic presentation to note that these primates weigh less than half a pound. Since the scale is electronic, it would be an easy matter to connect a recorder if the staff wished to keep a continuous record of weighings. Use of such devices can be made in conjunction with behavioral contingencies (e.g., food gathering) to ensure that weighing takes place regularly.

We are also evaluating different designs that will allow laboratory caged primates to be easily weighed without removing them from their home cages. This is especially important for larger, more aggressive animals who must otherwise be tranquilized in order to obtain accurate weights. The most promising procedure involves a universal base incorporating load cell technology. Each individual cage would be labeled with its empty weight and the primates could then be easily weighed by a subtractive process.

Acoustic stimulation is often a means of providing changes in the homes where other techniques may not be currently practical. We have installed speakers, a microphone, and playback systems in the cotton-topped tamarin exhibit in the San Francisco Zoo. At intermittent intervals, these primates hear brief segments of sound of varying kinds. Sometimes these are sounds of the same species recorded in the wild, sometimes sounds of conspecifics recorded in the laboratory, and sometimes they are other sounds indigenous to the cotton-tops' natural habitat. Zoo visitors can also hear the sounds and are encouraged by a recording to watch any changes in behavior and to listen to acoustic responses that the tamarins may make. Ken Gold and Chris Tromborg are currently evaluating the effects of the first weeks of use of this new system.

Current Research

Acoustic stimulation is one part of a number of exploratory efforts in the development of models that may be used to enrich the lives of primates in typical urban research facilities (Spinelli and Markowitz, 1985). Our most ambitious long-term plan is to develop an activity room that will enable monkeys to have daily periods of social activity in rich environments that allow them to exert control in a number of ways.

In the case of rhesus macaques, which are our current focus, this will require that we start with young animals and work carefully and be prepared to deal with inevitable changes in the expression of agonistic behavior as the animals mature (cf. Bernstein and Ehardt, 1985). The development of enrichment devices must be combined with the design of methods for painlessly managing the primates in these less-restricted environments. However, most of our time is being spent on the development of devices for use by individually caged monkeys. This is because it is unlikely that many laboratory facilities will be willing to go to the expense, in terms of both time and personnel, to develop the extensive controls needed to allow mature rhesus to engage in social activities. We are experimenting with devices ranging from commercial video and audio games through "pet" robots that the monkeys can acoustically control.

The most promising and relatively inexpensive devices include a matrix of hemp ropes that the macaques can use to produce musical notes and associated colors and a simple xylophone-like device with eight PVC wands that they can use to control computer-driven sounds and video displays. Having seen that the animals have some interest in playing with these devices, we will progress next to establishing a number of ways that they can use them to control other aspects of their environment. This will include delivering food to themselves, turning on one of several kinds of music that they choose, choosing to play specific games, selecting monkeys in other cages to play interactive games against, etc.

This work that we have begun at the University of California at San Francisco (UCSF) Animal Care Facility addresses a problem more critical in some ways than that in zoos which are, to their credit, slowly developing increasing numbers of complex environments. Almost every laboratory researcher to whom we have spoken has acknowledged some real sense of discomfort about the fact that they do not have richer living environments for their animals. They recognize that primates are in need of much more stimulation than is available in their barren lab cages in order to remain healthy representative members of their species. A number of researchers, especially those using positive reinforcement protocol in their studies, have emphasized that their subjects quite apparently look forward to the experimental sessions. Thus, these researchers recognize implicitly the importance of meaningful activities, in which it matters what the monkeys do.

At the outset of our efforts at UCSF, we invited principal investigators to tell us if they wished us to avoid working with their animals because it might interfere with their studies. Instead of requests that we forgo enrichment activities for any group, investigators lined up to ask that we consider working with their animals first. We are progressing in small increments to expose monkeys that were showing typical negative responses to humans and exhibiting excessive stereotypies to new opportunities and a much wider variety of stimuli. Researchers have expressed encouragement because of their perception that these monkeys have changed dramatically in their responses to humans, apparently welcoming company and enjoying more responsive environments.

Undoubtedly, some of this change in the primates' behavior is a function of increased habituation to the presence of humans who treat them kindly. Indeed, a useful improvement for many animal care facilities would be the scheduling of time for humans to spend positively interacting with monkeys that must be housed individually. However, limited personnel budgets and research designs that often make it unacceptable to tolerate the varieties of responses that individual humans will make in interacting with the primates mean that in many cases the best prospects for enrichment lie in providing devices that respond in standardized, measurable ways to the monkeys' requests.

Many of us who conduct research were trained in an era when the predominant emphasis was on careful control of the circumstances under which we conduct our experiments. Today, there is increasing emphasis on being certain that we have appropriately fit animals as well as controlled environmental conditions for testing. Our last brief suggestion is that providing animals with responsive living conditions can actually enhance rather than encumber research activities. In a direct sense, this can be accomplished by designing behavioral enrichment methods that are properly dovetailed with research needs. For example, if the research group is interested in relating motor coordination to cerebellar functioning, the apparatus left with the animals for use in their home cages may encourage fine motor responses. If the research involves inquiries about the visual system, the stimuli presented by enrichment devices may be selected so that upon entry to the testing situation the animal will already be proficient at visual discrimination problems.

In a less direct, but at least equally important manner, well-conceived environmental enrichment cannot help but produce more appropriate animal models for research. As indicated at the beginning of this chapter, animals that have learned that the environment is responsive to *their* actions are bound to be more representative of their species than subjects intentionally maintained in environments devoid of stimulating activity.

In closing, we would like to invite others to join us in the effort to make the quality of life better for captive animals and in helping ensure that the primates which we maintain are more adequate representatives of their species. Others will undoubtedly find widely contrasting avenues to developing richer lives for captive animals, some of which are bound to be better than those that we develop. We will look forward to incorporating some of these improvements for our own primates.

This is hard, time-consuming, and expensive work. It should not be undertaken without adequate budgets and trained personnel to develop equipment and evaluate the animals' responses to new regimens (Markowitz, 1982). But we believe that it is more than worth the effort and expense. It is time to stop denying that captive environments are inhumane and strive to make them as humane as possible. This must include attention to more than good nutrition and veterinary care. We would suggest that perhaps the single worst thing that one could do for animals who evolved in environments where what they did determined their lives, deaths, and relative status would be to put them in circumstances where nothing they did mattered.

Extremists who complain from a position of ignorance about the treatment of primates used in research and education have clouded over the need for making improvements in animal care by polarizing audiences. Instead of taking adversarial roles with respect to inappropriate criticisms, we need to get our houses in order and make our institutions more humane. Otherwise we may look back in envy upon the time when it was possible to study effectively vigorous representatives of varied primate species.

Conclusions

1. Environmental engineering for captive primates requires careful attention to behavioral opportunities as well as to traditional considerations of safety, reproductivity, physical appearance, and ease of maintenance.
2. Properly designed, responsive environments can enhance opportunities for the animals and simultaneously better serve educational and research purposes.
3. The choice is *not* between providing contingencies to encourage behavior and providing natural exhibits or captive quarters. Nature is full of contingencies. Many of these include dangers to which we are unwilling to expose captive animals. We must not allow this "protection" to leave the animals powerless.

Acknowledgement. This work was supported in part by a grant from the Academic Senate of the University of California, San Francisco.

References

Beach FA (1955) The descent of instinct. Psychol Rev 62:401–410.

Bernstein IS, Ehardt CL (1985) Sex differences in the expression of agonistic behavior in rhesus monkey (*Macaca mulatta*) groups. J Comp Psychol 99:115–132.

Cheney CD (1978) Predator–prey interactions. In: Markowitz H, Stevens VJ (eds) The behavior of captive wild animals. Nelson Hall, Chicago, pp 1–19.

Forthman-Quick DL (1984) An integrative approach to environmental engineering in zoos. Zoo Biol 3:65–77.

Foster-Turley P, Markowitz H (1982) A captive behavioral enrichment study with Asian small-clawed river otters (*Aonyx cinerea*). Zoo Biol 1:29–43.

Gay B et al. (1985) Redefining space allocation for caged nonhuman primates in biomedical research. Workshop presented by Association of Primate Veterinarians, American Society of Primatology.

Hutchins M, Hancocks D, Calip T (1978–1979) Behavioral engineering in the zoo: a critique. Intern Zoo News 25:18–23; 25:18–23; 26:20–27.

Lehrman DS (1953) Problems raised by instinct theories. Quart Rev Biol 28:337–365.

Markowitz H (1977) On natural zoos and unicorns. Keynote address, Conference of the Western Association of Zoological Parks and Aquariums, Seattle, WA.

Markowitz H (1982) Behavioral enrichment in the zoo. Van Nostrand Reinhold, New York.

Markowitz H (1983) Applications of learning theory to captive animal environments. 91st Annual Convention of the American Psychological Association.

Markowitz H (In press) About responsive environments for zoo animals. Zoo Biol.

Markowitz H, Stevens VJ (1978) Behavior of captive wild animals. Nelson Hall, Chicago.

Markowitz H, Stevens VJ, Mellen JD, Barrow BC (1982) Performance of a mandrill (*Mandrillus sphinx*) in competition with zoo visitors and computer on a reaction-time game. Acta Zool Pathol Antwerp 76:169–180.

Neuringer A (1969) Animals respond for food in the presence of free food. Science 166:339–341.

Spinelli J, Markowitz H (1985) The prevention of caging associated distress in laboratory animals. Lab Anim 14:19–28.

Yanofsky R, Markowitz H (1978) Changes in general behavior of two mandrills (*Papio sphinx*) concomitant with behavioral testing in the zoo. Psychol Rec 28:369–373.

40
Research Facility Breeding

DENNIS O. JOHNSEN and LEO A. WHITEHAIR

More than 30 distinct species of nonhuman primates are currently being bred in U.S. facilities to support biomedical and health research and testing programs conducted or sponsored in more than 45 governmental, academic, commercial, and nonprofit organizations. Domestic breeding of nonhuman primates on a large scale represents a relatively new development in this country that has occurred mainly since the early 1970s when sources of wild-caught animals began to be restricted. Most primates used in research at that time were wild caught, and domestic breeding of nonhuman primates for research purposes was estimated to be producing only about 2200 animals (Goodwin, 1975). Losses among imported animals were high as a result of gastrointestinal and respiratory diseases. Fortunately, the impending shortage of imported animals was anticipated, and substantial sums were invested in building up capacity for domestic breeding. Government grants and contracts for domestic primate breeding between 1972 and 1982 totaled well in excess of $20 million.

By 1982, as many as 6500 rhesus monkeys (*Macaca mulatta*) were probably being produced in breeding programs in the United States. This easily represented more than a fivefold increase in just 10 years. Critical needs for these animals in testing polio vaccine were met, and anticipated shortages were eased. However, the establishment of such domestic programs alone does not tell the whole story. The fundamental rationale for the government's role in supporting much of the primate breeding was to assure a stable source of high quality primates for research programs that could eventually be sold at full cost, thereby paving the way for gradually phasing out direct government support and subsidies. Because of the dramatic decrease in the supply of primates, caused primarily by the total ban on primate exports imposed by India in April 1978, and the need to recover the full cost of domestic breeding programs, the price of a 2-year-old rhesus monkey rose precipitously from approximately $50 in 1970 to well over $1,000 at present.

Accordingly, investigators have had to make difficult decisions about using fewer primates in research and seeking other models. On the positive side, powerful incentives were created to use animals more efficiently. This led to significant interest in reducing losses from disease, increasing breeding efficiency, and the sharing of primates through multiple-use protocols and mechanisms such as the Washington Regional Primate Research Center's Primate Information Clearinghouse. One indicator of how the quality of research primates has increased during this relatively short period is the fact that tuberculosis, once a common disease, is now relatively uncommon.

The macaque species clearly represent the most common group of nonhuman primates being bred. Among these, more rhesus monkeys are being produced than the other nonhuman primate species combined. As shown in Table 40.1, macaques are being bred in the United States at a number of locations, and at least 7900 live-born offspring were produced in 1984.

In reviewing the breeding of macaques, we will have reference to the rhesus monkey since the methods used to breed them are generally applicable to the other species as well. Four general types of breeding colonies are presently in use: free-ranging island colonies, semifree-ranging corral colonies, pen or run type single male harem colonies, and colonies where animals are cagemated in pairs. In terms of numbers, most breeding occurs on several colonies located on Morgan Island in South Carolina, Key Lois and Raccoon Key in the Florida Keys, and Cayo Santiago off the east coast of Puerto Rico. Corral breeding is generally carried out in large enclosures that are populated with multimale groups, females, and offspring; the populations in these corrals may exceed 100 or more animals. The most extensive corral breeding programs are located at the Delta, California, and Oregon Regional Primate Research Centers where this method is considered the most cost effective. Pen breeding is usually done in corn cribs or runs where there are from six to eight female breeders, their progeny including future breeders, and a single adult male breeder. Representative of such colonies are the NIH center at Perrine, Florida, and Hazleton Research Animal's Primate Center in Alice, Texas. Finally, the methods for caged matings are well established and meet special needs for pregnancies and fetal research.

Table 40.4 shows representative production statistics from each of the multianimal breeding systems. Reproduction is good in each system, and significantly more than one offspring can be expected per year for every two female breeders. There are a few clear advantages and disadvantages with each system. For example, in corral breeding there are usually more traumatic injuries. It is also difficult to reintegrate animals into their groups after they have been removed for more than a short time for hospitalization or other reasons. With island colonies, it is often difficult to monitor individual animals, and support costs may be high.

There are at least two major factors that have proven to be significant in improving these breeding programs. One is an amalgamation of previously diverse technical expertise, primarily that represented by the veterinarian and animal husbandry staff and that represented by primate behaviorists. This staffing will also increasingly benefit from the inclusion of geneticists. The other factor relates to computerization of data and information processing where great advances have occurred in recent years. Use of these new capabilities now permits breeding systems to be compared, evaluated, and ultimately to be refined and improved upon.

It is more difficult to make generalizations concerning the results of breeding other macaque species. Cynomolgus monkeys (*Macaca fascicularis*) have been bred because they are a potential substitute for the rhesus monkey. However, since wild-caught animals are still being imported, there has been no real urgency associated with breeding them. Experience to date indicates that cynomolgus monkeys do not seem to tolerate cold climates as well as rhesus monkeys. Experience with the other macaque species is somewhat anecdotal since it has been

TABLE 40.1. Principal U.S. institutions breeding nonhuman primate species of Asian origin for research purposes during 1984.

Rhesus Monkeys (*Macaca mulatta*)
- Hazleton Research Animals, Alice, TX
- Charles River Breeding Laboratories, Inc., Marathon Key, FL
- Litton Bionetics, Morgan Island, SC
- Delta Regional Primate Research Center, Covington, LA
- University of Puerto Rico, Sabana Seca, PR
- Oregon Regional Primate Research Center, Beaverton, OR
- National Institutes of Health, Perrine, FL
- Yerkes Regional Primate Research Center, Atlanta, GA
- Wisconsin Regional Primate Research Center, Madison, WI
- U.S. Navy, Pensacola, FL
- Gulf South Research Institute, New Iberia, LA
- New England Regional Primate Research Center, Southborough, MA
- University of Pittsburgh, Pittsburgh, PA
- California Regional Primate Research Center, Davis, CA
- Bowman Gray School of Medicine, Winston Salem, NC
- Centers for Disease Control, Atlanta, GA
- Coulston International, White Sands, NM
- Johns Hopkins University, Baltimore, MD
- Institute for Primate Research, Holloman AFB, NM
- University of Texas, Bastrop, TX
- Manheimer Foundation, Homestead, FL
- Yale University, New Haven, CT
- University of Illinois, Chicago, IL
- Eastern Virginia Medical College, Norfolk, VA

Estimated number of actual and potential breeders: 24,934

Live births: 5582

Pig-tailed macaque (*Macaca nemestrina*)
- Washington Regional Primate Research Center, Seattle, WA
- University of Colorado, Denver, CO
- Manheimer Foundation, Homestead, FL
- Yerkes Regional Primate Research Center, Atlanta, GA
- Pennsylvania State University, Hershey, PA

Estimated number of actual and potential breeders: 1536

Live births: 358

Crab-eating macaque (*Macaca fascicularis*)
- New England Regional Primate Research Center, Southborough, MA
- Bowman Gray School of Medicine, Winston Salem, NC
- Washington Regional Primate Research Center, Seattle, WA
- Hazleton Research Animals, Alice, TX
- California Regional Primate Research Center, Davis, CA
- Manheimer Foundation, Homestead, FL
- Delta Regional Primate Research Center, Covington, LA
- Gulf South Research Institute, New Iberia, LA
- Institute for Primate Research, Holloman AFB, NM
- University of Texas, Bastrop, TX
- Vanderbilt University, Nashville, TN

Estimated number of actual and potential breeders: 2232

Live births: 325

(Continued)

TABLE 40.1. (Continued).

Japanese macaque (*Macaca fuscata*)
Arashiyama West Primate Center, Dilley, TX
Oregon Regional Primate Research Center, Beaverton, OR

Number of actual and potential breeders: 600

Live births: 107

Stump-tailed macaque (*Macaca arctoides*)
Bowman Gray School of Medicine, Winston Salem, NC
Yerkes Regional Primate Research Center, Atlanta, GA
University of Illinois, Chicago, IL
University of Puerto Rico, Sabana Seca, PR
Wisconsin Regional Primate Research Center, Madison, WI
Manheimer Foundation, Homestead, FL
Brown University, Providence, RI
Pennsylvania State University, Hershey, PA

Estimated number of actual and potential breeders: 386

Live births: 44

Bonnet macaque (*Macaca radiata*)
California Regional Primate Research Center, Davis, CA

Number of actual and potential breeders: 149

Live births: 18

Taiwan rock macaque (*Macaca cyclopis*)
New England Regional Primate Research Center, Southborough, MA

Number of actual and potential breeders: 48

Live births: 4

Celebes black ape (*Macaca nigra*)
Oregon Regional Primate Research Center, Beaverton, OR
Manheimer Foundation, Homestead, FL
Yerkes Regional Primate Research Center, Atlanta, GA

Number of actual and potential breeders: 115

Live births: 9

Orangutan (*Pongo pygmaeus*)
Yerkes Regional Primate Research Center, Atlanta, GA

Number present: 34

Live births: 6

White-handed gibbon (*Hylobates lar*)
Yerkes Regional Primate Research Center, Atlanta, GA

Number present: 21

Live births: 2

Tree shrew (*Tupaia glis*)
Vanderbilt University, Nashville, TN

Number of breeders: 100

Approximate number of live births: 135

gained from a more restricted number of sites. Success similar to that realized in breeding the rhesus monkey has been seen in breeding pig-tailed, stump-tailed, Japanese, and bonnet macaques (*Macaca nemestrina*, *arctoides*, *fuscata*, and *radiata*, respectively). At the Washington Regional Primate Research Center, where the largest domestic breeding colony of pig-tailed macaques is located, there has been considerable concern with the impact of a fatal immune deficiency disease similar to human AIDS that causes retroperitoneal fibromatosis. Periodic clinical outbreaks of *Herpesvirus B* infections have had a negative effect on breeding performance in bonnet macaques at the California Regional Primate Research Center, which has the major colony of these animals (R. Hendrickson, 1985, personal communication). The breeding performance of Celebes black apes has also not been impressive. A viral disease that results in a fatal immunodeficiency disorder has also been responsible for the deaths of a number of these animals at the Oregon Regional Primate Resarch Center. Experience in breeding Taiwan rock macaques (*Macaca cyclopis*) has been similar, but Japanese macaques (*Macaca fuscata*) have bred exceedingly well in large outdoor corral settings such as the colony at the Oregon Regional Primate Research Center.

With respect to other species of nonhuman primates of Asian origin that are being bred for research use, no extensive amount of activity currently exists. Gibbons, in particular the white-handed gibbon, *Hylobates lar*, earlier offered a promising small ape model for certain types of cancer and communicable disease research. The methods for successfully pair breeding of these animals in enclosures that provided adequate room for exercise has been well described (Breznock et al., 1977). However, the very small number of animals in captivity and present restrictions on their importation make it unlikely that they will continue to be bred. With only one Regional Primate Research Center possessing a small number of orangutans, this is probably the case with *Pongo pygmaeus*, too. Significant attention will not be given here to other Asian species of possible research value including prosimians or the leaf monkey, *Presbytis entellus*. These animals have been successfully bred and maintained in captivity and do have demonstrated value as models for particular types of research.

Breeding of species of African origin in captivity for research purposes is currently an area of considerable activity. In the early 1970s, African nonhuman primates were not being extensively bred in the United States. However, largely because of the expansion of baboon breeding, African species now rank second in numbers of primates bred domestically. Baboons are equally or perhaps even more prolific than the rhesus monkey, are of a large and more desirable size for certain uses, and people often prefer to work with them more than with the rhesus monkey (Goodwin and Coelho, 1982). While figures concerning baboon breeding activity are presented in Table 40.2, the subject of baboon breeding will not be further addressed in this chapter since Dr. Goodwin treats it elsewhere in this volume.

Chimpanzees represent a quite different story. They can be expected to breed with significantly less prolificacy than the baboon and rhesus monkey and are expensive and often dangerous to maintain. Based on past experience, it is not unreasonable to expect that three years on the average will be required per breeding female chimpanzee to produce and rear an offspring to the age of 2 years. Considering maintenance costs for the mother, the baby, and the part-time presence of a male, the direct cost today for a 2-year-old chimpanzee can easily

TABLE 40.2. Principal U.S. institutions breeding nonhuman primate species of African origin for research purposes during 1984.

Baboon (*Papio* spp.)
- Southwest Foundation for Biomedical Research, San Antonio, TX
- Washington Regional Primate Research Center, Seattle, WA
- Manheimer Foundation, Homestead, FL
- Oregon Regional Primate Research Center, Beaverton, OR
- University of Illinois, Chicago, IL

Estimated number of actual and potential breeders: 3225

Live births: 425

Chimpanzees (*Pan* spp.)
- Primate Research Institute, Holloman AFB, NM
- Southwest Foundation for Biomedical Research, San Antonio, TX
- Coulston International, White Sands, NM
- Gulf South Research Institute, New Iberia, LA
- Yerkes Regional Primate Research Center, Atlanta, GA
- University of Texas, Bastrop, TX
- Lab. for Experimental Surgery and Medicine in Primates, Tuxedo, NY
- Primate Foundation of Arizona, Tempe, AZ
- University of Oklahoma, Norman, OK

Estimated number of actual and potential breeders: 1039

Live births: 104

African green monkey (*Cercopithecus aethiops*)
- Federated Medical Resources, Honeybrook, PA
- Bowman Gray School of Medicine, Winston Salem, NC
- Delta Regional Primate Research Center, Covington, LA
- Veteran's Administration Medical Center, Sepulveda, CA

Estimated number of actual and potential breeders: 1097

Live births: 102

Sooty Mangabey (*Cercocebus atys*)
- Yerkes Regional Primate Research Center, Atlanta, GA
- Manheimer Foundation, Homestead, FL

Estimated number of actual and potential breeders: 76

Live births: 13

Patas monkey (*Erythrocebus patas*)
- University of Puerto Rico, Caribbean Primate Research Center, Sabana Seca, PR
- Bowman Gray School of Medicine, Winston Salem, NC
- Delta Regional Primate Research Center, Covington, LA

Estimated number of actual and potential breeders: 271

Live births: 40

Gorilla (*Gorilla gorilla*)
- Yerkes Regional Primate Research Center, Atlanta, GA

Number present: 21

Live births: 2

Lemur (9 *Lemur* spp.)
- Duke University Center for Study of Primate Biology and History, Durham, NC
- Oregon Regional Primate Research Center, Beaverton, OR
- New England Regional Primate Research Center, Southborough, MA

Estimated number of actual and potential breeders: 319

Live births: 41

TABLE 40.2. (Continued).

Galago (*Galago* spp.)
Duke University Center for the Study of Primate Biology and History, Durham, NC
Oregon Regional Primate Research Center, Beaverton, OR
Wisconsin Regional Primate Research Center, Madison, WI
Estimated number of actual and potential breeders: 135
Live births: 111

exceed $30,000. One lesson learned from breeding chimpanzees is that animals isolated from their mothers and family social structure often grow up to be incompetent breeders. It is also true that female interbirth intervals may be extended, as they are in people, during periods of nursing. Accordingly, there may be a costly tradeoff between leaving infants long enough to assure proper social development if they are to become breeders and weaning them early enough so that pregnancies in the mother will occur sooner. In all likelihood, one of the most popular configurations for breeding chimpanzees is in an indoor and outdoor pen arrangement such as the University of Texas' facility in Bastrop where breeding is done in harem groups (Riddle et al., 1982). In situations where infants are removed from their mothers and nursery reared, experience at the Yerkes Regional Primate Research Center suggests that annual conception rates of 90% can be expected and that 80% of these pregnancies will result in live births (B. Swenson, 1985, personal communication). Infant mortality is typically between 5% and 20%; however nursery rearing of rejected infants can significantly reduce neonatal losses.

The United States has not imported chimpanzees for biomedical research since the implementation of the Convention on International Trade in Endangered Species in 1974 and will likely not import any in the future. Because of the relatively small numbers of behaviorally normal, experimentally "clean" animals, we are virtually assured that it will not be possible to breed chimpanzees here beyond the year 2000 unless positive steps are taken to meet that objective today. One approach under consideration is for NIH to include in its 1987 budget submission a funding request to support the implementation of a national chimpanzee management plan (Interagency, 1985). Its purpose is to assure the existence of a pool of breeders to produce indefinitely adequate numbers of chimpanzees for research and to meet future needs here for breeding stock. If such a plan is not strongly supported now, estimates indicate that there will be no clean chimpanzee breeding stock remaining in the United States by the turn of the century.

With respect to the breeding of other African species, African green monkeys (*Cercopithecus aethiops*), patas monkeys (*Erythrocebus patas*), and sooty mangabeys (*Cercocebus atys*) are being bred in significant numbers and in configurations similar to those used for macaque species (Kushner et al., 1982; Sly et al., 1983). Reports of success are mixed between the several colonies shown in Table 40.2. Current indicators suggest that they may not continue to be bred here in sufficient numbers to assure that domestic colonies will be self-sustaining. However, it should be noted that potential sources fo this species exist in feral populations on the Caribbean islands of Barbados and St. Kitts. The future for breeding gorillas and prosimians is less bright. These primates are not in

TABLE 40.3. Principal U.S. institutions breeding nonhuman primate species of American origin for research purposes during 1984.

Squirrel monkey (*Saimiri sciureus*)
University of South Alabama, Mobile, AL
Gulf South Research Institute, New Iberia, LA
Delta Regional Primate Research Center, Covington, LA
Hoffman-La Roche, Inc., Nutley, NJ
Stanford University, Palo Alto, CA
California Regional Primate Research Center, Davis, CA
Bowman Gray School of Medicine, Winston Salem, NC
New England Regional Primate Research Center, Southborough, MA
University of Puerto Rico, Sabana Seca, PR
Pennsylvania State University, Hershey, PA
National Institutes of Health, Poolesville, MD
Oregon Regional Primate Research Center, Beaverton, OR
Michigan State University, East Lansing, MI
Goucher College, Towson, MD

Estimated number of actual and potential breeders: 2248

Live births: 293

Cotton-topped tamarin (*Saguinus oedipus*)
Oak Ridge Associated Universities, Oak Ridge, TN
New England Regional Primate Research Center, Southborough, MA
University of Texas, Houston, TX
Wisconsin Regional Primate Research Center, Madison, WI
Texas A&M University, College Station, TX
National Institutes of Health, Poolesville, MD

Estimated number of actual and potential breeders: 571

Live births: 122

Tamarins other than cotton-tops (*Saguinus* spp.)
Centers for Disease Control, Atlanta, GA
Monel Institute, University of Pennsylvania, Philadelphia, PA
Oak Ridge Associated Universities, Oak Ridge, TN
Laboratory for Experimental Medicine and Surgery in Primates, Tuxedo, NY
Presbyterian-St. Luke's Hospital, Chicago, IL

Estimated number of actual and potential breeders: 368

Live births: 77

Common marmoset (*Callithrix jacchus*)
U.S. Army, Edgewood Arsenal, Edgewood, MD
Vanderbilt University, Nashville, TN
Texas A&M University, College Station, TX
Oak Ridge Associated Universities, Oak Ridge, TN
New England Regional Primate Research Center, Southborough, MA
Marmoset Breeding Farm, Riverview, FL

Estimated number of actual and potential breeders: 443

Live births: 125

Owl monkey (*Aotus trivirgatus*)
University of Missouri, Columbia, MO
Walter Reed Army Institute of Research, Forest Glen, MD
National Institutes of Health, Poolesville, MD
New England Regional Primate Research Center, Southborough, MA

Estimated number of actual and potential breeders: 390

Live births: 40

TABLE 40.1. Continued.

Cebus spp.
Gulf South Research Institute, New Iberia, LA
Manheimer Foundation, Homestead, FL
University of Puerto Rico, Sabana Seca, PR
Yerkes Regional Primate Research Center, Atlanta, GA
Number present: 305
Live births: 38
Ateles spp.
Gulf South Research Institute, New Iberia, LA
Number present: 200
Live births: 38
Callicebus spp.
California Regional Primate Research Center, Davis, CA
Number present: 36
Live births: 8

widespread demand for research use but are being successfully bred at the institutions indicated in Table 40.2.

Breeding of New World species does not generally reflect the type of success encountered in breeding macaques and baboons. The New World species most commonly bred and used in the United States is the squirrel monkey, *Saimiri sciureus*. Subspecies distinctions will not be discussed in this presentation, but their significance in breeding these and other New World species is probably important. The discovery that vitamin D3 is a dietary requirement was an important breakthrough in successfully maintaining and breeding these animals in captivity (Hunt et al., 1967). The principal institutions engaged in breeding squirrel monkeys for research purposes are shown in Table 40.3 with total numbers of animals produced in 1984. If the saying "less is better" has any place in this talk, it may well apply here. Colonies such as those at Monkey Jungle near Miami and the much smaller feral colony outside the cages at the Caribbean Primate Center in Puerto Rico offer two good examples (Fontaine and Hench, 1982). Data from the Puerto Rican colony show that nearly 100% of the free-ranging adult females deliver and carry young each year (B. Marriott, 1985, personal communication). More controlled approaches have produced less satisfying results, and a considerable amount of effort is being directed toward the improvement of methods that affect breeding performance in such settings.

A "state of the art" production unit for squirrel monkeys is located at the University of South Alabama Medical Center in Mobile. Animals are bred there in more than 40 modular pens, each providing about 500 ft^3 of space. They are housed in harem groups that basically consist of one male and nine females. Based on experience to date at this activity, colony breeding performance statistics for squirrel monkeys are depicted in Table 40.4. Some hard lessons have been learned in the breeding of squirrel monkeys. For example, it used to be a standard practice to wean offspring and to hold them for rearing in natal groups. Maternal performance and the survival of infants born to such animals were found to be poorer than those of wild-born animals. It is now considered better to leave future

TABLE 40.4. Breeding experience with selected species.

	Breeding systems	% adult females conceiving	% conceptions resulting in live birth	% survival of live birth to one year
Rhesus monkey	Island[1]	77	93	93
	Corral[2]	84	93	95
	Pen/harem[3]	67	86	90
Squirrel monkey	Pen/harem[4]	43	79	50
Cotton-topped tamarin	Caged pairs[5]	118	80	41
Owl monkey	Caged pairs[5]	95	70	64

[1–5]Based on information provided by: [1]Litton Bionetics, Inc., Morgan Island, SC; [2]Delta Regional Primate Research Center, Covington, LA; [3]Hazleton Research Animals, Alice, TX; [4]University of South Alabama, Mobile, AL; [5]New England Regional Primate Research Center, Southborough, MA.

breeders in family groups to provide the type of experience that they seem to require to perform well as breeders. Currently, one of the most pressing needs for research in squirrel monkeys, as in a number of captive primate species, is clearly the improvement of neonatal survival.

Captive breeding of tamarins represents variations on the same general theme. The cotton-topped tamarin, *Saguinus oedipus*, is a useful model for the study of human cancer of the colon, and the moustached tamarin, *Saguinus mystax*, is a preferred model for research on certain types of hepatitis. Several other *Saguinus* species and their various hybrids are also useful research animals. Table 40.3 shows the principal research facilities in the United States where research breeding of tamarins is occurring. Tamarins are most successfully bred in male–female family pairs with associated offspring. The Oak Ridge Associated Universities facility offers a good example of a tamarin and marmoset production unit. Some statistics that indicate what is reasonable to expect a breeding colony of tamarins to produce are presented in Table 40.4. A disease process called the "marmoset wasting syndrome" has represented a very serious threat to the survival of these animals, and a concerted effort to deal with it seems to have been helpful, although for reasons that presently are not completely understood. Success has been ascribed to better nutrition, provision of more space, an enriched environment, and improved husbandry and management. Breeding these animals seems to be more of an art than with other nonhuman primate species and is very much related to staff management and husbandry in a particular facility. One of the very useful research findings to emerge on this front was the discovery that ovulation is suppressed in females sharing a cage with a dominant female (Dawson and Dukelow, 1976).

The breeding of owl monkeys (*Aotus trivirgatus*), the common marmoset (*Callithrix jacchus*), and the cebid species is generally a more promising story. The owl monkey is useful primarily for malaria and eye research and for some time did not breed successfully in captivity. Several research breakthroughs have changed that pattern. The discovery of the acute and fatal sensitivity of these animals to infection with herpesviruses of other primate species, primarily the squirrel monkey, was important (Hunt and Melendez, 1966). This discovery showed the necessity of isolating owl monkeys from other primate species. The second discovery was the requirement of stimulating a more natural environment, which led to the adoption of reverse light cycles and provision of nest boxes. Finally, this was probably the first species where it was realized that a number of subspecies exist that were distinguishable only on the basis of karyotyping and that this could be an important consideration in establishing successful matings (Ma et al., 1976). Owl monkeys are bred in pairs like tamarins in cages provided with a nest box. The sites at which breeding of owl monkeys is presently being carried out and the number of live births in 1984 are shown in Table 40.3. Production statistics drawn from one of these colonies are shown in Table 40.4.

The common marmoset, *Callithrix jacchus*, is a species that has not really found a mission in the U.S. research laboratory. It has become a popular nonhuman primate model for toxicology studies in the United Kingdom, in no small part because of its fecundity, convenient size, and ease of handling. Breeding experience with this species in the United States, although not extensive, is similar. Several institutions in the United States are in the process of phasing out their colonies, and the trend is presently against this animal becoming well established here as a breeding resource for supplying research needs.

The situation with the larger South American species is similar. As Table 40.3 indicates, *Cebus*, *Ateles*, and *Callicebus* species are being bred for research purposes in the United States. The first two genera are being bred in very modest numbers and with relative ease. However, they are not available in sufficient numbers to be really useful, and their popularity, at least for research, is on the wane. Many of the animals on hand are simply being maintained, and the production of young is a by-product, not a primary objective.

In summary, it is both interesting and revealing to track the progress of breeding programs for meeting research needs since 1973. The differences between number of births in 1973 and 1984 are shown in Table 40.5. We have well-established resources available now for breeding rhesus monkeys, several other macaque species, baboons, and possibly chimpanzees on a self-sustaining basis. Methods for breeding a number of other species, including African green monkeys, patas monkeys, cebus and spider monkeys, owl monkeys, the common marmoset, and the cotton-topped and several other species of tamarins, have also been fairly well worked out. There is now a sufficient amount of fundamental knowledge to breed these animals in large numbers if interest, support, and adequate numbers of new breeders are provided. We must be concerned about this and assure that reasonable numbers of future breeders are retained since there may be few, if any, opportunities in the future to obtain breeders. If we are not, our endowment in breeding resources for these species may be irretrievably spent.

There are other actions that can be taken to help assure that these animals will continue to be bred here. First, it is in our best interest to support strongly efforts to conserve and manage wild populations. Establishing reserves and otherwise building walls around them may not be the only, or the best, way to do this. We must recognize that these animals have economic as well as altruistic value. Populations that are properly managed for sustained yield can be a source of long-term income and livelihood like the cropping of other forest products. More than $200 million annually is being spent on social forestry and reforestation programs in developing countries by donors such as the World Bank and the U.S. Agency for International Development. The focus of these programs is unfortunately on trees, and the value of animals mostly goes unrecognized. Beyond

TABLE 40.5. Nonhuman primate live births in the United States for species used in biomedical research: 1973[1] and 1984[2].

Species	Live births 1973	Live births 1984
Macaca mulatta	991	5582
Saimiri sciureus	185	293
Papio spp.	120	425
Macaca fascicularis	94	313
Macaca nemestrina	241	358
Aotus trivirgatus	2	40
Pan spp.	47	104
Cercopithecus aethiops	33	84
Macaca arctoides	68	44
Saguinus spp.	250	199
Cebus spp.	37	38
Macaca radiata	66	18
Galago crassicaudatus	64	53
Callithrix jacchus		75
Erythrocebus patas		40
Macaca fuscata		107
Tupaia glis		135

[1] Goodwin, 1975.
[2] Only those species listed whose total U.S. population in research-related activities was greater than 200 animals.

this, much field research is needed to establish how such habitats for nonhuman primates should be effectively managed and monitored. We need to be more active in this area than we presently are.

We also need to do more research closer to home. The biomedical research and the zoo communities have much to offer each other, both as complementary resources for genetic material and for carrying out research of mutual benefit. It is possible to offer several good examples of where this is occurring and why it should be happening more. It is satisfying that research on the technology of sperm and ova preservation, embryo transplantation and manipulation, in vitro fertilization, and artificial insemination is being carried out in the Primate Centers, several zoo programs, and other primate laboratories. There has also been substantial progress related to diseases that have threatened the existence of our nonhuman primate breeding programs. A prime example has been the very fine work that has been conducted on acquired immunodeficiency diseases.

It will be a real race to determine if the continuing availability of many of these primates in the United States and elsewhere can be assured. Rather than simply standing by and watching this race as the sand runs out of the glass, much effort needs to be given to managing better the fate of presently available and future breeding stocks such as is being attempted with chimpanzees. This needs to be coupled with relevant research and strengthened international participation that will require particular cultivation in developing countries where habitat destruction and the threat it poses to wildlife populations is the greatest. The challenge is indeed a formidable one, but the road to self-sustaining primate populations is not impassible.

Acknowledgment. The authors gratefully acknowledge the cooperation of those institutions and individuals indicated for providing much of the data and information contained in this report.

References

Breznock AW, Harrold JB, Kawakami TG (1977) Successful breeding of the laboratory housed gibbon (*Hylobates lar*). Lab Anim Sci 27:222–228.

Dawson GA, Dukelow WR (1976) Reproductive characteristics of free-ranging Panama tamarins (*Saguinus oedipus geoffroyi*). J Med Primatol 5:266–275.

Fontaine R, Hench KM (1982) Breeding New World monkeys at Miami's Monkey Jungle. Intern Zoo Yrbk 22:77–84.

Goodwin WJ (1975) Current status of primate breeding in the United States. In: Breeding simians for developmental biology, vol. 6. Laboratory Animals Ltd., Epping-Essex, pp 151–156 (Laboratory Animal Handbooks).

Goodwin WJ, Coelho AM Jr (1982) Development of a large scale baboon breeding program. Lab Anim Sci 32:672–676.

Hunt RD, Melendez LV (1966) Spontaneous *Herpes t* infection in the owl monkey (*Aotus trivirgatus*). Path Vet 3:1–26.

Hunt RD, Garcia FG, Hegsted DM, Kaplinsky N (1967) Vitamin D2 and D3 in New World primates: influence on calcium absorption. Science 157:943–945.

Interagency Research Animal Committee. National Chimpanzee Management Program, Chimpanzee Task Force, Primate Research Institute, Holloman Air Force Base, New Mexico, April 7–8, 1985 (draft). Available from IRAC, National Institutes of Health, Bethesda.

Kushner H, Kraft-Schreyer N, Angelakos ET, Wudarski EM (1982) Analysis of reproductive data in a breeding colony of African green monkeys. J Med Primatol 11:77–84.

Ma NSF, Jones TC, Miller AC, Morgan LM, Adams EA (1976) Chromosome polymorphism and banding patterns in owl monkeys (*Aotus*). Lab Anim Sci 26:1022–1036.

Riddle KE, Keeling ME, Alford PL, Beck TF (1982) Chimpanzee holding, rehabilitation, and breeding: facilities design and colony management. Lab Anim Sci 32:523–533.

Sly DL, Harbaughy SW, London WT, Rice JM (1983) Reproductive performance of a laboratory breeding colony of patas monkeys (*Erythrocebus patas*). Amer J Primatol 4:23–328.

41
Research Uses and Projections of Nonhuman Primates as Research Subjects

WILLIAM I. GAY

This paper discusses the use of nonhuman primates in medical research in the United States. It does not include data from the many excellent field studies of nonhuman primates that have been supported by government and private biomedical research agencies. These animals have not been removed from their natural habitat, and the studies have not altered their life expectancy.

The importance of nonhuman primates in various categories of medicine has been reviewed by King and Yarbrough (1985). The susceptibility of these animals to most infectious agents of humans and the similarity of their immune, reproductive, and nervous systems to their fellow primate, the human, have long been realized. As more obscure and difficult chronic diseases become national research priorities, continued dependence on a spectrum of primate species is inevitable.

Each research grant project (and subproject) supported by the National Institutes of Health is coded (catalogued) for the species of research animals used. Table 41.1 gives numerical data for the number of funded research projects that used nonhuman primates in 1983. This usage has varied very little in the past 5 years. Figure 41.1 is a bar graph comparing the same data.

Information on the types of animals used in research projects does not, of course, tell us how many animals are used in the various categories of medical research. These data will have to await a national survey expected to be conducted by the National Academy of Sciences. Meanwhile, the best information available on total use of nonhuman primates in research is shown in Table 41.2 (Greenhouse, 1984). The data must not be interpreted as numbers of animals used for experiments in one year, because they report animals on hand at the end of a year (inventory), as well as those purchased or raised. It is estimated that at least 50% of these primates are holdovers from a previous year. Dr. Dennis Johnsen covers the subject of primate breeding in Chapter 40 of this volume. The trend in numbers used is noted in the USDA table (Table 41.2). There is a percentage increase in the use of primates when compared with other large laboratory animals (Table 41.3), even though there is a decline in total numbers (Table 41.2). This trend indicate that these species are becoming established as vital to medical research.

Certain categories or disciplines of research are thought to be associated with extensive use of the nonhuman primates. We have obtained data from the NIH record system which show that all categories of research (Table 41.4) and all

TABLE 41.1. Numbers of NIH-funded research projects using nonhuman primates in fiscal year 1983.

Bid	Human projects	Total dollars	Nonhuman projects	Total dollars	Total	Total dollars
Aging	4	503,865	7	477,125	11	980,990
Allergy/inf. dis.	18	1,732,215	16	1,916,125	34	3,648,618
Arthritis/metab.	28	3,597,296	32	3,023,549	60	6,620, 845
Cancer	43	4,795,058	36	4,736,448	79	9,531,506
Dental	9	705,741	28	2,361,118	37	3,066,859
Environment.	5	1,076,748	26	2,342,563	31	3,419,311
Eye	37	4,690,162	118	12,517,378	155	17,207,540
Gen. med.	15	1,516,874	23	2,549,987	38	4,066,861
Child health	26	2,420,869	105	9,781,742	131	12,202,611
Heart/lung	65	9,128,299	112	19,051,143	177	28,179,442
Neurology	41	3,907,780	131	13,886,638	172	17,794,418
Res. resources	1	43,908	88	27,763,428	89	27,807,336
Res. services	0	0	8	1,627,211	8	1,627,211
Totals	292	34,118,815	730	102,034,733	1,022	136,153,548

disciplines of research (Table 41.5) use all common laboratory animal species. The categories of research for NIH are represented by our research Institutes; for example, Heart, Cancer, Eye, and Allergy (Table 41.4). When proposals are reviewed at NIH, they are reviewed for scientific content by NIH initial review groups or study sections organized by discipline, such as surgery, pathology,

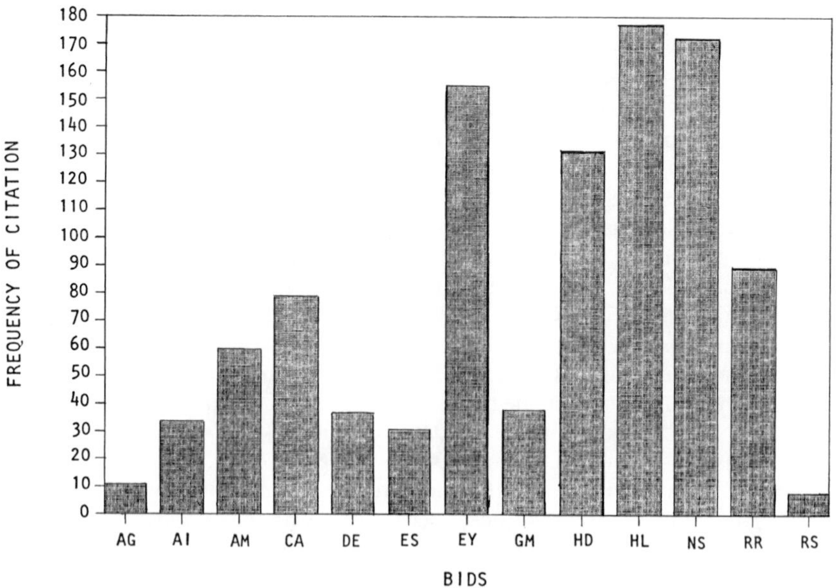

FIGURE 41.1. Bar graph of data in Table 41.1.

TABLE 41.2. Primates used in experimentation within the United States.

Fiscal year	Number of primates	Total number of animals[1]	Percent primates	Total number of reporting facilities[2]
1978	57,009	1,687,201	3.4	1072
1979	59,359	1,832,045	3.2	1211
1980	56,024	1,661,904	3.4	975
1981	57,515	1,658,439	3.5	1050
1982	46,388	1,576,556	2.9	1016
1983	54,926	1,680,242	3.3	1005

[1] Number does not include laboratory rats or mice.
[2] Number includes all reporting facilities, not just those using primates.
Source: Data from Animal Plant Health Inspection Service, U.S. Department of Agriculture, Report for 1978–1984.

TABLE 41.3. Trends in use of large laboratory animals as summarized by the Interagency Research Animal Committee from USDA Reports.

	Average			Percent	Change
	1973–1978	1979–1984	1973–1984	1973–1984	1973-1978–1979-1984
Registered research facilities	988	1129	1059	37	14
Total	1618	1747	1714	25	8
Dogs	188	187	188	3	−0.5
Cats	65	59	62	−14	−9
Primates	48	55	51	31	13
Guinea pigs	421	469	445	37	11
Hamsters	442	389	416	−4	−12
Rabbits	460	489	475	18	6
Wild animals	57	97	77	510	71

Source: Dr. Tom Wolfle, DRS/NIH, 1985, personal communication.

TABLE 41.4. Comparison of use of primates vs. other animal species in projects funded by NIH categorical institutes in fiscal year 1980.

Species	Allergy and infectious disease	Cancer	Eye	Heart/lung	Neurological diseases	Mental health
Artiodactyla[1]	41	87	9	308	27	6
Carnivores	50	169	145	1037	363	40
Lagomorphs[2]	308	352	216	561	123	25
Marine	—	—	—	4	—	—
Marsupials	—	1	2	2	6	—
Perissodactyla[3]	8	3	—	14	—	2
Primates	53	159	177	219	200	112
Rodents	817	2761	220	1155	920	334

[1] Ungulates (deer, pigs, ruminants).
[2] Rabbits.
[3] Horse, tapir, rhinoceros.

TABLE 41.5. Comparison of use of primates vs. other animal species by discipline in medical research.

	Total number applications	Rabbit	Dog	Cat	Primate	Farm Animal
Bacteriology and mycology	151	66	4	1	1	6
Biopsychology	186	2	4	12	27	1
Cardiovascular, renal, and pulmonary	587	78	213	48	14	54
Communicative sciences	283	3	4	32	29	1
Embryology/reproduction	572	84	15	6	49	86
Endocrinology	209	30	9	2	6	37
Immunobiology	696	182	14	1	7	27
Mammalian genetics	246	15	—	1	4	—
Metabolism/gen. med.	457	49	32	2	9	20
Molecular biology	224	5	—	—	—	2
Neurological sciences	767	69	18	127	48	10
Nutrition	171	3	2	1	5	6
Oral biology and medicine	283	20	13	7	24	17
Parasitology/trop. med.	211	3	5	3	7	6
Pathology	560	70	24	8	11	17
Pharmacology/therapeutics	423	37	41	15	6	4
Physiology	501	70	72	32	4	39
Radiation	201	6	15	2	2	2
Surgery	512	56	190	28	21	79
Toxicology	236	25	9	4	9	2
Virology	516	76	5	4	13	10
Totals	7992	949	689	337	296	426

Source: Data taken from research proposals assigned to NIH scientific review groups (study sections) organized by discipline.

genetics, or pharmacology (Table 41.5). Although these illustrations show numbers of projects funded (Table 41.4) or reviewed (Table 41.5) and not numbers of animals, a reasonable comparison can be made to show which category or discipline is most dependent on a species of animal.

Another measure of those disciplines of research that most frequently use primates is shown in Tables 41.6 and 41.7, which report numbers of articles in professional journals (O. A. Soave, FDA/DHHS, personal communication).

Although it is not possible, without a yet-to-be-initiated national survey, to determine the number of the various primate species used, we do know that the most frequently used species are the Old World species, which have been used for many years in research and which adapt to, and breed well in, a laboratory environment. The rhesus is the leading species as a laboratory subject, followed by the cynomolgus, the baboon, and the squirrel monkey. One way of estimating each species' popularity, or frequency of use, is through the records of the Primate Exchange Service operated by the Washington Regional Primate Research Center in Seattle (Table 41.8).

Recent research findings are showing a closer relationship between nonhuman primates and humans than previously thought. The importance of the nonhuman primate in research to improve human health has been reviewed in many articles and reports recently (King and Yarbrough, 1985; Hendrickson, 1984). In studying these reviews to assist in estimating trends in primate use, it was informative

TABLE 41.6. Citation of numbers of publications using primates in selected journals from sample years 1960, 1970, and 1980.

Journals		Citations	Primate	Rhesus
Amer J Physiol	1980	600	4	1
Amer J Physiol	1970	500	10	7
Amer J Physiol	1960	300	2	1
Exp'l Neurol	1980	140	5	3
Exp'l Neurol	1970	130	20	14
Exp'l Neurol	1960	115	0	0
Exp'l Eye Res	1980	160	4	2
Exp'l Eye Res	1970	145	2	1
Exp'l Eye Res	1960	125	1	1
J Endocrinol	1980	400	5	4
J Endocrinol	1970	325	4	2
J Endocrinol	1960	110	0	0
J Reprod Fertil	1980	150	5	3
J Reprod Fertil	1970	158	7	3
J Reprod Fertil	1960	120	0	0
Proc Soc	1980	350	4	2
Proc Soc	1970	725	8	8
Proc Soc	1960	875	1	?
Lab Anim Sci	1980	130	13	3
Lab Anim Sci	1970	115	53*	15
Lab Anim Sci	1960	105	5	4
J Neurophys	1980	225	18	10
J Neurophys	1970	95	3	2
J Neurophys	1960	75	2	1
J Pharm Expl Ther	1980	425	12	5
J Pharm Expl Ther	1970	240	5	3
J Pharm Expl Ther	1960	130	0	0
Fertil Steril	1980	230	3	1
Fertil Steril	1970	139	5	2
Fertil Steril	1960	70	2	2
Totals		7407	203	100
33 articles managerial			(170)	
1980		2810	73	34
1970		2572	117	56
1960		2025	13	9
7407/107 nonhumans primate or 2%				

Source: O. A. Soave, personal communication.

to separate discussions of past advances from promising and important research in progress. Papers reviewed by this author tend to have more reports in the "research in progress" category. This indicates a continued or increasing dependency on the nonhuman primate as a research subject.

Some of the important areas of research progress that are most frequently reported are:

1. RH disease, first identified on the red cells of rhesus monkeys in 1937, and resulting in reduction of the hazard to women giving birth to RH incompatible

TABLE 41.7. Frequency of citation of articles based on primate studies in journals representing various disciplines of biomedical research.

Discipline	Number of citations and percentages				
	1980	%	1981	%	Change
Nervous system	522	13	598	14.5	+1.5
Pharmacology	385	9	473	11	+2
Behavior	360	9	348	8	−1
Dental	206	5	215	5	0
Endocrine	199	5	207	5	0
Virology	190	5	139	3	−2
Reproduction	187	5	196	5	0
Cardiovascular	155	4	168	4	0
Metabolism	139	3	132	3	0
Primatology, general	128	3	130	3	0
Eye	123	3	221	5	+2
Toxicology	120	3	197	5	+2
Learning	104	2	85	2	0
Parasitology	89	2	87	2	0
Musculoskeletal	88	2	73	2	0
Blood, body fluids	81	2	52	1	−1
Genetics	78	2	61	1	−1
Digestive system	72	2	77	1	−1
Respiratory system	68	2	39	1	−1
Ecology	65	2	93	2	0
Immunology	62	2	82	2	0
Colony management	61	1	35	<1	—
Neoplasia	59	1	51	1	0
Bacterial infections	56	1	44	<1	<1
Embryology	53	1	51	<1	<1
Molecular biology	43	<1	42	<1	<1
Developmental biology	34	<1	45	<1	<1
Ear	30	<1	18	<1	<1
Disease general	24	<1	15	<1	<1
Gerontology	24	<1	34	<1	<1
Urology	24	<1	33	<1	<1
Biological clocks	20	<1	25	<1	<1
Radiation	20	<1	30	<1	<1
Integument	13	<1	27	<1	<1
Connective tissue	9	<1	9	<1	<1
Mycotic infections	6	<1	3	<1	<1
Total	3987		4135		

[1] 3% increase in number of citations.
Source: O. A. Soave, personal communication.

children, many of whom died of complications when their red blood cells reacted with the mother's serum.

2. Poliomyelitis was first transmitted to rhesus monkeys in 1909 and studied in rhesus, cynomolgus, and chimpanzees until 1955, when the Salk vaccine was introduced.
3. Hepatitis B had no animal model until it was discovered that the chimpanzee could be infected. Chimpanzee and marmosets were used in the development of a vaccine to control this serious infectious disease of humans.
4. Laser surgery in repair of injuries of the nerve sheath has been shown in baboons to be superior to conventional suturing of small nerve sheaths.

TABLE 41.8. Summary of primates placed through NIH-supported primate information clearinghouse, Washington Regional Primate Research Center, 1984.[1]

Macaca mulatta	33%
Papio spp.	18%
Saimiri sciureus	16%
Cercocebus	5%
Macaca fascicularis	5%
Callitrichidae	4%
Macaca nemestrina	3%
All others	16%

[1] 4030 recycled nonhuman primates; 3480 placed through new listings; 550 direct referral.
Source: Preliminary data from the Washington Primate Information Clearinghouse.

5. The basic studies of brain function that led to the Nobel Prize for Drs. Hubel and Wiesel have shown that eye disorders in children such as stabismus, cataracts, or amblyopia can, if untreated, result in primary damage to those developing areas of the brain involved in interpreting visual stimuli. Both monkeys and cats are born without fully developed visual cortices. Correct visual stimulation is necessary for development of this cortex, and, with early treatment of childhood eye disorders, this development can be assured.

Several research areas currently underway are at the forefront of research aimed at improving human health. Some of these research activities described below will also result in improved health for those primates we will meet on "The Road to Self-Sustaining Populations."

Simian Acquired Immunodeficiency Disease (SAIDS) is one special example of research likely to benefit humans and our fellow primates. The virus causing this condition in *Macaca* species is different from, but very similar to, that causing human AIDS. The pathology of SAIDS and AIDS is similar. It is believed that methods of control and prevention of these two diseases will be similar.

The National Institutes of Health is supporting research on SAIDS at four of the seven Regional Primate Research Centers it supports. Studies of the immune response in macaques, biochemical characterization of the SAIDS virus, and trials of an immunizing agent are all underway. It is possible that prevention methods for the primate disease will be developed before methods for the human disease.

Health care of primates at the NIH Regional Primate Research Centers is excellent. New or unusual diseases receive detailed attention. Several years ago, this led to the identification of leprosy in a sooty mangabey at the Delta Regional Primate Research Center. This disease had not been identified in any nonhuman primate species previously. It proved to be a transmissible disease, although not a serious threat to the primate colony. This is a unique opportunity for leprosy research, because it is now possible to obtain information on the immune response of a primate to leprosy and to evaluate new therapies. It also provides an animal model species in which an immunizing agent could be tested.

Celebes black macaques at the Oregon Primate Research Center have suffered from a naturally occurring diabetes similar to maturity onset diabetes in humans. This is serious as a disease for these primates as it is for humans.

Like their human counterparts, they develop antibodies against their insulin-producing cells.

Studies of these macaques are supported because of the opportunity to understand this illness and develop therapies for humans. This same research is essential to maintaining Celebes apes as self-sustaining populations in park collections.

Recent observations of aging female rhesus monkeys have shown menopausal changes similar to those in women. Researchers at the Wisconsin Regional Primate Research Center are studying this phenomenon to learn the role of the brain and endocrine system in initiating the menopausal change and understanding how its effects may be modulated.

Other chapters in this volume report on studies of the reproductive system, both for birth control and to overcome reproductive difficulties. These latter studies have contributed much to our ability to develop the present self-sustaining populations of primates in the United States.

Conclusion

The use of nonhuman primates in biomedical research is expected to continue for the rest of this century at a level no lower than present inventories. The supply of these animals will be largely from self-sustaining colonies within the biomedical community or their U.S. suppliers. In addition to reducing its dependence on wild primate populations, the biomedical community will continue to contribute to the preservation of natural populations through the new knowledge it will generate on behavior, environment, and reproduction.

References

Greenhouse D (ed) (1984) Trends in primate importants into the United States. ILAR News 27:6–12.

Hendrickson LJ (1984) Toward better health, the role of primates in medical research. In: Primate News 21 1:2–22.

King FA, Yarbrough CJ (1985) Medical and behavioral benefits from primate research. The Physiologist 28:75–89.

42
Approaches to Determining Colony Infections and Improving Colony Health

JAMES H. VICKERS

Introduction

In the middle and late 1950s, during the period of development of the polio vaccine in the race to stop the dreaded, crippling disease, approximately 200,000 primates per year, mostly Indian rhesus monkeys, were imported into the United States. During the 1960s with the licensing of the live, oral vaccine, these numbers decreased to approximately 50,000 (Vickers, 1983).

Improvements in the transportation of animals also contributed to the reduction in the numbers of animals needed for importation. Some of us working on polio programs worked in India, Africa, and Southeast Asia developing methods that reduced shipping losses and cross-infections between groups of animals. In the past, it was not uncommon to fly propeller-driven Constellations with seats removed carrying 2000 monkeys, making stops in the desert heat of the Middle East, the pouring rain of London, and frozen north of Newfoundland. The advent of jet aircraft made possible the rapid transportation of smaller shipments of animals with few layovers and greatly improved the condition of imported animals.

Beginning in 1973, the Indian government initiated an incremental reduction in the export of rhesus monkeys from 50,000 to 20,000 in 1974. Conservation was the main reason for this ban, because destruction of habitat by encroaching human populations and the effects of large-scale trapping during the 1960s and 1970s had greatly reduced the wild rhesus populations. The reduction continued until a complete ban was implemented in 1978. Subsequently, Bangladesh banned the export of rhesus monkeys, thereby eliminating the last wild source of these animals.

Great pressure from conservation groups is constantly being placed on the countries of Southeast Asia to ban the export of cynomolgus monkeys, and indeed, intermittent bans have occurred. The East African countries of Kenya, Tanzania, Uganda, Somalia, and Ethiopia have all suspended export of primate species at various times. Brazil, Peru, Colombia, and South Africa have already banned the export of other species of primates.

Today, requirements for polio vaccine safety testing have been reduced to approximately 2500 rhesus monkeys per year in the United States, and there is a continuing effort to improve testing methodologies. Thus, numbers should be substantially reduced again in the very near future. New vaccines or in vitro tests test yet to be developed may eliminate the need for lot to lot testing in monkeys altogether.

The FDA Experience

To meet the challenge of an uncertain and diminishing supply, the FDA began a program in 1973 to become self-sufficient in its requirements for rhesus monkeys by establishing domestic breeding colonies. At that time, it was not known how best to breed these animals in order to ensure high production at a reasonable cost and to provide for future generations of normal breeding stock that would perpetuate the colonies indefinitely (Vickers, 1983).

The program involved the establishment of rhesus monkey colonies at multiple locations with each colony producing 500 young per year for use in control testing and biomedical research, plus replacement breeding stock. With widely separated geographic locations, the effect of natural disasters of epidemic disease could be minimized, and a variety of breeding methods could be evaluated.

The first colony, established in 1974, was a free-ranging one on a 100-acre island off the southeast coast of the main island of Puerto Rico. This effort took advantage of an existing nucleus of 300 animals, supplemented with additional wild-caught Indian animals introduced in stages. The basic breeding concept was a good one, but the project developed programs pertaining to site location and management that prompted a decision to relocate the animals after 5 years. Every effort was made to maintain the social groups intact during the relocation process so that the colony could be reestablished with a minimum of disruption. The new site was on an island off the coast of South Carolina. The movement of the 1500 animals was highly successful. Minimal losses were sustained, and production rates at the current location have actually increased.

The second rhesus colony was established in the latter part of 1974, in Louisiana, using 22 half-acre corrals to hold the animals. Each corral contained approximately 50 breeding-age animals, essentially one social group. This colony was populated with young animals received from India. The decision to use young animals was predicated upon the extended breeding life of the animals and the increase in survival rate that is found when young, immature animals are imported rather than adults. The animals in this colony have now matured, and the corral breeding system has had the highest production rate (84% live births) of any breeding system used. The third colony, established in South Texas in 1977, utilized the harem method. One male and eight to ten females were held in a corn crib enclosure with a total of 180 such units used to accommodate the entire colony. This colony was developed utilizing young female animals from our other colonies and the animals obtained from several other smaller breeding colonies. The advantages of this system include short facilities set-up time and the ability to observe the animals closely. The disadvantages include labor-intensive maintenance and increased aggression between animals kept in relatively small enclosures.

The fourth colony, which began in 1977, was also free ranging on a 200-acre island in the Florida Keys. The breeding stock was transferred from another established colony in the Florida Keys that had been derived from animals born in India and imported several years earlier. These animals were selected from stock that was free from infection with herpes B virus, salmonella, shigella, and tuberculosis (Foster, 1975).

The establishment of domestic breeding colonies will ultimately lead to quality, standardized nonhuman primates for use in specific areas of biomedical research and quality control. Simultaneously, it will result in the conservation of natural populations throughout the world.

Health Management of the Corral-Type Colony

Every effort is made to protect the breeding colony animals from exposure to infectious or environmental agents that may be injurious to rhesus monkeys (Wolf, 1982–1985). This includes the use of personnel protective clothing that can be readily decontaminated. Normally only employees assigned to the project are allowed routine contact with the colony animals. Occasionally it is necessary to have other employees in the corral complex for roundups and repairs. These people must meet the necessary health requirements before they will be admitted to the colony area.

Surveillance and Health Care

All corral animals are inspected on a daily basis. This is done by walking through the corrals and watching the group movement. Injured, sick, obviously pregnant, and peripheral animals can be observed easily. When necessary an individual animal or a small portion of the group can be herded to the capture pen for more detailed examination.

Individual found injured or sick are caught for examination using the trapping chute and transfer cages. In the case of mild traumatic injury, it is generally considered best for the colony relationships to perform necessary treatment and return the animal to the group as quickly as possible. Cleaning, treatment, and suturing of a laceration and releasing the animal back into the corral will not disturb the animal's social position. Animals with more serious problems are removed and treated accordingly. Partial isolation can be accomplished in the holding pen area by setting up individual cages and racks. Better isolation can be obtained, however, by the transfer of the animal to the hospital facility. Reintroduction of individual animals may become a problem. The methods used to circumvent this problem include catching several animals to make a small group for reintroduction to the colony, releasing the returning animals at night, and forming new groups of chronically rejected animals.

Periodic Group Examinations

Periodic examinations of the colony are necessary to ascertain the health, nutritional, and reproductive status of the animals in the colony. Normally in an isolated, closed colony there are infrequent disease problems, so usually collection and examination of the entire colony twice yearly is adequate.

Activities during a roundup include:

1. tuberculin tests
2. physical examination including rectal palpation of females

3. tattooing and marking of infants
4. any surgery or treatment considered necessary
5. selection of juveniles for shipment
6. tetanus vaccination in animals to be retained as breeding stock

Roundups are done in the spring and fall. Summer roundups are difficult because infants are still small and very dependent on their mothers. Small infants can be separated from their mothers or injured during group manipulations. Also, rhesus monkeys seem quite susceptible to heatstroke. Even modest exertion and forced group activity can precipitate heatstroke. The spring roundup with special emphasis on pregnancy examinations also gives a very good analysis of the breeding status, number of pregnancies, and expected production for the colony. This examination would come about mid-gestation for most females. Likewise, the fall roundup would permit repeating routine tests and manipulations, confirming the reproductive status of each female, and tattooing the infants with their permanent number.

Daily Operational Procedures

Husbandry support is provided by the technical staff assigned to the corral colonies. These personnel enter each corral on a daily basis to observe the animals and distribute food. A quantity of Purina monkey chow sufficient to provide about 0.5 pounds per monkey per day is distributed either in a feeder or over the surface of the corral area while the personnel walk through the corral. During this inspection, the corral is examined for dead animals that may be hidden in clumps of grass, behind remnants of logs, or in depressions. These carcasses are removed, identified, labeled, and taken to the pathology department for examination. Cracked corn is scattered, which allows observation of the animals while they forage. The group is made to move about the corral. This provides closer inspection, and weak or injured animals lagging behind the main group are separated and removed for medical examination.

Additionally, during the birth season, a count of infants will be made on a daily basis and new births will be recorded. As infants are born, their mothers are captured, their numbers recorded, and the infant given a numbered eartag that will serve as its temporary physical identification. The number on the eartag correlates with a number in the data storage system that will be tattooed on the animal at the appropriate roundup period. The status of each corral on a daily basis is summarized and entered into the computer records daily. If an animal is relocated, this change is entered into the animal's permanent file.

The basis of this type of health program is preventive medicine. Detailed daily observation of animal activities, appearance, births, deaths, and illness provide necessary supplemental data so that appropriate actions can be taken. Routine veterinary care is available every day.

Any animal that dies or is born dead will be given a gross and histopathological examination of all major organ systems as well as parasitology and routine microbiological testing. Virus isolation is done as indicated. These data are reviewed for overall impact on the risk of the colony so that appropriate action can be taken.

Management of Infants

The management program for infants consists of natural rearing and weaning by the parents in the home corral. Reproductive efficiency does not suffer, and the infants survive better when there is normal socialization.

The bulk of the young animals are removed at 12 to 18 months of age. Infants retained for replacement stock are left in their family units. It is estimated that for the perpetuation of the colony 20% of the female infants and a smaller percentage of male infants be retained and raised in normal social groups to provide replacement stock. The colony-born replacement animals are now coming into production in this colony. It is anticipated that these animals will be successful breeders because they have inherent social positions and are already adapted to corral environment.

A nursery facility is available for rejected, weak, or clinically ill infants. These animals can be handled well in the nursery but often prove to be poor parents when not raised in social groups.

Quarantine Program

The general policy requires a 90-day quarantine period for all newly arrived nonhuman primates. The large investment in the existing colony should not be compromised by reducing the quarantine period. The quarantine facility is serviced by different personnel than the colony staff. Access to the quarantine facility is limited to authorized personnel. Quarantine procedures are in compliance with or exceed the requirements of the Animal Welfare Act of 1966 and the DHHS guidelines.

Upon admission, each new animal is assigned an identification number. This number is tattooed on the animal's chest for permanent identification. On admission each animal will be tuberculin tested according to standard procedures using 0.10 ml of KOT (containing at least 15 mg) injected intradermally in the upper eyelid with a 25-ga needle. Each test site is examined at 24, 48, and 72 hours after inoculation. Tuberculin testing is repeated every 2 weeks while the animal is in quarantine. Alternate eyelids are used in subsequent tests. Positive reactors are humanely sacrificed, necropsied, and cultured by the appropriate techniques. Animals with questionable reactions are isolated from their groups and retested within 2 weeks. If negative, this animal must have three additional negative tests at 2-week intervals before it can be returned to its original group. Positive reactors with disease extend the quarantine time for the group to be reinstituted from the time the animal was determined to be infected. All animals that die are given a complete necropsy and appropriate microbiological studies.

Shortly after arrival, a complete physical examination is made on each animal including a rectal palpation for the females. Additional tests performed in the quarantine period include the following and each is performed at least twice: (1) CBC, (2) Knotts test for microfilaria (once), (3) thin film examination for malaria, (4) stool culture for pathogenic enteric bacteria (three consecutive weekly tests), and (5) analysis of stool for helminths and protozoa by zinc sulfate and formol–ether concentration techniques. Intensive screening with three weekly fecal cultures are made for the detection of enteric pathogens. Tetanus toxoid is given to each animal. Special attention is given to adapting the animals to a standard diet and other nutritional supplements if found necessary. The canine teeth of male animals are cut level to the table surface of the molar teeth and filled, to reduce injury to other animals and personnel.

The genitalia of male animals are examined, and they are electroelaculated twice for semen evaluation. The volume of ejaculate is measured and the motility, total count, and live:dead ratios of spermatozoa determined. Sterile and anatomically defective males are not placed in the colony. Experience shows that males with counts of at least 3.5×10^8 spermatozoa/ml and with at least 60% motility are optimum for a breeding male.

Health Certification of Visitors and Staff

Visitors are restricted to those having valid reasons to inspect the colony area. Any visitors having contact with the colony animals will be required to provide written proof of:

1. negative tuberculin test and/or chest x-ray within 6 months of date of visit
2. proof of poliovirus vaccination at least 3 months prior to visit
3. proof of measles vaccination or past infection at least 3 months prior to visit
4. valid tetanus inoculation
5. history of no exposure to chicken pox, measles, mumps, rubella, yellow fever or other Arbovirus infections within 3 weeks of visit
6. visitors and staff exposed to the colony must be in overt good health with no signs of respiratory disease, gastrointestinal dysfunction, or cold sores

Harem Breeding Facilities

This colony utilizes the harem method with the basic enclosure being a corn crib type cage manufactured by Behlen Manufacturing Company. It consists of a circular sidewall 7.5 feet high of welded 2.5-gauge galvanized rods and a conical galvanized metal roof an additional 6 feet tall at the peak of the roof. The cage sits on a concrete block foundation and is secured by the use of five "soil anchors" providing an ability to withstand winds in excess of 100 mph. The diameter of the cage is approximately 126 ft². Multiple perches are provided to assure shelter from sun and rain. During December through March, a sheet metal screen will be placed on the northern exposure of each cage to provide shelter from cold winds. The floor of the cage consists of washed gravel 6–8 inches deep. The crib has a large swing-type door for access of personnel and a smaller guillotine door which will allow attachment of a chute for capturing the animals. Each cage is equipped with a lixit-type automatic waterer.

The colony has 180 enclosures set up in 12 rows. Adjacent enclosures are separated by a distance of not less than 20 feet. Three interconnecting 20 foot wide graded and graveled roadways in each unit provide all-weather access to any of the cages for daily care and observation. Buried water lines provide water to each cage. Security lights provide low-level illumination to the entire facility. Personnel entry into the cage is accomplished by use of a small, portable, lightweight anteroom that will be attached to the cage by means of harness snaps. Digital examinations, tuberculin testing, etc., will be done within the interior of the cage. A "capture gun" is available for emergencies.

Health Management of the Harem Colony

The health management program is essentially the same as described under the corral system, above. Quarantine, tuberculin testing, personnel health programs, and computerized record keeping are basic to colony management.

The Free-Ranging Colonies

Because of the nature of a free-ranging, island operation, medical and clinical intervention in the animals is infrequent (Taub, 1983–1985). All personnel understand, however, that any monkey showing signs of illness requires immediate medical attention, and every effort is made to capture such an animal. Daily observations are very effective in detecting ill or injured individuals. Veterinary care is provided at an on-site clinic.

Animals requiring more extensive treatment are transferred to the hospital facilities. After recovery, the animal is returned to its social group location and released.

Gross necropies are performed on all animals that die. Microscopic examination of tissues and isolation of organisms are conducted as necessary for diagnosis.

During the warm months, an insect control program (primarily mosquito, but also tick, chigger, and fly) is conducted. Once or twice a week, malathion is sprayed from a commercial fogger. This procedure was developed in conjunction with the County Mosquito Control Program. Malathion in concentrations suitable for fly control is also sprayed from a hand sprayer at appropriate areas around feeding/watering stations.

The health management for the attending staff and visitors of the free-ranging colony follow the basic programs as outlined for the corral and harem breeding colonies.

A registry of all visitors to the colony is maintained, visitor's name and title, date of visit, complete business address, and purpose of the visit.

Exporter Colonies in Countries of Origin

Over the past 25 years, there have been great improvements in the methods used by exporters in their capture and handling of monkeys in the countries of origin. With the ban on export of rhesus monkeys from India, Pakistan, and Bangladesh, the primary animal exported has been the cynomolgus monkey, *Macaca fascicularis*, from Southeast Asia. Several of the exporters from this area have greatly improved their methods and facilities.

The Ano Farm (J. Vacek, 1985, personal communication)

The Ano Farm was established on the island of Panay in the Central Philippines as a cynomolgus monkey conditioning and breeding facility in November 1978. It has the capacity to handle 600 monkeys and a regular stock of 300 young adults for export.

Utilizing advice from a number of internationally known primatologists, the Ano Farm has developed a program for sanitation and disease control.

When animals arrive from the trapping areas, they are given a general examination by a veterinarian, treated for internal and external parasites, tuberculin tested, and where applicable treated for disease or injury. Each animal is weighed, identified, and given an individual medical record.

Animals are quarantined for 6 weeks and the best animals selected for export. Mortality after shipment has averaged 2%.

An extensive sanitation program is conducted utilizing pressure sprays of halide-based disinfectants.

A dietitian provides a balanced diet utilizing fresh vegetables, and a monkey biscuit based on the NIH primate diet is baked on the premises.

Special programs to supply preshipment blood sera and measles-free animals are available.

SICONBREC (K. Hobbs, 1985, personal communication)

SICONBREC, also located in the Philippines, received animals from the trapping areas at its conditioning compound at Zamboanga.

Upon arrival, all animals are given a measles vaccination and checked for intestinal disorders. After a 2-week observation period, they are moved to the holding compound at Tanay and given a physical examination, weighed, tuberculin tested, checked and treated for parasites, and further treated as required by the veterinarian.

Screening programs for herpes B virus are available on request as well as preshipment sera checks for other viruses.

The holding facilities have four full-time veterinarians available, and protective clothing, masks, and foot baths with chloride-based disinfectants are used in disease prevention.

When holding areas are emptied, all cages are sprayed with quaternary ammonia compounds before removal to cage washing area where further cleaning and disinfection is performed. Waste materials are flushed into septic tanks for further processing.

Summary

The approaches to determining colony infections and improving colony health that we have felt the most important to our program are generally basic procedures that need continual enforcement. If colony managers grow lax in implementing these programs, the door is open for potentially disastrous disease problems. Those procedures recommended include:

1. Adequate quarantine of all animals prior to entrance into colony.
2. Observation of every animal every day. With the free-ranging colonies, as many animals as possible should be observed daily for signs of ill health.
3. Roundup and individualized examination of animals at least twice a year.
4. Twice yearly physical examination of all personnel in association with the colony including tuberculin testing and vaccinations.
5. Reduction of human contact to a minimum and those to have completed mandatory health requirements.

Areas for Research

As newer knowledge becomes available and our experience with colony management grows, there are important areas that should be addressed to reduce further the mortality encountered in these programs. Some of the troublesome areas that affect colony health now or in the future but appear solvable through research are as follows:

1. SAIDS: With the recent reports of Simian Acquired Immunodeficiency Syndrome in several large primate colonies, there is a need for methods to screen large colonies for retroviruses. Studies to develop accurate tests are essential. Methods for handling positive animals should be developed. Early detection of carriers is necessary.
2. Atypical mycobacteria: Atypical mycobacteria are more frequently being detected in primate colonies and were initially confused with avian, bovine, and human mycobacteria tuberculosis. Decisions for disposition of animals should be predicated on knowledge of clinical disease aspects of these agents.
3. Stillbirths and abortions: A number of animals are lost through stillbirths and abortions. Not enough effort has been placed on studies determining the cause of these losses. It is a difficult project to obtain good diagnostic material, but there is ample room for research in this important area.
4. Nutritional aspects of disease: Most large primate colonies are fed processed prepared food that may or may not follow the primate species' natural requirements. The effects on immune system and reproductive physiology are not fully known. The importance of nutrition and its effect on disease prevention is only recently receiving the attention it deserves. It follows that nutritional requirements of artificially reared primates is an important area for study.
5. Genetic surveillance of colonies: Most large primates are only a few years old and are only now producing second- and third-generation animals. The effects of inbreeding on the immune system and disease susceptibility is still unknown. The minimum size of the gene pool in the colony to prevent genetic diseases should be determined. A wealth of material for genetic study is available in these large colonies.

As the Hindu god Rama reached out his hand for help to the monkey king Hanuman over 3000 years ago, so we are still reaching out to the monkey for help in solving the problems of medical research. As we continue to accept this help, it should be our commitment to provide the most humane and disease-free environment for our research animals that current knowledge allows.

References

Boyd D (1981–1985) Technical progress reports. FDA Contract 223-81-1104.
Foster HL (1975) Establishment of a free-ranging rhesus monkey colony on Key Lois Island, Florida. Lab Anim Hdbk 6:107–117.
Taub D (1983–1984) Technical progress reports. FDA Contract 223-82-1101.
Vickers JH (1983) FDA regulatory use of nonhuman primates. Lab Anim 12:28–34.
Wolf RH (1982–1985) Technical progress reports. FDA Contract 223-82-1100.

43
Bacterial Infections of Nonhuman Primates

HAROLD M. MCCLURE, ANNE R. BRODIE,
DANIEL C. ANDERSON, and R. BRENT SWENSON

Introduction

A variety of bacterial infections can be expected to occur with considerable frequency in nonhuman primate colonies. These infections continue to be significant causes of morbidity and mortality and can have a serious impact on colony maintenance and management, domestic breeding programs, and research projects, if not actively pursued and controlled. Certain microbial agents, e.g., *Shigella*, *Salmonella*, pneumococcus and other streptococci, *Staphylococcus*, *Hemophilus*, and *Yersinia*, are encountered on almost a daily basis in nonhuman primate colonies. In addition, adequate microbial support will reveal numerous other bacterial pathogens that have been infrequently or rarely isolated in nonhuman primates, e.g., *Chromobacterium*, *Vibrio*, and *Listeria*. The correct clinical or postmortem diagnosis is often dependent upon the results of bacterial cultures. Adequate bacteriologic support is, therefore, required in order to provide the appropriate therapy for clinical cases and to arrive at a definitive diagnosis in necropsy cases. Information provided by bacterial cultures is particularly useful in the clinical situation in three important ways. First, if bacterial cultures are available for most of the cases of infectious disease, it gives the clinician a historical data base for the relative incidence of the more common infectious conditions that occur in the colony. Since identification of the etiologic agent may take from one day to several weeks, the data base from previous culture results is an invaluable source of reference data that is specific for the colony in question and often for the species in general. Second, since antibiograms are usually not available until 48–72 hours after submission, they are of limited value in initiating therapy in individual cases. However, the data base provided by previous antibiograms gives the clinician a good indication of the trends that exist in the colony relative to sensitivity patterns of microorganisms that are problems in the colony. Third, individual culture and sensitivity results are imperative in individual cases even though they are not available until therapy has been initiated, since they provide a rational basis for continuing or changing initial therapy. The availability of such information will improve the success rate for treatment of clinical disease and will, thereby, reduce colony mortality. The availability of a definitive diagnosis, with documented etiology, will also be an asset to colony management because it will dictate the need for isolation and/or quarantine of infected animals in order to prevent or control the spread of certain pathogens.

Materials and Methods

This publication is based primarily on our experience during the past 5 years (1980–1984) in a colony of approximately 1600 nonhuman primates (200 great apes and 1400 monkeys). During this period, we have done 8423 bacterial cultures from nonhuman primates and their environment. These include specimens from clinically ill animals, occasional surveys of selected groups of animals to determine carrier status, necropsy cases, and animal environment (e.g., cages, cage washer, water bottles, water supply, etc.). We routinely determine the genus and species of all isolates and do antibiograms on all isolates that are considered to be pathogens. Routine bacteriologic methods are used and are briefly summarized as follows for enteric cultures, blood cultures, urine cultures, vaginal cultures, CSF cultures, and cold enrichment culture for *Yersinia*.

Enteric Cultures

Feces or other intestinal tract material is inoculated on blood, MacConkey, Hektoen Enteric (HE), Salmonella–Shigella (SS) and Campylobacter agar plates and into selenite broth. Blood, MacConkey, HE, and SS agar plates and selenite broth are incubated at 35°C in an aerobic incubator. After 24 hours, all cultures are checked for pathogens, and these are subcultured to TSI slants and API identification strips (Analytab Products, Plainview, NY). Antibiograms and serology are done as needed. Small, colorless colonies are subcultured to TSI and held at room temperature for *Yersinia* isolation. *Campylobacter* plates are placed in a sealed plastic bag (e.g., a Zip-Loc storage bag), with *Proteus* growing on a chocolate plate to produce a microaerophilic atmosphere, and incubated at 42°C for 48 hours. Suspicious colonies are gram stained, and oxidase and catalase tests are done. If positive for all three, the organism is subcultured to another *Campylobacter* plate and a cephalothin (30 μg) and a nalidixic acid (30 μg) disc are dropped on the plate and read at 24 hours (89). Other differential tests done to identify the species of *Campylobacter* include culture at 25°, 35°, and 42°C; H_2S on TSI, and hippurate hydrolysis [55, 89].

Blood Cultures

Approximately 4 ml of blood are collected aseptically and inoculated into a vacutainer tube (B-D 4955) and a Roche septi-check blood culture bottle and incubated at 35°C. The vacutainer tube is vented after 24 hours and the septi-check "slide top" is placed on the septi-check bottle after 6 hours. Cultures are visually checked each day and are subcultured at 48 hours, 1 week, and 2 weeks. Positive cultures are gram stained and then worked up according to the results of the gram stain. Antibiograms are always done on isolates from blood cultures.

Urine Cultures

Urine, collected aseptically, is cultured on blood and MacConkey agar using a calibrated colony count loop (0.001 ml). After 24 hours incubation at 35°C, the number of visible colonies is multiplied by 1000 to determine the colony count. If the colony count is significant (> 10,000/ml), the organism is identified and antibiograms are done.

Vaginal Cultures

Material for vaginal culture is inoculated on blood, chocolate, phenylethyl alcohol (PEA), and MacConkey agar plates and into thioglycollate broth, and incubated at 35°C. Organisms are then selected for gram stain and further differential work up.

CSF Cultures

CSF or other body fluids are inoculated onto blood, chocolate, PEA, and MacConkey agar, onto a CDC anaerobe plate, and into thioglycollate broth. If positive, isolates are selected for further differentiation and identification. It should be noted that *Hemophilus*, a not-infrequent CSF isolate, will grow only on chocolate agar.

Cold Enrichment Culture for *Yersinia*

Yersinia organisms grow well on most laboratory media and are not difficult to isolate when cultures are taken from lesions that are not contaminated by other organisms. However, isolation from fecal material is much more difficult, as a result of their slow growth at 35°C, and can best be achieved by use of a cold enrichment technique. The cold enrichment technique used in our laboratory consists of the following:

One 5-ml tube of phosphate-buffered saline (see below) and one tube of cooked meat broth (BBL 21508) are inoculated with approximately 0.5 ml of feces. These are placed in the refrigerator at 4°C for 3–4 weeks. They are then subcultured to SS and MacConkey agar plates and incubated at room temperature for 48 hours. Any small, clear, colorless (non-lactose fermenting) colonies are transferred to TSI slants and held at room temperature for up to 72 hours. Organisms from any TSI slants with an alkaline slant/acid butt or acid/acid (with no gas) are transferred to a urea agar slant and held at room temperature for 48 hours. Urea positive cultures are transferred to an API strip and incubated at room temperature for 24–48 hours. Motility testing should also be done, since *Yersinia* is motile at 25°C but nonmotile at 35°C. Antibiograms are routinely done on all *Yersinia* isolates.

Phosphate-buffered saline: 1.08 g KH_2PO_4
8.36 g Na_2HPO_4
8.50 g NaCl

Add distilled water QS to 1000 ml. Does not have to be sterilized. Dispense into tubes and keep refrigerated until used.

The API system is used for identification of gram-negative bacilli and appears to provide a high degree of accuracy for the species identification of most isolates. Cold enrichment techniques are used for the isolation of both *Yersinia* and *Listeria*.

This presentation is based on a review of 8423 bacterial cultures that were done at the Yerkes Center during the past 5 years. These included 4455 enteric cultures (clinical and necropsy cases); 1533 blood cultures; 103 placental cultures; 304 urine cultures; 57 CFS cultures; 63 air sac cultures; 132 joint fluid cultures; 93 oral cavity or sputum cultures; 87 eye, ear, nose, and throat cultures; 94 vaginal cultures; 286 cultures of external lesions and/or abscesses; 1049 organ cultures at necropsy (nonenteric); and 167 miscellaneous cultures. The results of these cultures, with comments on literature reports concerning specific diseases produced, are discussed in the following paragraphs.

Results

Enteric Cultures

During the past 5 years, we have done 4455 enteric cultures. These included 1497 fecal cultures from clinical cases of diarrhea in the nonhuman primate colony, 595 cold enrichment cultures for *Yersinia* on animals with diarrhea; 417 fecal cultures from clinically normal animals to check carrier rates for *Shigella*, *Salmonella*, and *Campylobacter*; 200 cold enrichment cultures on clinically normal animals to check for *Yersinia* carriers; 1286 gastrointestinal tract cultures from necropsy cases (includes 638 cold enrichment cultures); and 460 cold enrichment cultures from the cecum and/or colon of wild rodents (217 mice; 137 rats), laboratory rats (100 albino rats), and other feral species (1 bird, 1 opossum, 1 mole, 1 shrew, 2 cats). The latter cultures were done to determine the incidence of *Yersinia* carriers in the wild and laboratory rodent population and in other feral species that may have access to the animal quarters. This large number of enteric cultures, 53% of the work load of our bacteriology laboratory, reflects, in large part, the frequent occurrence of diarrheal disease in nonhuman primates.

In a previous survey (prior to 1980), we reported the isolation of enteric pathogens from 29.5% of enteric cultures from 379 animals with diarrhea [95]. This recovery rate compared favorably with the reported isolation of enteric pathogens from 20% to 40% of human cases of diarrheal disease [132]. During the past 5 years (1980–1984) we have isolated enteric pathogens from 645 of 1497 (43.0%) cases of diarrheal disease in nonhuman primates. This improved success rate in isolation of enteric pathogens from cases of diarrheal disease is due in large part to our improved capability for isolating *Campylobacter* and the use of cold enrichment techniques for the isolation of *Yersinia*.

Enteric pathogens isolated from 1497 clinical cases of diarrheal disease in nonhuman primates during the past 5 years are listed in Table 43.1. As evident, multiple enteric pathogens were isolated from 145 of the 645 (22.5%) cases.

Campylobacter

Although *Campylobacter* cultures were not initiated in our laboratory until mid-1981, this species is now the most frequently isolated enteric pathogen. These organisms are also recognized as a frequent cause of diarrhea in humans [35, 86, 128, 148], many domestic and zoo animals [9, 62, 88, 123], and nonhuman primates [9, 88, 109, 151, 152]. *Campylobacter jejuni* is most prevalent in humans with 5% or less of the isolates being other species (e.g., *C. coli*, *C. fetus*, *C. faecalis*) [35, 72]. *Campylobacter laridis* has been proposed as the species designation for nalidixic acid-resistant thermophilic *Campylobacter* (NARTC) [86, 148]. A human strain of *C. jejuni* has been used to experimentally infect rhesus monkeys [37].

The pathogenesis of *Campylobacter* diarrhea has not been resolved; it is not clear whether diarrhea is related to an enterotoxin or to invasiveness of the organism [124]. It appears that both mechanisms may play a role in the production of diarrhea, depending up the characteristics of the isolate. The invasive proper-

TABLE 43.1. Enteric pathogens isolated in nonhuman primates during 1980–1984.

Organism	No. of cases
Shigella	129
Yersinia (Yer.)	62
Salmonella	6
Campylobacter	205
Enteropathogenic *E. coli* (EEC)	53
Yersinia + *Campylobacter*	16
Shigella + *Campylobacter*	67
Shigella + *Yersinia*	5
Shigella + *Yersinia* + *Campylobacter*	10
Aeromonas	26
Pseudomonas	18
Salmonella + *Campylobacter* + EEC	1
Campylobacter + *Pseudomonas*	1
Shigella + *Pseudomonas*	1
Shigella + *Campylobacter* + EEC	1
Vibrio + *Shigella* + *Campylobacter*	1
Shigella + EEC	3
Yersinia + Group A *Streptococcus*	1
Group B *Streptococcus*	1
Shigella + EEC + *Aeromonas*	1
Shigella + Yer. + *Aeromonas* + *Campylobacter*	1
Campylobacter + EEC	14
Salmonella + *Campylobacter*	2
Salmonella + *Shigella*	1
Campylobacter + *Aeromonas*	14
Shigella + *Aeromonas*	2
Yersinia + *Aeromonas*	2
Yersinia + *Campylobacter* + *Aeromonas*	1

diarrhea, depending upon the characteristics of the isolate. The invasive properties of some organisms have been demonstrated [28, 79, 107]; an enterotoxin has been identified [77, 133], and in some isolates, neither enteroinvasiveness nor an enterotoxin can be demonstrated [91]. It has been noted that only about one-third of patients with *Campylobacter* enteritis show a clinical picture indicative of intestinal invasion (blood, mucus, and leukocytes in feces) and that in the majority, the diarrhea is secretory in nature [133]. Gram's stain of fecal smears has been suggested as a method for making an early presumptive diagnosis of *Campylobacter* enteritis. Demonstration of gram-negative rods with typical *Campylobacter* morphology ("simple curve," "equivalent sign," or "seagull form") can be done in approximately 50% of the patients with *C. jejuni* infections [61]. Erythromycin is the drug of choice for treatment of *Campylobacter* enteritis [2, 16, 71]. Although treatment with erythromycin is reported to eradicate *C. jejuni* from the feces promptly, it apparently does not alter the natural course of the disease in humans when treatment is initiated 4 or more days after onset [2].

The species identification of 335 *Campylobacter* isolates from nonhuman primates with diarrhea in our colony include the following:

C. coli 197
C. jejuni 134
C. fecalis 3
C. fetus 1

An additional 95 *Campylobacter* isolates were obtained from either the colon or small intestine at the time of necropsy. These included 54 *C. jejuni* isolates and 41 *C. coli* isolates.

SHIGELLA AND SALMONELLA

Although *Salmonella* has been only infrequently identified in our colony, *Shigella* continues to be a significant and frequent cause of diarrhea. During the past 5 years, we had only ten *Salmonella* isolates from animals with diarrhea. These included three *S. derby*, three *S. typhimurium*, three *S. st. paul*, and one *S. litchfield*. In contrast, *Shigella* has been isolated from 221 clinical cases of diarrhea. The species identification of these *Shigella* isolates includes the following:

Shigella flexneri 4 177
Shigella flexneri 2 31
Shigella flexneri 3 10
Shigella flexneri 1 1
Shigella flexneri 6 1
Shigella boydii 1

An additional 44 *Shigella* isolates were obtained from either the colon or small intestine at the time of necropsy. These included 31 *Shigella flexneri* 4, seven *S. flexneri* 2, five *S. flexneri* 3, and one *S. boydii*.

It is well documented that clinically normal nonhuman primates may be carriers of *Shigella*, with carrier rates reported to range from 5% to 67% [43, 121, 159]. During the past 5 years, we have done bacterial surveys on fecal swabs from 417 clinically normal animals in order to determine the carrier rate for *Shigella* and *Salmonella* in our colony. Eighty-six of these were also checked for *Campylobacter*. Fourteen of 86 (16.3%) *Campylobacter* cultures were positive; ten *C. coli* and four *C. jejuni* were isolated. Thirty-nine of 417 (9.3%) animals were positive for *Shigella*; 34 *S. flexneri* 4, two *S. flexneri* 2, and one each of *S. flexneri* type 1, 3, and 6. *Salmonella* was isolated from one of the 417 cultures; one culture yielded *Y. enterocolitica*; and *Aeromonas hydrophila* was isolated from one.

Acute clinical disease may be precipitated in *Shigella* carriers by a variety of stressful situations. Shigellosis in nonhuman primates has been well characterized clinically and pathologically [94, 95, 102, 110]. Although usually considered a disease of the colon, experimentally infected animals do show lesions of the stomach and small intestine [74]; we have also encountered one clinical and three necropsy cases of *Shigella* gastritis in macaques of the Yerkes colony during the past 5 years. *Shigella* gastritis presents as a very acute, severe condition that results in very rapid dehydration and toxemia with a high mortality rate. In very acute, rapidly progressive cases of gastritis, cultures should be done to rule out or confirm *Shigella* gastritis.

Shigella has also been isolated from the oral cavity of macaques with gingivitis [5, 95, 97, 114]. During the past 5 years, we obtained gingival cultures from 73 macaques with gingivitis. *Shigella* was isolated from 25 (34%) of these cases; serotypes of the isolates included 19 *Shigella flexneri* 4 and six *Shigella flexneri* 2. Animals with this disease show swollen, hyperemic gums that may contain areas of necrosis. In more severe cases, there may be extensive ulceration with

TABLE 43.2. Antibiograms of selected pathogens.

	Percent susceptible				
	Staphylococcus aureus (99)[1]	*Shigella flexineri* (100)	*Yersinia enterocolitica* (100)	*Yersinia pseudo-tuberculosis* (60)	*Salmonella species* (10)
Ampicillin	53	97	23	97	100
Penicillin	53	0	0	77	0
Erythromycin	92	2	0	0	0
Tetracycline	98	91	38	73	40
Methicillin	98	0	0	0	0
Cephalothin	100	68	3	98	100
Lincomycin	67	0	0	0	0
Chloramphenicol	92	100	95	100	100
Kanamycin	95	82	100	100	100
Gentamicin	97	98	100	100	100
Polymixin	6	100	100	95	100
Trimethoprim + sulfamethoxazole	98	95	98	98	100

[1] Number of isolates checked.

gingival recession and exposure of tooth roots. This appears to be a specific disease syndrome in macaque species. Other isolates from the 73 cases of gingivitis included three *Eikenella corrodens*, one *Yersinia enterocolitica*, four group B Beta *Streptococcus*, one *Candida albicans*, one *Fusobacterium* species, three *Bacteroides* species, and one *Salmonella st. paul*. Antibiograms for *Salmonella* and *Shigella* isolates from our colony are shown in Table 43.2.

YERSINIA

Yersinia species continue to be isolated with considerable frequency from outdoor-housed animals in the Yerkes colony [101, 150]. These organisms are also being identified with increasing frequency in other facilities and in a variety of nonhuman primates [7, 12–14, 19, 129]. During the past 5 years, we isolated *Yersinia* organisms from 98 of 1497 (6.5%) clinical cases of diarrheal disease (often in conjunction with other enteric pathogens) and from 79 of 928 (8.5%) necropsy cases. Many of these 79 necropsy cases did not show lesions indicative of yersiniosis, and some of the isolates probably represent environmental (nonpathogenic) strains of *Yersinia*. At the present time, studies are being done to correlate biotype and serotype of the isolates with the presence of gross and/or microscopic lesions. Currently, *Y. enterocolitica* organisms that are salicin and esculin negative are considered pathogenic [150]. The pathogenicity of *Y. pseudotuberculosis* has been associated with autoagglutination, calcium dependency, and melibiose fermentation [153]. Overall, our clinical laboratory has identified 379 *Yersinia* isolates during the past 5 years. These included 212 *Y. enterocolitica*, 120 *Y. pseudotuberculosis*, 23 *Y. intermedia*, 19 *Y. kristensenii*, and 5 *Y. fredericksenii*.

Since all cases of yersiniosis occurred in outdoor-housed animals, wild rodents were suspected as the probable source of infection through fecal contamination of food or water. We subsequently trapped 542 wild rodents (394 mice and 148 rats) and cultured their cecal and/or colon contents by the cold enrichment technique. One hundred twenty-nine of the 542 (23.8%) rodents were culture positive for *Yersinia* organisms (14.2% of mice and 49.3% of rats). Organisms isolated from the rodents included 97 *Y. enterocolitica*, 11 *Y. intermedia*, nine *Y. pseudotuberculosis*, four *Y. fredericksenii*, four *Y. kristensenii*, two with both *Y. enterocolitica* and *Y. kristensenii*, one with *Y. pseudotuberculosis* and *Y. intermedia*, and one with *Y. pseudotuberculosis* and *Y. fredericksenii*. These observations indicate that a significant number of the wild rodents in the environment of our field station are *Yersinia* carriers and are the most likely source of infection in the nonhuman primates. Preliminary results of serotyping studies (done in collaboration with Dr. S. Toma, Yersinia Reference Laboratory, Ministry of Health, Toronto, Canada) indicate that many of the rodent isolates are of the same serotype as isolates from nonhuman primates. Additional studies are underway to compare further the serotypes and pathogenicity of rodent and nonhuman primate isolates.

Characteristics of yersiniosis in nonhuman primates include diarrhea (with or without blood), some degree of ulcerative enterocolitis, and mesenteric lymphadenitis. Hepatosplenic necrosis is also frequently observed. Imprint smears and sections of lesions show a characteristic overgrowth of gram-negative rods when examined microscopically [96, 101]. Antibiograms for *Yersinia* isolates from our colony are shown in Table 43.2.

Aeromonas

Aeromonas hydrophila has been incriminated as a cause of diarrhea and enteritis in humans [60, 81, 87, 131, 134, 156]. These organisms are oxidase positive and consequently are not members of the family Enterobacteriaceae; although not considered as part of the normal flora of the gastrointestinal tract, they have been isolated from fecal specimens of normal individuals [60]. Enterotoxic, cytotoxic, and hemolytic properties have been identified in culture filtrates of *A. hydrophila* isolated from patients with diarrhea [17, 44, 63].

Although the relationship between *A. hydrophila* and diarrheal disease in nonhuman primates is unclear at present, this organism should be considered as a potential enteric pathogen for these species. During the past 5 years, we have isolated *A. hydrophila* from 47 cases of diarrheal disease in nonhuman primates. In 26 of the cases, no other recognized enteric pathogens were isolated. In 21 cases, *Aeromonas* was isolated in conjunction with other enteric pathogens, such as *Shigella*, enteropathogenic *E. coli*, *Yersinia*, and *Campylobacter*.

Vibrio Cholerae

A *Vibrio cholerae* non-01 strain was recently isolated from a fecal culture of a 6-year-old, laboratory-born rhesus monkey in our colony. The animal was being maintained in an outdoor compound and presented with an acute, watery diarrhea. *Shigella flexneri* and *Campylobacter coli* were also isolated from the fecal culture. However, serologic studies on acute and convalescent serum samples showed a significant titer only to *V. cholerae* (from <1:10 to 1:160). *Campy-*

lobacter titers in both acute and convalescent samples were < 2; the *Shigella* titer was 1:40 in both samples.

In humans, non-01 *V. cholerae* is recognized as a cause of diarheal disease that may be indistinguishable from the epidemic form of cholera. Infected patients usually have a history of recent ingestion of shellfish or other seafoods or exposure to salt water [23, 31, 64, 75]. The source of infection in this rhesus monkey is unknown, and attempts to isolate the organism from other cases of diarrhea have been unsuccessful. However, this organism should be considered as a potential cause of diarrhea in nonhuman primates. Although nonhuman primates have been experimentally infected with *V. cholerae* [32, 33, 50, 51], this is apparently the first time that this organism has been isolated from a spontaneous case of diarrhea.

Fecal Leukocytes

Examination of Giemsa-stained fecal smears for the presence of leukocytes has been reported to be a reliable method for the detection of invasive bacterial enteric infections. The presence of fecal leukocytes has been stated to be more reliable in establishing infection than bacterial culture of the stool [54, 119]. One report regarding the association between diarrhea, *Shigella* infection, and the presence or absence of fecal leukocytes noted that the presence of cells in feces of monkeys with diarrhea was not predictive of whether or not *Shigella* would be isolated [59].

We routinely examine Giemsa-stained fecal smears of diarrhea cases for the presence of fecal leukocytes. During the past 5 years, fecal smears were examined from 817 cases of diarrhea; 664 (81.3%) of these smears were negative for fecal leukocytes and 153 smears were positive. Bacterial culture of the 664 fecal samples that were negative for fecal leukocytes revealed that 449 (67.6%) of these were negative for enteric pathogens; 215 (32.4%) were culture positive for a variety of organisms recognized as enteric pathogens. Of the 153 cases positive for fecal leukocytes, 66 (43.1%) were culture negative and 87 (56.9%) were culture positive for recognized enteric pathogens. Although enteric pathogens were often isolated from specimens that were negative for fecal leukocytes, there was a highly significant correlation between the presence of fecal leukocytes and the isolation of enteric pathogens (χ^2 value = 30.95). The results of our studies, with respect to fecal examinations for leukocytes, are summarized in Table 43.3 for selected enteric pathogens. It should be noted that these observations may have been affected by the following:

1. The association between the presence of fecal leukocytes and isolation of an enteric pathogen may have been reduced by the failure to use fresh feces from early in the course of the disease for the fecal smears.
2. Because of the high carrier rate for certain enteric pathogens in nonhuman primates, a positive culture/negative smear might occur in carriers whose diarrhea was not associated with the organism isolated.
3. The inclusion of organisms (as enteric pathogens) that may or may not be enteroinvasive will decrease the sensitivity of the fecal leukocyte test. Some organisms (e.g., *Aeromonas*) are probably not enteroinvasive; others (e.g.,

TABLE 43.3. Examinations for fecal leukocytes.

Bacterial isolate	Total cases	Fecal leukocytes	
		Positive	Negative
Shigella	65	35	30
Enteropath. *E. coli*	33	9	24
Aeromonas	19	1	18
C. jejuni	32	7	25
C. coli	66	9	57
Pseudomonas	13	7	6
Yersinia	25	4	21

EEC and *Campylobacter*) are believed to cause diarrhea by both enterotoxic and enteroinvasive mechanisms.

Blood Cultures

During the past 5 years, our bacteriology laboratory has done 1533 blood cultures; 896 of these were from clinical disease cases and 637 were cultures taken at the time of necropsy. A total of 233 of the 896 (26%) clinical cultures were positive; 26 (11.1%) of the culture-positive cases had two or more organisms isolated from the blood. A total of 341 of the 637 (54%) necropsy cases were culture positive; 98 of the 341 (28.7%) culture-positive cases had two or more organisms isolated from the blood. Septicemia, therefore, appears to occur with appreciable frequency in nonhuman primates, either in conjunction with a localized infection, umbilical infections in neonates, infected fight wounds, or gastrointestinal infections. Blood cultures should be routinely done as a diagnostic procedure in both clinical disease and necropsy cases; such cultures often reveal unsuspected septicemias.

Some of the more frequent, more important, and/or unusual blood culture isolates that have been identified from clinical cases in our laboratory during the past 5 years include the following: 18 *Campylobacter* isolates (2 NARTC, 5 *C. coli*, 3 *C. fetus*, 8 *C. jejuni*); 1 *Corynebacterium pseudotuberculosis*; 1 enteropathogenic *E. coli*; 15 *E. coli*; 5 *Hemophilus* species; 5 *Klebsiella pneumoniae*; 2 *Listeria monocytogenes*; 1 *Neisseria meningitidis*; 2 *Shigella flexneri*; 36 *Staphylococcus aureus*; 45 *Staphylococcus epidermidis*; 15 *Streptococcus pneumoniae*; 19 *Streptococcus viridans*, 28 other streptococci; 7 *Yersinia pseudotuberculosis*; and 7 *Aeromonas hydrophila*.

Bacterial isolates recovered from blood cultures taken at necropsy include the following: 23 *Aeromonas hydrophila*; 4 *Campylobacter* (1 *C. fetus*, 2 *C. jejuni*, 1 *C. intestinalis*); 2 *Corynebacterium pseudotuberculosis*; 100 *Escherichia coli*; 15 *Klebsiella pneumoniae*; 7 *Listeria monocytogenes*; 11 *Morganella morganii*; 19 *Proteus mirabilis*; 32 *Proteus vulgaris*; 1 *Salmonella typhimurium*; 1 *Shigella flexneri*; 38 *Staphylococcus aureus*; 24 *Staphylococus epidermidis*; 3 *Streptococcus pneumoniae*; 42 *Streptococcus viridans*, 56 other streptococci; 3 *Yersinia enterocolitica*; and 3 *Yersinia pseudotuberculosis*.

As evident from the above, a wide variety of organisms have been isolated from blood cultures of both clinical and necropsy cases. Although it may be difficult, at times, to determine precisely the relationship between bacterial isolates from the blood and clinical disease or death, we believe most of the above isolates were of etiologic significance. There is some question regarding the significance of *S. epidermidis* and *S. viridans* isolates. These organisms are often of significance in animals with indwelling vascular catheters or prostheses, intraventricular CSF shunts, or cardiac valvular lesions. Additional comments are offered in the following paragraphs on two of the above isolates: *Neisseria meningitidis* and *Listeria monocytogenes*.

Neisseria meningitidis

This isolate, *N. meningitidis*, Group C, from a blood culture of a chimpanzee, represents the second time that this organism has been isolated from chimpanzees in our colony. The previous isolate was from an ocular discharge of a chimpanzee with purulent conjunctivitis. Neither chimpanzee showed evidence of meningitis, and both recovered following treatment. Although these isolates were not the epidemic strain (Group A) of meningococcus, their presence in chimpanzees should be considered a potential health hazard to personnel. Although monkeys have been used in experimental meningococcal infections [104], we are not aware of any other instances in which these organisms have been isolated from spontaneous clinical diseases of nonhuman primates.

Listeria monocytogenes

Listeriosis has been infrequently reported in nonhuman primates [18, 57, 95, 103, 155, 164] but has become an endemic problem in our outdoor-housed macaque colonies. We have had a total of 15 cases, with 13 of these cases noted during the past 5 years. For the most part, cases have been documented in stillbirths and neonatal deaths. An occasional infection has responded to antibiotic therapy. Lesions consist primarily of focal hepatic necrosis and fibrinopurulent placentitis. Small gram-positive rods can usually be identified in areas of necrosis in both liver and placenta. Although this disease has been reported infrequently in other nonhuman primate colonies, it has become an endemic problem in our colony and should be considered as a potential cause of reproductive failure and neonatal death in nonhuman primate colonies. The 15 cases of listeriosis seen in our colony are summarized in Table 43.4.

CSF Cultures

During the past 5 years, 57 cerebrospinal fluid cultures have been done in our bacteriology laboratory. These have been done primarily on infant and juvenile great apes because of the relatively frequent occurrence of meningitis in nursery-maintained chimpanzees. CSF cultures are routinely done on young great apes that show lethargy, anorexia, and fever; this is done to provide an early diagnosis and to determine the antibiotic sensitivity of the bacterial isolate. This protocol, and the use of a commercially available pneumococcal vaccine in all great apes, has greatly decreased the death rate from meningitis in these animals.

Forty-seven of the 57 CSF cultures done during the past 5 years were culture negative. The ten positive cultures included five *Streptococcus pneumoniae* from four chimpanzees and one gorilla infant, two *Hemophilus influenzae* from chimpanzees, and three *Staphylococcus epidermidis* from a chimpanzee and a gorilla

TABLE 43.4. Listeriosis in Yerkes colony.[1]

Case number	Date diagnosed	Animal species	Age	Housing location
1	1-24-73	*M. nigra*	Stillbirth	A-2
2	12-16-76	*M. nigra*	3 days	S-7
3	4-18-80	*M. mulatta*	Abortus	A-1
4	6-18-81	*S. sciureus*	1 day	M-4,R-2
5	3-9-82	*M. mulatta*	Stillbirth	A-2
6	3-31-82	*M. arctoides*	3.5 years	S-6
7	4-2-82	*M. mulatta*	Stillbirth	BC-1B
8[2]	4-6-82	*M. mulatta*	Mother of case 7	BC-1B
9	4-6-82	*M. mulatta*	Stillbirth	A-2
10[2]	4-14-82	*M. mulatta*	Live infant	T-3
11	5-5-82	*M. mulatta*	Stillbirth	T-4
12	5-18-82	*M. arctoides*	Stillbirth	C-1
13	2-24-83	*M. mulatta*	Newborn	A-3
14	2-15-84	*M. mulatta*	Stillbirth	A-2
15	4-18-84	*M. mulatta*	2 years	A-1

[1] 10 of 10 isolates serotyped were serotype 1A.
[2] Nonfatal cases.

with intraventricular catheters. The gorilla was the same one that had previously had a *S. pneumoniae* meningitis. The catheter had been implanted to correct hydrocephalus that followed the earlier meningitis. The catheter in the chimpanzee had been implanted as part of an experimental procedure. Both animals with intraventricular catheters recovered following treatment. The two *Hemophilus* meningitis cases and three of the five *S. pneumoniae* meningitis cases also recovered following treatment.

Joint Cultures

During the past 5 years, bacterial cultures were done on 132 joint specimens from animals with arthritis. Ninety of these cultures were done on clinical arthritis cases and 42 were done at the time of necropsy. One hundred of these specimens were culture negative. *Staphylococcus aureus* was the most frequent isolate (11 cases); other, less-frequent isolates included *E. coli*, *K. pneumoniae*, *Proteus*, *Pasteurella multocida*, and a variety of streptococci. Based on these observations, septic arthritis is infrequently seen in our colony, and most of these cases are due to *S. aureus*. Most of the cases of arthritis that we see appear to be a reactive-type arthritis and may be related to enteric infections with *Shigella*, *Campylobacter*, or *Yersinia*. Some of the culture-negative cases had elevated *Mycoplasma* titers; however, significant increases in titers were not seen, and a limited number of cultures for *Mycoplasma* were negative.

Tuberculosis

Although tuberculosis appears to be a less-frequent problem in nonhuman primates than it was in previous years, a paper on bacterial diseases of nonhuman primates would be incomplete without some comment on this disease. This disease should always be considered a potential threat to nonhuman primate colonies and great efforts expended toward maintaining an active and effective tuberculosis surveillance program in both the animal colony and all personnel who have any contact with the nonhuman primates.

The apparent decreased incidence of tuberculosis in domestic nonhuman primates is probably related, in large part, to the fact that few nonhuman primates (especially rhesus monkeys) are currently being imported from the wild. Active tuberculosis surveillance programs also tend to eliminate infected animals and decrease the spread of disease in domestic colonies.

Only 29 cases of tuberculosis have been diagnosed in animals in the Yerkes colony during the past 19 years; no cases have been seen during the past 10 years (since 1975). Cases seen in our colony prior to 1975 included 23 rhesus monkeys, four chimpanzees, one stumptail macaque, and one orangutan. Affected animals included 14 cases from importers (in our colony for 1 week to 8 years), 12 cases in animals received from other domestic colonies (in our colony for 5 months to 6 years), and three cases in animals born in the Yerkes colony (18 months to 4 years of age). Twenty-seven of the 29 cases were due to *Mycobacterium tuberculosis*, one case was caused by an atypical *M. bovis*, and one case was culture negative. Twenty-two of the cases were detected by tuberculin tests; seven cases had multiple negative tuberculin tests, with their disease detected by radiography, tissue biopsy or at necropsy. At the time of necropsy, the disease was disseminated in six of the seven animals with negative tuberculin tests.

Mycobacteriosis

Mycobacteriosis, a term used to designate nontuberculous mycobacterial infection, is occurring with increasing frequency in nonhuman primates. Although most reports of mycobacteriosis in nonhuman primates have been due to infection with *M. avium* complex, other isolates include *M. kansasii*, *M. gordonae*, and *M. scrofulaceum* [38, 45, 80, 126, 135, 137, 140, 141, 154]. *Mycobacterium avium* complex infection in nonhuman primates is primarily a disease of the intestinal tract, with affected animals showing a chronic, refractory diarrhea, progressive weight loss, and eventual death. Gross lesions are usually limited to slight to moderate thickening of the wall of the small intestine and colon and variable degrees of lymphadenopathy of the mesenteric lymph nodes. Histologically, the lamina propria of the intestine and colon and the mesenteric lymph nodes show a diffuse infiltrate of large histiocytic cells that are usually filled with large numbers of acid-fast bacilli. Occasionally, focal granulomas may also be seen in the liver.

During the past 5 years, we have documented 11 fatal cases of mycobacteriosis in the Yerkes colony (Table 43.5). These occurred in a colony of outdoor-housed stump-tail macaques and were initially thought to be due to infection with *M. avium*. However, cultures of intestine and lymph nodes of these fatal cases (done by Dr. R. Chiodini, University of Connecticut) yielded mycobactin-dependent mycobacteria that were identified as *Mycobacterium paratuberculosis* [100]. Following identification of this organism in necropsy cases, fecal samples were collected for mycobacterial culture from 36 animals in the affected stump-tail macaque colony. Twenty-six of these 36 (72.2%) animals were culture positive for *M. paratuberculosis*. Fecal cultures from 105 nonhuman primates of various

TABLE 43.5. Paratuberculosis deaths in *M. arctoides* colony.

Case number	Date died	Origin of animal	Time in colony (months)
1	7-30-80	Other colony	19
2	3-5-81	Other colony	27
3	10-14-81	Colony born	34
4	6-7-82	Other colony	42
5	10-13-82	Other colony	46
6	4-5-83	Other colony	52
7	4-6-83	Colony born	31
8	5-18-84	Colony born	28
9	1-2-85	Other colony	73
10	4-30-85	Other colony	76
11	6-10-85	Other colony	78

species from other outdoor-housing facilities were all culture negative for *M. paratuberculosis* indicating that the infection has, to date, not spread to any other groups of outdoor-housed nonhuman primates.

Paratuberculosis in nonhuman primates is characterized clinically by chronic, refractory diarrhea and progressive weight loss that eventually results in death. The clinical and pathologic features of paratuberculosis are essentially identical to those reported for *M. avium* complex infections; a definitive diagnosis is dependent upon isolation and identification of the organism. Although paratuberculosis has been reported in rhesus monkeys, this represents the first time that *M. paratuberculosis* has been isolated from nonhuman primates. The diagnosis in the one earlier report was based on histologic features of the lesions [122]. The clinical and pathologic features of this disease in monkeys is essentially the same as that reported for paratuberculosis in the bovine species [20]. The recent isolation of a *M. paratuberculosis*-like organism from patients with Crohn's [21] is of interest and possible relevance to the occurrence of paratuberculosis in nonhuman primates.

Placental Cultures

Bacterial cultures are routinely done on all placentas that are available for examination. During the past 5 years, cultures were done on 103 placentas; 43 of these were culture negative, and a variety of organisms were isolated from the remaining 60 placentas. Isolates that were considered to be of pathogenic significance included six *Listeria monocytogenes*, six Beta streptococci, and one *Shigella flexneri*. Most other isolates, including such organisms as *E. coli*, *Proteus*, *Enterobacter*, *S. viridans*, *Staphylococcus*, etc., were considered to be of little or no etiologic significance. Beta hemolytic streptococci have been shown to colonize the female genital tract and to cause serious infections in neonatal humans [4, 36, 53, 69, 90]; listeriosis is a documented cause of fetal death in nonhuman primates [95, 103]; septic abortion caused by *Shigella* or *Salmonella* infection has been reported in nonhuman primates [146, 149]; and abortion in squirrel monkeys has been associated with *Yersinia* infection [14].

Vaginal Cultures

Vaginal cultures are obtained from animals that have aborted or given birth to stillborn infants, animals with a clinically evident vaginal discharge, and animals who have given birth to infants that are either unthrifty or clinically ill. During the past 5 years, vaginal cultures have been done on 94 nonhuman primates. For the most part, these cultures yielded a variety of and often multiple isolates whose clinical significance could not be determined with certainty. Isolates that were considered of potential clinical significance included Beta streptococci (29 isolates), *Shigella flexneri* type 4 (three isolates), one *Yersinia pseudotuberculosis*, and one *Salmonella typhimurium*. The most frequently isolated organisms included *E. coli* (48 isolates), *S. aureus* (29 isolates), *S. epidermidis* (26 isolates), *S. viridans* (24 isolates), diphtheroids (19 isolates), and *Proteus vulgaris* (15 isolates).

Air Sac Cultures

Most species of Old and New World nonhuman primates have submandibular air sacs that communicate directly with the trachea. In some species, such as the orangutan and, to a lesser degree, the chimpanzee, the air sacs are much more extensive and extend into the axillary spaces. Bacterial infections of the air sacs have been reported in the orangutan, chimpanzee, baboon, and owl monkey [3, 41, 46, 48, 84, 145], and we continue to see an appreciable number of air sac infections in our great ape colony, especially in orangutans and occasionally in chimpanzees.

During the past 5 years, we have done bacterial cultures on specimens from the air sacs of 63 animals; 54 of these were done on clinical cases and nine were specimens collected at necropsy. Multiple organisms are usually isolated from these cases, and isolates are primarily enteric organisms. The most frequently isolated organisms from the 63 cases noted above include *E. coli* (24 isolates), *Klebsiella pneumoniae* (19 isolates), *Pseudomonas* species (22 isolates), *Proteus* species (13 isolates), and *Pasteurella multocida* (11 isolates). A variety of other organisms such as *S. pneumoniae*, *Hemophilus* species, *S. aureus*, *Fusobacterium*, etc., were isolated from one to six times during this series.

Animals with air sac infections show varying degrees of distention of the air sacs with fluid and exudate. Impression smears of this material usually show large numbers of pleomorphic bacteria. Histologic examination of the air sac lining shows variable degrees of necrosis and inflammatory cell infiltrate. Aspiration pneumonia is a complication of air sac infection that may result in death of the animal.

Urine Cultures

Urine is cultured from all animals with clinical signs of a genitourinary tract infection. During the past 5 years, our bacteriology laboratory cultured 304 urine specimens; 289 of these were culture negative. Organisms considered to be clinically significant that were isolated from the 15 culture positive cases included: *E. coli* (six isolates), *S. aureus* (two isolates), *S. viridans* (two isolates),

Enterococcus (two isolates), and one isolate each of *Morganella morganii*, *Proteus*, and group B, Beta *Streptococcus*.

Pneumonia

Although seen less frequently than in previous years (when large numbers of animals were imported from the wild), bacterial pneumonia continues to be a significant cause of morbidity and mortality in nonhuman primates. Organisms that have been most frequently identified as a cause of pneumonia include staphylococci, streptococci (especially *S. pneumoniae*), *Hemophilus* species, *Klebsiella* species, *Pasteurella multocida*, *Escherichia coli*, and *Bordetella bronchiseptica* [42, 65, 78, 94, 95, 136].

Upper respiratory infections, sometimes followed by pneumonia, are seen with some frequency in nonhuman primates of the Yerkes colony. Pneumonia tends to occur most frequently during January through March, with the lowest incidence noted in September through November. Clinical cases are diagnosed by clinical signs and radiography; the causative agent for clinical cases is often not identified owing to the lack of appropriate material for culture. Occasionally, sputum cultures will reveal an organism that is believed to be the cause of the pneumonia. Gram-stained sputum or nasal exudate specimens should be examined to correlate culture results with organisms present within neutrophils in the smear. Blood cultures should always be done on animals thought to have pneumonia. In the great apes, more than 50% of the cases occur in animals that are less than 2 years old; most cases respond favorably to antibiotic therapy.

During the past 5 years, pneumonia has been diagnosed in 37 nonhuman primates that were presented for necropsy examination. These included 14 rhesus monkeys, eight squirrel monkeys, six pig-tailed macaques, four orangutans, three chimpanzees, one stump-tail macaque, and one hybrid macaque. The age distribution of these 37 cases included nine animals that died within the first week of life, seven animals that were 8 to 31 days of age, two animals that were 1 to 6 months old, and 19 animals that were 4 years of age or older. All four cases of pneumonia in the orangutan were associated with air sac infections; multiple organisms, including *Pseudomonas aeruginosa*, *Klebsiella pneumoniae*, *Proteus vulgaris*, *Escherichia coli*, *Pasteurella multocida*, *Clostridium* species, and *Bacteroides* species, were isolated from the lungs of these cases. Bacterial isolates from the other cases included *E. coli* (seven cases); *S. aureus* (five cases); *Corynebacterium pseudotuberculosis* (three cases); two cases each of *Enterobacter agglomerans*, *Klebsiella pneumoniae*, *S. aureus*, and *E. coli*; and one case each of *Corynebacterium* CDC Group F-2; *Hemophilus influenzae*; *Streptococcus pneumoniae*; *Aeromonas hydrophila*; *A. hydrophila* and *S. epidermidis*; *S. aureus* and *Enterobacter cloacae*; *Proteus vulgaris* and *Klebsiella pneumoniae*; *K. pneumoniae*, *S. aureus*, and *Acinetobacter lwoffi*; and *E coli*, *Pseudomonas* species, and *S. epidermidis*. Two pneumonia cases were not cultured, and one case was culture negative. Septicemia, caused by the same organism(s) as that present in the lungs, was documented in 13 of the 37 cases of pneumonia.

Wound Cultures

Bacterial infection of external wounds, usually inflicted during fights in group-housed animals, is a common problem in primate colonies. Prompt attention (bacterial culture, antibiotic sensitivity, and treatment) to such lesions will hasten recovery and often prevent serious, life-threatening infections that are prone to becoming septicemic. During the past 5 years, 286 cultures were done in our laboratory on infected fight wounds and/or other types of external lesions. The most frequently isolated organism was *Staphylococcus aureus*, which was recovered from 166 cases. Other frequent isolates included *S. epidermidis* (42 cases), *E. coli* (33 cases), Group B Beta *Streptococcus* (22 cases), *Proteus vulgaris* (27 cases), *Streptococcus viridans* (28 cases), *Proteus mirabilis* (17 cases), and *Aeromonas hydrophila* (10 cases). *Salmonella derby, Shigella flexneri* 4, and *Yersinia pseudotuberculosis* were each isolated from one case. A variety of other organisms that, other than various types of streptococci, were primarily gram-negative bacilli were isolated in a few instances from these cases.

Miscellaneous Bacterial Infections

Chromobacteriosis

Chromobacterium violaceum, the causative organism of chromobacteriosis, is widely distributed in nature and is usually considered to be nonpathogenic [24, 67]. Infection with this organism has been documented in humans [24, 67, 115, 116], a variety of domestic animals [70, 139, 163], and nonhuman primates [6, 47, 66, 70, 98]. One case of *Chromobacterium violaceum* infection has been documented in a macaque in the Yerkes colony [98]. This animal died 4 days after receipt from a primate facility in Florida.

Nocardiosis

Nocardiosis resulting from infection with *Nocardia* species has been documented in humans and a variety of animal species [56, 113, 117, 158], including nonhuman primates [1, 10, 42, 68, 85, 99]. These organisms have been associated with pulmonary, gastrointestinal, or skin infections. One case of pulmonary nocardiosis has been diagnosed in an orangutan in the Yerkes colony [99].

Moraxella

Moraxella nonliquefaciens, a part of the normal flora of the upper respiratory tract, has been associated with septicemia, pneumonia, bronchitis, and eye infections in human [11, 118, 130]. One case of *M. nonliquefaciens* speticemia has been documented in a rhesus monkey in our colony [95]. At necropsy this animal has superficial fight wounds and multifocal abscesses of the myocardium and kidneys.

Melioidosis

This disease, caused by *Pseudomonas pseudomallei*, is seen primarily in animals and humans of Southeast Asia [142]. Although not observed in our colony, the disease has been reported in various nonhuman primate species [15, 73, 111, 127, 144, 147]. This disease should be considered when examining animals from Southeast Asia.

Tularemia

Natural infection with *Francisella tularensis* has been reported in a squirrel monkey [29], tamarins, a marmoset, and a talapoin [112]. The disease has also been experimentally induced in macaques and African green monkeys [8, 22, 25, 52].

Leprosy

This disease, associated with acid-fast mycobacteria compatible with *M. leprae*, has occurred spontaneously in a chimpanzee [27, 82, 83] and a mangabey monkey [105, 106]. Suggested experimental transmission to a gibbon [157] and a chimpanzee [49] was reported several years ago. More recently, transmission to the mangabey monkey has been reported [92, 162].

Corynebacterium

A limited number of cases of *Corynebacterium* infection have been reported in nonhuman primates. These include *C. ulcerans* [39, 93], *C. equi* [143], and *C. pseudotuberculosis* [76]. We have, within the past few months, documented three cases of pneumonia in monkeys that were associated with *C. pseudotuberculosis* infection; all three animals also had *C. pseudotuberculosis* septicemia. *Corynebacterium* CDC Group F-2 was isolated from another case of pneumonia.

Other Bacterial Infections

Numerous other miscellaneous, infrequently encountered (or reported) bacterial infections have been reported in nonhuman primates. Some of these include tetanus [26, 125], leptospirosis [30, 138, 160, 161], brucellosis [120], actinobacillosis [108], treponematosis [40], borreliosis [34], and erysipelas [58].

Summary and Conclusions

Bacterial infections continue to be a frequent and significant cause of clinical disease and death in domestic nonhuman primates. Although enteric bacterial infections are the most common cause of morbidity and mortality in nonhuman primates, a variety of other bacterial infections are encountered with considerable frequency. Most commonly encountered bacterial infections are treatable and will respond favorably if promptly diagnosed. Consequently, bacteriologic support is required for the successful management of nonhuman primate colonies. Bacterial cultures should be utilized in the evaluation of clinical disease as well as at necropsy, as such is often required to provide appropriate therapy and/or to arrive at the correct diagnosis. In this paper, we have reviewed the literature with respect to reported bacterial infections of nonhuman primates and relate our experiences with regard to microbial isolates that have been encountered in the Yerkes primate colony during the past 5 years (1980–1984).

In conclusion, nonhuman primates, regardless of where they are maintained—in zoological collections or research laboratories—will experience spontaneous disease problems. Proper attention to and support of adequate diagnostic facilities and detailed evaluation of all cases of spontaneous disease will contribute substantially to our goal of self-sustaining populations.

Acknowledgments. This investigation was supported in part by NIH grant RR-00165 from the Division of Research Resources to the Yerkes Primate Research Center. The Yerkes Center is fully accredited by the American Association for Accreditation of Laboratory Animal Care.

References

1. Al-Doory Y, Pinkerton ME, Vice TE, Hutchinson V (1969) Pulmonary nocardiosis in a vervet monkey. J Amer Vet Med Assoc 155:1179–1180.
2. Anders BJ, Paisley JW, Lauer BA, Reller LB (1982) Double-blind placebo controlled trial of erythromycin for treatment of *Campylobacter* enteritis. Lancet 1:131–132.
3. Annual Report (1963) Cheyenne Mountain Zoological Park, Colorado Springs, Colorado, p 23.
4. Anthony BF, Eisenstadt R, Carter J, Kim KS, Hobel CJ (1981) Genital and intestinal carriage of group B Streptococci during pregnancy. J Inf Dis 143:761–766.
5. Armitage GC, Newbrun E, Polando V, Lodberg P, Anderson J (1978) Oral shigellosis and periodontal disease in macaques. I. Clinical and ultrastructural observations. J Dent Res 57(Suppl A):223.
6. Audebaud G, Ganzin M, Ceccaldi J, Merveille P (1954) Isolement d'un *Chromobacterium violaceum* à partir de lésions hépatiques observees chez un singe *Cercopithecus cephus* étude et pouvoir pathogène. Ann Inst Pasteur Lille 87:413–417.
7. Baggs RB, Hunt RD, Garcia FG, Hajema EM, Blake BJ, Fraser CEO (1976) Pseudotuberculosis (*Yersinia enterocolitica*) in the owl monkey (*Aotus trivirgatus*). Lab Anim Sci 26:1079–1083.
8. Baskerville A, Hambleton P, Dowsett AB (1978) The pathology of untreated and antibiotic-treated experimental tularemia in monkeys. Br J Exp Pathol 59:615–623.
9. Bauwens L, De Meurichy W (1984) The occurrence of thermophilic campylobacters in zoo animals. Acta Zool Pathol Antwerp 76:181–189.
10. Boncyk LH, McCullough B, Grotts DO, Kalter SS (1975) Localized nocardiosis due to *Nocardia caviae* in a baboon (*Papio cynocephalus*). Lab Anim Sci 25:88–91.
11. Bottone E, Allerhand J (1968) Association of mucoid encapsulated *Moraxella duplex* var. *nonliquefaciens* with chronic bronchitis. Appl Microbiol 16:315–319.
12. Bresnahan JF, Whitworth UG, Hayes Y, Summers E, Pollock J (1984) *Yersinia enterocolitica* infection in breeding colonies of ruffed lemurs. J Amer Vet Med Assoc 185:1354–1356.
13. Bronson RT, May BD, Ruebner BH (1972) An outbreak of infection by *Yersinia pseudotuberculosis* in nonhuman primates. Amer J Pathol 69:289–308.
14. Buhles WC Jr, Vanderlip JE, Russell SW, Alexander NL (1981) *Yersinia pseudotuberculosis* infection: study of an epizootic in squirrel monkeys. J Clin Microbiol 13:519–525.
15. Butler TM, Schmidt RE, Wiley GL (1971) Melioidosis in a chimpanzee. Amer J Vet Res 32:1109–1117.
16. Butzler JP, Skirrow MB (1979) *Campylobacter* enteritis. Clin Gastroenterol 8:737–765.
17. Chakraborty T, Montenegro MA, Sanyal SC, Helmuth R, Bulling E, Timmis KN (1984) Cloning of enterotoxin gene from *Aeromonas hydrophila* provides conclusive evidence of production of a cytotonic enterotoxin. Infect Immun 46:435–441.

18. Chalifoux LV, Hajema EM (1981) Septicemia and meningoencephalitis caused by *Listeria monocytogenes* in a neonatal *Macaca fascicularis*. J Med Primatol 10:336–339.
19. Chang J, Wagner JL, Kornegay RW (1980) Fatal *Yersinia pseudotuberculosis* infection in captive bushbabies. J Amer Vet Med Assoc 177:820–821.
20. Chiodini RJ, Van Kruiningen HJ, Merkal RS (1984) Ruminant paratuberculosis (Johne's Disease): the current status and future prospects. Cornell Vet 74:218–262.
21. Chiodini RJ, Van Kruiningen HJ, Merkal RS, Thayer WR Jr, Coutu JA (1984) Characteristics of an unclassified *Mycobacterium* species isolated from patients with Crohn's Disease. J Clin Microbiol 20:966–971.
22. Coriell LL, King EO, Smith MG (1948) Studies in tularemia. IV. Observations on tularemia in normal and vaccinated monkeys. J Immunol 58:182–202.
23. Dakin WPH, Howell DJ, Sutton RGA, O'Keefe MF, Thomas P (1974) Gastroenteritis due to non-agglutinable (non-cholera) vibrios. Med J Aust 2:487–490.
24. Dauphinais RM, Robben GG (1968) Fatal infections due to *Chromobacterium violaceum*. Amer J Clin Pathol 50:592–597.
25. Day WC, Berendt RF (1972) Experimental tularemia in *Macaca mulatta*: relationship of aerosol particle size to the infectivity of airborne *Pasteurella tularensis*. Infect Immun 5:77–97.
26. DiGiacomo RF, Missakian EA (1972) Tetanus in a free-ranging colony of *Macaca mulatta*: a clinical and epizootiologic study. Lab Anim Sci 22:378–383.
27. Donham KJ, Leininger JR (1977) Spontaneous leprosy-like disease in a chimpanzee. J Inf Dis 136:132–136.
28. Duffy MC, Benson JB, Rubin SJ (1980) Mucosal invasion in *Campylobacter* enteritis. Amer J Clin Pathol 73:706–708.
29. Emmons RW, Woodie JD, Taylor MS, Nygaard GS (1970) Tularemia in a pet squirrel monkey (*Saimiri sciureus*). Lab Anim Care 20:1149–1153.
30. Fear FA, Pinkerton ME, Cline JA, Kriewaldt F, Kalter SS (1968) A leptospirosis outbreak in a baboon (*Papio* sp.) colony. Lab Anim Care 18:22–28.
31. Fekety R (1983) Recent advances in management of bacterial diarrhea. Rev Inf Dis 5:246–257.
32. Felsenfeld O, Gyr K, Wolf RH (1976) Malnutrition and susceptibility to infection with *Vibrio cholerae* in vervet monkeys (*Cercopithecus aethiops*). J Med Primatol 5:186–194.
33. Felsenfeld O, Wolf RH (1972) Nonhuman primates in cholera experiments. In: Goldsmith EI, Moor-Jankowski J (ed) Medical primatology 1972, Part III. S Karger, Basel.
34. Fiennes R (1967) Zoonoses of primates. Cornell University Press, Ithaca NY.
35. Finch MJ, Riley LW (1984) *Campylobacter* infections in the United States. Arch Int Med 144:1610–1612.
36. Fischer GW, Weisman LB, Hemming VG, London WT, Hunter KW, Bosworth JM, Sever JL, Wilson SR, Curfman BL (1984) Intravenous immunoglobulin in neonatal group B streptococcal disease: pharmacokinetic and safety studies in monkeys and humans. Amer J Med 76:117–124.
37. Fitzgeorge RB, Baskerville A, Lander KP (1981) Experimental infection of rhesus monkeys with a human strain of *Campylobacter jejuni*. J Hyg, Camb 86:343–351.
38. Fleischman RW, du Moulin GC, Esber HJ, Ilievski V, Bogden AE (1982) Nontuberculous mycobacterial infection attributable to *Mycobacterium intracellulare* serotype 10 in two rhesus monkeys. J Amer Vet Med Assoc 181:1358–1362.
39. Fox JG, Frost WW (1974) *Corynebacterium ulcerans* mastitis in a bonnet macaque (*Macaca radiata*). Lab Anim Sci 24:820–822.
40. Fribourg BA, Mollaret HH (1969) Natural treponematosis of the African primate. Prim Med 3:110–118.

41. Giles RC Jr, Hildebrandt PK, Tate C (1974) Klebsiella air sacculitis in the owl monkey (*Aotus trivirgatus*). Lab Anim Sci 24:610–616.
42. Good RC, May BD (1971) Respiratory pathogens in monkeys. Infect Immun 3:87–99.
43. Good RC, May BD, Kawatomari T (1969) Enteric pathogens in monkeys. J Bact 97:1048–1055.
44. Goodwin CS, Harper WES, Stewart JK, Gracey M, Burke V, Robinson J (1983) Enterotoxigenic *Aeromonas hydrophila* and diarrhea in adults. Med J Aust 1:25–26.
45. Gribble DH (1972) Granulomatous enteritis and intestinal amyloidosis in nonhuman primates. Vet Pathol 9:81–82.
46. Gross GS (1978) Medical and surgical approach to laryngeal air sacculitis in a baboon caused by *Pasteurella multocida*. Lab Anim Sci 28:737–741.
47. Groves MG, Strauss JM, Abbas J, Davis CE (1969) Natural infections of gibbons with a bacterium producing violet pigment (*Chromobacterium violaceum*). J Inf Dis 120:605–610.
48. Guilloud NB, McClure HM (1969) Air sac infection in the orangutan (*Pongo pygmaeus*). Proc 2nd Int Congr Primatol 3:143–147.
49. Gunders AE (1958) Progressive experimental infection with *Mycobacterium leprae* in a chimpanzee: a preliminary report. J Trop Med Hyg 61:228–230.
50. Gyr K, Felsenfeld O, Wolf RH (1975) Intestinal absorption, exocrine pancreatic function and response to *Vibrio cholerae* infection in protein deficient patas monkeys (*Erythrocebus patas*). Trans Roy Soc Trop Med Hyg 69:247–250.
51. Gyr K, Felsenfeld O, Zimmerli-Ning M (1978) Effect of oral pancreatic enzymes on the course of cholera in protein-deficient vervet monkeys. Gastroenterology 74:511–513.
52. Hall WC, Kovatch RM, Schricker RL (1973) Tularemic pneumonia: pathogenesis of the aerosol-induced disease in monkeys. J Pathol 110:193–202.
53. Hammersen G, Bartholome K, Opperman HC, Lutz P (1977) Group B streptococci: a new threat to the newborn. Europ J. Pediat 126:189–197.
54. Harris JC, Dupont HL, Hornick RB (1972) Fecal leukocytes in diarrheal illness. Ann Intern Med 76:697–703.
55. Harvey SM (1980) Hippurate hydrolysis by *Campylobacter fetus*. J Clin Microbiol 11:435–437.
56. Hathaway BM, Mason KN (1962) Nocardiosis: study of fourteen cases. Amer J Med 32:903–909.
57. Heldstab A, Ruedi D (1982) Listeriosis in an adult female chimpanzee (*Pan troglodytes*). J Comp Pathol 92:609–612.
58. Hirsch DC, Boorman GA, Jang SS (1975) Erysipelas in a black and red tamarin. J Amer Vet Med Assoc 167:646–647.
59. Hirsch DC, Davidson JN, Beards LR, Anderson JH, Budd CP, Henrickson RV (1980) Microscopic examination of stools from nonhuman primates as a way of predicting the presence of *Shigella*. J Clin Microbiol 11:65–67.
60. Holmberg SD, Farmer JJ (1984) *Aeromonas hydrophila* and *Plesiomonas shigelloides* as causes of intestinal infections. Rev Inf Dis 6:633–639.
61. Ho DD, Ault MJ, Ault MA, Murata GH (1982) *Campylobacter* enteritis: early diagnosis with gram's stain. Arch Int Med 142:1858–1860.
62. Holt PE (1981) Role of *Campylobacter* spp. in human and animal disease: a review. J Roy Soc Med 74:437–440.
63. Hostacka A, Ciznar I, Korych B, Karolcek J (1982) Toxic factors of *Aeromonas hydrophila* and *Plesiomonas shigelloides*. Zbl Bakt Hyg I. Abt Orig A 252:525–534.
64. Hughes JM, Hollis DG, Gangarosa EJ, Weaver RE (1978) Non-cholera vibrio infections in the United States: clinical, epidemiologic, and laboratory features. Ann Int Med 88:602–606.

65. Hunt DE, Pittillo RF, Deneau GA, Schabel FM Jr, Mellett LB (1968) Control of an acute *Klebsiella pneumoniae* infection in a rhesus monkey colony. Lab Anim Care 18:182–185.
66. Johnsen DO, Pulliam JD, Tanticharoenyos P (1970) *Chromobacterium* septicemia in the gibbon. J Inf Dis 122:563.
67. Johnson WM, DiSalvo AF, Steuer RR (1971) Fatal *Chromobacterium violaceum* septicemia. Amer J Clin Pathol 56:400–406.
68. Jonas AM, Wyand DS (1966) Pulmonary nocardiosis in the rhesus monkey. Pathol Vet 3:588–600.
69. Jones DE, Kanarek KS, Lim DV (1984) Group B streptococcal colonization patterns in mothers and their infants. J Clin Microbiol 20:438–440.
70. Joseph PG, Sivendra R, Anwar M, Ong SF (1971) *Chromobacterium violaceum* infection in animals. Kajian Vet 3:55–66.
71. Karmali MA, Fleming PC (1979) *Campylobacter* enteritis in children. J Pediat 94:527–533.
72. Karmali MA, Penner JL, Fleming PC, Williams A, Hennessy JN (1983) The serotype and biotype distribution of clinical isolates of *Campylobacter jejuni* and *Campylobacter coli* over a three-year period. J Inf Dis 147:243–246.
73. Kaufmann AF, Alexander AD, Allen AM, Cronin RJ, Dillingham LA, Douglas JD, Moore TD (1970) Melioidosis in imported non-human primates. J Wildl Dis 6:211–219.
74. Kent TH, Formal SB, LaBrec EH, Sprinz H, Maenza RM (1967) Gastric shigellosis in rhesus monkeys. Amer J Pathol 51:259–267.
75. Kenyon JE, Piexoto DR, Austin B, Gillies DC (1984) Seasonal variation in numbers of *Vibrio cholerae* (non-01) isolated from California coastal waters. Appl Environ Microbiol 47:1243–1245.
76. Kim JCS (1976) *Corynebacterium pseudotuberculosis* as a cause of subacute interstitial nephritis in a chimpanzee. Vet Med Small Anim Clin 71:1093–1095.
77. Klipstein FA, Engert RF (1984) Purification of *Campylobacter jejuni* enterotoxin. Lancet 1:1123–1124.
78. Kohn DF, Haines DE (1977) *Bordetella bronchiseptica* infection in the lesser bushbaby (*Galago senegalensis*). Lab Anim Sci 27:279–280.
79. Lambert ME, Schofield PL, Ironside AG (1979) *Campylobacter* colitis. Br Med J 1:857–859.
80. Latt RH (1975) Runyon group III atypical mycobacteria as a cause of tuberculosis in a rhesus monkey. Lab Anim Sci 25:206–209.
81. Lautrop H (1961) *Aeromonas hydrophila* isolated from human feces and its possible pathological significance. Acta Pathol Microbiol Scand 51:299–301.
82. Leininger JR, Donham KJ, Meyers WM (1980) Leprosy in a chimpanzee: postmortem lesions. Int J Lepr 48:414–421.
83. Leininger JR, Donham KJ, Rubino MJ (1978) Leprosy in a chimpanzee: morphology of the skin lesions and characterization of the organism. Vet Pathol 15:339–346.
84. Lewis JC, Montgomery CA Jr, Hildebrandt PK (1975) Airsacculitis in the baboon. J Amer Vet Med Assoc 167:662–664.
85. Liebenberg SP, Giddens WE Jr (1985) Disseminated nocardiosis in three macaque monkeys. Lab Anim Sci 35:162–166.
86. Lior H (1984) New, extended biotyping scheme for *Campylobacter jejuni*, *Campylobacter coli* and "*Campylobacter laridis*." J Clin Microbiol 20:636–640.
87. Ljungh A, Popoff M, Wadstrom T (1977) *Aeromonas hydrophila* in acute diarrheal disease: detection of enterotoxin and biotyping of strains. J Clin Microbiol 6:96–100.
88. Luechtefeld NW, Cambre RC, Wang W-LL (1981) Isolation of *Campylobacter fetus* subsp. *jejuni* from zoo animals. J Amer Vet Med Assoc 179:1119–1122.

89. Luechtefeld NW, Wang W-LL, Blaser MJ, Reller LB (1981) *Campylobacter fetus* subsp. *jejuni*: background and laboratory diagnosis. Lab Med 12:481–487.
90. MacDonald SW, Manuel RF, Embil JA (1979) Localization of group B beta hemolytic Streptococci in the female urogenital tract. Amer J Obst Gyn 133:57–79.
91. Manninen KI, Prescott JF, Dohoo IR (1982) Pathogenicity of *Campylobacter jejuni* isolates from animals and humans. Infect Immun 38:46–52.
92. Martin LN, Gormus BJ, Wolf RH, Gerone PJ, Meyers WM, Walsh GP, Binford CH, Hadfield TL, Schlagel CJ (1985) Depression of lymphocyte responses to mitogens in mangabeys with disseminated experimental leprosy. Cell Immunol 90:115–130.
93. May BD (1972) *Corynebacterium ulcerans* infection in monkeys. Lab Anim Sci 22:509–513.
94. McClure HM (1975) Pathology of the rhesus monkey. In: Bourne GH (ed) The rhesus monkey, vol 2. Academic, New York.
95. McClure HM (1980) Bacterial diseases of nonhuman primates. In: Montali RJ, Migaki G (ed) The comparative pathology of zoo animals. Smithsonian Institution Press, Washington, DC.
96. McClure HM (1980) Diagnostic exercise (yersiniosis). Lab Anim Sci 30:513–514.
97. McClure HM, Alford P, Swenson B (1976) Nonenteric *Shigella* infections in nonhuman primates. J Amer Vet Med Assoc 169:938–939.
98. McClure HM, Chang J (1976) *Chromobacterium violaceum* infection in a nonhuman primate (*Macaca assamensis*). Lab Anim Sci 26:807–810.
99. McClure HM, Chang J, Kaplan W, Brown JM (1976) Pulmonary nocardiosis in an orangutan. J Amer Vet Med Assoc 169:943–945.
100. McClure H, Chiodini R, Anderson D, Swenson B (1984) Paratuberculosis in stump-tailed macaques (*Macaca arctoides*). Amer J Primatol 5:395.
101. McClure HM, King FA (1984) Yersiniosis: a review and report of an epizootic in nonhuman primates. In: Ryder OA, Byrd ML (ed) One medicine. Springer-Verlag, Berlin.
102. McClure HM, Guilloud NB (1971) Comparative pathology of the chimpanzee. In: Bourne GH (ed) The chimpanzee, vol 4. S Karger, Basel.
103. McClure HM, Strozier LM (1975) Perinatal listeric septicemia in a celebese black ape. J Amer Vet Med Assoc 167:637–638.
104. Mellins RB, Levine OR, Wigger HJ, Leidy G, Curnen EC (1972) Experimental meningococcemia: model of overwhelming infection in unanesthetized monkeys. J Appl Physiol 32:309–314.
105. Meyers WM, Walsh GP, Binford CH, Wolf RH, Gormus BJ, Martin LN, Gerone PJ, Baskin GB (1984) Comparative pathology of disseminated multibacillary leprosy. Comp Pathol Bull 16:1,5–6.
106. Meyers WM, Walsh GP, Brown HL, Fukunishi Y, Binford CH, Gerone PJ, Wolf RH (1980) Naturally-acquired leprosy in a mangabey monkey (*Cercocebus* sp.). Int J Lepr 48:495–496.
107. Michalak DM, Perrault J, Gilchrist MJ, Dozois RR, Carney JA, Sheedy PF (1980) *Campylobacter fetus* ss. *jejuni*: a cause of massive lower gastrointestinal hemorrhage. Gastroenterology 79:742–745.
108. Moon HW, Barnes DM, Higbee JM (1969) Septic embolic actinobacillosis: a report of 2 cases in new world monkeys. Pathol Vet 6:481–486.
109. Morton WR, Bronsdon M, Mickelsen G, Knitter G, Rosenkranz S, Kuller L, Sajuthi D (1983) Identification of *Campylobacter jejuni* in *Macaca fascicularis* imported from Indonesia. Lab Anim Sci 33:187–188.
110. Mulder JB (1971) Shigellosis in nonhuman primates: a review. Lab Anim Sci 21:734–738.
111. Mutalib AR, Sheikh-Omar AR, Zamri M (1984) Melioidosis in a banded leaf monkey (*Presbytis melalophos*). Vet Rec 115:438–439.

112. Nayar GPS, Crawshaw GJ, Neufeld JL (1979) Tularemia in a group of nonhuman primates. J Amer Vet Med Assoc 175:962–963.
113. Neu HC, Silva M, Hazen E, Rosenheim SH (1967) Necrotizing nocardial pneumonitis. Ann Int Med 66:274–284.
114. Newbrun E, Hoover C, Armitage G, Anderson J (1978) Oral shigellosis and periodontal disease in macaques. II. Microbiological findings. J Dent Res 57(Suppl A):351.
115. Nunnally RM, Dunlop WH (1968) Fatal septicemia due to *Chromobacterium janthinum*. Louisiana State Med Soc J 120:278–280.
116. Ognibene AJ, Thomas E (1970) Fatal infection due to *Chromobacterium violaceum* in Vietnam. Amer J Clin Pathol 54:607–610.
117. Peabody JW Jr, Seabury JH (1960) Actinomycosis and nocardiosis: a review of basic differences in therapy. Amer J Med 28:99–115.
118. Pedersen MM, Marso E, Pickett MJ (1970) Nonfermentative bacilli associated with man. III. Pathogenicity and antibiotic susceptibility. Amer J Clin Pathol 54:178–192.
119. Pickering LK, Dupont HL, Olarte J, Conklin R, Ericsson C (1977) Fecal leukocytes in enteric infections. Amer J Clin Pathol 68:562–565.
120. Pinkerton ME (1967) Bacteremia in wild baboons. Bact Proc 67:67.
121. Pinkerton ME (1968) Shigellosis in the baboon (*Papio* sp.). Lab Anim Care 18:11–21.
122. Pitcock JA, Gisler DB (1961) Paratuberculosis (Johne's Disease) in the monkey (*Macaca mulatta*). School of Aerospace Medicine, Report Number 61–86, Brooks Air Force Base, Texas.
123. Prescott JF, Munroe DL (1982) *Campylobacter jejuni* enteritis in man and domestic animals. J Amer Vet Med Assoc 181:1524–1530.
124. Price AB, Dolby JM, Dunscombe PR, Stirling J (1984) Detection of *Campylobacter* by immunofluorescence in stools and rectal biopsies of patientis with diarrhoea. J Clin Pathol 37:1007–1013.
125. Rawlins RG, Kessler MJ (1982) A five-year study of tetanus in the Cayo Santiago rhesus monkey colony: behavioral description and epizootiology. Amer J Primatol 3:23–39.
126. Renquist DM, Potkay S (1979) *Mycobacterium scrofulaceum* infection in *Erythrocebus patas* monkeys. Lab Anim Sci 29:97–101.
127. Retnasabapathy A, Joseph PG (1966) A case of melioidosis in a macaque monkey. Vet Rec 79:72–73.
128. Roop RM, Smibert RM, Krieg NR (1984) Improved biotyping schemes for *Campylobacter jejuni* and *Campylobacter coli*. J Clin Microbiol 20:990–992.
129. Rosenberg DP, Lerche NW, Henrickson RV (1980) *Yersinia pseudotuberculosis* infection in a group of *Macaca fascicularis*. J Amer Vet Med Assoc 177:818–819.
130. Rosett W, Heck DM, Hodges GR (1976) Pneumonitis and pulmonary abscess associated with *Moraxella nonliquefaciens*. Chest 70:664–665.
131. Rosner R (1964) *Aeromonas hydrophila* as the etiologic agent in a case of severe gastroenteritis. Amer J Clin Pathol 42:402–404.
132. Rudoy RC, Nelson JD (1975) Enteroinvasive and enterotoxigenic *Escherichia coli*: occurrence in acute diarrhea of infants and children. Amer J Dis Child 129: 668–672.
133. Ruiz-Palacios GM, Torres NI, Ruiz-Palacios BR, Torres J, Escamilla E, Tamayo J (1983) Cholera-like enterotoxin produced by *Campylobacter jejuni*: characterization and clinical significance. Lancet 2:250–253.
134. Sanyal SC, Singh SJ, Sen PC (1975) Enteropathogenicity of *Aeromonas hydrophila* and *Plesiomonas shigelloides*. J Med Microbiol 8:195–198.
135. Sedgwick C, Parcher J, Durham R (1970) Atypical mycobacterial infection in the pig-tailed macaque (*Macaca nemestrina*). J Amer Vet Med Assoc 157:724–725.

136. Seibold HR, Perrin EA Jr, Garner AC (1970) Pneumonia associated with *Bordetella bronchiseptica* in *Callicebus* species primates. Lab Anim Care 20:456–461.
137. Sesline DH, Schwartz LW, Osburn BI, Thoen CO, Terrell T, Holmberg C, Anderson JH, Henrickson RV (1975) *Mycobacterium avium* infection in three rhesus monkeys. J Amer Vet Med Assoc 167:639–645.
138. Shive RJ, Green SS, Evans LB, Garner FM (1969) Leptospirosis in barbary apes (*Macaca sylvana*). J Amer Vet Med Assoc 155:1176–1178.
139. Sippel WL, Medina G, Atwood MB (1954) Outbreaks of disease in animals associated with *Chromobacterium violaceum*. I. The disease in swine. J Amer Vet Med Assoc 124:467–471.
140. Smith EK, Hunt RD, Garcia FG, Fraser CEO, Merkal RS, Karlson AG (1973) Avian tuberculosis in monkeys. Amer Rev Resp Dis 107:469–471.
141. Smith EK, Ruebner BH (1970) Granulomatous enteritis and amyloidosis associated with *Mycobacterium avium* in *Macaca nemestrina*. Fed Proc 29:284.
142. Smith HA, Jones TC, Hunt RD (1972) Veterinary pathology. Lea & Febiger, Philadelphia.
143. Stein FJ, Stott G (1979) *Corynebacterium equi* in the cottontop marmoset (*Saguinus oedipus*): a case report. Lab Anim Sci 29:519–520.
144. Straus JM, Jason S, Lee H, Gan E (1969) Melioidosis with spontaneous remission of osteomyelitis in a macaque (*Macaca nemestrina*). J Amer Vet Med Assoc 155:1169–1175.
145. Strobert EA, Swenson RB (1979) Treatment regimen for air sacculitis in the chimpanzee (*Pan troglodytes*). Lab Anim Sci 29:387–388.
146. Swenson RB, McClure HM (1974) Septic abortion in a gorilla due to *Shigella flexneri*. Ann Proc Amer Assoc Zoo Vet, pp 195–196.
147. Tammemagi L, Johnson LAY (1963) Melioidosis in an orang-outang in North Queensland. Aust Vet J 39:241–242.
148. Tauxe RV, Patton CM, Edmonds P, Barrett TJ, Brenner DJ, Blake PA (1985) Illness associated with *Campylobacter laridis*, a newly recognized *Campylobacter* species. J Clin Microbiol 21:222–225.
149. Thurman JD, Morton RJ, Stair EL (1983) Septic abortion caused by *Salmonella heidelberg* in a white-handed gibbon. J Amer Vet Med Assoc 183:1325–1326.
150. Toma S, Wauters G, McClure HM, Morris GK, Weissfeld AS (1984) 0:13a, 13b, a new pathogenic serotype of *Yersinia enterocolitica*. J Clin Microbiol 20:843–845.
151. Tribe GW, Fleming MP (1983) Biphasic enteritis in imported cynomolgus (*Macaca fascicularis*) monkeys infected with *Shigella*, *Salmonella* and *Campylobacter* species. Lab Anim 17:65–69.
152. Tribe GW, Mackenzie PS, Fleming MP (1979) Incidence of thermophilic campylobacter species in newly imported simian primates with enteritis. Vet Rec 105:333.
153. Tsubokura M, Otsuki K, Kawaoka Y, Maruyama T (1984) Characterization and pathogenicity of *Yersinia pseudotuberculosis* isolated from swine and other animals. J Clin Microbiol 19:754–756.
154. Valerio DA, Dalgard DW, Voelker RW, McCarrol NE, Good RC (1978) *Mycobacterium kansasii* infection in the rhesus monkeys. In: Montali RJ (ed) Mycobacterial infections of zoo animals. Smithsonian Institution Press, Washington, DC.
155. Vetesi F, Balsai A, Kemenes F (1972) Abortion in Gray's monkey (*Cercopithecus mona*) associated with *Listeria monocytogenes*. Acta Microbiol Acad Sci Hung 19:441–443.
156. Von Graevenitz A, Mensch AH (1968) The genus *Aeromonas* in human bacteriology: report of 30 cases and review of the literature. N Engl J Med 278:245–249.
157. Waters MFR, Bakri BI, Rees RJW, McDougall AC (1978) Experimental lepromatous leprosy in the white-handed gibbon (*Hylobates lar*): successful inoculation with leprosy bacilli of human origin. Br J Exp Pathol 59:551–557.

158. Weed LA, Anderson HA, Good CA, Baggenstoss AH (1955) Nocardiosis: clinical, bacteriologic and pathological aspects. N Engl J Med 253:1137–1143.
159. Weil JO, Ward MK, Spertzel RO (1971) Incidence of *Shigella* in conditioned rhesus monkeys (*Macaca mulatta*). Lab Anim Sci 21:434–437.
160. Wilbert R, Delorme M (1927) Sur une spirochetose icterohémorragique du chimpanzee transmissible à l'homme. Ann Inst Pasteur Lille 41:1139–1155.
161. Wilbert R, Delorme M (1928) Note sur la spirochetose icterohémorragique du chimpanzee. CR Soc Biol (Paris) 98:343*345.
162. Wolf RH, Gormus BJ, Martin LN, Baskin GB, Walsh GP, Meyers WM, Binford CH (1985) Experimental leprosy in three species of monkeys. Science 227:529–531.
163. Woolley PG (1905) *Bacillus violaceus manilae* (a pathogenic microorganism). Johns Hopkins Med J 16:89–93.
164. Zwart P, Donker-Voet J (1959) Listeriosis BIJ in gevangenschap gehouden dieren. Tijdschr Diergeneeskd 84:712–716.

44
Mycotic Infections in Nonhuman Primates

George Migaki

Introduction

Although reports of spontaneous mycotic infections in nonhuman primates are relatively rare compared to those of infectious diseases caused by viruses, mycoplasmas, bacteria, protozoans, and helminthic parasites, one should be aware that fungi can contribute to poor health and even death in affected animals.

From a clinical standpoint, unlike other infectious diseases, most mycotic infections are not considered contagious, and there is little or no evidence of infection being spread from one animal to another, except for dermatophilosis and some of the dermatophytoses. Infection generally occurs by accidental contact with fungi in soil, on organic debris, or on vegetation where the organisms exist and grow as saprophytes (Emmons, 1962; Kaplan, 1973; Maddy, 1967). Therefore, natural sources or habitats of the fungi are important considerations in studying mycotic infections. Generally, inhalation of the fungal spores results in a primary pulmonary infection from which dissemination may occur to other parts of the body. Sometimes infection follows implantation of the fungus on abraded or denuded areas of the mucous membrane or skin. Many, though not all, fungi are secondary or opportunistic organisms; therefore, predisposing factors play a major role in the development of mycotic infections (Baker, 1971; Chick et al., 1975). Among the predisposing factors are conditions or circumstances associated with suppression of the animal's immune system such as prolonged use of broad-spectrum antibiotics or corticosteroids, the presence of preexisting diseases such as anemias or malignant neoplasms, coexisting septicemias or toxemias, deficiencies resulting from malnutrition, congenital defects of the immune system, or stresses associated with captivity (Chick et al., 1975; Emmons et al., 1977; Rippon, 1982).

Much of our current knowledge regarding mycotic infections in nonhuman primates has been gained from our experiences with mycotic infections in pets, livestock, laboratory animals (Ainsworth and Austwick, 1973; Migaki et al., 1978; Nielsen, 1970; Saunders, 1948), and humans (Baker, 1971; Chandler et al., 1980; Emmons et al., 1977; Rippon, 1982). Valuable information on mycotic infections in nonhuman primates has been documented by various investigators (Al-Doory, 1972; Fiennes, 1967; Griner, 1983; Kalter et al., 1968; McClure, 1975; McClure and Guilloud, 1971; Migaki, 1980; Migaki et al., 1982b).

This report is a review of the relative occurrence of spontaneous mycotic infections in nonhuman primates with emphasis on the recognition of the various disease processes manifested by these microbial agents. From a diagnostic standpoint, the diseases can be divided into (1) superficial or localized, those that are confined to the skin and cause relatively mild lesions, (2) deep or systemic, which cause severe granulomatous lesions, and (3) those caused by the actinomycetes. Nocardiosis, actinomycosis, and dermatophilosis are caused by actinomycetes, and although they are not fungi, they traditionally are grouped with mycotic infections because the lesions they produce are similar to those caused by fungi. Mycotoxicosis, resulting from the ingestion of food containing toxic fungal metabolites, is not included under this heading because the disease process is considerably different and should be discussed separately. Experimentally produced fungal diseases are not discussed except to indicate that nonhuman primates are susceptible and can be used to study further the pathogenesis and therapy of the disease.

Superficial Mycoses

Dermatophytoses, known also as ringworm, are caused by dermatophytes, which are parasites of the keratinized portion of the skin and adnexae and are caused by three genera: *Trichophyton*, *Microsporum*, and *Epidermophyton*. Infection generally results from direct contact with affected animals, an important epizootic and zoonotic consideration, and contaminated fomites. The gross appearance of the lesions varies considerably, but the principal clinical finding is localized loss of hair, frequently circular or ringlike, with the affected skin covered by fine scales or crusts. Although many genera and species of dermatophytes have been isolated from cutaneous lesions in nonhuman primates, most cases are due to the genus *Microsporum*. *Microsporum canis*, a common cause of ringworm in dogs and cats, has been isolated from scaly and alopecic areas of the skin of New World monkeys (Kaplan et al., 1958), as well as from a chimpanzee (*Pan troglodytes*) imported from Liberia (Klokke and de Vries, 1963), a rhesus monkey (*Macaca mulatta*) (Fig. 44.1) (Baker et al., 1971), and a capuchin monkey that was imported from Nicaragua and was responsible for causing infection in several humans who came in contact with it (Scully and Kligman, 1951). In an epizootic in Germany following importation of a gibbon (*Hylobates lar*) with cutaneous lesions from Thailand, the isolate was initially considered to be *M. audouini*, an anthropophilic dermatophyte, but later was identified as an atypical variant of *M. canis* (Seeliger et al., 1963). Another epizootic involved a gibbon (*H. Lar entelloides*) colony in Thailand in which the isolates were a variant strain of *M. canis* (Taylor et al., 1973). Other reports of *Microsporum* isolated from cutaneous lesions include *M. cookei* from a baboon (*Papio papio*) (Mariat and Tapia, 1966), *M. distortum* from a spider and three capuchin monkeys (Kaplan et al., 1957), and *M. gypseum* from a rhesus monkey (Koch and Jänisch, 1964). *Trichophyton mentagrophytes*, a common cause of ringworm in rodents, was isolated from several capuchin monkeys (*Cebus nigrivitattus*) in Austria that were severely debilitated from internal parasites, pneumonia, and enteritis (Bagnall and Grünberg, 1972). Other reports of *T. mentagrophytes* infection include three rhesus monkeys and a gibbon in India (Gugnani, 1971) and a green monkey that was

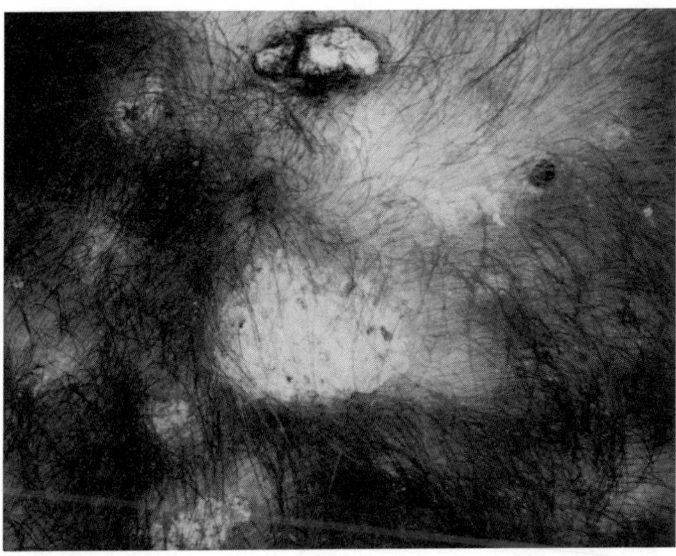

FIGURE 44.1. Dermatophytoses due to *Microsporum canis*. Skin with focal loss and thinning of hair and circular raised areas with crusts. AFIP neg. 85-8009. Courtesy of Dr. Henry J. Baker.

imported from Zaire (Hauck and Klehr, 1977). A *T. rubrum*-like organism was the cause of disease in several chimpanzees in Bratislava (Otcenasek et al., 1967) as was *T. gallinae* in a cynomolgus monkey imported from the Philippines (Gordon and Little, 1967). *Trichophyton violaceum* was isolated from an epizootic involving 60 baboons in Sukhumi in which the source of the infection was believed to be contaminated cages previously used by dogs (Voronin et al., 1948).

Deep Mycoses Caused by Yeasts

Candidiasis, known also as moniliasis or thrush, is caused by the genus *Candida*, the most common species being *C. albicans*. *Candida* appears to be a normal inhabitant of the mucous membrane, especially of the alimentary (Hunt et al., 1978) and genital tracts, and of the skin of nonhuman primates. Infection is endogenous rather than exogenous, the usual mode of infection for other mycotic agents. It has been reported that candidiasis and other mycotic infections may result from antibiotic therapy by direct tissue irritation, suppression of phagocytic capabilities of the host, inhibition of antibody synthesis, or suppression of naturally occurring candidacidal globulins (Seelig, 1966a, 1966b). Generally the lesions appear as whitish, creamy plaques caused mainly by the extensive growth of the candidal organism, thus producing a pseudomembrane. Candidiasis in nonhuman primates is probably more common than has been reported. At the London Zoo, *C. albicans* was isolated from lesions on the tongue of a woolly monkey, from a gorilla (*G. gorilla beringei*), and from a capuchin monkey in which the organism was recovered from the tongue, mouth, lungs, liver, and intestine (Fiennes, 1967). Esophagitis with formation of a pseudomembrane (Fig. 44.2) on the mucous membrane of the esophagus has been reported in a rhesus monkey (Kaufmann and Quist, 1969) and in a chimpanzee (Schmidt and Butler, 1970). Widespread lesions of candidiasis on the mucous membrane of the tongue, oral cavity, esophagus, colon, and hard keratin of nails have been described in a capuchin and five rhesus monkeys (Wikse et al., 1970). Most of these animals had been treated with antibiotics for diarrhea. *Candida albicans* was isolated from an enterocolitis in three spider monkeys that had received prolonged antibiotic and corticosteroid therapy (Patterson et al., 1974). *Candida albicans* also was isolated from both the penis and hand of a rhesus monkey with severe balanitis, paronychia, and onychia (Kerber et al., 1968). Infection of the nasal, pharyngeal, and intestinal mucosal surfaces and a pharyngeal lymph node was found in a capuchin (*C. apella*) that had been experimentally infected with *Schistosoma haematobium* (McCullough et al., 1977). An unusual finding was of *C. curvata* being isolated from lesions in the brain of a rhesus monkey in which the initial clinical sign was blindness (Herceg et al., 1977).

Coccidioidomycosis, known also as desert fever, is caused by *Coccidioides immitis*. The fungus grows as a saprophyte in the semiarid areas of the southwestern United States, Mexico, and Central and South America that have been identified as endemic areas. Inhalation of fungal spores is the usual means of infection, and epizootics may result from a severe dust storm originating from endemic areas. There are numerous reports involving nonhuman primates, but whether suppression of the animal's immune system is a necessary predisposing

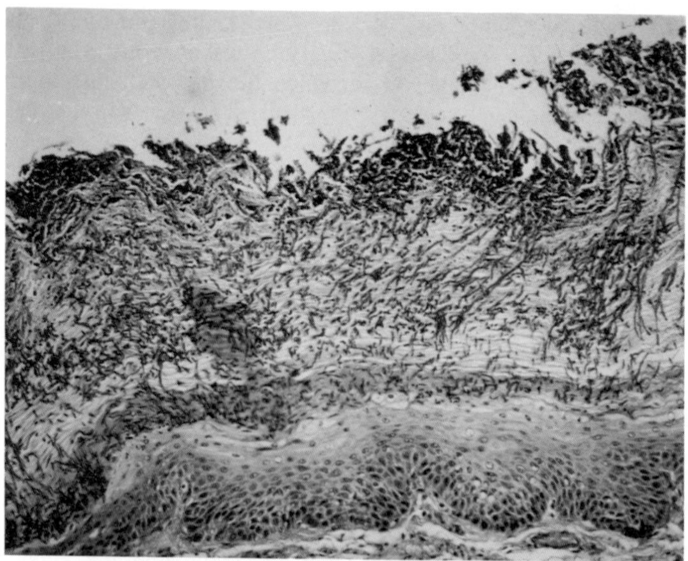

FIGURE 44.2. Candidiasis of the esophagus. Pseudomembrane, formed by large colonies of pseudohyphae and blastospores of *C. albicans*, on the superficial surface of the epithelium. PAS; ×160. AFIP neg. 85-8005.

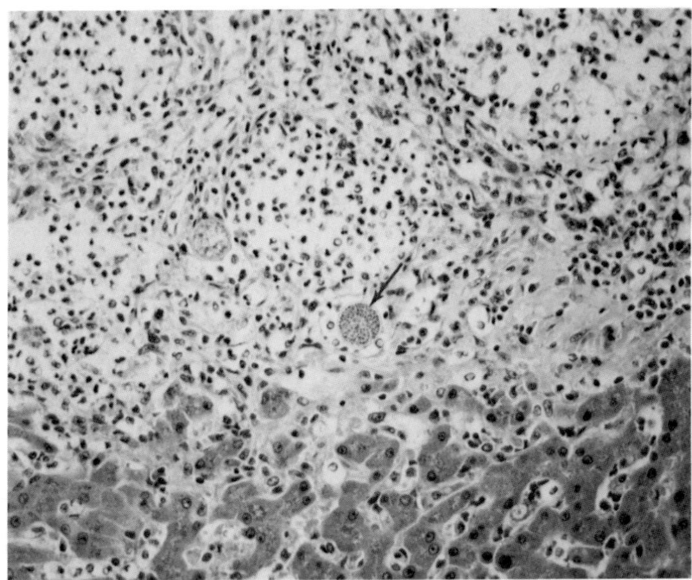

FIGURE 44.3. Coccidioidomycosis in the liver. Pyogranuloma containing *C. immitis* (arrow). H&E; ×250. AFIP neg. 85-8006.

factor is not known. There are differences in susceptibility among animal species, and vaccination may be necessary to provide protection to susceptible or compromised animals (Benirschke et al., 1981). Macaques, baboons, and gorillas appear to be most susceptible to *C. immitis* and the lesions generally are disseminated (Fig. 44.3). There are several reports from Davis, California, including that of a rhesus monkey with severe cavitary lesions in the lungs (Breznock et al., 1975). Another rhesus monkey, housed in an outdoor field cage, was found to have disseminated lesions involving the lungs, spleen, liver, and vertebrae, the latter being responsible for posterior paralysis (Castleman et al., 1980). Infection in a sooty mangabey (*Cercocebus atys*) imported from Sierra Leone was not recognized until 2 years after importation (Pappagianis et al., 1973). In addition to infection in the lung, there were "punched out" lytic lesions in many bones. Disseminated lesions in three gorillas have been reported in California; one having died at the Fresno Zoo and the other two at the San Diego Zoo (McKenney et al., 1944; Benirschke et al., 1981). A gelada baboon (*Theropithecus gelada*) died at the Toronto Zoological Gardens 1 month following importation from Norco, California (Rapley and Long, 1974). Lesions were found in the lungs, kidneys, lymph nodes, and esophagus. A 4-year-old colony-born baboon in San Antonio, Texas, had vertebral lesions resulting in lameness and a reluctance to move (Rosenberg et al., 1984).

Cryptococcosis is caused by *Cryptococcus neoformans*, which has a predilection for the brain and meninges, but other tissues also may be affected. Since *C. neoformans* is not dimorphic, infection may take place by inhalation or by direct contact on cutaneous surfaces from an infected animal. Although the source of infection is not always identifiable, a high correlation has been found between *C. neoformans* and old pigeon nests and droppings, which suggests that they provide a suitable substrate for growth. Infections are relatively common in nonhuman primates, and numerous cases have been described. Two marmosets (*Leontocebus geoffroyi*) purchased in Panama died 3 months later with disseminated lesions involving the lungs, spleen, and lymph nodes (Takos and Elton, 1953). Another disseminated case has been described in a patas monkey that had lesions on the left buttock, lungs, brain, thyroid, pancreas, adrenals, and spinal cord (Sly et al., 1977). Epileptiform seizures were observed. A rhesus monkey and a Formosan monkey (*Macaca cyclopis*) had disseminated infections with severe lesions in the brain and lungs (Garner et al., 1969). At the San Diego Zoo, cryptococcal meningitis has been described in several East Bornean proboscis monkeys (*Nasalis larvatus*) and in a Celebus crested macaque that also had a cryptococcal endometritis (Griner et al., 1983). It was believed that these animals became infected by inhaling dessicated fungal spores carried by the wind from soil containing avian feces. Also at the San Diego Zoo, a 12-year-old black lemur with a hepatoma was found to have a pulmonary infection, as did a purple-faced langur (*Presbytis senex vetulus*) with severe meningeal involvement (Fig. 44.4) and less severe lesions in the bronchi and tonsils (M. P. Anderson, 1979, personal communication). Seizures were noted several days before death. Systemic lesions were described in a 4-year-old lion-tailed macaque (*M. silenus*) in which an acute (overnight) sign of depression was the only clinical sign prior to death (Miller and Boever, 1983).

Blastomycosis, also known as North American blastomycosis, has been reported in many animal species, but no spontaneous cases have been reported in

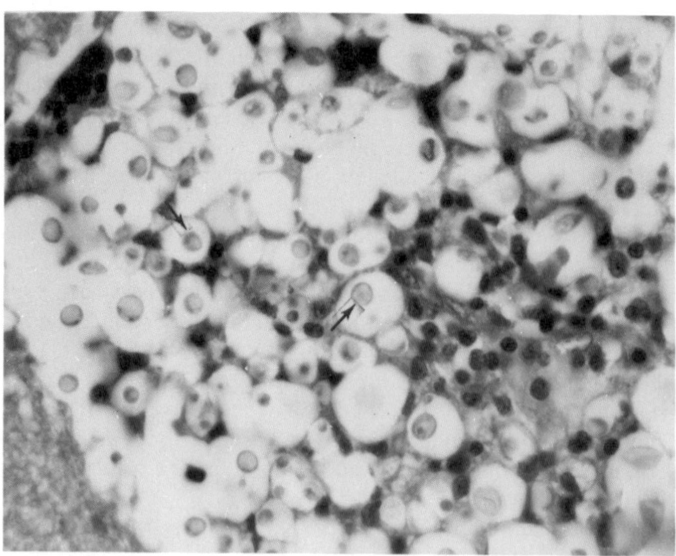

FIGURE 44.4. Granulomatous cryptococcal meningitis containing large numbers of *C. neoformans* (arrows). Note wide clear space (represents mucinous capsule) around each fungus. H&E; ×630. AFIP neg. 85-8007.

a nonhuman primate. However, they are susceptible to blastomycosis as typical cutaneous and systemic lesions were experimentally produced in rhesus monkeys (De Monbreun, 1935).

Paracoccidioidomycosis, caused by *Paracoccidioides brasiliensis*, is an endemic disease occurring only in Latin America. Although there are numerous reports of this disease in humans, it apparently is rare in animals. There is a single report involving an adult female squirrel monkey (*Saimiri sciureus*) that was imported to the United States from Bolivia (Johnson and Lang, 1977). At necropsy, approximately 8 months following importation, granulomatous lesions were found in the liver and colon.

Histoplasmosis is caused by *Histoplasma capsulatum*. Soil enriched by avian feces provides a favorable medium for growth of *H. capsulatum*. Infection generally results from inhalation of spores with the primary lesions in the lungs. The disease has been reported in many animal species, but there are few reports of spontaneous cases in nonhuman primates. In Kenya, systemic histoplasmosis has been reported in a *Cercopithecus neglectus* that had extensive lesions in the kidneys with less involvement of the liver and lungs (Frank, 1968). A granulomatous pneumonia, with granulomas in the liver and spleen, has been reported in a squirrel monkey that died 2 months after being purchased from a pet shop in Minneapolis (Bergeland et al., 1970).

African histoplasmosis, caused by *Histoplasma duboisii* and known also as large-form histoplasmosis, is a geographic disease in that the infection is indigenous to Africa. The organism in tissue sections can be distinguished from *H. capsulatum* by its larger size. Spontaneous cases have been reported in baboons (*Cynocephalus babium*) that were captured in French Guinea. Nodules were found on the skin of the buttocks, tail, and hands (Courtois et al., 1955; Mariat and Segretain, 1956). A baboon (*P. papio*) imported from Gambia and residing in England for about 4 years had elevated and ulcerated lesions on the skin of both buttocks, on the tail with involvement of the bone, and on both hands (Walker and Spooner, 1960).

Sporotrichosis, caused by *Sporothrix schenckii*, develops following a traumatic puncture wound of the skin resulting in localized subcutaneous nodules along superficial lymph vessels with dissemination to regional lymph nodes and possibly to visceral organs. A single case has been reported in a chimpanzee that had multiple nodules and ulcers on the skin of the face, eyelid, and forehead with enlarged sublingual and submaxillary lymph nodes (Saliba et al., 1968). The initial lesion was believed to be in the left nostril.

Geotrichosis, a relatively rare mycosis of animals, is caused by the genus *Geotrichum*, a fungus commonly found on vegetation and in soil. Six lowland gorillas at the New York Zoological Society developed watery diarrhea over a period of 4 days, and *G. candidum* was isolated from the feces (Dolensek et al., 1977). Feeding of contaminated hydroponically grown rye and barley grasses was believed to have been the cause of infection.

Deep Mycoses Caused by Hyphae

Zygomycosis is the term now used to identify the infections caused by the various species of zygomycetes (Chandler et al., 1980). This includes diseases formerly known by such names as mucormycosis, phycomycosis, and entomophthoromycosis. Fungi included in this category include the genera *Absidia*, *Mucor*, *Rhizopus*, *Cunninghamella*, *Basidiobolus*, *Mortierella*, *Conidiobolus* (*Entomophthora*), *Saksenaea*, and others. Like many of the other fungi, zygomycetes are ubiquitous soil inhabitants and are opportunistic organisms found throughout the world. Clinically, the lesions may be confined to the skin and subcutaneous tissue or may become generalized. Some of these fungi have a propensity to invade blood vessels causing thrombosis, infarction, or ulceration. Cutaneous zygomycosis has been described in two rhesus monkeys with multiple severe fight wounds on the skin resulting in metabolic acidosis that may have been a predisposing factor for the infection to occur (Baskin et al., 1984). Exposure of the traumatized and denuded skin to the fungal spores was the cause of the infection. Severe infection of the skin of the face and orbit of a rhesus monkey suspected of having diabetes mellitus was detected on arrival from India (Martin et al., 1969). Generalized mucormycosis has been described in a golden-bellied mangabey that died shortly after arriving at the Zoological Gardens of Antwerp (Kageruka and De Vroey, 1972). Another generalized case involved a mandrill (*Mandrillus sphinx*) that had lived in several zoological gardens in England (Lucke and Linton, 1965). Primary lesions were considered to be ulcers in the stomach (Fig. 44.5) and intestine with hematogenous spread to the lungs, heart, kidneys, liver, spleen, and lymph nodes. Gastric mucormycosis was found in a rhesus monkey with three discrete lesions of coagulative necrosis on the mucosa of the cardiac region of the stomach (Hessler et al., 1967). Extensive use of oral antibiotic therapy was believed to be a predisposing factor. Intestinal mucormycosis has been reported in two rhesus monkeys in which extensive ulcers of the cecum and ascending colon with focal necrosis in the liver were found in one

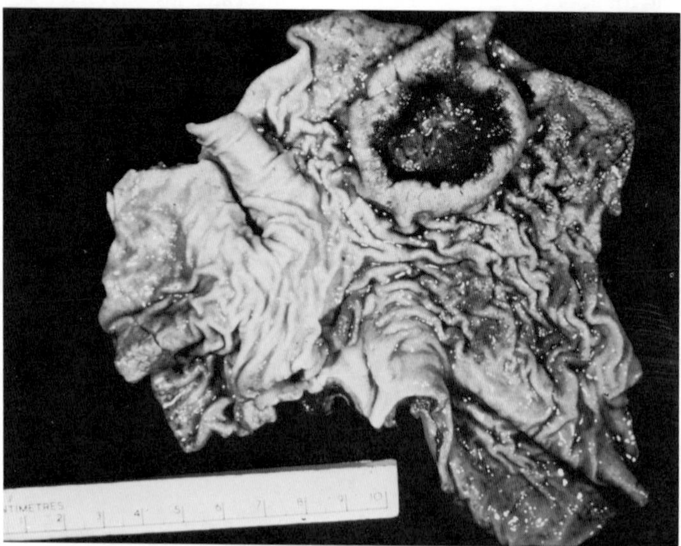

FIGURE 44.5. Zygomycosis. Ulcer in the stomach of a mandrill. AFIP neg. 79-3960. Courtesy of Dr. Vanda Lucke.

animal and an ulcerative colitis in the other animal (Gisler and Pitcock, 1962). Disseminated entomophthoromycosis has been described in a mandrill (*M. sphinx*) that had lived in a zoo for an undetermined length of time (Migaki et al., 1982a). *Entomophthora coronata* was isolated from cutaneous lesions of the nose and eyebrows of a wild free-living chimpanzee in Western Tanzania (Roy and Cameron, 1972).

Aspergillosis in nonhuman primates is relatively rare despite the fact that aspergillosis is a common respiratory infection in many animals species, especially avian. The disease is caused by the genus *Aspergillus*, usually *A. fumigatus*. Disseminated aspergillosis, including lesions in the lungs, liver, spleen, and kidneys, and tuberculosis as a dual infection suggesting that aspergillosis was primary, were found in various nonhuman primates at the London Zoo including a black mangabey, several wanderoo macaques (*M. silenus*), and a roloway monkey (*Cercopithecus roloway*) (Fiennes, 1967).

Paecilomycosis has been reported mostly in reptiles; relatively few cases have been reported in other animal species. A single case in nonhuman primates has been reported in a rhesus monkey with subcutaneous nodules over the right scapula, in the right thigh, and in the laryngeal wall (Fleischman and McCracken, 1977).

Infections Caused by Actinomycetes

Dermatophilosis, known also as streptothricosis, is caused by *Dermatophilus congolensis*. Infection occurs by direct contact with infected animals while insect bites, trauma, and prolonged wetness of the skin are considered to be predisposing factors. The disease generally is self-limiting to the superficial portion of the epidermis. On gross examination, there are discrete areas of the exudative dermatitis characterized by matted hairs, serofibrinous exudate, and crust formations on the surface of the skin, resulting in papillomatous masses composed of keratinaceous material. Multiple papillomatous lesions were reported on the skin of the legs, tail, back, ears, and head of recently imported owl monkeys (*Aotus trivirgatus*) (Fox et al., 1973; King et al., 1971; McClure et al., 1971), and a titi monkey (*Callicebus moloch*) (Fig. 44.6) (Migaki and Seibold, 1976).

Nocardiosis is caused by the genus *Nocardia*, most commonly *N. asteroides*. Infection may develop by contact on abraded skin, by inhalation, or by ingestion of contaminated feed since these organisms live as saprophytes in nature. Dissemination from the primary site of infection via blood and lymph vessels may result in secondary lesions in the brain and other organs. Disseminated pyogranulomatous lesions involving the liver, intestine, peritoneum, lung, and brain have been described in two rhesus monkeys and a pig-tailed monkey (*M. nemestrina*) (Liebenberg and Giddens, 1985). Ingestion of contaminated food with the primary site in the intestine was suspected. Another disseminated case has been described in a cynomolgus monkey with abscesses in the mandible, lung, kidneys, heart, liver, and brain (Sakakibara et al., 1984). The primary lesions were considered to be either the mandible or lung. Pulmonary nocardiosis without dissemination has been reported in an orangutan (*Pongo pygmaeus*) (McClure et al., 1976), in a vervet monkey (*C. aethiops*) (Al-Doory et al., 1969), and in two rhesus monkeys (Jonas and Wyand, 1966). *Nocardia caviae* was

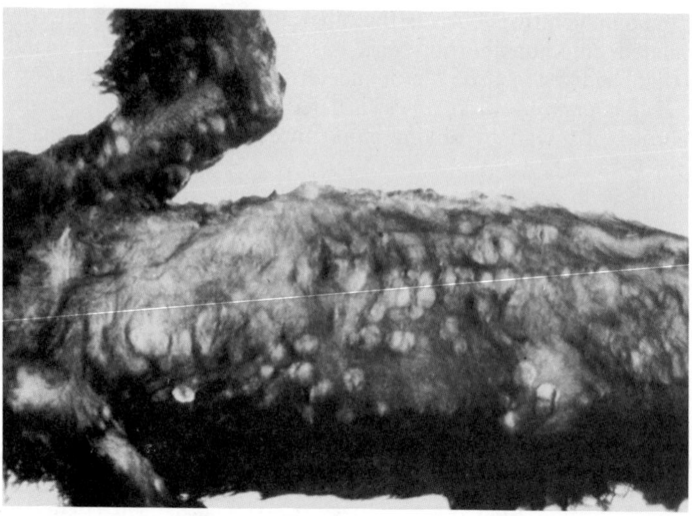

FIGURE 44.6. Dermatophilosis. Note widespread discrete raised nodules on the skin. AFIP neg. 71-12090. Courtesy of Dr. Herman Seibold.

isolated from a draining, multinodular swelling of the right hand of a baboon (*P. cynocephalus*) (Boncyk et al., 1975).

Actinomycosis is caused by the genus *Actinomyces*, a normal inhabitant of the flora of the alimentary tract. Damage to the mucous membrane provides the opportunity for infection to begin. Granulomatous abscesses, with suppurative centers containing "sulfur granules," were found along the sides of the nose of a white-throated cebus (*Cebus hypoleucus*) and a black spider monkey at the Philadelphia Zoological Garden (Weidman, 1935). At the London Zoo, a moor monkey (*Macaca maura*) died of an acute hemorrhagic enteritis in which *Actinomyces* was cultured from the intestinal wall; and *A. israelii* was the isolate from caseous lesions in the liver and pancreas of a gibbon (*H. moloch*) (Fiennes, 1967). *Actinomyces israelii* was the cause of multiple pyogranulomatous abscesses on the peritoneum and extensive fibrinous adhesions involving most of the abdominal organs of a drill (*M. leucophueus*) (Altman and Small, 1973).

Discussion

It should be reemphasized that precautions can be taken to lessen the possibility of an animal acquiring some of the mycotic infections. Special consideration should be given to the environment in which these animals are kept in captivity. Conditions that suppress the immune system such as certain preexisting diseases (e.g., diabetes mellitus, malignant neoplasms, malnutrition) and the prolonged use of antibiotics and steroids tend to increase susceptibility to mycoses, especially candidiasis, cryptococcosis, and zygomycosis. Vaccination and avoiding exposure to dust-containing arthrospores of *C. immitis* are appropriate preventive measures in endemic areas for coccidioidomycosis. Careful examination of animals and separation of those that are infected should be undertaken to avoid epizootics of those diseases occurring as the result of direct contact such as dermatophytosis and dermatophilosis.

Acknowledgments. This work was supported in part by Public Health Service grant RR00301-20 from the Division of Research Resources, National Institutes of Health, U.S. Department of Health and Human Services, under the auspices of Universities Associated for Research and Education in Pathology, Inc. The author thanks Chapman H. Binford, MD, and George D. Imes, DVM, for scientific review of the manuscript, Charmaine Goetz and Mary Ann Sonoda for technical assistance, and Catherine L. Wilhelmsen, DVM, for translation of articles published in German.

References

Ainsworth GC, Austwick PKC (1973) Fungal diseases of animals, 2nd ed. Commonwealth Agricultural Bureaux, Farnham Royal, Slough, England.

Al-Doory Y (1972) Fungal and bacterial diseases. In: Fiennes RNT-W (ed) Pathology of simian primates. S Karger, New York, pp 206–241.

Al-Doory Y, Pinkerton ME, Vice TE (1969) Pulmonary nocardiosis in a vervet monkey. J Amer Vet Med Assoc 155:1179–1180.

Altman NH, Small JD (1973) Actinomycosis in a primate confirmed by fluorescent antibody technics in formalin fixed tissue. Lab Anim Sci 23:696–700.

Bagnall BG, Grünberg W (1972) Generalized *Trichophyton mentagrophytes* ringworm in capuchin monkeys (*Cebus nigrivitatus*). Br J Dermatol 87:565–570.

Baker HJ, Bradford LG, Montes LF (1971) Dermatophytosis due to *Microsporum canis* in a rhesus monkey. J Amer Vet Med Assoc 159:1607–1611.

Baker RD (ed) (1971) Human infection with fungi, actinomyces, and algae. Springer-Verlag, Berlin.

Baskin GB, Chandler FW, Watson EA (1984) Cutaneous zygomycosis in rhesus monkeys (*Macaca mulatta*). Vet Pathol 21:125–128.

Benirschke K, Anderson MP, Oosterhuis JE, Nelson LS (1981) Coccidioidomycosis in gorillas and attempted prophylactic vaccination. Erkr Zootiere 24:119–122.

Bergeland ME, Barnes DM, Kaplan W (1970) Spontaneous histoplasmosis in a squirrel monkey. Primate Zoonosis Surveillance Report 1, January–February 1970, CDC, Atlanta, pp 10–11.

Boncyk LH, McCullough B, Grotts DD, Kalter SS (1975) Localized nocardiosis due to *Nocardia caviae* in a baboon (*Papio cynocephalus*). Lab Anim Sci 25:88–91.

Breznock AW, Henrickson RV, Silverman S, Schwartz LW (1975) Coccidioidomycosis in a rhesus monkey. J Amer Vet Med Assoc 167:657–661.

Castleman WL, Anderson J, Holmberg CA (1980) Posterior paralysis and spinal osteomyelitis in a rhesus monkey with coccidioidomycosis. J Amer Vet Med Assoc 177:933–934.

Chandler FW, Kaplan W, Ajello L (1980) Histopathology of mycotic diseases. Year Book Medical, Chicago.

Chick EW, Balows A, Furcolow ML (eds) (1975) Opportunistic fungal infections. CC Thomas, Springfield, IL.

Courtois G, Segretain G, Mariat F, Levaditi JC (1955) Mycose cutanée à corps levuriformes observée chez des singes africains en captivité. Ann Inst Pasteur 89:124–127.

DeMonbreun WA (1935) Experimental chronic cutaneous blastomycosis in monkeys. Arch Dermatol Syphilol 31:831–854.

Dolensek EP, Napolitano RL, Kazimiroff J (1977) Gastrointestinal geotrichosis in six adult gorillas. J Amer Vet Med Assoc 171:975–976.
Emmons CW (1962) Natural occurrence of opportunistic fungi. Lab Invest 11:1026–1032.
Emmons CW, Binford CH, Utz JP, Kwon-Chung KP (1977) Medical mycology, 3rd ed. Lea & Febiger, Philadelphia.
Fiennes R (1967) Zoonoses of primates. Cornell University Press, Ithaca, NY.
Fleischman RW, McCracken D (1977) Paecilomycosis in a nonhuman primate (*Macaca mulatta*). Vet Pathol 14:387–391.
Fox JG, Campbell LH, Reed C, Snyder SB, Soave OA (1973) Dermatophilosis (cutaneous streptothricosis) in owl monkeys. J Amer Vet Med Assoc 163:642–644.
Frank H (1968) Systemische Histoplasmose bei einem afrikanischen Affen. Dtsch Tierärztl Wochenschr 75:371–374.
Garner FM, Ford DF, Ross MA (1969) Systemic cryptococcosis in 2 monkeys. J Amer Vet Med Assoc 155:1163–1168.
Gisler DB, Pitcock JA (1962) Intestinal mucormycosis in the monkey (*Macaca mulatta*). Amer J Vet Res 23:365–367.
Gordon MA, Little GN (1967) *Trichophyton* (*Microsporum*) *gallinae* ringworm in a monkey. Sabouraudia 6:207–212.
Griner LA (1983) Pathology of zoo animals. Zoological Society of San Diego, San Diego.
Gugnani HC (1971) *Trichophyton mentagrophytes* infection in monkeys and its transmission to man. Hindustan Antibiot Bull 14:11–13.
Hauck H, Klehr N (1977) Meerkatzenfavus als Ursache für die Pilzinfektion eines Menschen. Z Allg Med 53:331–332.
Herceg M, Marzan B, Hajsig M, Naglic T, Huber I (1977) Pathomorphological observations of spontaneous candidal encephalitis in a monkey. Veterinarski Arkhiv 47:183–187.
Hessler JR, Woodard JC, Beattie RJ, Moreland AF (1967) Mucormycosis in a rhesus monkey. J Amer Vet Med Assoc 151:909–913.
Hunt RD, Anderson MP, Chalifoux LV (1978) Spontaneous infectious diseases of marmosets. Prim Med 10:239–253.
Johnson WD, Lang CM (1977) Paracoccidioidomycosis (South American blastomycosis) in a squirrel monkey (*Saimiri sciureus*). Vet Pathol 14:368–371.
Jonas AM, Wyand DS (1966) Pulmonary nocardiosis in the rhesus monkey. Pathol Vet 3:588–600.
Kageruka P, De Vroey C (1972) Generalized mucormycosis in the golden-bellied mangabey (*Cercocebus galeritus chrysogaster*, Lydekker). Acta Zoolog Pathol Antwerp 55:19–28.
Kalter SS, Al-Doory Y, Kuntz RE, Pinkerton ME (1968) Infectious diseases associated with the use of primates. In: Vagtborg H (ed) Use of nonhuman primates in drug evaluation: a symposium. University of Texas Press, Austin, pp 505–538.
Kaplan W (1973) Epidemiology of the principal systemic mycoses of man and lower animals and the ecology of their etiologic agents. J Amer Vet Med Assoc 163:1043–1047.
Kaplan W, Georg LK, Hendricks SL, Leeper RA (1957) Isolation of *Microsporum distortum* from animals in the United States. J Inf Dis 28:449–453.
Kaplan W, Georg LK, Ajello L (1958) Recent developments in animal ringworm and their public health implications. Ann NY Acad Sci 70:636–649.
Kaufmann AF, Quist KD (1969) Thrush in a rhesus monkey: report of a case. Lab Anim Care 19:526–527.
Kerber WT, Reese WH, Van Natta J (1968) Balanitis, paronychia, and onychia in a rhesus monkey. Lab Anim Care 18:506–507.
King NW, Fraser CEO, Garcia FG, Wolf LA, Williamson ME (1971) Cutaneous streptothricosis (dermatophiliasis) in owl monkeys. Lab Anim Sci 21:67–74.

Klokke AH, de Vries GA (1963) Tinea capitis in chimpanzees caused by *Microsporum canis* Bodin 1902 resembling *M. obesum* Conant 1937. Sabouraudia 2:268–270.

Koch HA, Jänisch W (1964) Eine Mikrosporum-gypseum-Enzootie bei Rhesusaffen (*Macacus rhesus*). Mykosen 7:86–89.

Liebenberg SP, Giddens WE (1985) Disseminated nocardiosis in three macaque monkeys. Lab Anim Sci 35:162–166.

Lucke VM, Linton AH (1965) Phycomycosis in a mandrill (*Mandrillus sphinx*). Vet Rec 77:1306–1309.

Maddy KT (1967) Epidemiology and ecology of deep mycoses of man and animals. Arch Dermatol 96:409–417.

Mariat F, Segretain G (1956) Etude mycologique d'une histoplasmose spontanée du singe africain cynocephalus babuin. Ann Inst Pasteur 91:874–891.

Mariat F, Tapia G (1966) Observations sur une souche de *Microsporum cookei* parasite du cynocephale (*Papio papio*). Sabouraudia 5:43–45.

Martin JE, Kroe DJ, Bostrom RE, Johnson DJ, Whitney RA (1969) Rhino-orbital phycomycosis in a rhesus monkey (*Macaca mulatta*). J Amer Vet Med Assoc 155:1253–1257.

McClure HM (1975) Pathology of the rhesus monkey. In: Bourne GH (ed) The rhesus monkey, vol II. Academic, New York, pp 346–348.

McClure HM, Guilloud NB (1971) Comparative pathology of the chimpanzee. In: Bourne GH (ed) The chimpanzee, vol IV. University Park Press, Baltimore, pp 167–172.

McClure HM, Kaplan W, Bonner WB, Keeling ME (1971) Dermatophilosis in owl monkeys. Sabouraudia 9:185–190.

McClure HM, Chang J, Kaplan W, Brown JM (1976) Pulmonary nocardiosis in an orangutan. J Amer Vet Med Assoc 169:943–945.

McCullough B, Moore J, Kuntz RE (1977) Multifocal candidiasis in a capuchin monkey (*Cebus apella*). J Med Primatol 6:186–191.

McKenney FD, Traum J, Bonestell AE (1944) Acute coccidioidomycosis in a mountain gorilla (*Gorilla beringeri*) with anatomical notes. J Amer Vet Med Assoc 104:136–140.

Migaki G (1980) Mycotic diseases in captive animals—a mycopathologic overview. In: Montali RJ, Migaki G (eds) The comparative pathology of zoo animals. Smithsonian Institution Press, Washington, DC, pp 267–275.

Migaki G, Seibold HR (1976) Dermatophilosis in a titi monkey (*Callicebus moloch*). Amer J Vet Res 37:1225–1226.

Migaki G, Voelker FA, Sagartz JW (1978) Fungal diseases. In: Benirschke K, Garner FM, Jones TC (eds) Pathology of laboratory animals, vol II. Springer-Verlag, New York, pp 1552–1586.

Migaki G, Toft JD, Schmidt RE (1982a) Disseminated entomophthoromycosis in a mandrill (*Mandrillus sphinx*). Vet Pathol 19:551–554.

Migaki G, Schmidt RE, Toft JD, Kaufman AF (1982b) Mycotic infections of the alimentary tract of nonhuman primates: a review. Vet Pathol Suppl 7:93–103.

Miller RE, Boever WJ (1983) Cryptococcosis in a lion-tailed macaque (*Macaca silenus*). J Zoo Anim Med 14:110–114.

Nielsen SW (1970) Infectious granulomas. In: Dobberstein J, Pallaske G, Stünzi H (eds) E Joest Handbuch Der Speziellen Pathologischen Anatomie der Haustiere. PP Verlag, Berlin, pp 590–632.

Otcenasek M, Dvorak J, Ladzianska K (1967) Trichophyton rubrum-like dermatophyte as a causative agent of dermatophytosis in chimpanzees. Mycopathol Mycol Appl 31:33–37.

Pappagianis D, Vanderlip J, May B (1973) Coccidioidomycosis naturally acquired by a monkey, *Cercocebus atys*, in Davis, California. Sabouraudia 11:52–55.

Patterson DR, Wagner JE, Owens DR, Ronald NC, Frisk CS (1974) *Candida albicans* infections associated with antibiotic and corticosteroid therapy in spider monkeys. J Amer Vet Med Assoc 164:721–722.

Rapley WA, Long JR (1974) Coccidioidomycosis in a baboon recently imported from California. Can Vet J 15:39–41.

Rippon JW (1982) Medical mycology, 2nd ed. WB Saunders, Philadelphia.

Rosenberg DP, Gleiser CA, Carey KD (1984) Spinal coccidioidomycosis in a baboon. J Amer Vet Med Assoc 185:1379–1381.

Roy AD, Cameron HM (1972) Rhinophycomycosis entomophthorae occurring in a chimpanzee in the wild in East Africa. Amer J Trop Med Hyg 21:234–237.

Sakakibara I, Sugimoto Y, Minato H, Takasaka M, Honjo S (1984) Spontaneous nocardiosis with brain abscess caused by *Nocardia asteroides* in a cynomolgus monkey. J Med Primatol 13:89–95.

Saliba AM, Matera EA, Moreno G (1968) Sporotrichosis in a chimpanzee. Mod Vet Pract 49:74.

Saunders LZ (1948) Systemic fungous infections in animals: a review. Cornell Vet 38:213–238.

Schmidt RE, Butler TM (1970) Esophageal candidiasis in a chimpanzee. J Amer Vet Med Assoc 157:722–723.

Scully JP, Kligman AM (1951) Coincident infection of a human and an anthropoid with *Microsporum audouini*. Arch Dermatol Syphilol 64:495–498.

Seelig MS (1966a) Mechanisms by which antibiotics increase the incidence and severity of candidiasis and alter the immunologic defenses. Bacteriol Rev 30:442–459.

Seelig MS (1966b) The role of antibiotics in the pathogenesis of *Candida* infections. Amer J Med 40:887–917.

Seeliger HPR, Bisping W, Brandt HP (1963) Über eine Microsporum Enzootie bei Kappen-Gibbons (*Hylobates lar*) verursacht durch eine Variante von *Microsporum canis*. Mykosen 6:61–68.

Sly DW, London WT, Palmer AE, Rice JM (1977) Disseminated cryptococcosis in a patas monkey (*Erythrocebus patas*). Lab Anim Sci 27:694–699.

Takos MJ, Elton NW (1953) Spontaneous cryptococcosis of marmoset monkeys in Panama. Arch Pathol 55:403–407.

Taylor RL, Cadigan FC, Chaicumpa V (1973) Infections among Thai gibbons and humans caused by atypical *Microsporum canis*. Lab Anim Sci 23:226–231.

Voronin LG, Kanfor IS, Lakin GF, Tikh NN (1948) Spontaneous diseases of lower monkeys, their prophylaxis, diagnosis, and treatment. In: Experimentation on the keeping and raising of monkeys at Sukhumi, Chapter 3. Acad Med Sci, Moscow.

Walker J, Spooner ETC (1960) Natural infection of the African baboon *Papio papio* with the large-cell form of *Histoplasma*. J Pathol Bacteriol 80:436–438.

Weidman FD (1935) Dermatoses of monkeys and apes. 9th Int Congr Dermatol 1:600–606.

Wikse SE, Fox JG, Kovatch RM (1970) Candidiasis in simian primates. Lab Anim Care 20:957–963.

45
The Pathoparasitology of Nonhuman Primates: A Review

JOHN D. TOFT, II

Introduction

Most people who have had more than cursory experience in the husbandry of nonhuman primate colonies will agree that parasitism is one of the most common disease entities that affects these animals. Numerous protozoal and metazoal genera have been described as infecting the members of all major nonhuman primate groups. Many of these are considered to be nonpathogenic, or at least their detrimental effects upon the host have yet to be eludicated. A large number, however, can produce lesions that result in serious debilitation and can create opportunities for secondary infections that may be fatal. This process appears to be exacerbated by the stress of capture and confinement.

This paper discusses the anatomic features of protozoan and metazoan parasitic infections of nonhuman primates. Since the paper is essentially a review of the literature, an extensive bibliography is included for those readers who wish to pursue the subject in greater detail. All tables in the text follow a published system [290]. Major nonhuman primate groups are classified according to a published taxonomy system [621]. Prosimians include the species in the families Tupaiidae; New World monkeys include the species in the families Callitrichidae and Cebidae; Old World monkeys include the species in the family Cercopithecidae; and the great apes include the species in the families Hylobatidae and Pongidae.

The numerous articles published previously that discuss the subject of nonhuman primate parasitology as a specific entity, or as a part of an overall discussion of general parasitology, systemic pathology, or the broad topic of nonhuman primate diseases, are outlined as follows: enteric and somatic protozoa [23, 72, 90, 338, 369, 370, 433, 443, 529, 604]; hemoprotozoa [2, 123, 135, 140, 236, 320, 372, 556, 797, 874, 928]; nematodes [650]; trematodes [147, 482]; cestodes [608]; acanthocephalans [770]; arthropods [92, 207, 281, 290, 302, 395, 405, 413, 457, 938]; pentastomids [788]; general parsitology [32, 339, 416, 427, 484, 487, 497, 499, 528, 530, 531, 791, 810]; systemic pathology [21, 97, 367, 429, 542, 572, 594, 774, 820, 846]; and nonhuman primate diseases [14, 34, 35, 112, 134, 312, 373, 407, 424, 431, 438, 440, 461, 469, 508, 570, 584, 585, 742, 746, 869].

In addition, there are many papers in the literature concerning parasitological surveys and parasite checklists for specific species of nonhuman primates. These are referenced as follows: rhesus monkey, *Macaca mulatta* [1, 220, 434, 435, 447, 680, 716, 741, 859, 912]; cynomolgus monkey, *Macaca fascicularis* [1, 45, 362, 393, 763, 912]; pig-tail monkey, *Macaca nemestrina* [912]; bonnet monkey, *Macaca radiata*, [912]; stump-tail monkey, *Macaca arctoides* [912]; Barbary ape, *Macaca sylvanus* [91]; Formosan rock macaque, *Macaca cyclopis* [490, 494]; African green monkey, *Cercopithecus* sp. [518]; silvered leaf monkey, *Presbytis cristatus* [3, 658, 659, 660]; chacma baboon, *Papio ursinus* [336, 578]; yellow baboon, *Papio cynocephalus* [485, 497, 616]; olive baboon, *Papio anubis*, syn. *P. doguera* [486, 488, 611]; baboons, *Papio* sp. [1, 573, 609, 610, 626, 823]; gelada baboon, *Theropithecus gelada* [493]; tamarin, *Saguinus fusciollis* [148]; owl monkey, *Aotus trivirgatus* [895]; squirrel monkey, *Saimiri sciureus* [584]; New World monkeys [121, 437, 492]; slender loris, *Loris tardigradus* [477]; tarsiers, *Tarsius bancanus* [61, 387]; chimpanzee, *Pan* sp. [284, 491, 613, 863]; and laboratory primates [712, 764, 911, 912].

References for additional reading on nonhuman primate parasitology, associated lesions, and related subjects outside the scope of this paper are as follows: pathology [290, 406, 427, 428, 742, 791]; necropsy techniques [910]; recognition of parasites [607]; comparative parasitology [127]; experimental parasitology [483, 643]; identification of parasitic eggs [421]; identification of parasites in tissue sections [128, 307, 848]; parasitic zoonoses [417, 432, 529, 931]; and parasite control [155].

Protozoan Parasites

Flagellates

The enteric flagellates and hemoflagellates described in nonhuman primates are listed in Table 45.1. All of the enteric flagellates are considered nonpathogenic except *Giardia lamblia*, which has been reported to cause diarrhea in monkeys [290, 742]. The ability of *Giardia* to cause disease is controversial, however, and the factors that govern its pathogenicity are not completely understood [529]. The majority of the hemoflagellates reported from nonhuman primates are also considered to be nonpathogenic. An exception is *Trypanosoma cruzi*, which is an important pathogen and is the cause of Chagas disease in humans. Many species of wild and domestic animals have been found to be infected with this parasite, including nonhuman primates, and it is speculated that most mammals are susceptible. In addition, since it is an important pathogen of humans, it has potentially serious public health implications and will be discussed in detail below.

South American Trypanosomiasis (Chagas Disease)

The cause of this disease, the hemoflagellate *Trypanosoma cruzi*, is distributed throughout South and Central America with extension into the southern and southwestern regions of the United States. Dr. Chagas made the original description of this parasite in *Callithrix penicillata* from Brazil [23, 109]. He also reported the first case of natural infection in the squirrel monkey [110]. Natural *T. cruzi* infection has been reported from numerous New World primate species (squirrel monkeys, marmosets, spider monkeys, cebus monkeys, and uakaris) [6, 20, 25, 32, 88, 159, 160, 212, 219, 290, 511, 559, 812]. *Trypanosoma cruzi* also

TABLE 45.1. Parasitic flagellates described from nonhuman primates.

Parasite genus-species	Location in host	Pro-simians	New World monkeys	Old World monkeys	Great apes	References
Hemoflagellates						
Trypanosoma cruzi	Blood, RE cells, muscle, heart, other tissues		X	X	X	6, 20, 23, 25, 32, 88, 109, 110, 159, 160, 212, 219, 290, 511, 559, 560, 782, 789, 812
Trypanosoma sanmartini	Blood		X			23, 212, 290, 324
Trypanosoma minasense	Blood		X			23, 109, 159, 161, 212, 290, 559, 560, 728
Trypanosoma rangeli	Blood		X			23, 212, 290, 342, 343, 528
Trypanosoma saimirii	Blood		X			23, 162, 212, 290, 728
Trypanosoma diasi	Blood		X			23, 162, 163, 212, 290
Trypanosoma lambrechti	Blood		X			23, 166, 212, 559, 560
Trypanosoma primatum	Blood				X	23, 72, 218, 290, 713, 778, 779, 936
Trypanosoma brucei	Blood			X		23
Trypanosoma perodictici	Blood	X				23, 37, 40, 41, 72, 388, 713, 779
Trypanosoma irangiense	Blood	X				23, 39, 41, 388
Trypanosoma sp.	Blood	X				136, 205, 213, 354, 357, 496, 620, 933
Enteric flagellates						
Trichomitus wenyoni	Cecum, colon			X		90, 290, 898
Trichomonas buccalis	Mouth			X		371, 742
Trichomonas tenax	Mouth			X	X	90, 290, 371, 391, 529
Trichomonas foetus	Intestine	X				61
Trichomonas hominis	Intestine				X	491, 494
Trichomonas sp.	Intestine	X	X			168, 387, 895
Pentatrichomonas hominis	Cecum, colon		X	X	X	290, 369, 371, 529, 680
Enteromonas hominis	Cecum			X	X	90, 290, 491, 530, 712
Retortamonas intestinalis	Cecum			X	X	90, 290, 529, 530

(Continued)

TABLE 45.1. (Continued).

Parasite genus-species	Location in host	Pro-simians	New World monkeys	Old World monkeys	Great apes	References
Chilomastix mesnili	Cecum, colon		X	X	X	90, 290, 369, 371, 485, 491, 530, 609, 654, 680, 712
Chilomastix tarsii	Intestine	X				61, 387, 685
Chilomastix sp.	Intestine		X			885
Hexamita pitheci	Cecum, colon			X	X	156, 290, 529
Hexamita sp.	Cecum			X	X	90, 290, 369, 896
Giardia lamblia	Anterior sm. intestine		X	X	X	90, 290, 491, 529, 530, 742
Giardia sp.	Intestine		X			168, 895

has been reported from Old World monkeys that originated in Asia (*Macaca* sp. and rhesus monkeys) [23, 290, 789, 791] and great apes (gibbon) [782]. Despite the reports in nonhuman primates other than New World monkeys, the question of whether or not *T. cruzi* exists outside the Western Hemisphere remains unanswered. A review of some of the earlier reports of this parasite in Asian monkeys concluded that infection was most probably acquired after the animals were in captivity [23]. Two additional cases have been reported involving rhesus monkeys that were members of colonies which were part of long-term research projects. Infection in these monkeys was considered also to have occurred during captivity, especially since they were housed outdoors in an endemic area of the United States [132, 436].

Two forms of *T. cruzi* are found in susceptible animals. The trypomastigote (trypanosomal) form occurs in the blood. The amastigote (leishmanial) form is found in groups in the cells of skeletal and cardiac muscle, the reticuloendothelial system, and other tissues. The life cycle is indirect with species of several genera of triatomid bugs belonging to the family *Reduviidae* (cone-nose bugs, assassin bugs, or kissing bugs) as biological vectors. For a detailed description of the morphology and life cycle of this parasite, the reader is referred to the following references: 290, 528, 530, and 791.

Infection with *T. cruzi* causes rather nonspecific clinical signs [290]. Generalized edema without necrosis or hemorrhage is said to be common. Anemia, hepatosplenomegaly, and lymphadenitis also can occur [290, 791]. Depression, anorexia, weight loss, and dehydration were seen in the case involving the gibbon [782]. Electrocardiographic patterns consistent with that of right bundle branch block are reported from the cebus monkey [88]. Also, a case of intrauterine death from congenital trypanosomiasis in marmosets has been reported [549].

The lesion mentioned most frequently in all of the naturally occurring cases of *T. cruzi* infection in nonhuman primates is myocarditis, which results in the destruction of myocardial fibers [88, 290, 511, 782]. Histopathologically, the infected myocardium contains numerous, randomly scattered, cystic structures of varying sizes, occupying individual myocardial fibers. These structures are pseudocysts that contain many individual circular to oval-shaped organisms from

1.5 to 4 μ in diameter, with a central nucleus and a prominent bar-shaped structure (kinetoplast). The kinetoplast is an intracellular organelle that is diagnostic for this parasite and aids in differentiating it from the intracellular forms of sarcocystis and toxoplasmosis. These organisms can also be seen filling the cytoplasm of large mononuclear reticuloendothelial (RE) cells, some of which can be multinucleated and have the characteristics of giant cells. Degenerating pseudocysts elicit a focal mononuclear cell inflammatory response, and there may be mild dystrophic mineralization of individual myocardial fibers in some of these areas [436].

The diagnosis of *T. cruzi* can be made by demonstrating and identifying the parasites in blood or other body fluids or in histological sections. Thin and thick blood smears should be stained with Giemsa preparations and examined for the presence of trypanosomes. Animal inoculation and xenodiagnosis can also be used to allow the organism to complete its life cycle. Xenodiagnosis involves the examination of known vectors for trypanosomes after they have been allowed to feed on suspected hosts [23, 290].

Because *T. cruzi* can cause serious disease in humans, all people involved with the care and use of nonhuman primates should exercise extreme care to avoid exposure either by accidental inoculation with trypanosomes or by contamination of mucous membranes or skin with infected material. This should be especially true for those persons working with New World monkeys or with any nonhuman primate maintained in an endemic area in the United States [23, 290].

Sarcodines (Ameba)

The parasitic sarcodines described in nonhuman primates are listed in Table 45.2. All are considered nonpathogenic except *Dientamoeba*, which sometimes can be pathogenic, and *Entamoeba histolytica*, which can cause severe enteric disease in humans and nonhuman primates. *Entamoeba histolytica* will be discussed in detail.

Amebiasis

The cause of this disease, *Entamoeba histolytica*, has a world-wide distribution and has been reported in New World monkeys (spider monkeys, cebus monkeys, woolly monkeys, howler monkeys, squirrel monkeys, and marmosets), Old World monkeys (rhesus monkeys, pig-tailed macaques, bonnet macaques, cynomolgus monkeys, colobus monkeys, proboscis monkeys, African green monkeys, baboons, and langurs), and the great apes (gibbons, orangutans, and chimpanzees) [10, 38, 51, 92, 172, 207, 222, 285, 290, 299, 329, 330, 369, 376, 423, 437, 445, 486, 491–494, 544, 545, 588, 599, 601, 661, 712, 742, 746, 747, 769, 791, 863, 929]. Infection is said to be common in Old World monkeys but uncommon or rare in New World monkeys obtained from their natural habitat [200, 742, 791, 869]. Young monkeys and New World monkeys are reported to sustain more severe lesions from infection with this parasite [46, 90, 207, 223, 704, 742, 869].

The morphology of this parasite has been discussed [90, 182, 290, 389, 529, 530, 748, 791]. Two races of *E. histolytica* are recognized, separated on the basis of size. The pathogenic race is the larger form; the trophozoites measure 20–30

TABLE 45.2. Parasitic sarcodines described from nonhuman primates.

Parasite genus-species	Location in host	Pro-simians	New World monkeys	Old World monkeys	Great apes	References
Ameba						
Entamoeba histolytica	Cecum, colon		X	X	X	3, 10, 38, 90, 168, 172, 207, 222, 285, 290, 300, 329, 330, 339, 369, 485, 486, 487, 491, 493, 494, 529, 530, 588, 599, 601, 610, 611, 616, 654, 661, 665, 684, 686, 687, 712, 742, 769, 791, 829, 863, 868, 869, 903, 939
Entamoeba hartmanni	Cecum, colon			X	X	57, 90, 290, 485, 491, 494, 529, 530, 611, 616
Entamoeba coli	Cecum, colon			X	X	90, 168, 290, 485, 486, 487, 491, 493, 494, 529, 530, 578, 616, 654, 680, 712, 741, 863
Entamoeba chattoni	Cecum, colon			X		90, 290, 444, 445, 485, 491, 529, 530, 616
Entamoeba gingivalis	Mouth			X	X	90, 290, 371, 529, 530, 742
Entamoeba polecki	Cecum, colon				X	491, 494, 611
Entamoeba sp.	Cecum, colon		X	X	X	486, 487, 491, 494, 573, 578, 895
Iodamoeba buetschlii	Cecum, colon		X	X	X	90, 207, 290, 369, 371, 485, 486, 487, 493, 494, 529, 530, 553, 611, 616, 712, 897
Iodamoeba wallacei	Intestine			X		719
Iodamoeba sp.	Cecum, colon		X	X		493, 494, 573, 578, 895
Endolimax nana	Cecum, colon		X	X	X	90, 207, 290, 369, 371, 485, 486, 487, 493, 494, 529, 530, 553, 611, 616, 654, 712, 741

TABLE 45.2. (Continued).

Parasite Genus-species	Location in host	Pro-simians	New World monkeys	Old World monkeys	Great apes	References
Endolimax sp.	Cecum, colon		X			895
Dientamoeba fragilis	Cecum, colon			X		90, 290, 371, 467, 486, 487, 529, 530

μm in diameter. The smaller nonpathogenic race has trophozoites that measure 12–15 μm. Only the pathogenic organisms ingest red blood cells; the presence of erythrocytes within trophozoites is helpful in distinguishing pathogenic from nonpathogenic amebae. Cyst forms are 10–20 μm in diameter and contain four nuclei and rodlike chromatin bodies.

These organisms reproduce by binary fission [90, 290, 530]. Prior to producing the cyst form, the amebae become round and small. A cyst wall is formed; the nucleus divides twice, resulting in four small nuclei. These nuclei divide upon rupture of the cyst wall. Thus, each original organism separates into eight small amebae. Each of these in turn develops into a trophozoite.

Infection with *E. histolytica* produces mild clinical signs or none at all. There is a great variability in virulence among strains of organisms [90, 290, 529, 791]. Pathogenicity is affected by the host species infected, the nutritional status of the host, environmental factors, and the bacterial flora present in the gastrointestinal tract [90, 290, 529, 567, 791]. *Entamoeba histolytica* usually lives in the intestinal lumen where it is nonpathogenic [529]. Only when it invades the mucosa does it become pathogenic and may lead to amebic dysentery [529]. Clinically, affected animals show the following signs: lethargy, weakness, dehydration, anorexia, vomiting, and severe diarrhea that may or may not be hemorrhagic or catarrhal [90, 223, 290, 379, 385, 461, 530, 588, 665, 704, 829, 868, 869, 890]. The gross and microscopic lesions associated with amebiasis in nonhuman primates have been described [57, 60, 90, 290, 300, 461, 588, 665, 704, 742, 791, 829, 869]. At necropsy, a mild to severe necroulcerative colitis can be seen. *Entamoeba histolytica* trophozoites can be found in wet smears from material from the colon of clinically ill animals or from the colonic contents overlying the lesions seen at necropsy [461]. Histologically, the colonic mucosa is necrotic and ulcerated down to the level of the muscularis mucosae; typical flask-shaped ulcers may be seen. These ulcers can be as small as a few millimeters or may become large and confluent and involve extensive areas of the colon. Trophozoites may be seen in or adjacent to the ulcers. Often, the host response is minimal unless secondary bacterial invasion has occurred [290, 461, 704, 791]. Extensive hemorrhage with neutrophilic and mononuclear inflammatory cell infiltrates has been described in lesions from New World monkeys [461]. Gastric amebiasis and death resulting from infection with *E. histolytica* has been reported in the silver leaf monkey, douc langurs, proboscis monkey, and colobus monkey [299, 330, 544, 545, 601, 661]. Some trophozoites may enter lymphatic channels or even the venules of the mesenteric vasculature. Most are filtered by the regional lymph nodes; a few, however, may be carried to distant parts of the body where they can produce the so-called amebic abscesses, particularly in the liver, lungs, or central nervous system [290, 300, 588, 665, 791]. Fatal amebiasis with abscess formation has been reported in a baboon [300], a chimpanzee [588], an orangutan [665], a group of spider monkeys [10], douc langurs [299], and several colobus monkeys [299, 544, 545].

The diagnosis of amebiasis depends upon the microscopic recognition of the causative organisms in the feces or in intimate association with typical lesions. These organisms are also common as nonpathogenic commensals in the digestive tract of nonhuman primates. Their presence in the feces of animals with clinical signs is not definitive evidence that protozoa are the cause of the gastrointestinal disease [290, 530, 791]. Wet-mount preparations may be used to examine the feces for trophozoites. This requires a fresh sample that must be placed immediately in a saline or buffer solution and examined while the preparation is still warm. The movement of the organisms can be seen. *Entamoeba histolytica* makes the most obvious kinds of progressive movement of all the intestinal amebae. Smears may be stained with Lugol's iodine solution, which identifies the nuclei and stains glycogen. Smears also may be fixed in Schaudinn's fluid and stained with Heidenhain's iron hematoxylin [90, 791]. Nuclear morphology can be used to distinguish pathogenic *E. histolytica* from nonpathogenic species [146].

Entamoeba histolytica causes amebic dysentery in humans; therefore, this organism in nonhuman primates poses a serious potential public health problem. The disease has been transmitted from laboratory primates to humans [48, 329]. These primates should be considered potential sources of infection, and proper care should be exercised in all phases of their management.

Sporozoans and Neosporans

The parasitic sporozoans and neosporans described from nonhuman primates are listed in Table 45.3. Infection with the coccidian parasites is considered essentially innocuous; there are no known lesions or diseases associated with their presence in the nonhuman primate gastrointestinal tract. Since *Cryptosporidium* sp. has been described recently in the gastrointestinal tract of monkeys, it will be discussed in greater detail. In addition, the malarial parasites, toxoplasmasids (sarcocystis and toxoplasma), piroplasmids (babesia and entopolypoides), and neosporins (encephalitozoon) will be discussed in detail.

Cryptosporidiosis

This disease, caused by the coccidian parasite *Cryptosporidium*, has been reported in the digestive tract of the rhesus monkey [138, 472], cynomolgus monkey, and bonnet monkey [906]. In one study, the organisms were seen in the epithelium of the common bile duct, the intrahepatic and pancreatic ducts, and the gallbladder of one monkey [472]. Histologic changes consisted of epithelial hyperplasia and mucosal inflammation. In another study, cryptosporidial organisms were found in the small-intestinal epithelium of seven rhesus monkeys and in both the small- and large-intestinal epithelium of one infant monkey. There were no lesions associated with the organisms in any of the monkeys except the infant. In this infant, the changes were characterized by atrophy of villi associated with many parasites in the brush border of the epithelium lining the villi and intestinal crypts [138].

TABLE 45.3. Parasitic sporozoans and neosporans described from nonhuman primates.

Parasite genus-species	Location in host	Pro-simians	New World monkeys	Old World monkeys	Great apes	References
Coccidian parasites						
Eimeria galago	Intestine	X				90, 290, 678
Eimeria lemuris	Intestine	X				90, 290, 678
Eimeria otolicni	Intestine	X				90, 290, 678
Eimeria pachylepyron	Intestine	X				139
Eimeria tupaiae	Intestine	X				604
Eimeria modesta	Intestine	X				604
Eimeria ferruginea	Intestine	X				604
Eimeria sp.	Intestine	X				168, 387
Isopora arctopitheci	Intestine		X			90, 290, 377, 378, 727
Isopora callimico	Intestine		X			217, 400
Isopora papionis	Intestine			X		576, 578
Isopora sp.	Intestine	X			X	90, 290, 485, 678, 722
Cryptosporidium sp.	Intestine			X		138, 472, 791, 906
Klossiella sp.	Kidney	X				97, 776
Haemogregarina cynomolgi	Blood			X		54, 290, 505, 525, 899
Piroplasmasid parasites						
Babesia pitheci	Erythrocytes			X		23, 290, 315, 385, 386, 449, 738, 837
Babesia sp.	Erythrocytes	X				72
Entopolypoides macaci	Erythrocytes			X		23, 269, 290, 332, 363, 485, 564, 565, 596
Theileria cellii	Erythrocytes			X		98, 99, 290
Malarial parasites						
Plasmodium cynomolgi	Erythrocytes			X		70, 143, 180, 236, 239, 243, 290, 316, 415, 563, 602, 702, 772, 773
Plasmodium knowlesi	Erythrocytes			X		80, 126, 243, 290, 318, 466, 502, 773, 797, 925
Plasmodium inui syn. (*P. shortti*)	Erythrocytes			X		70, 71, 186, 236, 237, 290, 353, 399, 415, 502, 795, 796, 816
Plasmodium coatneyi	Erythrocytes			X		236, 240, 241, 244, 290, 561, 658, 660, 662, 881
Plasmodium fieldi	Erythrocytes			X		236, 242, 290, 374, 881
Plasmodium gonderi	Erythrocytes			X		68, 290, 322, 733

TABLE 45.3. (Continued).

Parasite genus-species	Location in host	Pro-simians	New World monkeys	Old World monkeys	Great apes	References
Plasmodium fragile	Erythrocytes			X		180, 290, 703
Plasmodium siminovale	Erythrocytes			X		180, 181, 290
Plasmodium sp.	Erythrocytes			X		236, 290
Plasmodium brasilianum	Erythrocytes		X			165, 206, 211, 290, 337, 690, 835
Plasmodium simium	Erythrocytes		X			164, 165, 290, 291, 317
Plasmodium pitheci	Erythrocytes				X	290, 329, 353
Plasmodium malariae syn (*P. rodhaini*)	Erythrocytes				X	68, 69, 290, 321, 730, 731
Plasmodium reichenowi	Erythrocytes				X	65, 66, 71, 290, 321, 329, 714, 801
Plasmodium schweitzi	Erythrocytes				X	67, 70, 81, 290, 732
Plasmodium hylobati	Erythrocytes				X	71, 141, 290, 329, 729
Plasmodium eylesi	Erythrocytes				X	290, 882
Plasmodium jefferyi	Erythrocytes				X	290, 883, 884
Plasmodium youngi	Erythrocytes				X	244, 290
Plasmodium silvaticum	Erythrocytes				X	325, 542
Plasmodium girardi	Erythrocytes	X				84, 290, 319, 320
Plasmodium lemuris	Erythrocytes	X				290, 404
Plasmodium foleyi	Erythrocytes	X				319, 320
Hepatocystis kochi syn. (*H. simiae*)	Erythrocytes			X		2, 68, 290, 313, 314, 323, 452, 487, 488, 518, 524, 578, 586, 610, 652, 674, 823, 867
Hepatocystis semnopitheci	Erythrocytes			X		237, 238, 290, 465, 681, 714
Hepatocystis taiwanensis	Erythrocytes			X		70, 290
Hepatocystis foleyi	Erythrocytes	X				70, 84, 290
Hepatocystis sp.	Erythrocytes				X	174, 175, 573, 594
Sergentella anthropopitheci	Blood				X	173, 290

TABLE 45.3. (Continued).

Parasite genus-species	Location in host	Pro-simians	New World monkeys	Old World monkeys	Great apes	References
Toxoplasmasids						
Sarcocystis kortei	Striated muscle, oral cavity, heart, tongue, esophagus			X		23, 192, 290, 349, 361, 471, 555, 638
Sarcocystis nesbitti	Striated muscle, oral cavity, heart, tongue, esophagus			X		23, 290, 555
Sarcocystis sp.	Striated muscle, oral cavity, heart, tongue, esophagus	X	X	X		26, 168, 290, 369, 376, 573, 578, 623, 697, 834, 932
Toxoplasma gondii	Brain, lungs, liver, heart, kidney, lymph-nodes, blood, other tissues	X	X	X	X	12, 23, 36, 37, 81, 82, 119, 178, 183, 290, 381, 418, 470, 527, 578, 579, 581, 628, 633, 706, 707, 708, 709, 726, 783, 817, 831, 842, 858, 913, 934, 935
Neosporan parasites						
Encephalitozoon cuniculi	Brain, kidneys, heart, lungs, adrenals, other tissues		X			15, 79, 656
Microsporidian sp.	Intestine		X			781
Haplosporids						
Pneumocystis carinii	Lungs		X		X	117, 543, 679, 721, 820

A recent report documents severe intestinal disease in four juvenile macaques resulting from infection with *Cryptosporidium* sp. [906]. Two of the animals died naturally and the other two were humanely killed as a consequence of the illness, despite extensive therapy. Clinical signs in these young monkeys consisted of depression, dehydration, weight loss, and intractible diarrhea [906]. At necropsy, the animals were considered to be underweight for their age and dehydrated. The intestines were distended with gas and liquid. The mesenteric lymph nodes were enlarged in three of the four monkeys [906]. Histopathologically, the lesions in the small intestine consisted of blunting and fusion of villi, variation in height of the intestinal epithelium, necrosis of individual epithelial cells, and an increased mitotic index in the crypts. Variable numbers of organisms with both light microscopic and ultrastructural morphologic features consistent with *Cryptosporidium* sp. were seen adherent to enterocytes along the tips and side of villi, as well as within the crypts [906].

These cases indicate that cryptosporidial infection in young macaques can be a severe and potentially fatal disease. The cryptosporidia were associated with ultrastructural changes in the enterocytes that can result in malabsorption, as well as fluid loss in the infected host [906].

Malaria

This disease, caused by parasites in the family *Plasmodiidae*, genus *Plasmodium*, affects both humans and animals [32, 71, 290, 742, 791, 874]. Malaria is one of the most important hemoprotozoal parasitic diseases of primates in the tropical and semitropical regions of the world. Spontaneous infection is universal among nonhuman primates with the exception of a few species: rhesus monkey, tamarins and marmosets (*Callithrix* sp. and *Saguinus* sp.), and owl monkeys. The malaria parasites that infect the anthropoid apes are a different group of Plasmodia than those that afflict monkey species. They are also homologous to the malaria parasites of humans, and they are considered to be indistinguishable morphologically. Cross-infection to humans has been documented. In the natural host, these organisms do not produce severe disease—there are no outward signs of disease seen and usually no fever. There may be a slight anemia associated with a low-grade parasitemia in some animals, but they appear outwardly normal. In the aberrant host, infection with malarial parasites produces severe disease and debilitation that often leads to death.

Malaria parasites can be classified on the basis of the host infected—human, monkey, or anthropoid ape—or on the basis of the type of cyclic fever produced. Thus, quotidian malaria has a 24-hour cycle, tertian malaria cycles every 48 hours, and quartan malaria has a 72-hour cycle. The periodicity of the cyclic fever is determined by the length of time the organisms parasitize the host erythrocytes.

The life cycle of malaria parasites has been reviewed in detail [135, 140, 236, 290, 530, 556, 791, 874]. It is indirect with numerous mosquitoes in the genus *Anopheles* serving as biological vectors. Basically, the life cycle consists of two major phases: the sexual or sporogonic phase in the mosquito vector, and the asexual or shizogonic/gametogonic phase in the vertebrate host. The schizogonic phase is further divided into the exoerythrocytic or liver phase and the erythrocytic or blood phase.

The clinical signs reported in nonhuman primates infected with malaria consist of hepatosplenomegaly, fever, depression, listlessness, anorexia, and weight loss. Thrombocytopenia, leukopenia, progressive anemia, and reticulocytosis also have been reported. Diarrhea may be an accompanying symptom [290, 791, 816, 874]. Hematocrit, hemoglobin, mean corpuscular volume, and erythrocyte values are reported to be lower in infected nonhuman primates [186]. In general, young animals exhibit more severe symptoms than older animals [874]. Fever in infected nonhuman primates is less severe than in their human counterparts. The onset of fever coincides with the rupture of the parasitized erythrocytes and the release of toxic metabolic products into the bloodstream. Depending on the species of *Plasmodium* involved, this event can occur at 24-, 48-, or 72-hour intervals. Usually the natural host of a species of Plasmodia is asymptomatic.

Histopathology consists of functional and structural changes in the liver and pigment (hemazoin) deposition in the liver, spleen, and bone marrow. Hemorrhages in the brain, splenic rupture, and lower nephron (tubular) necrosis of the kidney have been reported [290, 542, 737, 791, 874].

The diagnosis of malaria depends on the demonstration and identification of the organisms in erythrocytes in thin or thick blood smears stained with Giemsa's or Wright-Giemsa stains [290, 791].

Malaria in most nonhuman primates is generally not fatal. However, it may cause debilitation, and overt disease can be precipitated by stress, concurrent disease, splenectomy, or immunosuppression. Infected primates may also serve as sources of infection for humans provided the required mosquito vectors are present. All people actively working with or caring for nonhuman primates should be alert to the possible existence and potential liabilities of malarial infection.

The Malaria of Old World Monkeys

Plasmodium knowlesi. This is the only known quotidian (24-hour) malarial parasite. It is distributed geographically throughout Southeast Asia, and the natural hosts include the cynomolgus monkey, leaf monkey (*Presbytes melalophus*), and the pig-tailed macaque.

Plasmodium knowlesi produces a virulent infection in the rhesus monkey that is almost always fatal and resembles acute *P. falciparum* infection of humans [70, 80, 126, 236, 243, 290, 318, 466, 502, 542, 773, 797, 874, 925].

Plasmodium cynomolgi. This is a tertian malarial parasite with a geographic distribution including the East Indies, Southeast Asia, and the Philippine Islands. The natural hosts include a wide variety of *Macaca* species including the cynomolgus monkey, Toque monkey, pig-tailed macaque, bonnet macaque, Formosan rock macaque (*Macaca cyclopis*), and several species of leaf monkeys (*Presbytes cristatus* and *P. entellus*). Infection in the rhesus monkey with this organism is not as severe as with *P. knowlesi* and usually consists of low-grade parasitemia of long duration. This organism is similar to *P. vivax* of humans and is also transmissible to humans [70, 143, 180, 236, 243, 290, 316, 542, 563, 602, 702, 773, 874].

Plasmodium gonderi. This tertian malarial parasite is distributed throughout west Africa and tropical central Africa and is the only simian *Plasmodium* found in Africa. Natural hosts include mangabeys (*Cercocebus* sp.) and drills (*Mandrillus* sp.). This parasite produces a high, chronic parasitemia in the rhesus monkey. Baboons (*Papio* sp.) and guenons (*Cercopithecus* sp.) are also susceptible. Humans have proven to be susceptible in experimental studies [68, 135, 290, 322, 337, 542, 733, 797, 874].

Plasmodium fieldi. *Plasmodium fieldi* is a tertian malarial parasite found on the Malay Peninsula. Natural hosts include the Asian species of *Macaca* including the cynomolgus monkey and the pig-tailed macaque. Infection with *P. fieldi* produces a severe disease in the rhesus monkey that is often fatal. Humans seem to be resistant to infection with this organism [242, 290, 374, 542, 874, 881].

Plasmodium fragile. This is also a tertian malarial parasite with a geographic range throughout southern India and on the island of Sri Lanka. Natural hosts include the Toque monkey (*Macaca sinica*) and the bonnet macaque. Infection in the rhesus monkey produces severe disease that often kills the host [135, 180, 290, 542, 703, 874].

Plasmodium siminovale. This tertian malarial parasite is found on the island of Sri Lanka. The natural host is the Toque monkey. Even though the parasitemia is not particularly severe, a pronounced anemia has been reported as accompanying this infection. This parasite is considered to be similar to *P. ovale* of humans [181, 290, 542, 874].

Plasmodium coatneyi. This species of malarial parasite causes a mild tertian malaria in susceptible hosts. Its geographic distribution includes the Malay Peninsula and the Philippine Islands. The natural nonhuman primate host is the cynomolgus monkey. It is quite closely related morphologically to *P. knowlesi*, and the infection in the rhesus monkey is similar to *P. knowlesi*, producing a severe and often fatal disease accompanied by severe anemia [240, 241, 244, 263, 290, 542, 874, 881].

Plasmodium inui. Plasmodium inui is a mildly pathogenic species that produces a quartan malaria in susceptible hosts. It is very widespread throughout Southeast Asia and extends from India to the Philippine Islands. The natural hosts include the Asian species of *Macaca*, cynomolgus monkey, and pig-tailed macaque. It has also been reported from members of the genus *Presbytis* and the Celebes black ape (*Cynopithecus niger*). This infection is frequently encountered in Asian monkeys and can persist for at least several years even in animals removed from endemic areas. The parasite is considered to be homologous to *P. malariae* of humans, and humans are susceptible to infection with this organism [70, 71, 236, 237, 290, 353, 399, 502, 542, 773, 795, 796, 874].

Plasmodium shorti. This quartan malarial parasite is found in India and on the island of Sri Lanka. Natural nonhuman primate hosts include the Toque monkey and the bonnet macaque. This organism has been transmitted experimentally to humans [70, 236, 290, 795].

The Malarial Parasites of New World Monkeys

Plasmodium simium. This is a tertian malarial parasite with a geographic distribution in the region of southern Brazil. The natural nonhuman primate hosts are howler monkey(*Alouatta* sp.) and woolly spider monkeys (*Brachyteles arachnoides*). The organism is similar to *P. vivax* of humans, and infection with this parasite has been reported in humans [135, 164, 290, 291, 317, 542, 874].

Plasmodium brazilianum. Plasmodium brazilianum is sometimes a markedly pathogenic species and is the commonest malarial parasite of New World monkeys. The natural nonhuman primate hosts include howler monkeys, spider monkeys (*Ateles* sp.), woolly spider monkeys, uakaris (*Cacajao* sp.), titis (*Callicecus* sp.), bearded sakis (*Chiroptes* sp.), capuchin monkeys (*Cebus* sp.), woolly monkeys (*Langothrix* sp.), and squirrel monkeys (*Saimiri* sp.) [2, 135, 211, 290, 337, 542, 690, 835, 874]. The geographic distribution ranges from Mexico throughout Central America and into South America down to Peru. *Plasmodium brazilianum* causes quartan malaria that can produce severe symptoms, and even adult monkeys have been known to die from this infection [542, 835]. Usually it is seen as an infection at equilibrium having a low parasitemia that may persist for several years [874]. This organism is considered to be the same as *P. malariae* of humans, and humans are susceptible to experimental infection. This species may actually be *P. malariae* introduced into the New World by early explorers and modified through numerous passages in wild monkeys [206, 290].

The Malaria of Anthropoid Apes

Plasmodium pitheci. This is a tertian malarial parasite that was originally described in 1907 from Borneo in an orangutan (*Pongo* sp.) [353, 542]. Orangutans are the only nonhuman primates in which this parasite has been reported, and there are few details about the organism and the disease it produces. It has proved to be noninfectious for gibbons and monkeys in limited experimental studies.

Plasmodium rodhaini. Plasmodium rodhaini is found in west to central tropical Africa where it causes quartan malaria in its natural nonhuman primate hosts, chimpanzees (*Pan* sp.) and gorillas (*Gorilla* sp.). There is no morphological difference between this organism and *Plasmodium malariae* of humans, and these two parasites are considered to be synonymous. This is the only malarial parasite that occurs as a natural infection in humans and nonhuman primates to any great extent. Infection is easily transmitted from humans to chimpanzees and vice versa [68, 69, 290, 321, 542, 730, 731, 874].

Plasmodium reichenowi. This mildly pathogenic species occurs in west, central, and east tropical Africa. The natural nonhuman primate hosts include chimpanzees and gorillas. *Plasmodium reichenowi* causes a mild quartan malaria in these species. This organism is very similar to *Plasmodium falciparum* of humans. There is only a slight morphological difference between the two organisms, but attempts to transmit *P. reichenowi* to humans have been unsuccessful [65, 66, 68, 70, 71, 290, 321, 542, 714, 801, 874].

Plasmodium schwetzi. Plasmodium schwetzi is a mildly pathogenic plasmodium species found in west Africa. The natural nonhuman primate hosts are chimpanzees and gorillas, and this parasite causes a mild tertian malaria in these species. The disease is often subclinical and not obvious unless the animal if splenectomized. *Plasmodium schwetzi* is very similar to *P. vivax* of humans. This organism can infect humans and has been transmitted from chimpanzees to humans via mosquitoes. The disease in humans consists of a mild febrile period followed by a spontaneous cure [67, 70, 71, 81, 290, 542, 732, 874].

Plasmodium hylobati, Plasmodium eylesi, Plasmodium jefferyi, Plasmodium youngi. These four closely related parasites are found in the East Indies. They produce a quartan malaria in gibbons (*Hylobates* sp.), which are the natural nonhuman primate hosts for these parasites. *Plasmodium hylobati*, *P. youngi*, and *P. eylesi* are reported to be pathogenic. A febrile response associated with the parasitemia has been seen in gibbons infected with *P. hylobati* and *P. youngi*. Details of the clinical disease and pathology have not been reported [141, 244, 290, 542, 729, 874, 882–884].

HEPATOCYSTOSIS

This disease is caused by parasites classified in the genus *Hepatocystis* in the family *Haemoproteidae*. The species reported to occur in nonhuman primates are listed in Table 45.3.

Hepatocystis kochi was formerly classified in the family *Plasmodiidae* as *Plasmodium kochi* and currently has several other synonyms including *H. joyeuxi*, *H. cercopitheci*, *H. bovilliezi*, and *H. simiae* [290]. These sporozoan parasites are distributed in India to the East Indies subcontinent and throughout the African continent south of the Sahara Desert. Nonhuman primates reported to be affected with *Hepatocystis* sp. include Old World monkeys (African green monkeys, guenons, mangabeys, baboons, patas monkeys, colobus monkeys, Formosan rock macaques, other *Macaca* species, and leaf monkeys) and great apes (gibbons) [2, 174, 175, 290, 313, 323, 487, 542, 557, 794, 867, 874, 884].

The incidence of *Hepatocystis* can exceed that of malaria [542] and has been reported from 42% to 56% in nonhuman primates obtained from west central Africa [68, 290, 524] and from 40% to 75% in species from east central Africa [290, 586, 674, 823, 867].

The life cycle of *Hepatocystis* is indirect with midges (*Culicoides* sp.) serving as the biological vector [290, 323, 542, 791]. *Hepatocystis kochi* is the only species in which the life cycle is completely known [313, 314, 530, 586]. The life cycle resembles that of *Plasmodium* sp. with the major exception that asexual schizogony does not take place in the host's erythrocytes. Schizogony in the liver produces grossly visible cysts called merocysts [290, 542, 874].

Hepatocystis infection in the nonhuman primate produces no cyclic fever or waves of parasites in the blood as occurs with malaria. Parasitemias are usually not too great, and evidently there is no adverse effect to the monkey's health.

Grossly, infected nonhuman primates have numerous, randomly scattered, grayish-white, translucent foci on the surface of the liver that correspond to the mature merocysts [290, 542, 867]. Histopathologically, there is no tissue reaction in the liver until the merocysts are formed. After the cyst develops, there is usually a neutrophilic exudate surrounding it. Following rupture of the cyst and release of the merozoites, a chronic granulomatous inflammatory reaction ensues with the infiltration of lymphocytes and macrophages. Healing results in fibrosis in and around the area where the cyst was located. These appear as white foci grossly [290, 542, 867].

Diagnosis is based on demonstration and identification of the parasite in thick blood smears or finding the typical hepatic lesions at necropsy and/or on histologic sections [290, 674, 867].

There are no public health considerations with this parasite since *Hepatocystis* sp. are not known to infect humans [290].

Toxoplasmosis

The cause of this disease, *Toxoplasma gondii*, is a cosmopolitan protozoan parasite. Spontaneous infection with *T. gondii* has been reported in New World monkeys (squirrel monkeys, spider monkeys, sakis, owl monkeys, uakaris, marmosets, tamarins, woolly monkeys, titi monkeys, howler monkeys, woolly monkeys, and cebus monkeys), Old World monkeys (rhesus monkeys, stump-tail macaques, cynomolgus monkeys, and baboons), great apes (chimpanzees), and prosimians (Malayan tree shrew, ring-tailed lemur, ruffed lemur, and slow loris) [12, 23, 36, 37, 81, 119, 178, 183, 290, 381, 418, 470, 527, 579, 581, 628, 633, 706–709, 726, 783, 817, 831, 842, 858, 913, 934, 935]. New World monkeys are reported to be more susceptible to this disease [23, 381, 429, 581, 742, 791,

913]; marmosets are very susceptible and may die within 5–6 days after contacting the disease [37]. There is some question as to whether or not the infection reported in the baboon (*Papio cynocephalus*) and the chimpanzee was acquired naturally since both animals had been inoculated intracerebrally with material from guinea pigs and rabbits shortly before death [23, 742]. Also, there is some doubt about the validity of the diagnosis regarding two fatal cases of toxoplasmosis in lemurs (*Lemur catta*) from Japan [23]. The diagnosis was based on the microscopic demonstration of the parasites, and the published photomicrographs are reported not to be entirely convincing [23].

The morphology of *T. gondii* has been reviewed previously [290, 303, 426, 530]. Toxoplasma tachyzoites (formerly called trophozoites) are crescent- or banana-shaped structures that measure 4–8 μ by 2–4 μ. One end is pointed, and the other is rounded and contains a centrally located nucleus. Tachyzoites can be found in various cells throughout the host and also in blood and peritoneal fluid. Initially, they occupy vacuoles in the host cells (current preferred terminology, group stage or colony) [303, 791]. As they multiply, a cyst forms around them. The encysted forms are known as bradyzoites (formerly called merozoites). The oocysts seen in the intestinal epithelial cells and feces of the cat measure 10 $\times$ 12 μ and are the smallest of the three common cat coccidia.

For an in-depth discussion of the life cycle of this interesting parasite, the reader is referred to the following references: 290, 302, 303, 426. Briefly, the life cycle consists of an enteroepithelial phase, which occurs only in the definitive host, and a extraintestinal or tissue phase, which occurs in all susceptible species (intermediate hosts). Asexual reproduction (endodyogony, endopolygony, and schizogony) and sexual reproduction (gametogony), which leads to the production of oocysts, occur in the intestinal epithelium of various domestic and feral members of the family *Felidae*. Unsporulated occysts are shed by these animals, the only known definitive hosts, and sporogeny occurs in the feces. In the extraintestinal cycle multiplication of tachyzoites occurs by endodyogony in all other tissues of the intermediate hosts, which include a wide variety of domestic and wild animals (including cats), birds, and some nonmammalian species [290, 302, 303, 791].

Infection with *Toxoplasma* can occur via transplacental transmission, consumption of tissue cysts, and consumption of oocysts. The organisms also can be spread mechanically and by insect vectors, such as cockroaches [290, 302, 303, 791, 879].

Nonspecific clinical signs reported in nonhuman primates infected with *T. gondii* include emesis, depression, diarrhea, anorexia, fever, cough, weakness, ocular and nasal discharges, pale mucous membranes, leukopenia, dyspnea, premature birth, abortion, and death [290, 429, 791]. Neurologic signs include circling, incoordination, paresis, and terminal convulsions [290, 429]. At necropsy, the most frequently observed abnormalities reported are cardiomegaly, myocardial necrosis, hemorrhagic lymphadenopathy, pulmonary edema, hepatic congestion, hepatocellular necrosis, splenomegaly, and splenic hyperplasia [36].

Histopathologically, focal hepatic necrosis, focal to diffuse necrotic lymphadenitis and splenitis, segmental intestinal lesions, interstitial and fibrinous pneumonia, and focal myocarditis have been reported. Necrotic foci and extracellular and/or intercellular tachyzoites are frequently found in conjunction with the inflammatory lesions. Lesions seen in the central nervous system include gliosis, focal hemorrhage, microscopic infarcts, and cellular degeneration [290, 429, 791]. *Toxoplasma gondii* cysts and free organisms have been noted in capillary endothelial cells and in the brain tissue, frequently with associated perivascular cuffing and cellular necrosis [429, 470, 581].

Diagnosis of toxoplasmosis depends on the demonstration and identification of the causative organism in smear preparations or in histopathologic sections, or by animal inoculation [23, 290, 429, 530]. Laboratory tests include the indirect fluorescent antibody test, the Sabin-Feldman dye test, and the hemagglutination test. Isolation of *T. gondii* itself is most reliable but is time consuming and expensive. Recognition of oocysts in fecal samples of cats is important for the prevention and control of both animal and human toxoplasmosis [290, 429, 791]. The following references should be consulted for information about preparing fecal samples and the morphological differences between the common feline coccidia oocysts [189, 191].

Because toxoplasmosis can occur in humans, reasonable care should be taken to prevent infection in those personnel responsible for the care and use of nonhuman primates. Feces from *Felidae* should be removed frequently (within 24 hours) and preferably incinerated or disposed of in some other way that will prevent contact of vectors and fomites with sporulated oocysts [290].

Sarcocystosis

This disease is caused by coccidian parasites commonly classified in the genus *Sarcocystis*. The cystic phase of this parasite has been described in skeletal muscle fibers and occasionally in cardiac or smooth muscle fibers in a wide variety of animals throughout the world [23, 225, 290, 461, 530, 532]. These cysts are common in the skeletal muscle of the tongue or the esophageal muscle of many nonhuman primates. *Sarcocystis kortei* and *S. nesbitti* have been described in the rhesus monkey, and other unnamed species have been reported in both Old and New World monkeys [23, 192, 270, 290, 361, 376, 452, 471, 534, 550, 555, 623, 638, 697, 791, 823].

It is now known that *Sarcocystis* has an obligatory two-host life cycle. The reader is referred to the definitive works published recently outlining the intricacies of the life cycle of this unique and interesting parasite [190, 302, 303, 743].

Lesions associated with spontaneous infections in nonhuman primates are rare [225, 290, 461, 791]. Inflammation characterized by infiltrates of lymphocytes, plasma cells, and eosinophils is associated with degeneration of the cysts within the muscle fibers. With time, there is a proliferation of fibrous connective tissue and resulting scar formation [791, 840]

Babesiosis

Two organisms are associated with this disease in nonhuman primates. *Babesia pitheci* has been reported from Old World monkeys (mangabeys, guenons, macaques, and baboons) [23, 290, 315, 385, 386, 449, 738, 837] and New World monkeys (marmosets) [385] . Its distribution and incidence in nature is unknown [290, 542]. The complete life cycle is unknown, but ticks are thought to be the biological vectors [23, 290, 542]. This babesial parasite is considered to be only slightly pathogenic in normal intact monkeys but can result in severe anemia and death after splenectomy. Marked poikilocytosis and anisocytosis are associated with the anemia [23, 290, 315, 449, 542, 837]. *Babesia pitheci* organisms are pyriform in shape and measure 2–6 μm long. Round, elliptical, oval, lanceolate, and ameboid stages have been also observed in peripheral blood smears [23, 290, 542].

The second babesia-like organism, *Entopolypoides macaci*, is a mildly pathogenic hemosporozoal parasite that has been described in Old World monkeys (cynomolgus monkeys, rhesus monkeys, baboons, and guenons) and great apes (chimpanzees) [32, 269, 290, 332, 363, 564, 565, 596, 791]. This organism does not have true pyriform stages, but early ring-shaped stages and ameboid stages with polypoid projections of cytoplasm similar to the true *Babesia* species have been seen. *Entopolypoides macaci* is smaller than *Babesia* and *Plasmodium* species parasites and is morphologically distinct [332, 542]. Parasitized erythrocytes are not enlarged, and pigment is not formed [542].

Fever, monocytosis, and anemia have been reported in parasitized nonhuman primates; however, infection with *E. macaci* appears to have little effect on the host [290, 542, 565]. Chronic, latent infections are known to occur, and splenectomy or immunosuppression will result in recurrence of the parasitemia and a marked increase in the intensity of hemolytic anemia and icterus. Under these conditions, the disease may be fatal [791]. There are indications that this organism is common in nonhuman primates [542].

Diagnosis of these two organisms depends on the demonstration and identification of the parasites within the host's erythrocytes [290, 791].

There is no public health significance associated with either *B. pitheci* or *E. macaci* infections because neither parasite has been reported in humans [290].

Encephalitozoonosis

The cause of this disease, *Encephalitozoon cuniculi*, is an obligate intracellular protozoan parasite that has been reported in a wide variety of vertebrate and invertebrate species [791]. Only a few cases have been reported in nonhuman primates, all in New World monkeys. Two of these cases involved squirrel monkeys, and one involved an unidentified microsporidian parasite in a dusky titi monkey (*Callicebus moloch*) [15, 79, 781].

Encephalitozoon cuniculi is a small, oval parasite that measures approximately $2.5 \times 1.5\mu m$. Division occurs by binary fission and produces two spores per sporont. The organisms can be distinguished from Toxoplasma and other parasites by their location, size, and positive staining characteristics with the gram stain and various silver impregnation methods. A coiled polar filament with four to five coils is a distinctive ultrastructural feature of *E. cuniculi* [791]. The life cycle of this parasite is not known completely at this time [791].

The signs described in nonhuman primates infected with *E. cuniculi* consist of nervous symptoms displayed by a 2-month-old squirrel monkey for approximately 1 month prior to death [79]. Lesions in this animal were focal granulomatous meningoencephalitis, hepatitis, and nephritis. Characteristic *E. cuniculi* organisms were seen by both light and electron microscopy. Also, a granulomatous encephalitis caused by *E. cuniculi* infection has been reported in a newborn squirrel monkey [15]. In the case involving the dusky titi, gram-positive, acid-fast microsporidial organisms having a polar filament with as many as seven coils were found in the jejunal epithelium. It was felt that this organism was one that normally infects arthropods rather than *E. cuniculi* and that the monkey became infected through ingestion of the arthropods that it was able to capture in its outdoor environment. No host response to the presence of the organisms in the intestinal epithelium was reported [781]. *Encephalitozoon cuniculi* infection also has been reported as the cause of death in two infant squirrel monkeys [656].

Diagnosis of *E. cuniculi* can be made by finding the parasites associated with the typical lesions during histopathological examination of the tissues or by demonstration of the organisms in the urine. Currently, an immunofluorescence test that detects antibodies against the organisms and an intradermal test have both proved reliable [429, 791].

The public health importance of this organism is unknown at this time. There is one report of a natural *E. cuniculi* infection in a human; however, its validity has been questioned by some authors [791]. Nevertheless, personnel working or caring for nonhuman primates should follow accepted personal hygiene practices, and because urinary excretion has been proposed as a possible mode of transmission, excrement from nonhuman primates should be handled with caution. Also, care should be taken to ensure that captive nonhuman primates are protected from exposure to species known to be carriers of this organism [290].

Ciliates

The parasitic ciliates described from nonhuman primates are listed in Table 45.4. *Balantidium coli* is the only species that has been associated with lesions of the intestinal tract.

Balantidiasis

The cause of this disease, *Balantidium coli*, has a world-wide distribution and has been reported in a number of nonhuman primate species including New World monkeys (howler monkeys, spider monkeys, and cebus monkeys), Old World

Table 45.4. Parasitic ciliates described from nonhuman primates.

Parasite genus-species	Location in host	Pro-simians	New World monkeys	Old World monkeys	Great apes	References
Balantidium coli	Cecum, colon		X	X	X	38, 90, 290, 331, 349, 368, 459, 461, 485, 486, 487, 491, 494, 529, 599, 611, 616, 694, 712, 791, 839, 863, 905, 929
Balantidium sp.	Cecum, colon			X		573, 578
Troglodytella abrassarti	Cecum, colon				X	82, 157, 284, 290, 463, 491, 529, 599, 624, 863
Troglodytella gorillae	Cecum, colon				X	90, 290

monkeys (rhesus monkeys, cynomolgus monkeys, and baboons), and great apes (orangutans, chimpanzees, and gorillas) [38, 90, 92, 285, 290, 294, 331, 349, 368–371, 437, 443, 459, 461, 487, 570, 599, 680, 687, 694, 723, 791, 839, 863, 905, 929]. The organism is usually nonpathogenic and is a common inhabitant of the cecum of nonhuman primates [90, 92, 290, 361, 461, 494, 611, 694, 712, 863]. Some have been reported to be symptomless carriers [137, 877].

Balantidium coli trophozoites are large, ovoid structures with a heavily ciliated outer surface [18, 89, 290, 351, 473, 530, 790, 791, 899]. This form measures 30–150 × 25–120 μm. Internal structures consist of a macronucleus and micronucleus, two contractile vacuoles, and numerous food vacuoles. Cyst forms are spherical to ovoid and measure 40–60 μm in diameter. Reproduction occurs by conjugation or by transverse binary fission. Infection occurs through ingestion of trophozoites or cysts [90, 290, 530, 791].

Infection with *B. coli* can cause severe ulcerative enterocolitis that can be fatal in great apes [53, 130, 285, 459, 742]. Signs of clinically ill animals are weight loss, anorexia, muscle weakness, lethargy, watery diarrhea, tenesmus, and rectal prolapse [90, 369, 598, 839]. At necropsy, lesions may resemble those seen in amebiasis and may consist primarily of an ulcerative colitis [90, 461, 791]. Histologically, the ulcers may be large and may extend down to the muscularis mucosae [290, 791]. There may be an accompanying lymphocytic infiltrate and, in time, coagulation necrosis and hemorrhage [290, 791]. Typical large *B. coli* organisms can be seen in masses associated with lesions in the tissues or in capillaries, lymphatics, or regional lymph nodes [90, 290, 791, 877].

Diagnosis depends on identification of the characteristic *B. coli* organisms associated with the typical colonic lesions [290, 461, 791]. Their presence as secondary invaders to a primary disease caused by other microorganisms should always be considered and must be ruled out [290].

Balantidium coli may cause diarrhea in humans; therefore, care should be taken in handling captive nonhuman primates to avoid infection [290].

Metazoan Parasites

Nematodes

The parasitic nematode genera described from nonhuman primates are listed in Table 45.5. Because nematodiasis is such a common occurrence in nonhuman primates, the majority of the genera listed will be discussed in detail.

Rhabditoids

Strongyloidiasis

This disease results from infection by the parasitic members of the genus *Strongyloides*. These small nematodes are prevalent in most tropical and subtropical areas, but their occurrence in the temperate zones is sporadic. Several species have been reported to affect nonhuman primates: *Strongyloides cebus* has been found in New World monkeys (cebus monkeys, woolly monkeys, spider monkeys, squirrel monkeys, and marmosets) [207, 290, 437, 461, 492, 541, 634, 650, 752, 791], *Strongyloides fulleborni* in Old World monkeys and great apes (rhesus monkeys, cynomolgus monkeys, guenons, baboons, and chimpanzees) [290, 345,

TABLE 45.5. Parasitic nematodes described from nonhuman primates.

Parasite genus-species	Location in host	Pro-simians	New World monkeys	Old World monkeys	Great apes	References
Strongyloidids						
Strongyloides fulleborni	Intestines			X	X	3, 284, 290, 345, 349, 360, 650, 741, 742, 764, 791, 863, 878, 921, 922, 929
Strongyloides cebus	Intestines		X			148, 290, 461, 541, 650, 791
Strongyloides stercoralis	Intestines				X	176, 290, 531, 541, 650, 791, 810, 922
Strongyloides papillosus	Intestines			X	X	290, 541
Strongyloides sp.			X	X	X	448, 570, 578, 654, 658, 905, 922
Pelodera strongyloides	Skin lesions			X		530, 742
Ancylostomatids						
Ancylostoma duodenale	Sm. intestine				X	38, 116, 290, 626, 650, 791, 929
Necator americanus	Sm. intestine		X	X	X	38, 75, 185, 280, 290, 486, 487, 491, 609, 650, 723, 742, 791, 841, 929
Globocephalus simiae	Sm. intestine			X		290, 916
Characostomum asimilium	Sm. intestine	X		X		290, 919
Necator sp.	Sm. intestine	X				387
Strongylids						
Oesophagostomum apiostomum	Colon, mesentery			X		3, 276, 290, 349, 350, 530, 650, 658, 742, 791, 921
Oesophagostomum bifurum	Colon			X	X	47, 276, 290, 339, 345, 485, 486, 487, 488, 493, 494, 530, 578, 609, 650, 742, 791, 863
Oesophagostomum aculeatum	Colon			X		276, 290, 362, 530, 650, 712, 742, 791, 836
Oesophagostomum stephanostomum	Colon			X	X	276, 280, 290, 491, 530, 609, 650, 712, 742, 791, 863
Oesophagostomum blanchardi	Colon				X	290, 650, 791, 919

TABLE 45.5. (Continued).

Parasite genus-species	Location in host	Pro-simians	New World monkeys	Old World monkeys	Great apes	References
Oesophagostomum sp.	Colon			X		1, 335, 488, 491, 573, 618, 654, 658, 660, 662, 922
Ternidens deminutus	Cecum, colon			X	X	9, 290, 335, 339, 349, 530, 578, 609, 626, 650, 712, 742, 764, 791, 836
Ternidens sp.	Cecum, colon			X		494
Trichostrongylids						
Molineus torulosus	Sm. intestine		X			201, 207, 290, 461, 509, 791, 926
Molineus vexillarius	Sm. intestine, stomach		X			148, 168, 201, 207, 290, 461, 689, 791
Molineus elegans	Sm. intestine		X			201, 207, 290, 461, 791
Molineus vogelianus	Sm. intestine	X				201, 290
Pithecostrongylus alatus	Intestine			X	X	290, 798, 851, 919
Trichostrongylus colubriformis	Sm. intestine			X	X	276, 290, 493, 509, 530, 578, 712, 810, 862
Graphidioides berlai	Intestine		X			290, 461, 919
Nematodirus weinbergi	Sm. intestine				X	290, 699, 919
Longistriata dubia	Sm. intestine		X			148, 207, 290, 461, 689
Nochtia nochi	Stomach			X		1, 58, 290, 339, 362, 508, 650, 742, 784, 791, 803, 851, 922
Tupaiostrongylus liei	Sm. intestine	X				203, 216
Tupaiostrongylus major	Sm. intestine	X				215
Tupaiostrongylus minor	Sm. intestine	X				215
Anoplostrongylus liei	Intestine	X				216
Hepatojarakus malayae	Intestine	X				216
Nycteridostrongylus petersi	Intestine, lungs	X				216
Trichostrongylus sp.	Sm. intestine			X		485, 486, 487, 491, 494, 616, 658

(Continued)

TABLE 45.5. (Continued).

Parasite genus-species	Location in host	Pro-simians	New World monkeys	Old World monkeys	Great apes	References
Metastrongylids						
Filaroides barretoi	Lungs		X			207, 290, 339, 584, 742
Filaroides gordius	Lungs		X			207, 290, 339, 742
Filaroides cebus	Lungs		X			63
Filaroides sp.	Lungs		X			148, 290
Filariopsis arator	Lungs		X			290, 919
Filariopsis asper	Lungs		X			207, 461, 865
Angiostrongylus costaricensis	Mesenteric arteries		X			802, 846
Angiostrongylus malaysiensis		X				535
Atractidids						
Probstmayria nainitalensis	Rectum			X		17
Probstmayria gombensis	Intestine				X	283, 284
Probstmayria gorillae	Intestine				X	284, 476
Probstmayria simiae	Intestine				X	284, 558
Ascaridids						
Ascaris lumbricoides	Sm. intestine			X	X	209, 290, 509, 530, 650, 676, 712, 723, 742, 791, 922
Ascaris sp.	Intestine			X		491
Subulurids						
Subulura distans	Stomach, sm. intestine	X		X		290, 346, 609, 626, 742, 922
Subulura malayensis	Colon			X		290, 922
Subulura jacchi	Sm. intestine		X			148, 168, 290, 554, 689, 742, 841, 922
Subulura perarmata	Cecum, colon	X				61
Subulura indica	Lg. intestine, cecum	X				477
Oxyurids						
Enterobius vermicularis	Lg. intestine				X	96, 131, 290, 362, 409, 618, 650, 718, 723, 742
Enterobius bipapillata	Lg. intestine			X	X	290, 409, 650, 723, 919
Enterobius brevicauda	Lg. intestine			X		290, 409, 485, 650, 759, 919
Enterobius anthropopitheci	Lg. intestine				X	209, 328, 409, 491, 650, 712, 723, 742
Enterobius buckleyi	Lg. intestine				X	290, 409, 650, 759, 919

TABLE 45.5. (Continued).

Parasite genus-species	Location in host	Pro-simians	New World monkeys	Old World monkeys	Great apes	References
Enterobius lerouxi	Lg. intestine				X	290, 409, 650, 759, 919
Enterobius pitheci	Lg. intestine			X		409
Enterobius parallela	Lg. intestine			X		409
Enterobius zakiri	Lg. intestine			X		409
Enterobius microon	Lg. intestine		X			290, 650, 919
Enterobius sp.	Lg. intestine	X		X	X	61, 387, 390, 448, 485, 658, 660, 662, 712, 922
Buckleyenterobius dentata	Lg. intestine			X		409
Trypanoxyuris trypanuris	Lg. intestine		X			409
Trypanoxyuris (Buckleyenterobius) atelis	Lg. intestine		X			85, 96, 290, 409, 759, 919
Trypanoxyuris (Buckleyenterobius) duplicidens	Lg. intestine		X			85, 409, 290, 919
Trypanoxyuris (Buckleyenterobius) lagothricis	Lg. intestine		X			85, 409, 290, 919
Trypanoxyuris (Enterobius) interlabiata	Lg. intestine		X			290, 409, 650, 759, 919
Trypanoxyuris minuta	Lg. intestine		X			290, 409, 410, 461, 650, 683, 841, 922
Trypanoxyuris sceleratus	Lg. intestine		X			207, 290, 409, 412, 461, 650
Trypanoxyuris brachytelesi	Lg. intestine		X			409
Trypanoxyuris callithricis	Lg. intestine		X			409
Trypanoxyuris tamarini	Lg. intestine		X			148, 168, 290, 410, 412, 450, 461, 689
Trypanoxyuris oedipi	Lg. intestine		X			410
Trypanoxyuris goeldii	Lg. intestine		X			410
Enterobius lemuris	Lg. intestine	X				409
Lemuricola nycticebi	Lg. intestine	X				411
Lemuricola malaysensis	Lg. intestine	X				28, 208, 411
Lemuricola contagiosus	Lg. intestine	X				411

(Continued)

TABLE 45.5. (Continued).

Parasite genus-species	Location in host	Pro-simians	New World monkeys	Old World monkeys	Great apes	References
Labatorobius scleratus	Lg. intestine		X			290, 919
Oxyuronema atelophorum	Lg. intestine		X			290, 339, 474, 919
Primasubulura jacchi	Lg. intestine		X			168
Primasubulura otolicini	Lg. intestine	X				232
Trypanoxyuris sp.	Lg. intestine		X			609, 689
Spirurids						
Chitwoodspirura serrata	Stomach, sm. intestine				X	290, 919
Spirura guianensis	Esophagus		X			148, 290, 623
Trichospirura leptostoma	Pancreas		X			26, 148, 149, 290, 461, 492, 649, 650, 806, 807
Protospirura (Mastophorus) muricola	Stomach		X	X		101, 290, 292, 488
Pterygodermatites nycticebi	Sm. intestine	X	X			536, 593, 856
Rictularia alphi	Sm. intestine		X			290, 597, 919, 922
Rictularia sp.	Sm. intestine		X			4, 591, 592, 930
Thelaziids						
Streptopharagus armatus	Stomach			X	X	290, 349, 578, 609, 836
Streptopharagus pigmentatus	Stomach			X	X	290, 339, 362, 485, 488, 494, 609, 712, 764
Streptopharagus baylisi	Stomach			X		485, 486, 487
Streptopharagus guptai	Rectum			X		809
Streptopharagus sp.	Stomach			X		485, 616
Gongylonema macroguber-naculum	Esophagus, stomach		X	X		290, 339, 461, 533, 547, 548, 609, 650, 791
Gongylonema pulchrum	Tongue, oral cavity, esopha-gus, stomach		X	X		290, 461, 531, 547, 548, 650, 791
Physocephalus sp.	Stomach			X		654, 421, 911
Metathelazia ascaroides	Lungs			X		187, 188, 290, 919
Thelazia callipaeda	Eyes			X		271, 290, 531, 810

TABLE 45.5. (Continued).

Parasite genus-species	Location in host	Pro-simians	New World monkeys	Old World monkeys	Great apes	References
Physalopterids						
Physaloptera tumefaciens	Stomach			X		290, 362, 494, 531, 650, 764, 836, 907, 921, 922
Physaloptera dilatata	Stomach		X			168, 290, 650, 919
Physaloptera masoodi	Stomach	X				477
Physaloptera sp.	Stomach	X				612, 654, 841, 911
Abbreviata caucasica	Stomach			X	X	73, 284, 290, 419, 485, 487, 531, 578, 609, 626, 742, 919
Abbreviata poicilometra	Stomach			X		290, 756, 799, 919
Abbreviata sp.	Stomach			X		616
Onchocercids						
Dirofilaria magnilarvatum	Subcutis, peritoneal membranes			X		13, 102, 290, 500, 580, 605, 693, 762, 838, 901
Dirofilaria corynodes syn. (*D. aethiops*, *D. schoutedeni*)	Subcutis			X		13, 102, 280, 290, 364, 507, 537, 580, 875, 885, 886, 887, 888
Dirofilaria immitis syn. (*D. pongoi*)	Subcutis, muscle, right ventricle				X	13, 102, 290, 580, 742, 873, 885
Dirofilaria repens syn. (*D. macacae*)	Subcutis			X		605, 754, 885
Dirofilaria sp.	Blood			X		561, 777
Edesonfilaria malayensis	Peritoneal cavity			X		290, 308, 531, 605, 635, 712, 920, 924
Loa loa	Subcutis, mesenteries, eyes		X	X		194, 196, 199, 280, 290, 514, 531, 609, 626, 646, 734, 755, 852, 872, 886
Macacanema formosana	Peritracheal connective tissue			X		43, 290, 531, 768
Meningonema peruzzii	Subdural space–medulla oblongata			X		648
Brugia pahangi	Lymphatic system	X		X		13, 102, 221, 290, 531, 660, 662, 766, 767, 810

(Continued)

TABLE 45.5. (Continued).

Parasite genus-species	Location in host	Pro-simians	New World monkeys	Old World monkeys	Great apes	References
Brugia malayi	Lymphatic system	X		X		13, 87, 102, 280, 290, 507, 531, 561, 605, 660, 662, 810, 900
Brugia tupaiae	Lymphatic system	X				642
Wuchereria kalimantani	Inguinal lymphnodes, testicles			X		561, 659, 660, 662
Filaria (?) *nycticebus*	Intestine	X				590, 886
Dipetalonema gracile	Peritoneal cavity		X			26, 52, 94, 148, 167, 168, 184, 210, 290, 310, 475, 508, 531, 537, 566, 568, 569, 580, 651, 742, 865, 885, 889
Dipetalonema caudispina	Peritoneal cavity		X			210, 290, 311, 580
Dipetalonema tenue	Subcutis, body cavity		X			102, 103, 290, 919
Dipetalonema barbascalensis	Peritoneal cavity		X			231
Dipetalonema petteri	Pleura and peritoneum	X				105
Protofilaria furcata	Thoracic cavity	X				114, 886
Parlitomosa zakii	Peritoneal cavity		X			210, 231, 461, 619, 886
Tetrapetalonema (Dipetalonema) *obtusus*	Periesophagal connective tissue		X			230, 290, 531, 580, 923
Tetrapetalonema (Depetalonemia) marmosetae	Subcutis, body cavity		X			210, 273, 290, 580, 742, 845, 885, 923
Tetrapetalonema (Dipetalonema) tamarinae	Peritoneal cavity		X			167, 210, 230, 290
Tetrapetalonema (Dipetalonema) atelensis	Connective tissue		X			290, 580, 742, 885, 919, 923
Tetrapetalonema (Dipetalonema) parvum	Connective tissue		X			207, 290, 580, 742, 885, 922, 923
Tetrapetalonema (Depetalonema) vanhoofi	Connective tissue		X			210, 461
Tetrapetalonema nicollei	Peritoneal cavity		X			210, 885
Tetrapetalonema dunni	Subcutis	X				603
Dipetalonema perstans	Subcutis, body cavity				X	102, 123, 197, 276, 290, 340, 348, 507

TABLE 45.5. (Continued).

Parasite genus-species	Location in host	Pro-simians	New World monkeys	Old World monkeys	Great apes	References
Dipetalonema vanhoofi	Peritoneal cavity			X	X	42, 100, 101, 290, 531, 609, 646, 669, 670, 671, 739, 742, 885, 923
Dipetalonema rodhaini	Subcutis, body cavity				X	290, 646, 671, 742, 885, 919, 923
Dipetalonema streptocerca	Subcutis, peritoneal cavity				X	122, 195, 197, 276, 290, 531, 646, 671, 742, 885, 923
Dipetalonema digitatum	Peritoneal cavity			X	X	102, 103, 114, 290, 605, 742, 758, 885, 889, 919, 923
Dipetalonema leopoldi	Subcutis				X	42
Dipetalonema gorillae	Subcutis				X	42
Tetrapetalonema papionis	Skin and skeletal muscle fascia			X		578
Dracunculids						
Dracunculus medinensis	Skin, subcutis, viscera			X		276, 290, 531, 609, 742
Onchocerca volvulus	Connective tissue				X	42, 646
Trichurids						
Trichuris trichiura	Cecum, colon		X	X	X	3, 75, 76, 276, 290, 339, 350, 461, 486, 487, 494, 650, 654, 688, 712, 742, 791, 836, 921, 922
Trichuris sp.	Cecum colon	X		X		61, 387, 390, 488, 491, 493, 494, 616, 658, 660, 662, 922
Capillaria hepatica	Liver		X	X	X	276, 290, 349, 350, 420, 507, 531, 742, 810, 854, 855
Capillaria sp.	Liver		X	X	X	922
Anatrichosomatids						
Anatrichosoma cutaneum	Nasal mucosa, skin			X		8, 74, 276, 290, 650, 712, 742, 791, 833
Anatarichosoma cynomolgi	Nasal mucosa			X		8, 129, 290, 650, 805
Anatrichosoma ocularis	Eye	X				282

349, 634, 650, 654, 712, 741, 742, 752, 764, 791, 863, 922, 929], and *Strongyloides stercoralis* and *Strongyloides* sp. in Old World monkeys (patas monkeys) and great apes (gibbons, chimpanzees, gorillas, and orangutans) [38, 171, 177, 222, 285, 290, 360, 504, 570, 650, 672, 687, 774, 791, 857, 861, 922]. Only adult females and larvae are found in the gastrointestinal tract of the host animal. Migrating larvae can be found in the lungs and other parenchymatous organs. Parasitic males have never been described [290].

The life cycle of *Strongyloides* sp. is complex and consists of both parasitic and free-living generations [116, 290]. The reader is referred to parasitology texts and referenced papers for a detailed discussion of this unique life cycle [30, 120, 272, 290, 540, 791, 811]. A variation in this life cycle, known as autoinfection, is a direct reinvasion of the host animal by filariform larvae that have developed during passage through the lower intestinal tract [171, 467]. This phenomenon results in hyperinfection of the infected host and is most responsible for sustained infections that result in clinical disease, severe damage to affected organs, and death [77, 95, 171, 290]. There is also evidence of intrauterine or transcolostral transmission [290, 589, 819]. Fatal cases of strongyloidiasis have been reported in the chimpanzee, gibbon, orangutan, patas monkey, and woolly monkey [50, 51, 171, 176, 228, 360, 422, 504, 526, 571, 636, 650, 677, 857, 861].

The disease in gibbons has been reported in detail [171]. Diarrhea, which may be hemorrhagic or mucoid, is the most common clinical sign described in infected animals [171]. Other common clinical signs are dermatitis, urticaria, anorexia, depression, listlessness, debilitation, vomiting, emaciation, reduced growth rate, dehydration, constipation, dyspnea, cough, prostration, and death [171, 274, 290, 461, 742, 791, 810]. Paralytic ileus is described in infected gibbons [171].

Gross lesions consist of catarrhal to hemorrhagic or necrotizing enterocolitis [171, 290, 461, 791]. There may be a secondary peritonitis associated with the enterocolitis [290, 428]. Pulmonary hemorrhage is the most common lesion outside the digestive tract [171, 290, 461, 791]. Histologic examination of the small intestine of the infected animal shows a multifocal erosive and ulcerative enteritis caused by adults, eggs, and rhabditiform larvae [171]. The mucosa contains numerous parasites, most of which are in intraepithelial tunnels or lumina of intestinal glands. These lesions may be infiltrated by neutrophils. Mononuclear cells and an occasional eosinophil can be seen in the lamina propria. Intestinal villi are short and blunt, and in severe infections bridging and loss of villi are seen [171]. In cases where autoinfection has occurred, changes in the small and large intestines in response to invasion by the filariform larvae range from a mild inflammatory cell response to severe, acute, or granulomatous or necrotizing enterocolitis. Larval invasion of the submucosal and serosal lymphatics results in a severe granulomatous endolymphangitis [171]. These changes are associated with various degrees of lymphatic obstruction and submucosal and serosal edema, fibrosis, or both. In the lungs, acute multifocal or diffuse hemorrhage is most common. Larval granulomas may be seen over the surface of the pleura. Filariform larvae also are seen in many tissues throughout the body, most commonly in the lymph nodes and liver [171]. Fatal strongyloidiasis has been described in lowland gorillas and chimpanzees [672].

This condition may be diagnosed by identification of typical larvae in the stool; by clinical signs; or by demonstration of parasitic adult females, eggs, and larvae at necropsy or at histologic examination [290].

Strongyloidiasis in nonhuman primate colonies is considered a potential public health problem. Infections by *S. fulleborni* that have been transmitted naturally from monkey to human have been reported [64, 290, 650, 742, 878]. Experimental infections in humans by *Strongyloides* sp. isolated from nonhuman primates also have been reported [49, 176, 272, 290, 650, 752].

Oxyurids

Oxyuriasis

This disease is caused by infection by nematodes in the family Oxyuridae. Commonly known as pinworms, these small nematode parasites inhabit the colon and cecum of nonhuman primate hosts. Genera described in nonhuman primates are *Trypanoxyuris* and *Oxyuronema* species found in New World monkeys [207, 410, 412, 461]; *Enterobius vermicularis* and other *Enterobius* species, found in Old World monkeys and great apes; *Enterobius anthropopitheci* in the chimpanzee [92, 96, 131, 222, 290, 362, 409, 570, 618, 650, 686, 687, 712, 718, 724, 742, 759, 774, 919, 922]; and several species in prosimian primates [28, 41, 106–108]. These parasites are considered cosmopolitan in geographic distribution but are more prevalent in temperate and cold climates. The life cycle is direct.

Most reports of oxyuriasis in nonhuman primates state that these infections are essentially innocuous [290, 461, 650, 791]. Clinical signs usually are limited to anal pruritis and irritation that may lead to self-mutilation, restlessness, and increased aggressiveness [131, 158, 200, 290, 461, 570, 650, 742, 774, 791]. Heavy pinworm infections are reported to be common in chimpanzees, and their coprophagic habits make constant reinfection inevitable [718]. Fatal cases of enterobiasis have been reported in chimpanzees [390, 439, 774, 775, 853], characterized by extensive ulcerative enterocolitis, peritonitis, and necrogranulomatous lymphadenitis involving the mesenteric lymphnodes. Numerous parasites with the morphologic characteristics of *Enterobius vermicularis* were associated with these lesions. There is also an early report of the death of a red spider monkey caused by an overwhelming pinworm infection [474].

Multiple intestinal polyps associated with immature male oxyurid parasites have been described in a male chimpanzee [848]. The gross and histologic characteristics of these lesions were identical to those produced by *Nochtia nochti* in the stomach and esophagus of Old World primates. It was thought that the lesion resulted from hypersensitivity to oxyurid infection in an aberrant host.

Adult oxyurids may be seen emerging from the anus. Perianal swabs or cellophane tape also can be used to recover the typical ellipsoid, asymmetrical pinworm eggs [276, 290].

Naturally infected nonhuman primates may be sources of infection in humans. Also, captive primates can acquire *E. vermicularis* infection from humans and then can act as reservoirs to reinfect them [290].

Strongylids

Oesophagostomiasis

This disease is caused by infection by nematodes in the genus *Oesophagostomum*. These parasites are known commonly as nodular worms and are considered to be the most common nematode parasite found in Old World monkeys and great apes

[42, 290, 442, 570, 645, 650, 712, 742, 791]. They have been described in baboons, mangabeys, guenons, macaques, chimpanzees, and gorillas [38, 47, 92, 285, 290, 339, 345, 350, 362, 367, 392, 438, 452, 486, 494, 578, 610, 618, 654, 712, 740, 774, 791, 618, 836, 863, 894]. They are rare in New World monkeys [207, 650, 922]. Their geographic distribution is widespread, almost universal. At least 11 different species have been proposed but not clearly defined [645, 922]. The species mentioned most frequently are *O. apiostomum*, *O. bifurcim*, *O. aculeatum*, and *O. stephanostomum* [276, 290, 531, 742, 764]. The life cycle is direct.

Infected monkeys usually are asymptomatic, and light infections usually go unrecognized [290, 742, 791]. Monkeys with severe infections may show general unthriftiness and debilitation characterized by increased weight loss and diarrhea; the mortality rate increases for this group [290, 742, 791]. Lesions seen at necropsy consist of the typical oesophagóstomum nodules, which are elevated, smooth 2–4 mm in diameter, and firm. They are seen most frequently on the serosal surface of the large intestine and cecum and in the mesentery supporting these organs [92, 290, 618, 645, 742, 791, 869] but also in ectopic sites, such as the peritoneal wall, mesentery of the small intestine, omentum, kidney, liver, lungs, or diaphragm [118, 645, 650, 791]. The nodules may be black or brown if there is associated hemorrhage; older nodules usually are white because of caseation of the contents. Viable worms may be seen in relatively young nodules; usually, however, the parasite is dead and surrounded by a mass of caseous debris. Older nodules may contain foci of mineralization [290, 350, 791]. The parasite and cell detritus usually are surrounded by a mantle of chronic inflammatory cells, mainly macrophages with scattered eosinophils and lymphocytes and plasma cells. Foreign-body giant cells sometimes are present in the cellular exudate. A fibrous capsule of various degrees of thickness and maturity, depending on the age of the nodule, surrounds the centrally located necroinflammatory mass [92, 290, 618, 791]. Sometimes ulcers form in the colonic mucosa at the point where the larval penetration occurred, and a migratory tract filled with inflammatory exudate connects the nodule in the wall with the intestinal lumen [290, 349, 350, 791]. Death of a chimpanzee from septicemia caused by bacterial invasion of esophagostomum nodules in the colon has been reported [892, 893, 894]. Rupture of the nodules may result in acute or chronic peritonitis with fibrous peritoneal adhesions [645, 791]. Adhesions may restrict intestinal motility and result in obstruction or rarely in ascites [290, 742, 791].

Oesophagostomum infection can be diagnosed by identifying the eggs in the feces. A problem arises, however, because the eggs of the different *Oesophaostomum* species cannot be differentiated from one another and also are indistinguishable from those of *Ternidens* and other hookworm species. The diagnosis of oesophagostomiasis based solely on typical eggs in the feces always should be questioned. Occasionally, adults are passed and can be identified. The postmortem diagnosis is based on typical nodular lesions or identification of adults, or both [290, 791].

This parasite has been reported to infect humans and therefore should be considered to have zoonotic potential [290, 347, 546]. Appropriate care in handling nonhuman primates should be exercised.

Ternideniasis

The cause of this disease, *Ternidens deminutus*, is a strongyle that is related to the oesophagostomes and hookworms [290, 650]. These parasites inhabit the cecum and colon and have been reported in Old World monkeys (macaques, guenons, and baboons) and the great apes (gorilla and chimpanzee) [9, 290, 339, 349, 531, 609, 626, 712, 742, 764, 836, 862, 919]. The morphologic features of the adult worms and their eggs are similar to those of *Oesophagostomum* [290, 742, 836]. The life cycle is direct and also is similar to *Oesophagostomum* [290]. There is little evidence of any lesions associated with this parasite; since it is a blood sucker, however, it can cause anemia and cystic nodules in the colonic wall [290, 531, 650].

This parasite can infect humans; it causes intestinal nodules. Infected captive animals should be handled with caution [290].

ANCYLOSTOMATIDS

Ancylostomiasis and Necatoriasis

The cause of these diseases, the hookworms usually found in humans, *Ancylotoma duodenale* and *Necator americanus*, are recorded occasionally in nonhuman primates, including monkeys, mandrills, baboons, gibbons, chimpanzees, and gorillas [38, 185, 222, 280, 290, 355, 358, 538, 570, 609, 647, 687, 718, 929]. Reports of their presence in South American monkeys are rare [650]. These parasites have a direct life cycle.

Clinical signs associated with heavy hookworm infection in nonhuman primates are similar to those produced by these parasites in humans and other animals and include anemia, eosinophilia, "pot-belly," dyspnea on exertion, and a general debilitation [290, 355, 358, 742]. Necropsy findings have included a general pallor of all tissues. The mucosa of the small intestine was thickened by a chronic inflammatory reaction. Small hemorrhages were seen throughout the intestinal mucosa, and large numbers of hookworms were attached to the mucosa [358].

The diagnosis of hookworm disease is based on finding eggs in the feces or mature worms in the bowel at necropsy [290]. Since hookworm eggs are morphologically identical to those of several species of strongyles that also infect nonhuman primates, diagnosis based on the eggs alone should be viewed with caution [650].

Because humans are the normal definitive host for these parasites, infected captive primates should be handled with caution [290].

TRICHOSTRONGYLIDS

Molineiasis

This disease is caused by trichostrongyles in the genus *Molineus*. These are small, slender, pale red worms that inhabit the upper digestive tract, duodenum, and sometimes the pyloric region of the stomach of nonhuman primates. Occasionally, they may involve the pancreas and mesentery [62, 148, 168, 201, 207, 290, 461, 509, 791, 919]. They are always found lying on the mucosa, never attached. Geographically, they are distributed throughout Central and South America, with one species occurring in Africa [201, 207, 290]. Species described in nonhuman primates include *M. vexillarius* in marmosets; *M. elegans* in squirrel monkeys, cebus monkeys, and howler monkeys; *M. vogelianus* in pottos; and *M. torulosus* in cebus monkeys, squirrel monkeys, and owl monkeys [148, 201, 207, 290, 461, 791, 926]. *Molineus torulosus* is the only species reported to be a specific pathogen [62, 290, 461, 509, 791].

The life cycle and method of transmission of these parasites are unknown. Diagnosis rests upon the identification of typical eggs in the feces or the presence of adult worms associated with typical lesions in the digestive tract [290, 531, 649].

Infection with *M. torulosus* has been reported to cause hemorrhagic or ulcerative enteritis, sometimes associated with diverticula of the intestinal wall [62, 290, 461, 509, 791]. Serosal nodules that involved the upper portion of the small intestine have been seen in capuchin monkeys [62]. These nodules communicated with the intestinal lumen through 1-mm reddish brown ulcers. Histologically, the nodules were composed of an intense granulomatous inflammatory response surrounded by a rim of proliferating fibrous connective tissue. The central portion contained a mass of nematode parasites and their eggs surrounded by eosinophilic debris. Neutrophils, histiocytes, and other chronic inflammatory cells were present adjacent to the worms. Chronic pancreatitis also was seen, with worms and eggs in inflamed pancreatic ducts [62, 791].

Nothing is known of the public health significance of this parasite [290].

Nochtiasis

This disease results from infection by trichostrongyles in the genus *Nochtia*. *Nochtia nochti* is a small, slender, bright red worm that has been described in the prepyloric region of the stomach of Asian macaques [58, 290, 508, 650, 746, 791, 803, 851]. Eggs are thin shelled and ellipsoid, typical of members of the Trichostrongyles. *Nochtia* eggs can be differentiated from those of the superfamily *Strongyloidea* because they are larger and more pointed and are embryonated when passed in the feces. Free parasites are not found in the feces or the gastrointestinal tract. The life cycle is direct [58, 290].

At necropsy, hyperemic, cauliflowerlike masses are seen protruding from the gastric mucosa at the junction of the fundus and prepyloric regions. Histologically, these masses are benign inflammatory polyps composed of hyperplastic fronds of gastric mucosa and inflammatory tissue. Adult worms and their eggs can be found deep at the base of the lesion [58, 290, 622, 650, 746, 791, 803, 851].

Diagnosis depends on identification of typical eggs in the feces of affected animals or on the finding at necropsy of characteristic gastric polyps containing the parasite [290, 791].

Nothing is known of the public health aspects of this nematode.

Metastrongylids

Angiostrongyliasis

The cause of this disease, the metastrongyle *Angiostrongylus costaricensis* normally is found in rats in South and Central America [304]. It also causes a clinical syndrome in humans, particularly in children who reside in this geographic region. This syndrome is characterized by an inflammatory granulomatous mass that usually is located in the wall of the appendix but that can extend to the ileum, the cecum and ascending colon, and regional lymph nodes [304]. Histologically, the granulomatous mass is composed of chronic inflammatory cells and nematode eggs. Adult parasites reside in the arteries of the intestinal wall and mesentery [304].

Similar parasitic granulomas in the wall of the small intestine have been reported from two mustached marmosets (*Saguinus mystax*) [802, 846]. The histomorphologic features of these lesions were identical to those described previously for *A. costaricensis* infection in humans. In addition to the chronic inflammatory cells, the granulomas contained numerous nematode eggs and many larvae. The eggs of this particular parasite are reported to hatch within the rat or monkey host, then migrate to the gut and pass out in the feces to complete the life cycle [304]. Adult parasites with morphologic features consistent with a diagnosis of *Angiostrongylus* sp. were found in the mesenteric arteries associated with the granuloma.

On the basis of the nonhuman primate species involved, the fact that this species originated in the geographical region where this parasite has been reported to occur, the gross and histological appearance of the lesions, and the finding of adult *Angiostrongylus* sp. in the mesenteric arteries intimately involved with the granulomas, these parasites were identified as *A. costaricensis*. People who work with marmosets should be aware of the presence of this parasite and look for additional cases to document further its occurrence in South American monkeys.

Pulmonary Nematodiasis

This condition is the result of infection with the metastronglyid lungworms in the genera *Filaroides* and *Filariopsis*. These parasites are most commonly seen in New World monkeys (marmosets, squirrel monkeys, cebus monkeys, and howler monkeys) [63, 148, 207, 290, 339, 461, 623, 650, 742, 791, 820, 865, 919].

In the live state, these parasites are very slender and fragile [650]. Adults are found in the terminal brochioles, respiratory bronchioles, and pulmonary aveoli [290, 461, 650, 791, 820]. The adult female is viviparious. They produce larvae that are coughed up, swallowed by the host, and passed in the feces. The remainder of the life cycle, and whether or not any intermediate hosts are required, is not known [461, 650, 820].

Gross lesions in the pulmonary parenchyma are subtle. The lung appears normal except for the presence of varying numbers of randomly located, small, elevated, subpleural nodules, which may be hyperpigmented and cause the pleura to bulge. Histopathologically, there are varying degrees of atelectasis and foci of chronic inflammatory cells infiltrating the affected alveolar spaces and intra-alveolar septae. Most infections are considered to be subclinical in nature, and although the parasite is common in certain species of New World monkeys, there is no evidence to suggest that the presence of lungworms has been the cause of death [461, 650, 820].

Diagnosis can be made by finding and identifying the typical lungworm larvae in the feces in the intact animal. In the dead animal, the presence of characteristic gross and histopathologic pulmonary lesions associated with metastrongylid parasites is also diagnostic.

The public health significance of these parasites is unknown [290].

Ascaridids

Ascariasis

This disease results from infection with members of the genus *Ascaris*, commonly known as roundworms. They are a common finding in the intestinal tract of nonhuman primates [650]. The specimens that have been recovered are reported to be indistinguishable from *Ascaris lumbricoides* in humans [19, 209, 844, 922]. Both Old World monkeys and great apes have been reported to be infected [38, 209, 222, 279, 349, 355, 468, 676, 712, 742, 813, 814, 922]. The life cycle of this parasite is direct.

Although roundworm infection in nonhuman primates is thought to be relatively innocuous and of little clinical significance [570, 650], fatal cases of ascariasis have been reported in both monkeys and great apes [349, 676, 813]. Death in the great apes was thought to be due to the presence of many worms, blockage of the bowel, and migration of the worms into the bile duct and liver.

Diagnosis of ascariasis is based on the presence of typical eggs in the feces or adults in the digestive tract at necropsy.

Since the ascarids reported in nonhuman primates are morphologically identical to *A. lumbricoides* in humans, cross-infection from infected animals to humans is possible. We could find no reports that documented such an occurrence; nevertheless, infected nonhuman primates should be considered a potential zoonotic threat and should be handled accordingly.

Spirurids

Trichospiruriasis

The cause of this disease, *Trichospirura leptostoma*, is a spirurid nematode that parasitizes the pancreatic ducts of several species of New World monkeys including marmosets, squirrel monkeys, and owl monkeys. Geographic distribution is confined to Central and South America [148, 149, 290, 461, 649, 650, 806, 807]. Male and female adult parasites measure up to 15 mm and 120 mm, respectively. The eggs are medium in size and are typically spirurid in that they are thick shelled and contain a larvae [534]. The complete life cycle is unknown [290, 461].

The parasite usually is found incidentally on histological examination. Infection usually causes little tissue destruction or inflammatory reaction. Tissue response apparently varies in proportion to the number of parasites present [461, 650]. Chronic pancreatitis in association with the worms has been described in marmosets (*Callithrix* sp.) [650, 807]. Acute pancreatitis in owl monkeys consisting of a patchy granulocytic interstitial infiltrate adjacent to intralobular ducts was thought to be associated with leakage of retained pancreatic secretions. Larger ducts containing cross sections of worms also contained granulocytes [649, 650].

Nothing is known about the public health significance of this parasite.

Pterygodermatitiasis

The cause of this disease, *Pterygodermatites nycticebi*, is a spirurid nematode that has been reported from prosimians (slow loris), New World monkeys (tamarins and marmosets), and the great apes (gibbons) [536, 593, 856]. Several reports in members of the family Callitrichidae refer to this parasite by a synonym, *Ric-*

tularia nycticebi [4, 591, 592, 930]. The life cycle of *P. nycticebi* is indirect with cockroaches serving as intermediate hosts [593].

Morbidity and mortality associated with infection of *P. nycticebi* have been reported in golden lion tamarins (*Leontopithecus rosalia*) [593]. Clinical signs in heavily infected animals included extreme weakness, passage of watery diarrhea that contained the adult parasites, anemia, leukopenia, and hypoproteinemia [593]. At necropsy, masses of *P. nycticebi* parasites were found throughout the gastrointestinal tract. Histopathologically, the anterior ends of the adult worms were embedded in the mucosa of the small intestine. Larvae were seen deeper in the submucosa. In a few cases, worms were seen in the tunica muscularis and the pancreatic ducts. There was severe clubbing of the small intestinal villi and randomly located foci composed of a necrotic pseudomembrane containing spirurid eggs, numerous yeasts, and pseudohyphae consistent with *Candida* sp. [593].

Diagnosis depends upon demonstrating and identifying the characteristic spirurid eggs, adult worms, or larvae in the feces, in the gastrointestinal tract at necropsy, or in histopathological slide preparations.

Nothing is known about the public health significance of this parasite.

Methods of control should be directed against the cockroach intermediate host through reducing populations and preventing consumption by susceptible hosts.

THELAZIIDS

Streptopharagiasis

This disease is caused by parasitic members of the genus *Streptopharagus*. These are thelaziid nematodes that have been described in the stomach of Old World monkeys and great apes [290]. *Streptopharagus armatus* is reported in the rhesus monkey, other macaques, guenons, patas monkey, baboon, and gibbon [290, 349, 836]. *Streptopharagus pigmentatus* has been reported in the rhesus monkey, cynomolgus monkey, guenon, baboon, and gibbon [290, 339, 362, 712]. The life cycle of *S. pigmentatus* has been reviewed recently [551]. Little is known about the anatomic effects of these parasites [290]; however, there is one report of the death of a baby chimpanzee as a result of a perforated esophagus secondary to the migration of *Streptopharagus* sp. larvae [127].

Gongylonemiasis

The cause of this disease, parasites in the genus *Gongylonema*, are small filiform, thelaziid nematodes that have been reported in many nonhuman primates including both Old and New World monkeys [16, 290, 461, 533, 650, 911, 919, 922]. The species most commonly mentioned are *G. macrogubernaculum* and *G. pulchrum*. They have a cosmopolitan geographic distribution. A characteristic feature of the adults is several rows of conspicuous oval to round cuticular bosses located at the anterior extremity [650]. The life cycle is indirect, with cockroaches or dung beetles serving as intermediate hosts [209, 650].

Infection with this parasite is asymptomatic. Its presence usually is recognized only histologically and is considered to be an incidental finding. Adults are found in tunnels in the stratum malpighii of the squamous epithelium of the esophagus, lip, tongue, and other parts of the buccal cavity. They have been recovered from bronchi and the stomach. There is little or no tissue reaction [290, 461, 650].

Gongylonema pulchrum has been reported to occur in humans [290].

PHYSALOPTERIDS

Physalopteriasis

This disease is caused by infection by members of the genus *Physaloptera*. Four species of physalopterids have been reported to occur in the upper gastrointestinal tract of nonhuman primates [290]. *Physaloptera tumefaciens* is common in the stomach of Asian macaques [290, 362]. *Physaloptera dilatata* is found in the stomach of New World monkeys (titi monkeys, bearded sakis, and marmosets) [168, 207, 290, 492, 919]. *Abbreviata caucasica* has been found in the esophagus, stomach, and small intestine of the rhesus monkey, baboon, and orangutan [290, 419, 531, 626, 742, 919]. *Abbreviata poicilometra* has been found in the stomach of mangabeys and guenons [290, 756, 799, 919].

The life cycle of the physalopterids is indirect; an arthropod intermediate host is required. The entire life cycle is not completely understood, and a second intermediate or paratenic host may be necessary [290, 810]. Lesions result from the attachment of the worms to the wall of the affected organ. Gastritis, esophagitis, enteritis, erosion, and ulceration of the mucosa at the point of attachment are seen at necropsy [290]. Hyperplastic gastric lesions and perforation of the stomach wall associated with *Physaloptera* sp. infection in cynomolgus monkeys have been described [287, 746].

Diagnosis depends upon identification of the ova in the feces or the presence of adult worms attached to the mucosa of the upper digestive tract [290].

The public health aspects of these parasites are unknown.

ONCHOCERIDS

Filariasis

This condition is caused by a variety of onchocercid or filarial nematodes that are commonly encountered parasites of nonhuman primates. The adult filarids are long, slender worms that inhabit various tissue sites in the host animal outside the gastrointestinal tract [290, 461, 531, 650, 742, 791, 889]. The length of the adult worm varies, depending on the species, from a few centimeters to as much as 30 cm. Female filariae are typically much larger than the males. The female worms produce small, primitive larvae called microfilariae that circulate throughout the peripheral blood or live in the skin of the definitive host [167, 168, 290, 461, 650, 742, 889]. The life cycle for these parasites is indirect. Obligatory intermediate hosts include an extensive variety of biting and blood- or lymph-sucking insects [461, 791]. The filarial worms reported from nonhuman primates are listed in Table 45.5.

Filariasis in New World Monkeys. At least 12 different species of filarid nematodes have been described from New World monkeys (marmosets, squirrel monkeys, cebus monkeys, spider monkeys, and owl monkeys). These include four species of *Dipetalonema* and seven species of *Tetrapetalonema* [52, 148, 168, 184, 207, 210, 230, 231, 290, 310, 311, 461, 508, 580, 742, 791, 845, 889, 923]. Mixed infections in the same animal are reported to be very common, with some animals containing as many as four different species at the same time [461]. These species live in the abdominal or thoracic cavities or in the subcutaneous tissues of the definitive host. The worms that locate in the subcutaneous tissues cause very little if any inflammatory response [461, 650]. Those species that are found in the serous cavities (*D. gracile* and *D. caudispina*) can cause a fibrinopurulent peritonitis or pleuritis with associated fibrinous adhesions that frequently results in entrapment of the worms [461, 650].

The filarid *Loa loa* has been reported from a spider monkey [872].

Filariasis in Old World Monkeys. *Dirofilaria corynodes* is reported to be the most prevalent filarial parasite of African Old World monkeys (vervets, mangabeys, colobus monkeys, and patas monkeys). These are large parasites that are found in the subcutaneous tissues of the trunk and lower extremities where their presence causes very little tissue reaction [650]. Two closely related species, *D. magnilarvatum* and *D. macacae*, have been reported from Asian Old World monkeys (cynomolgus monkeys) [650].

Macacanema formosana has been reported from Asian Old World monkeys (Taiwan macaque and cynomolgus monkey). This parasite commonly inhabits the peritracheal connective tissue and the diaphragm of the infected host [43, 290, 531, 650, 768].

Edesonfilaria malayensis has been described in Old World monkeys (cynomolgus monkeys and rhesus monkey) [290, 308, 531, 635, 712, 920, 924]. The adult worms usually are found free in the peritoneal cavity but have been reported from the subserosal connective tissue of the abdominal and thoracic cavities. In one report, they were associated with retroperitoneal masses composed of fibrous connective tissue and multiple foci of lymphoplasmocytic infiltrates. Numerous migratory tracts containing amorphous eosinophilic debris or adult *E. malayensis* worms were scattered throughout the masses [308]. In another report [635], six female adult *Macaca fascicularis* monkeys were found to be infected with *E. malayensis*. Clinical pathological findings in the infected animals included reduced values of hemoglobin and hematocrit, eosinophilia, elevated level of total protein, and a decreased A/G ratio. Gross lesions consisted of thickening of the connective tissues, hemorrhage, and adhesions of the serosa in the site occupied by the worms. Mechanical damage was seen occasionally in tissues adjacent to the location of the parasites such as the pancreas and iliopsoas muscle. Splenic nodules were seen in five of the six infected monkeys. Histopathologically, there was hemorrhage, fibroplasia of connective tissue, and proliferation of granulation tissue with infiltration of eosinophils, lymphocytes, and other inflammatory cells associated with the presence of worms in the tissues. The nodular lesions in the spleen consisted of a highly vascular network of large reticuloendothelial cells, reticulum fibers, eosinophils, and erythrocytes. Microfilariae were present in some of these lesions, and it was felt that the splenic nodules were most likely associated with their existence in the spleen [635].

Loa loa, normally a parasite of humans, has been reported from a variety of Old World monkeys (drills, baboons, mangabeys, and vervets) [199, 650, 755, 852]. Except for size, the worms described from both human and nonhuman primates are nearly identical morphologically. Another variation is the different circadian rhythm displayed by the microfilariae produced by the worms that infect nonhuman primates. These larvae circulate in the peripheral blood with a nocturnal periodicity [650]. Infection is usually asymptomatic, and significant lesions related to the presence of the adult *L. loa* in the subcutaneous tissues of nonhuman primate hosts has not been reported [650]. However, there has been a report of splenic lesions in drills infected with *L. loa*. Grossly, there were multiple nodules over the surface of the spleen resulting from the presence of granulomas that arose in the red pulp. Microscopically, the nodules were composed of fibrous connective tissue and numerous multinucleated giant cells, many of which contain disintegrating microfilariae within their cytoplasm. These lesions were attributed to the destruction of microfilariae within the spleen [198].

Brugia malayi and *Brugia pahangi* have been reported from a wide variety of Asian monkeys, particularly *Macaca* species [87, 500, 650]. *Brugia malayi* is also a parasite of humans. The adult parasites are found in the lymphatic and perilymphatic tissues of their nonhuman primate hosts. Symptoms and histopathology in the lymphatic system similar to that seen in human Malayan filariasis have not been reported in infected nonhuman primates [650]. Another species, *Brugia tupaidae*, has been described from the lymphatic system of prosimians (tree shrews) [650].

Meningonema peruzzii is a relatively recently reported filarid parasite from African Old World monkeys (vervets and talapoin monkeys) [648]. These worms were found only in the subarachnoid space along the dorsum of the brain stem at the level of the medulla oblongata. Female *M. peruzzii*, unlike most other filariae, are quadridelphic. Symptoms and lesions associated with infection by this parasite were not reported [648].

Filariasis in Great Apes. *Onchocerca volvulus*, a parasite of humans, has been reported from the gorilla [42, 650]. The parasite was located in a subcutaneous fibrous nodule morphologically similar to that formed by the parasite in the human host [42, 650].

Dipetalonema streptocerca and *D. rodhaini* are two filarid parasites reported from the chimpanzee. These two parasites, along with *O. volvulus* from the gorilla, are different from other filarids in that the microfilariae produced by the female remain in the dermis rather than circulating in the peripheral blood [646, 650].

Several other filarids have been reported from the great apes including *Dirofilaria pongoi* from the heart of an orangutan [650, 873] and *Dirofilaria immitis* in the abdominal cavity of another orangutan [650, 760]. *Loa loa* also has been reported from the chimpanzee and gorilla [42, 734]. *Dipetalonema vanhoofi*, a filarid parasite of the chimpanzee, inhabits the mesenteries and the connective tissue adjacent to the gall bladder, bile duct, liver, pancreas, and kidney; and the loose connective tissues and lymphatics surrounding the hepatic blood vessels [650, 671]. They also have been described from the periadrenal connective tissue [646, 650].

Diagnosis is based on demonstration and identification of the adult worms in the body cavities or subcutaneous connective tissues, or the characteristic microfilariae in the blood [111, 290, 791].

Several filarial nematodes affect humans (*Dirofilaria*, *Onchocerca*, *Loa*), but the public health significance for the majority of these species is unknown [276, 290].

Trichurids

Trichuriasis

This disease is caused by parasites in the genus *Trichuris*. Trichurid parasites are common inhabitants of the cecum and large intestine of nonhuman primates [276,

290, 531, 570, 650, 742]. These nematodes have a world-wide distribution but are more prevalent in the tropics and subtropics [290]. Nonhuman primates reported to be affected include New World species (howler monkeys, woolly monkeys, and squirrel monkeys) [76, 208, 365, 492, 650, 814], Old World species (rhesus monkeys, cynomolgus monkeys, Japanese macaques, Formosan macaques, African green monkeys, and baboons) [75, 290, 339, 362, 487, 494, 609, 712, 741, 764, 836, 922], and great apes (gibbons and chimpanzees) [222, 290, 359, 570, 686, 689, 712, 843, 863, 893]. These parasites are morphologically identical to and indistinguishable from *T. trichiura* in humans [290, 570, 650, 742]. The life cycle is direct.

Trichuriasis in nonhuman primates usually does not cause any significant clinical problems [570]. Light infections are reported to cause no apparent lesions; heavy infections, however, have been reported to cause anorexia, a gray mucoid diarrhea, and sometimes death [290, 339, 742, 843]. Fatal whipworm infections have been reported in two chimpanzees and a gibbon. Death of one chimpanzee was attributed to a severe parasitic enteritis; the second death was thought to be the result of a secondary bacterial infection resulting from the *Trichuris* infection [843, 893]. The death of a gibbon with chronic colitis caused by an overwhelming infection with *Trichuris* and oxyurid parasites also has been reported [359].

Diagnosis depends upon the identification of the characteristic double operculated eggs in the feces or adults in the cecum [290].

Because the trichurid species that affects nonhuman primates is morphologically similar to the whipworm found in humans, cross-infection from animal to human is possible [290, 570, 742]. Appropriate care in the handling and management of infected captive nonhuman primates is recommended.

Capillariasis

This disease results from infection with the cosmopolitan trichurid parasite *Capillaria hepatica*. It has been reported in the liver of a wide variety of mammalian hosts throughout the world including New World monkeys (squirrel monkeys, cebus monkeys, and spider monkeys), Old World monkeys (rhesus monkeys), and great apes (chimpanzee) [276, 290, 349, 350, 420, 507, 531, 742, 791, 810, 854, 855].

The anterior portion of these parasites is more slender than the posterior, but it is not as pronounced as in the whipworms. The eggs have bipolar opercula, and the shell contains many small perforations giving it a striated appearance. This feature is unique and is used to distinguish ova of *C. hepatica* from those of other trichurids [290, 650, 791].

The life cycle is direct and unique. Adult worms are found only in the hepatic parenchyma, and eggs are retained within the liver until the host dies or is killed. The eggs must be liberated from the liver either by decomposition of the original host or by passage through a predator or scavenger. Ingestion of infected liver tissue produces only spurious passage of the eggs in the feces. To become infective, the eggs must undergo embryonation under aerobic conditions. Infection occurs when embryonated eggs are ingested [290, 350, 650, 791].

The liver of infected animals reveals randomly placed white or yellow patches or nodules over the surface. Histopathologically, these foci are composed of adult *C. hepatica* and masses of eggs that are surrounded and infiltrated by proliferating fibrous connective tissue, chronic inflammatory cells, and foreign body giant cells. These lesions are ultimately converted to scar tissue, and the liver becomes cirrhotic. Fatal hepatitis has been reported in infected nonhuman primates [280, 290, 650, 742, 791].

Neither eggs or adult parasites will be found in the feces; therefore, diagnosis depends on demonstration and identification of the typical eggs and/or worms through liver biopsy or at necropsy [290, 791].

This parasite is pathogenic for humans, but because of the unusual life cycle of *C. hepatica*, infective nonhuman primates do not constitute a public health menace for persons caring for or working with them [276, 290].

Anatrichosomatids

Anatrichosomiasis

This condition is the result of infection with the anatrichosomatid parasites *Anatrichosoma cutaneum* or *Anatrichosoma cynomolgi*. These two species have been described from both Asian and African Old World nonhuman primates (rhesus monkeys, cynomolgus monkeys, patas monkeys, vervets, talapoin monkeys, mangabeys, and baboons) [8, 129, 142, 276, 290, 446, 585, 644, 712, 742, 805, 833] and great apes (gibbons) [74].

The adult worms are small and slender. The eggs are large, barrel shaped, have bipolar opercula, and unlike *Trichuris* and *Capillaria* contain a larva [290, 650].

The entire life cycle and method of transmission are not known, but the cycle is thought probably to be direct. The female worms migrate through the stratified layers of squamous epithelium forming tunnels in which the embryonated eggs are deposited [8, 290, 644, 650]. These tunnels are composed of epithelial cells and maintain their integrity. They are sloughed with the superficial keratin layers of the squamous epithelium and accumulate on the mucosal surface of the nares. Eggs are excreted from the host in the nasal secretions and less often in the feces [8, 290, 644, 650].

The original report of this parasite in nonhuman primates was from skin lesions on the extremities. Grossly, these lesions had the appearance of white, serpentine tracks on the palms and/or soles of the hands and feet [650, 833]. Since then they have been reported only from the stratified squamous epithelium of the external nares. Infection of the nares does not produce serious disease and is usually subclinical but is considered to be common in susceptible animals. Histopathologically, the affected epithelium is diffusely hyperplastic and parakeratotic, and there is a mild inflammatory infiltrate composed of leukocytes and plasma cells in the underlying lamina propria [8, 290, 644, 650, 791].

Diagnosis in the living animal can be made through the use of nasal mucosal scrapings or swabs that will reveal the characteristic eggs. In the dead animal, finding of the parasite in microscopic slides of the mucosa is considered to be diagnostic [8, 290, 650, 791].

This parasite has been reported in humans, where it causes a type of creeping eruption. Even though infection in humans is considered to be uncommon, those personnel who work with and care for nonhuman primates should handle those species known to be infected or susceptible to infection with proper caution [276, 290, 650].

A new species, *Anatrichosoma ocularis*, has been reported recently from the eye of a tree shrew (*Tupaia glis*) [282].

Trematodes

The parasitic trematodes described from nonhuman primates are listed in Table 45.6. The species most frequently mentioned in the literature are discussed below.

TREMATODIASIS

This disease in nonhuman primates can be caused by infection with a number of species of trematodes. Several of the more commonly encountered species will be discussed in detail.

GASTRODISCOIDIASIS

Gastrodiscoides hominis is a small, orange-red fluke that attaches to the mucosa of the cecum and colon [290, 339, 380, 791]. The parasite is distributed throughout the tropical orient and has been described in various *Macaca* species that range throughout this geographic area [276, 290, 339, 362, 380, 403, 570, 695, 712, 764, 791, 903, 940]. The life cycle is indirect with a snail serving as the intermediate host [276, 290, 791]. Diagnosis can be made by identifying the characteristic eggs in the feces or by finding the typical adult flukes in the lumen of the cecum or colon at necropsy [290, 791].

Infection usually is asymptomatic when the parasites are present in small numbers. Heavy infections produce a mucoid diarrhea and mild chronic colitis. Attachment of the flukes to the intestinal mucosa results in focal lesions characterized by hyperemia, loss of surface epithelium, and necrosis. Neutrophilic infiltrates may be associated with these lesions. The submucosa may be sclerotic because of proliferation of fibrous connective tissue and a lymphoplasmacytic cell infiltrate [276, 290, 296, 380, 403, 791].

This parasite has been reported to cause a mild diarrhea in humans, but because of the obligatory snail intermediate host in the life cycle, infected captive monkeys are not a direct health hazard for humans [276, 290].

WATSONIASIS

Watsonius watsoni, *W. deschieni*, and *W. macaci* have been reported to inhabit the intestinal tract of several Old World primate species (guenons, baboons, and cynomolgus monkeys) [276, 280, 290, 339, 362, 609, 742].

Adult trematodes of this genus are translucent, orange, and pear-shaped. The complete life cycle is not known but probably involves a snail intermediate host and is thought to be similar to that of *Fasciola hepatica* [276, 290, 742].

Watsonius watsoni and *W. deschieni* have been reported to be associated with diarrhea, severe enteritis, and death in monkeys [290, 339, 742]. Little else is known about the anatomic effects of these species [290].

Diagnosis can be made from the characteristic eggs in the feces or adults in the intestine at necropsy [290].

The public health considerations for these flukes are the same as described for *G. hominis*.

TABLE 45.6. Parasitic trematodes described from nonhuman primates.

Parasite genus-species	Location in host	Pro-simians	New World monkeys	Old World monkeys	Great apes	References
Plagiorchids						
Plagiorchis multiglandularis	Intestine			X		482
Lecithodendriids						
Novetrema nycticebi	Intestine	X				482
Odeningotrema apidion	Intestine	X				482
Odeningotrema bivesicularis	Intestine	X				482
Phaneropsolus bonnei	Intestine	X		X		482
Phaneropsolus lakdivensis	Intestine	X				147, 482
Phaneropsolus longipenis	Intestine	X			X	147, 482
Phaneropsolus orbicularis	Intestine		X			147, 148, 290, 461, 482, 689, 841, 917
Phaneropsolus oviformis	Intestine	X		X		147, 290, 482
Phaneropsolus simiae	Intestine			X		482
Phaneropsolus aspinosus	Intestine			X		657
Primatotrema macacae	Intestine			X		147, 290, 482
Dicrocoeliids						
Athesmia foxi	Bile ducts		X			26, 59, 147, 235, 275, 280, 290, 311, 333, 461, 482, 689, 742
Athesmia heterolecithodes	Bile ducts		X			482
Brodenia laciniata	Bile ducts, pancreas			X		147, 290, 482, 486, 487
Brodenia serrata	Pancreas			X		147, 290, 482
Concinnum brumpti syn. (*Eurytrema brumpti*)	Bile duct, pancreas				X	147, 280, 290, 482, 694, 701, 791, 828, 922
Controrchis biliophilus	Gall bladder, bile ducts		X			147, 461, 482
Dicrocoelium colobusicola	Bile ducts			X		147, 482
Dicrocoelium lanceatum	Bile ducts			X	X	147, 367, 482, 922
Dicrocoelium macaci	Bile ducts			X	X	147, 280, 290, 356
Euparadistomum cercopitheci	Gall bladder			X		482

TABLE 45.6. (Continued).

Parasite genus-species	Location in host	Pro-simians	New World monkeys	Old World monkeys	Great apes	References
Eurytrema pancreaticum	Pancreatic ducts			X		482
Eurytrema satoi	Bile ducts, pancreas			X	X	147, 280, 290, 482
Leipertrema rewelli	Pancreas				X	147, 290, 482, 760
Platynosomum amazonensis	Gall bladder, bile ducts		X			147, 148, 290, 461, 482
Platynosomum marmoseti	Gall bladder, bile ducts		X			147, 148, 290, 461, 482
Zonorchis goliath	Bile ducts		X			461, 482
Zonorchis microcebi	Bile ducts		X			482
Fasciolids						
Fasciola hepatica	Liver			X		147, 290, 339, 362, 482
Fasciolopsis buski	Duodenum, stomach			X		147, 276, 290, 361, 482
Opisthorchids						
Chonorchis sinensis	Bile ducts			X		482, 490
Opisthorchis felineus	Bile and pancreatic ducts			X		482
Heterophyids						
Haplorchis pumilio	Intestine			X		482
Haplorchis yokogawai	Intestine			X		147, 482
Metagonimus yokogawai	Intestine			X		147, 482
Pygidiopsis summa	Intestine			X		482
Microphallidids						
Spelotrema brevicaeca	Intestine			X		482
Echinostomatids						
Artyfechinostomum sp.	Intestine			X		147, 290, 482
Echinostoma aphylactum	Sm. intestine		X			147, 461, 482, 841
Echinostoma ilocanum	Intestine			X		48, 147, 276, 290, 482
Reptiliotrema primata	Intestine			X		147, 290, 482
Notocotylids						
Ogmocotyle ailuri	Sm. intestine			X		482
Ogmocotyle indica	Sm. intestine			X		147, 290, 339, 482, 490, 927

(Continued)

TABLE 45.6. (Continued).

Parasite genus-species	Location in host	Pro-simians	New World monkeys	Old World monkeys	Great apes	References
Troglotrematids						
Paragonimus westermani	Lungs, pleural cavity, diaphragm, body cavity, brain			X		147, 276, 288, 290, 362, 606, 742, 761, 810
Paragonimus africanus	Lungs			X		744
Achillurbainia sp.	Parotid gld.	X				655
Schistosomatids						
Schistosoma bovis	Mesenteric and abdominal veins			X		482
Schistosoma haematobium	Mesenteric, visceral, and abdominal veins			X	X	147, 170, 227, 276, 290, 482, 625, 627, 698
Schistosoma japonicum	Mesenteric and portal veins			X	X	147, 401, 482
Schistosoma mansoni	Mesenteric and abdominal veins		X	X	X	124, 147, 227, 276, 278, 280, 290, 367, 402, 482, 486, 487, 498, 507, 587, 698, 717, 822, 832
Schistosoma mattheei	Mesenteric and abdominal veins			X		49, 276, 482, 578, 609, 823
Schistosoma sp.	Mesenteric and abdominal veins			X	X	482
Diplostomatids						
Diplostomid mesocercariae	Visceral and pulmonary cysts		X	X		482
Neodiplostomum tamarini	Intestine		X			147, 148, 193, 290, 461, 482
Paramphistomids						
Chiorchis noci	Intestine			X		147, 290, 482
Gastrodiscoides hominis	Cecum, colon			X		147, 276, 290, 339, 362, 380, 403, 482, 660, 662, 658, 695, 712, 764, 791, 904
Watsonius deschiensi	Intestine			X		147, 276, 290, 482, 609, 742, 791
Watsonius watsoni	Intestine			X		147, 276, 280, 290, 339, 362, 482, 609, 742, 791
Watsonius macaci	Intestine			X		147, 276, 290, 362, 482, 791

Paragonimiasis

This disease, caused by the oriental lung fluke *Paragonimus westermanii*, has been reported in the cynomolgus monkey. Infection is directly associated with this animal's ingestion of infected raw crabs or crayfish as part of its dietary regimen [276, 288, 290, 362, 606, 742, 761, 810].

The adult flukes have a brown, plump, ovoid body with scalelike spines. The eggs are oval shaped, golden-brown in color, and have a partly flattened operculum at one end. The life cycle is indirect with snails and crabs or crayfish serving as intermediate hosts [276, 290, 791].

Adult flukes are found primarily in the lung but sometimes occur in ectopic sites such as the brain, liver, and other organs. The clinical signs reported in infected animals include coughing, wheezing, bloody or rusty-tinged sputum, moist rales, and progressive emaciation [290, 791].

At necropsy, lesions consist of focal areas of emphysema and soft, dark red to brown cysts that measure 2–3 cm in diameter and are randomly located throughout the pulmonary parenchyma. These cysts may be elevated above the lung surface, and pleural adhesions can sometimes be present. Two or more flukes occupy each cyst [290, 482, 791].

Histopathologically, the presence of the flukes provokes a leukocytic infiltration, and there is usually a mature fibrous capsule around the parasites that in turn are surrounded by a purulent exudate containing blood and groups of typical-appearing fluke eggs. Hemorrhage into the cyst often occurs, and this may lead to hemoptysis. Additional lesions described include hyperplasia of bronchial epithelium and submucosal glands and focal areas of inflammation in the lung parenchyma associated with groups of fluke eggs [290, 482, 791].

The diagnosis depends on the demonstration and identification of the typical eggs in the feces or the adult flukes in the pulmonary tissue at necropsy [290, 482, 791].

Paragonimus westermanii can affect humans; however, because of the obligatory molluscan and crustacean intermediate hosts in the life cycle, infected captive nonhuman primates are not a direct health hazard for humans [290, 482].

Schistosomiasis

Several species of schistosomatid flukes have been reported to infect nonhuman primates naturally. These include *Schistosoma mansoni* in New World monkeys (squirrel monkeys) [482, 832], Old World monkeys (mangabeys, patas monkeys, guenons, and baboons) [124, 227, 276, 278, 280, 290, 482, 487, 498, 587, 698, 791, 822], and the great apes (chimpanzees) [367, 402, 482, 717, 791]; *Schistosoma haematobium* in Old World monkeys (mangabeys, guenons, and baboons) [227, 276, 290, 482, 625, 627, 698, 791] and the great apes (chimpanzees) [170, 290, 482]; and *Schistosoma mattheei* in Old World monkeys (baboons) [49, 276, 482, 578, 609, 791, 823]. Although schistosomatids are considered to be extremely serious pathogens for humans, they are of little consequence in captive nonhuman primates and are usually found incidentally at necropsy [290, 487].

In the schistosomatids, both male and female forms are present and differ in appearance. They are usually found together in constant copulation with the long, slender female in the sex canal of the short, muscular male. The egg of *S. mansoni* is elongated-ovoid in shape, rounded at both ends, and bears a lateral spine. The egg of *S. haematobium* is also elongated-ovoid in shape, rounded at the anterior end, and bears a posterior terminal spine. Adult *Schistosoma mansoni* and *Schistosoma mattheei* inhabit the mesenteric veins, whereas *Schistosoma haematobium* adults are found in the pelvic or portal veins of susceptible hosts [290, 791].

The life cycle is indirect with snails serving as intermediate hosts [290, 791].

The reported clinical signs include pyrexia, hemorrhagic diarrhea or hematuria, and ascites [276, 290, 482, 791]. The principal pathologic effects are caused by the presence of eggs in the tissues [276, 290, 428, 482]. The eggs may be found almost anywhere in the abdominal or pleural cavities. The most frequently encountered lesion is thickening of the intestinal or urinary bladder walls caused by chronic inflammation. Microgranulomas surrounding typical schistosome eggs are also very common in the liver, brain, spleen, wall of the gastrointestinal tract and urinary bladder, and other organs. Continued insult can lead to stenosis of portions of the gastrointestinal tract, urinary bladder and other parts of the urogenital system, and cirrhosis of the liver [171, 276, 290, 428, 482, 698].

Diagnosis is made based on the finding and identification of the characteristic eggs in the feces or urine, the presence of adult schistosomes in the blood vessels at necropsy, or the finding of the typical lesions during histopathological examination of appropriate tissues [290, 482, 791].

Infected captive nonhuman primates are not of direct public health significant to humans because of the requirement for an obligatory molluscan intermediate host. However, because schistosomiasis is such an important and serious disease in humans, excreta from nonhuman primates should be decontaminated before disposal [290].

Athesmiasis

The cause of this disease, *Athesmia foxi*, is considered to be a moderately pathogenic fluke that inhabits the bile ducts of susceptible nonhuman primate species. It is a common finding in nonhuman primates obtained from South America and has been reported in a variety of New World monkeys (cebus monkeys, squirrel monkeys, tamarins, and titi monkeys) [59, 147, 235, 275, 280, 290, 311, 333, 461, 689, 742, 824].

The adult flukes are long and slender and measure 8.5×0.7 mm. Eggs are ovoid and golden-brown in color; they have a thick shell and are operculated [290, 791, 824].

The life cycle is indirect with a mollusk serving as a required intermediate host. However, because the method of infection of the vertebrate host is unknown, our knowledge about the life cycle of this particular fluke is incomplete [290, 791].

Infections in nonhuman primates are usually asymptomatic and most often considered an incidental finding. Aside from causing a moderate to marked distension of affected ducts, these parasites cause very little damage and do not invoke much of a host inflammatory response. Heavy infections can result in hyperplasia of the biliary epithelium and fibroplasia around eggs and the ducts. Extremely severe infections can result in a pronounced thickening of the bile ducts with resultant pressure and trauma to adjacent hepatic parenchyma leading to fatty degeneration of affected hepatocytes [59, 235, 290, 311, 461, 481, 742, 791, 824].

Diagnosis depends on the demonstration and identification of the adult flukes in the bile duct either at necropsy or on histopathological examination of liver sections, or by demonstration and identification of the characteristic eggs in the feces [290, 791].

Nonhuman primates infected with *A. foxi* do not pose any public health problems for humans. This parasite has not been reported from humans. Infected nonhuman primates are not a direct hazard to humans because of the need of a required molluscan intermediate host to complete the life cycle [280, 290].

Cestodes

Cestodiasis

This condition results from infection by one of any of the numerous tapeworm genera that have been described in the intestinal tract of nonhuman primates, including prosimians, New and Old World monkeys, and great apes [151, 202, 204, 222, 280, 290, 362, 365, 461, 486, 487, 578, 608, 654, 687, 712, 771, 774, 825, 826, 832, 836, 862, 908, 918]. Cestode genera and the primate group they parasitize are listed in Table 45.7. Life cycles for all the genera listed, except one, are indirect and require an arthropod intermediate host for completion of the cycle. *Hymenolepis nana* can complete its life cycle either through direct or indirect means [290, 461].

Although these parasites may be present in large numbers, clinical disease or enteric lesions are seldom associated with tapeworm infection. Diagnosis depends on the identification of characteristic eggs in the feces, passing of proglottids of adult worms, or the recovery of adult worms at necropsy [290].

Some tapeworm genera (*Hymenolepis*, *Raillietina*, *Bertiella*) rarely affect humans. Proper precautions in handling captive nonhuman primates, good personal hygiene by the caretakers, and care in disposing of bedding and feces of infected animals should be stressed in order to rule out accidental transfer of infection to humans [290].

Larval Cestodiasis

Nonhuman primates may serve as intermediate hosts for several species of tapeworm parasites and thus develop various larval forms of these parasites in their somatic tissues. The larval cestode species and the primate group they parasitize are listed in Table 45.7. The cestode larvae are classified as solid and bladder forms. The solid larvae are represented by the sparganum. The bladder larvae consist of cysticercus, coenurus, hydatid, and tetrathyridium.

Sparganosis

This term denotes infection with the elongate, nonspecific plerocercoid larvae of cestodes in the order Pseudophylloidea [461, 791]. The adult tapeworms belong to the genera *Diphylobothrium* and *Spirometra*, which are intestinal parasites of various carnivores, birds, and reptiles [608].

Spargana have been described in New World monkeys (squirrel monkeys and marmosets) [148, 204, 207, 290, 461, 608], Old World monkeys (rhesus monkeys, cynomolgus monkeys, vervets, baboons, and talapoin monkeys) [290, 495, 600, 609, 610], and prosimians (tree shrew) [771]. These larvae are solid with a

TABLE 45.7. Parasitic cestodes described from nonhuman primates.

Parasite genus-species	Location in host	Pro-simians	New World monkeys	Old World monkeys	Great apes	References
Anoplocephalids						
Bertiella studeri	Sm. intestine			X	X	32, 276, 290, 362, 461, 485, 486, 487, 494, 578, 608, 712, 791, 825, 836, 862, 908, 922
Bertiella mucronata	Sm. intestine		X		X	32, 151, 204, 280, 290, 608, 682, 689
Bertiella fallax	Sm. intestine		X			204, 280, 290, 608
Bertiella satyri	Sm. intestine		X	X		113
Bertiella okabei	Sm. intestine			X		765
Bertiella sp.	Sm. intestine			X	X	151, 480, 491, 570, 616, 654, 660, 764
Anaplocephala sp.	Sm. intestine				X	608, 753
Parabertiella sp.	Sm. intestine			X		3, 658, 660, 662
Moniezia rugosa	Sm. intestine		X			204, 280, 290, 461, 608
Thysanotaenia sp.	Sm. intestine	X				608, 918
Tupaiataenia quentini	Sm. intestine	X				771
Intermicapsifer sp.	Sm. intestine			X		539, 608
Atriotaenia megastoma	Sm. intestine	X	X			148, 168, 204, 290, 311, 461, 608, 623
Matheovataenia brasiliensis	Sm. intestine		X			478
Matheovataenia curzsilvai	Sm. intestine			X		582
Matheovataenia sp.	Sm. intestine		X			31, 461, 608
Davaineids						
Rallietina alouattae	Sm. intestine		X			204, 290, 608, 742
Rallietina demerariensis	Sm. intestine		X			32, 204, 290, 482, 841
Rallietina sp.	Sm. intestine	X	X			168, 290, 461, 608, 623
Paratriotaeniids						
Paratriotaenia oedipomidatus	Sm. intestine		X			207, 290, 461, 608, 689, 826
Dilepidids						
Dilepis sp.	Sm. intestine			X		608
Choanotaenia infundibulum	Sm. intestine			X		425

TABLE 45.7. (Continued).

Parasite genus-species	Location in host	Pro-simians	New World monkeys	Old World monkeys	Great apes	References
Hymenolepidids						
Hymenolepis nana	Sm. intestine		X	X	X	38, 276, 290, 349, 461, 584, 608, 742, 764, 791, 808, 880
Hymenolepis diminuta	Sm. intestine	X		X		276, 290, 349, 461, 608, 742, 764, 771, 791, 880
Hymenolepis cebidarum	Sm. intestine		X			204, 290, 623
Hymenolepis sp.	Sm. intestine	X			X	61, 387, 570, 764, 922
Vampirolepis sp.	Sm. intestine		X			461, 608
Mesocestoidids						
Mesocestoides sp. (*Tetrathyridium*)	Larva: peritoneal cavity			X	X	245, 290, 339, 608, 609, 712, 715, 751, 911
Taeniids						
Taenia crocutae (*Cysticercus*)	Larva: skeletal muscle			X		578
Taenia hydatigena (*Cysticercus tenuicollis*)	Larva: liver, peritoneal cavity			X		32, 280, 290, 339, 488, 506, 507, 608, 609, 810
Taenia solium (*Cysticercus cellulosae*)	Larva: brain, heart, muscle, subcutis			X	X	32, 280, 290, 507, 608, 609, 742, 745, 870, 876
Multiceps serialis (*Coenurus serialis*)	Larva: subcutis			X		32, 226, 290, 742
Multiceps brauni (*coenurus*)	Larva: subcutis, pleural and abdominal cavities, brain			X		32, 248, 290
Echinococcus granulosus (Hydatid cyst)	Larva: liver, lungs, peritoneal cavity	X	X	X	X	7, 32, 44, 55, 152, 179, 224, 276, 290, 309, 366, 397, 398, 408, 461, 501, 614, 615, 640, 663, 664, 691, 780, 810, 830
Diphyllobothriids						
Diphyllobothrium erinacei (sparganum)	Larva: subcutis, muscle		X	X		32, 148, 204, 280, 290, 509, 609, 742, 810

(Continued)

TABLE 45.7. (Continued).

Parasite genus-species	Location in host	Pro-simians	New World monkeys	Old World monkeys	Great apes	References
Spriometra reptans (sparganum)	Larva: subcutis		X			204, 207, 290
Spirometra sp. (sparganum)	Larva: abdominal cavity, subcutis, muscle	X		X		61, 290, 495, 578, 600, 610, 771, 911

scolex that contains a pseudosucker. They are white, ribbonlike, and of variable size and motility. They resemble the adult except they lack proglotids and mature genitalia. Spargana can vary from a few millimeters to several centimeters in length [608, 791, 847]. In nonhuman primates, spargana may be found in any part of the body: in retroperitoneal tissues, in abdominal or pleural cavities, or in subcutaneous and muscular tissues. They are commonly encased by a connective tissue capsule, and they do not incite much of an inflammatory response unless they die. These degenerating larvae may cause local inflammation and edema. Most infections in nonhuman primates are usually asymptomatic, and their presence is considered to be an incidental finding at necropsy [608, 791].

Diagnosis can be made in the live animal by radiography, which may reveal calcified nodules. Also, one may palpate mobile nodules in the subcutaneous tissue with localized edema. In the dead animal, diagnosis is made through demonstration and identification of the characteristic spargana larvae either grossly at necropsy or microscopically in histopathologic specimens [608, 791].

Cystricercosis

This condition is the result of infection with the larval form of various members of the family Taeniidae. Adult tapeworms of this family commonly parasitize birds and mammals [461, 608]. Cystericerci have been described in New World monkeys (squirrel monkeys and marmosets) [461], Old World monkeys (rhesus monkeys, baboons, mangabeys, patas monkeys, langurs, and vervets) [280, 290, 339, 608, 609, 742, 870, 876], great apes (gibbons and chimpanzees) [608, 745, 922], and prosimians (lemur) [86, 384, 608]. Cysticerci are oval, translucent cysts that contain a single invaginated scolex with four suckers. In those species that have them, a circle of hooks is present [461, 608, 847]. These cysts may be found in the abdominal or thoracic cavities, muscle, subcutaneous tissue, and central nervous system. Usually there is very little host inflammatory reaction to the presence of viable cysts. As the cysts enlarge, there may be compression of adjacent tissues. Dead cysts will provoke an intense chronic inflammatory reaction [461, 608, 870].

Symptoms in nonhuman primates are directly related to the tissue in which the cysticercus develops and the number present [608]. Involvement of the central nervous system can produce neurological disorders, but this appears to be less of a problem in infected nonhuman primates than in cerebral cysticercosis in the human patient [608, 870, 876].

Diagnosis depends upon the finding of the characteristic bladder-shaped structure in the tissues. Identification of the specific species involved is based on the characteristic hook size and structure [608].

Coenurosis

This condition is the result of infection with the larval form of the tapeworms *Multiceps multiceps* or *Multiceps serialis*, which are intestinal cestodes of dogs and related carnivores [290, 461]. Coenurosis has been reported in Old World monkeys (macaques, vervets, gelada baboon, and other baboons) [133, 226, 248, 290, 489, 501, 608, 700, 735, 742, 757, 860] and prosimians (lemur) [608].

The coenurus is a polycephalid larval form that produces both internal and external daughter cysts. The inner layer of the cyst wall is composed of germinal epithelium from which numerous scolices develop [290, 608, 847].

Coenuri have been described in the subcutaneous tissues, peritoneal cavity, liver, brain, and other organs of affected nonhuman primates [290, 461, 608, 757]. Clinical signs and histopathology depend on the number of coenuri present and their location. In general, infection in nonhuman primates has produced minimal symptoms and lesions [226, 501, 608, 700, 735, 860]. However, in those cases where there is involvement of the central nervous system, typical neurological symptoms are observed [608, 860].

Diagnosis can be made by radiography or the finding of a tumorlike mass in the subcutaneous tissues. Identification of the species of cestode is based on the hook structure of the scolex [290, 608].

Hydatidosis

This disease, also known as echinococcosis, is the result of infection by the larval stage of cestode parasites in the genus *Echinococcus*. The adult tapeworms are found in the intestinal tract of dogs, wolves, bush dogs, other members of the canine family, and related carnivores [290, 461, 608]. Hydatid cysts caused by *E. granulosus* have been described from a number of Old World monkeys (guenons, colobus monkey, mangabeys, mandrills, rhesus monkeys, other macaques, Celebes ape, and baboons) [7, 55, 152, 179, 200, 224, 290, 366, 397, 408, 501, 608, 614, 615, 663, 664, 691, 830], New World monkeys (marmoset) [461], great apes (chimpanzee, gorilla, and orangutan) [44, 309, 640], and prosimians (galago and lemurs) [290, 663, 780]. Recently, hydatid cysts from the tapeworm *E. vogeli* have been reported from a group of young great apes (gorillas, orangutans, and chimpanzees) [398].

Hydatid cysts are large, unilocular cysts. The inner layer of the cyst wall is composed of germinal epithelium from which numerous brood capsules develop. Multiple scolices then develop from the wall of the brood capsule. The cyst wall of *E. granulosa* is characteristically laminated and composed of a thick hyaline material [290, 461, 791, 847].

Hydatid cysts may be located in the abdominal cavity, liver, lungs, subcutis, or throughout the body [152, 179, 200, 290, 408, 461, 608, 615, 691, 791]. The size of the cyst and the amount of involvement and host reaction depends on its age and the location within the host. Abdominal distension or localized subcutaneous swellings are sometimes seen, but usually the presence of the cysts causes no clinical signs or ill effects and they are found incidentally at necropsy [152, 290, 366, 507, 608, 791, 830].

The gross appearance of the cyst is that of a varying sized, spherical mass, usually in the liver, but it may sometimes be embedded in the lungs or be free in the abdominal cavity. Rupture of pulmonary hydatid cysts and resulting anaphylactic shock have been suggested as the cause of death in several cases of echinococcosis in nonhuman primates [7, 608, 705]. Free scolices from ruptured cysts can implant in other tissues and produce additional cysts [608].

The diagosis of hydatidosis is usually not made until after the cyst reaches considerable size. Symptoms may mimic a neoplasm. Radiographs can be helpful in detecting the presence of pulmonary or calcified hepatic cysts. However, pulmonary changes can be mistaken for tuberculosis or neoplasia. Serological tests such as the Casoni intradermal skin test or tanned cell hemagglutination test are of value in the diagnosis of hydatid disease in nonhuman primates. Specific identification is based on the finding of detached scolices, or daughter cysts, in the cyst fluid. The hook is considered characteristic for the genus. If scolices are not present, the histomorphology of the cyst wall can be used as identifying criteria [152, 290, 430, 608, 791]. Abdominal ultrasonic scanning has been used successfully in diagnosing echinococcosis in gorillas [640].

Tetrathyridiosis

This condition results from infection with the larval stage of cestode parasites in the genus *Mesocestoides*. The adult tapeworms of this genus parasitize various birds and mammals [290, 608, 791]. It has been described in Old World monkeys (rhesus monkeys, guenons, cynomolgus monkeys, and baboons) [245, 290, 339, 608, 609, 712, 715] and great apes (gibbon) [751], but its occurrence in nonhuman primates is considered to be uncommon or even rare [290].

The tetrathyridial larva is flat and has an extremely contractile body. They may be confused with spargana. The anterior end is knotlike and contains an invaginated holdfast apparatus with four suckers. Length can vary from 2 to 70 mm,

TABLE 45.8. Parasitic acanthocephalans described from nonhuman primates.

Parasite genus-species	Location in host	Pro-simians	New World monkeys	Old World monkeys	Great apes	References
Moniliformis moniliformis	Sm. intestine				X	29, 290, 461, 742, 770, 791, 909, 922
Prosthenorchis elegans	Ileum, cecum, colon		X			26, 33, 115, 167, 168, 204, 219, 290, 311, 461, 552, 623, 689, 720, 742, 770, 791, 834, 841, 915
Prosthenorchis spirula	Ileum, cecum, colon	X	X			83, 115, 204, 290, 623, 689, 742, 770, 791, 841

depending on the species of cestode and species of host. The tetrathyridum is proglotted shaped in the monkey [290, 339, 608, 791].

These larvae usually are found free in the serous cavities of the body or are found encysted in various tissues. Tetrathyridum evokes little host response and is usually considered to be an incidental finding in nonhuman primates [245, 290, 339, 608, 791].

Diagnosis depends on demonstration and identification of the characteristic larval form in the body cavities or encysted in the host tissues [290, 608, 791].

Larval cestodes in nonhuman primates are of little public health importance to humans because infection can occur only by ingestion of the larval form. Of more importance to both humans and captive nonhuman primates is the possible ingestion of eggs passed by the infected definitive host. For this reason, feces from domestic and feral canids should be handled and disposed of with extreme care. Control of these parasites can only be accomplished through programs aimed at eliminating them from the definitive host [290].

Acanthocephalans

The parasitic acanthocephalans described in the alimentary tract of nonhuman primates are listed in Table 45.8. Those species most frequently encountered in nonhuman primates are discussed below.

Acanthocephaliasis

This disease in nonhuman primates is most frequently the result of infection with acanthocephalan parasites in the genus *Prosthenorchis*. These parasites are distributed throughout Central and South America and have been reported in a variety of New World monkeys. Prosimians, Old World primates, and great apes can become infected under laboratory or captive conditions [115, 204, 290, 742, 770]. The species involved are *P. elegans*, which inhabits the cecum or colon, and *P. spirula*, which favors the terminal ileum [115, 168, 204, 207, 290, 311, 461, 552, 583, 623, 632, 720, 742, 770, 791, 827, 834, 866].

The life cycle is indirect with cockroaches and beetles acting as the intermediate hosts [229, 290, 339, 461, 631, 770, 791].

Diagnosis depends upon identification of the characteristic thick-walled eggs or, more rarely, the worm itself in the feces. Conventional fecal flotation methods are ineffective as a means of demonstrating the eggs of these worms; fecal smears or sedimentation techniques must be used. At necropsy the finding of typical "thorny-headed worms" attached to the intestinal mucosa is considered diagnostic [168, 290, 461, 770, 791].

No distinctive symptoms accompany infection with acanthocephalans. Suspected cases must be confirmed by diagnostic methods. Clinical signs vary depending upon the severity of the infection. Diarrhea, anorexia, debilitation, and death all have been associated with acanthocephaliasis in New World monkeys. In cases of massive infection, there is often cachexia caused by secondary complications and perhaps pain, sometimes of sudden onset; death follows rapidly. Most often the parasite does not contribute directly to the animal's death, but rather produces lesions that allow secondary pathogens to become estab-

TABLE 45.9. Leeches described from nonhuman primates.

Parasite genus-species	Location in host	Pro-simians	New World monkeys	Old World monkeys	Great apes	References
Limnatus africana	Nasal cavities			X		125, 290, 506
Dinobdella ferox	Nasal cavities, pharynx			X		93, 276, 290, 295, 494, 696

lished, resulting in debilitation and the ultimate demise of the host [115, 168, 290, 461, 597, 718, 770, 791, 834].

Attachment of the proboscis of these parasites to the intestinal mucosa causes a pronounced, usually severe, granulomatous inflammatory response, and the nodules formed usually can be seen from the serosal surface. The proboscis often penetrates the mucosa and invades the muscular layers of the intestinal wall. If complete penetration of the intestinal wall occurs, a fatal peritonitis results. Adult parasites sometimes are found in the abdominal cavity. Severe infections can cause mechanical blockage of the intestinal tract, intussusception, or rectal prolapse. Under these circumstances, infected animals will be depressed and pass bright red blood and scanty feces [167, 168, 204, 290, 311, 461, 583, 623, 720, 770, 791, 834].

Histologically, a chronic, active inflammatory response is seen, with ulcers of the mucosa and granuloma and abscess formation in the intestinal wall associated with penetration of the proboscis and the resulting destruction of existing tissues. A focal suppurative to fibrinopurulent serositis also may be present in areas where the parasites approach penetration or actually rupture the intestinal wall [83, 290, 461, 770, 791]. A hepatic abscess and granulomatous myositis (diaphragm) associated with migration of an unidentified acanthocephalan has been reported from an adult bushbaby (*Galago crassicaudatus*) [78].

Infection with these parasites has not been reported in humans [290].

Annelida

The species of annelids that parasitize nonhuman primates are listed in Table 45.9.

DINOBDELLAIASIS

The cause of this condition is the leech, *Dinobdella ferox*, which is distributed geographically throughout southern Asia [276, 290, 820]. It is a frequent parasite of the nasal cavities of macaques that range throughout this region of the world.

Dinobdella ferox has been reported from several Old World monkey species (rhesus monkeys and Formosan macaques) [93, 290, 295, 494, 696, 820].

The life cycle of this parasite is direct. Adults are hermaphroditic and eggs are laid in cocoons that are attached to objects at the surface of a pond. After hatching, the immature leeches stay at the water's surface. Infection of the host occurs during drinking; the leech enters the body through the oral or nasal cavities, attaches to the mucosa of the upper respiratory tract, sucks blood for periods that

TABLE 45.10. Fleas described from nonhuman primates.

Parasite genus-species	Location in host	Pro-simians	New World monkeys	Old World monkeys	Great apes	References
Ctenocephalides felis	Hair, skin	X		X		281, 290, 609, 814
Tunga penetrans	Skin			X	X	280, 281, 290, 814
Pulex irritans	Hair, skin			X		290, 931
Ctenocephalides canis	Hair, skin			X		281, 814
Echidnophaga gallinacea	Skin	X				281, 814

may last a few days or many weeks, grows and matures, detaches, and drops out through the nostrils. The adult leeches are not parasitic [93, 290, 696].

Infection with a few parasites is usually asymptomatic, but heavy infection is reported to cause restlessness, epistaxis, anemia, weakness, asphyxiation, and sometimes death [290, 696, 820]. Histopathologically, the lesions are composed of a mild, focal, chronic inflammatory infiltrate, and increased mucus production involving the nasopharyngeal mucosa [290, 696, 820]. Diagnosis is based on recognizing and identifying the parasite in its typical anatomical location within the host [290, 696].

This leech presents some public health significance because it does attack humans; however, infection under laboratory conditions is improbable. Nevertheless, precautions should be taken when removing leeches from affected monkeys [290].

Arthropods: Insecta

The parasitic genera of Siphonaptera (fleas), Diptera (flies), and Mallophaga and Anoplura (lice) described from nonhuman primates are listed in Tables 45.10, 45.11, and 45.12, respectively. The most important members of these Orders will be discussed in detail below.

TABLE 45.11. Flies described from nonhuman primates.

Parasite genus-species	Location in host	Pro-simians	New World monkeys	Old World monkeys	Great apes	References
Cuterebra sp.	Skin, subcutis		X	X		281, 290, 792
Dermatobia hominis	Skin, subcutis		X			290, 584
Cordylobia anthropophaga	Skin			X		280, 281, 290
Alouattamyia sp.	Skin		X			207
Cochliomyia hominivorax	Skin			X		595

TABLE 45.12. Lice described from nonhuman primates.

Parasite genus-species	Location in host	Pro-simians	New World monkeys	Old World monkeys	Great apes	References
Anoplurans						
Pedicinus eurygaster	Hair			X		290, 479, 491
Pedicinus obtusus	Hair			X		290, 479, 491, 609
Pedicinus patas	Hair			X		290, 479
Pedicinus hamadryas	Hair			X		290, 609
Pedicinus mjobergi	Hair			X		290, 682
Pedicinus schaeffi	Hair			X	X	281, 290, 396, 460, 814
Docophthirus acionetus	Hair	X				214, 281, 396, 814
Phthiropediculus propitheci	Hair	X				281, 396, 814
Lemurphthirus galagus	Hair	X				281, 396, 814
Pediculus lobatus pseudohumanus	Hair		X			281, 396, 814
P. l. atelophilus	Hair		X			281, 396, 814
Harrisonia uncinata	Hair		X			281, 396, 814
Gliricola pintoi	Hair		X			281, 396, 814
Pediculus humanus friedenthali	Hair				X	281, 396, 814
Phthirus pubis	Hair				X	281, 396, 814
Phithirus gorillae	Hair				X	281, 396, 814
Pediculus humanus capitis	Hair		X		X	154, 281, 396, 814
Pediculus sp.	Hair		X			207, 736
Sathrax durus	Hair	X				214
Pedicinus longiceps	Hair			X		491
Mallophagans						
Trichodectes armatus	Hair		X			281, 814
Trichodectes colobi	Hair			X		281, 814
Trichodectes mjoebergi	Hair	X				281, 814
Trichodectes semiarmatus	Hair		X			281, 814
Trichodectes sp.	Hair		X			281, 814
Trichophilopterus babakotophilus	Hair	X				218, 814
Tetragynopus aotophilus	Hair		X			281, 814
Trichopilopterus ferrisi	Hair	X				281, 396
Eutrichophilus setosus	Hair			X		277, 281

TABLE 45.12. (Continued).

Parasite genus-species	Location in host	Pro-simians	New World monkeys	Old World monkeys	Great apes	References
Aotiella aotophilus	Skin		X			207, 396
Cebidicola armatus	Skin		X			207, 396
Cebidicola semiarmatus	Skin		X			207, 396

For detailed information regarding the morphology and life cycle of these parasites, the reader is referred to standard pathology or parasitology texts [32, 290, 428, 791, 811].

Flea Infestation

There is a relative paucity of information regarding the extent of flea infestation in nonhuman primates. The available reports concern fleas that, for the most part, are natural parasites of animals other than nonhuman primates (dogs, cats, and chickens). There is no suggestion about the importance of siphonapterids in nonhuman primates or of any potential role they may play in transmission of disease to humans [281].

Tunga penetrans (stick-tight, jigger, or chigoe flea) has been reported from Old World monkeys (guenons and baboons) [280, 281, 290, 791] and great apes (gorilla) [281, 814]. These parasites frequently invade the hard skin covering the ischial callosities where the female *T. penetrans* becomes firmly attached and penetrates into the epidermis that proliferates around the parasite. The implanted female fleas elicit severe irritation and pruritus, and secondary bacterial infections can occur particularly after removal of the parasite from the site of attachment [280, 281, 290, 352, 791].

Dermal Myiasis

The larvae (bots) of several species of flies in the families *Cuterebridae* and *Calliphoridae* are reported to infect nonhuman primates [281, 290].

New World monkeys (howler monkeys) are reported to be a natural host for *Cuterebra* sp. larvae [281, 290, 792]. Infection with these parasites produces dermal cysts or swellings, containing a central pore, primarily in the cervical region. A chronic inflammatory reaction occurs around these sites and a seropurulent exudate containing the dark feces of the larva may exude through the pore. Healing of these lesions is usually rapid after emergence of the larva. Secondary bacterial infections can occur, and these may be more severe than the primary infection [290].

Diagnosis depends on demonstrating and identifying the typical larvae from the characteristic dermal cysts [290].

Although these flies affect humans in the geographical locations where they normally occur, captive nonhuman primates are not considered to be a direct human public health hazard [290].

Pediculosis

Mallophaga

The mallophagans, or biting lice, are reported to be relatively rare on nonhuman primates and apparently are unimportant in regard on zoonoses [281]. Species of biting lice have been reported from prosimians (loris, indri, and mongoose lemur), New World monkeys (woolly spider monkeys, howler monkeys, and owl monkey), and Old World monkeys (colobus monkeys) [281, 814]. There is also a single report of infestation of rhesus monkeys with *Eutrichophilus setosus*, the porcupine-biting louse. These monkeys were housed in close proximity to a cage of porcupines, resulting in cross-infestation [277, 281]. There are no reports of Mallophaga infestations involving the great apes or humans.

Anoplura

Numerous species of anoplurans, or sucking lice, have been reported from a wide variety of nonhuman primates including prosimians (tree shrew, lemurs, and galagos), Old World monkeys (macaques, langurs, green monkeys, guenons, baboons, and colobus monkeys), New World monkeys (sakis, uakaris, howler monkeys, spider monkeys, marmosets, and tamarins), and great apes (gibbons, siamangs, chimpanzees, and gorilla) [234, 281, 396, 441, 814, 904]. There has been at least one report of a black spider monkey (*Ateles paniscus*) being infested with the human head louse. The infection was thought to have been the result of contact with an infected person, indicating that the Anoplura can be shared by humans and the New World monkeys, but not the Old World monkeys [154, 281].

Fiennes [281] regards the sucking lice as interchangeable among humans, the great apes, and the New World monkeys, with the possible exception of the marmosets and tamarins. Old World monkeys are not affected by the species of Anoplura that infect humans, the great apes, and New World monkeys. There are no reports of transmission of rickettsial diseases by lice from the great apes or New World monkeys to humans, or vice versa, though in theory such transmission would seem possible [281].

Arachnida

The parasitic genera of ticks and mites described from nonhuman primates are listed in Tables 45.13 and 45.14, respectively. The most important members will be discussed in detail below.

As for the parasitic insecta, the reader is referred to standard pathology or parasitology texts for detailed information regarding the morphology and life cycle of these parasites [32, 290, 341, 428, 811].

Tick Infestation

According to Fiennes [281], the problem of ticks on captive monkeys is not important because when engorged, the ticks drop off the host, and, under the conditions of captivity, reinfection does not occur [281]. Species of ixodid ticks have been reported from numerous nonhuman primates including prosimians (bushbabies), New World monkeys (spider monkeys), and Old World monkeys (rhesus monkeys, cynomolgus monkeys, baboons, colobus monkeys, bonnet

TABLE 45.13. Ticks described from nonhuman primates.

Parasite genus-species	Location in host	Pro-simians	New World monkeys	Old World monkeys	Great apes	References
Ixodids						
Rhipicephalus sanguineus	Skin	X		X		281, 290, 394, 395, 609, 862
Rhipicephalus appendiculatus	Skin	X		X		290, 394, 395, 486, 609
Rhipicephalus puchellus	Skin			X		488
Rhipicephalus haemaphysa-loides	Skin			X		850
Rhipicephalus evertsi	Skin			X		395
Rhipicephalus pravus	Skin			X		395
Rhipicephalus simus	Skin	X		X		395
Dermacentor auratus	Skin			X		850
Ixodes ceylonensus	Skin			X		850
Ixodes petauristae	Skin			X		850
Ixodes calvipalpus	Skin			X		281, 395, 814
Ixodes loricatus	Skin		X			144, 207, 281, 814
Ixodes schillingeri	Skin			X		281, 290, 394, 395, 814
Ixodes rasus	Skin			X		395, 488
Ixodes lemuris	Skin	X				395
Ixodes sp.	Skin	X				395
Amblyomma hebraeum	Skin	X		X		290, 394, 395, 578, 609
Amblyomma variegatum	Skin	X				395
Amblyomma sp.	Skin			X		395, 488, 850
Boophilus annulatus	Skin			X		281, 814
Hyalomma truncatum	Skin			X		395
Hyalomma sp.	Skin			X		395
Haemaphysalis wellingtoni	Skin			X		850
Haemaphysalis aculeata	Skin			X		850
Haemaphysalis cuspidata	Skin			X		850
Haemaphysalis kyasanurensis	Skin			X		850
Haemaphysalis minuta	Skin			X		850
Haemaphysalis leachii	Skin	X				395
Haemaphysalis lemuris	Skin	X				395

(Continued)

TABLE 45.13. (Continued).

Parasite genus-species	Location in host	Pro-simians	New World monkeys	Old World monkeys	Great apes	References
Haemaphysalis palmata	Skin			X		281, 395, 814
Haemaphysalis spinigera	Skin			X		290, 849, 850, 862
Haemaphysalis koningsbergi	Skin	X				11
Haemaphysalis hylobatis	Skin				X	11
Haemaphysalis bispinosa	Skin			X		637, 850
Haemaphysalis turturis	Skin			X		850
Haemaphysalis papuanakinneari	Skin			X		850
Haemaphysalis sp.	Skin			X		848
Argasids						
Ornithodorus talaje	Skin			X		144, 281, 290, 814
Argas reflexus	Skin			X		281, 814

macaques, langurs, and green monkeys) [281, 290, 394, 609, 814, 849, 850]. Argasid ticks have been infrequently reported from Old World monkeys (cynomolgus monkeys). In addition, the argasid tick, *Argas reflexus*, normally parasitic on pigeons and other avians, has been reported from an otherwise unidentified monkey [281, 814].

It appears that feral nonhuman primates are parasitized by ticks in most of the geographical areas in which they live and that they are infected by a variety of different species. The importance of ticks as parasites is their world-wide geographic distribution and their role as vectors of a wide variety of diseases, many of which are zoonotic. Since they can infect other animals and contaminate the premises, producing long-term difficulties in parasite control, procedures aimed at eliminating them from newly acquired animals are of primary importance [281, 791].

Most cases of tick infection are asymptomatic; however, heavy parasite loads can result in irritation, restlessness, weight loss, and anemia. Tick bites cause a local inflammatory reaction characterized by hyperemia, edema, and focal hemorrhage. Bite wounds may be involved with secondary bacterial infections [791].

Diagnosis is based on the signs and on the demonstration and identification of the specific species of tick on the host [290].

CUTANEOUS ACARIASIS

Scabies (mange)

The cause of this disease, *Sarcoptes scabiei*, the human itch mite, has been reported from Old World monkeys (cynomolgus monkeys and drills) [22, 290, 293, 305, 523, 891] and the great apes (gorillas, chimpanzees, orangutans,

TABLE 45.14. Mites described from nonhuman primates.

Parasite genus-species	Location in host	Pro-simians	New World monkeys	Old World monkeys	Great apes	References
Mesostigmates						
Pneumonyssus simicola	Lungs			X	X	1, 22, 24, 251, 254, 270, 286, 290, 301, 306, 326, 349, 393, 405, 413, 414, 451, 452, 456, 457, 522, 562, 570, 585, 629, 725, 821, 862, 871, 938
Pneumonyssus duttoni	Bronchi, trachea			X		145, 251, 253, 267, 290, 405, 413, 630
Pneumonyssus santos-diasi	Lungs			X		251, 290, 405, 413, 485, 488, 609, 938
Pneumonyssus longus	Lungs, bronchi, trachea			X	X	251, 253, 290, 405, 413
Pneumonyssus oudemansi	Lungs, bronchi, trachea			X	X	56, 249, 251, 253, 290, 405, 413
Pneumonyssus africanus	Bronchi			X		251, 290, 405, 413
Pneumonyssus mossambicencis	Lungs			X		251, 290, 405, 413, 485, 488, 578, 609, 938
Pneumonyssus congoensis	Trachea, lungs			X		233, 251, 290, 405, 413, 488, 609
Pneumonyssus rodhaini	Lungs and nasal fossae			X		246, 405, 413
Pneumonyssus vitzthumi	Lung, bronchi				X	56, 405, 413
Pneumonyssus vocalis	Laryngeal ventricles, vocal pouch			X		573, 577, 578
Pneumonyssus sp.	Lungs			X		405, 413
Rhinophaga dinolti	Nasal cavities, lungs			X		247, 290, 306, 405, 413, 653
Rhinophaga cercopitheci	Lungs, frontal sinuses			X		247, 268, 290, 405, 413, 821
Rhinophaga papionis	Lungs, nasal fossae			X		247, 268, 290, 405, 413, 488, 573, 574, 578, 609, 821
Rhinophaga pongicola	Maxillary sinuses, nasal fossae				X	250, 405, 413
Rhinophaga elongata	Nasal mucosa			X		488, 573, 574, 578
Pneumonyssoides stammeri	Large bronchiole, larynx, nasal cavities, sinuses		X			254, 290, 327, 405, 413, 871

(Continued)

TABLE 45.14. (Continued).

Parasite genus-species	Location in host	Pro-simians	New World monkeys	Old World monkeys	Great apes	References
Rhinophagus sp.	Mucosal and submucosal nasal tissues			X		450
Prostigmates						
Psorergates cercopitheci	Skin			X		27, 290, 517, 710, 793, 938, 939
Psorergates sp.	Skin			X		513, 517, 521
Demodex canis	Skin		X		X	281, 673
Demodex saimiri	Skin		X			520
Demodex sp.	Skin		X			383, 461, 668
Astigmates						
Sarcoptes scabiei	Skin			X	X	22, 281, 290, 293, 305, 334, 523, 675, 706, 718, 742, 864, 890, 891, 931
Sarcoptes pitheci	Skin			X		281, 673, 742
Prosarcoptes pitheci	Skin		X	X		257, 260, 265, 290, 673
Pithesaroptes talapoini	Skin			X		260, 265, 290
Cosarcoptes scanloni	Skin			X		265, 290, 804
Notoedres galagoensis	Skin	X				257, 259, 265, 290
Alouattalages corbeti	Skin		X			263, 290
Fonsecalges saimirii	Skin		X			148, 207, 263, 289, 290
Paracoroptes gordoni	Skin			X		257, 290, 512
Pangorillages pani	Skin				X	256, 290
Listrocarpus cosgrovei	Skin		X			148, 264, 290
Listsrocarpus hapalei	Skin		X			264, 290
Listrocarpus saimirii	Skin		X			264, 290
Listrocarpus lagothrix	Skin		X			264, 290
Rhyncoptes anastosi	Skin		X			261, 290
Rhyncoptes cebi	Skin		X			261, 290
Rhyncoptes cercopitheci	Skin			X		261, 290
Saimirioptes paradoxus	Skin		X			266, 290
Audycoptes greeri	Hair follicles		X			207, 266, 290, 515
Audycoptes lawrenci	Hair follicles		X			207, 266, 290, 515
Lemurnyssus galagoensis	Nasal cavities	X				258, 290

TABLE 45.14. (Continued).

Parasite genus-species	Location in host	Pro-simians	New World monkeys	Old World monkeys	Great apes	References
Mortelmansia brevis	Nasal cavities		X			148, 207, 252, 258, 290
Mortelmansia longis	Nasal cavities		X			148, 207, 252, 258, 290
Mortelmansia duboisi	Nasal cavities		X			148, 207, 258, 290, 492
Dunnalges lambrechti	Skin		X			207, 461, 515, 516
Rosalialges cruciformis	Skin		X			207, 461, 515, 516
Prosarcoptes sp.	Skin			X		804
Pithesarcoptes sp.	Skin			X		804
Kutzerocoptes sp.	Skin			X		517, 804
Trombiculidids						
Trombicula sp.	Skin			X		281, 848

gibbons, and siamiangs) [281, 290, 334, 706, 718, 742, 864, 890, 891]. A closely related species, *Prosarcoptes pitheci*, has been reported in Old World monkeys (African green monkeys and baboons) and New World monkeys in captivity (cebus monkeys) [257, 260, 265, 290, 673]. Two sarcoptiform species, *Dunnalges lambrechti* and *Rosalialges cruciformis*, have been reported from New World monkeys (marmosets and owl monkeys) [461, 515]. There appear to be no reports of *S. scabiei* infection in prosimians [281].

Signs associated with *S. scabiei* infection in nonhuman primates include intense pruritus, anorexia, weakness, weight loss, tremors, and emaciation. Gross lesions include thickening and scaling of the skin and severe alopecia. The severe itching can result in self-mutilation, with secondary hemorrhage and suppurative bacterial dermatitis [290, 334, 523, 791, 864, 891]. Death of a chimpanzee has been ascribed to a severe *S. scabiei* infection [675].

Histopathologically, the infected skin is characterized by hyperkeratosis, parakeratosis, and crusting. The epidermis contains burrows in which many parasites and eggs are seen [290, 305, 334, 742, 791].

The tentative diagnosis of scabies is based on the signs and lesions and confirmed by demonstrating and identifying the parasites and/or eggs in deep skin scrapings [290, 791].

Sarcoptes scabiei infections in nonhuman primates are transmissible to humans by direct contact [290, 334, 742, 891, 931]. Thus, infected nonhuman primates, or those suspected of being infected, should be handled with caution by those responsible for their care and management [290].

PULMONARY ACARIASIS

The cause of this condition is any one of at least ten species of lung mites in the genus *Pneumonyssus* that have been reported from the lower respiratory tract of Old World monkeys (rhesus monkeys, cynomolgus monkeys, pig-tailed macaques, patas monkeys, Celebes black apes, mangabeys, baboons, numerous members of the genus Cercopithecus, colobus monkeys, langurs, and proboscis monkeys) and the great apes (chimpanzees, gorillas, and orangutans) [24, 56, 233, 246, 247, 249, 251, 253, 270, 290, 326, 349, 405, 413, 456, 457, 585, 629, 630, 666, 742, 938]. Also, there has been one species of lung mite in the genus *Pneumonyssoides* that has been reported from the lungs, larynx, nasal cavities, and sinuses in New World monkeys (woolly monkeys and howler monkeys) [254, 290, 405, 413, 742, 871].

The most commonly encountered member of this genus, *Pneumonyssus simicola*, is found in the lungs of essentially 100% of imported rhesus monkeys [286, 290, 301, 326, 413, 414, 453, 454, 522, 725]. Also, this mite has been seen in the lungs of infant rhesus monkeys allowed to remain with their wild-caught parents after birth [519]. Reports of *P. simicola* infection in other macaque species are less frequent [24, 251, 254, 290, 393, 821]. Despite the common occurrence in feral *M. mulatta*, *P. simicola* has not been seen in laboratory-born monkeys taken from their mothers at birth [290, 464]. A high incidence of lung mite infection with species other than *P. simicola* has been reported from baboons [451, 452, 823]. The complete life cycle of this parasite is unknown [290, 301, 454].

The infection in the rhesus monkey is usually nonsymptomatic, and clinical signs are uncommon [290, 306, 413, 414, 456, 457, 791]. There have been reports of paroxysms of sneezing and coughing, but these may be the result of associated pulmonary disease [290, 375, 413].

Gross lesions are randomly located throughout the pulmonary parenchyma and consist of varying-sized pale spots or yellowish-gray foci that are usually flat or slightly umbilicated on the surface and contain translucent areas. Those located near the surface of the lungs elevate the visceral pleura. The lesions can resemble tubercules but are soft to the touch rather than firm. Adjacent lesions may become confluent. Bullous emphysematous lesions and hemorrhagic lesions may be seen in some cases. Many animals have fine, stringlike fibrous adhesions between visceral and parietal pleural surfaces and between all of the lung lobes. Under the dissecting microscope, the lesions present as pale, white, jellylike masses that have a small opening or slit in the center. These so-called "mite houses" can contain from 1 to 20 mites. The majority of these are females, but sometimes eggs, larvae, and male mites are also present. A characteristic golden-brown to black pigment permeates the lesions and surrounding pulmonary parenchyma [290, 326, 341, 349, 413, 414, 456–458, 503, 791].

Histopathologically, lung mite lesions are characterized by a localized bronchiolitis, peribronchiolitis, focal lobular pneumonitis, alveolar collapse or consolidation, and sometimes bronchiolectasis. There is thickening of the bronchiolar wall, loss of the lining epithelium, hyperplasia of the bronchiolar smooth muscle, and formation of peribronchiolar lymphoid aggregations. A pleocellular inflammatory cell exudate consisting of neutrophils, eosinophils, lymphoplasmacytes, and macrophages infiltrates the affected bronchiolar wall. There is little or no tissue necrosis or giant cell formation [290, 326, 341, 349, 413, 414, 456–458, 503, 791].

Macrophages, whose cytoplasm is laden with a golden-brown to blackish pigment and refractile crystals, are always present in and around the lesions and throughout the lung tissues. This pigment, which is not seen in the lungs of mite-free monkeys, does not contain carbon or melanin but is iron positive and birefringent under polarized light. The exact source of the pigment is not known, but it is felt that it probably results from the breakdown and excretion of the hosts' blood proteins by the mites [290, 326, 341, 349, 413, 414, 456–458, 503, 639, 791]. The immunological response to pulmonary acariasis has been reviewed by Kim [457].

Lung mite infection has been reported to be associated with pneumothorax [711] and pulmonary arteritis [510, 522, 914] in the rhesus monkey. There is also a report that describes extensive pleuritis and pericarditis associated with ruptured lung mite lesions [455].

Even though several earlier reports ascribe fatalities to *P. simicola* infection, it probably results in death only under conditions of massive infections. Such cases of massive infections and resultant death have been reported in the rhesus monkey, proboscis monkey, "lion macaque," pig-tailed macaque, douc langurs, and chimpanzee [14, 34, 298, 344, 357, 405, 457, 818]. The gross lesions and histopathology of lung mite infection in baboons and chimpanzees are similar to those described for rhesus monkeys [458, 570].

Diagnosis of lung mite infection in live monkeys is difficult. Thoracic x-rays or hematologic studies are of little value [290, 414, 791]. There has been some success in demonstrating lung mite larvae in tracheobronchial washings, but a negative finding is not conclusive proof that infection does not exist. Gross lesions are rather characteristic but must be differentiated from tuberculosis. Tissue sections containing the mites and/or the characteristic pigment and crystals are diagnostic [290, 414, 791]. Lung mites can be found in the feces of infected nonhuman primates [667].

There is no evidence that *P. simicola* infects humans. Therefore, there is no public health significance associated with lung mite infections in nonhuman primates [290].

Nasal Acariasis

Five species of nasal mites of the genus *Rhinophaga* have been described from the upper skull and olfactory mucosa of Old World monkeys (rhesus monkeys, baboons, *Cercopithecus* sp.) and great apes (orangutan) [249, 268, 405, 413, 450, 457, 573, 574, 653]. *Rhinophaga papinois* is found in the maxillary sinuses of the chacma baboon where it causes mucosal polyps [457, 547, 573]. In the lungs it causes pneumonitis and excessive mucus production. *Rhinophaga elongata*, also reported from the chacma baboon, is an extremely long mite that has been described in the apex of small mucosal nodules randomly distributed throughout the nasal cavity. The anterior third of the mite was embedded deeply in the nasal mucosa and in some cases in the adjacent bone. An inflammatory reaction and obstruction of the mucosal glands, which became greatly dilated, was associated with the presence of this mite [457, 573, 574]. *Rhinophaga dinolti* has been reported from the lungs and nasal cavities of the rhesus monkey. Lesions associated with the presence of this parasite in tissues have not been reported [247, 290, 306, 413, 653]. *Rhinophaga cercopitheci* has been reported from the lungs and frontal sinuses of several species of guenons (*Cercopithecus ascanius*, *C. mitis*). Lesions include pneumonitis and excessive mucus production [268, 290, 413, 821]. *Rhinophaga pongicola* has been reported from the maxillary sinuses and nasal fossae of an orangutan [250, 413].

TABLE 45.15. Pentastomids described from nonhuman primates.

Parasite genus-species	Location in host	Pro-simians	New World monkeys	Old World monkeys	Great apes	References
Linguatula serrata	Mesenteric lymphnodes, viscera		X	X		290, 493, 788
Porocephalus clavatus	Peritoneum, viscera		X			148, 290, 461, 623, 689, 750, 785, 787, 791, 800, 815
Porocephalus subulifer	Viscera	X		X		255, 290, 788, 791
Gigliolella brumpti	Mesentery	X				104, 290
Armillifer armillatus	Peritoneal cavity	X	X	X	X	153, 177, 207, 255, 280, 290, 339, 362, 382, 393, 461, 486, 641, 750, 788, 791, 815, 902, 937
Armillifer moniliformis	Viscera, peritoneal cavity			X		45, 262, 290, 788, 791
Porocephalus crotali	Peritoneal cavity		X			177, 207, 815
Nephridiacanthus sp. (juvenile)	Rectal wall			X		486, 487

LARYNGEAL ACARIASIS

A newly recognized mite, *Pneumonyssus vocalis*, has been reported from the mucosa of the laryngeal ventricles and vocal pouch of the chacma baboon where it elicits a mild local inflammatory response [457, 573, 577].

Pentastomids

The parasitic pentastomid nymphs described from nonhuman primates are listed in Table 45.15. They are discussed in detail below.

PENTASTOMIASIS

The parasites that cause this disease are considered to be highly aberrant arthropods [255, 290, 461, 785, 788]. Four genera have been described: *Linguatula*, which has a world-wide distribution; *Porocephalus*, found in both South America and Africa; *Armillifer*, which occurs in Africa, Asia, and Australia; and

TABLE 45.16. Anaplasmatids described from nonhuman primates.

Parasite Genus-species	Location in Host	Pro-simians	New World Monkeys	Old World Monkeys	Great Apes	References
Hemobartonella sp.	Erythrocyte		X			5

Gigliolella, found on the Island of Madagascar. The nymph form occurs in nonhuman primates, which serve as intermediate hosts in the pentastome life cycle [153, 177, 229, 290, 461, 788, 791, 937]. The adult forms of *Linguatula* are found in the nasal passages of dogs, other canids, domestic animals, and humans. Adults of the other two genera are found in the lungs and air sacs of various snakes [290, 461, 788, 791]. Pentastomid nymphs have been reported in a wide variety of nonhuman primates, including prosimians, New and Old World monkeys, and great apes [148, 150, 255, 262, 280, 290, 297, 355, 382, 393, 461, 493, 609, 623, 787, 788, 800, 937].

Infection with this parasite usually is asymptomatic. Dead nymphs act as foreign bodies and invoke an intense inflammatory response in the host. Fatal peritonitis has been reported in overwhelming infections with penetration of the intestinal wall by nymphs. When one infection follows another of considerable duration, there may be a lymphocytic response as a result of presensitization by the initial infection [150, 255, 290, 297, 461, 786, 787, 800, 902].

Diagnosis usually is based on an incidental finding at necropsy and hinges on the identification of the characteristic C-shaped nymph in the tissues [290]. Nymphs have been described in the lungs, liver, omentum, and serosa of the intestinal tract of nonhuman primates [290, 461, 623, 788, 791] but may be found in almost any tissue, including the brain [297, 788].

Pentastomids have been reported in humans in tropical Africa and Asia, but the parasite in captive animals is of no public health significance [290, 692, 749]. Infection of humans can occur only through the ingestion of eggs passed in the feces or in the saliva of the definitive host [290].

Acknowledgments. The portion of this chapter concerning parasites of the alimentary tract and pancreas is reprinted from Toft JD II (1982) The pathoparasitology of the alimentary tract and pancreas of nonhuman primates: a review. Vet Pathol 19 (Supp 7): 44–92. Used by permission of Veterinary Pathology and The American College of Veterinary Pathologists.

References

1. Abbott DP, Majeed SK (1984) A survey of parasitic lesions in wild-caught laboratory-maintained primates (rhesus, cynomolgus, and baboon). Vet Pathol 21:198–207.
2. Aberle SD (1945) Primate malaria. Natl Acad Sci–Natl Res Council Publ, Washington, DC.
3. Abrambulo PV III, Abass JB, Walker JS (1974) Silvered leaf monkey (*Presbytis cristatus*). II. Gastrointestinal parasites and their treatment. Lab Anim Sci 24:299–306.
4. Adams L (1980) Research scores breakthrough in answer to marmoset disease. Zoo Sounds 16:11.
5. Adams MR, Lewis JC, Bullock BC (1984) Hemobartonellosis in squirrel monkeys (*Saimiri sciureus*) in a domestic breeding colony. Lab Anim Sci 34:82–85.
6. Albuquerque RDR, Baretto MP (1969) Estudos sobre reservatorios e vectores silvestres de *Trypanosoma cruzi*, XXXII-infeccão natural do simio *Callicebus nigrifons* (Spix, 1823) pelo *T. cruzi*. Rev Inst Med Trop S Paulo 11:115–122.
7. Allen AM (1957) Pulmonary hyatid in a rhesus monkey. Arch Path 64:148–151.
8. Allen AM (1960) Occurrence of the nematode, *Anatrichosoma cutaneum*, in the nasal mucosae of *Macaca mulatta* monkeys. Amer J Vet Res 21:389–392.

9. Amberson JM, Schwarz E (1952) *Ternidens deminatus* Railliet and Henry, a nematode parasite of man and primates. Ann Trop Med Parasitol 46:227–237.
10. Amyx HL, Asher DM, Nash TE, Gibbs CJ Jr, Gajdusek DC (1978) Hepatic amebiasis in spider monkeys. Amer J Trop Med Hyg 27:888–891.
11. Anastos G (1950) The scutate ticks of Ixodidae of Indonesia. Entomol Amer 30(New Ser):1–144.
12. Anderson DC, McClure HM (1982) Acute disseminated fatal toxoplasmosis in a squirrel monkey. JAVMA 181:1363–1366.
13. Anderson RC (1957) The life cycles of dipetalonematid nematodes (Filaroidea: Dipetalonematidae): the problem of their evolution. J Helminthol 31:203–224.
14. Anderson S (1974) Diseases in freshly imported Proboscis monkeys (*Nasalis larvatus*). 18th Internationalen Symposium über die Erakrankungen der Zootier Acad Verlag, Berlin, pp 307–312.
15. Anver MR, King NW, Hunt RD (1972) Congenital encephalitozoonosis in a squirrel monkey (Saimiri sciureus). Vet Pathol 9:475–480.
16. Artigas P de T (1933) Sobre o parasitismo do *Saimiris sciureus* por um Gongilonema (*G. saimirisi*, *N. sp.*) e as possibilidades de infestacão humana. Rev Soc Paulista Med Vet 3:83–91. Cited by Dunn FL The parasites of Saimiri in the context of Platyrrhine parasitism. In: Rosenblum LA, Cooper RW (eds) (1968) The squirrel monkey, Chapter 2. Academic, New York, pp 31–68.
17. Arya SN (1981) A new species of the genus *Probstmayria* Ransom, 1907 (Nematoda:Atractidae) from the rhesus macaque, *Macaca mulatta*. Primates 22:261–265.
18. Auerbach E (1953) A study of *Balantidium coli* Stein, 1863 in relation to cytology and behavior in culture. J Morphol 93:405–445.
19. Augustine DL (1939) Some observations on some ascarids from a chimpanzee (*Pan troglodytes*) with experimental studies on the susceptibility of monkeys (*Macaca mulatta*) to infection with human and pig ascaris. Amer J Hyg 30:29–33.
20. Ayala FM (1961) Hallazgo de *Trypanosoma cruzi* Chagas, 1909 en el mono *Saimiri boliviensis* de la Amazonia Peruana. Rev Bras Malar Doenc Trop 13:99–105.
21. Ayers KM, Jones SR (1978) The cardiovascular system. In: Benirschke K, Garner FM, Jones TC (eds) Pathology of laboratory animals, chapter 1, vol I. Springer-Verlag, New York, pp 2–69.
22. Baker EW, Evans TM, Gould DJ, Hull WB, Keegan HL (1956) A manual of parasitic mites of medical or economic importance. Tech Publ Natl Pest Control Assoc, New York.
23. Baker JR (1972) Protozoa of tissues and blood (other than the Haemosporina). In: Fiennes RNT-W (ed) Pathology of simian primates. Part II. Infections and parasitic diseases. S Karger, Basel, pp 29–56.
24. Banks N (1901) A new genus of endoparasitic acarians. Geneesk Tidschr Nederl-Indie 41:334–336.
25. Barretto MP, Siqueira AF, Ferriolli F, Carvelheiro JR (1966) Estudos sôbre reservatorios e vectores silvestres do *Trypanosoma cruzi*. Rev Inst Med Trop S Paulo 8:103–112.
26. Baskin GB, Wolf RH, Worth CL, Soike K, Gibson SV, Bieri JG (1983) Anemia, steatitis, and muscle necrosis in marmosets (*Saguinus labiatus*). Lab Anim Sci 33:74–80.
27. Baskin GB, Eberhard ML, Watson E, Fish R (1984) Diagnostic exercise: skin lesions in sooty mangabeys. Lab Anim Sci 34:602–603.
28. Baylis HA (1928) Some further parasitic worms from Sarawak. Ann Mag Nat Hist (10)1:606–608; cited by Inglis WG, Dunn FL (1963) The occurrence of *Lemuricola* (Nematoda:Oxyurinae) in Malaya: with the description of a new species. Z Parasitenk 23:354–359.

29. Baylis HA (1929) A manual of helminthology, medical and veterinary. William Wood, New York. Cited by Ruch TC (1959) Diseases of laboratory primates. WB Saunders, Philadelphia.
30. Beach TD (1936) Experimental studies on human and primate species of strongyloides. V. The free-living phase of the life cycle. Amer J Hyg 23:243–277.
31. Beddard FE (1916) On two new species of cestodes belonging to the genera *Linstowia* and *Cotugnia*. Proc Zool Soc London, 87:695–706.
32. Belding DL (1965) Textbook of parasitology, 3rd ed. Appleton-Century-Crofts, New York.
33. Benirschke K (1979) Diagnostic exercise. Lab Anim Sci 29:33–34.
34. Benirschke K (1983) Occurrence of spontaneous diseases. In: Viral and immunological diseases in nonhuman primates. Alan R. Liss, New York, pp 17–30.
35. Benirschke K, Adams FD (1980) Gorilla diseases and causes of death. J Reprod Fertil Suppl 28:139–148.
36. Benirschke K, Low RJ (1970) Acute toxoplasmosis in woolly monkey (*Lagothrix* spp.). Comp Pathol Bull 2:3–4.
37. Benirschke K, Richart R (1960) Spontaneous acute toxoplasmosis in a marmoset monkey. Amer J Trop Med Hyg 9:269–273.
38. Benson RE, Fremming BD, Young RJ (1955) Care and management of chimpanzees at the Radiobiological Laboratory of the University of Texas and the United States Air Force, School Aviation Med. US Air Force Rept 55–48.
39. Berghe van den L, Peel E, Chardome M (1956) *Trypanosoma irangiense*, N. sp. parasite du singe de nuit *Protodicticus* [sic] *potto ibeanus* en Congo belge. Folia Scient Afr Cent 2:17
40. Berghe van den L, Chardome M, Peel E (1956) Etude d'un trypanosome du potto. Folia Scient Afr Cent 2:20.
41. Berghe van den L, Chardome M, Peel E (1963) Trypanosomes of the African lemurs, *Perodicticus potto ibeanus* and *Galago demidovi thomasi*. J Protozool 10:133–135.
42. Berghe van den L, Chardome M, Peel E (1964) The filarial parasites of the eastern gorilla in the Congo. J Helminthol 38:349–368.
43. Bergner JF Jr, Jachowski LA Jr (1968) The filarial parasite *Macacanema formosana*, from the Taiwan monkey and its development in various arthropods. Formosan Sci 22:1–68.
44. Bernstein JJ (1972) An epizootic of hydatid disease in captive apes. J Zoo Anim Med 3:16–20.
45. Bezubik B, Furmaga S (1960) The parasites in *Macacus cynomolgus* L. from Indonesia. Acta Parasit Pol 8:334–344.
46. Biagi FF, Beltran FH (1969) The challenge of amoebiasis: understanding pathogenic mechanisms. Int Rev Trop Med 3:219–239.
47. Bingham GA, Rabstein MM (1964) A study of the effectiveness of thiabendazole in the rhesus monkey. Lab Anim Care 14:357–365.
48. Bisseru B (1967) Diseases of man acquired from his pets. William Heinemann Medical Books, London.
49. Blackie WK (1932) A helminthological survey of southern Rhodesia. Mem London Sch Hyg Trop Med 9:91.
50. Blacklock B, Adler S (1922) A parasite resembling *Plasmodium falciparum* in a chimpanzee. Ann Trop Med Parasitol 16:99–106.
51. Blacklock B, Adler J (1922) The pathological effects produced by *Strongyloides* in a chimpanzee. Ann Trop Med Parasitol 16:283–290.
52. Blair WR (1904) Internal parasites in wild animals. Rep NY Zool Soc 8:129.
53. Blair WR (1912) Some common affections of the respiratory tract and digestive organs among primates. Zoologica 1:175–186, 1912. Cited by Ruch TC (1959) Diseases of laboratory primates. WB Saunders, Philadelphia.

54. Blanchard R, Langeron M (1913) Le paludisme des macaques (*Plasmodium cynomolgi* Mayer, 1907). Arch Parasitol 15:529–542.
55. Boever WJ, Britt J (1975) Hydatid disease in a mandrill baboon. JAVMA 167:619–621.
56. Böhm LK, Supperer R (1955) Zwei neue Lungenmilben aus Menschenaffen: *Pneumonyssus oudemansi* and *Pneumonyssus vitzthumi* (Ascarina, Halarachnidae). Oest Zool Z 6:11–29.
57. Bond VP, Bostic W, Hansen EL, Anderson AH (1946) Pathologic study of natural amebic infection in macaques. Amer J Trop Med 26:625–629.
58. Bonne C, Sandground JH (1939) On the production of gastric tumors, bordering on malignancy in Javanese monkeys through the agency of *Nochtia nochti*, a parasitic nematode. Amer J Cancer 37:173–185.
59. Bostrom RC, Slaughter LJ (1968) Trematode (*Athesmia foxi*) infection in two squirrel monkeys (*Saimiri sciureus*). Lab Anim Care 18:493–495.
60. Bostrom RE, Ferrell JF, Martin JE (1968) Simian amebiasis with lesions simulating human amebic dysentery. Abstr 51 19th Ann Meeting Amer Assoc Lab Anim Sci, Las Vegas.
61. Brack M, Niemitz C (1984) The parasites of wild-caught tarsiers (*Tarsius bancanus*). In: Niemitz C (ed) Biology of tarsiers. Gustav Fischer Verlag, New York, pp 77–84.
62. Brack M, Myers BJ, Kuntz RE (1973) Pathogenic properties of *Molineus torulosus* in capuchin monkeys, *Cebus apella*. Lab Anim Sci 23:360–365.
63. Brack M, Boncyk LH, Kalter SS (1974) *Filaroides cebus* (Gebauer, 1933)—parasitism and respiratory infections in *Cebus apella*. J Med Primatol 3:164–173.
64. Brannon MJC, Faust EC (1949) Preparation and testing of a specific antigen for diagnosis of human strongyloidiasis. Amer J Trop Med Hyg 29:229–239.
65. Bray RS (1956) Studies on malaria in chimpanzees: I. The erythrocytic forms of *Plasmodium reichenowi*. J Parasitol 42:588–592.
66. Bray RS (1957) Studies on malaria in chimpanzees: III. Gametogony of *Plasmodium reichenowi*. Ann Soc Belge Med Trop 37:169–174.
67. Bray RS (1958) Studies on malaria in chimpanzees: V. The sporogonous cycle and mosquito transmission of *Plasmodium vivax schwetzi*. J Parasitol 44:46–51.
68. Bray RS (1959) Pre-erythrocytic stages of human malaria parasites: *Plasmodium malariae*. Brit Med J 2:679–680.
69. Bray RS (1960) Studies on malaria in chimpanzees. VIII. The experimental transmission and pre-erythrocytic phase of *Plasmodium malariae*, with a note on the host-range of the parasite. Amer J Trop Med Hyg 9:455–465.
70. Bray RS (1963) Malaria infections in primates and their importance to man. Ergeb Mikrobiol Immunitätsfor Exp Therap 36:168–213.
71. Bray RS (1963) The malaria parasites of anthropoid apes. J Parasitol 49:888–898.
72. Bray RS (1964) A check-list of the parasitic protozoa of west Africa with some notes on their classification. Bull Inst for Afr Noire 26:238–315.
73. Brede HD, Burger PJ (1977) *Physaloptera caucasica* (= *abbreviata caucasica*) in the South African baboon (*Papio ursinus*). Arbeiten aus dem Paul-Ehreich-Institut und dem Georg—Speyer-Hause 71:119–122.
74. Breznock AW, Pulley TL (1975) Anatrichosoma infection in two white-handed gibbons. JAVMA 167:631.
75. Britz WE Jr, Fineg J, Cook JE, Miksche ED (1961) Restraint and treatment of young chimpanzees. JAVMA 138:653–658.
76. Brooks BA (1963) More notes on *Saimiri sciureus*. Lab Primate Newslett 2:3–4.
77. Brown HW, Perna VP (1958) An overwhelming strongyloides infection. JAVMA 138:653–658.
78. Brown RJ (1969) Acanthocephalan myositis in a bushbaby. JAVMA 155:1141–1143.

79. Brown RJ, Hinkle DK, Trevethan SP, Kupper JL, McKee AE (1972) Nosematosis in a squirrel monkey (*Saimiri sciureus*). J Med Primatol 2:114–123.
80. Brug SL (1934) Observations on monkey malaria. Riv Malar 13:121–142.
81. Brumpt E (1939) Les parasites du paludisme des chimpanzés. CR Sc Soc Biol 130:837–840.
82. Brumpt E, Joyeux C (1912) Sur un infusoire nouveau parasite du chimpanzé *Troglodytella abrassarti*, n. g, n. sp. Bull Soc Path Exot 5:499–503.
83. Brumpt E, Urbain A (1938) Épizootie vermineuse por acanthocéphales (*Prosthenorchis*) ayant sévi ala singerie due Museum de Paris. Ann Parasitol Hum Comp 16:289–300.
84. Bück G, Coudurier J, Quesnel JJ (1952) Sur deux nouveaux plasmodium observés chez un lémurien de Madagascar splénectomisé. Arch Institut Pasteur d'Algérie 30:240–243.
85. Buckley JJC (1931) On two new species of *Enterobius* from the monkey *Lagothrix humboldti*. J Helminthol 9:133–140.
86. Buckley JJC (1949) Cysticerci in liver and lung of ring-tailed lemur. Trans Roy Soc Trop Med Hyg 43:2.
87. Buckley JJC (1960) On *Brugia* gen. nov. for *Wuchereria* sp. of the "malayi" group, i.e. *W. malayi* (Brug, 1927), *W. pahangi* Buckley and Edeson, 1956, and *W. pateri* Buckley, Nelson and Heisch, 1958. Ann Trop Med Parasitol 54:75–77.
88. Bullock BC, Wolf RH, Clarkson TB (1967) Myocarditis associated with trypanosomiasis in a Cebus monkey (*Cebus albifrons*). JAVMA 151:920–922.
89. Burrows RB (1965) Microscopic diagnosis of the parasites of man VI. Yale University Press, New Haven.
90. Burrows RB (1972) Protozoa of the intestinal tract. In: Fiennes RNT-W (ed) Pathology of simian primates. Part II. Infectious and parasitic diseases. S Karger, Basel, pp 2–28.
91. Burton FD, Underwood C (1976) Intestinal helminths in *Macaca sylvanus* of Gibraltar. Can J Zool 54:1406–1407.
92. Butler TM (1973) The chimpanzee. In: Selected topics in laboratory animal medicine. Vol XVI: Aeromedical Review 1–73. USAF Sch Aerospace Med, Aerospace Med Div (AFSC), Brooks AFB, Texas, pp 1–81.
93. Bywater JEC, Mann KH (1960) Infestation of a monkey with the leech *Dinobdella ferox*. Vet Rec 72:955.
94. Caballero y CE, Peregrina DI (1938) Nemátodes de los mamiferos de México, I. An Inst Biol Univ Méx 18:169.
95. Cahill KM (1967) Thabendazole in massive strongyloidiasis. Amer J Trop Med Hyg 16:451–453.
96. Cameron TWM (1929) The species of *Enterobius* Leach, in primates. J Helminthol 7:161–182.
97. Casey HW, Ayers KM, Robinson FR (1978) The urinary system. In: Benirschke K, Garner FM, Jones TC (eds) Pathology of laboratory animals, chapter 3, vol I. Springer-Verlag, New York, pp 116–173.
98. Castellani A, Chalmers AJ (1910) Manual of tropical medicine. Ballière, Tindall and Cox, London.
99. Castellani A, Chalmers AJ (1913) Manual of tropical medicine, 2nd ed. Ballière, Tindall and Cox, London.
100. Chabuad AG (1952) Le genre *Dipetalonema* Diesing 1861: essai de classification. Ann Parasitol Hum Comp 27:250–285.
101. Chabaud AG (1955) Essai d'interprétation phylétique des cycles évolutifs chez les nématodes parasites de vertébres. Conclusions taxo-moniques. Ann Parasitol Hum Comp 30:83–126.
102. Chabaud AG, Anderson RC (1959) Nouvel essai de classification des filaires (superfamille des Filarioidea) II. Ann Parsitol Hum Comp 34:64–87.

103. Chabaud AG, Choquet MT (1953) Nouveau essai de classification des filaires superfamille des Filarioidea. Ann Parasitol Hum Comp 28:172–192.
104. Chabaud AG, Choquet MT (1954) Nymphes du pentastome *Gigliolella* (n. gen.) *brumpti* (Giglioli, 1922) chez un Lemurien. Riv Parassitol 15:331–336.
105. Chabaud AG, Choquet MT (1955) Deux nématodes parasites de lémurien. Ann Parasitol Hum Comp 30:329.
106. Chabaud AG, Petter AJ (1958) Les nematodes parasites de lémuriens Malgaches. I. Mém Inst Sci Madagascar Ser A 12:139–158. Cited by Inglis WG, Dunn FL (1963) The occurrence of *Lemuricola* (Nematoda:Oxyurinae) in Malaya: with the description of a new species. Z Parasitenkunde 23:354–359.
107. Chabaud AG, Petter AJ (1959) Les nematodes parasites de lémuriens Malgaches. II. Un nouvel oxyure: *Lemuricola contagiosus*. Mém Inst Sci Madagascar Ser A 13:127–132. Cited by Inglis WG, Dunn FL (1963) The occurrence of *Lemuricola* (Nematoda:Oxyurinae) in Malaya: with the description of a new species. Z Parasitenkunde 23:354–359.
108. Chabaud AG, Petter AJ, Golvan YJ (1961) Les nematodes parasites de lémuriens Malgaches. III. Collection recolteé par M et Mme Francis Petter. Ann Parasit Hum Comp 36:113–126. Cited by Inglis WG, Dunn FL (1963) The occurrence of *Lemuricola* (Nematoda:Oxyurinae) in Malaya: with the description of a new species. Z Parasitenkunde 23:354–359.
109. Chagas C (1909) Neue Trypanosomen. Vorläufige Mitteilung Arch Schiffsu Tropen-Hyg 13:120–122.
110. Chagas C (1924) Infection naturelle des singes due para (*Chrysothrix sciureus* L.) par *Trypanosoma cruzi*. CR Soc Biol 90:873–876.
111. Chalifoux LV, Hunt RD, García FG, Sehgal PK, Comiskey JR (1973) Filariasis in New World monkeys: histochemical differentiation of circulating microfilariae. Lab Anim Sci 23:211–220.
112. Chalmers DT, Murgatroyd LB, Wadsworth PF (1983) A survey of the pathology of marmosets (*Callithrix jacchus*) derived from a marmoset breeding unit. Lab Anim 17:270–279.
113. Chandler AC (1925) New records of *Bertiella satyri* (cestoda) in man and apes. Parasitol Cambr 17:421–425.
114. Chandler AC (1929) Some new genera and species of nematode worms, filarioidea, from animals dying in the Calcutta Zoological Garden. Proc US Nat Mus 75:1.
115. Chandler AC (1953) An outbreak of *Prosthenorchis* (Acanthocephala) infection in primates in the Houston Zoological Garden, and a report of this parasite in *Nasua narica* in Mexico. J Parasitol 39:226.
116. Chandler AC, Read CP (1961) Introduction to parasitology, with special reference to the parasites of man, 10th ed. John Wiley, New York.
117. Chandler FW, McClure HM, Campbell WG Jr, Watts JC (1976) Pulmonary pneumocystosis in nonhuman primates. Arch Pathol Lab Med 100:163–167.
118. Chang J, McClure HM (1975) Disseminated oesophagostomiasis in the rhesus monkey. JAVMA 167:628–630.
119. Chang J, Kornegay RW, Wagner JL, Mikat EM, Hackel DB (1980) Toxoplasmosis in a sifaka. In: Montali RJ, Migaki G (eds) The comparative pathology of zoo animals. Smithsonian Institution Press, Washington, DC, pp 347–352.
120. Chang PCH, Graham GL (1957) Parasitism, parthenogenesis and polyploidy: the life cycle of *Strongyloides papillosus*. J Parasitol Suppl 43:13.
121. Chapman WL, Crowell WA, Isaac W (1973) Spontaneous lesions seen at necropsy in 7 owl monkeys (*Aotus trivirgatus*). Lab Anim Sci 23:434–442.
122. Chardome M, Peel E (1949) La répartition des filaires dans la region de Coquilhatville et la transmission de *Dipetalonema streptocerca* par *Culicoides grahami*. Ann Soc Belge Med Trop 29:99–119.
123. Chase RE, Degaris CF (1938) Anomalies of venae cavae superiores in an orang. Amer J Phys Anthro 24:61–65.

124. Cheever AW, Kirschstein RL, Reardon LV (1970) *Schistosoma mansoni* infection of presumed natural origin in Cercopithecus monkeys from Tanzania and Ethiopia. Bull World Health Org 42:486–490.
125. Cheng TC (1964) The biology of animal parasites. WB Saunders, Philadelphia.
126. Chin W, Contacos PG, Coatney GR, Kimball HR (1965) A naturally acquired quotidian-type malaria in man transferable to monkeys. Science 149:865.
127. Chitwood M (1970) Comparative relationships of some parasites of man and Old and New World subhuman primates. Lab Anim Care 20:389–394.
128. Chitwood M, Lichtenfels JR (1972) Identification of parasitic metazoa in tissue sections. Exper Parasitol 32:407–519.
129. Chitwood MB, Smith WN (1958) A redescription of *Anatrichosoma cynomolgi* Smith and Chitwood 1954. Proc Helminthol Soc, Washington, DC 25:112–117.
130. Christeller E (1922) Über die Balantidienruhr bei den Schimpansen des Berliner Zoologischen Gartens. Virchows Arch 238:396–422. Cited by Ruch TC (1959) Diseases of laboratory primates. WB Saunders, Philadelphia.
131. Christensen LT (1964) Chimp and owners share worm infestation. Vet Med 59:801–803.
132. Cicmanec JL, Neva FA, McClure HM, Loeb WF (1974) Accidental infection of laboratory-reared *Macaca mulatta* with *Trypanosoma cruzi*. Lab Anim Sci 24:783–787.
133. Clark JD (1969) Coenurosis in a Gelada baboon (*Theropithecus gelada*). JAVMA 155:1258–1263.
134. Clarkson TB, Bullock BC, Lehner NDM, Manning PJ (1970) Diseases affecting the usefulness of nonhuman primates for nutrition research. In: Harris RS (ed) Feeding and nutrition of nonhuman primates. Academic, New York, pp 233–250.
135. Coatney GR, Collins WE, Warren McW, Contacos PG (1971) The primate malarias. US Govt Printing Office, Washington, DC, pp 1–366.
136. Coatney GR, Elbel RE, Kocharantana P (1960) Some blood parasites found in birds and mammals from Loei Province, Thailand. J Parasitol 46:701–702.
137. Cockburn TA (1948) *Balantidium* infection associated with diarrhoea in primates. Trans Roy Soc Trop Med Hyg 42:291–293.
138. Cockrell BY, Valerio MG, Garner FM (1974) Cryptosporidiosis in the intestines of rhesus monkeys (*Macaca mulatta*). Lab Anim Sci 24:881–887.
139. Colley FC, Mullin SW (1972) *Eimeria pachylepyron* sp. n. (Protozoa:Eimeriidae) from the slow loris in Malaysia. J Parasitol 58:110–111.
140. Collins WE (1974) Primate malarias. Adv Vet Sci Comp Med 18:1–23.
141. Collins WE, Contacos PG, Garnham PCC, Warren MSW, Skinner JC (1972) *Plasmodium hylobati*: a malaria parasite of the gibbon. J Parasitol 58:123–128.
142. Conrad HD, Wong MM (1969) *Anatrichosoma* sp. in Old World non-human primates. Program and Abstracts. No 52, p 41. 44th Ann Meeting Amer Soc Parasitologists, 3–7 Nov, Washington, DC. Cited by Orihel TC (1970) Anatrichosomiasis in African monkeys. J Parasitol 56:982–985.
143. Conran PB (1967) Monkey malaria. Abstr 38, 18th Ann Meeting Amer Assoc Lab Animal Sci, Washington, DC.
144. Cooley RA, Kohls GM (1944) The Argasidae of North America, Central America and Cuba. Amer Midl Natur Monogr 1.
145. Cooreman J (1946) Observations sur *Pneumonyssus duttoni* Newstead et Todd. Acarien parasite de la trachée de *Cercopithecus ascanius* Audebert au Congo belge. Rev Zool Bot Afr 39:331–335.
146. Copeland BE, Kimber J (1968) Nuclear size in diagnosis of *Entamoeba histolytica* on stained smears. Amer J Clin Path 38:664–668.
147. Cosgrove GE (1966) The trematodes of laboratory primates. Lab Anim Care 16:23–39.
148. Cosgrove GE, Nelson B, Gengozian N (1968) Helminth parasites of the tamarin, *Saguinus fuscicollis*. Lab Anim Care 18:654–656.

149. Cosgrove GE, Humanson G, Lushbaugh CC (1970) *Trichospirura leptostoma*, a nematode of the pancreatic ducts of marmosets (*Saguinus* spp.). JAVMA 157: 696–698.
150. Cosgrove GE, Nelson BM, Self JT (1970) The pathology of pentastomid infection in primates. Lab Anim Care 20:354–360.
151. Cram EB (1928) A species of the genus *Bertiella* in man and chimpanzees in Cuba. Amer J Trop Med Hyg 8:339–344.
152. Crosby WM, Ivey MH, Shaffer WL, Holmes DD (1968) *Echinococcus* cysts in the Savannah baboon. Lab Anim Care 18:395–397.
153. Cruz AA da, De Sousa L (1959) *Armillifer armillatus* in chimpanzee (*Pan satyrus verus*). Rev Cien Vet 54:21–24.
154. Cummings BF (1916) Studies on the Anoplura and Mallophaga being a report upon a collection from the mammals and birds in the Society's gardens. Proc Zool Soc London 1:253–295.
155. Cummins LB, Keeling ME, McClure HM (1973) Preventive medicine in anthropoids: parasite control. Lab Anim Sci 23:819–822.
156. Cunha A da, Muniz J (1929) Nota sobre os parasitas intestinaes do *Macacus rhesus* com a descripcão de una nova especie de Octomitus. Mem Inst Oswaldo Cruz Suppl 5:34–35.
157. Curasson GCM (1929) *Troglodytella abrassarti* infusoire pathogène du chimpanzé. Ann Parasitol Hum Comp 7:465–468.
158. Das KM (1965) Discussion. In: Ribelin WE, McCoy JR (eds) Pathology of laboratory animals. Charles C Thomas, Springfield, IL, pp 363–364.
159. Deane LM (1962) Infeccão natural sagüi *Callithrix jacchus* por tripanosoma do tipo *cruzi*. Rev Inst Med Trop S Paulo 4:225–229.
160. Deane LM (1964) Animal reservoirs of *Trypanosoma cruzi* in Brazil. Rev Bras Malar Doenc Trop 16:27–48.
161. Deane LM (1967) Tripanosomídeos de mamiferos de regiao Amazônico. Rev Inst Med Trop S Paulo 9:143–148.
162. Deane LM, Damasceno RG (1961) Tripanosomídeos de mamiferos de região Amazônico. II. Tripanosomas de macacos de zona de Salgado, estado do Pará. Rev Inst Med Trop S Paulo 3:61–70.
163. Deane LM, Martins R (1952) Sobre um tripanosoma encontrado em macaco da Amazonia e que evolui em triatomineos. Rev Bras Malar Doenc Trop 4:47–61.
164. Deane LM, Deane MP, Neto JF (1966) Studies on transmission of simian malaria and on a natural infection of man with *Plasmodium simium* in Brazil. Bull World Health Org 35:805–808.
165. Deane LM, Ferreira N, Sitonio JG (1968) Novo hospedeiro natural do *Plasmodium simium* e do *Plasmodium brasilianum*: o mono, *Brachyteles arachnoides*. Rev Inst Med Trop S Paulo 10:287–288.
166. Deane LM, Batista D, Ferreria Neto JA, DeSouza H (1970) Tripanosomídeos de mamiferos da região Amazônico. V. *Trypanosoma lambrechti* Marinkelle, 1968, em macacos de Estado do Amazonas, Brasil. Rev Inst Med Trop S Paulo 12:1–7.
167. Deinhardt JB, Devine J, Passovoy M, Pohlman R, Deinhardt F (1967) Marmosets as laboratory animals. I. Care of marmosets in the laboratory. Pathology and outline of statistical evaluation of data. Lab Anim Care 17:11–29.
168. Deinhardt F, Holmes AW, Devine J, Deinhardt Jean (1969) Marmosets as laboratory animals. IV. The microbiology of laboratory kept marmosets. Lab Anim Care 17:48–70.
169. De Korté WE (1905) On the presence of sarcosporidium in the thigh muscles of *Macaca rhesus*. J Hyg 5:451–453.
170. DePaoli A (1965) *Schistosoma haematobium* in the chimpanzee—a natural infection. Amer J Trop Med Hyg 14:561–565.

171. DePaoli A, Johnsen DO (1978) Fatal strongyloidiasis in gibbons (*Hylobates lar*). Vet Pathol 15:31–39.
172. Deschiens REA (1927) Sur les protozoaires intestinaux des singes. Bull Soc Path Exot 20:19–23. Cited by Miller MJ, Bray RS (1966) *Entamoeba histolytica* infections in the chimpanzee (*Pan satyrus*). J Parasitol 52:386–388.
173. Deschiens REA, Limousin H, Troisic J (1927) Eléments presèntant les caractéres d'un protozoaire sanguicole observés chez le chimpanzé. Bull Soc Path Exot 20:597–600.
174. Desowitz RS (1968) *Hepatocystis* sp. from a gibbon. Trans Roy Soc Trop Med Hyg 62:4.
175. Desowitz RS (1970) Observations on hepatocystis of white-cheeked gibbon (*Hylobates concolor*). J Parasitol 56:444–446.
176. Desportes C (1945) Sur *Strongyloides stercoralis* (Bavay 1876) et sur les *strongyloides* des primates. Ann Parasitol Hum Comp 20:160–190.
177. Desportes C, Roth P (1943) Helminthes rècoltés au cours d'autopsies pratiquées sur différents mammiferes morts á la ménagerie due Muséum de Paris. Bull Musée Hist Nat Paris 15:108–114. Cited by Dunn FL (1968) The parasites of Saimiri in the context of Platyrrhine parasitism. In: Rosenblum LA, Cooper RW (eds) The squirrel monkey. Academic, New York, pp 31–68.
178. Dickson J, Fry J, Fairfax R, Spence T (1983) Epidemic toxoplasmosis in captive squirrel monkeys (*Saimiri sciureus*). Vet Rec 12:302.
179. Dissanaike AS (1958) On hydatid infection in a Ceylon toque monkey *Macaca sinica*. Ceylon Vet 7:33–35.
180. Dissanaike AS, Nelson P, Garnham PCC (1965) Two new malaria parasites, *Plasmodium cynomolgi ceylonensis* subsp. nov. and *Plasmodium fragile* sp. nov., from monkeys in Ceylon. Ceylon J Med Sci 14:1–9.
181. Dissanaike AS, Nelson P, Garnham PCC (1965) *Plasmodium simiovale* sp. nov. A new simian malaria parasite from Ceylon. Ceylon J Med Sci 14:27–32.
182. Dobell C (1931) Researches on the intestinal protozoa of monkeys and man. IV. An experimental study of the *histolytica*-like species of *Entamoeba* living naturally in macaques. Parasitology 23:1–72.
183. Döbereiner J (1955) Toxoplasmose espontânea em macaco. Veterinaria 9:44–55.
184. Dodd K, Murphy E (1970) *Dipetalonema gracile* in a capuchin monkey (*Cebus capucinus*). Vet Rec 87:538–539.
185. Dollifus RP, Chabaud, AG (1955) Cing espéces de nématodes chez un atéle mort á la ménagerie du muséum. Arch Musée Natl Hist Nat 3:27–40. Cited by Dunn FL (1968) The parasites of Saimiri in the context of Platyrrhine parasitism. In: Rosenblum LA, Cooper RW (eds) The squirrel monkey. Academic, New York, pp 31–68.
186. Donovan JC, Stokes WS, Montrey RD, Rozmiarek H (1983) Hematologic characterization of naturally occurring malaria (*Plasmodium inui*) in cynomolgus monkeys (*Macaca fascicularis*). Lab Anim Sci 33:86–89.
187. Dougherty EC (1943) The genus *Filaroides* van Beneden, 1858, and its relatives. Preliminary note. Proc Helminthol Soc Washington, DC 10:69–74.
188. Dougherty EC (1952) A note on the genus *Metathelazia* Skinker, 1931 (Nematoda: Metastrongylidae). Proc Helminthol Soc Washington, DC 19:55–63.
189. Dubey JP (1973) Feline toxoplasmosis and coccidiosis: a survey of domiciled and stray cats. JAVMA 162:873–877.
190. Dubey JP (1976) A review of sarcocystis of domestic animals and of other coccidia of cats and dogs. JAVMA 169:1061–1078.
191. Dubey JP, Swan GV, Frenkel JK (1972) A simplified method for isolation of *Toxoplasma gondii* from the feces of cats. J Parasitol 58:1005–1006.
192. Dubin IN, Wilcox A (1947) Sarcocystis in *Macaca mulatta*. J Parasitol 33:151–153.
193. Dubois G (1966) Un néodiplostome (Trematoda: Diplostomatidae) chez le tamarin *Leontocebus nigricollis* (Spix). Rev Suisse Zool 73:37–42.

194. Duke BOL (1954) The transmission of loiasis in the forest-fringe area of the British Cameroons. Ann Trop Med Parasitol 48:349–355.
195. Duke BOL (1954) The uptake of the microfilariae of *Acanthocheilonema streptocerca* by *Culicoides grahamii*, and their subsequent development. Ann Trop Med Parasitol 48:416–420.
196. Duke BOL (1955) The development of *Loa* in flies of the genus *Chrysops* and the probable significance of the different species in the transmission of loiasis. Trans Roy Soc Trop Med Hyg 49:115–121.
197. Duke BOL (1956) The intake of the microfilariae of *Acanthocheilonema perstans* by *Culicoides grahamii* and *C. inornatipennis*, and their subsequent development. Ann Trop Med Parasitol 50:32–38.
198. Duke BOL (1960) Studies on loiasis in monkeys. III. The pathology of the spleen in drills (*Mandrillus leucophaeus*) infected with Loa. Ann Trop Med Parasitol 54:141–146.
199. Duke BOL, Wijers DJB (1958) Studies on loiasis in monkeys. I. The relationship between human and simian *Loa* in the rain-forest zone of the British Cameroons. Ann Trop Med Parasitol 52:158–175.
200. Dumas J (1953) Les animaux de laboratoire Paris (Ed. Médicales Flammarion, Paris). Cited by Ruch TC (1959) Diseases of laboratory primates. WB Saunders, Philadelphia.
201. Dunn FL (1961) *Molineus vexillarius* sp. n. (Nematoda:Trichostrongylidae) from a Peruvian primate *Tamarinus nigricollis* (Spix, 1823). J Parasitol 47:953–956.
202. Dunn FL (1962) *Raillietina* (*R.*) *trinitatae* (Cameron and Reesal, 1951), Baer and Sandars, 1956 (Cestoda) from a Peruvian primate. Proc Helminthol Soc Washington, DC 29:148–152.
203. Dunn FL (1963) A new trichostrongylid nematode from an oriental primate. Proc Helminthol Soc Washington, DC 30:161–165.
204. Dunn FL (1963) Acanthocephalans and cestodes of South American monkeys and marmosets. J Parasitol 49:717–722.
205. Dunn FL (1964) Blood parasites of Southeast Asian primitive primates. J Parasitol 50:214–216.
206. Dunn FL (1965) On the antiquity of malaria in the Western Hemisphere. Hum Biol 37:385–393.
207. Dunn FL (1968) The parasites of *Saimiri* in the context of Platyrrhine parasitism. In: Rosenblum LA, Cooper RW (eds) The squirrel monkey. Academic, New York, pp 31–68.
208. Dunn FL (1970) Natural infection in primates: helminths and problems in primate phylogeny, ecology, and behavior. Lab Anim Care 20:383–388.
209. Dunn FL, Greer WE (1962) Nematodes resembling *Ascaris lumbricoides* L., 1758, from a Malayan gibbon, *Hylobates agilis* F. Cuvier, 1821. J Parasitol 48:150.
210. Dunn FL, Lambrecht FL (1963) On some filarial parasites of South American primates with a description of *Tetrapetalonema tamarinae* n. sp. from the Peruvian tamarin marmoset, *Tamarinus nigricollis* (Spix, 1823). J Helminthol 37:261–286.
211. Dunn FL, Lambrecht FL (1963) The hosts of *Plasmodium brasilianum* Gonder and von Berenberg-Gossler, 1908. J Parasitol 49:316–319.
212. Dunn FL, Lambrecht FL, du Plessis R (1963) Trypanosomes of South American monkeys and marmosets. Amer J Trop Med Hyg 12:524–534.
213. Dunn FL, Lim BL, Yap LF (1968) Endoparasites patterns in mammals of the Malayan rain forest. Ecology 49:1179–1184.
214. Durden LA, DeBruyn EJ (1984) Louse infestations of tree shrews (*Tupaia glis*). Lab Anim Sci 34:188–190.
215. Durette-Desset M-C, Palmieri JR, Purnomo, Cassone J (1981) Two new species of *Tupaiostrongylus* Dunn, 1963 (Nematoda: Molineidae) from a tree shrew (*Tupaia tana*) of Indonesia. System Parasitol 3:237–242.

216. Durette-Desset MD, Chabaud AG (1975) Sur trois nematodes trichostrongylides parasites de Tupaiidae. Ann Parasitol 50:173–185.
217. Duszynski DW, File SK (1974) Structure of oocyst and excystation of sporozoites of *Isospora endocallimici* n. sp. from marmoset *Callimico goeldii*. Trans Amer Micr Soc 93:403–408.
218. Dutton JE, Todd JL, Tobey EN (1906) Concerning certain parasitic protozoa observed in Africa. Mem Lpool Sch Trop Med 21:87–97.
219. Eastin CE, Roeckel I (1968) *Trypanosoma cruzi* complicating *Prosthenorchis* infestation in the squirrel monkey (*Saimiri sciureus*). Abst 52 19th Ann Meeting Amer Assoc Lab Animal Sci, Las Vegas.
220. Eberhard ML (1981) Intestinal parasitism in an outdoor breeding colony of *Macaca mulatta*. Lab Anim Sci 31:282–285.
221. Edeson JFB, Wharton RH, Laing ABG (1960) A preliminary account of the transmission, maintenance and laboratory vectors of *Brugia pahangi*. Trans Roy Soc Trop Med Hyg 54:439–449.
222. Edsall G, Gaines S, Landy M, Tigertt WD, Sprinz H, Trapani RJ, Mandel AD, Benenson AS (1960) Studies on infection and immunity in experimental typhoid fever. I. Typhoid fever in chimpanzees orally infected with *Salmonella typhosa*. J Exp Med 112:143–166.
223. Eichhorn A, Gallagher B (1916) Spontaneous amoebic dysentery in monkeys. J Inf Dis 19:395–407.
224. Eisenbrandt DL, Floering DA, David TD, McKee AE (1978) Scanning electron microscopy of a cryofractured hydatid cyst. In: Scanning electron microscopy, vol II. SEM, AMF O'Hare, IL, pp 229–233.
225. Eisenstein R, Innes JRM (1956) Sarcosporidiosis in man and animals. Vet Rev Annot 2:61–78.
226. Elek SR, Finkelstein LE (1939) *Multiceps serialis* infestation in a baboon. Report of a case exhibiting multiple connective tissue masses. Zoologica: Sci Contrib NY Zool Soc 24:323–328.
227. Else JG, Satzger M, Sturrock RF (1982) Natural infections of *Schistosoma mansoni* and *S. haematobium* in Cercopithecus monkeys in Kenya. Ann Trop Med Parasitol 76:111–112.
228. Essbach H (1949) Strongyloidose beim Schimpansen. Beitr Path Anat 110:319–345. Cited by Ruch TC (1959) Diseases of laboratory primates. WB Saunders, Philadelphia.
229. Esslinger JH (1962) Hepatic lesions in rats experimentally infected with *Porocephalus crotali* (*Pentastomida*). J Parasitol 48:631–638.
230. Esslinger JH (1966) *Dipetalonema obtusa* (McCoy, 1936) comb. n. (Filarioidea:Onchocercidae) in Colombian primates, with a description of the adult. J Parasitol 52:498–502.
231. Esslinger JH, Gardiner CH (1974) *Dipetalonema barbascalensis* sp. n. (Nematoda: Filarioidea) from the owl monkey, *Aotus trivirgatus*, with a consideration of the status of *Parlitomosa zakii* Nagaty, 1935. J Parasitol 60:1001–1005.
232. Evans LB (1978) Fatal parasitism among free living bushbabies. J S Afr Vet Assoc 49:67–69.
233. Ewing HE (1929) Notes on the lung mites of primates (Acarina, Dermanyssidae), including the description of a new species. Proc Ent Soc Wash 31:126–130.
234. Ewing HE (1932) A new sucking louse from the chimpanzee. Proc Biol Soc Washington 45:117–118.
235. Ewing SA, Helland DR, Anthony HD, Leipold HW (1968) Occurrence of *Athesmia* sp. in the cinnamon ringtail monkey *Cebus albifrons*. Lab Anim Care 18:488–492.
236. Eyles DE (1963) The species of simian malaria: taxonomy, morphology, life cycle, and geographical distribution of the monkey species. J Parasitol 49:866–887.
237. Eyles DE, Warren McW (1962) *Plasmodium inui* in Sulawesi. J Parasitol 48:739.

238. Eyles DE, Warren McW (1963) *Hepatocystis* from *Macaca irus* in Java. J Parasitol 49:891.
239. Eyles DE, Coatney GR, Getz ME (1960) Vivax-type malaria parasite of macaques transmissible to man. Science 131:1812–1813.
240. Eyles DE, Fong YL, Warren McW, Guinn E, Sandosham AA, Wharton RH (1962) *Plasmodium coatneyi*, a new species of primate malaria from Malaya. Amer J Trop Med Hyg 11:597–604.
241. Eyles DE, Laing ABG, Dobrovolny CG (1962) The malaria parasites of the pig-tailed macaque *Macaca nemestrina nemestrina* (Linnaeus), in Malaya. Ind J Malar 16:285–298.
242. Eyles DE, Laing ABG, Fong YL (1962) *Plasmodium fieldi* sp. nov., a new species of malaria parasite from the pig-tailed macaque in Malaya. Ann Trop Med Parasitol 56:242–247.
243. Eyles DE, Warren McW, Fong YL, Sandosham AA, Dunn FL (1962) A malaria parasite of Malayan gibbons. Med J Malaya 17:86.
244. Eyles DE, Fong YL, Dunn FL, Guinn E, Warren McW, Sandosham AA (1964) *Plasmodium youngi* n. sp., a malaria parasite of the Malayan gibbon, *Hylobates lar lar.* Amer J Trop Med Hyg 13:248–255.
245. Ezzat MAE, Gaafar SM (1951) *Tetrathyridium* sp. in a Syke's monkey (*Cercopithecus albogulars*) from Giza Zoological Gardens, Egypt. J Parasitol 37:392–394.
246. Fain A (1952) Sur les acariens parasites du genre *Pneumonyssus* au Congo Belge. Description de deux espèces nouvelles chez le daman et le colobe. Rev Zool Bot Afr 45:358–382.
247. Fain A (1955) Deux nouveaux acariens de la famille Halarachnidae Oudemans, parasites des fosses nasales des singes au Congo Belge et au Ruanda-Urandi. Rev Zool Bot Afr 51:307–324.
248. Fain A (1956) *Coenurus* of *Taenia brauni* setti parasitic in man and animals from the Belgian Congo and Ruanda-Ufundi. Nature 178:1353.
249. Fain A (1957) L'acariase pulmonaire chez le chimpanzé et le gorille par des acariens du genre *Pneumonyssus* Banks. Rev Zool Bot Afr 56:234–242.
250. Fain A (1958) Un nouveau parasite de l'orang-utan *Rhinophaga pongoicola* n. sp. (Acarina-Halarachnidae). Rev Zool Bot Afr 58:323–327.
251. Fain A (1959) Les acariens du genre *Pneumonyssus* Banks, parasites endopulmonaires des singes au Congo Belge (Halarachnidae:Mesostigmata). Ann Parasitol Hum Comp 34:126–148.
252. Fain A (1959) Deux nouveaux acariens nasicoles chez un singe platyrrhinien *Saimiri sciurea* (L). Bull Soc Roy Zool Anvers 12:3–12.
253. Fain A (1961) *Pneumonyssus duttoni* Newstead et Todd (1906) est une espèce composite. Description des deux espèces du complex duttoni (Mesostigmata:Halarachnidae). Rev Zool Bot Afr 63:213–226.
254. Fain A (1961) Sur le statut de deux espèces d'acariens du genre *Pneumonyssus* Banks déscrites par H. Vitzthum. Désignation d'un neotype pour *Pneumonyssus simicola* Banks, 1901 (Mesostigmata:Halarachnidae). Z Parasitenk 21:141–150.
255. Fain A (1961) Le pentastomidés de l'Afrique Centrale. Ann Mus Roy Afr Centr Ser 8 Sci Zool 92:1–115.
256. Fain A (1962) *Pangorillages pani* g.n., sp. n. Acarien psorique du chimpanzé (Psoralqidae:Sarcoptiformes). Rev Zool Bot Afr 66:283–290.
257. Fain A (1963) Les acariens producteurs de gale chez les lemuriens et les singes avec une étude des Psoroptidae (Sarcoptiformes). Bull Inst Roy Sci Nat Belg 39:1–125.
258. Fain A (1964) Les Lemurnyssidae parasites nasicoles des Lorisidae africains et des Cebidae sud-américains. Description d'une espèce nouvelle (Acarina:Sarcoptiformes). Ann Soc Belge Med Trop 44:453–458.
259. Fain A (1965) Notes sur le genre *Notoedres* Railliet, 1893 (Sarcoptidae:Sarcoptiformes). Acarologia 7:321–342.

260. Fain A (1965) Nouveaux genres et espèces d'acariens Sarcoptiformes parasites (note préliminaire). Rev Zool Bot Afr 72:252–256.
261. Fain A (1965) A review of the family Rhyncoptidae:Lawrence, parasitic on porcupines and monkeys (Acarina:Sarcoptiformes). In: Naegele JA (ed) Advances in acarology, vol II. Cornell University Press, Ithaca, NY, pp 135–159.
262. Fain A (1966) Pentastomida of snakes—their parasitological role in man and animals. Mem Inst Butantan (São Paulo) 33:167–174.
263. Fain A (1966) Les acariens producteurs de gale chez les lémuriens et les singes. II. Nouvelles observations avec description d'une espèce nouvelle. Acarologia 8:94–114.
264. Fain A (1967) Diagnoses d'acariens Sarcoptiformes nouveaux. Rev Zool Bot Afr 65:378–382.
265. Fain A (1968) Étude de la variabilité de *Sarcoptes scabei* avec une revision des Sarcoptidae. Acta Zool Pathol Antverp 47:3–196.
266. Fain A (1968) Notes sur trois acariens remarquables (Sarcoptiformes). Acarologia 10:276–291.
267. Fain A, Schobbens S (1947) Lésions histopathologiques produites per l'acarien parasite *Pneumonyssus duttoni*. Newst et Todd. Rev Zool Bot Afr 40:12–16.
268. Fain A, Mignolet G, Bereznay Y (1958) L'acariase des voies respiratoires chez les singes due Zoo d'Anvers. Bull Soc Roy Zool Anvers 9:15–19.
269. Fairbairn H (1948) The occurrence of a piroplasm *Entopolypoides macaci*, in East African monkeys. Ann Trop Med Parasitol 42:118.
270. Fairbrother RW, Hurst EW (1932) Spontaneous diseases observed in 600 monkeys. J Path Bact 35:867–873.
271. Faust EC (1928) Studies on *Thelazia callipaeda*. Railliet and Henry, 1910. J Parasitol 15:75–86.
272. Faust EC (1933) Experimental studies on human and primate species of Strongyloides. II. The development of *Strongyloides* in the experimental host. Amer J Hyg 18:114–132.
273. Faust EC (1935) Notes on helminths from Panama. III. Filarial infection in the marmosets, *Leontocebus geoffroyi* (Pucheron) and *Saimiri örstedii* (Reinhardt) in Panama. Trans Roy Soc Trop Med Hyg 28:627.
274. Faust EC (1936) *Strongyloides* and strongyloidiasis. Rev Parasitol (Havana) 2:315–341.
275. Faust EC (1967) Athesmia (Trematoda:Dicrocoeliidae) Odhner, 1911 liver fluke of monkeys from Colombia, South America, and other mammalian hosts. Trans Amer Microsc Soc 86:113–119.
276. Faust EC, Beaver PC, Jung RC (1968) Animal agents and vectors of human disease, 3rd ed. Lea & Febiger, Philadelphia.
277. Fenstermacher R, Jellison WL (1932) Porcupine louse infesting the monkey. J Parasitol 18:294.
278. Fenwick A (1969) Baboons as reservoir hosts of *Schistosoma mansoni*. Trans Roy Soc Trop Med Hyg 63:557–567.
279. Fiennes RNT-W (1959) Report of the Society's pathologist for the year 1957. Proc Zool Soc London 132:129–146.
280. Fiennes RNT-W (1967) Zoonoses of primates. The epidemiology and ecology of simian diseases in relation to man. Weidenfeld and Nicolson, London.
281. Fiennes RNT-W (1972) Ectoparasites and vectors. In: Fiennes RNT-W (ed) Pathology of simian primates, part II. Infections and parasitic disease. S Karger, New York pp 158–176.
282. File SK (1974) *Anatrichosoma ocularis* sp. n. (Nematoda:Trichosomodidae) from the eye of the common tree shrew, *Tupaia glis*. J Parasitol 60:985–988.
283. File SK (1976) *Probstmayria gombensis* sp. n. (Nematoda:Atractidae) from the chimpanzee. J Parasitol 62:256–258.

284. File SK, McGrew WC, Tutin CEG (1976) The intestinal parasites of a community of feral chimpanzees, *Pan troglodytes schweinfurthii*. J Parasitol 62:259–261.
285. Fineg J, Britz WE Jr, Cook JE, Edwards RH (1961) Clinical observations and methods used in the treatment of young chimpanzees. Air Force Missile Development Center, Holloman Air Force Base, New Mexico, Report Number AFMCD-TR-61-12.
286. Finegold MJ, Seaquist ME, Doherty MJ (1968) Treatment of pulmonary acariasis in rhesus monkeys with an organic phosphate. Lab Anim Care 18:127–130.
287. Finkeldey W (1931) Pathologisch-anatomische Befunde bei der Oesophagostomiasis des *Javeneraffen*. Z Infektkr Haustiere 40:146–164. Cited by Ruch TC (1959) Diseases of laboratory primates. WB Saunders, Philadelphia.
288. Fischthal JH, Kuntz RE (1965) Six digenetic trematodes of mammals from North Borneo (Malaysia). Proc Helm Soc Washington, DC 32:154–159.
289. Flatt RE, Patton NM (1969) A mite infestation in squirrel monkeys (*Saimiri sciureus*). JAVMA 155:1233–1235.
290. Flynn RJ (1973) Parasites of laboratory animals. Iowa State University Press, Ames, IA.
291. Fonseca F da (1951) Plasmódio de primata do Brasil. Mem Inst Oswaldo Cruz 49:543–553.
292. Foster AO, Johnson CM (1939) A preliminary note on the identity, life cycle, and pathogenicity of an important nematode parasite of captive monkeys. Amer J Trop Med 19:265–277.
293. Fox H (1926) Scabies in a male drill. Rep Lab Comp Pathol, Philadelphia, pp 27–28.
294. Fox H (1928) Balantidium in the red howler (*Alouatta seniculus*). Rep Lab Comp Pathol, Philadelphia, p 27.
295. Fox JG, Ediger RD (1970) Nasal leech infestation in the rhesus monkey. Lab Anim Care 20:1137–1138.
296. Fox JG, Hall WC (1970) Fluke (*Gastrodiscoides hominis*) infection in a rhesus monkey with related intussusception of the colon. JAVMA 157:714–716.
297. Fox JG, Diaz JR, Barth RA (1972) Nymphal *Porocephalus clavatus* in the brain of a squirrel monkey, *Saimiri sciureus*. Lab Anim Sci 22:908–910.
298. Frank H (1962) Durch Milben verursachte tödliche Lungenerkrankung bei einem Affen. Berl Münch Tier Woch 76:135–137.
299. Frank H (1982) Pathology of amebiasis in leaf monkeys (*Colobidae*). Proc 24th Int Symp Dis Zoo Anim, pp 321–326.
300. Fremming BD, Vogel FS, Benson RE, Young RJ (1955) A fatal case of amebiasis with liver abscesses and ulcerative colitis in a chimpanzee. JAVMA 126:406–407.
301. Fremming BD, Harris MD Jr, Young RJ, Benson RE (1957) Preliminary investigation into the life cycle of the monkey lung mite (*Pneumonyssus foxi*). Amer J Vet Res 18:427–428.
302. Frenkel JK (1971) Protozoal diseases of laboratory animals. In: Marcial-Rojas RA (ed) Pathology of protozoal and helminthic diseases. Williams and Wilkins, Baltimore, pp 318–369.
303. Frenkel JK (1974) Advances in the biology of sporozoa. Z Parasitenk 45:125–162.
304. Frenkel JK (1976) *Angiostrongylus costaricensis* infections. In: Binford CH, Connor DH (eds) Pathology of tropical and extraordinary diseases, an atlas, vol II, section 9. Diseases caused by other nematodes; Chapter 10. Angiostrongyliasis. Armed Forces Institute of Pathology, Washington, DC, pp 452–454.
305. Fuerstenberg MHF (1861) Die Kraetzmilben der Menschen und Tiere. Englemann, Leipzig.
306. Furman DP (1954) A revision of the genus *Pneumonyssus* (Acarina:Halarachnidae). J Parasitol 40:31–42.
307. Gardiner CH (1982) Syllabus: identification of animal parasites in histologic section. Reg Vet Pathol, AFIP, Washington, DC, pp 1–71.

308. Gardiner CH, Nold JB, Sanders JE (1982) Diagnostic exercise. Lab Anim Sci 32:601–602.
309. Gardner MB, Esra G, Cain MJ, Rossman S, Johnson C (1978) Myelomonocytic leukemia in an orangutan. Vet Pathol 15:667–770.
310. Garner E (1967) *Dipetalonema gracile* infection in squirrel monkeys (*Saimiri sciureus*). Lab Anim Dig 3:16–17.
311. Garner E, Hemrick R, Rudiger H (1967) Multiple helminth infections in cinnamon-ringtailed monkeys (*Cebus albifrons*). Lab Anim Care 17:310–315.
312. Garner FM, Stookey JL (1968) Syllabus: diseases of nonhuman primates. Amer Reg Path, AFIP, Washington, DC, pp 22–28, 36–53.
313. Garnham PCC (1947) Exoerythrocytic schizogony in *Plasmodium kochi* Laveran: a preliminary note. Trans Roy Soc Trop Med Hyg 40:719–722.
314. Garnham PCC (1948) The development cycle of *Hepatocystis* (*Plasmodium*) *kochi* in the monkey host. Trans Roy Soc Trop Med Hyg 41:601–616.
315. Garnham PCC (1950) Blood parasites of East African vertebrates with a brief description of exoerythrocytic schizogony in *Plasmodium pitmani*. Parasitology 40:328–337.
316. Garnham PCC (1959) A new sub-species of *Plasmodium cynomolgi*. Riv Parasitol 20:273–278.
317. Garnham PCC (1963) Distribution of simian malaria parasites in various hosts. J Parasitol 49:905–911.
318. Garnham PCC (1963) A new sub-species of *Plasmodium knowlesi* in the long-tailed macaque. J Trop Med Hyg 66:156–158.
319. Garnham PCC (1973) Distribution of malaria parasites in primates, insectivores and bats. Symp Zool Soc London 33:377–404.
320. Garnham PCC, Uilenberg G (1975) Malaria parasites of lemurs. Ann Parasitol 50:409–418.
321. Garnham PCC, Lainson R, Gunders AE (1956) Some observations on malaria parasites in a chimpanzee, with particular reference to the persistence of *Plasmodium reichenowi* and *Plasmodium vivax*. Ann Soc Belge Med Trop 36:811–822.
322. Garnham PCC, Lainson R, Cooper W (1958) The complete life cycle of a new strain of *Plasmodium gonderi* from the drill (*Mandrillus leucophaeus*), including its sporogony in *Anopheles aztecus* and its preerythrocytic schizogony in the rhesus monkey. Trans Roy Soc Trop Med Hyg 2:509–517.
323. Garnham PCC, Heisch RB, Minter DM (1961) The vector of *Hepatocystis* (*Plasmodium*) *kochi*; the successful conclusion of observations in many parts of tropical Africa. Trans Roy Soc Trop Med Hyg 55:497–502.
324. Garnham PCC, Gonzales-Mugaburu L (1962) A new trypanosome in *Saimiri* monkeys from Colombia. Rev Inst Med Trop S Paulo 4:79–84.
325. Garnham PCC, Rajapaksa N, Peters W, Killick-Kendrick R (1972) Malaria parasites of the orang-utan (*Pongo pygmaeus*). Ann Trop Med Parasitol 66:287–294.
326. Gay DM, Branch A (1927) Pulmonary acariasis in monkeys. Amer J Trop Med 7:49–55.
327. Gebauer O (1933) Beitrag zur Kenntnis von Nematoden aus Affenlungen. Z Parasitenk 5:724–734.
328. Gedoelst L (1916) Notes sur la faune parasitaire du Congo Belge. Rev Zool Afr 5:1–90.
329. Geiman QM (1964) Shigellosis, amebiasis, and simian malaria. Lab Anim Care 14:441–454.
330. Geisel O, Krampitz HE, Willaert E (1975) Invasive amoebiasis caused by *Entamoeba histolytica* in a douc langur (*Pygathrix nemaeus* L. 1771). Berlin Muench Tieraerztl Woch 88:52–55.
331. Gisler DB, Benson RE, Young RJ (1960) Colony husbandry of research monkeys. Ann NY Acad Sci 85:758–768.

332. Gleason NN, Wolfe RE (1974) *Entopolypoides macaci* (*Babesiidae*) in *Macaca mulatta*. J Parasitol 60:844–847.
333. Goldberger J, Crane C (1911) A new species of *Athesmia* (*A. foxi*) from a monkey. Bull Hyg Lab 71:48–55.
334. Goldman L, Feldman MD (1949) Human infestation with scabies of monkeys. Arch Dermatol Syphilol 59:175–178.
335. Goldsmid JM (1974) The intestinal helminthzoonoses of primates in Rhodesia. Ann Soc Belge Med Trop 54:87.
336. Goldsmid JM, Rogers S (1978) A parasitological study on the chacma baboon (*Papio ursinus*) from the Northern Transvaal. J So Afr Vet Assoc 49:109–111.
337. Gonder R, Von-Berenberg-Gossler H (1908) Untersuchungen über Malariaplasmodien der Affen. Malar Lpz 1:47–56.
338. Goussard B, Collet J-Y, Garin Y, Tutin CEG, Fernandez M (1983) The intestinal entodiniomorph ciliates of wild lowland gorillas (*Gorilla gorilla gorilla*) in Gabon, West Africa. J Med Primatol 12:239–249.
339. Graham GL (1960) Parasitism in monkeys. Ann NY Acad Sci 85:842–860.
340. Grigorova O, Nesturch M (1934) Filariasis in a young chimpanzee. Trans Lab Exp Biol, Moscow 6:210–211.
341. Grinker JA, Karlin DA, Estrella PM (1962) Lung mites: pulmonary acariasis in the primate. Aerosp Med 33:841–844.
342. Groot H (1951) Nuevo foco de trypanosomiasis humana en Colombia. Anales Soc Biol Bogota 4:220–221.
343. Groot H, Renjifo S, Uribe C (1951) *Trypanosoma ariarii*, n. sp., from man, found in Colombia. Amer J Trop Med 31:673–691.
344. Grzimek B (1951) Tod durch Lungenmilben bei einem Schimpansen. Zool Gart Lpz 18:249.
345. Guilloud NB, King AA, Lock A (1965) A study of the efficacy of thiabendazole and dithiazanine iodide-piperazine citrate suspension against intestinal parasites in the *Macaca mulatta*. Lab Anim Care 15:354–358.
346. Gupta NK, Dutt K (1975) On three nematode parasites of the genus Subulura Molin, 1860 from India. Riv Parassitol 36:185–188.
347. Haaf E, Soest van AH (1964) Oesophagostomiasis in man in North Ghana. Trop Geogr Med 16:49–53.
348. Habermann RT, Menges RW (1968) Filariasis (*Acanthocheilonema perstans*) in a gorilla (a case history). Vet Med Sm Anim Clin 63:1040–1043.
349. Habermann RT, Williams FP Jr (1957) Diseases seen at necropsy of 708 *Macaca mulatta* (rhesus monkey) and *Macaca philippinensis* (cynomolgus monkey). Amer J Vet Res 18:419–426.
350. Habermann RT, Williams FP Jr (1958) The identification and control of helminths in laboratory animals. J Natl Cancer Inst 20:979–1009.
351. Habermann RT, Williams FP Jr, Thorp WTS (1954) Identification of some internal parasites of laboratory animals. Public Health Service Publ No 343, US Dept of Health, Education, and Welfare.
352. Haddow AJ, Williams MC, Woodall JP, Simpson DIH, Goma LKH (1964) 12 Isolations of Zika virus from *Aedes stegomyia africanas* (Theobald) taken in and above a Uganda forest. Bull World Health Org 31:57–69.
353. Halberstädter L, von Prowazek S (1907) Untersuchungen über die Malaria-parasiten der Affen. Arb Gesundh Amt, Berlin 26:37–43.
354. Hamerton AE (1932) Report on the deaths occurring in the Society's gardens during the year 1931. Proc Zool Soc London 103:613–638.
355. Hamerton AE (1933) Report on deaths occurring in the Society's gardens during the year 1932. Proc Zool Soc London 104:451–482.
356. Hamerton AE (1937) Report on the deaths occurring in the Society's gardens during 1936. Proc Zool Soc London 107:443–474.

357. Hamerton AE (1938) Report on the deaths occurring in the Society's gardens during the year 1937. Proc Zool Soc London 108:489–526.
358. Hamerton AE (1941–1942) Report on the deaths occurring in the Society's gardens during 1939–1940. Proc Zool Soc London 111:151–184.
359. Hamerton AE (1943) Report on the deaths occurring in the Society's gardens during 1942. Proc Zool Soc London 113:149.
360. Harper JS III, Rice JM, London WT, Sly DL, Middleton C (1982) Disseminated strongyloidiasis in *Erythrocebus patas*. Amer J Primatol 3:89–98.
361. Hartman HA (1961) The intestinal fluke (*Fasciolopsis buski*) in a monkey. Amer J Vet Res 22:1123–1126.
362. Hashimoto I, Honjo S (1966) Survey of helminth parasites in cynomologus monkeys (*Macaca irus*). Jap J Med Sci Biol 19:218.
363. Hawking F (1972) *Entopolypoides macaci*, a Babesia-like parasite in *Cercopithecus* monkeys. Parasitology 65:89–109.
364. Hawking F, Webber WAF (1955) *Dirofilaria aethiops* Webber, 1955, a filorial parasite of monkeys. II. Maintenance in the laboratory. Parasitology 45:378–387.
365. Hayama S, Nigi H (1963) Investigation on the helminth parasites in the Japan Monkey Centre during 1959–61. Primates 4:97–112.
366. Healy GR, Hayes NR (1963) Hydatid disease in rhesus monkeys. J Parasitol 49:837.
367. Healy GR, Myers BJ (1973) Intestinal helminths. In: Bourne GH (ed) The chimpanzee, vol 6. University Park Press, Baltimore, pp 265–296.
368. Hegner RW (1934) Specificity in the genus *Balantidium* based on size and shape of body and macronucleus with descriptions of six new species. Amer J Hyg 19:38–67.
369. Hegner RW (1934) Intestinal protozoa of chimpanzees. Amer J Hyg 19:480–501.
370. Hegner RW (1935) Intestinal protozoa from Panama monkeys. J Parasitol 21:60–61.
371. Hegner RW, Chu HJ (1930) A survey of protozoa parasitic in plants and animals of the Philippine Islands. Philippine J Sci 43:451–482.
372. Held JR (1969) Primate malaria. Ann NY Acad Sci 162:587–593.
373. Held JR, Whitney RA Jr (1978) Epidemic diseases of primate colonies. In: Chivers DJ, Ford EHR (eds) Recent advances in primatology, vol 4. Medicine. Academic, New York, pp 23–41.
374. Held JR, Contacos PG, Coatney GR (1967) Studies of the exoerythrocytic stages of simian malaria. I. *Plasmodium fieldi*. J Parasitol 53:225–232.
375. Helwig FC (1925) Arachnid infection in monkeys (*Pneumonyssus foxi* of Weidman). Amer J Pathol 1:389.
376. Henderson JD Jr, Webster WS, Bullock BC, Lehner NDM, Clarkson TB (1970) Naturally occurring lesions seen at necropsy in eight woolly monkeys (*Lagothrix* sp.). Lab Anim Care 20:1087–1097.
377. Hendricks LD (1974) A redescription of *Isospora arctopitheci* Rodhain, 1933 (Protozoa:Eimeriidae) from primates of Panama. Proc Helminthol Soc Washington, DC 41:229–233.
378. Hendricks LD (1977) Host range characteristics of the primate coccidian *Isospora arctopitheci* Rodhain 1933 (Protozoa:Eimeriidae). J Parasitol 63:32–35.
379. Herman CM, Schroeder CR (1939) Treatment of amoebic dysentery in an orangutan. Zoologica 24:339.
380. Herman LH (1967) *Gastrodiscoides hominis* infestation in two monkeys. Vet Med 62:355–356.
381. Hessler J, Woodard J, Tucek P (1971) Lethal toxoplasmosis in a woolly monkey. JAVMA 159:1588–1594.
382. Heuschele WP (1961) Internal parasitism of monkeys with the pentastomid, *Armillifer armillatus*. JAVMA 139:911–912.
383. Hickey TE, Kelly WA, Sitzman JE (1983) Demodectic mange in a tamarin (*Saguinus geoffroyi*). Lab Anim Sci 33:192–193.

384. Hill WCO (1951) Report of the Society's prosector for the year 1950. Proc Zool Soc London 121:641–650.
385. Hill WCO (1953) Report of the Society's prosector for the year 1952. Proc Zool Soc London 123:227–251.
386. Hill WCO (1954) Report of the Society's prosector for the year 1953. Proc Zool Soc London 124:303–311.
387. Hill WCO, Porter A, Southwich MD (1952) The natural history, endoparasites, and pseudoparasites of the tarsiers (*Tarsius carbonarius*) recently living in the Society's menagerie. Proc Zool Soc London 122:79–119.
388. Hoare CA (1932) On protozoal blood parasites collected in Uganda. Parasitology 24:210–224.
389. Hoare CC (1958) The enigma of host-parasite relations in amebiasis. Rice Inst Pamphlet 45:23–35.
390. Holmes DD, Kosanke SD, White GL (1980) Fatal enterobiasis in a chimpanzee. JAVMA 177:911–913.
391. Honigberg BM, Lee JJ (1959) Structure and division of *Trichomonas tenax* (O. F. Müller). Amer J Hyg 69:177–201.
392. Honjo R, Imaizumi K (1965) Diseases observed in monkeys. Bull Exp Anim 14:162–163.
393. Honjo S, Muto K, Fujiwara T, Suzuki Y, Imaizumi K (1963) Statistical survey of internal parasites in cynomolgus monkeys (*Macaca irus*). Jap J Exp Med 16:217–224.
394. Hoogstraal H (1956) African Ixodoidea. I. Ticks of the Sudan (with special reference to Equatoria province and with preliminary reviews of the genera *Boophilus*, *Margaropus*, and *Hyalomma*). US Navy Dept Washington, DC Res Rept NM 005 050.29.07.
395. Hoogstrall H, Theiler G (1959) Ticks (Ixodoidea, Ixodidae) parasitizing lower primates in Africa, Zanzibar, and Madagascar. J Parasitol 45:217–222.
396. Hopkins GHE (1949) The host associations of the lice of mammals. Proc Zool Soc London 119:387–604.
397. Houser WD, Paik SK (1971) Hydatid disease in a macaque. JAVMA 159:1574–1577.
398. Howard EB, Gendron AP (1980) *Echinococcus vogeli* infection in higher primates at the Los Angeles zoo. In: Montali RJ, Migaki G (eds) The comparative pathology of zoo animals. Smithsonian Institution Press, Washington, DC, pp 379–382.
399. Howard LM, Cabrera BD (1961) Simian malaria in the Philippines. Science 134:555–556.
400. Hsu C-K, Melby EC Jr (1974) *Isospora callimico*, n. sp. (Coccidia Eimeriidae) from Göldi's marmoset (*Callimico goeldii*). Lab Anim Sci 24:476–479.
401. Hsü HF, David JR, Hsü SYL (1969) Histopathological lesions of rhesus monkeys and chimpanzees infected with *Schistosoma japonicum*. Z Tropenmed Parasitol 20:184–205.
402. Hsü SYL, Hsü HF (1968) A chimpanzee naturally infected with *Schistosoma mansoni*: Its resistance against a challenge infection of *S. japonicum*. Trans Roy Soc Trop Med Hyg 62:901–902.
403. Hubbard GB, Butcher WI (1983) What's your diagnosis? Passengers. Trematodiasis and typhlitis caused by *Gastrodiscoides hominis*. Lab Anim 12:12, 14.
404. Huff CG, Hoogstraal H (1963) *Plasmodium lemuris* n. sp. from *Lemur collaris*. J Inf Dis 112:233–236.
405. Hull WB (1970) Respiratory mite parasites in non-human primates. Lab Anim Care 20:402.
406. Hunt RD, Jones TC, Williamson M (1970) Mechanisms of parasitic damage and the host response. Lab Anim Care 20:345–353.
407. Hunt RD, Anderson MP, Chalifoux LV (1978) Spontaneous infectious diseases of marmosets. Prim Med 10:239–253.

408. Ilievski V, Esber H (1969) Hydatid disease in a rhesus monkey. Lab Anim Care 19:199–204.
409. Inglis WG (1961) The oxyurid parasites (nematoda) of primates. Proc Zool Soc London 136:103–122.
410. Inglis WG, Cosgrove GE (1965) The pinworm parasites (Nematoda:Oxyuridae) of the Hapalidae (Mammalia:Primates). Parasitology 55:731–737.
411. Inglis WG, Dunn FL (1963) The occurrence of *Lemuricola* (Nematoda: Oxyurinae) in Malaya: with description of a new species. Z Parasitenk 23:354–359.
412. Inglis WG, Dunn FL (1964) Some Oxyurids (Nematoda) from neotropical primates. Z Parasitenk 24:83–87.
413. Innes JRM, Hull WB (1972) Endoparasites—lung mites. In: Fiennes RNT-W (ed) Pathology of simian primates. Part II. Infections and parasitic diseases. S Karger, New York, pp 177–193.
414. Innes JRM, Colton MW, Yevich PP, Smith CL (1954) Pulmonary acariasis as an enzootic disease caused by *Pneumonyssus simicola* in imported monkeys. Amer J Pathol 30:813–835.
415. Inoki S, Takemura S, Makiura Y, Hotta F (1942) Studies on *Plasmodium inui var. Cyclopsis* n. sp., new malaria parasite found in Formosan macaques (*Macaca cyclopsis*). Osaka Igakkuizassi 41:1327–1343.
416. Irving GW III (1972) Parasitology. In: Selected topics in laboratory animal medicine. Vol VIII. Aeromedical Review 2–72. USAF School of Aerospace Medicine, Aerospace Medical Division (AFSC), Brooks AFB, TX, pp 1–44.
417. Irving GW III (1972) Zoonoses of primates. In: Selected topics in laboratory animal medicine. Vol XIV. Aeromedical Review 13–72. US Air Force School of Aerospace Medicine, Aerospace Medical Division (AFSC), Brooks, AFB, TX.
418. Itakura C, Nigi H (1968) Histopathological observations on two spontaneous cases of toxoplasmosis in the monkey (*Lemur catta*). Jap J Vet Sci 30:341–346.
419. Jaskoski BJ (1960) Physalopteran infection in an orangutan. JAVMA 137:307.
420. Jensen JM, Huntress SL (1982) *Capillaria hepatica* infestation in a gelada baboon (*Theropithecus gelada*) troop. Amer Assoc Zoo Vet Ann Proc 48–49.
421. Jessee MT, Schilling PW, Stunkard JA (1970) Identification of intestinal helminth eggs in Old World primates. Lab Anim Care 20:83–87.
422. Johnsen DO, Gould DJ, Tanticharoenyos P, Diggs CL, Wooding WL (1970) Experimental infection of gibbons with *Dirofilaria immitis*. Trans Roy Soc Trop Med Hyg 64:937–938.
423. Johnson CM (1941) Observations on natural infections of *Endamoeba histolytica* in *Ateles* and rhesus monkeys. Amer J Trop Med 21:49–61.
424. Jones DM (1982) Veterinary aspects of the maintenance of orang-utans in captivity. In: Boer de LEM (ed) The orang-utan: its biology and conservation. Dr W Junk, The Hague, pp 171–199.
425. Jones ND, Brooks DR, Harris RL (1980) *Macaca mulatta*-a new host for *Choanotaenia* Cestodes. Lab Anim Sci 30:575–577.
426. Jones SR (1973) Toxoplasmosis: a review. JAVMA 163:1038–1042.
427. Jones TC, Hunt RD (1983) Diseases due to protozoa. Veterinary pathology, 5th ed. Lea & Febiger, Philadelphia, pp 719–777.
428. Jones TC, Hunt RD (1983) Diseases caused by parasitic helminths and arthropods. Veterinary pathology, 5th ed. Lea & Febiger, Philadelphia, pp 778–879.
429. Jortner BS, Percy DH (1978) The nervous system. In: Benirschke K, Garner FM, Jones TC (eds) Pathology of laboratory animals, chapter 5, vol I. Springer-Verlag, New York, pp 320–421.
430. Kagan IG, Allain DS, Norman L (1959) An evaluation of the hemagglutination and flocculation tests in the diagnosis of *Echinococcus* disease. Amer Trop Med Hyg 8:51–55.
431. Kalter SS (1980) Infectious diseases of the great apes of Africa. J Reprod Fertil Suppl 28:149–159.

432. Kalter SS, Heberling RL (1978) Health hazards associated with newly imported primates and how to avoid them. In: Chivers DJ, Ford EHR (eds) Recent advances in primatology, vol 4. Medicine. Academic, New York, pp 5–21.
433. Karr SL Jr, Wong MM (1975) A survey of sarcocystis in nonhuman primates. Lab Anim Sci 25:641–645.
434. Karr SL Jr, Henrickson RV, Else JG (1979) A survey for *Anatrichosoma* (Nematoda: Trichinellida) in wild-caught *Macaca mulatta*. Lab Anim Sci 29:789–790.
435. Karr SL Jr, Henrickson RV, Else JG (1980) A survey for intestinal helminths in recently wild-caught *Macaca mulatta* and results of treatment with mebendazole and thiabendazole. J Med Primatol 9:200–204.
436. Kasa TJ, Lathrop GD, Dupuy HJ, Bonney CH, Toft JD II (1977) An endemic focus of *Trypanosoma cruzi* infection in a subhuman primate research colony. JAVMA 171:850–854.
437. Kaufmann AF, Morris G, Richardson JH, Healy G, Kaplan W (1970) A survey of newly arrived South American monkeys for potential human pathogens. In: Primate zoonoses surveillance. Report No. 1. Centers for Disease Control, Atlanta, GA.
438. Keeling ME, McClure HM (1972) Clinical management, diseases and pathology of the gibbon and siamang. In: Rumbaugh DM (ed) Gibbon and siamang, vol 1. S Karger, Basel, pp 207–249.
439. Keeling ME, McClure HM (1974) Pneumococcal meningitis and fatal enterobiasis in a chimpanzee. Lab Animal Sci 24:92–95.
440. Keeling ME, Wolf RH (1975) Medical management of the rhesus monkey. In: Bourne GH (ed) The rhesus monkey, vol 2. Academic Press, New York, pp 11–96.
441. Kellogg VL (1913) Ectoparasites of the monkeys, apes and man. Science 38:601.
442. Kennard MA (1981) Abnormal findings in 246 consecutive autopsies on monkeys. Yale J Biol Med 13:701–712.
443. Kessel JF (1928) Intestinal protozoa of monkeys. Univ Calif (Berkeley) Publ Zool 31:275–306.
444. Kessel JF, Johnstone HG (1949) The occurrence of *Endamoeba polecki* Prowazek, 1912, in *Macaca mulatta* and in man. Amer J Trop Med 29:311–317.
445. Kessel JF, Kaplan F (1949) The effect of certain arsenicals on natural infections of *Endamoeba histolytica* and of *Endamoeba polecki* in *Macaca mulatta*. Amer J Trop Med 29:319–322.
446. Kessler MJ (1982) Nasal and cutaneous anatrichosomiasis in the free-ranging rhesus monkeys (*Macaca mulatta*) of Cayo Santiago. Amer J Primatol 3:55–60.
447. Kessler MJ, Yarbrough B, Rawlins RG, Berard J (1984) Intestinal parasites of the free-ranging Cayo Santiago rhesus monkeys (*Macaca mulatta*). J Med Primatol 13:57–66.
448. Keymer IF (1976) Report of the pathologist, 1973 and 1974. J Zool 178:456–493.
449. Kikuth W (1927) Piroplasmose bei Affen. Arch Schiffs Tropen Hyg 31:37–40.
450. Kim CS, Bang BG (1970) Nasal mites parasitic in nasal and upper skull tissues in the baboon (*Papio* sp.). Science 169:372–373.
451. Kim CS, Kalter SS (1972) Unilateral renal aplasia in an African baboon (*Papio* sp.). Folia Primatol 2:352.
452. Kim CS, Eugster AK, Kalter SS (1968) Pathologic study of the African baboon (*Papio* sp.) in his native habitat. Primates 9:93–104.
453. Kim CS, Bang FB, DiGiacomo RF (1972) Hemagglutination assay of antibodies associated with pulmonary acariasis in rhesus monkeys. Infect Immun 5:138–140.
454. Kim JCS (1974) Distribution and life cycle stages of lung mites (*Pneumonyssus* sp.). J Med Primatol 3:105–119.
455. Kim JCS (1976) Scanning electron microscopic studies of simian lung mites. J Med Primatol 5:3.
456. Kim JCS (1977) Pulmonary acariases in Old World monkeys. Vet Bull 47:249–255.
457. Kim JCS (1980) Pulmonary acariasis in Old World monkeys: a review. In: Montali

RJ, Migaki G (eds) The comparative pathology of zoo animals. Smithsonian Institution Press, Washington, DC, pp 383–394.

458. Kim JCS, Kalter SS (1975) Pathology of pulmonary acariasis in baboons (*Papio* sp.). J Med Primatol 4:70.
459. Kim JCS, Abee CR, Wolf RH (1978) Balantidiosis in a chimpanzee (*Pan troglodytes*). Lab Anim 12:231–233.
460. Kim KC, Emerson KC (1968) Descriptions of two species of Pediculidae (Anoplura) from great apes (Primates, Pongidae). J Parasitol 54:690–695.
461. King NW Jr (1976) Synopsis of the pathology of new world monkeys. In: First inter-American conference on conservation and utilization of American nonhuman primates in biomedical research. Scientific Publ No 317, Pan American Health Organization, Pan American Sanitary Bureau Regional Office of the WHO, Washington, DC, pp 169–198.
462. Kingston N, Cosgrove GE (1967) Two new species of *Platynosomum* (Trematode:Dicrocoeliidae) from South American monkeys. Proc Helminthol Soc Washington, DC 34:147–151.
463. Kirby H Jr (1928) Notes on some parasites from chimpanzees. Proc Soc Exp Biol Med 25:698–700.
464. Knezevich AL, McNulty WP Jr (1970) Pulmonary acariasis (*Pneumonyssus simicola*) in colony bred *Macaca mulatta*. Lab Anim Care 20:693–696.
465. Knowles R (1919) Notes on the monkey plasmodium and on some experiments in malaria. Ind J Med Res 7:195–202.
466. Knowles R, Das Gupta BM (1932) A study of monkey malaria and its experimental transmission to man. Ind Med Gaz 67:213–268.
467. Knowles R, Das Gupta BM (1936) Some observations on the intestinal protozoa of macaques. Ind J Med Res 24:547–556.
468. Kobayashi H (1925) On the animal parasites in Korea. Jap Med World 5:9–16.
469. Kohn DF, Haines DE (1982) Diseases of the Prosimii: a review. In: Haines DE (ed) The lesser bushbaby (Galago) as an animal model: selected topics. CRC, Boca Raton, FL, pp 285–301.
470. Kopciowska L, Nicolau S (1938) Toxoplasmose spontanée du chimpanzé. CR Soc Biol (Paris) 129:179–181.
471. Korté WF de (1905) On the presence of a sarcosporidium in the thigh muscles of *Macaca rhesus*. J Hyg 5:451–452.
472. Kovatch RM, White JD (1972) Cryptosporidiosis in two juvenile rhesus monkeys. Vet Pathol 9:426–440.
473. Kraschenninikow S, Wenrich DH (1958) Some observations on the morphology and division of *Balantidium coli* and *Balantidium caviae* (?). J Protozool 5:196–202.
474. Kreis HA (1932) A new pathogenic nematode of the family Oxyuroidea, *Oxyuronema atelophora* n. g., n. sp in the red-spider monkey, *Ateles geoffroyi*. J Parasitol 18:295–302.
475. Kreis HA (1945) Beiträge zur Kenntnis parasitischer Nematoden. XII. Parasitische Nematoden aus den Tropen. Rev Suisse Zool 52:551.
476. Kreis HA (1955) Beiträge zur Kenntnis parasitischer Nematoden. XVIII. Das Genus Probstmayria Ransom, 1907. Schweiz Arch Tierheilkd 97:422–433.
477. Krishnamoorthy RV, Srihari K, Rahaman H, Rajasekharaiah GL (1978) Nematode parasites of the slender loris, *Loris tardigradus*. Proc Ind Acad Sci 87B:17–22.
478. Kugi G, Sawada I (1970) *Mathevotaenia brasiliensis* n. sp., a tapeworm from the squirrel monkey, *Saimiri sciureus*. Jap J Parasitol 19:467–470.
479. Kuhn H-J, Ludwig HW (1967) Die Affenläuse der Gattung *Pedicinus*. Z Zool Syst Evolutions 5:144–256.
480. Kumar V, DeMeurichy W (1982) Efficacy of Yomesan against *Bertiella* sp (Anoplocephalidae/Cestoda) of a chimpanzee, *Pan schweinfurthii*. Riv Parassitol 43: 161–163.

481. Kumar V, DeMeurichy W, Van Peer L (1980) Microscopic pathology of liver of capuchin monkey (*Cebus albifrons*) infected with *Athesmia foxi* (Dicrocoelidae:Trematoda): a pictorial illustration. ACTA Zool Pathol Antverp 75:71–77.
482. Kuntz RE (1972) Trematodes of the intestinal tract and biliary passages. In: Fiennes RNT-W (ed) Pathology of simian primates, part II. Infectious and parasitic diseases. S Karger, Basel, pp 104–123.
483. Kuntz RE (1973) Models for investigation in parasitology. In: Bourne GH (ed) Nonhuman primates in biomedical research. Academic, New York, pp 167–201.
484. Kuntz RE (1982) Significant infections in primate parasitology. J Hum Evol 11:185–194.
485. Kuntz RE, Moore JA (1973) Commensals and parasites of African baboons (*Papio cynocephalus* L. 1766) captured in rift valley province of central Kenya. J Med Primatol 2:236–241.
486. Kuntz RE, Myers BJ (1966) Parasites of baboons (*Papio doguera* Pucheran, 1856) captured in Kenya and Tanzania, East Africa. Primates 7:27–32.
487. Kuntz RE, Myers BJ (1967) Microbiological parameters of the baboon (*Papio* sp): parasitology. In: Vagtborg H (ed) The baboon in medical research, vol 2. University of Texas Press, Austin, TX, pp 741–755.
488. Kuntz RE, Myers BJ (1967) Parasites of the Kenya baboon: arthropods, blood protozoa and helminths. Primates 8:75–82.
489. Kuntz RE, Myers BJ (1967) Primate cysticercosis: *Taenia hydatigena* in Kenya vervets (*Cercopithecus aethiops* Linnaeus, 1758) and Taiwan macaques (*Macaca cyclopis* Swinhoe, 1864). Primates 8:83–88.
490. Kuntz RE, Myers BJ (1969) A checklist of parasites and commensals reported for the Taiwan macaque (*Macaca cyclopis* Swinhoe, 1862). Primates 10:71–80.
491. Kuntz RE, Myers BJ (1969) Parasitic protozoa, commensals and helminths of chimpanzees imported from the Republic of the Congo. Proc 2nd Int Congr Primat, Atlanta GA, 1968, vol 3. S Karger, Basel/New York, pp 184–190.
492. Kuntz RE, Myers BJ (1972) Parasites of South American primates. Intern Zoo Yrbk 12:61–68.
493. Kuntz R, Myers BJ, Vice TC (1967) Intestinal protozoans and parasites of the gelada baboon (*Theropithecus gelada* Rüppel, 1835). Proc Helminthol Soc Washington, DC 34:65–66.
494. Kuntz RE, Myers BJ, Bergner JF Jr, Armstrong DE (1968) Parasites and commensals of the Taiwan macaque (*Macaca cyclopis* Swinhoe, 1862). Formosan Sci 22:120–136.
495. Kuntz RE, Myers BJ, Katzberg A (1970) Sparganosis and "proliferative-like" spargana in vervets and baboons from East Africa. J Parasitol 56:196–197.
496. Kuntz RE, Myers BJ, McMurray TS (1970) *Trypanosoma cruzi*-like parasites in the slow loris (*Nycticebus coucang*) from Malaysia. Trans Amer Micr Soc 89:304–307.
497. Kuntz RE, Myers BJ, Moore JA (1973) Parasitology. In: Kalter SS (ed) The baboon. Microbiology, clinical chemistry, and some hematological aspects. Primates in medicine, vol 8. S Karger, Basel, pp 79–104.
498. Kuntz RE, Huang T, Moore JA (1977) Patas monkey (*Erythrocebus patas*) naturally infected with *Schistosoma mansoni*. J Parasitol 63:166–167.
499. Kupper JL, Britz WE (1972) The squirrel monkey. In: Selected topics in laboratory animal medicine. Vol XVIII. Aeromedical Review 5–72. USAF School of Aerospace Medicine, Aerospace Medical Division (AFSC), Brooks AFB, TX, pp 1–16.
500. Laing ABG, Edeson JFB, Wharton RH (1960) Studies on filariasis in Malaya; the vertebrate hosts of *Brugia malayi* and *B. pahangi*. Ann Trop Med Parasit 54:92–99.
501. Lambert RA (1918) *Echinococcus* cysts in a monkey. Proc NY Path Soc 18:29–30.
502. Lambrecht FL, Dunn FL, Eyles DE (1961) Isolation of *Plasmodium knowlesi* from Philippine macaques. Nature 191:1117–1118.
503. Landois F, Hoepke H (1914) Eine endoparasitäre Milbe in der Lunge von Macacus rhesus. Zentralbl Bakteriol. Abt 1 73:384.

504. Lang EM (1966) The care and breeding of anthropoids. Symp Zool Soc London 17:113–125.
505. Langeron M (1920) Note additionnelle sur une hémogrégarine d'un macaque. Bull Soc Path Exot 13:394.
506. Lapage G (1962) Mönnig's veterinary helminthology and entomology, 5th ed. Williams and Wilkins, Baltimore.
507. Lapage G (1968) Veterinary parasitology. Oliver and Boyd, Edinburgh/London.
508. Lapin BA (1962) Disease in monkeys within the period of acclimatization and during long-term stay in animal houses. In: Harris RJC (ed) The problems of laboratory animal disease. Academic, New York, pp 143–149.
509. Lapin BA, Yakovleva LA (1960) Comparative pathology in monkeys. CC Thomas, Springfield, IL.
510. Lapin BA, Yakovleva LA (1963) Diseases of the cardiovascular system. In: Comparative pathology in monkeys. CC Thomas, Springfield, IL, pp 132–177.
511. Lasry JE, Sheridan BW (1965) Chagas' myocarditis and heart failure in the red uakari. Intern Zoo Yrbk 5:182–187.
512. Lavoipierre MMJ (1955) A description of a new genus of sarcoptiform mites and of three new species of Acarina parasitic on primates in the British Cameroons. Ann Trop Med Parasitol 49:299–307.
513. Lavoipierre MMJ (1955) The occurrence of a mange mite, *Psoregates* sp. (*Acarina*), in a West African monkey. Ann Trop Med Parasitol 49:351.
514. Lavoipierre MMJ (1958) Studies on the host-parasite relationships of filarial-nematodes and their arthropod hosts. I. The sites of development and the migration of *Loa loa* in *Chrysops silacea*, the escape of the infective forms from the head of the fly, and the effect of the worm on its insect host. Ann Trop Med Parasitol 52:103–121.
515. Lavoipierre MMJ (1964) A new family of acarines belonging to the suborder Sarcoptiformes parasitic in the hair follicles of primates. Ann Natal Mus 16:1–18.
516. Lavoipierre MMJ (1964) A note on the family Psoralgidae (Acari:Sarcoptiformes) together with a description of two new genera and two new species parasitic on primates. Acarologia 6:342–352.
517. Lavoipierre MMJ (1970) A note on the sarcoptic mites of primates. J Med Entomol 7:376–380.
518. Leathers CW (1978) The prevalence of *Hepatocystis kochi* in African green monkeys. Lab Anim Sci 28:186–189.
519. Leathers CW (1978) Pulmonary acariasis in a infant, colony-born rhesus monkey (*Macaca mulatta*). Lab Anim Sci 28:102–103.
520. Lebel RR, Nutting WB (1973) Demodectic mites of subhuman primates. I. *Demodex saimiri* sp. n. (Acari:Demodicidae) from the squirrel monkey, *Saimiri sciureus*. J Parasitol 59:719–722.
521. Lee KJ, Lang CM, Hughes HC, Hartshorn RD (1981) Psorergatic mange (Acari: Psorergatidae) of the stumptail macaque (*Macaca arctoides*). Lab Anim Sci 31:77–79.
522. Lee RE, Williams RB Jr, Hull WB, Stein SN (1954) Significance of pulmonary acariasis in rhesus monkeys (*Macaca mulatta*). Fed Proc 13:85–86.
523. Leerhoy J, Jensen HS (1967) Sarcoptic mange in a shipment of cynomolgus monkeys. Nord Veterinarmed 19:128–130.
524. Lefrou G, Martignoles J (1955) Contribution à l'étude de *Plasmodium kochi*. Plasmodium des singes africans. Bull Soc Path Exot 48:227–234.
525. Leger M, Bédier E (1922) Hémogrégarine du cynocéphale, *Papio sphinx* E. Geoffrey. CR Soc Biol 87:933–934.
526. Leibegott G (1962) Pericarditis verminosa (strongyloides) beim Schimpansen. Virchows Arch Path Anat 335:211–225.
527. Levaditi C, Schoen R (1933) Présence d'un toxoplasme dans l'encéphale du cynocephalus babuin. Bull Soc Path Exot 26:402–405.

528. Levine ND (1961) Protozoan parasites of domestic animals and of man. Burgess, Minneapolis.
529. Levine ND (1970) Protozoan parasites of nonhuman primates as zoonotic agents. Lab Anim Care 20:377–382.
530. Levine ND (1973) Protozoan parasites of domestic animals and of man, 2nd ed. Burgess, Minneapolis.
531. Levine ND (1976) Nematode parasites of domestic animals and of man. Burgess, Minneapolis.
532. Levine ND (1977) Nomenclature of *Sarcocystis* in the ox and sheep and of fecal coccidia of the dog and cat. J Parasitol 63:36–51.
533. Lichtenfels JR (1971) Morphological variation in the gullet nematode, *Gongylonema pulchrum* Molin, 1857, from eight species of definitive hosts with a consideration of gongylonema from *Macacca* spp. J Parasitol 57:348–355.
534. Lillie RD (1947) Reactions of various parasitic organisms in tissues to the Bauer Feulgen Gram and Gram-Weigert methods. J Lab Clin Med 32:76–88.
535. Lim-Boo-Liat (1974) New hosts of *Angiostrongylus malaysiensis* Bhaibulaya and Cross 1971, in Malaysia. Southeast Asian J Trop Med Pub Health 5:379–384.
536. Lindquist WD, Bieletzki J, Allison S (1980) *Pterygodermatites* sp. (Nematode: Rictulariidae) from primates in Topeka, Kansas Zoo. Proc Helminthol Soc Washington, DC 47:224–227.
537. Linstow von OFB (1899) Nematoden aus der Berliner zoologischen Sammlung. Mitt Zool Mus Berl 1(2)–1.
538. Linstow von OFB (1903) The American hookworm in chimpanzees. Amer Med 6:611.
539. Linstow von OFB (1912) Cestoda and cestodaria. In: Stilles and Hassall Index-catalogue of medical and veterinary zoology. Bull 85 Hyg Lab US Publ Health and Mar Hosp Serv.
540. Little MD (1962) Experimental studies on the life cycle of *Strongyloides*. J Parasitol Suppl 48:41.
541. Little MD (1966) Comparative morphology of six species of *Strongyloides* (Nematoda) and redefinition of the genus. J Parasitol 52:69–84.
542. Loeb WF, Bannerman RM, Rininger BF, Johnson AJ (1978) Hematologic disorders. In: Benirschke K, Garner FM, Jones TC (eds) Pathology of laboratory animals, chapter 11, vol I. Springer-Verlag, New York, pp 1000–1021, 1032–1050.
543. Long GG, White JD, Stookey JL (1975) *Pneumocystis carinii* infection in splenectomized owl monkeys. JAVMA 167:651–654.
544. Loomis MR, Britt JO (1983) An epizootic of *Entamoeba histolytica* in colobus monkeys. Amer Assoc Zoo Vet Ann Proc, p 10.
545. Loomis MR, Britt JO, Gendron AP, Holshuh HJ, Howard EB (1983) Hepatic and gastric amebiasis in black and white colobus monkeys. JAVMA 183:1188–1191.
546. Lothe DF (1958) An immature *Oesophagostomum* sp. from an umbilical swelling in an African child. Trans Roy Soc Trop Med Hyg 52:12.
547. Lucker JT (1933) *Gongylonema macrogubernaculum* Lubimov, 1931: two new hosts. J Parasitol 19:243.
548. Lucker JT (1933) Two new hosts *Gongylonema pulchrum* Molin, 1857. J Parasitol 19:248.
549. Lushbaugh CC, Humason G, Gengozian N (1969) Intrauterine death from congenital Chagas' disease in laboratory marmosets (*Saguinus fuscicollis labonotus*). Amer J Trop Med 18:662–665.
550. Lussier G, Marois P (1964) Animal sarcosporidiosis in the province of Quebec. Can J Publ Health 55:243–246.
551. Machida M, Araki J, Koyama T, Kumada M, Horii Y, Imada I, Takasaka M, Honjo S, Matsubayashi K, Tibat T (1978) The life cycle of *Streptopharagus pigmentatus* (Nematoda, Spiruroidea) from the Japanese monkey. Bull Nat Sci Mus (Ser A) 4:1–9.

552. MacKenzie PS (1979) Pathogenicity, identification and treatment of *Prosthenorchis elegans* infestation in squirrel monkeys (*Saimiri sciureus*). Prim Suppl 4:5–7.
553. MacKinnon DL, Dibb MJ (1938) Report on intestine protozoa of some mammals in the zoological gardens at Regent's Park. Proc Zool Soc London Ser B 108:323–345.
554. Magalhaes Pinto R (1970) Occurrence of *Subulura jacchi* (Marcel, 1857) Railliet and Henry, 1913 (Nematoda, Subuluroidea) in a new host: *Callithrix aurita coelestis* (M. Ribeiro, 1924). Atas de Sociedade de Biologia do Rio De Janeiro 13:143–145.
555. Mandour AM (1969) *Sarcocystis nesbitti* n. sp. from the rhesus monkey. J Protozool 16:353–354.
556. Manwell RD (1968) Simian malaria. In: Weinman D, Ristic M (eds) Infectious blood diseases of man and animals, vol II. Academic, New York, pp 78–88.
557. Manwell RD, Kuntz RE (1966) *Hepatocystis* in Formosan mammals with a description of a new species. J Protozool 13:670–672.
558. Maplestone PA (1931) Parasitic nematodes obtained from animals dying in the Calcutta Zoological Gardens. Pt 4–8. Rec Indian Mus 33:71–171.
559. Marinkelle CS (1966) Observations on man, monkey, and bat trypanosomes and their vectors in Colombia. Trans Roy Soc Trop Med Hyg 60:109–116.
560. Marinkelle CJ (1968) *Trypanosoma lambrechti* n. sp. aislado de micos (*Cebus albifrons*) de Colombia. Caldasia 10:155–165.
561. Masbar S, Palmieri JR, Marwoto HA, Purnomo, Darwis F (1981) Blood parasites of wild and domestic animals from South Kalimantan (Borneo), Indonesia. Southeast Asian J Trop Med Pub Health 12:42–46.
562. Masse R, Geneste M, Thiery G (1965) Acariose pulmonaire du singe traitement, prophylaxie. Rec Méd Vét Ecole Alfort 141:1227–1234.
563. Mayer M (1907) Über Malaria beim Affen. Med Klin 3:579–580.
564. Mayer M (1933) Über einen neuen Blutparasiten des Affen (*Entopolypoides macaci* n. g. n. sp.). Arch Schiffs Tropen Hyg 37:504–505.
565. Mayer M (1934) Ein neuer, eigenartiger Blutparasit des Affen (*Entopolypoides macaci* n. g. et n. sp.). Zentr Bakteriol Parasitenk Abt I Orig 131:132–136.
566. Mazza S (1930) Doble parasitismo por filarias en monos *Cebus* del norte. 5a Reun Soc Argent Pat Reg N 1929 2:1140.
567. McCarrison R (1920) The effects of deficient dietaries on monkeys. Brit Med J Feb 21 249–253.
568. McClure GW (1932) Nematode parasites of mammals with a description of a new species, *Wellcomia branickii* from specimens collected in the New York Zoological Park, 1930. Zoologica NY 15:1.
569. McClure GW (1934) Nematode parasites of mammals from specimens collected in the New York Zoological Park, 1932. Zoologica NY 15:49.
570. McClure HM, Guilloud NB (1971) Comparative pathology of the chimpanzee. In: Bourne GH (ed) The chimpanzee, vol 4. Behavior, growth and pathology of chimpanzees. University Park Press, Baltimore, pp 103–272.
571. McClure HM, Strozier LM, Keeling ME, Healy GR (1973) Strongyloidosis in two infant orangutans. JAVMA 163:629–632.
572. McClure HM, Chapman WL Jr, Hooper BE, Smith FG, Fletcher OJ (1978) The digestive system. In: Benirschke K, Garner FM, Jones TC (eds) Pathology of laboratory animals, chapter 4, vol I. Springer-Verlag, New York, pp 176–317.
573. McConnell EE (1977) Parasitic diseases observed in free-ranging and captive baboons. Comp Path Bull 9:2.
574. McConnell EE, Basson PA, Devos V (1971) Nasal acariasis in the chacma baboon. *Papio ursinus* Kerr 1792. Onderstepoort J Vet Res 38:207.
575. McConnell EE, DeVos AJ, Basson PA, DeVos V (1971) *Isopora papionis* n. sp. (Eimeriidae) of the Chacma baboon *Papio ursinus* (Kess, 1792). J Protozool 18:28–32.

576. McConnell EE, Basson PA, Thomas SE, DeVos V (1972) Oocysts of *Isospora papionis* in the skeletal muscle of Chacma baboons. Onderstepoort J Vet Res 39:113–116.
577. McConnell EE, Basson PA, DeVos V (1972) Laryngeal acariasis in the chacma baboon. JAVMA 161:678–682.
578. McConnell EE, Basson PA, DeVos V, Myers BJ, Kuntz RE (1972) A survey of diseases among 100 free-ranging baboons (*Papio ursinus*) from Krueger National Park. Onderstepoort J Vet Res 41(3):97–167.
579. McConnell EE, Basson PA, Wolstenholme B, DeVos V, Malherbe H (1973) Toxoplasmosis in "free-ranging" chacma baboons (*Papio ursinus*) from the Krueger National Park. Trans Roy Soc Trop Med Hyg 67:851–855.
580. McCoy OR (1936) Filarial parasites of monkeys of Panama. Amer J Trop Med 16:383–403.
581. McKissick GE, Ratcliffe HL, Koestner A (1968) Enzootic toxoplasmosis in caged squirrel monkeys. *Saimiri sciureus*. Pathol Vet 5:538–560.
582. Mendonca de MM (1983) *Mathevotaenia cruzsilvai* n. sp. (cestoda Anoplocephalidae), parasite of *Macaca irus* F. Cuvier, 1818. Bull Mus (Nat) d'Hist Nat A3:1081–1086.
583. Middleton CC (1966) Acanthocephala (*Prosthenorchis elegans*) infection in squirrel monkeys (*Saimiri sciureus*). Lab Anim Dig 2:16–17.
584. Middleton CC, Clarkson TB, Garner FM (1964) Parasites of squirrel monkeys (*Saimiri sciureus*). Lab Anim Care 14:335.
585. Migaki G, Seibold HR, Wolf RH, Garner FM (1971) Pathologic conditions in the patas monkey. JAVMA 159:549–556.
586. Miller JH (1959) *Hepatocystis* (= *Plasmodium*) *kochi* in the dog face baboon, *Papio doguera*. J Parasitol Suppl 45:53.
587. Miller JH (1959) The dog face baboon, *Papio doguera*, a primate reservoir host for *Schistosoma mansoni* in East Africa. J Parasitol 45:22–25.
588. Miller MJ, Bray RS (1966) *Entamoeba histolytica* infections in the chimpanzee (*Pan satyrus*). J Parasitol 52:386–388.
589. Moncol DJ, Batte EG (1966) Transcolostral infection of newborn pigs with *Strongyloides ransomi*. Vet Med 61:583–586.
590. Monnig HO (1920) *Filaria nycticebi* eine neue Filaria aus dem Nycticebus. Centralb Bakteriol 85:216–221.
591. Montali RJ, Bush M (1980) Diagnostic exercise. Lab Anim Sci 30:33–34.
592. Montali RJ, Bush M (1981) Rictulariasis in callitrichidae at the National Zoological Park. In: XXIII. Internationalen Symposiums über die Erkrankungen der Zootiere Halle/Saale. Akademie-Verlag, Berlin, pp 197–202.
593. Montali RJ, Gardiner CH, Evans RE, Bush M (1983) *Pterygodermatites nycticebi* (Nematoda:Spirurida) in golden lion tamarins. Lab Anim Sci 33:194–197.
594. Montgomery CA (1978) Muscle diseases. In: Benirschke K, Garner FM, Jones TC (eds) Pathology of laboratory animals, chapter 10, vol I. Springer-Verlag, New York, pp 841–853, 880–887.
595. Moore G, Myers BJ (1974) Parasites of non-human primates. In: Ann Proc Amer Assoc Zoo Vet, Washington, DC pp 79–86.
596. Moore JA, Kuntz RE (1975) *Entopolypoides macaci* Mayer 1934 in the African baboon (*Papio cynocephalus* L. 1776). J Med Primatol 4:1–7.
597. Moore JG (1970) Epizootic of acanthocephaliasis among primates. JAVMA 157:699–705.
598. Mooreman AE (1941) *Balantidium coli* and pinworm in a chimpanzee. J Parasitol 27:366.
599. Mortelman J, Vercruysse J, Kageruka P (1971) Three pathogenic intestinal protozoa of anthropoid apes: *Entamoeba histolytica*, *Balantidium coli* and *Troglodytella abrassarti*. In: Biegert J, Leutenegger W (eds) Proceedings of the Third International

Congress of Primatology, vol 2. Neurobiology, immunology, cytology. S Karger, Basel, pp 187–191.
600. Morton HL (1969) Sparganosis in African green monkeys (*Cercopithecus aethiops*). Lab Anim Care 19:253–255.
601. Muller R, Ruedi D (1981) Gastric amebiasis in a proboscis monkey (*Nasalis larvatus*). ACTA Zool Pathol Antverp 76:9–16.
602. Mulligan HW (1935) Description of two species of monkey *Plasmodium* isolated from *Silenus irus*. Arch Protistenk 84:285–314.
603. Mullin SW, Orihel TC (1972) *Tetrapetalonema dunni* sp. n. (Nematoda:Filaroidea) from Malaysian tree shrews. J Parasitol 58:1047–1051.
604. Mullin SW, Colley FC, Stevens GS (1972) *Coccidia* of Malaysian mammals: new host records and descriptions of three new species of *Eimeria*. J Protozool 19:260–263.
605. Mullin SW, Dondero TT Jr, Sivanandam S, Dewey R (1972) Filarial parasites of Malaysian monkeys. Southeast Asian J Trop Med Publ Health 3:548–551.
606. Murphy JC, Fox JG, Shalev M (1979) *Paragonimus westermani* infection in a cynomolgus monkey. JAVMA 175:981–984.
607. Myers BJ (1970) Techniques for recognition of parasites. Lab Anim Care 20:342–344.
608. Myers BJ (1972) Echinococcosis, coenurosis, cysticercosis, sparganosis, etc. In: Fiennes RNT-W (ed) Pathology of simian primates. Part II. Infectious and parasitic diseases. S Karger, Basel, pp 124–143.
609. Myers BJ, Kuntz RE (1965) A checklist of parasites reported for the baboon. Primates 6:137–194.
610. Myers BJ, Kuntz RE (1967) Parasites of baboons taken by the Cambridge Mwanza expedition (Tanzania 1965). East Afr Med J 44:322–324.
611. Myers BJ, Kuntz RE (1968) Intestinal protozoa of the baboon *Papio doguera* Pucheran 1856. J Protozool 15:363–365.
612. Myers BJ, Kuntz RE (1969) Nematode parasites of mammals (Dermoptera, Primates, Pholidata, Rodentia, Carnivora, and Artiodactyla) from North Borneo (Malaysia). Can J Zool 47:419–421.
613. Myers BJ, Kuntz RE (1972) A checklist of parasites and commensals reported for the chimpanzee (*Pan*). Primates 13:433–471.
614. Myers BJ, Kuntz RE, Vice TE (1965) Hydatid disease in captive primates (*Colobus* and *Papio*). J Parasitol Suppl 51:22.
615. Myers BJ, Kuntz RE, Vice TE, Kim CS (1970) Natural infection of *Echinococcus granulosus* (Batsch, 1786) Rudolph, 1805 in the Kenya baboon (*Papio* sp.). Lab Anim Care 20:283–286.
616. Myers BJ, Kuntz RE, Malherbe H (1971) Intestinal commensals and parasites of the South African baboon (*Papio cynocephalus*). Trans Amer Micr Soc 90:80–83.
617. Myers BJ, Kuntz RE, Kamara JA (1973) Parasites and commensals of chimpanzees captured in Sierra Leone, West Africa. Pro Helminthol Soc Washington, DC 40:298–299.
618. Mysorekar NR, Chakravarti RN, Chawla LS, Chhuttani PN (1966) Diseases of rhesus monkeys. III. Large intestine. J Assoc Phys Ind 14:583–587.
619. Nagaty HF (1935) *Parlitomosa zakii* (Filariinae). A new genus and species and its microfilaria from *Leontocebus rosalia*. J Egypt Med Assoc 18:483–496.
620. Naquira C (1963) Estudio preliminar sobre la infeccion celomica de *Triatoma infestans* por *Trypanosoma cruzi* y *Trypanosoma* sp. de mono. Biol Santiago 35:3–8.
621. Napier JR, Napier PH (1967) A handbook of living primates. Academic, New York.
622. Narama I, Tsuchitani M, Umemura T, Tsuruta M (1983) The morphogenesis of a papillomatous gastric polyp in the crab-eating monkey (*Macaca fascicularis*). J Comp Path 93:195–203.

623. Nelson B, Cogrove GE, Gengozian N (1966) Diseases of an imported primate *Tamarinus nigricollis*. Lab Anim Care 16:255–275.
624. Nelson EC (1932) The cultivation of a species of *Troglodytella*, a large ciliate, from the chimpanzee. Science 75:317–318.
625. Nelson GS (1960) Schistosome infections as zoonoses in Africa. Trans Roy Soc Trop Med Hyg 54:301–316.
626. Nelson GS (1965) The parasitic helminths of baboons with particular reference to species transmissible to man. In: Vagtborg H (ed) The baboon in medical research. University of Texas Press, Austin, TX, pp 441–470.
627. Nelson GS, Teesdale C, Highton RB (1962) The role of animals as reservoirs of bilharziasis in Africa. In: Wolstenholme, O'Conner (eds) Bilharziasis. Ciba Foundation Symposium Bilharziasis, Churchill, London, pp 127–149.
628. Nery-Guimaraes F, Franken AJ, Chagas WA (1971) Toxoplasmose em primates não humanos. I. Infeccões naturais em *Macaca mulatta* e *Cebus apella*. Mem Inst Oswaldo Cruz 69:77–96.
629. Newstead R (1906) On another new Dermanyssid Acarid parasitic in the lungs of the rhesus monkey (*Macaca rhesus*). Mem Lpool Sch Trop Med 18:45.
630. Newstead R, Todd JL (1906) On a new dermanyssid acarid found living in the lungs of monkeys (*Cercopithecus schmidti*) from the upper Congo. Mem Lpool Sch Trop Med 18:41–44.
631. Nicholas WL (1967) The biology of acanthocephala. Adv Parasitol 5:205–246.
632. Nielsen DH (1980) *Prosthenorchis elegans* infection in a primate colony. Amer Assoc Zoo Vet Ann Proc, pp 113–116.
633. Nigi H, Itakura C (1968) Spontaneous toxoplasmosis in *Lemur catta*. Primates 9:155–160.
634. Noda S (1962) Comparative studies on morphology of free-living stages of *Strongyloides* parasitic in monkeys. Jap J Parasitol 11:207–229.
635. Nonoyama T, Sugitani T, Orita S, Miyajima H (1984) A pathological study in cynomolgus monkeys infected with *Edesonfilaria malayensis*. Lab Anim Sci 34:604–609.
636. Nouvell J, Prot-Lassele J (1963) Shigella dysenteriae 2 chez des chimpanzé en captivité. Bull Acad Vét Fr 36:373–379.
637. Nuttal GHF, Warburton C (1915) The genus *Haemaphysalis*. In: Ticks, a monograph of the *Ixodoidea*, part III. Cambridge University Press, pp 349–550.
638. Offutt EP Jr, Telford IR (1945) *Sarcocystis* in the monkey. A report of two cases. J Parasitol Suppl 21:15.
639. Ogata T, Imai H, Coulston F (1971) Pulmonary acariasis in rhesus monkeys: electron microscopic study. Exp Mol Pathol 15:137.
640. O'Grady JP, Yeager CH, Esra GN, Thomas W (1982) Ultrasonic evaluation of echinococcosis in four lowland gorillas. JAVMA 181:1348–1350.
641. Ogunsusi RA, Mohammed AN (1978) The chimpanzee (*Pan troglogdytes*), a new host for nymphal *Armilliter armillatus* (Pentastomida:Porocephalida) in West Africa. Revue d'Elevage et de Med Vet Des Pays Trop 1:361–362.
642. Orihel TC (1966) *Brugia tupaiae* sp. n. (Nematoda:Filarioidea) in tree shrews (*Tupaia glis*) from Malaysia. J Parasitol 52:162–165.
643. Orihel TC (1970) Primates as models for parasitological research. In: Goldsmith EI, Moor-Jankowski J (eds) Medical primatology. S Karger, Basel, pp 772–782.
644. Orihel TC (1970) Anatrichosomiasis in African monkeys. J Parasitol 56:982–985.
645. Orihel TC (1970) The helminth parasites of nonhuman primates and man. Lab Anim Care 20:395–401.
646. Orihel TC (1970) Filariasis in chimpanzees. In: Bourne GH (ed) The chimpanzee, vol 3. University Park Press, Baltimore, pp 56–70.
647. Orihel TC (1971) *Necator americanus* infection in primates. J Parasitol 57:117–121.
648. Orihel TC, Esslinger JH (1973) *Meningonema peruzzii* gen. et sp. n. (Nematoda:

Filarioidea) from the central nervous system of African monkeys. J Parasitol 59:437–441.

649. Orihel TC, Seibold HR (1971) Trichospirurosis in South American monkeys. J Parasitol 57:1366–1368.
650. Orihel TC, Seibold HR (1972) Nematodes of the bowel and tissues. In: Fiennes RNT-W (ed) Pathology of simian primates. Part II. Infectious and parasitic diseases. S Karger, Basel, pp 76–103.
651. Ortlepp RJ (1924) On a collection of helminths from Dutch Guiana. J Helminthol 2:15.
652. Otsuru M, Sekikawa H (1968) A survey of simian malaria in Japan. Trans Roy Soc Trop Med Hyg 62:558–561.
653. Oudemans AC (1935) Kritische Literaturübersicht zur Gattung *Pneumonyssus*. Beschreibung dreier Arten, darunter einer neuen. Z Parasitenk 7:466–512.
654. Owen D, Casillo S (1973) A preliminary survey of the nematode parasites of some imported Old-World monkeys. Lab Anim 7:265–269.
655. Ow-Yang CK, Wah MJ (1975) A remarkable trematode from the parotid gland of *Tupaia glis*. Southeast Asian J Trop Med Publ Health 6:449.
656. Padovan D, Cantrell C (1983) Causes of death of infant rhesus and squirrel monkeys. JAVMA 183:1182–1184.
657. Palmieri JR, Krishnasamy M (1978) *Phaneropsolus aspinosus* sp. n. (Lecithodendriidae Phaneropsoline) from leaf monkey *Macaca fascicularis* (Raffles). J Helminthol 52:155–158.
658. Palmieri JR, Krishnasamy M, Sullivan JT (1977) Helminth parasites of the Old World leaf monkey *Presbytis* sp. from west Malaysia. Southeast Asian J Trop Med Publ Health 8:409.
659. Palmieri JR, Purnomo, Dennis DT, Marwoto HA (1980) Filarid parasites of South Kalimantan (Borneo) Indonesia. *Wuchereria kalimantani* sp. n. (Nematoda:Filarioidea) from the silvered leaf monkey, *Presbytis cristatus* Eschscholtz 1921. J Parasitol 66:645–651.
660. Palmieri JR, Purnomo, Lee VH, Dennis DT, Marwoto HA (1980) Parasites of the silvered leaf monkey, *Presbytis cristatus* Eschscholtz 1921, with a note on a *Wuchereria*-like nematode. J Parasitol 66:170–171.
661. Palmieri JR, Dalgard DW, Conner DH (1984) Gastric amebiasis in a silvered leaf monkey. JAVMA 185:1374–1375.
662. Palmieri JR, Van Dellen AF, Tirtokusumo S, Masbar S, Rusch J, Connor DH (1984) Trapping, care, and laboratory management of the silvered leaf monkey (*Presbytis cristatus*). Lab Anim Sci 34:194–197.
663. Palotay JL, Uno H (1975) Hydatid disease in four nonhuman primates. JAVMA 167:615–618.
664. Parker GA, Gilmore CJ, Roberts CR (1979) Diagnostic exercise. Lab Anim Sci 29:457–458.
665. Patten RA (1939) Amoebic dysentery in orang-utans (*Simia satyrus*). Aust Vet J 15:68–71.
666. Paulicki A (1872) Beiträge zur vergleichenden pathologischen Anatomie aus dem Hamburger Zoologischen Garten II. Grüne Psorospermienheerde in der Affenlunge. Magges Thierheilk 38:1.
667. Pavor ML (1965) Lung mites (*Pneumonyssus simicola*) in the feces of *Macaca mulatta*. Lab Primate Newslett 4:4.
668. Peddie JF, Larson EJ (1971) Demodectic acariasis in a woolly monkey. Vet Med/Sm Anim Clin 66:485–488.
669. Peel E, Chardome M (1946) Note préliminaire sur des filaridès de chimpanzés, *Pan paniscus* et *Pan satyrus* au Congo Belge. Rec Sci Méd Congo Belge 5:244.
670. Peel E, Chardome M (1946) Sur des filaridés de chimpanzés "Pan paniscus" et "Pan satyrus" au Congo Belge. Ann Soc Belge Med Trop 26:117–156.

671. Peel E, Chardome M (1947) Note complémentaire sur des filaridés de chimpanzés, *Pan paniscus* et *Pan satyrus* Congo Belge. Ann Soc Belge Méd Trop 27:241–250.
672. Penner LR (1981) Concerning threadworm (*Strongyloides stercoralis*) in great apes—lowland gorillas (*Gorilla gorilla*) and chimpanzees (*Pan troglodytes*). J Zoo Anim Med 12:128–131.
673. Phillipe J (1948) Note sur les gales du singe. Bull Soc Path Exot 41:597–600.
674. Pietrzyk J, Uminski J (1967) Malaria u malp *Cercopithecus* wywolana prezez *Hepatocystis kochi*. Zwierzeta Lab 2:72–79.
675. Pillers AWN (1921) Sarcoptic scabies (or itch) in the chimpanzee. Brit Vet J 77:329–333.
676. Pillers AWN (1924) *Ascaris lumbricoides* causing fatal lesions in a chimpanzee. Ann Trop Med Parasitol 18:101–102.
677. Pillers AWN, Southwell T (1929) Strongyloides of the woolly monkey (*Lagothrix humbodti*). Ann Trop Med Parasitol 23:129.
678. Poelma FG (1966) *Eimeria lemuris* n. sp., *E. galago* n. sp. and *E. otolicni* n. sp. from a galago *Galago senegalensis*. J Protozool 13:547–549.
679. Poelma FG (1975) *Pneumocystis carinii* infections in zoo animals. Z Parasitenk 46:61–68.
680. Poindexter HA (1942) A study of the intestinal parasites of the monkeys of the Santiago Island primate colony. Puerto Rico J Publ Health Trop Med 18:175–191.
681. Poisson R (1953) Sous-ordre des hémosporidies (Haemosporidiidea Danilewsky, 1889 emend.; Doflein, 1901). In: Grassé PP (ed) Traité de Zoologie anatomie-systématique biologie, vol I, fasc II. Masson, Paris, pp 798–906.
682. Pope BL (1966) Some parasites of the howler monkey of northern Argentina. J Parasitol 52:166–168.
683. Pope BL (1968) XV. Parasites. In: Malinow MR (ed) Biology of the howler monkey (*Alouatta caraya*). Bibl Primat. No 7. S Karger, Basel, pp 204–208.
684. Porter A (1945) Report of the honorary parasitologist for 1944. Proc Zool Soc London 115:384–386.
685. Porter A (1952) *Chilomastix tarsii* sp. n., a new flagellate from the gut of *Tarsius carbonarius*. Proc Zool Soc London 121:915.
686. Porter A (1953–1954) Report of the honorary parasitologist for the year 1952. Proc Zool Soc London 123:253–257.
687. Porter A (1954) Report of the honorary parasitologist for the year 1953. Proc Zool Soc London 124:313–316.
688. Porter A (1955) Summary of the report of the honorary parasitologist for the year 1954. Proc Zool Soc London 125:541.
689. Porter JA Jr (1972) Parasites of marmosets. Lab Anim Sci 22:503–506.
690. Porter JA Jr, Johnson CM, DeSousa L (1966) Prevalence of malaria in Panamanian primates. J Parasitol 52:669–670.
691. Powers RD, Price RA, Houk RP, Mattlin RH (1966) Echinococcosis in a drill baboon. JAVMA 149:902–905.
692. Prathap K, Law KS, Bolton JM (1969) Pentastomiasis. A common finding at autopsy among Malaysian aborigines. Amer J Trop Med Hyg 18:20–27.
693. Price DL (1959) *Dirofilaria magnilarvatum* n. sp. (Nematoda:Filarioidea) from *Macaca irus* Cuvier. I. Description of the adult filarial worms. J Parasitol 45:499–504.
694. Prine JR (1968) Pancreatic flukes and amoebic colitis in a gorilla. Abst 50 19th Ann Meeting Amer Assoc Lab Animal Sci, Las Vegas, 1968.
695. Prosl H, Tamer A (1979) The parasite fauna of the rhesus monkey (*Macaca mulatta*) and the Java ape (*Macaca irus*). Zentralbl Veterinärmed 26B:696–709.
696. Pryor WH Jr, Bergner JF, Raulston GL (1970) Leech (*Dinobdella ferox*) infection of a Taiwan monkey (*Macaca cyclopis*). JAVMA 157:1926–1927.

697. Pucak GJ, Johnson DK (1972) Sarcocystis in a Patas monkey (*Erythrocebus patas*). Lab Anim Dig 8:36–39.
698. Purvis AJ, Ellison IR, Husting EL (1965) A short note on the findings of schistosomes in baboons (*Papio rhodesiae*). Central African J Med 11:368.
699. Railliet A, Henry A (1909) Sur la classification des Strongylidae. I. Metastrongylinae. CR Soc Biol 66:85–88.
700. Railliet A, Marullaz M (1919) Sur un cénure nouveau du bonnet chinois (*Macacus sinicus*). Bull Soc Path Exot 12:223–228.
701. Railliet A, Henry A, Joyeux C (1912) Sur deux trématodes de primates. Bull Soc Path Exot 5:833–837.
702. Ramakrishnan SP, Mohan BN (1961) Simian malaria in the Nilgiris, Madras State, India. Bull Natl Soc Ind Malar Mosquito Dis 9:139–140.
703. Ramakrishnan SP, Mohan BN (1962) An enzootic focus of simian malaria in *Macaca radiata radiata* Geoffrey of Nilgiris, Madras State, India. Ind J Malariol 16:87–94.
704. Ratcliffe HL (1931) A comparative study amoebiasis in man, monkeys and cats, with special reference to the formation of the early lesions. Amer J Hyg 14:337–352.
705. Ratcliffe HL (1942) Deaths and important diseases. Rep Penrose Res Lab, pp 11–25.
706. Ratcliffe HL (1955) Causes of death in the animal collection. Rep Penrose Res Lab pp 6–16.
707. Ratcliffe HL (1962) Causes of death in the animal collection. Rep Penrose Res Lab Zool Soc Philadelphia, 1961, 6–18.
708. Ratcliffe HL (1963) Causes of death in the animal collection. Rep Penrose Res Lab Zool Soc Philadelphia, 1962, 13–24.
709. Ratcliffe HL, Worth CB (1951) Toxoplasmosis of captive wild birds and mammals. Amer J Path 27:655–667.
710. Raulston GL (1972) Psorergatic mites in patas monkeys. Lab Anim Sci 22:107–108.
711. Rawling CA, Splitter GA (1973) Pneumothorax associated with lung mite lesions in a rhesus monkey. Lab Anim Sci 23:259–261.
712. Reardon LV, Rininger BF (1968) A survey of parasites in laboratory animals. Lab Anim Care 18:577–580.
713. Reichenow E (1917) Parásitos de la sangre y del intestino de los monos antropomorfos africanos. Bol Real Soc Espan Hist Nat Secc Biol 17:312–332.
714. Reichenow E (1949–1953) Lehrbuch der Protozoenkunde, 6th ed, 3 vol. Fischer, Jena.
715. Reid WA, Reardon MJ (1976) *Mesocestoides* in the baboon and its development in laboratory animals. J Med Primatol 5:345–352.
716. Remfry J (1982) The endoparasites of rhesus monkeys (*Macaca mulatta*) before and after capture. Microbiologica 5:143–147.
717. Renquist DM, Johnson AJ, Lewis JC, Johnson DJ (1975) A natural case of *Schistosoma mansoni* in the chimpanzee (*Pan troglodytes*). Lab Anim Sci 25:763–768.
718. Rewell RE (1948) Diseases of tropical origin in captive wild animals. Trans Roy Soc Trop Med Hyg 42:17–36.
719. Reyes E (1970) *Iodamoeba wallacei* n. sp. of amoeba found in *Cercopithecus diana* L. (Mammalia-Primate) (Sarcodina-Endamoebidae) Bol Soc d Biol Concepcion 42:215–256.
720. Richart R, Benirschke K (1963) Causes of death in a colony of marmoset monkeys. J Path Bacteriol 86:221–223.
721. Richter CB, Humason GL, Godbold JH Jr (1978) Endemic *Pneumocystis carinii* in a marmoset colony. J Comp Path 88:171–180.
722. Rijpstra AC (1967) Sporocysts of *Isospora* sp. in a chimpanzee (*Pan troglodytes*, L.). Proc Kon Nederl Akad Wet 70:395–401.
723. Riopelle AJ, Daumy OJ (1962) Care of chimpanzees for radiation studies. Proc Int

Symp Bone Marrow Therapy and Chemical Protection in Irradiated Primates. Aug 15–18, 1962. Rijswijk, The Netherlands, pp 205–223.

724. Riopelle AJ (1967) The chimpanzee. In: UFAW Staff (ed) UFAW handbook on the care and management of laboratory animals, 3rd ed. Livingstone and the Universities Federation for Animal Welfare, London, pp 696–708.
725. Robertson OH, Loosli CG, Puck TT, Wise H, Lemon HM, Lester W Jr (1947) Tests for the chronic toxicity of propylene glycol and triethylene glycol on monkeys and rats by vapor inhalation and oral administration. J Pharmacol Exp Therap 91:52–76.
726. Rodaniche de ED (1954) Spontaneous toxoplasmosis in the whiteface monkey, *Cebus capucinus*, in Panama. Amer J Trop Hyg 3:1023–1025.
727. Rodhain J (1933) Sur une coccidie de l'intestin de l'ouistiti: *Hapale jacchus penicillatus* (Geoffroy). CR Soc Biol 114:1357–1358.
728. Rodhain J (1941) Notes sur *Trypanosoma minansense* Chagas: identité spécifique du trypanosome du saimiri: *Chrysothrix sciureus*. Acta Biol Belg 1:187–193.
729. Rodhain J (1941) Sur un Plasmodium du gibbon *Hylobates leusciscus* Geoff. Acta Biol Belg 1:118–123.
730. Rodhain (1948) Contribution á l'étude des Plasmodiums des anthropoides africains. Transmission du *Plasmodium malariae* de l'homme au chimpanzé. Ann Soc Belge Med Trop 28:39–49.
731. Rodhain J, Dellaert R (1943) L'infection á *Plasmodium malariae* du chimpanzé chez l'homme. Étude d'une premiére souche isolée de l'anthropoide *Pan satyrus verus*. Ann Soc Belge Med Trop 23:19–46.
732. Rodhain J, Dellaert R (1955) Contribution á l'etude de *Plasmodium schwetzi* E. Brumpt. II. Transmission du *Plasmodium schwetzi* á l'homme. Ann Soc Belge Med Trop 35:73–76.
733. Rodhain J, van den Berghe L (1936) Contribution á l'etude des Plasmodiums des singes africains. Ann Soc Belge Med Trop 16:521–531.
734. Rodhain J, van den Berghe L (1939) *Paraloa anthropopitheci* genre et espéce nouveaux de Filaroidea chez le chimpanzé du Congo Belge. Ann Soc Belge Med Trop 19:445.
735. Rodhain J, Wanson M (1954) Un nouveau cas de coenurose chez le babouin (*Theropithecus gelada* Ruppell). Riv Parassitol 15:613–620.
736. Ronald NC, Wagner JE (1973) Pediculosis of spider monkeys: a case report with zoonotic implications. Lab Anim Sci 23:872–875.
737. Rosen S, Hono JE, Barry KG (1968) Malarial nephropathy in the rhesus monkey. Arch Pathol 85:36–44.
738. Ross PH (1905) A note on the natural occurrence of piroplasmosis in the monkey (*Cercopithecus*). J Hyg 5:18–23.
739. Rousselot R (1956) Hepatite filarienne des anthropides. Bull Soc Path Exot 49:301–303.
740. Rousselot R, Pellissier A (1952) Pathologie du gorille. III. Oesophagostomose nodulaire a *Oesophagostomum stephanostomum* du gorille et du chimpanzé. Bull Soc Path Exot 45:568–574.
741. Rowland Eloise, Vandenbergh JG (1965) A survey of intestinal parasites in a new colony of rhesus monkeys. J Parasitol 51:294–295.
742. Ruch TC (1959) Diseases of laboratory primates. WB Saunders, Philadelphia.
743. Ruiz A, Frenkel JK (1976) Recognition of cyclic transmission of *Sarcocystis muris* by cats. J Infect Dis 133:409–418.
744. Sachs R, Voelker J (1975) A primate, *Mandrillus leucophaeus*, as natural host of the African lung fluke *Paragonimus africanus* in West Cameroon. Tropenmed Parasitol 26:205–206.
745. Sagartz JW, Tingpalapong M (1974) Cerebral cysticercosis in a white-handed gibbon. JAVMA 165:844–845.
746. Sakakibara I (1981) Naturally occurring diseases in cynomolgus monkeys. Jap J Med Sci Biol 34:263–267.

747. Sakakibara I, Sugimoto Y, Koyama T, Honjo S (1982) Natural transmission of *Entamoeba histolytica* from mother cynomolgus monkeys (*Macaca fascicularis*) to their newborn infants under indoor rearing conditions. Exper Anim 31:135–138.
748. Salis H (1941) Studies on the morphology of the *E. histolytica*-like amoebae found in monkeys. J Parasitol 27:327–341.
749. Sambon LW (1910) Porocephaliasis in man. J Trop Med Hyg 13:17–24, 212–217, 258–263.
750. Sambon LW (1922) A synopsis of the family Linguatulidae. J Trop Med Hyg 25:188–206, 391–428.
751. Sambon LW (1924) The elucidation of cancer. J Trop Med Hyg 27:124–174.
752. Sandground JH (1925) Speciation and specificity in the nematode genus *Strongyloides*. J Parasitol 12:59–80.
753. Sandground JH (1930) Notes and descriptions of some parasitic helminths collected by the expedition. Contrib Dept Trop Med Inst Trop Biol Med 5:462–486.
754. Sandground JH (1933) Report on the nematode parasites collected by the Kelley-Roosevelts expedition to Indo-China with descriptions of several new species. I. Parasites of birds. 2. Parasites of mammals. Z Parasitenk 5:542.
755. Sandground JH (1936) On the occurrence of a species of *Loa* in monkeys in the Belgian Congo. Ann Soc Belge Med Trop 16:273.
756. Sandground JH (1936) Scientific results of an expedition to rain forest regions in Eastern Africa. Bull Mus Comp Zool Harvard 79:343–366.
757. Sandground JH (1937) On a coenurus from the brain of a monkey. J Parasitol 23:482–490.
758. Sandground JH (1938) Some parasitic worms in the helminthological collection of the Museum of Comparative Zöology. 2. A redescription of *Tetrapetalonema digitata* (Chandler 1929) comb. nov., a filariid parasite of gibbon apes with an enumeration of its congeners. Bull Mus Comp Zool Harv 85:49.
759. Sandosham AA (1950) On *Enterobius vermicularis* (Linnaeus, 1758) and some related species from primates and rodents. J Helminthol 24:171–204.
760. Sandosham AA (1951) On two helminths from the orang-utan, *Leipertrema rewelli*, n.g., n. sp. and *Dirofilaria immitis* (Leidy, 1856). J Helminthol 25:19–26.
761. Sandosham AA (1954) Malaysian parasites. XV. Seven new worms from miscellaneous hosts. Stud Inst Med Res Fed Malaya 26:213–226.
762. Sandosham AA, Wharton RH, Warren M, Eyles DE (1962) Microfilariae in the rhesus monkey (*Macaca mulatta*) from East Pakistan. J Parasitol 48:489.
763. Sano M, Kino H, deGuzman TS, Ishii AI, Kino J, Tanaka T, Tsuruta M (1980) Studies on the examination of imported laboratory monkey *Macaca fascicularis* for *E. histolytica* and other intestinal parasites. Intern J Zoon 7:34–39.
764. Sasa M, Tanaka H, Fukui M, Takata A (1962) Internal parasites of laboratory animals. In: Harris RJC (ed) The problems of laboratory animal disease. Academic, New York, pp 195–214.
765. Sawada I, Kifune T (1974) A new species anoplocephaline cestode from *Macaca irus*. Jap J Parasitol 23:366–368.
766. Schacher JF (1962) Morphology of the microfilaria of *Brugia pahangi* and of the larval stages in the mosquito. J Parasitol 48:679–692.
767. Schacher JF (1962) Developmental stages of *Brugia pahangi* in the final host. J Parasitol 48:693–706.
768. Schad GA, Anderson RC (1963) *Macacanema formosana* n. g., n. sp. (Onchocercidae:Dirofilariinae) from *Macaca cyclopsis* of Formosa. Can J Zool 41:797–800.
769. Schiefer B, Loew FM (1978) Amebiasis and salmonellosis in a woolly monkey (Lagothrix). Vet Pathol 15:428–431.
770. Schmidt GD (1972) Acanthocephala of captive primates. In: Fiennes RNT-W (ed) Pathology of simian primates. Part II. Infectious and parasitic diseases. S Karger, Basel, pp 144–156.

771. Schmidt GD, File S (1977) *Tupaiataenia quentini* gen. et sp. n. (Anoplocephalidae:Linstowiinae) and other tapeworms from the common tree shrew, *Tupaia glis*. J Parasitol 63:473–475.
772. Schmidt LH, Greenland R, Genther CS (1961) The transmission of *Plasmodium cynomolgi* to man. Amer J Trop Med Hyg 10:679–688.
773. Schmidt LH, Greenland R, Rossan R, Genther C (1961) Natural occurrence of malaria in rhesus monkeys. Science 133:753.
774. Schmidt RE (1978) Systemic pathology of chimpanzees. J Med Primatol 7:274–318.
775. Schmidt RE, Prine JR (1970) Severe enterobiasis in a chimpanzee. Path Vet 7:56–59.
776. Schoeb TR (1984) *Klossiella* sp. infection in a galago. JAVMA 185:1381–1382.
777. Schultz AH (1939) Notes on diseases and healed fractures of wild apes, and their bearing on the antiquity of pathological conditions in man. Bull Hist Med 7:571–582. Cited by Ruch TC (1959) Diseases of laboratory primates. WB Saunders, Philadelphia.
778. Schwetz J (1933) Trypanosomes rares de la région de Stanleyville (Congo belge). Ann Parasitol Hum Comp 11:287–296.
779. Schwetz J (1934) Sur quelques trypanosomes rares de la région de Stanleyville (deuxieme note). Ann Parasitol Hum Comp 12:278–282.
780. Scott HH (1926) Report on the deaths occurring in the Society's gardens during the year 1925. Proc Zool Soc London 97:231–244.
781. Seibold HR, Fussell EN (1973) Intestinal microsporidiosis in *Callicebus moloch*. Lab Anim Sci 23:115–118.
782. Seibold HF, Wolf RH (1970) American trypanosomiasis in *Hylobates pileatus*. Lab Anim Sci 20:514–517.
783. Seibold HR, Wolf RH (1971) Toxoplasmosis in *Aotus trivirgatus* and *Callicebus moloch*. Lab Anim Sci 21:118–120.
784. Seibold HR, Wolf RH (1973) Neoplasms and proliferative lesions in 1065 nonhuman primate necropsies. Lab Anim Sci 23:533–539.
785. Self JT (1969) Biological relationships of the Pentastomida. A bibliography on the Pentastomida. Exp Parasitol 24:63–119.
786. Self JT (1972) Pentastomiasis host responses to larval and nymphal infections. Trans Amer Micr Soc 91:2–8.
787. Self JT, Cosgrove GE (1968) Penastome larvae in laboratory primates. J Parasitol 54:969.
788. Self JT, Cosgrove GE (1972) Penastomida. In: Fiennes RNT-W (ed) Pathology of simian primates. Part II. Infectious and parasitic diseases. S Karger, Basel, pp 194–204.
789. Seneca H, Wolf A (1955) *Trypanosoma cruzi* infection in the Indian monkey. Amer J Trop Med Hyg 4:1009–1014.
790. Sen Gupta PC, Ray HN (1955) A cytochemical study of *Balantidium coli* Malmsten, 1857. Proc Zool Soc India 8:103–110.
791. Shadduck JA, Pakes SP (1978) Protozoal and metazoal diseases. In: Benirschke K, Garner FM, Jones TC (eds) Pathology of laboratory animals, chapter 17, vol II. Springer-Verlag, New York, pp 1587–1696.
792. Shannon RC, Greene CT (1926) A bottfly parasite in monkeys. Zoopathologica 1:285.
793. Sheldon WG (1966) Psoregatic mange in the sooty mangabey (*Cercocebus torquates atys*) monkey. Lab Anim Care 16:276–279.
794. Shiroishi T, Davis J, Warren M (1968) *Hepatocystis* in white-cheeked gibbon *Hylobates concolor.* J Parasitol 54:168.
795. Shortt HE, Rao G, Qadri SS, Abraham R (1961) *Plasmodium osmaniae*, a malaria parasite of an Indian *Macaca radiata*. J Trop Med Hyg 64:140–143.
796. Sinton JA (1934) A quartan malaria parasite of the lower oriental monkey, *Silenus irus* (*Macaca cynomolgus*). Records Malaria Surv India 4:379–410.

797. Sinton JA, Mulligan HW (1932–1933) A critical review of the literature relating to the identification of the malaria parasites recorded from monkeys of the families Cercopithecidae and Colobidae. Rec Malar Surv Ind 3:357–380, 381–444.
798. Skrjabin KI, Shirhobalova NP, Shults RS (1954) Trichostrongylids of animals and man. In: Skrjabin KI (ed) Essentials of nematodology, vol 3. Acad Sci, USSR. (English edition, Israel Program for Scientific Translations, Jerusalem, 1960.)
799. Slaughter LJ, Bostrom RE (1969) Physalopterid (*Abbreviata poicilometra*) infection in a sooty mangabey monkey. Lab Anim Care 19:235–236.
800. Slaughter LJ, Dade AW, Chineme C, Andrews EJ (1974) Pentastoma larvae in a squirrel monkey. JAVMA 164:711.
801. Sluiter C, Wellengrebel N, Ihle J (1922) De dierlyke parasieten van den mensch en van onze hursdieren scheltema and holkema. Amsterdam.
802. Sly DL, Toft JD II, Gardiner CH, London WT (1982) Spontaneous occurrence of *Angiostrongylus costaricensis* in marmosets (*Saguinus mystax*). Lab Anim Sci 32:286–288.
803. Smetana HF, Oriehel TC (1969) Gastric papillomata in *Macaca speciosa* induced by *Nochtia nochti* (Nematoda:Trichostrongyloidea). J Parasitol 55:349–351.
804. Smiley RL, O'Connor BM (1980) Mange in *Macaca arctoides* (Primates:Cercopithecidae) caused by *Cosarcoptes scanloni* (Acari:Sarcoptidae) with possible human involvement and descriptions of the adult male and immature stages. Intern J Acarol 6:283–290.
805. Smith WN, Chitwood MB (1954) *Anatrichosoma cynamolgi*, a new trichurid nematode from monkeys. J Parasitol Suppl 40:12.
806. Smith WN, Chitwood MB (1967) *Trichospirura leptostoma* gen et sp n (Nematoda: Thelazioidea) from the pancreatic ducts of the white-eared marmoset, *Callithrix jacchus*. J Parasitol 53:1270–1272.
807. Smith WN, Levy BM (1969) The effects of *Trichospirura leptostoma* on the pancreas of *Callithrix jacchus*. Program and Abstracts No 7, p 45, 44th Annual Meeting Amer Soc Path, Washington, DC 1969.
808. Soave OA (1963) Diagnosis and control of common diseases of hamsters, rabbits, and monkeys. JAVMA 142:285–290.
809. Sood ML, Toong R (1973) *Streptopharagus guptai* n. sp. (Nematoda:Spiruridae) from the rectum of a rhesus macaque *Macaca mulatta* from India. Zoolog Anz 190:132–136.
810. Soulsby EJL (1965) Textbook of veterinary clinical parasitology, vol I. Helminths. FA Davis, Philadelphia.
811. Soulsby EJL (1968) Helminths, arthropods and protozoa of domesticated animals (Mönnig), 6th ed. Williams and Wilkins, Baltimore.
812. Sousa OE, Rossan RN, Baerg DC (1974) The prevalence of trypanosomes and microfilariae in Panamanian monkeys. Amer J Trop Med Hyg 23:862–868.
813. Stam AB (1960) Fatal ascaridosis in a dwarf chimpanzee. Ann Parasitol Hum Comp 35:675.
814. Stiles CW, Hassall A (1929) Key-catalogue of parasites reported for primates (monkeys and lemurs) with their possible public health importance. Hyg Lab Bull No 152 Public Health Serv, Washington, DC. Cited by Dunn FL (1968) The parasites of *Saimiri*; in the context of Platyrrhine parasitism. In: Rosenblum, LA, Cooper RW (ed) The squirrel monkey, chapter 2. Academic, New York, pp 31–68.
815. Stiles CW, Nolan MO (1929) Key-catalogue of primates for which parasites are reported. Hyg Lab Bull No 152, Public Health Serv, Washington, DC. Cited by Dunn FL (1968) The parasites of *Saimiri*; in the context of Platyrrhine parasitism. In: Rosenblum LA, Cooper RW (eds) The squirrel monkey, chapter 2. Academic, New York, pp 31–68.
816. Stokes WS, Donovan JC, Montrey RD, Thompson WL, Wannemacher RW Jr, Rosmiarek H (1983) Acute clinical malaria (*Plasmodium inui*) in a cynomolgus monkey (*Macaca fascicularis*). Lab Anim Sci 33:81–85.

817. Stolz G (1962) Spontaneous lethal toxoplasmosis in a woolly monkey. Schweiz Arch Tierheilkd 104:162–166.
818. Stone WB, Hughes JA (1969) Massive pulmonary acariasis in the pig-tail macaque. Bull Wildl Dis Assoc 5:20.
819. Stone WM (1964) *Strongyloides ransomi* prenatal infection in swine. J Parasitol 50:568.
820. Stookey JL, Moe JB (1978) The respiratory system. In: Benirschke K, Garner FM, Jones TC (eds) Pathology of laboratory animals, chapter 2, vol I. Springer-Verlag, New York, pp 72–113.
821. Strandtmann RW, Wharton GW (1958) A manual of mesostigmatid mites parasitic on vertebrates. Contrib No 4 Inst Acarology University of Maryland, College Park.
822. Strong JP, McGill HC Jr, Miller JH (1961) *Schistosomiasis mansoni* in the Kenya baboon. Amer J Trop Med Hyg 10:25–31.
823. Strong JP, Miller JH, McGill HC Jr (1965) Naturally occurring parasitic lesions in baboons. In: Vagtborg H (ed) The baboon in medical research, Univ of Texas Press, Austin, TX, pp 503–512.
824. Stunkard HW (1923) On the structure, occurrence and significance of *Athesmia foxi*, a liver fluke of American monkeys. J Parasitol 10:71–79.
825. Stunkard HW (1940) The morphology and life history of the cestode, *Bertiella studeri*. Amer J Trop Med 20:305–333.
826. Stunkard HW (1965) *Paratriotaenia oedipomidatis* gen. et sp. n. (Cestoda), from a marmoset. J Parasitol 51:545–551.
827. Stunkard HW (1965) New intermediate hosts in the life cycle of *Prosthenorchis elegans* (Diesing 1851), an acanthocephalan parasite of primates. J Parasitol 51:645–649.
828. Stunkard HW, Goss LJ (1950) *Eurytrema brumpti* Railliet, Henry and Joyeux, 1912 (Trematoda:Dicrocoeliidae), from the pancreas and liver of African anthropoid apes. J Parasitol 36:574–581.
829. Suldey EW (1924) Dysentérie amibienne spontanée chez le chimpanzé (*Troglodytes niger*). Bull Soc Path Exot 17:771–773.
830. Summers WA (1960) A case of hydatid disease in the rhesus monkey (*Macaca mulatta*). Allied Vet 31:141–143.
831. Sureau P, Raynaud JP, Lapeire C, Brygoo ER (1962) Premier isolement de *Toxoplasma gondii* à Madasgascar. Toxoplasmose spontanée et experimentale du *Lemur catta*. Bull Soc Path Exot 55:357–362.
832. Swellengrebel NH, Rijpstra AC (1965) Lateral-spined schistosome ova in the intestine of a squirrel monkey from Surinam. Trop Geograph Med 17:80–84.
833. Swift HF, Boots RH, Miller CP (1922) A cutaneous nematode infection in monkeys. J Exp Med 35:599–620.
834. Takos MJ, Thomas LJ (1958) The pathology and pathogenesis of fatal infections due to an acanthocephalid parasite of marmoset monkeys. Amer J Trop Med Hyg 7:90–94.
835. Taliaferro WH, Taliaferro LG (1934) Morphology, periodicity and course of infection of *Plasmodium brasilianum* in Panamanian monkeys. Amer J Hyg 20:1–49.
836. Tanaka H, Fukui M, Yamamoto H, Hayama S, Kodera S (1962) Studies on the identification of common intestinal parasites of primates. Bull Exp Anim 11:111–116.
837. Tanguy Y (1937) La piroplasmose du singe. Ann Inst Pasteur 59:610–623.
838. Taylor AER (1959) *Dirofilaria magnilvarvatum* Price, 1959 (Nematoda:Filarioidea) from *Macaca irus* Cuvier. II. Microscopical studies on the microfilariae. J Parasitol 45:505–510.
839. Teare JA, Loomis MR (1982) Epizootic of balantidiasis in lowland gorillas. JAVMA 181:1345–1347.
840. Terrell TG, Stookey JL (1972) Chronic eosinophilic myositis in a rhesus monkey infected with sarcosporidiosis. Pathol Vet 9:266–271.

841. Thatcher VE, Porter JA Jr (1968) Some helminth parasites of Panamanian primates. Trans Amer Micr Soc 87:186–196.
842. Thézé J (1916) Rapport sur les travaux de l'Institut d'Hygiene et de Bacteriologie. 1914–1915. Bull Soc Path Exot 9:449–469.
843. Thienpont D, Mortelmans J, Vercruysse J (1962) Contribution à l'étude de la Trihuriose du chimpanzé et de son traitement avec la methyridine. Ann Soc Belge Med Trop 2:211–218. Cited by Flynn RJ (1973) Parasites of laboratory animals. Iowa State University Press, Ames, IA.
844. Thornton H (1924) The relationship between the ascarids of man, pig and chimpanzee. Ann Trop Med Parasitol 18:99–100.
845. Tihen WS (1970) *Tetrapetalonema marmosete* in cotton-topped marmosets, *Saguinus oedipus*, from the region of San Marcos, Colombia. Lab Anim Sci 20:759–762.
846. Toft JD II (1982) The pathoparasitology of the alimentary tract and pancreas of nonhuman primates: a review. Vet Pathol 19(Suppl 7):44–92.
847. Toft JD II, Ekstrom ME (1980) Identification of metazoan parasites in tissue sections. In: Montali RJ, Migaki G (eds) The comparative pathology of zoo animals. Smithsonian Institution Press, Washington, DC, pp 369–378.
848. Toft JD II, Schmidt RE, DePaoli A (1976) Intestinal polyposis associated with oxyurid parasites in a chimpanzee (*Pan troglodytes*). J Med Primatol 5:360–364.
849. Trapido H, Work TH (1962) Non-human vertebrates and hosts and disseminators of Kyasanur Forest disease. Proc 9th Pacific Sci Cong 17:85–87.
850. Trapido H, Gorerdham MK, Rajagopalan PK, Rebello MJ (1964) Ticks ectoparasitic on monkeys in the Kyasanur Forest disease area of Shimoga District, Mysore State, India. Amer J Trop Med 13:763–772.
851. Travassos LP, Vogelsang EG (1929) Sobre um novo Trichostrongylidae parasito de *Maccacus* [sic] *rhesus*. Sci Med Ital 7:509–511.
852. Treadgold CH (1920) On a filaria, *Loa papionis* n. sp. parasitic in *Papio cynocephalus*. Parasitology 12:113–115.
853. Troisier J, Deschiens RE (1930) Deux cas d'oxyurose chez le chimpanzé traversée de la paroi intestinale jusgu au peritonine. Ann Parasit 8:562–565.
854. Troisier J, Deschiens RE, Limousin H, Delorme MJ (1928) L'infestation du chimpanzé par un nématode du genre *Hépaticola*. Ann Inst Pasteur, Paris 42:827–840.
855. Troisier J, Deschiens RE, Limousin H, Delorme MJ (1928) L'infestation du chimpanzé par un nématode du genre *Hépaticola*. Bull Soc Path Exot 21:211–222.
856. Tuggle BN, Beehler BA (1984) The occurrence of *Pterygodermatites nycticebi* (Nematoda:Rictulariidae) in a captive slow loris *Nycticebus coucang*. Proc Helminthol Soc Washington, DC 51:162–163.
857. Uemura E, Houser WD, Cupp CJ (1979) Strongyloidiasis in an infant orangutan (*Pongo pygmaeus*). J Med Primatol 8:282–288.
858. Uilenberg G, Ribot JJ (1965) Note sur la toxoplasmose des lemuriens (Primates:Lemuridae). Rev Elev Med Pays Trop 18:247–248.
859. Ulrich CP, Henrickson RV, Karr SL (1981) An epidemiological survey of wild caught and domestic born rhesus monkeys (*Macaca mulatta*) for anatrichosoma (Nematoda:Trichinellida). Lab Anim Sci 31:726–727.
860. Urbain A, Bullier P (1935) Un cas de cénurose conjonctive chez un gelada (*Theropithecus gelada*) (Ruppel). Bull Acad Nat Med 8:322–324.
861. Urbain A, Nouvel J (1944) Petite enzootic de strongyloidose observée sur des singes superieurs: gibbons á favoris blancs (*Hylobates concolor leucogenis* Ogilby) et chimpanzés (*Pan troglodytes* L.). Bull Acad Vét Fr 17:337–341.
862. Valerio DA, Miller RL, Innes JRM, Courtney KD, Pallotta AJ, Guttmacher RM (1969) *Macaca mulatta*: management of a laboratory breeding colony. Academic, New York.
863. Van Riper DC, Day PW, Finey J, Prince JR (1966) Intestinal parasites of recently imported chimpanzees. Lab Anim Care 16:360–362.

864. Van Stee EW (1964) Some observations on the clinical management of the chimpanzee. USAF Sch Aerospace Med (AFSC) Brooks AFB, TX, Tech Doc Rept SAM-TDR 64–45.
865. Van Thiel PH (1926) On some filariae parasitic in Surinam mammals, with the description of *Filariopsis asper* n. g., n. sp. Parasitology 18:128.
866. Van Thiel PH, Wiegand-Bruss CJE (1945) Présence de *Prosthenorchis spirula* chez les chimpanzés. Son rōle pathogéne et son dévelopment. Dans *Blattella germanica*. Ann Parasitol Hum Comp 20:304–320.
867. Vickers JH (1966) *Hepatocystis kochi* in *Cercopithecus* monkeys. JAVMA 149:906–908.
868. Vickers JH (1968) Gastrointestinal diseases of primates. In: Kirk RW (ed) Current veterinary therapy. III. Small animal practice. WB Saunders, Philadelphia, pp 393–396.
869. Vickers JH (1969) Diseases of primates affecting the choice of species for toxicologic studies. Ann NY Acad Sci 162:659–672.
870. Vickers JH, Penner LR (1968) Cysticercosis in four rhesus brains. JAVMA 153:868–871.
871. Vitzhum H (1930) *Pneumonyssus stammeri* ein neuer Lungenparasit. Z Parasitenk 2:595–615.
872. Vogel H (1927) Beiträge zur Anatomie der Gattungen *Dirofilaria* und *Loa*. Zbl Bakt I Orig 102:81.
873. Vogel H, Vogelsang EG (1930) Neue Filarien aus dem Orang-utan und der Ratte. Zbl Bakt Abt 1, Orig 117:480–485.
874. Voller A (1972) Plasmodium and hepatocystis. In: Fiennes RNT-W (ed) Pathology of simian primates. Part II. Infectious and parastic diseases. S Karger, New York, pp 57–73.
875. Vuylsteke C, Rodhain J (1938) *Dirofilaria schoutedeni* n. sp. de *Colobus polykomos uelensis*. Rev Zool Bot Afr 30:356.
876. Walker AE (1936) *Cysticercosis cellulosae* in the monkey. A case report. J Comp Path 49:141–145.
877. Walker EL (1913) Experimental balantidiasis. Philip J Sci Sect B 8:333–339.
878. Wallace FG, Mooney RD, Sanders A (1948) *Strongyloides fulleborni* infection in man. Amer J Trop Med 28:299–302.
879. Wallace GD (1972) Experimental transmission of *Toxoplasma gondii* by cockroaches. J Infect Dis 126:545–547.
880. Wardle RA, McLeod JA (1952) The zoology of tapeworms. University of Minnesota Press, Minneapolis.
881. Warren MCW, Wharton RH (1963) The vectors of Simian malaria: identity, biology, and geographical distribution. J Parasitol 49:892–904.
882. Warren MCW, Bennett GF, Sandosham AA, Coatney GR (1965) *Plasmodium eylesi* sp. nov., a tertian malaria parasite from the white-handed gibbon, *Hylobates lar*. Ann Trop Med Parasitol 59:500–508.
883. Warren MCW, Coatney GR, Skinner JC (1966) *Plasmodium jefferyi* sp. n. from *Hylobates lar* in Malaya. J Parasitol 52:9–13.
884. Warren MCW, Shiroishi T, Davis J (1968) A hepatocystis-like parasite of the gibbon. Trans Roy Soc Trop Med Hyg 62:4.
885. Webber WAF (1955) The filarial parasites of primates: a review. I. *Dirofilaria* and *Dipetalonema*. Ann Trop Med Parasitol 49:123–141.
886. Webber WAF (1955) The filarial parasites of primates: a review. II. *Loa*, *Protofilaria* and *Parlitomosa*, with notes on incompletely identified adult and larval forms. Ann Trop Med Parasitol 49:235–249.
887. Webber WAF (1955) *Dirofilaria aethiops* Webber, 1955, a filarial parasite of monkeys. I. The morphology of the adult worms and microfilariae. Parasitology 45:369–377.

888. Webber WAF (1955) *Dirofilaria aethiops* Webber, 1955, a filarial parasite of monkeys. III. The larval development in mosquitoes. Parasitology 45:388–400.
889. Webber WAF, Hawking F (1955) The filarial worms *Dipetalonema digitatum* and *D. gracile* in monkeys. Parasitology 45:401–408.
890. Weidman FD (1923) The animal parasites, their incidence and significance. In: Fox H (ed) Disease in captive wild mammals and birds: incidence, description, comparison. JB Lippincott, Philadelphia, pp 614–659.
891. Weidman FD (1923) Certain dermatoses of monkeys and an ape. Pemphigus, scabies, sebaceous cyst, local subcutaneous edema, benign superficial blastomycotic dermatosis and tinea capitis and circinata. Arch Dermatol Syphilol 7:289–302.
892. Weinberg M (1906) Kystes vermineux du gros intestin chez le chimpanzé et les singes inférieurs. CR Soc Biol, Paris 60:446–449.
893. Weinberg M (1907) Du rôle des helminthes, des larves d'helminthes, et des larves d'insectes dans la transmission des microbes pathogénes. Ann Inst Pasteur 21:417–442, 533–561. Cited by Ruch TC (1959) Diseases of laboratory primates. WB Saunders, Philadelphia.
894. Weinberg M (1908) Oesophagostomose des anthropoides et des singes inférieurs. Arch Parasitol 13:161–203.
895. Wellde BT, Johnson AJ, Williams JS, Langbehn HR, Sadun EH (1971) Hematologic, biochemical, and parasitologic parameters of the night monkey (*Aotus trivirgatus*). Lab Anim Sci 21:575–580.
896. Wenrich DH (1933) A species of *Hexamita* (Protozoa, Flagellata) from the intestine of a monkey (*Macacus rhesus*). J Parasitol 19:225–229.
897. Wenrich DH (1937) Studies of *Iodamoeba bütschlii* (protozoa) with special reference to nuclear structure. Proc Amer Phil Soc 77:183–205.
898. Wenrich DH, Nie D (1949) The morphology of *Trichomonas wenyoni* (Protozoa, Mastigophora). J Morphol 85:519–531.
899. Wenyon CM (1926) Protozoology: a manual for medical men, veterinarians and zoologists. 2 vol Balière. Tindall and Cox, London.
900. Wharton RH (1957) Studies on filariasis in Malaya: observations on the development of *Wuchereria malayi* in *Mansonia* (*Mansoniodes*) *longipalpis*. Ann Trop Med Parasitol 51:278–296.
901. Wharton RH (1959) *Dirofilaria magnilarvatum* Price, 1959 (Nematoda:Filarioidea) from *Macaca irus* Cuvier. IV. Notes on larval development in Mansonioides mosquitoes. J Parasitol 45:513–518.
902. Whitney RA Jr, Kruckenberg JM (1967) Pentastomid infection associated with peritonitis in mangabey monkeys. JAVMA 151:907–908.
903. Whitney RA, Johnson DJ, Cole WC (1967) The subhuman primate: a guide for the veterinarian. Edgewood Arsenal Maryland EASP 100–26.
904. Wigglesworth VB (1932) Exhibition of a new species of sucking louse from a chimpanzee. Proc Zool Soc London 1079.
905. Williams CSF, Murray RE, McGovney RM, Cockrell BY (1973) Adamantinoma in a spider monkey (*Ateles fusciceps*). Lab Anim Sci 23:273–275.
906. Wilson DA, Day PA, Brummer EG (1984) Diarrhea associated with *Cryptosporidium* sp. in juvenile macaques. Vet Pathol 21:447–450.
907. Windle DW, Reigle DH, Heckman MG (1970) *Physaloptera tumefaciens* in the stump-tailed macaque (*Macaca arctoides*). Lab Anim Care 20:763–767.
908. Witenberg GG (1964) Cestodiases. In: van der Hoeden J (ed) Zoonoses. American Elsevier, New York, pp 649–707.
909. Witenberg GG (1964) Acanthocephala infections. In: van der Hoeden (ed) Zoonoses. American Elsevier, New York, pp 708–709.
910. Wong MM (1970) Procedure in laboratory examination of primates with special reference to necropsy technics. Lab Anim Care 20:337–341.

911. Wong MM, Conrad HD (1972) Parasite nodules in the macaques. J Med Primatol 1:156–171.
912. Wong MM, Conrad HD (1978) Prevalence of metazoan parasite infections in five species of Asian macaques. Lab Anim Sci 28:412–416.
913. Wong MM, Kozak WJ (1974) Spontaneous toxoplasmosis in macaques: a report of four cases. Lab Anim Sci 24:273–278.
914. Woodard JC (1968) Acarous (*Pneumonyssus simicola*) arteritis in rhesus monkeys. JAVMA 153:905–909.
915. Worms MJ (1967) Parasites of newly imported animals. J Inst Anim Technic 18:39–47.
916. Yamaguti S (1954) Studies on helminth fauna of Japan. Part 51. Mammalian nematodes. V Acta Med Okayama 9:105–121.
917. Yamaguti S (1958) The digenetic trematodes of vertebrates. Vol I. 2 parts. In: Yamaguti S, Systema helminthum. Interscience, New York.
918. Yamaguti S (1959) The cestodes of vertebrates. Vol II. In: Yamaguti S, Systema helminthum. Interscience, New York.
919. Yamaguti S (1961) The nematodes of vertebrates. Vol III. In: Yamaguti S, Systema helminthum. Interscience, New York.
920. Yamaguti S, Hayama S (1961) A redescription of *Edesonfilaria malayensis* Yeh, 1960, with remarks on its systematic position. Proc Helminthol Soc Washington, DC 28:83–86.
921. Yamashiroya HM, Reed JM, Blair WH, Schneider MD (1971) Some clinical and microbiological findings in vervet monkeys (*Ceropithecus aethiops pygerythrus*). Lab Anim Sci 21:873–883.
922. Yamashita J (1963) Ecological relationships between parasites and primates. I. Helminth parasites and primates. Primates 4:1–96.
923. Yeh L-S (1957) On a filarial parasite, *Deraïophoronema freitaslenti* n. sp. from the giant anteater, *Myrmecophaga tridactyla* from British Guiana, and a proposed reclassification of *Dipetalonema* and related genera. Parasitology 47:196–205.
924. Yeh L-S (1960) On a new filarioid worm, *Edesonfilaria malayensis* gen. et sp. nov. from the long-tailed macaque (*Macaca irus*). J Helminthol 34:125–128.
925. Yokogawa S (1941) On the classficiation of the plasmodia found in the indigenous monkey (black-leg monkey) of Formosa found by us and previously reported. J Med Assoc Formosa 40:2185–2186.
926. Yorke W, Maplestone PA (1926) The nematode parasites of vertebrates. Blakiston, Philadelphia.
927. Yoshimura K, Hishinuma Y, Sato M (1969) *Ogmocotyle ailuri* (Price, 1954) in the Taiwanese monkey, *Macaca cyclopis* (Swinhoe, 1862). J Parasitol 55:460.
928. Young MD (1970) Natural and induced malarias in Western Hemisphere monkeys. Lab Anim Care 20:361–367.
929. Young RJ, Fremming BD, Benson RE, Harris MD (1957) Care and management of a *Macaca mulatta* monkey colony. Proc Anim Care Panel 7:67–82.
930. Yue MY, Jensen JM, Jorden HE (1980) Spirurid infections (*Rictularia* sp.) in golden marmosets, *Leontopithecus rosalia* (syn. *Leontideus rosalia*) from the Oklahoma City Zoo. J Zool Anim 11:77–80.
931. Yunker CE (1964) Infections of laboratory animals potentially dangerous to man: ectoparasites and other anthropods with emphasis on mites. Lab Anim Care 14:455–465.
932. Zarman V (1970) *Sarcocystis* sp. in the slow loris, *Nycticebus coucang*. Trans Roy Soc Trop Med Hyg 64:195–196.
933. Zaman V (1972) A trypanosome of the slow loris (*Nycticebus coucang*). Southeast Asian J Trop Med Publ Health 3:22–24.
934. Zaman V, Goh TK (1968) Isolation of *Toxoplasma gondii* from the slow loris, *Nycticebus coucang*. Ann Trop Med Parasitol 62:52–53.

935. Zaman V, Goh TK (1970) Isolation of *Toxoplasma gondii* from Malayan tree shrew. Trans Roy Soc Trop Med Hyg 64:462.
936. Ziemann H (1902) Über das Vorkommen von *Filaria perstans* und von Trypanosomen beim Schimpanse. Arch Schiffs Tropen Hyg 6:362.
937. Zumpt F (1961) The arthropod parasites of vertebrates in Africa south of the Sahara (Ethiopian region), vol 1 (Chelicerata). Pub 5th Afr Inst Med Res 9:1–457.
938. Zumpt F, Till WM (1954) The lung and nasal mites of the genus *Pneumonyssus* Banks (Acarina:Laelaptidae) with description of two new species from African primates. J Ent Soc So Afr 17:195–212.
939. Zumpt F, Till WM (1955) The mange-causing mites of the genus *Psorergates* (Acarina:Myobiidae) with description of a new species from a South African monkey. Parasitology 45:269–274.
940. Zwicker GM, Carlton WW (1972) Fluke (*Gastrodiscoides hominus*) infection in a rhesus monkey. JAVMA 161:702–703.

46
Overview of Simian Viruses and Recognized Virus Diseases and Laboratory Support for the Diagnosis of Viral Infections

SEYMOUR S. KALTER

Introduction

A phylogenetic relationship to the human is often suggested as the reason for selecting a particular simian host, such as the chimpanzee, for studies on human diseases. It is thought that the use of another primate permits the extrapolation of more meaningful data than those derived from other species of animals. However, there is more to selecting an appropriate nonhuman primate model than phylogenetic relatedness. Two widely separated species, the chimpanzee and, with a few exceptions, the marmoset, are highly susceptible to hepatitis A, while the rhesus monkey is not. Therefore, care in selection of the most appropriate model for the study of a particular disease is vital to the success of any research program.

Historically, the nonhuman primate has often been misused in medical research through failure to recognize these animals not as "test tubes" but as biological entities, with a full range of individual responses to internal and external influences. Fortunately, the scientific community is becoming more concerned about the use and abuse of these animals in research. An awareness that nonhuman primates differ among the species in their responses to human disease agents and to their own particular infectious agents is becoming apparent.

The influence of viruses and viral diseases on the health of nonhuman primate colonies and ultimately on the success of research programs has not been extensively studied. Investigators often fail to recognize infection, particularly latent infection and subclinical disease, and overt disease is frequently handled by disposal of the animal without attempt to discover the cause and epidemiology. Among the many viruses harbored by nonhuman primates, *Herpesvirus simiae* (B virus) has generated the most interest, and that because of its extreme danger to humans, not to other animals.

This report will attempt to provide an awareness of the influence of viruses and their diseases, as well as the importance of their detection, on the health of colonies necessary for "self-sustaining populations" of nonhuman primates.

Simian Viruses and Diseases of Monkeys and Apes

The presence of an extensive virus population (Hull, 1968; Kalter et al., 1980), counterparts of human viruses (Matthews, 1982), in nonhuman primates is now well established (Table 46.1). That many of these viruses are associated with a variety of diseases both in the host of origin and in alien hosts is also well recognized (Andrews, 1976; Hull, 1968; Kalter and Herberling, 1976; Kalter, 1983). Several of these simian agents are lethal for primates, both human and nonhuman; othersare oncogenic; and still others are capable of producing an assortment of clinical ailments. Nonhuman primates are equally susceptible to the viruses that cause infection/disease in humans, often with the same end result—death.

Nonhuman primate viruses, distinct from other animal viruses, have been recognized since Sabin and Wright (1934) described the presence of a herpesvirus (herpes B) in a fatal human case following a monkey bite. A number of investigators have since demonstrated antigenically distinct viruses in simian tissue (principally kidney cells) as well as in feces, throat washings, and spinal fluid (Herberling and Cheever, 1960; Hull et al., 1956; Hull, 1968; Kalter, 1960; Malherbe and Harwin, 1957). Although these viruses are antigenically distinct from the recognized human and other animal virus isolates, in general they conform to the biological criteria used to classify a virus (Kalter et al., 1980). The majority of the simian viruses may be included in existing virus families, notably Picornaviridae (picornaviruses, principally the enteroviruses). Adenoviridae (adenoviruses), and Herpesviridae (herpesviruses), and all are capable of infecting various species of primates including humans. The most notorious, however, are the herpesviruses, present in most simian species and associated with human and nonhuman primate fatalities as well as with oncogenesis. Retroviruses of nonhuman primates are well described and have also gained a certain amount of notoriety, not only because retroviruses are frequently responsible for tumor production in many animal species but also because recent findings have indicated that there is a close relationship between simian retroviruses responsible for simian acquired immunodeficiency syndrome (SAIDS) and the human disease AIDS. Other distinctive simian viruses exist but are found less frequently than the above viruses (Kalter et al., 1980).

In addition, viruses have been recovered from nonhuman primates that are either identical to or so closely related antigenically to the human agents that differentiation is impossible or of little consequence. These viruses are, in essence, primate (human and nonhuman) viruses. Included among these viruses are influenza viruses (Orthomyxoviridae), measles virus (Paramyxoviridae), reoviruses and rotaviruses (Reoviridae), monkeypox virus (Poxviridae), yellow fever virus (Togaviridae), rabies virus (Rhabdoviridae), and others. Several viruses originally isolated from nonhuman primates (respiratory syncytial virus, SV 5) have been reclassified as animal (human?) viruses rather than simian viruses. Unclassified viruses such as Marburg virus may or may not be simian in origin but are associated with diseases attributed to monkey contact. The various arboviruses (Togaviridae, Bunyaviridae, and other vector-transmitted viruses) are primate (human and nonhuman) viruses and generally not host restrictive. Disease occurs in both human and nonhuman primates as a result of infection with viruses derived from both human and nonhuman primates as well as from other animal sources. The full extent of infectivity of many of these viruses is not known.

A phylogenetic relationship to the human is often suggested as the reason for selecting a particular simian host (such as the chimpanzee) for studies on human

TABLE 46.1. Natural viral diseases of nonhuman primates.

Disease	Species	Virus	References
Adenovirus diseases			
Conjunctivitis–rhinorrhea	Patas monkey; *Macaca* sp.	SV17	Tyrrell et al. (1960); Bullock (1965)
Pneumoenteritis	African green monkey, baboon	V340, V404; V340	Kim et al. (1967); Eugster et al. (1969)
Conjunctivitis, upper respiratory infection, cough, nasal discharge	Rhesus	SV15	Landon and Bennett (1969)
Cough, skin lesions	Chimpanzee	Cytomegalovirus	Muchmore (1971)
Necrotizing pneumonia	Rhesus; stump-tailed macaque	SV11; SV15	Valerio (1971); España (1971)
Necrotizing viral, pancreatitis	Rhesus	"Adeno"; Ad. 31	Chandler et al. (1974); McClure et al. (1971)
Pneumonia	Rhesus; African green monkey, baboon, Bonnet monkey, cynomolgus, pig-tailed macaque	SV20; SV11	Moe et al. (1977); Boyce et al. (1978)
Viuria	Chimpanzee	Adenovirus	Asher et al. (1978)
Conjunctivitis–pneumonia	Rhesus	SV37	Vasileva et al. (1978)
Diarrhea	Rhesus	SV20, SV17, SV32	Stuker et al. (1979)
Aspiration pneumonia	Japanese macaque	"Adenovirus"	Umemura et al. (1985)
Arbovirus diseases			
Yellow fever	Red howler monkey; marmoset	Yellow fever	Balfour (1914); Anderson and Downs (1955); Laemmert and Castro-Ferreira (1945)
	Howler, marmoset		Vargas-Mendez and Elton (1953)
Kyasnur Forest disease	Langur, Bonnet monkey	Kyasanur Forest disease virus	Work and Trapido (1957); Webb (1969)
Dengue	"Monkeys" (Cercopithecus?)	Dengue	Monath et al. (1974)
Herpesvirus disease			
Salivary gland disease	Chimpanzee	Cytomegalovirus (CMV)	Vogel and Pinkerton (1955)
Generalized cytomegalovirus	Gorilla	CMV	Tsuchiya et al. (1970)
Genital malformations	Squirrel monkey	CMV	Ordy et al. (1981)
Herpesvirus	Rhesus	Herpes B (HBV, B virus)	Keeble et al. (1958)
B virus	Rhesus; cynomolgus; macaques	HBV	Pille (1960); Hartley (1964); Zwartouw et al. (1984)
Herpesvirus	Macaques	HBV	Roberts et al. (1984)
Herpesvirus (genital infection)	Rhesus	HBV	Zwartouw and Boulter (1984)
Respiratory disease	Bonnet monkey	HBV	España (1973)
Disseminated herpes	Rhesus	HBV	McClure et al. (1973)
Cerebral infarction	Rhesus	Herpesvirus	Daniel et al. (1975)

TABLE 46.1. (Continued).

Disease	Species	Virus	References
Oral lesions	Rhesus	Herpesvirus	Vizoso (1975)
Chickenpox	Apes; African green monkey	Varicella	Heuschele (1960); Lehner et al. (1984)
Exanthema (fatal)	Vervet	LVMV (Varicella)	Clarkson et al. (1967)
Herpesvirus	Patas monkey	PMHV (Varicella)	McCarthy et al. (1968)
Macular rash, pruritis	Apes	Varicella	McClure and Keeling (1971)
Chicken pox	Gorilla	Varicella-Zoster	White et al. (1972)
Varicella	Stump-tailed macaque;	Medical Lake Macaque (MLM);	Blakely et al. (1973);
	Patas	Delta herpes	Gard and London (1983)
Generalized rash, vesiculations, anorexia	Patas *Cercopithecus* sp. *Cercocebus* sp.	Delta herpes	Allen et al. (1974)
"Smallpox"	Chimpanzee	Varicella	Marennikova et al. (1974)
Generalized Herpes (fatal)	Patas	Delta herpes	Wolf et al. (1974)
Herpesvirus (fatal)	Marmoset;	*Herpesvirus tamarinus* (herpes M)	Holmes et al. (1964)
	Spider monkey	SMV (SMHV, herpes T)	Hull et al. (1972)
Herpesvirus	Marmoset; owl monkey; squirrel monkey	Marmoset herpes; herpes T	Melnick et al. (1964); Hunt and Melendez (1966); Daniel et al. (1967)
Pruritis	Owl monkey	Herpes T (HVT)	Emmons et al. (1968)
Encephalitis (resp. distress)	Owl monkey	Herpes T	Tate et al. (1971)
Herpesvirus	Marmoset; owl monkey; gibbon	HVT; herpes simples; *H. hominis*	Morita et al. (1979); Melendez et al. (1969); Smith et al. (1969)
Encephalitis (fatal)	Gibbon	*H. hominis*	Emmons and Linnette (1970)
Genital herpes	Chimpanzee	HSV-2	McClure et al. (1980)
Herpesvirus (generalized)	Gorilla	*H. hominis* (HSV-1)	Heldstab et al. (1981)
Encephalitis (fatal)	Gibbon	Herpesvirus	Ramsey et al. (1982)
Malignant lymphoma	Rhesus	Herpes-like	Stowell et al. (1971)
Lymphoma	Owl monkey	*H. hominis* (HVS)	Rabin et al. (1975)
Malignant lymphoma	Baboon	HVP (lymphotropic baboon herpesvirus)	Agrba et al. (1980)
Vesicular disease (fatal)	Macaques	"Herpes-type"	Lourie et al. (1971)

TABLE 46.1. (Continued).

Disease	Species	Virus	References
Lymphoma	Owl monkey	*H. Saimiri* (HVS)	Hunt et al. (1973)
Herpesvirus (fatal)	Patas, colobus	HVS-HSV	Loomis et al. (1981)
Generalized infection	Marmosets	*H. saimiri*	Benirschke (1983)
Myelitis and paralysis	African green monkey	SA8	Malherbe and Harwin (1957)
Pneumonia	Gelada	Herpes virus	Ochoa et al. (1982)
Orthomyxovirus diseases			
Grippe	Chimpanzee	"Influenza"	Mouquet (1926)
Death	Cebus	"Influenza"	Ratcliffe (1942)
"Influenza"	Cercopithecoid	"Influenza"	Panthier et al. (1949)
Fever, malaise	Baboon	A_2/Hong Kong/68	Kalter et al. (1969)
Respiratory disease	Gibbon	A_2/Hong Kong/68	Johnsen et al. (1971)
Papovavirus diseases			
Progressive multifocal leukoencephalopathy (PML)	Rhesus	SV40	Holmberg et al. (1977)
Paramyxovirus diseases in nonhuman primates			
Papular and desquamative exanthem	Orangutan	"Measles"	Fox and Weidman (1923)
Measles	Cynomolgus; rhesus; colobus; silver-leaf monkey (*Presbytis cristatus*)	Measles	Ruckle (1956); Shishido (1966); Potkay et al. (1966); Hall et al. (1972); Yamanouchi et al. (1973); Colman and Clarke (1975); Hime et al (1975); Remfry (1976); Montrey et al. (1980)
Giant cell pneumonia	Rhesus	Measles	Manning et al. (1968)
Measles (fatal)	Marmoset	Measles	Levy and Mirkovic (1971)
Endometritis, cervicitis, abortion	Rhesus	Measles	Renne et al. (1973)
Gastroenterocolitis	Marmoset	Measles	Fraser et al. (1978)
Encephalitis	Rhesus, pig-tailed macaque		Steele et al. (1982)
Coryza	Chimpanzee	"CCA"[1] (chimpanzee coryza agent)	Morris et al. (1956)
Pneumonia	Patas	Paraflu-3	Churchill (1963)
Upper respiratory infection	Gibbon	Paraflu-3	Martin and Kaye (1983)
Pneumonia	Chimpanzee	Paraflu-3	Jones et al. (1984)
Acute respiratory disease	Marmoset	Paraflu-1	Flecknell et al. (1983)
Mumps (parotiditis)	Rhesus	Mumps	Bloch (1937)
Gastroenterocolitis	Marmoset	"Measles"(?) SV5	Gibson et al. (1980)
Respiratory disease	Cynomolgus	Murayana	Nishikawa et al. (1977)

TABLE 46.1. (Continued).

Disease	Species	Virus	References
Picornavirus diseases			
Poliomyelitis (paralytic)	Chimpanzee	?	Goldman (1935); Müller (1935)
	Gorilla, orangutan	Polio 1	Guilloud and Kline (1966); Allmond et al. (1967); Guilloud et al. (1969)
Poliomyelitis	Chimpanzee	Polio	Howe and Bodian (1944); Melnick and Horstmann (1947)
	Colobus	Polio-1	Guillous and Kline (1966)
Myocarditis	Gibbon, chimpanzee	EMC	Schmidt (1948); Gainer (1967); Gaskin et al. (1980); Helwig and Schmidt (1945)
Myocarditis, pulmonary edema	Night monkey (*Aotes trivirgatus*), squirrel monkey	EMC	Roca-Garcia and SanMartin-Barber (1957)
Fever, diarrhea, respiratory disease	Chimpanzee	"Coxsackie-B5"	Kelly et al. (1978)
Diarrhea	Rhesus	SV6, SV48	Heberling (1972)
Neurologic	Rhesus, vervet	SV16	Kaufmann et al. (1973)
Hepatitis	Chimpanzee	HAV	Hillis (1967)
Poxvirus diseases			
"smallpox"	Cebus	"Smallpox"	Bleyer (1922)
Pox-infection	Orangutan	"Variola-like"	Gispen (1949)
Pox-infection (fatal)	Orangutan, chimpanzee, gorilla, "Cercopithecus," marmoset	"Monkeypox"	Peters (1966)
Pox-infection	Cynomolgus, rhesus; vervet, chimpanzee	"Monkeypox" Monkeypox (Variola?)	vonMagnus et al. (1959); Prier et al. (1960); Sauer et al. (1960); McConnel et al. (1962); Gispen and Kapsenberg (1967); Marennikova et al. (1972)
Subcutaneous tumors	Rhesus, baboon	? (YABA)	Bearcroft and Jamieson (1958)
Pox-like skin lesions	Rhesus (other macaques)	Tanapox (Yaba-like, Orteca, 1211)	Nicholas and McNulty (1968); Crandell et al. (1969); España (1968)
Molluscum contagiosum	Chimpanzee	Molluscum contagiosum	Douglas et al. (1967); Schmidt and Butler (1971)
Rabies (Rhabdovirus)			
Respiratory infection	Rhesus	Rabies	Boulger (1966)
Rabies	"Monkey," cebus, cynomolgus, squirrel monkey, "Micoleone" (golden lion marmoset)	Rabies	Richardson and Humphrey (1971)
	Chimpanzee		Miot et al. (1973)

TABLE 46.1. (Continued).

Disease	Species	Virus	References
Reovirus-rotavirus diseases			
Rhinitis	Chimpanzee	REO-1	Sabin (1959)
Diarrhea[2]	Chimpanzee; rhesus	Rotavirus; rotavirus (SA11)	Ashley et al. (1978); Stuker et al. (1980)
Retrovirus diseases			
Mammary tumor (adenocarcinoma)	Rhesus	MPMV (Mason Pfizer monkey virus)	Jensen et al. (1970)
Leukemia	Woolly monkey, gibbon	SSV/SSAV[3] GALV[3]	Theilen et al. (1971); Kawakami et al. (1972)
Leukemia-lymphosarcomas	Baboon, *M. arctoides*	? ("BILN")	Lapin et al. (1973)
Lymphoma	Baboon; stump-tailed macaque	Foamy virus 1	Rabin et al. (1976); Schnitzer (1981)
Pelvic endometriosis	Pig-tailed macaque		DiGiacomo et al. (1977)
SAIDS; lymphoma; lymphoma-retroperitoneal fibromatosis	Macaques	Type D (MPMV, SRV-1, SRV-2); HTLV-III (STLV-III)	Stowell et al. (1971); Manning and Griesemer (1974); Giddens et al. (1983); Henrickson et al. (1983); Letvin et al. (1983); Marx et al. (1984)
Miscellaneous diseases			
Marburg	African green monkey	Marburg	Martini et al. (1968)
Simian Hemorrhagic fever	Macaques	SHF	Lapin and Shevtsova (1966); Palmer et al. (1968); Allen et al. (1968)
Rubella	African green monkey	Rubella (Togaviridae)	Cabasso and Stebbins (1965)
Subacute Sclerosing Panencephalitis-like	Baboon	?	Kim et al. (1970)
Neurologic disease	Rhesus	Picornavirus	Kaufmann (1972)
Perinatal telencephalic leukoencephalopathy	Chimpanzee	?	Brack (1973)

[1] Respiratory syncytial virus.
[2] The significance of finding other viruses in stool samples, from both normal animals and those with diarrhea, emphasizes the need for further study.
[3] Closely related viruses—see text.

diseases. While this phylogenetic relationship appears to exist to a certain extent, variations in susceptibility to viruses among simian species must be considered. This difference in susceptibility is exemplified by certain marmoset species (with few exceptions) that phylogenetically are far removed from chimpanzees but are along with the chimpanzee both highly susceptible to hepatitis A.

Virus diseases of primates may be grouped according to the source of infection: natural, with which we are primarily concerned, or experimental. Experimental infectious disease studies also need consideration, particularly if they are not adequately controlled, allowing "in-house" epidemics to occur. Simian hemorrhagic fever was spread through several animal colonies by failure to sterilize adequately syringes or tattooing needles.

"Natural" infections in primate colonies have resulted from contact with rodents (gibbon ape leukemia) and humans (influenza, respiratory syncytial virus–croup). To develop and maintain primate colonies with minimal losses resulting from viruses, one must guard against the introduction of agents from the wild (newly imported animals) but also against the spread of a virus derived from the accidental infection of an animal in the colony by another animal (human?) source. A mechanism for monitoring the animals is necessary for early detection of an alien agent to prevent its spread.

With encroaching human populations, human contact with wild groups of animals needs consideration. These naive animals are highly susceptible to infectious agents by virtue of lack of previous contact (immunization). More often than not, such contact will have little effect on these groups of animals other than immunization. However, one bad experience, such as occurred with the introduction of measles into secluded human populations, can do much to endanger further an already endangered group of animals.

The full spectrum of virus diseases that occur among simian populations in the wild is unknown. Very few studies have been conducted on simian species in their native habitat. The majority of information available on virus diseases of nonhuman primates has been derived from captive animals or those in contact with humans or other species of animals. Trapping conditions are still deplorable in many instances and probably are the major source of infection and disease, but contact with other animal species does occur under natural conditions. Attempts to analyze data obtained from studies on wild populations of animals are clouded by the diverse conditions under which the different species live. The need for food and water results in migration patterns producing contacts impossible to determine. Even those animal populations that are stationary are in contact with other populations that are migratory. Under these conditions, precise origins of diseases are purely speculative.

In spite of these difficulties, serological studies and the isolation of viruses from animals immediately following capture have provided some information of value. Serological surveys furnish data regarding past infections; isolation data offer little in the way of past history, usually indicating only current infections. An important exception is the isolation of an agent from tissues, e.g., herpesviruses from ganglia, in the case of latent infections.

Antibody studies suggest widespread distribution of virus infections among wild populations. Clinical disease, however, has been reported rarely (Kalter and Heberling, 1971). With few exceptions, little is known about fatal virus diseases occurring in the wild. Animals that die are rarely found and rapidly disappear from the scene. Animals that are found, again with few exceptions, are rarely studied for specific cause of death, particularly with regard to viruses.

Verifiable reports of fatal virus diseases occurring in nonhuman primates in the wild are, therefore, few. High rates of fatalities among simians as a result of viruses have been reported, usually coincidental with human deaths, caused by yellow fever virus, Kyasanur Forest virus, and Marburg virus. The first two diseases were actually occurrences in the wild; the latter deaths were seen in the laboratory. Deaths caused by poxviruses (monkeypox), rabies, measles (rubeola), retroviruses (gibbon ape leukemia virus, SAIDS), picornaviruses (poliovirus, coxsackievirus), and others have been reported to occur as a result of natural infections. It is not clear what the impact of these viruses would be in colonies reared under controlled laboratory conditions.

Information on captive animals is more readily available than on those in nature. Review of documented captive animal disease outbreaks indicates that the majority are nosocomial. Mishandling, mixing of species or groups of animals, and breakdown in appropriate procedures account for most occurrences. As a result, disease caused by human agents spread by personnel either directly or indirectly occurs more frequently than diseases caused by simian viruses.

Only those virus diseases that occur naturally rather than experimentally will be discussed. It should be emphasized that the majority of these infections/diseases are more debilitating than fatal; nonetheless, the loss of time and the disruption of programs emphasize the need for proper handling procedures to sustain populations.

Viruses

In reviewing the major virus diseases observed in nonhuman primates as a result of "natural" infections, no attempt will be made to provide details on any particular disease or diseases. My intent is to emphasize the extensive occurrence of natural viral diseases that have been recognized and indicate the potential effect such occurrences could have on "self-sustaining" populations. I will also discuss measures necessary to prevent these diseases or minimize their spread. Serological assays of simian sera indicate widespread contact among simian species with all the following viruses and their various serotypes (Kalter and Heberling, 1971).

Adenoviruses (Adenoviridae)

This virus family (Adenoviridae) is the one most frequently isolated from simians. Over 20 serotypes are recognized, and one or another of these may be recovered from normal as well as sick animals at any time. As a result, association with disease is difficult to determine and requires careful laboratory support. Mainly present in the respiratory and intestinal tract of all simian species, adenoviruses may also be found in various tissues and conjunctival fluids. Cross-reactions with human adenoviruses are common (Willimzik et al., 1981), making differentiation between human and nonhuman strains difficult. As seen in Table 46.1, diseases similar to those occurring in humans are found in nonhuman primates as a result of the adenoviruses.

Arboviruses (Togaviridae, Bunyaviridae, etc.)

The use of nonhuman primates as sentinel animals for the detection of arboviruses indicates their susceptibility. A wide range in susceptibility to these viruses exists among primates. It might be assumed, because of the widespread prevalence and the large number of viruses included among the arboviruses, that infection/disease would be extensive. However, deaths as a result of arboviruses occur infrequently, and clinical disease, although it undoubtedly occurs, is also apparently infrequent (Table 46.1). It is noteworthy to recognize that monkeys play an important role in the maintenance of yellow fever in nature. This is a rare role for nonhuman primates to play in the epidemiology of infection.

Herpesviruses (Herpesviridae)

One of the most ubiquitous of virus families, the herpesviruses are also one of the most important in terms of infection/disease. As a natural infection, the herpesviruses are rapidly disseminated. In captivity, the herpesviruses are also responsible for infection, disease, and death. Undoubtedly, herpesviruses as a group are of greatest concern for nonhuman primates as well as to users of nonhuman primates. Herpesviruses are responsible for every conceivable type of clinical condition, extending from inapparent to mild localized disease to death. Any program designed to maintain self-sustaining populations must give thought to control of this group of viruses. A large number of herpesviruses are recognized, and although many are considered host specific, crossing of species barriers is common. When this occurs, it is frequently associated with a clinical pattern more severe than that seen in the original host. For this reason, B virus has probably become the simian virus of greatest concern among primate investigators.

Herpesviruses in their native hosts, while not innocuous, are probably of little concern in terms of colony management; however, in colonies of mixed populations or where human contact is not controlled, herpesviruses must be considered a major threat in maintenance of the colony. As Hunt (this volume) emphasizes, the cytomegaloviruses do not appear to be of any clinical consequence. However, until more is known about the full pathogenic potential of this virus(es), the frequency of occurrence in primates should make it suspect as a potential pathogen. There is also increasing concern regarding the oncogenic capabilities of the herpesviruses (Table 46.1).

Influenza Viruses (Orthomyxoviridae)

It is now apparent that all primates respond to infection with strains of influenza A viruses. This response may be clinical or subclinical. Clinical disease is similar to that seen in the human, with virus shed into the environment. Deaths appear to be associated with secondary bacterial infection. Serological studies indicate widespread distribution of antibody to influenza A, with greater numbers of positives seen to the newer strains (Kalter and Heberling, 1978). Vaccines apparently are effective but, in view of the lack of severe disease, the need for vaccination may be questioned. Although antibody to influenza virus B is detected, these seropositives are few, and it would appear that this virus is probably inconsequential as a disease factor. Epizootics in gibbons with type A (A_2/Hong Kong/68) have been reported (Johnson et al., 1971). Reports of isolated cases in nonhuman primates are also known (Table 46.1).

Parainfluenza Viruses (Paramyxoviridae

Several of the human parainfluenza viruses are responsible for disease in nonhuman primates. A certain amount of confusion exists in classification of these viruses because there have been isolations made from simians that suggest "true" simian viruses exists. Indeed, several have been found to be primate viruses that infect both human and nonhuman primates. Infection with the various parainfluenza viruses, mumps, measles (rubeola), and respiratory syncytial virus (chimpanzee coryza agent, CCA) has been reported. The exact relationship of two simian isolates, SV 5 and SV 41, to simians is still in need of study.

Papovaviruses (Papovaviridae)

Two virus genera, *Papillomavirus* and *Polyomavirus*, make up this virus family. The viruses included herein are of interest not so much for what they have done, but for their potential. Present in a number of different animal species including human and nonhuman primates, many of these viruses have the ability to produce tumors or transform cells. In spite of this recognized capability, only limited pathogenicity has been demonstrated. SV 40, one of the first simian viruses recognized (Rustigian et al., 1955), is present in *Macaca* sp. kidney cells. This virus has been widely distributed to humans, inadvertently incorporated into the poliovaccine. No adverse effects have been recorded as a result of this administration although some 30 years have passed since it was first given to humans. Holmberg et al. (1977) were able to isolate SV 40 following the immunological suppression of macaques with progressive multifocal leukoencephalopathy (PML). Inasmuch as these viruses are associated with latency (Norkin, 1982), are they activated when the individual is stressed or immunologically impaired?

Picornaviruses (Picornaviridae)

This family of viruses (Picornaviridae) originally included only the enteroviruses (polio-, coxsackie-, and echoviruses) but now includes the rhinoviruses and hepatitis A virus (HAV) derived from humans plus a number of animal viruses sharing the same biological characteristics (foot and mouth disease virus, encephalomyocarditis virus [see Hunt, this volume], and a number of insect viruses). Counterparts of the human enteroviruses are also present in various animal species including simians, rodents, cattle, swine, and others. These and the adenoviruses are the most numerous of the isolates encountered in routine primate virus isolation attempts. Antibody studies (Kalter, 1967) indicate that infection with the enteroviruses is widespread among nonhuman primates. In terms of disease, however, they are relatively unimportant although disease caused by one or another of these viruses has been recorded. Encephalomyocarditis virus evidently can be a major problem in primate colonies (Gaskin et al., 1980). Rhinoviruses do not appear to be highly pathogenic (Dick and Dick, 1974), but only limited information is available on this large group of viruses.

Poxviruses (Poxviridae)

A widely distributed virus group, most of these are restrictive to their natural host. Antigenic crossing among viruses in each genus makes differentiation of the viruses difficult. Several poxvirus genera exist in the family Poxviridae; infection of primates is generally with members of the vaccinia subgroup (Orthopoxvirus) (vaccinia–variola–monkeypox viruses). There is concern over poxvirus transmission, and human infection with monkeypox following contact with an infected chimpanzee has been reported (Mutombo et al., 1983). The extent of natural infection with this virus among nonhuman primates is not known. Unclassified Poxviridae (Yaba virus, Tanapox virus, molluscum contagiosum virus) are also known to be infectious for one or another nonhuman primate.

Reoviruses (Reoviridae)

Reoviruses isolated from nonhuman primates (SV 12 and SV 59) are so closely related antigenically to the recognized human reovirus types 1 and 2 that they are considered counterparts of the human types. In general, these viruses are of low pathogenicity although antibody surveys indicate widespread occurrence of all three reovirus types. Sabin (1959) did report the occurrence of a zoonoses of rhinitis in chimpanzees from which a type 2 reovirus was isolated. Reovirus association with diarrhea is suspected but not demonstrated. Another member of the family Reoviridae, the rotaviruses, are known to cause gastroenteritis in different animal species. It is apparent that all primates as well as other animal species may develop gastroenteritis following infection with one or another of the rotavirus types. SA 11, previously isolated from African green monkeys (Malherbe and Strickland-Cholmley, 1967) is a rotovirus antigenically related to an agent isolated from an abattoir. This virus has not been found in nonhuman primates with any frequency. Recently, however, Stuker et al. (1980) have reported a similar virus in macaques with diarrhea.

Retroviruses–Oncornaviruses (Retroviridae)

Tumors and tumor viruses as detailed by others in this volume (Hunt, Rabin and Benveniste) occur in nonhuman primates with great frequency. Although at one time it was thought that nonhuman primates did not develop malignancies, it is now recognized that these occur in simians as in humans. As indicated (see Herpesviruses), tumor induction is not restricted to the retroviruses. However, this family (Retroviridae) of viruses, which includes a number of RNA viruses, has been associated with oncogenesis in many animal species. Tumor induction in primates with retroviruses as a natural infection of primates (leukemia in gibbons, SAIDS in rhesus) has initiated extensive studies on the origin of these viruses, their relationship to each other, and their potential to induce tumors. It is still unclear what "diseases" the primate endogenous C-type retroviruses are capable of causing. They are apparently present in all vertebrate tissue (placentas) and as pseudotypes do cause metastatic disease (Kalter et al., 1975, 1977). Also, the relationship of such viruses as the AIDS–SAIDS viruses; the woolly monkey SSV–gibbon ape viruses (GALV) and other retroviruses to oncongenesis and their natural hosts is still unclear. The woolly monkey agent had been recovered from one monkey in contact with a gibbon ape that subsequently died of "leukemia."

Rabies (Rhabdoviridae)

This virus, frequently overlooked as a cause of deaths in nonhuman primates, does occur and is a disease that must be considered in colony management, particularly those that are "open" in which contact with wild animals is possible. The disease, similar to that seen in humans, has been reported in a variety of nonhuman primates (Richardson and Humphrey, 1971).

Miscellaneous Viruses

A number of diseases have been ascribed to one or another exotic virus (or suspected virus) either in nature or as the result of experimental studies. Infection resulting from experimental studies obviously does not concern those interested in colony management other than to suggest a possible etiology or cast suspicion when attempting to define an unknown outbreak. For example, lymphocytic choriomeningitis (LCM) virus (*Arenavirus*) has not been reported to cause natural disease of primates. However, the presence of this virus in rodents (natural host) as well as experimental studies (Armstrong et al., 1936) would strongly suggest that this disease could occur in any simian population.

Marburg disease is an infrequent simian disease of concern to colony managers. It is still not clear where or how this "one-time" outbreak originated. Neither is it known what role the African green monkey played in the spread of this disease other than that it served as the carrier of the agent bringing it into Frankfurt-Marburg, Germany, and into Belgrade, Yugoslavia (Martini and Siegert, 1971). It is important to recognize that the animals in Germany were used shortly after arrival and that there was no unusual mortality recorded. The animals in Belgrade were held for 6 weeks, and here the death rate was higher than usual (35%). It was originally believed that no unusual deaths among monkeys were observed in Uganda, the source of the monkeys. However, a recent report suggests that this was not the case (Smith, 1982). A retrospective report by Smith suggests that there were sick animals as well as humans in the area, but investigations into this relationship were not done.

Simian hemorrhagic fever (SHF) is another disease that has occurred unexpectedly, with high mortality rates (100% of animals with clinical disease), in primate colonies in the United States, Soviet Union, and England. This disease outbreak was primarily in *Macaca* spp., which appears to be the only genus susceptible to this virus. The epidemiology of the disease within colonies is not clear, but contaminated syringes as well as reuse of needles for tatooing without appropriate sterilization was suspected (Shevtsova, 1967; Lapin et al., 1967; Tauraso et al., 1968; Palmer et al., 1968, Wood et al., 1970).

These diseases and others that occur in areas from which laboratory primates are collected require consideration. Included also are such viruses as Lassa fever (*Arenavirus*), members of the Tacaribe complex such as Junin virus (Argentine hemorrhagic fever), and Machupo virus (Bolivian hemorrhagic fever). Other hemorrhagic fevers caused by various viruses, some classified and others not, include dengue (group B *Flavivirus*), Omsk hemorrhagic fever and Kyasanur Forest disease (Flavivirus). Crimean–Congo hemorrhagic fever viruses are in the family Bunyaviridae (*Nairovirus*), as is Rift Valley fever virus. Undoubtedly, there are others yet to be detected. Isolated reports suggest that these diseases do occur in monkeys and apes.

Mention has been made of chimpanzee infection with HAV (hepatitis A virus). Two unclassified hepatitis viruses—HBV (hepatitis B) and NANB (non A and non B hepatitis)—are infectious for chimpanzees, at least experimentally. Are simians independently susceptible to these viruses in nature or only as a result of human contact? The occurrence of HAV in chimpanzees was recognized when humans contracted the disease as a result of contact with infected animals (Hillis, 1961).

Gastroenteritis is a major problem in colony management. What is the etiology of this disease? It is evident that this syndrome is caused by many different viruses, some of which have been determined, others suspected, and still others yet to be determined. Thus, in addition to the rotaviruses, adenoviruses, reoviruses, picornaviruses, coronaviruses, calciviruses, astraviruses, orbiviruses, and others, one may ask, What is the role of other nonviral agents either in conjunction with viruses or acting independently in causation of diarrhea—parasites, bacteria, fungi, chemicals and other physiological insults?

Last we may question the influence of "slow" or exotic viruses on colony maintenance. Nonhuman primates have been useful in defining the etiology of several human diseases thought to be caused by exotic viruses. The majority of these diseases are considered members of the subacute spongiform virus encephalopathy group (scrapie and transmissible mink encephalopathy in animals, kuru and Creutzfeldt–Jakob's disease in humans). As suggested by Amyx et al. (1983), these and other human and animal diseases of this type may be caused by exotic viruses or conventional viruses. Included here are multiple sclerosis, Parkinson's disease, Alzheimer's disease, amyotrophic lateral sclerosis, Pick's disease, Huntington's chorea, and other similar diseases. Frauchiger and Fankhauser (1970) emphasize that naturally occurring diseases of the central nervous system are rarely, if ever, seen in nonhuman primates. On the other hand, reports are seen to the contrary! Is PML as reported by Holmberg et al. (1977) in rhesus monkeys an indication of simian exotic disease counterparts? Kim et al. (1970) have reported a subacute sclerosing panencephalitis-like syndrome in adult baboons; the etiology, however, was not determined. Similarly, Brack (1973) listed the occurrence of perinatal telencephalic leukoencephalopathy in chimpanzees.

Diagnosis and Detection

In attempting to define the role of viruses in self-sustaining populations of nonhuman primates, it is not only important clinically to recognize the existence of these viruses but to have methods for their detection. Diagnosis of virus diseases is necessary to maintain the well being of colony animals by prevention and control of the spread of these agents. Diagnosis is also helpful in order to apply specific therapy or preventive measures such as appropriate vaccines. In veterinary medicine as in human medicine, the clinician is faced with the dilemma of determining the etiology of zoonoses or the cause of an illness in an individual animal. Clinical acumen is important, but when specific etiologies of respiratory diseases, intestinal disturbances, neurological disease, or exanthems are desired, the support of a diagnostic laboratory is highly desirable (see Appendix A of this volume).

It is important to recognize that diagnostic methodologies have markedly improved during the past few years, and, as will be demonstrated herein, a serologic diagnostic method has now been developed to provide information within hours after the collection of appropriate specimens. A similar procedure for the rapid identification of an isolate is also available. What are the diagnostic procedures that are in use, and how are they applied? In essence, the laboratory procedures used for veterinary medicine, as would be expected, are the same as used in human medicine.

Approaches to the Laboratory Diagnosis of Viral Infections

The final diagnosis of an illness is completed by the combined efforts of the clinical and laboratory personnel. The clinician, recognizing a diagnostic problem, provides the laboratory with the necessary suitable specimens collected at the most appropriate time in relation to the illness (Kalter, 1971). The laboratory in turn considers the most appropriate test or tests for the type of specimen provided: (1) the direct examination of the specimen for the presence of a virus or its antigen(s); (2) serologic determination of antibody either as: (a) an early globulin (IgM), (b) present in a single serum specimen indicating past exposure, (c) present in high titer suggestive of recent infection, or (d) preferably, the detection of an antibody rise in a convalescent serum over that found in an acute serum sample; (3) the isolation of an agent, its identification, and the demonstration of its relationship to the illness by observing an antibody rise to the isolate in the convalescent serum, and (4) the significance of the findings in terms of specificity of reaction, cross-reactions of antigens, compatibility with clinical findings, etc. The cross-reactions among such viruses as the adenoviruses or herpesviruses is well known. It is apparent that the satisfactory completion of the above necessitates the availability of suitable reagents. Sensitive and specific reagents are the *sine qua non* of the laboratory's capabilities.

Methodologies

Most laboratories depend upon several procedures for the satisfactory completion of their mission. However, as indicated previously, the specimen(s) provided will determine the laboratory's approach to the problem.

Electron Microscopy

Direct visualization of a virus requires use of the electron microscope. The size of viruses precludes the use of the light microscope. Direct visualization requires a distinctive morphological charactetistic of the virus. Electron microscopy is of value in examination of material obtained from skin lesions (exanthems), feces (diarrhea, hepatitis), and certain biopsy specimens. This approach is of value in detecting viruses that are difficult to grow, or do not grow, in culture (HAV, rotaviruses). Limiting features are the requirement for large quantities of virus in the specimens as well as the inability to differentiate types within a virus family. Some specificity and sensitivity may be added by the application of an antibody (immunoelectron microscopy) to the system.

Serology

All too often, serological procedures are considered only in terms of classicial serological tests (complement fixation, neutralization, hemagglutination inhibition, etc.). Generally, serologic procedures, regardless of the test, require an acute and convalescent serum sample taken 2–3 weeks apart to detect antibody changes. A significant antibody rise (fourfold) is taken to indicate the host's response to the infectious agent. At times, additional serum samples must be collected to provide unequivocal information.

Serologic procedures are limited by the need to determine an antibody change. Exceptions occur where an antibody to an exotic agent is found or when IgM is present in an acute specimen.

Addition of serum to a procedure such a fluorescence microscopy when used for the direct visualization and identification of a virus or its antigens in a tissue(s) impression (immunofluorescence) is a rapid procedure. However, when used to detect antibody, the limitations inherent in any serologic test affect the results. The newer procedures that have been introduced such as enzyme immunoassays (EIA) or radioimmunoassays (RIA) are in essence similar to immunofluorescence. The virus or its antigens are detected by the presence of a homologous antibody. Detection of this reaction is done by labeling the antibody with an enzyme and detection of the enzyme-labeled antibody by a suitable substrate. In the RIA, instead of fluorescence or an enzyme, a labeled radioisotope (^{125}I) is used. The use of monoclonal antibody provides an enhanced specificity but in essence does not change the procedure.

Virus Isolation

Virus isolation studies are time consuming, often misleading and/or valueless, and more often than not are negative. Relating an isolate to the disease is nearly impossible unless a concomitant antibody rise to the agent is determined, and this, too, can be misleading because of contamination of the sample or unless Koch's postulates are established. Viruses are frequently encountered in body fluids or specimens that have no relationship to the illness under study. Laboratory expertise and cooperation with the clinician is required to resolve these findings.

To isolate an agent, it must be cultivated on a suitable medium—cell culture, animal, embryonate eggs. The choices are endless but the supply limited. Susceptibility varies, and maintenance of all the desired test systems is impossible. Host restrictions oftentimes require maintenance of a wide assortment of materials (cells, animals, etc.). Primary cells are considered preferable to secondary cells and newborn animals more desirable than older animals.

Newer Approaches

Attempts to unmask or free viruses from inhibiting substances, particularly fecal viruses by use of proteolytic enzymes, appear to be successful (Graham and Estes, 1980). Immunoassays (immunofluorescence, radioisotope, enzyme immunoassays), molecular hybridization, and other molecular procedures are under study as substitutes for the classical approaches.

A simple, rapid, and sensitive method currently under study in our laboratory provides serologic data or isolation identification within 3–5 hours (Heberling and Kalter, 1985). This method, the "dot immunobinding assay" (DIA), utilizes nitrocellulose membranes for the rapid adsorption of virus proteins. As seen in Figure 46.1, positive reactions are indicated by the presence of a dark dot (original test color is red). Differentiation between B virus and herpes simplex is also detected as reflected by differences in staining intensity of the dot. Comparisons with SN titers are indicated by the results detailed in the various columns. The small, open circles are for space indication purposes only. Serum determination or identification is done by addition of labeled serum (antiglobulin), which in turn is made visible by use of a substrate dye against the antiglobulin. The results are highly sensitive and specific. Field studies may be done because equipment needs are minimal. Highly infectious viruses, such as *H. simiae*, may be inactivated with a psoralen–UV source.

Failure to Determine Etiology

There are a number of reasons why the laboratory fails in its attempts to define the precise etiology. This occurs most frequently in virus isolation studies because (1) specimens are mishandled (agent inactivated before specimen arrives

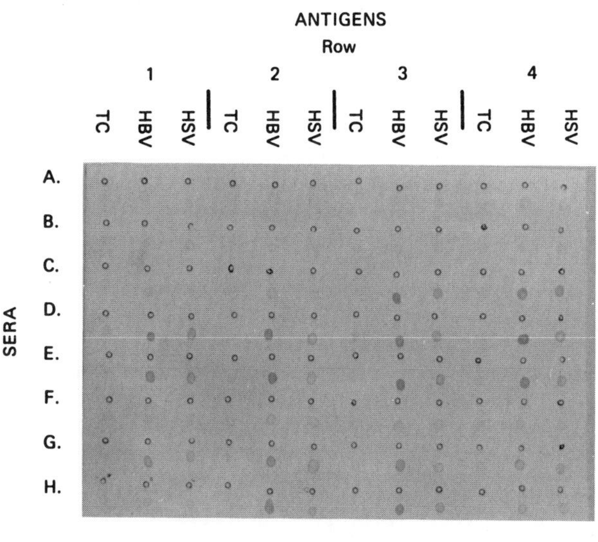

< 1:3 by SN: A1-C2, F1-F4, G3-H2, H4
1:3 by SN: G1, G2, H3
1:12-1:96 by SN: C3-E4

FIGURE 46.1. "DIA" testing of sera for antibody to B virus (HBV) and herpes simplex (HSV). Intensity of staining indicated the primary antigen. Open circles are for purposes of placement of antigen dots. Different rows indicate various serum titers (SN) used in test. T.C. is tissue culture control.

in laboratory), (2) specimens are taken at the wrong time in relation to illness, (3) an inappropriate specimen is collected, or (4) the laboratory uses the wrong procedures for isolation of the virus. Serologic studies are generally more rewarding but at times do not provide an answer because serum samples are not collected at appropriate times to demonstrate antibody, the selected procedure was inappropriate, or the laboratory procedure was not sufficiently sensitive to detect antibody.

Prevention and Significance of Disease in Self-Sustaining Colonies

It is apparent that natural virus diseases, with some few exceptions, have not been a major factor nor do they pose a threat to development of a self-sustaining primate colony. The diseases that do occur have been mainly those associated with the day-to-day operation of a colony and are usually restricted to an individual animal. However, the exceptions are devastating, and Benirschke (1983) emphasizes the occurrence of such spontaneous diseases in primate colonies and the need to eliminate them in order to create self-sustaining colonies. Those incidents that have caused untoward effects have been the result of breakdowns in colony management and the result of human error (simian hemorrhagic fever) or contact with humans (hepatitis A). Marburg disease is a unique situation, and it is still not clear what the epidemiological factors were with regard to that outbreak.

Thus, development and maintenance of self-sustaining colonies require programs in which the human element is carefully controlled. Recommendations suggesting such a program have been well described in the literature (Gerone, 1983; Kalter, 1973, 1983; Kalter and Heberling, 1976). Simply stated, prevention is the function of a well-trained staff who follow established procedures and recognize that infectious diseases, by definition, are caused by infectious agents. Yes, Virginia, there is a germ theory. The following recommendations will do much to protect an animal colony:

1. Obtain animals from reputable dealers with knowledge and records of sources, methods of trapping, maintenance following trapping, personnel contacts, and baseline samples for reference.
2. Institute appropriate quarantine measures with routine health examinations and collection of appropriate samples from staff as well as animals for reference and/or immediate study from staff as well as animals.
3. Maintain a carefully controlled separation of species and, if possible, separation of new groups.
4. Develop a program for the complete study of all animals that die or become ill. Have a comprehensive program available for determining etiology—bacteriology, mycology, parasitology, virology, and pathology.
5. Include vaccination and routine examination for infectious agents whenever possible and practical.
6. Maintain complete records on all animals following arrival in colony.
7. Minimize contact with human and other animal populations. Develop a procedure for control of human contacts.

Inasmuch as many of the human diseases that are transmitted to the captured animals result from inadequate and poorly maintained trapping facilities in the countries of origin, the development of self-sustaining colonies will eliminate this threat. However, a word of caution is necessary concerning established colonies. *These animals will need to be maintained under carefully controlled conditions because they will undoubtedly be more susceptible to infectious agents than their counterparts who have survived previous contact with infectious diseases.* Another problem of laboratory-bred animals that will need consideration is the possibility of latent infections. This type of infection as well as those agents (endogenous) passed vertically from mother to offspring will continue to be a potential threat to any colony.

Future Considerations

Although thoughts of alternate approaches to that of using an intact animal may be appealing to some, conceptualization and realization of an artificial, biologically intact robot does not appear to be in the immediate future. As already demonstrated, other approaches, such as mathematical, computerized, or test-tube biologies, will obviate some use of intact animals. However, it does not appear that any of these surrogates will be capable of immunological, neurological, endocrinological, or other biophysiological responses to a foreign antigenic stimulus. An animal model, preferably employing a nonhuman primate that will provide meaningful data, will continue to be necessary to further our understanding of human diseases. In this regard, expanded studies on species still available in sufficient numbers are needed to provide the most appropriate host. Of the several hundred recognized simian species, some few dozen are used in most research laboratories. The rhesus monkey was not selected because of recognized susceptibility—it was simply available!

The host response of these animals is extremely variable. Infection may not cause an overt pathologic response other than immunologic, but the animals may be shedding viruses in their saliva, feces, urine, and other body fluids that will be infectious for others.

A negative aspect to the continued use of animals, either intact or as contributors of the cells for culture, is the remote possibility of contributing an unwanted pathogen that has been residing in the animal as a latent infection or as an endogenous virus. Hsiung (1968) emphasizes the lack of information on the pathogenesis, persistence, and epidemiology of latent virus infections in primates. As emphasized by Hsiung (1968) and reiterated herein, the presence of viruses within "normal" primate tissues is extremely common. The source of these viruses is generally unknown as is their potential for causing disease.

Nowhere is this potential problem brought to our attention more emphatically than by the presence of retroviruses in primate tissues. For a number of years since their omnipresence in the placentas of vertebrates was described, attempts to associate pathogenic qualities have been fruitless (Kalter, 1983). The recognition that, with some few exceptions, these viruses are difficult to isolate (although readily observed by electron microscopy) and that adult tissues show little if any evidence of their presence has apparently escaped understanding. However, morphologically and biologically similar agents are now recognized to be associated with such primate diseases as AIDS, SAIDS, and various neoplastic diseases, and the possible relationship should be examined.

Thus, there is a realistic concern regarding the use of animal cells for direct employment as a source of vaccines or even for immortalization. As our molecular capabilities increase, what is the potential for removing that part of the genome responsible for carrying these endogenous agents? Molecular hybridization techniques utilizing various microbial cells could serve as a potential source of materials free of potential pathogens. Insertion of desired antigens would be valuable in eliminating the need for animal cells.

From the above, it would appear that there are a number of potential areas in need of exploration. Molecular studies should reduce the need for the large numbers of animals required for vaccine production. Complete elimination of animal colonies is inconceivable at present because no mechanism(s) exist to provide the information now offered by an intact animal. Colonies are needed in countries of origin as well as in those of utilization that provide animals of the highest quality. To do this, regulatory monitoring systems and support services must be provided. We have indicated a new rapid and specific virus diagnostic test that should do much to improve the quality of colony maintenance. Further study is required to expand and improve such procedures.

It is imperative that studies to determine the most appropriate simian species as models for specific human and nonhuman primate diseases be continued and supported. Gastroenteritis (diarrhea) is in need of a model system for understanding of the disease as well as prevention (vaccine development and evaluation), and as recently pointed out by Amyx et al. (1983), there is a need to establish primate models for the various exotic diseases now considered as possibly viral in etiology such as chronic encephalitis, Creutzfeldt–Jakob, schizophrenia, and hemorrhagic fever with renal syndrome. SAIDS–AIDS is also most appropriately studies in a nonhuman primate species, VandeBerg (1983) reiterated the need for nonhuman primate models and emphasized that little is known about the genetic aspects of viral diseases. Success with such diseases as poliomyelitis and hepatitis, in which the nonhuman primate has played such a positive role, should provide an impetus for expanded studies. Whitney (1983) cites several examples of nonprimate species serving as models in various research activities. Although this approach should be pursued, the immortal words of Koch should be kept in mind, "Meine Herren, vergessen Sie nie, dass die Mäuse keine Menschen sind" (Koprowski, 1958).

Acknowledgments. Studies reported herein were supported by NIH grant RR-00361 and WHO grant V4/181/38 (WHO Collaborating Center for Reference and Research in Simian Viruses). Special appreciation is extended to Dr. R. L. Heberling who has shared with me the trials and tribulations of maintaining this laboratory these past 18 years.

References

Agrba VZ, Lapin BA, Timanovskaya VV, Dzhochvliany MC, Kikosha LV, Churriov GN, Djatchenko AG (1980) Isolation of lymphotropic baboon herpesvirus (HVP) from oral swabs of hamadryas baboons of the Sukhumi monkey colony. Exper Pathol 18:269–274.

Allen AM, Palmer AE, Tauraso NM, Shelokov A (1968) Simian hemorrhagic fever. II. Studies in pathology. Amer J Trop Med 17:413–421.

Allen WP, Felsenfeld AD, Wolf RH, Smetana JF (1974) Recent studies on the isolation and characterization of Delta virus. Lab Anim Sci 24:222–228.

Allmond BW, Froeschle JE, Guilloud NB (1967) Paralytic poliomyelitis in large laboratory primates. Virologic investigation and report on the use of oral poliomyelitis virus (OPV) vaccine. Amer J Epidemiol 85:229–239.

Amyx HL, Asher DM, Gibbs CJ Jr, Gajdusek DC (1983) Future needs of primates in exotic diseases. In: Kalter SS (ed) Viral and immunological diseases in nonhuman primates. Alan R Liss, New York, pp 31–37.

Anderson CR, Downs WG (1955) The isolation of yellow fever virus from the livers of naturally infected red howler monkeys. Amer J Trop Med Hyg 4:662–664.

Andrews EJ (1976) Spontaneous viral infections of laboratory animals. In: Melby EC, Altman NH (eds) Handbook of laboratory animal science, vol III. CRC, Cleveland, OH, pp 119–178.

Armstrong C, Wooly JG, Onstott RH (1936) Distribution of lymphocytic choriomeningitis viruses in the organs of experimentally inoculated monkeys. US Pub Health Rep 51:298–303.

Asher DM, Amyx HL, Ludmila V, Asher S, Hooks JJ, Shah KV, Gibbs CJ Jr, Gajdusek DC (1978) Chronic adenoviral infection of the urinary tract in chimpanzees. J Supramol Struct Suppl 2:278(#703).

Ashley CR, Caul EO, Clarke SKR, Corner BD, Dunn S (1978) Rotavirus infections of apes. Lancet 1:477.

Balfour A (1914) The wild monkey as a reservoir for the virus of yellow fever. Lancet 1:1176–1178.

Bearcroft WGC, Jamieson MF (1958) An outbreak of subcutaneous tumors in rhesus monkeys. Nature 182:195–196.

Benirschke K (1983) Occurrence of spontaneous disease. In: Kalter SS (ed) Viral and immunological diseases in nonhuman primates. Alan R Liss, New York pp 17–30.

Blakely GA, Lourie B, Morton WG, Evans HH, Kaufmann AF (1973) A varicella-like disease in macaque monkeys. J Infect Dis 127:617–625.

Bleyer JG (1922) Ueber Auftreten von Variola unter Affen der Genera Mycetes und Cebus bei Vordringen einer Pockenepidemie im Urwaldgebiete an den Nebenflüssen des Alto Uruguay in Südbrasilien. Muench med Wochenschr 69:1009–1010.

Bloch O Jr (1937) Specificity of lesions of experimental monkeys. Amer J Pathol 13:939–944.

Boulger LR (1966) Natural rabies in a laboratory monkey. Lancet 1:941–943.

Boyce JT, Giddens WE, Valerio M (1978) Simian adenoviral pneumonia. Amer J Pathol 91:259–276.

Brack M (1973) Perinatal telencephalic leukoencephalopathy in chimpanzees. Acta Neuropathol 25:307–312.

Bullock G (1965) An association between adenoviruses isolated from simian tonsils and episodes of illness in captive monkeys. J Hyg 63:383–387.

Cabasso VJ, Stebbins MR (1965) Study of spread of rubella virus from inoculated to uninoculated contact *Cercopithecus* monkeys. J Lab Clin Med 65:612–616.

Chandler FW, Callaway CS, Adams SR (1974) Pancreatitis associated with an adenovirus in a rhesus monkey. Vet Pathol 11:165–171.

Churchill A (1963) The isolations of parainfluenza 3 virus from fatal cases of pneumonia in *Erythrocebus patas* monkeys. Brit J Exp Pathol 44:529–537.

Clarkson MJ, Thorpe E, McCarthy KE (1967) A virus disease of captive vervet monkeys (*Cercopithecus aethiops*) caused by a new herpes virus. Arch gesamte Virusforsch 22(1–2):219–234.

Colman PG, Clarke M (1975) Measles in monkeys. Vet Res 97:436.

Crandell RA, Casey H, Brumlow W (1969) Studies of a newly recognized poxvirus of monkeys. J Infect Dis 119:80–88.

Daniel MD, Garcia FG, Melendez LV, Hunt RD, O'Connor J, Silva D (1975) Multiple *Herpesvirus simiae* isolation from a rhesus monkey which died of cerebral infection. Lab Anim Sci 25:303–308.

Daniel MD, Karpas A, Melendez LV, King NW, Hunt RD (1967) Isolation of herpes-T virus from a spontaneous disease in squirrel monkeys (*Saimiri sciureus*). Arch gesamte Virusforsch 22:324–331.

Dick EC, Dick CR (1974) Natural and experimental infections on nonhuman primates with respiratory viruses. Lab Anim Sci 24:177–181.

DiGiacomo RF, Hooks JJ, Sulima MP, Gibbs CJ Jr, Gajdusek C (1977) Pelvic endometriosis and simian foamy virus infection in a pigtailed macaque. J Amer Vet Med Assoc 171:859–861.

Douglas JD, Tanner KN, Prine JR, Van Riper DC, Derwelis SK (1967) Molluscum contagiosum in chimpanzees. J Amer Vet Med Assoc 151:901–904.

Emmons RW, Gribble DH, Lennette EH (1968) Natural fatal infection of an owl monkey (*Aotus trivirgatus*) with herpes T virus. J Infect Dis 118:153–159.

Emmons RW, Lennette EH (1970) Natural *Herpesvirus hominis* infection of a gibbon (*Hylobates lar*). Arch gesamte Viruforsch 31:215–218.

España C (1968) A Yaba-like disease in primates. 17th Ann. Meeting Animal Care Panel Chicago, 1966. Cited by Hull RN, The simian viruses. Virol Monogr 2:29–30. Springer-Verlag, NY.

España C (1971) Review of some outbreaks of viral disease in captive nonhuman primates. Lab Anim Sci 21:1023–1031.

España C (1973) *Herpesvirus simiae* infection in *Macaca radiata*. Amer J Phys Anthro 38:447–454.

Eugster AK, Kalter SS, Kim CS, Pinkerton ME (1969) Isolation of adenoviruses from baboons (*Papio* sp.) with respiratory and enteric infections. Arch gesamte Virusforsch 26:260–270.

Flecknell PA, Parry R, Neelham JR, Ridley RM, Baker HF, Bowes P (1983) Respiratory disease associated with parainfluenza type 1 (Sendai) virus in a colony of marmosets (*Callithrix jacchus*). Lab Anim 17:111–113.

Fox H, Weidman FD (1923) Acute papular and desquamative exanthem in an orangutan. Arch Dermatol 7:462–464.

Fraser CEO, Chalifoux L, Sehgal P, Hunt RD, King NW (1978) A paramyxovirus causing fatal gastroenterocolitis in marmoset monkeys. Prim Med 10:261–270.

Frauchiger E, Fankhauser R (1970) Demyelinating diseases in animals, their prevalence to the pathogenesis of multiple sclerosis. In: Vinken PJ, Bruyn GW (eds) Handbook of clinical neurology. Elsevier, New York, p 664.

Gainer JH (1967) Encephalomyocarditis virus infections in Florida, 1960–1966. J Amer Vet Med Assoc 151:421–425.

Gard EA, London WT (1983) Clinical history and viral characteristization of Delta herpesvirus infection in a patas monkey colony. In: Kalter SS (ed) Viral and immunological diseases in nonhuman primates. Alan R Liss, New York, pp 211–212.

Gaskin JM, Jorge MA, Simpson CF, Lewis AL, Olson JH, Schobert EE, Wollenman EP, Marlowe C, Curtis MM (1980) The tragedy of encephalomyocarditis virus infection in zoological parks of Florida. Amer Assoc Zoo Vet Ann Proc, Washington, DC, pp 1–7.

Gerone PJ (1983) Biohazards and protection of personnel. In: Kalter SS (ed) Viral and immunological diseases in nonhuman primates. Alan R Liss, New York, pp 187–196.

Gibson SV, Soike KF, Wolf RH, Jirge SK, Baskin GB (1980) Fatal gastroenterocolitis in marmosets caused by two paramyxoviruses. Amer Assoc Lab Anim Sci, 31st Ann Session, Indianapolis, Abstract No 34.

Giddens WE Jr, Morton WR, Hefti E, Panem S, Ochs H (1983) Enzootic retropekritoneal fibromatosis in *Macaca* sp. In: Kalter SS (ed) Viral and immunological diseases in nonhuman primates. Alan R Liss, New York, pp 249–253.

Gispen R (1949) Smallpox reinfections in Indonesia. Ned T Geneesk 93:3686–3695.

Gispen R, Kapsenberg JG (1967) Cited in Vers Volksgezondh 140.

Goldmann KR (1935) Spontaninfektion spinalar Kinderlähmung bei Affen. Berl Munch Tierärztl Wschr 51:497–499.

Graham DV, Estes MK (1980) Proteolytic enhancement of rotavirus infectivity. Biologic mechanisms. Virology 101:432–439.

Guilloud NB, Allmond BW, Froeschle JE, Fitz-Gerald FL (1969) Paralytic poliomyelitis in laboratory primates. J Amer Vet Med Assoc 155(7):1190–1193.

Guilloud NB, Kline IC (1966) Paralytic poliomyelitis in the gorilla and orangutan. J Amer Phys Ther Assoc 46(5):516–518.

Hall WC, Kovatch RM, Herman PH, Fox JG (1972) Pathology of measles in rhesus monkeys. Vet Pathol 8:307–319.

Hartley EG (1964) Naturally occurring "B" virus infection in cynomologus monkeys. Vet Rec 76:555–557.

Heberling RL (1972) The simian adenoviruses. In: Fiennes RN T-W, Karger S (eds) Pathology of primates, part II. Basel, pp 572–591.

Heberling RL, Cheever FS (1960) Enteric viruses of monkeys. Ann NY Acad Sci 85:942–950.

Heberling RL, Kalter SS (1985) Dot-Elisa on nitrocellulose with psoralen inactivated herpesvirus. In: Abst Amer Soc Microbiol 85th Ann Mtg, Las Vegas, pp 21 (510).

Heldstab A, Rüedi D, Sonnabend W, Deinhardt F (1981) Spontaneous generalized *Herpesvirus hominis* infection in a lowland gorilla (*Gorilla gorilla gorilla*). J Med Primatol 10:129–135.

Helwig FC, Schmidt ECH (1945) A filter-passing agent producing interstitial myocarditis in anthropoid apes and small animals. Science 102:31–33.

Henrickson RV, Maul DH, Osborn BL, Sever JL, Madden DL, Ellingsworth LR, Anderson JH, Lowenstine L, Gardner MB (1983) Epidemic of acquired immunodeficiency in rhesus monkeys. Lancet 1:388–390.

Heuschele WP (1960) Varicella (chicken pox) in three young anthropoid apes. J Amer Vet Med Assoc 136:256–257.

Hillis WD (1961) An outbreak of infectious hepatitis among chimpanzee handlers at a United States Air Force Base. Amer J Hyg 73:316–328.

Hillis WD (1967) Etiology of viral hepatitis. Johns Hopkins Med J 120:176–185.

Hime JM, Keymer IF, Baxter CJ (1975) Measles in recently imported colobus monkeys (*Colobus guereza*). Vet Rec 97:392.

Holmberg CA, Gribble DH, Takemoto KK, Howley PM, España C, Osburn BI (1977) Isolation of simian virus 40 from rhesus monkeys (*Macaca mulatta*) with spontaneous progressive multifocal leukoencephalopathy. J Infect Dis 136:593–596.

Holmes AW, Caldwell RG, Dedmon RE, Deinhardt F (1964) Isolation and characterization of a new herpes virus. J Immunol 92:602–610.

Howe HA, Bodian D (1944) Poliomyelitis by accidental contagion in the chimpanzee. J Exp Med 80:383–390.

Hsiung GD (1968) Latent virus infections in primate tissues with special reference to simian viruses. Bact Rev 32:185–205.

Hull RN (1968) The simian viruses. Virol Monogr 2:1–66.

Hull RN, Dwyer AC, Holmes AW, Nowakowski E, Deinhardt F, Lennette EH, Emmons RW (1972) Recovery and characterization of a new simian herpesvirus from a fatally infected spider monkey. J Natl Cancer Inst 49:225–231.

Hull RN, Minner JR, Smith JW (1956) New viral agents recovered from tissue cultures of monkey kidney cells. I. Origin and properties of cytopathogenic agents SV1, SV2, SV4, SV5, SV6, SV11, SV12, and SV15. Amer J Hyg 63:204–215.

Hunt RD, Garcia FG, Barahona HH, King NW, Fraser CEO, Melendez LV (1973) Spontaneous *Herpesvirus saimiri* lymphoma in an owl monkey. J Infect Dis 127:723–725.

Hunt RD, Melendez LV (1966) Spontaneous herpes-T infection in the owl monkey (*Aotus trivirgatus*). Path Vet 3:1–26.

Jensen EM, Zelljadt I, Chopra H, Mason MM (1970) Isolation and propagation of a virus from a spontaneous mammary carcinoma of a rhesus monkey. Cancer Res 30:2388–2393.

Johnson DO, Wooding WL, Tanticharoenyos P, Karnjanaprakorn C (1971) An epizootic of A_2/Hong Kong/68 influenza in gibbons. J Infect Dis 123:365–370.

Jones EE, Alford PL, Reingold AL, Russell H, Keeling M, Broome CV (1984) Predisposition to invasive pneumococcal illness following parainfluenza type 3 virus infection in chimpanzees. J Amer Vet Med Assoc 185:1351–1353.

Kalter SS (1960) Animal "orphan" viruses. Bull WHO 22:319–337.

Kalter SS (1971) Collection and handling of specimens for detection of infection or disease. Lab Anim Sci 21:1015–1018.

Kalter SS (1973) Virus research. In: Bourne G (ed) Primates in biomedical research. Academic, Karger, Basel, pp 61–165.

Kalter SS (1983) Primate viruses and their significance. In: Kalter SS (ed) Viral and immunological diseases in nonhuman primates. Alan R Liss, New York, pp 67–90.

Kalter SS, Ablashi E, España C, Heberling RL, Hull RN, Lennette EH, Malherbe HH, McConnell S, Yohn DS (1980) Simian virus nomenclature. Intervirology 13:317–330.

Kalter SS, Heberling RL (1971) Comparative virology of primates. Bact Rev 35:310–364.

Kalter SS, Heberling (1976) Health hazards associated with newly imported primates and how to avoid them. In: Chivers DJ, Ford EHR (eds) Recent advances in primatology, vol 4. Academic, London, pp 5–21.

Kalter SS, Heberling RL (1978) Serologic response of primates to influenza viruses. Proc Soc Exp Biol Med 159:414–417.

Kalter SS, Heberling RL, Eichberg JW, Boenig DM, Cummins LB (1977) Pathologic and immunologic responses to the baboon endogenous virus and its pseudotype. In: Benhaelzen P, Hilgers J, Yohn DS (eds) 8th Int Symp on Comp Res on Leukemia and Related Diseases: Advances in Comparative Leukemia Research. Elsevier, Amsterdam, pp 100–102.

Kalter SS, Heberling RL, Hellman A, Todaro GJ, Panigel M (1975) C-type particles in baboon placentas. Proc Roy Soc Med 68:135–140.

Kalter SS, Heberling RL, Vice TE, Lief FS, Rodriguez AR (1969) Influenza (A_2/Hong Kong/68) in the baboon (*Papio* sp). Proc Soc Exp Biol Med 132:357–361.

Kalter SS, Ratner J, Kalter GV, Rodriguez AR, Kim CS (1967) A survey of primate sera for antibodies to viruses of human and simian origin. Amer J Epidem 86:552–568.

Kaufman AF (1972) The effects of spontaneous disease during quarantine, conditioning, and production of nonhuman primates for scientific use following importation. Pan Amer Health Org Sci Publ 235:39–48.

Kaufmann AF, Gary GW, Broderson JR, Perl DP, Quist KD, Kissling RE (1973) Simian virus 16 associated with an epizootic of obscure neurologic disease. Lab Anim Sci 23:812–818.

Kawakami TG, Huff DS, Buckley PM, Dungworth DL, Snyder SP, Gilden RV (1972) C-type virus—associated with gibbon lymphosarcoma. Nature New Biol 235:170–171.

Keeble SA, Christofinis GJ, Wood W (1958) Natural virus-B infection in rhesus monkeys. J Pathol Bacteriol 76:189–199.

Kelly ME, Soike K, Ahmed K, Iatropoulos MJ (1978) Coxsackievirus in an infant chimpanzee. J Med Primatol 7:119–121.

Kim CS, Kriewaldt FH, Hagino N, Kalter SS (1970) Subacute sclerosing panencephalitis-like syndrome in the adult baboon (*Papio* sp.). J Amer Vet Med Assoc 157:730–735.

Kim CS Sueltenfuss EA, Kalter SS (1967) Isolation and characterization of simian adenoviruses isolated in association with an outbreak of pneumoenteritis in vervet monkeys (*Cercopithecus aethiops*). J Infect Dis 117:292–300.

Koprowski H (1958) Counterparts of human disease in animals. Ann NY Acad Sci 70:369–382.

Laemmert HW Jr, Ferreira LdeC (1945) The isolation of yellow fever virus from wild-caught marmosets. Amer J Trop Med 25:231–232.

Landon JC, Bennett DG (1969) Viral induced simian conjunctivitis. Nature 222:683–684.

Lapin BA, Pekerman SM, Yakovleva LA, Dzhikidze EK, Shevtsova ZV, Kuksova MI, Danko LV, Krilova RI, Ya Akbroit E, Agrba VZ (1967) A hemorrhagic fever of monkeys. Vopr Virusol 12:168–173.

Lapin BA, Shevtsova ZV (1966) Studies on virus simian hemorrhagic fever. In: Int Cong on Microbiol Abst M 465–466.

Lapin BA, Iakovleva LA, Bukaeva IA, Indzhiia LV, Kokosha LV (1973) Induction of leukemia in monkeys with human leukemic blood. In: McNulty WP Jr (ed) International Congress of Primatology, 4th, Portland, Oregon, 1972, vol 4: Nonhuman Primates and Human Diseases. S Karger, Basel, pp 1–29.

Lehner NO, Bullock BC, Jones ND (1984) Simian varicella infection in the African green monkey (*Cercopithecus aethiops*). Lab Anim Sci 34:281–285.

Letvin NL, Eaton KA, Aldrich WR, Sehgal PK, Blake BJ, Schlossman SF, King NW, Hunt RD (1983) Acquired immunodeficiency syndrome in a colony of macaque monkeys. Proc Natl Acad Sci USA 80:2718–2722.

Levy BM, Mirkovic RR (1971) An epizootic of measles in a marmoset colony. Lab Anim Sci 21:33–39.

Loomis MR, O'Neill T, Bush M, Mondali RJ (1981) Fatal herpesvirus infection in patas monkeys and a black and white colobus monkey. J Amer Vet Med Assoc 179:1236–1239.

Lourie B, Morton WG, Blakely GA, Kaufmann AF (1971) Epizootic vesicular disease in macaque monkeys. Lab Anim Sci 21:1079–1080.

Malherbe H, Harwin R (1957) Seven viruses isolated from the vervet monkey. Br J Exp Pathol 38:539–541.

Malherbe HH, Strickland-Cholmley M (1967) Simian virus S all and the related O agent. Arch gesamte Virusforsch 22:235–245.

Manning JS, Griesemer RA (1974) Spontaneous lymphoma of the nonhuman primates. Lab Anim Sci 24:204–210.

Manning PJ, Banks KL, Lehner NDM (1968) Naturally occurring giant cell pneumonia in the rhesus monkey (*Macaca mulatta*). J Amer Vet Med Assoc 153:899–904.

Marennikova SS, Maltseva NN, Shelukhina EM, Shenkman LS, Korneeva VI (1974) A generalized herpetic infection simulating smallpox in a gorilla. Intervirology 2:280–286.

Marennikova SS, Shelukhina EM, Maltseva NN, Chimishkayar KL, Matsevich GR (1972) Isolation and properties of the causal agent of a new variola-like disease (monkeypox) in man. Bull World Health Org 46:599–611.

Martin DP, Kaye HS (1983) Epizootic of parainfluenza-3 virus infection in gibbons. J Amer Vet Med Assoc 183:1185–1187.

Martini GA, Siegert R (1971) Marburg virus disease. Springer-Verlag, New York/Heidelberg/Berlin, p 230.

Martini GA, Knauf HG, Schmidt HA, Mayer G, Baltzer G (1968) Uber eine bisher unbekannte, von Affen eingeschleppte Infektionskrankheit: Marburg Virus. Krankheit Deutsche med Wochenschr 93:559–571.

Marx PA, Maul DH, Osborn KG, Lerche NW, Moody P, Lowenstine LG, Henrickson RV, Arthur RV, Gilden RV, Sever JL, Levy SA, Minor RJ, Gardner MD (1984) Simian AIDS: isolation of a type D retrovirus and transmission of the disease. Science 223:1083–1086.

Matthews REF (1982) Classification and nomenclature of viruses. Fourth Report of the Virology Division, International Union of Microbiological Societies. Intervirology 17:4–199.

McCarthy KE, Thorpe E, Laursen AC, Heymann CS, Beale AJ (1968) Exanthematous disease in patas monkeys caused by a herpes virus. Lancet 2:856–857.

McClure HM, Chandler FW, Hierholzer JC (1978) Necrotizing pancreatitis due to simian adenovirus type 31 in a rhesus monkey. Arch Pathol Lab Med 102:150–153.

McClure HM, Keeling ME (1971) Viral diseases noted in the Yerkes Primate Center colony. Lab Anim Sci 21:1002–1010.
McClure HM, Olberding B, Strozier LM (1973) Disseminated herpesvirus infection in a rhesus monkey (*Macaca mulatta*). J Med Primatol 2:190–194.
McClure HM, Swenson RB, Kalter SS, Lester TL (1980) Natural genital *Herpesvirus hominis* infection in chimpanzees (*Pan troglodytes* and *P. paniscus*). Lab Anim Sci 30:895–901.
McConnell SJ, Herman YF, Mattson DE, Erickson L (1962) Monkey pox disease in irradiated cynomolgus monkeys. Nature 195:1128–1129.
Melendez LV, España C, Hunt RD, Daniel MD, Garcia FG (1969) Natural herpes simplex infection in the owl monkey (*Aotus trivirgatus*). Lab Anim Care 19:38–45.
Melnick JL, Horstmann DM (1947) Active immunity to poliomyelitis in chimpanzees following subclinical infection. J Exp Med 85:287–303.
Melnick JL, Midulla M, Wimberly I, Berrera-Ora JG, Levy BM (1964) A new member of the herpesvirus group isolated from South American marmosets. J Immunol 92:596–601.
Miot MR, Sikes RK, Silberman MS (1973) Rabies in a chimpanzee. JAVMD 162:54.
Moe JB, Schwartz LW, España C, Osborn BI (1977) Pneumonia in rhesus monkey fetuses induced by adenovirus SV-20. Amer Rev Resp Dis 115:287 (Abst.).
Monath TP, Calisher CH, Davis M, Bowen GS, White J (1974) Experimental studies of rhesus monkeys infected with epizootic and enzootic subtypes of Venezuelan equine encephalitis virus. J Infect Dis 129:194–200.
Montrey RD, Huxsoll DL, Hildebrandt PK, Booth BW, Arimbalam S (1980) An epizootic of measles in captive silvered leafmonkeys (*Presbytis cristatus*) in Malaysia. Lab Anim Sci 30:694–697.
Morita M, Iida T, Tsuchiya Y, Aoyama Y (1979) Fatal *Herpesvirus tamarinus* infection in cotton topped marmosets (*Saguinus oedipus*). Exp Anim 28:537–550.
Morris JA, Blount RE Jr, Savage RE (1956) Recovery of cytopathogenic agent from chimpanzee with coryza. Proc Soc Exp Biol Med 92:544–549.
Mouquet AD (1926) Maladie á allure grippale chez des chimpanzés. Bull Soc Ctr Méd Vét Paris 79:46–50.
Muchmore E (1971) Possible cytomegalovirus infection in man following chimpanzee bite. Lab Anim Sci 21:1080–1081.
Mukai N, Kalter SS, Cummins LB, Matthews VA, Nishida T, Nakajima T (1980) Retinal tumor induced in the baboon by human adenovirus 12. Science 210:1023–1025.
Müller W (1935) Spontane Poliomyelitis beim Schimpansen. Monatsschr Kinderheilkd 63:134–137.
Mutombo M, Arita I, Jazek Z (1983) Human monkeypox transmitted by a chimpanzee in a tropical rain-forest area of Zaire. Lancet 1:735–737.
Nicholas AH, McNulty WP (1968) In vitro characteristics of a poxvirus isolated from rhesus monkeys. Nature 217(5130):745–746.
Nishikawa F, Sugiyama T, Suzuki K (1977) A new paramyxovirus isolated from cynomolgus monkeys. Jpn J Med Sci Biol 30:191–204.
Norkin LC (1982) Papovaviral persistent infections. Microbiol Rev 46:384–425.
Ochoa R, Henk WG, Confer AW, Pirie GS (1982) Herpesviral pneumonia and septicemia in two infant gelada baboons (*Theropithecus gelada*). J Med Primatol 11:52–58.
Ordy JM, Rangan SRS, Wolf RH, Knight C, Dunlay VP (1981) Congenital cytomegalovirus effects on postnatal neurological development of squirrel monkey (*Saimiri sciureus*) offspring. Exp Neurol 74:728–747.
Palmer AE, Allen AM, Tauraso NM, Shelokov A (1968) Simian hemorrhagic fever. I. Clinical and epizootiologic aspects of an outbreak among quarantined monkeys. Amer J Trop Med Hyg 17(3):404–412.
Panthier R, Cateigne G, Hannoun C (1949) Adaptation to the egg and mouse of a strain of virus recently isolated from a case of influenza. Bull Inst Natl Hyg (Paris) 4:109–112.

Peters JC (1966) A monkeypox-enzooty in the Blijidorp Zoo. Intern Zoo Yrbk 6:274–275.

Pille ER (1960) B virus infection of monkeys. Vopr Virusol 5:542–547.

Potkay S, Ganaway JR, Rogers NG, Kinard R (1966) An epizootic of measles in a colony of rhesus monkeys (*Macaca mulatta*). Amer J Vet Res 27:331–334.

Prier JE, Sauer RM, Malsberger RG, Sillaman JM (1960) Studies on a pox disease of monkeys. II. Isolation of the etiologic agent. Amer J Vet Dis 21:381–384.

Rabin H, Neubauer RH, Pearson GR, Cicmanec JL, Wallen WC, Loeb NF, Valerio MG (1975) Spontaneous lymphoma associated with *Herpesvirus saimiri* in owl monkeys. J Natl Cancer Inst 54:499–502.

Rabin H, Neubauer RH, Woodside NJ, Cicmanec JL, Wallen WC, Lapin BA, Agrba VA, Yakoleva LA, Chuvirov GN (1976) Virological studies of baboon (*Papio hamadryas*) lymphoma: isolation and characterization of foamyviruses. J Med Primatol 5:13–22.

Ramsay E, Stair EL, Castro AE (1982) Fatal *Herpesvirus hominis* encephalitis in a white-handed gibbon. J Amer Vet Med Assoc 181:1429–1430.

Ratcliffe HL (1942) Deaths and important diseases. Repr Penrose Res Lab 11–25.

Remfry J (1976) A measles epizootic with five deaths in newly-imported rhesus monkeys (*Macaca mulatta*). Lab Anim 10:49–57.

Renne RA, McLaughlin R, Jenson AB (1973) Measles virus-associated endometritis, cervicitis, and abortion in a rhesus monkey. J Amer Vet Med Assoc 163:639–641.

Richardson JH, Humphrey GL (1971) Rabies in imported nonhuman primates. Lab Anim Sci 21(6):1082–1083.

Roberts JA, Prahalada S, Anderson JA, Lowestine LJ, Lerche NW, Henrickson RV (1984) Herpesvirus infection in macaques at the California Primate Research Center: a 14 year retrospective study of clinical and pathological findings. Lab Anim Sci 34: 500–501.

Roca-Garcia M, Sanmartin-Barberi C (1957) The isolation of encephalomyocarditis virus from *Aotus* monkeys. Amer J Trop Med Hyg 6:840–852.

Ruckle G (1956) Measles in humans and in monkeys: report of isolation from cynomolgus monkeys of an agent immunologically related to human measles virus. Fed Proc 15:610 (1990).

Rustigian R, Johnston PB, Reihart H (1955) Infection of monkey tissue with virus-like agents. Proc Soc Exp Biol Med 88:8–16.

Sabin AB (1959) Reoviruses—a new group of respiratory and enteric viruses formerly classified as ECHO type 10 is described. Science 130:1387–1389.

Sabin AB, Wright AM (1934) Acute ascending myelitis following a monkey bite, with the isolation of a virus capable of reproducing the disease. J Exp Med 59:115–136.

Sauer RM, Prier JE, Buchanan RS, Creamer AA, Fegley HC (1960) Studies on a pox disease of monkeys. Amer J Vet Res 21:377–380.

Schmidt ECH (1948) Virus myocarditis pathologic and experimental studies. Amer J Pathol 24:97–117.

Schmidt RE, Butler TM (1971) Case report: molluscum contagiosum in a colony born chimpanzee. Lab Prim Newslett 10(4)17.

Schnitzer TJ (1981) Characterization of a simian foamy virus isolated from a spontaneous primate lymphoma. J Med Primatol 10:312–328.

Shevtsova ZV (1967) Studies on the etiology of hemorrhagic fever in monkeys. Vopr Virusol 12:47–51.

Shishido A (1966) Natural infection of measles virus in laboratory monkeys. Jpn J Med Sci Biol 19(4):221–222.

Smith MW (1982) Field aspects of the Marburg Outbreak:1967. Prim Suppl 7:11–15.

Smith PC, Yuill TM, Buchanan RD, Stanton JS, Chaicumpa V (1969) The gibbon (*Hylobates lar*); a new primate host for herpesvirus hominis. I. A natural epizootic in a laboratory colony. J Infect Dis 120:292–297.

Steele MD, Giddens WE Jr, Valerio M, Mumi SM, Stetzer ER (1982) Spontaneous paramyxoviral encephalitis in nonhuman primates (*Macaca mulatta* and *M. nemestrina*). Vet Pathol 19:132–139.

Stowell RE, Smith EK, España C, Nelson VG (1971) Outbreak of malignant lymphoma in rhesus monkeys. Lab Invest 25:476–479.

Stuker G, Oshiro LS, Schmidt NJ (1980) Antigenic comparisons of two new rotaviruses from rhesus monkeys. J Clin Microbiol 11:202–203.

Stuker G, Oshiro LS, Schmidt NJ, Holmberg CA, Anderson JH, Glaser CA, Henrickson RV (1979) Virus detection in monkeys with diarrhea: the association of adenoviruses with diarrhea and the possible role of rotaviruses. Lab Anim Sci 29:610–616.

Suleman MA, Johnson BJ, Tarara R, Sayer PO, Ochieng DM, Muli JM, Mbete E, Tukei PM, Ndirangu D, Kago S, Else JG (1984) An outbreak of poliomyelitis caused by poliovirus type 1 in captive black and white colobus monkeys (*Colobus abyssinicus kikuynensis*) in Kenya. Trans Roy Soc Trop Med Hyg 78:665–669.

Tate CL, Lewis JC, Huxsoll DL, Hildebrandt PK (1971) Herpesvirus T as the cause of encephalitis in an owl monkey (*Aotus trivirgatus*). Lab Anim Sci 21:743–745.

Tauraso NM, Shelokov S, Allen AM, Palmer DE, Aulisio CG (1968) Two epizootics of simian haemorrhagic fever. Nature 218(5144):876–877.

Theilen GH, Gould D, Fowler M, Dungworth DL (1971) C-type virus in tumor tissue of a woolly monkey (*Lagothrix* sp.) with fibrosarcoma. J Natl Cancer Inst 47:881–889.

Tsuchiya Y, Jashiki O, Yamada H (1970) Generalized cytomegalovirus infection in a gorilla. Jpn J Med Sci Biol 23:71–73.

Tyrrell DAJ, Buckland FE, Lancaster MC, Valentine RC (1960) Some properties of a strain of SV17 virus isolated from an epidemic of conjunctivitis and rhinorrhoea in monkeys (*Erythrocebus patas*). Br J Exp Pathol 41:610–616.

Umemura T, Inagaki H, Goryo M, Itakura (1985) Aspiration pneumonia with adenoviral infection in a Japanese macaque (*Macaca fuscata fuscata*). Lab Anim Sci 19:39–41.

Valerio DA (1971) Colony management as applied to disease control with mention of some viral diseases. Lab Anim Sci 21:1011–1014.

VandeBerg JL (1983) Genetics of nonhuman primates in relation to viral diseases. In: Kalter SS (ed) Viral and immunological diseases in nonhuman primates. Alan R Liss, New York, pp 39–66.

Vargas-Mendez O, Elton NW (1953) Naturally acquired yellow fever in wild monkeys of Costa Rica. Amer J Trop Med Hyg 2:850–863.

Vasileva VA, Ivanov MT, Rumel NB, D'Yachenko AG, Kakubava VV, Danelyan GA (1978) Isolation and biological characterization of an adenovirus of rhesus macaques. Acta Biol Med Ger 37:1281–1287.

Vizoso AD (1975) Recovery of herpes simiae (B virus) from both primary and latent infections in rhesus monkeys. Br J Exp Pathol 56:485–488.

Vogel FS, Pinkerton H (1955) Spontaneous salivary gland virus disease in chimpanzees. Arch Pathol 60:281–288.

vonMagnus P, Anderson EK, Petersen KB, Birch-Andersen A (1959) A pox-like disease in cynomolgus monkeys. Acta Pathol Microbiol Scand 46:156–176.

Webb HE (1969) Kyasanur Forest disease virus infection in monkeys. Lab Anim Handb 4:131–134.

White RJ, Simmons L, Wilson RB (1972) Chickenpox in young anthropoid apes: clinical and laboratory findings. J Amer Vet Med Assoc 161:690–692.

Whitney RA (1983) Animals other than simian for the study of disease. In: Kalter SS (ed) Viral and immunological diseases in nonhuman primates. Alan R Liss, New York, pp 197–208.

Willimzik H-F, Kalter SS, Lester TL, Wigand R (1981) Immunological relationship among adenoviruses of humans, simians, and nonprimates as determined by the neutralization test. Intervirology 15:28–36.

Wolf RH, Smetana HF, Allen WP, Felsenfeld AD (1974) Pathology and clinical history of delta herpesvirus infection in patas monkeys. Lab Anim Sci 24:218–221.

Wood O, Tauraso N, Liebhaber H (1970) Electron microscopic study of tissue cultures infected with simian haemorrhagic fever virus. J Gen Virol 7:129–136.

Work TH, Trapido H (1957) Kyasanur Forest disease. A new virus disease in India. Summary of preliminary report of investigations of the Virus Research Centre on an epidemic disease affecting forest villagers and wild monkeys of Shimoga District, Mysore. Ind J Med Sci 11:340–341.

Yamanouchi K, Shishido A, Honjo S (1973) Natural infection of cynomolgus monkeys with measles virus. Exp Anim (Tokyo) 22:389–393.

Zwartouw HT, Boulter EA (1984) Excretion of B virus in monkeys and evidence of genital infection. Lab Anim 18:65–70.

Zwartouw HT, MacArthur JA, Boulter EA, Seamer JH, Marston JH, Charnove AS (1984) Transmission of B disease infection between monkeys especially in relation to breeding colonies. Lab Anim 18:125–130.

47
Virus-Associated Neoplastic and Immunosuppressive Diseases of Nonhuman Primates

HARVEY RABIN and RAOUL E. BENVENISTE

Introduction

Several primate colonies have experienced outbreaks of neoplastic disease accompanying immunosuppressive syndromes. These outbreaks possess the characteristics of an infectious disease, and in several cases viruses have been isolated and shown to have disease-inducing potential. The colonies involved have been at the California, New England, Oregon, and Washington Regional Primate Research Centers in the United States, and at the Institute of Experimental Pathology and Therapy in the U.S.S.R.

In this report, we will review the outbreaks at each of these centers and discuss the viruses isolated. Emphasis will be given to newer findings, especially in light of the possible analogy of some of the simian diseases to acquired immunodeficiency syndrome (AIDS) in humans.

Disease at the California Primate Research Center (CPRC)

From 1969 through 1975, 43 cases of spontaneous lymphoma were reported at the CPRC (Stowell et al., 1971; Manning and Griesemer, 1978). Forty-two of the cases were in rhesus monkeys (*Macaca mulatta*) and the other case in a stump-tailed macaque, *M. arctoides*. Although several types of viruses were detected in cultured tumor cells (adenoviruses, reoviruses, foamyviruses, and herpesviruses; Manning and Griesemer, 1978) and there was an indication that the condition was transmissible by inoculation of minced tumor tissue (Stowell et al., 1971), no etiologic agent was identified. There was also an indication that animals bearing lymphomas were immunosuppressed (Holmberg et al., 1978). Lymphomas were identified as T-cell, B-cell, or non-T, non-B-cell in origin (Holmberg et al., 1978).

Several years subsequent to the cases of lymphoma with accompanying immunosuppression, an outbreak of immunodeficiency disease, termed simian AIDS (SAIDS), was recognized at the CPRC (for review, see Gardner and Marx, 1985). The disease at the CPRC is characterized by generalized lymphadenopathy and histological, hematological, and functional evidence of immunodepression. The disease is generally fatal after several months, especially in younger animals. The outbreak was confined to a single large outdoor compound, and about 50% of the 64 animals housed in this compound have died. The only tumors reported were two cases of subcutaneous fibrosarcoma. Two other outbreaks of immunodeficiency have occurred in macaques at the CPRC, but only the 1969–1975 outbreak and the present one have included the presence of neoplastic disease (Henrickson et al., 1983).

That SAIDS at the CPRC appeared to have an infectious etiology requiring close contact was suggested by a study in which healthy monkeys were introduced into the corral where the disease was endemic. Some of these monkeys were placed in direct contact with the resident animals while others were placed within a cage separated from the resident monkeys by a 10-foot wide barrier. Only animals in direct contact with the resident monkeys developed disease (Henrickson, unpublished results as reported in Gardner and Marx, 1985). Transmission of disease with tissue homogenates from moribund animals (London et al., 1983) and with blood or filtered plasma (Gravell et al., 1984) further suggested the transmissible nature of the immunodeficiency disease. Transmission experiments using plasma from infected animals showed that the infectious portion of the plasma was filterable and banded in equilibrium sucrose gradients at 1.14 to 1.18 g/cm^3 (Gravell et al., 1984; Marx et al., 1984) suggesting the presence of a retrovirus. The experimentally induced disease at the CPRC was rapid, with monkeys dying between 5 and 21 weeks after inoculation. There appeared to be no neoplastic component associated with the disease at the CPRC except for the two cases of fibrosarcoma.

Disease at the New England Primate Research Center (NEPRC)

The immunosuppressive disease at the NEPRC has been mainly in Taiwanese rock macaques (*M. cyclopis*) but has also been observed in a few rhesus monkeys. In 1980–1981, mortality in Taiwanese rock macaques amounted to about one-third of the population; most of these deaths occurred in an outdoor facility containing gang cages. Location of housing constituted the largest risk factor for disease development. Analysis of necropsy records showed that causes of death in these macaques were noma, opportunistic infections (i.e., *Pneumocystis carinii*, cytomegalovirus [CMV] mononucleosis, and bacterial pneumonia), lymphoproliferative disease (including lymphoma), and diarrhea and wasting. Affected animals showed decreased proliferative responses to pokeweed mitogen (PWM), to concanavalin A (con A), and to xenogeneic cells in mixed lymphocyte cultures. Affected monkeys showed the presence of a large, immature-appearing abnormal monocyte in both the bone marrow and peripheral blood. Liver function tests also tended to be abnormal in affected animals (Letvin et al., 1983a). Detailed histologic study of 16 macaques confirmed the presence of CMV infection as well as lesions caused by SV40 virus and paramyxovirus (probably measles virus). Enterocolitis was seen in all animals, amyloidosis in several, and lymphodepletion, hemosiderosis, and erythrophagocytosis in half the animals (King et al., 1983). Lymphomas arising in macaques at the NEPRC were associated with evidence of immunodepression suggesting a possible link between lymphomas and lymphoproliferative disease and the immunosuppressive syndrome. This prompted transmission studies of lymphoma at the NEPRC (Hunt et al., 1983). Tumor homogenate from an adult male rhesus monkey was inoculated subcutaneously into a wild-caught female rhesus monkey. This recipient developed undifferentiated lymphoma that was surgically removed at

14 months. A 2-day-old male rhesus monkey also inoculated with the tumor homogenate died with an AIDS-like disease but without lymphoma at 14 months. In a second transmission series, tumor homogenate and blood from an adult female monkey with an undifferentiated retroorbital lymphoma were inoculated into an 8-month-old male rhesus monkey and a 10-month-old female rhesus monkey. The male was sacrificed at 5 months with an undifferentiated lymphoma, and the female recipient was sacrificed at 26 months with either a poorly differentiated or a mixed lymphocytic–histiocytic lymphoma. Recipient monkeys in these positive transmissions showed evidence of immunodeficiency disease as well as lymphoma. In other studies, the immunodeficiency syndrome was transmitted to five recipients by means of filtered (0.22 μm) tumor homogenates from two rhesus and one crab-eating macaque, *M. fascicularis* (Letvin et al., 1983b). Animals died as soon as a few weeks after inoculation or as long as 4 years afterwards. The transmission of the disease via cell-free filtrate suggested an infectious nature for the immunodeficiency disease. Its relationship with lymphoma, which also appeared to have a transmissible nature, is not as yet clear.

Disease at the Washington Primate Research Center (WPRC)

The major feature of disease at the WPRC is retroperitoneal fibromatosis (RF), which has been observed in 82 macaques from 1976 to 1983, including primarily pig-tailed macaques (*M. nemestrina*) but also *M. fascicularis*, *M. fuscata*, and *M. mulatta* (Giddens et al., 1983, 1985; Tsai et al., 1985a). RF at WPRC is characterized by an aggressive proliferation of highly vascular fibrous tissue subjacent to the peritoneum covering the ileocecal junction and associated mesenteric lymph nodes. Two RF syndromes have been recognized: localized, in which fibroproliferative lesions occur in solitary nodules either in the abdominal cavity or subcutaneously, and progressive, in which fibromatosis occurs throughout the abdominal cavity.

The case fatality rate of RF was 98% with a course of about 2 years. The leading cause of death was due directly to RF lesions in 43%, to enterocolitis in 36%, septicemia in 12%, amyloidosis in 5%, and malignant lymphoma in 2% (Giddens et al., 1985). Based on the occurrence of RF in colony-born and noncolony-born animals, the minimum incubation period for natural exposure is about 9 months. The incidence of RF was approximately 1% in all four species of macaques, and there were no differences in the incidence by sex (Tsai et al., 1985a). RF occurs predominantly in younger animals; the incidence is 5.7% in macaques 1–2 years old and 3.4% in animals 2–3 years old, but less than 1.0% in age groups of under 1 year and over 3 years. The predominant clinical signs in macaques with RF were recurrent diarrhea, weight loss, mesenteric lymphadenopathy, and opportunistic infections (predominantly bacterial pathogens). Most animals in the latter stages of illness showed marked immunodeficiency as evidenced by severe thymic and lymphoid atrophy, depressed mitogen (PHA, con A, and PWM) responses of peripheral blood mononuclear cells, and failure to mount normal antibody responses to the T-cell dependent antigen, bacteriophage ΘX174 (Stromberg et al., 1984; Tsai et al., in press b).

Experimental intraperitoneal inoculation of *M. nemestrina* and *M. fascicularis* in two separate experiments with suspensions of RF tissue resulted in the appearance of an immunosuppressive syndrome in 5 of 16 macaques and RF in 3 of 16 animals. Inoculation of RF tissue into nude mice, hamsters, and marmosets (*Saguinus labiatus*) did not result in disease (Giddens et al., 1985). These studies indicated that a transmissible agent might be involved in RF and immunosuppression in the macaque colony at WPRC.

Disease at the Oregon Primate Research Center (OPRC)

A colony of Celebes black macaques (*M. nigra*) at the OPRC has shown very high mortality with 47 out of 68 individuals dying from 1981 to 1983. Disease has been characterized by microbial and protozoan infections, anemia, diarrhea, low birth rate, and an apparent short life expectancy. The colony also has a high incidence of diabetes mellitus (Shiigi et al., in press). Laboratory findings on these monkeys show that they have a diminished proliferative response to con A and a reduced polyclonal immunoglobulin release response to PWM. Other findings suggest that there are more HLA DR-positive peripheral mononuclear cells in the diseased monkeys compared to controls and that diseased animals have lower percentages of OKT8-positive peripheral blood cells.

As is the case in the WPRC, a large number of animals in this colony have also been shown to have RF (W. McNulty, J. L. Palotay, and L. Olson, personal communication). Of 67 animals dying over the 6 years between 1978 and 1983, 17 had RF. Most of the RF-bearing monkeys also showed severe weight loss and persistent diarrhea.

Disease at the Institute of Experimental Pathology and Therapy, Sukhumi (IEPT), U.S.S.R.

Beginning in 1967, an outbreak of lymphoid disease, including lymphomas of various types, has been present at the IEPT (for review see Rabin, 1985). Disease is present predominantly in baboons (*Papio hamadryas*) and was first noted subsequent to the introduction into the colony of baboons that had been inoculated with human leukemic material.

The disease is characterized by marked splenomegaly and lymphadenopathy, gingivitis, dermatitis, increased skin pigmentation, and depression in mitogen (PHA and con A) responses. There is also a high frequency of apparently nonrandom chromosomal abnormalities involving chromosomes 1 and 20. The disease at IEPT is present in the main colony of about 900 animals and manifests itself primarily where there has been contact between diseased and healthy baboons. A colony of baboons living apart from the main colony has remained disease free. This, together with the observation that the disease can be transmitted from baboon to baboon (Falk et al., 1976), suggests an infectious etiology for the disease.

Virological Investigations—Type D Virus Isolates

The epidemiological findings at the various primate centers, together with the studies that showed disease could be transmitted experimentally with filtered biological fluids, suggested a viral etiology for SAIDS and RF. Cocultivation of peripheral blood lymphocytes from two monkeys (*M. cyclopsis*) with immunosuppressive disease at NEPRC with Raji cells (a human Burkitt lymphoma cell line) resulted in rapid (7 days) development of multinucleated giant cells (Daniel et al., 1984). Electron microscopic examination revealed retrovirus particles with charcteristic type D morphology. This initial isolate, termed retrovirus-D/New England, was readily passaged in Raji cells by means of cell-free filtered supernatants. Type D virus was also isolated from tissues (lymph node and spleen) and pharyngeal secretions of affected macaques. No isolates were obtained from the peripheral lymphocytes of 97 apparently healthy macaques.

Restriction enzyme patterns using DNA probe prepared from the prototype type D retrovirus, Mason Pfizer monkey virus (MPMV), showed that retrovirus-D/New England was a related type D virus (Daniel et al., 1984). A more detailed restriction endonuclease map of the two cloned viruses has since revealed that 46% of restriction sites are conserved showing that these viruses, although related, can be distinguished. Further molecular analysis has revealed that five separate isolates of retrovirus-D/New England obtained from three species of macaques (*M. cyclopis*, *M. fascicularis*, and *M. mulatta*) were indistinguishable from each other by restriction endonuclease analysis. It thus appears that one strain of type D retrovirus is present at the NEPRC.

Attempts were made to transmit the immunodeficiency disease to healthy, type D retrovirus-negative rhesus monkeys (Letvin et al., 1984). In these experiments, 11 macaques from 1 day to 1 year of age were inoculated. Three strains of virus were used at concentrations from 10^4 to 10^9 tissue culture infectious units. None of the inoculated animals developed progressive immunodeficiency disease, opportunistic infections, or neoplasms. Some animals did develop transient lymphadenopathy, fever, immunosuppression, and anemia. Most animals also developed persistent neutropenia and chronic viremia that lasted for several months. Two juvenile common marmosets (*Callithrix jacchus*) inoculated with the virus also failed to develop progressive immunodepressive disease.

At the CPRC, a virus was isolated by cocultivation of primary rhesus kidney cells with heparinized blood from a macaque with experimentally transmitted SAIDS. This isolate, after inoculation into juvenile rhesus monkeys, produced an acute form of SAIDS with a short latency period. Electron micrographs revealed the presence of type D retrovirus resembling MPMV. Competition radioimmunoassays (RIA) showed the CPRC isolate to have a major core protein (p27) that was indistinguishable from that of MPMV but to have an antigenically distinct envelope glycoprotein (gp70) (Marx et al., 1984).

In order to isolate the etiologic agent of RF at the WPRC, RF tissue from an immunodeficient rhesus monkey was cocultivated with mammalian cells known to support the replication of a wide variety of primate viruses. A retrovirus, which morphologically resembled a type D isolate, was obtained after cocultivation with canine, human, rhesus, bat, and mink cells (Stromberg et al., 1984).

Molecular hybridization experiments showed that this virus, initially termed SAIDS-D/Washington, was related to MPMV and to the endogenous type D virus of langurs, PO-1-Lu (Benveniste and Todaro, 1977). DNA hybridization experiments suggested the possibility that the RF isolate, because of its extent of hybridization to langur cellular DNA, may have arisen from a virus related to the endogenous langur type D virus. Competition RIAs showed complete cross-reactivity with MPMV p27. The WPRC isolate, like the CPRC isolate, did not compete in an RIA for MPMV gp70 showing that it too was only partially related to MPMV.

A separate report had described a virus associated with spleen cell cultures obtained from a macaque with RF at the WPRC (Hefti et al., 1983). This report described the virus as being heterogeneous and comprised of particles with morphology representing both type C and type D retroviruses. Unfortunately, this virus was lost in cultivation and is no longer available for study.

A total of 41 separate type D retroviruses have now been obtained from the WPRC. Twenty-one of these isolates were from RF tissues or plasma obtained from four species of macaques (*M. nemestrina*, *M. fascicularis*, *M. mulatta*, and *M. fuscata*). The additional 20 isolates have been obtained from macaques with recurrent diarrhea, significant weight loss, or lymphadenopathy. Molecular hybridization experiments reveal that the nonrepetitive cellular DNA of 11 RF tissues tested contain sequences homologous to retrovirus-D/Washington (R-D/W). This virus has not been isolated from 63 healthy feral *M. nemestrina* imported into the colony from Southeast Asia.

The association between RF and type D retroviruses suggested that this class of viruses might be responsible for RF in macaques. A biologically cloned virus, originally isolated from a rhesus monkey with RF (Stromberg et al., 1984), was therefore inoculated intravenously into six 1-year-old *M. nemestrina*. Four animals developed high titers of antibodies to the virus as measured by RIA to the major core antigen p27 and have remained virus-free and healthy 16 months after inoculation. The two other macaques became viremic (10^4 virus particles/ml of plasma) soon after inoculation. One of these animals died at 5 weeks experiencing weight loss, enterocolitis, candidiasis, and cryptosporidiosis. Histologically, the mesenteric lymph nodes revealed lymphoid depletion and capsular fibrosis, an early histological manifestation of RF. The other viremic animal had palpable abdominal nodes 15 weeks after inoculation; a biopsy at 26 weeks confirmed the presence of RF. These studies thus showed that in macaques not mounting a protective antibody response, RF could be induced after R-D/W inoculation (Benveniste et al., in press).

A preliminary screening of the healthy macaques at WPRC shows that the majority of the animals have antibodies to the type D virus (Benveniste et al., in press). This, together with the fact that feral animals imported into the colony are antibody and virus negative, suggests that the type D retrovirus is endemic in the colony at WPRC.

To establish better the relationship of R-D/W to MPMV, six structural proteins obtained from each virus have been purified and compared (Henderson et al., 1985). These proteins purified from each type D retrovirus included p4, p10, p12, p14, p27, and a phospho-protein designated pp18 for MPMV and pp20 for R-D/W. Amino acid composition and *N*-terminal amino acid sequence analysis show that these six proteins of R-D/W are distinct from the homologous proteins of MPMV but that these proteins from the two viruses share a high degree of amino acid sequence homology. These results agree with those obtained by liquid hybridization (Stromberg et al., 1984) and by restriction endonuclease analysis (Desrosiers et al., 1985) which suggest that these retroviruses are all distinct members of a closely related family.

The purified proteins from R-D/W were iodinated and used in RIAs to compare the various isolates to each other. Several of the structural proteins can be shown to be antigenically distinct among the retrovirus-D isolates and can therefore be used to type the isolates from the various regional primate centers (Benveniste and Arthur, unpublished observations).

The isolate from the WPRC has also been shown to transform various rodent fibroblast cell lines in vitro (Stromberg et al., 1984). Since the type D isolate at the CPRC is associated with SAIDS, the NEPRC isolates primarily with a transient lymphadenopathy, and the isolates from WPRC with retroperitoneal fibromatosis, it is possible that the retroviruses from the various primate colonies, which are molecularly distinct, are associated with different pathogenicities in vitro. Experiments utilizing recombinant hybrid molecular clones should resolve which portions of the virus are responsible for the different pathogenicities.

A similar type D virus isolate has been obtained from affected Celebes macaques at the OPRC by cocultivation of peripheral blood mononuclear cells with Raji cells (Shiigi et al., in press). Virus was isolated from five (three of which had RF) out of six macaques with reduced responses to con A and to PWM. Subsequent studies of Celebes macaques at the OPRC showed that virus could be isolated from monkeys with immunosuppressive disease and RF but not from monkeys without signs of disease (Marx et al., 1985). By membrane immunofluorescence, both immunosuppressed and healthy monkeys were seropositive. However, only sera from healthy monkeys showed virus neutralizing activity. Inoculation of plasma from Celebes macaques with immunosuppression and RF into two juvenile rhesus monkeys induced immunodepression, other signs of SAIDS and viremia, but not RF. At 14 months postinoculation, one recipient animal had recovered, while the other still showed clinical signs of disease and was virus positive. Virus isolates from the Celebes macaques were characterized by electron microscopy, RIA, restriction endonuclease fragmentation patterns, and by virus neutralization. RIA showed that the major core protein, p27, was identical with that of MPMV and the rhesus isolate from the CPRC. The Celebes isolate, however, could be distinguished from both MPMV and CPRC isolates in a competition RIA for p10. The virus could also be distinguished from MPMV and the CPRC isolate by restriction enzyme analysis. In neutralization assays using sera from monkeys infected with either the Celebes macaque virus or rhesus monkey virus from the CPRC, only homologous virus was neutralized even though endpoint titers were high (1:640 to 1:2580). The exact relationship of the type D isolates from the various primate centers is still under investigation, but preliminary studies reveal that the isolates from WPRC and the OPRC, both of which are associated with RF in macaques, might be closely related immunologically (L. Arthur, personal communication).

Virological Investigations—Lymphotropic Retrovirus and Herpesvirus Isolates

Lymphoma and lymphoproliferative disease have been components of the clinical and pathological findings at the NEPRC (Hunt et al., 1983; King et al., 1983). A serological investigation showed that 85% (11 of 13) of macaques with malignant lymphoma or lymphoproliferative disease at the NEPRC had antibodies that cross-reacted by membrane immunofluorescence with antigens of HTLV-I (Homma et al., 1984). In contrast, only 7 of 95 (7.4%) healthy macaques from the NEPRC or Taiwan had such cross-reactive antibodies. Radioimmunoprecipitation assays with selected macaque sera confirmed the immunofluorescence findings. More recent investigations of macaques at the NEPRC have led to the isolation and characterization of an HTLV-III-related virus, termed simian T-lymphotropic virus (STLV-III) (Daniel et al., 1985; Kanki et al., 1985). Virus was isolated from viably frozen splenocytes, fresh lymphocytes, or frozen serum from four rhesus monkeys. One of these animals had developed a malignant lymphoma 26 months after inoculation with lymphomatous tissue from a spontaneous case in another rhesus monkey. Tissue from the experimental case of lymphoma was inoculated into seven additional monkeys, all of which developed immunosuppressive disease. Pooled blood samples from three of these cases were inoculated into six other monkeys that also developed immunodeficiency disease. Cell-free filtered plasma from one of these animals was then inoculated into two other macaques which, in turn, developed a similar immunosuppressive condition. It was from frozen serum of these last two monkeys that virus was recovered. Both of these monkeys showed signs of brain lesions that histologically resembled lesions found in the brains of AIDS patients with encephalopathy (Shaw et al., 1985). The fourth monkey from which virus was isolated had a spontaneous immunosuppressive disorder. Virus was isolated on two occasions from this animal at an interval of 4 months.

This lymphotropic virus was isolated in fresh human T-cells or in the human T-cell tumor line, HUT-78. STLV-III induced the formation of pleomorphic and multinucleated giant cells in HUT-78 cells but failed to replicate in Raji cells. Electron microscopic examination of infected cells revealed budding and extracellular virus particles similar in appearance to HTLV-III (Daniel et al., 1985). As analyzed by radioimmunoprecipitation, STLV-III contained proteins of molecular weight = 24K, 55K, 120K, and 160K, similar to those seen in HTLV-III. Reciprocal cross-reactivity was observed between some of the HTLV-III and STLV-III proteins. Three of the macaque sera tested had antibodies cross-reactive with HTLV-I as well as with HTLV-III, suggesting either coinfection with both STLV-III and an HTLV-I-related virus or the presence of additional cross-reactive antigens. Type D virus was isolated from the spontaneous SAIDS case but not from the three experimental cases in this series (Kanki et al., 1985). In additional serologic surveys, antibodies cross-reactive with STLV-III were found in 28 of 67 (42%) wild-caught African green monkey sera as demonstrated by radioimmunoprecipitation analysis. Cross-reacting antibodies were not found in sera of 30 captive chimpanzees nor in sera of 60 captive baboons. Cross-reactivity was demonstrated, however, in 33–70% of HTLV-III-positive human sera (Kanki et al., 1985, 1985b).

At the IEPT in Sukhumi, U.S.S.R. (Lapin, 1976), baboons and macaques with lymphoma have been found to be positive for HTLV-I-related antibodies (Saxinger et al., 1984). Furthermore, serological evidence for transmission of virus from baboons with lymphoma to macaque, baboon, and owl monkey recipients was also obtained as was evidence for the presence of HTLV-I provirus in recipient monkeys. Evidence for the presence of HTLV-I-related viruses in macaques (Miyoshi et al., 1982. Komuro et al., 1984), African green monkeys (Yamamoto et al., 1983; Komuro et al., 1984), and chimpanzees (Komuro et al., 1984) in addition to baboons has also been reported. Results at Sukhumi as well as those at the NEPRC (Homma et al., 1984) raise the possibility that an HTLV-I-related virus may be involved in leukemogenesis at these two centers.

Evidence associating an Epstein-Barr(EBV)-related virus with leukemia at Sukhumi has also been obtained (Neubauer et al., 1979). Antibodies to this virus (*Herpesvirus papio*) were found in higher titers and directed against a wider number of viral antigens in diseased as opposed to control baboons. Moveover, *H. papio* DNA has been detected in the spleens of diseased baboons and in spleens of baboons in remission (Dyachenko et al., 1981). This virus has also shown the potential to induce an acute lymphoproliferative disease in adult cotton-topped marmosets (Deinhardt et al., 1978). EBV-related viruses have been found in cultures established from several leukemic or lymphoma-bearing nonhuman primates including those from an orangutan (Rasheed et al., 1977) and a rhesus monkey (Rangan et al., 1985). Several cases of lymphoma have occurred in immunosuppressed macaques used in transplantation studies, and an EBV-like virus was also associated with these tumors (Heberling et al., 1982).

Summary and Recommendations

Tumors and immunosuppressive disease have occurred concurrently in several species of macaques at primate centers in the United States and in baboons and macaques at the colony in Sukhumi, U.S.S.R. An outbreak of lymphoma occurred several years ago at the CPRC, and several cases of lymphoma and lymphoproliferative disease have been seen at the NEPRC. Cases of lymphoma have also been seen in other primate colonies such as the WPRC and in immunosuppressed macaques. RF has been observed in large numbers of macaques primarily at the WPRC and at the OPRC. In some colonies, the large numbers of tumors have been the principal feature of the disease outbreak, whereas in other colonies, the immunodepressive AIDS-like disease has been the main manifestation.

Several viruses have been isolated from animals at the various primate colonies. It will take further research to determine the precise etiologic role for each of these viruses, but some preliminary conclusions can be made. The type D retroviruses isolated at the CPRC appear to have the capacity to induce an acute form of immunosuppressive disease, neutropenia, viremia, and mortality and may play an important role in the AIDS-like disease at the CPRC. The type D retroviruses isolated at both the WPRC and the OPRC appear closely associated with RF and may be the etiologic agent of this disease at these centers. The role that these latter retroviruses play in immunosuppression is not yet clear nor is the role that SAIDS-D/California may play in the causation of RF. The recent isolation of the HTLV-III-related virus, STLV-III, at the NEPRC suggests that this virus may be an important factor in the disease at this center and possibly at other primate centers as well. In addition, if this lymphotropic virus is pathogenic in macaques, it will provide an important primate model for human AIDS. The roles that other lymphotropic viruses, such as the HTLV-I-related viruses and the EBV-related viruses, may play in the etiology of lymphoma remain to be determined.

In addition to the specific disease states that these various viruses may induce in a percentage of infected monkeys, it should be pointed out that the effects of chronic viremia on the host animal's general health, function, and reproductive capacity is not understood. Because primate colonies require satisfactory research animals and must also strive to be self-sustaining, attention must be paid to the possible long-range consequences of persistent retrovirus infection.

It is clear in most of these instances that disease spread requires contact among infected and uninfected monkeys. Epidemiologic studies and transmission experiments at the IEPT, the NEPRC, the CPRC, and elsewhere substantiate this idea. Managers of primate colonies should be alerted to the potential presence of these retroviruses and should monitor their colonies, note disease patterns, and take necessary steps to prevent further spread. Measures for controlling spread of virus have already been instituted at several colonies, and vaccines are being developed. It is also important to prevent the potential transmission of these viruses from primate to human. The disease-inducing potential of these viruses in humans is not yet understood. It would be well to regard these viruses, all of which replicate well in various human fibroblast and lymphocyte cells in vitro, with the utmost caution until more information indicating that they may be handled otherwise becomes available.

References

Benveniste RE, Todaro GJ (1977) Evolution of primate oncornaviruses: an endogenous virus from langurs (*Presbytis* spp.) with related virogene sequences in other Old World monkeys. Proc Natl Acad Sci USA 74:4557–4561.

Benveniste RE, Stromberg K, Morton WR, Tsai C-C, Giddens WE Jr (In press) Association of retroperitoneal fibromatosis with type D retroviruses. In: Salzman L (ed) Animal models of retrovirus infection. Academic, New York.

Daniel MD, King NW, Letvin NL, Hunt RD, Sehgal PK, Desrosiers RC (1984) A new type D retrovirus isolated from macaques with an immunodeficiency syndrome. Science 223:602–605.

Daniel MD, Letvin NL, King NW, Kannigi M, Sehgal PK, Hunt RD, Kanki PJ, Essex M, Desrosiers RC (1985) Isolation of T-cell tropic HTLV-III-like retrovirus from macaques. Science 228:1201–1204.

Deinhardt F, Falk L, Wolfe L, Schudel A, Nonoyama M, Lai P, Lapin B, Yakovleva L (1978) Susceptibility of marmosets to Epstein-Barr virus-like baboon herpesviruses. Prim Med 10:163–170.

Desrosiers RC, Daniel MD, Butler CV, Schmidt DK, Letvin NL, Hunt RD, King NW, Barker CS, Hunter E (1985) Retrovirus-D/New England and its relation to Mason-Pfizer monkey virus. J Virol 54:552–560.

Dyachenko A, Kokosha L, Lapin B, Yakovleva L, Agrba V (1981) Detection of baboon herpesvirus DNA in the tissues of hemoblastosis diseased and healthy monkeys from Sukhumi nursery USSR. J Sov Oncol 2:48–51.

Falk L, Deinhardt F, Nonoyama M, Wolfe LG, Bergholz C, Lapin B, Yakovleva L, Agrba V, Henle G, Henle W (1976) Properties of a baboon lymphotropic herpesvirus related to Epstein-Barr virus. Int J Cancer 18:798–807.

Gardner MB, Marx PA (1985) Simian acquired immunodeficiency syndrome. Adv Viral Oncol 5:57–81.

Giddens WE Jr, Morton WR, Hefti E, Panem S, Ochs H (1983) Enzootic retroperitoneal fibromastosis in *Macaca* spp. In: Kalter SS (ed) Viral and immunological diseases in nonhuman primates. Alan R Liss, New York, pp 249–253.

Giddens WE Jr, Tsai C-C, Morton WR, Ochs HD, Knitter GH, Blakely GA (1985) Retroperitoneal fibromatosis and acquired immunodeficiency syndrome in macaques: pathologic observations and transmission studies. Amer J Pathol 119:253–263.

Gravell M, London WT, Houff SA, Madden DL, Dalakas MC, Sever JL, Osborn KG, Maul DH, Henrickson RV, Marx PA, Lerche NW, Prahalada S, Gardner MB (1984) Transmission of simian acquired immunodeficiency syndrome (SAIDS) with blood or filtered plasma. Science 223:74–76.

Heberling R, Bieber C, Kalter S (1982) Establishment of a lymphoblastoid cell line from a lymphomatous cynomolgous monkey. In: Yohn D, Blakeslee J (eds) Adv in Compar Leuk Res. Elsevier/North Holland, New York, pp 385–386.

Hefti E, Ip J, Giddens WE Jr, Panem S (1983) Isolation of a unique retrovirus, MNV-1, from *Macaca nemestrina*. Virology 127:309–319.

Henderson LE, Sowder R, Smythers G, Benveniste RE, Oroszlan S (1985) Purification and *N*-terminal amino acid sequence comparison of structural proteins from retrovirus-D/Washington and Mason-Pfizer monkey virus. J Virol 55:778–787.

Henrickson RV, Maul DH, Osborn KG, Gardner MB (1983) Experimental transmission of simian acquired immunodeficiency syndrome (SAIDS) and Kaposi-like skin lesions. Lancet 2:869–873.

Henrickson RV, Maul DH, Lerche NW, Osburn KG, Lowenstine LJ, Prahalada S, Sever JL, Madden DL, Gardner MB (1984a) Clinical features of simian acquired immunodeficiency syndrome (SAIDS) in rhesus monkeys. Lab Anim Sci 34:140–145.

Henrickson RV, Arthur LO, Gilden RV, Sever JL, Levy SA, Munn RJ, Gardner MB (1984) Simian AIDS: isolation of a type D retrovirus and transmission of the disease. Science 223:1083–1086.

Holmberg CA, Osburn BI, Terrell TG, Manning JS (1978) Cellular immunologic studies of malignant lymphoma in rhesus macaques. Amer J Vet Res 39:469–472.

Homma T, Kanki PJ, King NW Jr, Hunt RD, O'Connell MJ, Letvin NL, Daniel MD, Desrosiers RC, Yang CS, Essex M (1984) Lymphoma in macaques: association with virus of human T lymphotrophic family. Science 225:716–718.

Hunt RD, Blake BJ, Chalifoux LV, Sehgal PK, King NW, Letvin NL (1983) Transmission of naturally occurring lymphoma in macaque monkeys. Proc Natl Acad Sci USA 80:5085–5089.

Kanki PJ, McLane MF, King NW Jr, Letvin NL, Hunt RD, Sehgal PK, Daniel MD, Desrosiers RC, Essex M (1985) Serologic identification and characterization of a macaque T-lymphotropic retrovirus closely related to HTLV-III. Science 228:1199–1201.

Kanki PJ, Kurth P, Becker W, Dreesman G, McLane MF, Essex M (1985b) Antibodies to simian T-lymphotropic retrovirus type III in African green monkeys and recognition of STLV-III viral proteins by AIDS and related sera. Lancet 1:1330–1332.

King NW, Hunt RD, Letvin NL (1983) Histopathologic changes in macaques with an acquired immunodeficiency syndrome (AIDS). Amer J Pathol 113:382–388.

Komuro A, Watanabe T, Miyoshi I, Hayami M, Tsujimoto H, Seiki M, Yoshida M (1984) Detection and characterization of simian retroviruses homologous to human T-cell leukemia virus type 1. Virology 138:373–378.

Lapin BA (1976) Epidemiology of leukemia among baboons of Sukhumi monkey colony. In: Clemmesen J, Yohn DS (eds) Comp Leuk Res (1975), Bibl Haemat No 43. S Karger, Basel, pp 212–215.

Letvin NL, Eaton KA, Aldrich WR, Sehgal PK, Blake BJ, Schlossman SF, King NW, Hunt RD (1983a) Acquired immunodeficiency syndrome in a colony of macaque monkeys. Proc Natl Acad Sci USA 80:2718–2722.

Letvin NL, King NW, Daniel MD, Aldrich WR, Blake BJ, Hunt RD (1983b) Experimental transmission of macaque AIDS by means of inoculation of macaque lymphoma tissue. Lancet 2:599–602.

Letvin NL, Daniel MD, Seghal PK, Chalifoux LV, King NW, Hunt RD, Aldrich WR, Holley K, Schmidt DK, Desrosiers RC (1984) Experimental infection of rhesus monkeys with type D retrovirus. J Virol 52:683–686.

London WT, Madden DL, Gravell M, Dalakas MC, Houff SA, Sever JL, Henrickson RV, Maul DH, Osborn KG, Gardner MB (1983) Experimental transmission of simian acquired immunodeficiency syndrome (SAIDS) and Kaposi-like skin lesions. Lancet 2:869–873.

Manning JS, Griesemer RA (1974) Spontaneous lymphoma of the nonhuman primate. Lab Anim Sci 24(1):204–210.

Manning JS, Griesemer RA (1978) Viruses detected in cells explanted into tissue culture from naturally occurring malignant lymphoma in rhesus monkeys. Curr Microbiol 1:157–162.

Marx PA, Bryant ML, Osborn KG, Maul DH, Lerche NW, Lowenstine LJ, Kluge JD, Zaiss CP, Henrickson RV, Shiigi SM, Wilson BJ, Malley A, Olson LC, Arthur LO, Gilden RV, Barker CS, Hunter E, Munn RJ, Heidecker G, Gardner MB (1985) Isolation of a new strain of SAIDS type D retrovirus from celebes black macaques (*Macaca nigra*) with immunodeficiency and retroperitoneal fibromatosis. J Virol 56:571–578.

Marx PA, Maul DH, Osborn KG, Lerche NW, Moody P, Lowenstein LJ, Henrickson RV, Arthur LO, Gravell M, London WT, Sever JL, Levy JA, Munn RJ, Gardner MB (1984) Simian AIDS: Isolation of a type D retrovirus and disease transmission. Science 223:1083–1086.

Miyoshi I, Yoshimoto S, Fujishita M, Tagachi H, Kubonishi I, Niiya K, Minezawa M (1982) Natural adult T-cell leukemia virus infection in Japanese monkeys. Lancet 2:658.

Neubauer RH, Rabin H, Strnad BC, Lapin BA, Yakovleva LA, Indzie E (1979) Antibody responses to *Herpesvirus papio* antigens in baboons with lymphoma. Int J Cancer 23:186–192.

Rabin H (1985) *In vitro* studies of Epstein-Barr virus and other lymphotropic herpesviruses of primates. In: Roizman B, Lopez C (eds) The herpesviruses, vol 4. Plenum, New York/London, pp 147–170.

Rangan SRS, Martin LW, Gormus BJ, Wang N (1985) Characterization of rhesus monkey (*M. mulatta*) lymphoid cellculture containing herpesvirus particles. Proc ASM p s-32.

Rasheed S, Rongey RW, Nelson-Rees WA, Rabin H, Neubauer RH, Bruszeski J, Esra G, Gardner MB (1977) Establishment of a cell line with associated Epstein-Barr-like virus from a leukemic orangutan. Science 198:407–409.

Saxinger WC, Lange-Wantzin G, Thomsen K, Lapin B, Yakovleva L, Li Y-W, Guo H-G, Robert-Guroff M, Blattner WA, Ito Y, Gallo RC (1984) Human T-cell leukemia virus: a diverse family of related exogenous retroviruses of humans and old world primates. Human T-cell leukemia/lymphoma viruses. Cold Spring Harbor Laboratory, New York, pp 323–330.

Shaw GM, Harper ME, Hahn BH, Epstein LG, Gajdusek DC, Price RW, Navia BA, Petito CK, O'Hara CJ, Groopman JE, Cho E-S, Oleske JM, Wong-Staal F, Gallo RC (1985) HTLV-III infection in brains of children and adults with AIDS encephalopathy. Science 227:177–181.

Shiigi SM, Wilson BJ, Malley A, Howard CF Jr, McNulty WP, Olson J (In press) Virus-associated deficiencies in the mitogen reactivity in Celebes black macaques (*Macaca nigra*). J Clin Immunopath.

Stowell RE, Smith EK, Espana C, Nelson VG (1971) Outbreak of malignant lymphoma in rhesus monkeys. Lab Invest 35:476–479.

Stromberg K, Benveniste RE, Arthur LO, Rabin H, Giddens WE Jr, Ochs H, Morton WR, Tsai C-C (1984) Characterization of exogenous type D retrovirus from a fibroma of a macaque with simian AIDS and fibromatosis. Science 224:289–292.

Tsai C-C, Giddens WE Jr, Morton WR, Rosenkranz SL, Ochs HD, Benveniste RE (1985a) Retroperitoneal fibromatosis and acquired immunodeficiency syndrome in macaques: epidemiologic studies. Lab Anim Sci 35:460–464.

Tsai C-C, Giddens WE Jr, Ochs HD, Morton WR, Knitter GH, Blakley GA, Benveniste RE (In press b) Retroperitoneal fibromatosis and acquired immunodeficiency syndrome in macaques: clinical and immunologic studies. Lab Anim Sci.

Yamamoto N, Hinuma Y, zur Hausen H, Schneider J, Hunsmann G (1983) African green monkeys are infected with adult T-cell leukemia virus or a closely related agent. Lancet 1:240.

48
Viral Diseases of Neonatal and Infant Nonhuman Primates

RONALD D. HUNT

Introduction

Infectious diseases and especially viral diseases of nonhuman primates have been studied since the advent of virology but have been the particular focus of attention of several groups during the past two decades [34, 37, 60, 68, 141]. Of particularly strong impact on viral primatology were the discoveries of *Herpesvirus saimiri* by our group in 1968 [103] and our subsequent demonstration of its oncongenicity in New World primates in 1969 [105]; and of Epstein-Barr virus (EBV) by Epstein and colleagues in 1964 [33] with the demonstration of its oncogenicity in cotton-top tamarins by Shope and colleagues in 1973 [137]. Another effect arose from the formation of the Special Virus Cancer Program (SVCP) in the early sixties by the National Cancer Institute (NCI). The SVCP target was basic viral research, relying heavily on nonhuman primates to uncover oncogenic viruses affecting human beings. Further, the creation of the Primate Centers program with its expanded use of nonhuman primates in diverse research projects required a more thorough understanding of primate diseases. Although each of these events heavily emphasized the oncogenicity of viruses, there also accrued considerable spin-off knowledge concerning other viruses of nonhuman primates. A number of laboratories have centered their attention on those viruses that naturally occur in nonhuman primates and have, for the most part, directed their activities toward resolving particular viral epidemics that trouble their own primate colonies.

Despite these efforts, our knowledge of viral infections of nonhuman primates is meager, especially with respect to diseases of the newborn. In part this stems from the facts that the number of primate research laboratories is few and that the populations being studied are small.

A substantial number of viruses recovered from nonhuman primates, primarily from *Macaca* spp., have been classified as simian viruses [67]. Some have been associated with spontaneous disease; most have not. These latter agents, however, must not be ignored as potential pathogens. Nonhuman primates have also been shown to be experimentally susceptible to many viruses of humans and lower animals. The finding of natural disease by these agents also would not be surprising.

This paper addresses viral diseases with particular emphasis on the part they play in major diseases and death in newborns and infants. Table 48.1 lists some

TABLE 48.1. Viral infections of human infants.

Herpesviruses
H. simplex (types 1 and 2)
H. varicella (chickenpox)
cytomegalovirus
Paramyxoviruses
measles (rubeola)
respiratory syncytial virus
mumps
parainfluenza viruses
Orthomyxoviruses
influenza virus
Togaviruses
Rubella (German measles)
Picornaviruses
enteroviruses
polioviruses
Coxsackieviruses (A and B)
ECHOviruses
Hepatitis viruses
hepatitis A (picorna-, enterovirus)
hepatitis B (proposed classification: Hepadnavirus)
non-A, non-B hepatitis
Poxviruses
smallpox (variola)

of the more important and classical viral infections that can affect the human newborn infant. Many of these viruses, or comparable simian viruses, are known to affect various species of nonhuman primates. Very little is known, however, concerning their role in disease of newborns in captivity, and virtually nothing is known of their importance to nonhuman primates in the wild. To a great extent, this stems from the populations being studied. If we use cytomegalovirus (CMV) as an example, the frequency of infection with this agent in human newborns is significant. At a minimum, 1% of all human infants are infected with CMV: of these, an estimated 10–20% will suffer overt disease [48]. Thus, CMV causes disease in at least one to two of every 1000 infants born, indeed a significant incidence in human beings. When translated to monkeys, 1000 births is a considerable number. CMV infection and possible death of one or two infant monkeys in 1000 births is very likely to be missed and although of the same frequency, would not be considered as significant a problem in monkeys as in humans.

With these opening remarks, let us now review in Table 48.1 those viral infections most likely to affect the newborn and infant nonhuman primate.

Herpesviruses

Herpesviruses are divided into three subfamilies [127]: *Alphaherpesviruses*, which are nonlymphotropic, most often associated with cytolytic disease, and usually latent in ganglia; *Betaherpesviruses* or the cytomegaloviruses (CMV), of which human CMV is the main prototype; and *Gammaherpesviruses*, which are lymphotropic, with some members associated with lymphomas and other neoplasms and latency detected in lymphoid tissues.

Of the over 30 simian herpesviruses [67], only a few have been associated with disease. In addition, human *H. simplex* and, rarely, *H. varicella-zoster* may cause disease in nonhuman primates [66]. All simian herpesviruses, however, should be considered as potential pathogens.

Alphaherpesviruses

Within this group, nine are known to affect various species of nonhuman primates. Two, *H. simplex* (types 1 and 2) and *H. varicella-zoster*, are viruses of human beings. With herpesviruses (as well as other virus groups), we must concern ourselves both with the infection in the natural or reservoir host and in alien species where the infection may behave in a different manner.

HERPESVIRUS SIMPLEX (HERPESVIRUS HOMINIS)

Human beings are the natural and reservoir hosts for *H. simplex*, of which there are two strains: *H. simplex* type 1, most usually associated with oral or conjunctival infection; and *H. simplex* type 2, usually associated with genital infection. Primary infection, occurring principally in youth or adolescence, generally escapes notice but may take the form of an acute gingivostomatitis of varying severity, which heals with no serious side effects. By adulthood, approximately 70% or more of all individuals have become infected as evidenced by the presence of serum-neutralizing antibodies [147]. The virus remains latent in ganglia, presumably for life, and some individuals suffer recurrent lesions precipitated by a variety of stimuli including fever ("fever blister"), colds ("cold sore"), emotional stress, menstruation, and certain foods and drugs. These lesions are physically and emotionally discomforting, but ordinarily heal in 7 to 10 days with no sequelae except in the case of recurrent keratitis, which can lead to blindness. Typically, primary or recurrent lesions are characterized by clusters of vesicles that rupture leaving erosions or ulcers. Hyperesthesia and neuralgia often precede recurrent lesions. Microscopically, ballooning degeneration, necrosis, intercellular edema, multinucleated giant cells, and typical intranuclear inclusion bodies may be seen.

Infection of the newborn between birth and 4 weeks of age may result in a much more serious disease with high mortality. The infection becomes disseminated with necrotizing lesions in most organs and tissues to include lung, liver, adrenal, lymph nodes, spleen, and brain. Even if the infant recovers, there is a high incidence of neurologic sequelae [117].

This pattern of mild, localized disease in adults and serious disseminated disease in neonates is shared by several other herpesviruses such as *H. suis* in swine and *H. canis* in dogs and is probably the case with simian alphaherpesviruses, although not well documented.

Several species of nonhuman primates are known to be naturally or experimentally susceptible to *H. simplex*, with the outcome of the infection dependent upon the species.

In the chimpanzee (*Pan troglodytes* and *P. paniscus*) [100], gibbon (*Hylobates lar*) [31, 140], and probably other apes, the infection is analogous to that in human beings, remaining localized and resolving. Recurrence of both oral and genital lesions has been described in chimpanzees [100]. Although not documented, presumably neonatal apes, like humans, would likely suffer disseminated disease. Fatal *H. simplex* encephalitis has also been reported in adult gibbons [119, 139, 140]. This may suggest a slightly less analogous host:virus relationship to that seen in humans, although *H. simplex* encephalitis can occur in adult human beings and is the most common cause of fatal encephalitis in humans. The experimental disease in cebus monkeys (*Cebus apella*, *C. albifrons*) also appears to be more analogous to *H. simplex* infection as it occurs in humans [36, 112, 135]. Localized lesions follow inoculation of such tissues as conjunctiva, skin, and cervix. Spread to superficial tissues may occur, but the disease is not serious, since spontaneous regression is the rule.

Natural infection in owl monkeys (*Aotus trivirgatus*) [104], tree shrews (*Tupaia glis*) [101], and lemurs (*Lemur catta*) [76] and experimental infection in tamarins (*Saguinus oedipus*) [19, 70] occur as epizootics, with rapid dissemination of lesions and high mortality. Although published observations of the disease in these species has concerned adults, presumably neonates would be equally or more susceptible.

Herpesvirus varicella-zoster

In human beings, varicella or chickenpox is one of the most frequent childhood infections [152]. It is characterized by a papulovesicular rash, histopathologically resembling the lesions of *H. simplex*. The disease is usually mild unless it is acquired as a congenital infection, when it is associated with a significant mortality owing to visceral dissemination. Zoster or shingles represents activation of latent varicella infection and is characterized by localized vesicular dermatitis, following the distribution of the peripheral or cranial nerves. Although associated with severe pain and hyperesthesia, it usually resolves, except in immunocompromised patients where there may be fatal dissemination.

Natural varicella has been recorded only in great apes, and then infrequently. It is described as a self-limiting disease in chimpanzee, gorilla, and orangutan [51, 94, 98, 150].

Simian Varicella

A number of viruses of simian origin share many characteristics with human varicella [44]. Different names have been coined for the various isolates. These include: Liverpool Vervet Monkey Virus, Delta Herpesvirus (Patas Herpesvirus), and Medical Lake Macaque Herpesvirus [11, 20, 40, 97, 123, 133]. All are closely related and can be considered a single agent, simian varicella, with strain variability. In contrast to varicella in humans, simian varicella is a severe generalized infection with high mortality. Susceptible species include African green monkeys or vervets (*Cercopithecus aethiops*), patas monkeys (*Erythrocebus patas*), pig-tailed macaques (*Macaca nemestrina*), Japanese macaque (*M. fuscata*), and cynomolgus monkeys (*M. fascicularis*). The disease described as varicella in chimpanzees by McClure and Keeling [98] may have been due to simian varicella. Simian varicella is characterized by a generalized vesicular dermatitis and disseminated systemic lesions typical of alphaherpesvirus infections [6, 106, 124, 125, 151].

Latent infection in ganglia occurs in animals that recover [62, 148]. Reactivation with generalized rash, in contrast to the dermatome distribution of herpes zoster, has been observed in two African green monkeys exposed to social or environmental stress [141].

Herpesvirus B (Herpesvirus simiae)

Several species of macaques serve as the natural or reservoir hosts for *Herpesvirus B*: *M. mulatta*, *M. fascicularis*, *M. arctoides*, *M. cyclopis*, and *M. radiata* [22]. The bulk of our knowledge of *Herpesvirus B*, however, comes from observations and studies of rhesus monkeys in which the infection is most analogous clinically, epidemiologically, and pathologically to *H. simplex* infection in human beings [72, 73, 129, 130]. Most evidence indicates that sexual activity is the chief means of spreading the disease [154, 155]. Infection may disseminate and cause death, although rarely [99]. This is perhaps more likely to occur in the cynomolgus monkey than in the rhesus. Presumably neonatal macaques would suffer more serious disease, as is the case with human infants; however, it has not been reported. This is surprising in view of the high incidence of *Herpesvirus B* in most nonhuman primate colonies.

The principal importance of *Herpesvirus B* is not its hazards to reservoir hosts, but rather that the virus can produce a fatal generalized disease in humans [128, 129, 131].

Herpesvirus T (Herpesvirus platyrrhinae)

This herpesvirus is carried as a latent viral infection by squirrel monkeys (*Saimiri sciureus*) [23, 59]. Based on evidence derived from circulating serum neutralizing antibodies, cebus monkeys (*C. albifrons*) and spider monkey (*Ateles* spp.) have also been suggested as likely reservoir hosts [54, 55]. Although morbidity is high, clinical disease is seen rarely in the reservoir host and has been documented only in the squirrel monkey. Here, as in *H. simplex* of humans or *Herpesvirus B* of rhesus monkeys, it is characterized by self-limiting vesicles and ulcers, most usually of the oral mucous membranes [79]. Fatal disease has not been seen in adults or neonates.

In marmosets, tamarins, and owl monkeys, *Herpesvirus T* causes an epizootic disease of high morbidity and mortality analogous to *H. simplex* in these species [54, 55, 59]. Latent infection is probably established in the few animals that recover; this could pose a threat if the disease should become reactivated.

Spider Monkey Herpesvirus

This virus was isolated from a young spider monkey with a fatal generalized herpesvirus infection distinguished by oral, labial, and dermal ulcers [56, 57]. Antibodies have been detected in adult spider monkeys, suggesting that they are the natural host for this virus and that latent infection poses a threat to infant monkeys.

SA 8

Originally, this virus was recovered from an African green monkey with myelitis and subsequently from a rectal swab of a healthy baboon [90, 91]. The serological incidence of infection in baboons is high. It has been shown that in neonatal baboons the virus causes a serious disease principally characterized by viral pneumonia with significant mortality [12, 28, 29, 114].

Tree Shrew Herpesvirus

Herpesviruses of uncertain classification have been recovered from healthy and diseased tree shrews [25, 26, 107]. Intravenous, but not intraperitoneal, inoculation has been shown to lead to a classical generalized herpesvirus disease. Recovered animals remain persistently infected. It has also been suggested that the virus may cause lymphoma [25].

Betaherpesviruses–Cytomegaloviruses

Betaherpesviruses or cytomegaloviruses are highly host specific. As such, we do not have to consider the diversity of host:virus interactions found in alphaherpesviruses. Cytomegaloviruses have been isolated from many different species of nonhuman primates including African green monkeys [10], owl monkeys [1], squirrel monkeys [120], marmosets [113], and rhesus monkeys [5].

In human beings, the incidence of cytomegalovirus infection increases with age, reaching 60–90% by adulthood as evidenced by complement-fixing antibodies. The vast majority of acquired infections are asymptomatic. Serious disease may, however, follow congenital infection and lead to abortion, stillbirth, or disseminated disease in neonates, with extensive necrotizing lesions in the liver and/or brain [17, 49, 117, 147]. Acquired infection in adults can lead to cytomegalovirus mononucleosis or cytomegalovirus hepatitis [49, 147].

None of these sequelae has been linked to natural cytomegalovirus infection in monkeys but should be anticipated, because they can be induced experimentally [115]. What has been seen in monkeys is reactivation of cytomegalovirus infection following immunosuppression. Under these circumstances, the virus causes a generalized disease characterized by necrotizing lesions and typical inclusion body-bearing megalocytes in many different organs and tissues. This has been a feature of the simian acquired immunodeficiency syndrome (SAIDS) [78]. It is also the case in human beings with AIDS and is of major importance in patients receiving immunosuppressive therapy.

Occasional inclusion body-bearing megalocytes encountered in the salivary gland or kidney in the course of routine histopathological examination are merely indicative of infection and are not of important pathological significance.

The virus is shed in the urine and/or saliva for protracted periods of time [145] allowing for rapid spread and a high incidence of natural infection.

Gammaherpesviruses

Gamma or lymphotropic herpesviruses have been identified in several species of great apes and Old and New World monkeys. None is recognized as causing any significant disease in its natural hosts in any way analogous to Epstein-Barr virus (EBV) infection in human beings.

Two viruses (*H. saimiri* and *H. ateles*) have been of particular importance in studying the oncogenicity of herpesviruses. *Herpesvirus saimiri* is carried by squirrel monkeys and *H. ateles* by spider monkeys. Both experimentally cause lymphoma in marmosets, tamarins, owl monkeys, and some other species. *Herpesvirus saimiri* also is suspected of having caused natural infection leading to lymphoma in owl monkeys [58, 61]. Since natural transmission among squirrel monkeys is probably by saliva, this may be similar to EBV transmission in humans. EBV has been shown to cause lymphoma in cotton-top marmosets [21, 32, 137].

Paramyxoviruses

Measles (Rubeola)

Measles is a highly contagious exanthematous disease of humans, principally children. In addition to the characteristic exanthematous rash, measles infection may result in giant cell pneumonia, enteritis, and encephalomyelitis. Secondary bronchopneumonia is an important complication. The rash is characterized by vesiculation and necrosis of epithelial cells, associated with intranuclear inclusion bodies and multinucleated cells. A peculiar giant cell known as the Warthin-Finkeldey lymphoid giant cell is one of the most characteristic pathologic features. These cells, often containing up to 100 small deeply basophilic nuclei and only rarely containing inclusion bodies, are found in lymph nodes, spleen, Peyer's patches, appendix, and tonsils. Primary pneumonia is also characterized by giant cells, but here the cells have fewer nuclei that are more leptochromatic and intranuclear, and cytoplasmic eosinophilic inclusion bodies are usually present. Inclusion bodies also may be found throughout the respiratory epithelium. Encephalomyelitis is characterized by demyelination, perivascular lymphocytic cuffing, and gliosis. Measles in humans is usually benign except as a congenital infection or when encephalitis is present. Measles virus also is associated with subacute sclerosing panencephalitis that occurs months to years after measles infection [60, 152].

Measles virus is infectious for several species of nonhuman primates including rhesus, cynomolgus, Taiwan macaque, pig-tailed macaque, baboons, African green monkeys, tamarins, squirrel monkeys, and chimpanzees [47, 60, 65, 69, 89]. Human beings serve as the source of infection. The disease, which mimics that seen in humans, is usually benign and is often unrecognized. Occasionally, however, it is fatal. In years past, many rhesus monkeys succumbed to severe measles giant cell pneumonia following transport from India. Measles pneumonia is encountered on occasion in captive populations of adult and neonatal macaques. Encephalitis may accompany the acute disease [143]. In other species, specifically tamarins, squirrel monkeys, and colobus monkeys, measles epizootics with exceptionally high mortality have been reported [2, 38, 87, 134]. These are largely unexplained, though possibly the result of variant viruses of greater virulence. We have inoculated squirrel monkeys, owl monkeys, and tamarins with measles virus recovered from infant rhesus monkeys that have died with giant cell pneumonia, with no resultant disease (unpublished observation).

Subacute sclerosing panencephalitis has been seen as a natural disease and induced in rhesus monkeys.

Respiratory Syncytial Virus

Respiratory syncytial virus (RSV) is one of the most important causes of childhood upper and lower respiratory disease [52]. Infection can be severe and fatal. The virus was originally isolated from chimpanzees with respiratory disease characterized by rhinorrhea, coughing, and sneezing [108]. Experimentally, upper and lower respiratory disease has been induced in adult owl monkeys [118] and cebus monkeys [121]. Squirrel monkeys and rhesus monkeys can be infected but without illness [9]. Antibodies to this virus have been found in rhesus, squirrel, and cebus monkeys and in gibbons, chimpanzees, orangutans, and gorillas [122]. The infection results in necrosis of nasal, tracheal, and bronchial epithelium with multinucleated giant cells that may contain intranuclear and intracytoplasmic eosinophilic inclusion bodies.

Mumps

Mumps is principally a disease of childhood. Although a generalized infection, the most distinctive feature is swelling of one or both parotid salivary glands. The infection is almost always benign; however, complications that include meningoencephalitis and abortion can occur [152].

Antibody to mumps virus has been identified in many different nonhuman primates to include apes and Old World and New World monkeys [65, 69]. Natural infection has not been associated with clinical disease. Mumps virus has been associated with hydrocephalus in children and has been experimentally induced in rhesus monkeys [86].

Parainfluenza

Several of the parainfluenza viruses are known to cause mild upper respiratory disease in nonhuman primates [27, 45, 95].

Orthomyxoviruses

Influenza

Influenza viruses affect individuals of all ages, but infection is most severe in the very young and the aged. Congenital malformations and death of fetuses and newborn infants have been associated with placental transmission of these viruses [117]. Several species of nonhuman primates can be infected including chimpanzees, gibbons, orangutans, baboons, rhesus monkeys, cynomolgus monkeys, squirrel monkeys, cebus monkeys, and owl monkeys, but the infection usually passes without clinical signs or is very mild [109, 110]. An exception was described by Johnsen et al. [63], who reported four deaths in a group of 36 gibbons experimentally or naturally infected with influenza A. Influenza virus has experimentally induced hydrocephalus in rhesus monkeys [85].

Togaviruses

Rubella (German Measles)

Rubella is a mild exanthematous systemic infection, predominantly of young children. The infection is of principal importance as a cause of abortion, stillbirth, and a variety of congenital anomalies when the virus affects pregnant women in the first trimester [117, 142]. Nonhuman primates of all classes have been found to have antibody to rubella, but spontaneous clinical disease has not been recognized. Attempts to induce disease or demonstrate the teratogenic effect of the virus in monkeys have failed [30, 136]; however, infection can be established.

Picornaviruses

Poliomyelitis (Infantile Paralysis)

Poliomyelitis infection in young children is most often unrecognized or clinically mild, characterized by fever, sore throat, gastrointestinal symptoms, headache, and malaise. Serious disease and paralytic poliomyelitis are more common in older children and adults [18]. Many species of nonhuman primates, including both New and Old World monkeys, are susceptible to the virus; however, spontaneous disease has been limited to the great apes (orangutan, gorilla, and chimpanzee) [64] and, in a single report by Suleman et al., black and white colobus monkeys [144]. Characteristic lesions are necrosis of anterior horn cells, perivascular cuffing with lymphocytes, and gliosis [60, 117].

Other enteroviruses such as the several Coxsackieviruses and ECHOviruses can cause a variety of syndromes in neonates and children, including myocarditis, hepatitis, gastroenteritis, and meningitis [117]. Neither these nor their simian origin counterparts have been recognized as important causes of natural disease in nonhuman primates. Simians are, however, susceptible, and there are isolated reports of natural disease. For example, a fatal illness has been reported in an infant chimpanzee, apparently caused by Coxsackievirus B5 [75]. Lesions included hepatic necrosis, myocarditis, and pneumonia. Experimentally, Coxsackievirus B4 has been associated with hepatitis and nephritis in squirrel monkeys [15, 16]. Enteroviruses SV2 and SV6 have been associated with diarrhea, but a causal relationship has not been demonstrated [56], and SV16 has been associated with an ill-defined neurologic disease in rhesus and African green monkeys [71].

Encephalomyocarditis virus, an enterovirus carried by rodents, can cause disease in many different species of animals. In nonhuman primates, the virus produces myocarditis characterized by an intense infiltration of polymorphonuclear and mononuclear cells. The disease has been recognized in gibbons, chimpanzees, baboons, owl monkeys, and squirrel monkeys [39, 50, 126].

Hepatitis Viruses

Although human infants can contract the various hepatitis viruses with varying consequences, none is an important cause of disease in nonhuman primates. In fact, the lack of suitable susceptible animal hosts has considerably hampered progress in understanding hepatitis. Through serologic and viral inoculation studies, natural or experimental hepatitis A infection has been recognized in chimpanzees, several species of macaques, African green monkeys, patas monkeys, baboons, tamarins, owl monkeys, and bushbabies; hepatitis B infection in gorillas, chimpanzees, gibbons, rhesus monkeys, and woolly monkeys; and non-A, non-B hepatitis in chimpanzees and *Saguinus mystax* and *S. labiatus* [3, 7, 35, 81–83, 93, 96, 138, 141, 153]. For the most part, natural and experimental infection is mild. Animals acquire infection from humans and in turn can transmit the disease to other primates including humans [77].

Pathologically, all three forms are similar to the disease in humans though less severe. There are degeneration and necrosis of hepatocytes (not massive necrosis), multinucleated hepatocytes, hypertrophy and hyperplasia of Kupffer cells, and lymphocytic and histiocytic inflammation, particularly in the portal tracts: A few neutrophils and eosinophils may accompany the cellular infiltrates [74, 116, 146]. Lesions compatible with human hepatitis are encountered in nonhuman primates with some frequency, but rarely are attempts made to establish an etiologic relationship. A recent example led to the suggestion that there may be hepatitis viruses of simian origin [88].

Poxviruses

In human beings, smallpox has essentially disappeared and is primarily of historical interest. Intrauterine and neonatal infection with smallpox or vaccinia has, however, resulted in a generalized necrotizing disease, intrauterine death, and abortion [80, 117]. Spontaneous smallpox has not been documented in nonhuman primates; however, monkeys can be infected with vaccinia and several simian poxviruses [42].

Monkeypox virus is immunologically related to smallpox virus [41]. Natural infection has been reported in macaques (*M. mulatta*, *M. fascicularis*), owl monkeys (*A. trivirgatus*), gorillas (*G. gorilla*), gibbons (*H. lar*), squirrel monkeys (*S. sciureus*), marmosets (*Callithrix jacchus*), and chimpanzees (*P. satyrus*). The infection is characterized by a generalized rash and pocks on mucous membranes in which there is epithelial necrosis leading to intraepidermal vesicles and pustules. Cytoplasmic inclusion bodies are present. Generalized fatal infection with necrotizing lesions in many visceral organs can occur [132, 149]. The disease has not been seen in the wild, but only in zoos and laboratories in Europe and North America. Human beings are susceptible, developing a disease resembling smallpox. All cases of human monkeypox, as of 1980, had been confined to the African continent. The source of infection has not been clearly established, although one example was associated with close contact with a chimpanzee [13, 42, 92, 111].

Or Te Ca Pox (an acronym coined from Oregon, Texas, and California where the disease has been described) or *Tanapox* (after the Tana River in Kenya) is a poxvirus disease of macaques and humans. Other species appear not to be susceptible. The virus is immunologically related to Yaba virus but not to smallpox or vaccinia [46, 102]. The lesions consist of multiple (but not diffuse), large, crusted elevations of the skin, especially on the face, arms, and perineum. Vesiculation is not seen. Microscopically, there is marked thickening of the epidermis and adnexae and hyperplasia of the epidermis somewhat similar to contagious ovine ecthyma. Individual cells become smaller, bear cytoplasmic inclusion bodies, and eventually become necrotic.

Yaba pox virus can infect humans, *M. fascicularis*, *M. arctoides*, *M. mulatta*, *M. nemestrina*, *Cercopithecus aethiops*, *Erythrocebus patas*, *Papio papio*, and probably other species [4, 8, 14, 43]. Infection is rare. In contrast to other poxes, the lesions are tumorlike masses in the subcutis, composed of proliferating, pleomorphic mononuclear cells (histiocytes) that contain irregularly shaped eosinophilic inclusion bodies. This resembles lumpy skin disease of cattle. The lesions eventually regress.

Prevention

In nonhuman primates, these infectious agents that may affect the newborn are contracted from one of three sources: human beings, another species of nonhuman primate, or latent infection in the same species. Vaccines, if available, can afford protection in the first two examples and should be employed. Measles (rubeola) and poliomyelitis vaccines have been shown to be effective and without side effects [84]. Vaccines have also been developed for protection against *H. simplex* and *Herpesvirus T* in owl monkeys and marmosets [24].* For other infections contracted from people, care should be exercised to prevent contact with individuals known to be suffering from an illness or in contact with it in their households. When the possibility exists of contracting disease from another species of nonhuman primate, strict separation of species should be enforced. In cases of neonatal disease caused by a virus carried by the same species (such as *H. simplex* in human infants), preventive measures are not easily instituted. Vaccines, under such circumstances, have not yet been developed. Recent progress in developing a CMV vaccine for human beings, however, is encouraging [48, 53].

Acknowledgment. Supported in part by NIH grant RR00168 from the Division of Research Resources, Animal Resources Program Branch.

References

1. Ablashi DV, Chopra HC, Armstrong GR (1972) A cytomegalovirus isolated from an owl monkey. Lab Anim Sci 22:190–195.
2. Albrecht P, Lorenz D, Klutch MJ, Vickers JH, Ennis FA (1980) Fatal measles infection in marmosets pathogenesis and prophylaxis. Infect Immun 27:969–978.

*Available from the New England Regional Primate Research Center, 1 Pine Hill Drive, Southborough, MA 01772 (617/481-0400).

3. Alter HJ, Purcell RH, Gerin JL, London WT, Kaplan PM, McAuliffe VJ, Wagner J-A, Holland PV (1977) Transmission of hepatitis B to chimpanzees by hepatitis B surface antigen-positive saliva and semen. Infect Immun 16:928–933.
4. Ambrus JL, Strandstrom HV, Kawinski W (1969) "Spontaneous" occurrence of Yaba tumor in a monkey colony. Experientia 25:64–65.
5. Asher DM, Gibbs CJ Jr, Lang DJ, Gajdusek DC (1974) Persistent shedding of cytomegalovirus in the urine of healthy rhesus monkeys. Proc Soc Exp Biol Med 145:794–801.
6. Ayres JP (1971) Studies of the Delta herpesvirus isolated from the patas monkey (*Erythrocebus patas*). Lab Anim Sci 21:685–695.
7. Barker LF, Maynard JE, Purcell RH, Hoofnagle JH, Berquist KR, London WT (1975) Viral hepatitis, type B, in experimental animals. Amer J Med Sci 270:189–195.
8. Bearcroft WGC, Jamieson MF (1958) An outbreak of subcutaneous tumours in rhesus monkeys. Nature (London) 182:195–196.
9. Belshe RB, Richardson LS, London WT, Sly DL, Lorfeld JH, Camargo E, Prevar DA, Chanock RM (1977) Experimental respiratory syncytial virus infection of four species of primates. J Med Virol 1:157–162.
10. Black PH, Hartley JW, Rowe WP (1963) Isolation of a cytomegalovirus from African green monkeys. Proc Soc Exp Biol Med 112:601–605.
11. Blakely GA, Lourie B, Morton WG, Evans HH, Kaufmann AF (1973) A varicella-like disease in macaque monkeys. J Infect Dis 127:617–625.
12. Brack M, Eichberg JW, Heberling RL, Kalter SS (1985) Experimental *Herpes neonatalis* in SA 8-infected baboons (*Papio cynocephalus*). Lab Anim 19:125–131.
13. Breman JG, Kalisa-Ruti, Steniowski MV, Zanotto E, Gromyko AI, Arita I (1980) Human monkeypox, 1970–79. Bull WHO 58:165–182.
14. Bruestle ME, Golden JG, Hall A III, Banknieder AR (1981) Naturally occurring Yaba tumor in a baboon (*Papio papio*). Lab Anim Sci 31:292–294.
15. Burch GE, Chu K-C, Soike KF (1982) Coxsackievirus B_4 nephritis in the squirrel monkey. Br J Exp Pathol 63:680–685.
16. Burch GE, Chu K-C, Soike KF (1983) Coxsackievirus B_4 hepatitis in monkeys. J Med 14:37–45.
17. Catalano LW Jr, Sever JL (1971) The role of viruses as causes of congenital defects. Annu Rev Microbiol 25:255–282.
18. Cherry JD (1976) Enteroviruses. In: Remington JS, Klein JO (eds) Infectious diseases of the fetus and newborn infant. WB Saunders, Philadelphia/London/Toronto, pp 366–413.
19. Cho CT, Liu C, Voth DW, Feng KK (1973) Effects of iodoxuridine on *Herpesvirus hominis* encephalitis and disseminated infections in marmosets. J Infect Dis 128:718–723.
20. Clarkson MJ, Thorpe E, McCarthy K (1967) A virus disease of captive vervet monkeys (*Cercopithecus aethiops*) caused by a new herpesvirus. Arch Ges Virusforsch 22:219–234.
21. Cleary ML, Epstein MA, Finerty S, Dorfman RF, Bornkamm GW, Kirkwood JK, Morgan AJ, Sklar J (1985) Individual tumors of multifocal EB virus-induced malignant lymphomas in tamarins arise from different B-cell clones. Science 228:722–724.
22. Daniel MD, Melendez LV, Hunt RD, Trum BF (1972) The herpesvirus group. In: Fiennes RNT-W (ed) Pathology of simian primates, part II. Infectious and parasitic diseases. S Karger, Basel/Munchen/Paris/London/New York/Sydney, pp 592–611.
23. Daniel MD, Karpas A, Melendez LV, King NW, Hunt RD (1967) Isolation of Herpes-T virus from a spontaneous disease in squirrel monkeys (*Saimiri sciureus*). Arch Ges Virusforsch 22:324–331.

24. Daniel MD, Barahona H, Melendez LV, Hunt RD, Sehgal P, Marshall B, Ingalls J, Forbes M (1978) Prevention of fatal herpes infections in owl and marmoset monkeys by vaccination. In: Chivers DJ, Ford EHR (eds) Recent advances in primatology, vol 4. Medicine. Academic, London/New York/San Francisco, pp 67–69.
25. Darai G, Koch H-G (1984) Tree shrew herpesvirus: pathogenicity and latency. Curr Top Vet Med Anim Sci 27:91–102.
26. Darai G, Zoller L, Matz B, Flugel RM, Moller P, Hofmann W, Gelderblom H, Delius H (1982) Tupaia herpesviruses: characterization and biological properties. Microbiologica 5:285–298.
27. Dick EC, Dick CR (1974) Natural and experimental infections of nonhuman primates with respiratory viruses. Lab Anim Sci 24:177–181.
28. Eichberg J, Kalter SS, Heberling RL, Brack M (1973) Experimental herpesvirus infection of baboons (*Papio cynocephalus*) and African green monkeys (*Cercopithecus aethiops*) and recovery of virus by tissue explants. Arch Ges Virusforsch 43:304–314.
29. Eichberg JW, McCullough B, Kalter SS, Thor DE, Rodriguez AR (1976) Clinical, virological, and pathological features of herpesvirus SA 8 infection in conventional and gnotobiotic infant baboons (*Papio cynocephalus*). Arch Virol 50:255–270.
30. Elizan TS, Fabiyi A, Sever JL (1969) Study of rubella virus as a teratogen in experimental animals: a short review. J Mt Sinai Hosp (NY) 36:108–112.
31. Emmons RW, Lennette EH (1970) Natural *Herpesvirus hominis* infection of a gibbon (*Hylobates lar*). Arch Ges Virusforsch 31:215–218.
32. Epstein MA, Achong BG (eds) (1979) The Epstein-Barr virus. Springer, Berlin/Heidelberg/New York.
33. Epstein MA, Achong BG, Barr YM (1964) Virus particles in cultured lymphoblasts from Burkitt's lymphoma. Lancet 1:702–703.
34. Espana C (1974) Viral epizootics in captive nonhuman primates. Lab Anim Sci 24:167–176.
35. Feinstone SM, Alter HJ, Dienes HP, Shimizu Y, Popper H, Blackmore D, Sly D, London WT, Purcell RH (1981) Non-A, non-B hepatitis in chimpanzees and marmosets. J Infect Dis 144:588–598.
36. Felsburg PJ, Heberling RL, Kalter SS (1973) Experimental corneal infection of the cebus monkey with *Herpesvirus hominis* type 1 and 2. Arch Ges Virusforsch 40:350–358.
37. Fiennes RNT-W (ed) (1972) Pathology of simian primates, part II. Infectious and parasitic diseases. S Karger, Berlin/Munchen/Paris/London/New York/Sydney.
38. Fraser CEO, Chalifoux L, Sehgal P, Hunt RD, King NW (1978) A paramyxovirus causing fatal gastroenterocolitis in marmoset monkeys. Prim Med 10:261–270.
39. Gainer JH (1967) Encephalomyocarditis virus infections in Florida, 1960–1966. J Amer Vet Med Assoc 151:421–425.
40. Gard EA, London WT (1983) Clinical history and viral characterization of Delta herpesvirus infection in a patas monkey colony. Monogr Primatol 2:211–212.
41. Gispen R, Verlinde JD, Zwart P (1967) Histopathological and virological studies on monkeypox. Arch Ges Virusforsch 21:205–216.
42. Gough AW, Barsoum NJ, Gracon SI, Mitchell L, Sturgess JM (1982) Poxvirus infection in a colony of common marmosets (*Callithrix jacchus*). Lab Anim Sci 32:87–90.
43. Grace JT Jr, Mirand EA (1963) Human susceptibility to a simian tumor virus. Ann NY Acad Sci 108:1123–1128.
44. Gray WL, Oakes JE (1984) Simian varicella virus DNA shares homology with human varicella-zoster virus DNA. Virology 136:241–246.
45. Grizzard MB, London WT, Sly DL, Murphy BR, James WD, Parnell WP, Chanock RM (1978) Experimental production of respiratory tract disease in cebus monkeys after intratracheal or intranasal infection with influenza A/Victoria/3/75 or A/New Jersey/7 virus. Infect Immun 21:201–205.

46. Hall AS, McNulty WP Jr (1967) A contagious pox disease in monkeys. J Amer Vet Med Assoc 151:833–838.
47. Hall WC, Kovatch RM, Herman PH, Fox JG (1971) Pathology of measles in rhesus monkeys. Vet Pathol 8:307–319.
48. Hamilton JD (ed) (1982) Monographs in virology, vol 12. Cytomegalovirus and immunity. S Karger, Basel/Munchen/Paris/London/New York/Sydney.
49. Hanshaw JB (1976) Cetomegalovirus. In: Remington JS, Klein JO (eds) Infectious diseases of the fetus and newborn infant. WB Saunders, Philadelphia/London/Toronto, pp 107–155.
50. Helwig FC, Schmidt ECH (1945) A filter-passing agent producing interstitial myocarditis in anthropoid apes and small animals. Science 102:31–33.
51. Heuschele WP (1960) Varicella (chicken pox) in three young anthropoid apes. J Amer Vet Med Assoc 136:256–257.
52. Hildreth SW, Hall CB, Menegus MA (1983) Respiratory syncytial virus. Clin Microbiol Newslett 5:93–95.
53. Ho M (1982) Cytomegalovirus. Biology and infection. Plenum, New York/London, pp 205–213.
54. Holmes AW, Caldwell RG, Dedmon RE, Deinhardt F (1964) Isolation and characterization of a new herpes virus. J Immunol 92:602–610.
55. Holmes AW, Devine JA, Nowakowski E, Deinhardt F (1966) The epidemiology of a herpes virus infection of New World monkeys. J Immunol 96:668–671.
56. Hull RN (1968) The simian viruses. Virol Monogr 2:1–66.
57. Hull RN, Dwyer AC, Holmes AW, Nowakowski E, Deinhardt F, Lennette EH, Emmons RW (1972) Recovery and characterization of a new simian herpesvirus from a fatally infected spider monkey. JNCI 49:225–231.
58. Hunt RD, Jones TC (1973) Nonhuman primates in viral oncology. Adv Vet Sci Comp Med 17:361–394.
59. Hunt RD, Melendez LV (1966) Spontaneous Herpes-T infection in the owl monkey (*Aotus trivirgatus*). Pathol Vet 3:1–26.
60. Hunt RD, Carlton WW, King NW Jr (1978) Viral diseases. In: Benirschke K, Garner FM, Jones TC (eds) Pathology of laboratory animals, vol II. Springer, New York/Heidelberg/Berlin, pp 1285–1365.
61. Hunt RD, Garcia FG, Barahona HH, King NW, Fraser CEO, Melendez LV (1973) Spontaneous *Herpesvirus saimiri* lymphoma in an owl monkey. J Infect Dis 127:723–725.
62. Iltis JP, Aarons MC, Castellano GA, Madden DL, Sever JL, Curfman BL, London WT (1982) Simian varicella virus (Delta herpesvirus) infection of patas monkeys leading to pneumonia and encephalitis. Proc Soc Exp Biol Med 169:266–279.
63. Johnsen DO, Wooding WL, Tanticharoenyos P, Karnjanaprakorn C (1971) An epizootic of A_2/Hong Kong/68 influenza in gibbons. J Infect Dis 123:365–370.
64. Jortner BS, Percy DH (1978) The nervous system. Poliomyelitis in nonhuman primates. In: Benirschke K, Garner FM, Jones TC (eds) Pathology of laboratory animals, vol I. Springer, New York/Heidelberg/Berlin, pp 344–347.
65. Kalter SS (1972) Serologic surveys. In: Fiennes RNT-W (ed) Pathology of simian primates. Part II. Infectious and parasitic diseases. S Karger, Basel/Munchen/Paris/London/New York/Sydney, pp 469–496.
66. Kalter SS (1982) Herpesviruses of nonhuman primates: their significance. Microbiologica 5:149–159.
67. Kalter SS (1983) Primate viruses—their significance. Monogr Primatol 2:67–89.
68. Kalter SS (ed) (1983) Monographs in primatology, vol 2. Viral and immunological diseases in nonhuman primates. Alan R Liss, New York.
69. Kalter SS (1985) Viral diseases of infant great apes. Monogr Primatol 5:57–106.
70. Kalter SS, Felsburg PJ, Heberling RL, Nahmias AJ, Brack M (1972) Experimental *Herpesvirus hominis* type 2 infection in nonhuman primates. Proc Soc Exp Biol Med 136:964–968.

71. Kaufmann AF, Gary GW, Broderson JR, Perl DP, Quist KD, Kissling RE (1973) Simian virus 16 associated with an episode of obscure neurologic disease. Lab Anim Sci 23:812–818.
72. Keeble SA (1960) B virus infection in monkeys. Ann NY Acad Sci 85:960–969.
73. Keeble SA, Christofinis GJ, Wood W (1958) Natural virus B infection in rhesus monkeys. J Pathol Bacteriol 76:189–199.
74. Keenan CM, Lemon SM, LeDuc JW, McNamee GA, Binn LN (1984) Pathology of hepatitis A infection in the owl monkey (*Aotus trivirgatus*). Amer J Pathol 115:1–8.
75. Kelly ME, Soike K, Ahmed K, Iatropoulos MJ (1978) Coxsackievirus in an infant chimpanzee. J Med Primatol 7:119–121.
76. Kemp GE, Losos GL, Causey OR, Emmons RW, Golding RR (1972) Isolation of *Herpesvirus hominis* from lemurs: a naturally occurring epizootic at a zoological garden in Nigeria. Afr J Med Sci 3:177–185.
77. Kessler H, Tsiquaye KN, Smith H, Jones DM, Zuckerman AJ (1982) Hepatitis A and B at the London Zoo. J Infect Dis 4:63–67.
78. King NW, Hunt RD, Letvin NL (1983) Histopathologic changes in macaques with an acquired immunodeficiency syndrome (AIDS). Amer J Pathol 113:382–388.
79. King NW, Hunt RD, Daniel MD, Melendez LV (1967) Overt Herpes-T infection in squirrel monkeys (*Saimiri sciureus*). Lab Anim Care 17:413–423.
80. Klein JO, Remington JS, Marcy SM (1976) An introduction to infections of the fetus and newborn infant. In: Remington JS, Klein JO (eds) Infectious diseases of the fetus and newborn infant. WB Saunders, Philadelphia/London/Toronto, pp 1–32.
81. LeDuc JW, Lemon SM, Keenan CM, Graham RR, Marchwicki RH, Binn LN (1983) Experimental infection of the New World owl monkey (*Aotus trivirgatus*) with hepatitis A virus. Infect Immun 40:766–772.
82. Lemon SM, LeDuc JW, Binn LN, Escajadillo A, Ishak KG (1982) Transmission of hepatitis A virus among recently captured Panamanian owl monkeys. J Med Virol 10:25–36.
83. Linnemann CC Jr, Kramer LW, Askey PA (1984) Familial clustering of hepatitis B infections in gorillas. Amer J Epidemiol 119:424–430.
84. Loomis MR (1985) Immunoprophylaxis in infant great apes. Monogr Primatol 5:107–112.
85. London WT, Fuccillo DA, Sever JL, Kent SG (1975) Influenza virus as a teratogen in rhesus monkeys. Nature 255:483–484.
86. London WT, Kent SG, Palmer AE, Fucillo DA, Houff SA, Saini N, Sever JL (1979) Induction of congenital hydrocephalus with mumps virus in rhesus monkeys. J Infect Dis 139:324–328.
87. Lorenz D, Albrecht P (1980) Susceptibility of tamarins (*Saguinus*) to measles virus. Lab Anim Sci 30:661–665.
88. Lucke VM, Bennett AM (1982) An outbreak of hepatitis in marmosets in a zoological collection. Lab Anim 16:73–77.
89. MacArthur JA, Mann PG, Oreffo V, Scott GBD (1979) Measles in monkeys: an epidemiological study. J Hyg (Cambridge) 83:207–212.
90. Malherbe H, Harwin R (1958) Neutropic virus in African monkeys. Lancet 2:530.
91. Malherbe H, Strickland-Cholmley M (1969) Simian herpesvirus SA8 from a baboon. Lancet 2:1427.
92. Manshande JP, Rutenda K-W (1983) Human monkeypox transmitted by a chimpanzee. Lancet 1:1110–1111.
93. Mao JS, Go YY, Huang HY, Yu PH, Huang BZ, Ding ZS, Chen NL, Yu JH, Xie RY (1981) Susceptibility of monkeys to human hepatitis A virus. J Infect Dis 144:55–60.
94. Marennikova SS, Maltseva NN, Shelukhine EM, Shenkman LS, Korneeva VI (1973) A generalized herpetic infection simulating smallpox in a gorilla. Intervirology 2:280–286.

95. Martin DP, Kaye HS (1983) Epizootic of parainfluenza-3 virus infection in gibbons. J Amer Vet Med Assoc 183:1185–1187.
96. Mathiesen LR, Mo/ller AM, Purcell RH, London WT, Feinstone SM (1980) Hepatitis A virus in the liver and intestine of marmosets after oral inoculation. Infect Immun 28:45–48.
97. McCarthy K, Thorpe E, Laursen AC, Heymann CS, Beale AJ (1968) Exanthematous disease in patas monkeys caused by a herpes virus. Lancet 2:856–857.
98. McClure HM, Keeling ME (1971) Viral diseases noted in the Yerkes Primate Center colony. Lab Anim Sci 21:1002–1010.
99. McClure HM, Olberding B, Strozier LM (1973) Disseminated herpesvirus infection in a rhesus monkey (*Macaca mulatta*). J Med Primatol 2:190–194.
100. McClure HM, Swenson RB, Kalter SS, Lester TL (1980) Natural genital *Herpesvirus hominis* infection in chimpanzees (*Pan troglodytes* and *Pan paniscus*). Lab Anim Sci 30:895–901.
101. McClure HM, Keeling ME, Olberding B, Hunt RD, Melendez LV (1972) Natural *Herpesvirus hominis* infection of tree shrews (*Tupaia glis*). Lab Anim Sci 22:517–521.
102. McNulty WP Jr, Lobitz WC Jr, Hu F, Maruffo CA, Hall AS (1968) A pox disease in monkeys transmitted to man. Clinical and histological features. Arch Dermatol 97:286–293.
103. Melendez LV, Daniel MD, Hunt RD, Garcia FG (1968) An apparently new herpesvirus from primary kidney cultures of the squirrel monkey (*Saimiri sciureus*). Lab Anim Care 18:374–381.
104. Melendez LV, Espana C, Hunt RD, Daniel MD, Garcia PG (1969) Natural *Herpes simplex* infection in the owl monkey (*Aotus trivirgatus*). Lab Anim Care 19:38–45.
105. Melendez LV, Hunt RD, Daniel MD, Garcia FG, Fraser CEO (1969) *Herpesvirus saimiri*. II. Experimentally induced malignant lymphoma in primates. Lab Anim Care 19:378–386.
106. Migaki G, Seibold HR, Wolf RH, Garner FM (1971) Pathologic conditions in the patas monkey. J Amer Vet Med Assoc 159:549–556.
107. Mirkovic R, Voss WR, Benyesh-Melnick M (1970) Characterization of a new herpesvirus for tree shrews. In: Proceedings of the 10th International Congress of Microbiology, Mexico City, pp 181–189.
108. Morris JA, Blount RE Jr, Savage RE (1956) Recovery of cytopathogenic agent from chimpanzees from coryza. Proc Soc Exp Biol Med 92:544–549.
109. Murphy BR, Sly DL, Hosier NT, London WT, Chanock RM (1980) Evaluation of three strains of influenza A virus in humans and in owl, cebus, and squirrel monkeys. Infect Immun 28:688–691.
110. Murphy BR, Hinshaw VS, Sly DL, London WT, Hosier NT, Wood FT, Webster RG, Chanock RM (1982) Virulence of influenza A viruses for squirrel monkeys. Infect Immun 37:1119–1126.
111. Mutombo MW, Arita I, Jezek Z (1983) Human monkeypox transmitted by a chimpanzee in a tropical rain-forest area of Zaire. Lancet 1:735–737.
112. Nahmias AJ, London WT, Catalano LW, Fuccillo DA, Sever JL, Graham C (1971) Genital *Herpesvirus hominis* type 2 infection: an experimental model in cebus monkeys. Science 171:297–298.
113. Nigida SM, Falk LA, Wolfe LG, Deinhardt F (1979) Isolation of a cytomegalovirus from salivary glands of white-lipped marmosets (*Saguinus fuscicollis*). Lab Anim Sci 29:53–60.
114. Ochoa R, Henk WG, Confer AW, Pirie GS (1982) Herpesviral pneumonia and septicemia in two infant gelada baboons (*Theropithecus gelada*). J Med Primatol 11:52–58.
115. Ordy JM, Rangan SRS, Wolf RH, Knight C, Dunlap WP (1981) Congenital cytomegalovirus effects on postnatal neurological development of squirrel monkey (*Saimiri sciureus*) offspring. Exp Neurol 74:728–747.

116. Popper H, Dienstag JL, Feinstone SM, Alter HJ, Purcell RH (1980) The pathology of viral hepatitis in chimpanzees. Virchows Arch (Pathol Anat) 387:91–106.
117. Potter EL, Craig JM (1975) Pathology of the fetus and the infant. Year Book Medical Publishers, Chicago.
118. Prince GA, Suffin SC, Prevar DA, Camargo E, Sly DL, London WT, Chanock RM (1979) Respiratory syncytial virus infection in owl monkeys: viral shedding, immunological response, and associated illness caused by wild-type virus and two temperature-sensitive mutants. Infect Immun 26:1009–1013.
119. Ramsay E, Stair EL, Castro AE, Marks MI (1982) Fatal *Herpesvirus hominis* encephalitis in a white-handed gibbon. J Amer Vet Med Assoc 181:1429–1430.
120. Rangan SRS, Chaiban J (1980) Isolation and characterization of a cytomegalovirus from the salivary gland of a squirrel monkey (*Saimiri sciureus*). Lab Anim Sci 30:532–540.
121. Richardson LS, Belshe RB, Sly DL, London WT, Prevar DA, Camargo E, Chanock RM (1978) Experimental respiratory syncytial virus pneumonia in cebus monkeys. J Med Virol 2:45–59.
122. Richardson-Wyatt LS, Belshe RB, London WT, Sly DL, Camargo E, Chanock RM (1981) Respiratory syncytial virus antibodies in nonhuman primates and domestic animals. Lab Anim Sci 31:413–415.
123. Riopelle AJ, Ayres JP, Seibold HR, Wolf RH (1971) Studies of primate diseases at the Delta Center. In: Goldsmith EI, Moor-Jankowski J (eds) Medical primatology 1970: Proceedings of the 2nd Conference on Experimental Medicine and Surgery in Primates, New York 1969. S Karger, Basel/Munchen/Paris/London/New York/Sydney, pp 826–838.
124. Roberts ED, Baskin GB, Soike K, Gibson SV (1984) Pathologic changes of experimental simian varicella (Delta herpesvirus) infection in African green monkeys (*Cercopithecus aethiops*). Amer J Vet Res 45:523–530.
125. Roberts ED, Baskin GB, Soike K, Meiners N (1984) Transmission and scanning electron microscopy of experimental pulmonary simian varicella (Delta herpesvirus) infection in African green monkeys (*Cercopithecus aethiops*). J Comp Pathol 94:323–328.
126. Roca-Garcia M, Sanmartin-Barberi C (1957) The isolation of encephalomyocarditis virus from *Aotus* monkeys. Amer J Trop Med Hyg 6:840–952.
127. Roizman B, Carmichael LE, Deinhardt F, de-The G, Nahmias AJ, Plowright W, Rapp F, Sheldrick P, Takahashi M, Wolf K (1981) Herpesviridae: definition, provisional nomenclature, and taxonomy. Intervirology 16:201–217.
128. Sabin AB (1934) Studies on the B virus. I. The immunological identity of a virus isolated from a human case of ascending myelitis associated with visceral necrosis. Br J Exp Pathol 15:248–269.
129. Sabin AB (1934) Studies on the B virus. III. The experimental disease in *Macaca rhesus* monkeys. Br J Exp Pathol 15:321–334.
130. Sabin AB, Hurst FW (1935) Studies on the B virus. IV. Histopathology of the experimental disease in rhesus monkeys and rabbits. Br J Exp Pathol 16:133–148.
131. Sabin AB, Wright AM (1934) Acute ascending myelitis following a monkey bite, with the isolation of a virus capable of reproducing the disease. J Exp Med 59:115–136.
132. Sauer RM, Prier JE, Buchanan RS, Creamer AA, Fegley HC (1960) Studies on a pox disease of monkeys. I. Pathology. Amer J Vet Res 21:377–380.
133. Schmidt NJ, Arvin AM, Martin DP, Gard EA (1983) Serological investigation of an outbreak of simian varicella in *Erythrocebus patas* monkeys. J Clin Microbiol 18:901–904.
134. Scott GBD, Keymer IF (1975) The pathology of measles in Abyssinian colobus monkeys (*Colobus guereza*): a description of an outbreak. J Pathol 117:229–233.
135. Sever JL (1973) Herpesvirus and cervical cancer studies in experimental animals. Cancer Res 33:1509–1510.

136. Sever JL, Meier GW, Windle WF, Schiff GM, Monif GR, Fabiyi A (1966) Experimental rubella in pregnant rhesus monkeys. J Infect Dis 116:21–26.
137. Shope T, Dechairo D, Miller G (1973) Malignant lymphoma in cottontop marmosets after inoculation with Epstein-Barr virus. Proc Natl Acad Sci USA 70:2487–2491.
138. Smith MS, Swanepoel PJ, Bootsma M (1980) Hepatitis A in non-human primates in nature. Lancet 2:1241–1242.
139. Smith PC, Yuill TM, Buchanan RD (1958) Natural and experimental infection of gibbons with *Herpesvirus hominis*. In: Annu Prog Rep SEATO Med Res Lab and SEATO Clin Res Cent. SEATO Laboratory, Bangkok, pp 258–261.
140. Smith PC, Yuill TM, Buchanan RD, Stanton JS, Chaicumpa V (1969) The gibbon (*Hylobates lar*): a new primate host for *Herpesvirus hominis*. I. A natural epizootic in a laboratory colony. J Infect Dis 120:292–297.
141. Soike KF, Rangan SRS, Gerone PJ (1984) Viral disease models in primates. Adv Vet Sci Comp Med 28:151–199.
142. South MA, Sever JL (1983) Congenital rubella syndrome—1982. Isr J Med Sci 19:921–924.
143. Steele MD, Giddens WR Jr, Valerio M, Sumi SM, Stetzer ER (1982) Spontaneous paramyxoviral encephalitis in nonhuman primates (*Macaca mulatta* and *M. nemestrina*). Vet Pathol 19:132–139.
144. Suleman MA, Johnson BJ, Tarara R, Sayer PD, Ochieng DM, Muli JM, Mbete E, Tukei PM, Ndirangu D, Kago S, Else JG (1984) An outbreak of poliomyelitis caused by poliovirus type I in captive black and white colobus monkeys (*Colobus abyssinicus kikuyuensis*) in Kenya. Trans Roy Soc Trop Med Hyg 78:665–669.
145. Swack NS, Hsiung GD (1982) Natural and experimental simian cytomegalovirus infections at a primate center. J Med Primatol 11:169–177.
146. Thung SM, Gerber MA, Purcell RH, London WT, Mihalik KB, Popper H (1981) Animal model of human disease. Chimpanzee carriers of hepatitis B virus. Amer J Pathol 105:328–332.
147. von Lichtenberg F (1984) Viral, chlamydial, rickettsial, and bacterial diseases. In: Robbins SL, Cotran RS, Kumar V (eds) Pathologic basis of disease, 3rd ed. WB Saunders, Philadelphia/London/Toronto/Mexico City/Rio De Janeiro/Sydney/Tokyo.
148. Wenner HA, Abel D, Barrick S, Seshumurty P (1977) Clinical and pathogenetic studies of Medical Lake macaque virus infections in cynomolgus monkeys (simian varicella). J Infect Dis 135:611–622.
152. Young NA (1976) Chickenpox, measles and mumps. In: Remington JS, Klein JO (eds) Infectious diseases of the fetus and newborn infant. WB Saunders, Philadelphia/London/Toronto, pp 521–586.
153. Zuckerman AJ (1980) Viral hepatitis in nonhuman primates. J Med Microbiol 13:Px (abstract 43).
154. Zwartouw HT, Boulter EA (1984) Excretion of B virus in monkeys and evidence of genital infection. Lab Anim 18:65–70.
155. Zwartouw HT, MacArthur JA, Boulter EA, Seamer JH, Marston JH, Chamove AS (1984) Transmission of B virus between monkeys in relation to breeding colonies. Lab Anim 18:125–130.

49
Acute Myocarditis in Golden Monkeys

CHANG YI

Introduction

The golden monkey is a rare and precious animal found only in China. It is valuable for both exhibition purposes and scientific study, but its chances of contracting infectious diseases are high.

Clinical Findings

The five golden monkeys in this study were recently caught from Qin Mountain of Shaanxi Province. They had been kept in quarantine for 6 months. Four animals ranged from 2 to 3 years and the fifth was a mature female. From 24–28 July 1979, four monkeys became ill and finally died. They showed similar symptoms: sudden onset of illness, anorexia, loose stools, weakness, dyspnea, and restlessness. Death occurred within 1 to 3 days. During the course of the illness, one female developed nuchal ridigity and another had hind limb paresis. On 26 August female 5 suddenly showed signs of anorexia, loose stools, poor activity, heart rate of 200/min, and temperature of 40°C. After 2 days she had black, tarry stools and marked dyspnea. The animal died on 30 August.

Necropsy Findings

All animals had cyanosis of the extremities, dehydration, pulmonary hemorrhage and edema, hemorrhage of thymus and pancreas, and ascites of a yellow, fibrinous nature. There was cardiomegaly with petechial hemorrhages. Females 2, 4, and 5 had pericardial effusion with a thick, yellowish, fibrinous liquid. Females 3 and 4 had, in addition, small amounts of light yellow fibrinous pleural fluid.

Histological Findings

Histologically, all animal had petechial hemorrhages, pulmonary edema, reticulocytosis of liver and spleen, and there was evidence of hepatitis. Serious pathological changes were found in the heart with interstitial edema scattered

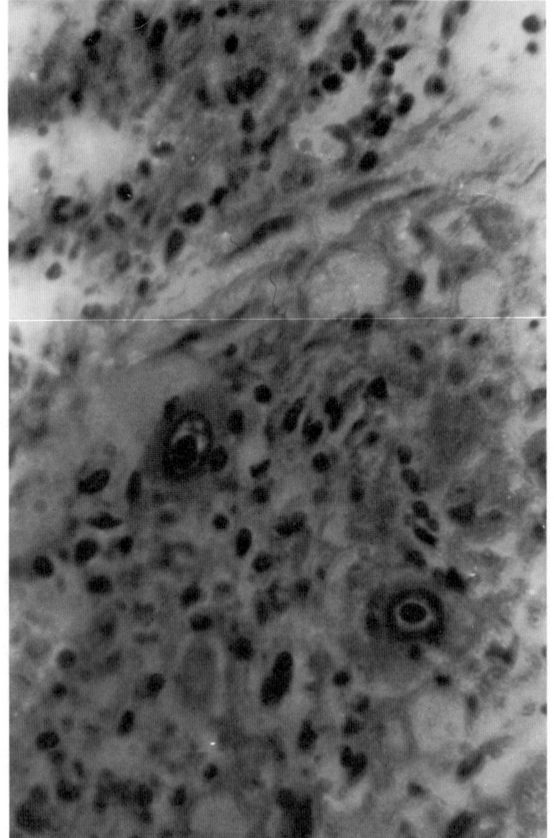

FIGURE 49.1. Myocardium of golden monkey with mononuclear cell infiltration, necrotic fibers, and inclusion bodies. H & E, ≈ 250×.

throughout with small amounts of mononuclear infiltration. There were areas of myocardial necrosis with many viral inclusions inside the cardiac cell nuclei (Fig. 49.1). Female 5 had slight interstitial myocarditis.

Discussion

There are three major types of common intestinal viral diseases: Coxsackievirus, poliomyelitis, and ECHOvirus infection. Among these, ECHOvirus does not affect primates, and poliomyelitis seldom causes myocarditis. It should be noted that perhaps because of the extensive use of poliomyelitis vaccine, the incidence of this disease in humans has dropped considerably. In contrast, myocarditis caused by Coxsackieviruses has greatly increased. Since this is a world-wide epidemic disease that affects both humans and animals, particular attention must be paid to it.

Myocarditis caused by Coxsackievirus type B is known to cause acute myocarditis. This disease generally occurs in the warmer season, between May and October. The rate of contracting the disease is fairly high among infants and young children. The younger the child, the more serious the pathological changes. This disease is also frequently associated with pericarditis. It can be contracted from patients' excretions and from flies. The clinical symptoms are sudden illness, anorexia, vomiting, spastic coughing, cyanosis and dyspnea, increased heart rate and signs of heart failure, cardiomegaly, rales in both lungs, and an enlarged liver. Those fortunate enough to survive recover gradually; however, one-third have central nervous system involvement.

Our group of golden monkeys possessed the above-mentioned characteristics, but we did not have the facilities to isolate the virus.

Since there is no effective cure for acute myocarditis, prevention is essential. The animal house and exercise playground must be kept clean and hygienic. Those who work with this species must wear quarantine clothes. It is also important to eradicate insect pests such as flies during summer and autumn. If there is a B type Coxsackievirus epidemic (e.g., upper respiratory infection, chest pain, aseptic meningitis), attention must be paid to the quarantine facilities, and injection of 4–6 ml placental globulin is given in order to increase resistance to the disease temporarily. If necessary, sick monkeys are isolated for approximately 2 weeks.

If myocarditis is confirmed, the sick animal must be kept away from outside disturbances and be provided with sufficient time to rest. Attention must also be paid to treating an animal suffering from heart failure with digitalis to be given immediately and ensure that sufficient oxygen is supplied. The use of intravenous solutions must be carefully monitored for amount and speed in order to prevent cardiac failure.

If a seizure occurs, a suitable amount of sedative might be used, but it must be carefully monitored to avoid respiratory depression. In view of the effusion and bleeding tendencies, one may consider using supplementary medicines such as corticosteroids and vitamins C and K.

Acknowledgment. Reprinted from Chinese Zoo Annual Magazine, November 1982, and translated by Ms. Mabel Lam, with permission.

50
Rearing and Intensive Care of Neonatal and Infant Nonhuman Primates

JOHN H. ANDERSON

Introduction

Hand rearing and intensive care of neonatal and infant monkeys are routine procedures in large, captive populations. Animals are orphaned as a result of neglect or loss of a critical parent and will usually require some form of hand rearing. Infectious agents, environmental stress, trauma, and inadequate nutrition are common problems that require intensive care. Maximizing the outcome of hand rearing and intensive care helps to reduce neonate and infant mortality, which is the leading cause of loss in primate populations.

The first section of this paper reviews the basic principles and techniques of hand rearing. The second section is a basic overview of intensive care providing some practical guidelines and techniques used for some of the commonly seen medical problems of neonates and infants.

A recent monograph, Clinical Management of Infant Great Apes (Graham and Bowen, 1985), provides an excellent source of information for those species. This manuscript references species other than the apes.

Hand Rearing

Hand rearing techniques have been developed and described for a number of species (Graham and Bowen, 1985; Breznock et al., 1979; Ausman et al., 1970; Cicmanec et al., 1979; Dronzek et al., in press; Hinkle and Session, 1972; Meier and Willis, 1984; Moore and Cummins, 1979; Ogden, 1979; Ruppenthal, 1979; Tardif et al., 1985; Valerio et al., 1970; Kaplan and Russell, 1973), most of which have been captive raised for biomedical research. The basic principles for each species are similar and provide methods that can be extrapolated to the threatened and endangered that may require captive propagation. Additionally, it is not unreasonable to anticipate care of a wildborn animal at a field research station where knowledge of practical techniques can enhance outcome.

Basic factors that need to be considered for maximizing successful hand rearing include housing, environmental temperature and humidity, adequate diet and feeding regimen, a proper surrogate, and socialization.

Housing and Environment

Warmth is one of the most important considerations for the neonate especially during the first 1–2 weeks of life. Surrounding air temperature should be maintained between 80° and 90°F depending on age and species. The younger the animal, the more heat it requires. This is especially true for the premature infant. In general, the beginning temperature for prematures and newborns should be 90°F, increased slightly if the animal cannot maintain body temperature, and gradually being reduced as the animal begins to thermoregulate.

A suitable environment can easily be provided by any number of methods. The basic requirement is a regulated heat source, a thermometer to monitor temperature, and a cage or container that will safely hold the animal and can be effectively cleaned. An enclosed cage or box placed on a heating pad can work very well, but precautions should be taken to ensure that the surface is not too hot. One large breeding facility of macaques uses this technique with two plastic guinea pig cages hinged together and provided with several ventilation ports (J. Wong, 1985, personal communication). Squirrel monkeys have been successfully reared by placing a small, circular open-bottomed wire mesh cage directly on a diaper-covered heating pad (C. Abee, 1985, personal communication). In an emergency, radiating heat lamps can be used, but it is important to thermostatically control these or at least recognize their limitations in terms of providing too much heat if not properly positioned and monitored. Measuring temperature can be done with any type of thermometer that is compatible with the cage. Poultry incubator thermometers or stick-on thermometers used for tropical fish tanks work very well, and the latter are not subject to damage by an active infant.

Many primate facilities use commercial human isolettes (Armstrong Baby Incubator, Model 500, Gordon Armstrong Co., Cleveland, OH). These are excellent and provide both heat and humidity, but they can be difficult to disinfect. An excellent commercially available system manufactured for small animals is a water heater warmer (Thermocare, P.O. Box YY, Incline, NV). This provides safe heat with humidity, is relatively easy to disinfect, and is reasonably priced. Heated surrogates have been used effectively (Dronzek et al., in press; Kaplan and Russell, 1973), but these are not adequate for premature or newborn infants or for infants that are hypothermic unless the surrounding air temperature is also warmed.

Elevated environmental humidity appears to be important especially in the more tropical species. A relative humidity range between 50% and 80% is generally used, the higher being used for marmosets and tamarins (Cicmanec et al., 1979; Dronzek et al., in press; Ogden, 1979). Like temperature, the higher humidity is most important for the premature and newborn, being gradually reduced as the animal matures.

Surrogates

A surrogate to cling to is of significant importance to the environment of the hand-reared neonate, and many types have been described (Dronzek et al., in press; Ogden, 1979; Ruppenthal, 1979; Kaplan and Russell, 1973; Herzog and Hopf, 1985). In general, a surrogate should allow an infant to cling in a position that mimics the position and shape of the real mother. Keeping this principle in

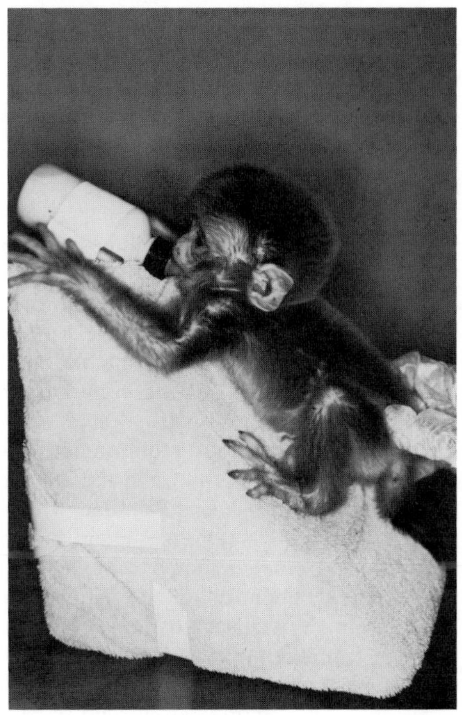

FIGURE 50.1. A towel-covered plexiglass frame surrogate can be used to condition a neonatal rhesus macaque to self-feed.

mind, innovative surrogates can be easily constructed for most species. The simplest design can be a wire mesh tube wrapped in soft terry cloth or other similar fabric. For macaques at the California Primate Research Center (CPRC), we use a towel-covered simple plexiglass frame that also has a bottle holder for self-feeding (Fig. 50.1). More complex surrogates contain internal heat sources (Dronzek et al., in press; Kaplan and Russell, 1973). One facility reports use of adult tanned skins in addition to the cloth covering (Dronzek et al., in press). Stationary surrogates appear to work well for macaques, *Aotus* (Kelly, 1985, personal communication), and *Saimiri* (Hennessy, 1985), if surrogate rearing is not prolonged. Movable or animate surrogates suspended from rubber bands or on mechanical rockers appear to enhance the outcome of *Saguinus* and *Callithrix* (Dronzek et al., in press; Ogden, 1979).

The complexity and outcome of prolonged surrogate rearing is beyond the scope of this paper. The surrogate appears to be critical in maximizing survival in the early hand-reared environment prior to socialization techniques.

Nutrition

An adequate nutritional regimen includes providing a proper diet using an appropriate feeding technique and feeding schedule. Published regimens are available for only a few species, but extrapolation of principles and techniques to other species will frequently lead to a successful outcome even though our knowledge of primate nutrition, in general, is still rudimentary when compared to many of our domestic species.

Nursing Formula

During the first 24 hours, the general practice is to feed the hand-reared neonate 5–10% dextrose in water. After this, several formulations have been used (Table 50.1). Ideally, the nursing formula should approximate milk for that species.

Accurate reported milk analysis is available for only a few of the many species of primates (Ausman et al., 1985). Formula selection, especially for the more exotic species, has been based largely on practical experience.

The use of commercially available human infant formula (Table 50.1) has worked well for many of the common laboratory species (*Saimiri*, *Macaca*, *Papio*, *Aotus*). Species such as *Saguinus* and *Callithrix* have demonstrated a higher caloric and protein requirement (Dronzek et al., in press; Ogden, 1979) during the neonatal period, and this has been remedied by adding sustagen to SMA or Similac or by using Primilac, which has been specifically formulated for primates (Table 50.1) (Dronzek et al., in press; Ogden, 1979). Weaning neonates to an adult-type diet as soon as possible is recommended in order to compensate for possible deficiencies in the formula (Ausman et al., 1985; Oftedal, 1980).

Palatability as well as nutritional content appears to be important, and several different formulations may have to be tried before one is accepted. As we learn more about each species, other formula requirements will become known. This is currently an area of much-needed research (Oftedal, 1980).

Feeding Techniques

Feeding and handling techniques vary with the species. As with designing a surrogate, a feeding and handling technique should try to approximate the natural situation in terms of nipple size, animal position when nursing, burping, and in some species (i.e., *Saguinus*) tactile stimulation such as rubbing the perineal area

Table 50.1. Milk formulas used for infant nonhuman primates.

Formula	kcal/oz	Genus
SMA	20	*Saimiri, Macaca*
Similac	20	*Saimiri, Papio, Macaca, Aotus, Cebus*
Enfamil	20	*Macaca, Papio, Hylobates*
Primilac	24.8	*Saguinus, Macaca*
SMA (1 Tblsp) Sustagen (1 tsp) 60 ml water	25	*Saguinus, Callithrix*
Whole cow's milk 120 ml Rice cereal 15 ml	19.9	*Varecia variegata, V. rubra*

[1] SMA: Wyeth Laboratories, Philadelphia, PA.
Similac: Ross Laboratories, Columbus, OH.
Enfamil: Mead Johnson and Co., Evansville, IN.
Primilac: Bio Serv, Inc., Frenchtown, NJ.
Gerber Strained Rice Cereal: Gerber Products Co., Fremont, MI.

to stimulate urination and defecation (Dronzek et al., in press). Pet-Nip nurser nipples (Borden, Inc., Pet-Ag Division, Hampshire, IL) are used for many species from *Aotus* to macaque size. The size of the hole in the nipple is important and may need to be enlarged, especially in the smaller species. As a general rule, if the neonate appears to be working too hard to obtain formula, enlarge the hole but not so much as to cause excessive delivery. Human-size nipples can be used for larger species such as *Papio*. A homemade nipple for *Callithrix* and *Saguinus* has been described (Dronzek et al., in press; Ogden, 1979) and appears to benefit by providing the ability to suckle more effectively.

Gavage feeding may be necessary for the premature or weakened neonate until a good sucking response can be elicited. Infant nasogastric feeding tubes can be easily passed into the stomach (Ruppenthal, 1979; Valerio et al., 1970) to administer the diet as well as to assess residual stomach content, which is of value in determining gastric emptying and if the proper volume is being administered (Ruppenthal, 1979). Weak infants as well as infants of smaller species will eat or lap diet dripped from a syringe or volumetric pipette (Ziegler et al., 1981).

The position of the infant during hand feeding should approximate its natural feeding position. The ideal situation is to mount the formula source on the surrogate (Fig. 50.1) and hand hold the animal on the surrogate to feed. This additionally enhances the initiation of self-feeding.

Feeding schedules and volumes are outlined in Table 50.2 and are generalized summaries from the references and personal experience. In general, during the first 1 to 2 weeks of life, neonates should be fed every 2–3 hours up to ten times daily depending on species. The caloric intake should be 150–250 kcal/kg per day during the first week and increasing for several weeks thereafter again depending on species. Volume per feeding is important. Excessive volume predisposes to gastric distention, vomiting, and aspiration. This is especially critical in weak neonates being gavage fed.

Conditioning infants to self-feed significantly reduces nursing care and probably allows for more normal feeding patterns to evolve. Conditioning involves placing them on the surrogate and directing them to the nipple (Fig. 50.1). Over a period of time, they will learn to feed themselves on their own schedule. Many infants are self-feeding by 2 weeks of age or earlier. Normal macaque neonates at our facility are self-feeding within 3–4 days of birth. Prematures may take up to several weeks (Ruppenthal, 1979). Remembering species-specific positions is important, and several good techniques have been described (Dronzek et al., in press; Kaplan and Russell, 1973). Formula should be changed several times over a 24-hour period since it rapidly becomes rancid in the heated neonatal environment (Ruppenthal, 1979).

Weighing infants daily and recording volume intake of formula are important to assess adequacy of feeding protocol and infant well being. After an initial 1–2 day period of adjustment, infants should gain weight and increase formula volume intake on a daily basis. If there is not a progressive increase in both, the dietary program and the health of the infant need to be evaluated.

Introduction to solid food and to an eventual adult-type diet should begin relatively early in a hand-rearing program, especially when formulas may only approximate the nutritional needs of an infant. In our macaque nursery, small pieces of banana or apple are offered to infants as early as the first week. While not much of the food item is consumed at this time, the animals at least begin to

TABLE 50.2. Birth weight and partial hand feeding guidelines for several species of nonhuman primates.

	Birth wt (gm)	Vol. per feeding 1st week (ml)	Interval between feeding (hours)	No. feedings per day 1st week	Begin self-feeding	Begin solid food	Foods used for diet change
Saguinus Callithrix	40–50	0.5–1.9	2	8–10	2–4 weeks	10–30 days	Yogurt, baby cereal, fruits, egg mash
Aotus	70–100	0.9–3.0	2–3	6–10	4–8 days	±14 days	Baby cereal, fruit, soaked monkey chow
Saimiri	100–115	1.0–3.0	2	8–10	4–20 days	2–4 weeks	Baby cereal, fruit, soaked monkey chow
M. mulatta	400–500	20–30	2–3	6–9	3–10 days	5–14 days	Baby cereal, fruits, soaked monkey chow
Papio sp.	800–950	40–60	2–3	5–8	8–14 days	10–20 days	Baby cereal, fruits, soaked monkey chow
Varecia variegata rubra	75–90	2.5–6.0	2	10		3–4 weeks	Fruit, vegetables, soaked monkey chow

mouth or lick at something other than formula. Some food items used are listed in Table 50.2. Fruits such as apple or banana, baby cereals, and yogurt are the most frequently used starter foods. Stone fruits may cause diarrhea and should be avoided. High protein monkey chow softened to a mush with formula or water should be instituted soon after the animals are eating other items. Feeding excessive amounts of fruit during this period is a common tendency that is important to recognize. Some species may require specific diet items during weaning transition and rapid growth. The weaning diet of leaf-eating species is a recognized researchable problem area (M. Loomis, 1985, personal communication). Again, monitoring weight gain is a valuable indicator of dietary adequacy.

A multivitamin supplementation added to the diet of nursery-reared animals is common practice (Ogden, 1979; Ruppenthal, 1979; Valerio et al., 1970). Pediatric liquid multivitamin preparations are frequently used. For New World animals it is important that vitamin D_3 be present for proper development (Cicmanec et al., 1979; Ogden, 1979; Oftedal, 1980; Abee, 1985).

Socialization

In order for the hand-reared neonate to develop into a reproductively successful animal, socialization with other members of its species at an early age is almost essential. The detrimental consequences of prolonged isolation and surrogate rearing for certain species are well documented (Hennessy, 1985; Ruppenthal and Sackett, 1979). Without going into great detail, it is worth mentioning several socialization techniques that have been successful. Nursery-reared rhesus macaque females at our facility develop into successful mothers after being placed into large peer groups at 3–6 months of age (unpublished observation). We have a fairly large nursery-reared population and always pair or group the animals by 3–4 weeks of age before moving them to larger peer groups. Recently, two separate breeding facilities of *Saguinus oedipus* reported successful introduction of surrogate-reared infants to family groups (Dronzek et al., in press; Tardif et al., 1985). The technique involved gradual introduction of 3–4-week-old infants and their surrogates to family groups. It will be important to identify successful socialization techniques for any of the endangered species that we elect to maintain in captive populations.

Additional Rearing Techniques

Avoiding the necessity to hand rear a neonate is obviously preferable to nursery rearing. Orphaned rhesus monkey neonates in our colony are frequently fostered to nursing females whose infants were lost because of stillbirths, medical problems, or management weaning (Brown et al., 1981). Marmoset triplets that traditionally have necessitated hand rearing of one infant are now managed by collaborative rearing (Hearn and Burden, 1979) or supplemental feeding (Ziegler et al., 1981) with minimal disruption of normal family socialization. Hand supplementation of an infant Titi monkey (*Callicebus moloch*) whose mother was lost from illness but whose father continued to care for it has been successful at our colony and eliminated the need to raise it in the nursery. This technique is also successful with *Aotus* (S. Kelly, 1985, personal communication).

Intensive Care

The principles of intensive care for nonhuman primates are similar to those used in humans. A wealth of information available in current pediatric texts (Graef and Cone, 1980; Orlowski, 1980) can be extrapolated and applied to many species. Veterinarians and other persons responsible for the care of these animals should have these references available. Additionally, the expertise of a human neonatologist or pediatrician can be extremely valuable. Many of these professionals enjoy providing assistance where it is needed.

The following section is a selected overview of techniques used to manage some of the commonly seen medical emergencies of neonates and infants. Included are resuscitation, guidelines and techniques for fluid therapy and patient management, and a brief outline of several commonly seen clinical conditions: prerenal failure, hypoglycemia, hypokalemia, and hypernatremia.

Resuscitation

Neonates most commonly require resuscitation at cesarean section or dystocia delivery. A newly delivered depressed newborn should be quickly rubbed dry and have its airways cleared of any secretions using cotton-tipped swabs or suction syringes. For very small infants, an IV catheter attached to a syringe can be used for aspiration. If respiration does not begin quickly and if the animal has poor muscle tone but a heart beat, assisted ventilation should be instituted. Slight anteroventral depression of the thoracic cavity may be all that is required, and this may be all that is posssible in the smaller species, i.e., neonates weighing less than 200–300 g. Doxapram hydrochloride (Dopram V injectable, A. H. Robins, Co., Richmond, VA) is used by some veterinarians, but in our experience its value is questionable and may cause bradycardia.

Infants that remain apneic require endotracheal intubation and positive pressure ventilation. Endotracheal intubation can be difficult, and practicing on anesthetized neonates is recommended. An instrument that significantly increases success is a Welch Allyn neonatal laryngoscope blade #635 (Welch Allyn, Inc., Skaneateles Falls, NY). This blade allows manipulation of the laryngeal anatomy while providing a tunnel to visualize the laryngeal opening in order to pass a 2.0 french endotracheal tube.

In an emergency, when a good laryngoscope is not available, cotton-tipped swabs can be substituted. Another technique is to use an 18–20 ga needle on a 1 or 3 ml syringe. Bend the tip of the needle making a 1–2 mm hook and use this as a laryngoscope. The hook can be placed between the tongue and the epiglottis and is used to pull the tongue forward in order to visualize the larnyx.

Positive pressure ventilation with a neonatal ambu bag (Penlon Infant Resuscitation Bag, Levingston Medical Products Co., San Jose, CA) is administered after intubation. Oxygen-enriched air aids success. Ventilation rates should be approximately 20–30/minute and care should be taken not to overexpand the lungs. Spontaneous respiration should occur within 5–30 minutes depending on the degree of depression and what anesthetic agents were used on the mother in the case of C-section. Cardiac rate should increase dramatically, but if it is significantly depressed or absent, cardiac resuscitation procedures should be followed with the use of sodium bicarbonate, isuprel, epinephrine, calcium gluconate, and atropine as needed.

Fluid Therapy and Patient Management

Fluid therapy is probably the most common treatment modality and is used for correction of dehydration and/or volume depletion, electrolyte imbalance, malnutrition, and blood loss. Fluid therapy is extremely important, generally not difficult, and can be extremely rewarding. The most common illnesses requiring fluid therapy are gastrointestinal diseases causing diarrhea and/or vomiting, prolonged inanition from maternal neglect, stress-related anorexia, and trauma. Fluid administration includes giving balanced fluid and electrolyte solutions, plasma, and blood. Going one step further is total parenteral hyperelimination (TPH), which involves the infusion of highly concentrated solutions of carbohydrates, lipids, amino acids, electrolytes, and other necessary nutrients. An excellent overview of TPH in neonatal exotic animals is available (Meier and Reichard, 1982) and will not be covered in this paper.

Patient evaluation, monitoring, and mangement are important during fluid therapy. It is essential to estimate needs in terms of maintenance, deficit, and loss and to calculate and regulate how much is being given. Fluid maintenance needs in the human newborn are roughly 60 ml/kg per day during the first 24 hours and change to 120 ml/kg per day. Deficit is estimated by weight loss and degree of dehydration and/or volume depletion. Loss is estimated by monitoring stool and urine volume as well as taking into consideration insensible water loss, which is influenced by maturity, respiratory rate, body and environmental temperature, relative humidity, and activity levels (Meier and Reichard, 1982). Loss in the infant with severe secretory diarrhea may double or even triple its maintenance needs. The following is a simplified example of estimating the volume requirements of a 500-g infant who is 10% dehydrated and has a profuse watery diarrhea. Maintenance needs are 120 ml/kg per day $\times$ 0.5 kg = 60 ml. Deficit is 10% $\times$ 500 g = 50 ml. Loss can be 60 to 120 ml/day. The requirements for 24 hours would be 60 + 50 + (60 to 120) = 170 to 230 ml.

Monitoring urine production is important in evaluation of renal perfusion and function. The normal human infant produces approximately 2 ml/kg per hour. Less than 0.5 ml/kg per hour suggests renal failure or volume depletion. A urinalysis is easy to perform with urine test strips (Bili-Labstix SG, Ames Division, Miles Laboratories, Inc., Elkhart, IN) and provides valuable information pertaining to physiological processes occurring during intensive care. Parameters that are measured include: occult blood, bilirubin, ketones, glucose, protein, pH, and specific gravity.

Blood glucose, BUN, plasma proteins, hematocrit, electrolytes, osmolarity, and blood gas are also parameters to monitor in order to provide adequate therapy. Blood glucose is critical and is analyzed with the use of whole blood strips (Dextrostix, Ames Division, Miles Lab, Inc., Elkhart, IN, or Chemstrip bG, Bio-Dynamics, Indianapolis, IN). Hypoglycemia is common in infants and neonates.

BUN is an indicator of renal function and perfusion and when elevated also increases serum osmolarity. Azostix whole blood sampling strips (Ames Division, Miles Laboratories, Inc., Elkhart, IN) can be used to determine if BUN is in the normal range or moderately elevated in the 20–60 mg/dl range.

Plasma proteins or total solids are a good indicator of the state of hydration and/or nutritional status. They are easily measured with a Goldberg refractometer (American Optical Co., Keene, NH). If they are low during fluid therapy (below $\pm$ 4.0 mg/dl), consider slowing the fluid rate or giving a transfusion. Hematocrit, like plasma proteins, is an indicator of dehydration, dilution, and also blood loss. Consider giving whole blood transfusions when the hematocrit is in the 10–15% range, especially if the plasma proteins are low or high.

The electrolytes Na, K, and Cl need to be maintained in normal balance. Hypokalemia, hyperkalemia, hypernatremia, hyponatremia, and hyperosmolarity are seen regularly in large colonies.

Techniques of Fluid Administration

Commonly used routes of fluid administration include intravenous, intraperitoneal, subcutaneous, and oral. Choosing a route is influenced by the patient needs, condition, anatomical size, and behavioral requirements. A subcutaneous or intraperitoneal route may be chosen over a theoretically better intravenous route in order not to disrupt maternal care. Several routes of administration may be used over a 24-hour period. IV therapy may be used for 8–12 hours during the day followed by SC administration for maintenance overnight. The choice of fluids is dependent on the condition being treated and the route of administration. Tables 50.3 and 50.4 are typical lists of fluids that should be available.

Intravenous Fluid Administration

Intravenous administration is frequently the preferred route and is extremely important for the critically ill infant that is severely volume depleted or dehyrated as well as those that require total parenteral hyperalimentation. The advantages of intravenous therapy are that deficits and imbalances can be corrected rapidly if necessary and that monitoring of patient status and regulation of administra-

Table 50.3. Common parenteral fluids and their content.

Fluid	Tonicity	Calories per liter	Ions (mEq) Na	K	Ca	Cl	Bicarbonate (mEq)
Lactated Ringer's	Isotonic	9	131	4	3	110	28
Dextrose 5% in half-strength saline	Hypertonic	170	77	0	0	77	0
Dextrose 2.5% in half-strength saline	Isotonic	85	77	0	0	77	0
Dextrose 2.5% in half-strength lactated Ringer's	Isotonic	89	65	2	1	54	0
Normal saline	Isotonic	0	155	0	0	155	0
Ringer's solution	Isotonic	0	147	4	4.5	155	0
Dextrose 50%	Hypertonic	1700	0	0	0	0	0

TABLE 50.4. Oral replacement of fluids and their content.

Solution	Calories per ounce	Carbohydrate source	Carbohydrate (g/100 ml)	mEq/liter		
				Na^+	K^+	Cl^-
Lytren (Mead Johnson)	9	Glucose	7.6	30.0	25.0	25.0
Pedialyte (Ross)	6	Glucose	5.0	30.0	20.0	30.0
5% glucose	6	Glucose	5.0			
10% glucose	12	Glucose	10.0			

tion are better controlled. Adequate restraint and sometimes Ketamine immobilization are frequently required for intravenous therapy. Fortunately, the anatomy of the monkey predisposes it to fairly comfortable restraint with the use of tape and a padded structure to tape it to (Fig. 50.2). We usually use a cross-shaped board padded with towels and gauze sponges or a rolled up sponge rubber pad covered with a towel. Infants can be restrained in this manner for up to 24 hours if necessary although we seldom restrain them for over 8 hours.

The commonly used IV sites are the saphenous veins over the posterior midline of each leg below the knee and the cephalic veins over the anteromedial surface of the arm. The femoral veins on the medial aspect of each thigh can also be used, especially if a cutdown technique is needed in a severely hypovolemic animal. Additional vascular sites commonly used in the premature human neonate are umbilical arteries. These are easily catheterized up to several hours postdelivery in macaques and are valuable for intensive care of the premature.

Numerous types of catheters are now available for intravenous therapy. For catheterization of peripheral vessels in infant macaques, we prefer the 24-ga teflon catheters (Jelco[tm], Criticon, Tampa, FL). With experience, these catheters are as easy to insert as butterfly infusion sets and avoid the problem of lacerating the vessels. An additional advantage is that an injection cap (BD Injection cap, Becton-Dickinson, Rutherford, NJ) with a heparin (100 μ/ml) lock can be put on the catheter and it can be taped in place overnight for intravenous therapy the following day or for bolus intravenous therapy. Longer catheters that can be advanced to larger central veins for total parenteral hyperalimentation or longer term maintenance are the 24-ga L-Cath[tm] (Luther Medical Products, Inc., Santa Ana, CA), which is inserted with a breakaway cannula, and the 22-ga Hydrocath (Cardiosearch, Tampa, FL), which is inserted over a wire. Both of these catheters are made of the apparently less thrombogenic polyurethane and also have wider lumens. The Hydrocath additionally has a hydrophilic hydromer coating that makes it slippery and nonabrasive, another quality that appears to inhibit thrombus formation.

The rate of fluid administration and fluid volume must be carefully calculated and monitored during intravenous therapy. Remembering blood volume of the patient, roughly 60–70 ml/kg, can help keep fluid volume administration in perspective. Hypovolemic shocky patients can have their blood volume equaled during the first hour of therapy in order to restore circulation and replace deficit. The administration rate should then be decreased and adjusted according to the calculated requirements. Pediatric microdrip IV sets with reservoirs or IV infusion pumps should be used in order to prevent overhydration.

FIGURE 50.2. The use of a rolled up, towel-covered pad, tape, and safety pin can comfortably restrain an infant primate for intravenous therapy. A 24-ga indwelling teflon catheter is secured in the saphenous vein, and a rectal probe monitors body temperature.

Intraperitoneal and Subcutaneous Administration

Intraperitoneal (IP) and subcutaneous (SC) fluid administration are frequently used techniques for rehydration and maintenance of nonhuman primate (Dronzek et al., in press; Ruppenthal, 1979; Abee, 1985). Practicality frequently dictates the use of these methods, but for the very small species they may be the only technique available to rehydrate them when parenteral therapy is indicated. The solutions used should always be isotonic and warmed to body temperature. The rate of fluid absorption is faster with IP administration, and this method is used for hypovolemic infants that are too small for IV therapy. The absorption rate for SC fluids is slower, and this technique is occasionally used for fluid therapy or maintenance overnight. The volume of fluid administered can be 2–10% of the infant's body weight given as one bolus or several doses over time. Several sites should be used for large bolus SC injections.

Oral Fluid Administration

Oral fluid therapy is very effective and frequently all that is required to rehydrate and maintain all but the most severely ill and shocky animals. Recent trends in human pediatrics advocate increased use of oral rehydration in place of intravenous for treatment of acute gastroenteritis (Tolia and Dubois, 1985). Several balanced electrolyte solutions are commercially available (Lytren Mead Johnson and Co., Evansville, IN; Pedialyte, Ross Laboratories, Columbus, OH) (Table 50.4). A balanced solution that can be prepared, if commercial solutions aren't available, consists of 4 tablespoons sugar, 1/2 teaspoon NaCl, 1/2 teaspoon baking soda, and 1/4 teaspoon KCl mixed with one liter of water. Most infants will drink these solutions when offered from a syringe or nurser. Gavage administration is effective, and calculated, multiple feedings should be given similar to nursing schedules.

Prerenal Failure

Acute prerenal failure from dehydration and hypovolemia is common. The objective of therapy is to restore urinary flow by reestablishing an effective circulating blood volume with fluid therapy. Persistent oliguria or anuria requires additional therapy with osmotic diuretics (mannitol, glucose) and/or furosimide (Inglefinger and Auner, 1980).

Hypoglycemia

Hypoglycemia is very common in the neonate and infant (Anderson et al., 1981), and it is important to recognize and correct it rapidly. The symptoms may include lethargy, depression, limpness, tremors, seizures, cyanosis, and anorexia. It can accompany many clinical problems including stress-related anorexia, maternal neglect, water deprivation, hypothermia, enteric disease, and trauma. Neonates, especially those that are unthrifty and/or premature, have very low glucose reserves and can succumb to hypoglycemia within a few hours of fasting. Maintaining adequate blood glucose levels is a critical factor in neonatal survival. Monitoring with whole blood sampling strips and supplementing with dextrose solutions are standard, easy-to-perform procedures.

Blood glucose values below 40 mg/100 ml should be cause for concern. Symptomatic animals are treated by administering glucose, 0.5–1.0 g/kg of body weight by IV push. This can be done by giving ±2.4 ml of 25% D/W per kilogram at a rate of 0.5 ml/minute. This is followed by glucose infusion rates of 5–15 mg/kg per minute (Lovejoy and Graef, 1980). Blood glucose should be periodically monitored using the sampling strips and should be kept between 45 and 125 mg/100 ml by adjusting the glucose infusion rate. Urine glucose measured with test strips is another useful guide. If urine glucose exceeds plus one, blood glucose is probably excessively elevated.

Oral supplementation is effective and frequently more practical for maintenance. This is done by estimating fluid requirements (100–300 ml/kg per day) and using 5–50% dextrose solutions as required. As a prevention of weaning hypoglycemia, a sugar-sweetened drink should be provided until the animal has adjusted (Anderson et al., 1981).

Hypernatremia

Hypernatremia (serum Na greater than 155–160 mEq/L) with dehydration and significant hyperosmolarity is a condition that may be seen in maternally neglected animals suffering with prolonged inanition, or in severe enteric disease with diarrhea, anorexia, and vomiting leading to loss of water in excess of solute and failure of water intake. Clinical symptoms are CNS in origin and range from lethargy to coma. The condition is serious since not only can it result in permanent brain damage and death if untreated, but rapid correction with water may result in cerebral edema, seizures, and death (Perkin and Levin, 1980).

Fluid administration must be carefully monitored so that serum Na does not drop too rapidly. A period of 48 hours should be allowed to reach normal Na values ($\pm$145 mEq/L). Rehydration should begin with normal saline for the first 4–8 hours and then decreasing to hypotonic half-strength saline. Glucose should be added as needed. If the patient shows initial neurologic improvement and then deteriorates, cerebral edema should be suspected even if serum Na concentrations and osmolarity remain greater than normal. Administration of hypotonic solutions should be stopped and osmotherapy with hypertonic saline or mannitol begun until signs of cerebral edema improve.

Hypokalemia

Potassium is the principal cation of the intracellular fluid. The serum K is an indirect measure of total body stores. The normal range in macaque infants and neonates is 2.5–4.5 mEq/L. Neonatal and infant diarrhea, especially that which is of a large secretory volume and/or of prolonged duration, frequently results in depletion. Serum values less than 2.0 mEq/L are cause for concern. Pronounced muscular weakness is the most apparent clinical sign. Characteristic ECG changes are usually present.

Potassium replacement is important in these cases and in animals with clinical signs is best accomplished with intravenous therapy. Concentration in the fluid should be increased to 30–50 mEq/L depending on severity and overall fluid requirements. Infusion rates should not exceed 0.5 mEq K/kg per hour. Continuous monitoring of ECG and frequent measurements of serum potassium concentration are essential in severely hypokalemic patients.

Oral K supplementation is effective, and Kaon Elixer (Adria Laboratories, Inc., Columbus, OH) can be mixed with oral self-feeding fluids or given by gavage. When by gavage, it should be diluted because of its tendency to cause vomiting.

Concluding Remarks and Researchable Problems

Neonatal and infant care of any species involves a multidisciplinary approach to optimize successful outcome. Nutrition, behavior, and medicine are essential components.

Adequate nutrition is the first essential step to successful captive rearing. Researchable areas include mother milk analysis and formula selection and then determining adequate diet during weaning and rapid growth. Diagnostic criteria to determine adequate nutrition need to be defined. Measurable physiological markers that correlate with normal or abnormal nutritional status would enhance our ability to select balanced diets. This would also provide a mechanism to compare wild and captive populations.

Socialization, like nutrition, is basic for normal development. Behavioral observation and analysis in both captive and wild populations provides valuable information for the development of rearing techniques. Behavioral research on the selected endangered species is essential.

Medical care optimizes the outcome of adequate nutrition and socialization. Preventive medicine, which includes nutrition and socialization, involves the methods used to keep animal populations free of disease. Monitoring of pathogenic organisms and knowledge of their biology in both captive and wild populations is important. Enteric pathogens are all too common in many captive populations and are responsible for a significant component of morbidity and mortality. Along with quarantine and good sanitation practices, other preventive measures are needed. Methods of immunization are continuing research areas in human medicine that should have application to nonhuman primates.

Optimal intensive care requires a knowledge of normal physiological parameters. Metabolic rates and requirements, urine output, electrolytes, serum chemistries, hematology, etc., should be established for each species.

While the principal issue of species survival is habitat protection and restoration, captive populations will play an important role in maintaining numbers as well as genetic diversity. Dedicated and skilled primatologists, veterinarians, clinicians, nutritionists, reproductive physiologists, and animal care technicians are an essential team that should be able to develop successful techniques for any species.

Acknowledgments. Many dedicated animal care personnel, primatologists, and clinicians have provided expertise and creativity for this manuscript. I thank them for the time and consideration they have given me. This work was supported by grant RR00169 from the California Primate Research Center, University of California, Davis, CA 95616.

References

Abee C (1985) Medical care and management of the squirrel monkey. In: Rosenblum LA, Coe CL (eds) Handbook of squirrel monkey research, Plenum, New York/London, pp 477–488.

Anderson JH, Rosenberg DP, Henrickson RV (1981) Post weaning hypoglycemia in infant rhesus monkeys. Abstract from the 32nd Annual Session, American Association for Lab Animal Science, Salt Lake City, Utah.

Ausman LM, Hages UC, Lage A, Heysted DM (1970) Nursery care and growth of Old and New World infant monkeys. Lab Anim Care 20:907–913.

Ausman LM, Gallivan DL, Nicolosi RJ (1985) Nutrition and metabolism of the squirrel monkey. In: Rosenblum LA, Coe CL (eds) Handbook of squirrel monkey research. Plenum, New York/London, pp 349–378.

Breznock AW, Porter S, Harrold JB, Kawakami TG (1979) Hand rearing infant gibbons. In: Ruppenthal (ed) Nursery care of nonhuman primates, 1st ed. Plenum, New York/London, pp 287–298.

Brown C, Samuels A, Brentson P, Henrickson RV (1981) Foster mothering: an alternative to nursery rearing. Abstract from the 32nd Annual Session, American Association for Lab Animal Science, Salt Lake City, Utah.

Cicmanec JL, Hernandez DM, Jenkins SR, Campbell SR, Smith JA (1979) Hand rearing infant Callitrichida (*Saguinus* spp. and *Callithrix jacchus*), owl monkeys (*Aotus trivirgatus*) and capuchins (*Cebus albifrons*). In: Ruppenthal GC (ed) Nursery care of nonhuman primates, 1st ed. Plenum, New York/London, pp 307–312.

Dronzek LA, Savage A, Snowdon CT, Whaling CS, Ziegler TE (In press) Techniques of hand rearing and reintroducing rejected cotton-top tamarin infants. Lab Anim Sci.

Graef JW, Cone TE (eds) (1980) Manual of pediatric therapeutics, 2nd ed. Little, Brown, Boston.

Graham CE, Bowen SA (eds) (1985) Clinical management of infant great apes. Alan R Liss, New York. (Monographs in Primatology Vol 5.)

Hearn JP, Burden FJ (1979) Collaborative rearing of marmoset triplets. Lab Anim 13:131–133.

Hennessy MB (1985) Effects of surrogate rearing on the infant squirrel monkey. In: Rosenblum LA, Coe CL (eds) Handbook of squirrel monkey research, 1st ed. Plenum, New York/London, pp 149–168.

Herzog M, Hopf S (1985) Some improvements of mother surrogates for squirrel monkeys: a technical note. Lab Prim Newslett 24:1–2.

Hinkle DK, Session HL (1972) A method for hand rearing of *Saimiri sciureus*. Lab Anim Sci 22:207–209.

Inglefinger JR, Auner E (1980) Renal disorders. In: Graef JW, Cone EF (eds) Manual of pediatric therapeutics, 2nd ed. Little, Brown, Boston, p 205.

Kaplan J, Russell M (1973) A surrogate for rearing squirrel monkeys. Behav Res Methods Instrum 5:379–380.

Lovejoy FH, Graef JW (1980) Management of the newborn, II. In: Graef JW, Cone TE (eds) Manual of pediatric therapeutics, 2nd ed. Little, Brown, Boston, p 134.

Meier JE, Reichard TA (1982) The use of total parenteral hyperalimentation in neonatal exotic animals. Proceedings, 1982 annual meeting American Association of Zoo Veterinarians, New Orleans, Louisiana.

Meier JE, Willis MS (1984) Techniques for hand rearing neonatal ruffed lemurs (*Varecia variegata* and *Varecia variegata rubra*) and a comparison of hand raised and maternally raised animals. J Zoo An Med 15:24–31.

Moore GT, Cummins LB (1970) Nursery rearing of infant baboons. In: Ruppenthal GC (ed) Nursery care of nonhuman primates, 1st ed. Plenum, New York/London, pp 145–151.

Oftedal OT (1980) Milk composition and formula selection for hand-rearing young mammals. Proceedings of the first annual Dr. Scholl nutrition conference. Chicago, pp 67–84.

Ogden JD (1979) Hand rearing *Saguinus* and *Callithrix* genera of marmosets. In: Ruppenthal GC (ed) Nursery care of nonhuman primates, 1st ed. Plenum, New York/London, pp 287–298.

Orlowski JP (ed) (1980) Pediatric intensive care. WB Saunders, Philadelphia (Pediatric Clinics of North America 27:3.)

Perkin RM, Levin DL (1980) Common fluid and electrolyte problems in the pediatric intensive care unit. In: Orlowski JP (ed) Pediatric intensive care. WB Saunders, Philadelphia, pp 567–586. (Pediatric Clinics of North America 27:3.)

Ruppenthal GC (1979) Survey of protocols for nursery rearing infant macaques. In: Ruppenthal GC (ed) Nursery care of nonhuman primates, 1st ed. Plenum, New York/London, pp 165–185.

Ruppenthal GC, Sackett GP (1979) Experimental and husbandry procedures: their impact on development. In Ruppenthal GC (ed) Nursery care of nonhuman primates, Plenum, New York/London, pp 269–284.

Tardif S, Lenhard A, Carson R, McArthur A (1985) Hand rearing of infant *Saguinus oedipus* with subsequent introdution into social groups. Amer J Primatol 8:368–369.

Tolia VK, Dubois RS (1985) Update on oral rehydration: its place in treatment of acute gastroenteritis. Pediat Ann 14:295–303.

Valerio DA, Darrow CC, Martin DP (1970) Rearing of infant simians in modified germ free isolators for oncogenic studies. Lab Anim Care 20:713–719.

Ziegler TE, Stein FJ, Sis RF, Coleman MS, Green JH (1981) Supplemental feeding of marmoset (*Callithrix jacchus*) triplets. Lab Animal Sci 31:194–195.

51
Prenatal and Neonatal Pathology of Captive Nonhuman Primates

NORVAL W. KING, JR. and LAURA V. CHALIFOUX

Introduction

It is appropriate in this volume where the overriding concern is the continued existence of self-sustaining populations of nonhuman primates that the issues of prenatal and neonatal mortality be addressed. For obvious reasons, there are essentially no published data regarding the incidence of fetal wastage and infant mortality in feral populations of nonhuman primates and the impact that these phenomena may have on the ability of natural populations to replenish themselves. Most of what we know regarding these topics stem from studies of relatively small, captive-breeding colonies where it is possible to compile accurate statistics regarding breeding histories and subsequent reproductive performance. Because these colonies are often quite different in terms of geographic location, types of breeding programs conducted, methods used to monitor reproductive cycles and detect pregnancy, and the resources available for pathologic evaluation of reproductive failure and infant mortality, the data derived from them may, at times, be quite different. Moreover, one should be reminded that the information published on these subjects spans at least five decades during which time there has been vast improvement in the manner in which nonhuman primates are maintained in captivity. This also has to be taken into account in comparing the data derived from present-day breeding programs where the breeding stock may be entirely composed of laboratory-born animals with those of earlier investigators whose colonies consisted solely of animals that were captured in the wild. And finally, one has to be extremely careful in applying information extrapolated from captive-breeding colonies to explain perceived problems of reproduction in natural populations.

Interest in the basic biology of the reproductive tract of nonhuman primates, particularly *Macaca mulatta* whose reproductive cycle is similar to that of the human female, dates back to the 1930s when such notable embryologists as Corner (1954), Hartman (1941), Hertig (1935), Heuser (1941), Lewis (1933, 1941), Ramsey (1949, 1954, 1966), Schultz (1937), and Wislockie (1938) and their colleagues defined the early developmental stages of human and monkey embryos and placentae. These hallmark investigations truly laid the foundation for what we know today regarding normal and abnormal enbryonic development and embryonic mortality of *M. mulatta* and the impact that these events have on the reproductive performance of this and other primate species. More recently, the investigations of Hendrickx and his colleagues have focused on the natural occurrence of embryonic and fetal mortality in captive *M. mulatta* as a baseline for evaluating the teratogenic effects of certain chemical compounds. The results of these studies have been comprehensively reviewed in a recent book chapter to which the reader is referred (Hendrickx and Binkerd, 1980).

Price et al. (1972, 1973) and Anver et al. (1973) reviewed the general subject of simian neonatology in a series of three papers based upon detailed, clinicopathologic observations on 84 infant monkeys. Their surveys included animals that were either stillborn or that died during the first 30 days of life out of a total of 225 born during a 21-month period at the New England Regional Primate Research Center (NERPRC), Southborough, Massachusetts. Rouse et al. (1981) and Kilborn et al. (1983) have also reported the results of retrospective studies of the causes of abortion and infant mortality in *Aotos trivirgatus* and *Saguinus oedipus*, respectively, at the NERPRC. The reader is referred to those papers for a detailed report on this subject.

In this report, we will present a summary of the gross and histopathologic findings in necropsies of 958 prenatal and neonatal deaths at the New England Regional Primate Research Center over the last 18 years. We will also identify those areas where additional research is needed if we are to effectively reduce this problem of major proportions in certain colonies of endangered species. For purposes of this presentation, the neonatal period is defined as the interval between birth and the first 30 days of postnatal life.

As a matter of background, the average daily census of nonhuman primates at the NERPRC over the past 18 years has been approximately 1400 animals. These are composed of approximately equal numbers of Old and New World species and include: *M. mulatta*, *M. fascicularis*, *M. cyclopis*, *M. arctoides*, *Saimiri sciureus*, *Cebus albifrons*, *A. trivirgatus*, *S. oedipus*, *S. fuscicollis*, and *Callithrix jacchus*. Of these, *M. cyclopis* and *S. oedipus* are currently listed as endangered and *M. arctoides* as threatened. We maintain relatively small but active breeding colonies of the following species, with the current number of breeding females listed in parentheses: *M. mulatta* (190), *M. fascicularis* (53), *M. cyclopis* (26), *M. arctoides* (14), *S. sciureus* (17), *A. trivirgatus* (26), and *S. oedipus* (31). The macaque breeding colonies are of two basic types: gang-caged harem colonies, and timed-mating colonies where individually caged females have their menstrual cycles monitored by daily vaginal swabs and are regularly placed in with the male for 5 days during the periovulatory period until pregnancy is detected. The latter colonies are maintained for investigators whose research requires fetuses of known age. All New World species are currently housed as breeding pairs, but this has not always been the case at this Center.

It is important to point out a major difference in the manner in which we rear infant Old World monkeys compared to most other primate colonies before one attempts to compare our experience with infant mortality in these species with that of other primate breeding facilities. We routinely remove all newborn macaque monkeys from their dams during the first day of life and hand rear them in a semiisolated nursery facility specifically designed for this purpose. The infants usually acquire some colostrum from their mothers prior to being transferred to the nursery where they are bottle fed a modified human infant formula for approximately 10–12 weeks. Infants reared in this way are monitored closely by the nursery personnel and are weighed daily for the first 8 weeks and weekly thereafter until they are transferred out of the nursery at approximately 6–8 months of age. Any sign of clinical illness is detected early, and appropriate treatment is administered. All infant New World monkeys, on the other hand, are left with their parents for rearing and are only transferred to the nursery if they become ill or are rejected or abused by their parents.

It has been our policy at the Center to autopsy all animals that die, including abortuses and stillbirths, and to examine their major organs and tissues histopathologically unless severe autolysis or mutilation precludes such an examination. In addition, all placentas from live as well as dead infants that are not eaten by the dam are routinely examined grossly and microscopically for evidence of lesions. The data presented herein represent a summary of a retrospective review of the pathologic records on this material. The data are presented in tabular form by individual species in Table 51.1. The total number of births at the NERPRC during the period of this study and statistics regarding infant mortality are presented by species in the companion paper by Johnson et al. (this volume). The discrepancies between the number of deaths listed in her paper and the number of necropsied prenatal and neonatal monkeys presented in this report are due to the problems of maceration, mutilation, and cannibalism that occur commonly in primate breeding colonies.

Viable Infants

The mean body weights and standard deviations for viable newborns of the following species born at the NERPRC are: *M. mulatta*, 517 (± 110) g; *M. fascicularis*, 366 (± 115) g; *M. cyclopis*, 565 (± 81) g; *M. arctoides*, 435 (± 21) g; *S. sciureus*, 100 (± 15) g; *A. trivirgatus*, 90 (± 10) g; and *S. oedipus*, 42 (± 5) g. The mean figures obviously are derived from infants that are either premature, term, or even, in some instances, postmature, but all of whom were viable.

Abortion and Stillbirth

The term *abortion* is generally defined as the expulsion of a fetus from the uterus prior to the age of extrauterine survivability. This is obviously a somewhat arbitrary definition because the developmental age at which a fetus can survive outside the uterus depends upon the kinds of life-support measures that are taken to sustain it. For lack of a better definition, we define an abortus as any fetus whose birth weight is less than that of the smallest infant of that species that we have successfully hand reared in our infant nursery. Currently, the upper weight limit for an abortus without the placenta of each of the following species is: *M. mulatta*, 300 g; *M. fascicularis*, 220 g; *M. cyclopis*, 330 g; *M. arctoides*, 250 g; *S. sciureus*, 80 g; *A trivirgatus*, 70 g; and *S. oedipus*, 30 g. The potential causes of abortion in nonhuman primates, as in most mammalian species, include genetic factors, developmental malformations of the placenta and/or embryo, ischemic lesions in the placenta, trauma, infection, and a variety of hormonal aberrations in the dam. Since many abortions occur very early in gestation, most of these probably go undetected unless pregnancy has been previously confirmed by radioimmunoassay measurements of maternal estrogen and progesterone

TABLE 51.1. Simian prenatal and neonatal deaths: gross and microscopic findings.

	Mm[1]	Mf	Mc	Ma	Aot	Sgo	Ss
Abortions							
No lesions found	13	2	9		23	9	
Placental infarcts					4		
Stillbirths							
No lesions found	24	7	5	3			2
Ruptured uterus	1						
Purulent placentitis	1	1	1		1		
Keratin squames, lungs	1	3	1			7	
Placental infarcts	1	1	1			2	
Abruptio placentae		2					1
Anencephaly		1					
Neonatal deaths							
No lesions found	126	52	2	9	25	147	127
Birth trauma/dystocia	6	15	4	3		19	18
Parental mutilation	4	3	1	1	15	37	18
Parental neglect						5	4
Septicemia	1						
Respiratory							
Immaturity						1	1
Pneumonia							
Purulent	1	1		2	2	5	
Adenovirus	1			1			
Unclassified		4	1		1	7	3
Keratin squames, lung		1					1
Gastrointestinal							
Gastroenteritis	1		1				
Gastric rupture		1					
Enteritis							
Purulent	1						
Viral		1					2
Intussusception							1
Colitis							1
Cardiovascular							
Congenital							
Interatrial septal defect		1					
Vestigial left vena cava	1						
Tricuspid atresia		1					
Transposition great vessels	2		1				1
Pericarditis					1		1
Central nervous system							
Hydrocephaly	3	1					
Purulent meningitis	1					4	
Congestion (severe)	4						
Hemorrhage	7		4	2	1	12	6
Encephalitozoonosis							4
Urogenital							
Toxic nephritis		1					
Purulent nephritis			1			1	
Bilobed kidney							1
Other							
Pleuritis, peritonitis, septicemia				1	1	1	
Dehydration	1						
Aspiration of milk or vomitus	2	2	1			1	
Umbilical hernia						2	
Gangrene, leg	1						

TABLE 51.1. (Continued)

	Mm[1]	Mf	Mc	Ma	Aot	Sgo	Ss
Thymic hypoplasia		1					
Accidental death		1				2	
Dwarfism/polydactyly						1	
Bacteria isolated							
Blood	8	6		1	5	4	25
Lung	4	1	1		1		4
Brain	3		1				2
Colon							1
Total animals with no lesions	163	59	16	12	48	156	129
Total animals with lesions	56	49	19	11	32	111	97
Total animals necropsied	219	108	35	23	80	267	226
Total necropsies; all species: 958							

[1] Abbreviations:
Mm - *Macaca mulatta* (rhesus monkey)
Mf - *Macaca fascicularis* (cynomolgus monkey)
M - *Macaca cyclopis* (Formosan rock macaque monkey)
Ma - *Macaca arctoides* (stumptail macaque monkey)
Aot - *Aotus trivirgatus* (owl monkey)
Sgo - *Saguinus oedipus* (cotton top tamarin)
Ss - *Saimiri sciureus* (squirrel monkey)

levels during the first 30 days of gestation. This is not done on a routine basis at the NERPRC, and therefore the true incidence of early abortion is not readily available. Since in most early abortions the conceptus is generally eaten by the dam, the chances of determining the precise cause are generally nil. Conceptuses expelled later in gestation may also be partially eaten by the dam and therefore be unsuitable for critical pathologic evaluation. As can be seen in Table 51.1, the two species in which unexplained abortion constitutes a major reproductive problem are *A. trivirgatus* with an incidence rate of 27 of 80 or 34% and *M. cyclopis* where the rate is 9 of 35 or 26%. Though placental infarction was associated with four of the 27 abortions in *A. trivirgatus*, the vast majority of the abortuses of these two species had no recognizable lesions to account for their expulsion. As already mentioned, placentas are often eaten by the dam and thus were not always available for examination, but from this survey it would appear that primary placental abnormalities are not major causes of abortion in our colonies. Multiple, small infarcts are observed with some frequency in the placental discs of healthy newborn monkeys and thus appear to be of little consequence in normal fetal development. It is only when the areas of infarction exceed the functional reserve of the bidiscoid, primate placenta that they have an adverse effect on fetal development and viability.

The term *stillbirth* is used to denote the delivery of a macerated or mummified fetus or for any infant for which there is direct clinical or pathologic evidence that the animal did not breathe postnatally. Macerated and mummified fetuses are not suitable for microscopic evaluation because of advanced postmortem autolysis, and, unless there are gross developmental abnormalities, the cause of death is generally not determined. For those near-term stillborns without advanced postmortem changes, the success rate in determining the cause of death varied with the species and the number of animals examined. In the rhesus monkey where we have the greatest numbers of stillbirths, the success rate was less than 20%. Placental lesions were the most common pathologic finding in stillborn macaques, and these included premature separation of the placenta, i.e., abruptio placentae, placental infarcts, placentitis, and rarely circumvallate placenta and placenta previa. Another lesion encountered in a small percentage of both New and Old World stillborns is the presence of large numbers of keratinized epithelial cells in the alveolar ducts of the lung resulting from aspiration of large quantities of amniotic fluid. This finding has been associated with fetal anoxia or other kinds of intrauterine distress that stimulates the fetus to increase its respiratory activity in utero. It is not a cause of intrauterine death per se but simply is a morphologic indication of some noxious, intrauterine event that caused the fetus to undergo gasping respiratory movements.

Neonatal Deaths

Neonatal death refers to the death of an infant that was born alive but that dies sometime between delivery and the first 30 days of life. Although early neonatal deaths may be difficult to distinguish from stillborns, the distinction is extremely important because this is the category of infant deaths for which there is the greatest opportunity for improvement. The basic distinction lies in determining whether or not the infant breathed postnatally. This may be determined directly if someone is present at the time of delivery, which in most instances is not the case, or by careful gross and microscopic evaluation of the lungs at the time of autopsy. We have found that examination of the lungs with a dissecting microscope prior to fixation can be useful in determining whether or not the lungs contain air at the time of autopsy. These findings should be correlated with the histopathologic findings also.

The causes of death of neonatal monkeys are many and varied and include factors that range from genetic or congenital to those that are acquired by or imposed on the infant during the early postnatal period. Our experience with infant macaques indicates that the majority of neonatal deaths in these species occur during the first day of life, and often the cause is not determined. Many of these occur within the first few hours after delivery during the night and thus the circumstances surrounding the death are not observed. Traumatic lesions including massive thoracic, abdominal, and intracranial hemorrhage incurred during the birth process, and multilation inflicted by the dam immediately after delivery account for only a small percentage of these, as do various congenital anomalies incompatible with extrauterine life. Clearly, more research is needed to elucidate the metabolic and physiologic abnormalities that contribute to the failure of certain newborn macaques to adjust to the extrauterine environment. Information of this kind will be essential to the development of effective intervention measures that could be used to reduce these unnecessary losses. At our Center, those infant macaques who survive the approximately 8–12-hour period between birth and transfer to the nursery have a greater than 95% chance of surviving the neonatal period and beyond. The small number of macaques that die in the nursery during the neonatal period usually do so from the consequences of birth trauma, such as intracranial hemorrhage, aspiration of milk or vomitus, or rarely from bacterial infections that either go undetected or fail to respond to antibiotic therapy.

The problem of neonatal mortality in our colonies of New World monkeys is clearly different and a more serious one. The differences relate to the fact that these infants are left with their parents for rearing. Since many of these deaths occur during the first 2 weeks of life rather than during the immediate postpartum period, parental neglect and/or abuse becomes important causative factors. The infants of these species are extremely small and totally dependent upon their parents for warmth and nutrition. Their thermoregulatory mechanism is poorly developed during the first several weeks of life, and thus they rely heavily upon the warmth derived from clinging to the body of a parent to ward off hypothermia. Similarly, they depend totally upon their dam for all their nutritional needs. For reasons that are not understood, captive New World primates will often abandon or worse yet severely mutilate their apparently healthy offspring during the first several weeks of life, and this is recognized as a major cause of neonatal mortality in these species. Though bodily mutilation is an easily recognized cause of death, benign neglect can be more difficult to recognize unless the parent–infant relationship is monitored carefully. This is obviously not feasible on a 24-hour basis, and for this reason a 12–18-hour period of abandonment at the bottom of a stainless steel cage may be sufficient to cause death from hypothermia or dehydration in these young neonates. Death from such causes is not associated with specific lesions in any particular organ system, and this no doubt accounts for the high number of animals listed in this category in Table 51.1. There is a pressing need for research aimed at addressing the issue of behavioral aberrations that underlie the serious problem of parental neglect and abuse in the captive colonies of New World primates if we are to deal effectively with this problem in parent-raised infants. In the meantime, pilot studies are underway to determine if the high rate of neonatal mortality in certain of these species such as the endangered *S. oedipus* can be drastically reduced by hand-rearing them as we do routinely with our macaques. Should this be the case, the long-term effects of hand rearing versus parent rearing on subsequent reproductive performance of these individuals also needs to be seriously assessed if we are to ensure that these colonies will be truly self-sustaining. Since neonatal infections such as pneumonia, meningitis, and enterocolitis occur more commonly in New World infants who are left with their parents than hand-reared macaques, it is likely that these infections are acquired from the adult animals. Hence, hand rearing infants of these species, if successful, could also have the secondary advantage of reducing the incidence of neonatal infectious diseases in such colonies.

Summary

It is clear that prenatal and neonatal mortality are significant clinical problems in captive-breeding colonies of nonhuman primates, but whether these occur in natural populations of these same species is not known. It is likely that the problem of *neonatal* mortality could be substantially reduced in captive-breeding colonies if an intensive research effort were directed toward understanding its manifold causes. On the other hand, we are less optimistic regarding the prospects of reducing the incidence of *prenatal* mortality since the likelihood of controlling its causes, over and above what we are currently doing, will be much more difficult. It is incumbent on those of us in the scientific community who are truly concerned about the future of self-sustaining populations of nonhuman primates, both captive and in the wild, to join forces in generating the knowledge needed to solve the immediate problems threatening their very existence and to assure ourselves that these remarkable species will have secure places to live and breed on earth forever.

Acknowledgment. Supported by NIH grant RR-00168 from the Division of Research Resources, Animal Resources Program Branch.

References

Anver MR, Hunt RD, Price RA (1973) Simian neonatology. II. Neonatal pathology. Vet Pathol 10:16–36.

Corner GW, Bartelmez GW (1954) Early abnormal embryos of the rhesus monkey. Contrib Embryol 35(231):1–9.

Hartman CG, Corner GW (1941) The first maturation division of the macaque ovum. Contrib Embryol 29(179):1–6.

Hendrickx AG, Binkerd PE (1980) Fetal deaths in nonhuman primates. In: Porter IH, Hook EB (eds) Human embryonic and fetal death. Academic, New York.

Hertig AT (1935) Angiogenesis in the early human chorion and in the primary placenta of the macaque monkey. Contrib Embryol 24(146):37–82.

Heuser CH, Streeter GL (1941) Development of the macaque embryo. Contrib Embryol 29(181):15–55.

Kilborn JA, Sehgal P, Johnson LD, Beland M, Bronson RT (1983) A retrospective study of infant mortality of cotton-top tamarins (*Saguinus oedipus*) in captive breeding. Lab Anim Sci 33:168–171.

Lewis WH, Hartman CG (1933) Early cleavage stages of the egg of the monkey (*Macaca rhesus*). Contrib Embryol 24(143):187–202.

Lewis WH, Hartman CG (1941) Tubal ova of the rhesus monkey. Contrib Embryol 29(180):7–14.

Price RA, Anver MR, Garcia FG (1972) Simian neonatology. I. Gestational maturity and extrauterine viability. Vet Pathol 9:301–309.

Price RA, Anver MR, Hunt RD (1973) Simian neonatology. III. The causes of neonatal mortality. Vet Pathol 10:37–44.

Ramsey EM (1949) The vascular pattern of the endometrium of the pregnant rhesus monkey (*Macaca mulatta*). Contrib Embryol 33(219):113–147.

Ramsey EM (1954) Venous drainage of the placenta of the rhesus monkey (*Macaca mulatta*). Contrib Embryol 35(238):151–173.

Ramsey EM, Harris JWS (1966) Comparison of uteroplacental vasculature and circulation in the rhesus monkey and man. Contrib Embryol 38(260):43–58, 90.

Rouse R, Bronson RT, Sehgal PK (1981) A retrospective study of etiological factors of abortion in the owl monkey, *Aotus trivirgatus*. J Med Primatol 10:199–204.

Schultz AH (1937) Fetal growth and development of the rhesus monkey. Contrib Embryol 26(155):71–97.

Wislocki GB, Streeter GL (1938) On the placentation of the macaque (*Macaca mulatta*), from the time of implantation until the formation of the definitive placenta. Contrib Embryol 27(160):1–66.

52
The Effect of Perinatal and Juvenile Mortality on Colony-Born Production at the New England Regional Primate Research Center

LORNA D. JOHNSON, ANDREW J. PETTO, DONALD S. BOY, PRABHAT K. SEHGAL, and MARY E. BELAND

Deforestation of primate habitats and local human predation, together with previous exportation of large numbers of primates for research, pets, and testing facilities, have created a crisis in animal populations in the wild. Of the colonies remaining in captivity, it is probable that those in research institutions and captive-breeding facilities offer the best hope for the reconstitution of sufficient numbers of animals to be relocated to natural habitats.

The New England Regional Primate Research Center (NERPRC) initiated a subdivision of reproductive biology to inventory the reproductive statistics of its colonies, to identify the reproductive problems, and to initiate proposals for research into reproductive failure. Records of all the animals living at the NERPRC from 1967 to the present have been reviewed, and vital statistics have been entered into the computer. This paper is a preliminary report on the effect of mortality rates of colony-born animals on the size of the reproductive population including the impact of spontaneous and experimental deaths in captive colonies of Old and New World monkeys.

Variables entered include species, animal identification, dates of births or arrival and death, type of death (stillborn, spontaneous, cesarean sacrifice, experimental sacrifice, or euthanasia), origin of animal (wild caught, colony born, or from another captive colony), sex, type of breeding (harem bred, time bred, or pair bred), type of rearing (parent or nursery), dam's and sire's identification and birth years, and generation number by dam and by sire.

There are four colonies of macaques at the NERPRC. Of these, *Macaca arctoides* is harem-bred, *M. cyclopis* was harem-bred previous to the last 2 years, and *M. fascicularis* and *M. mulatta* had both harem- and time-bred colonies. Harem-bred animals live in outdoor/indoor facilities, a male being grouped with several females. Time-bred animals live in individual cages indoors with controlled 12-hour light and dark cycles, humidity, and temperature. Menstrual cycles are monitored daily, and animals are bred during the periovulatory period. All newborn macaques are hand raised in the nursery.

Of the four colonies of New World primates, *Saimiri sciureus* is harem bred in outdoor/indoor facilities, and *Aotus trivirgatus*, *Callithrix jacchus*, and *Saguinus oedipus* are pair bred and raised in families indoors with controlled light, humidity, and temperature. Infants are raised by parents and siblings, unless they are abused or abandoned, in which case they are brought to the nursery for rearing. The only substantial numbers of nursery-reared infants are those of *S. oedipus*.

Nursery-raised infants are weaned to adult diets by 3 months of age. They are socialized in the nursery from 1 month of age in pairs and from 4 months of age in large play cages until puberty. The adult macaques are assigned to harem- or time-bred colonies, the squirrel monkeys to harem colonies, and the other New World primates to paired breeding.

The death and survival of colony-born infants of Old World primates at the NERPRC are shown in Table 52.1. The data are tabulated so that the rows "Transferred" and "Removed for experiments" show the percent of colony infants that were used for research purposes and were not available for future breeding. Deaths are those that occurred spontaneously in a captive colony. They are age-grouped as stillborn, neonatal (under 30 days), and as infants and juveniles from 30 days to 5 years. Survival includes adults that lived longer than 5 years and thus entered the reproductive group and juveniles that are presently alive, the survivors of which will be the future breeding replacement. The percent of juveniles is that proportion of the entire birth cohort from 1967 that are now alive as juveniles.

Macaca arctoides (the stump-tailed macaque) is now on the list of threatened animals. There have been no new arrivals in the colony since 1971. The stillbirth (19.4%) and the neonatal (11.2%) death rates are within the range of those reported by other captive colonies (Bernstein and Gordon, 1977; Harvey and Rhine, 1983; Hendrickx and Nelson, 1971; MacDonald, 1971; Price et al., 1972). Only 12.4% of colony-born animals survive as adults, and no juveniles remain. The colony suffers from the large number of infants that are transferred to investigators in eye research and from the aging of the surviving adults, six of whom are 14–15 years old.

Macaca cyclopis (the Taiwan rock macaque) is on the endangered species list. Approximately 17% were transferred or used in experiments up to 1977. No imported animals entered the colony from 1972 to 1985. The combined stillborn and neonatal death rate (21.4%) is approximately the same as that previously reported (Price et al., 1972). The high rate of juvenile deaths (30.6%) reflects the deaths from the 1977 to 1982 cohorts resulting from the spontaneous occurrence of simian acquired immunodeficiency diseases (SAIDS), which occurred in this colony in 1978 to 1980. Twenty-six percent of the colony infants lived to reproduce, and nine of the juveniles are still alive, reflecting the heavy toll of the SAIDS.

Of 607 *M. fascicularis* colony-born infants, 16% were removed from the colony for research. The stillbirth rate (23%) is slightly, but not significantly, higher than in the other macaque colonies. The stillborn and neonatal rates are within the range of those reported in other colonies (Hendrickx and Nelson, 1971; MacDonald, 1971; Price et al., 1972). The reproductive population is 22% of those born, and the number of live juveniles represents 25% of those born into the colony since 1967.

The *M. mulatta* colony has had 1076 colony-born infants. This primate is the most used in research investigations, and this is reflected in the high rate (46%) of transferred animals and experimental deaths. Stillborn and neonatal death rates are in the higher range of those reported (Bernstein and Gordon, 1977; Hendrickx and Binkerd, 1980; Hertig et al., 1971; Hird et al., 1975; Price et al.,

TABLE 52.1. Death and survival of colony-born infants of Old World primates, 1967–1983.

	Macaca arctoides		*Macaca cyclopis*		*Macaca fascicularis*		*Macaca mulatta*	
	No.	%	No.	%	No.	%	No.	%
Total births	170	100.0	173	100.0	607	100.0	1076	100.0
Transferred	66	38.8	14	8.1	50	8.2	209	19.4
Removed for experiments	16	9.4	15	8.7	45	7.4	290	27.0
Deaths								
Stillborn	33	19.4	32	18.5	141	23.2	181	16.8
Neonatal (< 30 days)	19	11.2	5	2.9	27	4.5	72	6.7
Infants and juveniles (< 5 years)	15	8.8	53	30.6	56	9.2	75	7.0
Survival								
Adults (5+ years)	21	12.4	45	26.0	136	22.4	124	11.5
Juveniles (< 5 years)			9	5.2	152	25.1	125	11.6

TABLE 52.2. Death and survival of colony-born infants of New World primates, 1967–1983.

	Saimiri sciureus		*Aotus trivirgatus*		*Callithrix jacchus*		*Saguinus oedipus*	
	No.	%	No.	%	No.	%	No.	%
Total births	414	100.0	203	100.0	100	100.0	462	100.0
Transferred	47	11.4	0	0	0	0	0	0
Deaths								
Stillborn	169	40.8	91	44.8	28	28.0	134	29.0
Neonatal (< 30 days)	65	15.7	20	9.9	17	17.0	133	28.8
Infants and juveniles (< 2 years)	66	15.9	50	24.6	19	19.0	54	11.7
Survival								
Adults (2+ years)	62	15.0	40	19.7	28	28.0	90	19.5
Juveniles (< 2 years)	5	1.2	2	1.0	8	8.0	51	11.0

1972). The colony has been heavily involved in fetal research for many years so that not only are many sacrificed at birth, but stillborn and neonatal deaths are also higher because of fetal manipulation. Thus, the adult population and breeding replacement from colony births had been low (11.5%). No animals have been imported from the wild since 1977. We have relied upon arrivals from other breeding colonies to replace those used in research.

The New World colony-born infants (Table 52.2) have not been used in terminal experiments, and with the exception of *S. sciureus*, there have been no transfers to other institutions. Thus, the low percent (15–28%) that reach reproductive age is largely due to the fact that these animals are difficult to raise and maintain in captivity. The greatest problem has been the extremely high perinatal death rate. Forty-five to 58% of these infants die at birth or within the first month of life. High stillbirth rates among *S. sciureus* (41%) and *A. trivirgatus* (45%) have been observed in other captive colonies (Hopf, 1967; Kaplan, 1977; Lehner et al., 1967; Rasmussen et al., 1980; Rosenblaum, 1972; Rouse et al., 1981; Santolucito and Sessions, 1971). Recently, it has been shown that many of the stillbirths in *S. sciureus* have been caused by cephalopelvic disproportion (Aksel and Abee, 1983). The high stillbirth rate among *A. trivirgatus* was shown to be due in part to breeding of animals with different karyotypes (Ma et al., 1976). The high perinatal mortality rates in *C. jacchus* and *S. oedipus* have been previously reported from other colonies (Evans, 1983; Gengozian et al., 1978; Hampton et al., 1970, 1978; Kilbourne et al., 1983; McIntosh and Looker, 1982; Ogden et al., 1978; Poole and Evans, 1982). The juvenile death rates in our colonies are also higher among New World primates (12–25%) than among Old World primates (7–9%), excluding the *M. cyclopis*. In part, this is due to the successful rearing of Old World macaques in the nursery with subsequent socialization. These nursery-raised animals have been shown to breed as successfully as the wild-caught parents. The percent of New World colony-born infants who survive to the reproductive age (15–28%) is similar to the survival of macaques (12–26%).

These mortality and survival rates represent the average over a period of 18 years. Stringent requirements for experiments involving sacrifice, attention to the known causes of stillbirth, hand-rearing of neonatal animals in the nursery, and the establishment of family groups among the New World species has greatly increased the survival rate of colony-born animals to the age of reproductive maturity. In our 1983 birth cohort of New World primates, 29–54% reach maturity and 48–88% of Old World primates are alive at 2 years of age with the exception of *M. arctoides*, which had two births, one stillborn and one transferred (Table 52.3).

At the NERPRC there is a new commitment to breeding captive animals, not only to replace those used in research but with the long-term plan to produce a surplus of animals that may be returned to a natural habitat. Until recently, there have been no breeding colonies in the sense that animals are allowed to mature to the reproductive age and are maintained only for breeding.

The Primate Centers were initiated 18–20 years ago as biomedical research institutions to provide primates to further understanding of human health and disease. At the time, animals were supplied by importation from the wild with no plans for the Centers to be self-sustaining in primate supply. Within the next 10 years, it became apparent that the Primate Centers would have to rely upon their

TABLE 52.3. Survival rate to 2+ years of the 1983 birth cohort.

Species	Total births	Percent survived
Old World primates		
M. cyclopis	8	87.5
M. fascicularis	57	68.4
M. mulatta	82	47.6
New World primates		
S. sciureus	17	29.4
A. trivirgatus	20	45.0
C. jacchus	13	53.8
S. oedipus	61	31.2

own breeding facilities to replace the animals to be used for research. At the present time, however, it has become imperative to breed these animals not only for research but, in addition, for conservation of the species. Collectively, the research and breeding institutions throughout the world are the only facilities that have colonies of wide enough gene diversity and of sufficient numbers to collaborate in reproduction for return to the wild in the future. Therefore, a concerted effort should be made to breed and rear captive animals for this purpose. The Primate Centers should have an important role in this endeavor. At the present time, funds are not specified for this purpose. Through conferences such as this one, we hope the attention of funding agencies will focus on the importance of such conservation.

At the NERPRC, the two colonies of endangered animals, *M. cyclopis* and *S. oedipus*, are of immediate importance. Since 1977 we have not used these animals in research projects other than those addressed to their own health and reproduction.

The *M. cyclopis* colony was small: Only 65 had entered the reproductive colony, and the colony had increased 25% by 1976. In 1980, an AIDS-like syndrome that occurred spontaneously in this colony (Letvin et al., 1983) may have accounted for a 70% stillbirth rate in that year and for the loss of 79% of juveniles that were born between 1977 and 1982 (Table 52.4). The impact of this loss of future breeding stock will lower the future birth rate from 1982 through 1987. The colony has been moved indoors, and since that time 80% of the 1983 and 1984 birth cohorts are still alive. In an endeavor to reestablish the colony, we requested and received as a gift four young colony-raised females from the Medical School of the National Taiwan University. After determining that none of the adult males in our colony had been infected with simian aids virus (STLV-III) (Kanki et al., 1985), we entered these females into breeding this year and one is now pregnant. Since the preparation of this manuscript a male *M. cyclopis* has been born and is alive in the nursery at two months of age.

The other endangered species, *S. oedipus oedipus* (cotton-topped tamarin), has not been involved in medical research since 1977. The gestation period of this species is approximately 144 days. Twins and triplets are more frequently born than singles, and estrus occurs during lactation. Thus, it could be expected that reproduction would be high. These animals, however, suffer from a perinatal mortality rate in excess of 50% because of stillbirth and neglect or abuse of their infants. The animals were pair bred from 1976 to 1980. When the first families

TABLE 52.4. The effect of AIDS-like syndrome on *M. cyclopis* breeding replacement.

Birth Cohort Year	Total live born		Infant and juvenile deaths		Live replacements	
	Female	Male	Female	Male	Female	Male
1977	5	2	5	2	0	0
1978	8	1	8	1	0	0
1979	2	9	1	6	1	3
1980	2	1	2	1	0	0
1981	4	2	3	2	1	0
1982	1	6	0	3	1	3
Total	22	21	19	15	3	6

were established, the older siblings were left with the family until puberty and shared in the raising of their younger siblings. These families and their offspring have been largely responsible for the increase in survival of infants to the reproductive age, and their offspring have now established families of their own. This method of rearing is the best hope for colony survival. The problem has been to get a family established because of the high perinatal death rate. In an effort to lower the neonatal death rate, we began raising neglected or abused newborns in the nursery. In 1984, we were able to raise 72.4% of these infants to weaning at 3 months of age, which is approximately the same survival as the parent-raised infants (68.6%) in that year (Table 52.5). Prior to 1982, the few nursery-raised infants had not produced successfully when returned to the colony. The nursery-raised animals are now socialized in large play cages connected to four home cages that include nesting boxes. The first of these socialized animals to be placed in breeding gave birth to live twins this year. It is hoped that the others who have only recently been paired will be successful. Since the preparation of this manuscript, two other cotton-topped tamarins raised in the nursery and socialized in play cages have given birth to two sets of twins who are alive at two and three months of age. If these animals are successful and if the families continue to produce as well as they have in the past 2 years, the survival of cotton-topped tamarins in captivity will be greatly increased.

If a commitment is to be made by research institutions toward production for conservation, particularly of endangered species, it would seem prudent at this time, when colonies of these animals are still available, to embark on several endeavors. First, within research institutions, animals must be set aside in breeding colonies where they are not used for any other purpose. These animals should be chosen after genealogy, karyotyping, and past reproductive performance have been determined. Selection of individuals that will produce effective stable populations in the future can be assessed from age and sex-specific fertility and mortality rates and genealogies with the aid of computers. These animals should be maintained in obstetrical units where pregnancy, delivery, and the postpartum period can be more closely monitored. The causes of perinatal death, whether they be hereditary, congenital, hormonal, nutritional, environmental, or pathological, must be investigated. Then, attention must be given to how to raise those animals destined for return to a native habitat.

Concomitant research must be done into methods of preservation of gametes, embryos, in vitro fertilization, and embryo transfer to surrogate mothers. These

TABLE 52.5. A comparison of survival to 3 months of age among nursery- and parent-reared cotton-top tamarins.

		Parent-reared			Nursery-reared		
		Total	Survived		Total	Survived	
Year of birth	Total live births	No.	No.	%	No.	No.	%
1980	24	13	8	61.5	11	3	27.3
1981	30	20	13	65.0	10	1	10.0
1982	50	25	14	56.0	25	10	40.0
1983	48	27	16	59.3	21	10	47.6
1984	64	35	24	68.6	29	21	72.4

methods will be of particular importance in endangered species of which only a few remain in captivity or the wild.

Second, between institutions, world-wide studbooks must be maintained, and exchange of animals, but preferably gametes, for maximizing the gene pools should be initiated. The zoological societies have a small number of animals of a large variety of species and together with anthropologists and conservationists have spear-headed the commitment toward preservation. The research institutions, however, have more animals of a few species. Thus, their commitment to this endeavor will decrease the time in which sufficient numbers of animals are produced for a viable colony in the wild. For example, in the future, it should be possible to return the cotton-topped tamarins to a natural habitat. It is estimated that over 700 of this species are now alive in research institutions. In collaboration with other groups of investigators involved in conservation, it should be determined how the young should be raised. In what environment and in what groups will they best be able to survive? Will they need halfway environments to raise the young, with or without their parents or families, before translocation? Where and how will areas be set aside to receive these animals?

The goal is defined. The task is Herculean, but the tools are at hand and the desire is expressed to give back that which we have taken from the wild.

Acknowledgment. Supported by NIH grant RR00168 from the Division of Research Resources, Animal Resources Program Branch.

References

Aksel S, Abee CR (1983) A pelvimetry method for predicting perinatal mortality in pregnant squirrel monkeys (*Saimiri sciureus*). Lab Anim Sci 33:165–167.

Bernstein IS, Gordon TP (1977) Behavioral research in breeding colonies of Old World monkeys. Lab Anim Sci 27:532–540.

Evans S (1983) Breeding of the cotton-top tamarin *Saguinus oedipus oedipus*: a comparison with the common marmoset. Zoo Biol 2:47–54.

Gengozian N, Batson JS, Smith TA (1978) Breeding of marmosets in a colony environment. Prim Med 10:71–78.

Hampton JK, Hampton SH, Levy BM (1970) Reproductive physiology and pregnancy in marmosets. In: Goldsmith EI, Moor-Jankowski J (eds) Medical primatology 1970, Proceedings of the Second Conference on Experimental Medicine and Surgery in Primates, New York, 1969. S Karger, Basel/Munchen/Paris/London/New York/Sydney, pp 527–535.

Hampton SH, Gross MJ, Hampton JK (1978) A comparison of captive breeding performance and offspring survival in the family *Callitricidae*. Prim Med 10:88–95.

Harvey NC, Rhine RJ (1983) Some reproductive parameters of the stump-tailed macaque (*Macaca arctoides*). Primates 24:530–536.

Hendrickx AG, Binkerd PE (1980) Fetal deaths in nonhuman primates. In: Porter IH, Hook EB (eds) Human embryonic and fetal death. Academic, New York, pp 45–69.

Hendrickx AG, Nelson VG (1971) Reproductive failure. In: Hafez ESE (ed) Comparative reproduction of nonhuman primates. Charles C Thomas, Springfield, pp 403–425.

Hertig AT, King NW, MacKey J (1971) Spontaneous abortion in wild-caught rhesus monkeys *Macaca mulatta*. Lab Anim Sci 21:510–519.

Hird DW, Hendrickson RV, Hendrickx AG (1975) Infant mortality in *Macaca mulatta*: neonatal and post-neonatal mortality at the California Primate Research Center, 1968–1972. J Med Primatol 4:8–22.

Hopf S (1967) Notes on pregnancy, delivery, and infant survival in captive squirrel monkeys. Primates 8:323–332.

Kanki PJ, McLane MF, King NW, Letvin ML, Hunt RD, Sehgal P, Daniel MD, Desrosiers RC, Essex M (1985) Serologic identification and characterization of a macaque T-lymphotropic retrovirus closely related to human T-lymphotropic retroviruses (HTLV) type III. Science 228:1199–1201.

Kaplan JN (1977) Breeding and rearing squirrel monkeys (*Saimiri sciureus*) in captivity. Lab Anim Sci 27:557–566.

Kilborn JA, Sehgal P, Johnson LD, Beland M, Bronson RT (1983) A retrospective study of infant mortality of cotton-top tamarins (*Saguinus oedipus*) in captive breeding. Lab Anim Sci 33:168–171.

Lehner NDM, Bullock BC, Feldner MA, Clarkson TB (1967) Observations of reproduction of laboratory-maintained squirrel monkeys. Lab Anim Newslett 6:1–3.

Letvin NL, Eaton KA, Aldrich WR, Sehgal PK, Blake BJ, Schlossman SF, King NW, Hunt RD (1983) Acquired immunodeficiency syndrome in a colony of macaque monkeys. Proc Natl Acad Sci 80:2718–2722.

Ma NS, Jones TC, Miller AC, Morgan LM, Adams EA (1976) Chromosome polymorphism and banding patterns in the owl monkey (*Aotus*). Lab Anim Sci 26:1022–1036.

MacDonald GJ (1971) Reproductive patterns of three species of macaques. Fertil Steril 22:373–377.

McIntosh GH, Looker JW (1982) Development of a marmoset colony in Australia. Lab Anim Sci 32:677–679.

Ogden JD, Wolfe LG, Deinhardt FW (1978) Breeding *Saguinus* and *Callithrix* species of marmosets under laboratory conditions. Prim Med 10:79–83.

Poole TB, Evans RG (1982) Reproduction, infant survival and productivity of a colony of common marmosets (*Callithrix jacchus jacchus*). Lab Anim 16:88–97.

Price RA, Anver MR, Garcia EG (1972) Simian neonatology. I. Gestational maturity and extrauterine viability. Vet Pathol 9:301–309.

Rasmussen KM, Ausman LM, Hayes KC (1980) Vital statistics from a laboratory breeding colony of squirrel monkeys (*Saimiri sciureus*). Lab Anim Sci 30:99–106.

Rosenblaum LA (1972) Reproduction of squirrel monkeys in the laboratory. In: Beveridge WIB (ed) Breeding primates. S Karger, Basel, pp 130–145.

Rouse R, Bronson RT, Sehgal PK (1981) A retrospective study of etiological factors of abortion in the owl monkey *Aotus trivirgatus*. J Med Primatol 10:199–204.

Santolucito JA, Sessions HL (1971) Reproductive performance in a group of squirrel monkeys during two consecutive seasons. Lab Prim Newslett 10:4–6.

53
Neoplasms and Proliferative Disorders in Nonhuman Primates

LINDA J. LOWENSTINE

Introduction

Understanding causes of morbidity and mortality in nonhuman primates is essential for maintaining these species in captivity and managing populations in the wild. Although it has been stated that neoplasia is relatively uncommon in nonhuman primates [21, 88], more recent reports and excellent review articles suggest that as the age of captive populations increases so does the incidence of neoplasia [70, 78, 107]. A survey of spontaneous neoplasms was conducted and the literature reviewed.

Materials and Methods

This survey was contributed to by eleven institutions: Chicago Zoological Society-Brookfield Zoo (BFZ); Bronx Zoo of the New York Zoological Society (BZ); Penrose Research Laboratory of the Philadelphia Zoological Society (PZ); Cincinnati Zoo (CZ); Oklahoma City Zoo (OCZ); Kansas City Zoo (KCZ); National Zoological Park (NZP); Litton Bionetics Research Laboratory (LBRL); Yerkes Regional Primate Research Center (YPC); Comparative Pathology Department of Johns Hopkins Medical School (JH); and Registry of Comparative Pathology of the Armed Forces Institute of Pathology (AFIP). In addition, the pathology records from the California Primate Research Center (CPRC) and the University of California at Davis Veterinary Medical Teaching Hospital (VMTH) were reviewed. Cases from the VMTH were zoo or pet animals. Cases from JH and AFIP represented a mixture of zoo and research animals. Attempts were made to include only spontaneous neoplasms, and cases previously reported in the literature are noted when known. The incidence of endometriosis was also investigated in cases from the CPRC and VMTH.

A total of 388 lesions (199 malignant) were reported from 363 individual nonhuman primates, representing 34 genera and at least 63 species (Table 53.1). The use of simple common names, e.g., "capuchin" or "marmoset," made assignment to an exact species impossible in some cases. Most common names were detailed enough, however, to permit identification of the genus and species using a standard of primate taxonomy [86]. Species divided into the broad categories of prosimians, New World primates, Old World primates, and apes are listed in

TABLE 53.1. Neoplasms in survey of 13 institutions maintaining or diagnosing diseases in nonhuman primates.

	Prosimians	New World	Old World: Macaques	Old World: Other	Apes	All
Total animals	41	42	204	49	27	363
Total tumors	42	46	219	52	29	388
Number malignant	19	20	127	21	12	199
Percent malignancy	45	43	58	40	41	51

TABLE 53.2. Prosimians in survey of neoplasms in nonhuman primates.

Species	Number
Lemurs	
Microcebus murinus	1
Cheirogaleus major	1
Lemur catta	7
L. mongoz	4
Tree shrews	
Tupaia sp.	2
Galagos	
G. crassicaudatus	9
G. senegalensis	2
Galago sp.	4
Galagoides demidovii	1
Lorises and pottos	
Loris tardigradus	1
Nycticebus coucang	4
Perodicticus potto	5
Total	41

TABLE 53.3. New World monkeys in survey of neoplasms in nonhuman primates.

Species	Number	Species	Number
Callitricidae		Cebidae	
Callithrix jacchus	1	*Alouatta* sp.	1
Callimico goeldii	1	*Alouatta villosa*	2
Cebulla pygmaea	1	*Ateles* sp.	2
Leontideus rosalia	1	*A. geoffroyi*	1
Saguinus mystax	1	*A. belzebuth*	1
Saguinus oedipus	7	*Cebus* sp.	2
		C. albifrons	2
Aotidae		*C. apella*	5
Aotus trivirgatus	4	*C. capucinus*	1
Callicebus moloch	1	*Lagothrix lagotricha*	2
		Saimiri sciureus	6
		Total	42

Tables 53.2 through 53.6. Macaques were separated from other Old World monkeys because they are commonly kept in research vivaria, and patterns of neoplasia, if distinct, might mask patterns in the other less numerous cercopithecines.

Neoplasms were tabulated by species, by organ system of origin, and by whether benign or malignant. Specific malignancies were listed for each group. The mean age of diagnosis or death was calculated for 179 known-age animals and the sex distribution for 274 animals for which data were given.

Organ System Distribution of Neoplasms in Nonhuman Primates

Prosimians (Tupaidae, Lemuridae, Galagoidae, Lorisidae)

Among prosimians, tumors of the skin and subcutis and those of liver and pancreas were most common, accounting for 43% of all lesions and two-thirds of malignancies (Table 53.7). Specific malignancies are listed in Table 53.8. Hepatocellular or biliary cancer was found in a mouse lemur (*M. murinus*), two greater galagos (*G. crassicaudatus*), and a potto (*P. potto*) and severe adenomatous hyperplasia of bile ducts in a Demidov's dwarf galago (*G. demidovii*). Pancreatic adenocarcinomas were found in a greater dwarf lemur (*C. major*) and a slow loris (*N. coucang*). Cutaneous and subcutaneous tumors were more numerous and included benign and malignant tumors of apocrine sweat glands in two galagos and a tree shrew (*Tupaia* sp.), a sebaceous adenoma and a basal cell tumor in two galagos, and a malignant mammary tumor in a tree shrew. Several mesenchymal tumors included a fibrosarcoma in a galago (previously in the literature [44]), a lipoma in a ring-tailed lemur (*L. catta*), a liposarcoma in the axillary region of a galago, and a myelolipoma in the perineal region of a potto (previously reported [9]). Multiple visceral lipomas and myelolipomas were found in the mediastinum and pulmonary hilus in two lemurs (*L. catta* and *L. mongoz*) from one zoo; these were extensive enough to have led to a gross diagnosis of malignancy.

A review of the literature on neoplasia in prosimians revealed that some tumors found in this study had been reported [107]. Chemicals such as benzopyrene and methylcholanthrene injected subcutaneously caused the development of fibrosarcomas in galagos and tree shrews, and other chemicals have produced epidermal and adnexal tumors as well as lipomas [78]. A more recent report listed 11 spontaneous neoplasms in nine *Tupaia glis* [47]. There were four tumors of the liver or pancreas, including two primary hepatocellular carcinomas and two pancreatic carcinoids (one had metastasized); three skin tumors, including an epidermoid carcinoma and two carinomas of apocrine sweat glands; three lymphoreticular malignancies (from which two different herpesviruses were isolated); and one mixed osteosarcoma and squamous cell carcinoma arising within the uterus. Liver and skin were also the sites affected in a review of lesions in zoo prosimians [45] in which two hepatomas in black lemurs (*M. macaco*) and a widespread ductal sclerosing adenocarcinoma of the mammary glands in a tree shrew were found. A review of the pathology of prosimians [6] found only a few tumors in the literature: an osteosarcoma of the hand of a ring-tailed lemur, a hepatoma in a black lemur, three unspecified neoplasms in lemurs, a pheochromocytoma in a ring-tailed lemur [95], mammary carcinomas in three ring-tailed lemurs [118], and spontaneous cholangiocarcinoma in a ring-tailed lemur

TABLE 53.4. Macaques in survey of neoplasms of nonhuman primates.

Species	Number
Macaca mulatta	151
M. fascicularis	16
M. arctoides	12
M. radiata	10
M. nemestrina	4
M. sinica	3
M. fuscata	2
M. sylvanus	2
M. nigra	2
M. maura	1
M. assamensis	1
Total	204

TABLE 53.5. Other Old World monkeys in survey of neoplasms in nonhuman primates.

Species	Number	Species	Number
Guenons, vervets, patas, etc.		Baboons	
Cercopithecus aethiops	3	*Papio papio*	7
C. (aethiops) sabeus	3	*P. hamadryas*	3
C. neglectus	2	*P. ursinus*	1
C. diana	1	*P. anubis*	2
C. mona	1	*P. cynocephalus*	3
C. albogularis	1	*Papio* sp.	5
C. mitis albogularis	1	*Mandrillus sphinx*	4
Allenopithecus nigroviridis	1	*M. leucophaeus*	1
Miopithecus talapoin	1		
Erythrocebus patas	3	Langurs	
		Presbytis obscurus	2
Mangabeys			
Cercocebus atys	4	Total	49

TABLE 53.6. Apes represented in survey of neoplasms in nonhuman primates.

Species	Number
Gibbons	
Hylobates lar	7
H. concolor	1
H. hoolock × *agilis*	1
Hylobates sp.	5
Orangutan	
Pongo pygmaeus	6
Chimpanzee	
Pan troglodytes	6
Gorillas	
Gorilla gorilla	1
Total	27

TABLE 53.7. Survey of neoplasia in prosimians by taxonomic group and anatomic location.

	Tupaiidae (tree shrews) (n = 2)	Lemuridae (lemurs) (n = 13)	Galaginae (galagos) (n = 16)	Lorisinae (lorises and pottos) (n = 10)
Skin and subcutis	2M[1]	1B	4B, 3M	1M
Musculoskeletal	0	0	0	0
Oral cavity	0	0	0	0
Gastrointestinal	0	1M	1B	2B
Liver and pancreas	0	2M	1B, 2M	2M
Respiratory	0	3B	1B	0
Hematopoietic/ lymphoreticular	0	1M	0	0
Urinary	0	3B	1B	0
Female reproductive	0	0	1B	1B, 1M
Male reproductive	0	2B	0	0
CNS/special senses	0	0	0	1M[2]
Endocrine	0	0	2B, 1M	0
Primary unidentified	0	0	0	2M
Totals	2 (2M)	13 (4M)	17 (6M)	10 (7M)

[1] B, benign; M, malignant; n, number of animals; total, number of tumors.
[2] This was an invasive intraorbital tumor without histologic diagnosis in a *Loris tardigradus*.

TABLE 53.8. Malignancies in prosimians reported in survey.

Species	Age (years)	Sex	Diagnosis	Contributor
Microcebus murinus	—	—	Heptocellular carcinoma (M)[1]	BZ
Cheirogaleus major	Ad	M	Pancreatic adenocarcinoma (M)	NZP
Lemur catta	10	F	Colonic adenocarcinoma	PZ
L. mongoz	17	M	Thymoma, malignant (I)	VMTH
Tupaia sp.	Ad	F	Papillary carcinoma, apocrine sweat glands, skin of abdomen	CZ
Tupaia sp.	Ad	F	Adenocarcinoma, mammary or apocrine (M)	AFIP
Galago crassicaudatus	7	M	Cholangiocarcinoma (also had bronchial adenoma)	NZP
G. crassicaudatus	Ad	F	Mass on chest (mammary ?) (M)	BFZ
G. crassicaudatus	Ad	F	Liposarcoma, axilla (M)	BFZ
G. crassicaudatus			Fibrosarcoma, skin (M)	CPRC[2]
G. crassicaudatus	5	M	"Liver cancer"	BZ
G. senegalensis	Ad	M	Thyroid carcinoma	NZP/AFIP
Nycticebus coucang	Ad	M	Gelatinous tumors liver and kidney (myxosarcomas?)	BFZ
N. coucang	16	F	Undifferentiated sarcoma, primary undetermined (M)	BFZ
N. coucang	5	F	Pancreatic adenocarcinoma (M)	PZ
Perodicticus potto	16	F	Myeloliposarcoma, perineum	AFIP[2]
P. potto	8	F	Ovarian carcinoma, solid (M) (? dysgerminoma)	AFIP
P. potto	10	M	Bile duct carcinoma (M)	PZ
Loris tardigradus	—	—	Invasive intraorbital tumor (sarcoma?)	BZ

[1] M, metastatic; I, invasive; Ad, adult.
[2] Previously reported in the literature.

[20]. In a recent paper describing neoplasms found in birds and mammals from the London Zoo, seven of 11 tumors reported from tree shrews and primates occurred in prosimians: an adenocarcinoma of the pancreas in a large tree shrew (*Lyonogale tana*), a malignant mixed mammary tumor and a thyroid adenoma in thick-tailed galagos (*G. crassicaudatus*), and a prostatic adenocarcinoma in a slow loris (*N. coucang*) in addition to three previously reported mammary carcinomas in ring-tailed lemurs [119]. One of the female lemurs with breast cancer also had a vaginal papilloma. Tumors reported in dwarf galagos (*G. demidovii*) were two bile duct carcinomas and a subcutaneous fibrous histiocytoma [10].

Thus, in all studies reported to date neoplasia in prosimians most frequently involves the integument, liver, and pancreas.

New World Monkeys (Callitricidae, Aotidae, Cebidae)

A total of 46 tumors were reported from 41 individuals (Table 53.9). Gastrointestinal lesions were the most common (26%), accounting for half of all malignancies (Table 53.10). Most were colonic adenocarcinomas in cotton-topped tamarins (*S. oedipus*), some of which are probably the same as reported [97]. Duodenal adenocarcinomas were found in a titi monkey (*Callicebus moloch*) and in a capuchin (*C. apella*). An esophageal squamous cell carinoma was reported in a squirrel monkey that also had a gastric leiomyoma, and an additional gastric leiomyoma was recorded in an owl monkey (*Aotus trivirgatus*). Similar tumors are in the literature, including several reports of colonic malignancies associated with chronic colitis ("marmoset wasting syndrome") primarily in cotton-tops [17, 22, 97]. A recent review of gastrointestinal neoplasms in all nonhuman primates [27] highlighted colonic and cecal adenocarcinomas in this species and recorded adenocarcinomas in the duodenum of a ring-tailed capuchin (*C. albifrons*), the ileum of a saddle-backed tamarin (*S. fuscicollis*), and the colon of a Goeldi's marmoset (*Callimico goeldii*). Also of interest in that paper were tumors of tooth gum origin in *Cebus* and *Ateles* that were found in this survey as well.

Tumors of the female reproductive tract, urinary system, and endocrine tissue accounted for 15%, 13%, and 11%, respectively of all tumors in platyrrhines. Most were benign. They included mainly gonadal tumors (an ovarian granulosa cell tumor in a spider and a squirrel monkey), lipid cell tumors or luteomas in two woolly monkeys (*Lagothrix lagothricha*), a congenital arrhenoblastoma in a young spider monkey (*A. belzebuth*), and an ovarian adenocarcinoma in a capuchin. A uterine leiomyoma was found in an aged squirrel monkey. Female reproductive tumors have been reported in past studies and have included fibromyomata and granulosa cell tumors in a squirrel monkey, ovarian adenocarcinoma in a capuchin (which might be the same individual as in this survey), a Brenner tumor in a black spider monkey, and a uterine hamartoma in a cotton-topped tamarin [107]. The ovaries of female cebids appear to have a developmental phase in which there is massive proliferation of interstitial cells [29]. This physiologic phenomenon may have contributed to the formation of the "luteomas" described in woolly monkeys. Tumors of the male gonads and accessory sex glands were infrequent in this survey and the literature. In this survey an interstitial cell tumor in a squirrel monkey, a seminoma in a howler monkey, and a prostatic adenocarcinoma in a 23-year-old squirrel monkey were the only

TABLE 53.9. Survey of neoplasia in New World monkeys by group and location.

	Callitrichidae (marmosets and tamarins) (n = 12)	Aotidae (owls and titis) (n = 4)	Cebidae (squirrels, capuchins, howlers, etc.) (n = 25)
Skin and subcutis	0	0	3B[1]
Musculoskeletal	0	0	0
Oral cavity	0	0	2M
Gastrointestinal	7M[2]	1B, 1M[2]	1B, 2M
Liver and pancreas	1B	0	1B, 1M[3]
Respiratory	0	0	0
Hematopoietic/ lymphoreticular	1M	1M	1B
Urinary	3B	1M	1B
Female reproductive	0	0	6B, 1M
Male reproductive	0	0	2B, 1M
CNS/special senses	0	0	0
Endocrine	0	1B	5B[4]
Primary unidentified	0	0	2M
Total	12 (8M)	5 (3M)	29 (9M)

[1] B, benign; M, malignant.
[2] All GI malignancies in callitricidae were colonic adenocarcinomas in *Saguinus oedipus* associated with chronlic colitis. An adenocarcinoma of Brunner's glands in a *Callicebus moloch* was also associated with chronic enteritis.
[3] A cholangiocarcinoma in a 25-year-old *Cebus albifrons* was associated with hepatic hemosiderosis. This animal also had a renal papillary cyst adenoma. Iron storage has been associated with hepatic tumors in humans and lemurs.
[4] A 14-year-old *Alouatta villosa* had multiple endocrine tumors: a thyroid chief cell adenoma; an adrenal pheochromocytoma; and an islet cell adenoma. Endocrine adenomas have been reported in wild *A. caraya* and *villosa*.

TABLE 53.10. Malignancies in New World monkeys in survey.

Species	Age (years)	Sex	Diagnosis	Contributor
Cebulla pygmaea	14	M	Myeloproliferative disease	AFIP
Saguinus oedipus	8	M	Colonic mucinous adenocarcinoma	AFIP[1]
S. oedipus	Ad	F	Colonic adenocarcinoma	AFIP
S. oedipus	Ad	F	Colonic adenocarcinoma	AFIP
S. oedipus	Ad	M	Colonic adenocarcinoma	AFIP
S. oedipus	Ad	M	Colonic adenocarcinoma	AFIP
S. oedipus	Ad	M	Colonic adenocarcinoma	AFIP
S. oedipus	Ad	F	Colonic adenocarcinoma	AFIP
Aotus trivirgatus	6	F	Renal adenocarcinoma	AFIP
A. trivirgatus	Ad	F	Myeloproliferative disease	AFIP
Callicebus moloch	4	M	Duodenal adenocarcinoma	BFZ/NZP
Cebus apella	?	?	Duodenal adenocarcinoma	AFIP
C. apella	Ad	M	Multiple tumors subcutis and viscera, primary and histopath undetermined	BFZ
Cebus sp.	Ad	M	Calcifying amyloblastoma	AFIP
Cebus sp.	40	M	Disseminated adenocarcinoma	AFIP
C. capucinus	Ad	M	Amyloblastic odontoma (I)	AFIP[1]
C. albifrons	Ad	F	Ovarian adenocarcinoma	AFIP[1]
C. albifrons	25	M	Cholangiocarcinoma, liver	AFIP
Saimiri sciureus	23	M	Prostatic carcinoma	YPC
S. sciureus	Ad	F	Esophageal squamous cell carcinoma	YPC

[1] Previously reported in the literature.

neoplasms of this organ system. In a review of the literature, only an additional seminoma in a wild howler monkey (*Alouatta caraya*) was found.

Both urinary tract and endocrine tumors in New World primates in this survey were primarily benign. These included an embryonal nephroma in a juvenile golden lion tamarin, a renal tubular adenoma in a young Goeldi's marmoset, and a hemangioma or vascular hamartoma in the kidney of a moustached tamarin (*Saguinus mystax*). A literature review and series of new cases of urinary tract tumors in nonhuman primates includes a renal carcinoma in a uakari (*Cacajao* sp.), a carcinoma in a capuchin (*C. apella*) and an owl monkey, a renal adenoma in a howler monkey (*A. caraya*), a hypernephroma in a capuchin (*C. a. fatuellus*), a papillary cystadenoma in a spider monkey (*A. geoffroyi*), and a renal papillary adenoma in a cotton-topped tamarin [60]. An additional case report of a renal papillary adenoma in a cotton-top has been published [11].

The endocrine tumors in this survey were a thyroid adenoma in a female squirrel monkey, a benign "adrenal tumor" in an adult female owl monkey, and multiple endocrine tumors (thyroid chief cell adenoma, a pheochromocytoma, and an islet cell adenoma) in a 14-year-old male mantled howler monkey (*A. villosa*). Interestingly, a 13-year-old female of this species from the same zoo died the preceding year of a thymoma which, although not invasive, had compressed trachea and esophagus. Two papers in the literature report the frequent occurrence of endocrine tumors, adrenal adenomas, in howler monkeys (*A. villosa* and *A. caraya*) in Latin America [52, 75]. Adrenal adenomas have also been reported from a *C. fatuellus* [107].

Although only two hematopoietic neoplasms (myeloproliferative disease in a pygmy marmoset [*Cebuella pygmaea*] and in an adult female owl monkey) were found in this study, there are several reports of both spontaneous and experimentally induced leukemias, lymphomas, and lymphoproliferative disorders in marmosets, tamarins, and owl monkeys. These are largely due to the sensitivity of these species to the oncogenic herpesviruses, *H. saimiri*, *H. ateles*, and Epstein-Barr virus [31, 35]. Affected species in one review included *Saguinus fusicollis*, *nigricollis*, and *oedipus*; *Cebus* sp. and *A. trivirgatus* [107]. Three black-tailed marmosets (*Callithrix argentata*) at the San Diego Zoo died of lymphoproliferative disease after being put in an exhibit with squirrel monkeys (the carriers of oncogenic *H. saimiri*) [45]. Owl monkeys have also been reported to develop spontaneous lymphomas and lymphoproliferative disease when placed in contact with squirrel monkeys [56, 91]. Myelogenous disorders are less common, but there are case reports of an eosinophilic myeloma in an owl monkey [16] and myeloproliferative disease in a cotton-top [67].

In contrast to the frequent occurrence in prosimians, only three tumors of the liver were recorded in this survey. Liver tumors were "lymphoid-like nodules" in the liver of a common marmoset (*C. jacchus*) that were not examined histologically, a fibroma in the liver of an adult female brown capuchin (*C. apella*), and a cholangiocarcinoma in a 25-year-old male *C. apella* that also had a renal papillary adenoma. A review article found liver tumors in platyrrhine monkeys only in marmosets and tamarins given aflatoxins and hepatitis virus [107], but there is a recent report of a cholangiocarcinoma in an older (25-year-old) male *C. albifrons* [13].

Although bone tumors are rare and none were reported from platyrrhines in this survey, osteosarcoma of the tibia has been reported in a squirrel monkey

[68]. A review article found only an osteoma in a capuchin (*C. fatuellus*) [107]. Skin tumors were infrequent: a hemangioma in a spider monkey and a brown capuchin and a papilloma on the eyelid of a spider monkey. Neoplasms of this organ system are also infrequent in New World primates in the literature although an epidermal squamous cell carcinoma was recently reported on a white-lipped tamarin (*S. fuscicollis*) [96].

A group of neoplasms not found in this survey but receiving three publications in recent years are carcinomas of the nasal and oral cavities and pharynx in common marmosets [2, 8, 81]. A total of 13 of these tumors have been reported from laboratory colonies in Britain (11) and Australia (2). This is interesting in light of the association of the oncogenic Epstein-Barr herpesvirus with nasopharyngeal carcinomas in humans, although antibodies to this virus were not found in one study [81].

In summary, inflammatory bowel disease, so common in New World monkeys [19], is an important factor in the large numbers of malignancies in the GI tract.

Old World Primates—Macaques

Tumors in macaques are listed in Tables 53.11 and 53.12. *Macaca* was the most frequently reported genus (204 individuals of which 151 were rhesus). Lymphoreticular and hematopoietic neoplasms (mainly lymphosarcoma) accounted for a quarter of all tumors and almost half (43%) of malignancies reported in

TABLE 53.11. Survey of neoplasms in macaques by anatomic location.

	Rhesus (*M. mulatta*) (n = 151)	Cynomolgus (*M. fascicularis*) (n = 16)	Other[1] (n = 37)
Skin and subcutis	17B, 16M	1B	4B, 1M
Musculoskeletal	3M	0	0
Oral cavity	1B, 4M	1M	0
Gastrointestinal[2]	5B, 12M	2M	3B, 5M
Liver and pancreas	3B, 2M	1B, 1M	3B, 1M
Respiratory	1M	0	2M
Hematopoietic/ lymphoreticular	47M[3]	0	8M[3]
Urinary	5M	1B	1M
Female reproductive	20B, 3M	2B, 2M	4B, 2M
Male reproductive	0	1B	0
CNS/special senses	3M	0	1M[4]
Endocrine	17B, 1M	4B, 1M	5B
Primary unidentified	2M	0	0
Total	162 (99M)	17 (7M)	40 (21M)

[1] See Table 53.3.
[2] This category includes two cases involving primarily the peritoneum: (1) a mesothelioma in a *M. nemestrina*; and (2) multiple hemangiosarcomas, hemangiomas, and fibromas of the peritoneum and gastric submucosa in a *M. fascicularis*.
[3] Forty-two of the lymphosarcomas in rhesus and the eight in "Others" (*M. arctoides*) were from one colony in which outbreaks of infectious lymphomas occurred.
[4] This was a malignant neuroblastoma in the abdominal cavity of a 1-day-old female *M. nemestrina*. Neuroblastomas are often tumors of infancy and childhood in humans and commonly occur in the abdominal cavity.

TABLE 53.12. Malignancies in macaques in survey of neoplasia in nonhuman primates.

Species	Age	Sex	Diagnosis	Contributor
Macaca arctoides	?	?	Leiomyosarcoma, lung, heart, kidney	LB
M. arctoides	9 1/2	F	Lymphosarcoma	CPRC[1]
M. arctoides	8 mo	M	Lymphosarcoma	CPRC[1]
M. arctoides	2 1/2	F	Lymphosarcoma	CPRC[1]
M. arctoides	2	F	Lymphosarcoma	CPRC[1]
M. arctoides	> 10	F	Lymphosarcoma	CPRC[1]
M. arctoides	5 1/2	M	Lymphosarcoma	CPRC[1]
M. arctoides	5	M	Lymphosarcoma	CPRC[1]
M. arctoides	3 1/2	F	Lymphosarcoma	CPRC[1]
M. fascicularis	7	F	Hepatocellular carcinoma	JH
M. fascicularis	Ad	M	Multiple hemangiosarcomas, hemangiomas, and fibromas in peritoneum and gastric submucosa	JH
M. fascicularis	4	M	Adrenal cortical carcinoma	LB
M. fascicularis	old	F	Squamous cell carcinoma, buccal pouch	VMTH[1]
M. fascicularis	?	?	Carcinoid of small intestine	AFIP
M. fascicularis	Ad	M	Ovarian carcinoma	AFIP
M. fascicularis	3	F	Ovarian carcinoma	PZ
M. fuscata	15 1/2	M	Gastroesophageal squamous cell carcinoma	AFIP/PZ
M. fuscata	?	?	Hepatocellular carcinoma	BZ
M. mulatta	Ad	M	Invasive osteofibroma, ribs	CPRC
M. mulatta	3	F	Multicentric fibrosarcoma, skin and subcutis	CPRC[1]
M. mulatta	3 1/2	F	Solitary fibrosarcoma, skin and subcutis	CPRC[1]
M. mulatta	> 9	M	Carcinoma, mandible (toothgerm?)	CPRC[1]
M. mulatta	> 8	F	Squamous cell carcinoma, skin of face	CPRC[1]
M. mulatta	16	F	Adenocarcinoma of jejunum	CPRC
M. mulatta	11	F	Gastric leiomyosarcoma	CPRC
M. mulatta	22	M	Adenocarcinoma of jejunum	CPRC
M. mulatta	6	M	Ependymoma, invasive, arising in the third ventricle-infundibular recess	CPRC
M. mulatta	26	M	Hepatocellular carcinoma and cholangiocarcinoma (2 distinct types)	YPC
M. mulatta	26	M	Carcinoma of small intestine	YPC
M. mulatta	Ad	F	Metastatic thyroid carcinoma	YPC
M. mulatta	25	M	Metastatic basal cell carcinoma	YCP
M. mulatta	29	M	Oral squamous cell carcinoma (M)	YCP
M. mulatta	Ad	F	Renal carcinoma	YCP
M. mulatta	24	F	Carcinoma of small intestine	YCP
M. mulatta	28	F	Colonic carcinoma (metastatic)	YPC
M. mulatta	29	M	Carcinoma of small intestine	YPC
M. mulatta	29	F	Lymphoma	YPC
M. mulatta	15	F	Squamous cell carcinoma, skin	YPC
M. mulatta	21	M	Basal cell carcinoma, skin	YPC
M. mulatta	Ad	M	Squamous cell carcinoma, skin of penis	YPC
M. mulatta	24	M	Carcinoma of small intestine	YPC
M. mulatta	24	M	Leiomyosarcoma, colon	YPC
M. mulatta	25	F	Mammary carcinoma	YPC
M. mulatta	21	M	Colonic carcinoma	YPC
M. mulatta	23	F	Bilateral lobular carcinoma, mammary glands	JH
M. mulatta	Ad	M	Squamous cell carcinoma, tongue	JH
M. mulatta	Ad	F	Transitional cell carcinoma, bladder	JH
M. mulatta	38	M	Invasive squamous cell carcinoma of the orbit	JH

TABLE 53.12. (Continued).

Species	Age (years)	Sex	Diagnosis	Contributor
M. mulatta	Old	F	Cervical epidermoid carcinoma, also papillomas of vagina and bladder and mammary gland intraductal carcinoma	JH
M. mulatta	Old	F	Ovarian adenocarcinoma	JH
M. mulatta	Ad	F	Metastatic squamous cell carcinoma, cervix	LB
M. mulatta	Ad	F	Mammary carcinoma	LB
M. mulatta	Ad	F	Metastatic mammary carcinoma	LB
M. mulatta	Ad	M	Mammary carcinoma	LB
M. mulatta	Ad	F	Mammary carcinoma	LB
M. mulatta	?	?	Malignant lymphoma	LB
M. mulatta	> 12		Colonic carcinoma	LB
M. mulatta	?	?	Esophageal squamous cell carcinoma	LB
M. mulatta	?	M	Myelogenous leukemia	AFIP
M. mulatta	?	?	Squamous cell carcinoma, skin over mammary gland	AFIP[1]
M. mulatta	?	?	Astrocytoma, grade 2, cerebrum	AFIP
M. mulatta	?	?	Adamantinoma, mandible	AFIP
M. mulatta	Ad	M	Histiocytic sarcoma, multicentric	AFIP
M. mulatta	6	M	Widely metastatic carcinoma, primary?	AFIP
M. mulatta	6	?	Undifferentiated sarcoma, elbow	AFIP
M. mulatta	9	?	Myxosarcoma, chest	AFIP
M. mulatta	6	F	Carcinoid of lung	AFIP
M. mulatta	7	M	Undifferentiated sarcoma, hand	AFIP
M. mulatta	> 9	M	Renal adenocarcinoma	PZ
M. mulatta	> 10	M	Renal carcinoma (metastatic hypernephroma)	PZ
M. mulatta	23	M	Renal carcinoma (hypernephroma)	PZ
M. mulatta	7 mo–> 9	35 F 7M	Lymphosarcomas (42 cases)	CPRC[1]
M. maurus	29	M	Squamous cell carcinoma, skin	PZ
M. nemestrina	5	F	Mesothelioma, abdominal cavity	YRPRC
M. nemestrina	1 day	F	Neuroblastoma, abdominal cavity	YRPRC
M. radiata	Ad	F	Pulmonary adenocarcinoma (type II cell)	CPRC[1]
M. radiata	> 12	F	Metastatic intestinal carcinoma	CPRC
M. radiata	> 12	F	Ovarian adenocarcinoma	CPRC[1]
M. sinica	> 11	M	Rectal adenocarcinoma invading to prostate	PZ
M. sinica	> 6	F	Ovarian carcinoma	PZ
M. sylvanus	?	?	Gastric adenocarcinoma	PZ

[1] These cases have been reported in the literature.

macaques. This was largely due to the presence in the survey of rhesus and stump-tails from one colony (CPRC) that had experienced outbreaks of infectious lymphoma apparently in conjunction with an epizootic of simian acquired immunodeficiency syndrome (SAIDS) [48–50, 74, 108, 112]. If all macaques from CPRC are excluded from the survey, lymphoreticular tumors become the fourth most common malignancy accounting for only 8% of malignancies of macaques. In addition to the outbreak at the CPRC, there are several other reports of hematopoietic neoplasms including plasmacytomas [51, 57, 72, 74, 117]. Tumors have been of both B- and T-cell origin and have arisen both in lymph nodes and extra nodal sites including the orbit [50, 76]. Although the exact etiology is unknown, their transmissible (infectious) nature is documented [57, 108]. Although fairly uncommon in most species, a thymoma has also been reported in a *M. arctoides* [107].

Tumors of the skin (17%), female reproductive tract (16%), endocrine glands (13%), and gastrointestinal tract (13%) were also relatively common in macaques in this survey. Neoplasms of the stomach and intestines were often malignant and included carcinomas and adenocarcinomas of both small and large colon (Table 53.11). Benign growths of the gastrointestinal tract were leiomyomas of esophagus and stomach in four macaques, a leiomyoma of the small bowel in one and of the large bowel in two animals, adenomatous polyps of small intestine and colon in two, and a salivary gland adenoma in a 26-year-old female rhesus. Also included were multifocal lesions of the peritoneal serosa in two macaques (Table 53.10). In a review of GI neoplasms in nonhuman primates, a similar array of lesions was reported [27]. In that study were proliferative lesions (papillomas) in the stomachs of several cynomolgous and stump-tails associated with the stomach worm *Nochtia nochtia*. Additional reports detail cases of esophageal squamous cell carcinoma in a rhesus [92] and papillomatous gastric polyps in a *M. fascicularis* [85].

Another proliferative disorder involving the intestinal tract not reported in this survey is the syndrome of retroperitoneal fibromatosis [40, 42]. This condition, involving the mesenteries and serosal surfaces, has been strongly associated with a type D retrovirus and a SAIDS-like syndrome [76, 109]. It has occurred endemically in primate center colonies of pig-tailed and Celebes macaques and has been responsible for many deaths in pig-tails in one colony [40, 42].

Both benign (21) and malignant (17) tumors of skin and subcutis, including mammary gland, were found in macaques in this survey. Among the malignancies were five squamous cell carcinomas of the face, penis, chest, and one unspecified site. A malignant basal cell tumor was reported in a rhesus and five benign basal cell tumors in a cynomolgous and four rhesus. Other benign tumors were calcifying epitheliomas in a rhesus and Barbary "ape," an epidermal inclusion cyst, an adnexal adenoma, and two cases of papillomas in rhesus. Mammary gland carcinomas were reported in seven rhesus, one of which was a male. One was an intraductal carcinoma in an old rhesus that also had cervical carcinoma and papillomas of vagina and bladder.

Mesenchymal skin tumors in the survey included benign lipomas (7), hemangiomas (3), and a fibroma. Malignancies were a myxosarcoma and two fibrosarcomas. The latter were in subadult female rhesus that had SAIDS [90]. The lesions were similar to "subcutaneous fibromatosis" in a recent report [115] and may be a variant of retroperitoneal fibromatosis. Similar experimentally produced lesions have also been called "Kaposi's-like skin lesions" [73].

Skin tumors in nonhuman primates have been reviewed in a case report of a trichoepithelioma in a Barbary ape [12]. Only 16 were found in the literature that did not include mammary neoplasms. Reported from macaques were squamous cell carcinomas, basal cell tumors, multiple papilliferous cystadenoma in a cynomolgous [65], and an unspecified carcinoma. There are also recent case reports of metastatic squamous cell carcinoma in a rhesus and a cynomolgous [53, 83]. A large survey of necropsies in nonhuman primates listed hemangiomas, lipomas, and a mast cell tumor [78]. Mammary neoplasia, while not common, has been reported in five macaques [107]. One had received oral contraceptives and another had been irradiated. Two intraductal mammary carcinomas in rhesus [30, 111] and a mixed mesodermal sarcoma have been reported [37].

Another skin tumor not found in this survey but reported in the literature is subcutaneous histiocytoma. The tumor is induced by Yaba and other closely related poxviruses and is usually self-limiting [106].

Female reproductive tract tumors were fairly common in macaques; most were benign [26, 33]. In contrast to New World monkeys, tumors of the uterus and cervix were more common than gonadal tumors. These were leiomyomas of uterus and cervix (16), a uterine hemangioma, and two cases with genital polyps. Malignancies were one squamous cell and two other carcinomas of the cervix. Benign ovarian tumors included two teratomas, a theca cell tumor, a granulosa cell tumor, a luteoma, a fibroepithelial polyp, and a cystadenoma. In addition, four ovarian carcinomas were reported (including a previously reported adenocarcinoma [15]). Recent review articles [78, 107] found tumors of the female genitalia among the most common in macaques. A fibroma of the uterus in a Japanese red-faced macaque (*M. fuscata*) was one of three neoplasms reported from macaques at the San Diego Zoo [45]. A high incidence of benign uterine tumors in macaques has also been highlighted in a study from the Russian primate center [70].

In literature reviews as well as in this survey, tumors of the male genital tract were uncommon [78, 107]. The one testicular tumor found in this survey was in an elderly *M. fascicularis* (*irus*), which was not examined histologically. A prostatic carcinoma was described in a rhesus that according to the authors brought the reported cases in this species to three [54].

Endocrine tumors were frequently reported in macaques. Only two of the 28 in the survey were malignant: an adrenal cortical carcinoma and a thyroid carcinoma. Benign tumors included eight adrenal cortical adenomas, five pheochromocytomas, five islet cell adenomas, three pituitary adenomas, and two thyroid adenomas. Similar tumors have been reported in reviews [78, 107]. Functional abnormalities (e.g., Cushing's syndrome) have not been documented with these tumors, and it seems that, as in human primates in which adrenal adenomas are found in 2% of adult necropsies, most are nonsteroid producing [98]. Pancreatic islet cell adenomas have also not been associated with endocrine imbalances although they are frequent findings. Such a tumor in a female lion-tail macaque was one of three neoplasms found in genus *Macaca* at the San Diego Zoo [45]. Pituitary adenomas are less frequently reported in the literature. A recent paper documented a thyrotroph adenoma and colloid goiter in a *M. fascicularis* that had no clinical, hematologic, or biochemical abnormalities [116]. Another pituitary tumor, a nonsecreting adenoma of the pars intermedia, was associated with galactorrhea in a male rhesus [18].

Additional neoplasms in macaques were scattered among other organ systems. Respiratory and musculoskeletal systems were represented by only three tumors each, all of which were malignant (Table 53.11). This is in keeping with the relatively rare incidence in reviews as well [41, 78, 107]. Recent reports of respiratory tumors in macaques have included a bronchial adenoma in a lion-tail from the San Diego Zoo [45], a clear cell carcinoma in a pig-tail [114], and the type II cell carcinoma in a bonnet monkey included in this survey [87]. Recent reports of bone tumors in macaques were an osteosarcoma of the long bone [102] and a mandibular myxoma [103] in rhesus.

Although several reviews found frequent tumors of the urinary tract in macaques [60, 78, 107] and although renal tumors have been reported from a family of rhesus [94], and a malignant nephroblastoma was reported from a cynomolgous [5], only seven were found in this study. Six were malignant. The benign tumor was a mesonephric adenoma of the bladder in a 15-year-old female *M. fascicularis* that also had endometriosis. Oral cavity neoplasms that are not frequent in the literature [107] were recorded from six macaques in this study. Three were squamous cell carcinomas.

Other Old World Monkeys Including Colobines, Guenons, and Baboons

Fifty-two tumors were found in 49 catarrhine monkeys from eight genera and at least 18 species (Table 53.5). Tumors of the urinary tract (19%), skin and subcutis (17%), female reproductive tract (14%), and liver and pancreas (14%) were the most common (Table 53.13). As in macaques, tumors of the male genitalia, respiratory, musculoskeletal, and nervous systems were infrequent. Interestingly, in contrast to macaques and New World monkeys, gastrointestinal tumors were infrequent. Most tumors recorded were benign. Although only three hematopoietic tumors were found, all were malignant. This made them the second most frequent malignancy, tied with liver and pancreas, and far behind urinary tract (Table 53.14). Seven of ten urinary tract tumors were malignant and accounted for 33% of malignancies in this group of nonhuman primates.

The malignant urinary tumors were five renal adenocarcinomas in baboons (*Papio* and *Mandrillus* spp.), bilateral renal adenocarcinomas in an Allen's

TABLE 53.13. Survey of anatomic location of neoplasms in Old World monkeys, excluding macaques.

	Guenons etc. (n = 17)	Mangabeys (n = 4)	Langurs (n = 2)	Baboons (n = 26)
Skin and subcutus	4B[1]	0	0	4B, 1M
Musculoskeletal	0	1B	0	1M
Oral cavity	0	0	1M	1B, 1M
Gastrointestinal	1B	0	0	1B
Liver and pancreas	2B, 1M	0	0	2B, 2M
Respiratory	1M	0	0	1B
Hematopoietic/	0	0	0	
lymphoreticular	3M	0	0	0
Urinary	2M	0	0	3B, 5M
Female reproductive	2B	1B, 1M	1M	2B
Male reproductive	0	0	0	1B
CNS/special senses	0	0	0	1B
Endocrine	1B	1B	0	2B
Primary unidentified	0	0	0	1M
Total	17 (7M)	4 (1M)	2 (2M)	29 (11M)

[1] B, benign; M, malignant.

TABLE 53.14. Malignancies in Old World primates, excluding macaques, in survey.

Species	Age (years)	Sex	Diagnosis	Contributor
Cercopithecus aethiops	—	—	Leukemia	LB
C. aethiops sabeus	14	F	Pancreatic adenocarcinoma	PZ[1]
C. aethiops sabeus	2	F	Lymphocytic leukemia	PZ
C. albogularis	Ad	F	Nasal carcinoma (M)[2]	BFZ
Allenopithecus nigroviridis	17	M	Bilateral renal papillary cyst adenocarcinomas	NZP[1]
Miopithecus talapoin	15	M	Myeloproliferative disease	BFZ
Erythrocebus patas	12	F	Renal adenocarcinoma	OCZ
Cercocebus atys	8	F	Uterine tumor metastatic to liver (adenocarcinoma)	PZ
Presbytis obscurus	2 1/2	M	Squamous cell carcinoma, oral cavity	PZ
P. obscurus	Ad	F	Ovarian papillary cyst adenocarcinoma	JH
Papio papio	23	M	Renal carcinoma	PZ
P. papio	23	M	Biliary adenocarcinoma, gall bladder, also had hepatic biliary cyst adenomas	PZ
Papio sp.	Ad	F	Squamous cell carcinoma, perineum	VMTH
Papio sp.	11	M	Squamous cell carcinoma, oral cavity	NZP/AFIP
Papio sp.	27	M	Disseminated adenocarcinoma, primary undetermined	AFIP
P. cynocephalus	18	M	Renal tubular adenocarcinoma	CPRC
P. cynocephalus	17	F	Pancreatic adenocarcinoma (also had uterine cyst adenoma)	PZ[1]
Mandrillus sphinx	15	F	Renal adenocarcinoma	PZ
M. sphinx	31	F	Renal adenocarcinoma	PZ
M. sphinx	Ad	F	Renal cell carcinoma	AFIP
P. ursinus	9	M	Giant cell tumor of bone (M)	AFIP/PZ[1]

[1] Previously reported in the literature.
[2] (M), metastasis.

swamp monkey, and a renal adenocarcinoma in a patas monkey. Benign lesions were renal adenomas in a guinea baboon (*P. papio*) and a drill (*M. leukophaeus*), both of which were 15-year-old males, and an unclassified "renal tumor" in a 4-year-old guinea baboon. Two had been previously reported in a review of primary renal tumors, but in that study most lesions had occurred in macaques [60]. A renal adenocarcinoma was identified in a silver leaf langur (*Presbytis cristatus*) at the San Diego Zoo [45].

Several skin tumors were diagnosed in monkeys of this group, but they were primarily benign (8/9). These included verrucae of the ischial callosities of an African green monkey, multiple basal cell tumors in a mandrill, a basal cell tumor in a Debrazza's monkey, hemangiomas in two young guinea baboons, lipomas in a patas monkey and an African green monkey, and a mast cell tumor in a male baboon. The one malignancy was a squamous cell carcinoma of the perineal region of a female baboon from California. A review of skin tumors in the literature revealed only a sarcoma in a *Cercopitheus* spp. and a squamous cell carcinoma of the sex skin in a *P. hamadryas* from the Russian primate center. A fibroma of the subcutis of the face and a basal cell carcinoma of the skin over the back have been reported in free-living chacma baboons. Although no mammary tumors were documented in this survey, a fibrosarcoma has been reported in a *C. aethiops*, a squamous cell epithelioma from a hamadryas baboon, and an adenocarcinoma in a female mandrill [107]. Additional skin tumors in the literature are Yaba virus histiocytomas, which occur in baboons as well as in macaques [14], and cutaneous papillomas (papova virus associated) in colobus monkeys [93]. A malignant fibrous histiocytoma has been reported in a yellow baboon [38].

Seven tumors of the female reproductive tract were found; five were benign. As in macaques, in contradistinction to New World monkeys, the tubular organs were more affected than the gonads. Benign tumors were fibromas or leiomyomas (in an adult sooty mangabey, a 17-year-old Debrazza's monkey, and an African green), one case of vaginal papillomas in an 18-year-old olive baboon that also had tracheal papillomas, and a uterine papillary cystadenoma in a 17-year-old yellow baboon (previously reported [106]). Malignancies were a uterine tumor of undefined histologic type that metastasized to the liver in a sooty mangabey and an ovarian adenocarcinoma in a spectacled langur. The literature also reports tumors in this group including an ovarian serous cystadenoma in a patas monkey, an ovarian sarcoma in a *P. anubis*, a granulosa cell tumor in a chacma baboon, a uterine myoma in a *Papio* sp., leiomyoma in *Cercopithecus* sp., and vaginal leukoplakia in a "*Cercocebus mona*" (*Cercopithecus*?) [107]. Three of five tumors reported from Old World monkeys other than macaques at the San Diego Zoo were uterine leiomyomas in a 26-year-old Allen's swamp monkey and 13-year-old silver leaf langur (which also had a renal adenocarcinoma), and an ovarian granulosa cell tumor in a 23-year-old bonneted langur (*P. pileatus*) [45]. An ovarian adenocarcinoma has been reported from an African green monkey [4].

In the parenchymatous organs of the digestive system (liver and pancreas) there were four benign tumors and three malignancies in this survey. These were a pancreatic acinar tumor in an 8-year-old male olive baboon and biliary or hepatic cystadenomas in a blue monkey (*C. mitis albogularis*), a mona monkey, and a 23-year-old male guinea baboon that also had an adenocarcinoma of the gall bladder. Pancreatic adenocarcinomas were reported from a green monkey and a 17-year-old female yellow baboon with the uterine cystadenoma, both of which previously appeared in the literature [107]. Other tumors in the literature have included a mixed cholangio- and hepatocellular carcinoma in a green monkey, a metastatic hepatoma in a sooty mangabey, an adenoma in a hanuman langur, a cystadenocarcinoma of the gall bladder and a bile duct hamartoma in chacma baboons [107], a pancreatic acinar cell adenoma in a mandrill [79], and a hepatoma in a baboon with cholelithiasis [66].

Few tumors were found in other portions of the digestive tract in this survey or in the literature. These were a squamous cell carcinoma of the oral cavity in an 11-year-old male baboon and a spectacled langur, fibroma on the lip of a hamadryas baboon, diffuse gastric adenomas in a 2-year-old male hamadryas baboon, and an omental lipoma in a mona monkey. In the literature are a lip fibroma and gingival squamous cell carcinoma in baboons, a gastric papillary adenoma, glandular polyps, and multiple adenomas all in hamadryas baboons, a gastric adenocarcinoma in a red-tailed monkey (*C. rufoviridis*), and a duodenal adenocarcinoma in a vervet [27]. A mandibular osteosarcoma and an amyloblastic odentoma in baboons are also in the literature [3, 99], and an adenocarcinoma of Brunner's glands was described in a yellow baboon [113].

As mentioned above, although only three hematopoietic tumors were in the survey, all were malignant: leukemia and lymphocytic leukemia in green monkeys and myeloproliferative disease in a talapoin. The infrequency of lymphoreticular tumors in this survey is in contrast to the numerous hematopoietic neoplasms (over 150) reported from hamadryas baboons from the Sukhumi primate center where an epizootic has been occurring since 1967 [70]. Six malignant lymphomas have also occurred in green monkeys at that facility. More recently, four cases of malignant lymphoma, one resembling Hodgkin's disease, in baboons (*P. anubis*, *P. cynocephalus*, and *P. anubis* × *cynocephalus*) have been reported from a U.S. primate research center [39]. Lymphoreticular neoplasms resembling Hodgkin's disease have been reported in a mangabey (*C. albigena*) in a review of tumors from the London Zoo [119]. In baboons a viral etiology (herpesvirus and/or C-type retrovirus) has been suggested and should be suspected for other primate hematopoietic malignancies.

Respiratory malignancies are infrequent in all nonhuman primates [41]. The findings of a nasal carcinoma in a Sykes monkey in this survey is especially interesting in light of a recent report of a bronchiogenic carcinoma in a Sykes monkey in east Africa [110] and a squamous cell carcinoma of the nasal cavity of a green monkey [33].

Endocrine tumors were infrequent. Islet cell adenomas were found in a young sooty mangabey and a yellow baboon, an adrenal lipoma in a hamadryas baboon, and a pheochromocytoma in the adrenal of a patas monkey. No adenomas were reported in a survey of 604 pairs of adrenals in young baboons [104]. A pituitary carcinoma in an olive baboon and thyroid adenomas in a *P. ursinus* and a *P. porcarius* (? *ursinus*) have been reported [107].

Other tumors in this survey were multiple osteochondromas involving the ribs in a 1-month-old sooty mangabey; a giant cell tumor of the forearm in a male chacma baboon (previously reported); and a cerebral axonal hamartoma in a baboon, probably reported in a survey of lesions in free-living chacma baboons in South Africa in which a meningioma was also reported [80]. Another CNS tumor in baboons was a medulloblastoma in a *Papio papio* [7].

In summary, although neoplasms in this group occur sporadically in all organs, the kidneys, integument, uterus, liver, and pancreas are most frequent except in colonies of baboons in which infectious hematopoietic tumors have occurred.

Apes (Hylobatidae and Pongidae)

This group was the least represented in this survey (only 29 tumors in 27 individuals) (Tables 53.6, 53.15). Hematopoeitic neoplasms were the most frequent, accounting for about one-third of all tumors and two-thirds of malignancies (Tables 53.15, 53.16). There were eight leukemias and lymphomas in gibbons (some of which may have appeared in the literature). This occurrence is well documented; hematopoeitic tumors in gibbons account for most of the reports of neoplasia in apes [24–26, 72, 105]. Several isolates of a type C retrovirus called gibbon ape leukemia virus (GaLV) have been obtained from these tumors [62–64]. There is also a report of a 13-year-old wild-caught female

TABLE 53.15. Survey anatomic location of neoplasms in apes by taxonomic group.

	Hylobatidae (gibbons) (n = 14)	Pongidae (orangutans) (n = 6)	Pongidae (chimpanzees) (n = 6)	Pongidae (gorillas) (n = 1)
Skin and subcutis	2B, 1M[1]	4B	3B	0
Musculoskeletal	0	0	0	0
Oral cavity	0	0	1M	0
Gastrointestinal	1B	1M	1B	0
Liver and pancreas	0	0	0	0
Respiratory	0	0	0	0
Hematopoietic/ lymphoreticular	8M	0	0	1B
Urinary	0	0	0	0
Female reproductive	1M	2B	1B	0
Male reproductive	0	0	0	0
CNS/special senses	0	0	0	0
Endocrine	1B	0	1B	0
Primary unidentified	0	0	0	0
Totals	14 (10M)	7 (1M)	7 (1M)	1 (0M)

[1] M, malignant; B, benign.

orangutan that developed subacute myelomonocytic leukemia. An Epstein-Barr-like herpesvirus was isolated from lymphocytes from this animal, and no retroviral activity (by reverse transcriptase) was found [36].

The single benign hematopoietic tumor in this survey was a splenic hemangioma. This was an incidental finding in a 29-year-old male lowland gorilla with atherosclerosis.

Nine skin tumors were recorded. The single malignancy was a pleomorphic sarcoma in the neck of a gibbon. The benign tumors were two lipomas in chimpanzees, a mast cell tumor on the eyelid of a orangutan, a hemangioma in a young chimpanzee (previously in the literature [78]), a xanthogranuloma in a white-handed gibbon, a trichoepithelioma and a sebaceous adenoma in a 24-year-old male orangutan, a fibroadenoma of the mammary gland in a 28-year-old female white-handed gibbon, and papillomas in an 18-year-old male orangutan. Lipomas in three chimpanzees and a mammary carcinoma in a 15-year-old orangutan were reported in one review [107]. A squamous papilloma in a chimpanzee has also been reported in another series [101].

Reproductive tract tumors found in this survey were uterine leiomyomas in a 45-year-old orangutan, a 44-year-old chimpanzee, and another chimpanzee older than 44 years; a granulosa cell tumor in a 26-year-old orangutan (which also had endometriosis); and a malignant granulosa cell tumor in a 38-year-old hybrid gibbon (*H. hoolock* × *agilis*). Leiomyomas have been reported in chimpanzees [78] (two of which are the ones in this survey); a uterine carcinoma in situ was also reported from a 23-year-old chimp, a teratoma of the ovary in a 11-year-old orangutan [107], and fibrothecomas in two chimps aged 37 and 48 years. The younger also had a Sertoli-Leydig cell tumor of the ovary [43]. Thus, ovarian tumors are not uncommon in apes, nor are benign uterine tumors, which seem to plague female cercopithecines and humans as well. As in all primates, male genital

TABLE 53.16. Malignancies in apes in survey of neoplasms in nonhuman primates.

Species	Age (years)	Sex	Diagnosis	Contributor
Hylobates sp.	—	M	Granulocytic leukemia	AFIP[1]
Hylobates sp.	3	M	Malignant lymphoma	AFIP
Hylobates sp.	—	F	Malignant lymphoma	AFIP
Hylobates sp.	—	M	Malignant lymphoma	AFIP
H. concolor	1	M	Acute lymphocytic leukemia	NZP/AFIP
H. lar	—	F	Lymphosarcoma	AFIP
H. lar	—	F	Lymphosarcoma	AFIP
H. lar	—	—	Lymphosarcoma	AFIP
H. lar	12	M	Pleomorphic sarcoma, neck metastatic to abdomen	AFIP
H. hoolock × *agilis*	38	F	Malignant granulosa cell tumor	NZP
Pongo pygmaeus	35	F	Esophageal squamous cell carcinoma	PZ
Pan troglodytes	—	—	Recurrent complex odontoma	AFIP[1]

[1] Previously reported in the literature.

tumors are rare (in contrast to the frequent occurrence of prostatic carcinoma in humans), although an interstitial cell tumor of the testis was reported on a lowland gorilla from the London Zoo [59] and a seminoma claimed for an orangutan [107].

Oral cavity and gastrointestinal tumors found in this survey were a papilloma in the pharynx of a white-handed gibbon, an esophageal squamous cell carcinoma in a 35-year-old orangutan, a recurrent complex odontoma of the maxilla in a chimpanzee, and multiple adenomas of Brunner's glands in a 44-year-old chimp. The last two have been cited in the literature along with jejunal polyposis [27]. An aged orangutan was reported to have a gastric squamous cell carcinoma [89]. Oral papillomas and proliferative lesions of the pharynx [27], a nasopharyngeal carcinoma, and a case of nasal polyposis have been reported in chimpanzees [1, 58].

Endocrine tumors in this survey were a thyroid adenoma in a 20-year-old gibbon and an adrenal cortical adenoma in an 8-year-old female chimpanzee. Another adrenal cortical adenoma has been reported in a chimpanzee and islet cell adenomatosis in another [78]. No tumors of liver and exocrine pancreas were found in this survey or in the literature.

With so few tumors and individuals in this group, no strong patterns of neoplasia emerge except hematopoietic neoplasms in gibbons.

Sex, Age, Incidence, and Proportion of Malignancies among Nonhuman Primates with Neoplasms and Proliferative Disorders

Females accounted for 56.6% of the 274 animals for which sex was recorded (Table 53.17). This slight excess was not unexpected and is probably not significant. Many primate breeding units are set up with an excess of females. In addition, neoplasms of the female reproductive tract were common while those of the male organs were not.

TABLE 53.17. Sex and age at diagnosis or death for nonhuman primates with neoplasms.

	Pro-simians	New World	Old World Macaques	Old World Others	Apes	All
Males	12	19	51	25	12	119
Females	14	16	90	21	14	155
Number known age	19	19	83	38	20	179
Range	2d–22y[1]	1m–40y	1d–38y	2w–31y	2m–45y	1d–45y
Mean	9.5y	10.3y	14.4y	10.6y	22.54y	12.4y
Number < 3y	1	4	8	6	6	23

[1] d, days; w, weeks; m, months; y, years.

The mean age at diagnosis or death was calculated for 179 known-aged animals in the survey (Table 53.17). In all groups, except the shorter lived prosimians, the mean age fell after the first decade of life. The ages given in the survey were often underestimated based on duration in captivity and were often given as a minimum age (e.g., "at least 14 years old"). Interestingly, several tumors were found in infant or juvenile animals (Table 53.18). Most were of types reported from younger age groups in other animals and humans [84, 98].

An important issue is whether or not neoplasms are a limiting factor in propagation. To answer this, one must have an appreciation for the incidence of neoplasia. This could not be documented for this survey as a whole, since the population

TABLE 53.18. Survey summary: neoplasms in infant and juvenile nonhuman primates (less than 3 years old).

Species	Age	Sex	Diagnosis	Contributor
Macaca assamensis	21m	F	Cutaneous hemangioma	PZ
M. nemestrina	1d	F	Neuroblastoma, abdominal cavity	PC
M. arctoides	24m	M	Lymphosarcoma	CPRC
M. arctoides	8m	M	Lymphosarcoma	CPRC
M. arctoides	28m	F	Lymphosarcoma	CPRC
M. fascicularis	1d	M	Hemangioma, skin of finger	CPRC
M. fascicularis	4m	M	Adrenal adenoma	LB
M. mulatta	12m	F	Calcifying epithelioma, skin	YPC
Papio papio	30m	M	Capillary hemangioma, subcutis	PZ
P. papio	2w	M	Hemangioma	PZ
P. hamadryas	31m	M	Diffuse gastric adenomata	PZ
Cercocebus atys	1m	M	Multiple osteochrondromas, ribs	CPRC
C. atys	7m	M	Islet cell adenoma	CPRC
Cercopithecus sabeus	1m	M	Subcutaneous lipoma	PZ
C. sabeus	24m	F	Acute lymphocytic leukemia	PZ
Presbytis obscurus	30m	M	Squamous cell carcinoma, oral cavity	JH
G. crassicaudatis	1y	M	Gastric adenoma	PZ
G. crassicaudatis	2d	M	Adrenal cortical adenoma	LB
Ateles belzebuth	1m	F	Arrhenoblastoma, ovary	PZ
Leontideus rosalia	9m	F	Embryonal nephroma	NZP
Ateles sp.	juvenile	M	Papilloma, eyelid	PZ
Callimico goeldii	24m	F	Renal tubular adenoma	NZP

TABLE 53.19. Incidence of neoplasia in nonhuman primate pathology cases, University of California at Davis Veterinary Medical Teaching Hospital.

Species	Number accessions	Number tumors (%)
Prosimians		
Lemur sp.	2	0
L. catta	4	0
L. mongoz	2	2 (100)[1]
New World monkeys		
Leontideus rosalia	1	0
Callimico goeldii	4	0
Callithrix jacchus	1	0
"Marmoset"	3	0
Callicebus moloch	2	1 (50)[2]
Aotus trivirgatus	1	0
Saimiri sciureus	16	1 (6.3)[3]
Ateles spp.	11	0
Lagothrix lagothricha	13	0
Cebus spp.	3	0
Apes		
"Gibbon"	8	0
Symphalangus syndactylus	1	0
Pan troglodytes	2	0
Gorilla gorilla	4	0
Primate (not otherwise specified)	14	0

Species	Number accessions	Number tumors (%)
Old World monkeys		
Cercocebus gallerita	2	0
C. torquatus	1	0
C. albigena	1	0
C. atys	1	0
Cercopithecus aethiops	1	0
C. sabeus	1	0
C. cephus	1	0
C. nictitans	3	0
C. mitis	1	0
C. neglectus	1	0
Miopithecus talapoin	1	0
Colobus sp.	5	0
Presbytis sp.	2	0
P. entellus	3	0
P. obscurus	1	0
Papio sp.	6	1 (16.7)[2]
Macaca sp.	3	0
M. fascicularis	1	1 (100)[2]
M. mulatta	3	0
Total	133	6 (4.5)

[1] 50% malignant.
[2] Malignant.
[3] Benign.

base (either animals necropsied or the total population at risk) was not available for most institutions. The incidence was calculated for neoplasia in nonhuman primate pathology cases at the VMTH (Table 53.19) and the CPRC (Table 53.20). The overall incidence was 4.5% and 2.4%, respectively. In the larger CPRC study, incidence in different groups could be calculated. Neoplasia was somewhat more frequent in prosimians (4/44 = 9%) and less frequent in New World monkeys (2/520 = 0.4%) than in macaques and other Old World monkeys (89/3432 = 2.6%). In macaques, the highest incidence (7.4%) occurred in stump-tails. This was a colony of stump-tails within which a subgroup in a single large outdoor cage suffered high mortality (44/54) during a 2-year period [49]. The eight lymphosarcomas and multiplicity of opportunistic infections in this group were accompanied by immunologic abnormalities suggestive of SAIDS.

A better appreciation for incidence was sought by reviewing the literature (Table 53.21). As in the VMTH and CPRC studies, the overall incidence was seldom more than 5% except for species with low numbers (in which a single tumor may give a spuriously high incidence) and in groups in which infectious hematopoietic neoplasms have occurred (e.g., the Sukhumi's hamadryas baboons).

TABLE 53.20. Incidence of spontaneous neoplasms in pathology accessions at the California Primate Research Center.

Species	Total pathology reports	No. with neoplasms (%)	% with malignancy
Prosimians			
Galago crassicaudatus panganiensis	41	4 (9.8)	25
Galago senegalensis	3	0	
New World monkeys			
Saguinus nigricollis	187	0	
S. fuscicollis	1	0	
Callicebus moloch	89	0	
Saimiri sciureus	227	1 (0.4)	0
Cebus apella	4	1 (25)	0
Ateles geoffroyi	6	0	
A. paniscus	6	0	
Old World monkeys			
Cercopithecus aethiops	32	0	
Miopithecus talapoin	3	0	
Cercocebus atys (fulliginosus)	56	2 (3.6)	0
Erythrocebus patas	2	0	
Presbytis entellus	19	0	
Papio papio	49	0	
P. anubis	9	1 (11)	0
P. cynocephalus	20	2 (10)	50
Macaca mulatta	2127	65 (3.0)[1]	76
M. fascicularis	329	3 (0.9)	0
M. radiata	539	6 (1.1)	50
M. arctoides	134	10 (7.4)[1]	80
M. nemestrina	113	1 (0.9)	100
Apes			
Hylobates lar	5	0	0
Totals	4001	97 (2.4)	65

[1] Species with outbreaks of lymphosarcoma and SAIDS.

TABLE 53.21. Incidence of neoplasms in necropsies of captive and free-living nonhuman primates in the literature.

Species	No. of necropsies	% with tumors	% Malignant	Source	References
Cebus albifrons	1	100	100	PC[1]	Siebold and Wolf (1973)
Cercopithecus aethiops	16	6.2	100	PC	Siebold and Wolf (1973)
Erythrocebus patas	158	0.6	0	PC	Siebold and Wolf (1973)
Papio sp.	18	11.1	0	PC	Siebold and Wolf (1973)
Macaca arctoides	134	2.2	0	PC	Siebold and Wolf (1973)
M. fascicularis	9	11.1	0	PC	Siebold and Wolf (1973)
M. mulatta	317	2.8	22	PC	Siebold and Wolf (1973)
Pan troglodytes	52	11.5	0	PC	Siebold and Wolf (1973)
Study total	1065	2.2	16.5	PC	Siebold and Wolf (1973)
New World spp.	870	0.3	100	PC	McClure (1979)
Macaques	938	4.8	49	PC	McClure (1979)
Baboons	26	7.7	0	PC	McClure (1979)
Misc. Old World spp.	72	0	0	PC	McClure (1979)
Apes	164	1.8	0	PC	McClure (1979)
Prosimians	106	1.9	50	PC	McClure (1979)
Study total	1276	2.0	44	PC	McClure (1979)
Tupaia glis belangeri	1650	0.4	100	Lab	Hofmann et al. (1981)
Baboons	3348	6.2	89	PC	Lapin (1982)
Macaques	9896	1.5	9	PC	Lapin (1982)
"Green monkeys"	408	2.6	5.8	PC	Lapin (1982)
Squirrel monkeys	50	10.0	100	PC	Lapin (1982)
"Red monkeys"	350	0.38	0	PC	Lapin (1982)
Capuchins	20	5.0	0	PC	Lapin (1982)
Study total	13,712	2.7	57	PC	Lapin (1982)
Prosimians	83	3.6	33	Zoo	Griner (1983)
New World spp.	301	2.3	42[2]	Zoo	Griner (1983)
Old World spp.	193	3.6	14.3	Zoo	Griner (1983)
Apes	31	0	0	Zoo	Griner (1983)
Study total	569	3.0	35	Zoo	Griner (1983)
Lemurs	126	2.4	?	Zoo	Benirschke et al. (in press)
Prosimians	167	3.6	0	Zoos	Benirschke et al. (in press)
Alouatta caraya	292	4.8	0	Wild	Maruffo (1967)
Papio ursinus	100	4.0	25	Wild	McConnell et al. (1974)

[1] PC, primate center.
[2] These were lymphoproliferative disease associated with *Herpesvirus saimiri*.

Many neoplasms reported have been life-threatening malignancies (Table 53.21). Although early reports found a low proportion of malignancies (e.g., 16%) [101], more recent reports gave rates of 35%, 44%, and 57% malignancy [45, 70, 78]. In this survey, malignancies accounted for 39–57% of the tumors depending upon the group of primates (Table 53.1). Also, benign tumors of the female reproductive tract are usually not life threatening, but limit reproductive ability and thus affect captive or wild propagation.

It has been stated that as captive populations age, the incidence of neoplasia will increase [78, 107]; this is true for both humans and domestic animals. Only two studies [47, 70] addressed the question of incidence and age directly (Table 53.22). In tree shrews, baboons, macaques, and green monkeys, neoplasia was most frequent in older age groups. A biphasic curve was noted for baboons

TABLE 53.22. Effect of age on the incidence (%) of neoplasms in nonhuman primates in the literature.

Species	Total necropsies	Percentage with tumors by age at necropsy (years)					Total	Reference
		< 1	1–4	4–12	13–20	> 20		
Tupaia glis	1650	0	0	37.5[1]	—	—	0.36	Hofmann et al. (1981)
Baboons	3348	0	3.0	24.2[2]	10.6	33.3	6.2	Lapin (1982)
Macaques	9896	0.08	1.25	1.9	3.55	15.2	1.5	Lapin (1982)
Green monkeys	408	0	0	4.29	6.25	50.0	2.6	Lapin (1982)

[1] Life expectancy for Tupaia 8–14 years.
[2] Peak age for lymphosarcoma occurring endemically in baboons in this primate center.

in the Russian colony. The younger peak was accounted for by infectious hematopoietic neoplasms.

Other Proliferative Disorders: Retroperitoneal Fibromatosis and Endometriosis

Retroperitoneal fibromatosis (RF) is an apparently unprovoked proliferation of fibrous tissue beneath the peritoneal covering of the mesenteric arteries, lymph nodes, intestines, and other viscera [40, 42]. The disease is endemic in one major primate research center where at least 80 cases have occurred in 9 years, 77 cases in pig-tailed macaques. A subcutaneous form of the disease has also been observed [45]. The incidence is unclear, but in one early abstract it was 3 of 900 (0.3%) in first generation, 4 out of 400 (1%) in second generation, and 4 of 30 (13.3%) in third generation pig-tailed macaques born in that facility; this was a total incidence of 1.2% for colony-born animals. Retroperitoneal fibromatosis has also been recognized in a group of Celebes macaques at another primate center [76]. There is a strong association between RF and a type D retrovirus, similar to the one associated with SAIDS at the CPRC [48, 76, 109]. Although RF has only been found at a few facilities, its presence could be important in small breeding units.

Endometriosis is a proliferative process in which endometrial tissue (epithelium and stroma) becomes implanted in the wall of the uterus (interna) or on the serosal surfaces of the abdomen (externa). This ectopic endometrium responds to cyclic changes in hormones and undergoes menstrual sloughing, causing peritonitis and adhesions [98]. Because the implants are under hormonal control, they are not considered to be neoplastic. Endometriosis can interfere with normal reproduction and may be life threatening because of the complications of peritonitis and adhesions (such as hydronephrosis). Endometriosis is most often reported from rhesus but can occur in any menstruating primate [28, 32, 77, 101].

An appreciation for the incidence of endometriosis was sought from the CPRC necropsy series and the literature (Table 53.23). As in neoplasia, the overall incidence is low, probably less than 10% of necropsies of female monkeys. However,

TABLE 53.23. Incidence of endometriosis in nonhuman primate necropsies.

Species	Cases of endometrosis	Reference or source
Macaca mulatta		
Laboratory (n = 38)[1]	12 (31)	DiGiacoma (1977)
Free ranging (n = 17)	2 (12)	DiGiacoma (1977)
M. mulatta (n = 63)[1]	21 (33.3)	McCann and Myers (1970)
M. mulatta (n = 317)[2]	6 (1.9)	Siebold and Wolf (1973)
M. mulatta (n = 2127)[2]	42 (2.0)	CRPC
M. fascicularis (n = 329)[2]	4 (1.2)	CPRC
M. radiata (n = 539)[2]	7 (1.3)	CPRC
Cercopithecus aethiops (n = 32)[2]	3 (9.4)	CPRC
Galago crassicaudatus (n = 41)[2]	1 (2.4)	CPRC

[1] Animals culled for reproductive problems.
[2] These numbers not corrected for either age or gender. If assume 50% females then the incidence doubles.

in the two studies where the population was composed of older females with poor reproductive histories, endometriosis was found in about a third of the cases.

The cause of endometriosis is not entirely understood. The incidence in rhesus is increased by irradiation and by surgical manipulation, specifically hysterotomy [28]. If cesarian sections or surgical embryo transfers become important in managing rarer populations of primates, care should be taken to prevent this sequel.

Summary

From the survey performed and from a review of the literature, it is apparent that spontaneous neoplasms occur in all groups of nonhuman primates. Neoplasms are most frequently reported from members of the genus *Macaca*, perhaps because of the frequency with which this species is kept in zoos and research vivaria and also because of the presence of infectious proliferative disorders (lymphosarcoma, retroperitoneal fibromatosis) within members of this genus.

As expected, neoplasms occurred more frequently in adult animals with the mean age in this survey varying from 9.5 years in prosimians to 22.5 years in apes. Recent studies have documented the increased incidence of tumors in older animals in tree shrews and several groups of Old World primates [47, 70]. Also there was no strong sexual bias although females accounted for about 56% of the neoplasms reported.

Each group of nonhuman primates had its own pattern of neoplasms based on organ systems most frequently affected. This is not unexpected when one remembers the different patterns observed between closely related domestic species such as dogs and cats [84], or between populations and genders of the same species, i.e., humans [98]. These patterns are summarized in Tables 53.24 and 53.25. They were a high incidence of neoplasms of skin and subcutis, liver, and pancreas in prosimians; a high incidence of neoplasms of the gastrointestinal tract in New World monkeys, especially cotton-topped tamarins, and a moderate incidence of tumors of the female reproductive tract, endocrine glands, and

TABLE 53.24. Summary of survey of neoplasms in the four taxonomic groups of nonhuman primates: organ system distribution of all tumors reported.

			Old World		
	Prosimians (n = 41)	New World (n = 42)	Macaques (n = 204)	Others (n = 49)	Apes (n = 27)
Skin and subcutis	11 (26.2)[1]	3 (6.5)	38 (17.4)	9 (17.3)	10 (34.5)
Musculoskeletal	0	0	3 (1.4)	2 (3.8)	0
Oral cavity	0	2 (4.3)	6 (2.7)	3 (5.7)	1 (3.5)
Gastrointestinal	4 (9.5)	12 (26.1)	28 (12.8)	2 (3.8)	3 (10.3)
Liver and pancreas	7 (16.7)	3 (6.5)	11 (8.5)	7 (5.7)	0
Respiratory	4 (9.5)	0	3 (1.4)	2 (3.8)	0
Hematopoietic/ lymphoreticular	1 (2.4)	3 (6.5)	55 (25.1)	3 (5.7)	9 (31.0)
Urinary	4 (9.5)	5 (10.9)	7 (3.2)	10 (19.2)	0
Female reproductive	3 (7.1)	7 (15.2)	33 (15.1)	7 (13.5)	4 (13.8)
Male reproductive	2 (4.8)	3 (6.5)	1 (0.5)	1 (1.9)	0
CNS/special senses	1 (2.4)	0	4 (1.8)	1 (1.9)	0
Endocrine	3 (7.1)	6 (13.0)	28 (12.8)	4 (7.7)	2 (6.9)
Primary unidentified	2 (4.8)	2 (4.3)	2 (0.9)	1 (1.9)	0
Totals	42	46	219	52	29

[1] Number (%).

urinary tract; a high incidence of hematopoietic neoplasms and also tumors of skin and subcutis in macaques, with a moderate incidence in female reproductive and gastrointestinal tract and in endocrine glands; a less clear pattern in other Old World species with urinary tract, skin and subcutis, and female reproductive tract all moderately affected; and finally, although the number of animals is small, tumors of the hematopoietic system in gibbons and of the skin and subcutis in all apes head the list in that group. Better reporting of more neoplasms in nonhuman primates will help further define these patterns.

The overall incidence of neoplasms is not great. They are present in less than 5% of most necropsy surveys, although reporting of neoplasms is sporadic and needs to be improved. In some species at some institutions the incidence of neoplasia has been higher, accounting for 10% or more of necropsy cases.

In addition to accumulation of more data on proliferative disorders, it is important to understand etiologic agents that may affect the pattern and occurrence. From the survey and the literature, it is apparent that infectious oncogenic agents (i.e., herpesviruses, papovaviruses, poxviruses, and type C, D, and E retroviruses) can enhance the incidence of neoplasia in certain groups of primates. While some are of little consequence (e.g., papovavirus warts in colobus monkeys and Yaba virus histiocytomas in macaques and baboons), others have had a more severe effect, e.g., type D and E retrovirus-related immunosuppression and lymphomas in Taiwanese rock macaques [23, 61, 71] and stump-tailed macaques [49], type C retrovirus leukemias in gibbons [26], *Herpesvirus saimiri* lymphoproliferative disease in black-tailed silver marmosets [45] and owl monkeys [56, 91], and type D retrovirus-related retroperitoneal fibromatosis in Celebes macaques [76].

TABLE 53.25. Summary of survey: distribution of malignancies (by organ system) among groups of nonhuman primates.

			Old World		
	Prosimians No. (%)	New World No. (%)	Macaques No. (%)	Others No. (%)	Apes No. (%)
Skin and subcutis	6 (32)	0	17 (13)	1 (5)	1 (8)
Musculoskeletal	0	0	3 (2)	1 (5)	0
Oral cavity	0	2 (10)	5 (4)	2 (10)	1 (8)
Gastrointestinal	1 (5)	10 (50)	19 (15)	0 0	1 (8)
Liver and pancreas	6 (32)	1 (5)	4 (3)	3 (14)	0
Respiratory	0	0	3 (2)	1 (5)	0
Hematopoietic/ lymphoreticular	1	2 (10)	55 (43)	3 (14)	8 (67)
Urinary	0	1 (5)	6 (5)	7 (33)	0
Female reproductive	0	1 (5)	7 (6)	2 (10)	1 (8)
Male reproductive	1 (5)	1 (5)	0	0	0
CNS/special sense	1 (5)	0	4 (3)	0	0
Endocrine	1 (5)	0	2 (2)	0	0
Primary unidentified	2 (11)	2 (10)	2 (2)	1 (5)	0
	19	20	127	21	12

The incidence of oncogenic agents in captive and free-living populations of nonhuman primates has been only partially investigated. Antibodies to *Herpesvirus saimiri* are present in virtually 100% of captive-born squirrel monkeys by 1 year of age [31]. Although squirrel monkeys are unaffected carriers, malignant lymphomas have occurred in other New World species in contact with them. Because of this virus and the equally dangerous cytolytic herpesvirus "T," which is also carried by squirrel monkeys, mixed exhibits of New World species should be discouraged. *Herpesvirus ateles*, another oncogenic virus of New World monkeys, has caused lymphosarcoma when inoculated into howler monkeys although spontaneous contact infections have not been documented. About 40% of spider monkeys in one study carried antibodies to *H. saimiri* that were probably directed against their own *H. ateles* [31].

Two different herpesviruses have been isolated and characterized from lymphoreticular neoplasms in tree shrews [47]. Whether or not these agents cause tumors or how widespread they are is unknown.

In gibbons, the incidence of infection with gibbon ape leukemia virus (GaLV) as detected by serology varied between colonies, ranging from zero to 67%. The highest incidence was in a small colony in which several animals developed leukemia. Gibbons can be antibody negative, persistently viremic shedders of virus and can act as a nidus of infection in a colony [63]. Antibodies to GaLV were not found in orangutans, chimpanzees, gorillas, macaques, woolly monkeys, and marmosets in one study [24].

The exact etiologic agent(s) of infectious lymphosarcomas in baboons and macaques is unknown. Several candidate viruses have included an Epstein-Barr-like herpesvirus in baboons [71] and a lymphocyte-associated herpesvirus in macaques [74] and several retroviruses, SAIDS-associated type D and type E or lente-like retrovirus and a simian T-cell lymphotrophic type C retrovirus STLV-I, related to HTLV-I of humans [23, 34, 51, 61, 69, 76]. Foamy viruses ubiquitous in primate tissues have also been isolated from lymphomas [100].

Antibodies to herpesviruses are widespread in macaques. Antibodies to the type D retrovirus (MPMV and SAIDS-related viruses) are also widespread and vary from zero to 48% of macaques tested in one survey [34]. Type D retroviruses are strongly associated with retroperitoneal fibromatosis, which can be a significant cause of morbidity and mortality if present in a colony. HTLV-1 has been associated with T-cell leukemias and cutaneous lymphoma in humans, but the association of the related simian TLV-1 to lymphomas in macaques is circumstantial [51]. Recent serologic surveys of prevalence of antibodies recorded a prevalence of 11–44% in macaques kept at a primate center in Japan, 3.5–57% of macaques in another Japanese study, but less than 1% in macaque sera from Europe in another survey [46, 55]. Antibodies to STLV-I/HTLV-1 have also been detected in 47% of African green monkeys and 14.5% chimpanzees in the Japanese study, and 60% of green monkeys and 5% of chimpanzees in the European survey [55, 82]. The incidence of antibodies to the newly described lente-like macaque virus ("STLV-III") has not yet been ascertained [61].

Thus, infection with potentially oncogenic viruses is not uncommon is some groups of Old World primates and apes. Further studies must be done to ascertain the prevalence in captive and wild primates and to better understand their role in infectious hematopoietic and fibrous neoplasms.

Further investigations are also warranted to determine the pathogenesis of the other common neoplasms in nonhuman primates. Possible etiologies can be hypothesized by comparison with other animal species and humans [84, 98]. Tumors of possible viral etiology, besides hematopoietic neoplasms, include mammary tumors in prosimians (retroviruses), liver tumors in prosimians (hepatitis-B-like agents), and chronic enteritis and colonic carcinomas in marmosets (paramyxo- or other viruses). Bacteria may also play a role in colitis in marmosets similar to *Campylobacter*-associated proliferative ileitis in hamsters and swine and Crohn's disease associated with intestinal cancers in humans. Multiple factors may be involved in the marmoset colitis/colonic carcinoma since not all colonies and species develop malignancies although enteritis is present [19]. Diet and environmental factors must also be considered in mammary tumors (high fat); skin tumors (solar radiation); gastrointestinal and pancreatic malignancies (high fat, low fiber, low vitamin A, C, E); and liver tumors (iron storage, aflatoxins). In some cases, e.g., urinary tract tumors, which occur in humans in 7–22% of human necropsies, the etiologies postulated in humans, such as cigar, cigarette, and pipe smoking, are unlikely in nonhuman primates. The pathogenesis of other tumors, e.g., those of the female reproductive tract, is undertermined in both human and nonhuman primates.

Although we now have a fairly long list of tumors in nonhuman primates, we know little about their pathogenesis or about their true incidence in captive or wild populations. More information is needed to understand and prevent morbidity and mortality caused by neoplasia and proliferative disorders in self-sustaining populations of nonhuman primates.

Acknowledgments. The author wishes to thank the following individuals who contributed to this survey: Dr. J. Ott-Joselin and Ms. P. Stout; Dr. E. Dolensek; Drs. K. Hinshaw and R. Snyder; Dr. S. Gosselin; Dr. M. Burton; Dr. R. Brannian; Dr. R. Montali; Dr. M. Valerio, Dr. A. Kincaid, and Dr. G. Migaki. Thanks are also due Drs. K. Osborn and S. Prahalada for their help in compiling cases from the CPRC, and Ms. J. Wall and M. Flores for preparation of tables and manuscript. This study was funded in part by the CPRC base grant from NIH-Division of Research Resources, RR00169.

References

1. Amyx HL, Salazar AM, Newsome DA, Gibbs CJ, Gajdusek DC (1982) Nasopharyngeal carcinoma with intracranial extension in a chimpanzee. J Amer Vet Med Assoc 181:1425–1426.
2. Baskerville M, Baskerville A, Manktelow BW (1984) Undifferentiated carcinoma of the nasal tissues in the common marmoset. J Comp Pathol 94:329–338.
3. Baskin GB, Hubbard GB (1980) Ameloblastic odontoma in a baboon (*Papio anubis*). Vet Pathol 17:100–102.
4. Baskin GB, Soike K, Jirge SK, Wolf RW (1982) Ovarian teratoma in an African green monkey (*Cercopithecus aethiops*). Vet Pathol 19:219–221.
5. Bennett BT, Beluhan FZ, Welsh TJ (1982) Malignant nephroblastoma in *Macaca fascicularis*. Lab Anim Sci 32:403–404.
6. Benirschke K, Miller C, Ippen R, Heldstab A (In press) The pathology of prosimians, especially lemurs. In: Cornelius CE, Simpson CD (eds) Advances in veterinary science and comparative medicine, 29.
7. Berthe J, Barneon G, Richer G, Mazue G (1980) A medulloblastoma in a baboon (*Papio papio*). Lab Anim Sci 30:703–705.
8. Betton GR (1983) Spontaneous neoplasms of the marmoset (*Callithrix jacchus*). Oral and nasopharyngeal squamous cell carcinomas. Vet Pathol 21:193–197.
9. Bingham G, Sembrat R, Migaki G (1976) Myeloliposarcoma in a potto (*Perodicticus potto*): a case report. Lab Anim Sci 26:473–477.
10. Brack M (1985) Tumors in dwarf galagos (*Galagoides demidovii*). Vet Pathol 22(4):334–346.
11. Brack M (1985) Renal papillary adenoma in a cotton-topped tamarin (*Saguinus oedipus*). Lab Anim 19:132–133.
12. Brack M, Martin DP (1984) Trichoepithelioma in a barbary ape (*Macaca sylvanus*): review of cutaneous tumors in nonhuman primates and case report. J Med Primatol 13:159–164.
13. Brown RJ, O'Neill TP, Kessler MJ, Andress D (1980) Cholangiocarcinoma in a capuchin monkey (*Cebus albifrons*). Vet Pathol 17:626–629.
14. Bruestle ME, Golden JG, Hall A, Banknieder AR (1981) Naturally occurring Yaba tumor in a baboon (*Papio papio*). Lab Anim Sci 31:292–294.
15. Bunton TE, Lollini L (1983) Ovarian adenocarcinoma in a bonnet monkey: histologic and ultrastructural features. J Med Primatol 12:106–111.
16. Chalifoux LV, King NW (1983) Eosinophilic myelocytoma in an owl monkey (*Aotus trivirgatus*). Lab Anim Sci 33:189–191.
17. Chalifoux LV, Bronson RT (1981) Colonic adenocarcinoma associated with chronic colitis in cotton top marmosets, *Saguinus oedipus*. Gastroenterology 80:942–946.
18. Chalifoux LV, MacKey JJ, King NW (1983) A sparsely granulated, nonsecreting adenoma of the pars intermedia associated with galactorrhea in a male rhesus monkey (*Macaca mulatta*). Vet Path 20:541–547.
19. Chalmers DT, Murgatroyd LB, Wadsworth PF (1983) A survey of the pathology of marmosets (*Callithrix jacchus*) derived from a marmoset breeding colony. Lab Anim 17:270–279.
20. Chang J, Wagner JL, Kornegay RW (1979) Spontaneous cholangiocarcinoma in a ring-tailed lemur (*Lemur catta*). Lab Anim Sci 29:374–376.

21. Chapman WL Jr (1968) Neoplasia in nonhuman primates. J Amer Vet Med Assoc 153:872–878.
22. Clapp NK, Henke MA, Holloway EC, Tankersley WG (1983) Carcinoma of the colon in the cotton-top tamarin: a radiographic study. J Amer Vet Med Assoc 183:1328–1330.
23. Daniel MD, Letvin NL, King NW, Kannagi M, Sehgal PK, Hunt RD, Kanki PJ, Essex M, Desrosiers RC (1985) Isolation of T-cell trophic HTLV-III-like retrovirus from macaques. Science 228:1199–1201.
24. Deinhardt F (1980) Biology of primate retroviruses. In Klein G (ed) Viral oncology. Raven, New York, pp 357–398.
25. DePaoli A, Gardner FM (1968) Acute lymphocytic leukemia in a white-cheeked gibbon. Cancer Res 28:2559–2561.
26. DePaoli A, Johnson DO, Noll WW (1973) Granulocytic leukemia in whitehanded gibbons. J Amer Vet Med Assoc 163:624–628.
27. DePaoli A, McClure HM (1982) Gastrointestinal neoplasms in nonhuman primates: a review and report of eleven new cases. Vet Pathol 19 (Suppl 7):104–125.
28. DiGiacomo RF (1977) Gynecologic pathology in the rhesus monkey (*Macaca mulatta*). II. Findings in laboratory and free-ranging monkeys. Vet Pathol 14:539–546.
29. Elwell MR, DePaoli A, Whitney GD (1980) Cytoplasmic crystalloids in the ovary of the woolly monkey. Vet Pathol 17:773–776.
30. Eydelloth RS, Swindle MM (1983) Intraductal mammary carcinoma and benign ovarian teratoma in a rhesus monkey. J Med Primatol 12:101–105.
31. Falk L, Desrosiers R, Hunt RD (1980) *Herpesvirus saimiri* infection in squirrel and marmoset monkeys. Cold Spring Harbor Conf on Cell Prolif 7:137–143.
32. Fanton JW, Hubbard GB (1983) Spontaneous endometriosis in a cynomolgus monkey (*Macaca fascicularis*). Lab Anim Sci 33:597–599.
33. Fincham JE, vanRensburg SJ, Kriek NP (1982) Squamous cell carcinoma in an African green monkey. Vet Pathol 19:450–453.
34. Fine DL, Arthur LO (1981) Expression of natural antibodies against endogenous and horizontally transmitted macaque retrovirus in captive primates. Virology 112:49–61.
35. Fleckenstein B (1979) Oncogenic herpesviruses of non-human primates. Biochim Biophys Acta 560:301–342.
36. Gardner MB, Esra G, Cain MJ, Rossman S, Johnson C (1978) Myelomonocytic leukemia in an orangutan. Vet Pathol 15:667–670.
37. Giddens WE Jr, Bielitzski JT, Morton WR, Ochs HD, Myers MS, Blakley GA, Boyce JT (1979) Idiopathic retroperitoneal fibrosis: an enzootic disease in the pigtail monkey (*Macaca nemestrina*). Lab Invest 40:294 (abstract).
38. Giddens WE Jr, Dillingham LA (1971) Primary tumors of the lung in nonhuman primates. Vet Pathol 8:467–478.
39. Giddens WE Jr, Tsai CC, Morton WR, Ochs HD, Knitter GH, Blakley GA (1985) Retroperitoneal fibromatosis and acquired immunodeficiency syndrome in macaques. Amer J Pathol 119:253–263.
40. Gleiser CA, Keeling ME, Raulston GL, Jardine JH (1981) A mixed mesodermal sarcoma in a *Macaca mulatta*. Vet Pathol 18:399–402.
41. Gleiser CA, Carey KD (1983) Malignant fibrous histiocytoma in a baboon. Lab Anim Sci 33:380–381.
42. Gleiser CA, Carey KD, Heberling RL (1984) Malignant lymphoma and Hodgkin's disease in baboons (*Papio* sp.). Lab Anim Sci 34:286–289.
43. Graham CE, McClure HM (1977) Ovarian tumors and related lesions in aged chimpanzees. Vet Pathol 14:380–386.
44. Griesemer RA, Mannig JS, Newman L (1973) Neoplasms in galagos. Vet Pathol 10:408–413.

45. Griner LA (1983) Pathology of zoo animals, chapters 37, 38, and 39. Zoological Society of San Diego, San Diego, pp 319–381.
46. Hayami M, Komuro A. Nozawa K, Shotake T, Ishikawa K, Yamamoto K, Ishida T, Honjo S, Hinuma Y (1984) Prevalence of antibody to adult T-cell leukemia virus-associated antigens (ATLA) in Japanese monkeys and other non-human primates. Int J Cancer 33:179–183.
47. Hofmann W, Moller P, Schwaier A, Flugel RM, Zoller L, Darai G (1981) Malignant tumours in Tupaia (tree shrew). J Med Primatol 10:155–163.
48. Holmberg CA, Henrickson R, Anderson J, Osburn BI (1985) Malignant lymphoma in a colony of *Macaca arctoides*. Vet Pathol 22:42–45.
49. Holmberg CA, Henrickson R, Lenninger R, Anderson J, Hayashi L, Ellingsworth L (1985) Immunologic abnormality in a group of *Macaca arctoides* with high mortality due to atypical mycobacterial and other disease processes. Amer J Vet Res 46:1192–1196.
50. Holmberg CA, Osburn BI, Terrell TG, Manning JS (1978) Cellular immunologic studies of malignant lymphoma in rhesus macaques. Amer J Vet Res 39:469–472.
51. Homma T, Kanki PJ, King NW, Hunt RD, O'Connell MJ, Letvin NL, Daniel MD, Desrosiers RC, Yang CS, Essex M (1984) Lymphoma in macaques: association with virus of human T lymphotrophic family. Science 225:716–718.
52. Houser RG, Hartman FA, Knouff RA, McCoy FW (1962) Adrenals in some Panama monkeys. Anat Rec 142:41–51.
53. Hubbard GB, Wood DH, Fanton JW (1983) Squamous cell carcinoma with metastasis in rhesus monkey (*Macaca mulatta*). Lab Anim Sci 33:469–472.
54. Hubbard GB, Easin RL, Wood DH (1985) Prostatic carcinoma in a rhesus monkey (*Macaca mulatta*). Vet Pathol 22:88–90.
55. Hunsmann G, Schneider J, Schmitt J, Yamamoto N (1983) Detection of serum antibodies to adult T-cell leukemia virus in non-human primates and people from Africa. Int J Cancer 32:329–332.
56. Hunt RD, Barahona HH, King NW, Fraser CEO, Garcia FG, Melendez LV (1973) Spontaneous *Herpesvirus saimiri* lymphoma in owl monkeys. Comp Leuk Res Leukemogenesis Bibl Hematol 40:351–355.
57. Hunt RD, Blake BJ, Chalifoux LV, Sehgal PK, King NW, Letvin NL (1983) Transmission of naturally occurring lymphoma in macaque monkeys. Proc Natl Acad Sci 80:5085–5089.
58. Jacobs RL, Lux GK, Spielvogel RL, Eichberg JW, Gleiser CA (1984) Nasal polyposis in a chimpanzee. J Allergy Clin Immunol 74:61–63.
59. Jones DM, Dixson AF, Wadsworth PF (1980) Interstitial cell tumour of the testis in a western lowland gorilla (*Gorilla gorilla gorilla*). J Med Primatol 9:319–322.
60. Jones SR, Casey HW (1981) Primary renal tumors in nonhuman primates. Vet Pathol 18:89–104.
61. Kanki PJ, McLane MF, King NW, Letvin NL, Hunt RD, Sehgal P, Daniel MD, Desrosiers RC, Essex M (1985) Serological identification and characterization of a macaque T-lymphotrophic retrovirus closely related to HTLV-III. Science 228: 1199–1201.
62. Kawakami TG, Buckley PM, dePaoli A, Noll W, Bustad LK (1975) Studies on the prevalence of type C virus associated with gibbon hematopoietic neoplasms. Bibl Haematol 40:385–389.
63. Kawakami TG, Sun L, McDowell TS (1977) Infectious primate type-C virus shed by healthy gibbons. Nature 268:448–450.
64. Kawakami TG, Sun L, McDowell TS (1978) Natural transmission of gibbon leukemia virus. JNCI 6:1113–1115.
65. Kim JC, Palazzo MC (1978) Multifollicular papilliferous cystadenoma in a cynomolgus monkey (*Macaca fascicularis*). J Med Primatol 7:189–191.

66. Kim JC, Rim BM (1980) Cholelithiasis and hepatic adenoma in an adult male baboon (*Papio* sp.). VM/Sac 75:257–259.
67. Kirkwood JK, James MP (1983) Myeloproliferative disease in a cotton-top tamarin (*Saguinus oedipus oedipus*). Lab Anim 17:70–73.
68. Knight JA, Wadsworth PF (1981) Osteosarcoma of the tibia in a squirrel monkey (*Saimiri sciureus*). Vet Rec 109:385–386.
69. Komuro A, Watanabe T, Miyoshi I, Hayami M, Tsujimoto H, Seike M, Yosheda M (1984) Detection and characterization of simian retrovirus homologous to human T-cell leukemia virus, type 1. Virology 138:373–378.
70. Lapin BA (1982) Use of nonhuman primates in cancer research. J Med Primatol 11:327–341.
71. Letvin NL, Eaton KA, Aldrick WR, Sehgal PK, Blake BJ, Schlossman SF, King NW, Hunt RD (1983) Acquired immunodeficiency syndrome in a colony of macaque monkeys. Immunology 80:2718–2722.
72. Lingeman CH, Reed RE, Garner FM (1969) Spontaneous hematopoietic neoplasms of nonhuman primates. Review, case report, and comparative studies. Natl Cancer Inst Monogr 32:157–170.
73. London WT, Sever JL, Madden DL, Henrickson RV, Gravell M, Maul DH, Dalakas MC, Osborn KG, Houff SA, Gardner MB (1983) Experimental transmission of simian acquired immunodeficiency syndrome (SAIDS) and Kaposi-like skin lesions. Lancet 2:869–873.
74. Manning JR, Griesemer RA (1974) Spontaneous lymphoma of the nonhuman primate. Lab Anim Sci 24:204–210.
75. Maruffo CA (1967) Spontaneous tumours in howler monkeys. Nature 213:521.
76. Marx PA, Bryant ML, Osborn KG, Maul DH, Lerche NW, Lowenstine LJ, Kluge JD, Zaiss CP, Henrickson RV, Shigi SM, Wilson BJ, Malley A, Olson LC, McNulty WP, Arthur LO, Gilden RV, Barker CS, Hunter E, Munn RJ, Heidecker G, Gardner MB (1985) Isolation of a new serotype of SAIDS type D retrovirus from Celebes black macaques (*Macaca nigra*) with immune deficiency and retroperitoneal fibromatosis. J Virol 56:571–578.
77. McCann TO, Myers RE (1970) Endometriosis in rhesus monkeys. Amer J Obstet Gynecol 106:516–523.
78. McClure HM (1979) Neoplastic diseases in nonhuman primates: literature review and observations in an autopsy series of 2176 animals. In: Montali R, Migaki G (eds) The comparative pathology of zoo animals. Smithsonian Institution Press, Washington, DC, pp 549–565.
79. McClure HM, Chandler FW (1982) A survey of pancreatic lesions in nonhuman primates. Vet Pathol 19(Suppl 7):193–209.
80. McConnell EE, Basson PA, DeVos V, Myers BJ, Kuntz RE (1974) A survey of diseases among 100 free-ranging chacma baboons (*Papio ursinus*) from the Kruger National Park. Onderstepoort J Vet Res 41:97–168.
81. McIntosh GH, Giesecke R, Wilson DF, Goss AN (1985) Spontaneous nasopharyngeal malignancies in the common marmoset. Vet Pathol 22:86–88.
82. Miyoshi I, Fujishita M, Taguchi H, Matsubayashi K, Meiva N, Taneoka Y (1983) Natural infections in nonhuman primates with adult T-cell leukemia virus or a closely related agent. Int J Cancer 32:333–336.
83. Morin ML, Renquist DM, Allen AM (1980) Squamous cell carcinoma with metastasis in a cynomolgus monkey (*Macaca fascicularis*). Lab Anim Sci 30:110–112.
84. Moulton JE (1978) Tumors in domestic animals, 2nd ed. University of California Press, Berkeley/Los Angeles/London.
85. Narama I, Tsuchitani M, Umemura T, Tsuruta M (1983) The morphogenesis of a papillomatous gastric polyp in the crab-eating monkey (*Macaca fascicularis*). J Comp Pathol 93:195–203.
86. Napier JR, Napier PH (1967) A handbook of living primates. Academic, New York.

87. Nicholls J, Schwartz LW (1980) A spontaneous bronchiolo-alveolar neoplasm in a nonhuman primate. Vet Pathol 17:630–634.
88. O'Gara RW, Adamson RH (1972) Spontaneous and induced neoplasms in nonhuman primates. In: Fiennes RNT-W (ed) Pathology of simian primates. S Karger, Basel/München/Paris/London/New York/Sydney, pp 190–238.
89. Ohgaki H, Hasegawa H, Kusama K, Sto S, Kawachi T, Masui M, Tanabe K, Kawasaki I, Hiramatsu H, Saito K (1984) Squamous cell carcinoma found in the cardiac region of the stomach of an aged orangutan. Gann 75:415–417.
90. Osborn KG, Prahalada S, Lowenstine LJ, Gardner MB, Maul DH, Henrickson RV (1984) The pathology of an epizootic of acquired immunodeficiency in rhesus macaques. Amer J Pathol 114:94–103.
91. Rabin H, Neubauer RH, Pearson GR, Cicmanec JL, Wallen WC, Loeb WF, Valerio MG (1975) Spontaneous lymphoma associated with *Herpesvirus saimiri* in owl monkeys. JNCI 54:499–502.
92. Rabin H, Neubauer RH, Gonda MA, Nelson Rees WA, Charman HP, Valerio MG (1978) Spontaneous esophageal carcinoma and epithelial cell line of an adult rhesus monkey. Cancer Res 38:3310–3314.
93. Rangan SR, Gutter A, Baskin GB, Anderson D (1980) Virus associated papillomas in colobus monkeys (*Colobus guereza*). Lab Anim Sci 30:885–889.
94. Ratcliffe HL (1940) Familial occurrence of renal carcinoma in rhesus monkeys (*Macaca mulatta*). Amer J Pathol 16:619–624.
95. Reichard TA, Ensley PK, Henrick MJ (1981) Pheochromocytoma in a ring-tailed lemur (*Lemur catta*). Amer Assoc Zoo Vet (Proc):44–45.
96. Richter CB, Buyekmihci N (1979) Squamous cell carcinoma of the epidermis in an aged white-lipped tamarin (*Saguinus fuscicollis leucogencys* Gray). Vet Pathol 16:263–265.
97. Richter CB, Lushbaugh CC, Swartzendruber DC (1980) Cancer of the colon in cotton-topped tamarins. In: Montali R, Migaki G (eds) The comparative pathology of zoo animals. Smithsonian Institution Press, Washington, DC, pp 567–571.
98. Robbins SL, Cotran RS, Kumar V (1984) Pathologic basis of disease. WB Saunders, Philadelphia/London/Toronto/Mexico City/Rio de Janeiro/Sydney/Tokyo.
99. Russell SW, Jenson FC, Vanderlip JE, Alexander NL (1979) Osteosarcoma of the mandible of a baboon (*Papio papio*): morphological and virological (oncornavirus) studies, with a review of neoplasms previously described in baboons. J Comp Pathol 89:349–360.
100. Schnitzer TJ (1981) Characterization of a simian foamy virus isolated from a spontaneous primate lymphoma. J Med Primatol 10:312–328.
101. Seibold HR, Wolf RH (1973) Neoplasms and proliferative lesions in 1065 nonhuman primate necropsies. Lab Anim Sci 23:533–539.
102. Sembrat RJ, Fritz GH (1979) Long bone osteosarcoma in a rhesus monkey. J Amer Vet Med Assoc 175:971–974.
103. Shaley M, Murphy JC, Fox JG, Wallstrom AC, Gottlieb LS (1980) Myxoma of bone in a nonhuman primate. Cancer 45:2573–2582.
104. Skelton-Stroud PN, Ishmael J (1985) Adrenal lesions in the baboon (*Papio* spp.). Vet Pathol 22:141–146.
105. Snyder SP, Dungworth DL, Kawakami TG, Callaway E, Lau DT-L (1973) Lymphosarcomas in two gibbons (*Hylobates lar*) with associated C-type virus. J Natl Cancer Inst 51:89–94.
106. Spencer AJ (1985) Diagnostic exercise: subcutaneous nodules in rhesus monkeys. Lab Anim Sci 35:79–80.
107. Squire RA, Goodman DG, Valerio MG, Frederickson T, Strandberg JD, Levitt MH, Lingeman CH, Harshbarger JC, Dawe CJ (1979) Tumors. In: Bernirschke K, Garner FM, Jones TC (eds) Pathology of laboratory animals. Springer-Verlag, New York/Heidelberg/Berlin, pp 1052–1283.

108. Stowell RE, Smith EK, Espana C, Nelson VG (1971) Outbreak of malignant lymphoma in rhesus monkeys. Lab Invest 25:476–479.
109. Stromberg K, Benveniste RE, Arthur LO, Rabin H, Giddens WE Jr, Ochs HD, Morton WR, Tsai CC (1984) Characterization of exogenous type D retrovirus from a fibroma of a macaque with simian AIDS and fibromatosis. Science 224:289–292.
110. Suleman MA, Tarara R, Mandalia KM, Weiss M (1984) A spontaneous bronchogenic carcinoma in a Sykes monkey (*Cercopithecus mitis stuhlmani*). J Med Primatol 13:153–157.
111. Tekeli S, Ford TM (1980) Spontaneous intraductal mammary carcinoma in a rhesus monkey. Vet Pathol 17:502–504.
112. Terrell TG, Gribble DH, Osburn BI (1980) Malignant lymphoma in macaques: a clinicopathologic study of 45 cases. JNCI 64:561–568.
113. Tsai C-C, Giddens WE Jr (1983) Adenocarcinoma of Brunner's glands in a baboon (*Papio cynocephalus*). Lab Anim Sci 33:603–605.
114. Tsai C-C, Giddens WE Jr (1985) Clear cell carcinoma of the lung in a pigtailed macaque. Lab Anim Sci 35:85–88.
115. Tsai C-C, Warner TFCS, Uno H, Giddens WE Jr, Ochs HD (1985) Subcutaneous fibromatosis associated with an acquired immune deficiency syndrome in pig-tailed macaques. Amer J Pathol 130:30–37.
116. Tsuchitani M, Narama I (1984) Pituitary thyrotroph cell adenoma in a cynomolgus monkey (*Macaca fascicularis*). Vet Pathol 21:444–447.
117. Uno H, Warner TF (1982) Plasmacytomas in rhesus monkeys. Arch Pathol Lab Med 106:278–281.
118. Wadsworth PF, Gopinath C, Jones DM (1980) Mammary neoplasia in ring-tailed lemurs (*Lemur catta*). Vet Pathol 17:386–388.
119. Wadsworth PF, Jones DM, Pugsley SL (1985) A survey of mammalian and avian neoplasms at the zoological society of London. J Zoo An Med 16:73–80.

54
Scaling and Anesthesia for Primates

CHARLES J. SEDGWICK

Introduction

Anesthesia is used in captive primates for two purposes—to render surgical patients insensitive to pain and to prevent biting when it is necessary to handle them. In either case, it is the work of the anesthesiologist to accomplish these purposes while maintaining the stability of vital organ systems. It is a further obligation to consider the potential impact on subtle primate social and behavioral systems of drug-induced incapacity. Removal of individuals from their stable social environments or otherwise changing their perceived status among conspecifics with various psychogenic chemicals can carry serious health risks for this special group of wild animals.

Scaling

Because of the vast range in body sizes that exists within the order Primates, it will be fruitful to comment here on the subject of physiological scaling. Scaling deals with the structural and functional consequences of differences in size among otherwise similar organisms (Schmidt-Nielsen, 1984). Scaling is a procedure born of physical laws. In a discussion of anesthesiology, it is customary to think only in terms of chemical laws. Thus, the activities and functions of water, salts, proteins, enzymes, oxygen, carbon dioxide, energy, products of protein, carbohydrate, and fat metabolism, and the pharmacodynamics of various drugs within an organism all seem to belong to the science of chemistry. It is often forgotten that physical laws are equally important since they determine how chemical processes are to proceed, what the time frame will be in which rates of diffusion and heat transfer will occur, and how moving fluids and gases will behave in various containments. When morphologically similar organisms within a group are also all nearly the same body size (weight, mass), it is customary to discount the physical laws that mediate chemical processes; but when they vary in body size by vast amounts, it is necessary to consider the physical side of the question carefully because this carries significant clinical ramifications. In zoological medicine there is concern for many morphologically similar groups of animals in which there is great variation in body size. Most notable of these is the order Primates.

To demonstrate how important scaling can be to primate anesthesiology, consider an extreme of difference in body size among primates, e.g., the difference between nursing-age infant tarsiers (*Tarsius* spp.), which may weigh 20g or less, and the largest adult male gorillas (*Gorilla gorilla*) at nearly 350 kg. There is a 17,500-fold difference in body size between these two examples. By scaling, the comparative aspects of clinical effects that are important to the application of anesthesiology, drug dosage, and monitoring may be furnished the anesthesiologist in a way that would be available from no other source save trial and error.

First, consider cardiac function. The stroke volume of any heart is proportional to its size (scale dependent for heart size). For the purposes of this discussion, the size of the mammalian heart is approximately 0.6% of the body size of the organism (Prothero, 1979). Yet, cardiac output must meet minimum tissue energy and oxygen requirements that are not determined by body size, but rather by a combination of factors that integrate body surface area and weight or mass, i.e., metabolic size (Fig. 54.1). Notice in Figure 54.1A that there is a 6:1 ratio between surface area and body size in the smaller animal represented graphically by a single cube. In Figure 54.1B, the animal has greater body size by a factor of 27

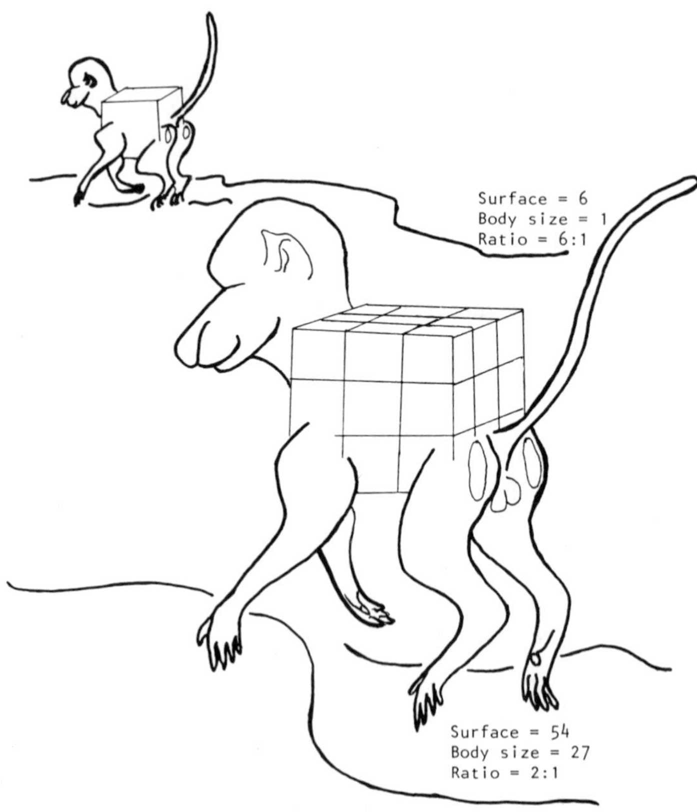

FIGURE 54.1. The concepts of metabolic size and body size.

(cubes), but the ratio between surface area and body size is only 2:1. With all environmental factors being equal, animal 1A would utilize energy and oxygen and dissipate heat at a rate three times greater than animal 1B. This means that the rate of energy and oxygen consumption in mammals increases with decreasing body size, and since the heart size that supplies this increased demand is a constant percentage of body size, the hearts of smaller mammals must beat more rapidly than those of larger mammals.

The resting heartbeat rate ($\int_h$) for any mammal can be computed as a scaleable (scale dependent) factor, but one that is based upon metabolic size rather than body size (Stahl, 1967). In considering morphologically similar groups of animals in which there is great diversity in body size, it is, therefore, necessary to be concerned with metabolic size just as it is with body size. Metabolic size is represented mathematically by the formula: body weight in kilograms to three-quarter power ($W_{kg}^{0.75}$ (Kleiber, 1932).

Here is an example of how this information can be helpful in the practice of anesthesiology. The calculated heartbeat frequency of an infant tarsier at rest (minimum energy cost or basal metabolic rate) is 640/minute while that of the adult gorilla is 56/minute. For a gorilla seen in a hypothetical anesthetic emergency where airway obstruction occurred and a life-threatening cardiac emergency developed, an anesthesiologist would probably have about 6–10 minutes (or the time required for from 300 to 500 heartbeats) in which to take corrective action to salvage the animal from the effects of hypoxemia and irreversible brain damage, or prevent death.

In 10 minutes the resting, normal tarsier's heart would beat about 6400 times. Obviously, the infant tarsier could not survive a serious deficit in cardiac output lasting 10 minutes. If brain tissue hypoxia reached pathological levels after 300–500 missed or defective heartbeats, the tarsier would succumb long before the 10 minutes it takes a gorilla's heart to toll away 500 beats. Instead, the infant tarsier with airway obstruction would be in critical clinical difficulty in from 30 seconds to about 1 minute, giving the clinician little time for corrective action. Scaling can demonstrate, theoretically, a principle that anesthesiologists should know from clinical experience, i.e., very small animals develop critical clinical emergencies much more rapidly than very large animals.

A second principle involves body heat and its generation and loss. All eutherian (placental) mammals, regardless of body size, have the same core body-temperature range (36–38°C). Core body temperature is the internal temperature of vital organs, such as the liver and kidneys for mammals in a nonhibernating state, and it is scale independent (nonscaleable) as a variable within the placental group of mammals (Schmidt-Nielsen, 1984). For example, rats have the same range for core body temperature as elephants (Kleiber, 1971). Apparently, a core body temperature set-point has been determined during phylogeny as a function of the anterior hypothalamus (Mountcastle, 1968). Pathological aberrations in this function resulting from environmental factors, such as anesthetics, can result in emergencies such as malignant hyperthermia.

Heat loss is scaleable (scale dependent). However, it is a function of metabolic size rather than body size as are energy utilization and oxygen consumption. The apparent absence of any obvious correlation between core body temperature and body size and the definite correlation that exists between heat loss and metabolic size provides an important principle relevant to the anesthesiology of primates, i.e., small individuals lose heat at a higher rate than large ones. This principle is particularly critical if subjects are anesthetized, endotracheally intubated, and forced to breathe cold, dry gases from compressed gas sources. Heat loss is scale dependent because heat production is scale dependent. A specific formula defines overall heat loss in terms of tissue conductance; it considers heat produced in the animal's core, transported through various thicknesses of different tissues to the body surface, radiating into the environment (Schmidt-Nielsen, 1984). However, for these purposes, it is not necessary to deal with variation in thickness or consistency of tissue. Rather, there is another more practical clinical consideration regarding heat loss in anesthetized subjects.

In the anesthetized, intubated primate subject, expired breath is saturated with water vapor that is lost through the endotracheal tube and carries away with it a great deal of heat from a subject's core. Because of this it is possible to compare, as a ratio of their total ventilation volume rates, heat loss of a very small intubated primate the size of the infant tarsier and that of an adult gorilla by comparing the rates of their respective minute ventilation volumes ($\dot{V}_T$) per unit of body weight. There are formulas for this calculation, and it is, therefore, scaleable (Loew and Ernst, 1981). The infant tarsier has a rate for the ratio between total pulmonary minute ventilation and body weight of 0.7 ml/minute per g of body weight while the gorilla has one of only 0.06 ml/minute per g. Therefore, the tarsier has a potential rate for heat and water vapor loss from pulmonary ventilation that is nearly 12-fold greater than that of the gorilla (0.7/0.06 = 11.6). It is tempting to suggest that neglecting to replace heat and water losses in an anesthetized 20-g primate for 1 hour of anesthesia would be a mistake comparable to subjecting a 350-kg primate to 12 hours of anesthesia without furnishing similar support.

Anesthetics

The time that clinically equivalent levels of anesthesia last, by comparison, in small and large animals following a single administration of the anesthetic dosage, appears to vary according to metabolic size. All aspects of the various responses are accelerated for the smaller individual and slowed for the larger one, and all dosage rates are increased for the smaller animal and reduced for the larger one. This can readily be appreciated when variation in body size between two members of a group is vast as it is between that of the tarsier and the gorilla. The smaller animal both recovers and succumbs more rapidly. For example, if pentobarbital is administered to equivalent levels of anesthesia, obtunding a specific source of surgical pain in a small and a large animal, the time period for persistence of the effect of the pentobarbital anesthetic is shorter for the small animal than for the large one. Experience of this kind suggests that anesthetic dosage is another factor that is scaleable to metabolic size more logically than to body size.

In most cases, chemical immobilization will be induced in primates using injectable agents intramuscularly. If the desired clinical procedure anticipated is a short one, such as physical examination or minor surgery requiring little analgesia for short periods of time, inhalation anesthetics and endotracheal techniques will be unnecessary. However, if there is a need to anesthetize a primate in order to perform surgical procedure, where it is impossible to foresee the extent of surgically induced pain and where multilevel anesthesia is desired, inhalation anesthetics and endotracheal techniques should be used. In some cases, where the patient's physical condition is critical, combinations of centrally and locally acting analgesics, relaxants, and sedative doses of general anesthetics may be used in balanced anesthesia.

The most common injectable anesthetic used in primates currently is ketamine hydrochloride. Its characteristics (shared with congeners phencyclidine and tiletamine) include analgesia, nearly normal pharyngeal and laryngeal reflexes, increased skeletal muscle tone, tachycardia, hypertension, and increased cerebrospinal fluid pressure (Dripps et al., 1977). Ketamine is a dissociative anesthetic. Subjects apparently experience profound analgesia while appearing to be awake, but oblivious to the environment.

Winters (1976) found that ketamine anesthesia induced electroencephalographic waves in cats that were suggestive of myoclonus. Wave frequencies were 1/7 to 1/2 (2–7/sec.) normal (15/sec.) with amplitudes three times (300 μV) those of awake (100 μV) subjects. For this reason, ketamine and its congeners are sometimes called cataleptic anesthetics. Extrapyramidal and reticular activating systems are variously unaffected or stimulated by ketamine, the net effect of which results in increased small motor neuron tone with increased skeletal muscle tension, tremors, myoclonic seizures, or convulsions.

In humans, ketamine, like its analogs, is hallucinogenic and may precipitate variable personality changes for undetermined periods of time (Dripps, 1977). Long-term personality change and hallucinogenic activity is more difficult to document in nonhumans, but in animals, such as the cat, it has long been recognized that subtle, pathological mechanisms might also be involved since characteristic behavioral and electrophysiological characteristics of a subject may fail to reappear subsequent to the administration of a single cyclohexylamine treatment (Adey et al., 1965).

While ketamine blocks the nonspecific thalamocortical system with deactivation of the cerebral cortex, it apparently fails to have much effect on the frontal lobes, hence corneal and palpebral brain reflexes persist (Green, 1979). Painful surgery on the eyelids or corneas should, therefore, not be attempted without providing supplemental anesthesia for ketamine.

Ketamine does not promote vomiting and neither does it prevent it, but it does frequently cause salivation. Ptyalism is not an indication for atropine administration with ketamine anesthesia since its administration could cause ventricular tachycardia or fibrillation in an already stimulated myocardium.

Dosage rates for ketamine have been suggested at between 5 and 20 mg/kg intramuscularly (i.m.) for primates (Green, 1979). This kind of prescription of dosage for such a diverse group of body sizes is hardly justified without qualification. The 5 mg/kg dosage rate is effective and usually will not cause severe myoclonus in very large apes, such as orangutans and gorillas, while dosages of 20 mg/kg in such subjects usually cause convulsions that must be treated with a muscle-relaxing ataractic such as diazepam. A 20 mg/kg intramuscular dosage is effective without causing severe myoclonic seizures only in the smallest of primates. Initial low dosages of ketamine (2 mg/kg for large primates and 10 mg/kg for small ones) often suffices for short procedures if the environment can be kept quiet, the light intensity low, and manipulations of the animal's body slow and gentle (such low dosages should not be counted upon to immobilize an excited physically exerted individual as might be found in a capture situation).

Nondirected, spontaneous movement should not necessarily be interpreted as inadequate ketamine anesthesia. With ketamine, it is often difficult to distinguish the difference between spontaneous movement because of underdosage and increased muscle tone caused by overdosage.

Intravenous (i.v.) injection of ketamine must be administered very slowly and carefully (reduce dosage 25%) to avoid an overdosage bolus effect resulting in respiratory depression, myoclonus, convulsions, or death. Many technicians routinely treat ketamine anesthetized subjects with phenothiazine derivative ataractics that are expected to obtund some of the signs of nervous system stimulation. However, this procedure is not always successful in diminishing such seizure activity (the phenothiazines are, themselves, central nervous system stimulants). Further, they may sometimes cause profound hypotension (due to alpha blockade and periperal vasodilation) in spite of the usual positive chronotropic effect of ketamine. Diazepam at a dosage rate of 0.5 mg/kg i.m. (0.3 mg/kg i.v.) in large primates and 1 mg/kg i.m. (0.5 mg/kg i.v.) in small primates is usually effective in temporarily ameliorating the myoclonal effects of ketamine (and other dissociative anesthetics).

Dissociative anesthesia may be converted to inhalation anesthesia by administration of the inhalant with a mask. It is possible to intubate a ketamine-anesthetized primate endotracheally without this, but the persistence of pharyngeal and laryngeal reflexes create a potential for trauma when endeavoring to manipulate the tube through the glottal opening. Mask induction with 3% halothane and 50% nitrous oxide in an appropriate flow of the carrier (oxygen) gas will usually facilitate intubation in about 5 minutes in the small primates and 7–10 minutes in the large ones. Once the primate relaxes from an anesthetic such as halothane in combination with nitrous oxide, the primate should be endotracheally intubated for precise control of anesthetic level and provision of ventilatory support (precision of inhalant anesthetic administration cannot be maintained adequately with a mask).

Anticholinergics such as atropine should be administered to treat bradycardia (sometimes associated with xylazine). Atropine should not be given to a primate without first auscultating and determining the heartbeat rate especially with ketamine anesthesia, which often has a profound chronotropic effect. Sinus tachycardia, often present with ketamine anesthesia, may be converted to ventricular tachycardia and bigeminal patterns or fibrillation by atropine. Atropine at 0.02 to 0.05 mg/kg is effective in the primate with the higher dosage (0.05) to be used in the smaller primates and the lower dosage (0.02) in the large primates. It may be administered intravenously, intramuscularly, or subcutaneously, but i.v. doses should be injected very slowly (over a period of 10 minutes) while auscultating the heart to preclude development of tachycardia. Scaling can be used to estimate the resting heartbeat rate, i.e., $\int_h = 241 \times W_{\text{kg}}^{-0.25}$. A rate that exceeds this by two- to threefold can justifiably be regarded as tachycardia.

Combination of ketamine with xylazine anesthesia seems to reduce the incidence of myoclonus and muscular tremors seen with ketamine when used alone. The centrally acting muscle-relaxing propensity of xylazine is undoubtedly responsible for this. Xylazine causes bradycardia in many instances when used in a variety of species. It may cause bradycardia in primates even when used in conjunction with ketamine.

Xylazine is not as effective by itself in primates as it is in some nonprimate species. Xylazine, by itself, provides relatively poor analgesia that lasts for a very short period. The dosage of xylazine is 0.5 mg/kg i.m. in small primates and 0.25 mg/kg i.m. in large primates when mixed with ketamine (ketamine dosages given previously are not reduced by this combination).

Another common anesthetic, fentanyl-droperidol, is a neuroleptanalgesic and is very effective in primates. A dosage of 0.3 ml/kg i.m. of the veterinary product is mentioned as the effective dosage by Green (1979), but the body size of the primate group tested was not mentioned, which should cause the anesthesiologist concern when using the agent in unfamiliar species, especially large ones. Neuroleptanalgesics are useful by having antagonistic agents available to counteract the narcotic portion of the mixture.

Discussion

It is common for clinicians working with primate anesthesia to have a great deal of experience with a single body size class for animals in this large, diverse group. When this is the case, it is important to remember that certain clinical procedures, physical evaluations, and drug and anesthetic dosage rates should not be extrapolated from the smallest classes to the largest ones based merely upon body size without considering metabolic size. Clinicians experienced with anesthesia in very small subjects as well as very large ones do not need intricate mathematical models to convince themselves that small subjects lose body heat more rapidly than do large subjects; it's a common clinical observation.

However, it is not uncommon to find clinicians who routinely deal with anesthesia for individuals of a large-size primate group occasionally trying to subject the much smaller animals to 1 or 2 hours of anesthesia without warming respiratory gases, warming intravenous fluids, or warming the subjects' body surfaces. Similarly, it is not uncommon to find technicians extrapolating physiological effective drug dosage rates from small-sized subjects to the large ones without allowing for a reduced metabolic rate seen in the larger ones.

Following are scaling equations selected from Schmidt-Nielsen (1984), Dripps et al. (1977), Loew and Ernst (1981), and Stahl (1967), which are useful for extrapolating physiological variables in the practice of anesthesiology for mammalian groups in which there is great diversity in body size.

Scale to body size
- tidal volume (ml) $= 6.2 \times W_{kg}^{1.01}$
- anatomical dead space volume (ml) $= 2.2 \times W_{kg}^{1.01}$
- blood volume (ml) $= 55 \times W_{kg}^{0.99}$
- heart size (g) $= 5.8 \times W_{kg}^{0.99}$

Scale to metabolic size
- oxygen consumption (ml/min) $= 11.6 \times W_{kg}^{0.76}$
- oxygen consumption (ml/min/kg) $= 11.6 \times W_{kg}^{-0.24}$
- lung ventilation rate (ml/min) $= 334 \times W_{kg}^{0.76}$

respiratory frequency (per min) $= 53.5 \times W_{kg}^{-0.26}$
heartbeat frequency (per min) $= 241 \times W_{kg}^{-0.25}$
cardiac output (dl/min) $= 2 \times W_{kg}^{0.75}$
cardiac output (dl/min/kg) $= 2 \times W_{kg}^{-0.25}$
daily energy requirement (kcal/24 hr) $= 70 \times W_{kg}^{0.75}$
Anesthetic and drug dosage

Nonscaleable Variables
core body temperature range
hemoglobin concentration range
erythrocyte size
blood pressure

References

Adey WR, Bell FR, Dennis BJ (1965) Effects of LSD-25, psilocybin, and psilocin on temporal lobe EEG patterns and learned behavior in the cat. Neurology 9:591–602.

Dripps RD, Eckenhoff JE, Vandam LD (1977) Dissociative anesthesia. Introduction to anesthesia. The principles of safe practice. WD Saunders, Philadelphia, pp 187–188.

Green CJ (1979) Animal anaesthesia. Lab Anim Handb 8. Laboratory Animals Ltd., London, pp 42–43.

Kleiber M (1932) Body size and metabolism. Hilgardia 6:315–353.

Kleiber M (1971) Body temperature and heat dissipation. Proceed Sympos Environm Requirements Lab Anim IER Publicat 71–02. Manhattan, 3–4 May, Kansas State University, p 171.

Loew HJ, Ernst EA (1981) The quantitative practice of anesthesia. Williams and Wilkins, Baltimore, p 19.

Mountcastle VB (1968) Regulation and control in physiology. Medical Physiology, vol 1. CV Mosby, St. Louis, pp 597–604.

Prothero J (1979) Heart weight as a function of body weight in mammals. Growth 43:139–150.

Schmidt-Nielsen K (1984) Scaling, why is animal size important. Cambridge Press, Cambridge, p 7.

Stahl WR (1967) Scaling of respiratory variables in mammals. J Appl Physiol 22:453–460.

Winters WD (1976) Effects of drugs on the electrical activity of the brain. Anesthetics, Ann Rev Pharmacol Toxicol 16:413–426.

55
Nutrition of Primates in Captivity

DUANE E. ULLREY

Introduction

The evolutionary development of the order Primates has led to an extant world population of as many as 203 species in 56 genera and 15 families (Nowak and Paradiso, 1983). They differ remarkably in size, from a mouse lemur (*Microcebus murinus*) weighing less than 100 g to the gorilla (*Gorilla gorilla*) weighing over 100 kg, and in natural dietary habits, from a galago (*Galago elegantulus*), a specialized gum eater, to the gelada (*Theropithecus gelada*), dependent on grass seeds, blades, and rhizomes. The morphologies of their digestive tracts range from a relatively simple tube, such as that of the angwantibo (*Arctocebus calabarensis*), to the much more complex arrangement of the forestomachs in langurs and colobids (subfamily Colobinae). These morphological features are related to the natural food supply, which is primarily invertebrates for the angwantibo and leaves, fruit, and flowers for the Colobinae. Many invertebrates are significant sources of protein and fat and require a relatively short and simple gut for their digestion. Leaves, flowers, and many fruits contain high concentrations of plant cell wall (cellulose, hemicellulose, and lignin), none of which can be digested by mammalian enzymes. Thus, primates that depend on leaves as an important source of nutrients have developed gastrointestinal compartments designed to harbor symbiotic microorganisms that can digest cellulose and hemicellulose.

Chivers and Hladik (1980) published a particularly useful discussion of the morphology of the gastrointestinal tract in primates and made comparisons with other mammals in relation to diet. They pointed out that mammals eating mostly animal matter (faunivores) have a simple stomach and colon and a long small intestine. Folivorous species have a complex stomach and/or an enlarged cecum and colon. Predominantly frugivorous species have an intermediate morphology, depending on the nature of the fruit and the other food eaten.

These species differences in dietary habits in the wild and in morphology of the gut lead one to conclude that all captive primates should not be fed the same way. Indeed, success in rearing some species has been very limited, in part because of inappropriate diet and husbandry. Most commercial primate diets were designed originally for omnivorous primates such as the macaques (*Macaca* spp.). Since such diets tend to be low in fiber and high in starch, their use as the only food for Colobinae is analogous to feeding ruminants an all-grain diet. The consequence may be rapid microbial fermentation, bloat, inflammation of the foregut, and diarrhea.

Despite such species differences, it is likely that all primates have about the same qualitative nutrient requirements for tissue metabolism. While few studies have been conducted to establish qualitative or quantitative need, a subcommittee of the National Research Council has attempted to define the nutrient requirements of nonhuman primates (NRC, 1978). Considerable extrapolation from data on other species was necessary, and for omnivorous primates, information on humans (NRC, 1980) can be useful. The items listed in Table 55.1 are probably required in the diet or must be synthesized by symbiotic microorganisms living in the gastrointestinal tract.

Table 55.2 is a listing of primates by family and briefly describes the typical environment in which they live and the dietary items they prefer. Presumably, when appropriately chosen, these natural foods provide all required nutrients. Foraging strategy in the wild is a complex process that has evolved to deal with a variety of environmental factors. Cant and Temerin (1984) have categorized these as (1) climate, (2) physical structure of the habitat, (3) predators, (4) co-consumers, (5) spatiotemporal distribution of food patches, (6) patch size, (7) arrangement of foods within a patch, (8) physical characteristics of food items, and (9) constituents of food items.

TABLE 55.1. Nutrients probably required in the diet of primates or that must be synthesized by gastrointestinal symbionts.[1]

Amino acids	Fatty acids	Minerals	Vitamins
Arginine	Linoleic	Calcium	A
Histidine		Phosphorus	D
Isoleucine		Sodium	E
Leucine		Potassium	K
Lysine		Chlorine	Thiamin
Methionine[2]		Magnesium	Riboflavin
Phenylalanine[3]		Sulfur	Niacin
Threonine		Iron	Pantothenic acid
Tryptophan		Copper	B_6
Valine		Zinc	Folacin
		Manganese	Biotin
		Cobalt	B_{12}
		Iodine	Choline
		Selenium	Myo-insitol
		Chromium	C[4]
		Molybdenum	
		Nickel	
		Tin	
		Vanadium	
		Silicon	
		Fluorine	
		Arsenic	

[1] Taurine and carnitine may also be important for the young.
[2] Probable that part of the requirement may be met by cystine.
[3] Probable that part of the requirement may be met by tyrosine.
[4] Certain prosimians may be able to synthesize vitamin C in their livers (J.I. Pollock, 1985, personal communication).

TABLE 55.2. Typical environment and dietary preferences of primates by family.

Tupaiidae: tree shrews (5 genera, 16 species)
- Forested areas of E Asia from India and SW China eastward through Malay peninsula to Borneo and Philippines
- Generally diurnal (*Ptilocercus* mainly nocturnal); arboreal and terrestrial
- Diet: insects, fruit, occasionally other animals, various kinds of plant material

Lorisidae: lorises, pottos, galagos (5 genera, 10 species)
- Forested areas of Africa south of the Sahara, S India, SE Asia, and E Indies (*Galago senegalensis*, open woodland, scrub, and grasslands with thickets)
- Noctural and arboreal
- Diet: insects, fruit, gums, shoots, young leaves, lizards, birds' eggs, snails (varies with species)

Cheirogaleidae: dwarf lemurs, mouse lemurs (4 genera, 7 species)
- Forested areas of Madagascar
- Nocturnal and arboreal
- Diet: fruit, flowers, gums, insects, insect secretions, nectar, spiders, frogs, lizards (varies with species)

Lemuridae: lemurs (3 genera, 9 species)
- Forested areas of Madagascar and Comoro Islands
- Primarily diurnal; arboreal, though *Lemur catta* partly terrestrial
- Diet: fruit, leaves, insects

Lepilemuridae: sportive and weasel lemurs (1 genus, 7 species)
- Forested areas of Madagascar
- Nocturnal and arboreal
- Diet: leaves and flowers

Indriidae: avahi, sifakas, indri (3 genera, 4 species)
- Forests and scrublands of Madagascar
- Avahi nocturnal, sifakas and indri diurnal; arboreal
- Diet: leaves, buds, fruits, nuts, bark, flowers

Daubentoniidae: aye-aye (1 genus, 1 species)
- Forests, mangroves, and bamboo thickets of Madagascar
- Nocturnal and arboreal
- Diet: fruit, insect larvae

Tarsiidae: tarsiers (1 genus, 3 species)
- Forests, mangroves, and scrub of Celebes, Sumatra, Java, Borneo, Philippines and nearby islands
- Nocturnal or crepuscular; mainly arboreal
- Diet: insects

Callitrichidae: marmosets, tamarins (4 genera, 15 species)
- Tropical forests of Central and South America
- Diurnal and arboreal
- Diet: insects, spiders, fruit, sap, buds, lizards, frogs, birds' eggs, snails, small birds (varies with species)

Callimiconidae: Goeldi's marmoset (1 genus, 1 species)
- Rainforests of the Upper Amazon
- Diurnal; arboreal, understory, and terrestrial
- Diet: fruit, insects, lizards, frogs

Cebidae: New World monkeys (11 genera, 31 species)
- Forest from NE Mexico to N Argentina
- Diurnal (Aotus nocturnal); mostly arboreal
- Diet: fruit, leaves, nuts, flowers, bark, shoots, gums, insects, spiders, small vertebrates, birds' eggs (varies with species)

TABLE 55.2. (Continued).

Cercopithecidae: Old World monkeys (12 genera, 85 species)
Africa, Gibralter, SW Arabian peninsula, SC and SE Asia, Japan, and E Indies
Nearly all diurnal; baboons mainly terrestrial, macaques arboreal and terrestrial, others mainly arboreal
Diet: fruit, leaves, seeds, roots, shoots, flowers, insects, spiders, snails, lizards, frogs, small mammals, birds (varies with species)
Hylobatidae: gibbons (1 genus, 9 species)
Rainforests of islands and mainland of SE Asia
Primarily diurnal and almost exclusively arboreal
Diet: siamang, 50% leaves, 40% fruit, 10% flowers, buds, insects; other species, mostly fruit but also leaves, shoots, flowers, insects, eggs, small vertebrates.
Pongidae: great apes (3 genera, 4 species)
Orangutan: forests of Sumatra and Borneo
Diurnal and primarily arboreal
Diet: fruit, vegetation, insects, small vertebrates, birds' eggs
Chimpanzees: tropical rainforest, forest-savanna, montane forest and swamp forest (pigmy) of central Africa
Largely diurnal; arboreal and terrestrial
Diet: fruit, leaves, flowers, seeds, stem, bark, gums, honey, insects, birds' eggs, small vertebrates, occasionally small antelope, baboons, other monkeys
Gorilla: tropical rainforest, montane rainforest, and bamboo forest of equatorial Africa
Diurnal and primarily terrestrial
Diet: leaves, shoots, stem, pith, roots, fruit, flowers, grubs
Hominidae: humans (1 genus, 1 species)
Ubiquitous on all continents in rainforests and deserts, from below sea level to high mountains
Primarily diurnal and terrestrial
Diet: anything that grows, swims, flies, walks, or crawls

In captivity, few of these items receive attention. Environment is highly controlled and food usually is offered in generous supply. There is little question, however, that the physical characteristics of food items and the constituents they contain are of particular concern. Because the feeding of folivores and insectivores poses special problems in this regard, they will receive particular attention. A related problem concerned with vitamin D deficiency in nursing infants will also be discussed.

Folivores

As pointed out by Milton (1984), plant-eating primates do not feed on plant parts at random but exhibit decided preferences. Selection of a particular food has generally been attributed to its concentration of nutrients or toxic substances and to the availability of that food in time and space. As animals and their food supply evolved together, each has developed systems to deal with the other. Plants have developed defenses to protect vital but potentially edible parts from predation. Such defenses may include increased proportions of fiber or elevated concentrations of phenolics and alkaloids. Primates do not produce tissue enzymes that can digest plant fiber, but various gastrointestinal modifications have evolved to accommodate bacteria that can digest cellulose and hemicellulose and detoxify secondary plant compounds. Thus, the structural polymers of plants become sources of energy through microbial fermentation to volatile fatty acids (VFAs), and potentially toxic substances are rendered harmless. The fermentation vat of colobines has a gross anatomical resemblance to the stomach of ruminants (Hill, 1952), and the Colobinae possess a highly evolved ruminantlike digestion (Bauchop and Martucci, 1968). The stomach has four parts: the sacculated and enlarged presaccus and saccus, the long tubus gastricus, and the short pars pylorica (Kuhn, 1964). Cellulolytic and methane-producing bacteria are present, and samples of gastric contents have a pH and VFA concentrations and proportions similar to those found in rumen fluid. Acetic acid predominates, followed by propionic and *n*-butyric acids. Lesser amounts of isobutyric, *n*-valeric, and isovaleric acids may also be present. It is probable that the presaccus and saccus also provide for nitrogen recycling and microbial synthesis of B vitamins and vitamin K.

Other folivorous primates, such as howlers (*Alouatta* spp.), have a simple stomach but a relatively capacious colon. In this genus, cellulolytic bacteria inhabiting the hindgut are likewise able to ferment cellulose and hemicellulose, yielding VFAs as energy sources for the host. It should be noted that there may also be considerable microbial fermentation in the cecum and colon of the colobines.

Considering natural dietary habits and gastrointestinal morphology, the dietary management of Colobinae should be similar to feeding programs for ruminants such as sheep or goats. Thus, at least two meals should be offered each day to ensure a regular supply of substrate for the microbial population in the foregut. Each meal should be similar to the last with minimum variation in fiber and readily fermentable carbohydrate. Dietary fiber concentrations should approach those in green leaves of natural browse, although the type of fiber should be carefully chosen. Because colobines appear not to ruminate (Ayer, 1948; Kuhn, 1964)—that is, they do not regurgitate and rechew their food—foods containing high concentrations of long lignin fibers can be troublesome. Lignin is a complex polymer found in plant cell walls that is not digested by the anaerobic bacteria found in the digestive tract. Lignin concentrations are higher in mature plant parts than in those that are young, and some plant species have higher lignin concentrations than other plant species in parts that are of the same phenological age. A lignin-related problem is the observation by Ensley et al. (1982) that even young *Acacia saligna* and *A. longifolia* leaves are unsuitable foods for hanuman langurs (*Presbytis entellus*) and douc langurs (*Pygathrix namaeus*) because phytobezoars form that obstruct the digestive tract with a long, ropy mass. Much of this mass appears to be lignin arising from the leaves, which contain about 25% of this constituent on a dry basis.

It is likely that colobines that have a choice would not select such high-lignin leaves. Lignin, after all, is a structural plant defense. However, field studies indicate that diets of *Colobus guereza* in eastern and western Africa typically contain greater than 50% leaf materials (Oates, 1977; Struhsaker, 1978; Gautier-Hion, 1983). Based on reported leaf composition from equatorial Africa (Hladik, 1978; McKey et al., 1981; Waterman and Choo, 1981; Baranga, 1982, 1983), diets of this species in the wild would often contain over 30% neutral detergent fiber (NDF) and 10% acid detergent fiber (ADF) even when primarily young leaves were selected. In captivity, colobines are often fed diets similar to those fed to primates that are not primarily folivorous. Such diets are typically based on commercial primate biscuits containing a low concentration of fiber (i.e., 2–5% crude fiber, 10–19% NDF, 5–7.5% ADF, dry basis). The high incidence of gastrointestinal disorders in captive colobines (Hill, 1964) illustrates the importance of improving their dietary management.

Because high-fiber primate biscuits are not commercially available, some zoos are using mixtures of a primate diet, fruits and vegetables, and browse to feed their colobines. In a study at the San Diego Zoo (Ullrey et al., 1982), the nutrient composition and digestibility of such a mixture fed to douc langurs (*Pygathrix nemaeus*), purple-faced langurs (*Presbytis senex*), and silvered leaf monkeys (*Presbytis cristatus*) was determined. The commercial biscuit used was Zu/Preem® Dry Primate Diet (product code 6990) manufactured by Hill's Pet Products, Topeka, Kansas. The fruit and vegetable mix contained apples, bananas, green beans, cabbage, carrots, oranges, spinach, and sweet potatoes. The browse used was freshly cut eugenia (*Syzygium paniculatum*) or hibiscus (*Hibiscus rosa-sinensis*). The morning feeding included primate biscuits and browse. The afternoon feeding included primate biscuits and fruits and vegetables.

The composition of dry matter consumed during the trial was as follows: 13–16% crude protein, 3–4% ether extract, 17–22% NDF, 9–13% ADF, 3–5% lignin, 52–59% starch and sugar, 5–7% ash, 0.6–0.9% calcium, and 0.3–0.4% phosphorus. The primate biscuit provided 39–56% of the dry matter, while fruits and vegetables and browse accounted for 40–51% and 4–17%, respectively. Although fruits and vegetables and browse were important sources of fiber (48–64% and 9–34% of ADF, respectively), the primate biscuit supplied 57–78% of the dietary crude protein, 38–78% of the calcium, and 61–78% of the phosphorus. The primate biscuit was also an important source of other minerals and vitamins. Thus, while this commercial primate biscuit did not contain as much fiber as might be desirable for colobines, it was, nevertheless, a very important source of nutrients. Where attempts have been made to eliminate such biscuits and to depend upon fruits and vegetables alone, the diet is frequently deficient in protein, calcium, vitamin D, and a number of other nutrients. Reduction or removal of some fruit items from fruit and vegetable mixtures would increase proportions of fiber and decrease the proportion of rapidly fermentable carbohydrate.

Higher fiber primate biscuits that should be more suitable for the Colobinae and that still provide appropriate levels of other nutrients are under development. Watkins et al. (1985) fed a high-fiber biscuit along with fruits and vegetables to black and white colobus (*Colobus guereza*). The high-fiber biscuit contained 25% crude protein, 34% NDF, and 12% ADF on a dry basis. Total diet dry matter consumed contained 16% crude protein, 25% NDF, and 9% ADF. Apparent digestibilities of diet fractions were as follows: dry matter, 87%; crude protein, 78%; NDF, 81%; ADF, 69%; gross energy, 85%. It is apparent that colobines can use cellulose and hemicellulose effectively and that high-fiber primate biscuits are potentially useful for such species.

The feeding of hindgut fermenters such as howlers (*Alouatta* sp.) is analogous to the feeding of horses or rabbits. Presumably, fiber digestion is not as efficient as it is in Colobinae, and howlers may accommodate by choosing very tender, young plant parts if they have an option. Milton (1980) reported that mantled howlers (*A. palliata*) on Barro Colorado Island in Panama preferred young leaves, fruit, and flowers of high nutritional quality. Leaves were selected that were high in protein and low in fiber. Fruits eaten generally had a soft pulp or aril and a high nonstructural carbohydrate content as compared to leaves. Flowers varied in their nutrient content, but low phenolic concentrations may have influenced selection in some cases. It is significant that mantled howlers (*A. palliata*), which eat considerable foliage, have a slower mean gut transit time (20 hours) than spider monkeys (4 hours), which are strongly frugivorous (Milton, 1984). Thus, howlers, like Colobinae, are adapted to digestion of fiber-containing diets and should benefit from a regular feeding schedule and a moderate-fiber diet of constant composition.

Insectivores

So little is known about the feeding of insectivorous primates that one is hesitant to include a discussion of this group. However, a recent observation at the National Zoo (M.S. Roberts and M.E. Allen, 1985, personal communication) may be so important for the captive rearing of young tarsiers (*Tarsius bancanus*) that a preliminary report may be in order. A group of tarsiers imported from the wilds of Borneo in 1983 produced young, but the two infants died shortly after birth with vertebral fractures. The primary food items for the tarsiers were crickets (*Acheta domesticus*). Reports on composition of such invertebrates are limited, but crickets, as commonly raised, are low in calcium ($< 0.2\%$, dry basis) in relation to the needs of vertebrates and may have a calcium to phosphorus ratio up to 1:8 (Allen and Oftedal, 1982). However, it has been found that the composition of crickets (including digestive tract contents) can be modified by feeding them a high calcium diet (Table 55.3; M.E. Allen and D.E. Ullrey, 1983, personal communication), and cricket calcium concentrations can be increased to 0.6–1.5%. This diet also supplies high levels of vitamins A, D, and E. When crickets fed this diet were offered to the above group of tarsiers, reproduction was successful and the young thrived. Other management changes may have contributed to this success, however.

Vitamin D Deficiency in Nursing Primates

In cold seasons and climates, subtropical and tropical primate species must be protected from low temperatures and are usually housed indoors, often without sunlight exposure. Skylights or windows may be used, but they are commonly made of materials that don't transmit the typical solar spectrum. In particular, most glass excludes wavelengths below 330 nm. As noted below, light in the range of 285–315 nm is especially important in the photobiogenesis of vitamin D. Thus, it is suspected that many of the persistent problems with rickets in nursing New World primates are due to a failure of photobiogenic conversion of provitamin D_3, 7-dehydrocholesterol, to previtamin D_3 and ultimately to vitamin D_3 itself. If solid foods containing vitamin D are not consumed or if mother's milk is low in this nutrient, then vitamin D deficiency might be expected.

Figure 55.1 illustrates a typical case of rickets in a Bolivian red howler (*Alouatta seniculus*) infant born in an exhibit without sunlight exposure. The mother's ration included a commercial primate diet plus various fruits and vegetables. The

TABLE 55.3. High calcium cricket diet.

Item	%
Ground yellow corn	6.8
Dehydrated alfalfa meal (17% crude protein)	10.0
Soybean meal, dehulled (48% crude protein)	28.7
Ground wheat	27.0
Calcium carbonate	20.0
Mono-dicalcium phosphate (18% Ca, 21% P)	2.0
Salt	0.5
Vitamin-trace mineral premix[2]	1.6
Selenium premix[3]	0.11
Vitamin E premix[4]	0.12
Vitamin A and D premix[5]	0.17
Corn oil	3.0
Total	100.00

[2] Contained the following per kg: 660,000 IU vitamin A, 132,000 IU vitamin D_3, 1100 IU vitamin E, 440 mg menadione sodium bisulfite, 660 mg riboflavin, 3520 mg nicotinic acid, 2640 mg D-pantothenic acid, 4 mg vitamin B_{12}, 22 g choline, 15 g zinc, 7.5 g manganese, 0.55 g iodine, 2 g copper, 11.9 g iron.
[3] Contained 200 mg selenium per kg.
[4] Contained 275,000 IU vitamin E per kg.
[5] Contained 30,000,000 IU vitamin A and 3,000,000 IU vitamin D_3 per kg.
Source: M.E. Allen and D.E. Ullrey, 1983, personal communication.

primate diet was not relished and there was some question about the amount consumed. The infant nursed exclusively for 5 months, ate very little solid food subsequently, and did not choose vitaming D-containing dietary items.

The hypothesis developed from consideration of the above and similar cases, plus current knowledge of vitamin D metabolism, is that howler monkeys, like other New World monkeys (Hunt et al., 1967), do not use vitamin D_2 effectively. Since they are strictly vegetarian (Braza et al., 1983), dietary intakes of vitamin D_3 sources in the wild are unlikely. Thus, they are dependent for their vitamin D supplies upon photobiogenic conversion of vitamin D precursor in the skin. Howler milk, like cattle and human milk (Hollis et al., 1981; Reeve et al., 1982), is probably low in vitamin D-active compounds and does not respond efficiently to increased levels of vitamin D added to the diet of the lactating female. As a consequence, nursing howler infants must have direct vitamin D supplementation, exposure to sunlight, or exposure to artificial light with a radiant spectrum in the range of 285–315 nm.

Rickets was a major medical problem among human infants at the end of the nineteenth century and the beginning of the twentieth. Suggestions as diverse as heredity and syphilis were proposed as causes of this debilitating disease (Norman, 1979). Ultimately, it was established that rickets could result from a lack of sunlight or a deficiency of a fat-soluble dietary factor. Early in this research it was noted that "a map of the incidence of rickets [was]. . . the practical equivalent of a map. . . of deficiency of sunlight" (Hess, 1929), and Huldschinsky (1919) demonstrated that rickets could be effectively treated by ultraviolet rays. At about the same time, Mellanby (1921) produced rickets in puppies when they were raised indoors on a high-cereal, low-fat diet. If cod-liver oil or butterfat were added to the diet, rickets were prevented. Mellanby suggested that the effective

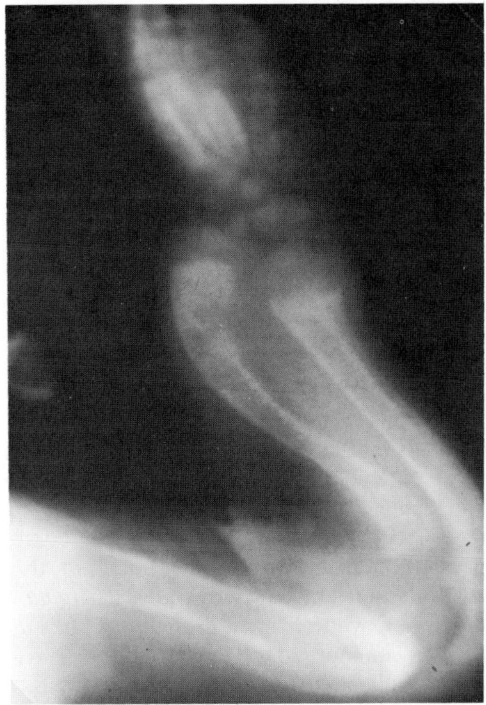

FIGURE 55.1. Radiograph of rickets in the forearm of a nursing Bolivian red howler monkey without sun or artificial UV light exposure.

factor was vitamin A. This proved not to be the case, and the distinction between vitamin A and the antirachitic factor came when McCollum et al. (1922) showed that vitamin A was destroyed by aerating and heating cod-liver oil to 100°C for 14 hours, while the antirachitic factor persisted. He named this factor vitamin D.

In succeeding years, the photobiogenesis of vitamin D_3, cholecalciferol, in cutaneous tissues was described. Provitamin D_3, 7-dehydrocholesterol, in the malpighian layer of the epidermis, is converted to previtamin D_3 by UV irradiation in the range of 285–315 nm, with maximum conversion at 297 (Holick et al., 1982) to 303 nm (Takada et al., 1979). Previtamin D_3, in turn, undergoes thermal isomerization to vitamin D_3 at a skin temperature of 37°C. Vitamin D_3 attached to a binding protein is transported in the blood plasma to the liver where it is hydroxylated to 25-OH-D_3. Further hydroxylation to 1,25-$(OH)_2$-D_3 takes place in the kidney. It is this compound that appears most active in promoting intestinal calcium and phosphate absorption and osteoclastic-mediated bone resorption. Other vitamin D_3 metabolites, with differing biological activities and significance, have been described by DeLuca and Schnoes (1983).

In 1924, Steenbock and Black noted that UV irradiation of a diet fed to rachitic rats resulted in a cure of the rickets, but the same diet without irradiation had no effect on the disease. Subsequently, it was discovered that the provitamin, ergosterol, present in plant tissues such as those in the rat diet, could be converted to vitamin D_2, ergocalciferol, by UV exposure. Still later it was determined that vitamin D_2 undergoes hydroxylation in the liver and kidney of mammals as does vitamin D_3.

Thus, it was established that certain vertebrates may have the option of deriving vitamin D need from exposure of the skin to the sun or by consuming vitamin D-active compounds from an irradiated vegetable diet. It should be noted that for this irradiation to be effective, the plant tissues should be dead. In the case of faunivores or omnivores, vitamin D supplies may also be derived from preformed vitamin D compounds in animal tissue, potentially both vitamin D_2 and D_3.

The evolutionary significance of the relative ability of 63 vertebrate species to bind vitamin D_2 or D_3 to plasma transport proteins has been investigated by Hay and Watson (1977). If vitamin D binding to plasma proteins correlates with the biological usefulness of the vitamin compounds, then their findings are relevant to the problem under consideration. All species of fish, reptiles, and birds that they studied bound vitamin D_2 with considerably less efficiency than vitamin D_3. This was true also for a mammalian monotreme, the Tasmanian echidna. Other mammals that bound vitamin D_2 with somewhat less efficiency (10–30%) than D_3 were Bennett's wallaby, European hedgehog, agoutis, lion, tiger, bactrian camel, and goat. The only primate examined that bound vitamin D_2 with less efficiency (15%) than vitamin D_3 was the large tree-shrew (*Tupaia tana*). No difference in binding affinity was noted in the thick-tailed bushbaby (*Galago crassicaudatus*), owl monkey (*Aotus trivirgatus*), white-throated capuchin (*Cebus capucinus*), white-fronted capuchin (*Cebus albifrons*), brown capuchin (*Cebus apella*), common marmoset (*Callithrix jacchus*), rhesus macaque (*Macaca mulatta*), olive baboon (*Papio anubis*), patas monkey (*Erythrocebus patas*), and white-handed gibbon (*Hylobates lar*). Other mammals, such as rats, pigs, and calves have recently been found to metabolize oral vitamin D_2 and D_3 differently, with significant discrimination against vitamin D_2 (Horst et al., 1981; Sommerfeldt et al., 1983). In any case, Hay and Watson (1977) speculated that vitamin D_2 may have a very limited role in vertebrates other than mammals. If true, this poses a particularly difficult problem for nonmammalian vertebrates that are nocturnal or insectivorous, as well as for nocturnal, insectivorous primates. Ergosterol is an essential sterol for many invertebrates (Bills, 1954; Gilmour, 1961), and it appears that irradiation of some invertebrates produces an antirachitic effect in rats, similar to vitamin D_2 (Rosenberg and Waddell, 1951). However, it is not clear whether invertebrates consumed as food would provide vitamin D_3-like compounds.

Some preliminary data on the spectral transmission characteristics of skylight materials that might be used in primate exhibits have been obtained through the assistance of S.E. Kaupp, National Marine Fisheries Service, Southwest Fisheries Center, La Jolla, CA, M.A. Strzelewicz of Michigan State University, and of R. Ruff, EG & G Gamma Scientific Inc., San Diego, CA. Polyurethane-fiberglass, acrylic-fiberglass, acrylic double skin, and Xcelite® double skin do not transmit light effectively below 370 nm. Vinyl transmits little light below 305 nm. Cellulose triacetate, plexiglass G-UVT (Rohm and Haas), and teflon do have potential to permit entry of biologically effective UV radiation into animal exhibits, but issues of physical strength and deterioration with prolonged solar exposure remain to be explored.

Preliminary tests on a variety of artificial light sources indicate that few provide effective energy levels within the desired 285–315 nm range. On the other hand, unless screened with cellulose triacetate, the Westinghouse sunlamp emits radiant energy of sufficient intensity at wavelengths below 285 to be potentially dangerous (Setlow, 1974).

Further research is urgently needed to solve these important problems. Studies are currently underway at the San Diego Zoo, and it is hoped that sufficient data will soon be available to provide practical guidelines for exhibit design that will prevent vitamin D deficiency in nursing primates.

References

Allen ME, Oftedal OT (1982) Calcium and phosphorus levels in live prey. Proc Reg Conf Amer Assoc Zool Parks Aquar, pp 120–128.

Ayer AA (1948) The anatomy of *Semnopithecus entellus*. Indian Publ House, Madras, India.

Baranga D (1982) Nutrient composition and food preferences of colobus monkeys in Kibale Forest, Uganda. Afr J Ecol 20:113–121.

Baranga D (1983) Changes in chemical composition of food parts in the diet of colobus monkeys. Ecology 64:668–673.

Bauchop T, Martucci RW (1968) Ruminant-like digestion of the langur monkey. Science 161:698–700.

Bills CE (1954) Vitamin D group. II. Chemistry. In: Sebrell WH, Harris RS (eds) The vitamins, vol 2. Academic, New York, pp 132–209.

Braza F, Alverez F, Azcarate T (1983) Feeding habits of the red howler monkeys (*Alouatta seniculus*) in the llanos of Venezuela. Mammalia 47:205–215.

Cant JGH, Temerin LA (1984) A conceptual approach to foraging adaptations in primates. In: Rodman PS, Cant JGH (eds) Adaptations for foraging in nonhuman primates. Columbia University Press, New York, pp 304–342.

Chivers DJ, Hladik CM (1980) Morphology of the gastrointestinal tract in primates: comparisons with other mammals in relation to diet. J Morphol 166:337–386.

DeLuca HF, Schnoes HK (1983) Vitamin D: recent advances. Ann Rev Biochem 52:411–439.

Ensley PK, Rost TL, Anderson M, Benirschke K, Brockman D, Ullrey D (1982) Intestinal obstruction and perforation caused by undigested *Acacia* leaves in langur monkeys. J Amer Vet Med Assoc 181:1351–1354.

Gautier-Hion A (1983) Leaf consumption by monkeys in western and eastern Africa: a comparison. Afr J Ecol 21:107–113.

Gilmour D (1961) Biochemistry of insects. Academic, New York.

Hay AWM, Watson G (1977) Vitamin D_2 in vertebrate evolution . Comp Biochem Physiol 56B:375–380.

Hess AF (1929) Rickets, osteomalacia and tetany. Lea and Febiger, Philadelphia.

Hill WCO (1952) The external and visceral anatomy of the olive colobus monkey (*Procolobus verus*). Proc Zool Soc, London 122:127–186.

Hill WCO (1964) The maintenance of langurs (Colobidae) in captivity: experiences and some suggestions. Folia Primatol 2:222–231.

Hladik A (1978) Phenology of leaf production in rain forest of Gabon: distribution and composition of food for folivores. In: Montgomery GG (ed) The ecology of arboreal folivores. Smithsonian Institution Press, Washington, DC, pp 51–71.

Holick MF, Adams JS, Clemens TL, MacLaughlin J, Horiuchi N, Smith E, Holick SA, Nolan J, Hannifan N (1982) Photoendocrinology of vitamin D: the past, present and future. In: Norman AW, Schaefer K, Herrath DR, Grigoleit HG (eds) Vitamin D, chemical, biochemical and clinical endocrinology of calcium metabolism. Walter de Gruyter, Berlin, pp 1151–1156.

Hollis BW, Roos BA, Draper HH, Lambert PW (1981) Vitamin D and its metabolites in human and bovine milk. J Nutr 111:1240–1248.

Horst RL, Littledike ET, Riley JL, Napoli JL (1981) Quantitation of vitamin D and its metabolites and their plasma concentrations in five species of animals. Anal Biochem 116:189–203.

Huldschinsky K (1919) Heilung von Rachitis durch künstliche Höhensonne. Dtsch Med Wochenschr 45:712.

Hunt RD, Garcia FG, Hegsted DM (1967) Vitamin D_2 and D_3 in new world primates: production and regression of osteodystrophia fibrosa. Lab Anim Care 17:222–234.

Kuhn JJ (1964) Zur Kenntnis von Bau und Funktion des Magens der Schlankaffen (Colobinae). Folia Primatol 2:193–221.

McCollum EV, Simmonds N, Becker JE, Shipley PG (1922) Studies on experimental rickets. XXI. Experimental demonstration of the existence of a vitamin which promotes calcium deposition. J Biol Chem 53:293–321.

McKey DB, Gartlan JS, Waterman PG, Choo GM (1981) Food selection by black colobus monkeys (*Colobus satanas*) in relation to plant chemistry. Biol J Linnaen Soc 16:115–146.

Mellanby E (1921) Experimental rickets. Medical Research Council, Special Reports, Series SRS-61.

Milton K (1980) The foraging strategy of howler monkeys: a study in primate economics. Columbia University Press, New York.

Milton K (1984) The role of food-processing factors in primate food choice. In: Rodman PS, Cant JGH (eds) Adaptations for foraging in nonhuman primates. Columbia University Press, New York, pp 249–279.

National Research Council (1978) Nutrient requirements of nonhuman primates. National Academy of Sciences, Washington, DC.

National Research Council (1980) Recommended dietary allowances. National Academy of Sciences, Washington, DC.

Norman AW (1979) Vitamin D: the calcium homeostatic steroid hormone. Academic, New York.

Nowak RM, Paradiso JL (1983) Walker's mammals of the world, vol 1, 4th ed. Johns Hopkins University Press, Baltimore.

Oates JF (1977) The guereza and its food. In: Clutton-Brock TH (ed) Primate ecology: studies of feeding and ranging behavior in lemurs, monkeys and apes. Academic, London, pp 276–319.

Reeve LE, Jorgenson NA, DeLuca HF (1982) Vitamin D compounds in cow's milk. J Nutr 112:667–672.

Rosenberg HR, Waddell J (1951) Nature of the provitamin D of the ribbed mussell, *Modiolus demissus* (Dillwyn). J Biol Chem 191:757–763.

Setlow RB (1974) The wavelengths in sunlight effective in producing skin cancer: a theoretical analysis. Proc Nat Acad Sci 1:3363–3366.

Sommerfeldt JL, Napoli JL, Littledike ET, Beitz DC, Horst RL (1983) Metabolism of orally administered [^{3}H] ergocalciferol and [^{3}H] cholecalciferol by dairy calves. J Nutr 113:2595–2600.

Steenbeck H, Black A (1924) The induction of growth-promoting and calcifying properties in a ration by exposure to ultra-violet light. J Biol Chem 61:405–422.

Struhsaker TTS (1978) Food habits of five monkey species in the Kibale Forest, Uganda. In: Chivers DJ, Herbert J (eds) Recent advances in primatology. Academic, London, pp 225–248.

Takada K, Okano T, Tamura Y, Matsui S, Kobayashi T (1979) A rapid and precise method for the determination of vitamin D_3 in rat skin by high-performance liquid chromatography. J Nutr Sci Vitaminol 25:385–398.

Ullrey DE, Robinson PT, Reichard TA, Whetter PA, Brockman DK (1982) Dietary studies with colobids at the San Diego Zoo. Unpublished manuscript.
Waterman PG, Choo GM (1981) The effects of digestibility-reducing compounds in leaves on food selection by some colobinae. Malaysian J Appl Biol 10:147–162.
Watkins BE, Ullrey DE, Whetter PA (1985) Digestibility of a high-fiber biscuit-based diet by black and white colobus (*Colobus guereza*). Amer J Primatol 9:137–144.

56 The Chinese Golden Monkey—Husbandry and Reproduction

JING-FEN QI*

Introduction

The golden monkey (*Rhinopithecus roxellanae*) is in the family Colobidae. It became known to the world in the latter part of the nineteenth century. In 1956, the golden monkey was first displayed in the Beijing Zoo, followed by zoos in Guangzhou, Chengdu, and Shanghai. In November 1979, a pair was loaned to Ocean Park (Hong Kong) from the Beijing Zoo for a 3-month period. This was the first time golden monkeys had been exhibited outside mainland China. It was reported that both the Lanzhou University and Wudu in Gansu had hand raised and bred the golden monkey successfully in April 1959.

The first pair at the Beijing Zoo were wild caught in Baoxing of Sichuan in June 1956. One died immediately after having been captured. It was reported that since 1963 the Beijing Zoo has kept 30 of the animals. Most came from Qinling of Shanxi, the others from Wenxian of Gansu, Baoxing, and Beichuan of Sichuan. Of the 30 golden monkeys, five were transferred to associate zoos in China (between 1964 and 1967), 15 died, and ten remained in the zoo. Two females from those remaining in the zoo were unable to reproduce (one was too old, the other too young), but the other four pair were able to breed.

Appearance

The golden monkey has a round head, short ears, and a long tail (equivalent to body trunk or even longer). There is a difference in body weight between the mature male and the female, the male always being bigger and heavier. The body weights of the four males in the Beijing Zoo vary from 7 to 10.5 kg. There is a pair of fleshy flaps on the male's upper lip. These flaps are difficult to see on the female.

The body fur is mainly golden in color. A patch of dark brown-colored hair is on the top of the head, and the coloration becomes brownish yellow around the cheeks and light yellow or white on the thorax and abdomen. Starting from the

*Translated from Wildlife 2:25–30, 1982, Beijing Zoo by Mabel Lam and reprinted by permission of the author.

neck area to the back, the shoulders and upper arms are more or less overlaid with long golden hairs, like a piece of golden straw cape, thus the name golden monkey or golden-thread monkey.

The face of the monkey is blue with golden hair on the side of the cheeks and the forehead extending to the center of the face, which divides the face into three blue circles—two eye circles and the protruding mouth circle. The connection point of these three circles is the upturned "snub" nose which is why the golden monkey is also known as the "snub-nosed" monkey. By watching the monkey sidewise, one can see the upturned nose tip and the subsided nose bridge forming a long and small gap, and at the same time the nose tip seems connected to the protruding superciliary ridge (particularly in the male), which makes the face look even more peculiar.

Husbandry

Environmental Conditions

The golden monkey display in the Beijing Zoo has a combined bedroom and exercise playground separated by an outside controlled sliding door. Total area of the bedroom is 8.22 m^2 (length 3.2 m, width 2.6 m, height 3 m) whereas the exercise playground is 12 m^2 ($4 \times 3 \times 2.5$ m). The area is large enough to rear a pair of mature animals and a youngster. Tree trunks and racks are installed in the bedroom and exercise playground for the animals to rest and play on. The floors of the bedroom and playground are made of concrete, sloping slightly for easy cleaning and prevention of water accumulation.

In Beijing it is unnecessary to install a heating system in the bedroom. The animals are free to stay outdoors or indoors overnight throughout the year. Even in the coldest season when temperatures drop to -18°C and wind velocity reaches grade 6–7, golden monkeys are strong enough to remain outdoors as long as wind shelters are provided. Sometimes urine forms icicles on the monkey's tail, but it still behaves with perfect composure, drinking the icy water from the drinking pan. In general, these animals are found in the central south, southwestern, and northwestern part of China. Although temperatures are not very low in those areas, the animals usually live in the wild at approximately 1400–3000 m above sea level. Thus, the animal has developed strong resistance to a cold climate.

Water and Food

A round drinking pan is provided inside the bedroom and the playground. Prior to drinking, the golden monkey squats in front of the pan, lowers its head, and immerses its mouth in the water.

The major diet includes coarse-fibered food such as browse and fruits. In order to satisfy nutritional demands, a small portion of other feeding materials must be fed to the animal.

Vegetative food

Browse: Mulberry, poplar, osier, elm and fresh apricot browse. Mulberry is the favorite browse. Dried oak browse is a substitute during winter.

Fruits and vegetables

Apple, pear, persimmon, peach, apricot, almond, grape, crabapple, banana, tangerine, Chinese hawthorne, cabbage, spinach, eggplant, cucumber, and carrot. During summer and autumn, more choices of fresh fruits and vegetables can be obtained whereas in winter and spring only apple and cabbage can be supplied.

Cereal crops

Millet bread, steamed bread, and biscuit

Oil crops

Peanut and sunflower seeds

Animal food

Hard-boiled egg and mealworm (larva of elytra species)

Supplemental food

Bone meal, calcium tablet, cod-liver oil, table salt, and clay (contains multielements such as iron, calcium, magnesium, manganese, cobalt, zinc, and sodium). Bone meal and table salt are mixed into the millet bread before feeding. Cod-liver oil and calcium tablet are only fed to pregnant and lactating females and to young animals. Clay is fed to the entire group of animals at appropriate intervals.

Method of Feeding

The golden monkey has a very good appetite. Therefore, the method of feeding and the amount of food must be strictly controlled to prevent harmful effects. They can be fed large amounts of browse with no harmful effect. However, if slightly overfed with cereal food, they become sick easily. Symptoms are dyspepsia, anorexia, poor activity, and distended stomach and abdomen. If serious, the stomach can rupture and the animal may inhale the vomited food and die. An experiment was done in 1980 to test the capacity of a golden monkey's stomach. A necropsy was performed on an aged male whose carcass weighed approximately 13 kg. The structure of the stomach wall was found not to be suitable for digesting cereal food. Therefore, the feeding of cereal must be strictly controlled to no more than 150 g per day.

A daily diet for a mature monkey is:

fruit 800 g
vegetable 200 g
millet bread and biscuit 150 g
peanut and sunflower seeds 20 g
hard-boiled egg ½ pc

This diet should be divided in half and fed at two intervals. Browse is fed three times a day.

Breeding

The Beijing Zoo successfully bred golden monkeys in captivity in 1964. A second baby was born in 1965. Eight females have conceived 17 times between 1973 and 1981. Among these 17 pregnancies, there were nine normal births and eight abortions or premature births. Births usually occur between March and May, particularly in April. Single births are the norm. The newborn baby weighs approximately 500 g. The umbilical cord detaches 3 days after birth. The young become sexually mature at age 6. Mating takes place throughout the year but more frequently during the months of August to November. Conception also frequently occurs in these 4 months. Copulation is usually seen in the morning and afternoon. It has been observed that a pair of golden monkeys copulated three times in one morning, each time lasting 1–2 minutes. Except for forced mating, the female will squat initially in front of the male when she is in estrus. She will raise her buttocks, lower her shoulders, position her body and lower or twist her head to face the male and start to vocalize. Mating will not be seen easily if a pair is kept together at all times, but if separated at night and put together during the day, mating can be seen as soon as they are released. After several years' observation and according to the records of two females' last menstrual dates, mating dates, and parturition dates, the Beijing Zoo calculated that the gestation period varies from 193.5 to 203 days (Table 56.1). The pregnant female gradually becomes less active during pregnancy, appears slow and prudent in motion, and becomes timid and reacts slowly. Prior to giving birth, the female usually leaves some of the food, and by the next morning one can see her holding the baby resting on the racks. Sometimes she licks up the secretions on the baby's fur. Birth usually takes place at night.

Baby Golden Monkeys

A newborn baby has long black hair on its head and back. Other parts are covered with a fine whitish-gray hair. Its entire face is whitish-gray and the eyes and nostril bridge are bluish-green in color. The upper and lower eyebrows and the rims of the lips are red (gradually becoming lighter and fading completely after 2 weeks). The palms of the four limbs are red, and the tips of the toes are pink.

TABLE 56.1. Female golden monkey's last menstrual period, mating, and parturition record.

Female monkey no.	Date of last menstrual period	Mating date (same cage)	Parturition date	No. of days between last menstrual period and parturition	Gestation (day)
6	1974.9.8	In same cage with male all the time	1975.4.9	212	—
6	1975.9.11	1975.9.17–9.26	1976.4.8	208	193–202
11	1974.10.24	In same cage with male all the time	1975.5.23	210	—
11	1976.9.26	1976.10.1–10.10	1977.4.24	208	194–204
Average				209.5	193.5–203

TABLE 56.2. Comparison between newborn and mature golden monkey's appearance.

Position	Mature golden monkey	Newborn baby
Body fur	Golden in color	Light brown in color
Head	Round shape	Long round shape
Face	1. Bluish	1. Whitish gray
	2. Has fine hairs that divide the face into 3 round circles	2. Entire face exposed
	3. Mouth part and superciliary ridge protrude distinctly	3. Mouth part slightly protrudes and superciliary ridge not distinct
Ear	Bear hairs	Bear no long hairs
Palm	Black in color	Red in color

The shape of a baby's head is different from a mature animal's. The top has a long round shape, flattened on two sides and longer at the back. The rear portion of the brain is large and protrudes backward distinctly forming a 90° angle with the short, small neck. Its four limbs look strong when compared to the long and slender body. The tail is long, equivalent to the trunk or perhaps even longer. Table 56.2 summarizes the appearance of the newborn golden monkey.

A baby basically feeds and sleeps in its mother's arms. Sometimes one can see the baby holding its mother's nipple in its mouth while sleeping. The limbs are not strong enough to walk. Therefore, when the mother descends to the ground, the baby clings against its mother's armpits and waist, and the mother's arms will give additional support. As the baby grows up, its arms become stronger. It no longer requires support from the mother. The mother just moves the baby's forelimbs under her armpits and the baby clings against her before moving. Four to 5 days after birth, the baby's neck becomes stronger and straightens. Its eyes are bright and piercing. The baby becomes active and does not need as much nursing and sleeping as before. Yet the mother always holds the baby's hind limbs tightly, hindering it from grabbing. At about 2 weeks of age, the baby begins to live independently. It plays and eats by the side of the mother.

Since maternal attachment is very strong, it is not easy to remove the baby from its mother. One has to take every opportunity (such as when the mother is caught for medical treatment) to capture the baby and get its body weight. Table 56.3 shows the body weights of various newborn golden monkeys.

Behavioral Observation

The golden monkey kept in captivity prefers to stay in high areas of the enclosure. They might occasionally descend to the ground for eating, drinking, grooming, scratching, hugging, mating, and giving birth. Their activities vary all the time and include jumping, climbing, and crawling. The mature animal sometimes hangs itself among the racks or tree trunks and swings its body back and forth. On the other hand, a young animal prefers to lift its arms and walk on the ground for a short distance.

The animals' special behavior is hugging each other with faces buried in each other's shoulder fur while closing their eyes, sometimes for long periods.

TABLE 56.3. Baby golden monkey weight record.

Monkey no.	Sex	Age	Body weight (gm)
3	M	68 days	1000
3	M	83 days	1075
3	M	174 days	1325
3	M	214 days	1450
79-2	Fm	6 months	1275
79-1	M	15 months	3150
77-1	M	3 years	5250

Although some males live in pairs, they seldom hug each other. It has been observed that a pair of mature animals that are able to reproduce and are affectionate hug frequently. Those less affectionate, timid, old, or immature seldom hug each other.

Proof of strong maternal love is evident in the behavior both parents exhibit to calm their infants. A baby was seen struggling in its mother's bosom, and the mother immediately lowered her head and moved her lips over the baby and the baby soon calmed. If the baby is separated from its mother by some distance and disturbed by strangers, the mother stares at the stranger and quickly holds back her baby, or the male comes up to the baby and opens his mouth toward the baby until the baby stops screaming or is away from danger.

When the female is pregnant, the male stays close. The author observed several times from nonexhibited areas that pregnant females became flustered when watched by outsiders. They appeared timid, moved their limbs aimlessly, and their eyes peered around. Yet they did not have the intention nor the ability to run away or hide. In one situation, the male came up, walked around the frightened female, and opened his mouth toward her. The female soon quieted down.

Male and female monkeys always look after each other. The males are strongly jealous of each other. In 1980 when the author was conducting a night observation, she observed a pair of affectionate monkeys separated from each other by their keeper because the female had just given birth to a baby a few days earlier. For safety purposes, the keeper kept the male alone in the bedroom at the western side of the exhibit. They sat opposite each other at a short distance on the rack in the playground. Except to descend to the ground for food, they did not change their position even while sleeping. One day the female holding the baby accidentally rested at the eastern side of the playground. A mature male kept at the eastern side of the bedroom immediately descended to the ground and walked close to the wire mesh. Suddenly a loud scream was heard from the western end. The author was startled and looked around; she then discovered that the female's spouse at the western end was running madly in and out of the area between his bedroom and the playground. When the author became curious, the male at the eastern end reacted the same, becoming hysterical and jumping up and down. Soon the female became scared and timid, and returned to her original position at the western end of the playground. Her spouse then quieted down. Thus, the author believes that golden monkeys may possibly fight with each other for partners, which has been mentioned in articles written on research done in the wild.

Summary

To keep golden monkeys in captivity, the selection of good feeding materials is very important. Browse must be fed in sufficient amounts, and the feeding of cereal foods must be strictly controlled.

The animals kept at the Beijing Zoo have been pregnant 17 times during the Zoo's 17 years of breeding history. Abortion and premature labor occurred eight times. Since abortion and premature labor occurred in 47% of conceptions, great attention should be drawn to this problem.

Five of the eight abortions and premature labors happened in exhibit areas. Another abortion occurred in a wild-caught pregnant female after a long transportation. Therefore, outside stress may possibly be one of the major factors that causes abortion and premature labor.

It has been customary to keep mature animals in pairs rather than in harems. Therefore, there is a possibility that the female might be forced by the male to mate with him even though she is pregnant. In this way, the female might suffer injury.

In order to breed the golden monkey, it is ideal to have both exhibit and breeding facilities, thus helping to prevent the pregnant female from disturbances.

In the selection of exhibit and breeding areas, it is recommended that the environment be clean with sufficient sunlight. These are high mountain primates. If they are moved to densely populated areas, they could suffer from infectious diseases. Also contamination from the surrounding environment may cause the beautiful color of the animal's body fur to become less glossy.

Two males and one female should not be kept in the same cage. It is also not wise to keep the male alone; otherwise, the male will develop a bad temper and gradually become aggressive.

The animals should not be captured or interfered with. It is important to have them live peacefully and have enough time to rest and sleep. It has been discovered that when they are interfered with by outside people or have exhibits lighted day and night, they will become tense, which results in a higher rate of illness and mortality.

The animals kept in the Beijing Zoo came from different locations and have different coloration. At the same time, it was also discovered that the color of the animal's body fur becomes darker with age. Therefore, a further study in the difference in coloration is needed.

57 Captive Status and Genetic Considerations

NATHAN R. FLESNESS

Introduction

The status of captive primate populations has been reviewed several times, as part of broader reviews of captive populations in general (Perry et al., 1972; Pinder and Barkham, 1978) and in specific reviews of captive primate populations (Stevenson, in press; Schmidt, in press).

The 1972 review by Perry et al. for the "rare" mammal species covered by International Zoo Yearbook (IZY) Census figures provided a grim result. They used simple and arbitrary criteria for self-sustaining status (population over 100, at least half captive bred), and IZY data through the calendar year 1970. They found *no* captive primate population that met their elementary criteria. They mention that the mongoose lemur, *Lemur mongoz mongoz*, might well have qualified if two key colonies had responded properly to the IZY questionnaire.

The 1978 review of captive breeding of all IUCN-listed species by Pinder and Barkham found a total of 26 species to be self-sustaining, using somewhat more complicated criteria. Just two of these were primates (mongoose lemur, *Lemur mongoz mongoz*, and lion-tailed macaque, *Macaca silenus*). This study used IZY data through the year 1975.

An update was performed by Stevenson (Stevenson, in press) just for primates. Using the then-available IZY data (International Zoo Yearbook Vols. 10–21), she was able to extend the analysis forward through calendar year 1979. By this time, she was able to find eight IUCN-listed primate species whose captive populations met the original Perry et al. criteria.

At the present moment, the most current published IZY survey is in Volume 23, which covers events through the end of calendar year 1981. This is only two additional years of data beyond Stevenson's review and provides a view of the status of populations as of 4 years ago. More current analysis seems needed.

Schmidt (in press) provides a more recent perspective. He used ISIS data for 1983 to provide a review of several aspects of the status of captive primate populations. Focusing on IUCN-listed taxa, he noted that only 20 such primate species have a captive population size over 30 and that only 15 of these were increasing in numbers.

I will provide an update on the status of captive primate populations from the most recently published ISIS reports, which permit carrying the evaluation forward through the end of 1984. To do this I use ISIS figures for 31 December 1984, with occasional addition from and extrapolation with IZY Volume 23 Rare Mammal Census figures for the end of 1981. For the sake of consistency, I have used taxa designated by IZY as IUCN-listed, rather than some other source (e.g., Mittermeier, 1982).

ISIS as a Data Source

The International Species Inventory System (ISIS) currently pools data from over 200 institutions in 14 countries. The information collected is similar to that of many zoological studbooks, but over 62,000 living specimens are now included in ISIS reports.

Of course, coverage of any one species is limited to participating institutions, and thus the system contains data on only a portion of the total global captive population of most taxa. This portion is larger than may be supposed, and in any case such partial data are still rather useful, given the timeliness of the reports.

To evaluate what fraction of the world's known captive primate populations are covered by ISIS, comparisons were made between the ISIS Species Distribution Report (ISIS, 1985) figures for 31 December 1984, and the most recently available IZY census figures, which are through 31 December 1981 (IZY, 1984). IZY-listed primate taxa total 5396 specimens. For these same taxa, ISIS lists 3126 specimens, or 58% of the (earlier) IZY totals. Additionally, ISIS provides information on a further 7125 living primate specimens of taxa not surveyed by IZY. Coverage similarities and differences are shown in Figure 57.1.

With some caution, one can use this approach to extrapolate from the more current ISIS census figures to make estimates about the rest of the world's captive population. Thus, populations are examined based on (1) just ISIS data, and (2) ISIS fraction captive bred extrapolated to the IZY census. When no census figures are available in IZY, extrapolation is based on the estimate that ISIS covers the same fraction (58%) of the world's zoological primates for species not listed in IZY, as for listed species.

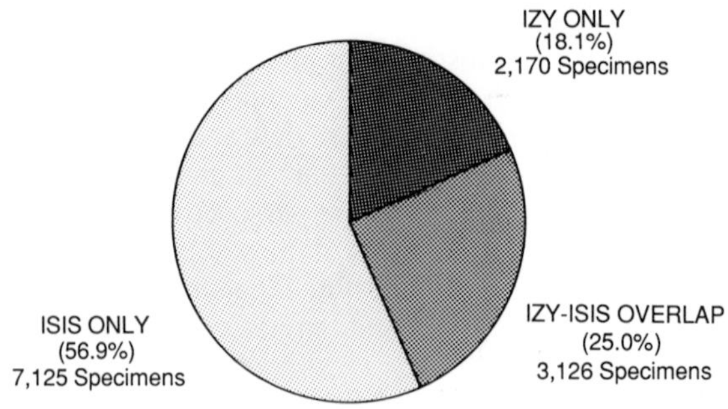

FIGURE 57.1. IZY and ISIS primate census, using latest available data (IZY, 1981, ISIS, 1984).

TABLE 57.1. IUCN-listed primate species meeting Perry et al. criteria.

		ISIS		IZY	
	ICUN	N[1]	CB	N	CB
Qualify on ISIS data alone:					
Lemur macaco (black lemur)	E	495	442	857	N/A
Callimico goeldi (goeldi's monkey)	R	136	121	141	127
Leontopithecus rosalia (golden lion tamarin)	E	394	384	295	N/A
Saguinus oedipus (cotton-topped tamarin)	E	242	193	508	387
Macaca silenus (lion-tailed macaque)	E	205	162	326	N/A
Pongo pygmaeus (orangutan)	E	341	228	378	222
Qualify using ISIS fraction captive-born + IZY census:					
Saguinus imperator (emperor tamarin)	I	64	39	107	43
M. sylvanus (barbary macaque)	V	62	52	765	N/A
Borderline cases:					
L. mongoz (mongoose lemur)	V	46	21	112	N/A
Cheirogaleus medius (fat-tailed dwarf lemur)[2]	V	61	58	39	N/A
Gorilla gorilla (gorilla)[3]	V	275	133	471	160

[1] N refers to the census, CB to the number censused that were identified as captive born.
[2] *Cheirogaleus medius* figures from ISIS (through 31 December 1984) suggest that extrapolation for the population held outside of ISIS would make the species just qualify. However, the IZY figures (through 31 December 1981) are considerably lower than the newer ISIS figures, so the extrapolation is in doubt.
[3] *Gorilla gorilla* registered with ISIS are 48% captive born, 48% wild born, and 4% "unknown." The latter reflects largely sloppy recordkeeping and reporting, especially since an international studbook does exist (Kirchshofer, 1980). This species is borderline by the Perry et al. criteria used here.

The taxonomy of many primate groups is not well understood. Importantly reproductive barriers may exist within taxa currently assigned species rank, such as *Aotus trivirgatus*, *Saimiri sciureus*, and *Ateles* sp. (Benirschke, 1983). There is also some uncertainty in the classification of some animals held in captivity, as a result of wild-caught individuals whose provenance was not established. In this paper, evaluations will be made primarily at the species level as a pragmatic decision. Given the likelihood that at least some "species" populations actually include individuals from two or more reproductively isolated populations, which will need to be maintained separately, the results to be presented below are too optimistic.

Following this approach, and using the simple Perry et al. criteria for purposes of consistency with previous studies, species captive populations were examined to see if they were over 100 in total size and more than 50% captive bred. For IUCN-listed species, the results are tabulated in Table 57.1 and shown as Figure 57.2, along with the results of the previous reviews.

IZY tabulates the census figures for 45 species of "rare" primates, 37 of which are listed by IUCN according to IZY. Of these latter, as shown in Figure 57.2, eight and possibly 11, now meet the minimal criteria set by Perry et al. ISIS data alone establish six species, and two more are included because their ISIS captive-bred fraction is over 50% and their IZY census is over 100. A further three species have very likely met the criteria, but data uncertainties leave them as borderline.

Thus, 8–11 species of IUCN-listed primates have captive populations meeting the most minimal criteria for likely success. This is only 22–30% of the 37

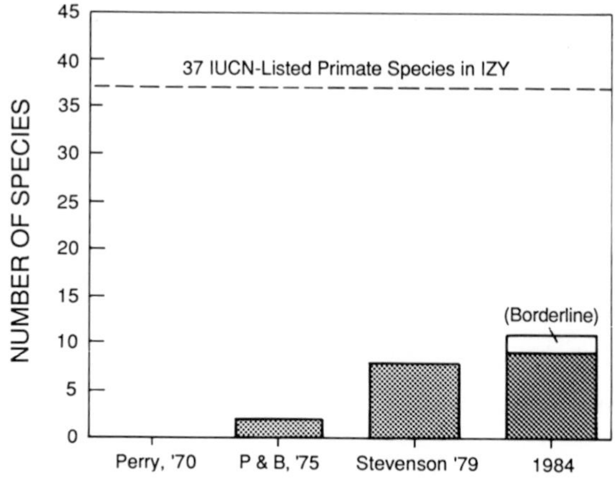

FIGURE 57.2. IUCN-listed primate species meeting Perry et al. criteria.

IUCN-listed primate species now held, according to IZY. This is hardly reason for self-congratulation, but some encouragement may be taken from the upward trend: about one species per year over the last decade. At this rate, it would take another quarter century to achieve minimum success with the IUCN-listed species now held. Unfortunately, it is quite unlikely that all of the less-successful captive populations will last this long. Eighteen species, or half of the IUCN-listed species held, have captive populations of ten or less (using IZY and ISIS data).

All captive primate populations should become self-sustaining, not least because many of those not now identified by IUCN as at risk will certainly become so in the near future (Soulé et al., in press). It is therefore important to extend this survey to primate taxa not (yet) listed by IUCN. Only eight such species are included in the IZY Volume 23 Census.

In the absence of IZY census data for these species, Stevenson (in press) used IZY birth figures for the calendar year 1979 in an effort to estimate the numbers of successful and unsuccessful captive populations. As she acknowledges, without at least corresponding census information, the birth figures are hard to interpret. She chose to simply identify those with a minimum annual "surviving" birth count of 30 or greater, and on this basis estimated that an additional 17 species are "successful."

Using the ISIS census and birth data through 1984 (ISIS Species Distribution Report, 31 December 1984), *all* primate species were surveyed to find populations meeting the Perry et al. criteria of census above 100, at least 50% captive bred. The results are shown in Figure 57.3, along with Stevenson's estimates from IZY birth reports. The data for species not IUCN listed is presented in Table 57.2.

Irrespective of IUCN status (combining Tables 57.1 and 57.2), ISIS-participant zoos have 24 species that fully meet these criteria. Three additional species are more than 50% captive bred by ISIS figures and have IZY census

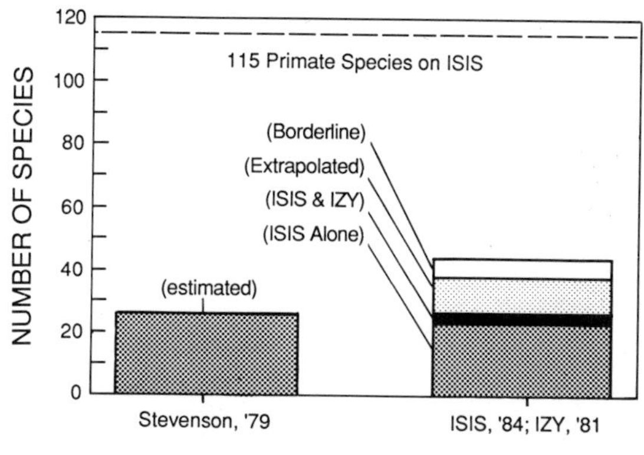

FIGURE 57.3. All primate species meeting Perry et al. criteria.

counts over 100. Then, using the 58% coverage figure for IUCN-listed species as the best estimator for ISIS coverage of nonlisted species, ISIS populations of 58 or greater that are at least one-half captive bred suggest a world population of 100 or more that is at least one-half captive bred. This extrapolation yields another ten species. A further six species are borderline.

These results are shown in Figure 57.3. Since there are 115 species of nonhuman primates held in ISIS-participant zoos, these 37–46 apparently successful species represent 32–40% of all primate species held.

Though rough and simplistic, this sort of analysis is still useful as a counting of milestones, in reasonably similar fashion to earlier reviews. These results suggest that the number of primate species meeting the Perry et al. criteria as of the end of 1984 is about one-third of the species held.

The reason for failure of so many of the others is not for lack of the ability to breed most species in captivity. Of the nine nonhuman primate families, six are represented by substantial numbers in captive zoological collections (Indriidae and Tarsiidae have together only 23 specimens on ISIS, all but one wild born, and there are no aye-ayes [Dabuentonidae] held). Zoological holdings of each of the six families are more than half captive bred (Table 57.3).

Overall, ISIS figures for 31 December 1984 show that 67.7% of the 10,296 primates reported were captive born. In spite of this, most populations have not reached adequate numbers for likely survival in captivity, and second or later captive generations are sometimes doing poorly, as with the mongoose lemur, *Lemur mongoz* (Schmidt, in press, and personal communication).

Criteria for Self-Sustaining Populations

The fairly cursory survey above produced results that are nonetheless fairly ominous. Roughly two-thirds of captive primate species failed to meet the simple arbitrary criteria defined by Perry et al. The only grounds for optimism are the fact that 15 years earlier, *all* primate species failed to meet the criteria.

TABLE 57.2. Species *not* IUCN-listed meeting Perry et al. criteria.

	ISIS		IZY	
	N	CB	N	CB
Qualify on ISIS data alone:				
Lemur catta (ring-tailed lemur)	579	529	N/A	N/A
L. variegatus (ruffed lemur)	388	363	250	N/A
Galago senegalensis (Senegal bushbaby)	142	101	N/A	N/A
Aotus trivirgatus (douroucouli)	142	84	N/A	N/A
Cebus apella (brown capuchin)	121	69	N/A	N/A
Saimiri sciureus (common squirrel monkey)	409	209	N/A	N/A
Ateles geoffroyi (black-handed spider monkey)	236	137	N/A	N/A
Callithrix jacchus (common marmoset)	185	161	N/A	N/A
Cercopithecus diana (diana monkey)	127	79	N/A	N/A
C. neglectus (DeBrazzas monkey)	126	98	N/A	N/A
C. patas (patas monkey)	127	93	N/A	N/A
Macaca fuscata (Japanese macaque)	217	187	N/A	N/A
M. nigra (black ape)	181	126	N/A	N/A
Papio hamadryas (hamadryas baboon)	105	70	N/A	N/A
P. sphinx (mandrill)	191	145	N/A	N/A
Colobus guereza (colobus monkey)	256	184	N/A	N/A
Hylobates lar (white-handed gibbon)	228	127	N/A	N/A
H. syndactylus (siamang)	122	76	N/A	N/A
Qualify if use ISIS captive-born fraction, IZY census:				
Microcebus murinus (lesser mouse lemur)	71	70	127	N/A
Qualify if extrapolate from ISIS census (assume ISIS includes less than or equal to 58% of world captive population):				
Cebuella pygmaea (pygmy marmoset)	71	58	66	N/A
Saguinus fuscicollis (brown-headed tamarin)	70	67	N/A	N/A
Cercocebus torquatus (sooty mangabey)	72	49	N/A	N/A
Cercopithecus aethiops (vervet)	59	34	N/A	N/A
C. mitis (blue monkey)	59	40	N/A	N/A
C. talapoin (talapoin monkey)	57	42	N/A	N/A
M. fascicularis (crab-eating macaque)	74	60	N/A	N/A
M. mulatta (rhesus macaque)	78	67	N/A	N/A
Papio cynocephalus (savanna baboon)	89	83	N/A	N/A
Theropithecus gelada (gelada baboon)	65	50	N/A	N/A
Borderline cases:				
Pithecia pithecia (pale-headed saki)	57	24	77	39
Pan troglodytes (chimpanzee)[1]	1770	844	N/A	N/A
Callicebus moloch (orabassu titi)	52	33	N/A	N/A
Cercopithecus talapoin (talapoin monkey)	57	42	N/A	N/A
M. nemestrina (pig-tailed macaque)	55	33	N/A	N/A
Presbytis entellus (entellus langur)	51	44	N/A	N/A

[1] *Pan troglodytes* data includes holdings in 20 research facilities. The zoo population alone is 412 animals, 154 captive born—too low a percentage (37%) to be viewed as borderline.

What about the criteria? They are of course flawed, as Perry and co-authors themselves recognized. For example, the deaths of a few wild-caught chimpanzees or gorillas would raise the proportion captive bred to over 50%, and leave the population size over 100 which would make these species certainly qualify instead of being borderline. Clearly, such deaths would not contribute to the population's long-term captive viability.

TABLE 57.3. Fraction captive born by family.

Primate family	ISIS census	No. captive born	Fraction captive born
Lemuridae	1682	1512	0.90
Indriidae	11	1	0.09
Lorisidae	183	323	0.57
Tarsiidae	12	0	0
Cebidae	680	1334	0.51
Callitrichidae	1401	1203	0.86
Cercopithecidae	2641	1919	0.73
Pongidae	2886	1471	0.51

Source: Data from ISIS Species Distribution Report for 31 December 1984, ISIS, 1985.

In spite of such flaws, it seems that the species lists developed using these criteria are not too misleading. Examination of the summarized 5-year birth and death rates (ISIS Species Distribution Report for 31 December 1984; ISIS, 1985) for the qualifying species (Table 57.1 and Table 57.2) determined that 41 of the 46 species had birth rates exceeding death rates (the exceptions were *Theropithecus gelada*, *Lemur mongoz*, *Saimiri sciureus*, *Pithecia pithecia*, and *Callicebus moloch*). The 5-year surplus of deaths over births is substantial only for the first two of these identified exceptions.

Ideally, a self-sustaining captive population would be one that met both demographic and genetic criteria.

Demographically, one would wish a Leslie-matrix type projection (e.g., Foose, 1980) based on the recent age and sex-specific fecundity and survivorship schedules as well as the present age distribution—all for captive-bred individuals—to predict a steady, significant population growth to achieve captive carrying capacity (of at least 250) in just a few generations, at which level it could be stabilized by coordinated management. There are several difficulties in carrying out the relevant demographic analyses for captive primate populations.

The first difficulty is whether or not the data exist at the holding facility—whether appropriate records are kept. This is a trivial, but critical, point. For example, 16.7% of living primate specimens registered with ISIS are of "unknown" origin, according to the facility now holding them. This figure is dropping, and only 5% of primate speciments arriving in ISIS institutions during 1984 were of "unknown" origin.

Assuming the data exist, the next difficulty is compiling it for all specimens. Ideally, a complete and current studbook is necessary. Ten primate species are officially covered by International Studbooks (IZY, Vol. 23), and nine of these have at least been published once, though some are badly out of date. This leaves 106 primate species held in ISIS-participant zoological collections without coverage by a published international studbook.

ISIS data are *not* globally complete but are current. They are a partial studbook for each taxon—for most taxa, the only source of information on age distribution, age-specific fecundity, and mortality, etc. A more sophisticated calculation of the relevant demographic parameters is necessary to utilize such partial data, taking into account the age classes through which each individual specimen is actually tracked, and therefore is at risk to death or birth while in the database. This could solve part of the data problem for ISIS institutions. The necessary computer software has been prototyped but is not finished at present.

A further difficulty and topic for research is the matter of doing demographic projections for small populations. The question of confidence intervals for demographic parameters and projections has not been resolved but is an area of active research (e.g., Polachek, 1985). For small enough populations, the data may lead to a deterministic projection of growth and success, but stochastic behavior may dominate and the population go extinct anyway. This is especially likely below the 50 individual range (Goodman, in press). Many projections have been made for zoological populations (Foose and Seal, in preparation) with perturbation modeling as a useful method for evaluating the confidence limits.

Genetically, the criteria for a captive self-sustaining population should be established in terms of the genetic diversity retained. For future adaptation to changing environments, whether captive, wild, or some highly modified "wild," more retained genetic diversity is better. Somewhat arbitrarily, recent discussion (Soulé et al., in press) has developed the genetic criteria of preserving 90% of the initial diversity for 200 years. This criterion is similar to an independently derived standard recently established for captive fish stocks (FAO, 1981).

There are several challenges, some useful research topics, in assessing the retained genetic diversity of captive primate populations. Like the demographic issues, the first questions are (1) the existence of the basic data—in this case usually pedigree information, and (2) compilation of the data from the many institutions concerned. The same solutions apply again.

If the complete pedigree data are available, one can proceed to try to evaluate the expected retained genetic diversity by (1) calculation of various descent probabilities (E. Thompson and M. Gilpin, personal communication), or (2) simulation. The first approach is new, the second has been applied to a couple of primate populations (J. Ballou and G. Mace, personal communication).

There is an alternative approach, which may not require complete pedigree data. This is the estimation or calculation of effective population size. This is a "fudge" factor, which attempts to convert the genetic dynamics of a real, messy, population, into something simple enough that the mathematics of theoretical population genetics applies to it.

There has been a lot of discussion of effective population size as a genetic criterion for captive populations (Soulé et al., in press, and references therein). It has also received extensive use in theoretical discussions of evolutionary processes. Unfortunately, it has not commonly been estimated for real populations. It is probably fair to say that no one presently knows how to calculate it exactly for the kind of populations we are talking about, though recent work by Lande and Barrowclough (in press) makes a good start.

What is well established is how to correct the census number of the population for sex ratio effects or progeny number variance effects (Crow and Kimura, 1970). These corrections typically reduce the population's effective size below that of the census number. Correction for overlapping generations, which all primate populations probably show, is more difficult and has not yet been combined simultaneously with the other corrections. Accurate practical application of this genetic evaluation technique promises to be a fruitful area for future research.

In the absence of completely explicit answers, and with only partial data available, considerable caution is necessary. Nonetheless, I have performed some very simple and crude calculations of effective population size for captive primate populations on ISIS using just the sex ratio correction and explicitly ignoring the complications of overlapping generations, progeny number variance, changing population sizes, and earlier history.

The procedure was to determine the number of living males that have been parents and the number of living females that have been parents, for each of the 115 species of primate on ISIS. These values were entered into the formula

$$\frac{4N_mN_f}{N_m + N_f}$$

to estimate a value for effective population size. The calculation, though crude, is worth making because no estimates have been made for almost all of these populations and because the results permit some preliminary evaluations.

Some confidence that the results are not too misleading comes from the fact that an independent calculation for one of these species, using the quite different approach of inbreeding-effective size, yielded an answer within about 5% (J. Ballou, personal communication) of that arrived at by this method.

There are further difficulties that relate to the quality of data available. Sometimes, the ISIS database is not comprehensive enough to trace an identified parent through potentially several moves to the current (or last) holding facility. These cases cause some uncertainty in the count of living breeders and thus uncertainty in the effective population size calculated.

The distribution of these crudely estimated effective population size values is not encouraging. Using ISIS data alone, only nine species of primates (Table 57.4) have estimated effective population sizes over 100. Restricting the chimpanzee (*Pan troglodytes*) data to zoological facilities only (excluding ISIS data for research facilities) would eliminate them from this category, leaving only eight species. Four of these are IUCN listed according to IZY. Several on this short list are taxa that may have biologically important subdivisions within the current species classification.

More than half of all species currently held in captivity (65) have an estimated effective population size of 25 or less in ISIS facilities, and about one-third (39) are estimated to be below 10. Extrapolation to non-ISIS populations would increase these estimates, but not by a great deal. These results unfortunately support those obtained from census and fraction captive-born figures. Very few captive primate populations can be viewed as adequate for the long-term.

TABLE 57.4. Captive primate populations with estimated effective population sizes of 100 or greater.

Species	IUCN	N_e (est.)
Lemur catta (ring-tailed lemur)		218
L. macaco (black lemur)	E	276
L. variegatus (ruffed lemur)		139
Saimiri sciureus (common squirrel monkey)		114
Leontopithecus rosalia (golden lion tamarin)	E	146
Saguinus oedipus (cotton-topped tamarin)	E	108
Hylobates lar (white-handed gibbon)		101
Pan troglodytes (chimpanzee)		404
Pongo pygmaeus (orangutan)	E	130

Source: Based on ISIS data as of 31 December 1984.

Neonatal Mortality, Inbreeding, and Outbreeding

Kurt Benirschke initially invited this paper to cover a somewhat different topic: specifically whether inbreeding or outbreeding depression was responsible for neonatal mortality in captive primate populations. I believe it is impossible to address this question generally at present. However, it is useful to look at the components of the question, and see what obstacles exist, and what answers are available.

Neonatal mortality *is* high in captive primates. ISIS-participant institutions have reported 13,687 primate births since 1974, when ISIS began. Of these, 730 were identified as stillborn, an additional 932 failed to survive the first day, and an additional 1054 died between day 2 and day 30. The stillbirth figures yield a perinatal mortality rate of 5.33%, which of course is an underestimate because of detection difficulties and variation in reporting practices.

Of the 12,957 live births reported, the 932 first-day deaths give a rate of 7.2% mortality on day one, followed by a further 8.1% over days 2 through 30. Thus, live births of primates average 15.3% neonatal or first month mortality. Counting the stillbirths, neonatal mortality over all 13,687 primate births reported to ISIS is 19.8%.

The figure of 15.3% neonatal mortality for nonhuman primates in (largely) zoological collections happens to be very close to the current 14% figure for neonatal mortality in human births in Western Africa, the highest regional rate recorded in a recent review of human infant mortality (Miller, 1985). This is also approximately the figure (14%) for human neonatal mortality world-wide over 30 years ago, for the period 1950–1955.

Since the early 1950s, human neonatal mortality has been reduced on the average world-wide to a present figure of 9%. The improved statistics for human primates are probably not related to any genetic issues, but rather to improved public health, especially prenatal care, and diet. It seems likely that a similar approach in nonhuman primates might yield similar results.

In the great apes, where it has been examined recently in more detail (Seal et al., 1985), infant (total first year) mortality averages 21.5% (data from Table III, Seal et al.). Since these same species have little inbreeding and are primarily first-generation captive-born and wild-caught individuals, inbreeding depression is improbable as a major cause of the current mortality rates in these species, though it may present problems in future decades as more captive-born specimens come of breeding age.

In published studies that have found statistical evidence linking mortality and inbreeding in a study of 18 species of captive primates (Ralls and Ballou, 1982) the *amount* of mortality (to age 6 months) was substantially higher than reported here, perhaps because they selected particularly complete sets of data. Mortality (including stillbirths) to age 6 months, averaged across the 18 species, was 34% in animals with no known inbreeding and increased by almost half again as much to 49% in "inbred" animals (average F was 0.16; J. Ballou, personal communication). This suggests that inbreeding depression in these species might account for as much as one-third of primate infant mortality, at least in these 18 species, but also suggests that the majority of infant mortality is *not* related to inbreeding.

Analysis of potential "outbreeding depression" is confounded by the profound lack of provenance data for wild-caught primates. In the great apes, this has led to confusion and management difficulties in orangutan, *Pongo pygmaeus* ssp., and to a lesser extent in gorilla, *Gorilla gorilla* spp. Regrettably, the same lack of provenance information typically applies to other species, obviating analysis of outbreeding and its consequences by these means.

Given the lack of adequate records during the time when these captive populations were established, unequivocal analysis will be difficult. However, chromosomal studies such as those of Benirschke (1983) and application of various molecular technologies may yet permit some partial analysis of this issue. In general, however, it would have been far better if just a little of the energy now being invested in discussions of subspecies problems, and possible outbreeding consequences, had been invested in ascertaining the actual place of capture of the population's wild-caught founders.

Conclusions

Captive primates in zoological collections are two-thirds captive born, and seven out of eight new arrivals during 1984 were also captive born. Nonetheless, two-thirds of captive primate species populations fail to meet the minimal criteria for long-term captive success established by Perry et al. For most primate species, captive breeding is a solved problem, but population growth and maintenance at an acceptable genetically effective population size have not been solved.

Neonatal mortality is high and could likely be lowered by improvements in basic management and husbandry, by analogy with similar improvements achieved for our own primate species.

Inbreeding-associated mortality may be a numerically significant fraction of current infant mortality in some species, but it is improbable as the dominant factor in infant primate deaths. Contribution of various possible forms of "outbreeding depression" is hard to evaluate because of earlier recordkeeping failures. Cytogenetic and molecular systematic work will be required to pursue this with current populations.

It is my opinion that many more species of primates could be maintained on a self-sustaining basis in zoological collections and that this would be very useful "conservation insurance" against increasingly likely disaster in the wild. I believe that each species will only be successful in the long run if emphasis is placed on coordinated *population* management, through programs such as the American Association of Zoological Parks and Aquariums Species Survival Plans (AAZPA, 1983) or the work of the Anthropoid Ape Advisory Panel of the British Federation of Zoos.

Acknowledgments. Most of the ISIS information used here was taken from the ISIS Species Distribution Report, published every 6 months by ISIS. Neonatal mortality figures and estimated effective population sizes were calculated from ISIS data. These data are available on a species-by-species basis as ISIS "studbook-like" reports. I would like to express thanks to the hundreds of people within zoological institutions who have provided data to ISIS and to IZY, to Larry Grahn who assembled the ISIS chimpanzee data, to the U.S. National Institutes of Health which funded that assembly, to Phil Garnatz and Paul Scobie who helped with the analysis, to Jon Ballou, Tom Foose, and Ulie Seal who commented on the manuscript, and to Minnesota Zoo Librarian Angie Norell who helped a great deal.

References

AAZPA (1983) Species survival plan. American Association of Zoological Parks and Aquariums, Wheeling, West Virginia.

Benirschke K (1983) The impact of research on the propagation of endangered species in zoos. In: Schoenewald-Cox C et al. (eds) Genetics and conservation. Benjamin Cummings, Menlo Park, CA, pp 402–413.

Crow JF, Kimura M (1970) An introduction to population genetics theory. Harper and Row, New York.

FAO/UNEP (1981) Conservation of the genetic resources of fish: problems and recommendations. Report of the expert consultation of the genetic resources of fish. FAO Fisheries Technical Paper 217.

Foose T (1980) Demographic management of endangered species in captivity. Intern Zoo Yrbk 20:154–156.

Foose T (1983) The relevance of captive populations to strategies for the conservation of biotic diversity. In: Schoenewald-Cox C et al. (eds) Genetics and conservation. Benjamin Cummings, Menlo Park, CA, pp 374–401.

Goodman D (In press) The minimum viable population problem. I. The demography of chance extinction. In: Soule M (ed) Minimum viable populations.

ISIS (1985) Species distribution report for 31 December 1984. International Species Inventory System, c/o Minnesota Zoological Garden, Apple Valley, Minnesota.

Kirchshofer R (1980) International Studbook for *Gorilla gorilla*. Frankfurt Zoological Society, Frankfurt, West Germany.

Lande R, Barrowclough G (In press) In: Soulé M (ed) Minimum viable populations.

Miller CA (1985) Infant mortality in the U.S. Sci Amer 253:31–37.

Mittermeier RA (ed) (1982) List of primates in the IUCN mammal red data book, IUCN/SSC/Primate Specialist Group Newsletter, No 2, pp 6–7. Stony Brook, NY.

Olney P (ed) (1983) Intern Zoo Yrbk 23:355–360.

Perry J, Bridgewater D, Horseman D (1972) Captive propagation: a progress report. Zoologica, Fall, 1972, 109–117.

Pinder NJ, Barkham JP (1978) An assessment of the contribution of captive breeding to the conservation of rare mammals. Biol Cons 13:187–245.

Polachek T (1985) The sampling distribution of age-specific survival estimates from an age distribution. J Wildl Manage 49:180–184.

Ralls K, Ballou J (1982) Effects of inbreeding on infant mortality in captive primates. Intern J Primatol 3:187–245.

Schmidt C (In press) A review of zoo breeding programs for primates. Intern Zoo Yrbk 24/25.

Seal US, Flesness N, Foose T (1985) Neonatal and infant mortality in captive-born great apes. In: Clinical management of infant great apes. AR Liss, Inc, pp 193–203.

Soulé M, Gilpin M, Conway W, Foose T (In press) The millenium ark: how long the voyage, how many staterooms, how many passengers? Zoobiology.

Stevenson MF (In press) Effectiveness of primate captive breeding. In: Harper D (ed) Symposium on the conservation of primates and their habitats. Vaughan Paper No 31, University of Leicester.

58
Incidence and Consequences of Inbreeding in Three Captive Groups of Rhesus Macaques (*Macaca mulatta*)

David Glenn Smith

Introduction

Nonhuman primates have proved useful as subjects for both behavioral and biomedical research because they are closely related to humans (Goodwin and Augistine, 1975) and provide animal models of human disorders (Cornelius and Rosenberg, 1983). The growing costs of acquiring nonhuman primates from their natural habitats, recent embargoes on exportation of feral animals for experimental purposes from some countries, such as India, the dwindling sizes of free-ranging nonhuman primate populations (Mason, 1979), and the high demand for, and severe shortage of, nonhuman primates available for biomedical research, have required the development of nonhuman primate breeding programs in the United States (Held, 1980).

The cumulation of inbreeding in these groups, however, threatens the continued productivity of such breeding programs (Smith, 1980a). In free-ranging groups of macaques, inbreeding is minimized by the emigration of all males to other groups soon after reaching sexual maturity and by periodic change in dominance rank that is closely correlated with reproductive success (Smith, 1981). Thus, young immigrant males rise to high rank and reproductive success and are replaced by new immigrant males before their own daughters become sexually mature. In captive groups in which emigration is suspended, older males of high rank are replaced by their own sons who, preliminary data indicate (Smith, 1982b, 1986), breed randomly with respect to paternal kinship. While avoidance of matrilineal inbreeding has been hypothesized (see, for example, Missakian, 1973) the studies upon which this hypothesis was founded were based upon observations of sexual activity that are now known to provide unreliable predictions of both relative reproductive success and paternity (Curie-Cohen et al., 1983; Stern and Smith, 1984).

Influence of Inbreeding upon Fitness

In the absence of migration, which increases genetic heterogeneity, inbreeding leads to increased homozygosity and results in a loss of fitness at loci where overdominance occurs or where rare alleles, whose frequencies are ordinarily maintained at low levels by recurrent mutation, are lethal or deleterious in homozygous form. The extent of this loss of fitness and the increase in mortality and morbidity and decline in fertility associated with such loss are, in theory, strict linear functions of the value of F_t, the average value of Wright's classic inbreeding coefficient for members of the tth generation (Cavalli-Sforza and Bodmer, 1971). $F_t = 1 - (1 - \Delta F)^t$ where ΔF is an inverse function of effective population size, N_e, or

$$\Delta F = \frac{1}{2N_e}.$$

Moreover, the value of F_1 provides an empirical estimate of ΔF.

Colony losses of some 8%, resulting from higher prereproductive mortality and lower fertility, for every 10% of inbreeding can be predicted based upon studies of the deleterious effects of inbreeding in various species (Dickerson, 1954; Falconer, 1960) including both human (Schull and Neel, 1965) and nonhuman (Ralls and Ballou, 1982) primates.

Influence of Inbreeding upon Research

As inbreeding accumulates within different captive groups housed at a breeding center, genetic variation within each group should decline while average differentiation (i.e., genetic subdivision) among all groups increases (Falconer, 1960). Similarly, the gene pool of the entire breeding stock at each breeding facility should become more homogeneous yet more unlike the gene pools at all other facilities. The contribution of intrinsic genetic differences, among research subjects, to differences in experimental results of biomedical research conducted at different facilities will also increase owing to this genetic subdivision. In other words, to the extent that factors under study are influenced by genes, the heritability of those factors will increase and obscure the causal relationships of interest.

It is to avoid this outcome that some have proposed the breeding of highly inbred strains of nonhuman primate species as standard subjects for biomedical research. Expectations for success of such plans are partially predicated on the belief that the small effective population sizes and local subdivision characteristic of free-ranging polygynous social groups of nonhuman primates have already promoted the elimination of many deleterious alleles that would otherwise lead to inbreeding depression and frustrate these breeding goals. For similar reasons, in fact, some breeders have questioned the need for breeding strategies designed to minimize close inbreeding in captive groups of nonhuman primates. An empirical study of the incidence and consequences of inbreeding in closed captive groups of nonhuman primates is needed to further address both of these issues.

Inbreeding in Three Groups of Rhesus Monkeys

We have used genetic data derived from the study of 28 genetic polymorphisms (Smith et al., 1985) in three groups of rhesus macaques (*Macaca mulatta*) at the California Primate Research Center (CPRC) to seek answers to the following questions that bear upon the development of colony management strategies at the CPRC.

1. To what extent has genetic subdivision of the rhesus groups at the CPRC alone already occurred?

2. What is the effective population size, N_e, of each group and the theoretical rate of random inbreeding, ΔF, associated with it?
3. Do members of rhesus macaque groups avoid consanguineous mating, as has been hypothesized for matrilineal relatives (Missakian, 1973), and how closely can the frequency of these matings for the first generation of potential inbreeding, F_1, be predicted by the value of N_e?
4. Is inbreeding in these groups associated with easily monitored deleterious effects?
5. What strategies for minimizing inbreeding in captive groups are the most successful and cost-effective?

Genetic Subdivision

Table 58.1 reports estimates of F_{ST}, a measure of genetic subdivision (Wright, 1965), among three of the captive rhesus groups at the CPRC that we have monitored most closely since 1977. Here, F_{ST} values for the founding female population (established in 1976) are compared with those based upon gene frequencies for the approximately 25 offspring born to each group in 1984. Since all three groups were established with adult and subadult animals originating from a heterogeneous source of primate importers in India, the F_{ST} value of 0.020 for the founding female members of the three groups is primarily due to male founder (i.e., sampling) effects. The F_{ST} value of about 0.049 for the 1984 offspring, therefore, represents an increase of 0.0285, but one that might be inflated as a result of sampling error.

Paternity Exclusion Analysis

We have also employed the genetic markers referenced above for paternity exclusion analysis (PEA). By this method (Smith, 1980b), all sexually mature males (i.e., possible fathers) lacking a genetic marker present in a particular offspring but absent in its mother are automatically excluded from paternity of that offspring. When all but one of the possible fathers in a given group are excluded, by at least one polymorphic system, from paternity of a given offspring, that sole

TABLE 58.1. Increase of genetic subdivision (F_{ST}) among three captive groups of rhesus macaques.[1]

System	1976 Founders	1984 Offspring
Transferrin	0.0159	0.0628
Prealbumin	0.0022	0.0419
Albumin	0.0173	0.0495
CA II	0.0475	0.0463
Gc	0.0225	0.0725
GPI	0.0101	0.0210
6PGD	0.0447	0.0813
Dia I	0.0005	0.0138
	0.0201	0.0486

[1] $F_{ST} = \dfrac{\sum_{i=1}^{k} (p_i - \bar{p})^2}{k \times \bar{p}\,(1 - \bar{p})}$, where $k = 3$.

TABLE 58.2. Distribution of number of paternity exclusions by year in three groups of rhesus macaques.

Number of non-excluded males	North Corral 2								
	1977	1978	1979	1980	1981	1982	1983	1984	Total
1	14	10	8	11	15	17	12	21	108
2	0	0	0	3	2	6	5	4	20
3	—	—	0	1	0	0	4	0	5
4	—	—	0	0	0	0	1	1	2
5	—	—	—	0	0	0	1	0	1
6	—	—	—	0	0	0	1	0	1
Total	14	10	8	15	17	23	24	26	137

Number of non-excluded males	North Corral 3								
	1977	1978	1979	1980	1981	1982	1983	1984	Total
1	19	14	17	7	15	15	5	17	109
2	2	0	0	1	2	6	3	4	18
3	0	0	0	0	3	1	3	1	8
4	0	0	0	0	0	0	4	0	4
5	—	—	0	0	0	0	5	1	6
Total	21	14	17	8	20	22	20	23	145

Number of non-excluded males	North Corral 7								
	1977	1978	1979	1980	1981	1982	1983	1984	Total
1	19	15	23	22	19	15	20	19	152
2	2	2	1	2	5	6	8	2	28
3	0	0	1	0	0	0	0	0	1
4	—	0	0	1	0	0	0	2	3
Total	21	17	25	25	24	21	28	23	184

nonexcluded male is declared the father. Males below the age of 4 years are regarded as possible fathers only in instances where all older possible fathers are excluded from paternity or when they exhibit sexual activity during the breeding season preceding the birth of the offspring in question. Table 58.2 shows the distribution of nonexclusions by year for each of the three rhesus groups.

THEORETICAL RATE OF INBREEDING

A theoretical estimate of the value of ΔF can be made if accurate estimates of reproductive success and of N_e, the inbreeding effective population size, can be made. Thus,

$$N_e = \frac{N_{t-1}\bar{K} - 1}{\dfrac{V_K}{\bar{K}} + \bar{K} - 1}$$

where $\bar{K} = m\bar{K}^m + (1 - m)\,\bar{K}^f$, $V_K = mV_K^m + (1 - m)\,V_K^f + m(1 - m)\,(\bar{K}^m - \bar{K}^f)^2$, m is the proportion of the adult members of the founding population, N_{t-1}

in number, which is comprised of males, $\bar{K}^m$ and $\bar{K}^f$ are the average numbers of offspring produced by males and females, respectively, in that group, and V_K^m and V_K^f are, respectively, the variances in those numbers (Crow and Kimura, 1970). These estimates, and those of

$$\Delta F = \frac{1}{2N_e},$$

which are based upon them, are given in Table 58.3 (Smith, 1985).

The values of N_e for the three groups range from about 16 to 20 and the average estimate of ΔF based upon these values is 0.029. This high expected value of ΔF results from the high variance in male reproductive success, which is illustrated in Table 58.4 by the marked differences, among founding males, in the numbers of their sexually mature sons and daughters presently in their group.

Empirical Estimate of Patrilineal Inbreeding

We have also used PEA to identify instances of inbreeding in the three rhesus groups and to estimate empirically the value of ΔF. The expected number of inbred offspring produced by a given male was estimated as the product of his relative reproductive success (column A in Table 58.5) during that breeding season and the number of animals born that year whose mothers are related to that male (i.e., who are potentially inbred to him). The sum of products over all possible fathers then gives the expected number of inbred offspring for that year in that group based upon the assumption of random mating with respect to paternal kinship.

When the remaining number of nonexcluded possible fathers, n, for a potentially inbred offspring was greater than 1, exactly $1/n$ of the offspring in question was assigned to each of the n nonexcluded possible fathers. This fractionation of

Table 58.3. Estimate of inbreeding effective population size (N_e) for three groups of rhesus macaques.[1]

	N^m_{t-1}	$\bar{K}^m$	V^m_K	m	N^f_{t-1}	$\bar{K}^f$	V^f_K	$\bar{K}$	V_K	N_e	$\Delta F = \frac{1}{2N_e}$
NC 2	2	25.50	42.25	0.0571	33	4	8.3	5.23	35.13	15.63	0.0320
NC 3	4	17.50	151.25	0.1081	33	4	8.3	5.46	41.32	16.71	0.0299
NC 7	4	25.50	151.25	0.0800	46	4	8.3	5.72	53.76	20.19	0.0248
		22.83	114.92	0.0817		4	8.3	5.47	43.40	17.84	0.0289

[1] $N_e = \dfrac{N_{t-1}\bar{K} - 1}{\dfrac{V_K}{\bar{K}} + \bar{K} - 1}$

where $\bar{K} = m\bar{K}^m + (1 - m)\bar{K}^f$, $V_K = mV_K^m + (1 - m)V_K^f + m(1 - m)(\bar{K}^m - \bar{K}^f)^2$, m is the proportion of the founding population, N_{t-1} in number, that is male, $\bar{K}^m$ and $\bar{K}^f$ are the average numbers of offspring produced by males and females, respectively, in that group, and V_K^m and V_K^f are, respectively, the variances in those numbers. The value $\bar{K}^f$ was estimated as $2R_0$ (see Table 58.7), and the value of V_K^f was computed from the distribution of live births to female founders since 1977, adjusted for $\bar{K}^f = 4.0$, as appropriate for a negative binomial distribution of progeny size (K) and averaged over all three rhesus groups. The observed (or hypothetical, for females) numbers of offspring ($N_{t-1} \times \bar{K}$) differ between males and females because the reproductive success of founding males, but not females, was almost completed by 1984.

TABLE 58.4. Distribution of numbers of potentially reproductive females and sexually mature males of founding patrilineages in the three rhesus groups in the fall of 1984.

Group	Founder of patrilineage	Number of female members born since 1976	Number of sexually mature males
NC 2	7820	30	6
	8241	10	4
	16684[1]	2	1
NC 3	3961	8	3
	4749	19	4
	7051	1	0
	7059	21	3
NC 7	11	10	3
	1243	9	2
	16888	20	1
	16898	2	0
	7618[2]	8	0
	7836[2]	0	0
	7838[2]	6	1

[1] The father of this male, whose mother was pregnant with him when placed in NC 2, is neither 7820 nor 8241.
[2] These three males were the original founders of NC 7 but were replaced with the other four males in the fall of 1977.

infants was required in only 97 of 466 cases (see Table 58.2) and is unlikely to have significantly distorted our estimates of relative reproductive success, given in column A, or the number of inbred offspring, given in column B, of Table 58.5. For only seven of the total of 132 offspring who were potentially inbred did the remaining nonexcluded possible fathers include males at least one of whom was unrelated to all others. Thus, while the true value of F is usually known, we recorded the status of these 132 offspring as simply either "inbred," "possibly inbred," or "noninbred."

About 45.9 inbred matings were expected while 42.8 were observed. The distribution of the numbers of observed inbred and noninbred matings and the distribution of the expected numbers of inbred and noninbred matings, with one degree of freedom, were statistically significantly different neither for the total data ($\chi^2 = 0.32$) nor for any of the three individual cages. We therefore conclude that captive rhesus monkeys can be expected to mate randomly with respect to paternal kinship, confirming conclusions based on earlier analyses of smaller subsets of these data (Smith, 1982b, 1986).

AVOIDANCE OF MATRILINEAL INBREEDING

An additional contribution to the values of F_t and $\triangle F$ is made by maternal inbreeding. The observed and expected rates of inbreeding in the matrilineage are given in Table 58.6. The number of inbred matings expected for each potential maternally inbred mating appears in the last column of Table 58.6 and is exactly equal to the relative reproductive success of the matrilineal male relative(s) during the breeding season preceding the birth of the infant in question. The sum of these values over all 40 matings that are potentially matrilineally inbred, 5.92, represents the total number of inbred matings expected under the null hypothesis that mating is random with respect to maternal kinship.

TABLE 58.5. Probability of paternally inbred matings assuming random mating with respect to kinship and observed number of inbred matings.[1]

Possible fathers	1980		1981		1982		1983		1984		Total	
	A	B	A	B	A	B	A	B	A	B	A	B
NC 2												
7820	0.567	1	1.33	1	0.654	1	1.171	1.2	1.200	1.5	4.922	5.7
8173	0.165	0	—	—	—	—	—	—	—	—	0.165	0
8175	0.900	1	0.835	1	0.522	0	0.242	0	0.200	0	2.699	2
8241	0.000	—	0.000	—	0.000	—	0.234	0.25	0.120	0	0.354	0.25
16684	0.000	—	0.000	—	0.000	—	0.000	—	0.000	—	0.000	—
17178	—	—	0.165	0	—	—	—	—	—	—	0.165	0
17274	—	—	0.995	1	3.390	3	1.102	2.58	2.300	3.5	7.787	10.08
17280	—	—	—	—	0.196	0	0.675	0.5	0.680	1	1.551	1.5
17298	—	—	0.000	—	—	—	—	—	—	—	0.000	—
17550	—	—	—	—	—	—	1.138	1.5	0.400	0	1.538	1.5
17714	—	—	—	—	—	—	0.174	0	—	—	0.174	0
18513	—	—	—	—	—	—	1.646	1.25	3.100	4	4.746	5.25
18551	—	—	—	—	—	—	0.161	0	0.160	0	0.321	0
Cage total	1.632	2(3)	3.325	3(5)	4.762	4(7)	6.543	7.28(13)	8.16	10(14)	24.422	26.28(42)
NC 3												
3961	—	—	0.000	—	0.000	—	0.113	0.40	0.650	0	0.763	0.40
4749	—	—	0.500	0	1.456	1	0.313	0.250	0.666	0.2	2.935	1.45
7051	—	—	0.075	0	—	—	—	—	—	—	0.075	0
7059	—	—	0.350	0	0.091	0	0.517	0	0.120	0	1.078	0
16953	—	—	0.075	0	0.136	0	—	—	—	—	0.211	0
17150	—	—	0.000	—	0.728	0	0.263	0.250	1.053	1.2	2.044	1.45
17155	—	—	0.500	1	0.368	0	0.050	0	0.585	0	1.503	1
17200	—	—	0.250	1	0.368	0	0.475	0.250	0.000	—	1.093	1.25
17204	—	—	0.150	1	0.368	1	0.317	0	0.172	0	1.002	2
17235	—	—	0.000	—	0.000	—	0.300	0	0.087	0	0.387	0
17638	—	—	—	—	0.360	0	0.425	0.250	2.232	1.2	3.017	1.45
18371	—	—	—	—	—	—	0.067	0	0.644	1	0.711	1
18414	—	—	—	—	—	—	0.200	0	0.436	0	0.636	0
18546	—	—	—	—	—	—	0.040	0	—	—	0.040	0
Cage total			1.900	3(7)	3.870	2(10)	3.080	1.4(9)	6.645	3.6(15)	15.495	10(41)

TABLE 58.5. (Continued).

Possible fathers	1980		1981		1982		1983		1984		Total	
	A	B	A	B	A	B	A	B	A	B	A	B
NC 7												
11	—	—	—	—	0.240	0	0.143	0	0.151	0.5	0.534	0.5
16888	—	—	—	—	0.428	0	0.536	0	0.545	0.25	1.509	0.25
16898	—	—	—	—	0.000	—	0.089	0	0.000	—	0.089	0
17179	—	—	—	—	0.238	0	0.000	—	0.000	—	0.238	0
17226	—	—	—	—	0.000	—	0.000	—	0.000	—	0.000	—
17466	—	—	—	—	0.470	0	0.625	1	0.390	1	1.485	2
17502	—	—	—	—	0.143	0	0.107	0	0.325	1.25	0.575	1.25
17700	—	—	—	—	0.000	—	0.393	1	0.696	1	1.089	2
18257	—	—	—	—	—	—	0.143	0	0.305	0.50	0.448	0.50
Cage total					1.519	0(12)	2.036	2(19)	2.412	4.50(18)	5.967	6.50(49)
Grand total	1.632	2(3)	5.225	6(12)	10.151	6(29)	11.659	10.68(41)	17.217	18.10(47)	45.884	42.78(132)[2]

[1] A, relative reproductive success × number of potentially inbred offspring; B, observed number of inbred matings. Total number of offspring who are potentially inbred (i.e., whose mother is paternally related to a sexually mature male in the group) is given in parentheses in rows labeled Cage total and Grand total.

[2] $\chi^2_{(1)} = 0.321$. Expected number of noninbred offspring = 132 − 45.88 = 86.12.

TABLE 58.6. Possible fathers and expected number of maternally inbred offspring based upon the null hypothesis of random mating with respect to kinship.

Cage number	Year of birth	Possible father who is related to offspring	Relation to offspring's mother	Random probability of inbreeding
2	1981	17274	full-sib	0.199[1]
3	1982	17150	full-sib	0.091
3	1982	16953	half-sib	0.136
3	1983	17200	half-sib	0.158
3	1983	18371	half-sib	0.017
2	1984	17280	half-sib	0.170
2	1984	18513	half-sib	0.310
2	1984	17274	half-sib	0.230
3	1984	17638	half-sib	0.248
3	1984	18371	half-sib	0.161
3	1984	17155	half-sib	0.065
3	1984	17235	half-sib	0.087
3	1984	17155	half-sib	0.065
3	1984	18371	half-sib	0.161
7	1984	17179	half-sib	0.196
7	1984	17700	half-sib	0.348
2	1980	16673	son	0.000
2	1981	17298	son	0.067
2	1982	17280	son	0.196
3	1981	17200	son	0.050
3	1981	17155	son	0.100
3	1981	17204	son	0.150
3	1982	17200	son	0.046
3	1982	16953	son	0.136
7	1982	17179	son	0.238
2	1983	18551	son	0.054
2	1983	17280	son	0.225
2	1983	18513	son	0.165
2	1983	17714	son	0.058
3	1983	18414	son	0.050
3	1983	17155	son	0.088
3	1983	18371	son	0.017
3	1983	17638	son	0.142
3	1983	17200	son	0.158
7	1983	17179	son	0.250
7	1983	17700	son	0.196
2	1984	17280	son	0.170
2	1984	18513	son	0.310
3	1984	18371	son	0.161
3	1984	17638	son	0.248
Total				5.917[2]

[1] Maternally inbred matings.
[2] $\chi^2_{(1)} = 4.77$, $P < 0.05$. Expected number of noninbred offspring = 40 − 5.9 = 34.1.

Thirty-nine of these 40 potential maternally inbred offspring could be unequivocally determined not to be inbred (none of the nonexcluded possible fathers were related to the offspring's mother). One of these 40 potential maternally inbred offspring was in fact inbred (all nonexcluded possible fathers were related to the offspring's mother), the issue of a full sib mating in NC 2 during 1981. Since the numbers of inbred and noninbred offspring expected (5.9 and 34.1, respectively) are statistically significantly different from those observed (1 and

39, respectively), we conclude that (1) captive rhesus macaques selectively avoid mating with matrilineal relatives but that (2) inbreeding in the maternal line contributes little to the theoretical values of F and ΔF. While avoidance of close inbreeding in the matrilines of rhesus macaques has been hypothesized (e.g., Missakian, 1973), we report here the first empirical confirmation, to our knowledge, of such avoidance.

An average value of the inbreeding coefficient, F, was estimated for each of the three groups by weighting $1/n$ by the appropriate inbreeding coefficient for each nonexcluded possible father, summing these weights, and dividing this sum by the total number of offspring born in each group. Then, $F_t = 1 - (1 - \Delta F)^t$ where t represents the number of generations that have expired since the first potentially inbred offspring was born. Since t is equivalent to 8.2 years (see estimate of the value T in Table 58.7) and the first potentially inbred offspring in NC 2, NC 3, and NC 7 were born in 1980, 1981, and 1982, respectively, the values of t for these three groups are 5/8.2 (or 0.61), 4/8.2 (or 0.49), and 3/8.2 (or 0.37). Solving for F_t, the contribution of paternal inbreeding to values of ΔF for NC 2, NC 3, and NC 7 are about 0.063, 0.023, and 0.028, respectively.

TABLE 58.7. Abridged life table for two groups of rhesus macaques, 1 January 1977–31 December 1982.[1]

	MYL	No. births	No. female deaths	q_x	l_x	$\bar{l}_x$	b_x^f	$\bar{l}_x b_x^f$
0	91.50	0	30	0.3279	1.000	0.7705	0.0000	0.0000
1	69.50	0	2	0.0288	0.6721	0.6548	0.0000	0.0000
2	57.92	0	1	0.0173	0.6433	0.6347	0.0000	0.0000
3	45.75	8	1	0.0219	0.6260	0.6151	0.0875	0.0331
4	37.16	38	2	0.0538	0.6041	0.5772	0.5113	0.2951
5	36.33	30	1	0.0275	0.5503	0.5366	0.4129	0.2216
6	37.00	30	0	0.0000	0.5228	0.5228	0.4054	0.2119
7	31.75	21	1	0.0315	0.5228	0.5071	0.3307	0.1677
8	31.08	26	1	0.0322	0.4913	0.4752	0.4183	0.1988
9	30.67	27	2	0.0652	0.4591	0.4265	0.4402	0.1877
10	36.92	28	1	0.0271	0.3939	0.3804	0.3792	0.1442
11	25.58	16	0	0.0000	0.3668	0.3668	0.3127	0.1147
12	17.50	11	0	0.0000	0.3668	0.3668	0.3143	0.1148
13	15.25	12	0	0.0000	0.3668	0.3668	0.3934	0.1443
14	11.17	5	1	0.0895	0.3668	0.3221	0.2238	0.0721
15	10.00	5	0	0.0000	0.2773	0.2773	0.2500	0.0693
16	1	0	0	0.0000	0.2773	0.2773	0.0000	0.0000
17	1	0	0	0.0000	0.2773	0.2773	0.0000	0.0000

$$R_0 = \Sigma \bar{l}_x b_x^f = 1.9753$$

$$R_1 = \Sigma \underset{x}{x} \bar{l}_x b_x^f = 17.1066$$

$$\sigma^2 = \frac{\Sigma \underset{x}{x^2} \bar{l}_x b_x^f}{R_0} - \left(\frac{R_1}{R_0}\right)^2 = 11.0049$$

$$\Sigma x^2 \bar{l}_x b_x^f = 169.8855$$

$$r = \frac{\ln R_0}{\frac{R_1}{R_0} - r^1 \frac{\sigma^2}{2}} = 0.083$$

$$T = \frac{\ln R_0}{r} = 8.2 \text{ years}$$

[1] See Keyfitz and Flieger (1971) and Smith (1982a) for definitions of parameters given above and methods for their estimation.

[2] Includes stillbirths.

Empirical Estimate of the Rate of Inbreeding and Effective Population Size

Despite the fact that macaques in these three groups avoid matrilineal inbreeding and mate randomly with respect to patrilineal kinship, the average value of ΔF among all three groups, 0.038, is somewhat higher than that predicted (0.029) based upon the value of N_e (see Theoretical Rate of Inbreeding, above, and Table 58.3). The correlation between reproductive success of fathers and their sons (Smith, 1985), especially in NC 2, undoubtedly contributes to this circumstance as does the rapid growth and increase in the rate of growth experienced by these groups in recent years (Smith, 1982a). Furthermore, since V_K can only be accurately estimated from knowledge of the lifetime reproductive success of each male and our estimates are yet incomplete, and our estimate of the value of t, applied to each of the three groups, is subject to some error, the difference between the theoretical and empirical values of ΔF is not disturbing.

The true value of N_e can be determined from the empirical values of ΔF, since

$$\Delta F = \frac{1}{2N_e}.$$

The values of N_e for NC 2, NC 3, and NC 7 are, therefore, 7.9, 22.1, and 17.9, respectively, and their average does not differ substantially from that predicted from population structure (see Table 58.3). These values represent between one-half and one-fourth the number of breeding adults in their respective groups and are somewhat lower than those estimated by Nozawa (1972) using other methods for free-ranging Japanese macaques.

Deleterious Effects of Inbreeding

Forty offspring of matings unequivocally known to be inbred were matched for age, sex, and cage history with peers who are known to be offspring of noninbred matings. These two groups were then compared for factors (mortality, fertility, and morbidity) known from studies of other species or predicted by theory of population genetics to influence fitness. Birth weight and infant growth rate were also studied because both can be easily computed from standard breeding colony records, have been shown to decline in inbred lines of other species (Falconer, 1960), and influence survival of rhesus offspring (Small and Smith, in press).

Birthweight and Infant Growth Rate

Infant growth rate was estimated by the coefficient for the linear regression of age upon weight at three intervals at which these animals were weighed during their first year of life. Birth weight was estimated as the y-intercept of this regression line. The description and justification of this method have been published elsewhere (Smith and Small, 1982). These data are provided in Figure 58.1 for each of the inbred and noninbred rhesus monkeys.

The apparent correlation between birth weight and infant growth rate identifies errors in estimation by the above method. This error was eliminated by excluding outliers either arbitrarily (i.e., those outside the ellipse in Fig. 58.1) or

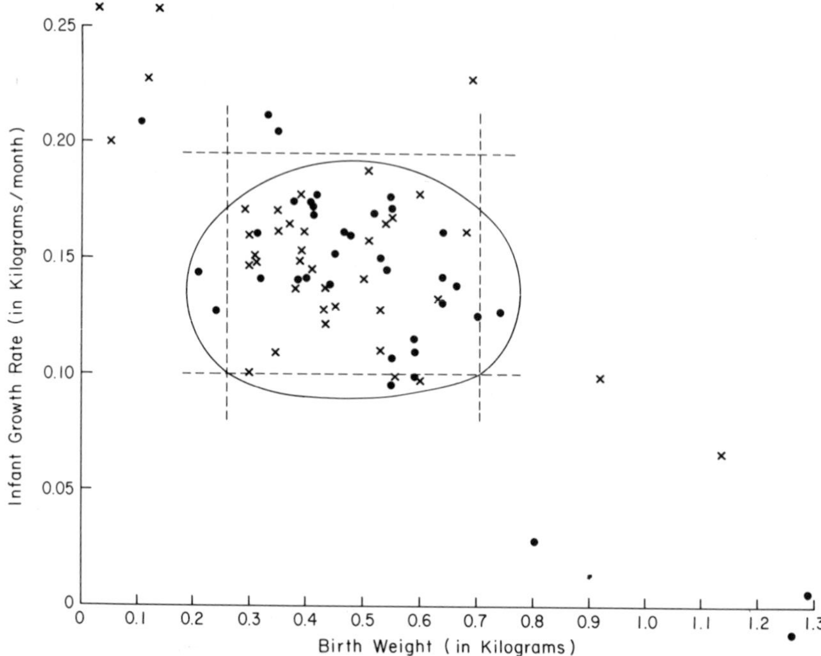

FIGURE 58.1. Birth weights and infant growth rates of inbred (x) and noninbred (•) rhesus monkeys with outliers (outside ellipse) and area of one standard deviation (dashed lines) indicated.

by eliminating those beyond one standard deviation of the mean values of both variables (i.e., those outside the dashed lines of the square in Fig. 58.1). Outliers with very low birth weights and very high growth rates were excluded because the three weights taken spanned a nonlinear growth spurt, resulting in a severe underestimate of birth weight. Outliers with very high birth weights and very low growth rates were excluded because weight loss occurred between at least two of the three weighings resulting in severe overestimates of birth weight. Approximately equal numbers of inbred and noninbred offspring, who, incidentally, tended to fall into the former and latter groups, respectively, were thus excluded from the subsequent analyses.

Table 58.8 indicates that the average birth weight, but not infant growth rate, of inbred offspring was statistically significantly lower than that for the noninbred offspring, an observation also reported for inbreds of other species (Falconer, 1960). Figure 58.2 shows the frequency distribution of both birth weight and infant growth rate in inbred and in noninbred offspring. That the bimodality of both distributions results from gender differences in birth weight and infant growth rate, none of which are statistically significant, is reflected in Table 58.9, which compares males and females separately for the inbred and outbred groups. Since the magnitude of the differences between the average birth weights of inbred and noninbred offspring does not decline, the concomitant decline of statistical significance of these differences is probably due to the

TABLE 58.8. Differences in birth weight and infant growth rate for inbred and noninbred offspring.

		Including outliers			
		Birth weight		Infant growth rate	
	n	Mean	S.D.	Mean	S.D.
Inbred	39	0.4410	0.2091	0.1533	0.0416
Noninbred	37	0.5288	0.2349	0.1396	0.0493
t		1.72[1]		1.31	

		Excluding outliers			
		Birth weight		Infant growth rate	
	n	Mean	S.D.	Mean	S.D.
Inbred	32	0.4411	0.1088	0.1451	0.0241
Noninbred	31	0.4971	0.1310	0.1457	0.0239
t		1.84[1]		0.10	

		Excluding observations 1 S.D. beyond means			
		Birth weight		Infant growth rate	
	n	Mean	S.D.	Mean	S.D.
Inbred	30	0.4319	0.1059	0.1483	0.0212
Outbred	26	0.5026	0.1090	0.1508	0.0211
t		2.45[2]		0.44	

[1,2]Statistically significant at the 0.05 and 0.01 levels of probability, respectively.

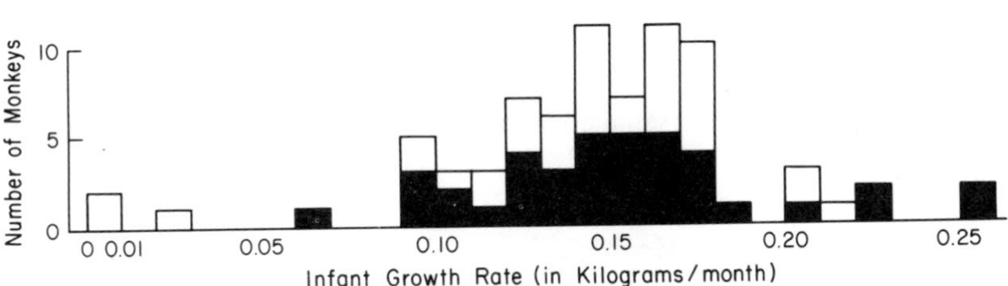

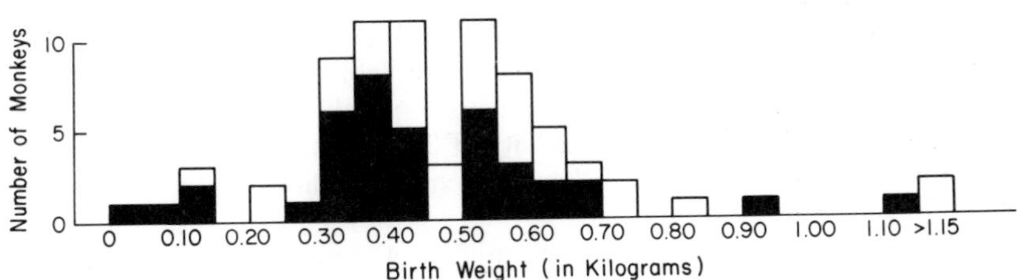

FIGURE 58.2. Distribution of birth weights and infant growth rates of inbred (■) and noninbred (□) rhesus monkeys.

TABLE 58.9. Comparison of average birth weights (BW) and infant growth rates (IGR) of male and female rhesus monkeys in the inbred and noninbred groups.

Including outliers

	Inbred group			Noninbred group			*t* for comparison of inbred and noninbred groups	
	n	BW	IGR	*n*	BW	IGR	BW	IGR
Males	18	0.483 ± 0.265	0.154 ± 0.045	17	0.551 ± 0.246	0.143 ± 0.043	0.79	0.73
Females	21	0.405 ± 0.143	0.153 ± 0.038	20	0.510 ± 0.230	0.137 ± 0.055	1.75[1]	1.08
t for comparison of males and females		1.12	0.04		0.53	0.35		

Excluding animals beyond 1 S.D. of means

	Inbred group			Noninbred group			*t* for comparison of inbred and noninbred groups	
	n	BW	IGR	*n*	BW	IGR	BW	IGR
Males	12	0.441 ± 0.101	0.152 ± 0.019	14	0.516 ± 0.114	0.149 ± 0.022	1.78[1]	0.37
Females	18	0.425 ± 0.111	0.146 ± 0.023	12	0.486 ± 0.105	0.153 ± 0.021	1.52	0.86
t for comparison of males and females		0.41	0.78		0.70	0.47		

[1] Statistically significant at the 0.05 level of probability.

higher variance relative to sample size when males and females are analyzed separately.

Mortality, Fertility, and Morbidity

Table 58.10 compares mortality (number of infant deaths), fertility (number of fertile seasons), and morbidity (number of animals taken inside to the hospital at least once for treatment of illness or trauma) experienced to date by inbred offspring with that of the noninbred offspring. While inbred offspring experienced both higher mortality and fertility, but not morbidity (as defined here), than noninbred offspring, as predicted by theory of inbreeding depression, these differences were not statistically significant. Since this lack of statistical significance might well be due to the paucity of data to date, we plan to continue monitoring vital events in these and subsequent inbred and noninbred offspring in these three rhesus groups.

The preliminary results of our ongoing studies of inbreeding in captive groups of rhesus macaques are consistent with the hypothesis that marked inbreeding depression can be expected unless countermeasures designed to preclude or minimize it are implemented. Moreover, owing to similarities in social structure

TABLE 58.10. Fertility, mortality, and morbidity of inbred and noninbred rhesus monkeys.

		No. breeding seasons experienced after age 2½		
Fertility	n	No. seasons pregnant		No. seasons not pregnant
Inbred	7	3[1]		9
Noninbred	7	7		7
χ^2			1.71, n.s.	
Mortality	n	No. deaths		No. surviving
Inbred	40	3		37
Noninbred	40	1		39
χ^2			1.05, n.s.	
Morbidity	n	No. monkeys removed to hospital at least once for disease or trauma		No. monkeys never removed to hospital
Inbred	36	7		29
Noninbred	30	6		24
χ^2			0.00, n.s.	

[1] One of these pregnancies was a liveborn offspring that died at the age of 15 months.

and mating patterns, there is little reason to expect different results for other captive-bred species of cercopithecoid monkeys. The development of inbred strains of nonhuman primates for biomedical research, which has already begun at some facilities, will, therefore, be more difficult and costly than some have anticipated, though not necessarily impossible. The results reported here also suggest the need for immediate strategies for preventing further genetic subdivision of colony- and zoo-bred nonhuman primates.

Colony Management Strategies for Minimizing Inbreeding

Control of Sex Ratio

Increasing the number of unrelated founding (or breeding) males will reduce the value of ΔF and increase genetic heterogeneity of breeding groups if a simultaneous increase in V_K^m occurs as seems likely from studies of interactions between demographic and social dynamics in these three groups (Stern and Smith, 1984). This follows from the well-known influence of the adult sex ratio upon the value of N_e, or

$$N_e = \frac{4N^m N^F}{N^m + N^F}$$

where N^m and N^F are the numbers of adult males and females in the breeding population. Since maintaining groups with low adult sex ratios is more economical and also improves success of PEA, however, it is desirable to achieve maximum genetic variation and equality of reproductive success among as few unrelated adult males as possible.

Selective Culling

Selective culling should be practiced to maximize the balance among numbers of members of founding matrilines and patrilines in each group since unique genes and gene combinations that become extinct cannot easily be replaced. Male members of lineages that are in jeopardy of extinction should be given priority as future breeding males while males produced by overrepresented lineages should be removed before reaching sexual maturity.

Culling practices should also ensure that a sufficient number of uniquely identifying genetic markers are carried by each breeding male in the group so that the relative reproductive success of all males or patrilines can be monitored. Thus, juvenile and subadult males homozygous for the most common alleles at most loci should be culled before reaching sexual maturity. No more than two or three subadult males, those of the appropriate lineages (those with the fewest breeding members) or those with the most unique genetic markers, should be allowed to reach sexual maturity in the group each year, and each of these males should be allowed to remain in the group only if his reproductive success is judged adequate by age 5 or 6. Older males should be removed, perhaps to indoor colonies for time-mated breeding, when their relative reproductive success declines, following several years of successful breeding (Smith, 1985). It is often found, in fact, that males who experience poor reproductive success in the context of social groups exhibit greater than average reproductive success under a time-mated breeding program.

Simulation of Male Emigration

The only practical and effective long-range solution to the inevitable loss of genetic variation and the increasing accumulation of inbreeding in captive groups of nonhuman primates is to develop strategies for exchanging males among groups. Infants that are exchanged between lactating mothers within the first 2 weeks of life are usually accepted by their foster mothers and seldom suffer injury at the hands of the other group members. Twenty-three of 32 such male infant swaps to foster mothers were successfully made during 1984 and 1985 among five captive rhesus groups at the CPRC (S. Smith, in press) and recent modifications of this procedure have since further improved the success of infant cross-fostering among social groups of rhesus macaques.

This foster infant program, to be continued on an annual basis for these seasonally breeding rhesus macaques, can provide for emigration rates not substantially lower than those experienced by free-ranging groups and can prevent marked genetic subdivision of the breeding groups at the CPRC. While this will counteract genetic subdivision and the accumulation of inbreeding in these groups for the short term, exchange of breeding stock among primate centers both inside and outside the United States will be required, in the long run, to sustain reproductive success of captive breeding stock at current levels. The CPRC recently acquired four male rhesus monkeys from mainland China that have been bred, in one outdoor group, to female rhesus monkeys of Indian origin. The hybrid offspring born to this group have been included in the foster infant program, and their survival and reproductive rates will be closely monitored and compared with that of their nonhybrid group peers to assess the success of this strategy.

The development of optimal management strategies for breeding genetically heterogeneous members of genetically well-defined species of nonhuman primates in captivity requires close cooperation among personnel trained in the study of primate behavior, genetics, demography, veterinary science, and medicine as well as business and administration. Breeding programs that are successful in the long run will be those that are closely monitored and based upon sound behavioral, demographic, and genetic data and principles.

Acknowledgments. This study was supported by National Institutes of Health grants PHS NO1-HD-9-2828 and RR-00169 and Faculty Research Grant D-1343 from the Regents of the University of California.

References

Cavalli-Sforza LL, Bodmer WF (1971) The genetics of human populations. WH Freeman, San Francisco.

Cornelius CD, Rosenberg DP (1983) Spontaneous diseases in nonhuman primates. Amer J Med 74:169–171.

Crow JF, Kimura M (1970) An introduction to population genetics theory. Harper and Row, New York.

Curie-Cohen M, Yoshihara D, Luttrell L, Benforado K, MacCluer JW, Stone WH (1983) The effects of dominance on mating behavior and paternity in a captive troop of rhesus monkeys. Amer J Primatol 5:127–138.

Dickerson GE (1954) Evaluation of selection in developing inbred lines of swine. Res Bull Mo Agricul Exp Sta, No 551.

Falconer DS (1960) Introduction to quantitative genetics. Ronald Press, New York.

Goodwin WJ, Augistine J (1975) The primate research centers program of the National Institutes of Health. Fed Proc 34:1641–1642.

Held JR (1980) Breeding and use of non-human primates in the U.S.A. Int J Study Anim Prob 2:27–37.

Keyfitz N, Flieger W (1971) Population: facts and methods of demography. WH Freeman, San Francisco.

Mason WA (1979) The role of primates in research. Res Resources Rep 3:11–14.

Missakian EA (1973) Genealogical mating activity in free-ranging groups of rhesus monkeys (*Macaca mulatta*) on Cayo Santiago. Behaviour 45:224–241.

Nozawa K (1972) Population genetics in Japanese monkeys. I. Estimation of the effective troop size. Primates 13:381–393.

Ralls K, Ballou J (1982) Effects of inbreeding on infant mortality in captive primates. Inter J Primatol 3:491–505.

Schull WJ, Neel JV (1965) The effects of inbreeding on Japanese children. Harper and Row, New York.

Small MF, Smith DG (In press) The influence of birth timing upon infant growth and survival in captive rhesus macaques (*Macaca mulatta*). Inter J Primatol.

Smith DG (1980a) Potential for cumulative inbreeding and its effects upon survival in captive groups of nonhuman primates. Primates 21:430–436.

Smith DG (1980b) Paternity exclusion in six captive groups of rhesus monkeys (*Macaca mulatta*). Amer J Phys Anthro 53:243–249.

Smith DG (1981) The association between rank and reproductive success of male rhesus monkeys. Amer J Primatol 1:83–90.

Smith DG (1982a) A comparison of the demographic structure and growth of free ranging and captive groups of rhesus monkeys (*Macaca mulatta*). Primates 23:24–30.

Smith DG (1982b) Inbreeding in three captive groups of rhesus monkeys. Amer J Phys Anthro 58:447–451.

Smith DG (1985) Use of genetic markers in the management of captive groups of rhesus monkeys at the California Primate Research Center, U.S.A. Jap J Med Sci Biol 38:44–48.

Smith DG (1986) Inbreeding in the maternal and paternal lines of two captive groups of rhesus monkeys. In: Taub DM, King FA (eds) Current Perspectives in Primate Biology. Van Nostrand Reinhold, New York, p 214.

Smith DG, Small MF (1982) Selection and the transferrin polymorphism in rhesus monkeys (*Macaca mulatta*). Folia Primatol 37:127–136.

Smith DG, Small MF, Ahlfors CE, Lorey FW, Stern BR, Rolfs BK (1985) Paternity exclusion analysis and its applications to studies of nonhuman primates. In: Cornelius CE, Simpson CF (eds) Advances in veterinary science and comparative medicine, vol 28. Academic, New York, p 1.

Smith, S (In press) Infant cross-fostering in rhesus monkeys (*Macaca mulatta*): a procedure for the long term management of captive populations. Amer J Primatol.

Stern BR, Smith DG (1984) Sexual behaviour and paternity in three captive groups of rhesus monkeys (*Macaca mulatta*). Anim Behav 32:23–32.

Wright S (1965) The interpretation of population structure by F-statistics with special regard to systems of mating. Evolution 19:395–420.

59
Hereditary Conditions of Nonhuman Primates

Kurt Benirschke

Introduction

In human medicine, a large number of conditions with a hereditary basis have been recognized. This has, of course, been possible because of the detailed knowledge of human genealogies, the detailed study of human diseases, and the long history of human medicine. In nonhuman primates (hereafter simply "primates"), this is not the case; only a few conditions are known to have a genetic basis. On first consideration, this might be inferred to be due to a relative paucity of deleterious mutations, but much more likely it is the result of less intensive study. Detailed genetic study of primates has commenced only in the recent one or two decades. Continuous breeding has an even shorter history, and few prospective inquiries have been made in the many species of primates now bred to detect potentially interesting aberrant genotypes. It is thus not unreasonable to expect that in the future much will be learned about genetic diseases of primates, some of which will doubtless be of great comparative interest, indeed have relevance to an understanding of human disease. An important consideration for the future of self-sustaining colonies of primates will be what decisions are to be made with such aberrant, perhaps lethal, genotypes. Should they be intentionally removed from the breeding stock, or should they be set aside for study, and how to do this bearing in mind the best interest of primates as well as their potential use as models? And it should be said at the outset that the full understanding of these conditions in the exploration of models not only will serve human medicine but it will also be essential in the long-term management of primate colonies. This aspect should not be lost sight of and is, of course, well supported from our knowledge of the management of human diseases.

For the purposes of this discussion, the topic will be considered under the headings of chromosomal disorders, known gene disorders and genetic polymorphisms, genetic basis of congenital anomalies, twinning events, and congenital anomalies with possible but undetermined genetic basis.

Chromosomal Disorders

The first major anomaly to be described was the autosomal trisomy in a chimpanzee. McClure and colleagues (1969) found trisomy of a small autosome in a 36-week-old *Pan troglodytes* and likened the condition to the human Down's syndrome. The female infant was growth retarded, had poor neonatal growth, frequent diarrhea and respiratory disorders, delayed neurologic development, partial syndactyly of toes with clinodactyly, prominent epicanthus, hyperflexibility of joints, a short neck with excessive skin folds, and a cardiac defect. This was later ascertained to be a patent ductus arteriosus, patent foramen ovale, and ventricular septal defect when the animal died at age 17 months (McClure, 1972). When the frozen cell strain was studied by quinacrine fluorescence methods, it was determined that the trisomic autosome had the staining characteristics of the human chromosome 21, and the homology with Down's syndrome was further strengthened (Benirschke et al., 1974). Maternal age (15) and paternal age (22) were not unusual for breeding chimpanzees, and the origin of the extra autosome could not be ascertained. In a 7-year-old Sumatran orangutan, Andrle and colleagues (1979) identified a similar trisomy. The retarded animal exhibited ear lobe, foot, and mouth anomalies, hypotonia, and inactivity but was maintaining herself in a troop of conspecifics at the Vienna Zoo. She had been born to 12- and 13-year-old parents and had eight apparently normal siblings. Rather more detailed chromosomal banding studies firmly established the similarity of this autosome 22 trisomy to the human Down's syndrome. A less detailed description of trisomy 22 exists in an apparently normal male gorilla (Turleau et al., 1972). These anomalies and genetic considerations have led various authors to suggest general homologies between human chromosome 21, ape chromosome 22, and mouse chromosome 16 and their respective trisomies as a complex anomaly known in humans as Down's syndrome (Polani and Adinolfi, 1980). The inversions associated with the process of speciation in orangutans and chimpanzees will be discussed elsewhere in this volume by Seuánez. Likewise the various complex chromosomal changes occurring coincident with speciation in *Aotus*, *Saimiri*, and *Ateles* are presented by Konstant in this volume. They have not been associated with aneuploidies at this time. Much less well documented are a few abnormalities summarized by Egozcue (1972b). Thus, the karyotypes purported to show XX/XY "mosaicism" in an unstudied baby male chimpanzee could be interpreted as XY or XX since they were not banded. This author also refers to reports of a trisomic (2n=53) howler monkey not further studied and a (2n =45 (instead of 46) normal male uakari without further description.

A clearly delineated 41,X anomaly was described in an imported rhesus monkey by Weiss et al. (1973). Considered to be a normal adult, she was found to have endocrine findings similar to oophorectomized animals. Fusion of vertebrae C_5C_6 was diagnosed roentgenographically. Hypoplastic uterus and tubes were found at laparotomy, and instead of gonads only a few tubular structures were seen in adnexal tissues. The condition was likened to the human Turner's syndrome. It must be emphasized that caution is in order when apparently X-O animals are first studied since complex sex chromosomal patterns exist in various species, including primates. Thus, Goeldi's monkey has 2n=48 in females and 2n=47 in males normally, a Y-autosome translocation accounting for this "anomaly" (Hsu and Hampton, 1970). Likewise, XX/"XO" sex determination systems are known from Bolivian and Peruvian owl monkeys and howler monkeys, with a Y/A translocation being the explanation (Ma et al., 1980).

In a colony of pig-tailed macaques, three female nursery animals were expected to have genetic anomalies based on their delayed development and phenotypes. Two had 43,XXX and one a mosaic 42,XX/43,XXX karyotypes (Ruppenthal et al., 1983). Perhaps better knowledge of neonatal phenotypes that is now developing in primate colonies will lead to more frequent and earlier recognition of chromosomal errors.

Bielert and colleagues (1980) found an unusually large but sexually underdeveloped adult chacma baboon among wild-caught animals and showed clearly that it possessed 42,XY as a karyotype. The infantile uterus and tubes were accompanied by streak ovaries, and the condition was likened to the XY gonadal dysgenesis of humans. They suggest that structural genes determining size may be located on Y and later (Bielert, 1984a,b) describe the female social status and reaction to conspecific males and females with and without hormonal therapy. It is interesting that this adult animal came from a wild troop while cytogenetic studies of 110 adult baboons captured in Kenya (*Papio cynocephalus*) yielded no chromosomal errors (Soulie and de Grouchy, 1981). Only some expected NOR-polymorphism and slight differences in secondary constrictions of chromosome 7 were seen.

Finally, three unusual animals have been described whose chromosomal assessment would need more refinements than was then possible in order to be acceptable as showing true anomalies. Interestingly, they are all langurs and thus suggest that better cytogenetic studies in these unusual genera are needed. Sharma and Gupta (1973) found a male *Presbytis entellus* with heteromorphic pair 21 and deduce an X to 21 translocation of a fragment from X. Egozcue (1971, 1972a) describes one hanuman langur male with possible mosaicism (2n=44 and 45) having two putative X chromosomes and a possible Y in one clone or, so interpreted, a fissioned metacentric. Structurally, it was a male with "poor spermatogenic activity." Bogart and Kumamoto (1978) found one of seven douc langurs to possess a mosaic karyotype with four of 17 metaphases having a reciprocal translocation between two large metacentrics. Their Giemsa-banded chromosomes are clear enough, but since none of the many abortuses this male produced was karyotyped, it remains speculative whether this anomaly was the cause of abortions. Parenthetically, it might be mentioned that no systematic study of primate abortuses has yet been undertaken to compare such findings with the aneuploidy rate of human conceptuses. The same authors found two of three crowned lemurs to be normal; a female, however, had three karyotypes in blood and skin cultures. One was normal, one had a polymorphic acrocentric pair (? due to NOR), and one a centric fission with deletion of a small acrocentric chromosome. This apparently normal animal never conceived.

Gene Disorders and Genetic Polymorphisms

Only few clearly defined genetic disorders have been identified in any primate species. Interestingly, one anomaly, the absence of one of the normally two lobes of the placenta, occurs in an organ whose genetic determination is virtually unknown. Most *Macaca* species possess two placental lobes while *Papio* with which hybridization has been described (Gray, 1971; Markarjan et al., 1974) has only one lobe. In a study of macaque placentas, Chez et al. (1972) described that 78% possessed two lobes but 22% had only one. From the genealogies available to these investigators, it appeared that this anomaly had a recessive inheritance.

In a prospective search for aminoacidopathies among primates, Shih et al. (1972) discovered arginase deficiency of red blood cells in crab-eating macaques, *M. fascicularis*. Breeding experiments clearly showed the enzyme defect to have an autosomal recessive inheritance with heterozygotes having intermediate enzyme values. The homozygotes were apparently normal, unlike the human counterparts with mental retardation, but their liver enzymes were normal, that of humans with this deficiency being markedly reduced (Michels and Beaudet, 1978). It was concluded that the red cell arginase levels represent a polymorphism maintained in the wild. This has been further supported by other studies in human patients and animals models (Weichert et al., 1976).

In another effort the authors used phenylalanine tolerance tests in order to find possible equivalents to the relatively common human condition phenylketonuria (PKU) (Jones et al., 1971). Altogether 174 animals of four different *Macaca* species had phenylalanine infusions and subsequent studies of phenylalanine and tyrosine levels. It was found that *M. mulatta* and *M. arctoides* cleared phenylalanine more quickly than *M. fascicularis* and especially *M. cyclopis*. Two male and one female of the latter species were potentially heterozygote deficient. Regrettably, the female was killed by a cage mate, and the male would not breed when raised to adulthood (M. L. Jones, 1985, personal communication). Thus, no potentially homozygote-deficient animals could be produced, but the study suggests that a model for PKU could exist in *M. cyclopis*.

Of course, many other polymorphisms exist in primates both intra- and interspecifically. Those concerned with blood group markers are described by Socha elsewhere in this volume. Keuppers and Ganesan (1977) found polymorphism of alpha$_1$-antitrypsin in *Macaca irus*, but in 200 individuals studied, the equivalent of the human deficiency was not encountered.

Hemoglobin changes have been used to study evolutionary relations (Buettner-Janusch and Hill, 1965). Silent alpha hemoglobin genes were detected in chimpanzee and gorilla (Boyer et al., 1971), but no diseases have been described in this system. From the study of transferrin polymorphism in several macaque species from different localities, Goodman and his colleagues (1965) infer that these animals might better be classified as "semispecies" than good species. Similar deductions were made by Nozawa et al. (1977) who studied broader serum electrophoretic protein polymorphisms in Asian macaques. These studies make reference to many other inquiries into genetic polymorphic states of primates. Less well understood is the occurrence of placental alkaline phosphatase in primates. Absent in various "lower species," it is abundant in humans, chimpanzees, and orangutans, but absent also in lowland gorillas (Doellgast and Benirschke, 1979; Doellgast et al., 1981). No pathologic state is associated with this deficiency, but its occurrence at this stage in the presumed evolutionary sequence of hominids remains puzzling.

Even more difficult to define is the relationship of spontaneous diseases to hereditary factors. Thus, a variety of primates have spontaneously developed diabetes mellitus; the so-called black Celebes ape (*Macaca nigra*) has been especially prone to do so (Howard, 1975). Whether this is so because of exposure in captivity to carbohydrate levels for which this species has not been selected or whether there truly exists a genetic error in insulin production is not elucidated. Howard (1975) concluded that it was due to inbreeding on Celebes Island. A somewhat similar situation exists in the excessive iron accumulation in tissues of a variety of lemur species. While it has been the experience of zoo pathologists that iron is accumulated by many species in captivity (Frye, 1982, *Procavia*; Randell et al., 1981, *Gracula religiosa*), among primates several species of *Lemur* are apparently more prone to do so. Wild captives are not known to contain an excess of iron stores, and juveniles in captivity quickly accumulate liver, duodenal, and lymphoreticular stores of this metal. Hepatomas and other tumors develop as well as cirrhosis. An approximately equal avidity for iron intake was found in young ruffed lemurs and rhesus monkeys (Gonzales et al., 1984). It is hypothesized that exposure to commercial pellets in captivity, containing manifold the iron available to these herbivorous prosimians in the wild, may be responsible for their developing a disease similar to human hemochromatosis. Whether these deductions are correct or not, it would seem that these species may be valuable for the understanding of human hemochromatosis for which no spontaneous and good model exists. The variability of susceptibility of different lemur species in particular is of interest with the recent description of a human juvenile form and HLA linkage (Editorial, 1984).

Finally, the inheritance of a blonde ("gold") pigmentation of hair was thought to be due to a recessive autosomal mode of transmission in troops of rhesus monkeys on Cayo Santiago (Pickering and van Wagenen, 1969). While not albino nor associated with any defect, the striking color change, referred to as "golden monkeys," was clearly abnormal, occurring in 1 of 10,000 births.

Genetic Basis of Congenital Anomalies

One might expect that defects caused by inbreeding in captive colonies would be a frequent event while in fact very few such examples have come to light. In the large colony of rhesus monkeys on Cayo Santiago, such effects have been specifically looked for but none was found (Rawlins and Kessler, 1983). Despite culling with presumed reduction of genetic heterogeneity, no defects other than the "golden monkey" previously referred to could be ascribed to inbreeding.

In ruffed lemurs at the London Zoo, a presumably recessive gene for hairlessness has been described (Goodwin, 1982). In this pedigree of *L. variegatus* (Fig. 59.1), where red ruffed and black and white ruffed lemurs were interbred, father × daughter mating resulted in four "naked lemurs" that were thought to be similar to the human BIDS syndrome (*b*rittle hair, *i*ntellectual impairment, *d*ecreased fertility, *s*hort stature; Jackson et al., 1974), but details of postmortem or their growth have not been published. In humans, alopecia and atrichia represent rare congenital defects with variable hereditary transmissions (Pinheiro and Freire-Maia, 1985), and it is not clear at this time whether the lemurine anomaly is truly equivalent to any of these conditions. Maintenance of this unusual gene pool would clearly be desirable for further study, but identification of heterozygotes would be of greatest interest for the long-term management of ruffed lemur colonies.

In black and white ruffed lemurs, the condition pectus excavatum was identified as due to an autosomal dominant gene brought into captivity by a wild-caught female with flat chest (Benirschke, 1980; Benirschke et al., 1985). The condition was found to be essentially the same as described in humans (Snyder and Curtis, 1934). No significant deleterious effect on reproduction, growth, pulmonary, or

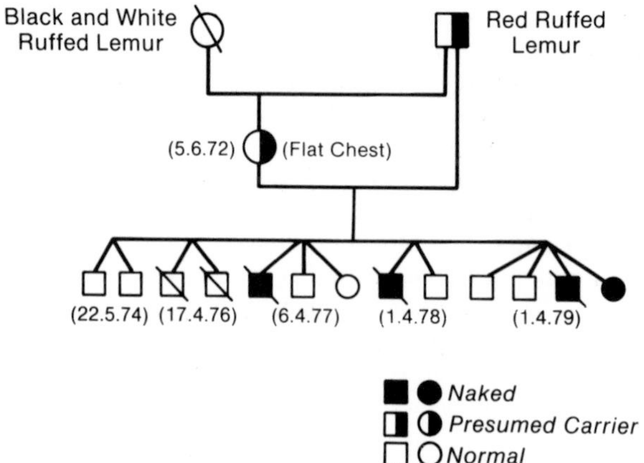

FIGURE 59.1. Pedigree of ruffed lemur family with autosomal recessive trait for hairlessness.

cardiac status could be identified, and it was concluded that elimination of this gene (easily done in a dominant) would serve no purpose for the success of this colony. While the mechanism by which this gene causes sunken chest could not be determined, it was troublesome that other skeletal anomalies occurred in the same genealogy (Benirschke et al., 1981; Benirschke et al., 1985). Whether they relate to the pectus gene is undecided but seems unlikely.

Diaphragmatic defects have been described in three primate species, but only in *Leontopithecus* is it possibly of hereditary origin. Hendrickx and Gasser (1967) found a left diaphragmatic hernia in one of 92 baboon fetuses dissected and refer to the infrequency of anomalies in primates in general. The defect was similar to that seen sporadically in humans. No evidence of heritable factors existed. A rather higher frequency of congenital anomalies was detected in a squirrel monkey colony (Stills and Bullock, 1981), which included a full-term male stillborn with a typical left diaphragmatic hernia. Likewise, Baker et al. (1977) found a similar defect in a stillborn female and refer to some consanguinity existing in their colony. The first indication of possible inheritance of diaphragmatic hernia came when Bush et al. (1980) found this defect in 11 of 130 golden lion tamarins. The founder population of *Leontopithecus* is small, and less than 300 are known to exist in toto. Hence it is not surprising that 10 of the 11 affected animals had some degree of consanguinity. Despite their efforts, the authors were unable to decide with clarity whether an autosomal recessive mode of inheritance or another system (dominant with incomplete penetrance or perhaps multifactorial inheritance with threshold) was at work. This is of interest not only for the maintenance of these tamarins but also because in humans this defect is on occasion familial (Gencik et al., 1982) but may have a heterogeneous pathogenesis. It is not surprising then, perhaps, that the trisomic chimpanzee was also found to have diaphragmatic hernia among other defects (McClure, 1972).

Twinning Events

There is good evidence from innumerable human studies that monozygotic (MZ) twinning is a sporadic event with uniform distribution throughout human races while fraternal or dizygotic (DZ) twinning has many determinants, among them strong genetic causes (Bulmer, 1970). It is dependent on maternal age (steadily rising to age 35) and particularly on race (highest frequency in the Nigerian Yoruba tribe, unusually low in Japanese). Of particular interest is its familial aspect, and recently it has been well demonstrated that its effect comes about by elevated FSH/LH secretion in mothers of DZ twins (Martin et al., 1984). In primates, there is a great variability in the frequency of twinning events. It is a regular phenomenon in Callitrichidae, the marmosets and tamarins, and from the sex distribution and other findings it is clearly DZ twinning. Triplets occur not infrequently while singletons are rare with a macerated fetus occasionally accompanying the singleton. Remarkably, all marmoset fetuses are connected vascularly in their placentas and thus they all are blood chimerae (Benirschke et al., 1962). The female XX/XY chimerae of marmosets, however, are not sterilized as the freemartin equivalent in cattle. The reason for this difference in embryonic development remains to be elucidated. Whether the true hermaphrodite with ovotestes described in an apparently normal female rhesus monkey was a "whole-body chimera" as occurs in some other species by fusion of early embryos of opposite sex is not known (Sullivan and Drobeck, 1966). Oocytes and presumed spermatogonia were found, and sex chromatin was demonstrated in neurons.

Next best studied is the frequent twinning process observed in some prosimian species (Pasztor and Van Horn, 1976). These authors found an approximately 20% DZ twinning rate in *Lemur catta*, and in two subspecies of *Galago* the rates were 9% and 27%. Dizygosity is derived from the sex distribution of the offspring and the fact that a majority of twins were delivered by a few females. In ruffed lemurs, DZ twins and higher multiple births occur regularly and without chimerism (Benirschke et al., 1985). Differences in FSH/LH levels of twinning and nontwinning lemurine species should be investigated.

In other primates, the incidence of twins is quite low. Taub (1983) finds only 0.43% of live births in free-ranging rhesus macaques to be twins. Koford et al. (1966) report in their review of the Cayo Santiago colony and that from Sukhumi that only four twin births occurred among 1748. Schultz (1956) reviews all records of primate twinning: *Alouatta*, 2; *Cebus*, 1; *Macaca*, 5; *Papio*, 6; *Cynopithecus*, 1; *Cercopithecus*, 4; *Nasalis*, 1; *Pan*, 4. At least two partially split macaques reported provide evidence that, on occasion at least, MZ twinning occurs. Aside from the double monsters described, monozygotic twinning in macaques has been further affirmed by the description of an acardiac monster with monochorionic placenta in a *Macaca fascicularis* (Hein et al., 1985). Kirchshofer et al. (1968) report on two gorilla and seven chimpanzee twins, and ten orangutan twin pairs have been observed (M.L. Jones, 1985, personal communication). Six of these were heterosexual pairs, and in one pair the sex was not recorded because they were stillborns. A set of male gorilla twin fetuses was described by Rosen (1972) and were thought to have been dizygotic because of some differences in measurements. Their placenta was not observed.

Congenital Anomalies with Possible but Undetermined Genetic Basis

Numerous congenital anomalies have been described in primates, and it appears that incidence and types vary widely. In general, they would appear to be less common than in humans. The first concern when an anomaly is encountered in primates is that it may have a hereditary basis. Indeed, not infrequently anomaly and inheritance are considered to be synonymous, which is surely incorrect.

Surveys of anomalies in wild monkeys are uncommonly conducted. McConnell and his colleagues (1974) examined the frequency of diseases in 100 free-ranging baboons (*Papio ursinus*) from the Kruger Park. The principal pathological findings concerned infectious agents, and few anomalies were encountered. These included two hamartomata of the brain stem, an accessory spleen, and a few minor deviations of normal tissue development. Major anomalies were not seen; perhaps such would not have survived to the age of capture, most being adults. Rawlins and Kessler (1983) studied the incidence of anomalies of rhesus monkeys on Cayo Santiago and differentiated "congenital" and "hereditary" features. Only two spontaneous anomalies were seen among 963 infants born, an anencephalic female and a blind male (0.2%). They consider these to be "congenital" (i.e., sporadic) but also state that "genetic error is the likely cause of the defect" (anencephaly), but then weigh possible other factors. Perhaps they were influenced by van Wagenen's (1972) report, who observed anencephalic offspring in one of two sets of twins sired by the same father. Koford et al. (1966) had previously observed three anomalies among 1003 animals of the same colony: right upper phocomelia, phallic hypoplasia (? anorchia), and hypoplasia of right face. Several minor anomalies, e.g., hypermastia, were observed as well, and all authors agree that these are probably underestimates of real incidence. Wilson and Gaven (1967) collected data on malformations of laboratory-bred primates and found a 0.44% incidence (13 in 2950). They comment on the likelihood that this is an underestimate but concede that the rate must be considerably lower than that found in human populations. Anencephaly, diaphragmatic hernia, anophthalmia, and other malformations resemble those of humans, but no hereditary traits were identified. They comment on the preponderance of cardiac anomalies detected in the Sukhumi (Russia) colony and the unusual frequency of limb malformations among Japanese macaques, and then describe the experimental thalidomide phocomelia model in rhesus monkeys. The most comprehensive review of pathologic and teratologic conditions in primates is that of Schultz (1956). He describes an amazing variety of skeletal, dental, and other malformations. While some of these may have been incurred through accidents during life, others clearly have their origin in disturbed embryogenesis, such as vertebral anomalies in a mandrill or cleft lip and palate in cebus monkeys. Most are assumed to be of sporadic nature. However, significant anatomical variations, for instance in the dentition or in the development of thumbs in spider monkeys (where the thumb normally is significantly reduced), must have a genetic basis. Specific breeding programs directed toward the identification of the nature of the inheritance have apparently not been undertaken.

The most frequent and striking congenital abnormalities of primates are the limb anomalies reported in many troops of Japanese macaques. In a review of the numerous studies in the Japanese literature, Yoshihro and colleagues (1979) report that more than 360 abnormal infants were seen in 22 troops since the first reported case in 1955. Males are slightly more frequently affected, and "absence deformities" (97%) outweigh all other types. Some troops have much higher rates of occurrence (up to 30%) with peaks occurring between 1969 and 1972. Stillborns have more anomalies than live births. Forelimbs are more common and remarkably, despite severe handicaps, many animals fare fairly well. The definitive cause of these anomalies has so far been elusive. Chromosome findings of severely malformed animals have been normal. Artificial breeding of affected animals has ruled out simple genetic mechanisms. The only suggestion of a hereditary factor is the finding that some females are more prone to deliver anomalous offspring. Since the anomalies followed about 2 years after provisioning by humans and is highest in provisioned troops, it is suspected that a teratogen is supplied with this food. This suggestion is enhanced by the easy production of limb anomalies by thalidomide. A virus etiology has been considered but is not proven. It may, of course, only be the case that cercopithecids have a much greater susceptibility to environmental teratogens in their limb development. This would be possibly a facile explanation for their frequent occurrence, as reported by Schultz (1956), the occurrence of phocomelia in a drill reported by Hill (1962), and that of two rhesus monkeys at the Seattle Primate Research Center (Morris, 1971). Minor anomalies of extremities, syndactyly of metacarpals associated with vertebral anomalies, have been reported in *Saguinus nigricollis* (Hetherington et al., 1975), but outside of macaques it seems to be uncommon.

Other anomalies, e.g., vertebral, dental, renal, palatine, etc., have all been described but appear to be sporadic. Perhaps only the underbite in langurs has a hereditary basis (Sirianni, 1979), but whether it should be classified as an anomaly is debatable.

Finally, a number of zoos have been concerned with recurrent episodes of apparent epilepsy in lowland gorillas. Perhaps because of the striking presentation in a rare animal, it has been considered to have a genetic basis. In view of the large amount of data accumulated in human epilepsy where some types of this disorder appear to be hereditary (Newmark and Penry, 1980), such considerations may be warranted and should be pursued more vigorously.

Conclusions

Compared with the abundance of genetic diseases and anomalies described in humans, very few such processes have been described in nonhuman primates. The reason for this discrepancy may be a truly lower frequency, but also less detailed observations have undoubtedly been made in primates. Perhaps because of a smaller foundation stock, more hereditary conditions have been observed in lemurs than in simians. Twinning is a specific genetic trait expressed in lemurs and callitrichidae most commonly, rarely in other forms. Some specific findings peculiar to certain species, e.g., hemosiderosis in lemurs, diabetes in black apes, diaphragmatic hernia in golden tamarins, susceptibility to phocomelia in Japanese macaques, and other conditions referred to, lend themselves to further inquiry. Not only would pathogenetic insights be helpful in understanding the respective human condition, they are essential for the establishment of self-sustaining populations of primates.

References

Andrle M, Fiedler W, Rett A, Ambros P, Schweizer D (1979) A case of trisomy 22 in *Pongo pygmaeus*. Cytogenet Cell Genet 24:1–6.

Baker CA, Hendrickx AG, Cooper RW (1977) Spontaneous malformations in a squirrel monkey (*Saimiri sciureus*) fetuses with emphasis on cleft lip and palate. J Med Primatol 6:13–22.

Benirschke K (1980) Pectus excavatum in ruffed lemurs (*Lemur* [*Varecia*] *variegatus*). 22nd Intern Symp Erkrank Zootiere. Akademie-Verlag, Berlin East, pp 169–172.

Benirschke K, Anderson JM, Brownhill LE (1962) Marrow chimerism in marmosets. Science 138:513–515.

Benirschke K, Bogart MH, McClure HM, Nelson-Rees WA (1974) Fluorescence of the trisomic chimpanzee chromosomes. J Med Primatol 3:311–314.

Benirschke K, Kumamoto AT, Bogart MH (1981) Congenital anomalies in *Lemur variegatus*. J Med Primatol 10:38–45.

Benirschke K, Miller C, Ippen R, Heldstab A (1985) The pathology of prosimians, especially lemurs. Adv Vet Sci Compar Med 30:167–208.

Bielert C (1984a) The social interactions of adult conspecifics with an adult XY gonadal dysgenetic chacma baboon (*Papio ursinus*). Horm Behav 18:42–55.

Bielert C (1984b) Estradiol benzoate treatment of an XY gonadal dysgenetic chacma baboon. Horm Behav 18:191–205.

Bielert C, Bernstein R, Simon GB, van der Walt LA (1980) XY gonadal dysgenesis in a chacma baboon (*Papio ursinus*). Intern J Primatol 1:3–13.

Bogart MH, Kumamoto AT (1978) Karyotype abnormalities in two primate species. *Pygathrix nemaeus* and *Lemur coronatus*. Folia Primatol 30:152–160.

Boyer SH, Noyes AN, Vrablik GR, Donaldson LJ, Schaefer EW, Gray CW, Thurmon TF (1965) Silent hemoglobin alpha genes in apes: potential source of thalassemia. Science 171:182–185.

Buettner-Janusch J, Hill RL (1965) Molecules and monkeys. Science 147:836–842.

Bulmer MG (1970) The biology of twinning in man. Clarendon Press, Oxford.

Bush M, Montali RJ, Kleiman DG, Randolph J, Abramovitz MD, Evans RF (1980) Diagnosis and repair of familial diaphragmatic defects in golden lion tamarins. J Amer Vet Med Assoc 177:858–862.

Chez RA, Schlesselman JJ, Salazar J, Fox R (1972) Single placentas in the rhesus monkey. J Med Primatol 1:230–240.

Doellgast GJ, Benirschke K (1979) Placental alkaline phosphatase in Hominidae. Nature 280:601–602.

Doellgast GJ, Wei SC, Robinson PT, Benirschke K (1981) Primate placental alkaline phosphatase. Fed Europ Biochem Soc Lett 135:61–64.

Editorial (1984) Idiopathic haemochromatosis in the young. Lancet 2:145.

Egozcue J (1971) A possible case of centric fission in a primate. Experientia 27:969–970.

Egozcue J (1972a) XX male *Presbytis entellus*? A retrospective study. Folia Primatol 17:292–296.

Egozcue J (1972b) Chromosomal abnormalities in primates. Med Primatol I:336–341.

Frye FL (1982) Iron storage disease (hemosiderosis) in an African rock hyrax (*Procavia capensis*). J Zoo Anim Med 13:152–156.

Gencik A, Moser H, Gencikova A, Kehrer B (1982) Familial occurrence of congenital diaphragmatic defect in three families. Helv Paed Acta 37:289–293.

Gonzales J, Benirschke K, Saltman P, Roberts J, Robinson PT (1984) Hemosiderosis in lemurs. Zoo Biol 3:255–265.

Goodman M, Kulkarni A, Poulik E, Reklys E (1965) Species and geographic differences in the transferrin polymorphism of macaques. Science 147:884–886.

Goodwin LG (1982) Annual report. Ruffed lemurs. J Zool 197:17.

Gray AP (1971) Mammalian hybrids. Farnham Royal: Commonwealth Agricultural Bureaux.

Hein PR, van Groeninghen JC, Puts JJG (1985) A case of acardiac anomaly in the cynomolgus monkey (*Macaca fascicularis*): a complication of monozygotic monochorial twinning. J Med Primatol 14:133–142.

Hendrickx AG, Gasser RF (1967) A description of a diaphragmatic hernia in a sixteen week baboon fetus (*Papio* sp.). Folia Primatol 7:66–74.

Hetherington CM, Cooper JE, Dawson P (1975) A case of syndactyly in the white-lipped tamarin *Saguinus nigricollis*. Folia Primatol 24:24–28.

Hill WCO (1962) Lobster-claw deformity in a drill (*Mandrillus leucophaeus* F. Cuv.). Bibl Primat, vol I. S Karger, Basel, pp 239–251.

Howard CF (1975) Diabetes and lipid metabolism in nonhuman primates. Adv Lipid Res 13:91–134.

Hsu TC, Hampton SH (1970) Chromosomes of Callitricidae with special reference to an XX/"XO" sex chromosome system in goeldi's marmoset (*Callimico goeldii*, Thomas 1904). Folia Primatol 13:183–195.

Jackson CE, Weiss L, Watson JHL (1974) "Brittle" hair with short stature, intellectual impairment and decreased fertility: an autosomal recessive syndrome in Amish kindred. Pediatrics 54:201–207.

Jones TC, Levy HL, MacCeady RA, Shik VE, Garcia FG (1971) Phenylalanine tolerance tests in simian primates. Proc Soc Exp Biol Med 136:1087–1090.

Kirchshofer R, Weisse K, Berenz K, Klose H (1968) A preliminary account of the physical and the behavioural development during the first 10 weeks of the hand-reared gorilla twins born in the Frankfurt Zoo. Intern Zoo Yrbk 121–128.

Koford CB, Farber PA, Windle WF (1966) Twins and teratisms in rhesus monkeys. Folia Primatol 4:221–226.

Kueppers F, Ganesan J (1977) Alpha$_1$-antitrypsin polymorphism in Malaysian *Macaca irus*. Biochem Genet 15:817–823.

Ma NSF, Renquist DM, Hall R, Sehgal PK, Simeone T, Jones TC (1980) XX/"XO" sex determination system in a population of Peruvian owl monkey, *Aotus*. J Hered 71:336–342.

Markarjan DS, Isakov EP, Kondakov GI (1974) Intergeneric hybrids of the lower (42 chromosome) monkey species of the Sukhumi monkey colony. J Hum Evol 3:247–255.

Martin NG, Olsen ME, Theile H, El Beaini JL, Handelsman D, Bathnagar AS (1984) Pituitary-ovarian function in mothers who have had two sets of dizygotic twins. Fertil Steril 41:878–880.

McClure HM (1972) Mongolism; Down's syndrome. Amer J Pathol 67:413–416.

McClure HM, Belden KH, Pieper WA, Jacobson CB (1969) Autosomal trisomy in a chimpanzee: resemblance to Down's syndrome. Science 165:1010–1012.

McConnell EE, Basson PA, De Vos V, Myers BJ, Kuntz RE (1974) A survey of diseases among 100 free-ranging baboons (*Papio ursinus*) from the Kruger National Park. Onderstepoort J Vet Res 41:97–168.

Michels VV, Beaudet AL (1978) Arginase deficiency in multiple tissues in argininemia. Clin Genet 13:61–67.

Morris LN (1971) Spontaneous congenital limb malformations in nonhuman primates: a review of the literature. Teratology 4:335–342.

Newmark ME, Penry JK (1980) Genetics of epilepsy. Raven, New York.

Nozawa K, Shotake T, Ohkura Y, Tanabe Y (1977) Genetic variations within and between species of Asian macaques. Japan J Genet 52:15–30.

Pasztor LM, van Horn RN (1976) Twinning in prosimians. J Hum Evol 5:333–337.

Pickering DE, van Wagenen G (1969) The golden mulatta macaque (*Macaca mulatta*): developmental and reproduction characteristics in a controlled laboratory environment. Folia Primtol 11:161–166.

Pinheiro M, Freire-Maia N (1985) Atrichias and hypotrichoses: a brief review with description of a recessive atrichia in two brothers. Hum Hered 35:53–55.

Polani PE, Adinolfi M (1980) Chromosome 21 of man, 22 of the great apes and 16 of the mouse. Devel Med Child Neurol 22:223–225.

Randell MG, Patnaik AK, Gould WJ (1981) Hepatopathy associated with excessive iron storage in Mynah birds. J Amer Vet Med Assoc 179:1214–1217.

Rawlins RG, Kessler MJ (1983) Congenital and hereditary anomalies in the rhesus monkeys (*Macaca mulatta*) of Cayo Santiago. Teratology 28:169–174.

Rosen SI (1972) Twin gorilla fetuses. Folia Primatol 17:132–141.

Ruppenthal GC, Caffery SA, Goodlin BL, Sackett GP, Vigfusson NV, Peterson VC (1983) Pigtailed macaques (*Macaca nemestrina*) with trisomy X manifest physical and mental retardation. Amer J Ment Defic 87:471–476.

Schultz AH (1956) The occurrence and frequency of pathological and teratological conditions and of twinning among non-human primates. Primatologia I:965–1014.

Sharma GP, Gupta CM (1973) X-autosome translocation in the Indian langur—*Presbytis entellus*. Curr Sci 42:576.

Shih VE, Jones TC, Levy HL, Madigan PM (1972) Arginase deficiency in *Macaca fascicularis*. I. Arginase activity and arginine concentration in erythrocytes and in liver. Pediat Res 6:548–551.

Sirianni JE (1979) Craniofacial morphology of the underbite trait in *Presbytis*. J Dent Res 58:1655.

Snyder LH, Curtis M (1934) The inheritance of "hollow chest." "Cobbler's chest" due to heredity—not an occupational deformity. J Hered 30:139–141.

Soulie J, deGrouchy J (1981) A cytogenetic survey of 110 baboons (*Papio cynocephalus*). Amer J Phys Anthro 56:107–113.

Stills HF, Bullock BC (1981) Congenital defects of squirrel monkeys (*Saimiri sciureus*). Vet Pathol 18:29–36.

Sullivan DJ, Drobeck HP (1966) True hermaphrodism in a rhesus monkey. Folia Primatol 4:309–317.

Taub DM (1983) Twinning among nonhuman primates. Amer J Primatol 4:357 (abstract).

Turleau C, deGrouchy J, Klein M (1972) Phylogenie chromosomique de l'homme et des primates hominiens (*Pan troglodytes*, *Gorilla gorilla* et *Pongo pygmaeus*). Essai de reconstruction du caryotype de l'ancetre commun. Ann Genet 15:225–240.

van Wagenen G (1972) Vital statistics from a breeding colony. J Med Primatol 1:3–28.

Weiss G, Weick RF, Knobil E, Wolman SR, Gorstein F (1973) An X-O anomaly and ovarian dysgenesis in a rhesus monkey. Folia Primatol 19:24–27.

Wiechert P, Mortelmans J, Lavinha F, Clara R, Terheggen HG, Lowenthal A (1976) Excretion of guanidino-derivates in urine of hyperargininemic patients. J Genet Hum 24:61–72.

Wilson JG, Gavan JA (1967) Congenital malformations in nonhuman primates: spontaneous and experimentally induced. Anat Rec 158:99–110.

Yoshihiro S, Goto S, Minezawa M, Muramatsu M, Saito Y, Sugita H, Nigi H (1979) Frequency of occurrence, morphology, and causes of congenital malformation of limbs in the Japanese monkey. Ecotoxicol Environm Safety 3:458–470.

60 Chromosomal and Molecular Characterization of the Primates: Its Relevance in the Sustaining of Primate Populations

HÉCTOR N. SEUÁNEZ

Introduction

Primates comprise a mammalian order of particular interest. Humans and our closest living relatives are included in it, and many extant and extinguished species of this order have supplied valuable data to reveal many insights into our descent. However, the survival of the extant primate species is now a challenge to us and to future generations. The extensive use of primates as animal models for experimentation, the indiscriminate destruction of forests and natural habitats, and uncontrollable poaching have contributed to the decline of many species that are presently endangered. Although several measures have been taken by conservationists to protect and to sustain primate populations, there is still a dearth of knowledge on the biology and behavior of many primate species and on how our limited understanding of these problems can be successfully applied for their protection. Chromosomes and DNA studies are just one of the many possible approaches to understanding how primate species are related and how close or distant they are from one another. If chromosome and DNA data are taken together with information from other fields of primate research, the possibility of implementing a consensus policy for conservation could eventually become a reality. In this report, I will comment on many aspects of chromosome and DNA studies that could be especially relevant, hoping to contribute to protecting and to sustaining primate populations.

Chromosome Studies in the Primates

Chromosome studies have been most valuable in characterizing primate species and for the understanding of specific divergence within the order. Since chromosomes are nuclear structures with definite morphological traits, as is evident with conventional and high-resolution banding techniques (Seuánez, 1979, 1984; Dutrillaux, 1979; Yunis and Prakash, 1982), the chromosome complement of a species exhibits a characteristic set of morphological attributes. Moreover, karyological comparisons have demonstrated that these attributes have been evolutionarily conserved within the primate order or are even shared with some species of other mammalian orders (Nash and O'Brien, 1982). Thus, tracing these morphological attributes may be illuminating in showing similarities, at the chromosome

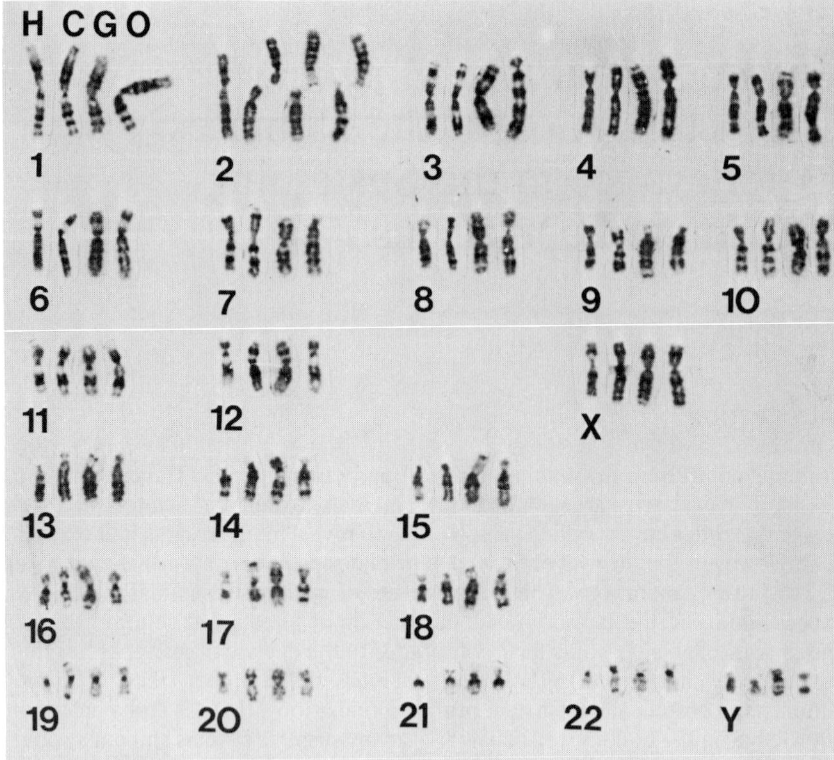

FIGURE 60.1. Karyological comparisons between human(H), chimpanzee(C), gorilla(G), and orangutan(O) chromosomes. Note morphological and G-banding similarities between species.

level, between different species (see Fig. 60.1 for a karyological comparison of humans and the great apes), as well as for the postulation of presumptive phylogenetic trees. These trees can be built on the assumption that chromosome change between species has operated on maximum parsimony pathways, the branches of which we may envisage as emerging from a common ancestor's karyotype through a minimum number of chromosome ancestor's karyotype through a minimum number of chromosome rearrangements. Recent advances in somatic cell genetics, based on cell hybridization and on the biochemical analysis of the hybrid clones, have succeeded in the assignment of many structural loci to primate chromosomes. These studies have shown that many linkage associations have been evolutionarily conserved in the primates and in other mammals and that many chromosomes of similar morphological attributes may contain, though not always, similar genes (O'Brien, 1984; O'Brien et al., 1985; Tables 60.1, 60.2). It is therefore likely that a similar macrostructural organization and gene content have been maintained in the primate chromosome complement for long periods of evolutionary radiation, except in cases where chromosome shuffling has been drastic, as it is in the case of the gibbon (Turleau et al., 1983).

TABLE 60.1. Chromosomal positions of homologous loci in humans and the great apes.[1]

Human chromosome	Marker	Chimpanzee chromosome	Gorilla chromosome	Orangutan chromosome
1p	PGM1	1	1	1q
1p	PGD	1	1	1q
1p	ENO1	1	1	1q
1p	AK2			1q
1p	FUCA			
1q	PEPC	1	1	1
1q	FH		1	1p
2q	IDH1	12(13)	12(11)	12(11)
2p	MDH1	13(12)	11(12)	11(12)
2p	ACP1	13(12)		11(12)
3	GPX1	2	2	(2)
4p	PGM2	3	3	3
5	HEXB	(4)	(4)	4
6p	GLO	5		
6p	MHC	5	5	5
6q	PGM3	5	5	5
6q	SOD2	5	5	5
6q	ME1	5	5	5
7q	GUSB	6	6	(10)
8p	GSR	7	7	6
9q	AK1	11	13	(13)
9p	ACO1	11	13	
9p	AK3	11		
10p	GOT1	8	8	7
11p	LDHA	9	9	8
11p	ACP2	9		
12p	GAPD	10	10	9
12p	TPI1	10	10	9
12p	LDHB	10	10	9
12q	PEPB	10	10	9
13q	ESD	(14)	14	(14)
14q	NP	15	18	(15)
15q	MPI	16		(16)
15q	PKM2	16	15	
15q	HEXA	16	15	
17q	TK	19	19	19
17q	GALK	19	19	
18q	PEPA	17	16	(17)
19	GPI	20	20	20
20p	ITPA	15(21)	18(21)	(21)
21q	SOD1	22	22	22
X	G6PD	X	X	X
X	HPRT	X		
X	GLA	X	X	X
X	PGK	X	X	

[1] The great ape cytological homolog and the syntenic homolog are the same in every case except human chr. 2p, 2q, and 20. In these cases and in those cases when no homologs have been mapped, the cytological homolog is in parentheses.

TABLE 60.2. Comparative gene mapping between humans, Old World and New World Monkeys, and the mouse lemur.

Human chrom.	Human loci	Gibbon chrom.	Rhesus chrom.	Baboon chrom.	African green monkey	Capuchin monkey chrom.	Owl monkey (*Aotus* K-VI) chrom.	Mouse lemur chrom.
1p	PGM1	5	1	1	1	15	12	3
1p	PGD	24	1	1	1	15	12	3
1p	ENO1	24	1	1	1	15	12	3
1p	AK2						12	
1p	FUCA	3			1			3
1q	PEPC			1	6			U4
1q	FH		1		6	U2	6	
1q	GUK1	5	1			U2		U5
2p	MDH1		15			4	2	4
2p	ACP1	19				4		4
2q	IDH1		9	12		14	16	
2	UGP2					4	2	4
3	GPX1	4	3		5			1
3p	ACY1					18		
4p	PEPS			5				
4p	PGM2		6	5		2	14	
5	HEXB		5					
6p	GLO			4			9	6
6p	HLA		2				9	
6q	PGM3	17	2			3	9	6
6q	SOD2	3	2	4			9	
6q	ME1			4		3	9	6
7q	GUSB		2	3		1		
7	MDH2		2				4	
8p	GSR		8		10			
9q	AK1	8			U1			10
9p	ACO1	8			U1		15	
9p	AK3					12		10
10q	GOT1	3						15
11p	LDHA	15	11	14	12	16	19	5
11p	ACP2	15	11			16		
12p	GAPD		12		13			7
12p	TPI1	U1	12		13	10	10	7
12p	LDHB	U1	12	11	13	10	10	7
12q	PEPB		12		13	10		7
12	CS		12		13			7
12	ENO2	U1						7
14q	NP	U2	7	7	U2		11	2
14q	CKBB	U2		7	U3			2
15q	MPI	6	7	7	U2	U1	11	2
15q	PKM2	6	7	7	U2	U1	11	2
15q	HEXA	6	7					2
15q	IDH2			7				
15q	SORD	6		7			11	2
15q	B2M						11	

TABLE 60.2. (Continued).

Human chrom.	Human loci	Gibbon chrom.	Rhesus chrom.	Baboon chrom.	African green monkey	Capuchin monkey chrom.	Owl monkey (*Aotus* K-VI) chrom.	Mouse lemur chrom.
16p	PGP					1		
19	GPI		19		25	8	25	U3
19	PEPD							U3
20q	ADA			10				
20p	ITPA		13	10				
21q	SOD1			3		9		
22	NAGA		13					
X	G6PD		X		X		X	X
X	HPRT	X						X
X	GLA	X	X		X		X	X
X	PGK	X			X		X	X

Chromosome studies are important for the characterization of a species beyond the limitations and constraints resulting from a purely morphological approach. This is because the taxonomic arrangements of the primates, as of most mammals, are entirely based on gross morphometric characteristics but not on the conception of species as natural populations of individuals among which gene flow normally occurs. It should not be surprising, then, if a taxonomic arrangement based on morphological attributes does not coincide with the genetic arrangement that might emerge from a chromosomal, biochemical, or molecular study. Since organic, chromosomal, and molecular evolution are not straightforwardly correlated, and in many cases they have been found to be uncoupled (Wilson et al., 1975; Wayne et al., in preparation), these contradictions are to be expected. Although we may initially describe the standard karyotype of a species, a substantial amount of chromosome change might be observed with a taxon considered to be a single species, when the definition of species is restricted to the pure realm of morphology. The opposite situation, that is, the identity of chromosome constitution between different species, has also been observed, as it is the case of the baboon (*Papio papio*) and the rhesus monkey (*Macaca mulatta*) (Finaz et al., 1978). This identity, however, does not question the inclusion of baboons and macaques in different taxa because their differences, at the morphological, biochemical, and molecular levels, as well as their geographic distribution and behavior, justify the inclusion of these species in two different genera (Napier and Napier, 1967). Thus, the identity of chromosome constitution is a mere consequence of the different rates at which evolution may be operative rather than direct proof of gene flow between macaques and baboons. However, while chromosomal identity may not be good evidence of gene flow, chromosomal diversity strongly suggests that gene flow may be normally impaired by the emergence of less fertile, infertile, or sterile offspring from two chromosomally different progenitors.

The observation of chromosomal diversity within a single taxon was clearly evident in *Aotus trivirgatus* (Ma et al., 1976). The analysis of some 300 specimens allowed the characterization of several karyomorphic populations, each with a different chromosome constitution. Detailed studies of the specimens permitted the establishment of a correlation with phenotypes and geographic distribution, thus leading to the postulation that *Aotus* included several species that have emerged through a process of rapid speciation (Hershkovitz, 1983). Chromosome change in *Aotus* has been considerably clarified by gene assignment because many linkage associations were found to be conserved or disrupted in the same way as predicted by the morphological study of chromosomes (Ma, 1984). It must then be concluded that the previous taxonomic arrangements of *Aotus*, which were based exclusively on the morphometric characteristics of the specimens, was limited and potentially misleading. Consequently, chromosome studies of all animals engaged in breeding programs must be carried out to avoid the production of potentially unfit individuals. Similar strategy should be applied whenever an area is repopulated because the indiscriminate mixing of two chromosomally different populations might actually reduce, rather than enhance, their survival in the wild.

Another clear example of chromosome variation within primate species is the case of the orangutan where two types of chromosome change were found to occur (Seuánez et al., 1976, 1979). One, corresponding to a polymorphic type of double inversion, was found to be present in both subspecies, the Bornean and the Sumatran orangutan. Another type of inversion, however, was found to be subspecies specific, and it allowed the distinction of subspecies by a simple karyotypic analysis (Fig. 60.2). Since the distinctions between Bornean and Sumatran orangutan might be difficult, when based on morphometric characteristics, chromosome studies are important for the maintenance of the two separate subspecies in breeding programs (Seuánez, 1982; de Boer and Seuánez, 1982). Indiscriminate breeding would result in the production of specimens genetically different from those normally existing in the wild and that would carry a chromosome inversion.

In our own laboratory, we have found additional evidence of intraspecific chromosome change in two platyrrhine species, *Cebus apella* and *Alouatta belzebul*.

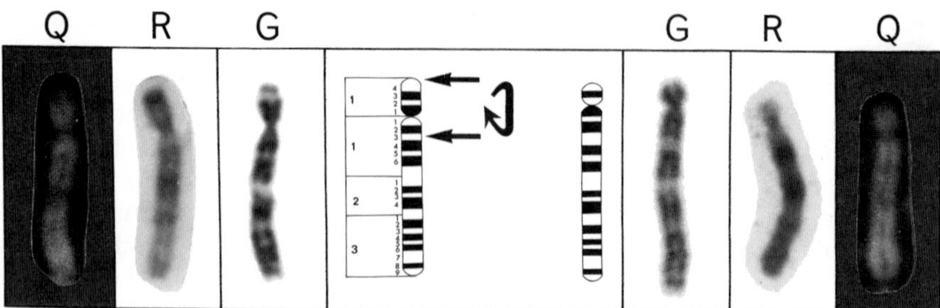

FIGURE 60.2. Chromosomal difference between the Sumatran (left) and the Bornean (right) chromosome 2 in the orangutan, as indicated by chromosome breaks and a pericentric inversion.

The tufted capuchin, *C. apella*, is one of the four species of the genus *Cebus*, and it is widely spread in the South American continent. Several taxonomic arrangements have been proposed to classify the different subspecies of *Cebus apella* (Cabrera, 1957; Mittermeier and Coimbra-Filho, 1981; Kinzey, 1980) that are found in different geographic regions. In recent years, we have undertaken a collaborative project with primate centers in Argentina and Brazil to study chromosome variation in this species and to establish a correlation between chromosome constitution, phenotypic characteristics, and geographic distribution. In a preliminary report, we analyzed 20 specimens from the three different regions of the South American coast where *Cebus apella* is distributed (Freitas and Seuánez, 1982). Our survey has now been extended to some other 100 specimens captured in the Paraguayan Chaco forest that are now being studied by our colleagues in Argentina (Matayoshi et al., personal communication). It also includes some 100 other specimens captured in the north of Brazil, at the Tucuruí dam reservoir (State of Pará), which have been studied by our colleagues in the Federal University of Pará (C. Barroso, personal communication). Furthermore, we have included six specimens of the very rare subspecies *Cebus apella xanthosternos* Wied, 1820, or yellow-breasted capuchins. This subspecies is particularly distinct for its intense yellow-orange pelage extending to its underparts, and its natural habitat, once extending along the coast of the States of Bahia, Espirito Santo, and Rio de Janeiro (Brazil), has now been greatly reduced by the destruction of the coastal forests.

Our preliminary results in this population survey suggest that *Cebus apella xanthosternos* is chromosomally distinct from other subspecies of *C. apella* in respect to the amount and position of constitutive heterochromatin in one chromosome pair. In fact, *Cebus apella*, as well as other species of the same genus, are characteristic in showing abundant heterochromatic material (Freitas and Seuánez, 1982; Dutrillaux et al., 1978; Couturier and Dutrillaux, 1981), especially in chromosomes 5, 9, 10, and 11 (Fig. 60.3). Of these regions, that of chromosome 9 is particularly evident in representing some 70% of the long arm, and extending to the long arm telomere. This peculiar chromosome pair has been found in the four species of the genus *Cebus* (Couturier and Dutrillaux, 1981), so it is not characteristic of any species within the genus. In *Cebus apella xanthosternos*, however, chromosome 9 shows a considerably smaller heterochromatic block, and its position is always intercalar and never terminal (Fig. 60.4); the heterochromatic region is in between two euchromatic regions, one proximal and another distal. Since this particular chromosome type has been found in five apparently unrelated specimens and in the homozygote condition, it is highly likely that it represents a subspecific chromosome trait absent in the other *Cebus apella* studied by us and in other species of the same genus. Here again, there is an association between a special phenotype and a characteristic chromosome trait, the significance of which is not yet known.

A more drastic difference in chromosome constitution has been found recently by us in *Alouatta belzebul* (Seuánez et al., unpublished observation). Two distinct subspecies, *Alouatta belzebul belzebul* and *Alouatta belzebul nigerrima*, showed marked chromosome differences in spite of having the same diploid chromosome number (2n=50). Such dissimilarities include differences in the number of acrocentrics and submetacentrics, differences in the G-band pattern of many chromosome pairs, and in the distribution of the nucleolar organizer regions

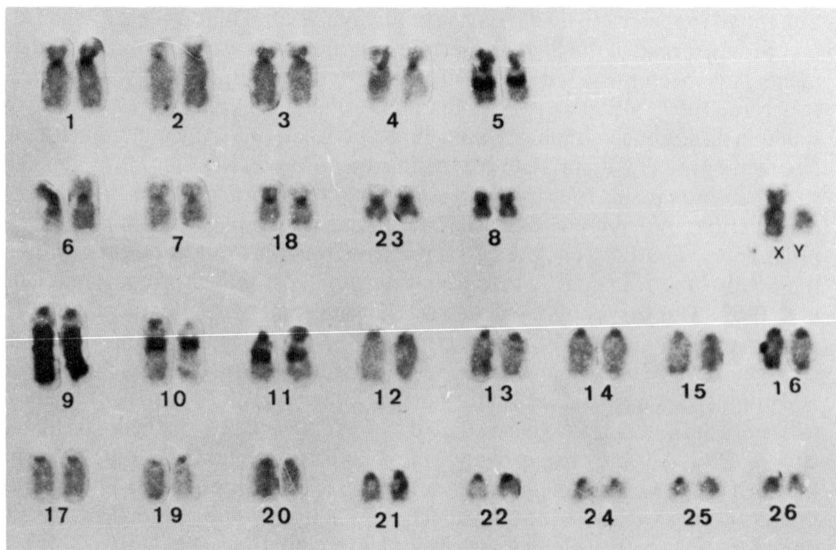

FIGURE 60.3. C-band karyotype of *Cebus apella*. Note abundant heterochromatic regions in chromosomes 5,9,10, and 11.

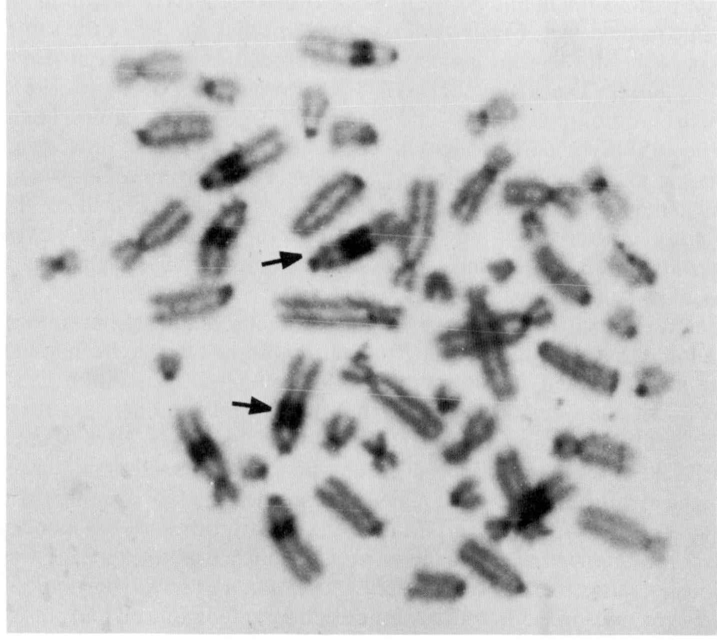

FIGURE 60.4. C-band chromosomes of *Cebus apella xanthosternos*. Note intercalar heterochromatic blocks in four chromosome pairs. Chromosome pair 9 is arrowed.

(NORs). While NORs are found in one chromosome pair in *A. belzebul belzebul*, they are present in three different chromosome pairs in *A. belzebul nigerrima* (see Figs. 60.5a, 60.5b, 60.6a, 60.6b, and 60.7 for a comparison of the karyotypes of the two subspecies). These findings point to substantial differences, at the chromosome level, between these howler monkey subspecies and, furthermore, question their inclusion in the same taxonomic arrangement. If species are conceived in terms of gene flow, it is highly likely that these chromosome differences might act as reproductive barriers between these groups and contribute to their genetic isolation rather than favor their cross-breeding.

The four examples mentioned in this report are illuminating in showing the vulnerability of morphometric taxonomy and in showing the importance of chromosome studies for the correct and precise characterization of species. Chromosome studies suggest that the amount of karyotypic change has been dissimilar among the different groups; it has been prominent in *Aotus* and *Alouatta* while it has been more conservative in *Cebus* and *Pongo*. This might be a consequence of different rates of chromosome change along the different primate lineages perhaps because karyological shuffling might occur more frequently in some groups with respect to others. These four examples of intraspecific karyotypic change point to the need of carrying out chromosome studies *within* species, looking at the largest possible number of specimens. One of the limitations of most reports in primate cytogenetics is the small number of specimens generally available for chromosome studies and the frequent extrapolations that result from the generalization of findings from one individual to the species level. Most reports on chromosome phylogenies in the primate order are affected by this bias and, moreover, consider the morphological attributes of chromosomes as the most important guidelines for the inference of phyletic affinities, even when morphological similarities do not coincide with data on gene assignment (Yunis and Prakash, 1982). Although it is not the intention of this writer to comment on all aspects of chromosome studies in the primates, extensive reviews in this field have been published in the recent years (Seuánez, 1979, 1984; Dutrillaux, 1979; deGrouchy et al., 1978), giving an overall picture of the karyological evolution of the order. Since many morphological characteristics of human karyotype, accounting for some 20% of the chromosome complement, have been conserved for 80 million years since the time humans and the domestic cat diverged from one another (Nash and O'Brien, 1982; O'Brien and Nash, 1982), the finding of chromosome similarities between primate species is a consequence of a relative karyotypical stability.

DNA Studies in the Primates

Molecular studies have been most useful in understanding primate evolution and have become a valuable complement to chromosome studies. While the vast majority of these studies is centered on the analysis, structure, and composition of nuclear DNA, recent work has now been focused on mitochondrial DNA because this material might be a better indicator of rapid evolutionary change, the rate of which is on the order of ten times higher than single copy nuclear DNA (Brown et al., 1982). The study of nuclear DNA may be based on single copy DNA (Hoyer et al., 1972; Kohne et al., 1972; Benveniste and Todaro, 1976), on

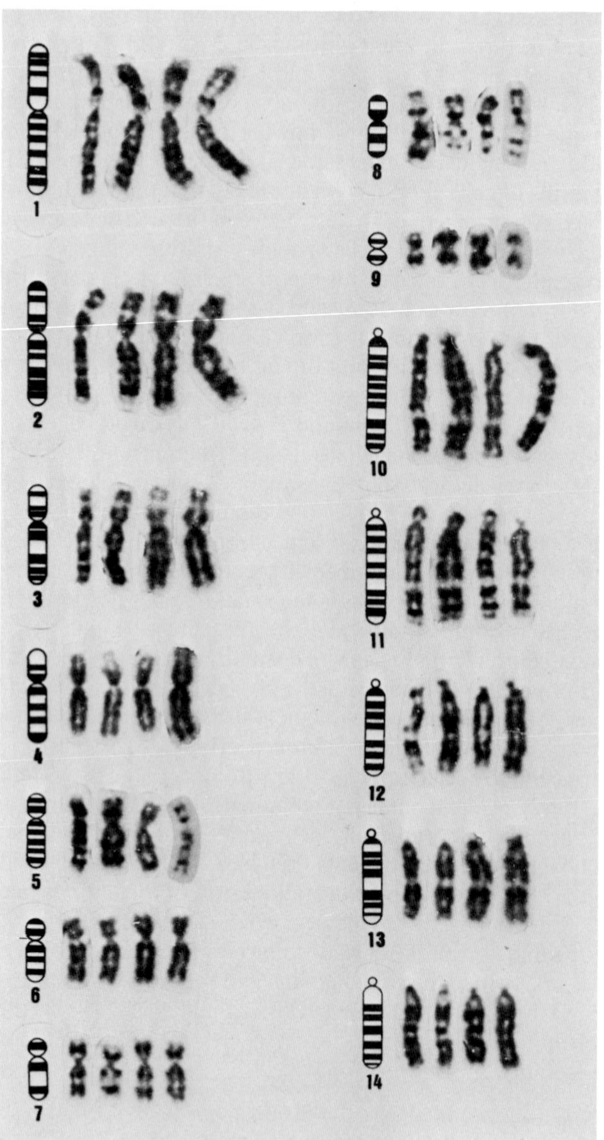

FIGURE 60.5. Karyotype of *Alouatta belzebul belzebul* with G-banding (2n=50). NOR staining occurs in chromosome 8, in two different regions.

repetitive DNA fractions (Gillespie et al., 1980; Mitchell et al., 1981; Arnheim et al., 1982), or on the analysis of isolated structural genes (Barrie et al., 1981; Jeffreys et al., 1982). It can also be based on the analysis of some externally acquired DNA sequences of viral origin, the endogenous retroviral DNA, which has been integrated into the primate genome and is present in multiple copies (Benveniste and Todaro, 1976). A comparative analysis of this material has

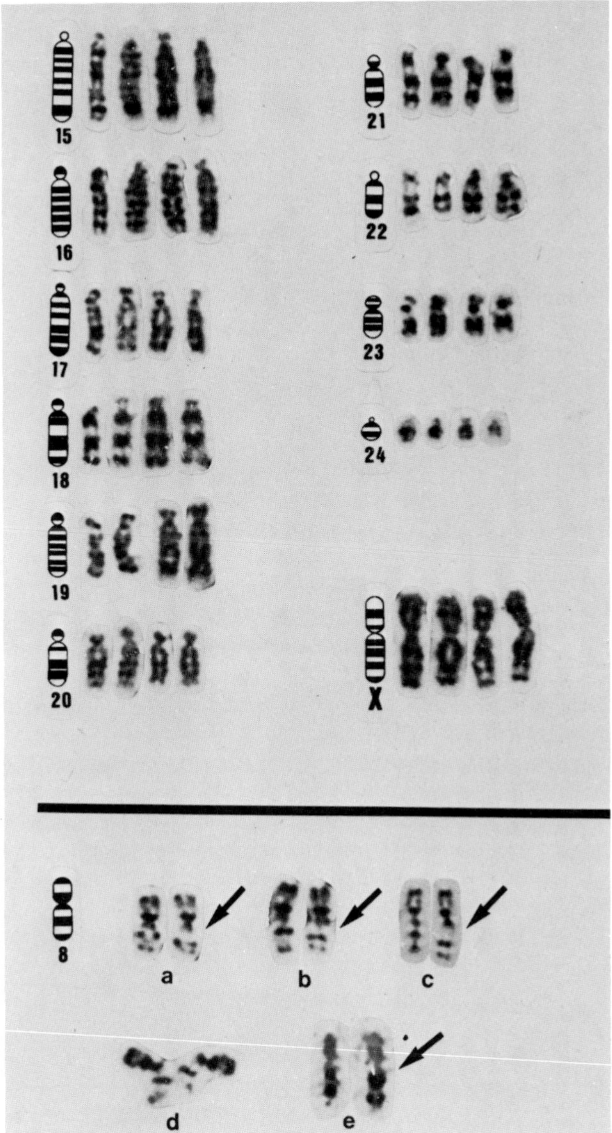

FIGURE 60.5. Continued.

supplied valuable information on primate phylogenies and has questioned the established view of human geographic origin.

A study of single copy DNA by hybridization experiments has permitted the estimations of the percentage of nucleotide replacements in the functional genome of the primates along different lines of descent because such percentage is a simple function of molecular mismatching when DNA from two different species is hybridized (Bonner et al., 1973). This approach allows for the calcu-

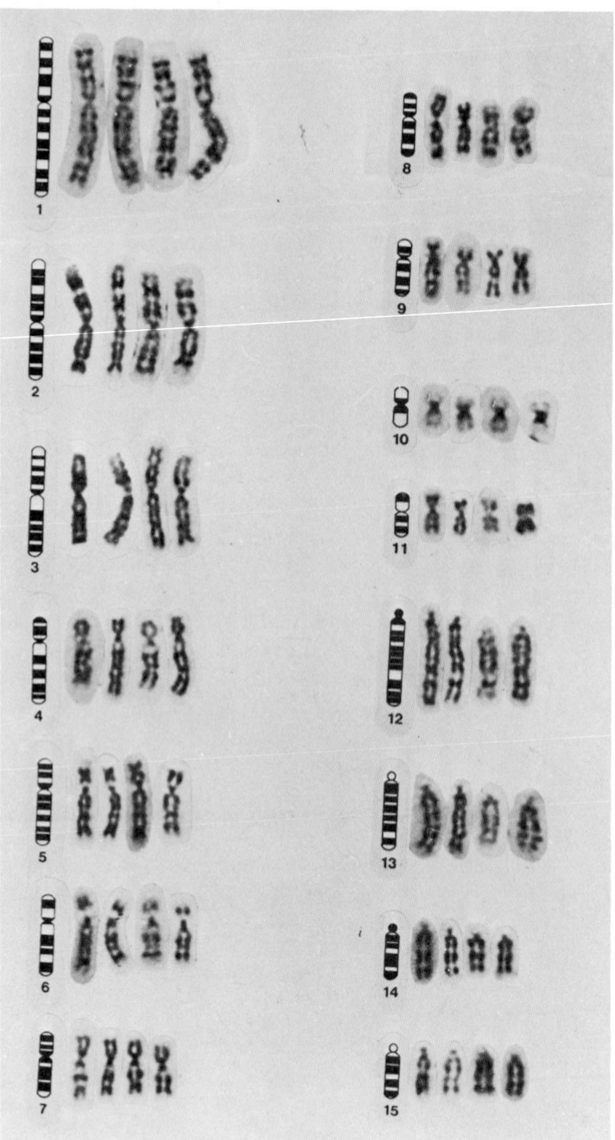

FIGURE 60.6. Karyotype of *Alouatta belzebul nigerrima* with G-banding (2n=50).

lation of molecular distances between species and for the postulation of phyletic trees as a result of the comparative analysis of these metrics. On the other hand, the study of repetitive DNA fractions has been most valuable in understanding primate evolution because a substantial proportion of the primate genome, as in all eukaryotes, is composed of reiterated sequences, most of which are apparently meaningless (Britten and Kohne, 1968). Since repetitive sequences are capable of

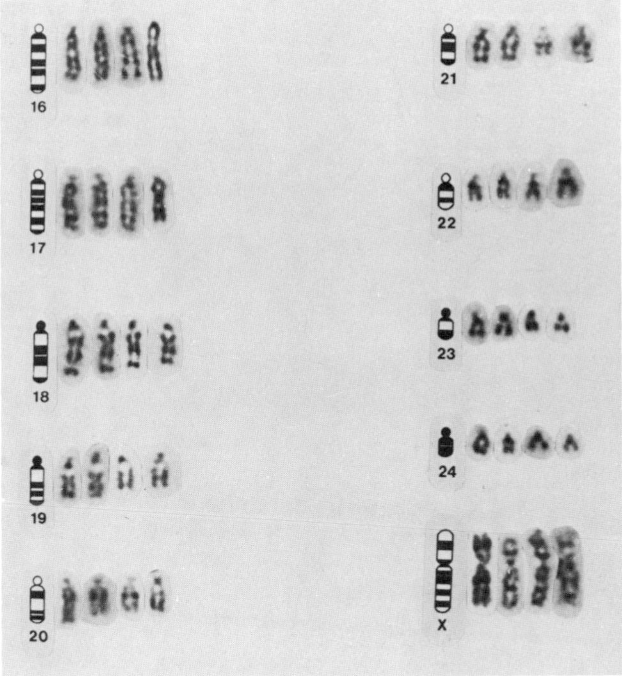

FIGURE 60.6. Continued.

reassociating at faster rates than single copy DNA, the kinetics of DNA reassociation has been a useful indicator of the proportion of repetitive DNA in the primate genome (Gummerson, 1972; Marx et al., 1979; Deininger and Schmid, 1979). Studies of this kind have shown that the emergence of repetitive sequence families in the primate genome has been discontinuous (Gillespie, 1977) because DNA amplifications have frequently occurred as independent events in different evolutionary steps, from original repeat units that were present in few or single copy in an ancestral genome. Some of these highly repetitive DNAs, formed by long tandem arrays of simple sequence repeats, have been isolated as "satellite" fractions by using specific gradients where they purify at the region of their buoyant density. In humans, four satellite DNAs were initially isolated (Corneo et al., 1973), though later studies showed that two of these were actually the same (Mitchell et al., 1979). These satellite DNAs were found to be shared with other nonhuman primates by experiments of DNA–DNA hybridization (Prosser et al., 1973) and in situ hybridization to great ape chromosomes (Jones et al., 1973; Mitchell et al., 1977; Gosden et al., 1977). The overall conclusion of these experiments is that human satellite DNAs have emerged at different times of the evolutionary divergence of the catarrhine primates and that they are not homogenous DNA fractions, as was originally thought. Satellite III DNA, for example, contains one ancestral component that was already present in our catarrhine ancestor some 24 million years before present, before the splitting of the

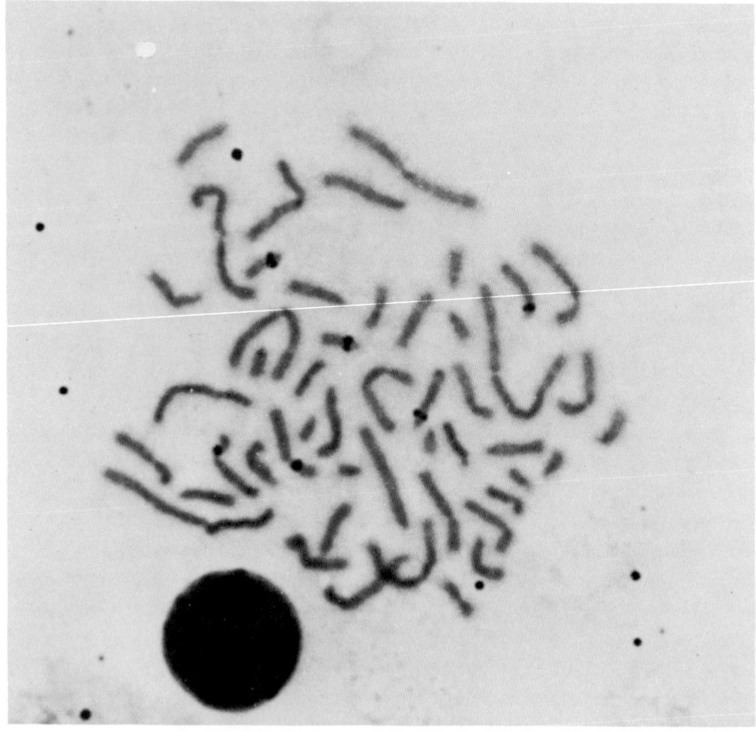

FIGURE 60.7. Chromosomes of *Alouatta belzebul nigerrima* with NOR staining in three chromosome pairs.

hominoid primates from the Old World monkeys. Another component, found in *male* satellite III DNA, is a 3.4 Hae III kb fraction that is exclusive of the human genome, the multiple copies of which are mostly located to the Y chromosome (Mitchell et al., 1979; Bostock et al., 1978). Satellite II DNA, which is related to satellite III, has apparently appeared in the human–great ape common ancestor or in the human genome after humans diverged from the common stock with the great apes (Mitchell et al., 1981). Satellite I DNA, which shows partial homology with satellite III but not with satellite II (Mitchell et al., 1979), is present in the genome of all hominoids, including the lesser apes (Mitchell et al., 1981). Thus, if DNA amplification occurred independently (Fig. 60.8), at different stages of primate evolution, the finding of a nonconservative chromosome distribution of these sequences in humans and the great apes is a logical consequence (Seuánez, 1974, 1984; Gosden et al., 1977; Seuánez et al., 1977). As in humans, repetitive DNA fractions have also been isolated from the nonhuman primates (Prosser et al., 1973; Prosser, 1974; Kurnit and Maio, 1974). In the African green monkey, for example, 20–25% of the genome was found to consist of a repetitive DNA fraction, or the α-component. In the chimpanzee, two different satellite DNAs were isolated, of which one showed close homologies with human satellite III DNA.

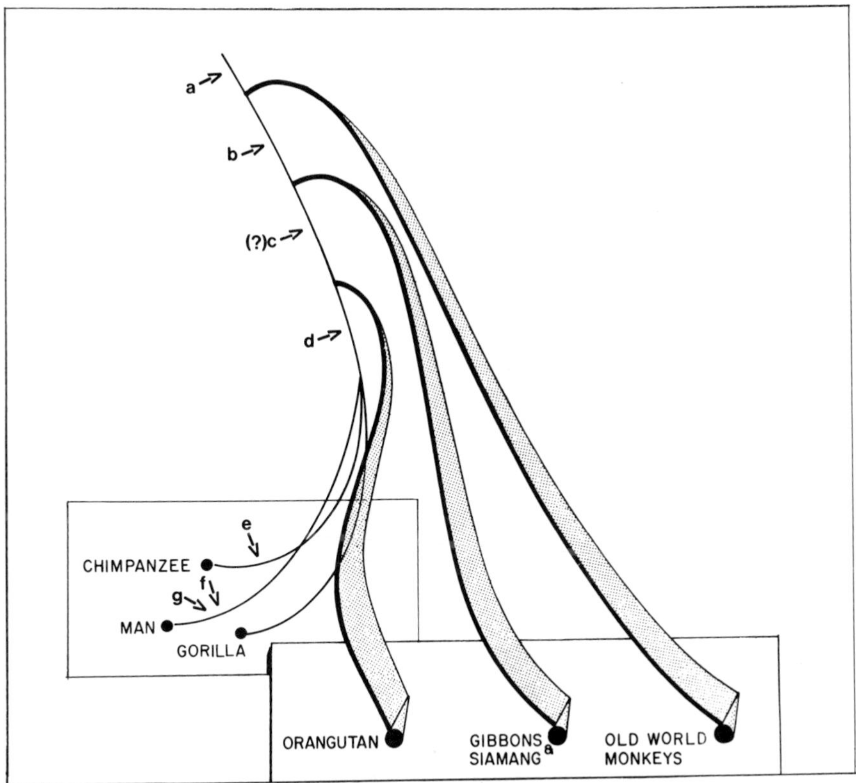

FIGURE 60.8. Evolution of human satellite DNAs and homologous sequences in the nonhuman primates. (a) Appearance of sequences homologous to the Hae III "ladder" of human satellite III DNA; (b) appearance of sequences homologous to human satellite I DNA; (c ?) probable appearance of sequences homologous to human satellite II DNA; (d) appearance of sequences homologous to the Hae III 2.1 kb male specific human DNA; (e) interspersion of the 2.1 kb sequences in the chimpanzee autosomes; (f) amplification of the 2.1 kb sequences in the human Y chromosome; (g) amplification of the Hae III 3.4 kb male specific sequence in the human Y chromosome.

The use of restriction endonucleases has further enhanced the isolation of repetitive DNAs when a basic sequence repeat is enclosed and delimited by two endonuclease recognition sites. In humans, for example, two male specific fractions were isolated (Cooke, 1976) of which one was identical with a male specific DNA previously purified by extensive hybridization of male and female DNA (Kunkel et al., 1976). This fraction, the Hae III 3.4 kb male specific DNA, was found to copurify with male satellite III DNA in a heavy iron gradient (Bostock et al., 1978) and to be exclusive of our own species. Contrary to this, the 2.1 Hae III kb fraction was found to be unrelated to any human satellite DNA and found to be preferentially amplified in the human Y chromosome after humans diverged from the great apes. However, its basic repeat unit was already present in the autosomes of humans, chimpanzees, and gorillas (Cooke et al., 1982).

Extensive endonuclease digestions of DNA of several organisms have shown a basic common pattern of cleavage consisting of a monomeric repeat unit of some 170 bp and multiples of this unit (Maio et al., 1977; Musich et al., 1977), showing repeat periodicities with one another and with the restriction pattern of the α-component of the African green monkey. In this cercopithecoid species, the restriction sites of the a-component fell between adjacent nucleosomes, suggesting that the repeat periodicities could have resulted from unequal crossing-overs in the nucleosome interstices (Musich et al., 1977). Sequence studies of the alfoid sequences in the African green monkey (Rosenberg, 1978) and humans (Manuelidis and Wu, 1978) have shown that the basic monomeric unit is complex, with little internal repetition, showing some 65% of interspecific similarity. Further studies have identified alfoid sequences along different primate orders, from the prosimians to humans (Grimaldi et al., 1981; Maio et al., 1981a, 1981b). One group of alfoid sequences, cleaved by Hind III and Eco RI endonucleases, have been found to be present in prosimians and therefore to be older than Old and New World monkeys. During the evolutionary divergence of the primates, however, nucleotide replacements and subsequent amplifications have altered these sequences that are dimeric in the cercopithecoids, tetrameric in the colobus monkey, and of higher complexity in the apes and humans (Maio et al., 1981a). Another group of alfoid sequences, cleaved by Kpn endonuclease, were found to be highly conserved and interspersed in the anthropoid genome but were absent in the prosimians (Maio et al., 1981b).

Other repetitive DNA multifamilies, which are present in the intermediate, or moderately repeated, fraction of the primate genome, include the ribosomal genes. The 18-28S cistrons comprise some 200–300 copies (Nelkin et al., 1980) while 5S rDNA is repeated some 7000 times in HeLa cells (Hatlen and Attendi, 1971). Endonuclease cleavage analysis of the 18-28S cistrons that form part of a transcribed DNA segment of 13 kb in length have shown a simple restriction pattern in the primates (Arnheim et al., 1982) with similar restriction maps among members of the same species. Although a few polymorphisms were identified with this restriction analysis and with sequence studies (Wilson et al., 1984) in limited regions of these cistrons, the overall picture from these reports is that intraspecific variations are considerably less frequent and significant than those observed between different primate species. Thus, it appears as if the whole set of multifamilies had evolved coincidentally, a phenomenon designated as "concerted evolution."

The chromosome distribution of the 18-28S sequences is multichromosomal in humans (Henderson et al., 1972), the great apes (Henderson et al., 1974, 1976), in some platyrrhines, such as *Cebus apella* (Freitas and Seuánez, 1982), *Callithrix jacchus* (Bedart et al., 1978), and *Leontopithecus rosalia* (Peixoto et al., 1981), and also in two prosimian species (Henderson et al., 1977; Table 60.3). In other primates, such as the gibbon (*Hylobates lar*), the rhesus monkey, and in a large number of cercopithecoid monkeys, the 18-28S cistrons are restricted to a single chromosome pair, clearly identified as a "marker chromosome" (Chiarelli, 1971) (Figs. 60.9, 60.10, 60.11). This situation is also observed in some platyrrhines, such as *Saimiri sciureus* (Lau and Arrighi, 1976), and the owl monkey, *Aotus trivirgatus* (Miller et al., 1977). A comparison between different species, at the chromosome level, has shown that the amplification of the 18-28S cistrons must have been independent from the process of karyotypic evolution of the primates (Henderson et al., 1977), contrasting with the more conserved

TABLE 60.3. Chromosome distribution of 18S and 28S ribosomal DNA genes in humans and the great apes.[1]

Humans[2]	Chimpanzee[3]	Pygmy chimpanzee[3]	Gorilla[3]	Orangutan[4]
2p	12	12	12	12
2q	13	13	11	11
9	11	11	13	13
13	14	14	14	14
14	15	15	18	15
15	16	16	15	16
18	17	17	16	17
21	22	22	22	22
22	23	23	23	23

[1] Chromosomes containing rDNA genes in each species are boxed. Chromosomes of the great apes are cytological homologs to human chromosomes.
[2] From Henderson et al. (1972).
[3] From Henderson et al. (1974, 1976).
[4] From Gosden et al. (1978).

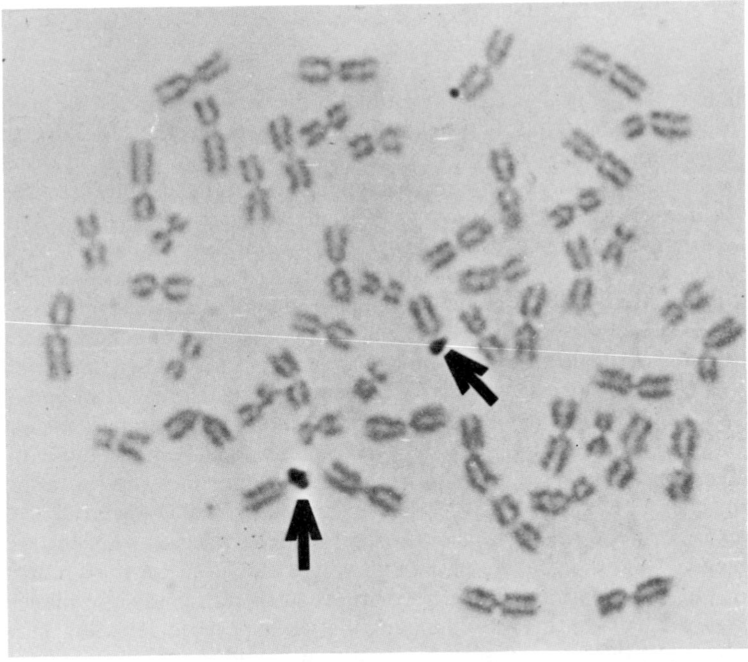

FIGURE 60.9. Silver staining in the active NORs of the siamang (*Symphalangus syndactylus*; 2n=50). Silver precipitation occurs in the short arm region of a pair of acrocentric chromosomes.

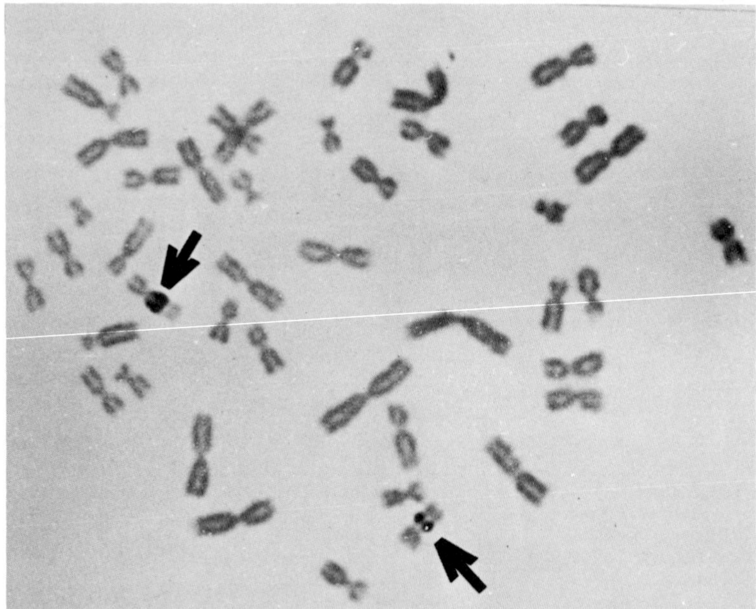

FIGURE 60.10. Silver staining of the active NORs in the gibbon (*Hylobates muelleri*; 2n=44). Note precipitation in the secondary constriction region of one pair of submetacentric chromosomes.

distribution of structural genes in the primate order or even between the primates and the domestic cat. Thus, the evolutionary conservation of the 18-28S cistrons and its concerted evolution in the primates is obviously unrelated to its disparent chromosome distribution. This paradoxical situation is evident between species such as humans with ten chromosomes containing 18-28S cistrons and the gorilla, where these sequences are located in only four chromosomes. It seems as if the 200–300 copies of these genes have substituted all previous copies in each evolutionary step by a hypothetical mechanism of molecular homogenesis. Of all possible mechanisms (Nelkin et al., 1980), the possibility of genetic exchange between nonhomologous chromosomes has been put forward (Arnheim et al., 1982) as the most logical alternative because nucleolar organizer regions are usually associated in metaphase chromosomes and in the interphase nucleus.

Other kinds of repetitive DNAs of considerable interest are those of the endogenous retroviruses that comprise ten to hundreds of copies in the mammalian genome (Benveniste, 1985). Retroviruses have been transferred between different vertebrate species because the viral particles released from one species can integrate into the genome of another where they are later transmitted in a regular Mendelian fashion. Thus, retroviruses act as transposable elements that overcome the barriers that reproductive isolation imposes to gene flow between individuals of different species. An interesting example of this genetic transposition is the interspecific viral transmission from the baboon ancestor to the domestic cat ancestor. This transmission occurred after the major radiation of

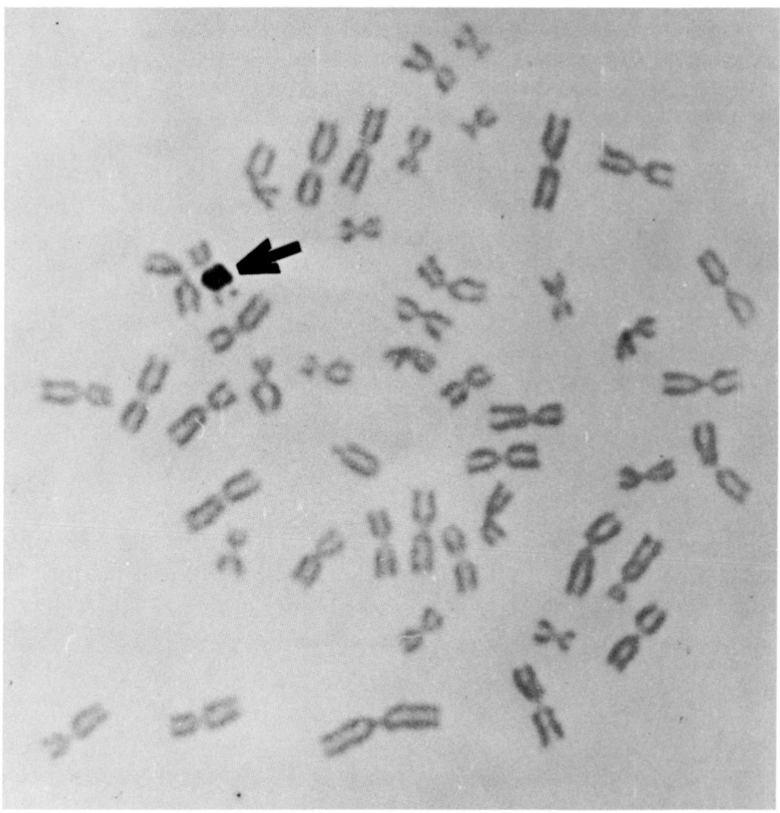

FIGURE 60.11. Silver staining of the active NORs in the siabon, an interspecific offspring of female siamang and a male gibbon. The siabon has a diploid number of 47 chromosomes, 25 comprising a siamang set and 22 comprising a gibbon set. Note that only the gibbon NOR is active in the hybrid.

the *Felidae* and prior to the speciation of the baboon and the gelada, after the common ancestor of these species diverged from the mangabeys, the macaques, and the mandrill (Benveniste, 1985).

Since endogenous retroviruses have a higher rate of evolutionary change than single copy cellular DNA, they could be useful indicators for the calibration of molecular clocks and for the inference of phyletic relationships between different species (Benveniste and Todaro, 1976). The baboon type C- endogenous retroviruses, for example, allow the distinction of different evolutionary rates between African and Asian primates, a finding that has led to the proposition that humans originated in Asia, contrary to the established idea on African origin. This is because all species of Old World primates contain sequence families related to the baboon virogene, which has been present in the catarrhine genome for the last 30 million years. During this period, however, the virogene sequences have diverged as functions of (1) the genetic distance of the carrier species to the baboon, and (2) the geographic distribution of the species independently of its genetic distance to the baboon. Thus, African primates showed more conserved molecular homologies with the baboon virogenes than their Asian counterparts, probably because the conservation of the retroviral homology would restrict the replication of the C-virus in their cells and would enhance their survival when exposed to the infectious viral particles released from baboons. Since human retroviral sequences were found to have diverged at a rate comparable to that of Asian primates, human emergence in the Asian continent has been proposed (Benveniste, 1985).

Finally, the analysis of mitochondrial DNA deserves a special comment in view of its potential value in evolutionary studies. This DNA component is present in multiple copies per cell; it is only maternally inherited, and it has a rate of evolutionary change some ten times higher than that of single copy cellular DNA (Brown et al., 1979). A common mitochondrial genome can, however, be shared by phylogenetically related species because of the crossing of the species barrier (Powell, 1983; Ferris et al., 1983), a finding that questions its reliability as indicator of evolutionary change. Nevertheless, a comparative study of mitochondrial DNA in the primates has allowed for the estimation of phylogenetic distances between them (Ferris et al., 1981), suggesting that humans are closer to the African apes than to the orangutan, but without resolving the chimpanzee–human–gorilla trichotomy. Further interpretation of these data has suggested that the African great apes might be closer to one another than to humans (Templeton, 1983), contrary to other reports, on a single copy DNA (Sibley and Ahlquist, 1984) and ribosomal DNA (Wilson et al., 1984), pointing to a larger molecular similarity between humans and chimpanzees.

Mitochondrial DNA studies have also been useful for the characterization of species and subspecies. Bornean and Sumatran orangutans, for example, show a 5% difference in their restriction patterns against 3.7% between the two chimpanzee species or some 1% within the same chimpanzee species (Ferris et al., 1981). This finding points to a molecular difference between orangutan subspecies that is coincident, though unrelated, to the chromosomal difference between Bornean and Sumatran orangutans. However, these findings have succeeded in characterizing subspecies independently from the rules and limitations of morphometric taxonomy and are clear examples for how chromosome and DNA studies can be complementary to the understanding of primate speciation and to the knowledge of extant species.

Summary

Much remains to be done, both at the chromosomal and molecular levels, to understand primate speciation, in spite of recent advances in the characterization of species and subspecies. Much could be done in favor of sustaining primate populations if genetic data are considered in the implementation of conservation policies. Since some primate populations are scarce, genetic programs should be devised to avoid inbreeding or the emergence of unfit individuals resulting from karyotypic abnormalities. Furthermore, primate genetic studies should be extended to the population level, relating chromosomal and molecular data to geographic distribution and morphometric characteristics. This approach will allow us to have a better understanding of primate populations and of how they can be helped to survive in a world of vanishing forests and dwindling natural resources.

Acknowledgment. This work was supported by CNPq/PIG no. 40/2404/82; FINEP conv. no. 4384073400; CEPG-UFRJ, and Fundação Universitária José Bonifácio.

References

Arnheim N, Krystal M, Schmickel R, Wilson G, Ryder O, Zimmer E (1982) Molecular evidence for genetic exchanges among ribosomal genes on nonhomologous chromosomes in man and apes. Proc Natl Acad Sci USA 77:7323–7327.

Barrie PA, Jeffreys AJ, Scott AF (1981) Evolution of the globin gene cluster in man and the primates. J Mol Biol 149:319–336.

Bedart MT, Ma NS, Jones TC (1978) Chromosome banding patterns and nucleolar organizing regions in three species of Callitricidae (*Saguinus oedipus*, *Saguinus fuscicollis* and *Callithrix jacchus*). J Med Primatol 7:82–97.

Benveniste RE (1985) The contributions of retroviruses to the study of mammalian evolution. In: MacIntyre RJ (ed) Molecular evolutionary genetics. Plenum, New York, pp 359–417.

Benveniste RE, Todaro GJ (1976) Evolution of type C viral genes: evidence for an Asian origin of man. Nature 261:101–108.

Bonner TI, Brenner DJ, Neufeld BR, Britten RJ (1973) Reduction in the rate of DNA reassociation by sequence divergence. J Mol Biol 81:123–135.

Bostock CJ, Gosden JR, Mitchell AR (1978) Localization of a male specific DNA fragment to a subregion of the Y chromosome. Nature 272:324–328.

Britten RJ, Kohne DE (1968) Repeated sequences in DNA. Science 161:529–540.

Brown WM, George MJR, Wilson AC (1979) Rapid evolution of animal mitochondrial DNA. Proc Natl Acad Sci USA 76:1967–1971.

Brown WM, Prager EM, Wang A, Wilson AC (1982) Mitochondrial DNA sequences in primates: tempo and mode of evolution. J Mol Biol 18:225–239.

Cabrera A (1957) Catálogo de Mamíferos de América del Sur, vol 1. Museo Argentino de Ciencias Naturales, Buenos Aires.

Chiarelli B (1971) Comparative cytogenetics in primates and its relevance for human cytogenetics. In: Chiarelli B (ed) Comparative genetics in monkeys, apes and man. Academic, London/New York, pp 276–304.

Cooke H (1976) Repeated sequence specific to human males. Nature 262:182–186.

Cooke H, Schmidtke J, Gosden JR (1982) Characterization of a human Y chromosome repeated sequence and related sequences in higher primates. Chromosoma 87:453–460.

Corneo G, Ginelli E, Zardi L (1973) Satellite and repeated sequences in human DNA. In: Pfeiffer RA (ed) Symposia Medica Hoechst. Schattauer Verlag, Stuttgart, pp 29–37.

Couturier J, Dutrillaux B (1981) Conservation of replication chronology of homologous chromosome bands between four species of the genus *Cebus* and man. Cytogenet Cell Genet 29:233–240.

de Boer LEM, Seuánez HN (1982) The chromosomes of the orangutan and their relevance to the conservation of the species. In: de Boer LEM (ed) Biology and conservation of the orangutan. W. Junk, The Hague.

Deininger PL, Schmid CW (1979) A study of the evolution of repeated DNA sequences in primates and the existence of a new class of repetitive sequences in primates. J Mol Biol 127:437–460.

de Grouchy J, Turleau C, Finaz C (1978) Chromosomal phylogeny of the primates. Ann Rev Genet 12:289–328.

Dutrillaux B (1979) Chromosomal evolution in primates: tentative phylogeny from *Microcebus murinus* to man. Hum Genet 48:251–314.

Dutrillaux B, Couturier J, Viégas-Péquignot E, Chauvier G, Trebbaue R (1978) Présence d'une heterochromatine abondante dans le caryotype de deux *Cebus*: *C. capucinus* et *C. nigrivittatus*. Ann Génét 21:142–148.

Ferris SD, Wilson AC, Brown WM (1981) Evolutionary tree for apes and humans based on cleavage maps of mitochondrial DNA. Proc Natl Acad Sci USA 78:2432–2436.

Ferris SD, Sage RD, Huang CM, Nielsen JT, Ritte U, Wilson AC (1983) Flow of mitochondrial DNA across a species boundary. Proc Natl Acad Sci USA 80:2290–2294.

Finaz C, Cochet C, de Grouchy J (1978) Identité des cariotypes de *Papio papio* et *Macaca mulatta* en bandes R,G,C at Ag-NOR. Ann Génét 21:149–151.

Freitas L, Seuánez HN (1982) Chromosome heteromorphisms in *Cebus apella*. J Hum Evol 10:173–180.

Gillespie D (1977) Newly evolved repeated DNA sequences in primates. Science 196: 889–891.

Gillespie D, Pequignot E, Strayer D (1980) An ancestral amplification of DNA in primates. Gene 12:103–111.

Gosden JR, Mitchell AR, Seuánez HN, Gosden C (1977) The distribution of sequences complementary to satellite I, II, and IV DNAs in the chromosomes of the chimpanzee (*Pan troglodytes*), gorilla (*Gorilla gorilla*) and orangutan (*Pongo pygmaeus*). Chromosoma 63:253–271.

Gosden JR, Lawrie S, Seuánez HN (1978) Ribosomal and human homologous repeated DNA distribution in the orangutan (*Pongo pygmaeus*). Cytogenet Cell Genet 21:1–10.

Grimaldi G, Queen C, Singer MF (1981) Interspersed repeated sequences in the African green monkey genome that are homologous to the human Alu family. Nucl Acids Res 9:5553–5568.

Gummerson KS (1972) The evolution of repeated DNA in primates. Thesis, Johns Hopkins University.

Hatlen L, Attardi G (1971) Proportion of HeLa cell genome complementary to transfer RNA and 5S RNA. J Mol Biol 56:535–556.

Henderson AS, Warburton D, Atwood KC (1972) Localization of ribosomal DNA in the human chromosome complement. Proc Natl Acad Sci USA 69:3394–3398.

Henderson AS, Warburton D, Atwood KC (1974) Localizing of rDNA in the chimpanzee (*Pan troglodytes*) chromosome complement. Chromosoma 46:435–441.

Henderson AS, Atwood KC, Warburton D (1976) Chromosome distribution of rDNA in *Pan paniscus*, *Gorilla gorilla beringei* and *Symphalangus syndactylus*. Comparisons to related primates. Chromosoma 59:147–155.

Henderson AS, Warburton D, Megraw-Ripley S, Atwood KC (1977) The chromosomal location of rDNA in selected lower primates. Cytogenet Cell Genet 19:281–302.

Hershkovitz P (1983) Two new species of night monkeys, genus *Aotus* (Cebidae,Platyrrhini). A preliminary report on *Aotus* taxonomy. Amer J Primatol 4:209–243.

Hoyer BH, van de Velde NW, Goodman M, Roberts RB (1972) Examination of hominoid evolution by DNA sequence homology. J Hum Evol 1:645–649.

Jeffreys AJ, Barrie PA, Harris DH, Fawcett DH, Nugent ZJ, Boyd AC (1982) Isolation and sequence analysis of a hybrid globin pseudogene from the brown lemur. J Mol Biol 156:487–503.

Jones KW, Prosser J, Corneo G, Ginelli E, Bobrow ME (1973) Satellite DNA constitutive heterochromatin and human evolution. In: Pfeiffer RA (ed) Symposia Medica Hoechst. Schattauer Verlag, Stuttgart, pp 45–61.

Kinzey W (1980) Distribution of some neotropical primates and the model of Pleistocene forest refugia. Proc Fifth Int Symp Assoc Trop Biol, Caracas. Colombia University Press.

Kohne DE, Chiscon JA, Hoyer BH (1972) Evolution of primate DNA sequences. J Hum Evol 1:627–644.

Kunkel LM, Smith KD, Boyer SH (1976) Human Y chromosome-specific reiterated DNA. Science 191:1189–1190.

Kurnit DM, Maio JJ (1974) Variable satellite DNAs in the African green monkey *Cercopithecus aethiops*. Chromosoma 45:387–400.

Lau YF, Arrighi F (1976) Studies of the squirrel monkey, *Saimiri sciureus*, genome. I. Cytological characterization of chromosomal heterozygosity. Cytogenet Cell Genet 17:57–60.

Ma, NSF (1984) Linkage and syntenic relationship in chromosomes of owl monkeys (with karyotypes I, III, V, VI, VII) homologous to man. In: O'Brien SJ (ed) Genetic maps, vol 3. Cold Spring Harbor, New York, pp 410–413.

Ma NSF, Jones TC, Miller AC, Morgan LM, Adams EA (1976) Chromosome polymorphisms in the owl monkey (*Aotus*). Lab Anim Sci 26:1022–1036.

Maio JJ, Brown FL, Musich PR (1977) Subunit structure of chromatin and the organization of eukaryotic highly repetitive DNA. Recurrent periodicities and models for the evolutionary origins of repetitive DNA. J Mol Biol 117:637–655.

Maio JJ, Brown FL, Musich PR (1981a) Toward a molecular paleontology of primate genomes. I. The Hind III and Eco RI dimer families of alfoid DNAs. Chromosoma 83:103–125.

Maio JJ, Brown FL, McKenna WG, Musich PR (1981b) Toward a molecular paleontology of primate genomes. II. The KPNI families of alfoid DNAs. Chromosoma 83:127–144.

Manuelidis L, Wu JC (1978) Homology between human and simian repeated DNA. Nature 276:92–94.

Marx KA, Purdom IF, Jones KW (1979) Primate repetitive DNAs: evidence for new satellite DNAs and similarities in non-satellite repetitive DNA sequence properties. Chromosoma 73:153–161.

Miller KC, Miller DA, Miller OJ, Tantravahi R, Reese RT (1977) Banded chromosomes of the owl monkey, *Aotus trivirgatus*. Cytogenet Cell Genet 19:215–226.

Mitchell AR, Seuánez HN, Lawrie S, Martin DE, Gosden JR (1977) The location of DNA homologous to human satellite III DNA in the chromosomes of chimpanzee (*Pan troglodytes*), gorilla (*Gorilla gorilla*) and orangutan (*Pongo pygmaeus*). Chromosoma 61:345–358.

Mitchell AR, Beauchamp SR, Bostock CJ (1979) A study of sequence homologies in four satellite DNAs of man. J Mol Biol 135:127–149.

Mitchell AR, Gosden JR, Ryder OA (1981) Satellite DNA relationships in man and the primates. Nucl Acids Res 24:3235–3249.

Mittermeier R, Coimbra-Filho AF (1981) Systematics: species and subspecies. In: Coimbra-Filho AF, Mittermeier R (eds) Ecology and behavior of neotropical primates, vol 1. Academia Brasileira de Ciências, Rio de Janeiro, pp 29–109.

Musich PR, Maio JJ, Brown FL (1977) Subunit structure of chromatin and the organization of eukaryotic highly repeated DNA: indications of a phase relation between restriction sites and chromatin subunits in African green monkey and calf nuclei. J Mol Biol 177:657–677.

Napier JR, Napier PH (1967) A handbook of living primates. Academic, London/New York.

Nash W, O'Brien SJ (1982) Conserved regions of homologous G-banded chromosomes between orders of mammalian evolution: carnivores and primates. Proc Natl Acad Sci USA 79:6631–6635.

Nelkin B, Strayer D, Vogelstein B (1980) Divergence of primate ribosomal RNA genes as assayed by restriction enzyme analysis. Gene 11:89–96.

O'Brien SJ (ed) (1984) Genetic maps. Cold Spring Harbor, New York, pp 1–581.

O'Brien SJ, Nash WG (1982) Genetic mapping in mammals: chromosome maps of the domestic cat. Science 216:257–265.
O'Brien SJ, Seuánez HN, Womak JE (1985) On the evolution of genome organization in mammals. In: MacIntyre RJ (ed) Molecular evolutionary genetics. Plenum, New York, pp 519–589.
Peixoto LIS, Ferrari I, Pedreira CM (1981) Estudo citogenético do mico-leão (*Leontopithecus rosalia*) do Brasil. Cienc Cult (Suppl) 33:645.
Powell JR (1983) Interspecific cytoplasmic gene flow in the absence of nuclear gene flow: evidence from *Drosophila*. Proc Natl Acad Sci USA 80:492–495.
Prosser J (1974) Satellite DNA in man and three other primates. Thesis, University of Edinburgh.
Prosser J, Moar M, Bobrow M, Jones KW (1973) Satellite sequences in chimpanzee (*Pan troglodytes*). Biochim Biophys Acta 319:122–134.
Rosenberg H, Singer M, Rosenberger M (1978) Highly reiterated sequences of SIMIANSIMIANSIMIANSIMIANSIMIAN. Science 200:394–402.
Seuánez HN (1979) The phylogeny of human chromosomes. Springer-Verlag, Berlin/Heidelberg/New York.
Seuánez HN (1982) Chromosome studies in the orangutan (*Pongo pygmaeus*): practical applications for breeding and conservation. Zoo Biol 1:179–199.
Seuánez HN (1984) Evolutionary aspects of human chromosomes. In: Roodyn D (ed) Subcellular biochemistry, vol 10. Plenum, New York, pp 453–535.
Seuánez HN, Fletcher J, Evans HJ, Martin DE (1976) A polymorphic structural rearrangement in two populations of orangutan. Cytogenet Cell Genet 17:327–337.
Seuánez HN, Mitchell AR, Gosden JR (1977) Constitutive heterochromatin in the Hominidae in relation to four satellite DNAs in man and homologous sequences in the great apes. Proc III Latin Amer Congr Genet, pp 171–178.
Seuánez HN, Evans HJ, Martin DE, Fletcher J (1979) An inversion in chromosome 2 that distinguishes between Bornean and Sumatran orangutans. Cytogenet Cell Genet 23:137–140.
Sibley CG, Ahlquist JE (1984) The phylogeny of the hominoid primates as indicated by DNA–DNA hybridisation. J Mol Evol 20:2–15.
Templeton AR (1983) Phylogenetic inference from restriction endonuclease cleavage site maps with particular reference to the evolution of human and the apes. Evolution 37:221–224.
Turleau C, Creau-Goldberg N, Cochet C, de Grouchy J (1983) Gene mapping of the gibbon. Its position in primate evolution. Hum Genet 64:65–72.
Wilson AC, Bush GL, Case SM, King MC (1975) Social structuring of mammalian populations and rate of chromosomal evolution. Proc Natl Acad Sci USA 12:5061–5065.
Wilson GN, Knoller M, Szura L, Schmickel RD (1984) Individual and evolutionary variation of primate ribosomal DNA transcription initiation regions. Mol Biol Evol 1:221–237.
Yunis JJ, Prakash O (1982) The origin of man: a chromosomal pictorial legacy. Science 215:1525–1529.

61 Considering Subspecies in the Captive Management of *Ateles*

William R. Konstant

Introduction

During the 1960s and early 1970s, thousands of spider monkeys (*Ateles* sp.) were exported from Central and South America to meet the needs of biomedical research, zoological exhibition, and the pet trade (Mack and Mittermeier, (1984). Today, this trade has come almost to a halt and institutions can no longer rely on a supply from the wild to replenish their stocks. In the wild, spider monkeys face a number of threats that could ultimately lead to the extinction of several subspecies. Foremost among these threats is the destruction of tropical forests. Whether the forests are cleared to provide land for settlement and agriculture or to supply lumber or fuel wood, their disappearance leads directly to the decline of species such as spider monkeys that depend on this habitat for their existence. Additional threats to *Ateles* include hunting of these large-bodied monkeys for food (Fig. 61.1) and their live capture to support local pet trades (Fig. 61.2). Taken in combination, these factors spell certain extermination for spider monkeys in many areas.

Although many zoos are actively engaged in breeding endangered wildlife and in several cases have been responsible for preserving species that have become extinct in the wild, current breeding programs for spider monkeys actually have little conservation value. A survey begun in 1984 by the World Wildlife Fund-U.S. Primate Program has shown that efforts to sustain distinct species and subspecies of *Ateles* in captivity are falling short for a variety of reasons (Konstant et al., 1985). Traditionally, zoos have considered spider monkeys common and replaceable, certainly not "priority species" such as the great apes, the golden lion tamarin, or the lion-tailed macaque. Consequently, little attention has been paid to their taxonomy or to breeding populations that are genetically distinct. Hybridization is common. Inbreeding depression is threatening the existence of a number of small, isolated colonies, and many collections remain misidentified. As a result, the entire zoological community is being presented with an inaccurate picture of *Ateles'* captive status.

If zoos wish to preserve the genetic diversity of *Ateles* in captivity, a great deal of work needs to be done. It will be necessary to compile as complete and accurate an inventory of captive populations as possible. Cytogenetic research will play an important role in this effort, helping zoos to distinguish between the different spider monkey taxa and thus establish homogeneous colonies. New

FIGURE 61.1. Adult female Guianan black spider monkey (*Ateles p. paniscus*) shot for food along the Brazil-Surinam border (photo by M. J. Plotkin).

FIGURE 61.2. Infant Mexican spider monkey (*Ateles geoffroyi vellerosus*) for sale in a market in the Sierra de Santa Martha, Veracruz (photo by E. Rodriguez Luna).

advances in the field of reproductive physiology, such as embryo transfer, might possibly help to build populations of some of the rarer subspecies quickly. Most important, cooperative breeding programs will be necessary to consolidate isolated individuals and unreproductive colonies and to provide for genetic exchange where none presently exists.

Endangered *Ateles* Subspecies

The greater part of the zoological community uses the taxonomy suggested by Kellogg and Goldman (1944), which recognizes four species and 16 subspecies of spider monkeys (Table 61.1). The Red Data Book (IUCN, 1982) lists *Ateles belzebuth, Ateles geoffroyi*, and *Ateles paniscus* as Vulnerable (14 subspecies) and *Ateles fusciceps* as Indeterminate (two subspecies). All are affected to some degree by the destruction of their tropical forest habitat, hunting for food, and live capture for pets. Based upon the most recent field information, the WWF-U.S. Primate Program now considers at least six *Ateles* subspecies to be endangered and in immediate need of some help if we are to prevent their extinction (Mittermeier, this volume).

Ateles fusciceps fusciceps

The brown-headed spider monkey is apparently restricted to the Pacific side of the cordillera of Ecuador (Kellogg and Goldman, 1944; Napier, 1976) and is probably very endangered. A WWF-U.S. Primate Action Fund project is now underway in the Cotacachi-Cayapas Ecological Reserve to determine this monkey's status there and to develop a management plan that would ensure its survival. At this time, there are no records of *Ateles fusciceps fusciceps* in captivity,

TABLE 61.1. Species and subspecies of *Ateles* (according to Kellogg and Goldman, 1944).

Genus	Species	Subspecies
Ateles	*belezbuth*	*belzebuth*
		hybridus
		marginatus
	fusciceps	*fusciceps*
		robustus
	geoffroyi	*azuerensis*
		frontatus
		geoffroyi
		grisescens
		ornatus
		pan
		panamensis
		vellerosus
		yucatanensis
	paniscus	*chamek*
		paniscus

and it is unlikely that there were ever many, based upon available trade records (Mack and Mittermeier, 1984). It is quite easily identified by its contrasting brown head and all black body.

Ateles belzebuth marginatus

This subspecies is endemic to Brazil, and its range is restricted to an area south of the Amazon, between the Rio Tapajos and the Rio Tocantins in the state of Pará (Kellogg and Goldman, 1944). It is thought to be protected only on the outskirts of one Amazonian national park (Rylands and Mittermeier, 1982). However, this subspecies has long been subjected to heavy hunting pressure, not only as a source of food, but also to provide bait to trap spotted cats (Mittermeier and Coimbra-Filho, 1977). In addition, recent flooding of its habitat resulting from the Tucurui hydroelectric project (Kingston, 1984) and reports of its absence in large areas of selectively logged forest within its range suggest that its present situation is critical (A. Johns, personal communication).

Fortunately, a small number of *Ateles belzebuth marginatus* can be found in a handful of Brazilian zoos (Konstant et al., 1985). Efforts are now underway to encourage these institutions to participate in genetic studies and a cooperative

FIGURE 61.3. *Ateles belzebuth marginatus* at the São Paulo Zoo, São Paulo, Brazil (photo by R. A. Mittermeier).

breeding program. White sideburns and triangular forehead patch on an all-black body distinguish this subspecies from other *Ateles*, and zoos should be alert to its possible presence in their collections, although it is unlikely that many were ever exported from Brazil (Fig. 61.3).

Ateles belzebuth hybridus

Ateles belzebuth hybridus can be found both in Colombia and Venezuela (Hernandez-Camacho and Cooper, 1976; Mondolfi and Eisenberg, 1979). However, it is known to be protected in only one Venezuelan reserve (Mondolfi and Eisenberg, 1979). Hunting pressure and habitat destruction are severe threats to its existence in northern Colombia (Scott et al., 1976).

In captivity, our survey has uncovered several colonies and isolated individuals of *Ateles belzebuth hybridus* that were either previously unreported or misidentified. Breeding success seems to be sporadic among these colonies, and they cannot at this time be considered sustainable. Information gathered concerning captive populations supports field data suggesting that there are two distinguishable forms of *Ateles belzebuth hybridus*, one lighter and one darker. Since this difference in coloration is apparently correlated with geographical distribution to some degree (Hernandez-Camacho and Cooper, 1976), there may be a need to recognize two subspecies instead of just one.

Ateles geoffroyi panamensis

The Panamanian or red spider monkey is found in Panama east of the Canal Zone (Cordillera de San Blas) and west through Chiriqui to central western Costa Rica (Kellogg and Goldman, 1944). A small population is protected on Barro Colorado Island, Panama. The subspecies is also protected in at least one Costa Rican national park (Konstant et al., 1985). Widespread habitat destruction has reduced its numbers drastically throughout its range.

It is unclear at present how many *Ateles geoffroyi panamensis* may be in captivity. Based upon primate trade data (Mack and Mittermeier, 1984), one would assume this subspecies to be somewhat common in North American collections. However, our survey has so far identified only a small number of individuals (Konstant et al., 1985). The characteristics that distinguish it are its dark brown head and limbs and contrasting red body (especially the abdomen).

Ateles geoffroyi azuerensis

This little-known subspecies at one time ranged from the Azuero Peninsula to the Burica Peninsula in southern Panama (Kellogg and Goldman, 1944). The most recent information indicates that it is restricted to the forested slopes of the western Azuero if it still exists at all (C. Skinner, personal communication). There is a record of a single *Ateles geoffroyi azuerensis* living in a North American zoo, but this is unconfirmed. Precise information on origin is essential for the identification of any unreported *azuerensis* in captivity, since the only museum specimens (two) ever collected were destroyed years ago and the subspecies remains poorly defined.

Ateles geoffroyi grisescens

This little-known subspecies presumably occurs in the valley of the Rio Tuyra and southeastward through the Serrania del Sapo of southeastern Panama and probably into the Cordillera de Baudo of northwestern Colombia (Kellogg and Goldman, 1944). It is not known to exist in any protected areas. Results of our survey suggest that there may be one or two *Ateles geoffroyi grisescens* in captivity (Konstant et al., 1985). More information is needed from field studies, both to determine this subspecies' status in the wild and to help identify it in captive collections. Published descriptions of its coloration do not appear to be in complete agreement with the available museum specimens.

Of these endangered *Ateles* subspecies, it appears that only one or two would benefit from any action that might be taken in captive collections at the present time. Several other subspecies are more common in captivity, and it is likely that attention paid to them now would reap significant benefits.

Ateles Subspecies in Captivity

A recent assessment of primate captive breeding suggests that only one (*Ateles geoffroyi*) of the four recognized spider monkey species is presently breeding at a sustainable rate (Stevenson, 1983). If that is true and if trends remain unchanged, we can expect only to preserve the Central American spider monkeys in zoos. This prediction is, however, flawed by some basic misconceptions.

At the present time, the zoological community relies heavily upon *International Zoo Yearbook* and ISIS statistics for information on the status of species in captivity. With regard to *Ateles*, however, neither source is close to being complete. In addition, there is currently no means of verifying the accuracy of reports. Consequently, we are probably underestimating total numbers of spider monkeys in captivity and perpetuating inaccurate information about the status of certain species and subspecies. Many members of the zoo community divide spider monkeys typologically into the general categories "blacks" and "goldens" and assign scientific names accordingly. Since several different subspecies easily fall into each color category, practical use of this dichotomy has resulted in two basic misconceptions about the captive status of *Ateles:* (1) The all-black spider monkeys in captivity are predominantly Amazonian taxa, and (2) all subspecies of the Central American *Ateles geoffroyi* are well represented in zoological collections.

Most zoos with all-black spider monkeys believe that they maintain either *Ateles paniscus paniscus* or *Ateles paniscus chamek*, this Amazonian species being commonly referred to as the "black spider monkey." There is, however, another all black subspecies, *Ateles fusciceps robustus* (= *rufiventris*) from northwestern Colombia and eastern Panama. Its common name, "brown-headed spider monkey" (a name which, in reality, only fits the endangered Ecuadorian subspecies, *Ateles fusciceps fusciceps*), is likely to be the reason that many zoos do not believe they maintain it. Our survey has identified repeated instances of *Ateles fusciceps robustus* being identified as one or the other of the *Ateles paniscus* subspecies. Prior to our survey, only two institutions reported *Ateles*

TABLE 61.2. *Ateles fusciceps robustus* in captivity (census as of July 1985).

Institution	Males/females	Total
North America		
Cleveland Metro	1/4	5
Erie Zoo	0/1	1
Glen Oak Zoo	0/1	1
Kings Island	0/1	1
Louisiana Purchase	0/1	1
Monkey Jungle	2/7	9
National Zoo	4/3	7[1]
Sedgewick County	0/1	1
Sequoia Park Zoo	1/7	8
Utica Zoo	1/2	3[1]
Europe		
Basel Zoo	1/3	4
Mulhouse	3/3	6
Australia		
Taronga Zoo	0/1(?)	1 (?)
Total		47 (48?)

[1]These individuals had been reported to ISIS an *International Zoo Yearbook* prior to the WWF–U.S. captive census. All others in this table were identified by the census.

fusciceps robustus in their collections: the National Zoo in Washington, DC, and the Utica Zoo in New York. Moreover, both colonies had been exhibiting effects of inbreeding (Ralls and Ballou, 1982). We can now confidently report at least six small breeding colonies of this subspecies here in the United States and Europe, in addition to a small number of isolated individuals in both regions (Table 61.2). While this news bodes well for sustainable management of Colombian spider monkeys, it suggests that the Amazonian *Ateles paniscus* spp. are even less secure in captivity than was originally believed (Stevenson, 1983) and that their status must be reevaluated.

The name "golden spider monkey" is usually used for *Ateles geoffroyi*, but in fact it is descriptive of only one or two of the nine subspecies recognized by Kellogg and Goldman (1944). The common name that best describes the entire group is "black-handed spider monkey," this being a trait that all *Ateles geoffroyi* subspecies appear to possess. The Nicaraguan spider monkey (*Ateles geoffroyi geoffroyi*) appears to be the most common of the Central American *Ateles* and indeed the most common spider monkey in captivity. It is endemic to Nicaragua and is not known to occur in any protected areas there. It is somewhat lighter in color than, but closely resembles, an adjacent subspecies, *Ateles geoffroyi frontatus*, known from at least two protected areas in Costa Rica. These two subspecies are somewhat difficult to distinguish, and there is certainly need for research into their taxonomic status. At the present time, these lighter colored Central American spider monkeys are the most secure in captivity, and there is no reason why they cannot be managed on a sustainable basis. As an example, the Birmingham Zoo in Alabama received an AAZPA silver breeding certificate in 1979 for breeding more than 25 *Ateles geoffroyi geoffroyi* since establishing its colony in 1957

FIGURE 61.4. Nicaraguan spider monkey (*Ateles g. geoffroyi*) colony at the Birmingham Zoo, Alabama. This is the most productive captive breeding group known to the author (photo courtesy of B. Truett).

(Fig. 61.4). The colony has produced at least an additional 56 infants since receiving the award and is a major supplier to other zoological collections (B. Truett, personal communication).

Goals for Captive Management and Future Research

Several very basic goals must be established if we intend to manage *Ateles* in captivity on a sustainable basis and reap the maximum conservation value from these efforts.

Most important, zoos must set as a priority conservation of those species and subspecies that are most endangered in the wild, lest they waste an important opportunity to conserve primate diversity. Of the six endangered *Ateles* subspecies listed earlier in this chapter, there is a reasonable chance that at least two can benefit from action taken by zoos. Three or four small breeding colonies of *Ateles belzebuth hybridus* have been identified in North America, South America, and Europe, and it is not unlikely that several more might be found. This represents a small gene pool from which to initiate a management plan, and its scattered distribution internationally may present some problems of coordination. Yet, inaction now can only lead to further population decline and eventual disappearance.

The same is true of *Ateles belzebuth marginatus*, with only a handful known to exist in several isolated Brazilian zoos, and it is difficult to consider any comprehensive management plan that would not consolidate the existing founder stock. Efforts are currently underway to accomplish this goal, and it is hoped that this spider monkey can be bred in captivity as successfully as other endangered Brazilian primates have been at institutions such as the Rio de Janeiro Primate Center (CPRJ-FEEMA).

Another major goal in spider monkey management is to overcome the effects of past practices in dealing with these species. At the simplest level, zoos should reexamine their collections and determine how certain they are of previous identifications. ISIS and *International Zoo Yearbook* reports should be updated accordingly. In addition, participating institutions should encourage nonparticipants to become part of the data pool. For some taxa, we are currently working with very small numbers, and the addition of only one or two colonies in a breeding program could spell the difference between sustainable growth and failure.

In the area of research, it is important that zoos contribute to a better understanding of *Ateles* taxonomy and reproductive physiology. Cytogenetic and biochemical analyses may provide answers to questions of species and subspecies validity. Preliminary work in this area began in the 1970s (Benirschke et al., 1980). Dr. Kurt Benirschke and his staff at the San Diego Zoo's Research Department have resumed work in this area with the cooperation of the WWF-U.S. Primate Program, conducting a project with dual emphasis on investigating the karyology of wild spider monkey populations and utilizing these findings to segregate captive populations. This research should help to give more conservation value to current captive breeding efforts.

Embryo transfer may prove a supplementary tool in propagation of *Ateles* as well. Although no research has yet been done on embryo transfer in *Ateles*, the incentive and direction now exist. Given the fact that the Central American *Ateles geoffroyi geoffroyi* (and possibly *Ateles geoffroyi frontatus*) currently exist in relatively large numbers in captivity, it is a likely candidate for research into transfer techniques that may eventually be applied to more endangered taxa. There is also the possibility that any known hybrid females, which should be removed from the current gene pool, can perhaps serve some useful purpose as surrogates and contribute to the conservation of genetically pure animals.

Summary

The spider monkeys (*Ateles*) consist of a number of species and subspecies that are either vulnerable or endangered in the wild. As a result of current management practices, only one or two of these taxa can be considered secure in captivity, and if nothing is done to change these practices, the situation is likely to become worse. Increasing awareness of these problems is now leading to increased interest in spider monkeys within the zoological community. Surveys and laboratory research combined can help to develop captive breeding programs that have greater conservation value than present efforts. Initial findings indicate that, in order to sustain *Ateles* in captivity, cooperation on an international level is likely to be necessary.

References

Benirschke K, Lasley B, Ryder O (1980) The technology of captive propagation. In: Soulé ME, Wilcox BA (eds) Conservation biology. Sinauer, Sunderland, MA.

Hernandez-Camacho J, Cooper RW (1976) The nonhuman primates of Colombia. In: Thorington RW, Heltne PG (eds) Neotropical primates: field studies and conservation. National Academy of Sciences, Washington, DC.

IUCN (1982) The IUCN mammal red data book. Part 1. Gland, Switzerland.

Kellogg R, Goldman EA (1944) Review of the spider monkeys. Proc US Natl Mus 96:1–45.

Kingston WR (1984) Tucurui hydro-electric project. Prim Eye 23:20-23.

Konstant WR, Mittermeier RA, Nash SD (1985) Spider monkeys in captivity and in the wild. Prim Conserv 5:82-109.

Kunkel LM, Heltne PG, Borgaonkar DS (1980) Chromosomal variation and zoogeography in *Ateles*. Intern J Primatol 1(3):223-232.

Mack D, Mittermeier RA (eds) (1984) The international primate trade. Traffic (USA), Washington, DC.

Mittermeier RA, Coimbra-Filho AF (1977) Conservation in Brazilian Amazonia. In: Prince Rainier, Bourne GH (eds) Primate conservation. Academic, New York.

Mondolfi E, Eisenberg JF (1979) New records for *Ateles belzebuth hybridus* in northern Venezuela. In: Eisenberg JF (ed) Vertebrate ecology in the northern neotropics. Smithsonian Institution Press, Washington, DC.

Napier PH (1976) Catalogue of the primates in the British Museum (Natural History). Part I. Families Callitrichidae and Cebidae. British Museum (Natural History).

Ralls K, Ballou J (1982) Effects of inbreeding on infant mortality in captive primates. Intern J Primatol 3(4):491-505.

Rylands AB, Mittermeier RA (1982) Conservation of primates in Brazilian Amazonia. Intern Zoo Yrbk 22:17-37.

Scott NJ Jr, Struhsaker TT, Glander K, Chiviri H (1976) Primates and their habitats in northern Colombia with recommendations for future management and research. First Inter-American Conference on Conservation and Utilization of American Nonhuman Primates in Biomedical Research. Scientific Publ No 317 PAHO, Washington, DC.

Stevenson MF (1983) Effectiveness of primate captive breeding. Proceedings of the Symposium on the Conservation of Primates and their Habitats, vol II. University of Leicester, Great Britain, pp 202-232.

62
Blood Groups of Apes and Monkeys

WLADYSLAW W. SOCHA and JAN MOOR-JANKOWSKI

Introduction

The study of red cell antigens of primate animals was initiated by Landsteiner and Miller in 1925. It was intensified in 1940 when it was recognized that not only apes' and monkeys' red cells contained antigens similar to those already known in humans, but that immunizations of laboratory animals with the red cells of monkeys resulted in a production of antibodies that defined a new, hitherto unrecognized allogenic property of the human red cells. The discovery of the Rh factor (Landsteiner and Wiener, 1940) was to be one of the most important chapters in the history of human blood groups.

The research initiated by the discovery made by Landsteiner and Wiener and continued until now in the Primate Blood Group Reference Laboratory at LEMSIP, New York University School of Medicine, embraced many species of anthropoid apes and Old and New World monkeys and resulted in the definition of a number of antigenic specificities and blood group systems (Socha, 1980). Some of them were found to be intimately related to the major blood group systems of humans or to blood group systems of other primate species (Socha and Moor-Jankowski, 1979). Structural and serological affinities among blood antigens and blood group systems of various primate species are probably exponents of the common evolutionary past (Socha and Ruffie, 1983).

The study of blood groups of nonhuman primates was a logical extension of the human serohematology, and, understandably, the first attempts to define specificities of primate red cells were carried out using the reagents originally prepared for typing human erythrocytes. The blood groups, such as A-B-O, M-N, Rh-Hr, Lewis, I-i, defined in that way in nonhuman primates were called *human-type blood groups* and were considered homologues of the human red cell antigens. The second category of red cell specificities were those detected by antisera specifically produced for typing primate animals and obtained either by immunizing laboratory animals with the red cells of apes and monkeys or, preferably, by iso- or cross-immunizations of primates. These so-called *simian-type blood groups* were believed to be primates' own specificities, some of which could be analogues of the human red cell antigens (Moor-Jankowski and Wiener, 1972).

Serological affinities among blood groups of various members of the order Primates point not only to common pathways of molecular evolution, but also—probably—to similar biological significance of red cell antigens in the species of this taxonomic entity. If this is so, all progress made during the last 80 years in detection and practical uses of the blood groups of humans should be equally applicable to nonhuman primates, at least those maintained in captivity.

Clinical Applications

The utmost importance of blood groups, and, in fact, the reason for their study, lies with the safety of blood transfusions between members of the same, or closely related, species. While in humans the blood transfusion is one of the basic tools of modern medicine, there are few veterinarian-primatologists who resort to this therapeutic method in their practice. Yet we can quote numerous cases in which sick animals or those incapacitated by experimentation were successfully treated and saved by administration of compatible homologous or even heterologous blood.

There is, for instance, the case of a leukemic chimpanzee maintained for a prolonged time by multiple homologous blood transfusions from donors provided by several primate laboratories tested by us for compatibility, and sent to the recipient's institution (Muchmore and Socha, 1976).

Freshly imported primates, and particularly chimpanzees, are often anemic, and some may not survive the quarantine without supportive transfusions.

Single or multiple blood transfusions are practiced by us routinely in chimpanzees who become anemic in the course of prolonged experimentation with hepatitis B, or non-A, non-B, or in vaccine testing. Transfusions of blood are required, as a rule, in primates used in bone marrow transplantation experiments, carried out by several research teams.

The importance of compatibility of the transfused blood, so obvious in human medicine, is not yet generally appreciated by practitioners dealing with primates. In most apes, which show the A-B-O red cell polymorphism and simultaneous presence of anti-A and anti-B isoagglutinins, even the first infusion of A-B-O incompatible blood will result in a transfusion reaction as severe and dramatic as that observed in human patients.

Reactions caused by the red cell incompatibilities other than A-B-O will come into play during multiple transfusions, where the agglutinating antibodies are formed as the result of previous nonmatched transfusions. The latter reactions will also occur when performing more than one transfusion in lower monkeys (e.g., macaques, baboons, etc.), which do not have A, B, or O antigens but have a number of simian-type antigens on their red blood cells. As we have seen in a series of cross-transfusions among intentionally mismatched baboons and rhesus monkeys, the half-life of transfused incompatible red cells may be so significantly shortened as to make the transfusion therapeutically worthless (Socha et al., 1982).

Exposure to red cell immunogens occurs naturally during incompatible pregnancies in humans and in various other species of mammals. The resulting erythroblastosis fetalis or hemolytic disease of the newborn may lead to perinatal death or to serious postnatal developmental disturbances. As for nonhuman primates, cases of erythroblastosis fetalis caused by feto-maternal blood group incompatibility were observed first by Gengozian et al. (1966) in marmosets.

Transplacental immunization caused by feto-maternal blood group incompatibility was, according to Soviet authors (Verbitsky et al., 1969; Verbickij, 1972), responsible for significantly increased frequency of "pathologic outcome" of pregnancies in the Sukhumi colony of hamadryas baboons. The same authors also reported successful experimental induction of erythroblastosis fetalis in the offspring of female baboons immunized during the pregnancy with red cells of their (incompatible) breeding mates. It is noteworthy that neither Sullivan et al. (1972) nor we (Wiener et al., 1975; Socha and Moor-Jankowski, 1976) were able to reproduce those results in baboons or macaques. Even though the newborn of the hyperimmunized macaque or baboon females had their red cells maximally coated with maternal antibodies, the infants did not show any clinical symptoms of erythroblastosis fetalis. There seems to be a protective mechanism that prevents the intravascular destruction even of heavily coated red cells. The nature of this mechanism is still unclear.

In anthropoid apes, on the other hand, the situation seems to resemble that in humans, as shown by cases of fetal erythroblastosis observed by us in captive chimpanzees (Wiener et al., 1977), and most recently orangutans (Socha and van Foreest, 1981). This latter case is of practical importance since the transplacental immunization was diagnosed early in pregnancy so that life-saving measures were undertaken to save the erythroblastotic infant. It is of particular interest that the blood group incompatibility that resulted in maternal immunization in the orangutans involved a red cell antigen closely related to, though not identical with, the human Rh_0 (D) specificity responsible for most erythroblastosis cases in humans.

To minimize the chances of incompatible pregnancies in chimpanzees, we routinely match the prospective breeding mates according to their blood groups. Because of the large number of blood types currently defined in these apes, however, it becomes increasingly difficult to find ideally compatible sires. We have, therefore, limited ourselves to compatibility for red cell antigens known for their high immunogenicity, namely the A-B-O, R-C-E-F, and V-A-B-D specificities.

Blood Groups in Genetics and Taxonomy

In humans, blood groups remain the most important and irreplaceable genetic marker despite the impressive progress in defining other new inheritable traits, such as plasma and red cell isozymes, factors of the major histocompatibility system H-La, serum groups, etc. The role of blood groups as chromosomal markers in nonhuman primates is even greater than in humans, because the knowledge of simian extraerythrocytic genetic systems is still either rudimentary or the polymorphisms of the systems are less developed than those of blood groups. This is well illustrated by a comparison offered by Chakraborty et al. (1979) for pig-tailed macaques: the two blood group systems known to occur in these macaques (the A-B-O system and the D^{rh} graded system) contribute more than one-half of the cumulative probability of exclusion of paternity obtained when 14 various genetic marker systems are evaluated.

The importance of blood groups as genetic markers was also assessed in chimpanzees (Socha, 1981). Here, the two complex chimpanzee blood group systems

(V-A-B-D and R-C-E-F) together with a set of unrelated red cell specificities offer altogether 67 polymorphic types, more than all the remaining polymorphic marker systems combined. The usefulness of blood grouping tests in solving problems of doubtful paternity and in assessing the pedigrees of captive chimpanzees was illustrated in examples quoted in a recent article (Socha, 1981). Sullivan et al. (1977) computed paternity exclusion probabilities in randomly chosen rhesus males when red cells of males, females, and offspring were tested with antisera of 21 specificities. The cumulative probability of exclusion by all those tests reached 73%.

Blood group testing occupies a special place in numerous population genetics studies in the field, in which comparative distributions of red cell antigens among troops of feral primates served for measuring the degree of genetic isolation (Brett et al., 1976; Froehlich et al., 1977; Socha et al., 1977). Hereditarily transmitted properties of red cell antigens may constitute, in some instances, useful tools for distinguishing among more or less closely related species, thus becoming one of the taxonomic parameters. This may be the case, for example, of some of the blood groups of macaques (Moor-Jankowski and Socha, 1979), baboons (Moor-Jankowski et al., 1973; Socha et al., 1977), gibbons (Moor-Jankowski and Wiener, 1972), and common and pygmy chimpanzees (Socha, 1984).

Current Status

Available information on blood groups of apes and monkeys is not evenly distributed among primate species. Understandably, the most abundant data were obtained for species most often employed for biomedical research.

Among apes, blood groups of the common chimpanzee (*Pan troglodytes*) have been the most thoroughly investigated. Table 62.1 summarizes the current status

TABLE 62.1. Blood groups of the common chimpanzee (*P. troglodytes*).

Blood group system	Blood group	Phenotype(s) observed
A-B-O	O	O
	A	A_1^{ch}, $A_{1,2}$, A_2^{ch}
M-N	M	v.O, v.A, v.B, v.D, v.AB, v.AD, v.BD
	MN^v or MN	V.O, V.A, V.B, V.D, V^q.O, V^q.A, V^q.B, V^q.D, V^{pq}.O, V^{pq}.A, V^{pq}.B, V^{pq}.D
		Rare forms: V^{qs}.O, V^{qs}.A, V^{qs}.B, V^{qs}.D
	Miltenberger	W^c, w^c
Rh-Hr	Rh-negative	rc_1, rc_2, rCc_1, rCF, rCF_1, rCF_2
	Rh-positive	Rc_1, Rc_2, RC, RCc_1, RCc_2, RCF, $RCFc_1$, $RCFc_2$, RCE, $RCEc_1$, $RCEc_2$, RCEF, $RCEFc_1$, $RCEFc_2$
		Rare forms: $R_{var}Cc_1$, $R_{var}CF$, $R_{var}CFc_1$, $R_{var}CEF$
I-i		i
Probably unrelated specificities		H^c, h^c, G_1^c, G_2^c, g.K^c, k^c (very rare), O^c, o^c, M^c, m^c, N^c, n^c, S^c, s^c, T^c, t^c
Secretor status		SecLes or SecnL

TABLE 62.2. Blood groups of pygmy chimpanzee (*P. paniscus*).

Blood group system	Blood group	Phenotype(s) observed
A-B-O	A	A_1
M-N	M	v.D
Rh-Hr	Rh-positive	$R^{ab}_{var}CE$
Unrelated specificities		g^c, H^c, K^c, N^c, n^c, O^c, o^c, T^c, t^c, S^c
Secretor status		Sec

of knowledge of blood groups of this species (Socha, 1981) based on records of approximately 500 captive animals, including numerous families.

There are close to 70 serological types identified so far in the common chimpanzee, most of which are associated with blood group systems known to exist in humans. It must be stressed, however, that homology among human and chimpanzee red cell specificities concerns only some of the membrane structures that are shared by humans and chimpanzees. The remaining specificities are exclusive attributes of either chimpanzee or human erythrocytes (Socha and Moor-Jankowski, 1979).

The total number of the pygmy chimpanzees (*Pan paniscus*) blood grouped thus far is far smaller than that of common chimpanzees, but even these limited data point to striking differences between *P. paniscus* and *P. troglodytes*. Table 62.2 summarizes the findings in a recent study of 14 *P. paniscus* (Socha, 1984). The peculiarity of the pygmy chimpanzee blood groups is their relative monomorphism in tests with isoimmune *P. troglodytes* sera, which is in striking contrast to the diversity of the phenotypes found in the common chimpanzee. In addition, there are several quantitative differences in antigens shared by both species of chimpanzees.

More thorough investigations of blood groups of *P. paniscus* as well as of other species of great apes has been hampered by lack of antisera produced specifically for the work on the red blood cells of these animals and the inherent limitations of the available reagents, i.e., reagents originally prepared for typing the red cells of humans and/or common chimpanzee.

Table 62.3 presents a summary of the blood group tests carried out on about 90 orangutans thus far investigated by us. Availability of the very potent anti-A and anti-B reagents made it possible to overcome the interference of strong species-specific antibodies in human sera that caused agglutination of all orangutan red cells. The anti-A lectin, free of such heteroagglutinins, allowed subgrouping of the orangutan A red cells.

All orangutan red blood cells react with anti-M sera but not with anti-N reagents. Further subdivision of the M type was made possible by the use of chimpanzee isoimmune sera of the so-called V-A-B-D specificities.

Of various human anti-Rh reagents only anti-Rh_0 was found to agglutinate orangutan red cells, though weakly and erratically. Recently we were able to clearly distinguish Rh-negative from Rh-positive orangutans with a set of cross-reactive chimpanzee isoimmune reagents of the so-called R-C-E-F specificities

TABLE 62.3. Blood groups of orangutan.

Presumable unrelated	Blood type	Phenotype(s) observed
A-B-O	A	A, $A_{1,2}$, A_2
	B	B
	AB	A_1B, $A_{1,2}B$, A_2B
M-N	m	v.O
	M	v^q.O, V^{pq}.O, V^q.B or V^{pq}.B
	Henshaw	He
Rh-Hr	Rh-negative	rc_1, sj^{Or}
	Rh-positive	RCc_1, SjOr (probably related to Rh_o(D)
I-i		i
Presumable unrelated specificities		K^c or k^c, O^c, T^c
Secretor status		SecLes or nSLes

TABLE 62.4. Blood groups of gorillas.

Blood group system	Blood type	Phenotype(s) observed
A-B-O	B	B
M-N	N	N
	MN	M_1N or M_2N; V.O
	Henshaw	He, he
Rh-Hr	Rh-positive	Rh_o^{Go}, $\overline{Rh_o}^{Go}$; RC, RCF, R
Probably unrelated specificities		G^c H^c
Secretor status		Sec

TABLE 62.5. Blood groups of gibbons.

Blood group system	Blood group	Phenotype(s) observed
A-B-O	A	A_1, A_2
	B	B
	AB	A_1B, A_2B
M-N	M	M^{Gi}
	N	N^{Gi}
	MN	$M^{Gi}N^{Gi}$
Rh-Hr	Rh-positive?	$R^{ab}c$ ($\overline{rh}^{Gi}$)
Xg	Xg(a+) Xg(a−)	(Sex linked)
Unrelated specificities		A^g, a^g, B^g, b^g, C^g, c^g
Secretor status		SecnL

(Socha and Moor-Jankowski, 1980). Somewhat related to the human Rh_0 and chimpanzee R^c antigen is the first orangutan isoimmune antibody, tentatively named anti-Sj^{Or}, obtained from a pregnant female bearing an erythroblastotic fetus (Socha and van Foreest, 1981).

Blood grouping tests have been carried out on lowland (*Gorilla gorilla*) and mountain gorilla (*Gorilla beringei*) without, however, establishing any signficant serological differences (Socha et al., 1973; Wiener et al., 1971). Table 62.4 gives a summary of these findings. As can be seen, although monomorphic with respect to the A-B-O and Duffy groups, gorillas display some diversity within specificities of the M-N and Rh-Hr systems detected either by antihuman and/or anti-chimpanzee antisera. Of the latter, isoimmune chimpanzee reagents of the R-C-E-F specificities enabled further subdivision of the Rh-positive gorillas into three types. Among a number of chimpanzee typing reagents of the unrelated specificities (not yet identified with any blood group systems) only two, anti-G^c and anti-H^c, were found to cross-react with gorilla red cells; neither detected individual differences in this species.

Gibbons display as many as five phenotypes of the A-B-O system and three M-N types (Table 62.5) when tested with the reagents produced for testing human blood (Wiener et al., 1963). They are also the only nonhuman primates whose red cells react with human anti-Xg^a reagent (Gavin et al., 1964). As in humans, the Xg^a of gibbons appears to be a sex-linked trait, but its occurrence was confirmed only in *Hylobates lar lar.*

Limited immunization experiments resulted in the production of specific gibbon isoimmune typing sera that defined three red blood cell antigens, A^g, B^g, and C^g (Moor-Jankowski et al., 1965; Wiener et al., 1966). Of these, the A^g appeared to be a species-specific trait of *H. lar lar*; the B^g, on the other hand, was detected only in *H. pileatus*, while C^g defined individual differences among white-handed gibbons.

The Old World monkeys constitute the largest group of nonhuman primates maintained in captivity. Of the eight macaque species so far investigated for their blood groups (Table 62.6), the crab-eating macaques (*Macaca fascicularis*) constitute by far the largest population for which the geographical distribution of the A-B-O groups has been reported (Terao et al., 1981). In addition to the A-B-O polymorphism detected by testing their saliva and serum, a number of polymorphic red cell specificities have been defined in macaques by isoimmune antisera produced in rhesus monkeys and found to cross-react type specifically with red cells of various species (Moor-Jankowski and Socha, 1979).

Knowledge of baboon blood groups is second only to macaques (Table 62.7). All four A-B-O groups have been observed, although group 0 was found only in Guinea baboons. Significant differences in the geographical distribution of the A-B-O groups among various populations of feral baboons have been reported (Brett et al., 1976; Socha et al., 1977), thus confirming the value of these traits for population and taxonomy studies. The greatest variety of blood types, however, is detected by means of hemagglutination tests using specific isoimmune baboon antisera as typing reagents. Most of the baboon antisera detect unrelated specificities, each of which represents, together with its recessive silent allele, a separate genetic system. Only recently a more complex blood group system, the so-called B^P graded blood group system of the baboon, has been described (Socha et al., 1983).

TABLE 62.6. Blood groups of macaques.

Species	Blood group system (phenotypes observed) A-B-O	Lewis	M-N	I-i	D^{rh}	Other
Rhesus monkey *Macaca mulatta*	B; A, AB, O } very rare	Les nL	M	i	D_1,D_2,D_3 and d	A^{rh}, a^{rh}, B^{rh}, b^{rh}, C^{rh}, F^{rh}, f^{rh}, G^{rh}, g^{rh}, J^{rh}, j^{rh}, L^{rh}, I^{rh}, M^{rh}, m^{rh}, N^{rh}, n^{rh}, O^{rh}, o^{rh}, P^{rh}, p^{rh}
Crab-eating macaque *M. fascicularis*	A B AB O } rare	Les	M	i	D_1, D_2, D_3, D_4, and d	A^{rh}, a^{rh}, B^{rh}, b^{rh}, C^{rh}, F^{rh}, f^{rh}, G^{rh}, J^{rh}
Pig-tailed macaque *M. nemestrina*	O A B AB	Les	M	nt	D_1, D_2, D_3, D_4, and d	M^{rh}, m^{rh}, N^{rh}, n^{rh}, O^{rh}, o^{rh}, P^{rh}, p^{rh}, A^{rh}, B^{rh}, b^{rh}, C^{rh}, F^{rh}, f^{rh}, G^{rh}, g^{rh}, J^{rh}, j^{rh}, L^{rh}, M^{rh}, m^{rh}, N^{rh}, O^{rh}, P^{rh}, p^{rh}
Stump-tailed macaque *M. arctoides*	B	nt	M	nt	D_2(var), D_3, and d	A^{rh}, C^{rh}, F^{rh}, N^{rh}, f^{rh}, M^{rh}, m^{rh}, O^{rh}, o^{rh}, P^{rh}, p^{rh}
Bonnet macaque *M. radiata*	A B AB	nt	M	nt	d	A^{rh}, B^{rh}, C^{rh}, G^{rh}
Barbary macaque *M. sylvanus*	A	nt	M	nt	D_2 and D_3	B^{rh}, C^{rh}, F^{rh}, M^{rh}, N^{rh}, O^{rh}, P^{rh}
Japanese macaque *M. fuscata*	B	nt	M	nt	d	B^{rh}, b^{rh}, C^{rh}, f^{rh}, f^{rh}, N^{rh}, n^{rh}, A^{P}, B_3^{P}
Celebes macaque *M. maurus (tonkeana)*	A	nt	M	nt	Tests inconclusive	F^{rh}, N^{rh}, B_3^{P}

The available information on blood groups of Old World monkeys other than baboons and macaques is limited to the groups defined by the human typing reagents, such as A-B-O, M-N, Lewis, I-i, and Duffy (Table 62.7). Specific isoimmune typing reagents have not been as yet produced for those species, while the available rhesus and baboon antisera proved to be unsuitable for testing their red cells.

Lack of the specific antisera has also limited the scope of information on the simian-type blood groups of the New World monkeys. A notable exception were marmosets, for which isoimmune blood grouping reagents were produced by Gengozian (1966) that detected individual differences among the animals tested.

Tests for the A-B-O blood groups carried out on saliva uncovered polymorphism in at least three of the five New World monkey species investigated (Table 62.8). Independent of their A-B-O types, as defined by saliva testing, are the

TABLE 62.7. Blood groups of other Old World monkeys.

Species	Blood group system (phenotypes observed)					
	A-B-O[1]	Lewis[2]	M-N	I-i	B^P	Other
Baboon (*Papio*) (various species)	A B AB O (rare)	Les	M	i	B_1, B_2, B_3, B_4, and b	A^P, a^P, C^P, c^P, G^P, g^P, N^P, n^P, S^P, s^P, T^P, t^P, U^P, u^P, V^P, V_1^P, v^P, L^P, l^P, Hu, hu, Ca, ca, M^P, m^P, O^P, o^P
Gelada (*Theropithecus gelada*)	O	nt	M	I	B^P	C^P, G^P, Ca[4]
Celebes ape (*Cynopithecus*)	A B O	Les	M	nt	nt	nt
Patas monkey (*Erythrocebus*)	A	Les, nL	nt	I i	nt	nt
Vervet monkey (*Cercopithecus pygerythrus*)	A B AB	Les nL		nt	nt	
Langur (*Presbytis entellus*)	B AB	nt	MN	nt	nt	nt
Drill (*Mandrillus*)	A	nt	nt	nt	nt	nt

A- or B-like specificities detected on the New World monkey red cells. For more details on this subject, the reader is referred to an earlier publication (Froehlich et al., 1977). Some of the New World species were found to be polymorphic with respect to the I-i system. All New World monkeys tested proved to be negative with anti-M and anti-N reagents, unlike the Old World monkeys, which were all M-positive.

TABLE 62.8. Blood groups of the New World monkeys.

Species	Blood group system (phenotypes observed)		
	A-B-O	Lewis	I-i
Marmoset (*Callithrix*) (various species)	A	nL	I i
Howler monkey (*Alouatta*)	B	nt	nt
Capuchin (*Cebus*) (various species)	O A B	nL	I i
Spider monkey (*Ateles*) (various species)	O A B	nL	I i
Squirrel monkey (*Saimiri*)	O A AB	nL	i

Conclusions

There are several specific areas in which blood groups come into play and are of importance in attempts to develop self-sustaining populations of captive primate animals. As potent immunogens, blood groups are part of self-nonself recognition mechanism and are of practical significance in:

1. Transplacental immunization due to materno-fetal blood group incompatibility, which was shown to contribute to pre- and perinatal mortality of great apes and possibly also of the Old World monkeys. Blood grouping tests are indispensable for selecting compatible breeding partners, while routine serological tests help monitor potentially incompatible pregnancies as one of the measures recommended for prevention of erythroblastosis fetalis.
2. Blood transfusions carried out for therapeutic purposes in apes and monkeys. Transfusion of incompatible blood may be life threatening in apes and therapeutically worthless in monkeys.
3. Experimental organ and tissues allo- and xenotransplantations performed with increasing frequency in the Old World monkey species considered to be promising, prospective sources of organs for human recipients. Blood group incompatibility is known to shorten the survival of the graft.

As normal traits with known and well-understood modes of inheritance, the blood groups of nonhuman primates, alone or in conjunction with other molecular characters, constitute valuable genetic markers that find numerous practical applications; for example:

1. In parentage investigations necessary for constructing pedigrees.
2. For assessing genetic structure of the monkey and/or ape colony as a means of inbreeding detection and monitoring.
3. For distinguishing among races, subspecies, and closely related species. Some of the blood groups are known to be species characteristics and as such constitute useful taxonomic parameters.

References

Brett FL, Jolly CJ, Socha W, Wiener AS (1976) Human-like ABO blood groups in wild Ethiopian baboons. Amer J Phys Anthro 20:276–289.

Chakraborty R, Ferrel R, Schull W (1979) Paternity exclusion in primates: two strategies. Amer J Phys Anthro 50:367–372.

Froehlich JW, Socha WW, Wiener AS, Moor-Jankowski J, Thorington RW (1977) Blood groups of the mantled howler monkey (*Alouatta palliata*). J Med Primatol 6:219–231.

Gavin J, Noades J, Tippett P, Sanger R, Race RR (1964) Blood group antigen Xg^a in gibbons. Nature (London) 204:1322–1323.

Gengozian N (1966) Formation of isohaemagglutinins in the marmoset, *Tamarinus nigricollis*. Nature (London) 209:722–723.

Gengozian N, Lushbaugh CC, Humason GL, Kniseley RM (1966) Erythroblastosis fetalis in the primate *Tamarinus nigricollis*. Nature (London) 209:731–732.

Landsteiner K, Miller CP (1925) Serological studies on the blood of the primates. J Exp Med 42:841–852.

Landsteiner K, Wiener AS (1940) An agglutinable factor in human blood recognizable by immune sera for rhesus blood. Proc Soc Exp Biol NY 43:223–224.

Moor-Jankowski J, Socha WW (1979) Blood groups of Old World monkeys, evolutionary and taxonomic implications. J Hum Evol 8:445–451.
Moor-Jankowski J, Wiener AS (1972) Red cell antigens of primates. In: Fiennes RNT-W (ed) Pathology of simian primates, Part I. S Karger, Basel, p 270.
Moor-Jankowski J, Wiener AS, Gordon EB (1965) Simian blood groups. The new blood factors of gibbon blood, A^g and B^g. Transfusion 5:235–262.
Moor-Jankowski J, Wiener AS, Socha WW, Gordon EB, Davis JH (1973) A new taxonomic tool: serological reactions in cross-immunized baboons. J Med Primatol 2:71–84.
Muchmore E, Socha WW (1976) Blood transfusion therapy for leukemic chimpanzee. Lab Prim Newslett 15:13.
Socha WW (1980) Blood groups of apes and monkeys: current status and practical applications. Lab Anim Sci 30:698–714.
Socha WW (1981) Blood groups as genetic markers in chimpanzees: their importance for the national chimpanzee breeding program. Amer J Primatol 1:3–13.
Socha WW (1984) Blood groups of pygmy and common chimpanzees. A comparative study. In: Sussman RL (ed) The pygmy chimpanzee. Evolutionary biology and behavior. Plenum, New York/London, pp 13–41.
Socha WW, Moor-Jankowski J (1976) Serological materno-fetal incompatibility in nonhuman primates. In: Ruppenthal GC (ed) Nursery care of nonhuman primates. Plenum, New York/London, pp 35–42.
Socha WW, Moor-Jankowski J (1979) Blood groups of anthropoid apes and their relationship to human blood groups. J Hum Evol 8:453–465.
Socha WW, Moor-Jankowski J (1980) Chimpanzee R-C-E-F blood group system. A counterpart of the human Rh-Hr blood groups. Folia Primatol 33:172–188.
Socha WW, Moor-Jankowski J, Ruffie J (1983) The B^p graded blood group system of the baboon: its relationship with macaque red cell antigens. Folia Primatol 40:205–216.
Socha WW, Rowe AW, Lenny LL, Lasano SG, Moor-Jankowski J (1982) Transfusion of incompatible blood in rhesus monkeys and baboons. Lab Anim Sci 32:48–56.
Socha WW, Ruffie J (1983) Blood groups of primates, theory, practice, evolutionary meaning. Alan R Liss, New York.
Socha WW, van Foreest AW (1981) Erythroblastosis fetalis in a family of captive orangutans. Amer J Primatol 1:326.
Socha WW, Wiener AS, Moor-Jankowski J, Jolly CJ (1977) Blood groups of baboons. Population genetics of feral animals. Amer J Phys Anthro 47:435–442.
Socha WW, Wiener AS, Moor-Jankowski J, Mortelmans J (1973) Blood groups of mountain gorillas (*Gorilla gorilla beringei*). J Med Primatol 2:364–369.
Sullivan P, Duggleby C, Blystad C, Stone WH (1972) Transplacental immunization in rhesus monkeys. Fed Proc 31:792.
Sullivan PT, Blystad C, Stone WH (1977) Immunogenetic studies on the rhesus monkey (*Macaca mulatta*). XI. Use of blood groups in problems of parentage. Lab Anim Sci 27:348–356.
Terao K, Fujimoto K, Cho F, Honjo SH (1981) Inheritance mode of human-type A-B-O blood groups in cynomolgus monkey. J Med Primatol 10:72–80.
Verbickij MS (1972) The use of hamadryas baboons for the study of the immunological aspects of human reproduction. In: Diczfalusy E, Standley CC (eds) The use of nonhuman primates in research in human reproduction. WHO Research and Training Centre on Human Reproduction, Karolinska Institutet, Stockholm, pp 492–505.
Verbitsky M, Volkova L, Kuksova M, Lapin B, Andreyev A, Gvazava I (1969) A study of haemolytic disease of the foetus and the newborn occurring in hamadryas baboons under natural conditions. Z Versuchstierk 2:136–145.
Wiener AS, Moor-Jankowski J, Gordon EB (1963) Blood groups of apes and monkeys. II. The A-B-O blood groups, secretor and Lewis types of apes. Amer J Phys Anthro 21:271–281.

Wiener AS, Moor-Jankowski J, Gordon EB, Daumy OM, Davis J (1966) Blood groups of gibbons: further observations. Int Arch Allerg Appl Immunol 30:466–470.

Wiener AS, Moor-Jankowski J, Gordon EB (1971) Blood groups of gorillas. Kriminalistik und forensische Wissenschaften 6:31–50.

Wiener AS, Socha WW, Moor-Jankowski J (1977) Erythroblastosis models. II. Materno-fetal incompatibility in chimpanzee. Folia Primatol 27:68–74.

Wiener AS, Socha WW, Niemann W, Moor-Jankowski J (1975) Erythroblastosis models. A review and new experimental data in monkeys. J Med Primatol 4:179–187.

63
The Mind of the Gorilla: Conversation and Conservation

FRANCINE PATTERSON

The gorilla has been known as an object of scientific study for less than 140 years and already the possibility of its extinction is a real danger. There are only about 240 mountain gorillas remaining in the world. Since none are reproducing in captivity, their situation can only be described as desperate. The lowland gorilla, though not so immediately endangered, is also threatened. In zoos, the number of gorillas decreases by about 3% every year. Because time is critical in the race against extinction, direct contact with two ambassadors of the gorilla species, Koko and Michael, can save valuable time in acquiring important data that can be used in the field, the laboratory, and the zoo (Patterson et al., 1985).

The Gorilla Foundation, the organization that holds trust over Koko and Michael, is dedicated to the preservation and welfare of gorillas. The Foundation's studies of the behavior and linguistic capabilities of lowland gorillas (*Gorilla gorilla gorilla*) have generated a wealth of information that is certain to affect the world's attitude toward the gorilla. Not only does this work offer an opportunity to replace derivative knowledge with direct interspecies communication, it gives an unprecedented view of the world from the perspective of a species other than Homo sapiens. Studies being carried out at the Gorilla Foundation have revealed not only the gorilla's intellectual ability, but also its sensitive emotional character. This work proceeds from the belief that only through an understanding of the gorilla and its problems can people take the necessary steps to reverse the gorilla's descent into extinction.

For the past 13 years, Koko has been living and learning in a language environment that includes American Sign Language (ASL) and spoken English. Koko, age 14, has acquired a vocabulary of over 500 signs and communicates in statements averaging three to six signs in length. Michael, a 12-year-old male, who joined her on the project about 9 years ago, has learned over 250 signs. Both of the gorillas initiate most of their conversations with humans and use their vocabularies in creative and original sign combinations to describe their environment, feelings, desires, and even what may be their personal past histories (Patterson and Linden 1981; Crail, 1981). They sign to themselves and to each other (Patterson 1980, 1984). They also comprehend spoken English and are being taught to read the printed word (Patterson 1978b; Tanner, 1984).

Conversations with gorillas resemble those with young children and in many cases need interpretation. In transcripts of deaf children's signing we find sentences like: "Poor no cry" (meaning they were poor and couldn't have anything and she cried) or "Perfect girl bed boy bear" (meaning Baby Bear's bed was perfect for the girl) (Livingston, 1983). We encounter similar utterances in signing with the gorillas, and we assign probable interpretations based on context and past use of the signs in question. Alternative interpretations are often possible.

There is also a broader issue that troubles the ape-language research waters—there is no agreement as to the definition of language. It would be safe to characterize language, however, as a shared set of symbols that are used openly or creatively to generate a potentially infinite number of statements in an orderly or rule-governed way. Other generally accepted features of language are that it can be used to give false information, ask questions, negate statements, and refer to things not present. There is considerable evidence that Koko and Michael use sign language to, among other things, create new meanings, ask and answer questions, express feelings, deceive, and refer to things displaced in time or space (Patterson and Linden, 1981; Patterson, 1980, 1978a).

But even if the gorillas' use of signs does not meet a particular definition of language, studies into that use *can* give us a unique perspective from which to understand more directly the physical and psychological requirements of the species. Some of what they tell us can be anticipated: "What do gorillas like to do most?" "Gorilla love eat good." Or, "What makes you happy?" "Gorilla tree." "What makes you angry?" "Work." Some responses, on the other hand, are quite unexpected: "How did you sleep last night?" (expecting "fine," "bad," or some related response). "Floor blanket" (Koko sleeps on the floor with blankets). "What happened?" (after an earthquake) "Darn darn floor bad bite. Trouble trouble."

The insights gained through discussions such as these do not support the traditional image of gorillas. These have been tragically misunderstood animals. They are stereotyped as bloodthirsty and destructive monsters, when in fact they are shy, placid, and unaggressive vegetarians. This misunderstanding is even more tragic in view of the danger of the extinction of gorillas.

Further, in scientific circles the gorilla has been considered the "dim-witted relative of the chimpanzee" (Maple and Hoff, 1982). In contrast to the chimp, the gorilla was not known to use tools or recognize its reflected image in a mirror. The research with Koko and Michael indicates that gorillas are aware of themselves and of their relationships, as sentient beings, with others. We have found the first evidence that gorillas have a self-concept. Koko has unequivocally demonstrated her self-awareness by attempting to rub away a dark spot of pigment on her gum, a spot that she had precisely located by looking into her mouth with a mirror (Fig. 63.1). Captured on videotape, this spontaneous experiment of nature has replicated laboratory experiments in which apes were marked while anesthetized—but without the methodological difficulties inherent in that technique. In the other experiments, however, gorillas never showed evidence of self-recognition (Suarez and Gallup, 1981; Ledbetter and Basen, 1982). Michael, too, has used a mirror to make self-directed grooming responses, attending to parts of his body he could only see in this way. In addition, both Koko and Michael have watched themselves sign and make faces in mirrors and have demonstrated self-awareness in their conversations: in their use of self-referents (*me, mine, myself*) and in their ability to describe themselves as the same as or different from others. For example, project research assistant Maureen Sheehan extensively questioned Koko about the differences between gorillas and humans (Patterson and Linden, 1981):

FIGURE 63.1. Using the mirror, Koko inspects the inside of her mouth.

M: What's the difference between you and me?
K: Head.
M: And how are our heads different?
(Koko beats on her head with her open hands quite hard, harder than a human would ever do.)
M: What else is different between us?
(Koko moves her hands up on her stomach, a gesture resembling blanket.)
M: Do you mean something about your stomach?
K: Stomach good that.
M: Oh, but what were you saying about blanket, different?
(Koko moves her hands up and down her torso, then pulls the hair on her belly. Maureen interpreted these new gestures as meaning "bodyhair.")
M: Now can you name something the same?
K: Eye.
M: Yes, that's right, we both have eyes.

Self-awareness is also shown by the gorillas when they talk about themselves in situations removed in space or time from the current one. This capacity for displacement is clearly and repeatedly demonstrated in conversations such as the following:

5 July 1984 with Barbara Hiller:
B: How about telling me about your birthday, was it a nice one?
(Koko's birthday was celebrated on July 4, the day before.)
K: Visit gorilla Koko-love.
B: What did you like best?
K: That. (To a velvet ribbon she received as a present.)
B: What else did you like?
K: Cat tiger.
B: Wow, you got a cat!
10 July 1984 with Francine Patterson:
F: What happened on your birthday?
K: Sleep eat.
F: Didn't something special happen on your birthday?
K: Old gorilla.
F: Yes, you're 13 years old and big.

Several months after her kitten "All Ball" died (he had slipped out the door and found his way to the road where he was hit by a car), the following conversation took place:

17 March 1985 with Francine Patterson:
F: How did you feel when you lost Ball?
K: Want.
F: How did you feel when you lost him?
K: Open trouble visit sorry.
F: When he died, remember when Ball died, how did you feel?
K: Red red red bad sorry Koko-love good.

Why have other studies with gorillas failed to produce positive self-concept results? One reason, perhaps, is that experiments of this type have been carried out under conditions stressful to the subjects. One critical methodological difference between the mirror studies with gorillas and those with chimpanzees was that the observer was next to the mirror in the former, and hidden behind a peephole in the latter. Since gorillas seem to be a great deal more sensitive than chimpanzees to changes in their environment, this factor in itself could explain the negative findings with gorillas. Think for a moment, would you be likely to examine yourself in a mirror while under the stern eye of a strange researcher? Wouldn't you more likely be preoccupied with your observer, or with anything of even the slightest interest, rather than with your own image in a mirror? In contrast, the work with Koko and Michael is carried out in a relaxed, informal, family atmosphere, where the gorillas are able to "be themselves."

Another important factor may be gorilla temperament. Primatologists have long reported that the gorilla is difficult to work with because of its aloofness, low level of motivation, and contrary nature. In the 1920s Robert Yerkes, noted for his work with the great apes, characterized gorillas as reserved, independent, obstinate, and negativistic, characteristics that he felt could profoundly discourage work with them (Yerkes and Yerkes, 1929).

Such gorilla stubbornness and negativism have been encountered and documented in this work, but certain findings indicate that this is evidence of the gorilla's intelligence and independence rather than of its stupidity. And it is just this ornery independence that seems to spark episodes of humor and verbal playfulness. A characteristic incident involved Koko and assistant Barbara Hiller. Koko was nesting with a number of white towels and signed, "That red," to one of the towels. Barbara corrected Koko, telling her that it was white. Koko repeated her statement with additional emphasis, "That RED." Again Barbara stated that the towel was white. After several more exchanges, Koko picked up a piece of red lint, held it out to Barbara, and, grinning, signed "That red."

Koko's intelligence has been tested (Fig. 63.2), and her scores have been in the low normal range for humans (Patterson, 1979b), indicating that there is probably much more going on in the mind of a gorilla than we once might have guessed. And the key to discovering the gorilla's full potential may be language. In the work with Koko and Michael not only sign language but also spoken and written English are used. As in work with hearing children who are language disabled, sign has been just one tool, rather than the only means of communication. Early research had shown that the great apes could comprehend spoken words, and the project has been designed to take advantage of every opportunity to expose the gorillas to language. Tests have shown that the gorillas understand English as well as they understand sign. In one standardized language comprehension test (the first ever administered to apes) called the Assessment of Children's Language Comprehension, novel phrases corresponding to sets of pictures were given to the gorillas under conditions in which the tester did not know the correct answers. Koko's performance was twice as good as might have been expected by chance, and there was no significant difference in her performance whether the instructions were given in sign only or English only (Patterson, 1978b).

Although they once used a speech synthesizer to produce English word sounds, when the project was at Stanford University, the gorillas now converse exclu-

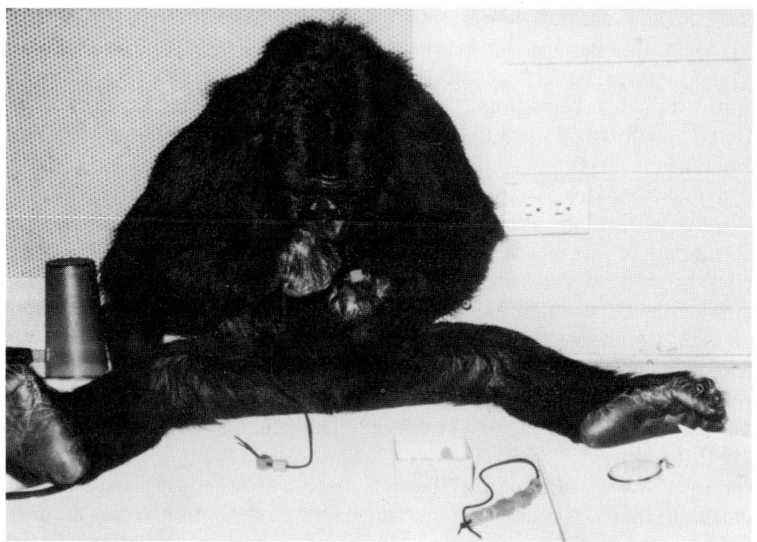

FIGURE 63.2. Koko works on a task from the Stanford Binet Children's Intelligence test.

sively in sign. (A new system is being planned.) Koko's working sign vocabulary has increased from approximately 12 signs after 1 year of training to about 500. Her emitted vocabulary—those signs she has used correctly on one or more occasions—is about 1000. Her recognition vocabulary in English is several times that number of words. Michael's working sign vocabulary is about 250 signs, and he has used about 400 different ones. Most of the signs were learned by direct instruction: through the molding of the gorillas' hands into signs or through imitation. But Koko and Michael have both used the language in diverse ways not explicitly taught. In a very real sense, the work has involved the mapping of skills, rather than the teaching of skills.

This mapping is being done through observations in relatively unstructured and uncontrolled situations and through rigorous tests. The best possible linguistic and cognitive performances are likely to be given in the informal setting, with support coming from tests. The approach has been to give Koko and Michael vocabulary instruction but no direct teaching of any other language skill.

The gorillas have taken these basic building blocks of conversation (signs) and, on their own, added new meaning through a process similar to inflection, called modulation. They do this through changes in motion, hand location, hand configuration, facial expression, and body posture. The sign *bad*, for instance, can be made to mean *very bad* by enlarging the signing space, increasing the speed and tension of the hand, and exaggerating facial expression. Deaf people use this feature of sign language creatively, for example, to exaggerate a point by signing *long* from toe to shoulder instead of from wrist to shoulder (Klima and Bellugi, 1979). Koko, like humans, has exploited sign language in similar ways, as when she signed *thirsty* from the tip of her head to her stomach, instead of down her throat.

The gorillas have been observed to use these kinds of variations to mark relations of size (e.g., small versus large *alligator* sign), number (*bird* versus *birds* by repeating the sign), location (scratch-on-back), possession (*Koko's-baby* signed simultaneously), manner, degree, intensity or emphasis (*tickle* signed with two hands), agent or object of an action (*you-sip* signed by moving the "f" hand toward the intended agent), negation (negating the *attention* sign by changing its location), to express questions (through eye contact and facial expression, Fig. 63.3), and as a form of word play akin to wit or humor (simultaneously signing *sad frown* when asked to "smile" for the camera).

A conversation with Koko was documented on film that involved this kind of creativity with the sign *rotten*. Koko demonstrated the standard form of the sign in an exchange of insults after her companion called her a "stinker." Koko then inflected the sign by using two hands (perhaps meaning "really rotten") and in the same sequence, brought the sign off her nose toward her companion, conveying the idea "you're really rotten." Koko's use of *rotten* in this conversation also demonstrates her grasp of the connotation of a word rather than its denotation or concrete or specific meaning. Thus, the meaning of the gorillas' signs are not necessarily identical to the most common meanings, and in some cases they are entirely different.

To interpret the gorillas' conversations correctly, one must discover what the signs mean to them. A number of techniques have been employed to accomplish this. Word-association games have been included in their activities (Patterson, 1984). The gorillas are asked to tell their teachers what they think of when a

FIGURE 63.3. By holding eye contact and the sign *pink*, Koko asks a question about the color of her smock.

given word is mentioned. Examples of stimulus-response pairs from sessions with Michael include: Teacher "hand," Michael "foot," and similarly, afraid-hug, dead-stink, and hungry-eat (Patterson and Peterson, 1984). The gorillas have also simply been asked to give definitions. Examples from data on Koko follow.

2/12/84
What's an insult?
  Think devil dirty.
What's a stove?
  Cook with.

7/12/84
What's an injury? (Voiced only.)
  There bite (to a cut on her hand).

7/13/84
What is crazy?
  Trouble surprise.

2/8/85
When do people say darn?
  Work obnoxious.

4/21/83
What can you think of that's hard?
Rock . . .work.

2/9/84
What's a smart gorilla?
Me.

There are also words that Koko does not understand, and she sometimes corrects her companions when they apply them to her: On Christmas evening in 1984 Koko picked up the telephone, listened to the dial tone, then signed "Ron" and handed the receiver to Barbara Hiller. When another companion commented, "She's a goofball!" Koko responded, "No, gorilla."

Table 63.1 shows a group of signs that Koko applies as invectives, many in ways that were not taught—she has apparently seized and used connotations with negative or insulting impact. One interesting sequence was captured on film. Koko had ripped the leg off a rag doll (which had earlier lost its other leg to Michael). The pieces (including both legs) were collected and placed in front of Koko who was then scolded for destroying the doll. Koko's reaction was "bad bite" then "you dirty bad toilet." (Patterson, 1978a)

Some of the gorillas' teachers have been in the habit of using the English word *stinker* to refer to them when they have been stubborn or mischievous. Now both gorillas use the sign *stink* (which the gorillas use most frequently to refer to flowers) in ways that translate as *stinker*. This interpretation has been confirmed by direct questioning:
Barbara Weller with Michael (12 November 1980):

B: What do I mean when I call you sweet gorilla?
M: Good.
B: What do I call you when you bad?
M: Stink.

In the case of the word *stink*, both Koko and Michael have given definitions that have expressed meanings at these two levels.

Francine Patterson with Koko:
F: What is stink?
K: Rotten devil know.

TABLE 63.1. English glosses for signs Koko uses as insults, expletives, and derogatives.

Bad	Nut
Big-trouble	Obnoxious
Bird	Old
Blew-it	Rotten-lousy
Darn	Skunk-stink
Devil	Stubborn-donkey
Dirty	Stupid
False-fake	Toilet
Frown	Trouble
Mad	Unattention

F: What else does stink mean?
K: Nice all leaf (Koko refers to flowers and plants as *stinks*).
Barbara Weller with Michael:
B: What stink mean?
M: Pull-out-hair think bad.
B: Can you think of another meaning for stink?
M: Good flower.

Another way Koko and Michael have created novel meanings for basic vocabulary signs is through an unusual coining process in which they employ signs whose spoken equivalents match or approximate the sounds of English words for which no signs have been modeled.

For example, Koko uses a modulated *knock* sign to mean *obnoxious*. This indicates that she knows:

1. That the sign *knock* is associated with the spoken word *knock*.
2. That *knock* sounds like the spoken word *obnoxious*.
3. That the sign *knock* can therefore be applied semantically to mean someone who is being obnoxious.

Other examples include the substitution of the sign *tickle* for *ticket*, *skunk* for *chunk*, and *lip stink* for *lipstick*.

A few years ago, some of Koko's companions were in the habit of using the expression "you blew it" in English with Koko, and she began to counter with a forceful and aggressive blow, which after some time was deduced to be Koko's way of conveying "you blew it" to her teachers. This interpretation has been confirmed by directly asking Koko how she signs "you blew it."

In some cases, signs have been repeatedly demonstrated that are difficult or impossible for Koko to form. Her solution has often been to make substitutions based on the sound of the corresponding English word: *knee* for *need*, *red* for *thread*, *lemon* for *eleven*, and *bird* for *word*.

The gorillas also communicate new meanings by making up their own entirely new signs. The intended meaning of some of the gestures has been obvious (e.g., *nailfile*, *eyemakeup*, or *pocket*) because of their iconic form, but the meaning of others has had to be worked out, over time, from records of the situations in which they occurred. For example, Koko invented a sign for a game which translates as *walk-up-my-back-with-your-fingers*.

Koko and Michael have generated dozens of novel gestures, with meanings ranging from *eyeglasses* to *runny-nose*. Table 63.2 lists a sample of Koko's invented signs. Some of Michael's invented words such as *hit-in-mouth*, *hit-in-nose*, *unlisten*, *pull-out-hair*, and *pull-out-teeth* are very similar to signs young deaf boys create (e.g., *hit-in-eye*; *cut-on-eye*) (Livingston, 1983).

Data on such inventions indicate that the gorillas, like human children, take initiative with language by making up new words and by giving new meanings to old words. On the next level, there is evidence that Koko and Michael can generate novel names by combining two or more familiar words. For instance, Koko signed "bottle match" to refer to a cigarette lighter, "white tiger" for a zebra, and "eye hat" for a mask. Michael has generated similar combinations, such as "orange flower sauce" for nectarine yogurt and "bean ball" for peas. Samples of the speech of 2- to 5-year-old children contain parallel phrases: "barefoot head"

TABLE 63.2. English glosses for a subset of the signs invented by Koko.

Above	Nailfile
Away	Nailpolish
Barrette	Obnoxious
Bite	Pocket
Blew-it	Runny-nose
Bracelet	Scarf
Clay	Sip
Eyeglasses	Stethoscope
Eyemakeup	Thermometer
Frown	Tickle
Headband	Unattention
Log	Walk-up-back

for a bald man and "giraffe bird" for an ostrich (Chukovsky, 1963). These all appear to be expressions of original ideas, not mere imitations of adult statements. Other examples in the samples of the gorillas' signing are "elephant baby" for an Pinocchio doll and "bottle necklace" for a canned soda six-pack holder.

Critics have commented that such phrases are merely the pairing of two separate aspects of what is present. Many of the above examples, however, cannot be explained in this way—when Koko signed "bottle match," neither a bottle nor a match was present. Currently, novel objects are being introduced during videotaped sessions in order to elicit such new names under controlled conditions for further study.

The gorillas have applied such new descriptive terms to themselves as well as to novel objects. When angered, Koko has labeled herself a "red mad gorilla," and in one conversation, that took place with Barbara Hiller, Koko, who had been reduced to drinking water through a thick rubber straw from a metal pan on the floor because of her constant nagging for drinks, referred to herself as a "sad elephant" (Patterson, 1980).

Intrigued by examples of language use such as these, the project research team went on to obtain empirical evidence for the gorillas' metaphoric capacity. In one experiment, the gorillas were asked to assign various descriptive words to pairs of colors. The results revealed that they assigned *warm* to red and *cold* to blue, *hard* to brown and *soft* to blue-grey, *sad* to violet and *happy* to orange, and *loud* to yellow and *quiet* to light green. Koko indicated her answers either by pointing or by verbal description (such as "orange that fine," when asked which color [orange or violet] was happy). Ninety percent of the gorillas's responses were identical to those of three project research assistants who took the same test (Patterson, 1980).

The gorillas have also been asked to represent feeling states such as love, hate, happiness and anger with paints on canvas. Given free choice of ten or more colors, the gorillas produced works with contrasting color and form. Asking them to paint emotions seemed a reasonable request because the gorillas had earlier demonstrated some primitive representational ability in their drawings and paintings done after models or from memory. Both Koko and Michael titled these works appropriately. Figure 63.4 is an example of Michael's representational art: Asked by his teacher to draw the Gorilla Foundation's black and white dog named

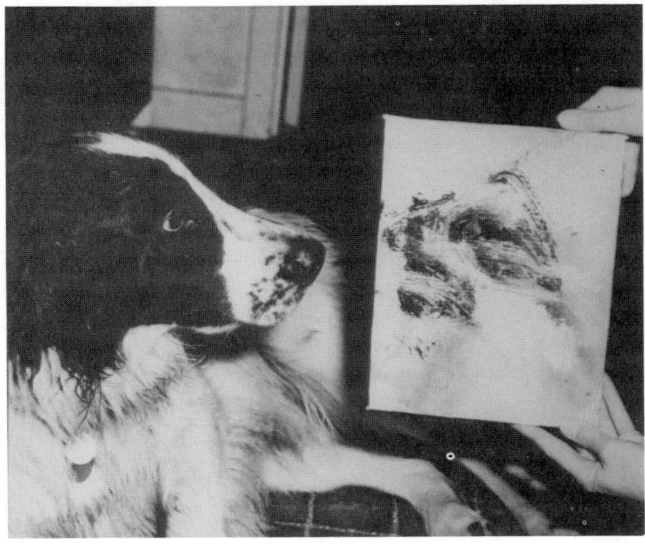

FIGURE 63.4. The dog Apple and Michael's painting titled "Apple chase."

Apple, he titled the work "Apple chase," and the black and white painting bears a resemblance to the dog's head.

Another creative aspect of the gorillas' language behavior is humor. Humor, like metaphor, requires a capacity to depart from what is strictly correct, normal, or expected. For example, when asked to demonstrate her invented sign for *stethoscope* for the camera, Koko did it on her eyes instead of on her ears (Fig. 63.5). Asked to feed her chimp baby, she put the nipple to the doll's eye and signed "eye."

Appreciation of this kind of wit is sometimes dependent on recognizing the sign behind the distortion. A skeptic might see this as simple error, but in the case of signs that the gorillas themselves invent, such as *stethoscope*, this is not likely, and there are consistencies that run across the gorillas' humorous use of signs.

During the past several years, we have noticed Koko giving an audible chuckling sound at the results of her own and her companions discrepant statements or actions. She discovered that when she blew bugs on her companions, a predictable shrieking and jumping response could be elicited. Originally, she laughed at this outcome, but now she chuckles in anticipation of the prank as well. Accidents and unexpected actions by others can also cause Koko to laugh. Chuckles were evoked, for instance, by a research assistant accidentally sitting down on a sandwich and by another playfully pretending to feed M&M's to a toy alligator.

Koko has also made verbal "jokes." On 30 October 1982 Barbara Hiller showed Koko a picture of a bird feeding its young.

K: That me (to the adult bird).
B: Is that really you?
K: Koko good bird.

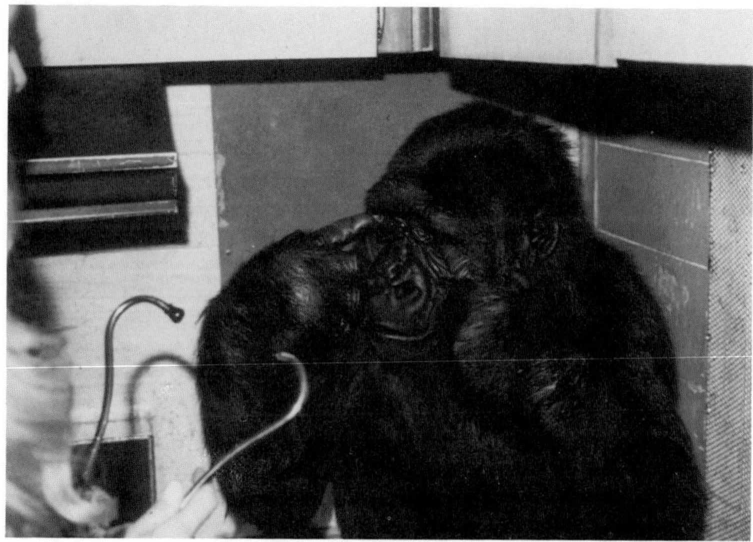

FIGURE 63.5. Koko places her fingers to her eyes when asked to identify a stethoscope. The correct response, her invented the sign for the instrument, would be to place the index fingers to the ears.

B: I thought you were a gorilla.
K: Koko bird
. . . .
B: Can you fly?
K: Good. (*Good* can mean *yes*.)
B: Show me.
K: Fake bird, clown. (Koko laughs.)
B: You're teasing me. (Koko laughs.)
B: What are you really?
(Koko laughs again and after a minute signs)
K: Gorilla Koko.

Researchers in developmental psychology have found that the earliest form of humor in young children, incongruity-based humor, relies on similar principles of discrepancy applied to objects, actions, and verbal statements.

Asked outright what she thought was funny, Koko signed "that nose" to a toy bird's mouth, and "hat" to a rubber key she placed on her head. Children between the ages of 3 and 6 may laughingly apply different names to familiar objects or make statements about nonexistent things or properties.

In stark contrast to the gorillas' ability to express humor is their ability to communicate their thoughts and feelings about death. This is, perhaps, especially significant in view of the species' fight against extinction. The first evidence appeared about 7 years ago, when one of Koko's teachers asked her, "When do gorillas die?" and she signed, "Trouble, old." The teacher also asked, "Where do gorillas go when they die?" Koko replied, "Comfortable hole bye." When asked

"How do gorillas feel when they die—happy, sad, afraid?" she signed "Sleep." Koko's reference to holes in the context of death has been consistent and is puzzling since no one has ever talked to her about burial, nor demonstrated the activity. That there may be an instinctive basis for this is indicated by an observation at the Woodland Park Zoo in Seattle, Washington. The gorillas came upon a dead crow in their new outdoor enclosure, and one dug a hole, flicked the crow in, and covered it with dirt (Hancocks, 1983). This is just one of many areas in which our knowledge of not only how gorillas live, but also how they think and feel is being expanded.

The research of the Gorilla Foundation has several direct implications for the cause of conservation. It is now possible to learn an unprecedented amount about the psychological and physical needs of the great apes. We have a new tool. We can ask them directly how they would like to be treated.

Jane Goodall has employed this new tool, using Koko as an informant. She wanted to confirm for native observers at the Gombe Stream Reserve in Tanzania that apes were more comfortable in the presence of humans crouching or sitting than with people who were standing. She asked us to question Koko about her preference. On several occasions we asked Koko, "Do you like people to stand up or sit down when they're with you?" Koko clearly and emphatically responded "down"—signing with both hands—to each inquiry with the exception of one, when she stressed her point even more by prostrating herself (Patterson, 1979a).

The Foundation's work may also increase our understanding of communication among gorillas in the wild. The first step in developing such an understanding, recognition of semantically significant gestures and sounds, becomes easier as we become more familiar with gorillas as communicators. Koko and Michael share a number of communicative gestures with wild gorillas. Two of these gestures, for example, communicate invitations to play. One involves crouching, bowing, and slapping the ground with the hands. The other, which seems to mean "hurry come play" is expressed by one open hand shaken vigorously. Heretofore, field researchers have not always recognized the significance of semantic gestures observed in wild gorillas, because they were unfamiliar with the animals' communicative habits. A study of communication among feral gorillas by an investigator familiar with Koko's and Michael's signs would be considerably enriched.

But perhaps our most interesting findings of the last 13 years relate to how astonishingly like us gorillas are: Self-aware, even self-conscious, they are capable of aesthetic appreciation, vanity, embarrassment, humor, pride, empathy, jealousy, and grief. As each new development occurs, new similarities are discovered: For instance, as far as we know, only humans and gorillas have a natural inclination to pet animals (Fig. 63.6). In the wild, Dian Fossey observed the gorilla Tuck petting a small antelope trembling in the brush (Fossey, 1983). Koko has adopted a number of pets, from her kittens to tiny tree frogs (Cohn, 1983).

It surprises most people that a gorilla could be so gentle with such tiny helpless creatures, but Koko is not the first, and this seems to be consistent with gorilla temperament. In the 1930s a gorilla named Toto was reported to have adopted a kitten (Hoyt, 1941). Humans are not excluded from Koko's gentleness. She is especially careful with infants and aged people.

It is possible that this highly sensitive nature has contributed to the gorilla's poor captive breeding record. Incompatibility of mates, a high level of male

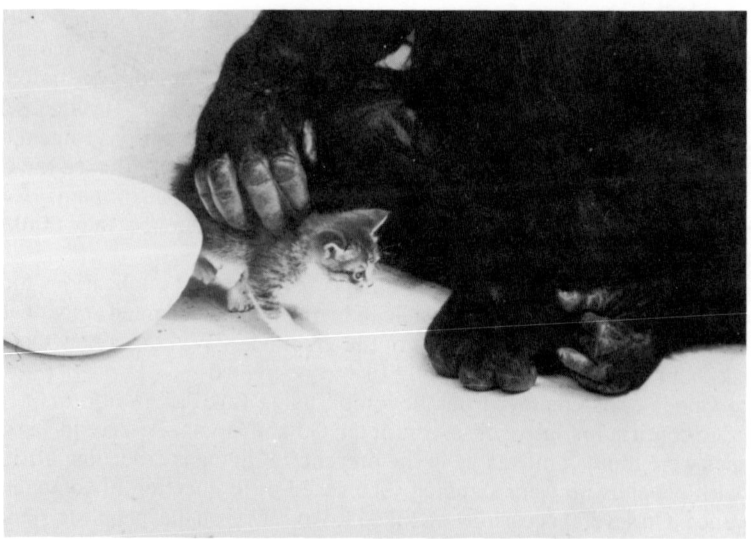

FIGURE 63.6. Koko pets her kitten All Ball.

sterility, and poor maternal skills are all facets of this problem—stress may be a factor in all, especially the latter two. Stress is known to figure in human infertility, and recent studies with chimpanzees at Ohio State University have shown that autonomic responses such as heart rate increase greatly when the apes are shown pictures of strangers (Berntson and Boysen, in press). It seem conceivable that constant exposure to a viewing public with strange faces could contribute to infertility through stress. In our experience, gorilla behavior is disrupted by the intrusion of strangers, even at a distance.

Maternal behavior can also be disrupted by procedures imposed in captive settings. Typically, gorilla mothers have been separated from the group at parturition time because zoo officials fear that the male may harm the infant. Separation is stressful, and in humans isolation can result in child abuse; a direct parallel has been found in captive gorillas. On the other hand, social pressure by curious group members actually helps the mother to give the infant proper care. Removing the valuable infants because of fear of incompetent mothering, without giving the mother a chance, is also an unfortunate practice that only serves to perpetuate the problem.

To summarize the material presented on Koko and Michael: With an emotional and expressive range far greater than previously believed, they have revealed a lively and sure awareness of themselves as individuals. Koko expressed it particularly well on one occasion when she was asked whether she was an animal or a person. Her signed response was "fine animal gorilla."

Koko and Michael are changing the way we view the world: they are forcing a reexamination of traditional thought about animals. This, in turn, should generate greater respect and concern not only for gorillas, but for other wildlife as well. Public sympathy and commitment are of immeasurable help in the politics of survival of all endangered species, especially in view of the limited resources that have so far been available to conservationists.

The Gorilla Foundation is currently seeking to establish the first gorilla preserve. This would be a private place, away from the public, spacious and natural. We are actively looking in Hawaii, where the climate closely approximates that of the gorilla's native habitat. Such a sanctuary would make it possible to learn more about how gorillas relate to their natural environment, further assisting efforts to promote the survival of the endangered gorilla in the wild.

References

Bernston G, Boysen S (In press) Cardiac correlates of individual recognition in the chimpanzee. J Comp Phys Psych.
Chukovsky K (1963) From two to five. University of California Press, Berkeley.
Cohn R (1983) Gorilla's pet. Gorilla 7(1):3.
Crail T (1981) Apetalk and whalespeak: the quest for interspecies communication. JP Tarcher, Los Angeles.
Fossey D (1983) Gorillas in the mist. Houghton Mifflin, Boston.
Hancocks D (1983) Gorillas go natural. Anim Kingd 86(1):10–16.
Hoyt M (1941) Toto and I. Lippincott, New York.
Klima ES, Bellugi U (1979) The signs of language. Harvard University Press, Cambridge, MA.
Ledbetter DH, Basen JA (1982) Failure to demonstrate self-recognition in gorillas. Amer J Primatol 2:307–310.
Livingston S (1983) Levels of development in the language of deaf children. Sign Lang Stud 40:193–285.
Maple T, Hoff MP (1982) Gorilla behavior. Van Nostrand Reinhold, New York, p 208.
Patterson F (1978a) Conversations with a gorilla. Natl Geogr 154(4):438–465.
Patterson FG (1978b) Linguistic capabilities of a young lowland gorilla. In: Peng FC (ed) Sign language and language acquisition in man and ape: new dimensions in comparative pedolinguistics. Westview Press, Boulder, CO, pp 161–201.
Patterson F (1979a) Talking gorillas as informants. Gorilla 2(2):1–2.
Patterson FG (1979b) Linguistic capabilities of a lowland gorilla. Thesis, Stanford University, California.
Patterson FG (1980) Innovative uses of language by a gorilla: a case study. In: Nelson K (ed) Children's language, vol 2. Gardner, New York, pp 497–561.
Patterson F (1984) Gorilla language acquisition. Natl Geogr Soc Res Rep 17:677–700.
Patterson F, Linden E (1981) The education of Koko. Holt, Rinehart and Winston, New York.
Patterson F, Peterson T (1984) Free word association by a gorilla. Gorilla 7(2):7.
Patterson F, Share E, Cornwall C (1985) Interspecies communication and conservation. Prim Conserv 5:39–40.
Suarez SD, Gallup GG (1981) Self-recognition in chimpanzees and orangutans but not gorillas. J Hum Evol 10:175–188.
Tanner J (1984) Koko: reading report. Gorilla 8(1):4–6.
Yerkes RM, Yerkes AW (1929) The great apes. Yale University Press, New Haven, CT.

64
Translocation of Primates

SHIRLEY C. STRUM and CHARLES H. SOUTHWICK

Introduction

Active primate conservation and management is becoming increasingly important to the long-term survival of populations and species as habitats are destroyed and animals exploited. The future of many species may depend on modifying current directions and exploring new survival possibilities. Translocation is an established but little used conservation and management technique for primates that may have great future potential.

Translocation involves capturing wild animals from one part of their natural distribution and moving them to another part. Several other techniques resemble or overlap with translocation, and it will be helpful to define these as they have been used in the literature (for a review see Konstant and Mittermeier, 1982). *Reintroduction* places individuals (either captive on wild caught) into habitat that is suitable but currently lacking that species. *Introduction* attempts something more daring—putting individuals into habitats where that species has never naturally occurred. Finally, *rehabilitation* strives to recondition individuals so that they can resume a "natural" existence.

Although translocation is a common technique in both the conservation and management of nonprimate species (Hamilton and King, 1969; Brambell, 1977; Campbell, 1980; Hanks et al., 1981; Hitchens, 1984; Hall-Martin and Hillman, in press), translocation of primates is rare. Konstant and Mittermeier (1982) concluded, after a careful review of the literature on introduction, reintroduction, and translocation of neotropical primates, that reintroduction and translocation are preferable to introductions and rehabilitation efforts. Unfortunately, there were no follow-up data in any of the cases of translocation that they examined.

This paper briefly reviews two recent primate translocations, baboons in Kenya and rhesus monkeys in India. The results suggest some general conclusions about the factors that contribute to successful primate translocations. The general merits of translocation as a conservation and management technique are also considered.

Gilgil Baboons

A population of baboons near Gilgil, Kenya, has been studied since 1970 and intensively studied since 1972 (e.g., Harding, 1976; Strum, 1975, 1981, 1982, 1983, 1984). The baboons live in a high-altitude savanna that has been used for ranching domestic stock since the 1920s (Blankenship and Qvortrop, 1974). In addition to cattle, sheep, and baboons, there has been a large complement of other wildlife in the area. In fact, the additional water and pasture provided for the cattle were responsible for the expansion of wildlife populations.

Changes in land use, from ranching to settled agriculture, began in 1977. Agriculture and the increase in the human population caused conflicts between baboons and people. Our research into techniques to control crop raiding by baboons provided some initial solutions (Strum, 1984, in press; Musau and Strum, 1984; Oyaro and Strum, 1984), and Debra Forthman-Quick's investigation of taste aversion among the baboons (in press a, in press b) demonstrated that this method could have great potential for primates. But initial solutions that excluded the baboons from the farming area became untenable as more and more land was converted to agriculture. Without a refuge area where baboons could range and feed on natural foods, crops became a nutritional necessity rather than a foraging bonus (Strum, in press).

The decision to move three focal troops was an attempt to save valuable research animals for future study while simultaneously exploring the feasibility of translocation as a management and conservation tool. In September 1984, the troops were moved from Gilgil to the Laikipia Plateau, a distance of about 120 miles.

By far the most difficult part of the translocation was finding a suitable site (O'Bryan, Strum, and Else, in preparation). We considered ecological factors such as availability of water, sleeping places, and the nature of the vegetation and terrain. But other factors played an equally important role: the distance from agriculture and human settlement and general accessibility.

Negotiations were complicated by the negative attitudes of locals toward baboons and toward scientists. Finally, after a year-long search, two ranches were selected.

We used standard baboon trapping procedures employed in Kenya since the mid-1950s. These were modified on the basis of our long-term understanding of the baboons. The team of people included Baboon Project staff and the Institute of Primate Research, Nairobi, who helped with the capture and took biomedical samples.

The baboons were habituated to rectangular metal traps placed within their respective home ranges. The traps were wired open and baited with food for 2 to 3 weeks. When all the animals readily entered the traps, they were set.

Each troop was captured in its entirety, except for one male who escaped and could not be recaptured. The baboons were sedated twice, first when removed from the traps and again for the collection of biomedical samples. They were, however, fully awake for the rest of the translocation, which included several days in individual cages at a holding area (for details on methods of capture, holding, and release see O'Bryan, Strum, and Else, in preparation).

Travel to the release site took one day. All the females and immatures were released that evening near water and good sleeping sites. Two methods were used to help prevent the troop from traveling very far. First, the subadult and adult males were held captive at the release site for several days. They thus became a focal point for the rest of the troop. It was unlikely that the females would go very far without the males. By the time the males were released, the females had carved out an area of familiarity within the otherwise alien terrain, and female conservatism could act as a check on male proclivities to wander further. Second, the troop was provisioned in one area for several weeks. This served to orient the troop twice a day, as well as supplement their diet.

The translocation of each troop had unique aspects. Two troops were released near each other and the third, 30 miles away. The new sites shared some characteristics with Gilgil, the baboon's original home: the same altitude, savanna-type ecosystem, rocky sleeping sites, and some of the same baboon foods, and all were located on ranchland. The sites differed from the baboons' original home in that there was less rainfall, higher ambient temperatures, a more arid-adapted flora, a larger number of wild predators, and a different array of other wildlife species. Water was available from rivers or dams rather than from troughs. At one site the vegetation was very dense while visibility was similar to Gilgil for the other two troops.

The baboons in all three groups responded extremely well to the capture, processing, transport, and release. There was no mortality and only one serious injury (an infant was injured in the holding area by a subadult male in an adjacent cage) among the 131 baboons captured.

The troops remained cohesive, and integrated social units from the start and operated much as they had before the translocation. Holding out the males seemed to work very well; the troops did not wander far from the release sites immediately following the translocation. The animals were provisioned for 8 weeks but even on the first day, the baboons foraged on local familiar and unfamiliar "natural" foods. Within 3 weeks from the day of release, provisioning was more of an aid to orient the monkeys than a dietary necessity.

During the next 3 months, the baboons explored the area around the release site, systematically following the distribution of baboon foods and returning to areas of abundant food. Within the first 3 months, the troops had clearly already made their initial adjustment to their new home.

Two troops suffered no mortality, but six immatures disappeared from the third troop. We assumed that these animals died of starvation during the 3 weeks following their release since local baboons were dying as a result of the drought and we were unable to provision this troop. Provisioning then went into effect until the rains came several weeks later and no further mortality occurred.

Local baboons reacted both aggressively and with curiosity to the translocated troops. Intertroop aggression was short-lived, and the majority of troop contacts were vigilant and peaceful. Two factors may have contributed to the easy integration of the translocated troops into the local populations. The drought may have left indigenous baboons little extra energy and few resources to protect against the intruders. In addition, the local baboons were not habituated to the observers. Our presence with the Gilgil baboons may have given the introduced troops an advantage. Yet the one troop that was not continuously followed by us also became integrated into its population. Within several weeks after translocation, a male had already transferred in from the indigenous population.

Six months after the release, all troops had established a home range that included the release area and adjacent areas containing seasonally available foods; the daily routine was usually predictable. More new foods were added to the diet. Occasionally the baboons would explore new areas either on their own or by following local troops. Nine months after their release, two of the troops appeared to have a large and relatively stable home range.

Reproduction was not disrupted by the translocation among more than 50 females. There was one miscarriage and one stillbirth, but other pregnant females gave birth to healthy offspring within the first 2 months, and female cycling appeared normal.

The only deaths in this period were two old adult males (one disappeared and was found later and one died from disease) and one yearling. Local males transferred into the translocated troops, and by the end of the 9-month period, translocated males began to emigrate into indigenous troops. Thus, male movement patterns appeared normal within 9 months.

Subgroup Translocation of Rhesus Monkeys in India

In September 1983, M. Farooq Siddiqi and Charles Southwick (Southwick et al., 1984) translocated a subgroup of 20 rhesus monkeys from a group of 130 individuals in Aligarh District, India, 30 km away to a canal bank habitat that formerly contained rhesus. Despite the serious decline of rhesus in India in recent decades, some areas have extremely large groups creating problems of agricultural damage and public nuisance. At the same time, relocating some of the animals, it was hoped, would conserve and help restore the total population of rhesus in India. This was viewed as a pilot study on the restoration of rhesus monkeys to suitable habitat.

The 20 rhesus, representing a cohesive subgroup of four adult males, six adult females, five infants, and five juveniles within the larger group, were selectively captured with nets at their sleeping site in a mango grove. The individuals were caught, transported without sedation, and released together on the same day in the new location. There were no outbreaks of aggressive behavior, no injuries, and no mortality as a result of the trapping or translocation.

The monkeys were provisioned several times daily at their new location with rice, grams (a soybean-type of legume), and mixed fruits and vegetables. Their new home differed from their previous location in that there were more natural foods, better cover, and a different array of human foods (bordering fields of wheat, sugar cane, and grams instead of a mango grove). The new location also had greater distances from human habitations and no roadside vehicle traffic. Pedestrians and domestic animals walked along the canal bank, but neither they nor the nearest villagers fed the monkeys in the first few weeks. The closest other rhesus were 6–7 km, outside the normal range of the translocated group, and there was no direct contact between them.

Initially, the group was unsettled and moved somewhat restlessly along the canal bank. Three juveniles dispersed and were gone for several months, but by February they had returned. The remaining monkeys explored the area on either side of the release site, covering 2–4 km per day, as if they were looking for something. They usually returned to the provisioning site, however, which was the original release point.

The group was provisioned for 5 months; this food supplied the majority of their daily needs. By the third month, however, the group moved 2 km from the original release point to settle next to a village and bridge crossing the canal. Here they subsisted on a combination of natural foods and handouts from the villagers, much as they had in their original location. They lodged in an open grove of banyan and jamun trees that also contained several old abandoned buildings associated with a former canal inspection house. This grove could probably have met the majority of their nutritional needs. If the villagers had not adopted the group, however, the monkeys would likely have raided nearby field crops and orchards.

As with the baboon study, the most difficult part of the rhesus translocation was finding a suitable habitat where local people would permit the release of the monkeys. Suitable rhesus habitats in ecological terms are very abundant in northern India, but most villagers and local people do not want monkeys nearby.

Once settled into a fixed range and stable behavior pattern, the group prospered. All six adult females gave birth to healthy infants during their first birth season (June and July 1984), so obviously the trapping and translocation the previous September had not disrupted the sequence or success of the normal mating season. Eight months after release, three other rhesus joined the translocated group. Within 8 months, the local villagers claimed that this group was now "their monkeys," brought to them by Hanuman, and they should not be disturbed.

A large-scale translocation of rhesus, for the purpose of establishing a breeding population within the United States, was undertaken by the FDA in the fall of 1979 when 1500 monkeys were moved from islands off the southwestern coast of Puerto Rico to Morgan Island, South Carolina (Vickers, this volume). The initial moves were made in September, and all were complete by December 1979. The move involved extensive behavioral study of the monkeys and ecological analysis of the habitats. Great care was taken to keep social groups intact throughout the periods of trapping, transportation, and release. The translocation went very well with no deaths, serious injuries or illnesses. No excessive aggression or behavioral disruptions occurred. Excellent reproduction occurred in the first season after translocation.

Discussion

The baboon and rhesus experiments demonstrate that translocation can be successful and can be a useful tool with some primates, at least. Combined with the small amount of data in the literature (Kawai, 1960; Morrison and Menzel, 1972; Angst, 1973; Bailey et al., 1974; Wilson and Elicker, 1976; Konstant and Mittermeier, 1982; Vickers, this volume), these results suggest factors important to the success of primate translocation.

1. *The composition and nature of the group*: Individuals or artificially created groups composed of strangers (whether or not these strangers have been formed into a social group before moving them) do not do as well after release as intact groups or subgroups. Translocated *individuals* usually disperse and are often repelled by members of the indigenous population. Artificial social groups usually disintegrate after release. Furthermore, Kawai (1960) found that the behavioral plasticity, so evident in natural groups of Japanese monkeys, was lacking in composite groups whose extreme conservatism prevented them from

adapting to new situations. By contrast, intact social units retained their integrity and adapted quickly and were also able to insinuate themselves into an existing population of conspecifics. Not only do these groups do better in the new environment, but the social continuity of the group apparently allows individuals to cope with the trauma of the translocation. Kawai (1960) and Morrison and Menzel (1972) conclude from their own experiments that the adaptability of primates depends on the existence of a stable, normal social group.

2. *The release site*: The availability of food, water, sleeping sites, and the presence of other species, both nonhuman and human, must be assessed carefully. Although the new location should be as good and as familiar as possible, the translocation of rhesus from Cayo Santiago to Desecheo in the 1960s (Morrison and Menzel, 1972) demonstrated that primates can survive even when placed in a habitat that is impoverished by their own standards. In the successful cases, extensive surveys were conducted by individuals familiar with the behavior, ecology, and biology of the species. At even the best release sites, provisioning was useful during the initial period of adjustment.

3. *Procedures*: Moving intact social units into appropriate habitats is still a risky business. The procedural issues are as important as the biological ones. Capturing, transporting, releasing, and monitoring are all fraught with difficulties. Translocation is a labor-intensive operation whose success depends on careful planning by people who understand the animals. Fortunately, some suggestions for future efforts are offered by the previously successful projects.

In summary, successful translocations have depended on moving intact social units to appropriate areas, although these need not have been part of the previous range of that species. When the moves were well planned and the procedures reflected an understanding of the species' behavior, ecology, and biology, the chances for success increased. Provisioning played a pivotal role in the initial adjustment of the translocated animals. Failures, by contrast, usually violated one or all of these provisos by using individuals or artificially created social groups and/or placing them into appropriate locations with no monitoring and no provisioning. These findings support the suggestions of Konstant and Mittermeier (1982) that translocations and reintroductions should use intact, socially cohesive groups.

Conclusions

There are a number of reasons why primates should be translocated. These range from managing pest populations to reintroducing species into newly protected parts of their range. The recent Gilgil baboon and Indian rhesus translocations have given added weight to earlier results supporting the feasibility of translocation (as cited above). But how realistic a tool is translocation? To answer that question, we must consider both the costs and the benefits of the technique.

A major cost is manpower. Translocation is a labor-intensive operation regardless of whether 20 or 131 animals are moved. Furthermore, trained personnel are required at every step. The behavior, ecology, and biology of the species must be at least partially understood for appropriate sites to be selected and successful procedures to be constructed.

The monetary considerations are difficult to delineate because of the extensive preliminary work required. In the Indian rhesus case, the actual translocation costs were about $150 per animal. The Gilgil baboons were more expensive to move, roughly $500 per individual, but this figure includes a year of preliminary work and a 6-month follow-up study (see O'Bryan, Strum, and Else, in preparation, for details).

Other considerations are more difficult to evaluate. In addition to trained personnel and money, any translocation project needs to fit into the political and economic realities of the countries where the primates reside. In both Kenya and India, while suitable habitat was available, public relations were critical to each stage of negotiations. In India the bureaucracy provided many obstacles and the local people had an ambivalent attitude toward the monkeys. They liked them for cultural–religious reasons but could not afford the economic losses caused by crop raiding. Yet, when the translocated group took up residence near a village, the villagers adopted them, both protecting and providing for the group. The attitude of the local people was critical to the ultimate success of this translocation.

In Kenya, although governmental agencies were easily convinced about the project, the attitude of locals toward baboons was unambiguously hostile. Baboons were seen as vermin rather than wildlife. Most potential release sites were usually eliminated because the local people, both black and white, were not receptive to the idea. Additional incentives were needed to secure the willingness of local landowners to house the project.

Thus, translocation requires trained personnel, knowledge of the animals, money, and a favorable attitude, or convincing incentive, for its success. On the other hand, primates show good survivorship, seem able to adapt with minor changes in behavior, are nonmigratory so their movements are easy to control after translocation, and the continuity of their social world seems to reduce the trauma of the move. The return on investment seems good. What is more, costs are always relative. Given two possibilities, moving the animals from one wild location to another or moving them from the wild into captivity, the translocation alternative may be the least expensive and most effective in the long run. Bringing animals into captivity incurs many of the same costs as translocation. They have to be captured, transported, and released. Preliminary research on behavior, ecology, and biology is as critical to a captive operation as to a translocation, if not moreso. But captive animals need to be housed and fed indefinitely, while translocated primates will be self-sufficient and self-sustaining soon after release.

Finally, if the purpose of the exercise is to conserve and preserve a species or population, or to rapidly expand an existing population, translocation seems preferable to captive propagation. It has the advantage of maintaining the animals in as "natural" a state as possible.

Translocation may, therefore, be an important future direction in primate management and conservation. The initial work suggests some basic guidelines:

1. Translocation should involve intact social units. Moving these groups into areas previously occupied by the species (i.e., reintroduction) increases the chances of success without jeopardizing any existing populations.
2. Groups should not be moved into national parks or other protected areas that already have a healthy resident population of that species.

3. Either declining or expanding populations are prime candidates for translocation.
4. The potential of introduction, rather than reintroductions, should be explored with populations of nonendangered primate species. It may be possible to augment existing suitable habitats to include areas where primates do not naturally occur.

Since the majority of primates live in forests in Third World countries, the current rate of forest destruction may make translocation appear unrealistic. Yet many instances can be cited where areas of tropical forests are belatedly safeguarded as parks (IUCN, 1982; McNeely and Miller, 1984). In these, some or all resident primate species may have disappeared while the same species may exist precariously in other parts of their range. Under these circumstances, translocation/reintroduction of remnant or vulnerable groups into safer areas seems preferable to the other alternatives: letting them die, trying to protect them in situ against insurmountable odds, or placing them into captivity.

In some cases, like the rhesus monkey, overexploitation has left large amounts of suitable habitat without primates. Attitudes are changing in the Third World both for economic and aesthetic reasons (Western, in press) creating opportunities where translocation can play an important role.

It seems to us that the weight of the evidence, no matter how scanty, is that translocation has been an underutilized tool in safeguarding the future of primates. It is a costly technique, but then conserving wild animals is an expensive undertaking and translocation may be among the least expensive of the alternatives, in the longer term.

Acknowledgments. The baboon translocation has been sponsored by the Office of the President, the Institute of Primate Research, Nairobi, and the National Museums of Kenya. Funding for the work was provided to Strum by the National Geographic Society and the University of California, San Diego, with additional logistic support from Ker and Downey Safaris, Kenya, and the Morjaria family. This paper benefited greatly from discussions with David Western.

The rhesus translocation was supported by USPHS grant RR-01245 to the University of Colorado as part of a broader study of rhesus populations. Field arrangements were handled by Dr. M. F. Siddiqi. We are grateful to many officials in the Government of India in New Delhi and the Government of Uttar Pradesh in Lucknow who granted permission and provided advice for this translocation project.

References

Angst W (1973) Pilot experiments to test group tolerance to a stranger in the wild *Macaca fascicularis*. Amer J Anthro 38:625–630.

Bailey R, Baker R, Brown D, vonHildebrand P, Mittermeier R, Spousel L, Wolf R (1974) Progress of a breeding project for nonhuman primates in Colombia. Nature 248: 453–455.

Blankenship L, Qvortrup S (1974) Resource management on a Kenya ranch. Ann S Afr Wildl Mang Assn 4:185–190.

Brambell MR (1977) Reintroduction. Intern Zoo Yrbk 17:112–116.

Campbell S (1980) Is reintroduction a realistic goal? In: Soulé ME, Wilson BA (eds) Conservation biology. Sinauer, Sunderland, MA, p 263.

Forthman-Quick D (In press a) Controlling primate pests: the feasibility of conditioned taste aversion.

Forthman-Quick D (In press b) The effects of consumption of human foods on the activity budgets of two troops of baboons, *Papio anubis*, at Gilgil, Kenya.

Hall-Martin A, Hillman K (In press) An assessment of translocation operations as a means of conserving African rhinoceros species.

Hamilton P, King JM (1969) The fate of the black rhinoceros released in Nairobi National Park. E Afr Wildl J 7:73–83.

Hanks J, Densham WD, Smuts GL, Jooste JF, Joubert SCJ, le Roux P, Milstein P (1981) Management of locally abundant mammals—the South African experience. In: Jewell PA, Holt S (eds) Problems in management of locally abundant wild mammals. Academic, New York, p 21.

Harding RS (1976) Ranging patterns of a troop of baboons (*Papio anubis*) in Kenya. Folia Primatol 25:143–185.

Hitchens PM (1984) Translocation of black rhinoceros from the Natal Game Reserves 1962–1983. Lammergeyer 33:45–48.

IUCN Commission on National Parks and Protected Areas (1982) United Nations list of national parks and protected areas. IUCN publication, Gland, Switzerland.

Kawai M (1960) A field experiment in the process of group formation in Japanese monkeys and the releasing of the group at Ohirayama. Primates 2:181–255.

Konstant W, Mittermeier R (1982) Introduction, reintroduction and translocation of neotropical primates: past experiences and future possibilities. Intern Zoo Yrbk 22:69–77.

McNeely JA, Miller K (1984) National parks, conservation and development: the role of protected areas in sustaining society. Smithsonian Institution Press, Washington, DC.

Morrison JA, Menzel EW (1972) Adaptation of free ranging rhesus monkey group to division and transplant. Wildl Monogr 31.

Musau JM, Strum SC (1984) Response of wild baboon troops to the incursion of agriculture at Gilgil, Kenya. Intern J Primatol 5:364.

Oyaro HO, Strum SC (1984) Shifts in foraging strategies as a response to the presence of agriculture in a troop of wild baboons at Gilgil, Kenya. Intern J Primatol 5:3:71.

Southwick CH, Siddiqi MF, Johnson R (1984) Subgroup relocation of rhesus monkeys in India as a conservation measure. Amer J Primatol 6:423.

Strum SC (1975) Primate predation: interim report on the development of a tradition in a troop of olive baboons. Science 187:755–757.

Strum SC (1981) Processes and products of change: baboon predatory behavior at Gilgil, Kenya. In: Teleki G, Harding R (eds) Omnivorous primates. Columbia University Press, New York, p 255.

Strum SC (1982) Agonistic dominance in male baboons: an alternative view. Intern J Primatol 3:175–202.

Strum SC (1983) Why males use infants: In: Taub D (ed) Primate paternalism. Van Nostrand Reinhold, New York, p 146.

Strum SC (1984) The Pumphouse Gang and the great crop raids. Anim Kingd 87:36–43.

Strum SC (In press) A role for long-term primate field research in source countries.

Western D (In press) The origins and development of conservation in East Africa.

Wilson M, Elicker J (1976) Establishment, maintenance and behavior of freeranging chimpanzees on Ossabaw Island, Georgia, USA. Primates 17:451–474.

65
Conservation Program for the Golden Lion Tamarin: Captive Research and Management, Ecological Studies, Educational Strategies, and Reintroduction

Devra G. Kleiman, Benjamin B. Beck, James M. Dietz, Lou Ann Dietz, Jonathan D. Ballou, and Adelmar F. Coimbra-Filho

Introduction

The future conservation of most threatened species will require not only the preservation and management of critical habitats but also scientifically managed propagation programs for captive animals by zoos. Zoos will undoubtedly have primary responsibility for the preservation and protection of genetic diversity through the maintenance of viable captive populations (or their deep-frozen equivalents). However, they should also have a role to play in supporting and contributing to the preservation of natural habitats through research and public education on environmental issues. Conservation programs by zoos, by international and national conservation organizations, and by governments should converge, as the size of critical habitats and refuges becomes smaller and the amount of land available to zoos and their involvement with endangered species becomes greater.

A successful conservation program involves a management plan and an educational strategy. However, basic research is essential for the development of scientifically based recommendations on preservation and management, whether the management plan is within a zoo or field context, and for educational programs. Traditionally, the fields of psychology, physiology, behavior, reproduction, and genetics have been emphasized in zoo research. The results of such research can lead to improved management of a captive population. Field research in similar disciplines as well as in ecology not only improves the preservation of a single species, but also often results in the protection of a portion of the natural habitat within a refuge or national park.

Basic research also provides the foundation upon which the educational elements of a conservation program must be built. Creating a public constituency sensitive to and supportive of conservation programs cannot be accomplished without the attractive presentation of research information that is appropriate for the target audience. There is an enormous need to improve the public image of conservation in developing and developed countries. Systematically developed education programs, based on the presentation of results from scientific research, are thus ultimately indispensable to successful long-term conservation.

Additionally, conservationists in developed countries must recognize that the support and conduct of future conservation programs ultimately needs to be in the hands of trained professionals from the less-developed countries. Thus, scientists from developed countries should include in their overseas budgets support for training in conservation biology to ensure that a cadre of professional wildlife managers and biologists can continue programs initiated from outside.

One major area where zoo conservation interacts with programs of maintaining and preserving natural habitats involves the use of reintroductions, translocations, and introductions of endangered species (see Konstant and Mittermeier, 1982, for definition of terms). These techniques are being developed and may be used to increase the chances of survival of a natural population, mainly when the wild population and its critical habitat are exceedingly vulnerable.

Konstant and Mittermeier (1982) have summarized the limited attempts to translocate, reintroduce, and introduce various primate species within South America. Many appear to have been unsuccessful, although the results have often been difficult to evaluate. One major flaw has been the paucity of long-term monitoring of individuals to determine the impact of the release on the local fauna and flora or the effectiveness of the preparation of the animals prior to the release (most are not prepared). Konstant and Mittermeier (1982) recommend the use of these techniques in only limited cases, especially because of the potential damage that the introduction of animals may have on an already established fauna and flora.

This paper summarizes a conservation program that includes reintroduction of captive-born animals into the wild, in the context of a broad strategy to save a species from extinction. In this case, the reintroduction created local interest in and support for habitat and species protection, fostered the development of techniques to prepare (rehabilitate) animals for a return to the wild, and repopulated areas devoid of the species, while paving the way for an increase in genetic diversity of the captive and wild population.

History of Golden Lion Tamarin Conservation Efforts

The golden lion tamarin was historically found in the coastal forests of the states of Rio de Janeiro and Espírito Santo, south of the Rio Doce. The original range is now reduced to remnant and scattered forests within the state of Rio de Janeiro. Available habitat is probably less than 2% of the original forest inhabited by golden lion tamarins (Hershkovitz, 1977; Coimbra-Filho and Mittermeier, 1977; Kleiman, 1981).

The golden lion tamarin's precarious condition has resulted primarily from the near total destruction of the Atlantic coastal rainforests of southeastern Brazil. Intensive logging, initially for the commercially valuable tree species and then for other human activities such as farming and cattle ranching, has contributed to the nearly complete deforestation of this region. More recently, forest destruction has intensified, with charcoal and wood being used as alternative fuels to petroleum both for home consumption and to fuel ceramics factories. Finally, a relatively new threat to the remaining forest has arisen from the construction of condominiums and weekend cottages near beaches with some of the only remaining golden lion tamarin habitat. These condominiums are being developed for the wealthier inhabitants of Rio de Janeiro, a city within a 2-hour drive of the major remaining golden lion tamarin population.

In addition to habitat destruction, golden lion tamarins have traditionally been captured for pets and for exhibition in zoos both within and outside of Brazil. It has been illegal to capture and/or export wild golden lion tamarins since the late 1960s, thus international trade in this species has declined significantly within the past 15 years. However, individuals still occasionally appear for sale in the markets of major cities such as Rio de Janeiro.

Adelmar Coimbra-Filho and his colleagues have been working within Brazil for nearly 20 years to ensure the survival of wild golden lion tamarins. These efforts have resulted in (1) the creation of the Poço das Antas Biological Reserve for *Leontopithecus rosalia* in 1974 and (2) the development of a breeding facility for endangered primates endemic to Brazil (Rio de Janeiro Primate Center, CPRJ-FEEMA)(Coimbra-Filho and Mittermeier, 1977).

International zoo involvement with the preservation of golden lion tamarins was initiated in the United States in the late 1960s when the Wild Animal Propagation Trust (WAPT—now disbanded) convened a conference entitled *Saving the Lion Marmoset* (Bridgwater, 1972). This conference reviewed the status of the species in captivity and in the wild and recommended research areas that needed immediate attention for the future reproduction and management of the species in captivity. Shortly thereafter, Marvin Jones (1973) published the first International Studbook for the golden lion tamarin, which contained information on natality, mortality, and animal pairings for this species in captivity. The existence of the Studbook permitted the development of an aggressive program of management for the captive population.

Following the 1972 WAPT Conference, the National Zoological Park, Smithsonian Institution, initiated long-term studies on the reproduction, social behavior, and husbandry of this species in captivity. The initiation of this captive research and management program occurred simultaneously with Coimbra-Filho's efforts to protect remaining habitats in Brazil and develop a captive breeding program for endangered Brazilian primates. These two efforts were the basis for the Golden Lion Tamarin Conservation Program as it exists today, with its five major elements.

Golden Lion Tamarin Conservation Program

The Captive Population: Cooperative Research and Management

When the National Zoological Park initiated a research program with golden lion tamarins in the early 1970s, it was estimated that there were fewer than 80 individuals of this species remaining in captivity. At that time all projections suggested that the captive population was destined for extinction since the mortality rate of the captive animals was exceeding natality (see Kleiman, 1977a; Kleiman and Jones, 1977; Kleiman et al., 1982).

The major goal of the NZP research program was to find methods to improve the poor reproductive performance and survivorship of animals in captivity. Thus, studies of reproduction, social behavior, and nutrition were initiated with the aim of applying the results of the studies to the captive management of the species. Simultaneously, the major biomedical problems were being defined and physiological norms established. The results of these studies have been summarized and published in a variety of periodicals (see Bush et al., 1980, 1982; Hoage, 1977, 1982; Kleiman, 1977b, 1981, 1983; Kleiman et al., in press; Kleiman et al., 1982; Montali et al., 1983) and have enabled us to manage better this captive population, resulting recently in a phenomenal growth in numbers.

Some of the major behavioral findings include confirmation that (1) golden lion tamarins breed best in captivity when maintained in monogamous pairs and nuclear family groups, and (2) monogamy appears to be maintained by both physiological and behavioral suppression of maturing offspring (Kleiman, 1978, 1979). Both parents and older juveniles participate in the rearing of infants within the family group, and juvenile experience with younger siblings appears to be important for later sexual and parental competence (Hoage, 1977, 1982). Among the behaviors exhibited by family groups that are critical to the successful development of young are carrying infants and the active sharing of food with the young during and after weaning, a behavioral attribute found in few mammals (Brown and Mack, 1978; Hoage, 1977, 1982).

Additionally, a major finding with a powerful impact on captive management has been that females, including related females, exhibit incredibly high levels of aggression toward each other (Kleiman, 1979). Deaths of several daughters (and one pregnant mother) have occurred while young females were still living with their families, prior to and during the time of puberty. Aggression toward daughters by mothers occurs even before sexual maturity, when the daughters are likely to be reproductively inactive.

Coimbra-Filho and colleagues have also contributed to our understanding of golden lion tamarin biology, with studies of taxonomy, reproduction, distribution, and ecology (Coimbra-Filho, 1969, 1977; Coimbra-Filho and Mittermeier, 1973, 1977, 1978; Coimbra-Filho and Maia, 1979a, 1979b; Rosenberger and Coimbra-Filho, 1984). Thse studies have provided considerable groundwork for the ecological research and the reintroduction.

In 1974, D. G. Kleiman assumed responsibility for the International Golden Lion Tamarin Studbook and began to manage the captive golden lion tamarin population at an international level. In the late 1970s as the captive population of golden lion tamarins began to grow, it first became possible to send pairs to institutions that had not previously been involved in the golden lion tamarin breeding program. Concurrently, an international Cooperative Research and Management Agreement (CRMA) was drawn up and signed by the owners and new holders of golden lion tamarins. The standards for golden lion tamarin management and maintenance were defined, and both owners and holders agreed to adhere to decisions on population management made by an elected management committee. The agreement also included a prohibition on permitting golden lion tamarins to enter commercial trade.

Major decisions by the management committee in recent years have involved providing animals for exhibit and breeding to institutions wishing to join the golden lion tamarin consortium and developing a long-term management plan to ensure the long-term viability of the captive population through demographic and genetic management. The latter has involved (1) the removal of a certain number of individuals from the breeding pool via vasectomy, hormone implants, and single-sexed groupings, (2) altering the genetic composition of the captive population by pairing animals to maximize the contribution from individual founders that are underrepresented in the captive gene pool, and (3) developing mechanisms for placing a ceiling on the total size of the captive population (see Kleiman et al., 1982; Ballou, 1985).

Recent studies in genetics and physiology have also affected the captive breeding program. In the late 1970s a diaphragmatic hernia was discovered in an animal born at the National Zoo (Bush et al., 1980). Careful screening of living and dead animals suggested that the diaphragmatic hernia was present in between 5 and 10% of the captive population. Although it initially appeared that the defect could have been the result of a single recessive gene, recent analyses of the pedigrees of affected individuals suggest that this problem is polygenic or not genetic at all. Moreover, attempts to breed for this trait have generally had negative results (Kleiman, Ballou, and Evans, unpublished observations). The diaphragmatic hernia has now been found in the majority of lineages of captive golden lion tamarins and thus is not a characteristic that selective breeding could easily remove from the population.

Since the average inbreeding coefficient for the captive population is not particularly high (Ballou, 1985), the diaphragmatic condition is unlikely to be the result of inbreeding. However, when inbreeding has occurred in golden lion tamarins, it has resulted in reduced viability of offspring, as is the case with numerous other primate species (Ralls and Ballou, 1982).

From a captive population of fewer than 80 outside of Brazil in 1971, there were approximately 370 animals by the end of 1983. Recently, the population has been increasing at a rate of 20–25% per year (Ballou, 1985). The major management efforts are currently being devoted to slowing population growth and balancing the contribution of the various founders to the population.

The Wild Population: Field Research on Population Size and Behavioral Ecology

By 1981 it was clear that the captive population was secure and that there were animals surplus to the needs of the breeding program. However, the wild population's status was still precarious. In 1974, the Poço das Antas Federal Biological Reserve in the state of Rio de Janeiro had been established, and a director (D. Pessamilio) was appointed in 1977. A survey in 1980 by K. M. Green (unpublished observations) suggested that although golden lion tamarins existed in the Reserve, there were major problems to be solved before protection and long-term conservation of the golden lion tamarin could be guaranteed.

Shortly thereafter, Kleiman initiated discussions with officials from IBDF (Instituto Brasileiro de Desenvolvimento Florestal), World Wildlife Fund–U.S., and with Adelmar Coimbra-Filho, Director of the Rio de Janeiro Primate Center (CPRJ-FEEMA), to determine the feasibility of developing a formal collaborative program for golden lion tamarin conservation involving species preservation in the wild and captivity, and possibly the release of captive-born golden lion tamarins into suitable natural habitat in Brazil.

It was clear from initial discussions that deforestation continued to erode the remaining natural habitat and that only tiny isolated and unprotected populations of tamarins existed outside the Poço das Antas Reserve. Green (unpublished observations) estimated that fewer than 100 individuals were likely to exist in the Reserve and emphasized the degraded nature of the habitat, its extreme vulnerability as a result of human development, and the need for rapid implementation of and support for a management plan to increase the carrying capacity of the Reserve and protect it from further degradation.

The small size and uneven habitat quality of Poço das Antas suggested that a viable natural population of golden lion tamarins could *not* be sustained in the long-term without remedial action. Thus, it became evident that we needed to develop methodologies for (1) genetic exchange between wild and captive populations through reintroduction, translocation, and pinpoint genetic intervention, (2) increasing carrying capacity of the Reserve through rehabilitating degraded tropical forest habitats, and (3) developing a conservation education strategy that would insure public support for the protection and expansion of remaining forest blocks outside the Reserve.

Although reintroduction of captive-born animals to the wild was considered an aim of our collaborative conservation program, the necessity for and feasibility of a release needed to be evaluated prior to any reintroduction plans. Release of captive-born animals required prior knowledge of (1) the behavioral ecology of the species, including home-range sizes, group size and composition, feeding and foraging patterns, habitat preferences, and group movements, (2) the status of the wild population within the Poço das Antas Reserve, and (3) availability of suitable habitat. Additionally, (4) assurance that the release of captives would in no way jeopardize the safety of the remaining natural population was needed.

Studies of the behavioral ecology and status of golden lion tamarins were initiated in the Poço das Antas Reserve in 1983 by James Dietz. The relevant data have been collected by Dietz and Brazilian students through a trap, mark, and release program, with selected tamarins being outfitted with radio collars in order to follow the movements of individuals and family groups. Table 65.1 presents a summary of the trapping and handling that has occurred in the field through June 1985.

Table 65.2 provides some information on the size and composition of free-living golden lion tamarin groups in the wild. For purposes of this table, adults were considered to be individuals weighing more than 500 g, juveniles and subadults weighed between 200 and 500 g, and infants weighed less than 200 g. There tends to be a sex ratio of 1:1, with no more than two adults of each sex per group. The trapping program has provided insufficient data to confirm whether or not wild golden lion tamarins are strictly monogamous (Kleiman, 1977b), but it does appear that only one litter is raised at a time, thus suggesting that only a single female breeds within a group.

TABLE 65.1. Captures and handling of wild and reintroduced golden lion tamarins, November 1983 to June 1985.

	Wild tamarins	Released tamarins	Total
Captures (trapped)	169	53	222
Individuals captured	101	28	129
Chemical immobilizations	153	47	200
Number of individuals outfitted with radio transmitters	21	12	33

TABLE 65.2. Sex and age composition of captured groups of golden lion tamarin, October 1983 to October 1984.

	Adults			Subadults and juveniles			Infants			Total			
Group	M[1]	F	?	M	F	?	M	F	?	M	F	?	Σ
PW	1	1	3						2	1	1	5	7
P20	2	1			1					2	2		4
RS	1	2		1	1					2	3		5
NW	2	2		2	1					4	3		7
AR	2	3		3						5	3		8
RN	2	3		1	2					3	5		8
OS	2	1	1	2						4	1	1	6
P10	2	3		1			2			4	4		8
VA	2	2		2			1	1		3	5		8
CO	2	2		3	2				2	5	4	2	11
Totals	18	20	4	12	10		3	1	4	33	31	8	72
Mean (n = 10)	1.8	2.0	0.4	1.2	1.0		0.3	0.1	0.4	3.3	3.1	0.8	7.2

[1] M, male; F, female.

Home-range size for individual groups is approximately 40 hectares within the Reserve, with home-range overlap of adjacent groups averaging about 10%. Golden lion tamarins are dependent on the use of holes in trees for shelter at night, and thus good tamarin habitat includes sufficient holes for denning (see Coimbra-Filho, 1977), especially since each tamarin group depends on several tree holes within its home-range (Dietz, unpublished observations). Tamarins also depend on bromeliads as a source of food with high protein content (e.g., small vertebrates and insects). Good tamarin habitat thus additionally includes forest with a profusion of bromeliads.

Systematic censuses of the wild population have resulted in estimates of the overall numbers and distribution of remaining free-living golden lion tamarins. Table 65.3 summarizes the most recent estimates of golden lion tamarin numbers in currently protected sites. There may be an additional 100 animals existing precariously in tiny isolated forest patches. These data clearly indicate that golden lion tamarins continue to be severely threatened in the wild. The 5000-hectare Poço das Antas Reserve with its slightly more than 100 individuals is the only long-term protected plot of land in Brazil for this species. All addi-

TABLE 65.3. The approximate numbers of golden lion tamarins remaining in the wild in protected habitat.

Available forest	Approximate size of area (ha)	Estimated no. groups	Estimated no. individuals[1]
Poço das Antas Reserve	5000	25	153
Campos Novos Naval Reserve (São Pedro de Aldeia)	800	12	90

[1] Based on mean group size of 7.2 individuals.

tional animals existing in the remaining southeastern coastal rainforests of Brazil are inhabiting vulnerable habitat that is being deforested at unprecedented rates.

Preliminary studies on the genetics of the captive and wild population, through the biochemical analysis of blood proteins, suggest that free-living golden lion tamarins may be less genetically diverse than the captive population (Forman et al., in press). These tentative findings suggest that inbreeding may be a major problem with remaining wild golden lion tamarins and that an exchange of genetic material is likely to be required in the future to maintain a vigorous and viable wild population.

The Habitat: Protection, Management, Preservation, and Restoration

The mechanisms whereby one can rapidly increase the carrying capacity of currently degraded habitat for golden lion tamarins while paving the way for a natural regrowth of the forest is of basic research interest as well as being of major conservation importance. The Poço das Antas Reserve is a patchwork of primary and secondary forest and pasture land (Ferreira et al., 1981). Estimates by Green (unpublished observations) of the amount of forest cover suggest that less than 40% of the entire 5000-hectare reserve is forested, with perhaps 10% in undisturbed climax forest.

At the beginning of the field studies project we initiated a major protection and restoration effort for the Poço das Antas Reserve. Much of the support for the Reserve's rehabilitation has come from contributions by zoos that hold and own golden lion tamarins outside of Brazil, from IBDF, and from the World Wildlife Fund–U.S.

Major problems that were addressed and remedied between 1981 and 1983 included (1) making and placing signs indicating the existence of the Reserve and its protected status; (2) hiring and training guards to patrol and enforce antihunting regulations within the Reserve; (3) providing additional vehicle fuel to enable the Reserve's guards to get to and from work and to monitor the Reserve's boundaries; and (4) eliminating most squatters involved in slash-and-burn agriculture and periodic hunting.

More recently, we constructed fire breaks to improve fire control since the Poço das Antas Reserve has been exposed to continued degradation through fires that annually burn nearly 20% of its area.

Long-term management and restoration of this Reserve must involve the development of innovative techniques for the rehabilitation of tropical forests. One of our first efforts was to distribute lime over a 500-hectare area to elevate the pH of the soil in degraded areas and to encourage natural regeneration of woody species where grasses have been dominant for many years. Also, several experimental plots have been established to determine which species will regenerate most rapidly given different soil types. From these efforts we hope to improve our understanding of how rapidly a tropical forest with structural complexity and species diversity can regenerate through several treatments, including selective plantings (Dietz and Pessamilio, unpublished observations).

Despite these major gains, there are several remaining problems within the Poço das Antas Reserve that retard the creation of additional habitat for golden lion tamarins. For example, a railroad track still bisects the Reserve and trains pass through several times a day. Moreover, a dam has recently been completed which, when operational, may flood 15–20% of the Reserve. Additionally, there is a 2000-hectare plot of land adjacent to the Reserve that is owned by the Government of Brazil and may be turned over to the Department of Agriculture for development. Any intensive farming operation bordering the Poço das Antas Reserve would further jeopardize the remaining habitat through the additional human impact and application of insecticides. We have encouraged annexation of this block of land to the Reserve by IBDF, with limited success to date.

Finally, the Poço das Antas Reserve may be too small to maintain a genetically viable population of golden lion tamarins in perpetuity. Thus, we need to identify and protect additional blocks of forest that are already inhabited by golden lion tamarins or that can support the species. We are currently encouraging the protection of any significant blocks of forested land by public and private owners.

Conservation Education

The long-term preservation of species or habitats cannot be accomplished without public and professional education at every level, which results in community support for conservation. One of our major educational goals has involved the training of students and young professionals in up-to-date techniques of wildlife management and conservation biology. As biologists and educators from a developed country, we must encourage future environmental protection and conservation of uniquely Brazilian ecosystems through the identification, involvement, and support of as many interested Brazilian students as are available and logistically possible. Essential to our whole program is the training of a cadre of Brazilian wildlife ecologists, behaviorists, and conservation educators to maintain the long-term continuity of this and future conservation programs.

Second, L. A. Dietz is coordinating our public education programs and educational research efforts. We are attempting to obtain local, national, and international attention to foster increased public interest in and support generally for environmental protection and conservation. Finally, we hope to serve as a catalyst for other much-needed conservation education programs in Brazil by contributing to the development of education as a conservation tool. Each step of the program is being documented and its effectiveness evaluated.

The golden lion tamarin ecological studies in the Poço das Antas Reserve have provided the information necessary to define the major threats to the survival of the golden lion tamarin and thus establish specific objectives for the education program:

1. To reduce deforestation in the lowland areas around the Reserve
2. To assure the permanent conservation of at least a part of the privately owned forests in the area
3. To reduce fires in forests and cleared areas in the region.
4. To reduce commerce of golden lion tamarins
5. To reduce hunting within the limits of the Reserve

Because changing human opinions and behaviors is a slow process, this education program must be ongoing. Thus, we have encouraged major community involvement in the planning and implementation of the program. We have found that the local community is more interested in implementation if it has invested in the development of a project.

Our efforts have been concentrated in the three municipalities surrounding the Poço das Antas Reserve. The area (2292 km²) has a total population of 89,000 inhabitants (1980 census) with a relatively low density of 39 inhabitants per square kilometer. The principal economic activities are agriculture and fishing. We have worked with the entire population, more specifically the Reserve staff, local authorities, teachers, students, landowners, and other adult residents.

The education level of this target population is relatively low for the state of Rio de Janeiro. Forty-one percent of the population have had no formal instruction (1980 census), and only a fourth of the population over 10 years old have completed fourth grade. The large landowners (over 50 hectares) have a higher level of education and generally live in metropolitan areas outside the three target municipalities.

In January and February 1984 with the help of 30 local volunteers, we conducted 519 interviews with a sample of the population of the rural and urban communities of the municipality of Silva Jardim (in which the Reserve is located). We collected information concerning knowledge, attitudes, and behaviors regarding the golden lion tamarin, the Reserve, and the conservation of local flora and fauna that helped us to plan appropriate educational strategies and materials. The initial results will be compared with similar interviews conducted in 1986, from which we can evaluate the effectiveness of our educational activities and the conservation education program in general.

The following are responses to some questions from the first round of interviews:

1. 92% listen to radio and 76% watch television although much of this rural area has no electricity. The use of radio and TV to transmit our conservation program, therefore, became a priority.
2. 58% recognized the golden lion tamarin from a photo, but they knew very little regarding the animal's habits, or that it is endangered and close to extinction.
3. In answer to the question "What is the major problem in your municipality?" the majority of those interviewed responded "deforestation" (28%) while other common responses were "roads" (12%), a local dam (10%), and "unemployment" (7%).
4. 94% of respondees indicated that they had never been harmed or had had property damaged by wild animals.
5. Of the landowners interviewed, 74% responded that they wanted to "leave alone" or protect the wildlife on their properties.
6. 76% did not know if lion tamarins were beneficial to humans. The benefits that were mentioned included "cure of diseases," "beauty," or "happiness."
7. To the question "Does the forest provide you any benefits?" 88% responded "yes." The benefits most often mentioned were "pure air," "firewood," "wood," and "watershed protection."

In general, we found that we did not have to confront negative attitudes about the golden lion tamarin or the forest. Instead, we needed to increase the level of knowledge and global thinking, principally concerning the long-term consequences of human actions on the local environment and the interrelations of humans, fauna, habitat, and the well being of all.

In the communities we first talked with local authorities and invited them to visit the Reserve where we explained our goals, the problems we were facing, and the help we needed. They have been fully supportive. These first contacts have also led to the formation of a group of young people interested in conservation whose ideas and interests largely contributed to the educational materials we have produced. We tested prototypes of all materials with members of the target population before final production and use.

Materials we have produced include:

1. *Press releases*: An informed press can help tremendously in communicating to the public the necessity for conservation and environmental protection.
2. *30-second public service messages for television and radio*: The Brazilian Roberto Marinho Foundation helpfully donated air time on national and local stations.
3. *Video copies of the television news reports and public service messages*: These and other films of Brazilian flora and fauna were provided for local showings.
4. *A pamphlet that serves as the cover for a school notebook*: It contains a story about the lion tamarin and the forest, the problem of deforestation, and some suggestions for solutions, and is distributed in the schools.
5. *A factual pamphlet*: Current scientific information about the golden lion tamarin, the Poço das Antas Biological Reserve, and the conservation of the Atlantic Coastal Forest has been prepared and is distributed to students, teachers, the press, and others interested in the subject.
6. *An educational poster*: The Reserve guards and local students distribute the poster to public places in the region.
7. *Short slide-tape programs*: These are presented in various locations to different target audiences.
8. *A package of materials "Profit While Conserving Nature"*: This explains to landowners the benefits (financial and otherwise) of establishing "Private Wildlife Refuges" on their properties.
9. *A golden lion tamarin logo*: This logo (prepared by the National Zoo's Office of Graphics and Exhibits) is used on all project materials and is now a symbol for the Poço das Antas Reserve.
10. *T-shirts, buttons, and stickers*: These are presented as recognition for those who contribute to the conservation of the lion tamarin and the forests, as prizes for school activities related to conservation, and for sale by local conservation groups to finance their activities.
11. *Other existing materials*: Posters, pamphlets, and books produced by the Brazilian Foundation for Nature Conservation (FBCN) and by the World Wildlife Fund are also used, as well as the children's book "Artes e Manhas do Mico-Leão" by Yves Hublet.

The activities we have employed in the education program have included:

1. *Lectures* for local authorities, conservation groups, farmers, high school and university students, and other interested people.
2. *Training courses* for Reserve guards and local teachers.
3. *Promotion of the formation of conservation groups*.

4. *Press events*: We have capitalized on the tremendous interest in this project shown by the Brazilian press.
5. *A traveling exhibit* that has now been presented in six communities to approximately 4500 people.
6. *A children's play* based on the book "Artes e Manhas do Mico-Leão" presented in three municipalities to a total estimated audience of 1500. All the actors are young local residents.
7. *Special classes on conservation of local wildlife in local schools*: With the help of a team of local young people we have given over 150 presentations reaching all the students of our target population (approximately 7000).
8. *Educational field trips to the Reserve*: The guards are Reserve guards and members of local conservation groups. Participants have an opportunity to see the reintroduced tamarins free in the forest and to learn about the forest by walking along a nature trail planned to stimulate interest and observation.
9. *School essay contests* on conservation themes.
10. *A parade* with the theme "The Natural Resources of Our Municipality" organized by local teachers in Silva Jardim in 1984 to commemorate the anniversary of the municipality.
11. *Visits to local landowners* who still have forested land appropriate for lion tamarins.

In all these activities, we attempt to present the golden lion tamarin as a symbol for the conservation of its habitat. Conserving an area of forest for a golden lion tamarin, we will at the same time conserve nearly all of the elements of that ecosystem.

The final results from our education efforts are not yet known, but a large part of the target population now clearly recognizes the golden lion tamarin and knows that it is endangered and why. They also are more aware of the existence of the Poço das Antas Reserve. Many landowners are interested in establishing private wildlife reserves. Also, we currently have more requests for internships, lectures, and educational materials than we can fill, and there has been a dramatic increase in the number of golden lion tamarins donated to the Reserve or reported to IBDF as being held illegally in captivity. The challenge now is to make conservation not just something currently in vogue, but a subject of continuing community action.

We hope that the ongoing educational research in Brazil, including the evaluation of different conservation education media in terms of their effect on the attitude of the local population, will lead to the development of a model for conservation education programs using the most efficient and effective techniques.

Reintroduction

Results from 1984

The chronology of the 1984 release (through 30 June 1985) is presented in Table 65.8. Fifteen captive-born animals were sent to Brazil in November 1983, of which four died during a 6-month quarantine and preparation period in the Rio de Janeiro Primate Center (CPRJ). One litter of two was born during quarantine, and a wild-born animal was added to the group being prepared for release.

During May and June 1984, the animals were moved from CPRJ to the Poço das Antas Reserve. They were held there in large cages at the release site for 12 to 29 days to acclimate to the local environment. Between May and July 1984, 14 animals were released into the wild, including one family group of eight animals and three adult pairs. The release site contained no lion tamarins although the forest appeared suitable. By 30 June 1985, 11 of the released animals had died or had been removed (rescued).

Causes of death or debilitation included predation, exposure, disease, starvation, snakebite, social conflict, and disappearance. Most losses occurred shortly after release. The major cause of mortality was an apparent disease that affected a family group in February 1985 and resulted in the death of five animals. The symptoms included severe diarrhea, lethargy, and dehydration and may well have been caused by a virus. However, during this same period one female produced a single offspring and another had twins. The twins currently survive.

Starvation could be identified as contributing to a death in only one case. Released animals were provisioned with gradually decreasing amounts of banana, orange, and ground meat. Food provisioning was discontinued in March 1985.

Released animals have been monitored almost daily for a full year to document their adaptation to a wild existence. The long-term regular monitoring of individuals is unique for a primate release study.

Preparation

The preparation phase has involved differential training in various aspects of foraging and feeding as well as in locomotion. Additionally, in order to choose the best reintroduction candidates in the future we have studied social behavior and activity to develop personality profiles for individual animals that can later be correlated with differential survivorship of the released tamarins.

Before the first release, we quarantined animals at the Rio de Janeiro Primate Center for 6 months because the potential for disease transmission, from the captives to the wild population (or the reverse), was a major concern. This also enabled us to observe the process of adaptation as our animals, captive-born in the Northern Hemisphere, responded to the changes in climate, light, and diet within Brazil.

Feeding

We reasoned that the most serious deficit of captive-born tamarins would be in locating and harvesting natural foods. Captives are typically fed finely cut produce, diced marmoset diet, and immobilized prey all in a bowl at a predictable place in their cage at a predictable time. Wild tamarins eat whole fruits and invertebrates and small vertebrates that are cryptic and capable of rapid escape or self-defense. These food items are distributed widely and quite unpredictably in space and time in the forest. Prey is frequently embedded (Parker and Gibson, 1977) within crevices, rotten wood, rolled leaves, and bromeliads. We used a training protocol designed to replace gradually the expectation of finding cut foods in a traditional place with the tendency to search for food that is spatially distributed and hidden. Once the animals began to forage for hidden food, we presented likely sites that were empty, thus countering any expectation that all likely sites would contain a pay-off.

TABLE 65.4. Protocol for training golden lion tamarins to search for food that is distributed in space and time and hidden.

Condition	Traditional	Distributed	Embedded	Empty embedded	Total sites (with food)
1	1	—	—	—	1
2	1	1	—	—	2
3	1	1	1	—	3
4	1	2	1	—	4
5	1	2	2	—	5
6	1	1	2	1	5(4)
7	—	1	2	1	4(3)

Table 65.4 shows this protocol. "Traditional" is a bowl of cut food at the established site. "Distributed" is a similar bowl at a randomly selected site in the cage. An "Embedded" site is one of a large set of pseudo-naturalistic "puzzle-boxes" that contain the same types and amount of foods as a traditional and distributed site, but the food is not visible. Embedded sites were also placed randomly in the cage.

Each condition in the protocol was presented in the morning and afternoon for 5 days. We scored the animals' use of the feeding sites for the first hour after the morning and afternoon feeding. We noted every time an animal visited a site (defined as touching the container or food) and took food from the site. We also noted which animals first extracted food from an embedded site. This paper documents the performance of the family group from Kings Island, Ohio, on the first six conditions of the protocol.

As shown in Table 65.5, under Condition 1 with just a bowl at a traditional site, each animal in the group makes about 8 "Visits" to the bowl and "Takes" 10.5 pieces of food in the hour following feeding by the keeper. By adding a distributed site (another bowl at a random place), the numbers of "Visits" and "Takes" increase slightly. Under Condition 3, where we add an embedded site, the number of "Visits" nearly doubles while the number of "Takes" rises only slightly. Adding a second distributed site (Condition 4) makes little difference, but a second embedded site (Condition 5) is accompanied by substantial increases in "Visits" and "Takes." Finally, in Condition 6, we drop a distributed bowl and add an empty site; "Visits" increase even further and "Takes" decrease. Note that the total amount of available food is the same under all conditions: It simply becomes more distributed and more embedded. By adding extra feeding sites, especially

TABLE 65.5. Feeding behavior of a family group of golden lion tamarins on six conditions of the training protocol.

Condition	$\overline{X}$ Visits/hr	$\overline{X}$ Takes/hr	Takes/visits
1 (T)	8.1	10.5	1.30
2 (T+D)	9.9	11.5	1.16
3 (T+D+E)	19.2	12.0	0.62
4 (T+D+D+E)	19.0	13.2	0.69
5 (T+D+D+E+E)	20.9	15.7	0.75
6 (T+D+E+E+EE)	22.7	14.0	0.61

embedded sites, the number of "Visits," a measure of *foraging effort*, increases dramatically.

While the number of "Takes" increases somewhat with the number of feeding sites, the ratio of "Takes" per Visit, a measure of *foraging efficiency*, drops by more than half by Condition 6. The animals are not only expending more foraging energy but are receiving a smaller relative pay-off. This would appear to be suitable preparation for life in the wild where foraging efficiency is much lower than in captivity. Parenthetically, the use of distributed and embedded food sites is an effective means to increase the activity of primates and other omnivorous scavenge-hunters (Hamilton, 1973) in captivity.

We looked at which animals actually opened and ate first from the embedded sites. The alpha pair of the family opened only 8% (6 of 80) of the embedded sites, while the six offspring opened 89% (71 of 80; three went unopened). There was a nonsignificant correlation of +0.47 between individual ages and frequency of opening embedded sites by each member of the family (Spearman Rank Correlation, $n = 8$, $P > 0.05$), but it was clear that adults were much less apt to open embedded sites.

The frequency of opening embedded sites before release is strikingly correlated (+0.86, Spearman Rank Correlation, $n = 6$, $P < 0.05$) with time to death or rescue after release: Animals that were more adept at opening embedded sites survived longer.

Once the tamarins had completed our feeding protocol, we introduced whole fruits. We were unable to get natural fruits in sufficient quantity, but we chose cultivated analogs when available, and we chose cultigens such as banana and papaya that are found on remnant plantations in the Reserve. We also provided quail eggs in man-made "nests."

Table 65.6 shows the performance of the adult pairs pooled and of the family group in exploiting these foods. The family group was more successful. One might expect that a group of eight is more likely to discover accidentally and exploit such foods, but the family was housed in twice the space as each adult pair. Further, we biased the results in favor of the faltering adults by hiding the fruits and eggs less well and by cutting small windows in the rinds and shells. We even cracked the eggs and added jelly to get some adults to eat eggs. This exemplifies the dilemma of having to choose between rigorous science and preparation for survival that we encountered frequently in this program. In this case, however, the difference between adult pairs and an age-graded family is still evident.

TABLE 65.6. Success of a family group and three adult pairs of golden lion tamarins in exploiting whole fruits and quail eggs.

Adult pairs		
Fruits:	66.7%	(360/540)
Eggs:	63.4%	(104/164)
Combined:	65.9%	(464/704)
Family groups		
Fruits:	90.3%	(140/155)
Eggs:	71.1%	(32/45)
Combined:	86.0%	(172/200)

As additional preparation, we provided as many natural food items and forest features as we could. Our ongoing field study on wild tamarins indicated that bromeliads are important sources of food and water. The captives quickly learned to hunt in and drink from bromeliads and to explore thoroughly hollow logs and decaying wood. Insects (especially orthopterans), frogs, and lizards were taken readily.

Avoiding Danger

The captives displayed so little selectivity that we wondered how they would avoid dangerous animals and noxious foods. They learned quickly not to seize bees and biting ants and never sampled mushrooms that grew profusely in their cages. Thorns were negotiated without difficulty. Large over-flying birds and birdlike silhouettes caused our captives to give alarm chirps and escape to the tree core and lower strata: We suspect that this response is genetically "hard-wired." We saw two opportunistic encounters with wild nonvenomous snakes: One of the snakes was eaten and the other successfully defended itself against five mobbing tamarins and escaped. We think that captive tamarins will try to eat small to medium-sized reptiles and amphibians and may not discriminate between dangerous and nondangerous species. One of the reintroductees was killed following an apparent encounter with a snake; we did not see the bite, but it occurred during a tamarin intergroup encounter when it was unlikely that the victim was foraging.

We presented a large toad (*Bufo marinus*) to the family group. They approached cautiously at first, but within 2.5 minutes two of the animals had bitten the toad squarely on the parotid sacs. We literally had to wrest the toad from yet a third tamarin. One animal went into convulsions and barely recovered after 4 hours; another was ataxic for 2 hours. Both frothed, cried, and vomited while the others watched closely. Nonetheless, they seemed eager to get to the toad on the next day. We placed the toad in a jar and for 30 minutes scored the number of times each tamarin "Visited" the jar and the number of 1-minute intervals each animal was at the jar. After 30 minutes, we presented a group of lively grasshoppers, a highly favorite food, in the same jar for 30 minutes and used the same scoring

TABLE 65.7. Response of a family group of golden lion tamarins to a toad and a favored food (grasshoppers) 24 hours after two family members (asterisks) had become seriously ill from biting the toad.

	Toad		Grasshoppers	
Animal no.	Visits	Intervals	Visits	Intervals
1	1	2	1	2
5	17	18	22	26
6*	3	6	4	4
7*	6	5	3	10
8	9	15	6	10
9	9	16	15	18
M	4	8	3	6
R	0	0	0	0
Total	49	70	54	76

TABLE 65.8. Golden lion tamarin reintroduction: chronology of events between November 1983 and June 1985.

Event	Date	Deaths (or removals)	Births	Wild born added	Total alive
Captive-borns to Brazil	XI 83				15
During quarantine in Rio (CPRJ)	to VII 84	4	2	1	14
After release	to VI 85	11 (incl. inf.)	3	1	7

Causes of death (or removal)

Predator	Exposure/ starvation	Snake	Social conflict	Disappear	Medical
1	1	1	1 (remove)	2	5 (viral?)

system. After a break we again presented the jar full of grasshoppers for 30 minutes and then the toad for 30 minutes. Table 65.7 shows each individual's total number of "Visits" to and intervals at the jar for 60 minutes with the toad and 60 minutes with grasshoppers. The affected family members and the observers showed as much interest in the fateful toad as they did to the tasty grasshoppers. Of course, the experiment is flawed since the jar decreased olfactory cues, and olfaction may be more crucial in identifying danger than in finding insects. However, it appears that neither the affected nor the observing animals learned much from this near-fatal encounter. We must entertain the nonadaptionist hypothesis that captive and wild tamarins learn little about reptiles and amphibians: They try to eat the small ones, and if they are unlucky they may die. The area of recognition and avoidance of danger is one where we know little and where we are methodologically handicapped because of wanting to avoid potentially fatal experiments.

Locomotion

During preparation, we induced the tamarins to jump, climb, and hang to develop locomotor skills. However, after bringing the tamarins to the "halfway house" enclosures in the forest, we were struck by their reluctance to use natural vegetation of various textures, diameter, and flexibility. Older animals, particularly, preferred to stay on the wood frame and wire mesh of the cage. We had to dismantle the cages soon after the release to force some tamarins to enter and use trees. Once they moved into natural vegetation, another deficit became obvious: the animals were unable to plot a cognitive route through the forest between themselves and an incentive. Their movements were characterized by false starts, fruitless retracing of pathways to dead ends, and, finally, descent to and travel across the ground. At best, travel by the reintroductees was slow and hesitant. At worst they got disoriented and lost. Some simply sat, appearing to give up, and had to be rescued. Two perished on the ground, one taken by a feral hunting dog and one, as noted above, likely killed by a snake.

Our preparation program for 1985 has been amended to include profuse natural cage furnishings, especially small diameter branches and vines. Further, the entire network is knocked down and reassembled twice each week to force the animals to move through a novel three-dimensional array as opposed to using familiar routes.

Age

The poorer performance of adults as compared to younger animals in exploiting embedded food sites and whole fruits and eggs, and the adults' apparently greater deficits in locomotor ability, suggested that we look at the relationship between age at reintroduction and survivorship. Using days to death or first rescue as the measure of survivorship, this correlation is -0.58 for the 13 captive-born reintroductees. If we use days to actual death (ignoring rescue), it is -0.56. Both negative correlations are statistically significant (Spearman Rank Correlation, $n = 13$, $P < 0.05$). Three of four captive-born adults released as pairs were dead within only 17 days of release. We interpret these results to mean that young tamarins are more likely to have the vitality and behavioral flexibility to survive the dramatic environmental changes between life in a zoo cage and life in the Brazilian forest.

Future reintroductions will include age-graded family groups and no adult pairs. Although we believe that our preparation program does confer a postintroduction survival advantage, we will test this belief in 1985 by releasing a group with no formal training. We will rescue and train them if necessary, but this is another systematic step in forging a true cost-effective science of reintroduction which, hopefully, will be applicable to other primates and indeed to a wide variety of vertebrates.

Summary and Conclusions

Of an original 15 captive-born animals transported to Brazil and released into the wild, there are currently three individuals still alive. One pair has successfully reproduced and reared offspring; in our minds successful reproduction constitutes successful reintroduction. We not only have begun repopulating suitable but empty tamarin habitat but also have infused new genetic material into the Poço das Antas Reserve.

Although the reintroduction of captive-born animals was a major goal of the golden lion tamarin program in 1984, the methodology that we are developing for reintroduction can eventually be used to translocate wild individuals into established home-ranges and saturated habitats at a time of our choosing. This will eventually lead to a more outbred population.

It is important to state that the long-term conservation of golden lion tamarins was not necessarily dependent on the release of captive-born animals. If support had been available since the species was identified as endangered for the other elements of the program, i.e., habitat protection and restoration, public and professional education, field research and censusing of the wild population, and a strong management program for both the captive and wild population, the reintroduction aspect of this program might not have been pertinent. As it was, the reintroduction of captive-born animals to the wild acted as a springboard for the entire conservation program and as a result, the ultimate chances for survival of the golden lion tamarin have been dramatically improved.

Acknowledgments. This research program has profited tremendously from the support of numerous individuals and institutions, over nearly a 15-year time span. The following individuals could well be considered co-authors for each of the separate elements of the program: (1) captive research and management: Ron Evans, David Mack, Lynn Rathbun, Robert Hoage, Ken Green, Melissa Ditton, G. Maliniak; (2) field studies: Laurenz Pinder, Carlos Peres; (3) habitat restoration: Dionizio Pessamilio; (4) education: Elizabeth Nagagata; (5) reintroduction: Inês Castro, Beate Rettberg, Vera Cruz, Rosa de Sá.

Additionally, J. Block, M. Bush, L. Phillips, and R. Montali of the NZP staff have been very helpful, as has staff of the Office of Graphics and exhibits.

Supporting institutions include: Smithsonian Institution International Environmental Sciences Program and Educational Outreach Program; World Wildlife Fund–U.S.; National Geographic Society; Friends of the National Zoo; Wildlife Preservation Trust International; Frankfurt Zoological Society; National Institutes of Mental Health (Grant #27 241); Roberto Marinho Foundation; Instituto Brasileiro de Desenvolvimento Florestal, and Fundação Brasileira para a Conservação da Natureza and a host of zoos and zoological societies that have generously given of their time and money.

In Brazil, we are grateful for the assistance of A. Pissinatti, R. da Rocha e Silva, and C. Padua Valladares (CPRJ-FEEMM) and the observers and educators of Silva Jardim.

Finally, the following individuals have provided generous support, inspiration, and thoughtful input since the beginnings of the golden lion tamarin research: T. H. Reed, John F. Eisenberg, R. A. Mittermeier, Jeremy Mallinson, George Rabb, and Warren Thomas.

References

Ballou JD (1983, 1985) International Studbook for the Golden Lion Tamarin *Leontopithecus rosalia*. National Zoological Park, Smithsonian Institution, Washington, DC.

Bridgwater DD (ed) (1972) Saving the lion marmoset. Wild Animal Propagation Trust, Wheeling, WV.

Brown K, Mack DS (1978) Food sharing among captive *Leontopithecus rosalia*. Folia Primatol 29:268–290.

Bush RM, Montali RJ, Kleiman DG, Randolph J, Abramovitz MD, Evans RF (1980) Diagnosis and repair of familial diaphragmatic defects in golden lion tamarins. J Amer Vet Med Assoc 171:866–869.

Bush RM, Custer RS, Whitla JC, Smith EE (1982) Haemotologic values of captive golden lion tamarins (*Leontopithecus rosalia*): variations with sex, age and health status. Lab Anim Sci 32:294–297.

Coimbra-Filho AF (1969) Mico-leão *Leontopithecus rosalia* (Linnaeus 1766) situação atual de espécie no Brasil (Callitrichidae-Primates) An acad Brasil Cienc 41 (Supl):29–52.

Coimbra-Filho AF (1977) Natural shelters of *Leontopithecus rosalia* and some ecological implications (Callitrichidae: Primates). In: Kleiman DG (ed) The biology and conservation of the Callitrichidae. Smithsonian Institution Press, Washington, DC, pp 79–89.

Coimbra-Filho AF, Maia A de A (1979a) O processo da muda dos pêlos em *Leontopithecus r. rosalia* (Linnaeus 1766) (Callitrichidae, Primates). Rev Brasil Biol 39:83–93.

Coimbra-Filho AF, Maia A de A (1979b) A sazonalidade do processo reprodutivo em *Leontopithecus rosalia* (Linnaeus 1766) (Callitrichidae, Primates). Rev Brasil Biol 39:643–651.

Coimbra-Filho AF, Mittermeier RA (1973) Distribution and ecology of the genus *Leontopithecus* Lesson, 1840 in Brazil. Primates 14:47–66.

Coimbra-Filho AF, Mittermeier RA (1977) Conservation of the Brazilian lion tamarins *Leontopithecus rosalia*. In: Prince Rainier, Bourne GH (eds) Primate conservation. Academic, New York, pp 59–94.

Coimbra-Filho AF, Mittermeier RA (1978) Reintroduction and translocation of lion tamarins: a realistic appraisal. In: Rothe H, Wolters HJ, Hearn JP (eds) Biology and behaviour of marmosets. Eigenverlag H Rothe, Göttingen, pp 41–48.

Ferreira LM, Poupard JP, Rocha SB (1981) Plano de Manejo Reserva Biologica de Poço das Antas. Technical Document No. 10, Instituto Brasileiro de Desenvolvimento Florestal (IBDF) and Fundação Brasileira Para a Conservação da Natureza (FBCN). Editora Gráfica Brasiliana Ltda, Brasilia.

Forman L, Kleiman DG, Bush RM, Dietz JM, Ballou JD, Phillips L, Coimbra-Filho AF, O'Brien SJ (In press) Genetic variation within and among lion tamarins. Am J Phys Anthropol.

Hamilton W (1973) Life's color code. McGraw-Hill, New York.

Hershkovitz P (1977) Living New World monkeys (Platyrrhini) with an introduction to primates, vol 1. University of Chicago Press, Chicago.

Hoage RJ (1977) Parental care in *Leontopithecus rosalia rosalia*: sex and age differences in carrying behavior and the role of prior experience. In: Kleiman DG (ed) The biology and conservation of the Callitrichidae. Smithsonian Institution Press, Washington, DC, pp 293–305.

Hoage RJ (1982) Social and physical maturation on captive lion tamarins *Leontopithecus rosalia rosalia* (Primates: Callitrichidae). Smithson Contribs Zool No 354:1–56.

Jones M (1973) International studbook for the golden lion tamarin, 1972. Unpublished.

Kleiman DG (1977a) Progress and problems in lion tamarin *Leontopithecus rosalia rosalia* reproduction. Intern Zoo Yrbk 17:92–97.

Kleiman DG (1977b) Characteristics of reproduction and sociosexual interactions in pairs of lion tamarins (*Leontopithecus rosalia*) during the reproductive cycle. In: Kleiman DG (ed) The biology and conservation of the Callitrichidae. Smithsonian Institution Press, Washington, DC, pp 181–190.

Kleiman DG (1978) The development of pair preferences in the lion tamarin (*Leontopithecus rosalia*): male competition or female choice? In: Rothe H, Wolters HJ, Hearn JP (eds) Biology and behaviour of marmosets. Eigenverlag H Rothe, Göttingen, pp 203–208.

Kleiman DG (1979) Parent-offspring conflict and sibling competition in a monogamous primate. Amer Nat 114:753–760.

Kleiman DG (1981) *Leontopithecus rosalia*. Mammal Spec No 148:1–7.

Kleiman DG (1983) The behavior and conservation of the golden lion tamarin *Leontopithecus r. rosalia*. In: Thiago de Mello M (ed) A primatol Brasil. An 1° Congr Bras Primatol, Belo Horizonte, pp 35–53.

Kleiman DG, Hoage RJ, Green KM (In press) Behavior of the golden lion tamarin, *Leontopithecus rosalia rosalia*. In: Coimbra-Filho AF, Mittermeier RA (eds) Ecology and behavior of neotropical primates, vol 2.

Kleiman DG, Jones M (1977) The current status of *Leontopithecus rosalia* in captivity with comments on breeding success at the National Zoological Park. In: Kleiman DG (ed) The biology and conservation of the Callitrichidae. Smithsonian Institution Press, Washington, DC, pp 215–218.

Kleiman DG, Ballou JD, Evans RF (1982) An analysis of recent reproductive trends in captive golden lion tamarins *Leontopithecus r. rosalia* with comments on their future demographic management. Intern Zoo Yrbk 22:94–101.

Konstant WR, Mittermeier RA (1982) Introduction, reintroduction and translocation of neotropical primates: past experiences and future possibilities. Intern Zoo Yrbk 22:69–77.

Montali RJ, Gardiner CH, Evans RF, Bush M (1983) *Pterygodermatites nycticebi* (Nematoda: Spirurida) in golden lion tamarins. Lab Anim Sci 33:194–197.

Parker S, Gibson K (1977) Object manipulation, tool use and sensorimotor intelligence as feeding adaptations in cebus monkeys and great apes. J Hum Evol 6:623–641.

Ralls K, Ballou JD (1982) Inbreeding and infant mortality in primates. Intern J Primatol 3:491–505.

Rosenberger AL, Coimbra-Filho AF (1984) Morphology, taxonomic status, and affinities of the lion tamarins, *Leontopithecus* (Callitrichinae, Cebidae). Folia Primatol 42:149–179.

66
Before We Pilot the Ark

H. SHELDON CAMPBELL

"There is no rush to cure extinction!" We all took that down when Jared Diamond said it at the opening of this conference. Could it be because few people comprehend extinction that they do not rush to cure it?

In poetry we often feel in a few lines the full impact of an otherwise abstract idea. Take, for example, these four lines about the abstraction, extinction:

> They left the skyways and the wooded lands,
> Forsook the winds, the glory of the sun,
> And time entombed them in his silent sands;
> Their race was over and they had not won.[1]

Even at the intellectual level extinction remains a difficult abstraction. I received a call not long ago from a local high school student who wanted my advice on a class assignment. He was supposed to pick an animal in the zoo and do a behavioral study of it. His ambition knew no limits! What he wanted was to study not just any animal, but "one of the really extinct ones."

Encounters like that bring to us the full realization that extinction is not fully understood or felt as an idea. There was, of course, little or no comprehension of the term in western nations until the latter half of the nineteenth century. Large parts of the globe remained unexplored and unexploited, and Christian ideas of creation did not admit extinction. Dinosaur bones were explained away, either as not bones at all or as the remains of antediluvian creatures that could not be accommodated on the ark or, more likely, that had fallen victim to the wrath of God.

With Darwinism and the concept of natural selection, extinction became a comprehensible idea, but one that few people thought about in the years before World War II and that, even now, has penetrated only a little into the public consciousness—witness that high school student, or take a Masai tribesman standing guard over his cattle. What thought does he give to the approaching extinction of the large mammals with which he has lived for centuries?

---

[1]A. Kulik.

The Necessity for the Ark

If, then, we are going to pilot the ark, we need first to establish and keep established the necessity for the ark. We must, in other words, achieve *on a global basis* a far greater comprehension of what extinction means than we have now.

But even comprehension is not enough. What we must achieve also in a large number of people throughout the world is a state of caring, the kind of gut feeling about extinction that is expressed in Kulik's poem. One can comprehend without caring. Our problem here is that the approaching extinction of many species cannot be classified as good news, and with the world in the state it is right now, when, to quote from Marc Connelly's *Green Pastures*, "Everything that's fastened down is coming loose," most people don't want to see, hear, or read any more bad news.

A High Value on the Ark

What people care for most they spend money on, for in our society, value is easily measured by the dollars we spend:

We value freedom and spend billions for atom bombs and rockets.
We value emotional highs and spend billions on alcohol and drugs.
We value personal security and spend billions on law and order.
We value entertainment and spend billions for rock music concerts, baseball games, and video tapes.
We value health and spend billions for cancer research and vitamin pills.

In that monetary scale of values, what is left for the ark? We note that with the possible exception of the value set on freedom, most of those values are ego centered—their value is to us as living, breathing individuals. And in valuing the prevention of extinction through whatever means—captive breeding or preserving natural habitats—we must be future conscious, concerned with a world in which we personally will be extinct.

Who Goes into the Ark?

Even as we develop the necessity for the ark and begin to move it into our scale of high values, we are faced with the problem of choosing which plants and animals will go into it. We cannot take them all—there are not enough dollars in the world to handle that. If a choice narrows down to one between sea slugs and mountain gorillas, which one do we choose—and why? Evolution does not care. What are the determinants—and who are the determiners?

Do we look at economic values? Our sensibilities and sentimentalities? Similarity to us? Do we use triage based on the availability of space, facilities, and personnel? Do we choose on the basis of doing what costs us the least? For, as Conway, Rabb, and others have pointed out, no matter what animals we choose for captive breeding, the costs will be enormous. Right now the Zoological Society of San Diego is spending $250,000 a year (less, it is true, than the annual salary of a baseball star) to maintain eight California condors of the 18 in captivity for a last-ditch breeding program that, if it works, will probably extend from here to eternity. But let us say it goes only 20 years. That means 5 million to save one species of bird.

Who Will Pilot the Ark?

Finally, if we succeed in establishing necessity, value, methods of choice, and an assured method of financing, who keeps the ark moving in the right direction? What is the right direction? Where is our future Ararat? Are the pilots going to be scientists accompanied by economists and propagandists? Or do we need to develop a priesthood of the ark whose chants and incantations can, over the centuries ahead, keep it afloat?

67
Who Will Pilot the Ark?

RUSSELL A. MITTERMEIER

I would like to start off by answering the question in the title of this plenary session, "Who Will Pilot the Ark?" I think that the best answer to this question is that all of us here have to work together to pilot the ark, and we have to recognize that the ark is not just what we have managed to bring into captivity, but rather our entire planet.

Looking specifically at the primates, which are the subject of this meeting, I think that all of us, regardless of whether we are field biologists, biomedical researchers, or members of the zoo community, have the common goal of maintaining the current diversity of the Order Primates, with its 55 genera and 200 species, and I think that we have to realize that there is no single simple solution to conserving primate diversity. Conservation of wild populations must remain our first priority, but there is no doubt in my mind that captive colonies are going to play an increasingly important role in primate conservation in the future. This conference has made a tremendous contribution to the whole cause of primate conservation by bringing us all together to work out whatever differences we might have and to determine how we can best work together to ensure that our ark remains filled.

However, I think that it is necessary to remember that our ark is in constant danger of sinking because of shortage of fuel. The total resources available for conservation are pathetically limited—just as an example, World Wildlife Fund–U.S. has annually only about 2.5 million dollars available for all projects worldwide, for all species and not just primates. That may sound like a lot of money, but it wouldn't even rate as a small development project of the kind that is often at the root of our conservation problems. I am not going to enter into the whole issue of influencing development and aid agencies because that is not within the scope of this conference. However, something that *is* within our scope is how we can all contribute to conservation of wild populations of primates and their habitats. We very much welcomed Roger Short's "Call of the Wild" appeal, and, needless to say, we have some more ideas on how you can spend your money for conservation.

What I would like to do right now is present some recommendations from the IUCN/SSC Primate Specialist Group. For those of you who might not be familiar with our group, it is one of the specialist groups of the Species Survival Commission of the International Union for Conservation of Nature (IUCN) and has some 150 members in 30 different countries, representing much of the existing field expertise on wild populations of nonhuman primates. In recognition of the importance of the captive component of primate conservation, we have also incorporated a captive section into the group, and it now has about 30 members and is growing rapidly. We have also included a captive section in our newsletter/journal, *Primate Conservation*, and have made special efforts to increase circulation of this publication to the zoo community.

The recommendations of the Primate Specialist Group are as follows:

In order to achieve maximum results in primate conservation and to reach our common goals of self-sustaining populations and maintenance of the current diversity of the Order Primates, we call upon the zoo community to:

1. *Use their great expertise in conservation education to increase public awareness here at home of the problems facing primates and their habitats, and also to make this expertise available to enable us to develop more effective conservation programs in the developing countries*;
2. *To assist in developing training programs in which zoo staff teach essential zoo technologies to our colleagues and counterparts from the developing countries.* One excellent example of this kind of activity is the Training Course held every year at the Jersey Wildlife Preservation Trust, and there is definitely a need for many more training programs of this kind in other zoos. However, training activities can also be carried out by sending veterinarians, curators and other zoo staff to work in the countries where primates occur, and in such cases it is a two way street since zoo staff members will also get first hand field experience that will enable them to develop more effective captive management programs. An excellent example of this kind of program is the one that is currently carried out by the Brookfield Zoo in Chicago;
3. *To take a more active role in supporting field programs on key ecosystems and key endangered species of primates and other animals, perhaps starting with species that zoos are focusing on in the captive setting.* In this regard, we were delighted with Don Lindburg's suggestion this morning for an AAZPA Field Conservation Fund, which we think would make a major contribution to *in situ* conservation, and we can also point to the excellent field conservation program (Wildlife Conservation International) of the New York Zoological Society. Of course, it is often pointed out that the New York Zoological Society is a wealthy organization, but it is important to note that there is need for all levels of activity and that even the smallest zoo can become active in field projects. A few hundred dollars can go a long way in many developing countries, and can accomplish more than one might ever imagine in furthering conservation.

Zoos should also recognize that field projects involve a long term commitment and that many of them may have to be carried out for years, or even indefinitely, in order to be effective. At the same time, however, such long term projects are sure to provide information and opportunities that will greatly enhance captive management of endangered species.

As an example of the kind of activities that we have in mind, some projects that would be especially appropriate for our host institution, the San Diego Zoo, would be:

- Conservation of the red-ruffed lemur (*Varecia variegata rubra*) on the unprotected Masoala Peninsula;
- Conservation of the lion-tailed macaque (*Macaca silenus*) in south India;
- Conservation of the pygmy chimpanzee (*Pan paniscus*) in Zaire;
- Conservation of the mandrill (*Mandrillus sphinx*) in Gabon; and
- Improved management of the Sapo National Park in Liberia, where the San Diego Zoo has already been active and where there are at least a dozen species of primates.

Finally, we would also call on all field primatologists, in other words most of our own members, to work with the zoo community, and to help zoos develop and find support for the programs that we have just recommended to them. The regional Action Plans that the

Primate Specialist Group is now preparing should identify the priorities and provide the guidelines for developing field programs for zoos, and we, the members of the IUCN/SSC Primate Specialist Group, stand ready to participate in public awareness and fund-raising activities that will help to make such programs possible.

This meeting represents a great step forward in identifying our common goals in primate conservation. It is now up to all of us to follow up on what we have started here, to ensure that all primate species alive today are still with us as we enter the next century.

68
The Road to the Ark from the Zoo's Perspective

WARREN D. THOMAS

Since that symbolic time when primeval man stood erect and made his impact on this planet over a million years ago, the world and its resources have been shrinking and rapidly disappearing. The wilderness and its wildlife have been in a crisis state for hundreds of years, but only recently have we become aware of it. The world as we know it is in a period of dramatic change, and we can only speculate as to what it will be like in the next century or two.

A worldwide effort is in progress to protect the environment and the creatures therein. Unfortunately, much of this comes too little and too late, and for many of the life forms in the wild, there is virtually no hope. The zoo's role in this drama is questionable. Regardless of all the conservation efforts and programs of all the zoos combined, there is no way that we can guarantee that we will save a single animal—but we are morally bound to try.

The zoo of the past was usually a lovely park setting, with animals held in compounds to amuse and amaze the viewer. The zoo of today, however, must face a heavy burden with staggering responsibilities. With some species, if we are not collectively successful, they will simply cease to exist. It is no longer possible to sit back and let nature take its course. We must use every device available to improve the prospects of survival.

Most zoos are now heavily engaged in research, from a behavioral standpoint as well as reproductive manipulation. We are just beginning to scratch the surface in terms of artificial insemination, frozen semen, embryo banks, and embryo transfers, all high-tech means of improving the reproductive potential of animals.

Detractors, who are against any animal ever being kept in captivity, criticize us by saying that even if we are successful, these animals in captivity will be markedly changed from the original wild stock. That may be true. Change is inevitable. We do, however, submit that it is better to have a modified individual to be able to put back into a safe, wild situation than none at all.

Zoos are moving away from the idea of showing as many different types of animals as possible toward showing fewer types and making a major commitment to each one. Continuing this trend of specializing in lesser animal numbers, our resources will not be dissipated in so many directions.

Cooperation between zoos has now taken on a strong new direction as characterized by the Species Survival Plan programs. Using the golden lion tamarin program as a model, there has been a substantial proliferation of these cooperative programs considering rare and endangered animals as total captive populations. The management of each species is abrogated to an elected committee with a Species Manager and/or Studbook keeper. These programs give us new hope for the establishment of our ultimate goal of self-sustaining, vigorous, captive populations.

The level of cooperation between zoos has markedly changed. Not too many years ago, zoos sending animals on breeding loan to other zoos was uncommon. Today it is more the rule than the exception. It is now common for a major zoo to have as many as 200–300 animals out on loan to other institutions and to have a like number on loan to them. The cooperation between zoos that is now very good will, by necessity, get even better. The trend is to increase the numbers of animals under these cooperative efforts. This will probably extend, in time, to even more common species to ensure their long-term viability. In addition to studbooks, careful records are now maintained and decisions on breeding are made by well-substantiated information rather than individual whim. Some zoos that have been parochial and local must become national and international in their scope of endeavor because of the very nature of the animals entrusted in their care.

There is a great need for additional field biology. We must have many more conservation plans funded and directed by a consortium of zoos. The zoo community can and must become a cohesive group, working in harmony, to accomplish these enormous, expensive projects. The sum total of all of our efforts must be to address ourselves to the traditional reasons for having a zoo.

More and more we must give emphasis to those programs that can support a continued future for the species, with the idea in mind that whenever possible these captive reserves are to be drawn upon for reintroduction into safe wild environments—not necessarily the same location where the animal was originally obtained. We may have to look for alternative locations compatible with the given animal.

The natural world is a legacy, to be held in trust, for the generations to follow. Our efforts must not be in vain so that future generations will appreciate the wealth of the wild as we know it.

69
Who Will Save the Children?

DENNIS A. MERITT, JR.

This is perhaps a strange subtitle to "Who Will Pilot the Ark?" In reality, the primate populations of 1985 are under study, are being investigated, probed, observed, and scrutinized intensely. Whether in nature, in a laboratory, or in a zoological garden this holds true. We are concerned, motivated, and enthusiastic about our work. If we were not, we would not be here today. But how do we ensure the future for the Order Primates of which we are members? What can we do to make certain that future generations of aye-aye, mountain gorillas, and muriqui are there, living, breathing, behaving, and not a series of skins and skulls in a museum collection? Perhaps our encounter this week gives a clue to the future and provides direction.

In my recollection, no one has attempted to bring together before, in one forum, field biologists, conservationists, researchers, zoological professionals, students, and a lot of other participants. Dr. Benirschke is to be thanked and congratulated for accomplishing what some had characterized as an impossible task. Accolades must also go to the Morris Animal Foundation and the Zoological Society of San Diego for enthusiastically supporting his idea. First attempts like this conference are always difficult, not without some frustration and pain, and, as we have seen, not without controversy, but they always are a learning experience.

Early in the week a friend and colleague reflected and shared with me his thought that what was apparent to him was how misinformed some of us were. How poor the communication was between the various disciplines represented here. Perhaps the key lies in the latter, communication. We have gathered here this week from various specialties with different areas of expertise and interests to report, more properly to share and communicate our results, our ideas, our hopes, and our beliefs for the Order Primates.

While we may not necessarily agree on all points discussed and raised nor the techniques used, at least we have shared time, space, and ideas. Hopefully and ideally, we have listened to each other as well. This is a first but nevertheless valuable step. Roger Short, struck by the events of the moment, reflected and commented emphatically, "What is so striking is how little we actually know about any given primate species." I might add, how little we know about the interaction between the animals and their environment whether in Africa, Asia, the Orient, or the Neotropics.

How then will we proceed for the future? What will the plan be? Who will ensure the primate generations to come, who will save the children? Some of us will still be here, perhaps a bit older, ideally wiser, the sages of tomorrow. But we will need help, an enormous amount, from those as yet unidentified, but nonetheless to come.

I feel obligated to share with these future colleagues, our progeny if you will, some random but useful thoughts. I leave it to others to share their thoughts as they apply to the future role of zoos, funding, the formation of an international body to provide direction and guidance, and the special requirements of medical research. These are all important and necessary considerations, some requiring further detailed discussion, others immediate action.

We must always remember our obligation to train and educate the nationals in the countries where we work as well as people in our own nation. It is our obligation to provide information that is not readily available to them, information and resources that we may take for granted. It is also critical that we make our results known and available in the country of origin where we work and to publish our results in their language or journals if at all possible; Dan Janzen in Costa Rica can serve as a model in this respect for each of us.

As we preach and pronounce while guests in foreign countries, let us not forget how recent our own conservation ethic is, nor how badly we have permanently scarred our own landscape here in North America. Let us also remember that it is hard to have a conservation ethic, perhaps impossible, when you are hungry and a primate may represent your only meal, the single largest source of protein available to you. Put yourself in the place of any of a number of people in less-developed countries.

Let us also proceed with caution and thought, lest we confuse allies and benefactors with exotic terminology, buzz words, descriptive adjectives, and phraseology, without clear definition. Consider how we implicate and indict with the use of "slash-and-burn agriculture" without giving our audience a chance to know there are degrees of effort. Clearly and without argument there is a difference between the practice of slash-and-burn agriculture for subsistence farming for a family or an Indian village and the larger scale multinational approach practiced by those investing in converting land to cattle ranching. Yes, we all know there is a difference, but do those outside of our own group assembled here know?

Similarly we must define what we mean by research. Clearly, there are as many interpretations of this word as there are ears to hear it being vocalized. We must be specific and detailed in describing research and its aims and objectives. Some of us have even coined new terms to avoid the use of the word, using blanketing phrases such as "scientific studies." How much simpler might it be if we took the time to define clearly and simply what we mean by research. By the nature of the work each of us does we have an obligation, a responsibility, to share our knowledge with others.

There are decided advantages to knowing the animal one is to study in the field prior to arriving at the study site. This goes beyond what can be read in the literature or what can be gleaned from a more experienced colleague. Captive animals provide this unique opportunity to experience the living, breathing creature. We are allowed some insight by sharing time with them. Zoos are a unique resource in many ways. Is it not better to arrive in the field with some idea of what the animal eats, how it handles its food, when it is most likely to be active? Zoos provide these as well as a host of other opportunities on a daily basis.

While I might go further, I will not, and I close with these thoughts. There is a reason for optimism. Our gathering here for this meeting is optimistic. The remarks of Russ Mittermeier as well as others earlier in the week give us reason for hope. The success of other organized programs with multifaceted approaches is clear. Alison Jolly's calm, stimulating, and heartening approach is refreshing and thought-provoking, amid many doomsday projections. David Chivers has also given us a new perspective on forest life and utilization. There are many reasons to be optimistic, not the least of which is our own individual commitment to what we believe in. While we may not save all the members of our order, we will save as many as we possibly can. With our actions today and those of colleagues who will follow in our footsteps, there will be someone to save the primate children of the future. So long as we cooperate, communicate, and continue to educate, this will be a reality.

70
The Importance of an Interdisciplinary Approach: Getting the Conservation Act Together

JEREMY J.C. MALLINSON

Introduction

Having helped Gerald Durrell for over a quarter of a century in the development of the Jersey Wildlife Preservation Trust and having been privileged to have studied various aspects of natural history and the animal kingdom in Africa, India, and South America, I believe that such a gestation has provided me with a realistic insight as to some of the complexities that we are confronted with, and an exposure to the jigsaw puzzle of components that somehow or other we are going to have to place together carefully if we are going to manage to attain our conservation objectives by sustaining species diversity. Regrettably, we are not just dealing with the animal kingdom, but with what more and more seems to be the irresponsibility and irrationality of the human race. Therefore, the ingredients of my paper embody some of my thoughts as to how at least something that I know a reasonable amount about—how captive breeding programs can really aid species preservation—providing we can get our conservation act together now.

In a recent paper, presented at a Primate Symposium held at Bielefeld University, West Germany, the author underlined the importance of an interdisciplinary approach in the establishment of self-sustaining populations. Also, in order to provide a fuller awareness as to some of the practical difficulties that arise in our attempts to secure captive reservoirs for endangered primate species, some of the existing problems were highlighted. Although some aspects of this paper were covered at the Bielefeld Symposium, a number of the points raised are hereby addressed and expanded upon (see Mallinson, in press).

As stated on a number of previous occasions, the long-term future of animal species in captivity increasingly relies upon national and international cooperation and coordination (Mallinson, 1980). However, in the final analysis, "survival cocktails" will materialize only if the necessary ingredients of what have been termed as captive breeding professionals, field workers, conservationists, scientists, educationists, academics, and others genuinely interested in animal conservation are willing to cooperate wholeheartedly. Without such a marriage of disciplines and endeavors, many species will evaporate prematurely from the survival reservoirs for endangered species that the modern zoo and other breeding centers at present have the opportunity to help secure for posterity (see Mallinson, 1982a, 1984a).

At this stage of the evolution of captive breeding *aiding conservation*, it is important for us not just to dwell on what the captive breeding community has achieved to date, but rather to consider what the real potentials are, providing that we can now get our conservation act together. It is, therefore, the main objective of this paper to provide an awareness as to the diverse components involved that have resulted in a number of the conservation goals being achieved to date, to examine the type of coordinated strategy that could be of importance for an interdisciplinary approach.

Wildlife Preservation Trust

It is difficult to present a paper at such an international conference on the road to establishing self-sustaining populations without first making reference to the conservation role of the Jersey Wildlife Preservation Trust (JWPT) and its sister organization, Wildlife Preservation Trust International (WPTI).

The Wildlife Preservation Trusts have a great number of ambitious captive breeding programs that entail a wide range of responsibilities, and the Trust has always attended to these by:

1. Conducting biological research on the maintenance and breeding of animals in captivity
2. Placing its own animals with other breeding facilities in order to build up their numbers
3. Becoming actively involved with national and international bodies devoted to the cooperative management of threatened species held in captivity, or to animal conservation in general.

and more recently by:

4. Collaborating with governments of countries in which many endemic species are threatened
5. Training people from these and other countries on the breeding of animals in captivity at the International Training Centre for Conservation and Captive Breeding in Jersey (see Waugh, 1983)
6. Supporting field research on the species in its care.

It can be seen from such diverse activities that the WPT has a multifaceted approach to captive breeding for conservation. It is from such past and present experiences that the author can address some of the problems that have arisen and, as at the Bielefeld primate symposium, go on to present an outlook on the future role of cooperative breeding programs for primate conservation.

Existing Problems

It should first be recognized that no single approach may be applied to the diverse array of species requiring captive propagation programs and that each individual species may require a different consideration. If an animal species is going to flourish in a captive environment, it is of vital importance to have knowledge of its life-style in the wild, so that proper provision can be made to provide better physical and social captive conditions. To neglect to take into account such environmental, social, and nutritional requirements could be considered as detrimental as it is to hang a masterpiece in a dark corner or fail to exercise a high-spirited race horse.

In our attempts to cultivate a good cross-pollination between the qualities of both wild and captive conditions, it has been found that the fundamental knowledge gained can sometimes present some major managerial problems that may be difficult to overcome. For instance, in the wild state gorillas usually live in social groups, each group member having an opportunity to contribute to the gene pool if group composition allows breeding opportunities for both males and females. If breeding success is not possible, emigration ordinarily occurs, resulting in dispersal and outward gene flow from the natal group to some other social unit (Fossey, 1982). Taking these factors into consideration, one of the chief hurdles that now confronts owners of gorillas that are breeding in captivity is just how can one best attempt to emulate the social groupings and transfer options that are open to gorillas in the wild (see Mallinson, 1983).

As Georgina Mace (1983) so succinctly records, all the available evidence suggests that inbreeding reduces individual survival, though the effect is difficult to quantify. However, a breeding program based on inbreeding would result in the highest possible rate of loss of genetic variation from the population, so if conservation aims are to preserve, as far as possible, the original gene pool, inbreeding of any sort should be studiously avoided. In order to comply with such principles of genetic management, certain moves that may be considered exemplary as far as increasing genetic variation may prove to be both stressful and counterproductive as far as the gorilla individual is concerned. For example:

1. In captive conditions, accommodation varies greatly from one location to another, whereas in the wild state, the environment remains the same.
2. Medical care, diets, house temperature, and outside climatic conditions also often differ considerably.
3. Although social groupings are becoming more common, there are still considerable differences between the background and make-up of one group and another. On some occasions, requirements necessitate the removal of an individual from a social group in a seminaturalistic surrounding to a more sterile environment, in order to make up a solitary pair.
4. The problems that have arisen with hand-reared animals as opposed to mother-reared individuals; as well as the advantages shown by specimens that have had infant handling experience.
5. Professional staff play a paramount role in providing satisfactory management for the animals that come under their care. In some cases, when gorillas are moved from one country to another, they not only miss conspecifics but also the security of the people who have been in charge of them; for people, like gorillas, vary greatly. It should also be recognized that some animals may initially experience a degree of uncertainty and stress when spoken to in an unaccustomed language.

A further existing problem that should be borne in mind is that although the management of animal species has always varied considerably between one collection and another, one does find that in the majority of cases primates, especially anthropoid apes, are confined to their inside accommodation for over two-thirds of the 24 hours. Often in connection with this factor, insufficient thought has been given to either the inside space or the furnishings that are made available for the long periods concerned. Therefore, it should be recognized that in comparison to a primate living in the wild, its environment in captivity becomes dramatically restricted and emaciated; however, most authorities acknowledge the fact that it is not necessarily the size of the accommodation in which primates are kept that really counts, but rather just how the space available is utilized. As Hediger (1969) records, "within certain limits the quality of the space is much more important than the overall size." It should also be appreciated that when a reintroduction program is considered to be viable, it is of the utmost importance for the stock to have experienced similar social and environmental conditions to those of wild populations (see Mallinson, 1982b).

It can be deduced from the examples presented that although the best possible use should be made of the available captive gene pool and textbook-type moves be considered, very careful attention must be given to the multitude of components that go to make up the qualities that can result in a satisfactory transfer of primates from one captive environment to another.

Cooperative Efforts

The Species Survival Plan (SSP) of the American Association of Zoos and Aquariums (AAZPA) makes it clear that cooperation without coordination is insufficient to preserve wild animals in captivity over long periods of time, and that captive populations fragmented among many small collections can be preserved only if they are managed scientifically as a whole (see Conway, 1982).

An increasing number of zoological collections are at long last no longer attaching a price tag to the rare species they have represented, but rather make such stock available to be included in cooperative breeding programs (see Mallinson, 1984b).

The Conservation and Animal Management Committee of the National Zoo Federation, the Common Management of Species Group (see Mallinson, 1984c), and the Anthropoid Ape Advisory Panel all work to promote cooperation and coordination with breeding programs within the quarantine barrier of the British Isles and Ireland. In the United States some 32 species are now covered by the Species Survival Plans of AAZPA, which promote interinstitutional cooperation in North America on an unprecedented scale.

In 1981, the Australian Association of Zoo Directors decided to place various species in their collections under common management for breeding and genetic management. In the same year, the election took place of a seven-person International Management Committee (see Kleiman, 1984), which acts as the decision-making body the International Co-operative Research and Management Agreement (RMA). The subsequent combined work of field researchers, the Rio De Janeiro Primate Center, international zoo personnel, and various conservation funding agencies probably represents the finest example to date of what can be achieved when various disciplines working for conservation of a species cooperate and coordinate their efforts in order to secure its survival (see Mallinson, 1984d).

In October 1983, at the 38th Annual Conference of the International Union of Directors of Zoological Gardens (IUDZG) held in Melbourne, Australia, delegates unanimously adopted a resolution recognizing the increasing importance of scientifically developed programs for the management of species, resolving that the Union requires its members to make every effort to cooperate in the development and coordination of such programs. The resolution dealt with the selection of species, the importance of a unified record system, species management, the problems surrounding ownership, the curtailment of breeding, when required, and recognizing that a successful breeding program may create an excess in number (see IUDZG, 1983).

In September 1984, a combined statement on behalf of Amsterdam, Antwerp, Copenhagen, and Rotterdam Zoos was presented at the 39th Annual Conference of IUDZG by the Director of the Amsterdam Zoo, Dr. Bart Lensink. The statement recognized the important regional communication concerning species management that was already well established in both the British Isles and North America, but pointed out some of the difficulties that presented themselves in the adoption of a clear world-wide philosophy, when language barriers sometimes inhibited the essence of free discussion and the economic realities that were attached to some considerations that would affect the free movement of animals from one location to another. As the conservation and captive breeding principle had been accepted by the IUDZG Melbourne Resolution, the four zoos, who have all expressed a will to work together intend to form a bridge between Continental Europe and the two English-speaking regions of the British Isles and North America. It is considered that, in practical terms, such a combined statement represents a significant stride forward in promoting such regional communications.

Future Strategy

It has been acknowledged for some time that it is of the utmost importance to look at total populations, both nationally and internationally; as well as to establish which institutions will participate in a propagation plan, and the number of spaces that can be produced for a particular species. The conclusion of the Melbourne Resolution recognized that the support of conservation organizations at local, national and international levels is vital to the long-term success of species conservation. It recommended that IUDZG members should continue to develop the closest possible links with local and national conservation organizations.

The task now in front of us is just how best are we going to provide the links for such a complex chain? The author considers that it is important to establish a framework within which regions can integrate their efforts on an international basis, and that this in turn will go to represent a more interdisciplinary approach to the long-term future of the breeding programs concerned (see Fig. 70.1).

The role of species managers has been well defined in the Species Survival Plan of the AAZPA (SSP/AAZPA, 1984). In order to further such management strategies for rare and endangered primates, it is advocated that species coordinators should be appointed by national bodies in each of the geographical regions, e.g.,

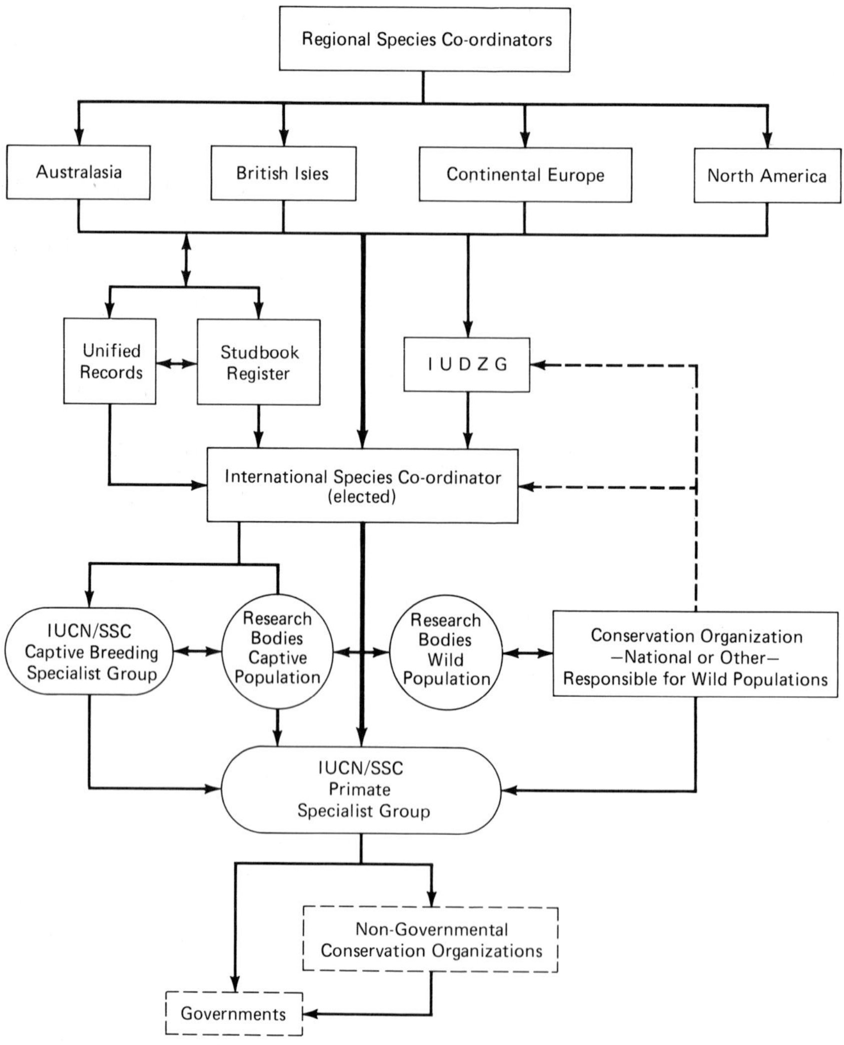

FIGURE 70.1. A coordinated approach to the development of viable captive primate populations.

1. Australasia (Australia and New Zealand)
2. British Isles and Ireland—CMSG
3. Continental Europe (Belgium, Germany, Netherlands, Scandinavia)
4. North America—AAZPA/SSP

and that these representatives should keep in direct contact with each other, as well as with:

1. A unified record system (e.g., ISIS)
2. The international studbook or world register holder, if relevant
3. The International Union of Directors of Zoological Gardens.

The national coordinators should elect one of their number to act as their representative (perhaps for an initial 2- or 3-year period) in order for them to develop formal links with:

1. IUCN/SSC Captive Breeding Specialist Group
2. IUCN/SSC Primate Specialist Group

In turn, these international conservation specialist groups should be responsible for linking up with research bodies and keeping themselves up to date as to the status of wild populations. With the combined knowledge as to how a species is doing in both the wild as well as in captivity, the specialist group, through the IUCN Species Survival Commission, could make their conservation recommendations to the relevant governmental departments of those countries where the primate species is endemic, as well as to international nongovernmental conservation organizations.

Although it is recognized that some of the organizations in the flow chart have neither the infrastructure nor probably the will to be active participants in information exchange, let alone decision making, it is important to debate just how best we can link our efforts so that we can develop the machinery to transmit data that can be used for species conservation in the most positive fashion. As J. Knowles (personal communication) pointed out, "Can we, with the people and budgets available to us, manage all the endangered species, the excessive paperwork and bureaucracy?"

Summary

It is evident from what has been recorded that captive breeding programs have made great progress in recent years. However, if we are going to attain our conservation goals by the establishment and maintenance of self-sustaining populations, we are going to have to adopt a far more interdisciplinary approach as well as a framework that can link up all those working for a particular species survival.

The absence of such frameworks undoubtedly handicaps our real conservation potentials, for how often have we heard "If only we had been aware that such and such a person, or such an organization had previously been, or were now actively involved with the conservation of a particular species"? How much unnecessary duplication of work and financial expenditure could have been avoided if proper communication between all parties concerned with conservation was commonplace. Perhaps the recent plunder of the golden headed lion tamarins (*Leontopithecus chrysomelas*) from the State of Bahia, Brazil, which has resulted in an estimated 50% of the remnant wild population being illegally removed and exported to both Europe and Asia, could have been minimized or prevented had there been a well-lubricated channel of communication between all the disciplines that have been involved, during the last 2 years, in trying to conserve this now critically endangered species (see Mallinson, 1985).

Regrettably, one is aware of animal conservationists working in a given geographical region who are unwilling to divulge and pass on knowledge that would aid the conservation of a species after it has either walked, flown, or swum out of their particular theater of interest and responsibility. Also, it has been established that some specialists, and even some conservation bodies, instead of electing to work together to attain a specific conservation goal, have preferred to work in isolation; and through such short-sightedness, some erosions of animal populations have been recorded as well as some important conservation projects unnecessarily handicapped.

Because of the ever-increasing number of species joining the floundering ranks of the vulnerable, rare, and endangered, the limited space in suitable captive breeding centers, and the escalating cost of maintaining and developing breeding programs, it is of the utmost importance for the available resources and space to be assessed by the international zoo and research communities, and criteria for species selection to be arrived at as soon as possible.

In conclusion, the author believes that in order to develop future long-term management strategies and to establish captive reservoirs of rare and endangered species, it is important for all those who are directly involved with the management of conservation programs to recognize what we are neither expert in nor knowledgeable about. In some cases, we shall have to adopt a degree of humility in order to take on board a far more interdisciplinary approach to our responsibilities. In the future, we should strive to develop the framework of a well-fabricated coordinated approach and maintain good communication with all the disciplines that are able to contribute to an acceleration of our progress along the road to the establishment of self-sustaining populations of the world's vanishing primates. Like proficient omelette makers or mixers of good cocktails, it is our responsibility to gather up and nurture the diverse ingredients that will result in the "bonanza omelettes" and "survival cocktails" that should culminate in the type of long-term conservation goals that this international conference has been convened to secure.

Acknowledgments. To Ged Caddick for carrying out the artwork for Figure 70.1. To Roger Wheater for supplying data on the combined statements made at the 39th Annual Conference of IUDZG with regard to the promotion of regional communication concerning captive breeding programs. To John Knowles for his valuable comments on both the text and the flow chart. To Dr. Lee Durrell for so succinctly summarizing the way Wildlife Preservation Trust has attended to its wide range of responsibilities in connection with the great number of JWPT's/WPTI's ambitious programs.

To the organizers of the Bielefeld Symposium on captive propagation and conservation of primates, held at the University's Center for Interdisciplinary Research, which provided me with so much food for thought.

References

Conway WG (1982) The species survival plan: tailoring long-term propagation species by species. In: AAZPA Annual Conference Proceedings, Phoenix, Arizona, USA, pp 6–11.

Fossey D (1982) Reproduction among free-living mountain gorillas. Amer J Primatol Suppl 1:97–104.

Hediger H (1969) Man and animal in the zoo: zoo biology. Routeledge & Kegan Paul, London.

IUDZG (1983) Policy statement. Proc 38th Ann Conf Intern Union Directors Zool Gardens, Melbourne, Australia.

Kleiman DG (1984) The behavior and conservation of the golden lion tamarin *Leontopithecus r. rosalia*. In: de Mello MT (ed) A Primatologia No Brasil, Sociedade

Brasileira de Primatologia, Anais do 1° Congresso Brasileiro de Primatologia, Belo Horizonte, pp 35–53.

Mace GM (1983) The present status and future management of populations of great apes in the United Kingdom. A report for the Anthropoid Ape Advisory Panel of Great Britain and Ireland.

Mallinson JJC (1980) National and international zoo co-operation: an example from the Jersey Wildlife Preservation Trust. Intern Zoo Yrbk 20:179–180.

Mallinson JJC (1982a) Co-operative breeding programmes outside the United States. In: AAZPA Annual Conference Proceedings, Phoenix, Arizona, USA, pp 52–57.

Mallinson JJC (1982b) Cage furnishings and environments for primates with special reference to marmosets and anthropoid apes. Nat Fed Zool Gardens Great Britain and Ireland Newsl Pullout Supp No 12.

Mallinson JJC (1983) An update on gorilla breeding in the British Isles with special reference to the work of the Anthropoid Ape Advisory Panel. In: Harper D (ed) Proceed Sympos Conservation of Primates and Their Habitats. University of Leicester, Vaughan Paper No 1, pp 233–266.

Mallinson JJC (1984a) The breeding of great apes at the Jersey Wildlife Preservation Trust and a look into the future. Zoo Biol 3:1–11.

Mallinson JJC (1984b) Survival reservoirs for endangered species: the conservation role of a modern zoo. Biologist 31(2):79–84.

Mallinson JJC (1984c) Regional joint management of species group—British Isles and Ireland. IUCN/SSC Primate Specialist Group Newsletter 4:47–48.

Mallinson JJC (1984d) Golden lion tamarin's survival hangs in balance. Oryx 18:72–78.

Mallinson JJC (1985) Plunder of the golden lions: Brazil's endangered primates. Country Life CLXXVII(4565):368–369.

SSP/AAZPA (1984) Species survival plan guide of the American Association of Zoological Parks and Aquariums.

Waugh DR (1983) The Wildlife Preservation Trust Training Programme. Dodo J Jersey Wildl Preserv Trust 20:12–16.

71 Recommendations of Workshops

A. Strategies for the Extremely Endangered Primates

George Rabb, Katharine Milton, A. H. Harcourt, Dennis A. Meritt, Jr., Alison Jolly, and Russell A. Mittermeier

B. Artificial Breeding

John P. Hearn

C. Virus Diseases

Seymour S. Kalter

A. Strategies for the Extremely Endangered Primates

It was indicated at the outset that no definition for extremely endangered existed, although it was suggested that species with a total population of 1000 or an effective population size of 500 or less be considered extremely endangered. Despite this handicap, we addressed some of the issues. First came some sobering statistics from Russ Mittermeier. The most critical figure was his estimate that one of every seven primate species would be seriously in danger of extinction by the turn of the century. He then enumerated ten taxa in critical shape now, starting with the monotypic family represented by the aye-aye and proceeding down the taxonomic hierarchy: the indri, muriqui, golden-rumped tamarin, lion-tailed macaque, cotton-top, mountain gorilla, etc.

He then touched on four major objectives for conservation action:

1. Better knowledge of status of species from field surveys
2. Improvement of prospects for survival in the wild—reserves, ecology, field station "guard function"
3. Enhancing education, including training, public awareness
4. Utilization of all avenues, including captive breeding, to assure survival

Sandy Harcourt discussed the mountain gorilla conservation program, more especially that in Rwanda. He emphasized the cooperation among conservation organizations in this effort. The building of a public awareness was praised and the taking over of the educational function by the Education Ministry was cited. The effect of tourism was seen as providing valuable leverage in promoting recognition of value to maintaing the animals.

Results of conservation effectiveness of the studies:

1. Contrast of Zaire and Rwanda in reproductive success (unmanaged versus "managed")
2. Education component, actually changing attitudes, including recognition by locals of watershed value of park
3. Tourism supports park

General issues: justification of this effort questionable in view of small population versus all gorillas, site occupies but 0.5% of land, etc., but effects noted above counter these objections. He urged us to remember conservation is not just inside protected areas, and we must give attention to fundamentals such as human demography and compatible usage or development.

Kathy Milton described the situation of the woolly spider monkey, a species down to perhaps 300 animals in pockets in southeast Brazil. She spoke to the objectives of her field study in a private reserve:

1. Determining population size
2. Nature of the forest
3. Diet and activity pattern
4. Social organization

Some points of interest in her study: (1) Fusion–fission pattern of group behavior found—multimale mating seen; but cohesive social pattern elsewhere. This is perhaps diet related, which in turn may be related to nature of forest. (2) Puzzle of distribution—20 animals/35,000 hectares in one reserve versus 40–45 animals/422 hectares in study area.

She spoke against ill-prepared and ill-considered moves of animals to captivity, citing learning needs of this species in relation to diet, making reintroduction difficult at best.

Dennis Meritt followed with an account of the successful mounting of a captive breeding program for Geoffroy's tamarin, starting with salvaged specimens from the markets of Panama City. It was pointed out that we should perhaps use such a species for testing reintroduction techniques.

Alison Jolly remarked on the particular opportunities for conservation among the Madagascan lemurs and the desirability of achieving more competence in captive maintenance and breeding of potential flagship species. She cited that only 12 of the 23 species of lemurs are kept in captivity, and we are thereby denying ourselves knowledge and research opportunities as well as the educational value the animals have as ambassadors of their species.

In the general discussion, George Rabb mentioned the two major dimensions of biological and of social-political factors that have to be considered in reviewing the appropriate course of action for any particular species, and how both in-situ and ex-situ conservation actions have generalized pros and cons that must be addressed on a case-by-case basis.

The general discussion ranged over several areas that had not been well covered in the talks and previous discussions. Among these, the notion that we needed teams of wide talents, starting with human population demographers, to fully address the conservation problems involved. It was pointed out that at one extreme of our concerns was the development activity of government and international agencies, where there are signs of recognition that their biologically ignorant aid and assistance programs were too often counter the interests of Western countries as well as the affected countries of the Third World.

Disappointment was expressed that the session had not dwelled on the need for improving techniques for assessing the population biological problems and dealing with the fundamentals of the minimum viable populations concept. Dr. Rabb indicated that the conference program was otherwise structured, that there has been attention to the small population issues at genetic, ecological, and behavioral levels, at least in the zoo community, and that certainly much more should be done on the biology of extinction, where we have only the pilot Lovejoy study on Brazil's rainforests.

Nate Flesness tried to provoke the panel into identifying a particular population status as the critical point for major action. The panel respectively declined. Warren Thomas pointed out that what the conservation question comes to is who is willing to take initiative. He agreed it was important to conserve in situ and pointless to conserve without environmental conservation, but we don't know the future environment, and hedging bets with captive populations seems warranted.

There was dismay expressed at the posing of a dichotomy of protected reserves versus captive breeding, when dual or multiple strategies might be best. It was evident that there was a general desire to have a plan to coordinate the efforts of field researchers, zoos, and conservation activist organizations. This will probably have to take place in a true workshop session devoted to planning, where we might achieve integration of the action planning of the Species Survival Commission Primate Specialist Group, individual and cooperative zoo conservation and management programs for captive breeding, public education tactics, and identification of resources necessary.

The panel thus contributed to the stimulating dialogue characteristic of the conference. This should be followed up with a more focused effort toward a comprehensive conservation strategy recognizing the captive and field data base needs, the human equation from local populations to international agency level, and the contributions needed institutionally and professionally. We must have such a strategy if we are to move beyond the present levels of conservation activity in the field and of management capacity in zoos.

B. Artificial Breeding

The Workshop considered the current status of research in reproductive physiology relevant to the artificial breeding of exotic species. The possible applications to primate conservation were considered. In this chapter, some general points are noted by way of clarifying the questions. The priority species for research, priority techniques for application, and the collaborative links on which future development of the field depends are then reported.

1. In the next 50 years the human population of the world will double to 10 billion. Much of this increase will be in developing countries, resulting in restricted environments for wildlife and isolated populations. It is necessary to plan now ways of overcoming these restrictions. Artificial breeding technology is one area that may assist.

2. Animals in captivity and in the wild will require management as one stock. Wherever possible, natural breeding should be given priority. Artificial breeding may be required to overcome cases of infertility, incompatibility, or consanguinity.
3. Researchers are often castigated for exploiting animals for human benefit. It is encouraging to see that the human is now becoming the experimental animal for a new technology in reproduction that can assist conservation of animals.
4. Much of the research required is long-term and should be initiated immediately in order to provide results within the time available. The work requires a fundamental approach if it is to succeed. It should be remembered that 20 years and many millions of dollars were required to develop artificial breeding technology for cattle and for humans. There is still only 15% success in in vitro fertilization in humans. Application of this technology to rare primates will not take as long or be as expensive, but it cannot be immediately applied without appreciation of species differences.

Priority Species

Virtually all primate species will be endangered by the turn of the century. It is difficult, therefore, to attach immediate priorities. Of greatest importance at present are species such as the mountain gorilla and other great apes; the smaller apes in the forests of Southeast Asia; and the rarer Callitrichid monkeys in the diminishing forests of South America. Any studies on the reproductive biology of these species would be helpful, and, where possible, more common but closely related species could be studied to develop artificial breeding techniques.

Priority Techniques

There have been great advances in the development of artificial breeding technology. The following areas require additional work:

1. Maturation, freezing, and transfer of sperm. It is still only possible to routinely carry out these procedures in few species.
2. Maturation, freezing, and transfer of oocytes. This technology has yet to be developed but is greatly needed to allow collection and transfer of female genetic material.
3. Collection, freezing, and transfer of embryos. This field is developing rapidly but is still only routine in few species.
4. There is an urgent need for new noninvasive methods for monitoring reproductive function. These need to focus on analysis of urine and the development of ELISA assays that can be used in the field. In addition, applications of ultrasound technology would assist.
5. There are a number of other new technologies that are developing but are not yet practical for any application in the conservation of primates. These include sexing of embryos, in vitro growth of oocytes, in vitro fertilization, cloning, and chimerism.
6. A greater effort is needed in research on neonatal survival and rearing. Too many primates are lost unnecessarily, and improvements in artificial rearing techniques would reduce these losses.

Collaborative Links for Future Development

The future will depend on much closer research collaboration between workers in field, zoos, and laboratories. No single one of these components has the facilities, the expertise, or the funds to solve the problems confronting primate conservation. The following points need particular attention:

1. Population genetics. Close contact must be maintained between primate breeding programs and the ISIS genetic advice service. It is hoped that ISIS will be able to provide early warning of species that face major problems, and therefore crisis intervention and management can be planned.
2. Closer links should be forged between zoos with captive primate populations and laboratories with the facilities to study reproductive physiology. In particular, small zoos are unlikely to obtain the sophisticated facilities required for such study.
3. Far closer collaboration is encouraged between field workers and scientists in research laboratories and zoos. Field data are essential for the proper development of captive conditions. In addition, field data in reproduction will greatly assist in assessing the reproductive capacity of animals, both in the field and in captivity.
4. In future, with populations declining or becoming isolated, the need for artificial reproductive intervention may be necessary. The transfer of gene material between captive populations, across national boundaries, or between small populations in captivity and in the wild would lead to improved genetic status of the populations concerned.

C. Virus Diseases

There is obvious concern regarding the effect of viruses and virus diseases on the future of "self-sustaining colonies." This concern was not only apparent to the participants of this workshop but was also noted throughout the conference by other speakers, not part of our panel.

An extensive list of monkey and ape virus diseases was provided along with mechanisms for minimizing their occurrence. The etiologies of these diseases, which occur either in individual animals or as major zoonoses, were viruses of simian and nonsimian (principally human) origin. A few diseases could be prevented by vaccines, a number of which were developed in primates, but many could not be. Some viral diseases are not preventable in nature because their epidemiology in nonhuman primates is not understood or because vector control or vaccination, when a vaccine exists, is impractical; notably:

1. Vector-transmitted disease such as yellow fever, Kyasanur Forest disease, and dengue fever, from which both human and nonhuman primates suffer. There is little that can be done to prevent such diseases from occurring other than vector control. The relationship of monkeys to the maintenance of yellow fever in nature was discussed.
2. Diseases (simian hemorrhagic fever, Marburg) that have made rare appearances without clearly defined epidemiologic origins. These diseases occurred without warning and were considered "new" because there were no previous indications of their existence. It is highly probable that good husbandry with

appropriate control over the handling of the animals would have prevented these outbreaks. At the present time, SAIDS appears to be in this group as well.

Although simian viruses per se do account for diseases of various clinical types in monkeys and apes, their pattern of disease appears to be individual rather than zoonotic. As listed by Kalter and described in detail by Hunt, *it was predominantly the human viruses responsible for widespread outbreaks* that would be of major concern to colony management. Rabin reviewed the current SAIDS situation and the unique EBV-like virus disease at the Sukhumi primate center in the U.S.S.R. He also indicated concern for the future of colonies as a result of infection with these viruses. Both Kalter and Vickers described the maintenance and management techniques necessary to minimize the spread of viral agents among primate colonies. Mechanisms for diagnosing virus diseases as well as a rapid, specific diagnostic test were described.

Several suggestions, problems, and questions regarding the future of primates, particularly within the framework of self-sustaining colonies, were discussed:

1. It would appear that we are "between a rock and a hard place" in terms of recommendations for future management policies. Inasmuch as the major viral disease problems of nonhuman primates are human related, how do we reconcile the rigid seclusion of a colony advocated by those concerned with infections with the intimacy demonstrated by others?

   What effect will suggested "tourist" programs have on the health of colonies in the wild and in captivity? What is the possibility of such programs defeating their goal of conserving species? It was suggested that this may be a good immunizing mechanism, but there is no guarantee that infection without serious disease would occur. It could be analogous to those human situations where measles in a naive adult population causes a highly fatal rather than a "childhood" type of disease.
2. Procedures to monitor the occurrence and prevention of disease are highly desirable, even essential. Zoos would be an excellent source for detecting infections and collation with overt disease.
3. The role of viruses in diseases of the newborn in captivity, let alone in the wild, is not well understood, nor is their role in abortions and stillbirths nor their effect on breeding.
4. Little is known about diseases occurring in the wild compared with diseases of captivity. The susceptibility of different species is not well understood nor is the influence of geographic and ecologic factors.
5. Careful evaluation of simian viruses in alien species (species other than the original host) is necessary. Most isolates have been derived from rhesus, African green monkeys, chimpanzees, baboons, and squirrel monkeys. Are there others?
6. What is the role of nutrition on viral diseases? The role of genetics?
7. What is the significance of retroviruses? Is their omnipresence in all vertebrates of consequence? Is SAIDS the tip of the iceberg in relation to the entire potential of retroviruses? What is the potential danger to other species? To humans?
8. There is a need for model systems for studies on cancer, exotic diseases, and gastroenteritis. Closer collation among all disciplines is also needed. There is a lack of collated studies; very few investigators have their sick or dead

animals completely studied for the precise etiology of the illness (bacteriology, mycology, parasitology, virology). Many investigators do not understand that "pathology" *does not* provide etiology.

9. What systems may be substituted for the nonhuman primate in biologic testing? Source of tissues? Molecular hybridization?
10. What mechanisms are necessary to protect colonies from infectious agents, particularly those colonies that are "self-sustaining"? What will be the effect of infectious diseases on naive (virgin) populations?
11. There is a continued need for the development and evaluation of diagnostic procedures as well as for rapid procedures.
12. A realistic appraisal of arbovirus infections is needed along with studies to ascertain their possible danger to captive colonies.
13. Virus vaccines still need to be developed.

All members of the panel expressed a fear that the majority of investigators using nonhuman primates in their research were their own worst enemy. Unless investigators markedly change their pattern of thinking, self-sustaining colonies of simians are an unreachable dream. It was agreed that continued studies on primates were necessary. As Koch so aptly stated the situation: "Meine Herren, vergessen Sie nie, die Mäuse keine Menschen sind."

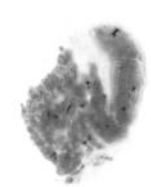

72
Strategies for the Conservation of Highly Endangered Primates

RUSSELL A. MITTERMEIER

Introduction

Of the 200 living species of primates currently recognized, about one in every seven are already considered endangered. Some of these (e.g., the golden-rumped lion tamarin, *Leontopithecus chrysopygus*; Perrier's diademed sifaka, *Propithecus diadema perrieri*) are literally on the verge of extinction, and most of the others could easily be extinct by the turn of the century if action on their behalf is not taken in the very near future. Some of the world's most endangered primate species are listed in Table 72.1, which begins with the only endangered primate family, the Daubentoniidae, and also includes endangered genera, species, and subspecies. This list is far from complete and is in need of constant modification, but it serves to indicate the large number of primates that may no longer be with us in a few decades if adequate conservation measures are not instituted now.

In developing a program for the conservation of an endangered species, several steps have to be followed, and these can be broken down into three broad categories and several subcategories that are discussed briefly in the remainder of this article.

Locating Remaining Populations of the Target Species and Determining Basic Geographic Distribution

Before any conservation can begin, it is essential to know where the species of concern is found and approximately how many individuals remain. This usually involves survey work of the most basic kind and seems very obvious, but the fact remains that we still know remarkably little about wild populations of even some of the best known endangered primates.

In carrying out surveys like these, it is necessary to pay particular attention to locating populations of the target species in existing protected areas, since these are the places in which the species will be most likely to survive, and in obtaining some indication of population numbers in such protected areas. This will help in determining whether or not the park or reserve in question is likely to be adequate for the long-term survival of the species of if it merely harbors a disappearing population remnant.

TABLE 72.1. The world's mosts endangered primates.[1]

Genus/species	Range	Comments
Endangered families		
Daubentoniidae		
D. madagascariensis	Formerly the eastern forests of Madagascar	A monotypic family; highly endangered by habitat destruction and killed on sight by villagers as evil omen; no population estimates available
Endangered monotypic genera		
Brachyteles		
B. arachnoides	Formerly the southeastern portion of Brazil's Atlantic forest region	A monotypic genus; highly endangered by habitat destruction and poaching; 300–400 remain
Allocebus		
A. trichotis	Eastern forests of Madagascar	A monotypic genus; known from four specimens, three of them collected in the last century; possibly extinct?
Indri		
I. indri	Eastern forests of Madagascar	A monotypic genus; endangered by habitat destruction within its small and patchy range; no population estimates available
Varecia		
V. variegata	Eastern forests of Madagascar	A monotypic genus; endangered by habitat destruction and hunting; no population estimates available
Simias		
S. concolor	Mentawai Islands off Sumatra, Indonesia	A monotypic genus; endangered by habitat destruction and hunting and the most endangered of the four Mentawai primates; population estimated at 10,000
Endangered polytypic genera		
Leontopithecus *L. rosalia* *L. chrysomelas* *L. chrysopygus*	Southeastern section of Brazil's Atlantic forest region in states of Rio de Janeiro, Bahia, and São Paulo	All three species highly endangered by habitat destruction and illegal live capture within their very small ranges; only a few hundred *rosalia* and *chrysopygus* survive in the wild; *chrysomelas* also depleted, but no recent population estimates available
Rhinopithecus *R. roxellanae* (including *roxellanae*, *brelichi*, and *bietii* as subspecies) *R. avunculus*	China and Vietnam	All four taxa endangered by habitat destruction and perhaps also hunting; *R. roxellanae brelichi* down to 500 individuals; *R. roxellanae bietii* to about 200; *R. roxellanae* to 3700–5700; *R. avunculus* known from only a handful of museum specimens, possibly extinct?

TABLE 72.1. (Continued).

Genus/species	Range	Comments
Endangered species and subspecies		
1. Species/subspecies numbering only in the hundreds		
Hapalemur simus	Now known only from the humid forest east of Fianarantsoa on the east coast of Madagascar	Although no population estimates are available, this species has been considered extremely rare and on the brink of extinction
Propithecus diadema (including *diadema*, *candidus edwardsi*, *holomelas*, and *perrieri* as subspecies)	Eastern forests of Madagascar	Although no population estimates are available, the species as a whole is considered highly endangered because of habitat destruction and perhaps some hunting, and all subspecies must exist at low population sizes; *P. d. perrieri* is the rarest
Macaca silenus	Western Ghats of south India	Highly endangered by habitat destruction and occasional hunting; population estimated at 670–2000
Colobus badius gordonorum	Uzungwa Mts. and Magombera Forest Reserve in Tanzania	Endangered by habitat destruction and heavy hunting pressure; Magombera Forest Reserve, thought to be its main stronghold, has a population of only 300 animals
Gorilla gorilla beringei	Virunga Volcanoes of Zaire, Rwanda, and Uganda; Bwindi Forest in Uganda	Highly endangered by habitat encroachment and poaching; total population in the Virungas 255 animals, in Bwindi 95–130
2. Species/subspecies numbering in the low thousands		
Macaca nemestrina pagensis	Mentawai Islands off Sumatra, Indonesia	Endangered by habitat destruction and hunting in very restricted range; population estimated at 39,000
Cercocebus galeritus galeritus	Tana River, Kenya	Endangered, very small population in fragmented habitat; total numbers estimated at 1200–1700
Colobus badius rufomitratus	Tana River, Kenya	Endangered, very small population in fragmented habitat; total numbers estimated at 1400–2000
C. b. kirkii	Zanzibar	Very small population in fragmented habitat; total numbers estimated at 1700
C. b. preussi	Cameroon	Endangered by habitat destruction and hunting; population estimated at less than 8000
Presbytis potenziani	Mentawai Islands off Sumatra, Indonesia	Endangered by habitat destruction and hunting; population estimated at 46,000
Hylobates klossii	Mentawai Islands off Sumatra, Indonesia	Endangered by habitat destruction and hunting in very restricted range; population estimated at 36,000
H. moloch	West Java	Endangered by habitat destruction in already fragmented habitat; total numbers estimated at 2400–7900

TABLE 72.1. (Continued).

Genus/species	Range	Comments
Pan troglodytes verus	West Africa from southern Senegal to western Nigeria; now largely restricted to Guinea, Sierra Leone, Liberia, and Ivory Coast	Endangered by habitat destruction, hunting, and live capture for export; total in known habitats estimated at 1500; in potential habitats 15,700
P. paniscus	Central Zaire east of the Zaire River	At risk from habitat destruction and hunting in parts of its range; numbers in known habitats about 2200; in potential habitats 13,000
Pongo pygmaeus (including *pygmaeus* and *abelii* as subspecies)	Borneo and Sumatra	Threatened by habitat destruction and occasional live capture; 5000–15,000 still thought to exist in Sumatra, 3500 in Sabah, and about 250 in Sarawak; no published estimate for Kalimantan; recent unpublished information indicates that the total population may be higher than previously believed
Gorilla gorilla graueri	Eastern Zaire	Endangered by habitat destruction and hunting; total numbers estimated at 4000
G. g. gorilla	Western Africa (including Cameroon, Gabon, Central African Republic, Congo Republic, Equatorial Guinea, Nigeria, and Angola)	Threatened by habitat destruction and hunting; total numbers unknown, but recent estimate for Gabon 35,000+7000 indicates that total population may be higher than previously believed
Species/subspecies for which population estimates not available, but known to be of conservation concern		
Mirza coquereli	Forests of western Madagascar	May be endangered
Lepilemur mustelinus (including *mustelinus, edwardsii, microdon,* and *septentrionalis, ruficaudatus, leucopus,* and *dorsalis* as subspecies)	Forest of western and southern Madagascar	May be endangered
Lemur macaco	Coastal areas of northwestern Madagascar and islands of Nosy Be and Nosy Komba	Probably endangered or vulnerable
L. mongoz	Northwestern Madagascar and Moheli and Ndzouani in the Comores	Endangered both on the Madagascan mainland and the Comores
L. rubriventer	Sparsely distributed in eastern forests of Madagascar	Poorly known, but probably endangered

TABLE 72.1. (Continued).

Genus/species	Range	Comments
Callithrix flaviceps	Atlantic forest region of eastern Brazil in southeastern Espirito Santo and adjacent parts of Minas Gerais	Endangered by widespread habitat destruction
C. aurita	Atlantic forest region of eastern Brazil in São Paulo and adjacent parts of Minas Gerais and Rio de Janeiro	Endangered by widespread habitat destruction
C. geoffroyi	Atlantic forest region of eastern Brazil in Espirito Santo and Minas Gerais	Endangered by widespread habitat destruction
Saguinus oedipus	Northwestern Colombia	Endangered by widespread habitat destruction and live capture for export
Saguinus bicolor bicolor	Vicinity of Manaus in Brazilian Amazonia	Endangered by habitat destruction in a very small range
Callimico goeldii	Widely but very sparsely distributed in upper Amazonia	A rare species about which very little is known
Callicebus personatus (including *personatus*, *melanachir*, and *nigrifrons* as subspecies)	Atlantic forest region of eastern Brazil from southern Bahia to São Paulo	Endangered by widespread habitat destruction and hunting in some areas
Saimiri oerstedii	Western Panama and southern Costa Rica	Endangered by widespread habitat destruction
Chiropotes satanas satanas	Lower Brazilian Amazonia	Endangered by habitat destruction and heavy hunting pressure in its restricted range
Cacajao calvus calvus	Upper Brazilian Amazonia, between the Rio Japurá, the Rio Solimoes, and the Rio Auatí-Paraná	Possibly endangered
Cebus apella xanthosternos	Atlantic forest region of eastern Brazil, restricted to small area in southern Bahia	Endangered by widespread habitat destruction and heavy hunting pressure
Alouatta fusca (including *fusca* and *clamitans* as subspecies)	Atlantic forest region of eastern Brazil from southern Bahia to Rio Grande do Sul	Endangered by widespread habitat destruction and hunting; the northern subspecies *fusca* may be close to extinction
Lagothrix flavicauda	Cloud forest region of the northern Peruvian Andes	Endangered by habitat destruction and hunting
Ateles geoffroyi azuerensis	Azuero Peninsula of Panama	Endangered by habitat destruction and hunting; possibly on the verge of extinction?

TABLE 72.1. (Continued).

Genus/species	Range	Comments
A. fusciceps fusciceps	Pacific slope forests northern Ecuador	Endangered by habitat destruction and hunting; possibly on the verge of extinction?
A. belzebuth hybridus	Disjunct distribution in northern Colombia and northern Venezuela	Endangered by habitat destruction and hunting
Cercocebus galeritus sanjei	Uzungwa Mts., Tanzania	A recently discovered subspecies quite restricted in range
Cercopithecus erythrogaster	Southern Nigeria	Endangered by habitat destruction and hunting
C. erythrotis erythrotis	Fernando Po in Equatorial Guinea	Thought to be endangered
Papio leucophaeus	Cameroon, eastern Nigeria, and Fernando Po	Endangered by heavy hunting pressure
Colobus satanas	Cameroon, Equatorial Guinea (including Fernando Po), Gabon, and Congo Republic	Endangered by habitat destruction
C. badius bouvieri	Congo Republic	Possibly endangered
C. b. pennanti	Fernando Po	Possibly endangered
Presbytis aygula	West Java	Endangered by habitat destruction
Pygathrix nemaeus	Laos, Cambodia, Vietnam, perhaps Hainan in China	A monotypic genus; poorly known and possibly endangered or vulnerable
Hylobates pileatus	Thailand, Laos, and Cambodia	Endangered by habitat destruction and hunting in Thailand; status unknown in Laos and Cambodia

[1] Species are ranked first according to taxonomic uniqueness, and then according to degree of depletion of wild populations.

During the course of such basic survey work, it should also be possible to identify the major threats to the endangered primate species and to begin thinking about ways to improve protection of surviving populations.

Improving Prospects for Survival in the Wild

Once one knows where the species still occurs, a number of steps can be taken:

1. Conduct preliminary ecological and behavioral studies to determine the basic requirements of the species, with the eventual goal of developing management plans to ensure the survival of the species and its habitat.
2. Improve protection in existing parks and reserves since many of these exist mainly on paper and are protected inadequately or not at all. This should usually be done by working in collaboration with governmental and non-

governmental conservation agencies in the country where the work is being conducted and frequently will require infusion of outside funding. It is also important to recognize the importance of privately protected parks and reserves as well, since these are often in more pristine condition than government protected areas.

3. Establish new protected areas if none exist within the range of the species or if existing protected areas appear inadequate. This can sometimes be done right after initial survey work, but more often requires more in-depth ecological studies to determine the most appropriate sites for protected areas. Again, this has to be done in conjunction with governmental and nongovernmental agencies and can be a time-consuming and expensive process.
4. Develop long-term research programs and field stations to provide a constant scientific presence in the most important areas for the primate species or community of concern. This not only serves to gather longitudinal data on the biology of the animals and their habitat, but also functions as a very effective deterrent against poaching and habitat encroachment. Indeed, the international prestige that such a program may communicate to local people often serves to protect a species much more effectively than a large guard force.
5. Determine the genetic makeup of existing populations, especially in the case of species with small, fragmented, widely separated populations (e.g., the muriqui, *Brachyteles arachnoides*; the golden-rumped lion tamarin, *Leontopithecus chrysopygus*). The eventual goal of such studies would be to develop an effective management plan for the entire remaining population and, if necessary, to develop a translocation program to ensure that the genetic diversity of the species is maintained. Needless to say, such manipulation could only occur after the ecological and behavioral requirements of the species had been carefully studied and the areas in which the remnant populations survive had been adequately protected.

Increasing Public Awareness of the Importance of the Species and Its Habitat

Without the support of local people in the countries in which endangered primates occur, programs on their behalf are unlikely to be successful. Education programs and public awareness campaigns should be an integral part of any broad primate conservation program and should focus first on people in the immediate vicinity of the protected area(s) in which the species occurs and second on government officials who are responsible for making and implementing conservation policy. Fortunately, primates are usually attractive and appealing creatures that are often ideal as symbols of public awareness campaigns. Using them as so-called "flagship" species can result in improved protection of other species as well, and they are often the best vehicles for conveying the importance of tropical forest conservation as a whole. Species that have been used effectively as flagship species include the mountain gorilla (*Gorilla g. beringei*), the muriqui (*Brachyteles arachnoides*), the golden lion tamarin (*Leontopithecus rosalia*), and the yellow-tailed woolly monkey (*Lagothrix flavicauda*).

Training Nationals from Countries in Which the Endangered Species Occur

Training of students and researchers from the developing countries in which the vast majority of primates occur is essential for the long term. Foreign researchers can only make a limited contribution to ensuring the survival of endangered wildlife; in the final analysis, success will depend on initiatives by people living in the countries where these animals are found. Conservation-oriented training of promising young students from these countries should be a high priority and should include everything from local training of guards and park wardens up to Ph.D. level training abroad.

Develop Captive Breeding Programs for Selected Endangered Species

Finally, it is necessary to determine if captive breeding is a viable option for the species in question. If it appears both viable and politically feasible in the broader context, efforts should be undertaken to see that such a program begins as soon as possible. The first question that should be asked is whether or not there are individuals of the species in captivity, and, if so, could they be better managed. The golden lion tamarin (Kleiman, this volume) is an excellent example of a species that was reasonably well-represented in captivity and has benefited greatly from a carefully managed program that has now been underway for more than 12 years. The spider monkeys (*Ateles* spp.) and the cotton-topped tamarin (*Saguinus oedipus*) are examples of species that are abundant in captivity but are not yet managed effectively. All endangered species currently in captivity should be rigorously managed, and the Studbook and SSP Programs of the AAZPA and other organizations provide an excellent framework for making this possible.

If the endangered species in question is not yet represented in captivity or exists only in very small numbers, one must ask whether it is justified to remove individuals from the wild to develop a captive program. If it is deemed appropriate to capture wild individuals, these should be taken from unprotected forests and preferably ones that are already slated for destruction; animals should never be removed from parks or reserves unless these areas are totally inadequate for the survival of the species. Other factors (e.g., politics in the country in which the species occurs; long-term commitment to breeding the species by a number of zoos or other institutions; ease of breeding and maintaining large captive populations, etc.) will also have to be considered before a final decision is made.

An example of an endangered primate for which captive breeding would not be appropriate at this time is the mountain gorilla. This animal occurs only in protected areas, and several international organizations are helping to improve protection of these parks and reserves. The governments of the three countries (Rwanda, Uganda, Zaire) in which mountain gorillas occur have shown considerable interest in its conservation, and the Rwandan government has already made a major commitment to it. Efforts to remove animals from the wild would send all the wrong messages to dedicated conservationists in these African countries.

On the other hand, captive breeding appears to be a very viable option for several endangered *Callithrix* species from Brazil's Atlantic forest region (e.g., *C. geoffroyi*, *C. aurita*, *C. flaviceps*, *C. kuhli*). These animals breed very well in captivity and can still be obtained from unprotected forest fragments that would otherwise be destroyed in a few years' time. Furthermore, the lead in developing captive breeding programs for these species is being taken by a Brazilian institution, the Rio de Janeiro Primate Center, in collaboration with foreign zoos, and only captive-born animals are being shipped out of Brazil. Given the rate at which these adaptable monkeys breed in captivity, it is quite possible that there may be more of them in captive colonies than in the wild by the turn of the century.

Summary

Needless to say, the five major steps outlined above do not have to be undertaken sequentially. Some survey work is obviously necessary before any other field activities can begin, but ecological studies at one or two sites can be undertaken while surveys are underway, and efforts to improve protection of parks and reserves, to develop new reserves, to increase public awareness, and to train local people can and should be carried out simultaneously with long-term field studies. Indeed, a successful conservation program for an endangered species should include ongoing efforts in all these categories. Captive breeding can also be carried out hand-in-hand with conservation of wild populations and should not be viewed only as a last-ditch effort for species that are on the brink of extinction. Indeed, in the future, we would hope that collaborative efforts between zoos and field workers will increase, and that carefully planned programs tailor-made to the needs of the endangered species in question will prevent primates from approaching the brink of extinction and will help to ensure that the current diversity of the Order Primates is maintained.

References

Ali R (1982) Update on the status of India's lion-tailed macaque. IUCN/SSC Primate Specialist Group Newsletter 2:21.

Davies G, Payne J (1982) A faunal survey of Sabah. Kuala Lumpur, Malaysia, World Wildlife Fund-Malaysia.

Harcourt AH (1983) Conservation and the Virunga gorilla population. Afr J Ecol 21:139–142.

Harcourt AH (1984) Conservation of the Virunga gorillas. IUCN/SSC Primate Specialist Group Newsletter 4:36–37.

Kappeler M (1981) The Javan silvery gibbon (*Hylobates lar moloch*)—habitat, distribution, numbers. Ph.D. dissertation, University of Zurich.

Marsh C (1978) Problems of primate conservation in a patchy environment along the lower Tana River, Kenya. In: Chivers DJ, Lane-Petter W (eds) Recent advances in primatology, vol 2. Conservation. Academic, London, pp 86–87.

Poirier F (1983) The golden monkey in the People's Republic of China. IUCN/SSC Primate Specialist Group Newsletter 3:31–32.

Rijksen HD (1978) A field study on Sumatran orang utans (*Pongo pygmaeus abelii* Lesson 1827). H. Veenam & Zonen B.V., Wageningen, Netherlands.

Rodgers WA (1981) The distribution and conservation status of colobus monkeys in Tanzania. Primates 22(1):33–45.

Silkiluwasha F (1981) The distribution and conservation status of the Zanzibar red colobus. Afr J Ecol 19:187–194.

Tattersall I (1982) The primates of Madagascar. Columbia University Press, New York.

Tattersall I (1983) Studies of the Comoro lemurs: a reappraisal. IUCN/SSC Primate Specialist Group Newsletter 3:24–25.

Teleki G, Baldwin L (1979) Known and estimated distributions of extant chimpanzee populations (*Pan troglodytes* and *Pan paniscus*) in Equatorial Africa. Special report to the IUCN/SSC Primate Specialist Group.

Tutin C, Fernandez M (1982) Preliminary results of a gorilla and chimpanzee survey in Gabon. IUCN/SSC Primate Specialist Group Newsletter 2:19.

World Wildlife Fund (1982) Forest pack, No 5, Launch Edition. The World Wildlife Fund Tropical Forests and Primates Campaign, October 10, 1982. World Wildlife Fund, Gland, Switzerland.

73
Research Needs in Captive Primate Colonies

KURT BENIRSCHKE

Introduction

When the speakers for this conference were invited to participate, they were also admonished to highlight areas that demand attention for research. Specifically, they were challenged to address questions whose resolution, accomplished by targeted research, would materially aid in the creation of self-sustaining populations of captive primates. The following, then, are the most important points brought out in presentations and their discussions; however, most papers here reproduced will raise additional topics whose pursuit is deemed important and which might serve as guideposts to funding agencies.

Some aspects discussed are so elementary that they need hardly be mentioned. They fall under the category of expected animal husbandry. Thus, it is obvious that attention must be paid to immunization schedules, to surveillance of parasite loads—better, to attempts at eradicating parasites—to dental prophylaxis and therapy, to increasing awareness of virus diseases and their avoidance by monitoring keeper staff and mixing different taxa. For captive populations in particular, the avoidance of boredom through better design of enclosures, "psychotherapy," or an increase in appropriate furniture and varied diets is a challenging need. There were three overriding concerns mentioned so often, however, that they need brief and special discussion here. They are perinatal mortality, nutrition, and genetics.

Perinatal Mortality

It is apparent that nonhuman primates have an excessive perinatal mortality in captivity, presumably much higher than that occurring in wild populations. The reasons for this destructive feature are not yet apparent and need immediate and critical attention. It is also in this area that a combined approach by many different scientific disciplines may be helpful. It is apparent from the few detailed pathologic inquiries that exist into perinatal mortality that more often than not the true etiology and pathogenesis of perinatal deaths are elusive. The same was true a few decades ago in human medicine, and, as a result, a new discipline was born, perinatology. In many medical centers, there exists highly developed expertise to deal with extremely small human infants. Their help and equipment are often available to us, because many of these colleagues delight in being able to be of service to infant primates. Their technology, such as ultrasound, radiology, and incubators with waterbeds, should be developed in the care of perinatal problems of primate colonies. A better pathologic appraisal of the deaths but also a more thorough investigation of the social and behavioral environment of the failing mother are needed. For it is apparent that "mismothering" is a frequent cause of neonatal death. Indeed, Nadler's survey revealed that over 80% of infant gorillas are pulled for hand-rearing! In this vein, the discussion ensuing the paper delivered by Dr. Williams should be of great interest. Dr. Short related that ewes that reject their young would immediately accept them when a vaginal balloon was inserted and inflated. Dr. Hodgen indicated that vaginal palpation in macaques may have a similar effect. These simple measures could be of great benefit if systematically explored. Behaviorists' input in solving these problems is needed. But also it will be of great importance to anticipate more accurately the births of primates in order to rearrange the social structure, if needed, to provide appropriate shelter—e.g., waterbeds—and give additional overall attention to the parturient when needed. Toward this end the development of techniques that allow pregnancy diagnosis and perhaps the staging of pregnancies is essential. One can anticipate that "dipstick" methods for urinary hormone analysis are on the horizon and that they would find a welcome addition to the presently sparse armamentarium of colony managers. Perhaps the increased use of video equipment and better records on early rearing experience will provide clues as to causes of mismothering and perinatal deaths.

Nutrition

We have been complacent in the feeding of primates because of the availability of primate biscuits that we thought were able to fulfill the needs of all nonhuman primates. The realization of complex stomachs in leaf-eating primates, greatly variable milks of different species, the requirement of vitamin D_3, and other relatively recent findings suggest that the great heterogeneity of primates is not satisfied by feeding one staple diet. Above all, even if nutritionally adequate, its application two or three times a day may lead to satiety but also to boredom in animals that normally spend an enormous amount of time searching for food. We have recognized here that little is known of the normal nutritional content of the browse animals ingest in the wild, the fiber content needs, the manner in which nocturnal primates acquire vitamins, the need for specific wavelengths of light, and many other aspects of a well-balanced nutrition. We have recognized the great need to investigate the nutrients consumed in the wild, how animals acquire essential vitamins and minerals, and what these are. In short, the whole complex topic of more adequate nutrition needs thorough investigation and then computerization, say with ISIS, so as to make it universally available quickly. We have not paid enough attention to the fact that nonhuman primates are a very diverse group, having presumably very diverse needs. While we have eliminated rickets because of the discovery of the vitamin D_3 needs in South American primates, we still believe that all diarrheas must have an infectious origin. In lemurs, for instance, it may equally well be the result of too high a CHO content in the biscuits, but then we don't know what the normal stools of wild lemurs are like, to cite just one example.

Genetics

Given the fact that few new primates will be imported—possibly none in the not too distant future—it is essential that the extant population of captive primates be managed appropriately. First of all, this requires attention to pedigrees that are available for some endangered species and to some extent through the ISIS files. It is imperative, however, to consider the need for similar information on *all* primates in captivity, not only those now listed as endangered. This would perhaps be accomplished easiest if ISIS were more widely used and were more securely funded. Some species, perhaps most, would benefit from expanded use of karyology. This has been most obviously demonstrated by the finding of different chromosome numbers in *Aotus* and different chromosomes in *Pongo*. It is now evident that *Ateles* has many different karyotypes, which must be understood lest there develop innumerable hybrids in captivity. We have learned that an urgent need for precise genetic delineation of chimpanzee subspecies exists. Such an RFP (request for proposal) should include study of DNA (including mitochondrial DNA) and other taxonomically decisive features of the few animals now available with known origin and, as R. Short has pleaded, a search for a possible new form of chimpanzee in Gabon described in older reports needs to be encouraged. Likewise, modern molecular biology must be brought to bear promptly on deciphering the relationship of mountain and the two forms of lowland gorillas, not a difficult task when specimens are available. Towards this goal, O. Ryder has included in this volume a precis for the adequate collection of this material in the field and its transfer to the laboratory. The captive population of Geoffroy's tamarins detailed by D. Meritt should be considered as a possible species to study exchange of gene flow between captive and wild species, as could also be undertaken with lemurs. Finally, greater attention needs to be paid to heritable traits in captive reproductive groups. Not only will this benefit the field of "animal models," but also such information is ultimately needed for the benefit of captive primates; witness the pectus excavatum found in ruffed lemurs as a dominant trait.

There is perhaps another need for research in captivity, and this refers to the less glamorous species. Much emphasis is paid to the highly visible animals such as gorilla, orangutan, woolly spider monkey, and those animals identified with human applied research potential, the macaques, baboons, squirrel monkeys, etc. There is a wealth of species whose detailed study would greatly add to our knowledge of evolution, physiology, and many other disciplines, which are at least as scarce and perhaps even more endangered, e.g., the Madagascan prosimians. The great diversity of their biology as detailed by Jolly and by Pollock testifies to the greater interest scientists should pay to these taxa. Living primarily in zoological gardens, they are underinvestigated and at least as much in need of greater insight as are their more visible successors. Their seasonality, reproductive strategies employed, and nutritional components all should be fascinating topics for investigation. Here is a problem where different levels of society need greater collaboration than has been at work in the past.

Finally, with the recognition of retroviruses in humans, as well as some nonhuman primate species, has come the awareness that much more attention needs to be paid to their impact on self-sustaining primate populations, a point well made in Rabin and Benveniste's contribution.

74
Researchable Problems in the Natural Realm

DONALD G. LINDBURG

Indulge with me in a bit of fantasy. The year is 1990, and the occasion is the Morris Animal Foundation's Fifth Annual Conference on self-sustaining populations. I had the good fortune of attending the first conference in this series, held in San Diego in 1985. At that time I was a graduate student at Berkeley. I now have degree in hand and will leave shortly for South India to study the highly endangered lion-tailed macaque. As the first grantee of the new field conservation program of the American Association of Zoological Parks and Aquariums, I've been asked to give a report on my plans to this conference. Funding of the AAZPA field program, as many of you known, has become available through the charging of a small conservation tax on admissions at zoos all over the United States.

I am pleased to see that Kurt Benirschke is in attendance at this conference. Some here will remember that he organized the first one in this series back in 1985. Kurt is now Director-Emeritus of Research at the San Diego Zoo but remains very involved in conservation causes in his operation of the grants program for the AAZPA. It is to him that I will present my final report on the lion-tailed macaque when I return from India 2 years hence.

At the time of our 1985 meeting, the most recent field surveys indicated about 1000 lion-tails left in the wild. Scientific colleagues in India have been monitoring this population very closely over the years and currently report a state of decline. I am most fortunate to have the financial resources enabling me to join up with Indian primatologists in the search of information on the failure of this rare primate to achieve self-sustaining status.

The captive population also continues to grow, and in fact the numbers are so good that there is some thought of possible reintroduction into the wild, provided suitable habitat remains available. At the same time, managers of the captive population can expect their efforts to improve with additional information from the wild state on diet and nutrition, genetics, behavior, and reproduction.

As I noted a moment ago, we have reason to believe that current census figures from the wild are firm. We also know a great deal about the population structure of this species. The publication in 1981 of the manual on "Techniques for the Study of Primate Population Ecology" (Subcommittee on Conservation of Natural Populations, 1981) and its dissemination by the World Wildlife Fund to scientists in the Third World has made it possible to obtain solid information on lion-tails and on many other endangered primates. Today, we also have the benefits of satellite mapping, which provides us with detailed information on the loss of habitat. Even so, a closer look at the causes of decline is required, and I am here today to report on a few of the investigations we plan to undertake. My program draws inspiration from the coalescing views of field biologists, zoo conservationists, and biomedical researchers in evidence at the 1985 conference. Despite many different specialties and philosophical perspectives, one sensed at that time a unanimity in forging ahead with programs that would secure the future for some of nature's primate diversity. The time taken by knowledgeable individuals to join in discourse over a cup of coffee or a beer with me and other neophyte primatologists in attendance was a most gratifying aspect of that conference.

Not only are the numbers of lion-tails left in the wild alarmingly low, but like many of the world's endangered species, the remaining population is fragmented. Lion-tails are obligate rainforest primates, and that forest which remains exists only in small patches. Charles Southwick suggested to me the importance of assessing the diversity of food plants remaining in some of these patches. Insofar as there is seasonality in the appearance of fruit in the tropics, the possibility of food scarcity at certain seasons could place resident lion-tail troops under nutritional stress. Remedial action could take the form of translocation for deprived troops, or even short- and long-term measures to subsidize existing food sources.

Competition with other species sharing the same foods, for example with hornbills and squirrels, will be investigated. In the past, primatologists have in many cases focused their attention almost exclusively on the primates themselves, with only casual attention to other life forms in the same community. At the San Diego conference, David Western (this volume) made a strong case for an ecosystems approach, i.e., careful attention to the place of particular species within a community of life forms, as essential to understanding the reasons why populations flourish or decline.

Our plans call for study of the availability of food in time and space and how lion-tails are able to locate food. In her studies of neotropical primates, Katherine Milton (this volume) pointed out the complexity of food distribution in tropical forest habitats, and our poor understanding of the ways in which animals such as spider monkeys know when and where to find it. It would seem that a rather detailed mental map of the forest is required, and this may entail information that is acquired over the lifetime of individuals, perhaps even across generations. This prospect should give us pause in advocating reintroductions into the wild, for we have to ensure that captive-reared candidates are up to the task of learning to harvest these food sources.

In noting the fragmentation of populations such as the lion-tailed macaque, many individuals at the 1985 conference wondered about the genetic consequences of that situation. We know very little about genetic diversity in the wild. David Glenn Smith (this volume) spoke to us about inbreeding depression in captive populations, leading us to wonder about a future in which certain individuals may have to be transferred between wild troops, many of which now exist as genetic isolates. We are confident that this can be done for lion-tails, since other macaques and several baboon troops have previously been trapped or darted in order to collect the necessary samples.

Time permits me to discuss only one further point, and that is reproductive studies. In the first field reports on lion-tails, Green and Minkowski (1977) stated that females in the wild first reproduce at 5 years of age and appear not to survive much beyond the age of 9. They further suggested that interbirth intervals approached 20 months. Therefore, a breeding female in the wild would contribute only two or three offspring during her lifetime, on the average. I should stress that this was a first assessment of reproductive potential in the wild, based on visual estimates of age. It is at odds with reports for other wild populations of macaques. There are nevertheless hints from other, more recent reports, that offspring production in this species is on the low side (e.g., Kumar and Kurup, 1981). The difficulty with these assessments is that up to now, reproductive potential has been based on what can be seen with the naked eye—sex skin swelling, episodes of mating, estimates of age, and the appearance of infants. If it should prove to be the case that fecundity is low in certain wild populations, we lack the techniques for determining why that should be the case.

I am happy to report that in the captive sector a new technology has recently been developed that will now enable us to ascertain *when* and *if* a female primate in the wild is fertile. For several years now, it has been more or less routine to assess fertility in captive subjects by measuring ovarian steroid excretion in small samples of urine. Unfortunately, urine is virtually impossible to collect in the wild state. The only known example is a study with Amboseli vervet monkeys (Andelman et al., 1985), which habitually rest low in acacia trees at night. Field workers in this instance would arise in the wee hours of the morning, and with containers held over their heads, would dash about under acacia trees in an attempt to intercept the urine stream of voiding females. That a reasonable measure of success was obtained is an indication of the insightful and determined efforts of these investigators. But, alas, it is not an approach that has broad application.

So, we come to the breakthrough that I mentioned a moment ago. Many field workers have informed me that there is one biologically relevant substance that is easily collected in almost every case, and that is feces. It has been known for a long time that as much as 95% of specific steroids in at least some mammals are excreted through the gut. Sam Wasser (personal communication), at the University of Washington, and Bill Lasley (see earlier work on avian species; Czekala and Lasley, 1977) at U.C.-Davis have been working diligently on the problems of steroid extraction from primate fecal matter and in adjusting sample values for metabolic clearance rates. Not only do we now have a technique for measuring ovarian activity in individual lion-tail females, but we have it in the form of a field kit. I will be in the exhibit room afterward, for any who might be interested in examining our fecal field analyzer.

Finally, I want to conclude with a few more words about the funding of my project. That, too, began at the memorable 1985 conference in San Diego, with Roger Short's reference (this volume) to the "call of the wild." It seems that zoos, which, for decades upon decades have taken thousands upon thousands of animals from the wild, have finally realized that a debt is owed. They have discovered that the public is more than willing to add a few cents to each admission dollar to return something to the wild. They have discovered, furthermore, that their credibility as conservers of wildlife has increased immeasurably more than could ever be realized from support of all their captive breeding programs. I understand as well that natural history museums are about to follow suit. Sensitized to the fact that their shelves are stocked with thousands upon thousands of specimens, collected to satisfy human curiosity, they too have heard the "call of the wild" and are taking steps to provide financial support of conservation activities. On this happy note I conclude my report. I'm off to the field. So long, Kurt, I'll see you in a couple of years.

Acknowledgments. I thank my wife, Linda, for suggesting to me that a futuristic approach would have potential as a way of "summing up" several of the points relevant to research on wild populations.

References

Andelman SJ, Else JG, Hearn JP, Hodges JK (1985) The non-invasive monitoring of reproductive events in wild vervet monkeys (*Cercopithecus aethiops*) using urinary pregnanediol-3-glucuronide and its correlation with behavioural observations. J Zool London 205:467–477.

Czekala NM, Lasley BL (1977) A technical note on sex determination in monomorphic birds using faecal steroid analysis. Intern Zoo Yrbk 17:209–211.

Green S, Minkowski K (1977) The lion-tailed monkey and its South Indian rain forest habitat. In: Prince Rainier, Bourne GH (eds) Primate conservation. Academic, New York.

Kumar A, Kurup GU (1981) Infant development in the lion-tailed macaque, *Macaca silenus* (Linnaeus): the first eight weeks. Primates 22:512–522.

Subcommittee on Conservation of Natural Populations (1981) Techniques for the Study of Primate Population Ecology. National Academy Press, Washington, DC.

Appendix A
Collection and Handling of Animal Specimens for Detection of Viral Infections[1]

SEYMOUR S. KALTER

The laboratory detection of an infectious agent is dependent upon the availability of appropriate specimens. These specimens are limited in number and consist of: blood (probably the most useful for the detection of viremia and the presence or absence of antibody), throat swabs or washings, feces or rectal swabs, urine, spinal fluid, skin scrapings, tissue biopsy, or the appropriate tissue(s) collected at necropsy. Depending upon the clinical problem, the specimens to be collected should be:

1. *From sick animals*: If there are clinical indications (cough, rash, diarrhea, etc.), obtain those specimens most appropriate to the "disease"—throat swabs or washings from animals with respiratory symptoms, skin scrapings or biopsy from skin lesions, stool samples from animals with diarrhea. Blood (acute phase) samples are always collected for serological studies (detection of antibody) as well as for virus isolation studies. Early blood samples are most important for baseline information and antibody changes. Convalescent blood samples are collected 2–4 weeks after onset for comparison with the acute phase serum to detect any significant antibody increase. Single serum samples have value for epidemiological surveys (antibody indicates past infection), detection of IgM (appears early in certain infections); high titer of antibody may have some significance, particularly in the case of exotic diseases.

2. *From dead animals*: Abnormal tissue findings indicate areas to collect. "Normal" tissue should be obtained along with that considered to be pathologic. Viable tissue is desired for virus isolation studies. Such tissue provides the laboratory with the most susceptible host cells.

Tissues and specimens for laboratory studies *must not* be placed in formalin or any other preservative. Tissues for pathologic examination must be collected separately!

---

[1]Those primate facilities without virological laboratory support may use the NIH and WHO Collaborating Center for Reference and Research in Simian Viruses, P.O. Box 28147, San Antonio, Texas 78284. Attn.: S. S. Kalter.

Collection and Handling of Specimens

Isolation of a virus is dependent upon the availability of viable tissue (specimen) or, in the case of swabs or washings, the maintenance of the specimen in such a manner as to preserve whatever viruses may be therein. For studies requiring a virology laboratory, collect and handle specimens as follows:

1. *Serum samples for serology*: Collect aseptically 1.0–10.0 ml of blood depending upon the size of the animal. Allow to clot (1–4 hours at room temperature), rim to free clot, refrigerate approximately 1 hour, centrifuge lightly, remove serum to a sterile tube, and either send to laboratory to store frozen (−20°) or refrigerated until convalescent serum is collected. Vacu-tainers are ideal for collection of blood samples. Save clot (or heparinized blood) if virus isolation studies are desired (see below for handling).
2. *Virus isolation samples*: Blood or tissue should be collected aseptically, placed in individual labeled containers with transport medium (Hanks Balanced Salt Solution or other buffered preparations as used in cell cultures plus 0.5% bovine serum albumin [virus antibody free], plus antibiotics), and then placed on *wet ice* (not frozen) and sent directly to the diagnostic lab. Most viruses will remain viable 2–4 days if kept cold. If a delay is contemplated, freeze specimens either in CO_2–alcohol mixture or in liquid nitrogen. Store at −70°C or lower until shipped. Small (approximately "pea" size) samples are all that is necessary. Biopsy materials are handled as a tissue after collecting aseptically.
3. *Swabs*: Swabs are best placed in 0.5–1.0 ml of a transport medium, then rotated vigorously in the fluid, and the fluid expressed from the swab against the inside of the tube. The swab should be sterilized and discarded.

   Throat swabs should be obtained by carefully swabbing the posterior pharynx, tonsillar area, or visible lesions with sterile swab moistened in transport medium.

   Rectal swabs are obtained by carefully inserting swab, rotating with care so as not to puncture the bowel wall, removing the feces-stained swab, and then treating as described for throat swabs in transport medium. Feces are placed in a sterile, dry container. Handle these specimens as indicated above.
4. *Shipping of specimens for virus isolation*: The collected samples, *each in its own separate labeled container*, are placed on wet ice and shipped by rapid transport to designated laboratory. Do not freeze unless specimens are to be kept (or delayed) for long periods of time, then use CO_2–alcohol bath or liquid nitrogen. Store at −70°C (or lower) if necessary until shipped. When shipped on dry ice, seal containers to prevent CO_2 from inactivating virus in specimen. Shipping of serum specimens for serologic testing *does not* require freezing or refrigeration (gamma globulins are relatively stable), but should be sent by the most rapid transport available. Tissue specimens are sent on wet ice in order for the laboratory to cultivate the "virus" containing tissue. Frozen tissue will not grow in culture although the virus may survive therein.

Seal all containers against leakage. Wrap to protect against breakage. Label all specimens for future identification. Call laboratory to advise of problem.

Appendix B
The Collection of Samples for Genetic Analysis: Principles, Protocols, and Pragmatism

OLIVER A. RYDER

Only approximately one-half of all mammal species have been karyotyped, and, for many of these, relatively few individuals only have been sampled, and from limited portions of the species' range. We stand largely ignorant of the intraspecific chromosomal variability within many taxa for which both scientific and conservation interests exist (Ryder, in press).

Electrophoretic variation in enzymatic and otherwise identifiable proteins can provide useful insights into systematics and population genetics aspects of a wide variety of organisms. Such investigations have been undertaken but only for a relatively small number of vertebrate species, and complete studies of single species or even subspecies are conspicuous by their rarity (Ryder et al., 1981; Smith et al., 1982).

However, many practical decisions necessary to properly manage endangered animal populations for conservation purposes in situ and ex situ could benefit from or require a more thorough knowledge of the systematics of the populations of interest and in context of higher taxonomic categories (subspecies, species, genera).

Field biologists have the greatest opportunity to collect data and samples for laboratory analysis that can be expected to shed light on outstanding systematic questions. Similarly, zoo and colony managers may have in their collections individual specimens or groups of specimens that could be employed usefully for addressing systematic and other biological research questions of conservation importance. All too often, advantage of such opportunities is not taken.

Recognizing that situations in the field and even in laboratory settings often pose practical problems for the collection of specimens appropriate for genetic analysis, a listing of the types of specimens useful for genetic analysis, as well as protocols for obtaining, storing, and shipping such samples, is hoped to provide a stimulus for increased numbers of collaborative studies involving field biologists, conservationists, and laboratory investigators.

Living tissue in the form of whole blood with anticoagulant such as heparin or ACD (acid, citrate, dextrose) in the form of tissue biopsies is required for chromosome studies and may, under certain circumstances, be indefinitely propagated and stored for future retrospective studies. Living tissue must be quickly transported to a laboratory facility properly equipped to sustain, propagate, process, and store the samples in a useful condition. The time frame within which living tissue samples must be delivered to such a laboratory rarely exceeds 1 week and should always be accomplished as rapidly as possible. They must never be frozen.

For small species the amount of blood that can safely be taken may limit the extent of possible studies. A rule of thumb indicates that a blood sample up to 2% of body weight may safely be taken as a single sample. For larger species, blood sample volumes in excess of those required for chromosomal and electrophoretic analysis may be processed to yield purified proteins and nucleic acids (DNA) useful for systematic studies. A sample protocol for blood sample collection and storage from a large ungulate (equid) is included at the end of this discussion.

Living tissue in the form of a sterile skin biopsy may be propagated in tissue culture and the proliferating cells utilized for a wide variety of genetic and biochemical investigations (Ryder and Benirschke, 1984). Of paramount importance for the successful collection of skin biopsies is a thorough and complete cleaning and disinfection of the sampling site. All too often, samples arrive in an advanced state of contamination with bacteria and fungi rendering the sample completely useless. Skin biopsy samples depend on sustenance from the tissue culture fluid prior to their explantation to culture flasks. Because only small numbers of cells are required to proliferate a cell culture, it is important to sample only a small piece of tissue on the order of 3 × 3 × 3 mm. Skin biopsy samples may be refrigerated but, like blood samples, must be kept from extremes of heat and cold. For this reason, shipment of samples should be performed in well-insulated containers. A sample protocol for collection and shipment of skin biopsy samples is attached.

Blood and other tissue samples preserved by freezing are useful for electrophoretic survey, DNA extraction, and other analyses. Biological activity within frozen tissues is preserved as an inverse function of temperature; the colder the better. Storage in liquid nitrogen (−196°C), at −70°C in ultracold freezers, or on dry ice (solid CO_2) are optimal. Shipment of frozen samples should include sufficient dry ice to last well beyond the anticipated length of journey. Two kilograms of dry ice tightly packed in a styrofoam insulated container will keep frozen tissue for 24 hours at best. As much as 5 kg of dry ice may be required to ship samples safely when the journey will take several days.

The availability of nitrogen vapor freezers with essentially zero-volume liquid nitrogen have been very successfully exploited by field expeditions to store frozen samples over at least 1 month's time, thus allowing for extensive sampling of field specimens for genetic analyses and safe transport back to the laboratory.

Many powerful new analyses may be performed on relatively small amounts of DNA. Although DNA samples may be obtained from samples of frozen tissue, when field conditions make collection, storage, and transport of frozen or living tissue impossible, useful systematic information can be expected to be ultimately derived by preservation of a DNA sample. In the most simple instance, mincing of organ tissue into 80–95% ethanol preserves DNA for later purification although proteinaceous components are denatured. Field purification of DNA may be considered as well. Once purified, DNA may be stable over millenia, particularly if kept in proper preservative or in a very low-humidity environment.

As opportunities to sample wild populations for genetic and systematic studies decrease with the increasing rarity of wildlife, it is no longer excusable to let occasions for specimen collection pass by. Important conservation and management decisions require input of genetic and systematic data.

Skin Biopsies for Tissue Culture

General Principles

Tissue cultures must be sterile; bacteria or fungi kill the cultures. The fresher the sample is, the better it will grow. *No antiseptic* (such as mercurochrome) should be used before biopsy; soap and alcohol are best.

Procedure

1. Clip the hair in the area to be sampled, if excessive.
2. Moisten with alcohol.
3. Use single-edged razor blade and shave, if necessary.
4. Use alcohol-drenched sponge to clean this area very, very well.
5. Open a package with dry sponge, scalpel, and forceps, preferably with teeth (sterile suture removal kits exist in medical supply houses).
6. Get the recipient bottle ready; it should contain tissue culture fluid with antibiotics.
7. Grasp piece of skin from center of shaved area with forceps and cut off a full-thickness small piece of skin, about the size of a bean or pea; place immediately into the recipient bottle and close tightly.
8. *Label* bottle with name, sex, case number, and ship immediately in prepared container. It can be cooled but must never be frozen or heated.
9. Discard scalpel in safe place. Should you contaminate the knife or scissors accidentally before biopsy by touching hair, etc., then use a new instrument. Alternatively, dip into alcohol and flame it.

Note: it is imperative that only a very small piece be biopsied. Often this is easiest in the loose skin behind the earlobe or on the back. The area does not matter. Large pieces do not do well in the small amount of fluid, and they are not needed.

Mail to: Dr. Oliver A. Ryder
Research Department
San Diego Zoo
PO Box 551
San Diego, CA 92112

Instructions for Collecting and Marking of Grevy's Zebra Horse Blood Samples

1. Five blood samples are required from each animal; use two yellow-stoppered tubes containing anticoagulant solution (ACD), two empty red-stoppered (no additives) tubes, and one empty green-stoppered tube (heparinized).
2. Bleed the animal and, using a large diameter needle (18 ga), slowly inject blood into each of the five tubes specified in step 1 above. IMPORTANT: The green-stoppered tube must be gently inverted repeatedly for about 3 minutes. This is to dissolve the heparin anticoagulant present in the green-stoppered tubes.

3. Immediately label the tubes with the following information: species (i.e., Grevy's zebra), the individual's name and any other identification, sex, and the date the samples were taken.
4. Pack the samples in the insulated box. Do not refrigerate or chill the samples. Mail the samples to: Dr. O. A. Ryder, San Diego Zoo Hospital, Old Globe Way, San Digeo, CA 92103. Use Postal Service Express Mail or air couriers, such as Federal Express.

References

Ryder OA (In press) Genetic investigations: tools for supporting breeding programs. Intern Zoo Yrbk 25.

Ryder OA, Benirschke K (1984) Value of frozen tissue collection for zoological parks. In: Collection of frozen tissues: value, management, field and laboratory procedures, and directory of existing collections. Association of Systematic Collections, Lawrence, KS, pp 6–9.

Ryder OA, Brisbin PC, Bowling AT, Wedemeyer EA (1981) Monitoring genetic variation in endangered species. In: Scudder, Reval (eds) Evolution today. Proceedings of the Second International Congress on Systematic and Evolutionary Biology. Hunt Institute for Botanical Documentation. Carnegie Mellon University, Pittsburgh, PA, pp 417–424.

Smith MW, Aquardo CF, Smith MH, Chesser RK, Etges WJ (1982) Bibliography of electrophoretic studies of biochemical variation in natural vertebrate populations. Texas Tech Press, Lubbock, p 105.

Index